CAMBRIDGE LIBRARY COLLECTION

Books of enduring scholarly value

Technology

The focus of this series is engineering, broadly construed. It covers technological innovation from a range of periods and cultures, but centres on the technological achievements of the industrial era in the West, particularly in the nineteenth century, as understood by their contemporaries. Infrastructure is one major focus, covering the building of railways and canals, bridges and tunnels, land drainage, the laying of submarine cables, and the construction of docks and lighthouses. Other key topics include developments in industrial and manufacturing fields such as mining technology, the production of iron and steel, the use of steam power, and chemical processes such as photography and textile dyes.

The International Exhibition of 1862

Replete with detailed engravings, this four-volume catalogue was published to accompany the International Exhibition of 1862. Held in South Kensington from May to November, the exhibition showcased the progress made in a diverse range of crafts, trades and industries since the Great Exhibition of 1851. Over 6 million visitors came to view the wares of more than 28,000 exhibitors from Britain, her empire and beyond. Featuring explanatory notes and covering such fields as mining, engineering, textiles, printing and photography, this remains an instructive resource for social and economic historians. The exhibition's *Illustrated Record*, its *Popular Guide* and the industrial department's one-volume *Official Catalogue* have all been reissued in this series. Volume 2 continues with further illustrated examples of British design and innovation, featuring exhibits that one might find not only at the Victorian factory, workshop or farm, but also in the home.

Cambridge University Press has long been a pioneer in the reissuing of out-of-print titles from its own backlist, producing digital reprints of books that are still sought after by scholars and students but could not be reprinted economically using traditional technology. The Cambridge Library Collection extends this activity to a wider range of books which are still of importance to researchers and professionals, either for the source material they contain, or as landmarks in the history of their academic discipline.

Drawing from the world-renowned collections in the Cambridge University Library and other partner libraries, and guided by the advice of experts in each subject area, Cambridge University Press is using state-of-the-art scanning machines in its own Printing House to capture the content of each book selected for inclusion. The files are processed to give a consistently clear, crisp image, and the books finished to the high quality standard for which the Press is recognised around the world. The latest print-on-demand technology ensures that the books will remain available indefinitely, and that orders for single or multiple copies can quickly be supplied.

The Cambridge Library Collection brings back to life books of enduring scholarly value (including out-of-copyright works originally issued by other publishers) across a wide range of disciplines in the humanities and social sciences and in science and technology.

The International Exhibition of 1862

The Illustrated Catalogue of the Industrial Department

VOLUME 2: BRITISH DIVISION 2

ANONYMOUS

CAMBRIDGE
UNIVERSITY PRESS

University Printing House, Cambridge, CB2 8BS, United Kingdom

Published in the United States of America by Cambridge University Press, New York

Cambridge University Press is part of the University of Cambridge.
It furthers the University's mission by disseminating knowledge in the pursuit of education, learning and research at the highest international levels of excellence.

www.cambridge.org
Information on this title: www.cambridge.org/9781108067294

This edition first published 1862
This digitally printed version 2014

ISBN 978-1-108-06729-4 Paperback

THE

ILLUSTRATED CATALOGUE

OF THE

INTERNATIONAL EXHIBITION.

SHEEP SHEARING.

DESIGN FOR A WALL MOSAIC, BY C. W. COPE, R.A., BEING ONE OF A SERIES ILLUSTRATING SCIENCE, ART, AND INDUSTRY, AND FOR DECORATING THE PERMANENT BUILDINGS FOR INTERNATIONAL EXHIBITIONS.

THE INTERNATIONAL EXHIBITION *of* 1862.

THE ILLUSTRATED CATALOGUE

OF THE

INDUSTRIAL DEPARTMENT.

BRITISH DIVISION—VOL. II.

PRINTED FOR HER MAJESTY'S COMMISSIONERS.

Printed for Her Majesty's Commissioners by

Clay, Son, & Taylor, Bread Street Hill. | Edmund Evans, Raquet Court.
Clowes & Son, Stamford Street. | Petter & Galpin, La Belle Sauvage Yard.

Spottiswoode & Co. New Street Square.

CONTENTS.

VOLUME II.

CONTENTS.

CLASS X.

CIVIL ENGINEERING, ARCHITECTURAL, AND BUILDING CONTRIVANCES.

SUB-CLASS A.—*Civil Engineering and Building Contrivances.*

[2226]

ALGER'S PATENT FURNACE COMPANY (Limited), 4 *Victoria Street, Westminster.*—Model of an elliptical blast furnace, now erected at Stockton-on-Tees, and in blast.

The advantage which the elliptical blast furnace possesses over other furnaces, consists in its combining large capacity with that degree of narrowness which insures the horizontality of the lines of equal temperature from the tuyères upwards. Thus the whole of the ore arrives in a uniform state of preparation for fusion at a melting zone, possessing perfect uniformity of temperature. The quality of the iron is improved, the descent of the charge more uniform, and there being *two* openings for tapping, one at each end of the ellipse, the furnace is more under control, and bridging, or scaffolding, is greatly diminished. There is a saving of one-third in the cost of construction, one-third in the fuel; less labour is required, and there is, besides, a saving in the blowing. Any ordinary blast furnace can be altered to the elliptical form at a small expense.

[2227]

ALLEN, EDWARD ELLIS, 5 *Parliament Street, S.W.*—Corrugated fibrous sheets for roofing, partitions, &c.

[2228]

ALLEN, H., 17 *Percy Street, W.*—Double model: a lift for manufactory; ditto for private house.

1. Model of a lift for raising or lowering invalids to the different floors of a house with ease and comfort.

2. The same adapted for raising or lowering goods at a factory or warehouse.

[2229]

ARBRUTH, G. B., & THOS. SCOTT, 18 *Parliament Street.*—Models of iron armour for ships and forts.

[2230]

ARCHITECTURAL POTTERY COMPANY, THE, *Poole, Dorset.*—Mosaic, tesselated, and white glazed tiles; patent glazed bricks; and orange-tree tile-tubs.

[2231]

ASHBY, ROBERT, 34 *Smith Street, Chelsea, S.W.*—Model for fireproof building.

[2232]

ATKINSON, W., 1 *Victoria Street, London.*—Portland cement.

BOWER, GEORGE, *St. Neots, Huntingdonshire.*—Patented vertical gas apparatus and combined purifier, for private use, and for exportation.

GAS APPARATUS AT BLENHEIM.

Of the numerous inventions which have so pre-eminently distinguished the present age, none has contributed in a greater degree to the comforts of civilised life than that of illumination by gas.

In these days of its almost universal adoption in our cities and towns, it is quite superfluous to dilate on its numerous advantages, which must be manifest to all, though their full force can perhaps be appreciated only by those who can remember the sombre appearance formerly presented by the streets of our large towns at night, as contrasted with their present brilliant aspect.

There is, however, a still more extended field for the operations of gas lighting, and much yet remains to be done in our villages, and in the mansions and private residences of the nobility and gentry. It is believed that this is due to certain misapprehensions and not unnatural prejudices which have hitherto existed on the subject, and that when these can be effectually removed, gas lighting will no longer be, in a great measure, confined to cities and towns, but its advantages will be as widely appreciated and embraced as they undoubtedly deserve to be. Its non-introduction into many of our villages has been mainly owing to the belief that it would prove a commercial failure, whereas it has been most conclusively shown by experience, that any compact village of a thousand inhabitants may be lighted with gas, so as to pay a good per-centage on the original capital embarked.

It is only within a comparatively recent period that the prejudices of private gentlemen, as to the advisability or practicability of introducing gas into their dwellings, have been partially removed. By some, danger was contemplated; by others, it was regarded as a nuisance, or as too complex in its management and manufacture; others, again, shrank from it on the score of economy, as involving fearful outlay in plant, and large cost of maintenance; while the possessor of the ornamental domain imagined in such plant an unsightly structure, emitting dense smoke and noxious vapour, giving to the mansion the appearance of a manufactory, and altogether inconsistent with that picturesqueness and quiet which are so generally and justly appreciated in country life. These suppositions are, however, an entire fallacy, for it may be confidently stated, that science has completely removed all ordinary chance of danger, or possibility of nuisance; that on the score of economy, in regard to the cost of apparatus, and the method and expense of making the gas, much has been done to reduce and overcome objections; whilst, by judicious arrangements, and the use of a portable apparatus (such as the one exhibited, which, from its compactness, can be placed in any out-building), nothing calculated to offend the eye, or the most fastidious taste, can be objected to. For works of greater magnitude, a low and secluded position (hidden it may be by trees and shrubbery), is usually chosen; and the requisite buildings may be so designed as to combine the ornamental with the useful.

The above engraving represents the patented apparatus as erected and fixed by the exhibitor for the palace of his Grace the Duke of Marlborough, at Blenheim.

Mr. Bower's inventions, designed for the purpose of removing the objections here alluded to, have been extensively adopted in various parts of Great Britain and the Continent, the following being a list of 100 gas works selected out of a great number of cities, towns, villages, factories, public buildings, and private establishments which have been lighted by him during the past few years. It may be mentioned as a proof of their general applicability, that the necessary apparatus for lighting the railway tunnel now in course of construction under Mont Cenis, has recently been supplied by him for the Italian Government.

BOWER, GEORGE—*continued.*

CITIES, TOWNS, AND VILLAGES.

Bourn............Lincolnshire, remodelled.
Beaufort.........Breconshire.
Bishop's Castle.Salop.
Bolsover.........Derbyshire.
Caistor ....Lincolnshire.
Casale...........Piedmont.
Crickhowell.....South Wales.
Coalville.........Leicestershire.
Collingham ...Nottinghamshire.
Donnington ...Lincolnshire.
Dokkum ......The Netherlands.
Gefle ....Sweden.
Higham Ferrars.Northamptonshire.
Hatfield.........Hertfordshire.

Harlingen ......The Netherlands.
Irthlingboro' ..Northamptonshire.
King's Cliffe...Northamptonshire.
Kimbolton......Huntingdonshire.
March...........Cambridge, remodelled.
Middleham ...Yorkshire.
Milverton ......Somersetshire.
Mountsorrel ...Leicestershire.
Oakham.........Rutland, remodelled.
Purmerende...The Netherlands.—Plant only.
Potton ....Bedfordshire.
Quorndon ......Leicestershire.
Redbourn ......Hertfordshire.
Repton ....Derbyshire.

Red Hill ......Surrey.
Saffron Walden.Essex, remodelled.
San Sebastian.Spain.
Sandy...........Bedfordshire.
St. Ives ....Hunts, remodelled.
St. Neots ......Hunts, remodelled.
Stevenage ......Hertfordshire.
Swineshead ...Lincolnshire.
Spa..............Belgium.
Syston ....Leicestershire.
Wellingborough.Norths., remodelled.
Whittlesea ...Cambs., remodelled.
Whitwick ......Leicestershire.
Wigston ......Leicestershire.

FACTORIES, RAILWAY STATIONS, AND COLLIERIES.

Abersychan	South Wales	Iron Works.
Allen, J., Esq.	Ivy Bridge, near Plymouth	Paper Factory.
Barrow, R., Esq.	Staveley	Three Coal Pits.
Bolckow and Vaughan	Middlesboro'-on-Tees	Iron Works.
Bayer and Co.	Sourabaya, Dutch East Indies	Sugar Works.
Bolckow and Vaughan	Eston, near Stockton	Iron Works.
Brown, Humphrey, Esq.	Tewkesbury	Silk Factory.
Brymbo Iron Company	Near Wrexham	Iron Works.
Cocker and Sons	Hathersage, near Sheffield	Needle Factory.
Castelli, M.	Serravelle, Italy	Cotton Factory.
Combe and Co.	Wolvercott, near Oxford	Paper Mill.
Ebbw Vale Company	Near Newport, Monmouthshire.	Iron Works.
Gordon, J., Esq.	33, New Broad Street, London	For Export.
Harrison, Ainslie, and Co.	Lyndal, Lancashire	Iron Works.
Hepburn and Sons	Dartford	Tannery.
Italian Government	Montcalier	State Railway Station.
Lansdorff, The Count	St. Petersburg	Cotton Mill.
Moscow Sugar Factory Company	Moscow	Sugar Works.
Mancardi, S., Esq.	Turin, Piedmont	Silk Mill.
Midland Railway Company	Toton, Nottinghamshire	Siding and Station.
Metallic Works Company	St. Petersburg	General Metal Works.
Milne, H. B., Esq.	Gefle, Sweden	Saw Mill for Gas from saw dust.
National Silk Spinning Company	Novara, Piedmont	Silk Mill.
Russian Spinning Company	St. Petersburg	Cotton Mill.
Rignon and Co.	Savigliano, Piedmont	Silk Mill.
Sampson Mill Company	St. Petersburg	Cotton Mill.
Schipoff and Co.	Moscow	Cotton Mill.
Smith and Sons	Watford, Hertfordshire	Paper Mill.
Gilineshoff, S. W., Esq.	St. Petersburg	Spinning Mill.
Shute, Thomas, Esq.	Watford, Hertfordshire	Silk Mill.
Saunders, Thomas, Esq.	Dartford	Paper Mill.
Towgood, Messrs., Brothers	St. Neots	Paper Mill.
Towgood, Alfred, Esq.	Helpstone Peterborough	Paper Mill.
Treschow, T., Esq.	Lorwig, Norway	Iron Works.
Tzarskoe Railway Company	St. Petersburg	Gardens and Railway Station.
Taylor and Robinson	Rastrick, Huddersfield	Woollen Mill.
Tunnel under Mount Cenis	Bardoneche	Italy.
Vitale, Placido, Esq.	Messina	The New Theatre.
Whalley and Hardman	Kirkham, near Preston	Cotton Factory.

PRIVATE ESTABLISHMENTS.

Marlborough, His Grace the Duke of, Blenheim Palace, Woodstock, Oxfordshire.
Westmoreland, Right Honourable the Earl of, Apthorpe Park, Wansford, Lincolnshire.
Roden, Right Honourable the Earl of, Tollymore Park, County Down, Ireland.
Macclesfield, The Right Honourable Lord, Sherborn Castle Oxfordshire.
Ashby, Captain, Nazeby Woolleys, Northamptonshire.
Bolckow, J., Esq., Marton Hall, near Middlesborough.
Benyon, Rev. E. R., Culford Hall, Bury St. Edmunds.
Benyon, R., Esq., M.P., Englefield, Reading.
Fothergill, R., Esq., Abernant House, Aberdare.
Gwyn, Howell, Esq., Dyffryn House, Neath.
Harter, Rev., Cranfield Court, Newport Pagnell.
Haigh, G. H., Esq., Grainsby Hall, near Louth, Lincolnshire.
Ledeboer, J., Esq., Macassar, Dutch East Indies.
Newton, Geo., Esq., Croxton Park, Cambridgeshire.
Pearce, J. D. M., Esq., M.A., Craufurd College, Maidenhead.
Roden, R. B., Esq., Ponty Moil, South Wales.
Stevens, Rev. T., St. Andrew's College, Bradfield, near Reading.
Vaughan, John, Esq., Middlesboro'-on-Tees.
Vansittart, S. H., Esq., Bisham Abbey, Maidenhead.

In reference to illuminating gas it is noticeable, that though many articles, such as oil, peat, resin, &c., readily produce it, the exhibitor and patentee fearlessly asserts, that, after innumerable experiments, and the trial of almost every conceivable plan for generating artificial light, our coal mines furnish the best and most economical materials for the purpose. For, as the illuminating power of gas depends on the relative proportions of its constituents—carbon and hydrogen—and as these are found to exist in coal in the proper proportions, it is obvious that taking into consideration its low cost, and the simplicity of the means required to convert the volatile portion of it into gas, coal is the best substance for the purpose. It is true, for instance, that water,

BOWER, GEORGE—*continued.*

which is costless, will produce an illuminating gas when decomposed, and its hydrogen liberated and carbonised, but the cost of decomposition for the production of hydrogen—one only of the elements of which it is composed—is actually greater than that of highly illuminating gas produced from coal; and notwithstanding the many attempts to supersede it, it may be emphatically stated, that nothing whatever can in any commercial sense compete with coal, even giving it a range of price far beyond what it is at present.

Having established this fact, it became necessary to devise a cheap, simple, and economical apparatus, and the result has been the exhibitor's inventions, patented in 1852, 1859, and 1860; the last of which (the vertical retort), in connection with the combined purifying apparatus, has been pronounced far superior to all others, in every essential property requisite for the manufacture of gas on a small scale.

These features may be noted in the articles exhibited, viz., the vertical retort or gas generator, with its appurtenances, and the combined hydraulic main, condenser, and purifier, being equal in their conjoint capacity to a power representing about twenty lights. (The gasholder cannot be exhibited, for want of space.)

The gas generator consists of a conical retort, set vertically in an iron case, lined with fire-brick; both ends of such retort are open, the top being surmounted by a hopper, for the purpose of charging it with coal, to be afterwards closed by a luted plug, and the bottom provided with a luted door, having a false bottom or diaphragm projecting about six inches into the retort. This door, when closed, is retained in its position by means of a lever, having a swing catch and wedge; the fire-grate being arranged around the retort, so as to bring the fire itself into immediate contact with its outer surface. The mode of operation is, to heat the retort to a bright red, the bottom door being then luted, and raised to its proper position by means of the lever and catch; the retort is next filled, by the use of the hopper, with the necessary charge of dry coal or cannel, and the top closed with the luted plug. After the lapse of three or four hours, the gas will be extracted, the wedge and catch may be removed, and the door lowered by the lever to the pair of horizontal bars; and being removed, the coke falls out leaving the retort free for renewed operations.

The combined purifying apparatus consists of the hydraulic main, condenser, and purifier, united in one vessel or case, the base of which forms the hydraulic main; and a receptacle for the products, separated from the gas by the condenser, which condenser is so formed that the gas passes around the purifying vessel, in a space, the inner surfaces whereof are exposed to the water, forming the lute or seal for the purifying lid; and the outer surfaces are exposed to the atmosphere.

The purifier is provided with a cover, and four tiers of shelves, or perforated plates, which, when in operation, are covered with lime, and through which the gas percolates.

The gas is brought from the retort by means of a pipe into the hydraulic main, thence passing up and down the spaces forming the condenser, into the purifier, and from thence it is conveyed into the gasholder, ready for use.

The principal advantages of these arrangements for private works on a small scale are:—

1. They occupy but little space, are very simple, and require but little labour or skill to manage.

2. No bricks are required to set the retort, further than the few sent with the apparatus, and these are moulded of suitable shapes.

3. The retort being set vertically, and surrounded by fire, in immediate contact with it, requires less fuel than if set horizontally, and the fire may be lighted and permitted to go out with impunity, the same as an ordinary shop or hall stove.

4. It is adapted for common coal as well as cannel, and may also be adapted for the generation of gas from wood, peat, or oil, in situations where coal is difficult to be obtained.

5. In this arrangement of retort, by merely removing the top and bottom covers, when red-hot, the current of air that passes through, will remove all deposits of carbon from the interior.

6. The whole of the apparatus for removing impurities from the gas by condensation and purification, which by the ordinary process consists of three or four separate and cumbrous vessels, is effectually combined in the limits of one solid base.

7. The retort, when worn out, can be replaced, without requiring a skilled workman to fix it.

[2233]

BALE, T. S., *Mount Pleasant, Newcastle, Staffordshire.*—Mosaic and ornamental floor, wall tiles, and glazed bricks.

[2234]

BARNETT, S., 23 *Forston Street, Hoxton.*—Diving apparatus.

[2235]

BARRETT, HENRY, 12 *York Buildings, Adelphi.*—Model of fireproof flooring.

[2236]

BASFORD, WILLIAM, *Elgreave Street, Burslem.*—Front-facing brick, in connection with walls or fronts of cottages, and other buildings; roof and floor tiles, &c.

[2237]

BEART'S PATENT BRICK COMPANY, *Arsley, and King's Cross, London.*—Bricks and agricultural drain pipes.

[2238]

BELLMAN & IVEY, 14 *Buckingham Street, Fitzroy Square, W.*—Specimens of various imitations of scagliola marble.

[2239]

BETHELL, JOHN, 38 *King William Street, London, E.C.*—Specimens of creosoted woods.

BROWN-WESTHEAD, MARCUS, *Manchester.*—Patent hoist governor and patent safety railway platform lift.

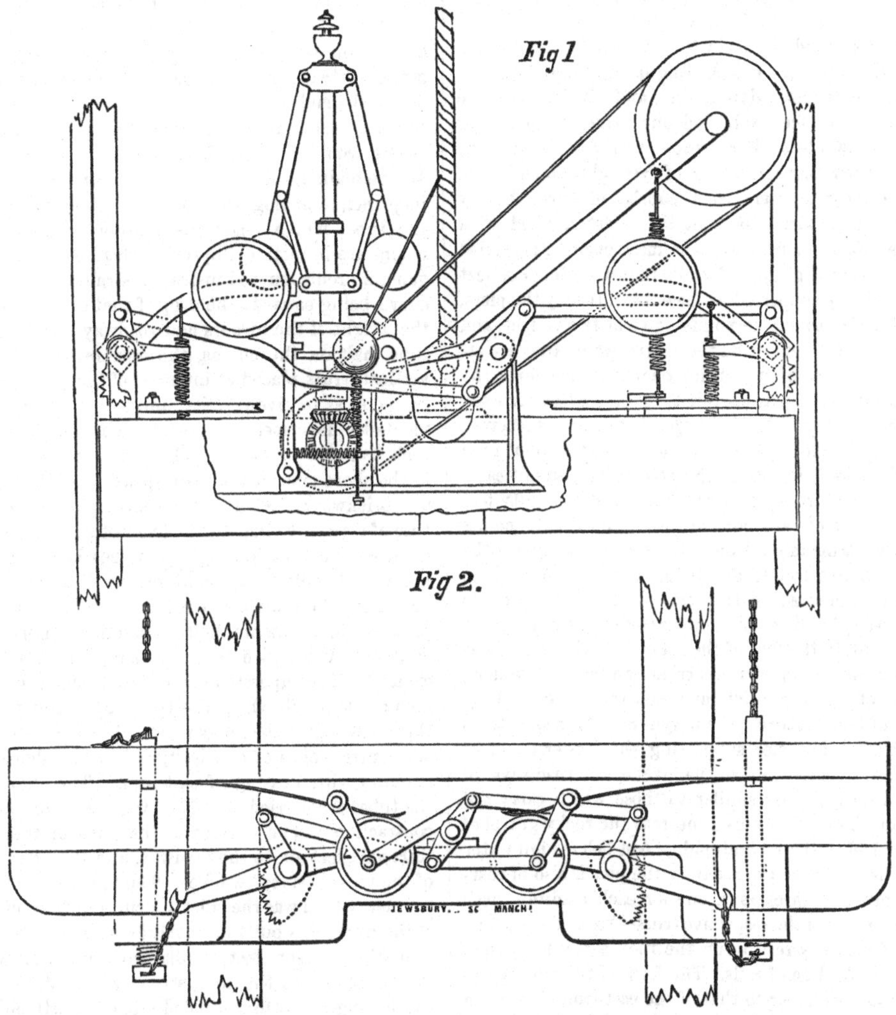

Fig. 1 represents a patented apparatus used in the northern districts of England for preventing accidents to life and injury to property in hoisting and lifting machinery. The machine is attached to the upper side of the ascending and descending chamber or cage. It is simple in construction and easily kept in order, and has already proved itself in many instances to be a secure and efficient apparatus for the purpose intended. It is evident that no matter what the cause of the cage or box travelling at too great a speed, the expansion of the governor from an excessive rate of motion must, per force, operate upon the other portions of the machine, and thus permit the cams or wedges to be instantly projected against the uprights or guides of the shaft or well-hole, and so arrest the box. A small weighted lever falls if the rope of suspension is severed, and thereby catches the box, or cage, within a few inches, and that instantaneously. This is a double security in the event of the rope breaking, as the machine, in that case, does not depend upon the action of the governor alone. It has been proved that the greater the weight of the cage, or bóx and its contents, the more safely it is secured; for the heavier the load the greater will be the resistance of the cams, as they will take a much firmer hold of the guides or uprights. Some alteration of the plan, as shown n Fig. 1, makes this invention equally applicable to the cages used in coal and other mines, where it is essential that the momentum of the cage should be *gradually* reduced in the event of over-speed. This invention combines the peculiar and important advantage, viz., the springs to insert the cams are not brought into action except at the required time, thereby preserving their elasticity unimpaired. The cost of application to ordinary hoists ranges from £25 to £30, including royalty.

Fig. 2 is a plan for preventing accidents to railway platform lifts. These hoists are extremely dangerous in consequence of the liability of one or more of the suspending chains breaking. By this invention the platform is evenly and simultaneously arrested at its four corners in the event of one or all of the suspension chains being fractured, by too great a load, or from the alternations of temperature affecting unequally the nature of the metal employed in the manufacture of the chains. It is, moreover, advantageous in its application, as the platform can be readily adjusted or repaired whenever requisite without having to stay the platform. Cost of application, including royalty, £75 to £85.

Brunel, Isambard, *Duke Street, Westminster.*—Models of Saltash and Chepstow bridges, designed by late Mr. Brunel.

Model of Bridge on the South Wales Railway over the Wye at Chepstow.

This bridge was designed by the late Isambard Kingdom Brunel, Esq., D.C.L., F.R.S., Engineer of the Railway. It is constructed for a double line, and consists of three side spans of 100 feet each, and one principal span over the river of 300 feet. Each roadway over the land openings is carried between a pair of wrought-iron girders. The intermediate piers each consist of three hollow cylinders of cast-iron, six feet in diameter, filled with concrete. Each roadway over the main opening is carried between a pair of girders of similar construction, 300 feet long, which are supported at the extremities by the piers, and at four intermediate points—two at twelve feet, and two at sixty-two feet—from the centre, where they are attached to two sets of suspension chains, which form the tensional parts of a pair of rigid trusses. In these trusses, the tension delivered by the suspension chains is received by a straight tube of plate-iron, of circular section, nine feet in diameter; which is supported at its extremities on the superstructure of the piers, with its centre fifty feet above the line of rails, and at two intermediate points by vertical struts raised from the suspension chains at the points sixty-two feet from the centre of the span where the girders are attached to them. Rigidity is given to the structure by diagonal chains extending from the top of each strut, to the foot of the other. Each tube where it rests on the eastern pier is carried by a system of rollers, to allow of the expansion and contraction caused by changes of temperature. The weight of iron in each truss is 460 tons. The pier supporting the western end of these trusses consists, up to the level of the roadways, of six cast-iron cylinders similar to those which carry the land spans, and which here penetrate the rocky gravel of the river bed to the rock at eighty feet below high water spring tides. Above the roadway the pier is also of cast-iron, forming two archways, one for each roadway, over which the tubes of the respective trusses rest. The eastern pier is of masonry resting on the rock a few feet below the level of the line of rails. The form of this pier above the roadway is similar to that of the cast-iron pier at the western end. The bridge was commenced in April, 1850, and was completed in three years. On the successive completion of each tube, it was temporarily rendered rigid with chains, and being placed at right angles to the river, one end was supported on pontoons, and the other on a rolling truck. The pontoons were then drawn across the river by warps from the opposite shore. The tube was next lifted into its place on the top of the piers by chain purchases, and the rest of the truss was then completed. The operation of floating was rendered difficult by the great rise and fall of the tide, which is forty-two feet at spring tides. The contractors for the iron work were Messrs. Finch and Willey, of the Windsor Foundry, Liverpool. The total cost of the bridge was £77,000.

Model of the Royal Albert Bridge on the Cornwall Railway over the Tamar at Saltash.

This bridge was also designed by the late Mr. Brunel. It is for a single line, and consists of two spans of 455 feet each over the river, and seventeen land openings of spans varying from ninety to seventy feet. The land openings—of which there are ten on the Cornwall, and seven on the Devonshire side—form curved viaducts leading to the main spans. Throughout the structure, at a level of one hundred feet above high water, the rails are laid on a ballasted platform of planks carried on cross girders between pairs of plate-iron girders. In the viaducts, the ends of the girders rest on piers of limestone masonry, each pier consisting of two square pillars which spring from a common base, and are united at the top. In the main spans the girders are supported by trusses, in principle analogous to those at Chepstow; but here the tubes which resist the tension of the suspension chains are in section elliptical instead of circular, and in general profile, curved instead of straight, the rise of the curve, being equal to the drop of that of the chains; thus the weight of the girders and roadway rests half on the tube, half on the chains, the girders being carried by vertical struts, placed at intervals of forty feet, diagonally braced so as to give rigidity, and by intermediate attachments to the suspension chains. The weight of iron in each truss is 1,070 tons. The substructure of the piers at the shore ends of the main spans is of granite masonry and brickwork. That of the centre pier consists at the base of a granite pillar thirty-five feet in diameter, resting on a rock foundation eighty-six feet below high water mark, and built to a height of ten feet above it, from which rise four hollow octagonal columns of cast-iron, built up in segments bolted together internally, and which carry the girders on an entablature above their capitals. The superstructure of each pier consists of an archway through which the trains pass, and over which the ends of the tube are carried. The superstructure of the centre pier is of cast-iron, and of the shore piers of masonry with a casing of cast-iron. The shore ends of the tubes are carried on rollers, to allow of expansion and contraction. The centres of the ends of the tube are thirty-six feet above the roadway, and the extreme depth of the truss is sixty-two feet. The lower part of the centre pier, which was the chief difficulty in the construction of the work, was built in a cofferdam or cylinder of plate-iron, thirty-seven feet in diameter and ninety feet in length, closed at the top, strongly stayed throughout, and having its bottom divided into compartments, which were kept clear of water partly by a supply of compressed air, partly by pumping. This cylinder was correctly placed on the rock through the mud which was there, thirteen feet in depth, and which being loaded with shingle ballast, assisted to keep out the water. Each truss was put together on the Devonshire shore of the river. Docks were formed, and pontoons prepared with wooden framings to carry the truss. Warps were led from these pontoons to various points on shore, and to vessels moored in the stream. The operation of floating in each case was performed without delay or accident, and the ends of the tube placed on the piers which had been built up to receive them. The truss was then lifted by hydraulic presses, the piers being built up underneath. The total cost of the whole work was £225,000. It was commenced in the beginning of 1853, and was opened on May 3rd, 1859, by H. R. H. the late Prince Consort, Warden of the Stanneries, by whose gracious permission it was called the Royal Albert Bridge.

These models were made for the late Mr. Brunel by Mr. Salter, of Hammersmith, and are both to the scale of ten feet to one inch.

[2240]

BOWER, GEORGE, *St. Neots, Huntingdonshire.*—Patented vertical gas apparatus and combined purifier, for private use, and for exportation. (*See pages* 2, 3, 4.)

[2241]

BROOKE, EDWARD, *Field House Fire Clay Works, Huddersfield.*—Glazed sewer tubes, fire-bricks, furnaces, retorts, glass pots, &c.

[2242]

BROWN, JOHN, *Chapel Field, Norwich.*—Models of patent for rendering windows, &c., wind and water-tight.

Windows and doors are by this patent rendered impervious to draft, dust, and other annoyances, without interfering with perfect ventilation. These results are essential to health and comfort.

By this invention, when fitted to shop windows and show-cases, jewellery, silver and plated goods; cutlery, books, lace, and other articles liable to injury from the effects of gas, dust, or damp, are completely protected.

The agent of John Brown is T. Burton, 35, Wellington Street, Strand, W.C.

[2243]

BROWN, R., *Surbiton, Surrey.*—Italian and other roofing tiles; ornamental bricks, red, green, black, and white; ornamental ridge, &c.

[2244]

BROWN-WESTHEAD, MARCUS, *Manchester.*—Patent hoist governor, and patent safety railway platform lift. (*See page* 5.)

[2245]

BRUNEL, ISAMBARD, *Duke Street, Westminster.*—Models of Saltash and Chepstow bridges, designed by the late Mr. Brunel. (*See page* 6.)

[2246]

BUNNETT & CO., *Deptford, Kent.*—Patent revolving iron shutters, and ornamental brass sashes, &c.

[2247]

BURGESS, THOS. H., 4 *Upper Marsh, Lambeth.*—A stand, which will admit of boots being made without sitting.

[2248]

BURT & POTTS, 38 & 65 *York Street, Westminster.*—Patent water-tight wrought iron window and frame.

[2249]

CARTWRIGHT, J. M., & CO., *Swadlincote, Burton-on-Trent.*—Fire-bricks and arches for locomotive engines.

[2250]

CHALMERS, JAMES, *London* (*late of Montreal, Canada*).—Drawings of proposed channel railway, connecting England and France.

[2251]

CHAPMAN, J. W., *Park Road, Richmond, Surrey.*—Plans of estates, &c.

[2252]

CHAPPUIS, P. E., 69 *Fleet Street, London.*—Patent reflectors for diffusing daylight in dark places, and reflecting artificial light.

[2253]

CHRISTMAS, R., & JONES, 28 *Lord Street, Birkenhead.*—Castellated circular turret, random rubbed; white quartz.

[2254]

Clark & Co., *Gate Street, Lincoln's Inn Fields.*—Patentees and manufacturers of revolving shutters in steel, iron, and wood, for shop fronts, private houses, fireplaces, &c.

REVOLVING SHUTTERS FOR SHOP FRONTS.

Clark and Co.'s patent self-coiling revolving safety shutters, in steel, iron, and wood, are adapted to close, with security and ease, every description of opening, as exhibited in shop front, Class X. The steel shutter is both thief and fire-proof, and as cheap as the ordinary wood shutters.

STEEL SHUTTER FITTED TO FIRE-PLACE.

WOOD SHUTTER ADAPTED TO BAY WINDOW.

[2255]

Clark, Edwin, *Great George Street, Westminster.*—Model of Clark's patent hydraulic graving docks, Victoria Docks.

MODEL OF HYDRAULIC LIFT, GRAVING DOCKS (CLARK'S PATENT), AS ERECTED AT THE THAMES GRAVING DOCKS, VICTORIA DOCKS.

This model represents a plan for docking vessels, patented by Mr. Edwin Clark, and carried out on a large scale at the works of the Thames Graving Dock Company, where it may be seen in daily use.

The system is entirely novel, and differs from an ordinary graving dock in that, instead of the vessel being floated into a pit, and the water pumped out or allowed to run out with the tide, the vessel is raised bodily out of the water, cradled upon a shallow pontoon, on which it is afterwards floated away to any place convenient for its repair. The apparatus for these enormous lifts consists of a series of hydraulic presses contained in and supported by cast-iron columns sunk into the ground in two parallel rows, the space between being sufficient for the vessel to pass through.

From the cross-head of each ram the ends of a pair of girders are suspended; these girders pass across the dock, and form a platform, on which the vessel and pontoon are lifted.

The pumping power is a small steam engine placed near the presses, the communication between it and the presses being through wrought-iron pipes. The engine does not pump direct into the hydraulic cylinders, but into an intermediate valve-chest, by which the raising power is regulated, and the uniform rise of the whole ship and pontoon secured.

The pontoons are large, shallow vessels, constructed of wrought-iron framing and shell, and are divided into several water-tight compartments, in each of which is a valve; they are made of various sizes, corresponding with the weight of the vessels they are intended to carry. The seven pontoons now in use vary from 160 to 320 feet in length, draw from 3 feet to 6½ feet when loaded, and carry vessels of from 500 to 3,000 tons.

The hydraulic rams will safely raise a dead weight of 6,000 tons, but can be adapted to lift any weight.

The peculiarities of this system are the raising the vessel to the level of the workshops and repairing-yards, and keeping it high and dry there in full light, exposed to the drying influences of the air; while, from the vessel being carried above the pontoon, its bottom is more accessible.

The blocking or shoring the vessel, under this system, is most effectually and rapidly performed, the operation being simply the drawing in of blocks fitted to the side of the vessel, which blocks are carried on the wrought-iron transverse girders. The pontoon, being highly elastic longitudinally, accommodates its shape to the keel of the ship, whatever be its form; thus insuring a perfect bearing throughout.

Each pontoon in itself forms a complete graving dock, and one hydraulic lift is sufficient for a great number of pontoons. The cost of a graving dock complete is, therefore, little more than the cost of the pontoon, which, for all ordinary vessels, varies from £600 to £10,000; and the rapidity of an operation is so great, that at least six vessels can be docked and set afloat in an ordinary working day.

The Thames Graving Dock Company, during the three years of their practical working, have most successfully docked upwards of 400 vessels, weighing 220,000 tons.

[2256]

CLARKE, GEORGE, *Manufactory, South Crescent Mews, Burton Crescent, London.*—Clarke's improved fire escape, in use by the Royal Society for the Preservation of Life from Fire.

The great utility and importance of this machine consists in its extreme cheapness, combined with simplicity of construction, any person being able to work it after a few hours' practice. Its use has now become so apparent that no city, town, or village, and even large manufacturing premises, should be without one.

The main ladder of the escape reaches a height of 33 feet, and can instantly be applied to a secoud-floor window. Under the escape is a canvas trough, protected from flaming by a copper gauze.

The upper ladder folds over, and can easily be raised by levers to the position represented; and by adding an additional ladder, which is the work of a few minutes, will reach the height of 70 feet.

In cases where gardens are in the front of the houses, and the gates are not of sufficient width to admit the escape, the upper ladders unship by means of shifting levers, and can be used separately.

IMPROVED FIRE ESCAPE.

PRICES :—

Fire escape reaching 42 feet £45
Ditto, reaching 50 feet 50
Ditto, reaching 60 feet, with copper gauze and shifting levers, with all late improvements 63
Ditto, reaching 70 feet . . . £73 10s.
Ditto, above 70 feet . . . £105

From the Report of the Royal Society for the years 1860 and 1861, it appears that their fire escapes, manufactured solely by Clarke, were used successfully at no less than 103 fires, and 155 human beings were rescued from the flames.

[2257]

CLARKE, JOHN VIZETELLY, 251 *High Holborn, W.C.*—Gas regulators and apparatus.

[2258]

CLERK, FRANCIS NORTH, *Mitre Works, Wolverhampton, Staffordshire.*—Metal roofing and galvanized fittings for roofs and buildings.

[2259]

CLIFF, JOSEPH, & SON, *Wortley, near Leeds.*—Clay retorts, fire-bricks, sanitary pipes, chimney-tops, terra-cotta ornaments, &c. (*See page* 12.)

[2260]

COLLA, J. G., 18 *Parliament Street.*—Ornamental tiles, &c.

[2261]

COOKEY, E., & SON, *Frome Selwood.*—Valves for regulating the flow of gas in gas manufactories.

[2262]

CORY, WILLIAM & SON, *Commercial Road, S.*, Owners.—J. H. Adams, 1 Grove Hall Terrace, Bow, E., Engineer.—A float, with machinery for discharging screw colliers and other vessels with great rapidity, in the stream.

This vessel is fitted with Sir William Armstrong's hydraulic cranes, and is provided with other machinery and apparatus for screening the coals when required, and depositing them in barges without breakage. Two steam colliers of the largest dimensions may be discharged at once. By a suitable arrangement of the hatchways and holds of the steamers, three cranes may be worked on each steamer at the same time, and each crane can discharge 60 tons of coals per hour. The owners are prepared to undertake to discharge steamers not exceeding 1,200 tons cargo, in ten hours, night or day. They have similar machinery on fixed buildings in operation at the Victoria Docks.

Builders and owners of steamers can obtain from Mr. Adams the requisite particulars for the adaptation of their vessels.

Cory, William, & Son—*continued.*

A FLOAT, WITH MACHINERY FOR DISCHARGING SCREW COLLIERS AND OTHER VESSELS WITH GREAT RAPIDITY, IN THE STREAM.

CLIFF, JOSEPH, & SON, *Wortley, near Leeds.*—Clay retorts, fire-bricks, sanitary pipes, chimney tops, terra cotta ornaments, &c.

Cliff's patent enamelled clay retort is especially adapted for the use of gas works, by its smooth interior, and freedom from fire-cracks, by which the adhesion of carbon is prevented.

Its excellence is attested by the experience of the leading gas engineers of the day.

Cliff's Wortley fire-bricks are the most durable bricks manufactured for forge purposes, and the linings of blast-furnaces and glass-works.

The following designs in terra cotta chimney tops have proved themselves the most efficient wind guards introduced.

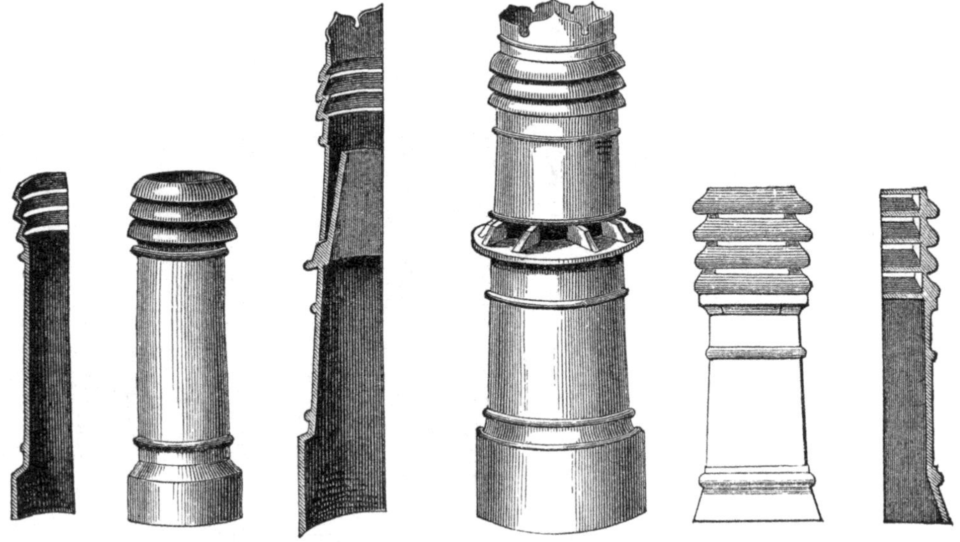

TERRA COTTA CHIMNEYS.

JOSEPH CLIFF and SON are the largest manufacturers of the patent salt-glazed socketed drain pipes in the kingdom. These pipes are made, in circular, up to 54 inches diameter, and in egg-shape, to 36 × 24, for sewer and water culverts. They can be shipped in any quantity, and in all sizes, from 2-inch to 36-inch diameter, at the ports of London, Liverpool, or Hull.

London depot, No. 4 Wharf, inside Great Northern Goods Station, King's Cross, N. M. B. Newton, agent, who will have pleasure in attending to any correspondence or appointment during visitors' stay in London.

[2263]

COSSER, FREDK. C., 145 *York Road, Lambeth.*—Models of railway carriage signal, and steam engine; improved chimney-pot and ventilator, for curing smoky chimneys.

[2264]

COUNCIL OF ARCHITECTURAL MUSEUM, *South Kensington, W.*—Specimens for which prizes have been awarded to artist-workmen.

[2265]

COWEN, JOS., & CO., *Blaydon Burn, Newcastle-on-Tyne.*—Patent fire-clay gas retorts, fire-bricks, tiles, &c.

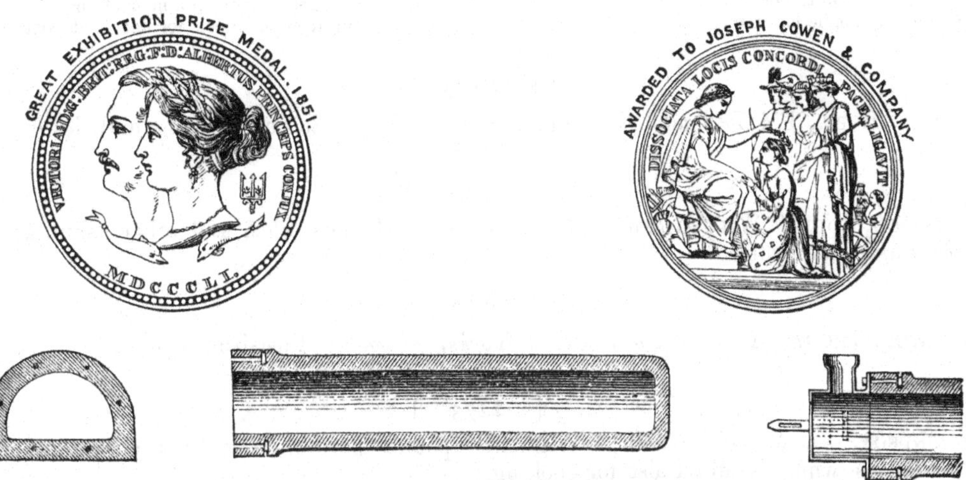

Obtained the Prize Medal at the Exhibition of 1851.

Joseph Cowen and Co. are manufacturers of patent fire-clay gas retorts, fire-bricks, tiles, bearers, and all descriptions of fire-clay goods used in gas-works, blast furnaces, potteries, chemical works, &c. The Prize Medal obtained by them was the only one awarded at the Great Exhibition for fire-clay goods. They have at all times a large stock of the ordinary size of fire-bricks and tiles, and can ship them in any quantity at a day's notice. Patent fire-clay retorts are made by this firm of any size or shape, and to fit existing mouthpieces. These retorts are well adapted for small gas-works, as they can be used without an exhauster. Drawings of retort settings may be obtained by application.

Mead and Bell, 13, Cliff-street, New York, are Jos. Cowen and Co.'s agents for America.

[2266]

CRESSWELL, JOHN, 100 *Islington, Birmingham.*—Patent self-folding shutters.

Cresswell's Patent Shutters are (at present) unequalled for strength, neatness, and convenience.

They require no latches, bars, bolts, or other fastenings, but when closed are perfectly secure, and cannot by any accident be left unfastened.

It will be seen upon the first inspection that, from the simplicity of the construction, there being no springs, gearing, pulleys, or complicated apparatus, the chances of getting out of order are the most remote; certainly less than those of any other shutters now in use. When constructed in iron, they are perfectly fire-proof.

For bay windows they are particularly suitable; the hitherto almost *insurmountable* difficulties attending the inclosing of a bay window with shutters are entirely removed. The patentee observes, that "his shutters are the only ones without objectionable features for the bay window."

The adoption of these shutters for bed-room windows, particularly in the country, where security is required, is suggested, as they may be fixed at small cost, and without interfering with the existent window dressings. They are especially adapted for show-glasses, book, museum, or other cases, where safety and protection from dust are essential.

J. Cresswell solicits inspection of his models and of work already executed, and will be happy to give any further information that may be required, and furnish estimates, &c. Any orders he may receive will have his best personal attention. He is prepared to furnish the trade with the sheets ready for fixing, and to grant licences to parties desiring to manufacture their own.

[2267]

DODMAN, GEORGE, 4 *Back South Parade, Manchester.*—Patent hoist safe, on eccentric principles.

[2268]

DOULTON, HENRY, & Co., *Lambeth.*—Stoneware pipes, and other articles, for sanitary and building purposes.

[2269]

DOWNING, GEORGE FRAS., 122 *King's Road, Chelsea, S. W.*—Floor-cloth.

The following are exhibited :—

1. Model of double straining frame for floor-cloth.
2. Section of built floor-cloth roller.
3. Ratchet handle for ditto.
4. Stand for ditto.
5. Various trucks for moving floor-cloths.
6. Hanging iron and hook for floor-cloth battens, &c.

[2270]

DUNCAN, ROBT., 174 *Trongate, Glasgow, and at Bowling.*—Self-acting time and tide-gauge.

[2271]

EASTWOOD, JOHN & WM., *Belvidere Road, Lambeth.*—Bricks, tiles, and other manufactures, plain and ornamental.

[2272]

EDINGTON, THOMAS, & SONS, *Phœnix Iron Works, Glasgow.*—Cast-iron pipes.

[2273]

EDMUNDSON, J., & Co., *Dublin.*—Wigham's patent portable gas apparatus for private residences, which is suited also for cooking.

By means of this apparatus, noblemen's or gentlemen's mansions, mills, manufactories, farm buildings, railway stations, &c., even in the most remote districts, can be supplied with brilliant light, at an expense much less than from oil or candles. Any servant can manage the apparatus with ease, and it can be removed from one residence to another, if found requisite. All information as to price, &c., can be procured by application to the sole agents for the patentee, J. Edmundson & Co., Gas Engineers, 34, 35, 36, Capel Street, Dublin.

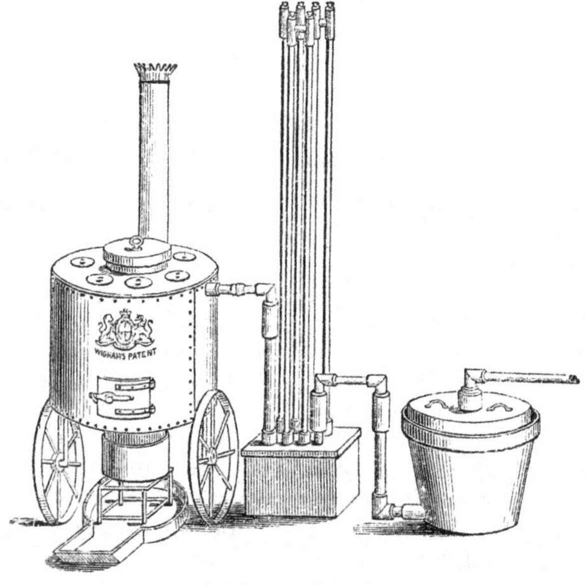

WIGHAM'S PORTABLE GAS APPARATUS.

[2274]

EDWARDS, GEORGE HENRY, 90 *Aldersgate Street, E.C.*—Patent fastening for sash-lines, instantaneously connected and disconnected.

[2275]

EFFERTZ, PETER, 40 *Brown Street, Manchester.*—A diving bell.

[2276]

ELKIN, W. H., 27 *Belvedere Road, London, S.*—An improved window, which can be cleaned or repaired without danger.

This improvement, which admits of the sashes being turned inside out with the greatest ease and safety, also provides for replacing broken sash-lines, without taking down the beads, effecting a saving thereby of more than double its cost. Fittings, according to size and character of window, at from 2s. to 6s. per set, may be obtained from the patentee.

Old windows can be altered to the above principle at a small cost.

[2277]

ERRINGTON, J. E., *V.P. Institution C.E.*, 13 *Duke Street, Westminster.*—Viaduct across the valley of the Lune.

[2278]

FAIRFAX, BRYSON, & Co., *Birmingham, Liverpool, and London.*—Model of Maillefert's aerostatic tubular diving-bell.

[2279]

FAYLE & Co., 31 *George Street, Hanover Square.*—Architectural building blocks, facing and fire-bricks, &c.

[2280]

FIELD, W., 13 *Parliament Street.*—Pulpit for Westminster Abbey.

[2281]

FISHER BROTHERS & Co., *the Hayes Fire Clay Works, Stourbridge.*—Fire-bricks, gas retorts, and glass-house pots.

FISHER BROTHERS & Co are manufacturers of fire-bricks, gas retorts, and glass-house pots; and proprietors of best glass-house pot and crucible clay.

[2282]

FISON, C. O., *Stowmarket, Suffolk.*—Kiln-tiles for drying malt, &c. (used by Messrs. Bass & Co.); white bricks, pantiles.

Improved Kiln Tiles. The excellence of these tiles will be certified by maltsters throughout England, who prefer them to a metal drying floor. Their advantages are:—1st. No stoppage of the holes, as in ordinary tiles. 2nd. The greatest possible draught. 3rd. They are made of a very fine and durable earth. Price 9d. and 10d. each.

Drying tiles for wool, cotton, &c., 12-inch . . . . 6d. & 9d. each.

Perforated tiles for paper-makers, 12-inch . . . . 9d. & 10d. „

Best White Suffolk Bricks.

Red and White Pantiles, new waterproof pattern; requiring no mortar. The white are invaluable, where a cool roof in summer is important.

[2283]

FITZ-MAURICE, HON. MAJOR, *Conway Lodge, Hyde Park Gate.*—Patent apparatus for making oil and coal gas.

[2284]

FRANK, JOHN CHRISTIAN, E.B.A., 1 *Quadrant Grove, Haverstock Hill.*—Standing white colour; improvement in oils; improvement in metallurgy.

[2285]

FREEN & Co.—Specimens of cement.

Glover, George, & Co., *Ranelagh Works, Ranelagh Road, Pimlico, London; Offices, 22, Parliament Street, Westminster, and 15, Market Street, Manchester.*—Standard gasometers, similar to those deposited at the Exchequer.

GLOVER, GEORGE & Co.—*continued.*

CONTENTS OF DIAGRAM.

I. MESSRS. GEORGE GLOVER & Co's FIVE-FEET GAS-HOLDER; being a copy of the five-feet gas-holder belonging to the set of national standards constructed from Mr. George Glover's designs, and under his superintendence, with proper balances, indices, and apparatus, as the "Sales of Gas Act" requires.

Under this Act it was found impossible to adopt, as a standard measure, any of the gas-holders hitherto in use. The material of which they are made is very liable to corrosion when in contact with gas and water. To retard corrosion, paint is used. The paint in the inner surface of the bell diminishes its measuring capacity, and its renewal from time to time aggravates the evil. The coating of paint softens, swells, frequently rises in blisters, falls off in flakes, or crumbles away. In its ascent from the cistern the painted surface of the bell brings with it a quantity of water, which adheres to it in the form of a film, and numerous drops which adhere especially to the inner surface of the flat cover. These occasion further diminution of capacity, whilst the evaporation of the water on the outer surface of the bell lowers the temperature, diminishes the volume of gas contained, and causes error in the testing of meters.

Their measuring part is not a truly cylindrical vessel. They not only differ from each other to the extent of 3 or 4 per cent. in their measuring capacity, but the various divisions into feet, and the subdivisions of the feet, differ in the same gas-holder.

Their scales are not engraved upon the bell, and they can easily be tampered with. These circumstances precluded their adoption as standard measures for gas.

The essential properties of the national standards deposited at the Exchequer, and the derived standards for London, Edinburgh, and Dublin, which have also been constructed from Mr. George Glover's designs, are these:—

1. The metal of which they are made is an anti-corrosive alloy, which resists the chemical action of the constituents of coal-gas and water.

2. The surface of the bell readily parts with water.

3. The bell, or the measuring part of the instrument, is a truly cylindrical vessel, and sufficiently rigid to resist change of form from the application of any ordinary forces.

4. It has a correct scale engraved upon it, to indicate its capacity in cubic feet, and the subdivisions of the cubic feet into minute fractional parts.

5. It is correctly balanced, and a part of the counterpoise suspended by a cord, passing over a spiral, preserves its equipoise at varying depths of its immersion in the water in the cistern.

6. The sides of the bell are maintained vertical in its ascent and descent.

7. The taps are lined with the anti-corrosive alloy; and the density of their rubbing-surfaces is so varied, that the friction is reduced to a minimum, and their soundness and durability is thus secured.

8. The different parts of the gas-holder are so perfectly adapted to each other, that, when put together as a whole, the instrument works easily, steadily, and correctly.

The daily use of the national standards for more than twelve months has shown them to be adapted to the purpose for which they were made, and has fully justified the opinion expressed by the Astronomer Royal, in his Report to the Lords Commissioners of Her Majesty's Treasury, that they were capable "of being applied to the verification of gas-measurers of every class and of gas-meters of every class," "entirely fitted to maintain the character of our national standards," and "as accurate as it is possible for human skill to make them."

II. MR. GEORGE GLOVER'S DIRECT TRANSFERRER.—The graduation of the gas-holders was a matter of difficulty, and involved nice scientific considerations. No method was known by which the cubic foot bottle or unit of measure could be used directly in the graduation of gas-holders, or their division into multiples and decimal parts of a cubic foot. The indirect method of applying it by what was termed a "transferrer," failed to give satisfactory results, and the simple and direct method by which the volume of air defined by the contents of the bottle was transferred into the gas-holder, being graduated, was found perfectly adapted to the purpose: this direct transferrer, invented by Mr. George Glover, is now generally used in the graduation of gas-holders for testing meters.

III. THERMOMETERS.—Thermometers are used of a peculiar construction, with elongated bulbs, by which sufficient delicacy of indication is insured. One is let into a hollow column in connection with the outlet of the standard gas-holder; and another is similarly situated in connection with the outlet of the instrument being tested. On one side of the thermometer is a scale for temperature; on the other there is a scale for corrections, arising from dilatation or contraction, occasioned by variations of temperature and moisture in the gas.

IV. THE PRESSURE GAUGES.—These have no joinings; they are made of one piece of glass tube, of large bore, and they have an enamelled scale, which is easily read.

V. TESTING TABLE.—Messrs. George Glover and Co.'s testing table is truly levelled, and so constructed as to be easily maintained in a horizontal position.

The same principles which have been successfully applied in constructing the national standard gas-holders, Messrs. G. Glover and Co. apply in the manufacture of their patent dry gas meter. *See* "Illustrated Catalogue," Class XXXI.

VI. THE PHOTOMETER.—Next in importance to correct standard measures for gas is the correct measurement of its illuminating power. This is determined by means of the photometer, in connection with which minute quantities of the gas under examination can be measured by Messrs. G. Glover and Co.'s patent dry gas meter, the dial of which is modified for this purpose, and the $\frac{1}{500}$th part of a cubic foot is measured each second.

VII. THE PNEUMATOMETER.—Illustrating the extreme accuracy in the measurement of gas by means of their patent dry gas meter, by a suitable modification of the index, it is used as a pneumatometer for measuring the capacity of the chest. This enables the experimental physiologist to determine with precision the quantity of carbon discharged from the lungs at each expiration, and the minute variations which occur in the same individual at different periods of the day and different seasons of the year. It is valuable to the physician and surgeon in the diagnosis of disease, in testing recruits for the army, and applicants for life assurance.

[2286]

GARRETT BROTHERS, *Tunstall, Staffordshire; Paddington, London.*—Ridging, roofing, and flooring tiles; plain and ornamental pavements, &c.

[2287]

GIBBS, G., *Brentford.*—Iron breakwaters, &c.

[2288]

GIBBS & CANNING, *Tamworth.*—Glazed stoneware sewerage pipes, &c.; fire-bricks, and terra cotta.

[2289]

GIBSON & TURNER, *Ball's Bridge, Dublin.*—Models of bridges.

[2290]

GILKES, WILSON, & Co., *Middlesbro'-on-Tees.*—Model of the Beelah Viaduct, Westmoreland.

Model to a scale of 1 inch to a foot of a railway viaduct over the river Beelah, on the line of the South Durham and Lancashire Union Railway, Westmoreland.

This viaduct is the lightest and cheapest combination of cast and wrought iron that has ever been adopted. It is much cheaper than stone, and for rapidity of construction is unequalled. The whole structure, 1,000 feet long, and 200 feet high, was erected in four months.

[2291]

GLOVER, GEORGE, & Co., *Ranelagh Works, Pimlico.*—Standard gasometers, &c. (*See page* 16.)

[2292]

GRAY, JAMES, M.D., *Glasgow.*—Coating to preserve iron, wood, and stone.

[2293]

GREENWOOD, JOHN, 10 *Arthur Street West, London Bridge.*—Patent india-rubber stops to make air-tight joints.

[2294]

HARTLEY, T. H., *Esher Street, Westminster.*—Sculptured specimens of marble-work.

[2295]

HAWKSHAW, JOHN, and WILLIAM HENRY BARLOW, 33 *Great George Street, Westminster.*—Model of suspension bridge proposed to be erected at Clifton.

[2296]

HEINKE BROTHERS, 79 *Great Portland Street, London.*—Sub-marine helmet, dress, and diving apparatus.

Obtained First Class Medals at the Great Exhibition, 1851, *and at the Paris Exhibition,* 1855.

Heinke Brothers have effected improvements in this apparatus, by which the diver is enabled to remain any length of time under water. It is now an invaluable aid in the recovery of property from wrecks; in subaqueous engineering, and in pearl and sponge diving. This firm are submarine engineers to the English, French, Russian, Spanish, Portuguese, Sardinian, Canadian, Peruvian, Brazilian, and Indian Governments.

Extract from the Report of the International Jury on the Paris Exhibition of 1855, *relative to Mr. E. Heinke's Diving Apparatus:—*

"The principal improvement which he has introduced consists in enabling the diver to remain under water when an accident occurs, such as the breaking of a glass, which would otherwise have allowed the water to penetrate into the dress."—Vol. ii., page 41.

[2297]

HELPS, ARTHUR, *Vernon Hill, Bishop's Waltham.*—Clays, various; terra cotta, brick, and tileware.

[2298]

Hemming, Samuel C., & Co., 21 *Moorgate Street.*—Samples of iron buildings and iron roofing.

INTERIOR OF AN IRON CHURCH.

[2299]

HOLLAND, W., *St. John's, Warwick.*—Apparatus for raising and lowering window sashes.

[2300]

HOOD & SON, SAMUEL, 68, *Upper Thames Street, and West London Iron Works, Notting Hill, London.*—Wrought-iron sashes, staircase, and baluster.

A cast-iron circular staircase, which can be made of any radius without strings or plates.

Staircase balusters, with adjusting caps, to suit various bevils. Wrought iron sashes and casements.

[2301]

HOWIE, JOHN, *Hurlford Fire Clay Works, N.B.*—Fire-bricks, troughs and mangers, chimney cans, vases, fountain, &c.

[2302]

INGHAM & SONS, WILLIAM, *Wortley, near Leeds.*—Fire-bricks, gas retorts, sanitary tubes, and terra cotta.

[2303]

JACKSON, R. W., *Greatham Hall, Durham.*—Model of West Hartlepool harbour and docks.

[2304]

JAMIESON, ROBERT, *Glasgow.*—Permeating timber, prevents dry rot; coating stone, wood, iron, &c., prevents decay.

[2305]

JENN, JOSEPH, JUN., 38 *Whittlebury Street, Euston Square.*—Jelly or cake moulds.

[2306]

JONES, WILLIAM, *Springfield Tileries, Newcastle, Staffordshire.*—Terra metallic ridging, roofing, and paving tiles, red, blue, and buff.

[2307]

Kennedy, Lieut.-Colonel J. P., *Torrington Square.*—Elements essential to railway success, illustrated by Baroda works.

[2308]

Knight, Bevan, & Sturge, 155 *Fenchurch Street, and Belvedere Road, Lambeth, and Northfleet, Kent.*—Portland cement.

A block of solid Portland cement weighing five tons, and samples of their manufacture are shown; also two large blocks composed of nine parts of shingle to one part of cement for breakwaters, shown in the open court of the eastern annexe, weighing respectively eight, and three-and-a-half tons. This cement, from its very superior hydraulic properties and relative cheapness to stone, is an article of great commercial importance. It is principally employed for blocks for breakwaters, harbour works, concrete work, stuccoing, plastering, bridges, cisterns, aqueducts, sewers, tanks, reservoirs, flooring, and paving, &c. &c.

[2309]

Laidlaw & Son, *Edinburgh and Glasgow*—Gas meters and fittings.

[1310]

Lawrence Brothers, *City Iron Works, Pitfield-street, London, N.*

Working Models.

Improved Warehouse Lifts, so constructed that when used for lowering goods, only the brake is required, the cage returning to the upper floors for a fresh load by means of a balance-weight. This Lift can be worked by one man from any floor.

Warehouse Crane, with expanding jib, for loading carts in streets where the pavements are wide.

Diving Bell, with signal apparatus and safety-valve, to prevent accidents from the breaking of the air hose. This model shows the apparatus as used by Messrs. H. Lee & Son, at the Dover Pier.

Travelling Crane, the traversing motions being worked from the crab. The bevilled wheels ordinarily used are dispensed with.

Lawrence's Patent Sluice, in which the pressure of the water is made to raise the sluice. The mode of construction is shown in the engraving.

A. The paddle, the top fitting the chamber, B.

C D. Small sluices or valves, connected together by a rod, so that when C is opened, D is closed.

F. Rod from paddle, A, to machinery, K.

G. Rod from valves C and D to machinery, K.

H. Passage to high level. On turning the handle of machinery, K, the valve, C, is opened, and D closed; the water in the chamber, B, immediately runs off to the low level, and the high level water, passing through channel H, presses against the piston plate attached to the paddle, A, and forces the sluice up, the rising being regulated by a brake. To lower the sluice, the machinery is reversed, and the rod, G, lowered, closing C, and opening D. The chamber, B, immediately fills with the high level water, and the sluice is forced down, closing the culvert.

Six sluices of large size, constructed as the model, are now at work at the Lavender Entrance of the Commercial Docks, Rotherhithe, and at the same Docks a similar plan has been adopted for opening the sluices on the gates of the Old Entrance Lock.

Improved Hydrant, or Fire-Cock, worked by a screw. All the parts are so arranged that they are not liable to get out of order or leak, even under very high pressure. This hydrant is used by H.M. War Department at many of their establishments.

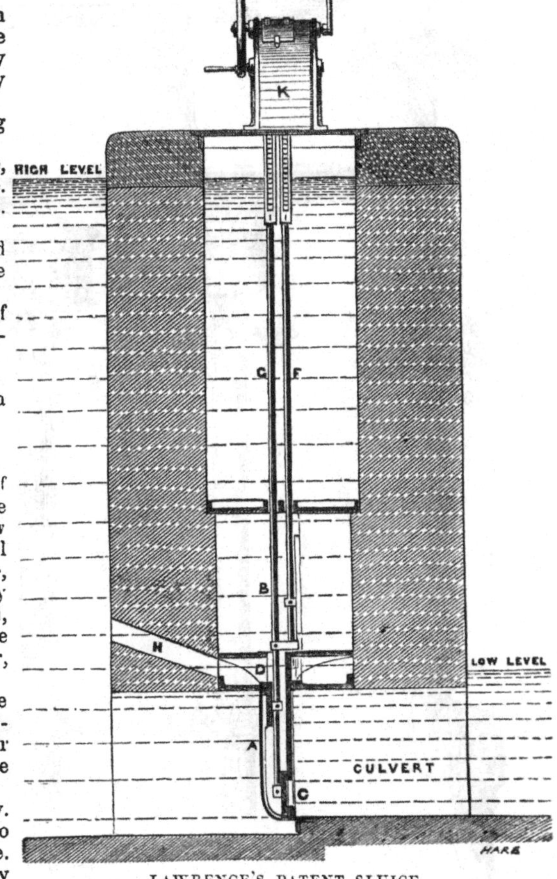

Lawrence's Patent Sluice.

[2311]

Lee, Son, & Smith, 16 *Upper Ground Street, Blackfriars, S.*—Limestone, lime; Portland and Scott's cement; Scott's plaster.

[2312]

Lucas, A., & Sons, *Fire-brick Works, near Gateshead-on-Tyne.*—Fire-bricks' goods.

[2313]

MACFARLANE, WALTER, & Co., *Saracen Foundry, Glasgow.*—Architectural cast-iron appliances, pipes, gutters, cresting, finials, &c.

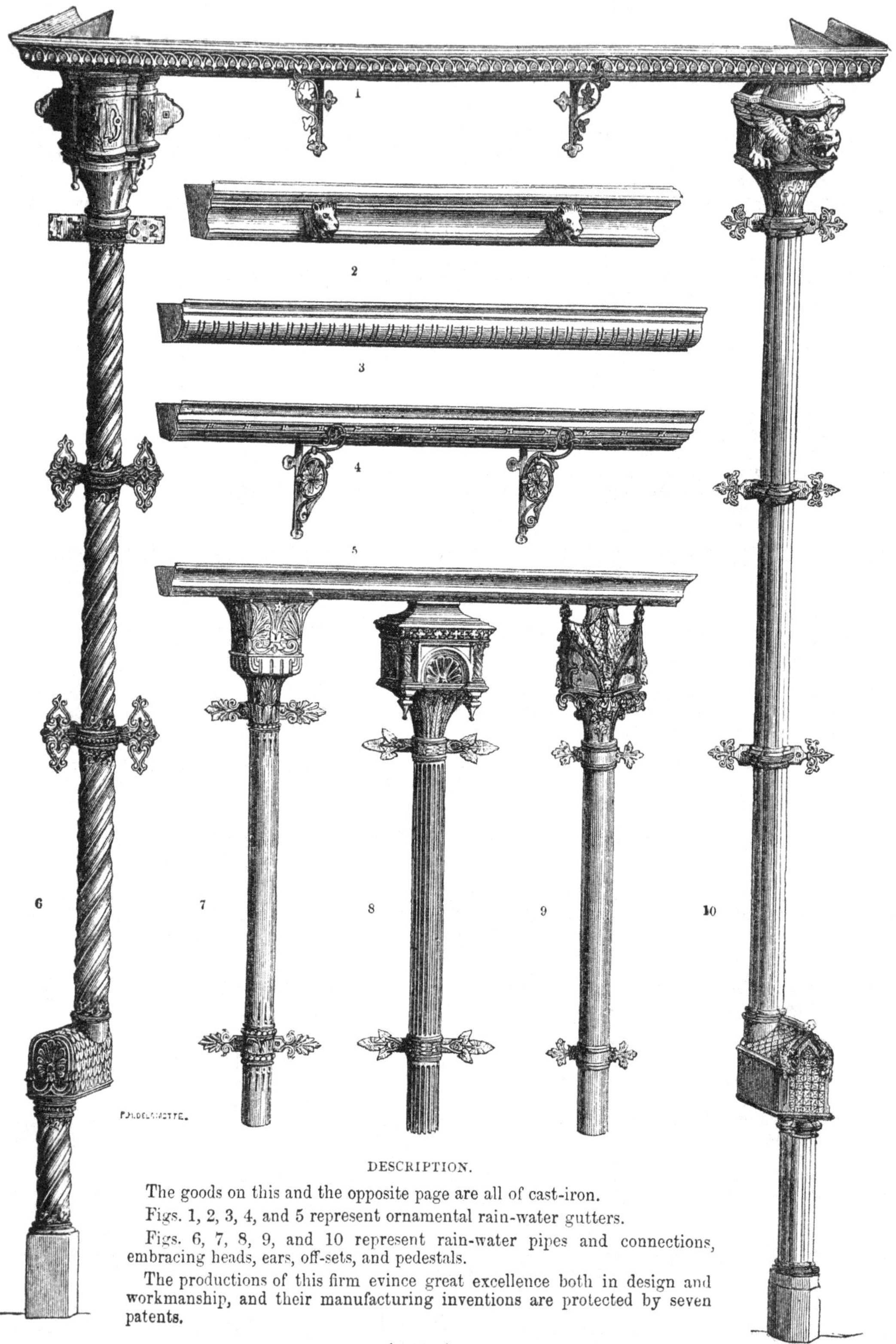

DESCRIPTION.

The goods on this and the opposite page are all of cast-iron.

Figs. 1, 2, 3, 4, and 5 represent ornamental rain-water gutters.

Figs. 6, 7, 8, 9, and 10 represent rain-water pipes and connections, embracing heads, ears, off-sets, and pedestals.

The productions of this firm evince great excellence both in design and workmanship, and their manufacturing inventions are protected by seven patents.

MACFARLANE, WALTER, & CO.—*continued.*

DESCRIPTION.

Figs. 1, 5, and 7.—Cresting for roofs, walls, balconies, gallery fronts, &c.
Fig. 2.—Weathervane.
Fig. 3.—Ridge-plate and cresting.
Figs. 4 and 6.—Bannerets.
Figs. 8 and 9.—Crosses.
Figs. 10, 11, and 12.—Ornamental ridge-plates.
Figs. 13 and 14.—Finials.

The decorative treatment of architectural cast-iron work is a matter worthy of consideration, and the examples exhibited by W. MACFARLANE and Co. illustrate their views on this subject.

[2314]

Macintosh, John, 40 *North Bank, Regent's Park.*—Samples of telegraphic cables.

[2315]

Maclaren, Robert, & Co., *Eglington Foundry, Glasgow, N.B.*—Four cast-iron pipes.

[2316]

Macneill, Sir John, LL.D., F.R.S., 23 *Cockspur Street.*—Model of bridge over the Boyne.

BRIDGE OVER THE BOYNE.

Malleable Iron Lattice Bridge over the river Boyne, near the town of Drogheda, on the line of the Dublin and Belfast Junction Railway, completed in 1855.

Total length	1750 feet.
Height of under side of girders above high water	90 "
Span of centre openings	264 feet.
Span of side openings	138 " 8 in.
Span of stone arches	60 "

Weight of iron in lattice work about 700 tons.

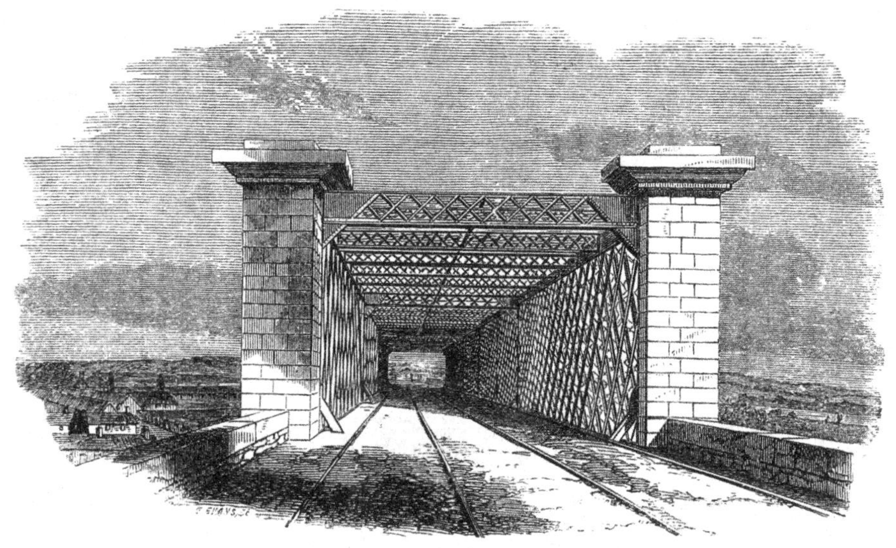

[2317]

Maw, George, *Benshall Hall, Shropshire.*—Manufactures of clay, &c.

[2318]

Maw, George, *Broseley, Shropshire.*—Collective series of artificial productions, illustrating the clay manufactures of the Shropshire Coal Field. (*See pp.* 26, 27.)

[2319]

M'LINTOCK, WILLIAM, 38 *Kirk Street, Gorbals, Glasgow.*—Hydraulic (Arden) lime; bituminous shale, containing paraffin oils.

[2320]

MEARS, GEORGE, & Co., 267 *Whitechapel Road.*—Bell 15 cwt., and self-acting apparatus for striking ditto.

[2321]

MOREWOOD & Co., *Dowgate Dock, London, and Lion Works, Birmingham.*—Galvanized iron roofs, farm buildings, &c.

MOREWOOD & Co. manufacture patent galvanized tinned iron and galvanised iron, plain or corrugated, curved, and in tiles, of all gauges. Black or painted corrugated iron, galvanized or black-cast gutters, pipe, &c. They are also the makers of Morewood's patent continuous galvanized iron roofing, which is cheaper than felt. Full particulars may be obtained on application.

Mining sheds, engine sheds, farm sheds, and every description of galvanized farm building, are constructed by these exhibitors.

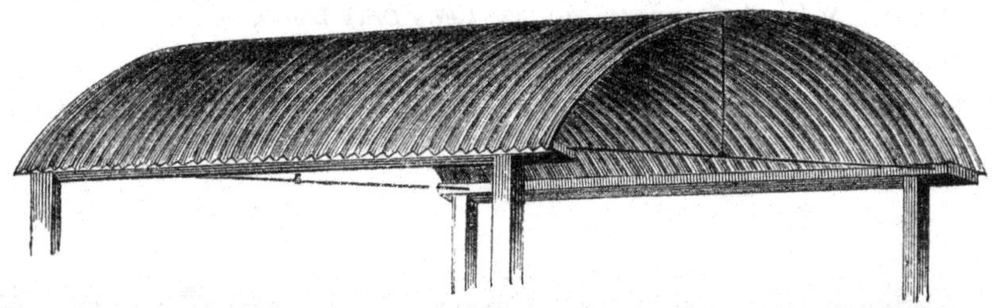

Specimens of the following are exhibited:—

Patent galvanized tinned iron and galvanized iron, plain and corrugated.

PATENT CONTINUOUS GALVANIZED IRON ROOFING.—This roofing is cheaper than felt when mixed complete. Full particulars may be learned on application, and estimates will be supplied for roofs, churches, and every description of galvanized iron building.

[2322]

MORTON, FRANCIS, & Co., *Liverpool.*—Patented improvements in permanent railway fences; iron telegraph poles; galvanized corrugated iron roofs, buildings, &c. (*See pages* 28 *to* 30.)

[2323]

MURRAY, JOHN, 7 *Whitehall Place.*—Cellular wine bin, or *porte-bouteilles.*

[2324]

NORMAN, RICHARD & NATHAN, *Burgess Hill, Sussex.*—Specimens of ornamental bricks, tiles, &c. &c.

[2325]

PAINE, MRS., *Farnham, Surrey.*—Artificial stone; terra cotta bricks, pipes, &c., made from soluble silica (patent).

[2326]

PART, J. C., 186 *Drury Lane.*—Martin's cement and plaster of Paris.

Obtained Prize Medal in 1851.

PART'S MARTIN'S CEMENT is the best internal cement in use, and can be painted upon within twenty-four hours of its application. A saving of 45 per cent. in the bare cost of material will be effected by using this cement. Manufactured only by J. Cumberland Part, 186 Drury Lane, London, and at Derby.

PLASTER OF PARIS.—Coarse, fine, and superfine.

[2327]

PATENT BITUMENIZED WATER GAS AND DRAINAGE PIPE COMPANY, 14A *Cannon Street, E.C.*—Inoxidable pipes, one-fourth weight of iron pipes.

[2328]

PEAKE, THOMAS, *The Tileries, Stoke-upon-Trent, and City Road Basin, London.*—Terrometallic bricks, tiles, &c.

[2329]

PELD, B., *Birkenhead.*—Model of iron bridge.

MAW, GEORGE, *Shropshire, Broseley.*—Collective series of architectural productions, illustrating the clay manufactures of the Shropshire Coal Field, classified and arranged by GEORGE MAW, F.S.A., F.L.S., on behalf of the undermentioned exhibitors.

BURTON, Messrs. JOHN AND EDWARD, *Ironbridge.*

COALBROOK DALE COMPANY, *Lightmoor Works, Coalbrook Dale.*

DAVIS, Messrs. GEORGE & Co., *Broseley.*

DOUGHTY, Mr., *Jackfield, near Broseley.*

EVANS, Mr. ROBERT, *Jackfield, near Broseley.*

EXLEY, Mr. WM., *Jackfield, near Broseley.*

LEWIS, Mr. G. W., *Jackfield, near Broseley.*

MADELEY WOOD COMPANY, *Madeley Wood Fire Brick Works.*

MAW, Messrs., & Co., *Benthall Works, Broseley.*

SIMPSON, Messrs., W. B., & SONS, 456 *West Strand. London Agents for* MAW & Co.

THORN, Mrs., *Broseley.*

CHIMNEY PIECE, COMPOSED OF ENAMELLED TILES AND STONE.

Division A, roofing materials.

Common, plain, and ornamental flat roofing tiles, unglazed, glazed, and enamelled, of various patterns and colours. (Arrangements designed by Mr. Digby Wyatt.)

Pantiles.

Roof crestings, plain, flanged, and with fixed and loose ornaments, brown, black, glazed, and enamelled. Ventilating roof crest tiles.

Various hip and gutter tiles, and flanged hip crestings.

MAW, GEORGE—*continued.*

Division B, *paving materials.*

Illustrations of the revival of pictorial mosaic, consisting of a pictorial mosaic pavement, 13′ 3″ × 10′ 9″. Subject, "Apollo, and the Four Seasons," designed by Mr. M. Digby Wyatt, and manufactured by Maw and Co. For sale. Can be adapted to any size to suit particular dimensions.

Facsimile copy of head, from ancient Roman pavement at Bignor, in Sussex. Maw & Co.

Examples of tesselated pavements. Maw & Co.

Examples of geometrical mosaic pavements. Maw & Co.

Examples of encaustic tile pavements, and of combinations of plain and encaustic tiles. Maw & Co.

Moresque mosaics for wall linings. Maw & Co.

Various combinations of enamelled Majolica tiles, for wall, cornice, bath, and fire-place linings. Maw & Co.

Box of loose examples of various mosaics and tiles, manufactured by Maw and Co.

Examples of common square and hexagonal paving tiles, loose, and in combinations. Maw & Co.

Drawings of mosaic and encaustic tile pavements, executed by Maw & Co.

Stable paving bricks. John and Edward Burton.

Malt kiln, flue tiles, and bearers.

Division C, *draining materials.*

Flanged and unflanged sanitary tubes, from four to twelve inches in diameter, with bends, junctions, traps, &c.

Agricultural draining pipes and horse-shoe pipes.

Gutter bricks, of various sizes.

Flood-bolt bricks, for irrigation. Mrs. Thorn.

Eave spouting bricks and cornice, with inner and outer returns, manufactured by Mr. Exley.

Division D, *fire-bricks, furnace materials, stove fittings, &c.*

Common fire-bricks, of various forms, arch bricks, bull-heads, pin bricks, soap bricks, &c.

Various fire-clay blocks, used in the construction of iron furnaces.

Fire lumps, of various sizes.

Fire squares, of various sizes.

Fire-place chucks, cooking-apparatus-slabs, and oven slabs, of various forms and sizes.

Grate backs, sundry patterns. Coalbrook Dale Company.

Gas retorts, retort bricks and covers.

Arnott stove-pot linings. Coalbrook Dale Company.

Fire-clay kiln bars. Madeley Wood Company.

Cundy's hot air stove fittings. Coalbrook Dale Company.

Rebated flue bricks, key bricks, and various materials used in porcelain works. Messrs. Burton, and Madeley Wood Company.

Division E, *bricks and materials used in the construction of walls.*

Common, pressed, and moulded bricks and blocks for walls, arches, copings, window jambs, gables, chimneys, plinths, cornices, &c.

Boosey or manger bricks.

Various glazed and enamelled arch and other bricks. Maw & Co.

Log bricks for pit shafts.

Penneystone Mount bricks, made from the refuse of ironstone pits. Mr. Doughty.

Division F, *accessaries to the decoration of buildings and various articles not included in the other divisions.*

Various terra cotta architectural decorations.

Various examples of enamelled terra cotta decorations, from the designs of Mr. Digby Wyatt, including pillar caps, chimney tops, round columns, arch bricks, &c. Manufactured by Maw & Co.

Examples of terra cotta balustrading. Mrs. Thorn, and Coalbrook Dale Company.

Various terra cotta chimney tops.

Flower borders, edgings, tiles, and returns.

Hot-house and vinery squares and channels.

Sundry terra cotta vases, flower pots, stands and pedestals, orchid pots, mignionette boxes, flower and orange tree boxes, &c. Designed by Mr. Kremer. Manufactured by the Coalbrook Dale Company.

Step bricks.

Chimney piece, composed of enamelled tiles and stone. Manufactured by Maw & Co. Designed by Mr. M. Digby Wyatt. The stonework executed by Mr. Richard Yates, builder, Shiffnal. For sale. *See Engraving.*

Orange tree, and mignonette boxes, the former having a space between the slate lining and tiles to keep the soil cool, composed of Majolica tiles, set in electro-bronzed framing. Manufactured by Maw & Co., from designs by Mr. M. Digby Wyatt. For sale.

Division G, *raw materials.*

Various specimens of clays and other materials from the Shropshire Coal Field, in the native and burnt state; also made up into squares to show their relative shrinkages.

Section of Shropshire Coal Field, showing the disposition of clay beds, and other materials used in the manufacture of the series. Mr. G. Maw.

Morton, Francis, & Co., *Liverpool.*—Patented improvements in permanent railway fences; iron telegraph poles; galvanized corrugated iron roofs, buildings, &c.—Manufacturers of every description of fire-resisting iron roofs and buildings for agricultural and colonial purposes; iron roofs for docks, railways, shipbuilding yards, &c.; cranes, weigh-bridges, wire ropes, electric telegraphs, &c.

Francis Morton & Co.'s patented improvements in the construction of permanent efficient fences for home and foreign railways, parks, pleasure-grounds, &c., have obtained the Silver Medals and highest commendations of all the principal Agricultural Societies of the United Kingdom.

Fig 10.

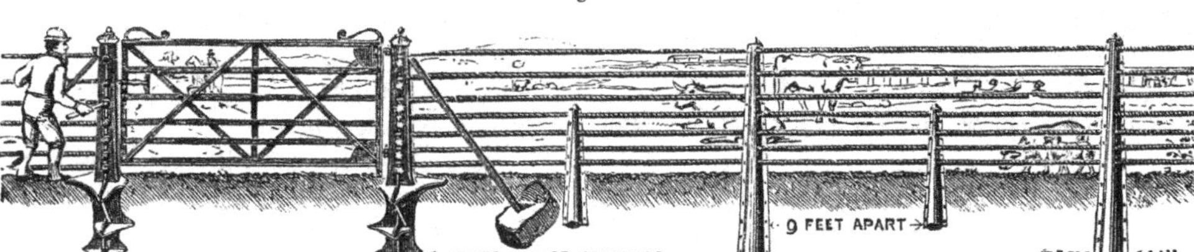

Fig. 5.

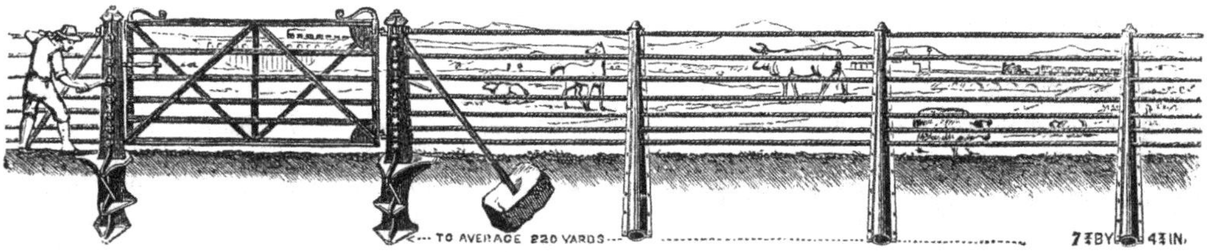

1.—Strained Cable Fence, Strands, and Posts, all galvanized; fitted with Francis Morton's patent winding straining pillars, and galvanized tapered oval iron posts—the strongest, most rigid, and durable form of iron fence known. Price 1s. 10d. per yard (see Fig. 5) and 2s. 2d. per yard. (See Fig. 10.)

Fig. 8.

FRONT VIEW.

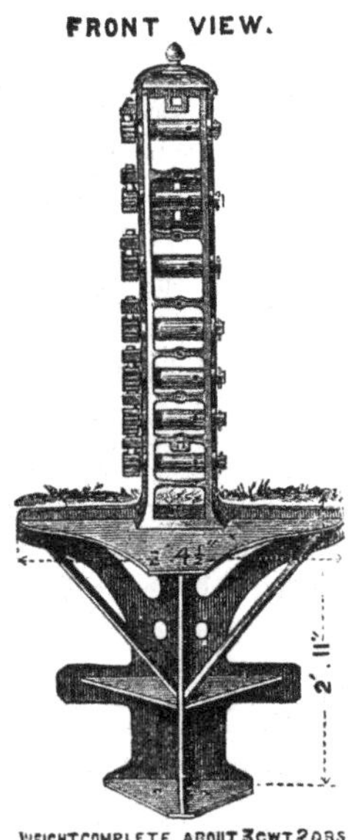

2.—Examples of Patent Winding Straining Pillars, in cast and galvanized hammered iron. (Specially patented for India.) These powerful fencing pillars are self-acting, and specially adapted for straining every description of fencing wire, and areprepared for hanging field and level-crossing gates. They supersede skilled labour, and save more than half the cost of fixing wire fencing by any other method. Each pillar is entirely complete in itself, will stretch, and always keep perfectly rigid, without expense, from 400 to 500 yards of fence. Price 38s. to 45s. each. (See Fig. 8.)

3.—Patent Winding Straining Brackets (specially patented for India), for attaching to wood main posts, are extremely portable for long inland transport; save two-thirds the cost of fixing by the ordinary mode; while, like the patent winding straining pillars, they double the value and efficiency of the fence. (See Drawing.)

4.—Examples of Patent Galvanized Tapered Oval Iron Fencing Posts (see Fig. 9) are the most complete and permanently efficient iron fencing posts in use. These galvanized iron posts are perfectly inflexible in every direction; they cannot be bent, broken, or thrust aside by trespassers, or by the heaviest cattle. They are self-fixing, requiring no stone or wood blocks to fasten them.

Fig. 9.

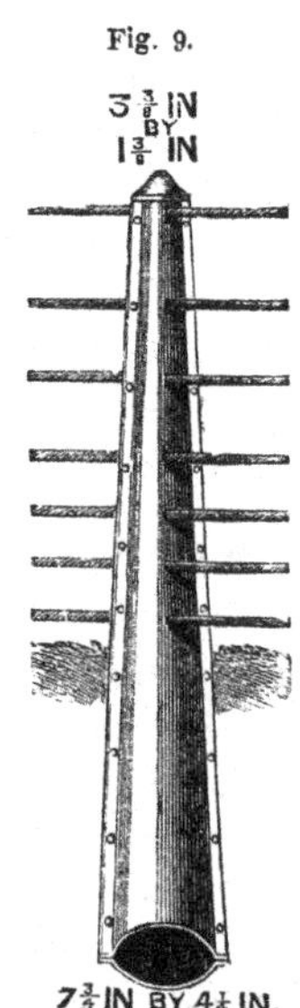

5.—Model of Wrought Iron Side Stay, for curves for bends in wire fencing. These, when fitted in the exhibitors' patent galvanized oval iron fencing posts, are concealed below the ground, thus removing a great defect and eye-sore hitherto inseparable from strained fencing.

6.—Examples of Best Prepared Galvanized Signal Cords, as originally applied and manufactured by them in 1846.

7.—Examples of Best Prepared Galvanized Fencing Strands, which have stood the test of 16 years' wear on railways, and are still in good condition.

8.—Examples of Wire Ropes for collieries, mines, railways, &c.

Drawings.—Illustrations of the various applications of the exhibitors' patent strained cable fences and galvanized iron manufactures.

MORTON, FRANCIS, & CO.—*continued.*

9 & 10.—HALF-SIZE MODELS OF PATENT GALVANIZED OVAL TAPERED IRON TELEGRAPH POLES, specially patented for India. They possess great facility of transport, great strength and rigidity when fixed, with simplicity of erection, superior electric action, reduced cost; and they save in fixing one-half the usual labour. (See Figs. 6 and 7.) *For further description, see next page.*

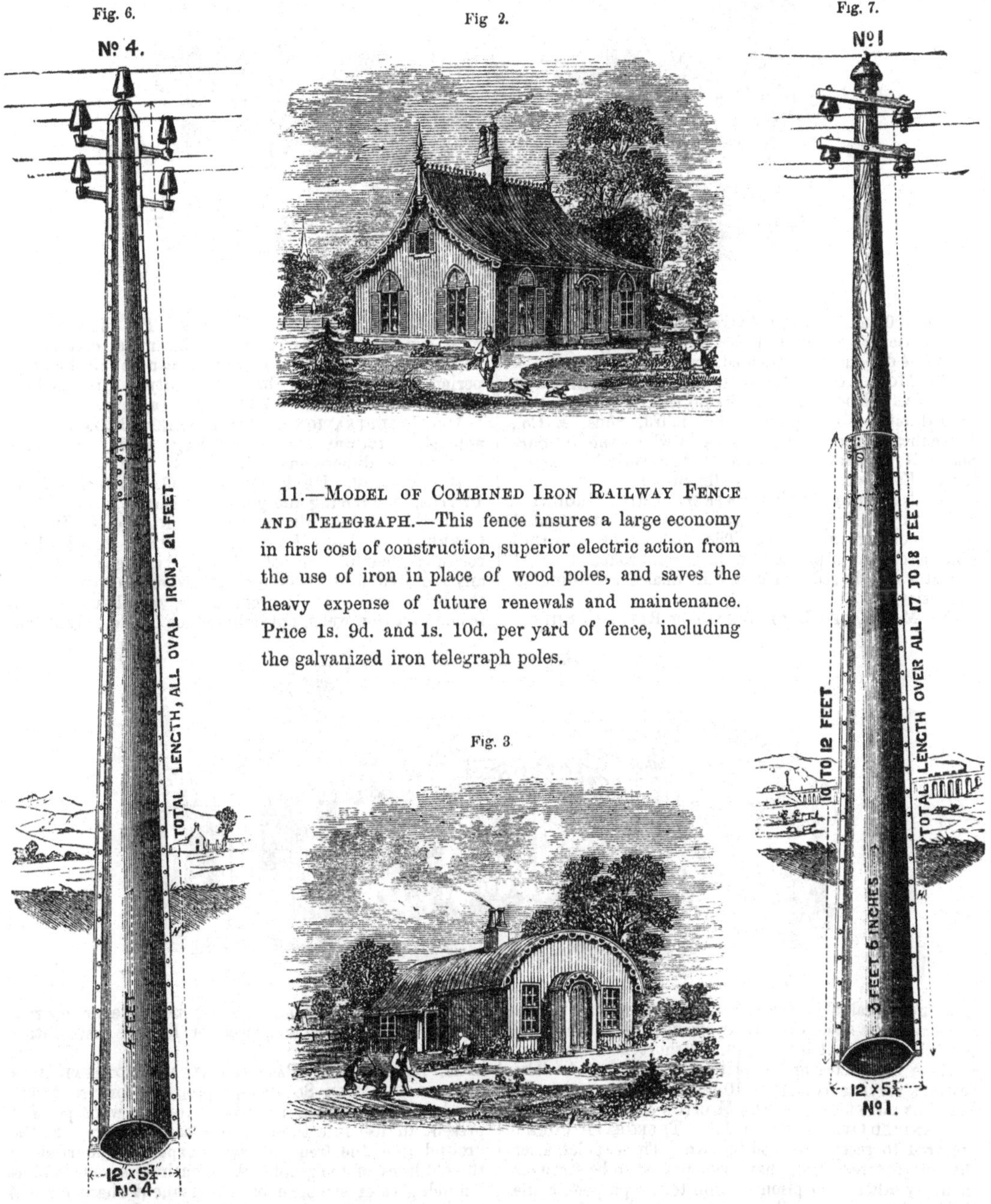

11.—MODEL OF COMBINED IRON RAILWAY FENCE AND TELEGRAPH.—This fence insures a large economy in first cost of construction, superior electric action from the use of iron in place of wood poles, and saves the heavy expense of future renewals and maintenance. Price 1s. 9d. and 1s. 10d. per yard of fence, including the galvanized iron telegraph poles.

12.—MODEL OF "GARDENER'S OR FARM BAILIFF'S COTTAGE OF GALVANIZED CORRUGATED IRON."—Price, erected complete, with five rooms and entrance porch, £120. (See Fig. 3.)

13.—MODEL OF "ORNAMENTAL SHOOTING LODGE AND COUNTRY HOUSE" OF GALVANIZED CORRUGATED IRON.—Price, erected complete, with seven rooms, £350 to £400. (See Fig. 2.)

NOTE.—These iron buildings are easy of conveyance and erection where carriage and labour are very expensive, and, as constructed by Francis Morton & Co., possess all the comfort of stone or brick buildings.

MORTON, FRANCIS, & CO.—*continued.*

14.—DRAWING OF "FARM YARD," wholly covered with Francis Morton and Co.'s galvanized corrugated iron fire-resisting roofs. These iron roofs are economical in first cost, and are permanent. They save all cost of future repairs and maintenance, are not injured by violent winds, removable without injury, and, above all, are fireproof. Crops thus housed are protected from the ravages of vermin, wet, hail-storms, fire, and waste, and are thereby brought to market in the finest condition. Price of galvanized corrugated iron roofs to cover an area of 100 feet square, delivered and erected complete on proprietor's wood wall plates and uprights, £370. All other dimensions estimated for.

Fig. 1.

15.—MODEL OF "GALVANIZED CORRUGATED IRON SHIP-YARD ROOF," 400 feet by 80 feet, as now erecting by exhibitors for Messrs. John Laird, Sons, & Co., Birkenhead, over their new dock in which the armour-plated Government war vessel "Agincourt" is being built. Estimates according to dimensions.

16.—DRAWING OF "VOLUNTEERS' DRILL GROUND," 300 feet by 70 feet, covered with galvanized corrugated self-supporting iron roofs. The method of construction here shown provides a large unbroken covered area at a greatly reduced cost. Estimates according to dimensions.

17.—MODEL OF "DOCK WHARF OR RAILWAY SHED," covered with galvanized iron tiles. The lightest and most convenient form in which this metal can be exported or used in this country; it can be laid by inexperienced workmen without difficulty. Price: Tiles 3 feet by 2 feet, £10 per 100, packed in cases.

18.—ILLUSTRATION OF "GALVANIZED IRON ROOFS," applied to railway stations of large span. Estimates according to dimensions. See Fig. 4.

19.—VARIOUS EXAMPLES OF "GALVANIZED IRON PLATES," for roofing and general purposes.

20. EXAMPLE OF FRANCIS MORTON & CO.'S "HEAVY CORRUGATED IRON FLOORING PLATES," for bridges, roadways, floors of fireproof buildings, &c., prepared to bear any required load. These plates, when substituted for the brick arches of fireproof floors, or when used for inside partition walls, not only reduce inside weight, but save much valuable space, and impart great lateral stiffness to the structure; their strength is not limited, as in plain wrought-iron plates, to short bearings only, as these heavy corrugated plates are prepared to take bearings varying from 4½ to 10 feet.

Fig. 4.

21. EXAMPLE OF FRANCIS MORTON & CO.'S "PATENT GALVANIZED OVAL TAPERED IRON TELEGRAPH POLE," prepared to receive a wood top, &c. These poles, after the most severe testings, have been proved to be stronger than any other description of iron telegraph pole, while their lightness especially recommends them for long overland transport. The weight of these poles is only from 95lbs. to 100lbs. each, if made throughout entirely of iron, and 17 feet long. These poles are all manufactured of iron in its most enduring form, both as regards strength and its power to resist deterioration; they are also free from the slightest flexibility, no wear and tear can arise at the joints, and having no loose or separate parts, nothing can be misplaced or lost in distribution overland.

22. EXAMPLE OF "PATENT GALVANIZED OVAL IRON TELEGRAPH POLE SOCKETS," prepared for use where wood poles are abundant. By putting the wood poles at present in use into these sockets as they decay at the ground line, the frequent heavy expenses incurred by the old lines of telegraph will be entirely saved; and as a much greater strength of a permanent character will be obtained by this means at the point where the strain is severest and decay the most rapid, the duration of the present wood poles will be more than doubled, and for the future a cheaper description of wood pole may be used, besides preventing accidents occasioned by storms, &c. These sockets are free from the uncertainties inseparable from cast-iron.

[2330]

PERKINS, M. A., 6 *Francis Street, Regent Square, London.*—Warming and ventilating buildings by hot water.

[2331]

PILLAR, S. J., 91 *Newman Street, Oxford Street.*—Model bridges, pocket umbrella, &c.

[2332]

PORTER, J. T. B., & Co., *Lincoln, and* 7 *John Street, Adelphi, London.*—Gas works, for private use; drawings of ditto, for towns, villages, &c.

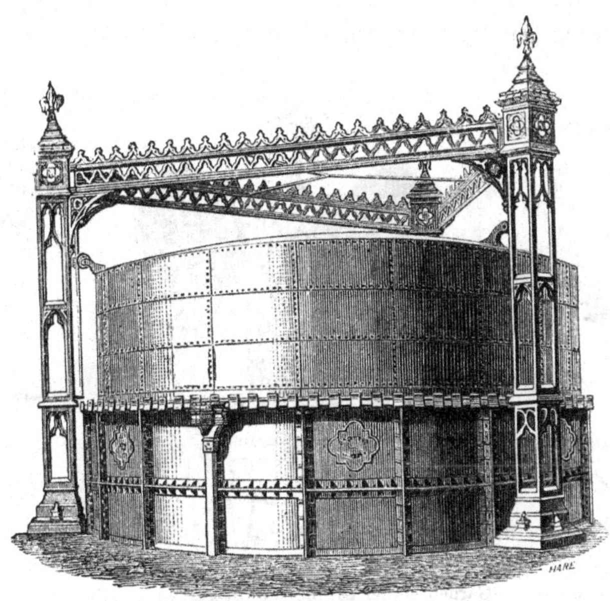

SELF-ACTING GAS-HOLDER, FRAMEWORK, PILLARS, &c.

The exhibitors are manufacturers of patent portable coal gas works, by which any villa, farmhouse, mansion, country residence, railway station, church, or other isolated building where 10 lights and upwards are required, can be lighted by gas made upon the premises, the gas thus produced being cheaper than that obtained from ordinary public gas works, and so pure that it will not injure the most delicate fabric, or tarnish the most costly gilding. The cost of gas varies with the price of coal, from 2*s.* to 4*s.* per 1,000 cubic feet—prices one-sixth the cost of candles or oil for equal amounts of light. The attention required by these works is trifling, and need not incur additional labour, as any groom, under-gardener, or other man-servant, can work the apparatus quite efficiently, in addition to his ordinary duties. Any kind of bituminous coal can be used, and the refuse of the household heap is available, the small coal being adapted to the purpose. The two sets of apparatus exhibited are for 10 and 50 lights respectively.

An apparatus for the production of gas from cannel coal, peat, or peat mixed with oil is also exhibited as particularly adapted to those countries where ordinary bituminous coal is scarce, and cannel coal may be obtained at less than £5 per ton.

Drawings of improved gas works for towns and villages are exhibited, showing the arrangement of the various portions of small public gas works.

The above engraving is an illustration of a self-acting gas-holder, framework, pillars, &c., designed and erected by Messrs. Porter & Co. for Messrs. Hodges & Co., at their distillery, Lambeth; showing the application of design to the purposes of an ordinary gas-holder.

Messrs. Porter & Co. have erected and furnished from 100 to 200 gas works to which they can refer, and have also been honoured with prizes and medals from the Highland and Agricultural societies of Scotland and other societies in England, for their small gas works adapted to country residences and farm buildings.

[2333]

Ramage, Robert, 55a *Holywell Street, Millbank, Westminster, S.W.*—Glass and metal patent and other ventilators, &c.

[2334]

Rennie, G., & Son, 6 *Holland Street, Blackfriars.*—Model of system of docking, in connection with floating gravingdocks.

[2335]

Reynolds, William, *Sheffield.*—Artificial stone; metallic mortar for building, painting, and plastering.

[2336]

Rosher, F. & G., *Ward's Wharf, Blackfriars, and Chelsea, London.*—Garden ornaments of artificial stone, and garden edgings in various materials.

The following are exhibited:—Garden border edging tiles in Terra-cotta, Terro-metallic, and Red ware.

Articles in Artificial Stone.

Pair of lions.
Eagle.
Maltese vase.
Antique vase (Roman).
Acanthus vase.
Tulip vase.
Greyhound and leveret.
Triton boy.
Boy with dolphin.
Octagon gate terminal.
Trusses.
Specimens of balustrading.

[2337]

Scott, M., 26 *Parliament Street.*—Models of timber breakwater; submarine foundations; and a diving apparatus.

[2338]

Siebe, Augustus, 5 *Denmark Street, Soho, London.*—Diving apparatus, as manufactured for Her Majesty's Board of Admiralty, &c. (*See page* 33.)

[2339]

Siliceous Stone Company, Patent, *Cannon Row, Westminster, and Works at Ipswich.*—Articles in artificial stone; specimens of natural stone indurated by Ransome's process.

[2340]

Simmons, George, 7 *New Palace Yard, Westminster.*—Simmons' patent gas and water connector, full size. (*See page* 34.)

SIEBE, AUGUSTUS, 5 *Denmark Street, Soho, London.*—Diving apparatus, as manufactured for Her Majesty's Board of Admiralty, the Board of Ordnance and Crown Colonies; Imperial Navy of France and Ponts et Chaussées; Imperial Navy of Russia, Sweden, Turkey, America, and other maritime powers. Spring weighing machines, sportsman's stilyard, rotatory pump and syringe, self-pressure cocks, paper knotting machines, &c.

Obtained First-class Medals at the International Exhibitions of 1851, 1855.

The art of diving and remaining under water for a lengthened period is a subject which has occupied the attention of scientific men since the earliest records of history; but it was not until the early part of the 18th century that it began to assume a practical form in the shape of the cumbersome diving bell, which, although appearing in various forms in the hands of different inventors, was found inapplicable to the removal of wrecks and deep sea diving. It was not until the exhibitor (about 1830), in conjunction with Mr. Deane, invented the first diving equipment, professionally known as the "open diving helmet," that operations under water could be carried on with any degree of success. As the invention was not patented, and was in great demand, many imitators soon entered the field, but without introducing any new feature; it was left for the exhibitor to complete what he had begun by inventing the close helmet and dress, in 1837—the principle now generally adopted, by which all danger of water entering the dress or helmet was removed. This he speedily followed up by adding the segmental neck screw, by means of which the head of the helmet can be removed by an eighth of a turn; also, the safety valve, to prevent water entering the dress in case of accident to the pipe; also by strengthening the pipes with a cylindrical coil of wire; adding water cistern to prevent the heating of the air pump cylinders, and many minor improvements; by the aid of which, the late Sir Chas. Pasley, C.B., was enabled, from 1839 to 1844, to carry on successfully the submarine operations to clear the anchorage at Spithead, and remove the wrecks of the "Royal George" and "Edgar," sunk respectively in 1782 and 1711. Although some hundreds of the exhibitor's improved close helmet have been twenty years in use, it is satisfactory to state that no death is recorded to have taken place from any cause connected with the apparatus.

[2341]

SLACK & BROWNLOW, *Manchester.*—Self-acting cistern filter.

[2342]

SMITH, ARCHIBALD, *Princes Street, Haymarket.*—Door spring for swing door; weather-tight casement fastening and water bar.

[2343]

SPARKES, J., 308 *Regent Street.*—Upright bench, for a working shoemaker; self-acting prismatic ventilator, for hall.

[2344]

STEPHENSON, WILLIAM, & SONS, *Newcastle-on-Tyne.*—Fire-clay, gas retorts, and fire-bricks, &c.

[2345]

STEWART, D. Y., & CO., *Glasgow.*—Cast-iron pipes.

[2346]

STUTTER, C., *Woolpit, Suffolk.*—White and red facing bricks, stable clinkers, and other kiln goods.

[2347]

SZERELMEY, NICHOLAS CHARLES, *Laboratory, New Palace, Westminster, Pannonia Leather Factory, Park Road, Acre Lane, Clapham.*—Arabian, zopisso, and granitic preserving and indurating compositions. (*See page* 34.)

SIMMONS, GEORGE, 7 *New Palace Yard, Westminster.*—Simmons' patent gas and water connector, full size.

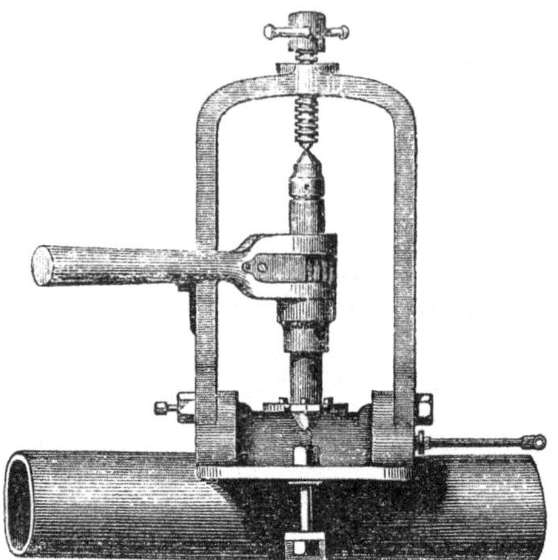

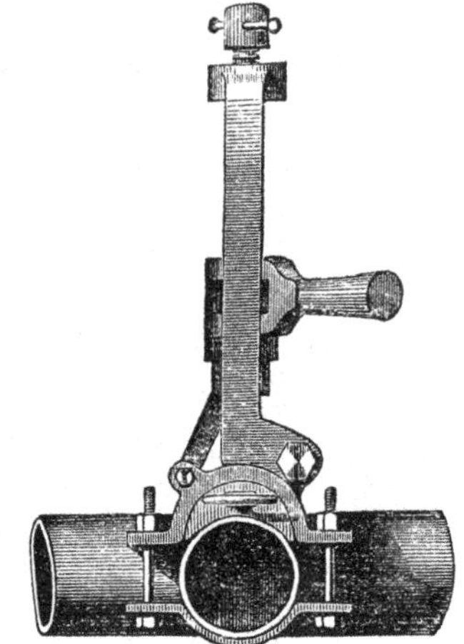

The escape of gas occasioned by the present mode of connecting services to mains, is costly to the gas companies, highly dangerous to the workman engaged, and hazardous to the property in the neighbourhood.

The gas companies now lose about one-fifth of the gas delivered from their works; and when it is remembered that gas under only one inch pressure escapes at the rate of 5,000 cubic feet per hour from a hole made to receive an inch and a half service, it is easily understood what a large proportion of this loss must be attributed to the service laying. The fire in Wood Street, City, February 27th, 1859, resulted from this operation, when property to the extent of upwards of £100,000 was destroyed.

At a meeting of the Metropolitan Association of Medical Officers of Health, held at Whitehall, Dr. Aldis stated "That the use of this machine would avert the risk of human life, as he could fairly testify, having seen it in operation in Horseferry Road, Westminster. One workman had been knocked down seven times in attaching service pipes, from the pernicious effects of the escape of gas on the present objectionable system."

This machine is the only connector whereby
can be avoided. It is so simple and compact
most unskilful hand can perform the work in
minutes, without loss or danger of breaking
Dr. Lankester, remarking upon this machine at
meeting, said "that simplicity indeed was
highest recommendations, as it was in other i
and the one before them would tend greatly t
the supply of gas and water."

At present, no services can be attached
charged with water; but by the use of this ma
inhabitants may secure to themselves a contin

It is useful for putting in temporary valves
piece is required, or where a main is being re
the gas or water in the same.

It is also convenient for steam pipe connecti

Connectors in stock with tools complete for
services, $\frac{3}{8}$ and $\frac{1}{2}$ inch . . . .
Ditto, $\frac{3}{4}$ and 1 inch . . . .
Ditto, $1\frac{1}{4}$, $1\frac{1}{2}$ and 2 inch . . . . .
Any other size can be made to order.

J. COWDY, Manufacturer, 3A Bond Court, Walbrook, London.

SZERELMEY, NICHOLAS CHARLES, *Laboratory, New Palace, Westminster, Pannonia Leather Factory, Park Road, Acre Lane, Clapham.*—Arabian, zopissa, and granitic preserving and indurating compositions.

1. *Wooden Railway Sleepers and Building Timber*, prepared so as effectually to resist dry rot and other decay.

The sleepers exhibited have been severely tested during the last ten years; some have been buried in the ground that period, some have been immersed in the sea for four years and a half. None of them exhibit the slightest indications of decay. The application of the process presents no difficulty, is remarkably cheap, and can be carried on in any part of the world without great expenses.

2. *Zopissa Composition for preserving iron and wooden ships and vessels.*—This valuable composition, in the case of wooden vessels, supersedes the use of copper sheathing, tar, and paint; it effectually closes the pores of the wood, excluding the air, and preventing the absorption of water. It forms a smooth bronze brown enamel surface, preventing the ravages of worms (*Torredo navalis*) and the attachment of barnacles. In iron vessels it completely prevents rust both within and without, and effectually closes the joints of the plates. It will last three times as long as all paints hitherto invented and used on iron vessels.

3. *Granitic Composition*, to be used as a paint for preserving iron from rust, and timber from decay. Is applicable to painting carriages, doors, chairs, corrugated iron roofs, iron houses, viaducts, bridges, railings, gutters, tanks, water and gas pipes, shutters, iron and wooden fences, telegraph posts, iron guns and shot, whether in the open air, under ground, or otherwise. It never requires renewal, is applicable and effective in all parts of the world, and under every change of climate.

4. *Bricks* composed of sand and chalk, or sand and lime, or pure chalk. These bricks are made without burning, they are stronger and cheaper than ordinary bricks, and can be made with great rapidity by a machine, which will turn out about 16,000 or 18,000 a day. The bricks are the invention of Mr. N. C. Szerelmey, and are manufactured for this country only by Messrs. Bodmer Brothers, Newport, Mon., and 2, Thavies' Inn, Holborn, London.

5. *Silicate Zopissa Composition*, for preserving public and private buildings of stone, brick, stucco, or cement statuary and other similar works of art, from atmospheric and other corroding and destroying influences.

This composition will at once arrest the progress of decay or chemical change, penetrate the surface, fill in and consolidate it, and, by its cohesive powers, permanently seal it from the action of free gases, atmospheric air, and damp.

It has been successfully applied, amongst other buildings, to the inner courts of the New Palace of Westminster, the principal entrance of the Bank of England, the whole of the Kennington and Regent Square Churches, the Gresham Club House, City, the interior of St. Paul's Cathedral, &c. &c. &c.

[2348]

TAYLOR, WILLIAM J., 5 *Church Street, Chelsea.*—Specimen of plastering, for external purposes, in Portland cement.

Improved and patented method of finishing Portland cement for walls of buildings and other erections, without the use of surface colour.

[2349]

THORN & Co., *Grosvenor Row, Pimlico, S.W.*—Atmospheric bells; Trinidad asphalte; specimen stone of old Westminster Bridge.

[2350]

TOD & M'GREGOR, *Clyde Foundry.*—Meadowside building yard and graving-dock, Glasgow. (For Engraving, *see page* 36.)

Model of private graving dock and basin, showing also the various workshops connected therewith, designed for Messrs. Tod and M'Gregor by Messrs. Bell and Miller, Civil Engineers, Glasgow. The illustration on the following page represents the graving dock, dockyard, and premises. The large tidal basins, with wharves and quays, on the rivers Clyde and Kelvin, adjoining the building yard, have a depth of water sufficient to admit vessels of the largest tonnage for repairs, &c.

The dock is 500 feet long, 80 feet wide, with 20 feet water at spring tides. It is entirely built of squared masonry, freestone, and granite. The gates are of malleable iron, weighing upwards of 60 tons, and are of peculiar construction, hanging on pivots without the support of quadrant rollers. The bearing is not in the usual manner of hollow quoins, but a flat surface on heel-posts of planed cast-iron, shutting upon a polished face of the granite quoin stone—iron to granite, without the intervention of any softer material, and perfectly water-tight.

The tides on the Clyde fall only eight feet at springs, leaving ten to twelve feet water on the dock sill; this renders necessary a heavy pumping engine of 250 horse-power, working two 52-inch pumps, which empties the dock in two-and-a-half hours, without waiting for the ebb.

The platform is kept clean, by the discharge from the pumping engine through the chambers in the masonry behind the gates.

The tidal basins and wharves have together 1,070 feet of quays constructed along the banks of the Clyde and Kelvin, with room for 650 feet additional. The whole water frontage is 2,400 feet.

The dockyard contains a complete arrangement of buildings and machinery, steam-hammers, &c., for repairing, entirely independent of the works in the building-yard adjoining. The total ground occupied by the building-yard and dockyard is twenty acres. There is sufficient accommodation to admit seven vessels of 3,000 tons each, repairing and fitting out at one time, besides those building on the stocks.

In addition to variouscranes from five to twenty tons, there is a moveable steam crane capable of lifting eighty tons, for boilers and heavy machinery.

[2351]

TUPPER & COMPANY, 61A *Moorgate Street, London, and Birmingham.*—Galvanized iron manufactures connected with building and architecture.

[2352]

TURNER, W., & GIBSON, J. W., *Dublin.*—Balance rolling bridges for railways over water and public roads; iron roofs, &c.

[2353]

VAVASSEUR, HENRY, & Co., *Sumner Street, Southwark, London.*—Galvanized, corrugated, and plain sheet iron, &c. (*See page* 37.)

[2354]

VIGNOLES, C., *F.R.S.*, 21 *Duke Street, Westminster.*—Models and drawings of Bilbao railway, Spain. (*See pages* 40 *and* 41.)

[2355]

VIEILLE MONTAGNE ZINC COMPANY, *Manchester Buildings, Westminster.* R. G. FISHER, Architect. F. BRABY & Co., Manufacturing Agents, 358–360 *Euston Road, London.*—Models showing the use of zinc for roofing purposes. (*See page* 38.)

Tod & M'Gregor, *Clyde Foundry.*—Meadowside building yard and graving-dock, Glasgow.

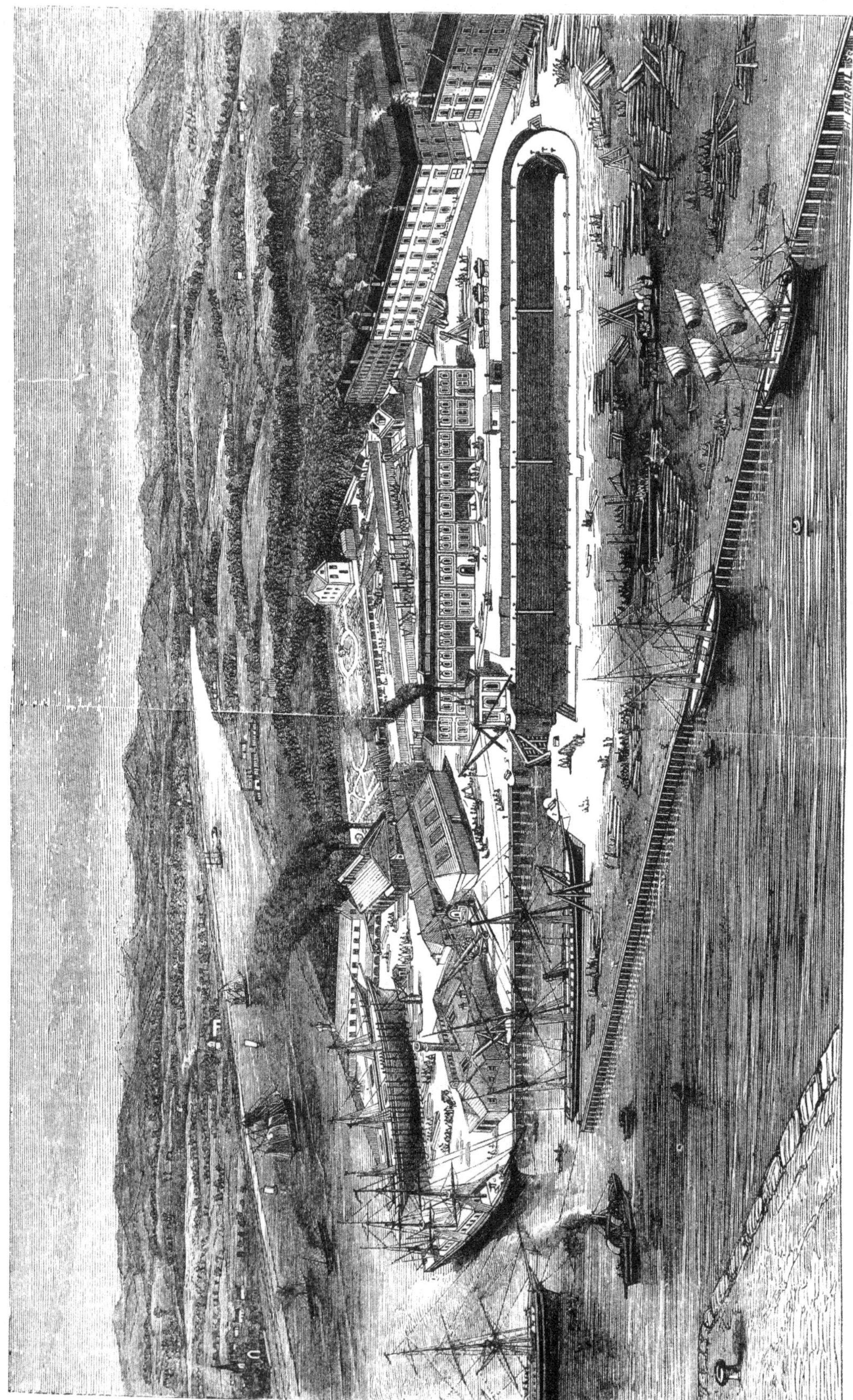

MEADOWSIDE BUILDING YARD AND GRAVING-DOCK, GLASGOW.

VAVASSEUR, HENRY, & Co., *Sumner Street, Southwark, London.*—Galvanized, corrugated, and plain sheet iron, &c., for building and roofing purposes, tanks, and cisterns.

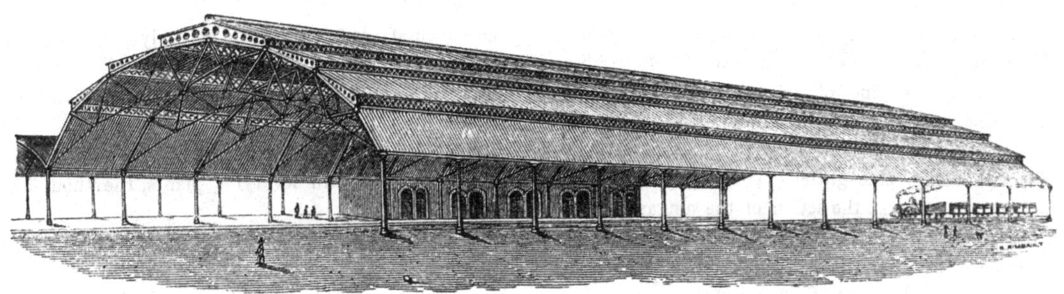

AMSTERDAM STATION ON THE DUTCH-RHENISH RAILWAY.

No. 1. A sheet of galvanized iron, No. 16 gauge, corrugated, with a 10-inch flute, used in covering the Amsterdam station on the Dutch Rhenish Railway.

No. 2. A sheet of galvanized iron, No. 20 gauge, corrugated, with a 5-inch flute, as used in the construction of the Palace of Industry, Amsterdam.

No. 3. Specimen of galvanized iron roofing, used in covering No. 1 Slip in H.M. Dockyard, Portsmouth.

No. 4. A sheet of galvanized iron, No. 24 gauge, corrugated, and curved with a 3-inch flute.

No. 5. Case containing galvanized iron fittings for iron buildings.

No. 6. Galvanized iron tank.

No. 7. Galvanized iron cistern.

No. 8. Specimen of galvanized iron coffee-spouting.

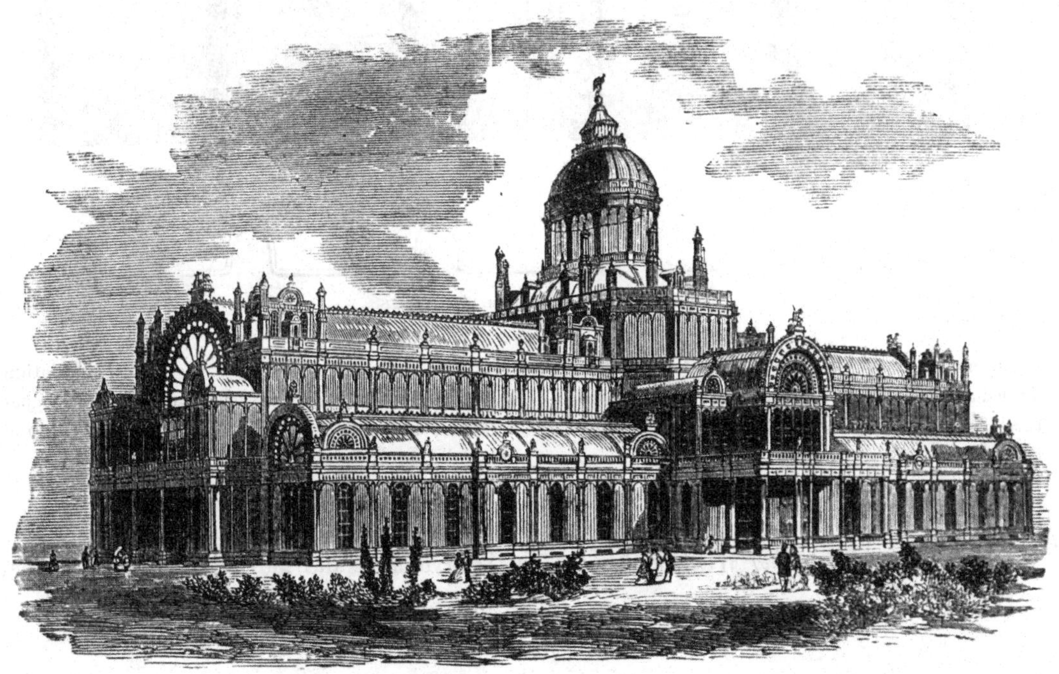

THE PALACE OF INDUSTRY, AMSTERDAM.

Messrs. HENRY VAVASSEUR & Co. are contractors for every description of iron buildings.

Vieille Montagne Zinc Company, *Manchester Buildings, Westminster.* R. G. Fisher, Architect. F. Braby & Co., Manufacturing Agents, 358–360 *Euston Road, London.*—Models showing the use of zinc for roofing purposes.

Models of zinc roofs, showing an economic system of framework, but having also due regard to strength. Corrugated zinc sheets, in extensive use for railway buildings. Sheet zinc, of superior quality, for roofing, each sheet bearing the stamp of "Vieille Montagne—F. Braby & Co."

O G moulded zinc gutter. This form of gutter is well adapted to resist the action of the sun, and may be fixed either by screws or spikes, through zinc tubes, or by the ordinary brackets. It also gives an ornamental and architectural finish to the eaves of the building.

Semi-circular zinc gutter, for farm buildings and out-houses.

Rain-water pipes and heads of various forms.

Zinc water-balls for cisterns, much cheaper than copper, but equally efficient.

A zinc cistern.

Ornamental clock-case, made entirely of hammered zinc.

Roll and verandah caps, for making good the joints of sheets in covering roofs.

Braby's Italian-formed corrugated zinc.

Zinc wire—the thin for tying plants, the thick for laundries.

Zinc mouldings of various designs.

A zinc casement.

Speaking pipe and circular elbow.

Zinc ridging—durable, and far cheaper than lead or slate.

Zinc sash-bars for skylights, conservatories, garden hand-frames, church and cottage windows.

Casement bars.

MODELS FOR ROOFING PURPOSES.

The above shows section and elevation of Italian-formed zinc, as used for the verandah of the Horticultural Society's conservatory, and on the refreshment rooms of the Exhibition, &c.

Zinc friezes and frets for lamps, verandahs, ventilation and decorative purposes; replacing either lead or copper for these applications, and being cheaper, lighter, and more elegant in appearance.

[2356]

Walker, John, 32, *King William Street, City.*—Corrugated and galvanized iron in roofs, churches, &c.

[2357]

Walker, C., & Sons, *Little Sutton Street.*—Gas valves, water valves, hydrants, regulating columns, &c.

[2358]

CENTRAL COTTAGE IMPROVEMENT SOCIETY, *Cottage Architectural Museum*, 37 *Arundel Street, Strand.*—Models and plans of labourers' cottages.

President: HIS GRACE THE DUKE OF MARLBOROUGH.
Vice-President: HIS GRACE THE ARCHBISHOP OF CANTERBURY.

"AS THE HOME, SO THE PEOPLE."

The objects of this Society are, to furnish plans and specifications for suburban dwellings for artisans; also, for agricultural labourers—for village lodging-houses—and suggestions for the general improvement of existing dwellings, cottages, gardens, &c.; and to establish Auxiliary Societies, some of which are now in course of formation.

The Society have published four designs for cottages:—No. I. has a kitchen, and a second room, to be used either as a sitting or extra bed-room, with two good bedrooms on second floor. No. II. is similar to No. I., but slightly larger, and one bed-room up-stairs has a movable partition for the better separation of children, if desired. No. III. is similar to No. I., with the addition of a convenient scullery or wash-house. No. IV. is more commodious, to meet the requirements of the artisan class.

No. I. has been built for £162; No. II., £168; No. III., £175; and No. IV., at £220 per pair.

The Museum is open every week-day from 12 till 4, free.

The Museum is particularly appropriated to the collection of models of cottages, plans and specifications thereof, books and papers of every description bearing on the subject of labourers' dwellings, with a view to place the matter before the public in a compendious and practical form.

Plans with specifications, 5s. each, to be obtained of the Secretary.

Annual subscription, £1; life subscription, £10. Donations also received.

THE EXHIBITION MODEL COTTAGES, nearly opposite the Eastern Dome, on the Grounds of the South Kensington Museum.—By the kind permission of the Lords of the Committee of Council on Education, the Council of the Central Cottage Improvement Society have been enabled to erect a pair of cottages on the artisan's plan, No. IV., for the inspection of visitors to the International Exhibition, and to which their attention is earnestly invited.

The special aim of the Society is practically to demonstrate to the satisfaction of the public the possibility of erecting dwellings, so much needed by the working classes, for a sum returning a fair rate of interest, in the neighbourhood of large towns, and remunerative (in agricultural districts) in many senses, particularly by improving the condition of the labourer.

Economy of space and material, compactness, solidity, convenience, and salubrity have been carefully studied; and the Council believe that no plan for cottages in pairs can be devised more appropriate to these conditions, or more suited to English tastes and habits. The cost of erection, by contract, is £215, and this closely corresponds with the estimate of the Society, and the cost of those already built in various localities.

In conclusion, the Council sincerely hope that this plan will be approved by the public, and materially assist in procuring better dwellings for the humbler classes.

Admission free to season ticket-holders, and by orders, which can be obtained in the Exhibition, Class 10, No. 31.

DESCRIPTION OF PLAN.—Ground Floor —A sitting-room, bed-room, or work-room, as may be required, 12 ft. by 10 ft., containing a press bed and two cupboards; kitchen, 12 ft. by 9 ft., fitted up with oven and boiler range; a pantry, and store-closet under the stairs; wash-house, 8 ft. 6 in. by 7 ft. 9 in., has a fire-place, fire-clay oven, copper, sink, and a dresser.

Up-stairs are two convenient bed-rooms, 12 ft. by 10 ft. and 12 ft. by 9 ft., with fire-places, stoves, and store-closets. All the rooms are well ventilated.

[2234]

BARNETT, SAMPSON, 23 *Forston Street*, *Hoxton*.—Diving apparatus maker to the Royal Navy and the various maritime powers.

RECOVERING THE GUNS AND STORES FROM THE "ROYAL GEORGE."

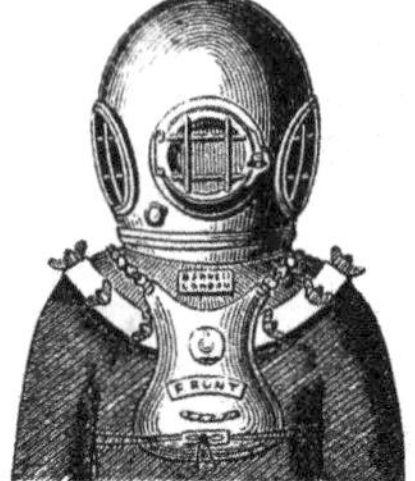

FRONT VIEW.

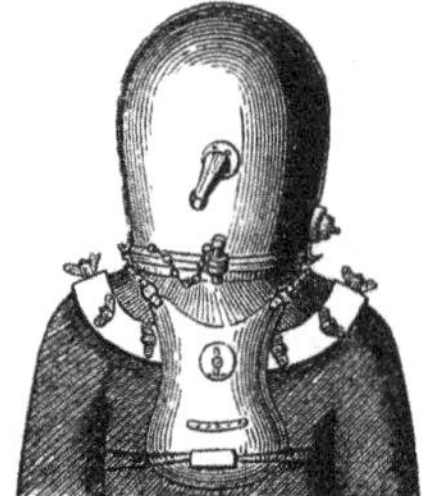

BACK VIEW.

A. Diver equipped in water-tight dress, copper helmet, with glass eyes, boots with leaden soles, &c.

B. Air tube for supplying the diver.
C. Signal or life line.
D. Attendants at signal line.
E. The 3-barrel atmospheric air engine.
F. Ladder line for use in thick water.
G. Rope ladder for ascending and descending.
H. Weight to steady the ladder.
I. Diver stopping a leak under the water-line.

Simplicity in diving apparatus is most essential, as professional divers are seldom to be obtained when their services are of immediate importance, and they often require so much for salvage, as materially to lessen the value of what they recover.

The apparatus here shown can be immediately understood and used by any ordinary labouring man, after reading the book of directions supplied with it; thus rendering it valuable in remote parts, where the services of the practised diver cannot be obtained.

Since the introduction of the tight dress and jointed helmet, the diving apparatus has come into very general use, and by its aid a vast amount of property has been recovered. The use of the apparatus is now generally understood, and its value universally acknowledged. Those who, in urgent cases, have descended with merely a rope to depend upon, will be surprised at the simplicity of an apparatus, by the use of which they are enabled to see and breathe with perfect freedom, and to remain under water until hunger compels them to ascend. The celebrated diver, Mr. Deane, has frequently remained under water for five hours consecutively, with the apparatus made for him by the exhibitor. No vessel ought to go to sea without one of these machines, for if it only once recovers an anchor it more than repays its whole cost. Vessels and crews have often been saved by stopping a leak from the outside, under the water-line, as shown in the illustration. This apparatus is extensively used in pearl and sponge diving.

The patent improvements of S. Barnett are the result of thirty years' experience in the manufacture of diving apparatus. While retaining all that was valuable in former apparatus, they possess the following advantages :—

1. Should the diver wish to raise himself without signalling the attendant, he can do so by simply placing his finger on the valve, which afterwards rights itself.

2. Without assistance, he can open his own helmet, which is so constructed that the front eye can never be lost, or become tight.

3. The indicator always denotes the depth the diver is at.

4. The condensing box secures a more continuous stream of air.

5. The tight dress is free from the inconveniences presented by the old loose dress in sitting and lying down.

At an official trial at H.M. dockyard, Portsmouth, by order of the Lords of the Admiralty, in presence of their officers, by the government diver, these improvements were fully tested on all their points.—See report of the *Times*, Nov. 30, 1861, under "Naval Intelligence."

The apparatus, complete, consisting of the treble-barrel atmospheric engine, with duplicate working parts, the helmet and tubes, two sets of waterproof clothing, six sets of under-clothing, &c. &c., securely packed for exportation, in one case, delivered in London, price £100.

Orders can be sent through any merchant or agent, or direct to the manufacturer.

Sub-Class B.—*Sanitary Improvements and Constructions.*

[2368]

Askew's Patent Window Sash and Improved Ventilation Company.—A reversible ventilating window, of which both sides can be cleaned from within.

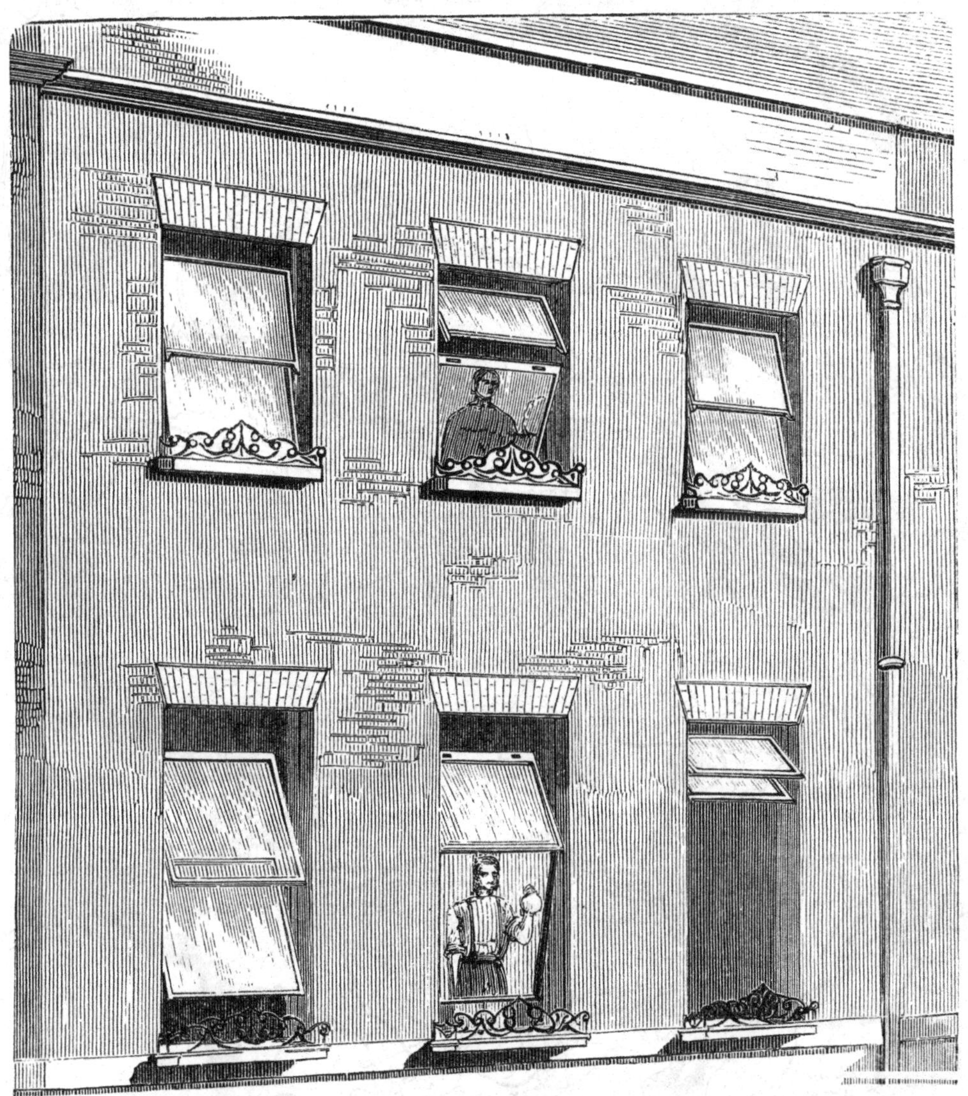

A REVERSIBLE WINDOW, OF WHICH BOTH SIDES CAN BE CLEANED FROM WITHIN.

The accompanying sketch shows some of the chief advantages derivable from the employment of this patent, which may be briefly enumerated as follows:—

1. The outside of the glass may be readily cleaned from the interior of the room, thereby effectually preventing the frightful accidents continually occurring to domestic servants and others, from standing or sitting on the outside sills.

2. A perfect system of ventilation, allowing the admission of air, even in windy weather, without the evils arising from a downward draught, rendering the invention peculiarly applicable to hospitals, barracks, and other large buildings, as well as to private dwellings.

3. Its extreme simplicity and non-liability to derangement, and the readiness with which it can be applied to existing window sashes.

4. The entire exclusion, at pleasure, of all draughts, not only at the sides, but also at the meeting rails of the sashes.

It has already been used at Pembroke House; at the private residence of the Right Hon. W. F. Cowper, M.P., Chief Commissioner, Board of Works, 17, Curzon Street, May Fair; and at the residence of Mr. King, 19, Percy Street, Bedford Square; Messrs. Mappin Brothers, King William Street, London Bridge; Messrs. Dakin and Co., St. Paul's Churchyard; Messrs. Parkins and Gotto, Oxford Street; Mr. Cox, Southampton Row, Russell Square; Mr. Edgley, 3, Serjeant's Inn, Fleet Street; Mr. Tilbury, Ferdinand Street, Kentish Town; Mr. Magotti, 76, Seymour Street; and at St. Thomas's Hospital; and in every case the inventor has given unqualified satisfaction.

Applications to be made to the Secretary, at the Offices of the Company, 9 Adam Street, Adelphi, W.C.

Vignoles, C., F.R.S., 21 *Duke Street, Westminster.*—Models and drawings of Bilbao railway, Spain.

INDEX MAP OF MODEL OF THE PASSAGE OF

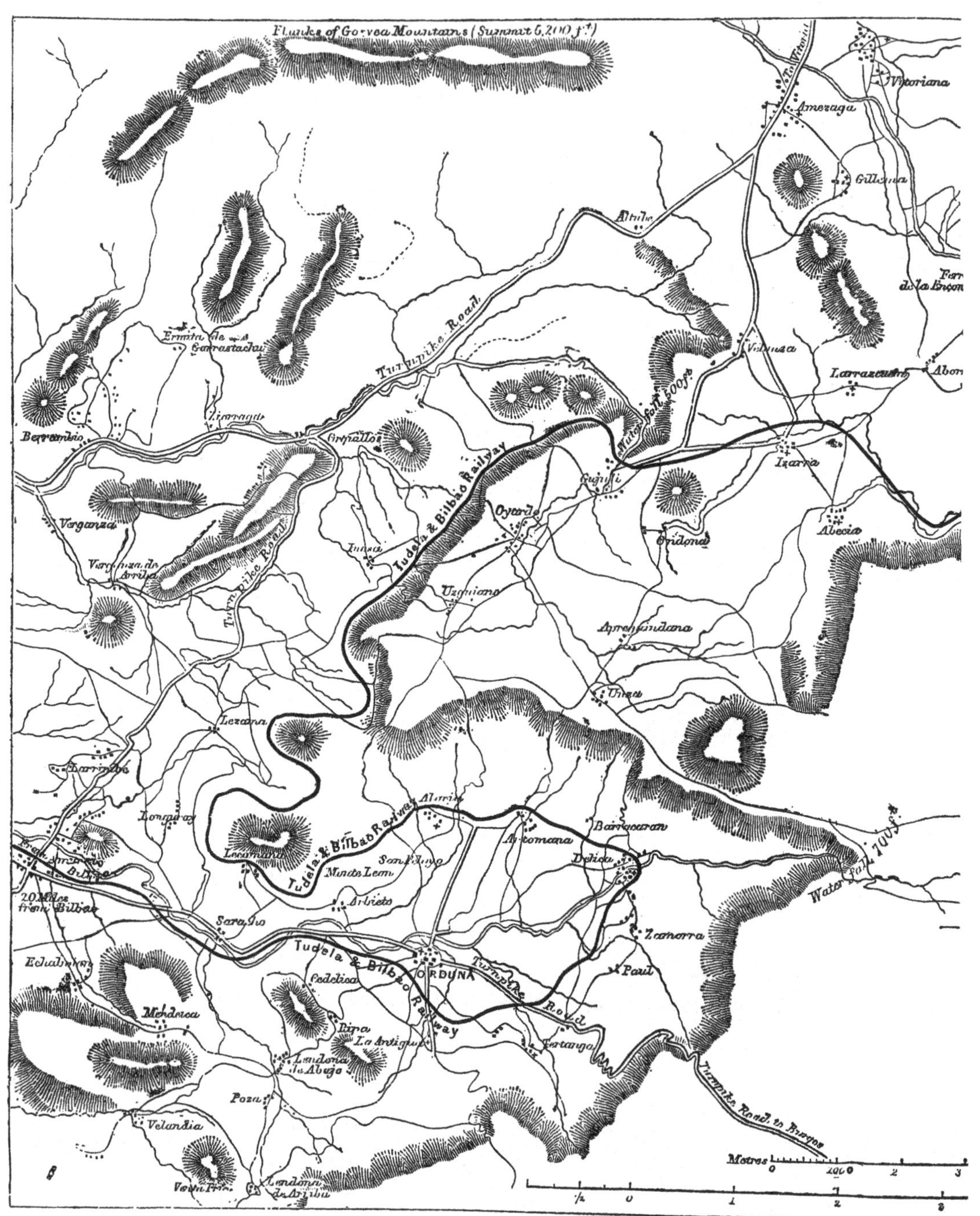

MODEL OF THE PASSAGE OF THE TUDELA AND BILBAO RAILWAY ACROSS THE CHAIN OF THE

DESIGNED AND EXHIBITED BY CHARLES

Henry Montague Mathews, } District Engineers.
Christopher Bennison, }
Thomas Vignoles Croudace, Assistant Engineer.

Horizontal Scale of Model, $\frac{1}{5000}$;
Vertical Scale of Model, $\frac{1}{2000}$;
Minimum Radii of Railway Curves,

VIGNOLES, C.—*continued.*

THE TUDELA AND BILBAO RAILWAY.

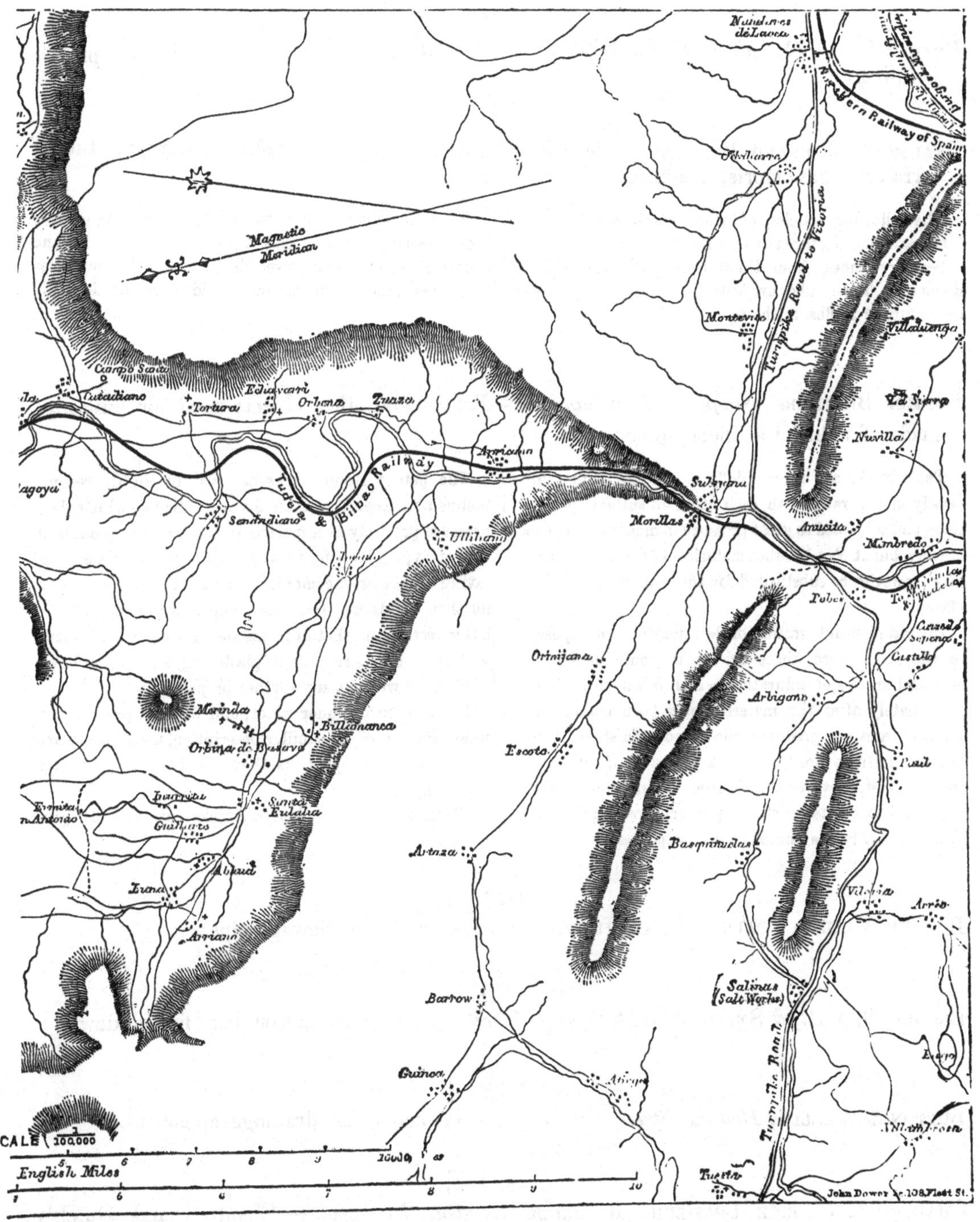

CANTABRIAN PYRENEES, THROUGH THE BASQUE PROVINCES, IN THE NORTH OF SPAIN, 1862.

VIGNOLES, F.R.S., ENGINEER-IN-CHIEF.

2⅗ inches per mile—1 inch to 416⅔ feet.	HENRY VIGNOLES, Principal Resident Engineer.
” 1 inch to 166⅔ feet.	PERCIVAL SKELTON, Artist.
807 yards; Steepest Gradient, 1 in 70.	STEPHEN SALTER, Modeller.

[2369]

BAZALGETTE, JOSEPH WILLIAM, *Spring Gardens.*—Drawings of the metropolitan main drainage, sewers, and intercepting works.

[2370]

BEAGLE & Co., 71 *Cannon Street, West, London.*—Patent ventilator, for public and private buildings.

[2371]

BEAUMONT, EDWARD BLACKETT, *Darfield Pottery, Barnsley, Yorkshire.*—Sanitary tubes: terra cotta gas retorts, fire-bricks, filters, &c.

The following articles, of which specimens are exhibited, are manufactured at this Pottery, viz.:—

Sanitary tubes, from three inches to four feet in diameter. These tubes are tested, when required, to bear a pressure of 200lbs. to the square inch.

Fire bricks, gas retorts, vases, and terra cotta ware of every description. This clay is peculiarly adapted for the construction of chemical vessels, and for other purposes where resistance to the action of acids is required.

[2372]

BODMER BROTHERS, *Newport, Monmouthshire.*—Bricks, and other objects used for building, made of unburnt artificial stone.

These bricks are made chiefly of sand and lime, intimately incorporated with each other in suitable proportions, and subjected to great pressure in moulds. Furnace cinders, burnt clay, or other materials of a similar nature, may, however, be substituted for the sand, with excellent effect.

Instead of disintegrating or deteriorating on exposure to the atmosphere, these bricks, in consequence of a chemical process of induration, which commences almost immediately after the materials have been compressed, improve, and are gradually converted into stone. They absorb very little moisture, and are capable of withstanding any frost, however severe, after having once become indurated to a certain extent: properties which numbers of burnt clay bricks cannot be said to possess.

The patent stone bricks can also be highly recommended on account of their accurate shape—which they preserve precisely as imparted to them by the moulds of the press; and owing to which their use effects a great saving of mortar or cement. They are also commendable for their handsome appearance and pleasing colour; the latter resembling that of freestone. Partition walls made with these bricks require no plastering, and can be hung with paper without any further preparation.

For structures under water, and for coal pits, sewers, wells, and works of a similar description, these bricks are pre-eminently adapted, as they can be made perfectly water-tight.

Black and other coloured bricks can also be produced.

[2373]

BOURNE VALLEY POTTERY COMPANY, *Nine Elms, Vauxhall.*—Sewage pipes.

[2374]

BROOKS, B. & R., & SMITH, J., 154 *Goswell Street.*—New invented sash bars for windows.

[2375]

BURTON & WALLER, *Holland Street, Southwark.*—Gas and water drainage apparatus.

[2376]

CHANTRELL, GEORGE FREDERIC, 6 *Hatton Garden, Liverpool.* — Chantrell and Dutch's water-closet, &c. (*See page* 45.)

[2377]

CHEAVIN, S., *Pen Street, Boston.*—Patent double action rapid belt water purifier; damp proof paints and cement, &c.

[2378]

CLIFF, JOHN, & CO., *Imperial Potteries, Princes Street, Lambeth, London, S.*

Chemical vessels, pharmaceutical apparatus, &c. &c. to order.

Brown and white stoneware of all descriptions. Illustrated price lists will be sent on application.

Obtained large Medals at the Exhibitions of 1851 *and* 1855.

"The Jury noticed with great commendation the care and attention bestowed by these exhibitors on chemical and other apparatus."—(*Extract. See Jurors' Report, Class* 27, *p.* 583.)

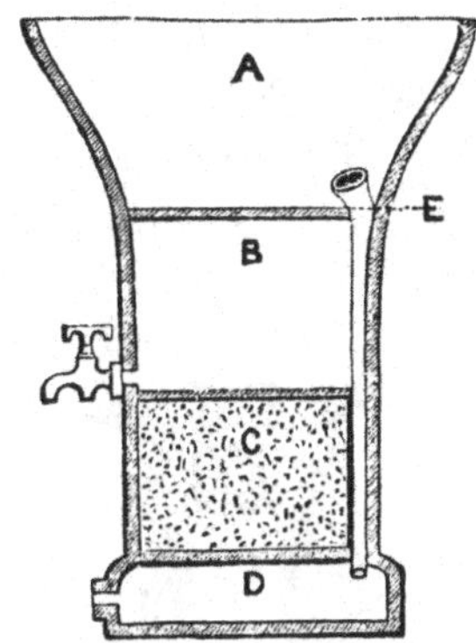

PATENT FILTER BY ASCENSION.

The large jar exhibited in Paris by this firm is in use in the present Exhibition building as a sherry butt, in the cellars of the Refreshment Department, and is the largest stoneware vessel in the world in actual service.

GREAT EXHIBITION OF 1851.

(*Extract. See Jurors' Report, Class* 25, *p.* 541.)

"Stephen Green & Co., Lambeth. This firm exhibits some very remarkable specimens of stoneware, of great size, designed for the use of breweries, distilleries, &c., and which, on account of their hardness of glaze and other qualities, are of great value in many processes of chemical manufacture." The Jury have awarded a Prize Medal.

(*Extract. See Jurors' Report, Class* 27, *p.* 583.)

"Although by the decision of the constituted authorities the Medal which has been awarded to Messrs. Stephen Green & Co., in Class 27, has been withdrawn in favour of the similar honour awarded by the Jury of Class 25, the author of the present report cannot pass on to other exhibitors without giving some account of the objects which chiefly attracted the attention of his Jury. These are the large jar, the condensers, the air-tight stoppers, and the acid pump, exhibited within the Building; and the whole apparatus of the retort placed outside. The condensers are not only large, but perfect, and the spherical stopper and valve are so ground as to be perfectly air-tight, and must be regarded as an admirable and most useful contrivance. The jar is perhaps the largest piece ever manufactured in this ware." The Jury noticed with great commendation the care and attention bestowed by these exhibitors on chemical and other apparatus.

Imperial Potteries, Lambeth, London.

[2379]

Cooke, William, Civil Engineer, 26 *Spring Gardens.*—Ventilating and sanitary appliances; inexpensive, and of general utility.

Apparatus for effecting ventilation without dust or draught. This invention is self-acting, simple, and inexpensive, is always in its place, gives no trouble, is not liable to damage or derangement, admits an unceasing and imperceptible supply of pure air without dust or draught, and may be used with safety in sick rooms and sleeping apartments during the night. When out of use it is out of sight.

It is equally applicable to apartments, buildings, and carriages.

[2380]

Dale, Thomas, *Manager, Great Yarmouth Water Works.*—Improved service-box for supplying water-closets, and preventing waste.

[2381]

Danchell, F. Hahn & Co., 38 *Red Lion Square.*—Filtering, water-softening, and water-testing apparatus.

Water Purifying Apparatus.—The following are exhibited:—

Cistern Filters.—To be placed direct into house cisterns, and capable of yielding from two quarts to two gallons of water in a minute, according to size.

Fountain Filters.—To be connected either with the service-pipe direct from the main, or with the supply-pipe from the cistern, and capable of yielding from two quarts to two gallons per minute, according to size.

Portable Household Filters of stoneware, from one to ten gallons size.

Portable Table Filters of porcelain, earthenware terra cotta, &c., from one to four gallons size.

Self-Regulating Apparatus for softening water to be placed in cisterns, and constructed for softening from 100 to 100,000 gallons of water per diem.

Water-Testing Apparatus, requiring no knowledge of chemistry, to ascertain the presence in water of any deleterious substances in solution. Arranged for domestic use, hydraulic engineers, sanitary officers, and others.

Having the contract for supplying the Exhibition with filtered water, numerous very novel designs in Vase and Fountain Filters will be found in use in various parts of the building.

For information on the subject of purification of water with reference to the above articles, see a "Treatise on Water, its Impurities and Purification," by F. Hahn Danchell; published by Renshaw, 356 Strand.

G. Kent, Sole Manufacturer, 199 High Holborn.

Chantrell, George Frederic, 6 *Hatton Garden, Liverpool.*—Chantrell & Dutch's water-closet, combining slate cistern, basin, and trap with patent flushing apparatus.

This apparatus is strong, durable, efficient, and exceedingly cheap, and prevents waste of water. Price—with thirty-gallon slate cistern, measuring-box, solid double valve with vulcanised india-rubber washers, overflow and air pipe, flushing pipe and connections, improved hopper basin and trap, self-acting motion and ball-valve for main supply, complete and ready for fixing—£3. A liberal discount allowed to the trade.

It may be readily fixed and examined, by merely unscrewing the lower valve-seat, by any ordinary skilled workman.

Large-sized cisterns and other descriptions of cisterns in proportion.

The inventors, after many years' practical experience in sanitary matters, find that a cistern is indispensable; for with any one charge system, should the closet be used when the main supply is off (which is very often the case), it becomes foul. The self-acting principle in this apparatus (as shown) being strong, and free from complications, always insures a thorough flushing of the closet at the time of use.

A, cistern with double bottoms, forming measuring chamber, B, between which the double valve, C, acts, which is held in the position shown by the seat of the closet; the latter is weighted at the back, and, when used, merely drops half an inch in front, acting upon the lever, D, closing the outlet to E, opening the inlet from A, charging B. When the seat is free, the valve returns to its former position; the water accumulated in B flushes the closet (or urinal, to which it is also adapted). E, flushing pipe; F, air and overflow pipe.

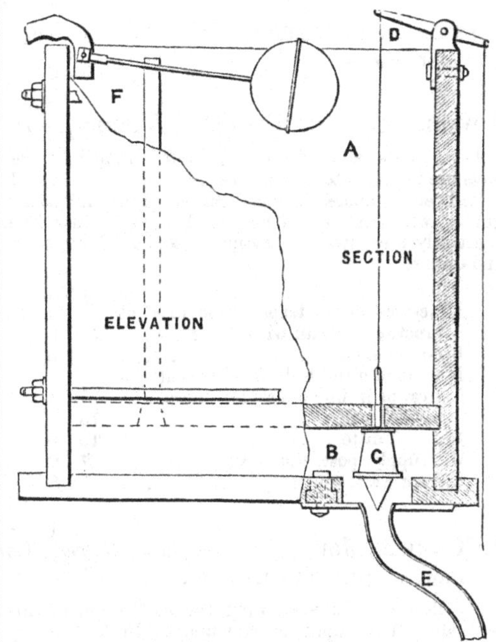

[2382]

Edwards, Frederick, & Son, 49 *Great Marlborough Street, London, W.*—Models and drawings of an improved method of constructing chimneys and ventilation.

[2383]

Field & Allen, 27 *Frederick Street, Edinburgh.*—Articles for housebuilding purposes.

[2384]

Finch, John, 11 *Adam Street, Adelphi.*—Patent Porcelain Bath, designed by his late Royal Highness Prince Albert (Rufford and Finch, patentees).

Obtained a Medal at the Great Exhibition, 1851, *"for Baths, &c." Gold Isis Medal of the Society of Arts,* 1850.

Patent Porcelain Bath designed by His late Royal Highness the Prince Consort. This Bath is patronised by H. R. H. the Duke of Cambridge for Her Majesty's war department; by the Emperor of the French, the Emperor of Russia, Lord Palmerston, &c., and is extensively used in public institutions and private houses.

Patent Porcelain Housemaid's and Kitchen Sinks.

These are made from the suggestions of Miss Florence Nightingale, and are not liable to the complaint made of the ordinary sink in her valuable "Notes on Nursing." "The ordinary oblong sink is an abomination. That great surface of stone, which is always left wet, is always exhaling into the air. I have known whole houses and hospitals smell of the sink."

[2385]

Gotto, Frederick, Architect, *Leighton Buzzard, Bedfordshire.*—Gotto's self-discharging effluvia trap.

[2386]

Jennings, George, *Holland Street, Blackfriars Road.*—Domestic, sanitary, and building appliances, tending to comfort and health.

[2387]

Key, E., *Sharrington, viâ Thetford, Norfolk.*—Models of country cottages.

[2388]

Keynsham Blue Lias Stone and Cement Company, 6 *Martin's Lane, Cannon Street.*—Samples of blue lias lime.

[2389]

KITE, C., 20 *Liverpool Street, King's Cross.*—Improved chimney tops, ventilators, and stable requisites.

[2390]

LIPSCOMBE, FREDERICK, & Co., 233 *Strand, near Temple Bar.*—Patent self-cleansing charcoal water filters.

[2391]

LOVEGROVE, JAMES, *Town Hall, Hackney, N.E.*—Traps to prevent effluvia from drains and gulleys.

Lovegrove's patent drain traps and ventilating valves have been applied to noblemen's mansions, public buildings, dwelling-houses, stables, garden paths, and street gullies, with complete success, and in every case have effectually prevented the escape of foul air from sewers and drains.

	s.	d.
Nine-inch outlet traps, to be fixed at junction of drain with sewer	15	0
Six-inch ditto	12	6
Nine-inch ditto, to be fixed beneath an area, or in line of drain, if more convenient	15	0
Six-inch ditto	12	6
Air supply, post, and valve	7	6
Cistern waste pipe traps	7	6
Garden and yard sinks, with valve and cesspit complete	9	6
Iron stable traps, 11×11 ditto	15	0
Rain pipe, closet valve } 4-inch P traps, with valve }	5	6
Iron area or yard sink, 9×4, with valve	7	6

Architects and surveyors should specify the drains to be trapped and ventilated with Lovegrove's traps and ventilating valves.

Orders for fixing under the instructions of the inventor must be forwarded to J. Lovegrove, Civil Engineer, Surveyor, Town Hall, Hackney, N.E.

The traps may be obtained at Jennings' Depôt, 5, Holland Street, Blackfriars, S.E.

[2392]

M'KINNELL, JOHN, 15 *Langham Street, London.*—Patent ventilator, for buildings of all kinds, ships, and carriages.

Models are exhibited showing the application of this ventilator. 1. In upper apartments. 2. In floors where the joists run from wall to wall. 3. Where girders intervene. 4. Where the fresh air is supplied horizontally, and the vitiated discharged at the ridge, as in the Royal Chapel and the Queen's School, Windsor Park. 5. In ships, as it has been adopted by Her Majesty's Emigration Commissioners.

Ventilators in various forms.

Economic Gas Regulators.

[2393]

MOORE, JOSIAH, 81 *Fleet Street.*—Ventilators for houses.

[2394]

NIXON, T., *Kettering, Northampton.*—Greenhouse.

[2395]

PIERCE, WILLIAM, 5 *Jermyn Street, London.*—Huthnance's patent heating apparatus, for drying rooms, &c.; sanitary improvements in stove grates for hospitals, cottages, &c.

Obtained Prize Medal at the Exhibition of 1851.

[2396]

PRITCHARD, WILLIAM, 3 *Ware Street, Kingsland Road, N.E.*—Patent life-protecting machine for cleaning windows.

The object of this apparatus is to secure to domestic servants immunity from the risks to life and limb incurred in window cleaning. It is at once simple, effective, and inexpensive. It can be applied to any window in town or country, at the very trifling cost of £2 and upwards.

[2397]

RIDDELL, JOSEPH HADLEY, 155 *Cheapside, London.*—Patent portable cooking stove, and patent slow combustion heating stove.

[2398]

ROBSON, W., *Newcastle.*—Firebricks, &c.

[2399]

ROSSER, SAMUEL EGAN, *Percy Chambers, Northumberland Street, Strand, London, W.C.*—Warming, ventilating, and desiccating apparatus.

[2400]

SILICATED CARBON FILTER COMPANY, *Bolingbroke Gardens, Battersea, London.*—Silicated carbon filters, for universal application (Dahlke's patent).

[2401]

SMITH, GEORGE, & Co., *Sun Foundry, Glasgow.*—Patent composite grave monuments and tablets, and ornamental drinking fountains in iron. Sanitary structures, such as baths and dry-deodorising closets and stable-fittings.

NO. 14. ORNAMENTAL DRINKING FOUNTAIN FOR WALL PURPOSES. HEIGHT, 6 FEET.

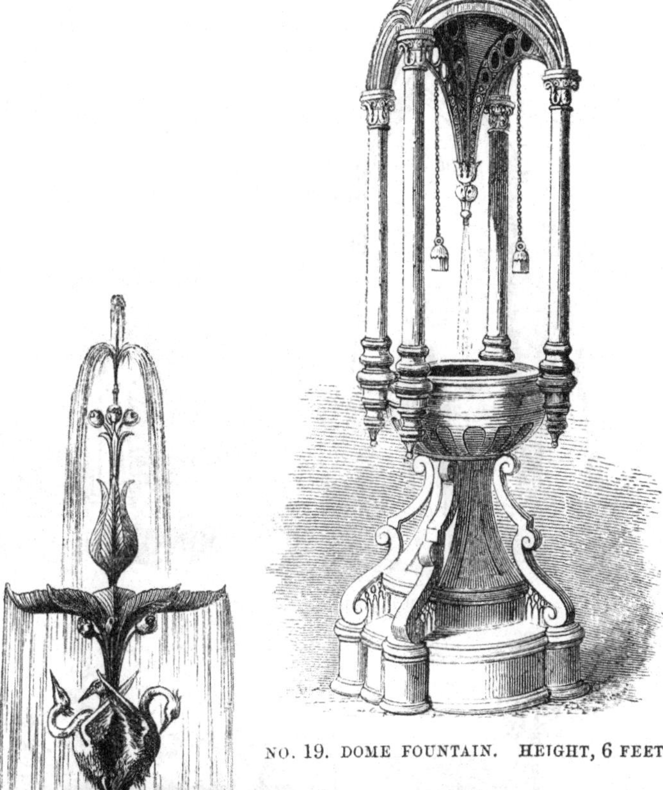

NO. 19. DOME FOUNTAIN. HEIGHT, 6 FEET.

Manufacturers of rain-water goods, and all kinds of cast-iron fittings for plumbers' and architectural purposes, and patentees of GEORGE SMITH & Co.'s patent baths, bath stands, and lavatories, dry-deodorising closets, commodes, urinals, patent composite grave monuments and tablets; also cast-iron plain and ornamental fountains, cattle troughs, and their registered stable fittings, which were selected by the late Prince Consort for Holyrood Palace stables, and for which first premium was awarded at the Royal Highland and Agricultural Shows.

NO. 12. ORNAMENTAL FOUNTAIN. HEIGHT, 10 FEET. DIAMETER OF TROUGH, 7 FEET.

Smith, George, & Co.—*continued.*

Patent composite grave monuments and memorial structures. In these monuments panels of marble, stone, or slate, upon which inscriptions can be engraved either before or after erection, are combined with ornamental cast-iron framework capable of the simplest or most elaborate designs, at extremely moderate cost.

NO. 11. GRAVE MONUMENT. 7 FEET HIGH.

NO. 25. MEMORIAL OBELISK. 14 FEET HIGH.

EMBLEMATIC TOMB RAILING.

SMITH, GEORGE, & CO., *continued.*

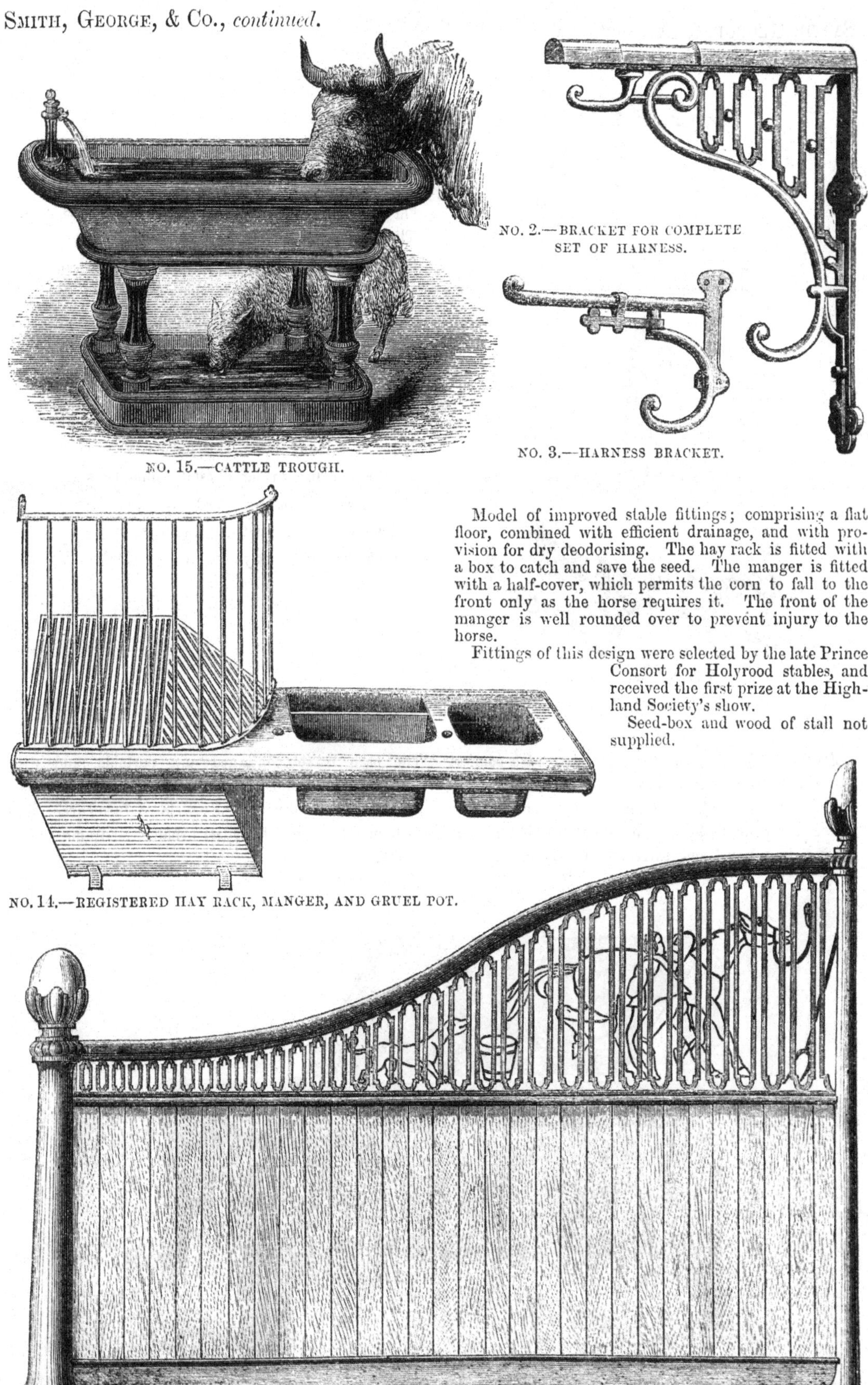

NO. 15.—CATTLE TROUGH.

NO. 2.—BRACKET FOR COMPLETE SET OF HARNESS.

NO. 3.—HARNESS BRACKET.

Model of improved stable fittings; comprising a flat floor, combined with efficient drainage, and with provision for dry deodorising. The hay rack is fitted with a box to catch and save the seed. The manger is fitted with a half-cover, which permits the corn to fall to the front only as the horse requires it. The front of the manger is well rounded over to prevent injury to the horse.

Fittings of this design were selected by the late Prince Consort for Holyrood stables, and received the first prize at the Highland Society's show.

Seed-box and wood of stall not supplied.

NO. 14.—REGISTERED HAY RACK, MANGER, AND GRUEL POT.

NO. 5.—ORNAMENTAL STALL DIVISION.

Smith, George, & Co., *continued.*

PATENT LAVATORY.

Patent Lavatory, in the bed-room or dressing-room, combines in one the various appliances required for ablutionary purposes. The stand itself is converted into a foot, sponge, or sitz bath, and the basin is fixed to a hollow swivelling pillar. The framework is filled in with ornamental glass, and is used as a mirror; and when used as a sponge bath, a curtain prevents injury to the walls and furniture.

The Patent Lavatory is also made without the glass screen, and in this form is admirably adapted for use in large establishments, such as barracks, hospitals, schools, &c. &c. These articles are supplied without the glass or porcelain basins, and may be made portable, if required.

Patent Egyptian Bath; shaped to give comfort to the user, and to economise water; and having provision for raising and maintaining the temperature, for giving a vapour bath, and for supplying hot or cold showers in a novel and extremely refreshing manner. The bath is self-contained and portable, and with the folding cover appears, in the bed-chamber or dressing-room, as an ornamental article of furniture. This bath may be had without the ornamental frame, cover, or interior fittings.

PATENT EGYPTIAN BATH.

SMITH, GEORGE, & Co., *continued.*

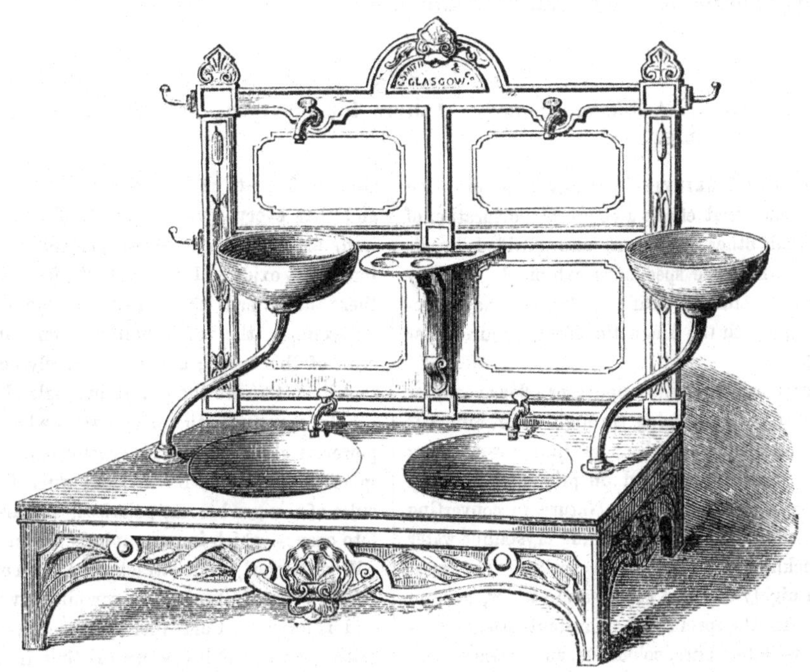

PATENT LAVATORY, ADAPTED FOR USE AS A WASH-HAND AND FOOT RANGE.

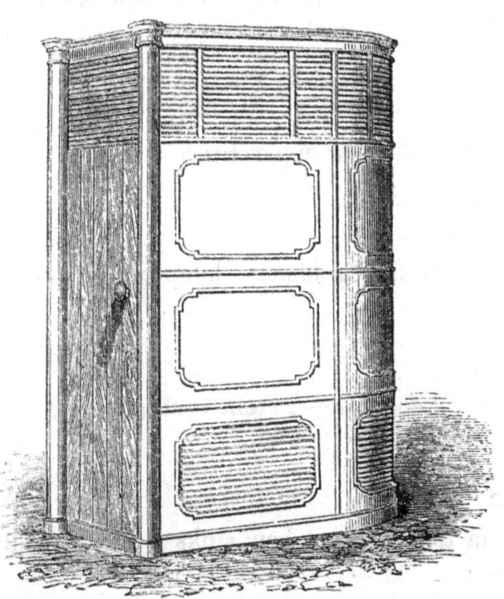

PATENT DRY DEODORISING COMMODE, IN IRON STRUCTURES, WITH VENTILATING VENETIAN PANELS, AND CAN BE SUPPLIED IN RANGES.

PATENT DRY DEODORISING COMMODE FOR THE BED-CHAMBER CLOSET.

The shell or frame is of cast iron and painted imitation wood; the working details are of the simplest construction, and do not get out of order. It is the most satisfactory appliance yet introduced for working out the system of dry deodorisation a system recommended by the best authorities, and which must gradually work its way to general adoption.

[2402]

SOWOOD, THOMAS, *Blue Boar Court, Manchester.*—Models of apparatus for curing smoky chimneys, ventilation, and heating buildings.

[2403]

SPENCER, THOMAS, 32 *Euston Square.*—Filters for purifying water with a new compound magnetic iron oxide.

THE MAGNETIC PURIFYING FILTER is the only one known to science that effects a chemical purification of water. Several other fluids are also purified by its agency, as shown in the specimens exhibited. Already some of our greatest scientific authorities have pronounced that, " with the Magnetic Filter, impure water is impossible."

No matter how chemically impure or offensive water may originally be, in passing through these Filters it becomes as pure and sparkling as the purest spring water. The change is effected on principles precisely analogous to those exercised by Nature, in converting impure surface water into the refreshing crystalline water we find trickling from a natural spring. The most impure and highly coloured bog or drain water, or even sewer water (see the specimens), is instantaneously rendered by these filters pure, colourless, and tasteless.

In the limited space at command, it is impossible to describe adequately the philosophical principles brought into practice by this discovery of Mr. Spencer.

We may convey some idea by stating, first, that this gentleman has discovered magnetic oxide of iron—loadstone in fact—to be Nature's chief agent of purification, and that every rock or substratification that contains iron, also contains a small per centage of this now important oxide. Moreover, that where it most abounds, there the water is the purest. In the Malvern district, for example, the rocks contain from ten to fifteen per cent. of this oxide; and it is scarcely necessary to add, that its waters are the purest in England. Mr. Spencer has also expounded the principles by which this purifying power is governed, viz., magnetic oxide attracts atmospherical oxygen to its surface; when there, the molecules of this gas become polarised, and are thus resolved into ozone—which important body is polarised oxygen.

When formed, ozone attracts the carbon of moist organic matter with avidity, and by combining with it, carbonic acid is formed. Consequently, the deleterious organic matter, and mephitic gases existing in impure water are decomposed and converted by means of the magnetic oxide into healthful and refreshing carbonic acid. Perhaps the greatest practical feature of this invention, is the mode by which Mr. Spencer converts ordinary ores into this now most important oxide.

[2404]

TAYLOR, J. J., 52 *Spring Gardens, Manchester.*—Portable gas apparatus.

[2405]

TAYLOR, JOHN, Jun., 53 *Parliament Street.*—Patent facing blocks, damp proof course, roof tiling, &c. (*See page* 53.)

[2406]

TENWICK, JOHN, *Albion Foundry, Clarendon Street, Landport.*—Patent ventilators for sewers, &c.; patent cesspools and gratings.

[2407]

TYE & ANDREW, *Brixton Road.*—Patent effluvia trap for kitchen sinks; also a means for flushing drains.

[2408]

UNDERHAY, E. G., *Crawford Passage, Clerkenwell, London, E.C.*—Underhay's patent regulator water-closets, high-pressure valves, and basin apparatus. (*See page* 54.)

[2409]

WARNER & SONS, JOHN, *Crescent, Cripplegate, London.*—Ship and portable water-closets, sanitary contrivances, flushing apparatus for high pressure. (*See page* 55.)

TAYLOR, JOHN, Jun., 53 *Parliament Street.*—Patent facing blocks, damp proof course, roof tiling, smoke consuming and ventilating grates, and other sanitary building appliances.

The inventor, in the course of his professional practice as an architect, has had his attention particularly directed to the following too frequent defects in house construction:—

1st. The heat in summer and the coldness in winter of a slate roof, and the want of a tiled roof that shall be as light, and laid to the same pitch as slates. 2nd. Wet penetrating brick walls, and the difficulty of preventing it except by undue thickness, or the aid of cement, paint, &c. 3rd. Damp rising up the walls from the foundations (the fruitful source of unhealthy dwellings), and the want of sufficient air beneath the floors for the prevention of dry rot, &c. He has been enabled to invent and successfully bring into use the following:—

TAYLOR'S PATENT TILING FOR ROOFS.—Slate is generally applicable for roofing, as it admits of being laid to a flat pitch and is light, but is so absorbent of heat that rooms in the roof become unbearable.

Plain tiling has not this objection, but must be laid to a steeper pitch, is much heavier, being nearly of double thickness, and requiring greater strength of timber.

Pan-tiling is lighter, but so pervious to weather, as to be only suitable for sheds and similar buildings.

TAYLOR'S PATENT FACING BLOCKS.—The defects of ordinary brickwork are—

1st. It absorbs moisture into its entire substance.

2nd. The through-joints admit wet into the interior.

3rd. A wall one-brick thick, although strong enough, is not stiff enough, there being no vertical bond.

4th. No one-brick wall can be fair inside and outside.

Concrete has not been used with success in walls, as it requires to be retained as in a trench, and its external surface cannot resist the action of the weather.

TAYLOR'S PATENT DAMP-PROOF COURSE.—In the construction of foundations, three essentials have been hitherto partially effected by as many separate means.

1st. Damp prevented rising up the walls, by a layer of asphalte, sheet-lead, slates in cement, &c.

2nd. The introduction of air by air-bricks at intervals.

3rd. Strengthening and bonding by the use of rough York stone, &c.

In the patent damp-proof course, these effects are combined.

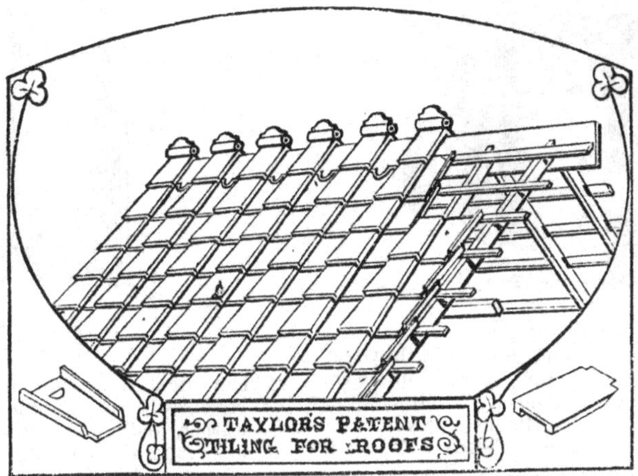

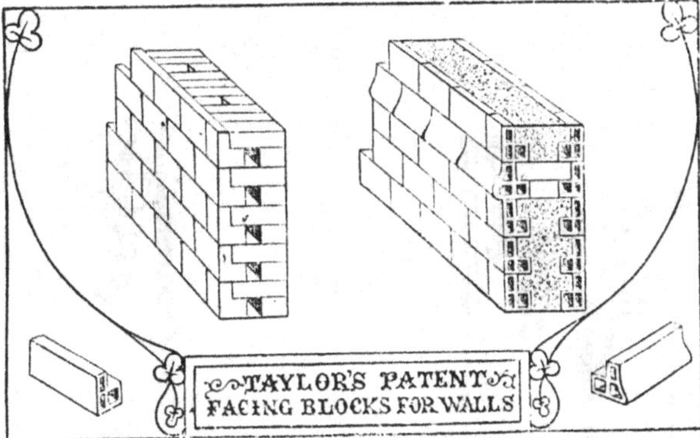

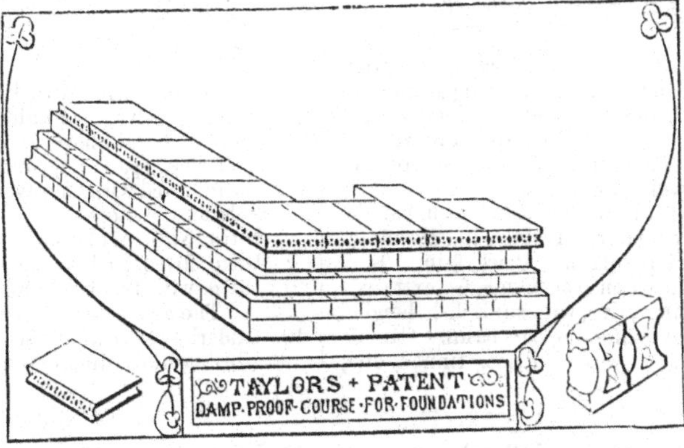

Constructed specimens of each of the above can be seen in the Court of the Eastern Annex, Class IX.

These inventions are now being extensively used in churches for the Ecclesiastical Commissioners, the Herbert Hospital now erecting for the War Office, and barrack huts at Hounslow, Colchester, &c.; also farm buildings and labourer's cottages for the Crown, and in a vast variety of villas and other buildings.

For all further information, supply, &c., apply to the offices, No. 53 Parliament Street, W.

The patent tiles may be laid to as flat a pitch as slates; their weight is 656 lbs. per square. Countess slating is 640 lbs. per square; plain tiling is 1,624 lbs. per square; pan-tiling is 840 lbs. per square. Thus it appears that it is as light as slating, and less than half the weight of the ordinary tiling. It is thoroughly rain and snow-proof, extremely pleasing in appearance, and combines all the advantages of slates and tiling without the drawbacks attending them. Price the same as plain tiling.

The patent walls have—

1st. A dry area, or space, immediately within the external face, preventing the absorption of moisture, and rendering them cool in summer and warm in winter.

2nd. All the through-joints are intercepted.

3rd. The wall is strengthened by the vertical bond effected by the facing block.

4th. The work is fair inside and outside. Concrete for walls is retained by the facing block as in a trench, which also protects it from the action of the weather.

1st. Damp rising is completely prevented, by a highly vitrified and non-absorbent material having an air space through the joints.

2nd. Air is supplied through the perforations, securing a circulation beneath the surface of the walls.

3rd Strengthening and bonding are effected by the use of an imperishable material, capable of sustaining 600 feet of vertical brickwork upon each superficial foot. These are economically combined in the one article, with a saving of one course of brickwork in height in the building.

UNDERHAY, F. G., *Crawford Passage, Clerkenwell, London, E.C.*—Underhay's patent regulator water-closets, high-pressure valves, and basin apparatus.

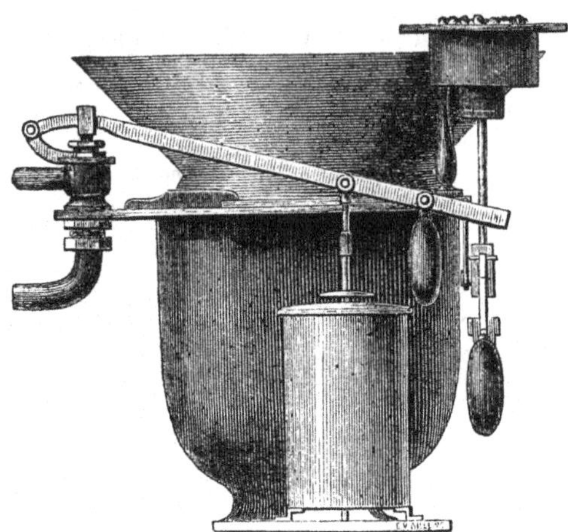

PATENT PAN CLOSET.

These closets (Underhay's make) are in use at the Great International Exhibition (Galleries); Horticultural Gardens, Kensington; Houses of Parliament; Windsor Castle; Crystal Palace; Grosvenor Hotel, Pimlico, &c. &c.

1. Regulator pan closet, complete, as above	£2	10	6
Ditto, best quality	2	13	6
2. Regulator valve closet, with flat plate and white basin	3	10	0
Ditto, best quality	4	4	0
Extra for sunk dish	0	4	6
Ditto, fancy basin	0	4	6

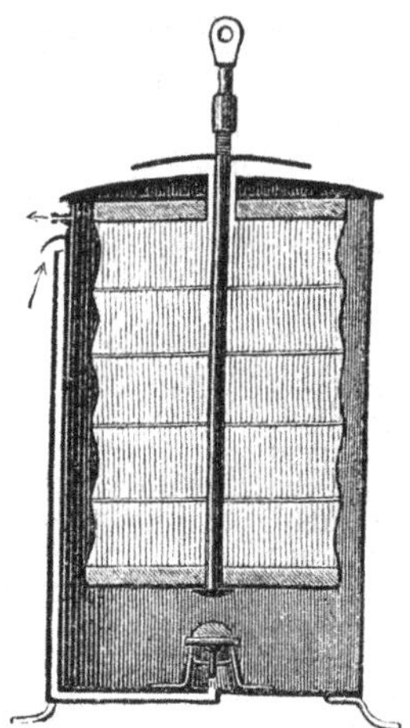

PATENT REGULATOR.

3. The patent regulator consists of a cylindrical metal vessel, with an internal diaphragm of prepared leather, and valve at the bottom. On lifting the diaphragm the regulator becomes charged with air through side tube, and on depressing it the air is expelled through the small orifice above side tube, by the size of which the time elapsing in emptying the regulator is determined. Price 10s. 6d.

4. Underhay's patent lever valve and regulator, with sunk dish, handle, and weight. This apparatus can be used with any kind of closet, and can be had fitted complete on an iron frame, dispensing with all trouble in fixing, as it then only requires screwing to the floor (as above). Price 25s.
5. Patent lever valve, with tinned end, sunk dish, handle, rods, weight, and regulator. Price, ¾ in., 17s. 6d.
6. Ditto, with union and regulator. Price, ¾ in., 19s. N.B. Nos. 5 and 6, fitted on iron frame, 6s. extra.
7. Patent lever self-closing valve, with sunk dish, handle, and weight, for supplying hopper basin. Can be screwed to back wood-work. Price, tinned end, ¾ in., 7s. 6d.

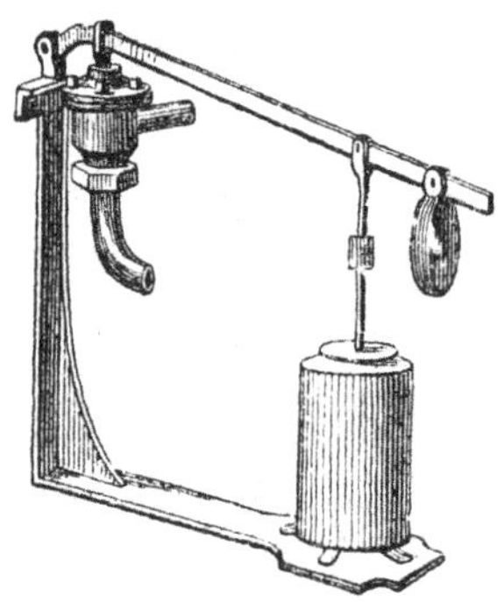

8. Iron frame, fitted with Underhay's patent regulator, ¾ supply valve, and union, which can be attached to old closets (as above). Price £1.

9. Underhay's patent apparatus for washhand basin, fitted with engraved ivory knobs for cold and waste water	£1	15	0
Ditto, with regulator	2	5	0
Ditto, fitted for hot, cold, and waste water	2	12	6
Ditto, with two regulators	3	12	6

The above prices are exclusive of basins.

10. Underhay's patent self-closing valves (flush with basin when fixed)	0	3	9

---

11. Underhay's patent equilibrium ball valve *for very* high pressures.

This valve will work equally well under high or low pressures, requiring but a small ball and comparatively short rod. The water continues to run *full on* till the cistern is *nearly full.*

Round shank, price, with copper ball, ¾ in.	0	7	7

12. Improved extra strong round-way screw-down bib and stop cocks; these cocks cannot leak between the spindles and cap, are very durable, easily re-washered, and specially calculated for high service and constant pressure.

Bib, round shank, price, ¾ in.	0	5	0
Stop, price, ¾ in.	0	5	0

WARNER & SONS, JOHN, *Crescent, Cripplegate, London.*—Ship and portable water-closets, sanitary contrivances, flushing apparatus for high pressure.

Obtained a Prize Medal in 1851.

JOHN WARNER & SONS, bell and brass founders to Her Majesty, hydraulic engineers, and manufacturers of fire engines, ships' pumps, patent brass and iron pumps, garden engines, lamps, urns, brazery goods, plumbers' work, water-closets, steam and gas cocks, lead, tin, and copper pipe, imperial standard weights and measures.

No. 148½.—Warner's patent pan closet, with regulating valve for high or low pressure. Any number of these closets can be fixed to one cistern.

No. 145½.—Warner's spring valve closet, on cast-iron frame, with vulcanised india-rubber valve and patent supply valve attached. Any number of these closets can be attached to one cistern.

No. 68.—Warner's brass lift and force pump for house purposes.

No. 383 and No. 384.—Warner's screw-down stop and bib cocks for high pressure.

No. 298.—Round shank bib cock.

No. 287.—Stop cock.

Closets and cocks in great variety.

No. 316½.—Patent equilibrium ball valve, with copper ball and rod, for the supply of cisterns at high or low pressure.

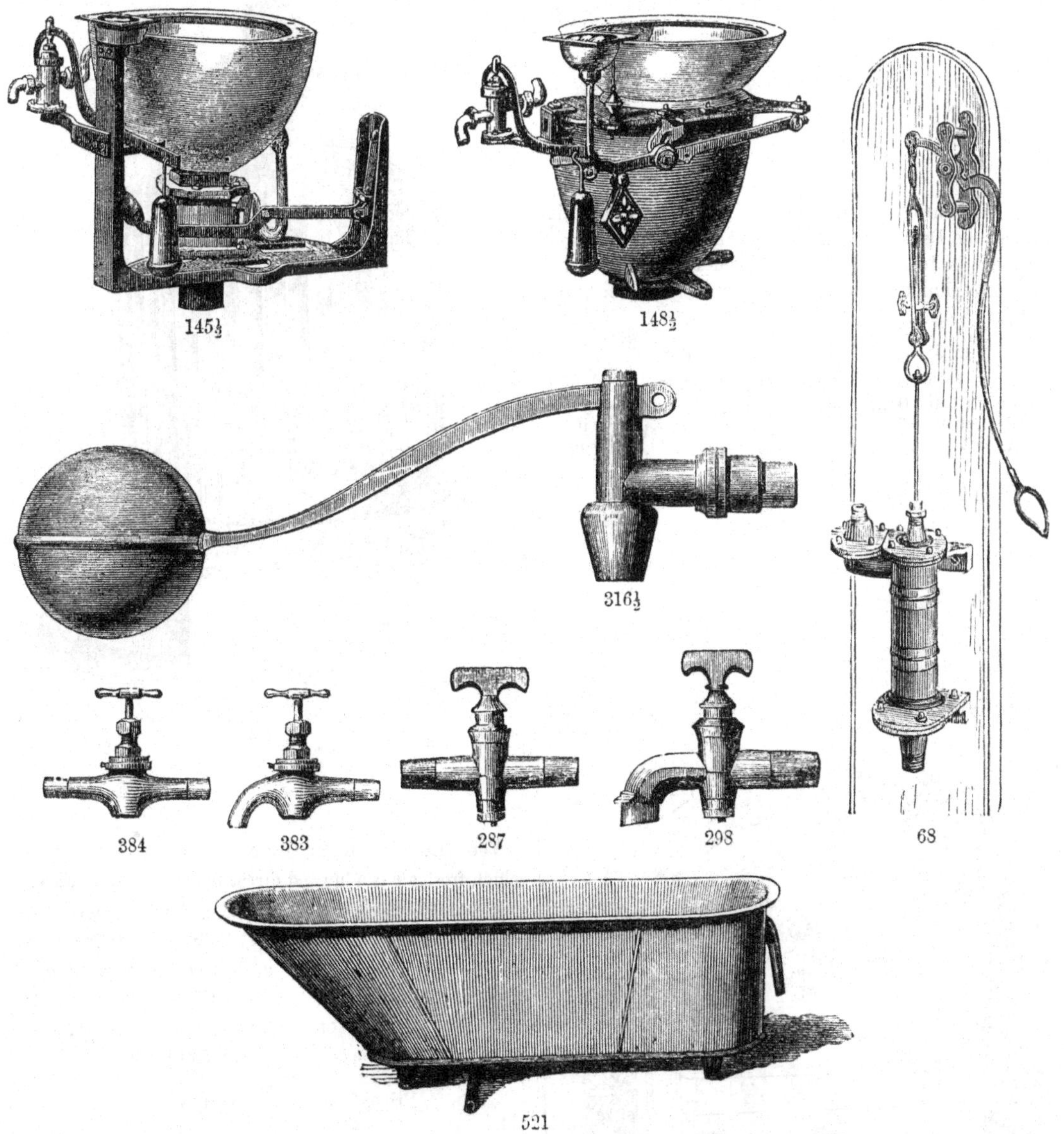

IMPROVED SHAPE ALBERT PATTERN ENAMELLED WHITE MARBLE BATH.

Baths of JOHN WARNER & SONS' manufacture can be had made in copper, tinned iron, or zinc, to any size; also, dome boilers, gas boilers, or tinned copper coils, for supplying baths, nurseries, or bed-rooms with hot water.

WOODCOCK, W., 26 *Great George Street, Westminster.*—Close stoves, open fire-places.

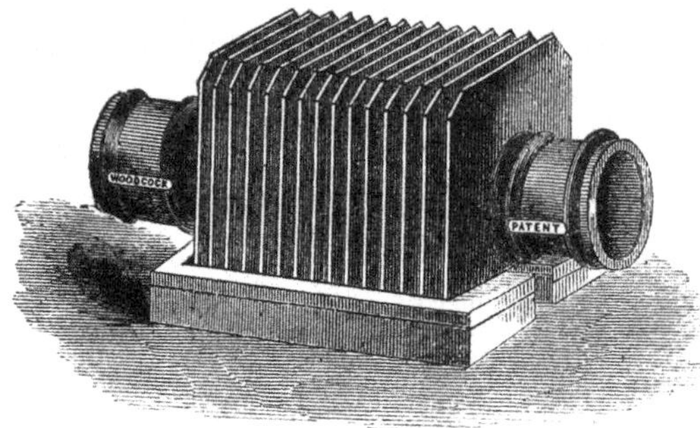

Hot Water Battery, for greenhouses, halls, &c. Advantages:—The condensing into a space of 14×15 inches the power of 25 feet of 4-inch pipe, inducing circulation of the air, and the ability to produce at will a perfectly dry, partially damp, or saturated atmosphere.

Price £2 5 0 each.

"The Gurney Stove," used in the Houses of Parliament, the Department of Science and Art, in numerous cathedrals, including St. Paul's and York Minster, and in many hundreds of churches, public buildings, and private houses.

	£	s.	d.	Capable of warming	
D	9	0	0 ...	15,000 cubic ft. of air	Or double the quantity if in daily use.
C	16	0	0 ...	30,000 "	
B	26	10	0 ...	70,000 "	
A	35	0	0 ...	120,000 "	

Ventilating Fresh Air Grate—Is set in a chamber, to which fresh air is admitted direct from the outer source, and warmed before passing through the front openings of the grate into the room. It is to a great extent a smoke consumer, *prevents all draughts from doors and windows,* and consumes about half the usual quantity of coal. By means of open spandrils, or otherwise, this stove can readily be adapted to any front required.

Price, from £2 2 0 upwards.

[2410]

WEST HARTLEPOOL HARBOUR COMPANY, *West Hartlepool.*—Model of harbour, docks, &c.

[2411]

WOODCOCK, W., 26 *George Street, Westminster.*—Close stoves, open fire-places. (*See page* 56.)

[2412]

WOODWARD, JAMES, *Swadlincote, Burton-on-Trent.*—Terra cotta chimney-tops, glazed sewerage-pipes, garden-edgings, &c.

SUB-CLASS C.—*Objects shown for Architectural Beauty.*

[2423]

BLANCHARD, MARK HENRY, 74 *Blackfriars Road, London.* — Patent articles in terra cotta.

1. Portion of a flight of patent terra cotta fireproof stairs, possessing great advantages over the ordinary stone stairs, in being fireproof, more durable, stronger, and, even with all the additional decorating, cheaper than stone.

The capability of this material for decoration of the most elaborate character is well exemplified in the grand staircase of the Turkish Baths, Victoria Street, Westminster.

2. A variety of useful articles and ornamental work, including tracery for window heads.

These will be found very economical, as regards price and durability, when compared with stone work.

The exhibitor desires to call particular attention to his patent reversible tesselated pavement, which is unsurpassed in elegance, cheapness, and durability.

He also directs the notice of contractors and others to his patent mile, distance, fencing, and telegraph posts, which are imperishable in any soil or climate.

Those who adopt them will find them to effect a great saving in time, labour, and cost, and to possess great advantages over wood.

[2424]

BOUCNEAU, A., 48 *Warren Street.*—Three statuary French and Italian marble chimney pieces: style, Louis Quatorze; Louis Seize; Italien.

[2425]

THE COUNCIL OF THE ARCHITECTURAL MUSEUM, 18 *Stratford Place, W.*—Architectural and decorative carvings in stone and wood.

[2426]

CLAY, C., 21 *Sidmouth Street, Regent Square.*—Inlaid marble table.

[2427]

EARP, THOS., 1 *Kennington Road, Lambeth.*—Marble, alabaster, and stone reredos; stone drinking-fountain; alabaster chimney-piece; oak lectern.

[2428]

EDWARDES, BROS., & BURKE, 142 *and* 144 *Regent Street.*—Three sculptured statuary chimney-pieces; mediæval monument. (*See page* 58.)

[2429]

FIELD, W. B., *Parliament Street.*—A marble column.

EDWARDES, BROS., & BURKE, 142 *and* 144 *Regent Street; and* 29, 30, *and* 31 *Warwick Street,* 17 *Newman Street, Oxford Street; Carrara, Brussels, and Invernettie, N.B.*—Three sculptured statuary chimney-pieces; mediæval monument.

A STATUARY MARBLE CHIMNEY-PIECE, CINQUE CENTO.

[2430]

FORSYTH, JAMES, 8 *Edward Street, Hampstead Road, London.*—Working model of statuary marble font, with carved oak cover. Executed for the Earl of Dudley, and fixed in Witley Church. Designed by Mr. S. W. DANKES, Architect. Richly carved oak bench ends for choir of Chichester Cathedral. Designed by Mr. WILLIAM SLATER, Architect.

MODEL OF MARBLE FONT IN WITLEY CHURCH.

JACKSON & SONS, GEORGE, 49 *Rathbone Place, Oxford Street.*—Specimens of *carton pierre* enrichments for architectural purposes,

Chimney-piece executed in *carton pierre.* The peculiar advantage of this mode of execution over wood is, that the material is not liable to shrink through heat. Greek candelabrum, executed under the direction of C. R. Cockerell, Esq., R.A; griffins, with candelabrum between; Louis XVI. door and over-door; various mouldings; ovals; centre flowers for ceilings; room cornices; compartment of ceiling, &c., showing the advantages of the material for lightness, sharpness of detail, and relief.

MAGNUS, GEORGE EUGENE, 39 *and* 40 *Upper Belgrave Place, S.W.*—Enamelled slate bath, billiard-table, chimney-pieces, door-way, stoves, &c.

Obtained the Prize Medal in 1851, *and two First-class Medals at the Paris Exposition,* 1855. *Also, Medal of the Society of Arts.*

1. HALL TABLE—representing Irish green, Verona, Sienna, and Genoa green marbles; lapis lazuli circle in back, with raised shield and ciphers.

2. HOT-WATER COIL CASE.—Black, with inlaid mosaic pilasters; Verona caps and bases.

3. STAIRCASE (portion of)—representing Sienna marble, with bronzed metal stringing and balusters, serpentine newel, and hand-rail; treads, risers, and soffit in grooves, and moveable.

4. BATH.—Enamelled pale green, with porphyry casing and capping, step, and riser.

5. LARDER FITTINGS.—Enamelled pale green wall-lining, and pink granite shelves and brackets.

6 to 13. CHIMNEY-PIECES—representing various marbles, and fitted with suitable grates for dining-rooms, drawing-rooms, libraries, &c.

14. PEDESTAL AND VASE.—Porphyry, inlaid black, with ormolu mounting.

15. PEN TRAYS.—Two carved pen trays, enamelled *vert de mer*.

16. CABINET WITH SLATE TOP.—The top of Sienna, inlaid with Florentine sprig, &c. &c.

17. PILASTER.—Ground of Fior di Persico, inlaid with lapis lazuli, Sienna, and other marbles, and bird centre.

18. PEDESTAL STOVE.—Porphyry, with fire-stone interior.

19. BILLIARD TABLE.—Magnus's patent, with rich Florentine mosaic subjects inlaid in maroon panels, on Verona ground.

20. PAIR OF SLATE DOORS.—Verona, with malachite sunk panels, enriched with ormolu mountings; the centre of malachite, with Florentine subject; the architrave mouldings of serpentine; wall-lining of rich Sienna, surmounted by entablature of Irish green, with serpentine moulding, and white figures and scrolls represented in relief.

21. FOUR ALTAR TABLETS.—Enamelled black, with sunk gilt letters and illuminated capitals.

22. SLATE CIRCLE, with Florentine bird and gilt moulding.

23. ILLUMINATED CLOCK, with enamelled slate case; lapis lazuli and various marbles, with rich metal mountings.

24. ETRUSCAN VASES (*three*) for Etruscan chimney-piece.

MAGNUS'S ENAMELLED SLATE obtained the medal of the Society of Arts, the prize medal of the Great Exhibition, 1851, and two first-class medals at the Paris Exposition. It is patronised by her Majesty the Queen, by the Empress of the French, by the princes of India, has been used in the seraglio at Constantinople, and in most of the Continental palaces. It is largely employed in Government buildings, in clubs, first-class hotels and railway stations; and is recommended and extensively used by our best architects in the mansions of the nobility and gentry. It is also adapted to houses of less pretensions, being handsomer, more durable, and not nearly half the cost of marble.

ARTICLES MADE:—

Chimney-pieces.
Cabinet stoves.
Baths and fountains.
Console slabs and brackets.
Hall tables.
Sideboards.
Hot water coil cases.
Columns and pilasters.
Vases.
Pedestals for statues and busts.
Altar tablets.
Clock dials.
Chess and other table tops.
Washstand and dressing table tops.
Wall linings.
Door-plates and handles.
Dairy and larder fittings, including wall lining to ditto.
Billiard tables, &c. &c. &c.

Inferior imitations of these beautiful productions are now being made. Architects are requested, in order to protect the public, themselves, and the inventor, to observe that the name of "MAGNUS" is on the under side of each piece.

Plain slate works of all descriptions in a very superior style, and at low charges.

Shippers and merchants will find MAGNUS'S Enamelled Slate more suitable for export than marble. Its great strength, ten times that of vein marble and statuary, renders it safe from breakage. For the frames and legs of billiard tables it is the only material that will withstand the effects of climate. (See Reports of the Juries of the Great Exhibition, p. 571.)

MITCHELL, J., *Walton Street, Brompton.*—A marble chimney-piece.

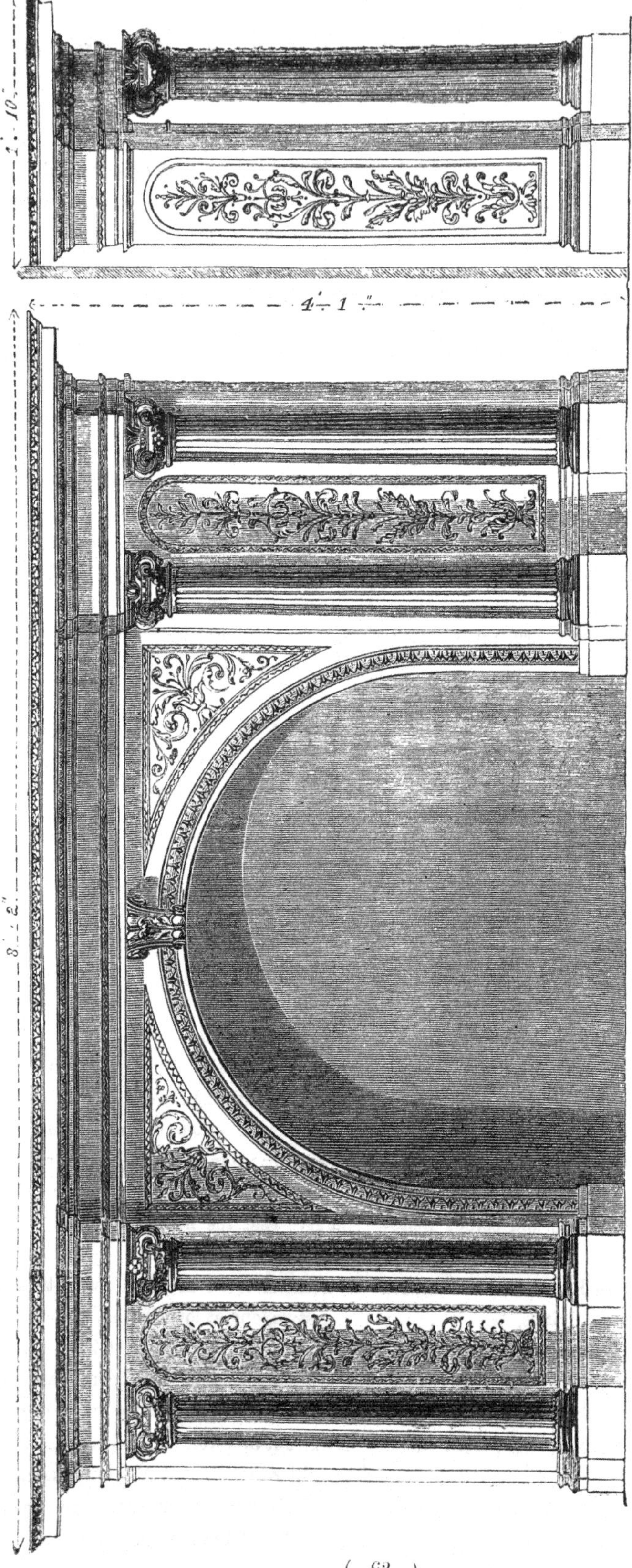

IN PURE STATUARY MARBLE, PRICE TWO HUNDRED GUINEAS.

GEORGE MITCHELL has a large selection of well-designed marble chimney-pieces and monuments, always ready for fixing and exportation. He undertakes to restore and to keep in repair marble and stone work, and will send directions for cleaning and restoring. An illustrated book and a priced catalogue will be sent post free on application.

[2431]

Georgi, Gustavo, 18 *Homer Street, Lambeth.*—Scagliola imitation of Florentine mosaics; ornamental models for marble chimney-pieces, ceilings, &c.

[2432]

Gomm, Henry J., 18 *Royal Street, Lambeth.*—Portion of a Caen stone chimney-piece, illustrating the "Cock and Jewel."

[2433]

Harmer, J. M., 10 *Thornhill Bridge Place, Caledonian Road.*—Models of architectural ornaments.

[2434]

Jackson & Sons, George, 49 *Rathbone Place, Oxford Street.*—Specimens of *carton pierre* enrichments for architectural purposes. (*See page* 60.)

[2435]

Magnus, George Eugene, 39 *and* 40 *Upper Belgrave Place, S.W.*—Enamelled slate bath, billiard-table, chimney-pieces, door-way, stoves, &c. (*See page* 61.)

[2436]

Mitchell, J., *Walton Street, Brompton.*—A marble chimney-piece. (*See page* 62.)

[2437]

Nesfield, W. E., *Bedford Row, W.C.*—A drinking fountain, mediæval style.

[2438]

Palmer, John Earle, *Guildhall, Swansea.*—Model of a font, in Maltese stone.

[2439]

Palmer, J. E., *Swansea.*—Model of a font.

[2440]

Poole, Henry & Son, 11 *Great Smith Street, Westminster.*—Marble mosaic pavement; incised and inlaid surface decoration in alabaster and stone.

Marble mosaic pavement, manufactured for the choir of Chichester Cathedral, from the designs of Mr. William Slater, architect. The portion exhibited comprises the richest part of the pavement. The central compartment will be placed immediately in front of the communion table. The entire work is composed of different kinds of marble, English and foreign. The design of the central portion shows a combination of conventional floreated ornament, with geometrical patterns. The floreated border is composed of Verde di Prata, Griotte, &c. The central cross, with the circle around it, is also much enriched with floreated ornaments in Italian, English, and Irish marbles. A considerable portion of the pavement is composed of rich geometrical patterns, so arranged as to give varied alternations of colour.

Doorway in Steetley Stone, designed by Mr. William Slater, architect, is to form the entrance to a mortuary chapel, now being built at Sherborne, for G. D. W. Digby, Esq.

The following marbles are employed in the shafts:—Red Devonshire spar, Irish green, Italian Sienna (yellow), and Staffordshire alabaster. The carving is executed by Mr. Samuel Poole.

Specimens of surface decoration:—

Staffordshire alabaster, incised and filled in with cement.

Inlaid Staffordshire alabaster of various tints.

Inlaid English stones of various colours.

Marble Mosaic.

Inlaid marble pavements, &c. &c.

[2441]

Pulham, James, *Broxbourne.*—Architectural and garden decorations, fountains, vases, figures, flower baskets, candelabra, &c., in terra cotta.

[2442]

Richardson, E., *Harewood Square, London.*—Mural monuments, &c.

[2443]

ROBERTSON & HUNTER, *Polished Granite Works, Wellington Road, Aberdeen.*—Drinking fountains of Aberdeenshire granites.

The exhibitors manufacture fountains, pedestals, columns, chimney-pieces, table tops, vases, curling stones, and all kinds of mural and other monumental erections.

[2444]

SERPENTINE MARBLE COMPANY, J. C. GOODMAN, 5 *Waterloo Place, Pall Mall.*—Vases, font, and pedestals.

[2445]

STANLEY, W., *Brighton Cottages, Earl Road, Old Kent Road, Camberwell.*—Enamelled slate, stone, and marbled glass.

[2446]

THOMAS, J., 32 *Alpha Road, Regent's Park.*—Carved chimney-piece in statuary marble, and ornamental grate.

[2447]

WESTMINSTER MARBLE COMPANY, THE, *Earl Street, Westminster.*—A sculptured marble chimney-piece.

Class XI.

MILITARY ENGINEERING, ARMOUR AND ACCOUTREMENTS, ORDNANCE AND SMALL ARMS.

Sub-Class A.—*Clothing and Accoutrements.*

[2466]

Carter, Lieut.-Col., *Monmouth.*—New accoutrements and boots for the soldier, the sportsman, and the tourist.

Obtained the large silver medal of the Society of Arts in 1847, *for his suspension of a knapsack.*

The weight of the knapsack falls equally on both shoulders, while the chest and arms are free. By means of the straps, the gun, the fishing rod, the artist's easel, &c., can be carried on the shoulders, where they ride the lightest, without the necessity of holding them with the hand, thus freeing the arms entirely for walking; and by means of the iron bars, being immediately under the shoulders, a considerable amount of weight can also be most conveniently carried. The basket can be increased in size, if desired, either for fish or game.

[2467]

Cattanach, William, *Sporran Maker Bankfoot, viâ Perth.* — Highland dress purses or "sporrans," with improvements.

[2468]

Firmin & Sons, 153 *Strand, London, and* 2 *Dawson Street, Dublin.*—Metal buttons and military ornaments.

[2469]

Holmes, Thomas, 15 *Princess Terrace, Regent's Park, and* 22 *John Street, Edgware Road.*—Improved self-acting cartouch box and military gaiter.

[2470]

Mackenzie, Captain J. D., *R.E. Office, Devonport.*—Light volunteer knapsack.

[2471]

Mitchell, H., 39 *Charing Cross.*—Photographs of British war medals, &c.

[2472]

Munn, Major, *Throwley, Kent.*—Cartouch-box, compact, light, and waterproof.

[2473]

Troubridge, Colonel Sir Thomas, Bt., C.B., 8 *Queen's Gate, W.*—Volunteer or tourist valise, suspended by a metal yoke.

SUB-CLASS B.—*Tents and Camp Equipages.*

[2486]

CLARKE, WILLIAM HENRY, 3 *Vernon Place, Bloomsbury.*—Models of ambulance waggons, &c.

[2487]

COTTON, CHARLES PHILIP, 8 *Lower Pembroke Street, Dublin.*—Model of improved tent.

[2488]

EDGINGTON, BENJAMIN, *Duke Street, London Bridge.*—Military tent, with stove, and models of other tents.

Obtained a prize medal in 1851.

MODELS AND DRAWINGS OF MILITARY TENTS AND MARQUEES.—Marquees, suitable for military purposes; also for agricultural and horticultural exhibitions, dinners, public meetings, &c.

IMPROVED MILITARY OR TRAVELLING TENT.—The two porches, and the complete ventilation, are improvements of great value—it can be erected with ease by two men; while its peculiar shape offers most effectual resistance to wind and rain.

[2489]

EDGINGTON, FREDERICK, *Thomas Street, Gloucester Place, Old Kent Road, London.*—Marquees and tents.

The exhibitor manufactures flags of all nations in silk or bunting; cricket marquees and tents; rick cloths; waggon cloths; engine and machine covers; sacks, ropes, &c.

Monster new marquees and tents may be had from the exhibitor on hire. Handsome and capacious marquees lined, floored, lighted, and tastefully decorated. For the erection of these marquees experienced workmen are sent to all parts of the kingdom.

The marquee shown in the illustration, 200 feet long by 40 feet wide, was erected at Lord Brownlow's, Great Berkhampstead, in October, 1860, upon the occasion of the Herts rifle contest.

	£	s.	d.
6 feet square.. ...	5	0	0
8 " "	6	10	0
10 " "	8	10	0
12 " "	10	10	0

The second illustration shows a square tent, of novel construction, without a pole in the centre, whereby an unbroken space is secured. This invention claims novelty as well as utility; it is light, portable, very strong, easy of erection, extremely simple, pretty, and cheap.

[2490]

EDGINGTON, JOHN, & Co., 17, *Smithfield Bars, E.C.*—Marquee; rick-cloth; temporary ball-room; three travelling tents; two garden tents.

[2491]

PICHLER, F., 162 *Great Portland Street.*—Folding and self-supporting tent.

[2492]

RHODES, MAJOR G. (S. W. SILVER & Co., 34 *Bishopsgate Street*).—Portable waterproof tents.

[2493]

Turner, Geo., *Northfleet, Kent.*—Improvements in the construction and fittings of tents and marquees (patented 1855).

Turner's Patent Tents and Marquees.—Adapted for military camps and hospitals; anglers, sportsmen, emigrants, tourists, gold diggers, &c.; also for railway and mining operations abroad.

These tents are suited to all climates; are perfectly waterproof; resist both heat and cold; are provided with every needful appliance for sleeping and cooking; are thoroughly ventilated; of great strength and stability; can be pitched and struck with ease and expedition; and stowed away compactly for transport. The chimney forms the support to the roof, and when packed occupies less space than a pole.

The hammocks are suspended at a sufficient distance from the ground, to keep the occupant dry, and out of the influence of those night damps which act so fatally upon man.

The ventilation is easily controlled by means of the sliding hood; and the screw pegs possess the greatest holding power—one weighing only 1 lb. being capable of a resistance of from 700 lbs. to 800 lbs. in ordinary turf.

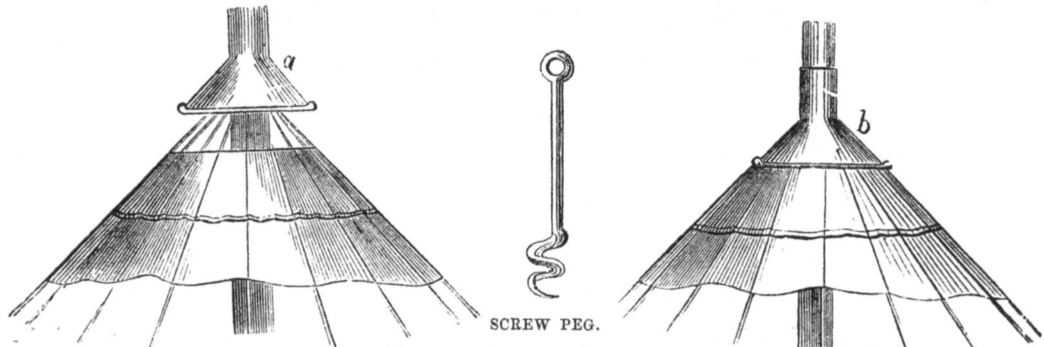

SCREW PEG.

(*a*) SHOWING THE VENTILATOR OPEN. (*b*) SHOWING THE VENTILATOR CLOSED.

For full sheets of engravings of tents and marquees for all purposes, pamphlets with particulars, reports of military officers on trials, testimonials, and prices, application should be made to the patentee, Northfleet, Kent, or McNeill & Moody, 23, Moorgate Street, London, E.C.

[2494]

Unite, John, 130 *Edgware Road, Paddington, W.*—Model of rick-cloth marquee, 30 ft. by 15 ft.; and round tent.

SUB-CLASS C.—*Arms, Ordnance, &c.*

[2505]

ADAIR, COLONEL SHAFTO, 7 *Audley Square.*—Military model of London and the adjacent country, in relief.

MILITARY MODEL OF LONDON.

Colonel ADAIR, A.D.C. to the Queen, Designer and Exhibitor.

Messrs. WILDE & Sons, New Cross, S.E., Modellers.

This model shows the defensive capabilities of London by forts, redoubts, and continuous lines, on an area of 22 × 14 miles.

The octagon of defence, = 50 miles 430 yards, is included within lines drawn through Woolwich, Anerley, S.E.; Kingston, S.; Twickenham, S.W.; Horsington Hill, W.; Harrow, N.W.; Hendon, N.; Stamford Hill, N.E.; East Ham, E.; measuring—

	M.	Yds.		M.	Yds.		M.	Yds.		M.	Yds.		M.	Yds.
S.E.	7	176	S.W.	3	264	N.W.	4	1584	N E.	7	1232			
S.	10	528	W.	4	1056	N.	4	1232	E.	7	1408	=	50	430

Each face of this polygon represents a line of battle, of which the works at the angles give *points d'appui*, and the intermediate works supports.

The scale is of six inches to the mile, with vertical augmentation to give appreciable relief. The ground slopes outwards, at a favourable angle for manœuvre and artillery fire. Sixty-two roads permit the sorties, which an interior railway system facilitates; and streams supply means of inundation, as in the marshes of the Lea and of the Brent.

The forts and permanent works are on the German trace, as best adapted to defence by direct fire, and by sorties.

For it is assumed that the fire of breech-loading ordnance, and of volunteer infantry will supply the principal defensive power; wherefore, in order to obtain effective fire, no re-entering angle, or angle of defence, should be less than a right angle.

It is also desirable that the works should be of a simple trace, but formidable, from a wide front of fire.

The forts are adapted to prolonged resistance; the redoubts secured from insult; the lines completely swept by flanking fire; and observed, in reverse, by the permanent works.

The slopes of the country are followed, so as to give low angles of depression from ramparts secured from enfilade.

The unit of calculation for construction and armament = 600 yards.

The forts and large redoubts are constructed in brick.

The lines are in earth, with a scarp in concrete; the main ditch has a lunette, and wide ramps for sortie. Mortar batteries are constructed with parados. The lowest command = 22 feet.

The casemated batteries are recessed, so that a rolling projectile would clear in descent the angle included between the terre-plein, and the face of the casemate.

The flanks of the bridge-heads on the W. and S.W. fronts are traced on sides of a triangle, whose base coincides with the mid-stream line.

The armament is calculated on the regulated war scale, less the difference between the mean service ranges of rifled and smooth-bored guns, multiplied into the relative rapidity of fire from breech-loading and muzzle-loading ordnance. This difference, on equivalent fronts, equals a deduction of $\frac{7}{16}$ from the received proportion.

All guns on ramparts are mounted on garrison carriages, in Haxo casemates, turned in concrete, and stepped in the splay.

Each casemated flank to be supplied with a turn-table, and built on a modification of the Haxo casemate, allowing the gun to traverse through 90°. Lead concrete to be used against injury from cone of blast.

The fire of casemated flanks grazes the line of defence, the ditch being swept by carronades, with a drop in front.

The re-entering angles in the bastioned lines, form *places d'armes* for sortie, under cover of the shoulder angle.

Nine points for forts on an inner line of defence are indicated on Telegraph Hill, Forest Hill, Tooting Common, the Ridgeway, in Richmond Park, near Mortlake, near Ealing, on Hanger Hill, Whembley.

Volunteer alarm-posts, stations for fire-brigades, and police picquets, are marked by red flags.

The lines having been carefully traced in advance of towns and villages, the amount of house property to be purchased is small.

100 yards are allowed in depth for works, and 300 yards additional for a military zone, on which few buildings now exist, and none could hereafter be constructed.

EXTERIOR SIDES.		LINES OF			GUNS.			
		Manœuvre.		Musketry.	Position.	Flank.	Mortars.	Carrons.
		M	Yds.	Yds.				
S.E.	Anerley	*7	588	11,900	476	99	48	†98
S.	Kingston..........	6	940	8,000	60	32	30	24
—	—	4	1360	5,900	85	16	—	16
S.W.	Twickenham....	2	1430	———	151	—	—	—
W.	Hanwell	3	1170	3,900	61	14	—	13
—	—	3	920	6,100	9	26	—	‡20
N.W.	Harrow..........	5	500	10,800	61	41	12	36
N.	Hendon	5	500	11,900	115	30	15	18
N.E.	Stamford Hill. .	8	620	20,700	173	52	24	49
E.	East Ham.......	8	920	10,400	131	40	—	36
	For Redoubts...				—	—	—	40
		54	148	89,600	1322	350	129	350

* Excluding Woolwich lines. † Including Woolwich lines.
‡ Inundation of Brent.

ESTIMATE.	ACRES.		
Land—Forts ...	1,090		
— Works ...	4,935		
— Lines ...	8,896		
	—— 14,921 ...		£1,492,100
Works—Forts ...		£450,000	
— Works ...		893,875	
— Casem. and Mort. Bats.		67,000	
Lines—Artillery ...		249,750	
— Musketry ...		623,200	
		———	... 2,283,825
Guns, 1,801—excluding 350 carronades ...			360,200
			£4,136,125

[2506]

ADAMS, ROBERT, 76 *King William Street, City, E.C.*—Patent breech-loading guns, rifles, and revolvers.

[2507]

AKRILL, ESAU, *Beverley, Yorkshire.*—Gilby's patent breech-loading and self-priming rifles.

Honourable mention at the Paris Exhibition, 1855.

The remaining stock of these rifles (of which specimens are exhibited), along with the sole right of manufacture, is to be sold on moderate terms; and also an entirely new breech-loading cannon, believed to be the most simple, rapid, and effective system of breech-loading for cannon yet invented. It may be seen by applying to E. Akrill, or to John Gilby, the inventor and patentee, Beverley, Yorks.

[2508]

ALLEN, JOHN WILLIAM, 22 & 31 *West Strand, London.*—Portable camp bucket, canteen, &c.

[2509]

ARMSTRONG, Sir WILLIAM GEORGE, C.B., *Newcastle-on-Tyne.*—Breech-loading and muzzle-loading ordnance, with projectiles; 2 fuses.

[2510]

BAKER, FREDERICK T., 88 *Fleet Street, E.C.*—Sporting guns and rifle.

[2511]

BAYLISS, E. & SON, *St. Mary's Square, Birmingham.*—Military implements.

[2512]

BIELEFELD, CHARLES, 21 *Wellington Street, Strand.*—Gun-wads, and cartridges.

[2513]

BIRMINGHAM SMALL ARMS TRADE, *Birmingham.*—Military rifles and guns, and pistols of various descriptions. (*See page* 7.)

[2514]

BLAKELY, T. W., 34 *Montpellier Square, London, S.W.*—Blakely's patent cannon.

BLAKELY ORDNANCE.

[2515]

BRAZIER, JOSEPH, *The Ashes, Wolverhampton.*—Sporting and military gun locks; implements; breech-loader barrels, actions, &c.

BIRMINGHAM SMALL ARMS TRADE, *Birmingham.*—Military rifles and guns, and pistols of various descriptions.

SPECIMENS OF MILITARY ARMS manufactured by the Birmingham Small Arms Trade for the English and Foreign Governments :—

1. *Small bore* (·451) rifle, with wind gauge sight.
2. *Small bore* (·451) rifle, plainer quality.
3. *Short Enfield* rifle, ·577 bore, with sword bayonet.
4. *Navy* rifle, ·577 bore, with sword bayonet.
5. *Enfield* rifle, ·577 bore, with bayonet.
6. *American* infantry rifle, ·580 bore, with bayonet.
7. *Italian* infantry rifle, ·702 bore, with bayonet.
8. *Brazilian* infantry rifle, ·584 bore, with bayonet.
9. *Spanish* infantry rifle, ·568 bore, with bayonet.
10. *Portuguese* infantry rifle, ·577 bore, with bayonet.
11. *Engineer's* rifle (Lancaster's patent), ·577 bore, with sword bayonet.
12. *Sergeant's* rifle, ·577 bore, with bayonet.
13. *Artillery* rifle, ·577 bore, with sword bayonet.
14. *Constabulary* carbine, ·656 bore, with bayonet.
15. *Cavalry* rifled carbine, ·577 bore.
16. *Cavalry* rifled pistol, ·577 bore, 10-inch.
17. *Cavalry* rifled pistol, ·577 bore, 8-inch.

BENTLEY and PLAYFAIR, Birmingham, Manufacturers.

21. Double-barrel fowling-piece.
22. Breech-loading double-barrel fowling piece, with front action locks.
23. Erskine's patent breech-loading double-barrel fowling-piece, with eccentric action to slide barrels forward.
24. Small bore, ·451 rifle, with wind gauge sight and movable shade.
25. Small bore ·451 rifle, full stocked, wind gauge sight, with shade.

COOPER and GOODMAN, Birmingham, Manufacturers.

26. Cooper's patent breech-loading rifle.
27. Cooper's patent breech-loading rifle.
28. Case containing a section of Cooper's patent breech-loading rifle, and parts with cartridges for the same.

ISAAC HOLLIS and SONS, Birmingham, Manufacturers.

29. Small bore, ·451 rifle, with wind gauge sight, and "Aston's" pattern rifling, warranted not to foul.
30. "Hay" pattern rifle, with 36-inch ·577 bore barrel and wind gauge sight.
31. Double-barrel fowling-piece, with laminated steel barrels.
32. Double-barrel breechl-oading fowling-piece.
33. Isaac Hollis and Sons' patent solid trigger-guard.
34. Isaac Hollis and Sons' improved sight shades and protectors.

KING and PHILLIPS, Birmingham, Manufacturers.

35. One rifled patent breech-loading cavalry carbine.
36. One rifle corps carbine, with patent breech-loading action.
37. One pistol hand breech-loading gun, sixteen squared barrel.

JOSEPH SMITH, Birmingham, Manufacturer.

38. Breech-loading shot gun, chain twist barrels.
39. Double gun, steel barrels.
40. Double gun, stub Damascus barrels.
41. Single rifle, ·451 bore, wind gauge sight.
42. Pair of under and over pistols, and implements.
43. A set of best gun implements.
44. A set of second quality gun implements.

C. P. SWINBURN and SON, Birmingham, Manufacturers.

45. Jacob's army double rifle, with sword, sighted up to 2000 yards.
46. Bailey's patent breech-loading military rifle, with bayonet.
47. Bailey's patent breech-loading rifled carbine for cavalry.
48. Small bore rifle, with patent lock and quadrant sight.
49. Swinburn's small bore rifle, with Newton's patent sight.
50. Case containing skeleton action, showing the principle of Bailey's patent breech-loader, with all its parts.

TIPPING and LAWDEN, Birmingham, Manufacturers.

51. Breech-loading double rifle.
52. Breech-loading single rifle.
53. Muzzle-loading shot gun.
54. Patent repeating pistol, in case.
55. Patent repeating pistol, in case.
56. Patent repeating pistol, in ornamental case.
57. Patent repeating pistol, in ornamental case.
58. Dressing case, fitted with patent repeating pistol.

THOMAS TURNER, Birmingham, Manufacturer.

59. Turner's patent rifle musket, with bayonet, ·452 bore.
60. Turner's patent rifle, plain, iron mounted, with concentric fore sight, with shade.
61. Turner's patent rifle, windage back sight, concentric fore sight, with shade.
62. Turner's patent rifle, best full stocked, with Turner's improved windage fore sight.
63. Turner's patent rifle, half stocked, pistol hand-octagon barrel, with Turner's improved windage fore sight.

JAMES WEBLEY, Birmingham, Manufacturer.

64. Wilson's patent breech-loading military rifle, ·577 bore, with bayonet.
65. Wilson's patent breech-loading short rifle.
66. Wilson's patent breech-loading rifled cavalry carbine.

B. WOODWARD and SONS, Birmingham, Manufacturers.

67. Breech-loading double gun.
68. Best muzzle-loading double gun.

PRYSE and REDMAN, Birmingham, Manufacturers.

69. Gun barrels in different stages of manufacture.

WM. TRANTER, Birmingham, Manufacturer.

70. Tranter's patent double action revolver, gilded.
Tranter's patent double trigger revolver, gilded.
Tranter's patent military gun locks, made by machinery, to interchange.

[2516]

BREECH-LOADING GUN COMPANY, *Great Portland Street, London.*—Guns, rifles, rests, slings, stadiometers; military percolators (patented).

The following are exhibited, viz.:—Single and double rifles, both military and sporting, secured by "Leetch's" and "Sturrock's" patents. They are rifled on "Scott's" patent cylindric principle; no special cartridge or wad is required; and they are the only breech-loaders which take the Government ammunition, with the skin, or any other whole cartridge. Sturrock's registered sitting rifle rests; the stadiometer, for judging distance, has been adopted by Government; King's registered rifle slings; Herr Mott's porous charcoal military water filters; Lefauchaux and muzzle-loading shot guns. All communications must be addressed to "The Secretary."

[2517]

BRIDER, GEORGE, 30 *Bow Street, Covent Garden, London.*—Implements for breech and muzzle-loading firearms.

[2518]

BRINE, LIEUT., R.N., *Army and Navy Club.*—Model of Crimean monument; specimens, &c.

[2519]

BRITTEN, BASHLEY, *Redhill.*—Projectiles for rifled cannon.

[2520]

BROWN, CAPTAIN, *Abbey Mills House, Romsey.*—Artificial parchments; compressed gunpowder cartridges; solid paper tubes; self-lubricating ramrod.

[2521]

BROWN, JOHN, 8 *Shelley Terrace, Stoke Newington.*—Repeating pistol, to fire fourteen times without reloading.

[2522]

BURNETT, CHARLES J., *Edinburgh.*—Various firearms; anti-Crimean elongated projectile, chain-shells, and other projectiles.

[2523]

CALISHER & TERRY, *Whittall Street, Birmingham, and Norfolk Street, Strand.*—Terry's patent breech-loading rifles and pistols. (*See page* 9.)

[2524]

CLINTON, LORD ARTHUR PELHAM, R.N., & HART, GEORGE W., *Southsea.*—Shot proof port, or embrasure, fitted for batteries or ships of war.

[2525]

COLLINSON, T. B., LT.-COL., *Chatham.*—Model of proposed land defences for Plymouth: models of parts of Preston and Colchester barracks.

[2526]

CULLING, CHARLES, *Downham Market, Norfolk.*—Patent safety gun.

[2527]

DAW, GEORGE H., 57 *Threadneedle Street, London.*—Patent central fire breech-loading gun and cartridge. (*See page* 10.)

[2528]

DONNELLY, Capt., R.E., *South Kensington Museum.*—Rolling drawbridge, requiring neither counterbalance, weight, nor extra length.

CALISHER & TERRY, *Whittall Street, Birmingham, and Norfolk Street, Strand.*—Terry's patent breech-loading rifles and pistols.

TERRY'S PATENT BREECH-LOADING RIFLES AND PISTOLS.

"THE TIMES," *July 22nd*, 1858.

"A breech-loading rifle carbine, the invention of Mr. Terry, of Birmingham, has been under test on board Her Majesty's ship *Excellent*, under the superintendence of Captain Hewlett, C.B., from May 10th until the present time; during which time 1,800 rounds have been fired from it with unprecedented accuracy at various ranges, without cleaning the weapon; which, notwithstanding, gives no recoil; in proof of which Captain Hewlett gave the inventors the following certificate, which is fixed on the stock of the gun:—

'This is to certify that I have seen 1,800 rounds fired from this rifle without cleaning.

'July 20, 1858. 'R. H. HEWLETT.'

The rifle missed fire but twice in the 1,800 rounds, and, whether discharged by officer or man, 86 per cent. were 'hits.' Yesterday the rifle was taken to the camp at Browndown, and its capabilities exhibited before the troops and Instructors in Musketry of the 15th Foot (Lieutenant Cuthbert) and Royal Marine Light Infantry (Major Lowder). The practice at 700 and 800 yards was marvellous, notwithstanding a very powerful wind, and will be continued to-day. Its advantages over the old pieces are, 3lbs. less in weight and five shots to one in time of firing, giving it the advantages of a revolver with a tremendous range, and no necessity for cleaning out under about a couple of thousand rounds."

The Terry rifle has, since the publication of the above, been supplied to the Sydney Government for the whole force of the mounted police; to the rifle corps at Queensland, Australia; to the whole of the free rifle corps, Adelaide, South Australia; to the mounted police of Auckland, New Zealand; to several companies of volunteers at the Cape of Good Hope, Bombay, Madras, Kurrache, India; to the whole of Her Majesty's 18th Hussars stationed at Brighton; and to a mounted volunteer corps at Shanghai, China.

The Adelaide free rifles have accepted many challenges, and have won every prize for which they have contended; including the gold cup and other prizes at Melbourne.

The Terry rifle is much prized by many sportsmen in England, and has been supplied to His Majesty the King of Denmark.

DAW, GEORGE H., 57 *Threadneedle Street, London.*—Patent central fire breech-loading gun and cartridge.

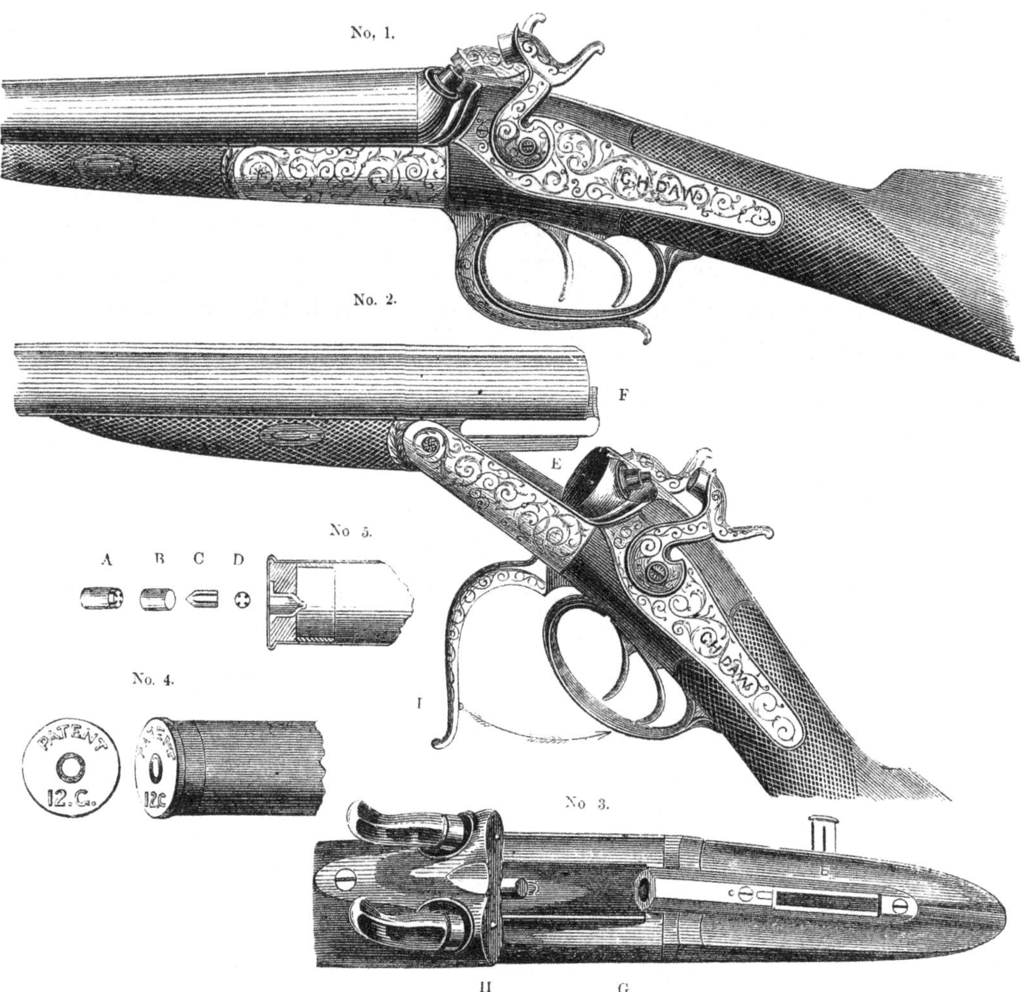

The extreme simplicity, safety, additional strength, and uniformity in shooting, of the above valuable patented invention, renders it the most perfect breech-loader ever yet produced. The parts are made by novel and patented machinery, insuring a degree of cheapness and accuracy never before attained.

No. 1. Gun complete.

No. 2. Gun opened ready for loading.

No. 3. Part of stock, with hinged fore-part, showing connection for barrels.

No. 4. Form of cartridge, with cap in the centre, and below the surface.

No. 5. Section of cartridge, showing brass cup, with communication hole in centre, and direct into the charge.

A Percussion cap, with brass anvil inside, ready to be placed in brass cup, as shown in Nos. 4, 5.

B Percussion cap.

C Brass anvil, with grooves for communicating the flame to the powder. The conical end is placed towards the fulminate, and receives the blow of the piston from the fall of the hammer.

D Bottom of anvil, showing the grooves and front part, which rest against the shoulder inside the cup, for resisting the blow of the piston.

E Piston points, for exploding percussion caps.

F Self-acting steel slide, for drawing out exploded cartridge cases.

G Socket for receiving and fixing steel bolt on the barrel lump at breech end of barrels.

H Steel bolt for locking and fastening the barrels.

I Lever connected with steel bolt for opening or closing the breech.

((Extract from *Bell's Life*, Nov. 8, 1861.)

"We have carefully examined the weapon ourselves, and we earnestly recommend it to the attention not only of our sporting readers, but also to the whole of the gun-making fraternity."

(*The Field*, Dec. 21, 1861.)

"The gun which is represented above has been tried in our presence with complete success by Mr. Daw, the well-known gunmaker, of Threadneedle Street, London. ....We therefore do not imagine that there will be the slightest tendency in this gun to get out of order, and, as far as we have been able to try it, we have the highest opinion of its merits."

The principle illustrated in the gun exhibited is applicable to every description of firearm.

[2529]

DOUGALL, J. D., 23 *Graham Street, Glasgow.*—Breech-loading firearms.

Patent lockfast breech-loaders, the mechanism of which interlocks the barrels and stock by a powerful lever and eccentric rod.

[2530]

DRYDEN, CHARLES, 10 *Denmark Street, Soho.*—Gun locks.

[2531]

DU CANE, Capt. EDMUND FREDERICK, 13 *Victoria Road, Kensington, London.* — Self-balanced iron shutters; rolling drawbridges, &c., without counterweights.

[2532]

EBRALL, SAMUEL, *Shrewsbury.*—A pair of double sporting guns and rifles; a double breech-loading gun, a double rifle.

[2533]

FAIRMAN, JAMES, 23 *Jermyn Street.*—1 double breech-loading gun, with sliding barrels; 1 double breech-loading gun, with drop barrels.

[2534]

FAWCETT, PRESTON, & Co., *Liverpool.*—Gun on carriage and boat-slide combined, for land and sea service.

Fawcett, Preston, & Co., are the designers and manufacturers of the rifled gun and wrought-iron carriage, for mountain service and mule transport.

This gun carries a 7lb. shell or 9lb. solid shot, and gives, with an elevation of 5°, a range of 1,800 yards, and is also specially adapted for a boat gun, for use afloat or ashore. The exhibitors have adapted to the carriage above illustrated a wrought-iron boat slide of simple construction, which, by the addition of a pair of wheels on board, is converted into a land carriage, and can be run ashore, obviating the inconvenience of carrying a separate carriage specially for land service in the boat, and lessening the difficulty of transferring the gun, from the boat slide to the land carriage in the ordinary way.

Fawcett, Preston, & Co. also manufacture all descriptions of heavy ordnance and field artillery complete, with carriages, limbers, &c. &c., in brass, steel, or iron, smooth-bored or rifled. They are also licensees under Blakely's patent.

[2535]

FAWCUS, GEORGE, *Alma Place, North Shields.*—Civil and military scaling ladders.

Key-bolts connect the ladders (in lieu of lashings); cleats, at the joinings on the side pieces, meet and support the ends, forming a continuous smooth surface for the hand to grasp and slide along.

[2536]

FOWKE, CAPT. FRANCIS, R.E., *Park House, South Kensington.*—Collapsing canvas boat pontoons; fire engine for military purposes. *See* SHAND & MASON, *Class VIII.*

[2537]

FOX, CAPT. & LIEUT.-COL. A. LANE, *Grenadier Guards, Park Hill House, Clapham, S.*—Model illustrating the parabolic theory, for the range of projectiles in vacuo.

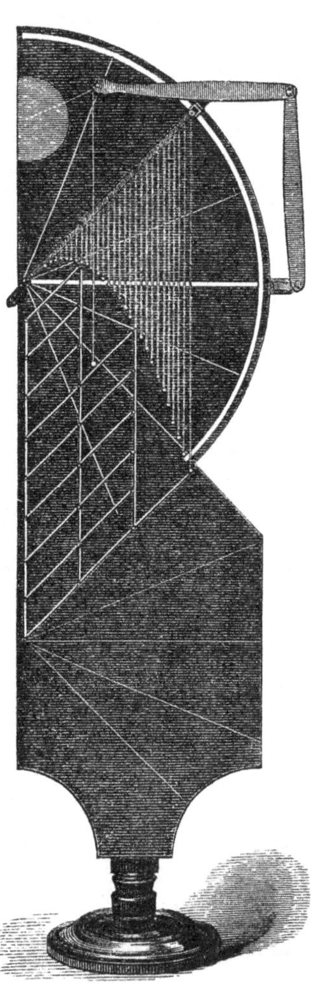

MODEL ILLUSTRATING THE PARABOLIC THEORY.

The three forces which combine to influence the flight of a projectile in the air are—

1. The velocity caused by the exploded gunpowder producing a movement of transition, in continuation of the axis of the piece; 2. The force of gravitation drawing the bullet to the ground; and, 3. The resistance of the air.

The parabolic theory deals only with the first two forces, viz., the movement of transition and the movement of gravitation.

In the model of the parabolic theory of projection in vacuo, the movable bar, to which the wires with white beads at their ends are attached, represents the line of fire. This bar can be set to any angle of elevation or depression. The movable bar is divided into 30 equal parts, representing the points at which the bullet, if influenced by the force of transition only, would arrive at the end of each successive second of time. In the present instance, it is supposed to move at the small velocity of 321·6 feet, or 107 yards, per second. From each of the thirty points on the movable bar a wire is suspended, with a white bead at the end; these wires increase in length, as the squares of the times according to the fall of gravitation, at each successive second of time. By this means a uniform curve is produced throughout the parabola, each bead representing one second of the actual flight in vacuo. The parallelogram apparatus is movable on pivots, by which it may be adjusted to any angle at which the movable bar may be set, the three points of the parallelogram always coinciding with the beads indicating the tenth, twentieth, and thirtieth second of time; thereby demonstrating, at all angles of elevation or depression, the operation of a compound force. To show how the range on any plan may be obtained from the impetus, another contrivance has been added to this model. The impetus is the height to which the ball would ascend if fixed vertically. At the distance of four times the impetus a protractor is fixed, and a movable arm is used to hold a thread, with a weight at the end, upon the surface of the board. To use it, mark off on the protractor an angle equal to the elevation, prolong the line till it touches the line of fire, let fall a vertical line upon the plane, and the intersection will be the range upon the plan.

For further particulars, see the "Journal of the Royal United Service Institution," volume v., page 497.

[2538]

GARDEN, R. S., 29 *Piccadilly.*—Punt gun on Prince's breech-loading principle.

[2539]

GIBBS, GEORGE, *Corn Street, Bristol.*—Breech and muzzle-loading double guns; sporting and target rifles.

[2540]

GLADSTONE, HENRY, & CO., 22 *Lawrence-Pountney Lane, London.*—"Capt. Haye's seamless skin cartridge."

[2541]

GISBORNE & BOLTON, 3 *Adelaide Place, E.C.*—Electrographic target and signal apparatus.

[2542]

GRAINGER, JAMES, 60 *Vyse Street, Birmingham.*—Gun, pistol, rifle, and military locks.

[2543]

GREENFIELD, JOHN, & SON, 10 *Broad Street, W.*—Portable Minié bullet compressing machine; selection of bullets, bullet-moulds, &c.

[2544]

GREENER, WILLIAM, *Rifle Hill Works, Birmingham.*—Rifle artillery, double guns, rifles, &c.

Awarded First Class Medals, Exhibition, 1851; *New York,* 1853; *and Paris,* 1855, *Two Silver Medals.*

WILLIAM GREENER is the inventor of the present improved system of gunnery, as attested by a public Parliamentary grant, session 1858.

Double guns, of the very highest quality, made also for exportation, at £6, £8, £12, £15, to £35 each.

Double rifles, from £8 to £40 each.

Single guns, from £3 to £15.

Single rifles, £3 to £25.

Elephant, lion, and tiger shell rifles.

Harpoon and sealing guns. Punt and all other descriptions of guns for wild fowl shooting.

The exhibitor contracts for every description of naval and military small arms to any extent.

W. GREENER, if required, can furnish these arms as perfect in their shooting as any rifle ever made, the "Whitworth" not excepted.

[2545]

HALE, WILLIAM, 6 *John Street, Adelphi.*—Hale's war rockets, comet shells, and apparatus for directing their flight.

[2546]

HARRINGTON, JOSIAH, 6 *Lansdowne Terrace, West Brixton, S.*—A self-priming musket.

[2547]

HEMMING & CO., 21 *Moorgate Street.*—Electric targets, iron roofing for churches, and other buildings.

[2548]

HODGES, E. C., 6 *Florence Street, Islington.*—Improved breech-loading actions.

[2549]

HOLLAND, HARRIS J., 98 *New Bond Street.*—Breech-loading guns and rifles, of various recent patents.

[2550]

JACKSON, RICHARD, 30 *Portman Place, Edgware Road, London, W.*—Rifle, muzzle-loader.

[2551]

JAMES, COLONEL SIR HENRY, *Ordnance Survey Office, Southampton.*—Maps, books, and instruments of the Ordnance Survey, with specimens of photozincography.

[2552]

JEFFERY, ALFRED, *Limehouse, E.*—Muzzle-loading rifled ordnance projectiles, showing application of Minié rifle principle to cannon.

[2553]

JEFFRIES, GEO., *Golden Ball Street, Norwich.*—Patent portable cartridge machine; patent breech-loading gun.

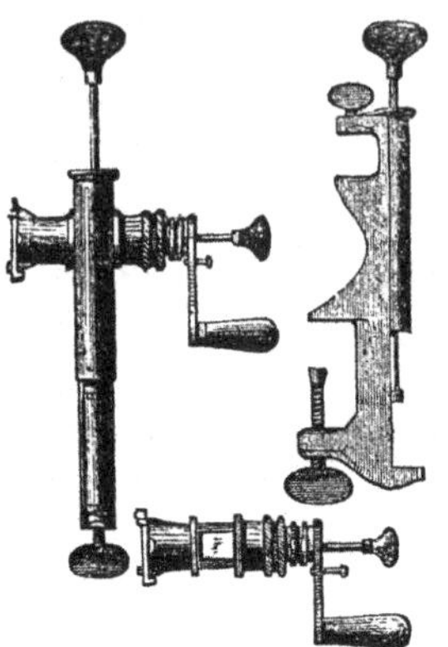

IMPROVED MACHINE FOR FILLING CARTRIDGES FOR BREECH-LOADING FIREARMS.

These machines have been extensively used for the last two years, and have just had several important improvements; the cross-box, or turnover tool, being made removable, so that one box can be changed for another. They are suitable for different sized cartridges, viz., 12 or 16 guage. This arrangement admits of the entire apparatus being fitted in an ordinary gun-case. It is of great importance to gentlemen travelling, and is the only patent machine in the kingdom. It is in general use amongst gentlemen and gunmakers in England, Ireland, and Scotland, and is sold by all respectable gunmakers.

The exhibitor is the sole maker and patentee.

[2554]

JONES, JOHN, *Serjeant Major, Brompton Barracks, Chatham.*—Iron-band gabion, sap-roller, field suspension-bridge, floating-bridge, field bedstead, ambulance litter, rafters for stabling and hutting, field trip.

IRON-BAND GABION FOR FIELD AND GARRISON WORKS.—Is formed of ten bands of galvanized sheet iron, 20-gauge, each three and a quarter inches wide, and seventy-seven inches long, fixed, in a circle, on twelve deal or other pickets, three feet long. It is constructed by two men in five minutes by passing the bands, basket-fashion, down the sides of the pickets. After many trials and reports of its efficiency, it supersedes, by an order from the Secretary of State for War, dated 11th December, 1860, the old cumbersome wicker gabion. Its weight with the pickets is 29 lbs. It is very convenient for carriage, and is very easily repaired in sieges.

SUSPENSION BRIDGE FOR ALL ARMS AND GUNS.—Is formed by any number of bands, from one to four, laid on each other according to the strength, width, and length of the bridge required, and connected together in their lengths by bolts and nuts through small holes at their ends. When thus connected, they form long strips, and so act as bearers to the superstructure or planking. The ends of the bearers are formed into loops, and fastened with bolts and nuts; through these loops is passed a horizontal pole placed behind two stout baulks of timber sunk in the ground, and hauled taut. In this way the bridge is steadied, and rendered fit for the service required of it.

SUSPENSION BRIDGE FOR INFANTRY.—Is formed as above, but without planking or superstructure of any kind, by placing the bearers close together, and weaving them at intervals with the gabion pickets. The pickets so woven, besides preventing a lateral separation of the bands, serve as cleats to keep the men from slipping in crossing.

FLOATING BRIDGE FOR ALL ARMS AND GUNS.—This is formed in the same way as the two preceding, using a greater or less number of bands according to the nature of the presumed weight to pass over it, as cavalry, infantry, artillery, &c., and is supported on piers of boats, pontoons, trestles, or other contrivances.

These three bridges, subjected to examination by a permanent committee of Royal Engineers, have been commended for their ingenuity, and a printed description of them has been ordered by the Secretary of State for War to be circulated to officers commanding Royal Engineers.

SAP-ROLLER.—Is made in the same manner as the gabion, except that the outer circle has two lengths of bands connected together, and the inner only one; the intervals between the two circles being filled with fascines.

FIELD BEDSTEAD.—AMBULANCE LITTER.—These can be formed in the field of rough timber or any other chance material, using the iron-bands in place of sacking or planking.

RAFTERS FOR STABLING AND HUTTING.—In the construction of these field services, the bands can be used for the rafters.

TRIP FOR CHECKING INFANTRY AND CAVALRY.—This is formed by laying the bands singly on the ground three or four feet apart, edge-wise and buttoned, and connecting them by wooden toggles attached to wire or rope three or four feet long. Thus connected, they are laid in rows, parallel and chequerwise, at any distance that may be considered best; the parallel rows being held in their places by rope or wire. This kind of obstacle would, on service, be found to occasion much more confusion than crows-feet, trous-de-loup, &c.

The price of the bands is determined by the nature of the contract entered into by the War Department.

The iron-band, primarily, is for the construction of gabions; but it can be adapted to the various field purposes above described, without, in any way, impairing its efficiency for gabions; and its use in these various forms enables the equipment of the Engineer Department to be considerably reduced; thus saving an immense expenditure to the country both in the provision of *materiel* and in transport.

[2555]

JOYCE & CO., *Upper Thames Street.*—Percussion caps, gun wadding, cartridges, &c.

[2556]

LANCASTER, ALFRED, 27 *South Audley Street, Grosvenor Square.* — Improved Lancaster rifle, specimens of fine sporting firearms.

[2557]

LANCASTER, CHARLES WILLIAM, 151 *New Bond Street, London.*—Breech-loading guns and rifles, for sporting purposes. Military rifles, oval bore. New method of iron-plating ships of war.

OVAL BORE CANNON: WROUGHT-IRON AND CAST-IRON SHELLS.—This gun illustrates the application of this system to cast-iron service guns. The specimen shown has fired 604 rounds of wrought-iron shell at high angles of elevation.

ADVANTAGES—Great range and accuracy when the elongated shell is used. Power to use molten iron in shell, and at the same time the service round shot and shell, as well as grape and canister, may be used without damage to the rifling, and with great precision, the range being equal to the usual service gun.

OVAL BORE RIFLES: MILITARY PATTERN.—The system of construction followed in these rifles may be described as follows:—The inside of barrel is cut by proper machinery in a spiral form, the difference between major and minor axis being 012 of an inch. This rifle is adopted in Her Majesty's service, being the arm of the Royal Engineers, and using the usual Enfield rifle ammunition, the system, as proved by actual service during the Indian Mutiny, gives the highest results as a military weapon and arm of the first precision. The trials just conducted at Woolwich, by order of the War Office, have resulted in the complete success of these rifles.

BREECH-LOADING SHOT GUNS.—The special advantages of this system consist in the absence of any pin to the cartridge, a perfectly central fire, and extreme simplicity of mechanical arrangement.

PROTECTING THE BOTTOMS OF IRON SHIPS.—Copper sheathing by this method is used to protect the portion of the iron hull below the water line. A layer of bitumen is placed upon the iron and screw rivets at the junction of the plates to prevent the disruption of the copper sheathing. The bitumen interposed between the copper and iron completely prevents any galvanic action.

[2558]

LANG, JOSEPH, 22 *Cockspur Street, London.*—Guns; rifles; new improved revolvers and other pistols; air guns; percussion walking-stick gun and rifle.

[2559]

LEETCH, JAMES, 68 *Margaret Street, Regent Street, London, W.*—Breech-loading firearms, empty cartridges, models, &c.

[2560]

LEWIS, GEORGE EDWARD, 32 & 33 *Lower Loveday Street, Birmingham.*—Sporting guns; pistols; military and sporting rifle.

[2561]

LONDON ARMOURY COMPANY, *Bermondsey.*—Long Enfield rifles, machine-made; Kerr rifles, ditto; Kerr's revolving pistols, Adams' ditto.

[2562]

LOVELL, MAJOR J. W., *Brompton Barracks, Chatham.*—Sap shield.

[2563]

LUCAS, W. H., 109 *Victoria Street.*—Model of a self-adjusting rolling bridge for forts, requiring no counterweights.

[2564]

MACINTOSH, JOHN, 40 *North Bank, Regent's Park.*—Breech-loading firearms, ordnance, and cartridges.

[2565]

MANTON, J., & SON, 4 *Dover Street, Piccadilly.*—Best guns and rifles.

[2566]

MARRISON, ROBERT, *Great Oxford Street, Norwich.*—Self-extracting breech-loader, vertical cylinder cartridge chargers; specimen of ornamental engraving.

[2567]

MERSEY STEEL & IRON COMPANY, *Liverpool.*—Gun blocks; guns; armour-plates; and the first battery-plate ever broken.

[2568]

MILLER & PEARER, *Glasgow.*—Brass cannon.

[2569]

MONT STORM, WILLIAM, 3 *Rood Lane, E.C.*—Patent breech-loading military and sporting firearms.

MONT STORM'S BREECH-LOADING ARMS.—Breech-loading arms may be divided into twelve different systems or "species," and there are various "varieties" of at least eleven of these species. The twelfth species (Mont Storm's "SELF-SEALING CHAMBER SYSTEM") is of comparatively recent development, and its plan is adapted to universal application to every style and class of both military and sporting arms, or the ready conversion of the present muzzle-loading arms to breech-loaders. Some of its many points of merit may be enumerated as follows :—

It has a chamber, but no lever—lateral, vertical, or other—to catch in the accoutrements, dress, or bridle-rein.

It is confined to no special ammunition.

The charge may be varied, but the arm cannot be overloaded.

The explosion is within a solid chamber; the recoil is upon a solid breech.

The connection between the stock and barrel is strong, graceful, and "FIXED:" thus it is adapted for the use of the bayonet for infantry.

The opening and closing of the chamber is effected, with unprecedented ease and rapidity, with the mere finger and thumb, even when the weapon and the soldier are lying upon the ground; and in the case of cavalry in action, the left hand remains entirely free, to govern the reins. It is a perfect MUZZLE-LOADER. The force of the explosion, irrespective of special ammunition, CLOSES the joint, in contrast to its effect in other breech-loading arms: thus there is no escape of gas.

It cannot stick fast, or clog, by rust or powder dirt.

There is no sliding or abrasion of one surface upon another, in opening and closing the breech: thus there is no friction or wear.

In the insertion of the cartridge, the ball constitutes the handle or ramrod—an important feature.

It cannot be fixed, accidentally or purposely, till the chamber is locked in place; and the locking device is solid, "self-acting," and INFALLIBLE of operation.

It is extremely simple, involves no delicate parts, and cannot easily get out of order.

There are no specialities of lock, stock, barrel, or mountings; thus there are no mysteries in its repair, it is of economical construction, and any approved species of "SELF-ACTING PRIMER" may be applied to it.

These arms may be thoroughly and quickly cleaned without the application of WATER.

Though these arms are only now about to be brought before the public in this country, they have received more approbation and praise, at numbers of public trials before Governmental and military authorities in America, than any arm hitherto known.

[2570]

MOORE & HARRIS, *Military Contractors, Great Western Gun Works, Birmingham.*—Sporting guns, breech-loaders; English volunteer prize rifles, &c.

GUNS, No. 5,035 AND No. 5,050.

Bar and back-action sporting breech-loading double guns, with improved method of holding the breech ends of the barrels down to the breech-piece. One of the short-comings of the ordinary breech-loader being, that the hook-shaped contrivance for holding the barrels down, is not immediately under the strain with which it has to contend in the discharge of the gun; the centre of the breech-piece being cut away, all the strain acts upon the attaching of the turnpin (in the case of the breech piece), and upon a hook between the two barrels (in the other direction). The guns exhibited are constructed upon an improved, more perfect, and much simpler plan; the loop upon each barrel receiving the end of a steel tumbler-shaped bolt, by turning the lever to guard when the gun is closed, which bolt turns under the surface of the breech-piece, and is consequently much safer; the parts are less subject to wear either by use or the vibration of shooting. The weight of these guns does not exceed that of the ordinary muzzle-loader of same calibre.

Rifles, double and single, are also made on this principle, to which its advantages are quite as valuable and appropriate.

Price of improved breech-loading double guns, complete in case, with all implements, and leather cover, forty-five pounds (£45).

DOUBLE GUN, No. 5,052.

Bar double gun, so constructed that all the detonating is completed on the breeches, protecting the joint of false breech from the flame of cap or corrosion; and the side lock-screw passing through the iron, renders a well-proportioned and properly-striking hammer compatible with the above arrangement.

A very desirable feature in this gun is, that the part of the stock which is first destroyed by use, in the ordinary gun, is perfectly protected; and also, none of the sharp edges (which present themselves round the detonating of the regular bar gun), when the barrels are taken from the stock.

Price complete, with implements in case, and leather cover, thirty-five pounds—(£35).

Specimens of gun-barrel tubes, in the rough, oxidated, to show the variety of twists and figures produced by the manner of working Rose's patent iron; also, specimens for machinery made by the same process as the gun-barrel iron, but on a larger scale, showing more than four thousand square rods of iron and steel, as seen by the presented ends of the specimens; and which would be maintained and seen by microscopic power if drawn down to the sizes of the smallest specimens. The extent to which it will endure friction, without increasing temperature, is one of its valuable qualities for machinery and gunnery. Specimens of the same for cable-links.

[2571]

MORTIMER & SON, *Edinburgh.*—Single rifle in case, improved rifling and sight, breech-loading gun, double rifles.

[2572]

MURCOTT & HANSON, 68 *Haymarket, S. W.*—Samples of four patents for breech-loaders, and patent for firing explosive compounds.

[2573]

NEWTON, WILLIAM EDWARD, 66 *Chancery Lane.*—1. New gabion. 2. Improved riveting for interior slopes of field batteries. 3. Portable camp fire-place.

[2574]

PAPE, WILLIAM R., 36 *Westgate Street, Newcastle-on-Tyne.*—Sporting guns and rifles.

This exhibitor was, for two successive years, the winner of the great gun trials held in London in 1858 and 1859.

Those guns and rifles have won the approval of the most celebrated and experienced sportsmen at home and abroad, who have pronounced them the best sporting weapons of the present day.

The barrels are made from W. R. Pape's improved laminated steel, at his works, Newcastle-on-Tyne.

[2575]

PARFREY, Y., *Victoria Road, Pimlico.*—Breech-loading double gun, sliding action, drawing its own cartridges.

[2576]

PARSON, WILLIAM, *Swaffham, Norfolk.*—Six improved double guns.

[2577]

PARSONS, P. M., 9 *Arthur Street West, London Bridge.*—Patent breech-loading firearms. (*See pages* 18, 19.)

[2578]

PATON, EDWARD, *Perth.*—Pair of double breech-loaders, double breech-loading rifle, single long-ranged rifle.

[2579]

POTTER, JOHN, *Lynn, Norfolk.*—Breech and muzzle-loading double guns, and machine for compressing rifle bullets.

Parsons, P. M., 9 *Arthur Street West, London Bridge.*—Patent breech-loading firearms.

Parsons' Patent Breech-Loading Fire Arms.—This system of breech-loading is applicable both to small arms and ordnance, and consists mainly in constructing the plug to close the breech of a spherical form, and combining it with such other parts as are necessary to carry out its application in different ways, according to the particular purpose for which it is used. It also consists of improved methods of making the joint at the breech gas-tight and self-acting.

Figs. 1, 2, and 3 are sections of the breech portion of hand rifles, with the parts in the position they would be at the time of the discharge; the dotted lines in each figure show the position of the plug when the breech is

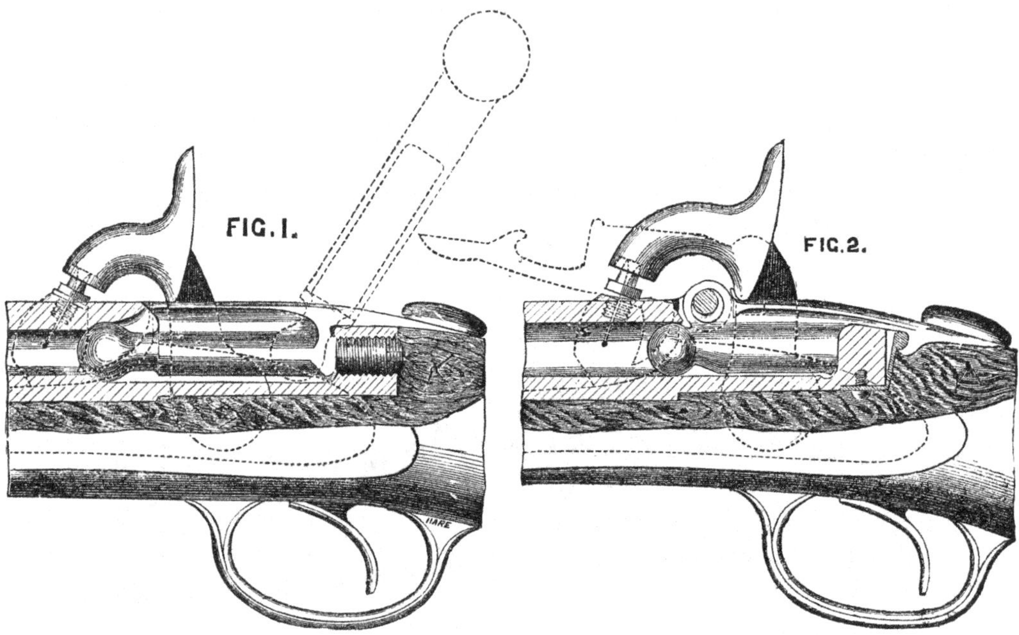

opened for the insertion of the cartridge. In Fig. 1, the plug consists of a bolt terminating in a knob of a spherical form, which slides in the chamber, and by which it is retained in its place. In opening the breech, the tail end of the plug is lifted up until it is released; the spherical knob, at the same time, turns on its spherical end in its seat, in the manner of a ball and socket; it is then drawn back to the position shown by the dotted lines. The plug can be removed from the chamber instantaneously when required, by a peculiar movement in one particular position, without removing any pins or screws, for the purpose of cleaning, or, if necessary, to render the arm for the time unserviceable; although it is impossible for it to escape from its place by accident. In Fig. 2, the plug is in one piece with the lid, which is hinged to the breech; in opening the breech, in this arrangement, the

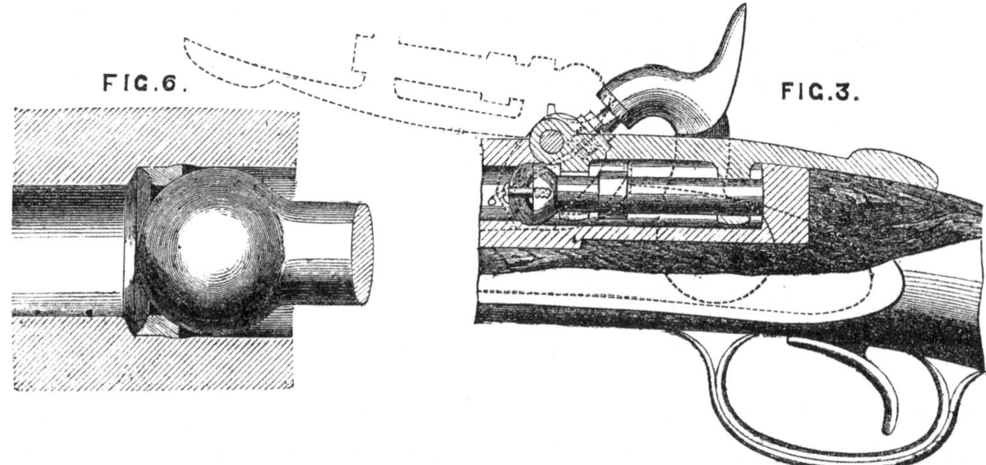

tail end of the plug is lifted, and the spherical end turns in its seat, as before, sufficient play being allowed in the hinge joint for that purpose; it is then turned back on its hinge joint, as shown by the dotted lines: in this arrangement, the gas valve for maintaining a tight joint, to be afterwards described, is not applied, it being intended to use a cartridge with a greased felt wad at its base, which supplies its place. In Fig. 3, the breech is opened in the same manner as in Fig. 2, except that the requisite play is allowed between the lid and the plug. In all these arrangements, the hammer is so arranged that when it falls on the nipple at the discharge of the piece, it also covers the lid, and thereby prevents all possibility of the plug rising out of its place.

PARSONS, P. M.—*continued.*

Figs. 4 and 5 show the application of this system to a cannon. In this, a complete sphere is secured in a chamber in the breech of the gun by a nut; both the sphere and the nut have a cylindrical hole through them corresponding to the bore of the gun. The sphere is capable of turning freely on its own centre in its seat, and motion is given to it by the lever. When in the position shown at Fig. 4, the charge is introduced through it and the nut; it is then turned a quarter of a turn to the position shown at Fig. 5, by which the aperture through it is brought across the bore, and the breech thereby closed. The lever is so arranged that it covers the touchhole, and thereby prevents the gun being discharged, except when the plug is in the proper position and the breech closed. The lever is also so constructed that it can be removed out of its place almost instantaneously, and the gun thereby rendered useless for the time, if necessary.

Two different methods of securing a self-acting gas-tight joint at the breech are employed, according to the form of the plug used. When it is of a conical form, or a sphere turning in its seat, as in Figs. 1, 4, and 5, a ring of copper or other soft metal is fitted in between the plug and the breech, as shown on an enlarged scale at Figs. 6 and 7, which the force of the explosion drives or wedges in between the spherical plug and its seat in the breech of the gun. When a flat surface is employed to close the breech, a hollow spherical cup, or

FIG. 4.

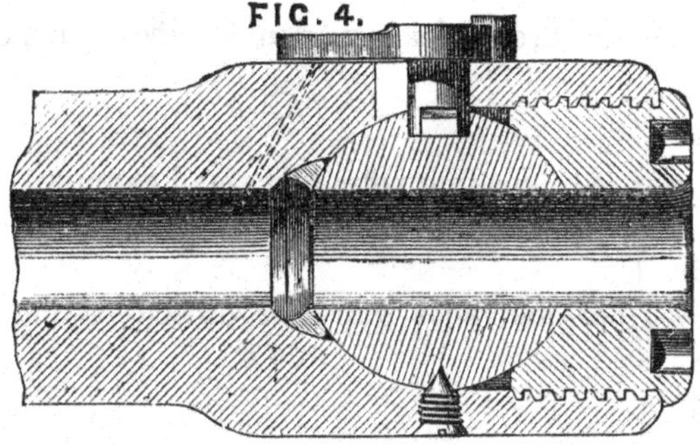

FIG. 5.

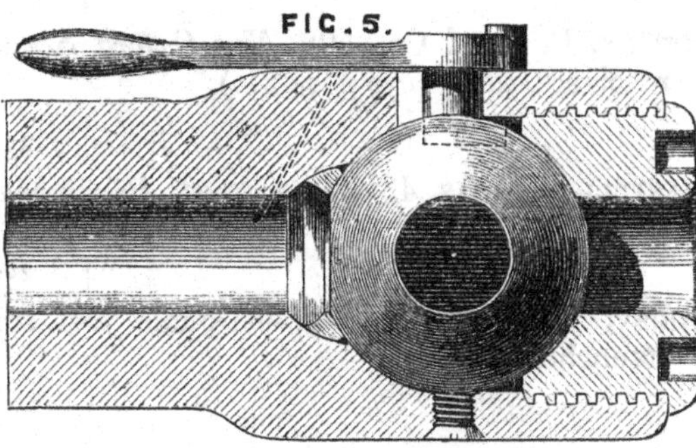

disc of copper, is attached to the plug, with its convex side presented to receive the force of the explosion, and its periphery resting against the base of the plug and the interior of the barrel or breech, as shown at Fig. 8. On the discharge of the piece, the force of the explosion tends to flatten and spread out the cup or disc, and its periphery is thereby forced into close contact with the interior of the breech, and the joint thereby made good.

The advantages claimed for these breech-loading rifles are, that they are easily and quickly loaded; perfectly gas-tight; so simple and strong that they cannot get out of order; so secure that it is impossible an accident can occur from their use, even through the greatest carelessness; and that they admit of being manufactured at small cost by self-acting machinery, and so that all the parts of any rifle may be interchangeable with those of any other. The arrangements shown at Figs. 2 and 3 have probably a slight advantage in facility of loading, but all the other points are strongly developed in the arrangement shown at Fig. 1: in fact, it is hardly possible to conceive a more secure or simple breech-loading arrangement, as the moving parts are reduced to a minimum—viz., to one—and that a simple bolt of hardened steel, which is proof against injury.

The advantages claimed for the arrangement adapted to ordnance are, that the surfaces which make the breech joint are perfectly protected from injury; and as in the

FIG. 8.

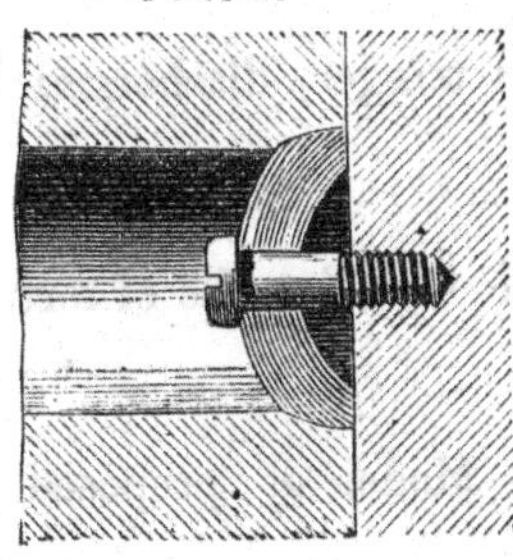

FIG. 7.

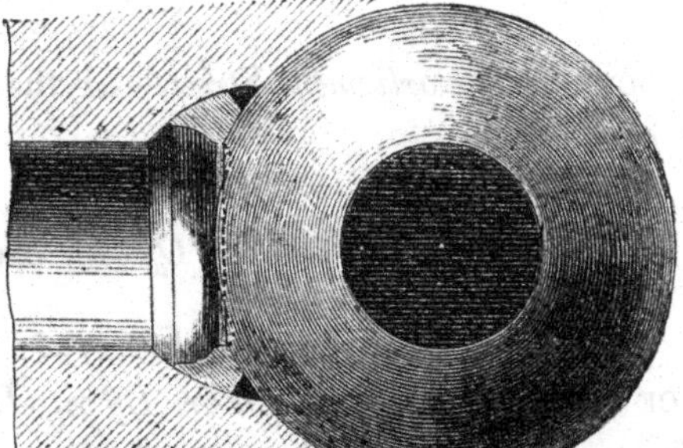

working they slide upon each other, they are constantly scraped clean, and free from deposit and other extraneous matter, and the joint is made tight by the explosion of the charge itself, and does not depend upon the proper tightening up of a screw or wedge. The opening and closing of the breech is effected by one simple movement of the lever, and the gun can consequently be loaded and discharged with the greatest rapidity, and with but a small fraction of the force necessary where a screw, wedge, or other mechanical appliance, is employed to force up the plug in opposition to the explosive power. The plug not having to be lifted out of its place in opening the breech, its weight is not on this account confined to any limit; it is, therefore, made five or six times as heavy as those that require removal in working the gun, by which a sufficient amount of inertia in it is secured, and the strain on the breech screw thereby very materially reduced. The strength of the breech is also not impaired by having a large slot cut through it. There is no chance of injury occurring to the breech screw from neglect in properly tightening up, as it is never unscrewed, except when required to remove the plug for the purpose of examination or cleaning. When guns of this description are used in casemate batteries, or between deck on board ship, the smoke from the discharge can be prevented from escaping at the breech by inserting the shot into the nut before opening the breech.

[2580]

PRINCE, F. W., 15 *Wellington Street, London Bridge.*—Improved breech-loading cannon and small arms.

[2581]

PURSALL, W., & Co., 45 *Hampton Street, Birmingham.*—Percussion caps; military and sporting ammunition.

[2582]

REEVES, CHARLES, *Charlotte Street, Birmingham.*—Swords, field and dress; military rifles, and rifles to load either at breech or muzzle.

[2583]

REILLY, E. M., & Co., 502 *New Oxford Street.*—Guns, breech-loaders, double rifles, patent revolvers, &c. (*See page* 21.)

[2584]

RESTELL, THOMAS, 43 *Broad Street, Birmingham.*—Breech-loading rifles and small cannon.

[2585]

RIGBY, WM. JNO., *Dublin.*—Double and single rifles and shot guns; a breech-loading staunchion gun, &c.

[2586]

RICHARDS, W., & Co., *Birmingham and London.*—Breech-loading rifles, guns, cannon, &c.

[2587]

SCHLESINGER, JOSEPH, *George Street, Birmingham.*—Patent needle firearm breech-loader.

[2588]

SCOTT, M., 26 *Parliament Street, Westminster.*—Sunken, but movable barrier, to exclude enemies' ships from ports.

[2589]

SCOTT, W. & C., 95 *Bath Street, Birmingham.*—Muzzle and breech-loading guns, rifles, and military small arms.

[2590]

SCOTT & ARBUCKLE, 18 *Parliament Street.*—Hythe position trigger, discharged by pressure instead of pull.

[2591]

SECRETARY OF STATE FOR WAR, *War Office, London.*—Models of works of fortifications.

[2592]

SECRETARY OF STATE FOR WAR, *War Office, London.*—Model of proposed barrack at Colchester.

[2593]

SECRETARY OF STATE FOR WAR, *War Office, London.*—Plan of Netley Hospital.

[2594]

SECRETARY OF STATE FOR WAR, *War Office, London.*—Plan for Herbert Hospital (Woolwich).

[2595]

SECRETARY OF STATE FOR WAR, *War Office, London.*—Plan for Regimental Hospital, for 60 and 120 men.

Reilly, E. M., & Co., 502 *New Oxford Street;* Branch Manufactory, 315 *Oxford Street.*—Guns, breech-loaders, double rifles, patent revolvers, &c.

Specimens of the following are exhibited:—

Improved Breech-Loaders, Double and Single, Sporting Guns and Rifles.—These guns are substantial, durable, and possess all the latest improvements, viz., "double grip," with lever over guard; "sliding action," to withdraw cartridge cases; "the lockfast," eccentric patent, or "vertical action," central fire, &c. They load with cartridges containing the entire charge, powder, shot, and primer, all in one. Flattering testimonials to the excellence of these guns have been received by the makers, from gentlemen in all parts of the country.

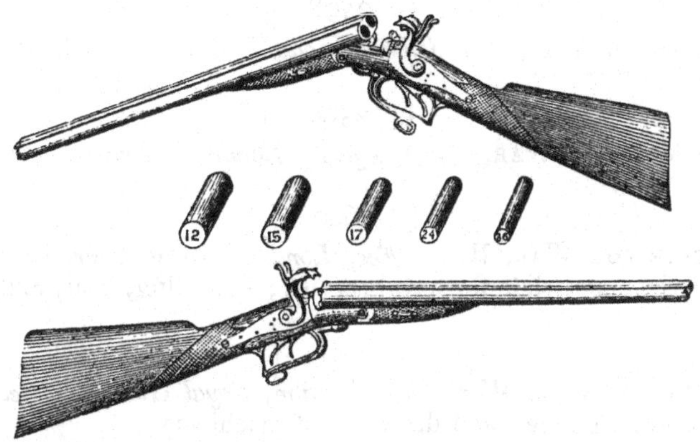

Breech-Loaders, of improved construction, Damascus barrels for the twelve or fifteen cartridge, good locks, chequered wood fore-end, well stocked, &c., plain finishing and engraving, from £21 0 0

Ditto with the latest improvements, lever over guard, the workmanship and shooting guaranteed, superior quality, Damascus barrels, finest locks, &c., made on any of the above-mentioned superior and approved systems, highly finished, &c. £31 10 0

Double Rifles on the same Principle.—Breech-Loaders of various calibres, 12 to 36 bore, for India, Africa, and the Colonies, also as pea rifles for rabbit and sea-fowl shooting, the prices commencing at . . 20 guineas

Double Fowling Pieces, Superior Double Rifles, &c.—Finest London manufacture.

Double-Barrelled Guns, fine Damascus barrels, 12 to 16 bore, superior locks, handsomely stocked, well engraved and highly finished . . . . 12 to 20 guineas.

The best Gun, with first-class Damascus or laminated steel barrels, very superior locks, first-class finishing, &c. £25 0 0

With Brazier's best locks, very superior finishing and engraving, in best oak case and apparatus complete 31 10 0

Double Guns in pairs, the barrels to interchange, fitted in double case complete, 30, 40, and 50 guineas the pair.

Superior Double-Barrel Rifles, of various calibres, from 12 to 40 gauge, the latest improved systems of grooving, flush and long-range sighting, carefully regulated for perfect accuracy . . . . 20 to 35 guineas

All the work is done under the immediate supervision of the exhibitors, and they undertake the repair and restoration of guns.

Every article of their own manufacture is guaranteed for efficiency and safety, having been subjected to the severest proof in the rough as well as finished state, and thoroughly tested for correct shooting.

On the premises they have a range of fifty yards for trials of shooting, and every facility for testing their weapons.

Orders by letter will be attended to punctually, and the goods, if ready, forwarded by same evening's railway train

Revolvers on the latest approved systems, the Ordnance pattern, improved double action lockworks, for quick or slow and accurate firing, direct action lever rammers, bolted cylinders, &c. &c. Likewise Adams's patent, recently perfected and simplified. Deane's, Harding's, Tranter's, Colt's, and others in endless variety.

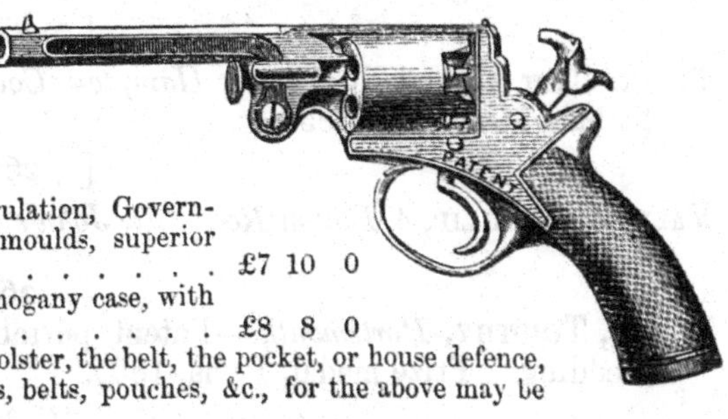

A revolver of the pattern shown in the illustration, regulation, Government No. 54 bore, in best oak case, with conical moulds, superior fittings, and apparatus complete, price £7 10 0

The same revolver, highly finished and engraved, in mahogany case, with best apparatus £8 8 0

Excellent self-acting revolvers, of various sizes, for the holster, the belt, the pocket, or house defence, with cases and apparatus, from five guineas. Holsters, belts, pouches, &c., for the above may be obtained from the exhibitors.

[2596]

SECRETARY OF STATE FOR WAR, *War Office, London.*—Models of ambulances.

[2597]

SECRETARY OF STATE FOR WAR, *War Office, London.*—Models of 12-pounder Armstrong gun, carriage and limber, forge waggon, rocket carriage, &c.

[2598]

SECRETARY OF STATE FOR WAR, *War Office, London.*—Model of ballistic gun and pendulum.

[2599]

SECRETARY OF STATE FOR WAR, *War Office, London.*—Armstrong rifled 100-pounder cannon.

[2600]

SECRETARY OF STATE FOR WAR, *War Office, London, Royal Carriage Department.*—Armstrong guns; carriage and slide for naval service; travelling, boat, and field carriages.

[2601]

SECRETARY OF STATE FOR WAR, *War Office, London, Royal Gun Factories.*—Rifled ordnance and their details, tools, gauges, and drawings of machinery.

[2602]

SECRETARY OF STATE FOR WAR, *War Office, London, Royal Laboratory.*—Cartridges, fuzes, projectiles, tubes, &c.

[2603]

SECRETARY OF STATE FOR WAR, *War Office, London, Royal Small Arms Factory.*—Specimens illustrating the manufacture of the Enfield rifle and small arms, &c.

[2604]

SECRETARY OF STATE FOR WAR, *War Office, London, Chemist.*—Fuses to be fired by magneto-electricity and electro-magnetism.

[2605]

SMITH, GEORGE, 40 *Davies Street, Berkeley Square.*—Sporting guns and rifles.

This exhibitor was for many years in the well-known establishment of J. Purdey. He holds the office of honorary armourer to the National Rifle Association. His improved breech-loading double-barrelled rifle for deer-stalking has no superior in power and correctness.

[2606]

SYLVEN, THOMAS, 33 *Leicester Square.*—A muzzle-loader altered to breech-loader; barrels and action; breech-loading gun.

[2607]

TREEBY, TWY, 1 *Westbourne Terrace Villas, Paddington, W.*—Self-indicating target, saves the lead, and requires no marker.

[2608]

TRULOCK & HARRISS, 9 *Dawson Street, Dublin.*—Breech and muzzle-loading sporting guns and rifles, single and double.

[2609]

TYLER, CAPT., Royal Engineers, *Hampton Court.*—Sheet iron gabion, prepared and sent to Chatham in August, 1853.

[2610]

VALLANCE, PHILIP, 4 *Bolton Road, St. John's Wood.*—Telescopic rifle sights.

[2611]

WHITE, TIMOTHY, *Portsmouth.*—Patent portable barracks and buildings; improved system of bedding. Prize medal, Paris, 1855.

[2612]

WHITWORTH RIFLE AND ORDNANCE COMPANY, *Sackville Street, Manchester.*—Rifled ordnance, gun carriages, small arms, and ammunition.

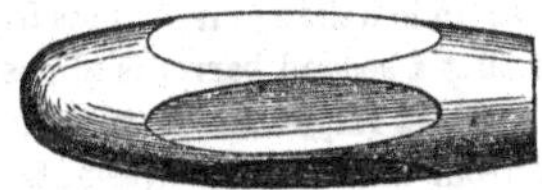

THE WHITWORTH ORDNANCE COMPANY, Sackville Street, Manchester, sole manufacturers of the Whitworth Ordnance, exhibit rifled ordnance, ranging in size from the 1-pounder to the 70-pounder gun—

- 1-pounder muzzle-loading rifled cannon, mounted on carriage.
- 6-pounder, ditto, ditto.
- 6-pounder breech-loading rifled cannon, without carriage.
- 12-pounder brass rifled field piece.
- 32-pounder rifled ship's cannon.
- 70-pounder, ditto.
- Projectiles of various weights, solid shot, and shell of all sizes from 1-pounder to 70-pounder.
- One of the flat-fronted projectiles which were fired through the armour plates and side of the *Trusty* during official trials at the Nore.

The bores of all these cannon are hexagonal in the cross section.

Rifling.—The pitch is in all cases equal to 20 times the diameter of bore.

Material of cannon may be steel-iron, brass, wrought or cast iron; but one of the two first-mentioned metals is preferred.

Projectiles.—Solid projectiles are usually cast, and then planed. One man will mould 200 of the 12-lb. shot per day; one man can plane the same number per day; or the projectiles can be cast so as not to require planing. Hollow shot are treated in a similar manner, and then filled in the same manner as the ordinary spherical shells. No special fuse is required, as the flash of the explosion ignites a fuse in the front, placed and used like the ordinary simple time fuse.

Ranges.—The average ranges obtained from the 12-pounder rifled cannon, with a 12-lb. shot and 1¾ lbs. powder, are at point blank 380 yards; at 1°, 900 yards; at 5°, 2,600 yards; at 10°, 4,500 yards; at 20°, 7,000 yards; at 35°, 10,000 yards, or nearly 6 English miles.

The charge of powder usually employed is equal in weight to one-sixth the weight of shot.

Penetration through iron armour-plates has been in all cases successfully effected, when hard metal flat-fronted shot have been used. As a general rule, the calibre of the guns employed should be slightly in excess of the thickness of metal to be pierced, and they will then send their hard metal flat-fronted projectiles through armour-plates of forged or rolled wrought iron placed either upright or inclined to the perpendicular. Flat-fronted projectiles may be made to pass through water to a considerable distance, and they will then retain great penetrating power.

The drill of the gunners with the rifled muzzle-loading cannon is similar to that practised for the smooth-bore cannon.

The Whitworth cannon is equally applicable for the use of solid shot, hollow shot, shell, tubular and flat-fronted shot.

The tangent back-sight is elevated by a rack and pinion, the latter having a micrometer wheel for finer readings than the divisions on the tangent stem allow.

Gun Carriages.—These are made of the simplest form of construction, and of the fewest number of parts; so that they may be constructed wholly by ordinary machinery.

THE WHITWORTH RIFLE COMPANY, Sackville Street, Manchester, are the manufacturers of the Whitworth rifles, a case of which is exhibited, containing sporting and military rifles of various weights and lengths.

The bore is hexagonal, being 45″ measured across the flats, and 49″ across the corners of the hexagon. The pitch, or rifling turn, is one in 20 inches.

The average figure of merit at 500 yards is four inches.

The cartridges contain the powder charge, lubricating wad, and projectile, arranged in proper order, so as to be pushed down into the barrel by the ramrod at one operation, and without reversing the cartridge.

The front-sight can be moved across the barrel by screw, and the slide of the *back-sight* is raised and adjusted with great nicety by means of a double-acting rack and pinion.

[2613]

WILKINSON & SON, 27 *Pall Mall.*—Best guns, rifles, and swords for real service; defensive armour.

[2614]

WILLIAMS, A. H., & BOCCIUS, G., 135 *Fenchurch Street, London.*—Patent percussion cap holder for military, naval, and sporting uses.

[2615]

WOODWARD, JAMES, 64 *St. James's Street, S. W.*—Guns and rifles, muzzle and breech-loading.

[2616]

WOOLLCOMBE, R. W., 14 *St. Jean d'Acre Terrace, Stoke, Devonport.*—Cannon and projectiles for projection with cycloidal rotation.

[2617]

WYLEY, ANDREW, 21 *Barker Street, Handsworth, Birmingham, and Rose Lodge, Belfast.*—Patent automatic breech-loader, self-cocking, self-capping, using any ammunition.

The breech B has a more or less conical lip entering three-quarters of an inch or more, so that escape is impossible. The nipple is placed in the axis of the breech, and usually screwed from inside, with or without a cartridge piercer of steel or platinum. The cock works in a slot in the middle of the stock; there is no tumbler; and the trigger, or, as above shown, a small catch connected to the trigger by a link, engages in bents cut in the circular head of the cock. Owing to the increased leverage, the pull of the trigger is very light, and yet the bents as deep, and the gun as safe, as with the common tumbler lock.

Segments of screw threads project from the breech, on opposite sides, and lock into hollow threads in the breech case A. The last is about twice the length of the breech, and is screwed to the barrel, and fixed by two screws to the break-off D, by which great strength is imparted, while the barrel and breech-case is easily removed for cleaning.

Fig. 4 (half scale) shows part of the priming tube, containing 40 or 50 caps, pushed forward by a spiral spring. The last is outside, and therefore much more effective, acting on the cap-driver by a pin or pins travelling in a slot in the tube. These pins are retained in a transverse slot, while the caps are being dropped down the tube. When filled with caps, the tube is passed up the hole in the stock from the butt end, the projecting pins running in a grove on one side. When the tube is home, it is turned one-quarter round, which frees the pins from the transverse slot, and allows the spring to act; the same motion locking the tube against the butt plate, the flat handle acting as a spring catch to retain it there. When the breech is drawn back in the chamber, by a simple contrivance, the leading cap is reversed, and brought into a position to be taken off by the nipple.

The action of loading is as follows:—The breech being unlocked by striking up the lever, it is drawn away from the barrel. It thus forces back the cock, sliding over the face of the latter, and sweeping out the exploded cap, which falls through the opening F. The nipple then enters a fresh cap, and the breech being brought up by the half-round projection in front of the break-off, naturally swivels upon it as far as the pins (*p*) will permit, into the position shown in Fig. 2; when it may be loaded alike with cartridge or loose charge. The above is all done by one continuous motion in half a second. The charge being introduced, and the above motion reversed, the gun is ready for firing; when proper cartridges are used, in less than four seconds from the previous discharge, so that a second barrel is almost superfluous.

A most important feature in this gun, distinguishing it from all previous attempts at self-cocking and self-capping, is, that it can be cocked, uncocked, and capped by hand, as easily as the common weapon, so that if its priming arrangement were totally destroyed, it is still a breech-loader of the most perfect kind. The only part liable to fouling is about the rear of the breech, from the cap flash; but as this is quite accessible, it admits of easy cleaning, and the two pins (*p*) being loosened, the breech can be removed. But even if by carelessness allowed to rust ever so much, it does not interfere with the action.

For rapid firing, a stiff paper cartridge is used, as shown in the section, Fig. 1; the ball projects fully two-thirds of its length, and thus expands in the grooves as in a muzzle-loader. The cartridge may be long enough to contain the ball as at V, having, in this case, the lubrication before it. A cartridge like the last is used for shot (*w*). These project from the breech, so as to be easily withdrawn by hand, after firing. (*u*) is a cartridge like the Enfield; but with the paper envelope of the ball continued in front, and enclosing the lubrication (*l*). When it is desirable to use the loose charge, a metal tube is fixed in the breech to supply the place of the thick cartridge.

The back sight in rifles, H, Fig. 5, slides vertically in two holes in the front of the cheeks of the breech-case, at the usual distance from the eye. The leaf has a simple traverse, so as to act as wind-gauge. An interchangeable supplementary sight K, for distances up to 2,000 yards, or more, is carried in the fore part of the stock. I, Fig. 2, is a point-blank sight, which can be used without lowering the main sight, in case of a surprise or snap shot. L is a spirit level, which is slid over the legs of the sight for long ranges. In non-military rifles, the foresight is placed on the end of a blade spring, and retracts within a strong sheath. A fine sight is thus less liable to injury from rough usage, or the common practice of leaning the gun against a wall.

WYLEY, ANDREW—*continued.*

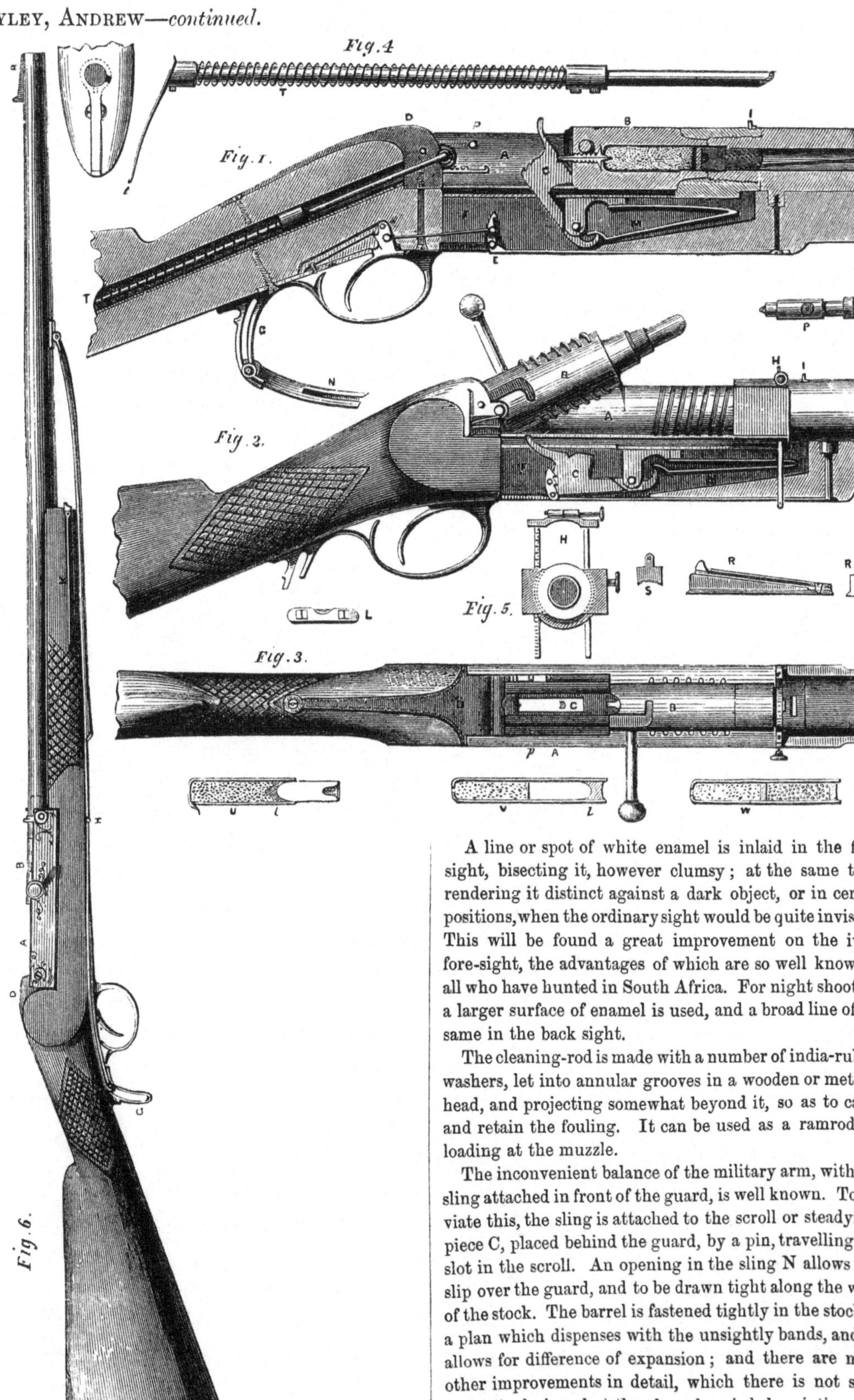

A line or spot of white enamel is inlaid in the fore-sight, bisecting it, however clumsy; at the same time, rendering it distinct against a dark object, or in certain positions, when the ordinary sight would be quite invisible. This will be found a great improvement on the ivory fore-sight, the advantages of which are so well known to all who have hunted in South Africa. For night shooting, a larger surface of enamel is used, and a broad line of the same in the back sight.

The cleaning-rod is made with a number of india-rubber washers, let into annular grooves in a wooden or metallic head, and projecting somewhat beyond it, so as to catch and retain the fouling. It can be used as a ramrod for loading at the muzzle.

The inconvenient balance of the military arm, with the sling attached in front of the guard, is well known. To obviate this, the sling is attached to the scroll or steadying-piece C, placed behind the guard, by a pin, travelling in a slot in the scroll. An opening in the sling N allows it to slip over the guard, and to be drawn tight along the wood of the stock. The barrel is fastened tightly in the stock by a plan which dispenses with the unsightly bands, and yet allows for difference of expansion; and there are many other improvements in detail, which there is not space to particularise; but the above hurried descriptions will enable those conversant with the subject to judge how far this arm fulfils the conditions of a perfect breech-loader.

Class XII.

NAVAL ARCHITECTURE, SHIPS' TACKLE, &c.

Sub-Class A.—*Ship Building for purposes of War and Commerce.*

[2646]

Aston, James J., 4 *Middle Temple Lane, London.*—Working model boat, fitted with Aston's patent disc propeller.

[2647]

Aylin, J., R.N., *Wilton, near Brough, Yorkshire.*—Wedge-armed anchor, shackle, &c.

[2648]

Basire, James, 4 *King Street, Westminster.*—Models of Brown and Harfield's patent capstans, and C. Langley's unsinkable ship. (*See page* 2.)

[2649]

Bethell, John, 38 *King William Street, London, E.C.*—Models of a new mode of building ships.

[2650]

Browning, Henry, *Avon Cottage, Clifton Wood, Bristol.*—Patent composition for the preservation of ship's bottoms against the action of water or atmosphere.

[2651]

Burden, William, *Hay Well, Great Malvern.*—Oblique paddle-wheel, without back lift, illustrating a new theory of motion.

[2652]

Burnett, Charles J., *Edinburgh.*—Fan propellers, with shields and accompaniments.

The models exhibited are illustrative of improved steam ship propellers, with diagonal shields, offered to Admiralty 16th August, 1860. The current of water driven out centrifugally by a fan (or screw) is deflected into a straight line, by unhinging on a diagonal surface. An amount of power is thus rendered available for propulsion, which would otherwise be, and in the ordinary screw propellers is, lost, or worse than lost, in consequence of its action on unequal, constantly varying masses of water below and above it, the unequal resistance of which strains and shakes both screw axle and ship.

[2653]

Caird & Co., *Greenock.*—Models of steam-ships and marine engines. (*See page* 3.)

BASIRE, JAMES, 4 *King Street, Westminster.*—Models of Brown and Harfield's patent capstans, and C. Lungley's patent unsinkable ship.

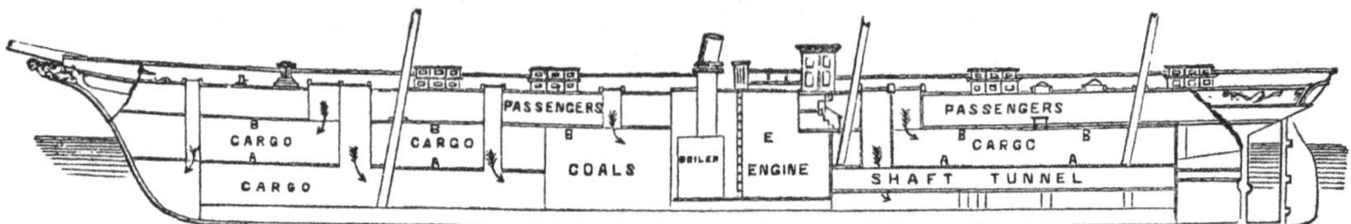

No. 1. Model of unsinkable and fireproof ship "Briton," belonging to Union Steam Ship Company, built by, and upon the plan patented by, Mr. Charles Lungley, of Deptford Green Dockyard.

No. 2. Longitudinal section of same, with fittings, showing how the invention may be applied to all ships. The lower deck, marked A, is made of iron, water-tight, and fitted with water-tight trunks, to communicate with the upper deck, so that access can be had at all times distinct from the other decks. By this plan, if the vessel's bottom is torn out, the water can only get into the space under the lower deck, the trunks preventing it going into the other holds. The deck, marked B, is also made of iron, as well as the trunks. These decks give the means of fitting iron storerooms and divisions for other purposes; and the more fittings put in, the more secure is the ship against fire or combustion, which is localised. E is engine and boiler space, which is inclosed by iron walls, so that if water gains access it is confined, and can be pumped out by separate pumps. The model is made to show the principle of the invention, which can be adapted to every kind of merchant, transport, or war ship.

No. 3. Model of river steamer, with paddles, the invention of James Basire. These wheels have shown the following results, as compared with the ordinary wheels —18 per cent. greater speed, and wheels with feathering floats—16 per cent. greater speed. The rims form a series of continuous floats, each being at such an angle as will best free the water from them. They are less liable to get out of order, and form less resistance to the atmosphere. This model has been saturated and coated with the zopissa composition, for preserving iron and wooden ships against rust and decay, invented by N. C. Szerelmy. (See Class X.)

No. 4 is model of Messrs. Miller and Knill's, of 39 Pudding Lane, City, patent marine steam governor. This invention is intended to supply a want that has been long felt in the merchant steam navy, especially in screw steamers, in providing them with a cheap, simple, and effective governor; and although there are several descriptions in use, they are all more or less complicated and expensive. It is well known where a steamship is in a heavy sea, that there is great danger to the engines, when by the motion of the ship the screw is only partly immersed, and the speed of the engine is greatly increased. This governor has now been in use some time, and with the most favourable results, as the annexed will show:—

"Dundee, Perth, and London
"Shipping Company's Office,
"Dundee, 8th February, 1862.

"I hereby certify that this company's screw steam ship 'Queen' was fitted with Miller and Knill's 'Patent Marine Governor,' in March last; that she has since been employed in the Mediterranean and coasting trades, and that it has been found to answer the purpose it was intended to serve so well, that they have resolved to supply their other steamers with it as soon as possible.

"The 'governor' prevents the engines from racing in heavy weather, lessens the risk of a 'breakdown,' and allows them to be driven at greater speed with safety, so that the vessel now makes her passages in less time and with less consumption of fuel than formerly.

(Signed) "THOS. COUPER, Manager."

No. 5. Soul's engine-room telegraph, as now fitting in the British navy. It is not uncommon to see the signals on the dial of a telegraph, arranged altogether at variance with the condition of the engines. An engine moving full speed ahead has to pass in succession, and sometimes rapidly, through all the different degrees of speed, and even to stop, before it can commence a retrograde motion. And again, in moving backwards, it has to pass in the same order through the different periods, before it can be brought to advance. An arrangement of signals is

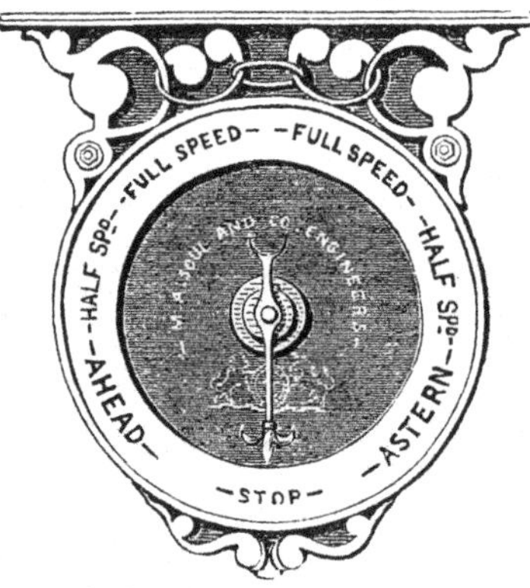

therefore required and provided by this telegraph to correspond with similar conditions. The attention of the engineer is directed to the telegraph dial by the sound of the alarum. On seeing a retrograde direction of the hand, he knows at once that the speed has to be slackened; he follows the degree indicated successively by the telegraph with the starting lever or wheel until the hand comes to a stop. It is therefore evident, that the hand upon the dial has to pass through the same degrees as the starting lever or wheel of the engine. The motion of the handle on deck is transmitted as usual, by means of a series of shaftings and tooth-wheels, where such are necessary. The alarum hammer is directly attached to a spring, and is acted upon by a wheel having on its circumference a series of evolute cogs or projections, the same in number as there are signals upon the dial. To insure a quick release of the hammer, in order that such may strike the bell effectively, a peculiarly shaped cam is mounted upon it, which, after being raised by the before-mentioned wheel to the extent requisite, allows the alarum hammer to fall, and re-act, without meeting any hindrance which might impede its free action, should the wheel not be moved quick enough. Mr. Soul's invention dispenses, it will be readily seen, with the use of several springs and hammers, without diminishing the number of signals required. There is this to recommend it—easy adjustment of working parts, and the reduction of the whole mechanism to a form more simple than any now in actual use.

No. 6. New system of slide valve for marine and land engines, locomotives, &c. Invented and patented by E. Plainemaison, Engineer of the North of France Railway Company. Agents: H. P. Burt & Co., 2 Charlotte Row, City. The object of this invention is to reduce the pressure of steam upon the working parts of the slide valve, thereby greatly diminishing the friction and wear and tear, and greatly increasing the facility of reversing the gear.

CAIRD & CO., *Greenock.*—Models of steam-ships and marine engines.

The following are the descriptions of the models exhibited:—

1st. A small model in steel to a ½-inch scale of a pair of oscillating engines, of 160 horse-power collectively, suitable for merchant paddle steamers of any size.

2nd. A small model in steel to a ½-inch scale of a pair of inverted cylinder direct acting engines, of 400 horse-power collectively (originally designed by J. T. Caird, Greenock), suitable for merchant screw steamers of any size.

3rd. A pair of horizontal direct acting screw engines, of 10 horse-power collectively, fitted with improved variable expansion gear. Engines of this class are suitable for war vessels of any size.

4th. A model to a ¼-inch scale of Royal Mail Company's steam-ship "Atrato," built, engined, and equipped complete, by Caird and Co., 1853. Gross tonnage = 3,466.

5th. A model to a ¼-inch scale of the North German Lloyd's screw steam-ship "Hansa," built, engined, and equipped complete, by Caird and Co., 1861. Gross tonnage = 2,991.

[2654]

CALLEY, SAMUEL, *Brixham, Devon.*—Ship's worn sheathing; patent compositions for metals and wood; metallic paints, ochres, &c.

The exhibitor manufactures patent composition for ships' metal sheathing, iron ships, iron, wood, and other surfaces; and also the celebrated Torbay iron ore and metallic paints, and mineral ochres. Prices and testimonials may be obtained on application at the works.

[2655]

CAMPBELL, ROBERT F., 8 *Brook Street, Hyde Park.*—Apparatus for management of vessels. New mechanical motion.

[2656]

CARR, THOMAS, *New Ferry, near Birkenhead.*—Models of two patent steering apparatuses.

[2657]

CLAY, JOHN, 82 *Castle Street, Edgeley.*—Ship and propellers.

[2658]

CLIBBETT, WILLIAM, *Appledore, Devon.*—Half model of barque.

[2659]

Clifford, Charles, 49 *Fenchurch Street, and Temple.*—Improved systems of unlashing, lowering, and releasing ships' boats, from vessels stationary or under weigh, without possibility of canting, by one of the crew sitting in the boat.

"The means of lowering boats evenly, and of readily disengaging the tackles, are *desiderata* wanting throughout the naval service."—Parliamentary Report on the Loss of the *Amazon*, 1852.

Upwards of 1,500 boats have been fitted on this system, 350 being for the Royal Navy.

The *third* "MAN OVERBOARD" picked up during the first voyage of H.M.S. *Shannon* (under Captain Sir William Peel, V.C.), through the celerity with which boats were lowered by Clifford's gear; ship under full sail.

Captain Vaughan, C.B., says:—"I was below, at the mess-room table, when I heard the cry of 'Man overboard;' but in three minutes from that time the boat was manned, lowered, and the man picked up. I lowered the boat myself, *single-handed.*"

Of the various forms of fatal accidents to which mankind is liable, drowning in sea voyages is in by far the largest proportion. The majority of instances of sailors and others falling overboard, and lost before help can reach them, either never come to our knowledge, or pass unheeded in the crowd of events that daily press upon our notice.

It requires a catastrophe like the loss of the *Amazon*, or the *Birkenhead*, the *Queen Victoria*, the *Austria*, the *Pomona*, or the *Royal Charter*, with all its attendant horrors, to bring us to think of and appreciate the perils incident to navigation.

During the space of only a few months of 1859-60, not less than 1,484 persons lost their lives at sea from the destruction of six ships; and it was officially stated that upwards of 1,000 men are annually lost from American ships alone, by falling or being washed overboard, while the numbers lost from British ships are probably equally large.

One of the chief causes of this lamentable loss of life is the want of any means for lowering the boats speedily and safely in case of accident to the ship. On the occasion of the loss of the *Amazon*, the Parliamentary report stated the supply of boats was ample, but "that the means of lowering boats evenly, and readily disengaging the tackles, &c., are *desiderata* wanting throughout the naval service;" and that "it may be expected some useful means for supplying these defects may be devised." Clifford's system accomplishes these *desiderata*, and by it a boat laden with any crew can be instantly and safely lowered, even if the ship is moving rapidly through the water. It has been approved and adopted by the Admiralty, and every naval department of the Government, by the surveyors at Lloyd's Register of British and Foreign Shipping, by the Institution of Civil Engineers, and most of the leading Steam Companies. After repeated competitive trials, it is the only plan made compulsory in all ships chartered by H.M. Emigration Commissioners, the Council of India, and the Marine Board of Melbourne. The Committee of the Royal National Life Boat Institution, consisting of some of the first naval men of the country, passed a vote of thanks to its inventor on account of the number of lives it has saved, a list of which the Journal of the Institution, Jan. 1862, gives in the following words:—

"We think we shall be rendering a service to the great cause of humanity, by giving every possible publicity to the list of lives saved by this invention, as in most of the instances we record the men have fallen overboard in heavy gales, and when the ship was moving rapidly through the water; the officers in command stating their firm belief that but for it they would have been lost; and also that the lowering and disengaging the boat being the result of the single act of one man only, is the chief cause of its great success. In some cases the entire crews of ships when foundering or wrecked, in collision and suddenly sunk, or on fire, owe their preservation to it. From H.M. ships *Shannon, Racoon, Princess Royal, Archer, Trafalgar, Emerald, Diadem, Chesapeake, Mersey, Calypso, Ganges*, 20 men were saved by it. H.M.S. *Perseverance*, from a vessel run into at night, and entirely sunk in less than ten minutes, took off 15 men and 1 boy. From the troopships *Lady M'Naghten, Australasian, John Duncan, Dutchman, Kate, Clara*, 10 men. From ships chartered by H.M. Emigration Commissioners *Commodore Perry, Washington Irving, Aloe, Black Eagle, Transatlantic, Ebba Brahe, Medway, Omega, Rodney, Blundell, Admiral Boxer, Champion of the Seas, Hoogley*, 17 persons (2 being women). From merchant steam ships (Royal

CLIFFORD, CHARLES—*continued.*

Mail) *Tasmanian, Queen of the South, Duke of Richmond, Duke of Rothsay, Queen,* 8 men. The *John Masterman, Rodney, Merchantman,* by it lowered down their boats and took away the entire crews of three ships that were about foundering at sea, or on fire, and which had lost their own boats when attempting to lower them by the ordinary tackling; by it also, on the memorable occasion of the fire of the troop-ship *Sarah Sands,* 'the life-boats filled with the women and children were lowered in perfect safety,' the *Times,* in its account, stating that 'for once in the case of a conflagration at sea the boats were lowered in safety.' The official report of the chief officer of the *Pomona* to the Board of Trade, when she foundered off Malta, was, that the only people saved, 18 in number (2 being women and 1 a child) 'are indebted to Clifford's lowering apparatus for their lives.' Thus we have certain accounts of more than 100 people being saved, probably not half of what have really occurred."

The committee appointed by Admiralty order to report upon this apparatus, expressed its unanimous conviction "that no captain, whether in the Queen's or mercantile navy, should be permitted to put to sea without it."—*Times, Dec.* 11, 1856.

In the House of Commons, Admiral Berkeley said "that in every trial which had been made of it, its use had been attended with complete success, and he hoped to see it universally adopted."—*Times, March* 18, 1857.

C. Clifford is prepared to unlash, lower, and entirely disengage from any ship, either stationary, under weigh, or going at any speed, in a gale, or in smooth water, a boat laden with a full crew, against any other invention or crew in the world, for any sum to £100, to be given for placing a life-boat on an exposed part of our coast. As hundreds of our best seamen are annually lost through the want of such means of instantly lowering a boat—which Parliament has decided to be "wanted throughout the naval service"—it is hoped some one will be found with sufficient spirit and humanity practically to test this challenge.

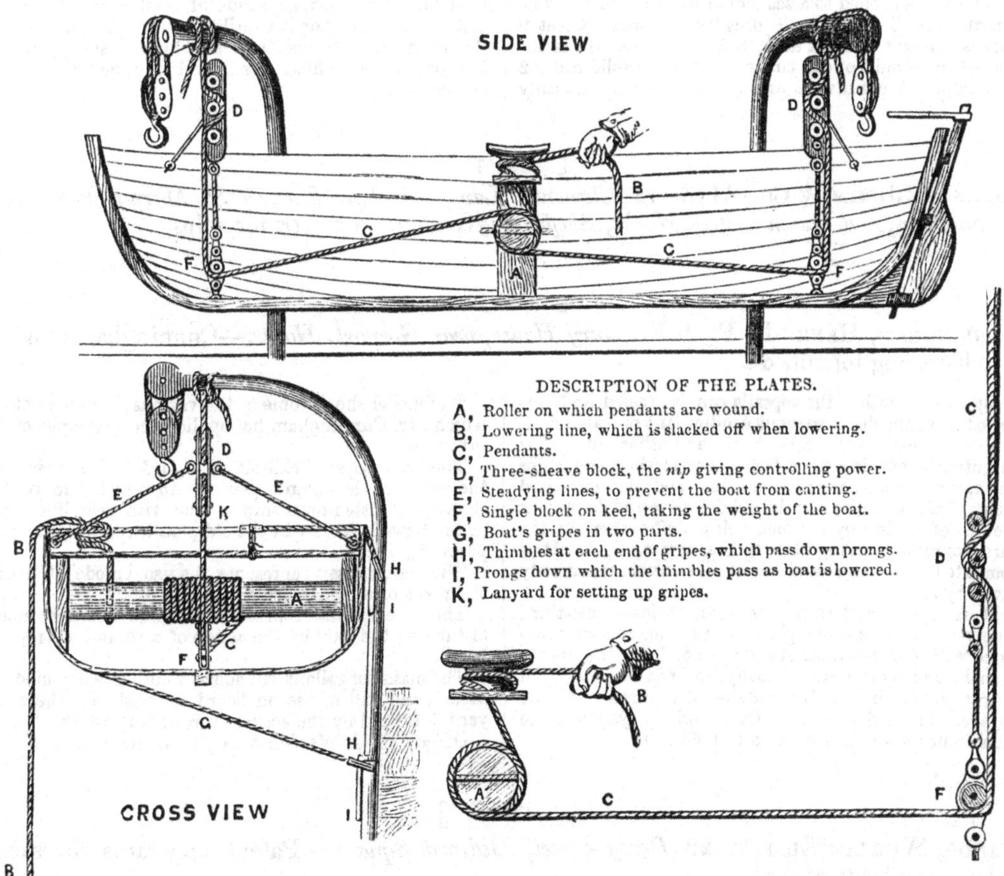

DESCRIPTION OF THE PLATES.

A, Roller on which pendants are wound.
B, Lowering line, which is slacked off when lowering.
C, Pendants.
D, Three-sheave block, the *nip* giving controlling power.
E, Steadying lines, to prevent the boat from canting.
F, Single block on keel, taking the weight of the boat.
G, Boat's gripes in two parts.
H, Thimbles at each end of gripes, which pass down prongs.
I, Prongs down which the thimbles pass as boat is lowered.
K, Lanyard for setting up gripes.

On slacking off the lowering line B, the roller A revolves, and the pendants CC are unwound evenly as the boat descends into the water, when the pendants being tapered at the ends, overhaul themselves, and the boat is perfectly free. The controlling power is obtained by the blocks DD, which act like a sailor's "turn and a half" in the boat on each pendant; the *nip* of the blocks exists only when they sustain the weight of the boat, and ceases when it reaches the water. This power in the block to decrease the weight of the boat, and thus enabling the man attending the lowering line to control the descent, whatever the weight may be, but yet allowing all to run free the moment the lowering line is let go, is its chief feature, and that which befits it for the purpose to which it is here applied, and for which it was specially designed.

INSTRUCTIONS FOR LOWERING.—One of the boat's crew takes charge of the lowering line B, and with one round turn on the cleet, slackens it off slowly. The lashings release themselves by the thimbles passing down the prongs II. When the boat reaches the water, the lowering line is let go, the pendants overhaul themselves, and the boat is perfectly free.

Thus by this one simple act of the one man, the boat is unlashed, lowered, and released from the ship.

[2660]

COMMISSIONERS OF IRISH LIGHTS.—Fastnet Rock Lighthouse, off Cape Clear, S.C. Ireland.

[2661]

COMMISSIONERS OF NORTHERN LIGHTHOUSES, *Edinburgh.*—Lighthouse apparatus and models.

[2662]

CORPORATION OF TRINITY HOUSE, THE, *London.*—Models of lighthouses, and of a light vessel.

[2663]

COUCHMAN, JOHN WILLIAM, *Tottenham Green, Middlesex.*—1. A model of a new principle of street making. 2. A model of an iron combination bridge. 3. A model of a floating battery.

CONCHMAN'S FLOATING BATTERY.—A design for a 12-gun floating battery by John William Conchman, Engineer; modelled to a scale of 5 feet to 1 inch. The length over all is 150 feet; breadth of beam 38 feet 9 inches; height of main deck, 8 feet at sides. It is intended to be built of oak timber, of ribs in solid order 2 feet thick; the gunways and sides to 2 feet vertically below water level, to be covered with 3 inch iron plates; and the flush deck with 2 inch ditto, secured with bolts with mushroom-shaped heads of steel 8 inches in diameter. The interior is ventilated through gangways on the flush deck; by apertures at stem and stern; and by the port-holes, which open to the under side of the domed roofs.

[2664]

COULSON, JUKES, & Co., 11 & 12 *Clement's Lane, London, E.C.*, 7 *St. Mary's Row, Birmingham, and Queen's Steel Works, Sheffield.*—General ironwork for ships.

[2665]

CUNNINGHAM, HENRY D. P., R.N., *Bury House, near Gosport, Hants.*—Cunningham's patent self-reefing topsail, &c.

By this invention, the topsails can be reefed and unreefed from the deck, without sending any one aloft. It is also applicable to topgallant sails and other sails. This invention is now in use on board several thousand ships belonging to the mercantile marine, and also on board many of H. M. ships; and the old defective and dangerous method of reefing by the men going aloft and out on the yards, is rapidly giving place to the new method. It is computed that many hundreds of lives have been already saved by it.

It has been found that sails wear, at least, one-third longer than on the old plan; ships, too, can be navigated with fewer regular seamen, and, from the ability to make and shorten sail so easily, sail can be carried on longer, thus considerably abridging the duration of the voyage. Ships fitted on the Cunningham system make much quicker voyages than on the old plan.

The model shows some of the various arrangements by which Mr. Cunningham has applied the principle of his invention.

The fore topsail exhibits the yard turned round by the action of the chain topsail tye in which the yard is shown, and fitted for a ship of war with reef lines, &c., in the topsail, to reef in the old plan if required for purposes of exercise.

The main topsail represents the usual mode of fitting the yards of merchant ships.

The mizen topsail represents one of the earliest modes of turning the yard by the action of a wound up rope or band.

The main topgallant sail shows another early mode of fitting, and still in use on board some ships, where the yard is turned by the action of wound up ropes.

Office: G. C. Warden & Co., 12 London Street.

[2666]

DANDO, WILLIAM ELBERT, 29 *Percy Street, Bedford Square.*—Patent apparatus for safely lowering boats at sea.

[2667]

DAY, WILLIAM, & Co., *Bow Road, London, E.*—Patent marine cements and compositions, for coating the insides and outsides of ships.

[2668]

DENNY, WM., BROTHERS, *Dumbarton.*—Sectional model of a screw steamer, in a glass case; models of two screw ships.

[2669]

DUNCAN, ROBERT, 174 *Trongate, Glasgow, and at Bowling.*—Slip cradle or carriage, and self-acting time and tide gauge.

DUNCAN'S SELF-RELIEVING SLIP CRADLE.—An improved method of relieving the cradle from underneath vessels while on slip for repairs; entirely doing away with the raising of vessels by wedge, hydraulic, or other means.

Also, a new method of adjusting bilge blocks on the arms of the cradle; by which the same blocks can fit vessels of almost any shaped bottom.

Robert Duncan, Engineer, 174 Trongate, Glasgow, and at Bowling, on Clyde.

[2670]

DUNLOP, DAVID, *Hurlet, Glasgow.*—Angulated invulnerable steam-ram, propelling either way, sweeping enemies from decks by machinery.

[2671]

EDDY, C. W., *Sutton, Loughborough.*—Armour-plated steamers, submarine shell and ram, and other naval inventions.

[2672]

ELIOT, EDWARD J., 7 *Southampton Row, Russell Square, W.C.*—An improved hydraulic apparatus for raising sunken vessels.

[2673]

ELLIS, GEORGE, 4 *Collier Street, N.*—War ships, safety ports.

[2674]

ESCOTT, R. A., *Old Charlton, Kent.*—Sections of guns and shots, showing rifling.

[2675]

FORMBY, ROBERT, *Liverpool.*—Patent apparatus for working ships' pumps by water power.

[2676]

FRYER, F. A., 3 *Leadenhall Street, London.*—Horizontal patent propeller direct-acting steam engines, and steam-steering safety propellers. (*See page* 8.)

[2677]

FULLER, GEO. L., C.E., 69 *Lombard Street, London, E.C.*—Model of floating-ship lift for open waters, Mackelcan's patent.

[2678]

FYFE, T., 46 *Leicester Square.*—Patent rigging; horizontal ship; patent knapsack; method of ventilating ships' holds.

[2679]

GITTINS, RICHARD, 28 *New Street, Dorset Square, London.*—Model of a new invention for propelling steam ships.

[2680]

GRANTHAM, J., 31 *Nicholas Lane, E.C.*—Plan for preserving iron ships from concussions, &c.

[2681]

GRAY, JOHN WILLIAM & SON, 114 *Fenchurch Street, City, and Margaret Street, Limehouse, E.*—Patent engines, patent ship's pumps, anchor dropper, lightning conductor, deck lights; closets, night life buoys, and brass work.

The exhibitors are engineers, and workers in copper and brass, and patentees and manufacturers of the following apparatus, viz.:—

Gray's spherical steam-engine and ordinary engines.

Agricultural and locomotive engines.

Spherical and other pumps.

Ship's side ports, with lifting and securing apparatus.

Deck illuminators and ventilators.

Portable fire-engines; patent anchor stoppers.

Apparatus for distilling sea-water; steam gauges and fittings.

Signal lanterns; ship's cooking apparatus; engineers' tools, &c.

They are also agents for Sir W. Snow Harris' patent lightning conductors, as applied in the royal navy; for Hosmer's patent self-acting house and street cleaning apparatus; and for Grimaldi's patent revolving steam-boilers for ship and land purposes. These boilers possess six times the evaporating power of ordinary boilers.

FRYER, F. A., 3 *Leadenhall Street, London.*—Horizontal patent propeller direct-acting steam engines, and steam-steering safety propellers.

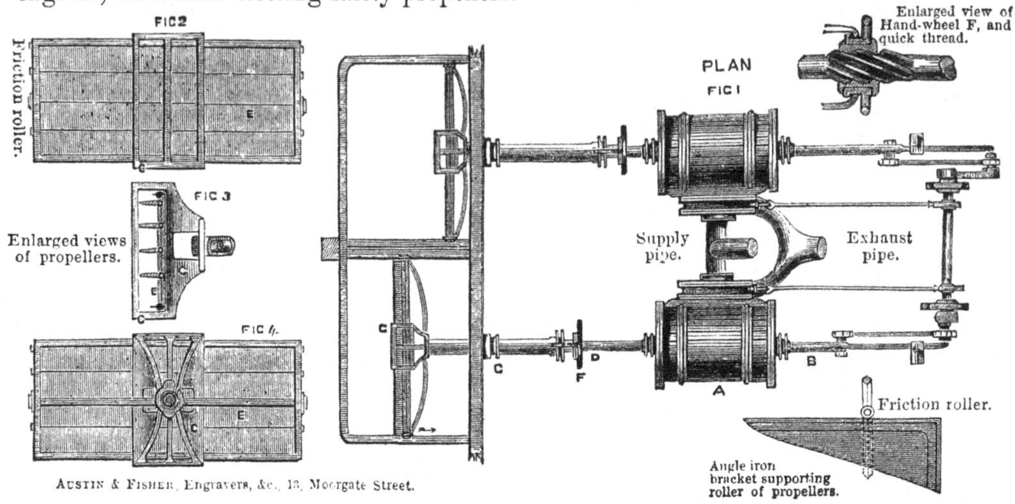

The stroke of propellers can be multiplied by mechanical contrivances.

The above engraving represents Fryer's patent marine steam propellers. Fig. 1 is a plan of the general arrangement of the machinery. Fig. 2 a front view. Fig. 3 a section, and Fig. 4 a back view of the propeller. A the cylinder. B the piston rod, giving direct motion to the propeller through stuffing box C. D shows a quick thread worm for reversing the propeller E, by means of the hand wheel at F. G the frame supporting the floats when making the forward stroke; in making the backward stroke, the floats turn edgewise, offering little resistance to the water. The arrow shows the direction in which the propellers are supposed to be moving.

FRYER'S HORIZONTAL PATENT PROPELLER DIRECT-ACTING STEAM ENGINES, STEAM STEERING SAFETY AND LEVER PROPELLERS.—These new propellers and steam engines combine so many valuable properties with admirable simplicity, that it is hoped they will commend themselves to the mercantile community and the shipping interest in such a manner as to give full development to the wonderful power of steam, both for the speed, safety, convenience, and economy of vessels and canal boats. The inventor has long deplored that so mighty an agent for steering vessels with security, even in a storm, should never before have been employed, notwithstanding the enormous annual loss of lives and property.

Among many advantages may be enumerated the following:—

Economy.—It is calculated that the cost of these engines and propellers, where so much superfluous machinery is entirely abolished, will not much exceed half of those now in use, both for screw and paddle-wheel steamers; this, therefore, is a consideration of the very first importance, in addition to which, there is a considerable saving in fuel, and the facility for working the steam expansively, from the proximity of the cylinders.

Space.—The diameter of the cylinder or cylinders represent the amount of space required for these engines, both in height and width; and being fastened on a bed plate close to the keel of the vessel, are always considerably below the load water mark, besides being perfectly protected from shot, shell, &c.

Speed.—From the immense propelling power brought to bear in a direct manner upon the water, the opinions of several practical engineers have been given, that twenty-five statute miles per hour can be attained. There is also very little resistance given on the return stroke of the propellers.

Steam Steering.—Particular attention is invited to this part of the invention, as showing an entirely novel application of steam power, by means of which its full power can be concentrated at any moment on either side of the bow or stern of the vessel, so as, in two or three minutes, to turn her completely round on her centre, without the use of the rudder; an invaluable power to a vessel in a storm, as thereby she can turn her head to the wind, and thus make for the open sea. The want of this power has been seriously felt by the "Great Eastern," on the occasion of a recent voyage to America, in a peculiarly disagreeable manner.

General Remarks.—It is surprising that, after the lapse of so many years since the introduction of steam vessels, so few as about 2,000 only should be registered in the United Kingdom, against upwards of 34,000 sailing vessels to the present date, as shown by the official lists; thereby proving the immense scope there exists for the introduction of steam into vessels of every class, including collier vessels and canal boats, and to which the merits of this invention, it is hoped, will greatly tend. The propellers being entirely under water, are not exposed to the violence of the waves as paddle wheels or the screw propeller is, nor can it foul as the latter does. These engines also act as ballast; they can be constructed of any strength, and are equally applicable for canal boats as ocean steamers. There are many other advantages in connection with this patent; and in consequence of the magnitude of the subject, the patentee purposes granting licenses to responsible parties for working the same in the most liberal spirit. All communications on the subject to be addressed to the under-mentioned, who will forward detailed circulars and tracings of the drawings.

Frederick A. Fryer, sole agent for the patentee, 3 Leadenhall Street, London, E.C.

Modellists: Messrs. Lewis & Sons, 5 Wych Street, Strand, London, W.C.

[2682]

GREEN, Messrs. RICHARD & HENRY, *Blackwall Yard.*—Models of a 51-gun screw frigate; a clipper ship; boat-lowering apparatus, &c.

[2683]

GRIFFITHS, ROBERT, 69 *Mornington Road, London.*—Two screw propellers, and a model of a frigate, with portable armour-plates. (*See pages* 10, 11.)

[2684]

HALE, WILLIAM, 6 *John Street, Adelphi.*—Gun and rocket boat, with apparatus attached for firing Hale's rockets.

[2685]

HALL, ROBERT, 37 *Princes Stairs, Rotherhithe, and* 58 *Paradise Street, S.E.*—Ship's figure head.

The exhibitor designs and carves all descriptions of figures, ornaments, &c., for the decoration of ships, public buildings, &c. He executes, in all kinds of wood, shields, crests, and other heraldic decorations, from authentic drawings. The specimen exhibited is the figure head of the "Algerine."

[2686]

HALL, J. & J., *Arbroath and Dundee.*—Half model of vessel.

[2687]

HEWITT, WILLIAM, 3 *Brislington Crescent, Bristol.*—Feathering screw propeller.

[2688]

HIGGINS, ARTHUR, 10 *St. Vincent's Parade, Clifton.*—A trader for narrow rivers, with new arrangement of rudder.

[2689]

HORNSEY, WILLIAM, *West Front, Southampton.*—Patent marine engine room, telegraphs and gongs. (*See page* 11.)

[2690]

IMRAY, JOHN, *Bridge Road, Lambeth.*—Hirsch's patent propeller, with boss for altering pitch. (*See page* 12.)

Griffiths, Robert, 69 *Mornington Road, London.*—Two screw propellers, and a model of frigate, with portable armour-plates.

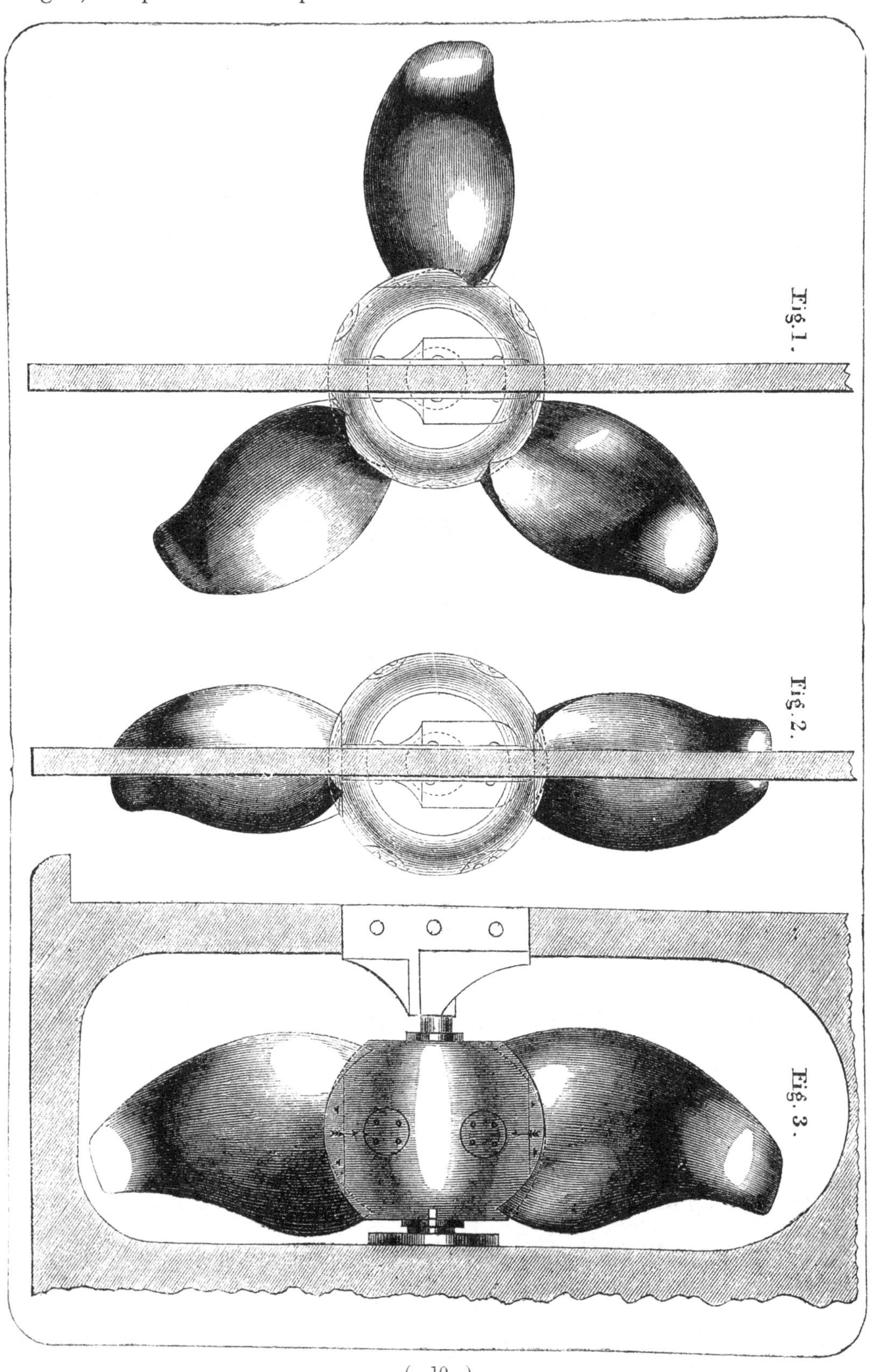

GRIFFITHS, ROBERT—*continued.*

Griffiths's Improved Patent Screw Propeller, patent 319, Feb. 20th, 1858.

Fig. 1, end view of a three-bladed screw propeller.

Figs. 2 and 3, side and end view of a two-bladed screw propeller.

LIST OF PRICES :—

All sizes of two-blade propellers in iron, complete, with centre bored ready for keying on screw shaft, from 7 feet diameter to 14 feet inclusive, price 15s. 6d. per foot, calculating on square of the diameter of the propeller. Thus a screw, of 10 feet diameter, will amount to £77 10s.

For all sizes of propellers, above 14 feet diameter, 20s. per foot. Screw propellers with three blades will be 25 per cent. above the price of two-bladed propellers.

Spare or extra blades one-sixth the cost of screw for each for two-bladed propellers, and one-eighth for three-bladed propellers. All screw propellers having their blades fixed to the boss for extra security, with gun-metal bolts and nuts, 7½ per cent. on the above price.

Gun-metal screw propellers, of all sizes, at per lb., varying according to the state of the metal market.

Patent right, 5s. per nominal horse-power of engines.

Patent Screw Propeller Manufactory, London Works, near Birmingham.

London Address, 69, Mornington Road, N.W.

A model of a frigate, with portable armour plates, by Robert Griffiths ; scale, ¼-inch to a foot.

HORNSEY, WILLIAM, *West Front, Southampton.*—Patent marine engine-room telegraphs and gongs.

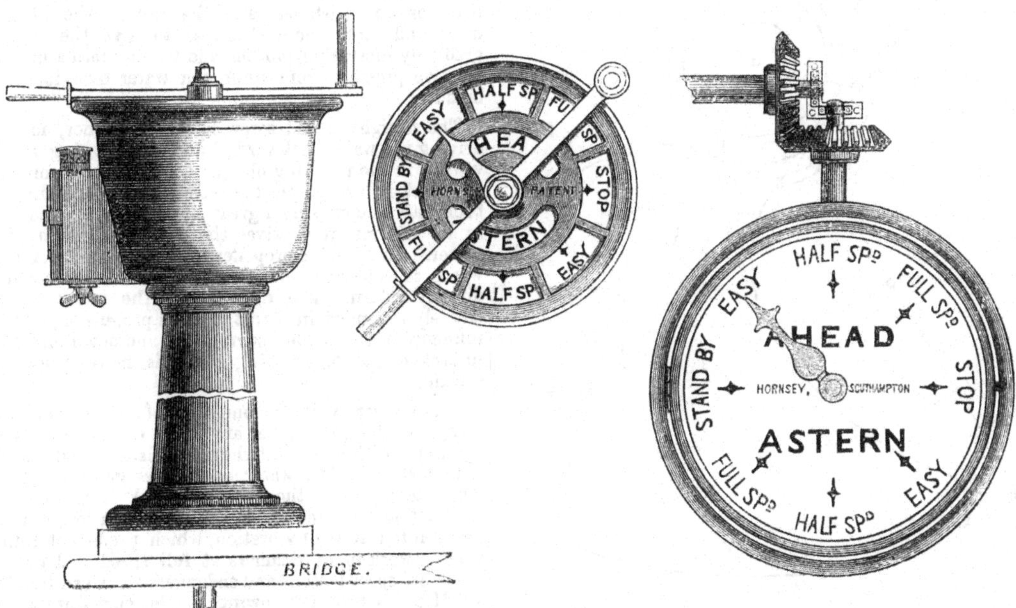

These telegraphs were selected as the standard for the Royal Navy in 1858, by Captain Halstead.

They have been supplied and fixed by the patentee to several of Her Majesty's line-of-battle ships and frigates, and the steam ram "Defence," and are extensively adopted by mercantile and Royal Mail steamers.

The engine-room dial being fixed in the engine-room, within view of the starting gear, is connected by shafting and bevil-wheels to the brass columns on the bridge. The dial on the top of the columns is divided in the same manner as the engine-room dial, and glazed with embossed glass segments for illuminating at night. A gong in the interior of the apparatus in the engine-room is sounded at every move of the pointer, which cannot move from one division to another in either direction without attracting the attention of the engineer by striking the gong.

[2691]

JAMIESON, ROBERT, C.E., *Glasgow.*—Preservative compositions for coating and permeating materials for marine architectural purposes.

[2692]

JECKS, ISAAC, *Great Yarmouth.*—Ship with iron passage, to allow missiles to pass entirely through her.

IMRAY, JOHN, *Bridge Road, Lambeth.*—Hirsch's patent propeller, with boss for altering pitch.

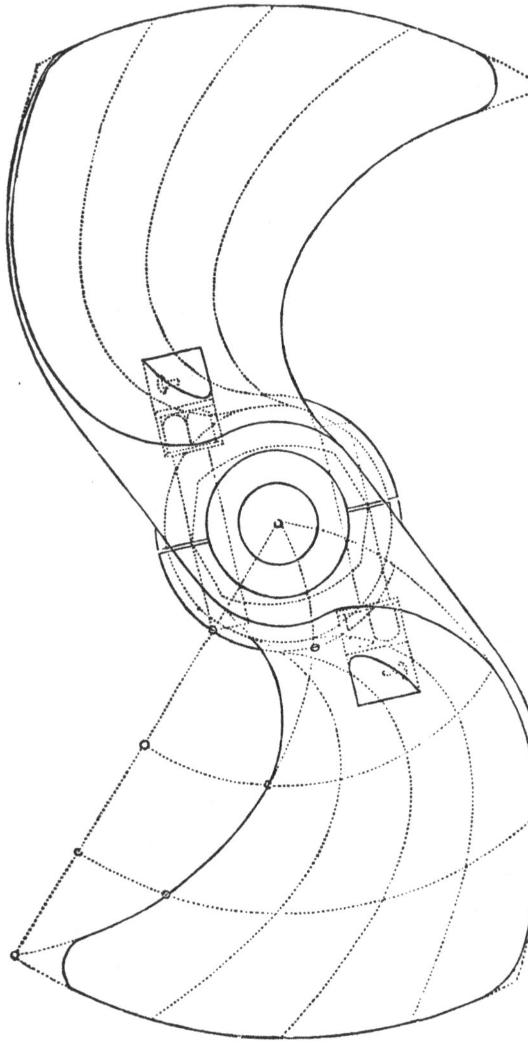

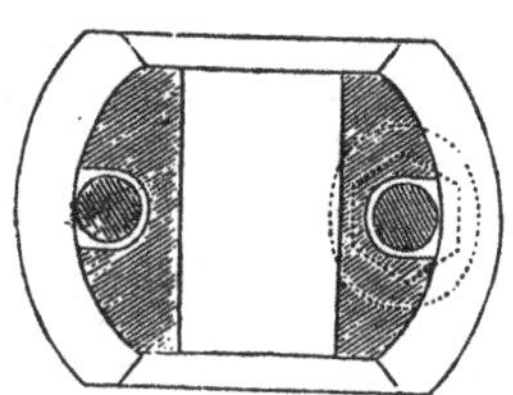

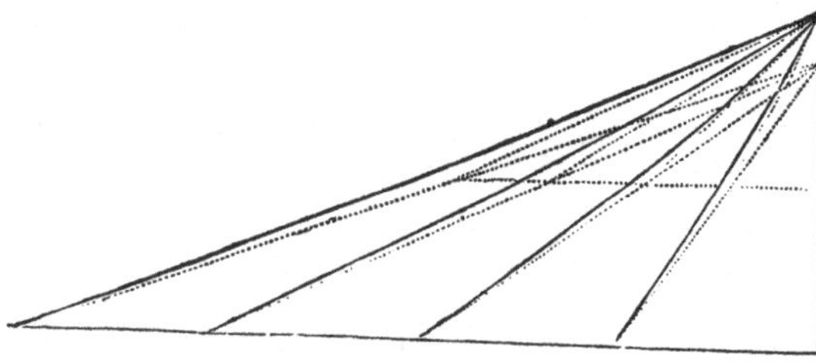

The blades of this new propeller are curved in such a manner as to secure certain advantages and avoid certain evils.

The object of a propeller being to convert the rotatory motion of its surface into a pressure directed in the line of its axis, that is the best propeller which converts a given rotatory power into the greatest longitudinal pressure, consistently with facility of steerage and absence of vibration. The extent to which these objects are attained by Hirsch's propeller may be best understood by comparing it with an ordinary straight-bladed screw, the form of propeller found best until Hirsch's propeller was tried.

The *slip* of a screw represents the yielding of the fluid medium, the resistance of the fluid to this yielding being a measure of the propelling force. When all this force is applied at once, the water suddenly put in motion at the front edge is scarcely acted on by the rest of the blade. In Hirsch's propeller the front or entering edge is so inclined that it cuts the unbroken water with little or no resistance, and the rest of the blade is more and more inclined, so as to give the water a gradually increasing motion, and to maintain a uniform reactive pressure from unbroken water over its whole breadth.

The straight-bladed screw acts like a fanner, not only driving the water backwards, but also throwing it outwards; in the resulting oblique action it loses much of its effect in propelling the vessel, and by breaking the backwater and causing a great divergence and eddying in its streams, it deprives the rudder of much of its power. In Hirsch's propeller the blades are curved inwards, so as to drive the water in an unbroken column directly astern. The reaction of the water is thus entirely expended in direct forward propulsion, and the influence of the rudder, surrounded and acted on by the unbroken and fast-moving fluid, is more quick and certain.

And farther, while the one blade of an ordinary screw is moving along the upper arc of its course, it displaces the water with ease; but the other blade, moving at the same instant in deep water, encounters great resistance, which tends to lift the vessel and jerk it to one side. This action, repeated twice in every revolution, puts the vessel into a state of vibration, which renders it impossible to work many steamers at full speed, and even at moderate speed loosens and endangers the stern-framing. In Hirsch's propeller, owing to the curvature of the blades, the successive parts of their surfaces are brought gradually into action in all parts of their revolution, and their force is thus divided and delivered easily and gradually without vibration, and with proportionally less expenditure of power.

In the trials on the Australian postal steamer *Western*, while a four-bladed screw of the ordinary kind gave a speed of ten knots per hour, Hirsch's propeller, two-bladed, gave eleven knots, with a saving of power, reduction of vibration, and increased facility of steering, so marked as to excite the surprise of all on board.

When it has been desired to separate the blades from the boss, or to alter their pitch, the patentee has successfully applied an arrangement, represented in the diagram, which is highly approved by engineers and practical men, on account of its superior neatness, simplicity, and strength, as compared with all other arrangements for the same purpose.

Models and drawings of Hirsch's patent screw propeller may be seen, and particulars obtained, on application to Mr. John Imray, Engineer (agent for the patentee), 65, Westminster Bridge Road, Lambeth, S., London.

[2693]

JONES, JOSIAH, JUN., *Liverpool.*—Models of Jones's patent angulated iron-cased ships, and of other vessels.

[2694]

KING, J. CHARLES, 12 *Portland Road, Regent's Park, W.*—Design of a ship of steel, cast in sections.

[2695]

KIRKALDY, DAVID, 4 *Corunna Street, Glasgow.*—Specimens of coloured engineering drawing, H.M.S. "Persia;" also photographs, engravings, &c., from exhibitor's drawings.

Photograph of the drawing of the *Persia*, price £2 2s.
Proof prints of *Arabia* and *La Plata* . . . 1 10s.
Transfer prints of ditto ditto ditto . . . . 1 15s.
Experiments on wrought iron and steel . . . 0 10s.

Any of these will be sent free on receipt of post-office order.

D. Kirkaldy exhibits, in addition to the above, photographs from the drawings of the *Louis XIV.*, *Europa*, &c.; and engravings of the screw propeller and lines of the *Europa*. His coloured drawings of the engines of H.M.S. ship *Hector* is exhibited by Messrs. R. Napier and Sons.

[2696]

LAIRD, JOHN, SONS, & Co., *Birkenhead.*—Models of several classes of ships.

[2697]

LORDS OF THE ADMIRALTY, *Whitehall.*—Models of ships, &c., furnished by the Admiralty.

[2698]

MCGREGOR, JAMES, *Beechwood, Partick, near Glasgow.*—Model of screw steamer.

[2699]

MARE, MESSRS. C. J. & Co., *Millwall.*—A model of the armour-plated war frigate "Northumberland," and the Government steam-transport "Himalaya."

[2700]

MITCHELL, C., & Co., *Newcastle-on-Tyne.*—Paddle and screw steamers, for river and sea navigation.

[2701]

MULLEY, WILLIAM ROBINSON, *Lockyer Street, Plymouth.*—Model of an auxiliary and reserve rudder.

[2702]

PALMER & SWIFT, *Langbourn Chambers, Fenchurch Street.*—Patent hydraulic marine propellers, regulated independent of the engine.

[2703]

PATTERSON, WILLIAM, JUN., Ship Builder, *Bristol.*—Models of screw and paddle steamships, merchant ships, and yachts.

[2704]

PEARSE, M., & Co., *Stockton-on-Tees.*—Model of Government troop steamer for the Lower Indus.

[2705]

PERETTE, AUGUSTE, 25 *Curzon Street, May Fair, London.*—Centrifugal and centripetal propeller (combined); small boiler for the epuration of oil.

[2706]

PILE, SPENCE, & Co., *Dockyard, West Hartlepool.*—Case of models of steam vessels for various purposes; patent graving dock. (*See pages* 14, 15.)

Pile, Spence, & Co., *Dockyard, West Hartlepool.*—Case of models of steam vessels for various purposes; patent graving dock.

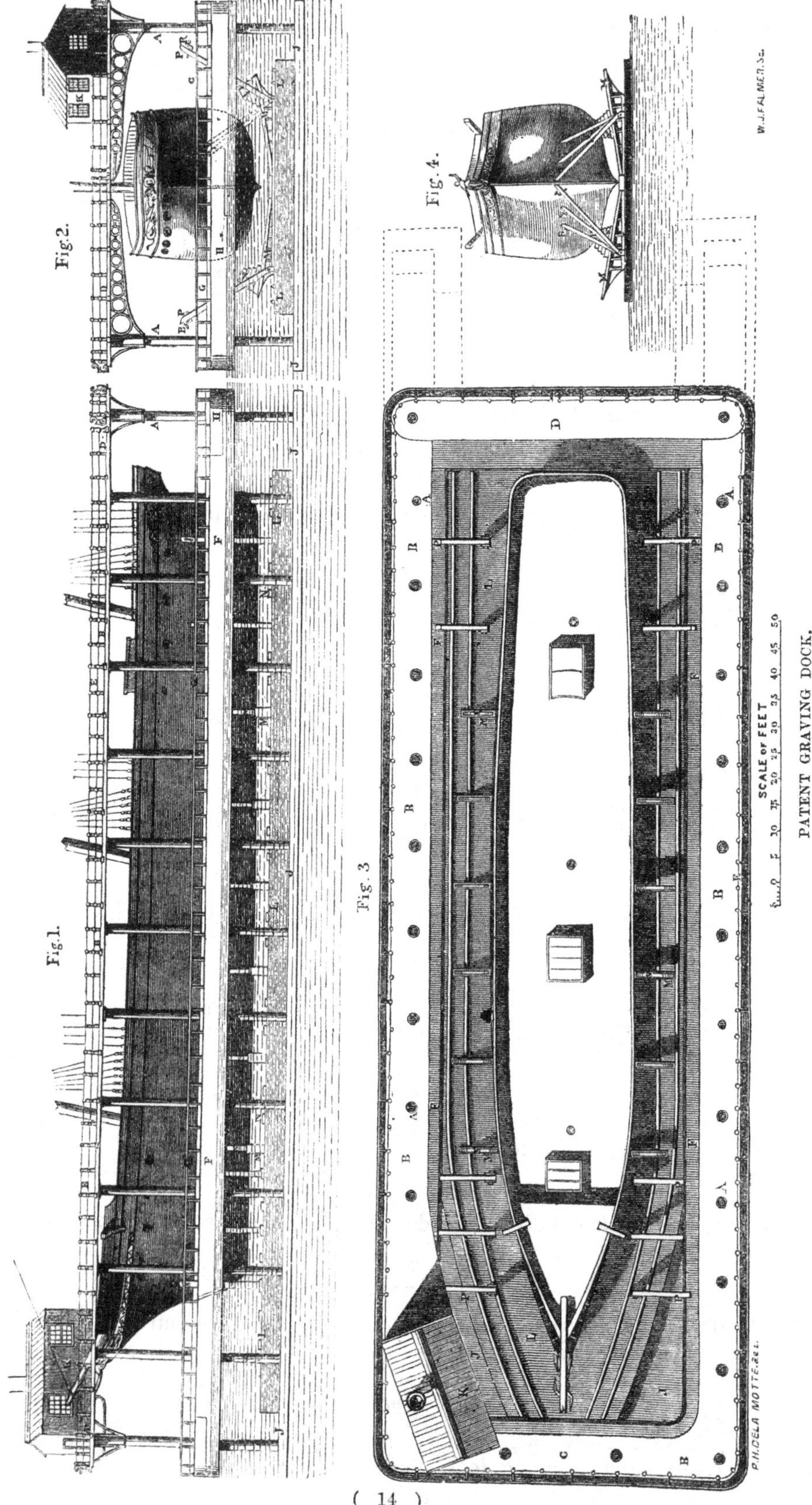

PILE, SPENCE, & CO.—*continued.*

The object of this invention is to facilitate the raising and lowering of ships or vessels out of the water, for the purpose of repair or inspection.

Fig. 1, a longitudinal elevation; Fig. 2 is a front elevation of the dock entrance; Fig. 3 is a plan corresponding to Fig. 1; and Fig. 4 is an end view of a floating pontoon having a vessel supported thereon. The improved floating dock consists of a series of columns, A, arranged at equal distances asunder in two parallel lines. The columns, A, are pillars of wrought-iron, the lower extremities of which are firmly fixed to the bottom pontoon or lift, and act as air tubes to admit the air into the pontoon or lift as the water is pumped out. On the upper extremities of the columns, A, is built a platform, B, which is carried completely round the dock, as shown in the plan, Fig. 3, of the engraving. The columns, A, serve also as guides for the floating pontoon, F, which extends from end to end, and from side to side of the dock.

A series of tubular apertures corresponding to the number of the columns, A, are made in a vertical direction through the pontoon, F. These openings encircle the columns, A, and sufficiently large to admit of the pontoon rising and falling easily. The outer end, H, of the pontoon is formed in two parts, and these are made to swing back when required, by means of a rack and pinion, or other mechanical contrivance. The floating pontoon, F, has pendant from its lower side a series of chains, I, the lower ends of which are secured to the submerged pontoon, J, by means of which chains the submersion of the pontoon or lift is regulated to any depth. This elevating pontoon, J, is constructed of iron, thoroughly water-tight; it carries the columns, A, which are securely fixed thereon. It is so arranged that it may be partially filled with water, so as to give it a greater specific gravity than the surrounding fluid, in order that it may be submerged with facility; upon discharging this water from the pontoon, sufficient buoyant power is imparted to it to lift a vessel out of the water.

In addition to the pontoon, J, there is a secondary pontoon, L, which is constructed so as to be easily attached to it; this pontoon is made to any required size, according to the weight of the vessel to be lifted, and is attached to the pontoon, J. Upon this secondary pontoon, L, the cradle, M, and chock-blocks, N, for preventing the ship from heeling over, are arranged. Prior to the vessel being docked for examination or repair, the pontoon, L, is secured to the lower pontoon, J, as shown in the end view, Fig. 2, of the accompanying plate. The vessel is then floated into the dock, and the pontoon, J, is raised by pumping air into, or water out of, the interior thereof, the vessel being kept meanwhile equidistantly from the columns, A. When the pontoon, L, touches the keel of the vessel, the blocks, N, are brought beneath the hull, in order to keep the ship in an upright position. The blocks, N, are drawn down the inclined surface of the cradle by means of the chains, O, which are carried away below the cradle, and on to windlasses fitted for the purpose on the platform of the pontoon, F. The bow and stern of the ship is further steadied and supported by means of the shores, P, which are jointed to the pontoon, L, so that they may be readily thrown back out of the way when it is desired to release the vessel. When the ship is floated over the cradle, M, and rests upon the blocks, N, the shores, P, are brought up against the bow and stern of the vessel by the chains, Q, which may be actuated in manner similar to the chains, O. Or the chains may be made fast to eyes screwed into the cradle, M, the slack of the several chains being taken up on spindles actuated by means of the winch handles, R.

When it is desired to remove the ship from the dock, the pontoon, L, is cast off from the pontoon, J, and she is floated out thereon. To float the ship from off the pontoon, L, after repairs, the pontoon and ship are again brought into the dock and placed over the elevating pontoon, J; water is then let in to both pontoons, J and L, and they sink accordingly, leaving the ship floating on the surface. The elevation, Fig. 4, shows the ship floating upon the pontoon, L, and free from the dock.

[2707]

PORT OF DUBLIN CORPORATION, *Ballast Office*, 21 *Westmoreland Street*, *Dublin.*—Lighthouse models.

[2708]

PROCTER, SAMUEL, *Churwell*, *Leeds.*—Model auxiliary screw; three-masted schooner; inverted cylinders, inclosing slides to screw.

[2709]

RANDOLPH, ELDER, & CO., *Glasgow.*—Models of vessels.

[2710]

RENNIE, GEORGE, & SONS, 6 *Holland Street*, *Blackfriars*, *and Greenwich.*—Models of ships; model of a floating graving dock. (*See page* 16.)

[2711]

RICHARDS, JOHN, *Iron Works*, 27 *Hill Street*, *Milford.*—Model of iron ship; anchors, cable.

[2712]

RICHARDSON, C. J., 34 *Kensington Square.*—Drawings of projecting shields for ships.

[2713]

RICHARDSON, DUCK, & CO., *South Stockton Iron Ship Yard*, *Stockton-on-Tees.*—Models of iron screw steamers.

RENNIE, G., & SONS, 6 *Holland Street, Blackfriars.*—Models of ships; model of a floating graving dock.

FLOATING DOCKS.—Messrs. George Rennie and Sons have lately constructed two floating docks on their patent, of which the accompanying engraving shows the general appearance, for the Spanish Government, capable of lifting vessels of from 5,000 to 6,000 tons dead weight, such as H.M.S. *Warrior* would be at her light draught.

Dimensions as follows:—

	No. 1.	No. 2.
Length	320 feet.	350 feet.
Breadth	105 „	105 „
Depth of base or floating chamber . . .	11 ft. 6 in.	12 ft. 6 in.

Floating docks of similar construction are suitable for localities where masonry graving docks are difficult and expensive to execute, or where there is but little rise and fall of tide; and will be found to be of service now that vessels of iron are so much in use both in the royal and mercantile navies.

The model exhibited is about one-third of the length of the floating dock constructed for the Spanish Government for the arsenal of Ferrol. The ends are both open, so that no gates are required, and merely the sides are closed in, against which the shores of the ships rest when docked.

The vessel and dock are lifted by the buoyancy of the lower compartment till the vessel is out of the water.

The engine and pumps are placed on the upper part of the side walls, for pumping the water out of the several chambers of the base or lower compartment. The tops of the sides are used as buoyant chambers, to prevent the possibility of the dock sinking altogether through carelessness in handling.

MODEL OF A FLOATING GRAVING DOCK.

ARRANGEMENT FOR DOCKING SEVERAL VESSELS AT ONCE BY MEANS OF FLOATING DRY DOCKS.—The arrangement exhibited shows three shallow flat horizontal slipways, radiating from a common centre. This system is intended for places where there there is but little rise and fall of tide, as in the Mediterranean; and is now being carried out at the Spanish Royal Arsenal at Carthagena, in conjunction with the floating dock lately constructed by Messrs. George Rennie and Sons for the Spanish Government.

In order to dock a vessel by this means, it is first raised out of the water by the floating dock; the floating dock, with the vessel on it, drawing about ten to eleven feet of water, is then to be hauled into a shallow basin, which is so arranged that the way on the base of the floating dock is level with the ways of the slips in it. The floating dock is then lowered by admission of water into the base or floating chamber till it rests on the bottom of the basin. The vessel is then hauled off, and can be repaired at leisure. This operation can be repeated as often as desired with the same floating dock, until the slipways are occupied with the number of vessels they are capable of containing.

To place the vessels in the water again, the operation is simply reversed.

The model shows only three slipways; but this number may be increased so as to obtain the required accommodation.

In case of repairs of a simple description, or such as will take a short time, or when merely an examination of the bottom of a vessel is required, the operation of hauling-off is not necessary, the vessel being merely lifted out of the water by the floating dock, examined, and afterwards allowed to float again by submerging the dock.

[2714]

RICHARDSON & CO., J. WIGHAM, *Low Walker, Newcastle-on-Tyne.*—Models of ships and steamers.

[2715]

ROBERTS & CO., RICHARD, 10 *Adam Street, Adelphi.*—Models of screw steamers, windlass and screws.

[2716]

ROBERTSON, A. J., *Hattonburn, Kinross.*—Medels of ships.

[2717]

ROGERSON & CO., JOHN, *Newcastle-on-Tyne.*—Model of floating dock, and ferry steamer.

[2718]

ROSE & CROWDER, *Wapping.*—Parallel lift-dock, for repairing ships in tideless waters.

[2719]

RUSSELL, J. SCOTT, *London.*—Models of ships built on the wave principle since 1851.

[2720]

SADLER, WILLIAM, *Tredegar Place, Bow Road, Middlesex.*—Frigates, floating batteries, and gun-boats.

[2721]

SAMUDA, BROTHERS, *London.*—Models of steam-vessels "Leinster," "Victoria," "Tamar," armour-cased frigates, &c.

[2722]

SHARPE, BENJAMIN, *Hanwell Park, Middlesex.*—Shot proofing for ships and batteries; gunnery instruments.

[2723]

SIMONS, WILLIAM, & CO., *London Works, Renfrew.*—Models and plans of iron ship-building and marine architecture and engineering.

[2724]

SIMPSON, ROBERT, *Dundee.*—Models of clipper ship, screw steamer, and swift river steamer.

[2725]

THAMES IRON WORKS AND SHIP-BUILDING COMPANY, *Blackwall.*—Models of iron-cased frigate "Warrior," and other vessels.

[2726]

THOMPSON, HERBERT LEWIS, 47 *Parliament Street, Westminster.*—Models and drawings showing improved construction of iron ships.

[2727]

TOD & MCGREGOR, *Glasgow.*—Model of a screw steamer.

Model of screw steamer, *City of New York,* of the Inman line, 2,560 tons, 550 horse power, built by Tod and McGregor, Glasgow.

[2728]

TOVELL, G. R., *Ramsey, Isle of Man.*—Models of ships and vessels (Tovell & Miller's patent).

[2729]

TRUSS, T. S., 53 *Gracechurch Street, London.*—Patent swift propeller.

These propellers are constructed to produce currents, the action of which will cause the surrounding water to act upon the vessel to propel her.

Without altering the angle of placement of the blades, few or many revolutions may be employed to accomplish a given speed.

With a given power, these propellers accomplish a greater speed than any others now in use, and the greater the draught of water the greater the speed.

They are simple in construction, and not liable to foul.

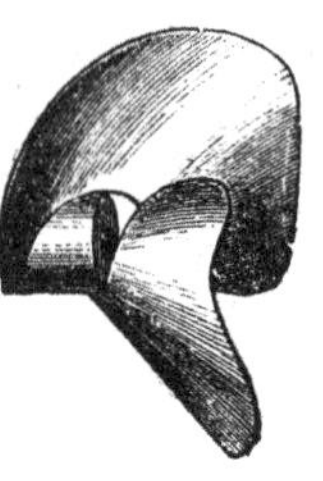

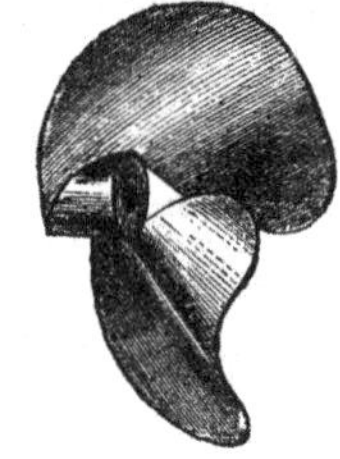

[2730]

VERNON, THOMAS, & SON, *Liverpool.*—Model, Woodside landing stage; caisson, Malta dockyard; and plans of first-class merchant ships.

[2731]

VINES, RICHARD, 3 *Great College Street, Camden Town.*—Newly-invented transverse floats, for propelling steam vessels without backwater.

[2732]

WALKER, WILLIAM HAMMOND (Messrs. TODD, NAYLOR, & Co.), *Liverpool.*—Floating hydraulic ship lift.

[2733]

WATSON, GEORGE, 50 *Lower Shadwell, E.*—Watson's boat lowering and disengaging apparatus.

[2734]

WRIGHT, JOSEPH WILLIAM, 4 *Cumberland Place, Old Kent Road, S.E.*—Paddle-wheels of an improved construction.

SUB-CLASS B.—*Boat and Barge Building, and Vessels for Amusement, &c.*

[2743]

AYLING, EDWARD, 50 *Lower Fore Street, Lambeth.*—Racing oars and sculls.

[2744]

BIFFEN, WILLIAM, *Middle Mall, Hammersmith.*—Models of boats. Great benefits in carriage stowage, &c.

[2745]

CANNON, HENRY, 14 *Blackwall, Middlesex.*—Model of "Peterboat," used in the whitebait fishery.

[2746]

CORYTON, JOHN, 89 *Chancery Lane.*—Ship lifeboat and main rigging, on vertical wave-line system.

[2747]

HALKETT, PETER ALEXANDER, 142 *High Holborn.*—New life-boat; a Franklin expedition knapsack boat; cloak boat. (*See page* 19.)

Halkett, Peter Alexander, 142 *High Holborn*.—New life-boat; a Franklin expedition knapsack boat; cloak boat.

The principle of these inventions consists in making a curved cylinder of india-rubber cloth serve for the sides and ends of a boat, by becoming distended when inflated with air. The diameter of the cylinder being large in proportion to the size of the boat, a very rigid construction is made, and no wood or framework of any kind is required. The centre part within the cylinder, or bottom of the boat, is also formed of india-rubber cloth. Before describing the life-boat marked No. 1, the inventor will refer to the smaller boats. No. 3, the boat cloak, was first invented for the purpose of crossing rivers in exploring, travelling, &c. It does not weigh more than 7 lbs. Mr. Galton, the well-known African traveller, recommends this form of the invention. In the "Art of Travel," he says:—"The inflated india-rubber boat is an invention which has proved invaluable to travellers. They have been used in all quarters of the globe, and are found to stand every climate. They stand a wonderful deal of wear and tear. For the general purposes of a traveller, I should be inclined to recommend as small a mackintosh boat as can be constructed, such as the cloaks that are convertible into boats."

Sir John Franklin, having seen one of these boat cloaks, asserted that if he had had such a boat at the Coppermine River, in his expedition in 1819–21, the disasters and loss of life which then happened would have been obviated. He was, at the time he saw the boat, preparing for his last Arctic expedition, and he expressed a desire to have one to take with him. The inventor recommended him to take, instead of a cloak boat, a knapsack boat, as shown in No. 2, which was the first of the kind made. This is constructed of strong canvas. It weighs no more than a regulation knapsack, and will contain three or four men. They have been used in many climates; in South Africa; by explorers in Central America, in India, and China, and in all the Arctic searching expeditions. It was the means of rescuing two of the men who were with the amiable and gallant Frenchman, M. Bellot, on the floating ice, when he lost his life; and had there been a few more minutes to spare before the ice made so rapidly away, this devoted officer's life would have been saved by its means. They have been well spoken of by many of the distinguished officers commanding those expeditions. Sir R. M'Clure writes, in his despatches to the Admiralty: "I cannot refrain from noticing the excellence of Halkett's boats. These admirable little articles were inflated on board, and with the greatest facility carried upon a man's shoulders over ice, which, from its excessive roughness, no other boat could possibly have been got across without being smashed. By their means a large party were relieved, who were without tents, clothing, fuel, provisions, or in any way provided to withstand the severities of a Polar night, with the thermometer at 8 deg. *minus*." And in another place: "It is impossible to recommend these boats too highly upon a service of exploring, where every article of weight is objectionable. Their whole fitting is but 25 lbs." Dr. Rae (Nov., 1847) wrote: "The boat was found most useful. It carried two men, with a quantity of stores, weighing upwards of 2 cwt., without being in the least degree overloaded. Although in constant use for upwards of six weeks on a rocky coast, it never required the slightest repair." In another letter, April, 1852: "Remembering that the one I had with me in 1846–7, at Repulse Bay (where it had undergone much rough usage), had been left at York Factory, I sent for it, and had it brought more than two thousand miles in the winter to Bear Lake. During the summer season of 1851, it was in constant use for setting nets and other purposes along the Arctic coast. It is now still in very serviceable condition."

No. 1. The life-boat is made of No. 1 canvas, and roped as necessary. It is thirty or thirty-five feet long. In proportion to the number it carries it is a very cheap boat. It can be rolled up and stowed away, so as to take no more room than a spare sail. Inside the cylinders, which are of canvas, and in compartments, there are india-rubber cloth cylinders of the corresponding shape. These can be inflated by ten or twelve men in about six minutes. The canvas takes the shape, by distension, of a boat large enough to contain more than one hundred men, and floating at ease a weight of many tons. Such a boat is strong enough to suffer no damage from concussions, either against the ship's side or on a beach. It can neither overturn nor sink. It is propelled by eighteen oars, or paddles, and though slow in speed, might, the men relieving each other, make constant progress at the rate of three miles per hour, carrying, besides its crew, provisions and water for a fortnight.

In addition to this invention of the inflated life boat, a sail-ladder, when the wind is not too high at the time to use it, has been designed for getting the passengers rapidly and easily into the boat, especially the women and children, who often in cases of fire or wreck are found very difficult to be got into boats. The ladder, made of canvas, is six or eight feet wide "puckered," or folded into various rows of steps, with ropes passing down on each side of the rows. It has been found that even while such a canvas ladder is tossed about with a very violent motion, men, boys, women, and children can go down and up the steps with great facility. The feet can hardly go in any other direction than into the "pucker," and they are very firmly retained in the hollow made by the pucker. In the middle of the breadth of canvas a plain unpuckered surface, some two feet wide, is retained, in order that those who choose may slip down into the boat, and that light articles may be hastily and safely sent down into it from the ship's deck.

The inventor never received any profit or benefit from the use of these boats, and he did not retain a monopoly of their manufacture. It is not with any motive of self-interest that he proposes the application of this principle for ship's life boats, but solely with a view of directing the attention of the public, and especially of members of Parliament, to the fact that such appliances for saving life can be efficiently carried out, that he exhibits plans and drawings of boats such as one he had constructed, and which carried more than a hundred persons. He believes that in this invention the difficulty is practically surmounted of an emigrant ship carrying with it the means of sending away at once all the people on board in case of wreck or fire.

The exhibitor begs to refer persons desirous of further information respecting the boats to S. Matthews and Son, successor to Charles Macintosh & Co., 58, Charing Cross; and to Mr. W. Wellby, who made the large boat, and witnessed the trials made with it, and whose address will be given upon application to Mr. Matthews, or to Mr. Weir, 142, High Holborn.

ROYAL NATIONAL LIFE-BOAT INSTITUTION, 14 *John Street, Adelphi, W.C.*—Life-boat on her transporting carriage; models of life-boats, and of other life-saving apparatus; gold and silver medals of the institution; large wreck chart of the British Isles for 1861, barometer model indicators, &c.

Patroness—HER MOST GRACIOUS MAJESTY THE QUEEN.

President—VICE-ADMIRAL HIS GRACE THE DUKE OF NORTHUMBERLAND, K.G., F.R.S.
Chairman—THOMAS BARING, Esq., M.P., F.R.S., V.P., Chairman of Lloyd's.
Deputy-Chairman—THOMAS CHAPMAN, Esq., F.R.S., V.P.
Secretary—RICHARD LEWIS, Esq. Inspector of Life-boats—Captain J. R. WARD, R.N.

DRAWINGS OF THE LIFE-BOAT OF THE ROYAL NATIONAL LIFE-BOAT INSTITUTION.

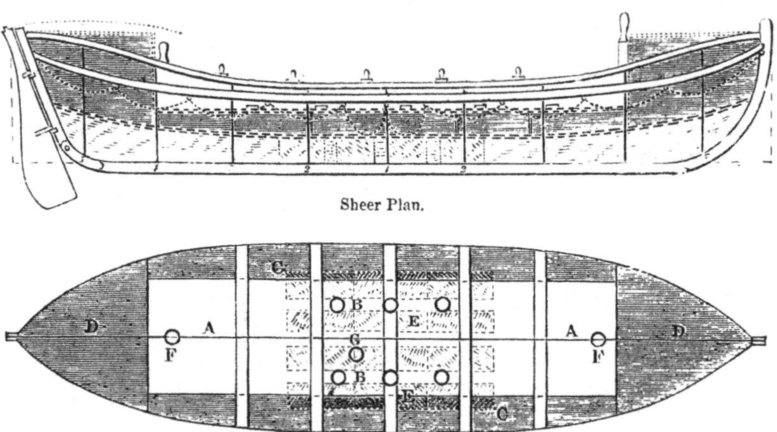

Sheer Plan.

Deck Plan.

This life-boat possesses in the highest degree all the qualities which it is desirable that a life-boat should possess:—1. Great lateral stability. 2. Speed against a heavy sea. 3. Facility for launching and for taking the shore. 4. Immediate self-discharge of any water breaking into her. 5. The important advantage of self-righting if upset. 6. Strength. 7. Stowage-room for a number of passengers.

During the past two years (1860–1) the life-boats of the National Life-boat Institution have been instrumental in rescuing the crews of the following vessels:—

Vessel	Saved
Schooner *Ann Mitchell*, of Montrose	1
Schooner *Jane Roper*, of Ulverstone	6
Brig *Pallas*, of Shields	3
Ship *Ann Mitchell*, of Glasgow ...	9
Smack *John Bull*, of Yarmouth ...	5
Schooner *Catherine*, of Newry ...	4
Barque *Niagara*, of Shields... ...	11
A Barge of Teignmouth	2
Brig *George and James*, of London...	8
Brig *Zephyr*, of Whitby	6
Coble *Honour*, of Cullercoats ...	3
Schooner *Eliza*, of North Shields ...	7
Barque *Oberon*, of Liverpool... ...	15
Brigantine *Nancy*, of Teignmouth...	9
Smack *Wonder*, of Teignmouth ...	2
Brig *Scotia*, of Sunderland	6
Sloop *Three Brothers*, of Goole ...	5
Sloop *Charlotte*, of Woodbridge ...	5
Brig *Ann*, of Blyth	8
Sloop *Hope*, of Dublin	3
Schooner *Druid*, of Aberystwyth ...	5
Barque *Vermont*, of Halifax, U.S. ...	16
Schooner *William Keith*, of Carnarvon	2
Brig *Flying Fish*, of Whitby ...	5
Smack *Elizabeth Ann*, of Lyme Regis...	3
Steam Dredge, at Newhaven ...	9
Schooner *Admiral Hood*, of Rochester	6
Schooner *Susan and Isabella*, of Dundee	5
Schooner *Rose*, of Lynn	3
Brig *Prodroma*, of Stockton ...	11
Brig *Eliza*, of Middlesborough ...	7
Brigantine *Freia*, of Konigsberg ...	6
Brigantine *Diana*, of Fredrikshamn	7
Brig *Gloucester*, of South Shields ...	7
Brig *Lovely Nelly*, of Seaham ...	6
Brigantine *Nugget*, of Bideford ...	5
Schooner *Prospect*, of Berwick ...	6
Sloop *Thomas and Jane*, of St. Ives...	3
A Fishing-boat of Whitburn ...	4
Brig *Arethusa*, of Blyth	8
Schooner *Dewi Wyn*, of Portmadoc	8
Flat *Cymraes*, of Beaumaris... ...	2
Schooner *William*, of Morecambe ...	5
Smack *Gipsy*, of Newry	4
Schooner *Margaret Anne*, of Preston	4
Brig *New Draper*, of Whitehaven ...	8
Schooner *William*, of Liverpool ...	5
Lugger *Nimrod*, of Castletown ...	3
Brig *Providence*, of Shields	8
Brig *Mayflower*, of Newcastle ...	8
Schooner *Village Maid*, of Fleetwood	4
Barque *Guyana*, of Glasgow ...	19
Brig *Roman Empress*, of Shields ...	10
Brig *San Spiridione*, of Galaxide ...	2
Schooner *Voador du Vouga*, of Vianna	8
French Brig, *La Jeune Marie Thérese*	6
Barque *Perseverance*, of Scarborough	5
Schooner *Elizabeth*, of Bridgewater	4
Ship *Danube*, of Belfast	17
Schooner *Hortensia*, of Hanover ...	4
Schooner *Oregon*, of Stonehaven ...	4
Brig *St. Michel*, of Marans	8
Spanish Barque *Primera de Torrevieja*—Saved vessel and 1 of the crew	1
Schooner *Harrell*, of Penzance—Saved vessel and crew	4
Brig *Anne*, of Plymouth—Saved vessel and crew	8
Schooner *Betsey*, of Peterhead—Saved vessel and crew	6
Barque *Frederick*, of Dublin ...	1
Barge *Peace*, of London	2
Lugger *Saucy Lass*, of Lowestoft ...	11
Schooner *Fly*, of Whitby—Saved vessel and crew	4
Smack *Adventure*, of Harwich ...	10
Pilot cutter *Whim*, of Lowestoft ...	7
Barque *Undaunted*, of Aberdeen ...	11
Wrecked boat on Blackwater Bank, on the Irish Coast	1
Schooner *Skylark*, of Folkestone ...	6
Brig *Lively*, of Clay, Norfolk ...	5
Barque *Robert Watson*, of Sunderland	5
Schooner *Auchincruive*, of Grangemouth	6
Schooner *Friends*, of Lynn	4
Schooner *Eliza Anne*, of Dublin ...	5
Brig *Content*, of Sunderland... ...	5
Smack *Ellen Owens*, of Cardigan ...	3
Schooner *Epimachus*, of Amsterdam	5
Total... ...	498

For these and other life-boat services the Institution has voted £1,893 as rewards. It has also granted rewards amounting to £515 10s. for saving 373 shipwrecked persons, by shore-boats and other means, making a total of 871 persons saved from a watery grave during the last two years.

The number of lives saved by the life-boats of the society, and other means, since its formation, is upwards of 12,200; for which services, 82 gold medals, 704 silver medals, and £15,250 in cash have been granted as rewards. The Institution has also expended since its establishment nearly £60,000 on life-boat establishments.

Contributions are received by all the bankers in the United Kingdom; and by the Secretary, Richard Lewis, Esq., at the Institution, 14 John Street, Adelphi, London, W.C.

Royal National Life-boat Institution—*continued.*

THE LIFE-BOAT OF THE ROYAL NATIONAL LIFE-BOAT INSTITUTION GOING OFF TO A WRECK.

[2748]

HAMLEY, JOHN ISAAC, 16 *Capland Street, Lisson Grove.*—Model of life-boat.

[2749]

HAWKESWORTH, AMORY, & ANNERSLEY, GEORGE, 65 *Lincoln's Inn Fields.*—Model of the Hartlepool Seamen's Association Life-boat, in use since 1853.

[2750]

HUTCHINS, WILLIAM, *Croom's Hill, Greenwich.*—Self-righting, indestructible pneumatic life-boat, and shot-proof ship's cutter.

[2751]

JORDESON, THOMAS POWDITCH, *Eastcheap.*—Patent life-boat, and apparatus for converting ships' boats into life-boats.

[2752]

LEARWOOD, THOMAS, *Truro, Cornwall.*—Life-boat, propelled without oars through the surf; cannot fill; self-righting.

[2753]

PRESTON, LIEUT. THOMAS, R.N., *Lowestoft.*—Double rudders, less liable to accident; propeller, blades unship from arms.

[2754]

PYM, JOHN, 4 *Laurence Pountney Hill, London.*—Double sheer hulk for raising sunken vessels (Pym's patent).

[2755]

RICHARDSON, HENRY THOMAS, *Aberhirnant, Bala, N.W.*—Model of "Richardson's patent iron tubular life-boat."

[2756]

ROYAL NATIONAL LIFE-BOAT INSTITUTION, 14 *John Street, Adelphi, W.C.*—Life-boat on her transporting carriage, models of life-boats, and of other life-saving apparatus, &c. (*See pages* 20, 21.)

[2757]

SEARLE, EDWARD, *Stangate, Lambeth.*—Model of state barge.

The exhibitors hold the appointment of boat builders to Her Most Gracious Majesty the Queen, H.M. the Emperor of the French, H.M. the Emperor of Austria, H.R.H. the Prince of Wales, H.R.H. the Prince of Prussia, H.S.H. Prince Edward of Saxe Weimar, H.H. Ismael Pacha, H.H. Prince Duleep Singh, the Lords Commissioners of the Admiralty, the Right Honourable the Board of Ordnance, the Honourable Board of Conservators of the River Thames, the Universities of Oxford and Cambridge, the Eton and Westminster Schools, the "Guards," "Leander," and other distinguished Clubs, and most of the leading Amateurs.

[2758]

STEVENS, WILLIAM, *Trinity Square, Tower Hill.*—Model, ships and boats, and every requisite for fitting and rigging.

[2759]

THOMPSON, NATHAN, 21 *Rochester Road, N.W.*—Models of Thompson's new patented system for building boats by machinery.

[2760]

TWYMAN, HENRY, 26 *Hardres Street, Ramsgate.*—Lugger life-boat, built with air-tight compartments.

[2761]

WATSON & DAVISON, 5 *Munster Square, Regent's Park.*—Patent safety rowlocks.

[2762]

WENTZELL, ANDREW, *Lambeth, and Crystal Palace.*—Improved models of boats (various kinds) for speed and pleasure.

SUB-CLASS C.—*Ships' Tackle and Rigging.*

[2774]

ADCOCK, JOHN, *Marlborough Road, Dalston, London.*—A marine "odometer;" or, improved ship's log; various modifications.

[2775]

BERREY, CAPTAIN GEORGE A., 32 *Fenchurch Street, London.*—A "sphereometer," for facilitating the practice of great circle sailing.

[2776]

BIRT, J., 5 *Wellclose Square.*—Model of the mortar and rocket life-saving apparatus.

[2777]

BLAKENEY, J. W., & Co., *Hull, Glasgow, and Sunderland.* (*See page* 24.)

[2778]

BROWN, J. H., *Adelaide Place, London Bridge.*—A floating buoy for saving ship's papers when wrecked.

[2779]

BROWN, LENOX, & Co., *Billiter Square, E.C., and Millwall, Poplar, E.*—Malleable cast iron blocks and sheaves.

[2780]

DANBY, JAMES F., 11 *Cantelowes Road, Camden Square, N.W.*—Model of "Danby's patent anchor."

[2781]

GIFFORD, WM. J., *Wellington,* and 39 *Devonshire Street, Queen Square, London.*—A model gaffyard rig.

The following is exhibited:—A model of a full-rigged vessel, viz., a three-masted steamer, in a glass case, showing the application of the new system of rigging and sail-making, called the "GAFFYARD RIG." The general object of this system is the perfection of the art of sailing, more especially "close-hauled" (*i.e.*, obliquely against the wind), with complete command of the course and position at any angle with the wind, or point of the compass. Some of the improvements and modifications in which this system differs from others, may be understood by the following particulars:—

1. The tension of the canvas is equalised everywhere; and at any part liable to undue or irregular strain, the sail is secured by an interposed breadth of a cloth manufactured expressly for the purpose. It may also be secured by "cord-bands."

2. The sails are "bent head and foot."

3. A more perfect "set" of the sail is thus obtained, with this essential peculiarity, that the curving, or "bellying," is in the perpendicular and not in the horizontal direction. The surface is therefore straight, in the horizontal sense.

4. The sails are to windward instead of to leeward of the masts and gear, the result being that when "close up," the masts are sheltered from the wind; and secondly, the wind passes off the sails without impediment at any angle.

5. The yards, as well as the gaffs, are "fore and aft." This new system may, in some sense, be viewed as a combination of the two common rigs known as the "square" and the "fore and aft" rigs; and hence the term "gaffyard" rig is applied to it.

BLAKENEY, J. W., & Co., *Hull, Glasgow, and Sunderland.*

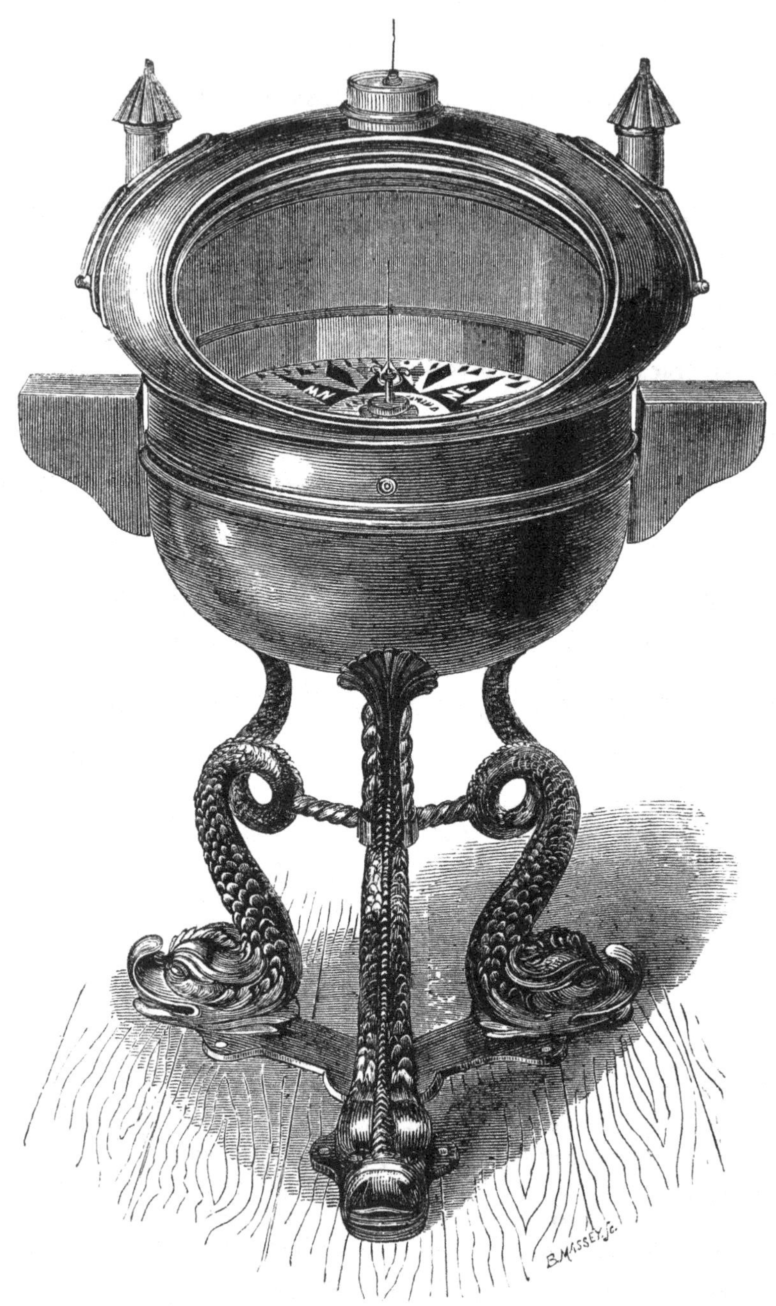

IMPROVED ANTI-VIBRATION STEERING COMPASS FOR STEAMERS.

[2782]

GLADSTONE, THOMAS MURRAY, 30 *Parliament Street, London, S.W.*—Two models of a "patent anchor," an iron and a wood stock.

[2783]

GODDARD, JOHN MAYNARD, 9 *Ship Street Lane.*—Specimens in the manufacture of ships' blocks.

[2784]

HAWKS, CRAWSHAY, & SONS, *Newcastle-on-Tyne.*—Model of Trotman's patent anchor, 95 cwt., supplied to H.M. frigate "Warrior." (For Illustration, *see* page 28.)

INTERNATIONAL EXHIBITION,
PARIS, 1855.
GRANDE MEDAILLE 1RE CLASSE,
À J. TROTMAN, 42 CORNHILL,
LONDON.

The model of Trotman's anchor, 95 cwt., made for Her Majesty's iron-cased frigate "Warrior," and exhibited by Messrs. Hawks, Crawshay, & Sons, Newcastle-on-Tyne.

The Lords Commissioners of the Admiralty nominated as an anchor committee the Honourable Admiral Sir M. Stopford, K.C.B., &c.; Admiral George R. Mundy, C.B., &c.; the late Admiral Charles Hope, C.B.; and other naval officers; with whom were asociated Duncan Dunbar, Esq., the chairman of the General Shipowners' Society, and five gentlemen of "Lloyd's Classification Committee"—"to investigate and determine, by a series of practical proofs and tests, the relative merits of different descriptions of anchors." Their unanimous report, dated 1st of February, briefly recapitulates the order of merit, as follows:—

ART. 29.—"The committee here beg to recapitulate the order in which they consider the anchors to stand, together with their relative per-centage of inferiority or superiority to the Admiralty anchor, the value of which being taken as the standard or unit:

Trotman	1·28 or 28 per cent.	superior to Admiralty anchor.
Rodgers	1·26 or 26 ,,	
Mitcheson	1·20 or 20 ,,	
Lenox...	1·13 or 13 ,,	
Porter	1·09 or 9 ,,	
Aylen	1·09 or 9 ,,	
Admiralty	1. = the standard."	

The Royal yacht, "Victoria and Albert;" also the French and Russian imperial yachts, "La Reine Hortense," "Le Prince Jerome," &c. &c.; the steam-ship "Great Eastern;" the ships of the Cunard Company, Peninsular and Oriental Company, the Royal Mail Company, the Messageries Imperiales, the Austrian Lloyd's, and the mercantile marine generally, are supplied with Trotman's improved anchors at about one-third less weight than would be required for ordinary anchors—a desideratum which the anchor committee deem "of vast importance to the shipping interest."—*Vide paragraph 9 Official Report.*

The distinguishing feature peculiar to Trotman's anchor is the palm being set at an acute angle to the line of strain, and differing from that of the arm: in action, it is found to bite the ground instantaneously as a ploughshare, and by reason of the vibratory motion of the arms, the pressure of the upper arm on the shank imparts increased penetration to the lower arm in the ground; or in other words, the heavier the strain, the more tenacious the holding properties. It possesses other advantages besides strength, and holding-power more than doubled; viz., freedom from fouling the cable—increased efficiency with reduced weight, affording very material relief to ship's bows in a head sea—facility of transport to or from ships by means of boats—convenience of stowage—elasticity of form, which enables it to sustain sudden strains or jerks at short stay-peak, and concussion, when let go on hard or rocky bottoms.

Comparison suggests the following conclusions:—the angle of the palm and arms of Porter's and other anchors being identical; the ordinary anchor, likewise, in action is a mere scraper, accumulating, as it were, the loose surface, instead of biting and retaining its fulcrum of resistance in unbroken ground; its form rigid and inflexible, a mass of iron, one-third of which is never available, and really mischievous—as the upper arm is ever liable to be fouled or hooked by the cables of other ships in crowded anchorages—presenting always a dangerous projection to ship's bottoms, in shoal water, tidal harbours, and rivers. The principle of Trotman's anchor obviates these objections. It is flexible in its parts, each contributing its portion of duty to the whole, and adapting itself to every emergency.

The following are some of the eminent firms licensed by the patentee to make Trotman's anchors, viz.—Messrs. Hawks, Crawshay, & Sons, Newcastle-on-Tyne and London; Wood & Co., Liverpool, Chester, and London; John Abbot & Co., Gateshead-on-Tyne; Pow & Fawcus, North Shields; Robert Wight & Son, Sunderland and Seaham; N. Hingley & Sons, Netherton and Cradley Iron-works; Henry P. Parkes, Dudley and Liverpool.

Patentee's office, 42, Cornhill, London.

[2785]

HERBERT, GEORGE, *Dartford.*—Bury, Beacon, Telegraph Station Battery, moored from centre of gravity; ship's motion metre, self-registering.

[2786]

HOLSGROVE & REED, *Sunderland.*—Model of Reed's patent anchor; specimens of iron ship knees and forgings.

[2787]

HOLTUNG, WILLIAM, *Church Street, Walmer.*—Model of balista, for making communication with stranded vessels.

[2788]

HUNTER, SAMUEL, 22 *Grey Street, Newcastle-on-Tyne.*—Model of a new anchor.

[2789]

JEULA, HENRY, *Lloyd's, E.C.*—Martin's patent anchor: immediate hold, immense power, no fouling, easy tripping and fishing, great lightness.

[2790]

LAING, JAMES, 2 *M'Vicar's Lane, Perth Road, Dundee.*—Helixameter, for experimenting on the screw propeller. Combined screw pump for ships, and compound ventilator for ships.

HELIXAMETER.—The purpose of this machine is to make an experimental investigation into the properties of the screw as a propeller, more particularly in consideration of the "pitch." The tables and diagrams accompanying the machine are constructed on experimental data ascertained by its use, and on examining these some very remarkable properties of the screw will be observed. The most prominent of these are, that the greatest possible thrust is obtained, and that neither positive nor negative slip occurs when the pitch is equal to the diameter.

COMBINED SCREW PUMP FOR SHIPS.—The objects in the construction and action of this pump are, small first cost, easy keep, and certain action under any probable circumstances, besides giving with small power a large and continuous discharge of water.

COMPOUND VENTILATOR FOR SHIPS.—This system produces both a downward and upward current of air acting in combination. The ventilators are made either of metal as fixtures, or of cloth as wind-sails, to be used at pleasure. A thorough ventilation is obtained by this system.

[2791]

LONGHURST, JOHN, *Ticehurst, Sussex.*—Breakless cable chain.

[2792]

MACDONALD, JOHN, 13 *Henry Street, Vauxhall.*—Compass, with accompaniments, for longitude and latitude; also, ship's lamps.

[2793]

MARTIN, CLAUDE, 10 *Bath Place, Hatcham.*—Improved Porter's anchors, and a *bombe-mitraille* (unloaded).

[2794]

MOORE, C., *Swansea.*—A spherical indicator, for nautical and astronomical purposes.

[2795]

PARKES, HENRY PERSHOUSE, *Chain and Anchor Works, Tipton, Staffordshire.*—Chains and anchors of various descriptions.

[2796]

PEACOCK, GEORGE, F.R.A.S., *Starcross, Devon.*—Refuge buoy-beacon, granulated cork poncho-mattress; life and treasure preserver; unfoulable anchor.

[2797]

RETTIG, EMIL, *London.*—Martin's anchor (patented), with holding power double that of ordinary anchors, and inability to foul.

[2798]

RICH, WILLIAM, 14 *Great Russell Street, Bloomsbury.*—Improved kite for carrying a line or man, &c., on shore from stranded vessels.

[2799]

ROGERS, M. D., *St. Leonard's Road, Poplar.*—Models of boat lowering gear, chain cable, stopper, controller, and windlass.

Wood and Rogers' patent suspending and detaching apparatus for ships' boats. The advantages of this method will be obvious from the following observations:—

The boat being suspended from four points instead of two, prevents the possibility of canting while persons are getting in or in lowering; and it may be detached, however fast the vessel may be going through the water, with ease and safety.

This apparatus is so compact, that the boat may be filled with provisions, &c., there being no ropes to foul or kink, to the danger of persons in the boat.

The suspension is so secure, that the boat may hang to the davits without gripes or lashing.

The detaching is effected by simply raising a lever.

Kendall and Rogers' patent chain cable controller, for windlass to be fixed on deck. The advantage of this is, that it prevents the chain from riding, in paying out or heaving in, which it effectually does by avoiding the necessity of fleating. It is simple, and not expensive.

Rogers' patent chain cable stopper is intended to secure any link in heaving up, and to assist in paying out chain; to give cable in bringing up ships, or at the chain cable locker.

[2800]

ROGERS, WILLIAM, *Waterloo Street, Swansea.*—A ship's steering apparatus.

[2801]

ROYAL HUMANE SOCIETY, 4 *Trafalgar Square, W.C.*—Models of apparatus, used for rescuing and recovering persons apparently drowned or dead.

[2802]

SAMUEL, DAVID A., 3 *Cedar Place, West Ham, Essex.*—Steering apparatus.

This is a model of a steering apparatus on a new principle. This new method will be much quicker in its action, and more safe at sea than the old chain principle. Three turns of the wheel will bring the helm from port to starboard, and a single person will be able to steer in the heaviest sea, without fear of being overpowered by the wheel, as often occurs in very rough weather. It is also well adapted for river purposes.

[2803]

SMITH, ROBERT, 23 *Fish Street Hill.*—Solid cork life buoy, jacket, belt, and waterproof cork socks.

***Obtained honourable mention at the Paris Exhibition,* 1855.**

The buoys and belts invented by R. Smith are celebrated for their durability and buoyancy.

The socks, invented in October, 1861, are superior to any yet offered to the public.

[2804]

SOLOMON, JOSEPH, 22 *Red Lion Square, London.*—Sphereometers, for facilitating great circle sailing, obviating abstruse calculations.

[2805]

STONE, JOSIAH, *Deptford, London.*—Side lights, blocks, copper nails, rullocks, boat fittings, &c.

[2806]

TENWICK, JOHN, *Albion Foundry, Clarendon Street, Landport.*—Patent steering apparatus.

[2807]

TROTMAN, J., 42 *Cornhill.*—Model of Trotman's anchor used on board H.M. yacht "Victoria and Albert."

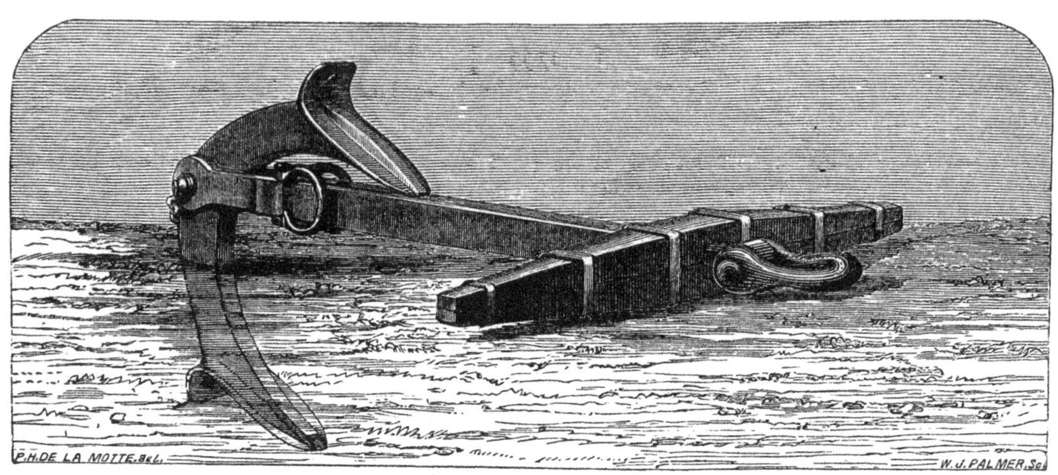

TROTMAN'S ANCHOR.

[2808]

TYLOR, J., & SONS, *Warwick Lane, Newgate Street, London.*—Apparatus for distilling fresh from sea water.

[2809]

WALKER, THOMAS, & SON, *Oxford Street, Birmingham.*—Ship logs, and sounding machines.

[2810]

WARD, CAPTAIN J. R., R.N., *New Brentford.*—Cork life-belt, for use of life-boat men, and ship's crews.

CORK LIFE-BELT FOR USE OF LIFE-BOAT MEN.

These life-belts possess buoyant power averaging 26 pounds, being double that of ordinary cork life-belts.

The advantages of this belt are—

1. It secures great buoyancy without inconvenience to the wearer.

2. It is perfectly flexible.

3. It affords great protection to the body against injury when in the water.

4. It is very strong and durable, is little liable to injury, and is readily repaired if injured.

5. Being tightly secured round the waist, it cannot slip upwards or downwards, but is always in the best position for preserving the equilibrium of the wearer.

These life-belts have already been the means of saving a large number of lives. They are supplied to the life-boats' crews of the National Life-boat Institution on the coasts of the United Kingdom, and to the crews of coast-guard stations.

Manufacturer, Mr. J. Birt, Jun., maker of life-saving apparatus, 5 Wellclose Square, London.

[2811]

WATSON, THOMAS, 49 *Rupert Street, W.*—Application of friction break to ships' capstans (working model).

This invention brings the capstan under the entire control of one man, who is entirely out of danger, and prevents the necessity of the men having to walk backwards. At an inquest held on the bodies of some men killed by one of these accidents on board H.M.S. "Nile," in January, 1861, the jury expressed a hope that something might be invented which would prevent the recurrence of such accidents in future.

[2812]

WEST, JOHN GEO. & CO., 92 & 93 *Fleet Street.*—Ships' and boats' binnacles, and patent liquid compasses.

[2813]

WIGHT, ALEXANDER, 14 *Lansdowne Crescent, W.*—Compound iron cable.

[2814]

WOOD & CO., *Liverpool, London, and Stourbridge.*—Chains, cables, and anchors.

PRINTED BY PETTER AND GALPIN, BELLE SAUVAGE WORKS, LUDGATE HILL, LONDON, E.C.

Class XIII.

PHILOSOPHICAL INSTRUMENTS, AND PROCESSES DEPENDING UPON THEIR USE.

[2845]

Ackland, William, 19 *Church Row, Newington Butts.*—Dividing engine, and instruments divided thereby.

[2846]

Adie, Patrick, 395 *Strand.*—Patent semicircle; patent diastameter; patent theodolite level; patent level; patent surveying compasses; standard barometer; eidograph (Wallace's).

[2847]

Adie, Richard, 55 *Bold Street, Liverpool.*—A gold disc steam and vacuum gauge; an alcohol hermetic barometer; a double telescope.

[2848]

Aldous, W. Lens, 47 *Liverpool Street, King's Cross.*—Microscopic drawings of the human breath, and other curiosities.

[2849]

Alison, Dr. S., 80 *Park Street, W.*—Differential double stethoscope; sphygmoscopes; stethogoniometer; and hydrophone, used in chest diseases.

[2850]

Allan, Thomas, C.E., 1 *Adelphi Terrace, Strand, W.C.*—Mechanical or automatic recording telegraph; electro-magnetic engine; submarine cables, &c.

[2851]

Bagot, Charles E., M.D., *Claremont Mall, Dublin.*—Nephelescope, for viewing the upper strata of clouds.

[2852]

Bailey, J. W., 162 *Fenchurch Street, London.*—Sextants; artificial horizons; theodolites; levels; prismatic compasses; drawing instruments.

Pillar sextant, on counterpoise stand.
Bell metal and box sextants.
Gravatt's dumpy level.
Transit and Everest's theodolites.
Prismatic compass.
Artificial horizons, mercurial and dark glass.
Universal compass (a new method of describing an ellipse), which comprises triangular and hair compasses, pen, and pencil joints; and forms a complete set of instruments.

[2853]

BAKER, CHARLES, 244 *High Holborn.*—Microscopes, and their appliances; surveying, engineering, and drawing instruments; ivory and box rules, &c.

The following are exhibited:—

Binocular and other microscopes, achromatic object-glasses and apparatus, and materials for mounting preparations: also surveying, levelling, and drawing instrument of all kinds.

	£	s.	d.
No. 1.—Highly finished compound microscope, with mechanical and secondary stages, and all the latest improvements, with two Huyghenian eye-pieces	20	0	0

No. 1 MICROSCOPE.

	£	s.	d.
No. 1 A.—A microscope of the same size, but without secondary stage	13	10	0
No. 1 B.—A small microscope, having mechanical stage, &c.	11	10	0
No. 1 B.—With plain stage	7	15	0
No. 2.—Two sizes smaller, with mechanical stage and one eye-piece	8	15	0
No. 2.—With plain stage	6	15	0
No. 3.—A student's ditto, with slow motion, object glasses, case, and apparatus complete	6	15	0
No. 3.—Ditto, without slow motion	5	15	0
No. 4.—A student's microscope, complete ...	4	15	0
No. 5.—Educated microscope, complete ...	3	3	0

ACHROMATIC OBJECT GLASSES, OF LARGE ANGULAR APERTURE.

	Degrees.	£	s.	d.
3 inch	10	1	15	0
2 „	12	1	10	0
Ditto ...	15	1	17	6
1½ inch	20	1	17	6
1 „	23	1	17	6
⅔ inch	30	2	5	0
½ „ with adjustment	60	3	0	0
4/10 „ „	70	3	5	0
¼ „ „	75	3	5	0
¼ „ „	95	3	15	0
⅛ „ „	115	5	5	0
⅛ „ „	125	6	6	0
German, ½, 1, & 1½ inch	... ...	1	2	6
Ditto, ¼, ⅓, ½ „	... ...	1	5	0

	£	s.	d.
A first-rate 6-inch transit theodolite, bright or bronzed, with vertical and horizontal circles, and verniers divided on silver to 20 min. The clamping and tangent adjustments of the most approved make. An inverting and erecting eye-piece, with tripod, stand, and mahogany case	22	10	0

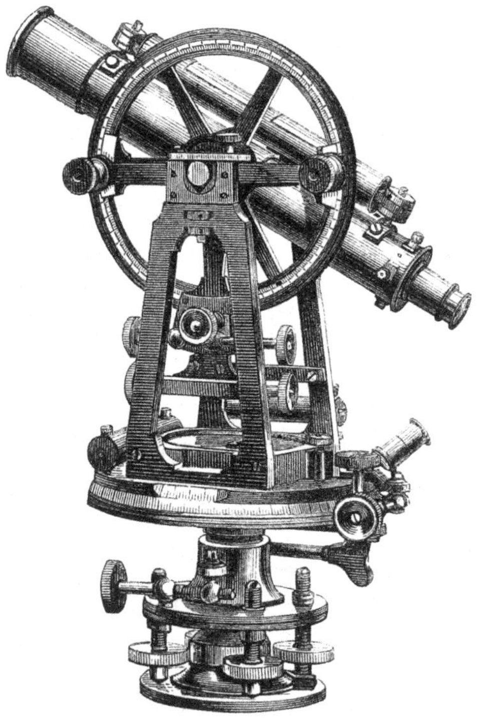

TRANSIT THEODOLITE.

	£	s.	d.
6-inch ditto, without vertical arc	21	0	0
6-inch bronzed or bright theodolite, with all the latest improvements	20	0	0
Ditto, ditto, 5 inch	18	10	0
Gravatt's improved 14-inch dumpy level, bronzed or bright, complete	10	10	0
Ditto, ditto, with silver ring compass ...	11	10	0
12-inch ditto	11	0	0
Ditto, without compass	10	0	0
Station staves, from ... ... 35s. to	2	5	0
Mining dials, from ... ... 6*l.* 10s. to	10	0	0
Pocket sextants, from	3	3	0
Engine-divided protractors, from	0	12	0
Improved German silver beam compass ...	1	10	0
Proportional ditto, fully divided	1	2	6
Tubular compasses	1	10	0
Napier ditto	1	5	0
Pillar ditto	1	0	0

[2854]

BARNETT, JOHN, 3 *Whitehall Street, Tottenham.*—Microscopic preparations.

[2855]

BEALE, PROFESSOR, F.R.S., *London.*—Microscopes for class demonstration; microscopical specimen of animal tissue, nerves, vessels, &c.

[2856]

BELLHOUSE, WILLIAM DAWSON, 1 *Park Street, Leeds.*—Medical galvanic apparatus, containing primary, secondary, and combined currents.

[2857]

BENHAMS & FROUD, 40, 41, & 42 *Chandos Street, Charing Cross.*—Platina apparatus for chemical uses.

[2858]

BESTALL, WILLIAM, 1 *Victoria Cottage, Royal Road, Kennington Park.*—Apparatus for showing the beautiful phenomenon of polarized light.

[2860]

BLIGH, JOHN, 30 *Charles Street, Berkeley Square, W.*—Sensitive thermometer, and self-acting ventilator.

[2861]

BOLTON & BARNITT, 146 *Holborn Bars.*—Chemical, galvanic, and pneumatic apparatus.

[2862]

BRAHAM, JOHN, *Bristol.*—Spectacles from earliest dates; patented anti-ophthalmioscopic, rifle, and sporting spectacles; helical spring eye-glasses.

The exhibitor's improved pantescopic and anti-ophthalmioscopic spectacles, spherical eye-preservers, rifle and sporting spectacles, and helical spring eye-glasses (protected by letters patent of Her Majesty the Queen and the Emperor of the French, granted August, 1861), may be obtained, wholesale and retail, from himself and from licensed agents in all towns of the United Kingdom.

[2863]

BRETT, JOHN WATKINS, 2 *Hanover Square.*—Submarine telegraph cables successfully established by the inventor; Roman type-printing telegraph.

[2864]

BRITISH AND IRISH MAGNETIC TELEGRAPH COMPANY, *Liverpool.*—Telegraph instruments, insulators, and apparatus; submarine telegraph cables. (*See page* 4.)

[2865]

BRITISH ASSOCIATION, *Kew Observatory.*—Philosophical instruments.

[2866]

BROWN, DAVID STEPHENS, *Eton Lodge, Ashby Road, Islington, London.*—Self-acting sympiesometer; portable barometer; sensitive sympiesometer.

[2867]

BUCKINGHAM, JAMES, Civil Engineer, *Walworth Common, London.*—Refracting telescope, equatorially mounted, 20 inches aperture; portable ditto, 5 inches aperture.

1. Equatorial Refracting Telescope (in the Nave), 28½ ft. focus and 20 inches aperture; believed to be the largest existing; all the clamps and slow motions in right ascension and declination are at the eye-end, where also the declination circle is read.

2. Portable Equatorial (in the North Gallery), 7 ft. focus, $5\frac{1}{10}$ in. aperture, circles 11¼ in. diameter, graduated on platina, adjustable to latitudes 30° to 60°, having entire motion in azimuth to facilitate placing in position: can be used as a transit.

3. Very delicate Level, without adjustment, divided to seconds of space.

4. Micrometer, divided on platina, with new method of illuminating by prisms.

5. Object Glasses of 2¾, 3¾, $5\frac{1}{10}$, 9, and 20 in. clear aperture; and one 8¼ in. on the dialytic system.

None of these instruments are for sale; they are exhibited only to show the convenience and novelty of the fittings. The object glasses, which are free from chromatic and spherical aberration, were made for the exhibitor by William Wray, Optician, 1 Clifton Villas, Upper Holloway, N.

BRITISH AND IRISH MAGNETIC TELEGRAPH COMPANY, *Liverpool.*—Telegraph instruments, insulators, and apparatus; submarine telegraph cables.

The following patented inventions are exhibited :—

THE ACOUSTIC TELEGRAPH INSTRUMENT. Sir C. T. and E. B. Bright's patents, 1855—1860. The acoustic instrument conveys signals to the ear instead of to the eye of the operator. Two bells are used of different tone, and muffled so as to prevent prolonged vibration. A single conducting wire only is employed, a relay being used to connect up a local battery on a positive current being received with one bell, or if a negative current with the other bell. This apparatus is used at all the principal stations of the Company. It is not liable to get out of order, and utilizes both currents, besides saving the cost of a writing clerk, required at each instrument when visual signals are used.

THE MAGNETIC TELEGRAPH INSTRUMENT. Henley's patent, 1848. In this invention the magneto-electric current is applied in place of a galvanic battery. It is used extensively by the Company for railway telegraphs and other purposes, and is peculiarly applicable to hot climates, where it is found very difficult to keep galvanic batteries in order.

THE NEEDLE TELEGRAPH. Edward Highton's patent, 1848. This is a simple form of visual telegraph, requiring only one wire, and is extensively used by the Company for railway telegraphs and other purposes.

THE TRANSMITTING INSTRUMENT. Sir C. T. and E. B. Bright's patent, 1852. This is a relay, so constructed that when connected to a single wire, at an intermediate point, it will act as a relay to transmit a positive or negative current at will in either direction.

RESISTANCE COILS, for testing the position of a fault in telegraph conductors from a distant station. Sir C. T. and E. B. Bright's patent, 1852.

VACUUM LIGHTNING PROTECTOR. Ditto, ditto, ditto. In this invention the resistance of the air to the passage of electricity, between the points of the conductor, may be exactly regulated by partial exhaustion of the chamber in which the points are placed, so as to allow lightning to pass freely from the wire to earth without entering the telegraph instrument or cable, etc., to which this apparatus is connected. The points are at the same time preserved from dust.

MERCURIAL RELAY AND RECORDING APPARATUS. Sir C. Bright's patent, 1862. In the relay a fine stream of mercury, or other conducting fluid is employed; contact being made by a magnetic needle or its arm passing through the stream. The use of a metallic spring or pin is thus dispensed with, rendering the relay exceedingly sensitive and speedy in its action.

APPARATUS for compensating for the variations in force of currents sent and different length of pauses on long lines of telegraph, and RECORDING APPARATUS used for determining the variation of currents, showing the rise and fall in the wire, and registering any currents of terrestrial magnetism. Sir C. Bright.

SHACKLES and SWIVEL INSULATORS, for use round sharp curves and where great stretches of wire are required. Sir C. T. and E. B. Bright's patents, 1852 and 1858. Employed extensively in London and elsewhere for the overhouse telegraphs.

SPECIMENS of the Company's two Irish cables, containing six wires each, laid 1853—4, and collection of other cables. Specimens of concrete on cable.

ISULATORS ON ARMS, of different lengths. Sir C. T. and E. B. Bright's patent, 1852. The wires are so arranged as to avoid contact if one or more break.

[2868]

BURROW, W. & J., *Great Malvern.*—Malvern landscape and target telescopes. (*See page* 5.)

[2869]

BURTON, EDWARD, 47 *Church Street, Minories, London.*—Optical and mathematical instruments.

[2870]

BUSS, THOMAS ODEMCY, 3 *Upper East Smithfield, Tower Hill.*—Hydrometers and saccharometers for fluids.

[2871]

BUTTERS, THOMAS E., 4 *Belvedere Crescent, Lambeth.*—Parallel glasses for mathematical instruments.

[2872]

CAMERON, PAUL, Mathematical Instrument Maker, *Glasgow.*—Marine compass; marine barometer; instruments for determining ships' position at sea.

BURROW, W. & J., *Great Malvern.*—Malvern landscape glasses and target telescopes.

This drawing represents a "BURROW'S LANDSCAPE GLASS," containing twelve lenses constructed of the purest glass, in such combinations as to produce high power and accurate definition with wide and brilliant field.

These glasses are achromatic, and are made in two sizes, price 6 guineas and 3½ guineas respectively, sling included.

The 6 guinea glass will show hits on a target at 500 yards, distinguish colours on the race-course at a mile, and define objects in a landscape and ships at sea at 15 to 20 miles. Size 4½ in. × 4¾. Diameter of object-glasses, 2 inches.

The 3½ guinea glass will do the same at 400 yards, ¾ of a mile, and 10 to 12 miles. Size 3 in. × 4¼. Diameter of object-glasses, 1½ inches.

References to noblemen, distinguished officers in the army and navy, sportsmen, travellers, scientific men, racing judges, &c., who use these glasses in preference to field glasses of the ordinary make.

BURROW'S TARGET TELESCOPE for rifle practice will show hits on the target at the long ranges. Length 7 inches; weight 3 ounces. Price, 25*s.*; covered Russia, 30*s.*—For centres at 1000 yards, 3 guineas.

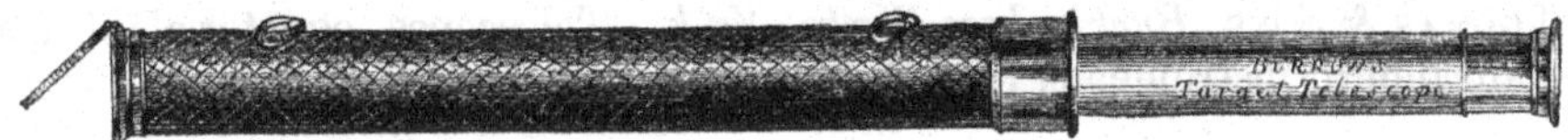

[2873]

CASARTELLI, JOSEPH, 43 *Market Street, Manchester.*—Microscopes; telescopes; mining and surveying instruments.

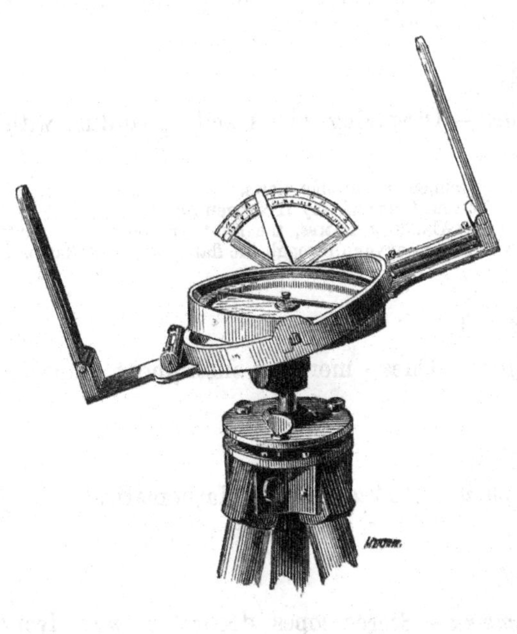

	£	s.	d.
1. Astronomical telescope, 3½ inches aperture, equatorially mounted, with movable axis for setting to different latitudes, suitable for either a pillar or portable stand	50	0	0
2. Astronomical telescope, 3½ inches aperture, mounted on a firm portable stand; fine and coarse horizontal and vertical movement	40	0	0
3. Large size, first-class microscope, stand, two eye-pieces, pliers, forceps, &c. &c. . .	20	0	0
4. A new arrangement of circumferenter, or miner's dial, of first-class quality, with sights mounted on movable plate for sighting up or down inclines, with quadrant attached, showing the degree of inclination, &c. (*See illustration*)	14	14	0
5. Ditto, as above, smaller size . . .	14	0	0
6. Circumferenter or miner's dial, with tangent screw adjustment of first-class quality	13	10	0
7. Best 14 inch dumpy level, with compass	14	14	0
8. Portable anemometer (invented by J. Dickenson, Esq., Government Inspector of Mines) for showing the velocity of air current in coal mines, by which can be computed the quantity of air, in cubic feet, passing per minute through the airways	2	10	0
8. Steam-engine indicator, for taking diagrams, showing the working of the engine, and computing the power exerted	6	6	0
9. Steam-engine indicator as above, small size	5	5	0
10. Vacuum gauge for indicating the state of the vacuum in the condenser of the steam-engine	3	10	0
11. Patent steam-pressure gauge . . .	3	5	0

[2874]

CASELLA, LOUIS P., 23 *Hatton Garden.*—Mathematical, philosophical, surveying, reliable, popular, and standard and meteorological instruments.

[2875]

CHADBURN BROTHERS, *Nursery, Sheffield.*—Spectacles, telescopes, microscopes, reading-glasses, optical lenses, &c.

[2876]

CHANCE BROTHERS, *Birmingham.*—Dioptric sea lights and lanterns.

[2877]

CHATTERTON, T., 14 *King's Terrace, Bagnigge Wells Road.*—Barometers.

[2878]

CLARK, GEORGE, 30 *Craven Street.*—Improvements in manufacturing, connecting, laying, and raising electro-telegraphic cables.

[2879]

COOK, JAMES EDGAR, 22 *Mearns Street, Greenock.*—Damp and water resisting mirror—patented.

[2880]

COOKE, THOMAS, & SONS, *Buckingham Works, York.*—Telescopes, equatorials, transit instruments, altazimuth instruments, theodolites, levels, &c.

[2881]

CHEYNE, J. B., & MOSELEY.—Recording apparatus, to serve as a check upon signal-men, engine-drivers, and others.

[2882]

COX, FRANCIS B., 50 *Camden Street, Birmingham.*—Box and ivory rules; Carrett's, Hawthorn's, and Routledge's engineers' rules; English and foreign rules.

[2883]

COX, FREDERICK J., 22 *Skinner Street, London.*—Dissolving views and apparatus, with various methods of illumination.

VERTICAL DISSOLVING VIEW APPARATUS, 1½ inch condensers, oxy-hydrogen jets, clockwork motion to lime cylinders, gas-bags, and generator; price 40*l.*

APPARATUS FOR EDUCATIONAL PURPOSES, 3½ inch condensers, oxy-hydrogen jets, and gas generator; price 22*l.*

Specimens of suitable views.

Various forms of Oxy-Hydrogen Jets.

OXY-CALCIUM LAMPS, the light produced by passing a stream of oxygen through the flame of a spirit lamp.

[2884]

CRABBE, REV. GEORGE, *Merton Rectory, Thetford.*—Cheap meridian instrument, showing solar time at noon within one second.

[2885]

CRONMIRE, J. M. & H., 10 *Bromehead Street, Commercial Road East.*—Mathematical instruments.

[2886]

CUTTER, WILLIAM G., *Crystal Palace, Sydenham.*—Stereoscopes decorated with ivory debuscopes, and folding reflectors for designers.

[2887]

CUTTS, J. P., SUTTON, & SON, Opticians to Her Majesty, 43 *Division Street, Sheffield.*—Optical, mathematical, and philosophical instruments.

Specimens of coloured and white crown glass, best plate glass, achromatic flint glass, and metals used in the construction of achromatic object glasses, for telescopes, microscopes, and cameras.

Crown and flint discs used in the construction of the best quality of achromatic object glasses for telescopes.

A block of achromatic flint lenses, ground and polished by hand to the required radius.

A block of white crown lenses, ground and polished by hand to the required radius.

An achromatic object glass of best quality, composed of flint and crown glass, the inner surfaces of which are cemented together with Canada balsam.

Lump of plate glass.

Rough plate glass of various thicknesses and diameters, rounded ready for grinding by machinery to the curves required.

Flint glass cast in the form of a plano-convex lens, for the purpose of being ground to the radius of the tool in which it was cast. When ground and polished, they are mounted in brass cells with a plano-convex lens of plate glass, and form the condensers for a phantasmagoria lantern.

Lump of Brazilian pebble, from which slabs are cut and ground into spectacle eyes.

Specimens of extra white and coloured glass, cut and rounded for spectacle eyes to be ground by machinery.

Tray containing specimens of white and coloured spectacle eyes, ground and finished ready for mounting into spectacle frames.

Double convex and plano-convex lenses in different stages of manufacture. The exhibitors can grind any diameter from $\frac{1}{16}$ to 16 in., and any focus from $\frac{1}{40}$ to 72 in. They produce yearly over 10,000 dozen lenses.

Tray containing a variety of double convex and plano-convex lenses from $\frac{1}{10}$ in. focus and $\frac{1}{18}$ in. diam. to 3 in. focus $2\frac{1}{2}$ in. diam., suitable for microscope and telescope work.

Samples of perspectives and telescopes with four lenses.

Portable achromatic telescopes with mahogany, leather, and whalebone bodies.

Portable achromatic telescopes with bronzed mounts and draws for deer-stalking.

Portable achromatic telescopes, with German silver mounts and draws.

Portable achromatic telescopes, with silver-plated mounts and draws, fancy wood or leather bodies, and lenses of warranted quality.

Achromatic day and night ship telescopes, with wood, leather, and corded bodies.

Improved taper achromatic navy, yacht, and masthead telescopes, with leather bodies and shades, and short draws.

Double screw taper clip stand, suitable for the above, to screw into mast, &c.

Achromatic telescopes, on brass stands, with and without rack ajustment, for astronomical observation.

Opera and marine glasses of all kinds.

Provers (pillar and holding) for linen and cloth.

Simple and compound microscopes, with the latest improvements. Magic lanterns and sliders.

Sets of lenses, in brass mounting, for phantasmagoria lanterns.

Stereoscopes, patent achromatic, with and without rack adjustment. Sextants and quadrants.

Barometers, aneroid and metallic, and stands for ditto.

Pocket compasses and sun-dials.

Magnets—horseshoe, straight bar, and parallel, with brass wheel. Needles only, for ships' compasses.

Samples of boxwood and ivory.

Scales and rules, parallels, &c.

Samples of every description of spectacle in iron, steel, horn, shell, silver, and gold.

Horn box readers, burners, and magnifiers.

Horn, brass, German silver, mahogany, and rosewood picture glasses, from $1\frac{3}{4}$ in. to 13 in. diam. Any focus can be supplied.

Shell and horn double fiddle pattern microscopes, fitted with plano-convex lenses.

Full catalogue and prices of articles exhibited and manufactured by the exhibitors may be had on application at their works.

[2888]

DALLMEYER, J. H., 19 *Bloomsbury Street, London, W.C.*—Telescopes, microscopes, photographic lenses, apparatus, &c. (*See pages* 8 & 9.)

[2889]

DANCER, J. B., *Manchester.*—Binocular and monocular microscope; microscopic photographs; micrometers; equatorial telescope; dissolving view lantern.

The exhibitor is a manufacturer of the following instruments, &c., of which he has always a large stock on hand, viz. :—

Monocular, binocular, and dissecting microscopes.

Photographic micrometers.

Astronomical telescopes, with equatorial mountings and plain stands. Tourists' telescopes. Opera-glasses. Spectacles and eye-glasses in gold, steel, and shell.

He also manfactures and repairs philosophical instruments in general, and supplies the public and the trade with microscopic photographs and other objects.

[2890]

DARKER, WILLIAM HILL, 9 *Paradise Street, Lambeth.*—Illustrations of action of polarized light on crystalline and other bodies.

[2891]

DAVIS, E. & J., 1 *Albion Street, Leeds, and Derby.*—Two coal dials; three anemometers; two pressure gauges; one vacuum gauge, oil tap, cistern, and steam thermometers; indicator; flax tester.

Hedley's dial, with angular motion, for coal pits and mines. Vernier dial for ditto.

Davis' pressure gauge in skeleton.

Ditto, ditto, commercial ditto.

Ditto, vacuum ditto.

Anemometers for regulating the ventilation of coal pits.

Cistern and steam thermometers.

Level. Self-closing oil tap.

Testing machine for yarns and fibrous substances.

DALLMEYER, J. H., 19 *Bloomsbury Street, London, W.C.*—Telescopes, microscopes, photographic lenses, apparatus, &c.

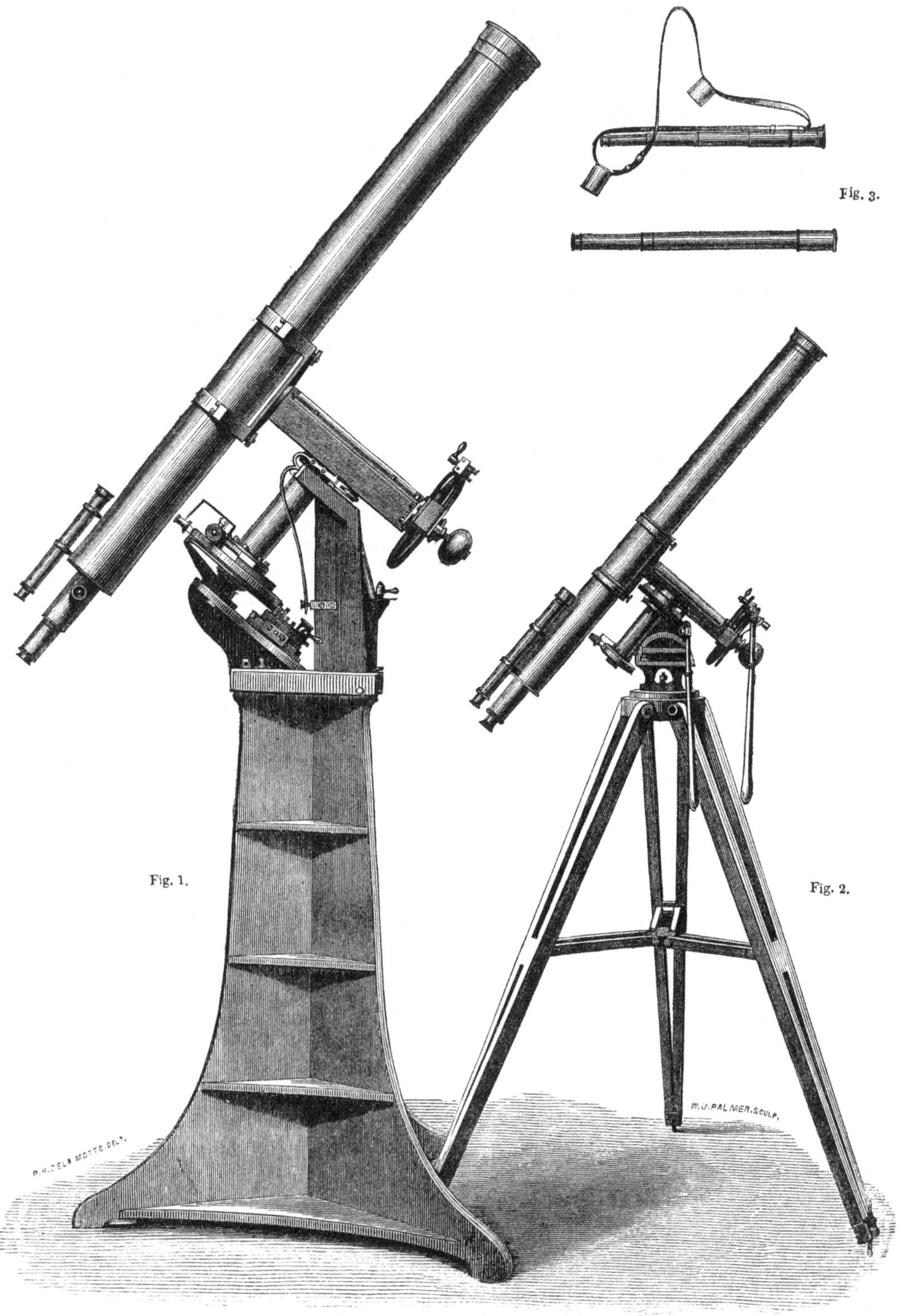

Fig. 1. Fixed equatorial, with clock-work. Fig. 2. Universal portable equatorial. Fig. 3. Terrestrial telescopes, naval, reconnoitring, deer-stalking, etc. (Exhibited in the Nave.)

Dallmeyer, J. H.—*continued.*

The engraving is a representation of the general form of Dallmeyer's microscopes. They are manufactured of four different dimensions. Each of these instruments is composed of several distinct parts, and the simple stand of each size forms the basis of a complete instrument to which any of the other parts may be added subsequently.

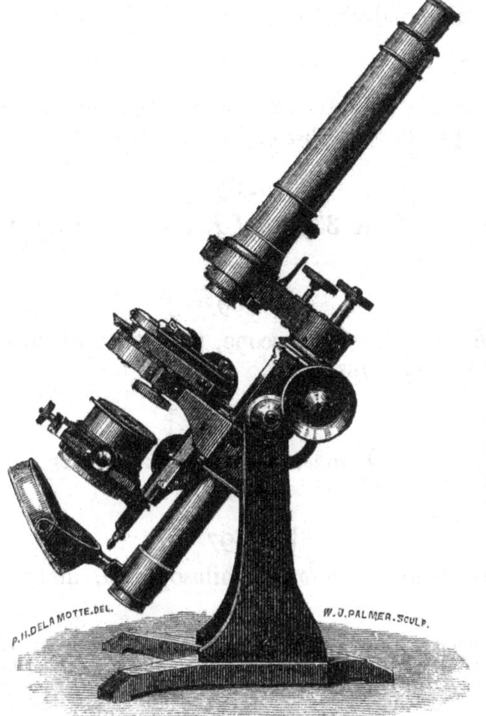

No. 1A Microscope.

Achromatic Object-Glasses.

The first of a new series of objectives was exhibited at the Soirée of the London Microscopical Society, in March, 1860.

Fig. 1.

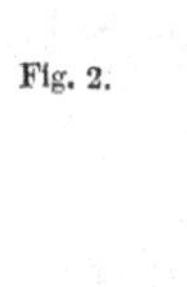

Fig. 2.

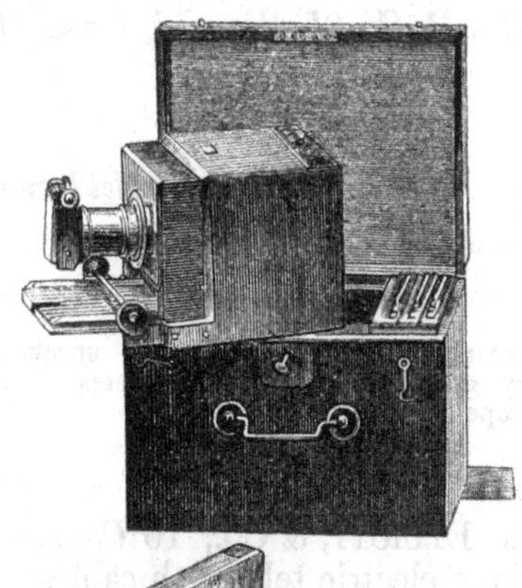

Fig. 3.

Fig. 1.—The new Dallmeyer Triple Achromatic Lens, and a Bellows Camera.

Fig. 2.—A pair of quick-acting portrait Lenses, specially constructed for taking album pictures, attached to a camera and shutter of new design.

Fig. 3.—A pair of Dallmeyer's new Stereoscopic Lenses, and a No. 1 Triple Achromatic Lens. Also a travelling Stereoscopic Camera, with which, on removing the central partition, full-sized pictures on plates $7\frac{1}{4} \times 4\frac{1}{2}$ can be taken. (Exhibited in Class XIV.—No. 3069.)

[2892]

DE GRAVE, SHORT, & FANNER, 59 *St. Martin's-le-Grand.*—Hydrostatic assay balances, scales and weights for diamonds, bullion, &c.

[2893]

DE LA RUE, WARREN, Ph.D., F.R.S., *Observatory, Cranford.*—Lunar, solar, and planetary photographs; photographs of the solar eclipse of July 18th, 1860. (*See page* 11.)

[2894]

DENT & Co., 61 *Strand, and* 34 & 35 *Royal Exchange.*—Ship and azimuth compasses and dipleidoscopes.

[2895]

DESVIGNES, MR., *Lewisham, S.E.*—Mimoscope, patented: an instructive philosophical toy, affording endless variety and amusement.

[2896]

DUNCAN, CHARLES STEWART, *Monmouth Road, Bayswater* — Ratan deep-sea electric telegraph cable.

[2897]

ELLIOTT BROTHERS, 30 *Strand, London.*—Philosophical, mathematical, optical, and surveying instruments.

The model of a new magneto-electric propeller, invented and exhibited by Harry Whiteside Cook, Esq., of Rossall Hall School, is intended merely to demonstrate the principle.

In practice the compound core would of course be flexible. The battery is intended to be carried by the propeller.

The force employed is that known as the axial force. The electric contact being broken at the moment this power reaches its maximum, the momentum imparted is sufficient to carry the propeller and its train within the influence of a second magnetizable portion of the core, when the contact is again renewed and again broken. The result of this alternation is found to produce a continuous and indefinitely rapid rate of progression.

[2898]

FORBES, R. C., 95 *Warwick Street, Liverpool.*—Artificial horizon for observing altitudes in hazy weather.

[2899]

FRITH, PETER, & Co., *Sheffield and London.*—Optical instruments.

The following are exhibited :—

IMPROVED SPECTACLES.

Concave, convex, and meniscus spectacle lenses. Military, marine, and tourists' telescopes. Achromatic microscope and telescope objectives.

MICROSCOPES.

Woollen and linen provers. Twin photographic stereoscopic view lenses. Camera lucida right angle and compass prisms. Riflemen's telescopes. Exhibition opera-glasses, lenses, &c. Astronomical and surveying instruments made to order.

[2900]

GLASS, ELLIOTT, & Co., 10 *Cannon Street, London;* Manufactory, *East Greenwich.*—Submarine electric telegraph cables.

[2901]

GODDARD, JAMES THOMAS, *Whitton, near Hounslow.*—Cloud mirror and sunshine recorder.

[2902]

GOWLAND, GEORGE, *Liverpool.*—Vertical and semi-vertical compasses, with circular magnets; sextants with artificial horizon, and binocular glasses.

[2903]

GREEN, SAMUEL, & SON, 7 *Helmet Row, Old Street, London.*—Pocket compasses and sundials.

De la Rue, Warren, Ph.D., F.R.S., ***Observatory, Cranford.*****—Lunar, solar, and planetary photographs; photographs of the solar eclipse of July 18th, 1860.**

A series of Astronomical Photographs, comprising photographs of the several phases of the total eclipse of July 18, 1860, taken at Rivabellosa, near Miranda de Ebro, in Spain; photographs of the moon in her different phases, taken at Mr. De la Rue's observatory at Cranford; photographs of a lunar eclipse; photographs of Jupiter and of Saturn and the moon, taken together, just after the occultation of Saturn by the moon, also obtained at Cranford.

Nos. 1 to 31 inclusive show (as seen direct, that is, not inverted) the several phases of the total solar eclipse, the irregularity of the moon's edge being very apparent in the several pictures. These pictures were obtained by means of the Kew Heliograph, an instrument contrived by Mr. De la Rue for taking sun-pictures, at the suggestion of Sir John Herschel to the Royal Society, whose property it is. The Kew instrument was transported to Bilbao, in Spain, on the occasion in question, by H.M. steam ship Himalaya, and thence to Rivabellosa, over the Cantabrian Pyrenees,* with the co-operation of Mr. Vignoles, C.E. Mr. De la Rue's party, consisting, besides himself, of Mr. Beckley, Mr. Reynolds, Mr. Downes and Mr. E. Beck, formed one section of the Himalaya expedition organised by the Astronomer Royal, Mr. Airy.

The pictures having the greatest interest are those taken during the totality, in which may be seen the luminous prominences. These prominences, it is now known, belong to the sun, and it may be regarded as certain that they project at all times beyond the solar surface; but they only become visible during a total solar eclipse, because on all ordinary occasions their light is less bright than that of our own atmosphere illuminated by the sun's rays. A paper on the results of the photographic expedition to Spain has been read by Mr. De la Rue to the Royal Society, as the Bakerian lecture of the present year. It is shown in this paper that the phenomena depicted by the photographs completely establish the view that the luminous prominences really belong to the sun, and that they are not occasioned by any action of the moon's edge on light coming originally from the sun.

Nos. 32 to 45 inclusive show the different phases of the moon, and bring prominently under view the wonderful craters which cover the greater portion of the surface of our satellite especially in the upper or southern hemisphere (the pictures are as seen in an inverting telescope). Conspicuously visible is the crater Tycho, from which radiate a series of furrows like lines of longitude on a globe; lower down on the right is Copernicus. These pictures, about eight inches in diameter, are enlarged by means of a camera from negatives 1 inch in diameter, as seen in No. 53. It must be borne in mind that, during the taking of these pictures, the moon is in motion with respect to the observer, so that the telescope has to be made to follow her motion very exactly indeed to secure such perfect pictures, which are still sharp although magnified eight times linear.

Nos. 46 to 49 are photographs of the lunar eclipse of February 27th, 1858; No. 46 being the moon just before the commencement. It will be observed how indistinct is the boundary of the earth's shadow in 47, 48, and 49.

Nos. 50 and 51, photographs of Jupiter, enlarged from the original negatives.

No. 52, a photograph of the moon and Saturn, taken together just after the occultation of that planet by the moon on May 8th, 1859; the planet Saturn is surrounded by a black circle to indicate its position on the plate.

No 53, an original lunar negative.

The several lunar and planetary photographs were taken with a reflecting telescope of 10 feet focal length and 13 inches aperture, made by Mr. De la Rue, and erected at his observatory at Cranford, Middlesex.

No. 55, a moving model to illustrate the phenomena of the total solar eclipse.

No. 56, stereoscopic view of the moon, 2 inches in diameter.

No. 57, stereoscopic view of the moon, 8 inches in diameter.

These stereoscopic views are produced by placing in the stereoscope two views of the moon, taken under different circumstances of libration. By observing the position of the crater Tycho in the collection of photographs, it will be seen that it is sometimes nearer to, sometimes further from, the moon's edge; by selecting two pictures we can obtain such as have the proper stereoscopic relation.

No. 58, a stereoscopic view of Saturn, produced by reducing two hand-drawings made at an interval of four years. The exact coincidence of these drawings in the stereoscope is an evidence of their accuracy.

No. 59, solar spots, printed by the ordinary typographical press from a copper block, produced by means of light and electro-metallurgy, by M. Paul Pretsch, from an original negative, taken at Cranford, on a scale of 3 feet to the sun's diameter. The printing block is absolutely untouched by the graver.

* The district may be seen in Mr. Vignoles' model, exhibited in the department of "Civil Engineering," which does not, however, quite extend to Rivabellosa.

[2904]

GRIFFIN, JOHN JOSEPH, F.C.S., 119 *Bunhill Row, London, E.C.*—Chemical and philosophical instruments, and their applications.

[*Obtained a Prize Medal in Class X. at the Exhibition of* 1851.]

Gas-burners, constructed to produce great heat, without light, for chemical use :—

A.—GAS BURNERS WITHOUT BELLOWS.—Three sizes. Each gives a single flame, when a solid substance is to be heated, and a number of small flames when liquids are to be boiled or evaporated. Suitable jackets or furnaces, made of fire-clay and iron, are provided, to concentrate and economize the heat.

B.—BLAST GAS FURNACES.—These consist of multiple blowpipe gas burners, blowing machines, and fire-clay furnaces filled with flints. The heat they produce will melt silver, gold, copper, cast iron, nickel, cobalt, and wrought iron. A large size will fuse 25 lbs. of cast iron in about an hour. They produce no smoke.

C.—GAS FURNACE FOR TUBES.—Will heat tubes of glass, porcelain, or iron, up to 36 inches long, to bright redness in a few minutes. The heat can be easily reduced and regulated for the whole length, or at any part, in order to adapt it to combustions in organic analysis.

Collection of graduated apparatus for centigrade testing, containing everything necessary for the preparation of the test liquors, or their use in volumetric analysis. In a mahogany cabinet.

Assortment of graduated instruments for testing in the arts, namely—alcalimeters, acidimeters, hydrometers, thermometers, alcoholometers, saccharometers, eudiometers, and liquid measures.

Collection of chemical apparatus and tests, suitable for a physician, or for a hospital laboratory, containing everything necessary for the detection of poisons, the analysis of urine, the testing of medicines, and the performance of many other chemical operations that occur in medical practice. In a mahogany cabinet.

Collection of chemical apparatus and tests suitable for a travelling engineer, a miner, or a naval or military officer; comprehending whatever is necessary for the qualitative analysis of minerals, ores, or chemical and medical compounds generally; being the collection described in the "*Admiralty Manual for Scientific Enquiry.*" In a strong mahogany cabinet.

Colonial Chemistry.—A collection of apparatus for testing cane juice, to determine the exact quantity of lime necessary to clarify it. Also, Twaddell's hydrometers, and Baume s hydrometers, saccharometers, and alcoholometers, graduated at 84° Fahr., to suit West Indian liquor lofts.

Collection of apparatus and tests for the examination of minerals, ores, and chemicals, by the blowpipe. In a portable cabinet.

Elementary collection of chemical apparatus and preparations for the use of young chemists. In portable cabinets; several sizes.

Collection of chemical apparatus, as required by each pupil for the Oxford and Cambridge Middle Class examinations.

Miscellaneous instruments for chemical researches and demonstrations, namely :—

Portable furnaces, for general chemical use; assay furnaces and assay tools.

Graham's dialyser, for effecting chemical analysis by liquid diffusion.

Still for determining the quantity of alcohol in wines; alcoholometers for testing it.

Retorts of cast iron, with movable heads, for distilling coal oils, &c.

Specimens of evaporating basins, crucibles, flasks, funnels, beakers, cut filters, test-papers, cheap balances, simple and compound blowpipes, water baths, air baths, pneumatic troughs for water and mercury, gasometers, and many patterns of supports for chemical apparatus.

Galvanic batteries, pattern cells of different constructions.

Air-pump, on Tate's plan, which has two pistons in one cylinder, and which dispenses with valves between the cylinder and the receiver, and thus gains power and accuracy. Small size, horizontal position, for screwing to a table.

Another air-pump, on Tate's plan, large size. The cylinder is placed in a vertical position, and the exhaustion is rapidly effected by a circular motion, regulated by a fly-wheel. Suitable for manufacturing purposes, or for rapid action when many experiments are to be made during a lecture.

The spectroscope, for optical experiments in chemistry, and several instruments for demonstrations and researches in other branches of chemical physics.

[2905]

GRUBB, THOMAS, *Dublin.*—Great equatorial (achromatic), 12 inches aperture; improved and perfect system of equipose throughout.

[2906]

HART, WILLIAM D., 7 *North College Street, Edinburgh.*—Electrical apparatus.

[2907]

HELY, ALFRED AUGUSTUS, 26 *Upper Albany Street, Regent's Park.*—Pocket reflecting telescope for astronomical purposes.

[2908]

HENLEY, W. T., 46 *St. John Street Road.*—Magneto-electric alphabetical telegraph.

This instrument is now coming much into use with private firms that have branch establishments; also with colliery owners; and with mill proprietors in Manchester and other large towns in the manufacturing districts, as a means of communication between their mills and warehouses. It is well adapted for railway purposes. Any person that can read can use it without any previous knowledge of telegraphy. It requires no battery, and has no complicated machinery to get out of order. The only attention required is the application of a little oil once a month. Gentlemen engaged in business can communicate with their partners or confidential clerks at their other houses of business with the greatest facility, without the necessity of imparting their secrets to a telegraph clerk.

The instrument is represented in fig. 1, with case complete, and in fig. 2 with the case removed. The apparatus consists of two parts : that in which the current is induced, and by which it is transmitted; and that for receiving and indicating the signals. The first part is in the form of a

HENLEY, W. T.—*continued.*

Fig. 1.

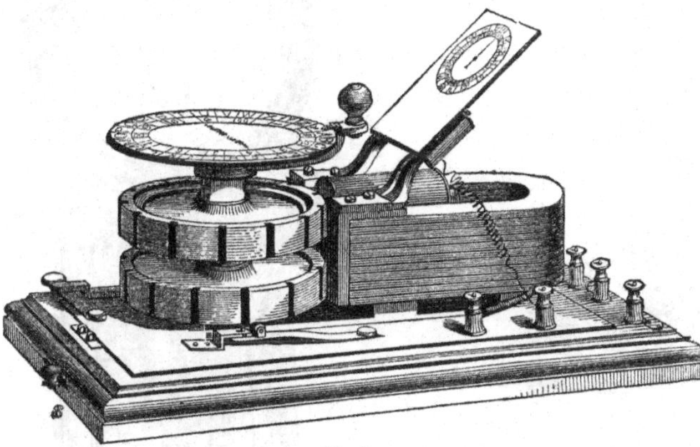

Fig. 2.

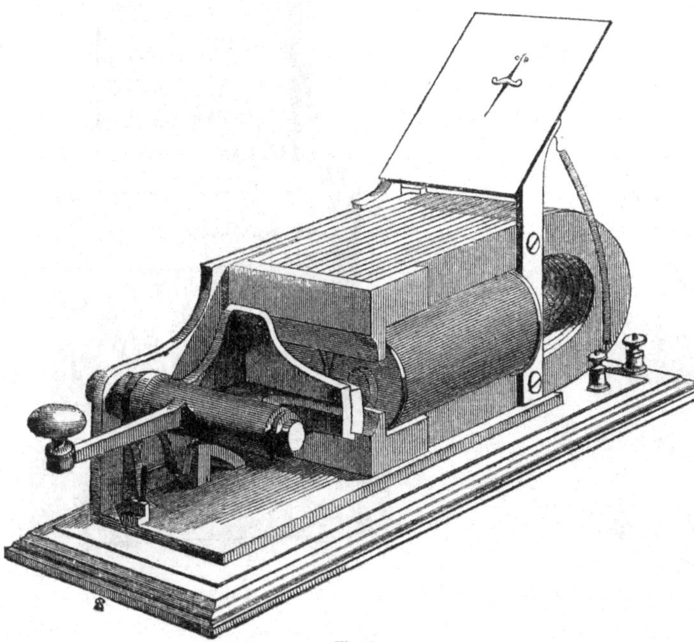

Fig. 3

horseshoe permanent magnet, with a temporary magnet or armature surrounded with coils of insulated wire placed within its poles, and a double wheel of brass revolving on a fixed axis. These wheels have on the periphery twenty-six pieces of iron, thirteen on each wheel. They are formed by turning a ring of soft iron to fit each wheel, attaching it firmly by twenty-six screws, and dividing it into thirteen pieces by the wheel-cutting engine. These pieces of iron, as they pass the ends of the permanent and temporary magnets, make a connection between the poles of each in such a way as to reverse the polarity of the latter completely at the passing of each piece, and consequently inducing a current of electricity alternately in opposite directions.

The receiving part consists of a dial with the letters of the alphabet in a circle, with a revolving index or pointer and a toothed wheel of a peculiar construction. This wheel is propelled step by step by the action of a magnetic needle, the upper end of which is formed into a pair of pallets, which act on the inclined teeth of the wheel, and the lower end oscillates in a slot formed in two pieces of iron fixed on the poles of a small electro magnet, through the coils of which the current received from the distant stations passes.

In the transmitting apparatus a dial is fixed outside of the case, on the stud on which the double wheel revolves. This has letters and figures corresponding to the receiving dial. The wheels are revolved by a handle with a knob and pointer moving round the edge of the stationary dial; and as one revolution of the wheels causes twenty-six currents of electricity to be transmitted, these will cause the wheel and index on the receiving dial to make one revolution also. It therefore follows that the two will keep time together, and whatever letter the pointer attached to the revolving wheels is made to stop at will be indicated on the receiving dial. The springs seen on the base in fig. 2 are, one for making a short circuit when receiving, so that the current may not have the resistance of the sending coils to overcome; the other is for putting the hand right on the receiving dial if it should go wrong. As these instruments have only the one moving piece of machinery, without any multiplying power cranks, cams, or other complications, and have no break pieces or current reversers (the circuit always remaining unbroken) they cannot get out of order, and the magnetism will remain the same for many years.

Fig. 3 represents a machine on a similar principle, arranged as a needle telegraph, and is also used for working the Morse printing telegraph. It has a magnet and armature as in figs. 1 and 2, but instead of rotating wheels in front of the

HENLEY, W. T.—*continued.*

Fig. 4.

Fig. 7.

Fig. 5.

Fig. 8.

Fig. 6

HENLEY, W. T.—*continued.*

Fig. 9.

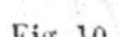

Fig. 10.

magnet it has a segment of a wheel, which is made to oscillate by a lever, the motion being limited by india-rubber stops. Every depression of the lever causes the needle on the dial to be deflected; the upward motion of the lever bringing it back again.

Figs. 4 and 5 show a modification of the letter-showing telegraph represented in figs. 1 and 2. In this case the armature is placed in a different position, and only one wheel is used, working on the upper side of the magnet. In this instrument a ratchet is used for the handle to drop into for the purpose of stopping opposite each letter.

Fig. 6 is a letter-showing magnetic telegraph. It is worked by a lever or pedal. Reverse currents are produced by its motions, all the downward movements inducing them in one direction, and the opposite by allowing the lever to recover its position. It has to be depressed thirteen times to produce one revolution of the hand round the receiving dial. This instrument is somewhat slow in its action from this circumstance; but it is in use in several places, and gives satisfaction owing to its non-liability to get wrong.

Fig. 11.

The operator, keeping his eye on his own dial, knows that any letter he causes the hand to pause at will be in like manner indicated on the distant instrument. All the letters and figures on the dial marked with a dash are those at which the hand stops when the lever is pressed down. Those without dashes by its upward movement.

Fig. 7 represents a relay, or instrument by which a small amount of power is made to bring into action a greater quantity from another source. It consists of an electro-magnet with semicircular pieces fixed to the poles, within which a soft iron needle or cross-piece (rendered magnetic by a magnetic needle, to which it is fixed) oscillates when the current, either voltaic or magneto-electric, is made to pass through the coils. A piece of gold fixed to the moving magnetic needle making contact with another piece in an adjusting screw, and completing the circuit battery either for printing or for transmission to another station. This instrument is used either with or without reverse currents; a permanent magnet, seen at one end of the sketch, is used for adjusting instead of a spring.

A time-keeper, usually termed a regulator, is shown in fig. 8; it has apparatus for transmitting alternate reverse currents of electricity for actuating the companion shown in fig. 9, or for other purposes. The currents are sent every second, and pass through the coils shown at the back of fig. 9.

This might be many miles distant from the regulator. It has a spring with fusee and chain; once winding keeps it going for six months. The train, with the escapement, remains stationary until the current passes throughout the coils, when the deflection of the magnetic needle causes the escape wheel to move half a tooth, the seconds-hand on the dial moving one second division. The next current, being in the reverse direction, deflects the needle the opposite way, and causes the escape wheel to move another half tooth, and thus the two time-keepers work accurately together. A clock with a very large dial may be used in this way, as the power of the spring does all the work of moving the hands, the current only being required just to liberate the escapement at the proper time.

Fig. 10 shows a very delicate galvanometer, fitted up as a differential one, the needle axis pivoted in jewels with levels, adjustments, &c.

Fig. 11, a ditto not differential, with needle suspended by silk fibre; either of these will be affected by an extremely feeble current.

[2910]

HETT, A., 4 *Albion Grove, Islington.*—Injected microscopic preparations.

[2911]

HICKS, JAMES, 8 *Hatton Garden.*—Meteorological instruments.

[2912]

HIGHLEY, SAMUEL, F.G.S., F.C.S., &c., Philosophical Instrument Maker, 70 *Dean Street, Soho, London, W.*—Educational microscope and philosophical instruments.

£1 5s. £5 5s. £2 12s. 6d. £13 13s. £3 3s. £7 10s. 6d. 12s. 6d.

MICROSCOPES.—Naturalists' pocket-lenses, on telescope stand, and universal motion support, for the examination of minerals, &c. Improved Quekett's pocket dissecting microscope. Dr. Lionel Beale's pocket clinical microscope, lecturers' demonstrating microscope, and lamps.

Highley's "Educational Microscope," "Hospital students' microscope," and large microscope with complete mechanical motions.

A series of cheap achromatic object-glasses, comprising 3-inch, 2-inch, 1-inch, ½-inch, ¼-inch objectives.

Accessory apparatus and appliances for the microscope.

An improved bi-prism spectroscope, and accessory apparatus for observations on gases, vapours, Gladstone's "absorption spectra," and stellar phenomena.

Universal polariscope, for table or lecture-room demonstrations on polarized light.

Universal electro-magnet, in glazed lantern, for magnetic, diamagnetic experiments, &c.

Bohnenberger's electroscope; astatic galvanometer; Melloni's apparatus for illustrating the athermic and diathermic character of minerals, and the refraction and polarization of heat. An educational astronomical telescope, &c.

[2913]

HINTON, W., 21 *Greville Street.*—Improved barometers.

[2915]

HOOPER, WILLIAM, 7 *Pall Mall East, S.W.*—Submarine telegraph cables insulated with india-rubber.

[2916]

HORNE & THORNTHWAITE, 121, 122, & 123 *Newgate Street.*—Rhumkorff's induction coil; spectrum apparatus; meteorological instruments; chemical apparatus; microscopes, &c.

[2917]

HUDSON & SONS, *Greenwich.*—Animal, vegetable, and fossil tissues and structures, and minerals for microscopic use.

[2918]

HUGHES, JOSEPH, *Queen Street, Ratcliff, London, E.*—Nautical, optical, and surveying instruments.

[2919]

JACKSON & TOWNSON, 89 *Bishopsgate Within, London.*—Chemical and scientific apparatus for general and special purposes.

REVENUE STANDARD STILL, by authority of the Board of Customs, for the alcohol test of wines; and by the Board of Inland Revenue for the analysis of beer.

WORKING LABORATORY for qualitative and quantitative analysis; bottles with enamelled labels.

PORTABLE CASE of BLOWPIPE and APPARATUS. APPARATUS for VOLUMETRIC ANALYSIS, arsenic determinations, benzole estimations, and chemical research generally.

CONDENSERS for DISTILLATION, various kinds of drying-baths.

FURNACES for ASSAYING, organic analysis, and general purposes.

NEW SPECIFIC GRAVITY BOTTLE, for weighing alcohols or ethers, at any increment of temperature, without removing the liquid.

[2920]

JOHNSON, HENRY, Inventor, 39 *Crutched Friars.*—Volutors for tracing spiral curves; deep-sea pressure-gauges, for recording the pressure or density of sea-water at various depths; deep-sea thermometers.

Invented by HENRY JOHNSON, 39 *Crutched Friars;* manufactured by F. HOFFMANN, 32 *Wilmington Square, Clerkenwell.*

THE VOLUTOR.

Papers on this instrument were read by the Rev. Dr. Booth, F.R.S., before the Mechanical Section of the British Association, at the meeting held at Leeds in 1858, and the meeting held in Oxford in 1860.

This instrument has been contrived for the purpose of facilitating, by means of mechanical arrangements, the drawing of volutes, an operation that requires much time and care, and also the drawing of other spiral curves.

During the action of the instrument a band is wound round its centre, which regulates the pencil, and thus a continuous curve is traced with a radius varying in length at every point.

The radius of the curve is increased during each revolution by the circumference of the axis, and a circular cylinder will be found convenient as an axis for tracing spirals whose radii increase in arithmetical or other proportions.

In drawing volutes a flat band may be used, wound round a cylinder, so that each coil encloses the preceding one, and increases the diameter of the axis; but a grooved cone appears to be more convenient, as the proportions of a cone and the distances between the grooves may be more readily adapted to the curves required.

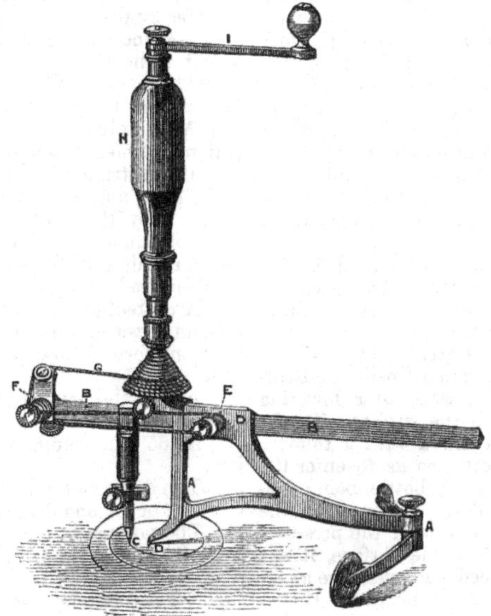

A stand with wheels, A, moved round the central point O, supports a horizontal arm or bar B, and which moves through a horizontal tube, D, on the stand. To the horizontal arm is fixed the tracing pencil C, pressed down by a vertical spiral spring.

A steel rod rises through a perforated grooved cone (or other axis) and its handle H, and is furnished with a small winch handle I, by which the stand is made to revolve, while the cone is held still by its handle H; and a band G, one end of which is attached to the cone and the other to one end of the horizontal arm, is wound round the cone, and the end of the arm is gradually drawn towards the centre, and the curve is traced by the pencil.

In volutes commenced at the extremity of the radius vector the band is attached to the base, and adjusted to the groove selected for the first curve, and is wound round the cone approaching the apex as the instrument revolves.

When tracing volutes commencing at the centre and receding from it, a movable set of pulleys F should be fixed with a screw on to the outer end of the horizontal arm.

One end of the band being fastened to the pulleys, and the other attached to the apex of the cone, it will be wound round the cone approaching the base, and the pencil will recede from its position in the centre, tracing the curve as the outer end of the arm is drawn by the band towards the centre.

When variations of radius less than the circumference of the axis at each revolution are required, the effect of the band wound round the centre may be modified by passing it over some of the pulleys, one set of which, E, is fixed on the stand, and the other set, F, is movable, and may be screwed on to either end of the arm. The effect varies according to the number of lines of band over which it is distributed; as, for instance, when the band is passed over one pulley the effect is distributed over two lines, and the radius varies in a revolution one-half of the circumference of the axis; when the band is passed over two pulleys the effect is distributed over three lines, and the radius varies in a revolution one-third of the circumference of the axis, &c.

When tracing the fillet of a volute the cone should be turned round until the band is tightened, after the pencil has been placed in its proper position.

The size of the axis is thus slightly altered, and a proportionate distance maintained between the curves.

In drawing a parallel curve, as the size of the centres must coincide, it will be necessary to alter the length of the band to suit the position of the pencil.

THE DEEP-SEA PRESSURE-GAUGE.

A paper on this instrument was read by James Glaisher, Esq., F.R.S., before the Mathematical and Physical Science Section of the British Association, at the meeting held at Manchester in 1861.

In deep-sea soundings the pressure of water is too great to admit of accurate measurement by the compression of any highly elastic fluid confined in a small portable instrument.

For a long period water was considered incompressible, but it has been found to possess a slight degree of elasticity, sufficient to render its compression in a vessel avail-

JOHNSON, HENRY—*continued.*

able as an indication of the compression or density of the water into which it is lowered.

In the year 1762, December the 16th, Mr. Canton communicated to the Royal Society the results of his experiments on the compressibility of water—*Philosophical Transactions*, vol. lii., page 640.

He took a small glass tube of about two feet in length, with a ball at one end of it of an inch and a quarter in diameter, and filled the ball and part of the tube with water exhausted of air, and left the tube open that the ball, whether in rarefied or condensed air, might always be equally pressed within and without. He placed the ball and tube under the receiver of an air-pump, and could see the degree of expansion of the water answering to any degree of the rarefaction of the air; and also placed the ball and tube into the glass receiver of a condensing engine, in which he could see the degree of compression answering to any degree of condensation of the air.

In this way he found by repeated trials, when the temperature was about 50° Fahrenheit, and the barometer about a mean height, that the water expanded and rose in the tube, by removing the weight of the atmosphere, one part in 21,740, and that it was as much compressed under the weight of an additional atmosphere.

More recently, Mr. Perkins found, when subjecting water to great pressure, a diminution in volume of $\frac{6}{100}$th parts under a pressure of 1120 atmospheres, equal to one part in 18,666 per atmosphere.

The experiments of Mr. Perkins, exhibited at the Adelaide Gallery, appeared to be intended as a demonstration of the fact of progressive compression, rather than a basis for minute calculation.

The effect of pressure of water at great depths is illustrated by a very interesting experiment made by Rear-Admiral Sir James Clarke Ross, who, after lowering several bottles which returned to the surface with the corks reversed, lowered a bottle fitted with a tube; a cork being suspended in the bottle so as to enter the tube in the event of the water in the bottle, being condensed under heavy pressure, and expanding upon the raising of the bottle and the diminution of the pressure.

Upon the return of the bottle to the surface, it was found that the cork had been forced some distance along the tube, and the compression of the water in the bottle, and its subsequent expansion, were thus demonstrated.

In experiments conducted with a pressure-gauge made of metal, it was found that air-bubbles adhered to the inner surface of the pressure-gauge, and materially affected the results.

This difficulty is avoided in the instrument now exhibited, which is composed of glass, so that the absence of air-bubbles may be ascertained by inspection before any experiment is made.

The instrument consists of a cylindrical glass vessel with a long neck or stem finely graduated; within which are placed a flat elastic ring to act as an index, and an elastic stopper.

When used, the pressure-gauge should be well rinsed with *warm* water, to prevent the adhesion of air to its inner surface, and then filled to the top of the stem with sea-water boiled to free it from air.

In the event of this water being poured in while warm, it will be necessary to fill up the stem after the water has cooled down to the temperature of the atmosphere, so that the stopper may be inserted without confining any air beneath it. A small vent, or grooved needle, affording a passage for the escape of superfluous water, should be pushed in with the stopper, which should be slightly lubricated to prevent excessive friction, until the lower end of the stopper is coincident with the zero, or top line of the graduated scale, marked 2000, when it will also touch the flat elastic ring.

The vent should then be withdrawn and the stem will remain tightly closed by the stopper.

When lowered into water of greater density, the water in the pressure-gauge is compressed by external pressure until of equal density with the surrounding water, and the elastic stopper and the elastic ring are pressed along the tube towards the cylinder.

When raised, as the external pressure diminishes, the water in the pressure-gauge expands, and gradually presses back the elastic stopper, the elastic ring remaining as an index to mark the extreme compression.

When the water attains the temperature of the atmosphere the stopper will have returned to its original position, less a small difference arising from friction.

The volume of water in the cylinder and stem is considered as consisting of 2000 parts, of which the cylinder contains nine-tenths, or 1800 parts or degrees, and the stem one-tenth, or 200 degrees, and which are numbered 1801 to 2000.

The graduated scale on the stem may easily be read to one-tenth of a degree, or $\frac{1}{20000}$th part of the whole volume of water.

For the compression of one part in 20,000 of *boiled sea-water* a pressure is required of 15·8 lbs. avoirdupois per square inch, equal to the pressure of a depth of 35·446 feet, or nearly six fathoms.

This amount of pressure, which is the result of several experiments, and which is confirmed by the observations of Mr. Canton, appears to be a fair basis for the compilation of tables of comparison of depth and pressure.

The instruments should, however, be attached to sounding lines, and the indications compared with the depths shown by the lead. The results would form a table of comparison of depth and pressure of practical use in determining depths when strong currents render the use of the lead uncertain.

A correction will be required for the variation in volume of water with change of temperature, and which is not uniform, being greater at high temperatures, as, for instance—

At 86° the volume is for this object estimated at..	20,000	parts.
At 65° the volume is contracted to ..	19,932·5	„
The difference for 21° being ..	67·5	parts,
or for one degree 3·21 parts.		
The volume at 65° of	19,932·5	parts,
is contracted at 31° to	19,880	„
The difference being for 34° ..	52·5	parts,
or for one degree 1·55 parts.		

A series of experiments will be made to determine the correction required on account of friction.

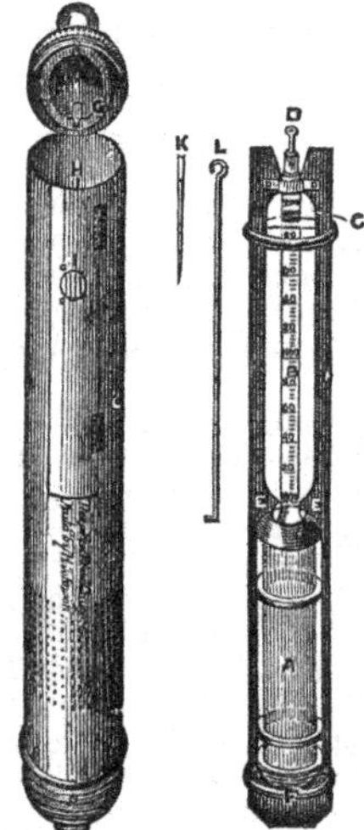

References—
A. Cylinder.
B. Stem with graduated scale.
C. Flat elastic ring or index.
D. Elastic stopper.
E. Metal frame lined with caoutchouc.
F. Caoutchouc rings preventing concussion.
G. Caoutchouc rings at top and bottom of the case, securing the frame in position.
H. Metal hook on door fastening down the top of case.
I. Clasp to door, let in to avoid projection.
K. Vent or grooved needle inserted with stopper.
L. Hook used to draw up the elastic ring.

The following table shows the volume of water for each degree of temperature, from 31° to 86° Fahrenheit.

JOHNSON, HENRY—*continued.*

Variation in the volume of Sea Water, boiled to free it from air, with change of temperature. Thermometer 67·5° *Fahr. Barometer* 29·92.

Deg.	No. of Parts.	Deg.	No. of Parts.	Deg.	No. of Parts.
Fahr.		Fahr.		Fahr.	
86°	20000·0*	64°	19930·0	42	19888·0
85	19996·0	63	19927·5	41	19886·7
84	19992·5	62	19925·0	40	19885·5
83	19989·0	61	19922·5	39	19884·5
82	19985·5	60	19920·0	38	19883·5
81	19982·0	59	19917·5	37	19883·0
80	19978·5	58	19915·0	36	19882·5
79	19975·0	57	19913·0	35	19882·0
78	19971·5	56	19911·0	34	19881·5
77	19968·0	55	19909·0	33	19881·0
76	19964·7	54	19907·0	32	19880·5
75	19961·5	53	19905·0	31	19880·0
74	19958·25	52	19903·0	30	19880·0
73	19955·0	51	19901·0	29	19880·0
72	19951·5	50	19899·0	28†	19880·0
71	19948·0	49	19897·0	27	19880·0
70	19945·0	48	19895·0	26	19880·0
69	19942·5	47	19894·0	25	19880·0
68	19940·0	46	19892·5	24	19880·0
67	19937·5	45	19891·0	23	19880·0
66	19935·0	44	19890·0	22	19880·0
65	19932·5	43	19889·0		

* The volume at 86° being considered as unity, and divided into 20,000 parts.

† A gentle motion kept up to equalize the temperature of the sea water has prevented its freezing at 28·5°.

THE DEEP-SEA THERMOMETER.

A paper on this instrument was read by James Glaisher, Esq., F.R.S., before the Mathematical and Physical Science Section of the British Association, at the meeting held at Manchester in 1861.

This instrument is intended to be used simultaneously with the Deep-sea Pressure-Gauge for the purpose of

in which the proportion of brass, the more dilatable metal, is two-thirds, and of steel one-third.

Upon one end of a narrow plate of metal about a foot long, *a*, are fixed three scales of temperature, *h*, which ascend from 25° to 100° Fahrenheit, and which are shown more clearly in the drawing detached from the instrument.

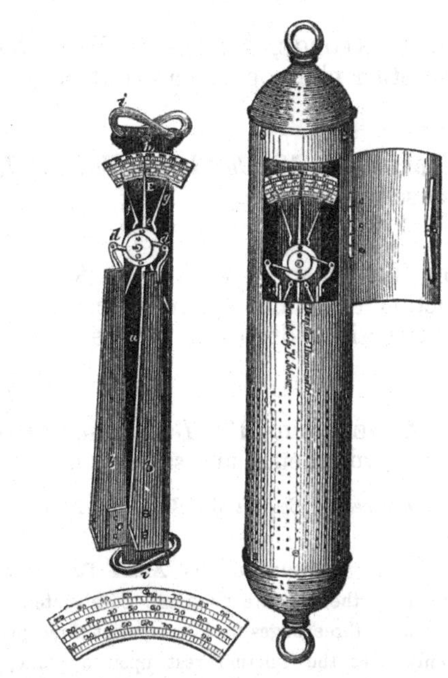

Upon one of these scales the present temperature is shown by the pointer E, which turns upon a pivot in its centre. The register index *g* to the maximum temperature, and the index *f* to the minimum temperature, are moved along the other scales by the pin upon the

[2921]

JOHNSON, W., Manufacturer, 188 *Tottenham Court Road.*—Spectacles cut from solid steel; invisible spectacles; process of manufacture.

[2922]

KIESSLER & NEU, 29 & 49 *Spencer Street, Goswell Road.*—Analytical balances, &c.

[2923]

KNIGHT, GEORGE, & SONS, 2 *Foster Lane, London, E.C.*—Chemical, electrical, galvanic, and other philosophical apparatus.

[2924]

KULLBERG, V., 12 *Cloudesley Terrace, London, N.*—Self-registering mariner's compass, or course indicator.

[2925]

LADD, WILLIAM, 11 & 12 *Beak Street, Regent Street, W.*—Focimeter for lighthouses; induction coils, and apparatus connected therewith; compound microscopes of improved construction; ditto with magnetic stage; a triple-barreled air-pump.

[2926]

LADD & OERTLING, 192 *Bishopsgate Street Without.*—Bullion, chemical, and assay balances; metal hydrometers and saccharometers. (*See page* 21.)

[*Obtained the Council Medal in Class X. of the Great Exhibition of* 1851, *and the First Class Medal of the Paris Exhibition of* 1855.]

The BULLION, CHEMICAL, and ASSAY BALANCES represented in the opposite page are constructed upon the system of three edges working against three planes. Not only does the fulcrum rest upon a plane, but the pans also are suspended by inverted planes, upon knife-edges, affixed to the ends of the beams. The advantages obtained by this system are twofold: first, it admits of the balance being adjusted to the greatest point of sensibility without diminishing its precision and constancy; and, secondly, when the balance is not in actual use, the pans are resting upon supports entirely independent of the beam. The bullion balances are made of sizes varying in length from 24 to 60 inches. In the chemical and assay balances the knife-edges as well as the planes are of agate, in order to protect the most important parts of the instrument against the fumes of the laboratory and the effects of damp climates.

Ladd and Oertling are manufacturers of balances to her Majesty's Exchequer, the Bank of England, the Assay-offices of the Royal Mint, &c.

[2927]

LAING, JAMES, 2 *McVicar's Lane, Perth Road, Dundee.*—Motoroscope: a new optical instrument giving motion besides relief to the individual objects of the stereoscope.

The object of this instrument is to give *motion* to stereoscopic figures, besides relief; that is, to give motion to the individual figures in a stereoscopic view; such as to show a carpenter in the act of sawing, or a machine in action. The relief and the motion combined, certainly presents to the eye one of the most extraordinary optical delusions that has yet been produced by only apparent phenomena.

[2928]

LANKESTER, DR. EDWIN, 8 *Saville Row.*—An ozonometer for registering the hourly variations of ozone.

[2929]

LEWIS, JOSEPH, *Dublin.*—Lewis's patent automaton register and pentagraph, applied to photo-printing and printing surfaces.

[2930]

LOWE, RIGHT HONOURABLE R., 34 *Lowndes Square.*—Spectacles which magnify without glass or any other refracting medium.

[2931]

MACDONALD, DR., 4 *Coburg Place, Kennington Lane.*—Instrument to facilitate finding the longitude at sea.

LADD & OERTLING, 192 *Bishopsgate Street Without.*—Bullion balances, &c.

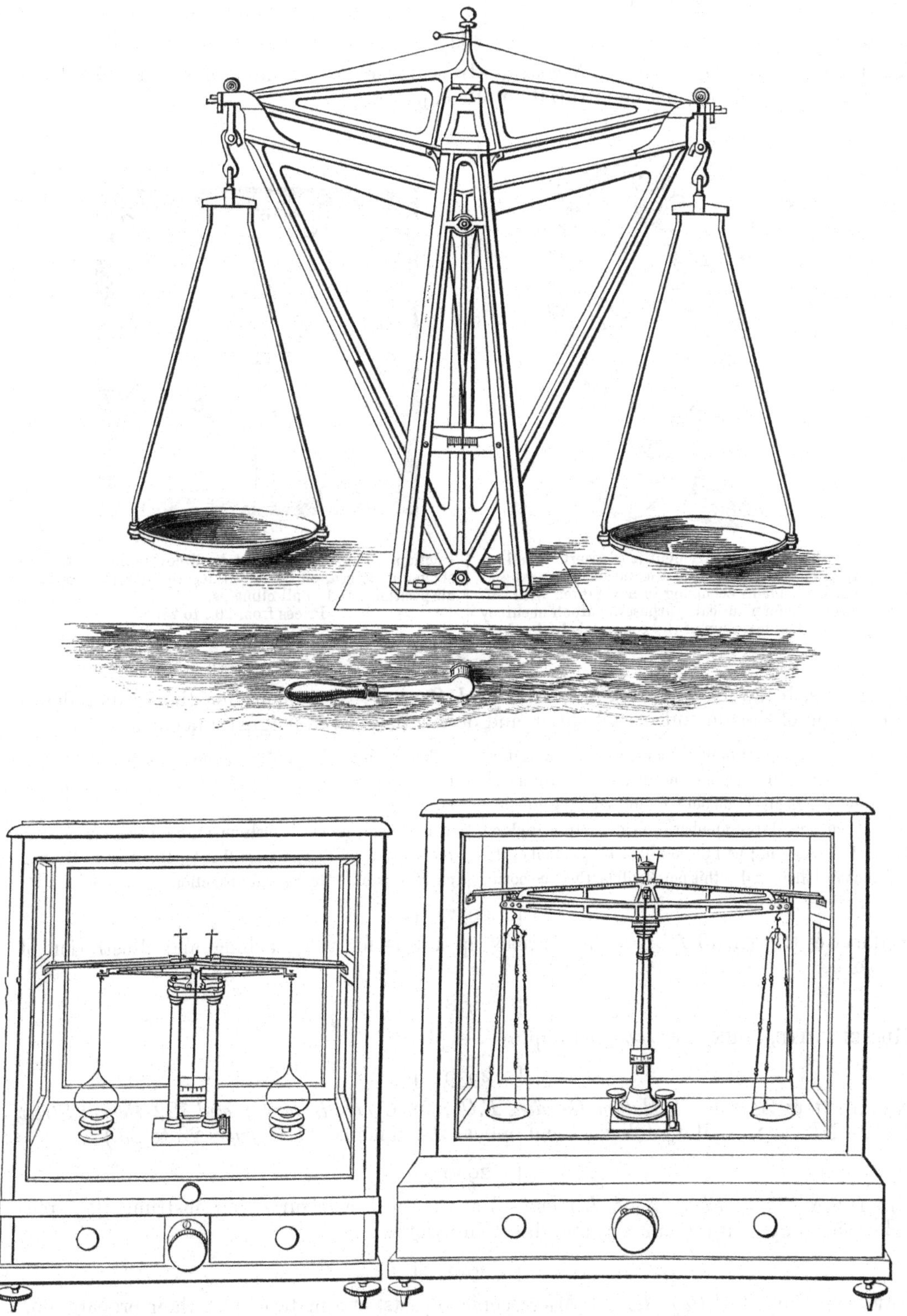

BULLION, CHEMICAL, AND ASSAY BALANCES.

[2934]

MINCHIN, HUMPHRY, M.B., 56 *Lower Dominick Street, Dublin.*—Galactoscope—for measuring the transparency of milk.

[2935]

MOORE, CHARLES, *Quay Parade, Swansea.*—Indicator for ascertaining nautical and astronomical problems, and magnetic variation of compasses.

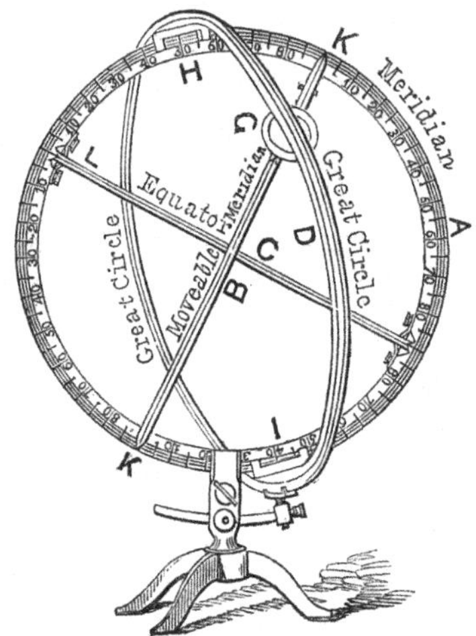

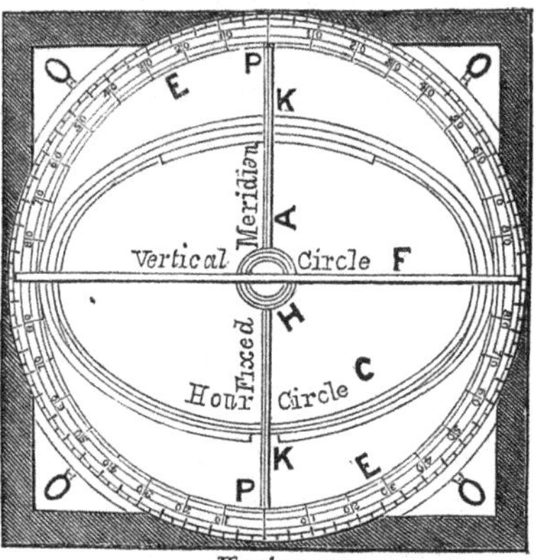

The purposes of this instrument is to ascertain and indicate approximate solutions of the various nautical and astronomical problems occurring in navigation, with sufficient accuracy for nautical purposes. Much accuracy depends upon the deviation of the compasses by local attraction. This is immediately discoverable by this instrument at sea, on board all descriptions of vessels, in all positions, and in all climates.

Prices from 20*l.* to 25*l.*

[2936]

MORTIMER, JOHN, *Pippinford Park, Maresfield, Sussex.*—Instrument for the readier determination of the amount of inclination and declination of the magnetic needle.

A COMPASS for determining the amount of inclination and declination of the magnetic needle, and the true north.

Obtain the smallest amount of dip and the greatest amount (90); mark the number of degrees traversed on the small dial for this; half these will be the amount of the apparent variation, and at this point will be the true north.

Two such needles with opposite poles, poised horizontally and placed in a line will detect any amount of local attraction.

The small globular instrument, on being arranged to the latitude of the place, will show the true declination and the point of terrestrial attraction.

[2937]

MURRAY & HEATH, 43 *Piccadilly.*—Various apparatus for the teaching and illustration of science.

[2938]

MUSSELWHITE, JOHN, *Devizes.*—An improved syphon.

[2939]

NEGRETTI & ZAMBRA, 1 *Hatton Garden, E.C.; 59 Cornhill, E.C.; and* 122 *Regent Street* (late NEWMAN).—Meteorological and optical instruments. (*See pages* 23 to 25.)

[2940]

NEWTON & CO., 3 *Fleet Street, London.*—Mathematical and surveying instruments; philosophical apparatus; lanterns and dissolving views.

[2941]

NORMAN, JOHN, 178 *City Road.*—Microscopic objects, and materials for their preparation.

The exhibitor manufactures and prepares all kinds of microscopic objects, and supplies slips, cells, and all the requisites for mounting. Established 1846.

Negretti & Zambra, 1 *Hatton Garden, E.C.;* 59 *Cornhill, E.C.; and* 122 *Regent Street* (late Newman).—Meteorological and optical instruments.

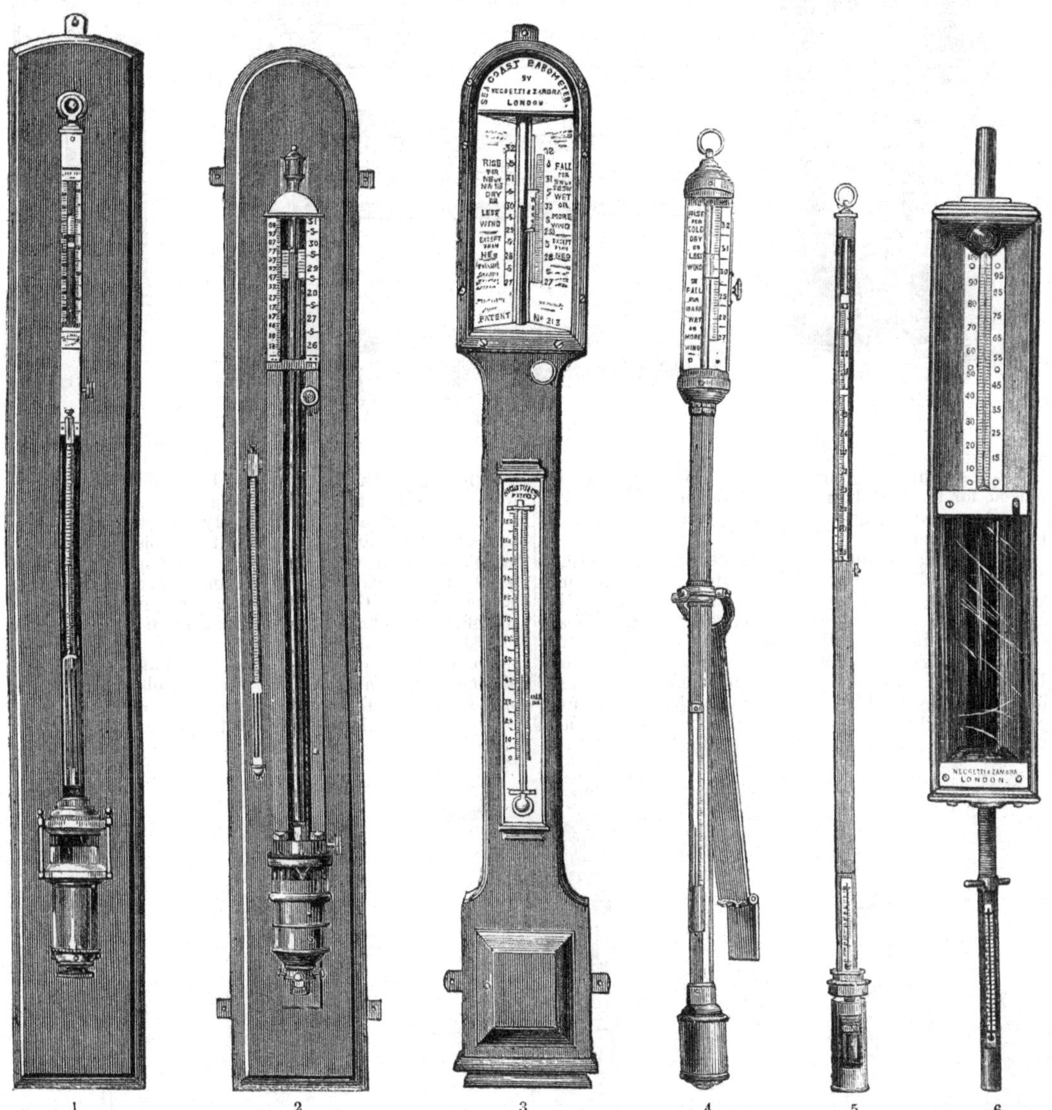

1 2 3 4 5 6

Standard Meteorological Instruments.

No. 1.—Standard barometer on Fortin's principle, reading from an ivory zero point in the cistern, to insure a constant level, with mercury boiled in the tube. The barometer tube is enclosed and protected by a casing of brass throughout its whole length; the upper portion of which has two longitudinal openings opposite each other; on one side of the front opening is the barometrical scale of English inches, divided to show, by means of a vernier, $\frac{1}{500}$th of an inch; on the opposite side is sometimes a scale of French millimètres, reading also by a vernier to one-tenth of a millimètre.

No. 2.—Negretti and Zambra's modification of Newman's Standard Barometer, in which a greater amount of light is admitted to the reading surfaces of the instrument.

No. 3.—The Fishermen's and Life-Boat Station Barometer made by Negretti and Zambra, especially for the Board of Trade, Royal Life-Boat Institution, and British Meteorological Society, to be fixed at all the principal seaports, fishing and life-boat stations on the British coast.

No. 4.—Board of Trade Standard Marine Barometer, as made by Negretti and Zambra for Her Majesty's Government, with spare tube to replace in case of accidents.

No. 5.—Negretti and Zambra's Patent Portable Mountain Barometer. To make an observation the instrument is suspended vertically, and the cistern unscrewed until the surface of the mercury is brought exactly level with the extreme end of the ivory zero point. The reading is then taken by the scale on the limb and vernier. To make the instrument portable, it is inclined until mercury from the cistern fills the tube, the cistern must then be screwed up as far as it will go.

No. 6.—Actinometer (Sir John Herschel's) for ascertaining the absolute heating effect of the solar rays, in which *time* is considered one of the elements of observation. See "Report of Royal Society on the Physics and Meteorology."

Negretti & Zambra—*continued.*

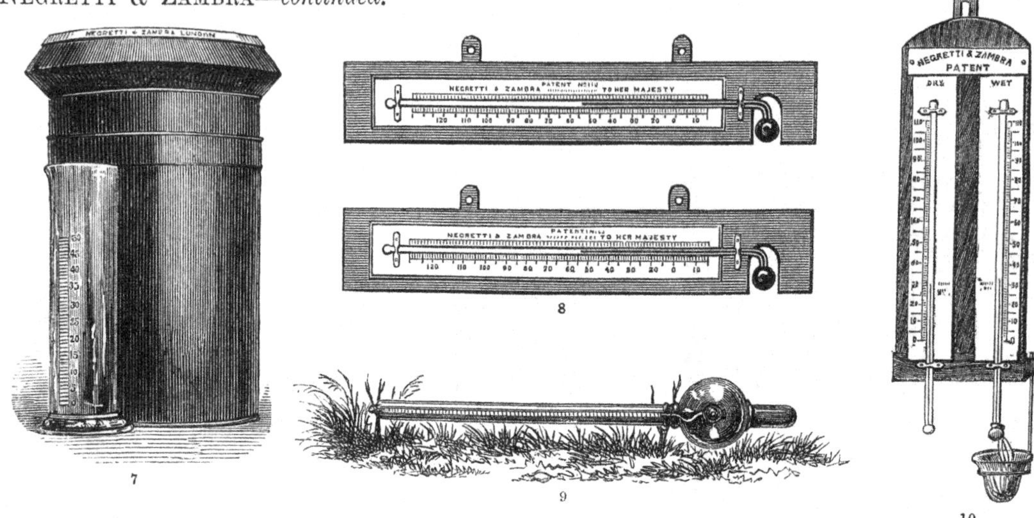

7 8 9 10

No. 7.—Glaisher's Rain Gauge. This gauge is arranged for the reception of the water only which falls upon its receiving surface, and for the prevention of loss by evaporation.

No. 8.—Negretti and Zambra's Patent Self-registering Maximum and Minimum Thermometers. The only instrument of the kind adapted for transmission to India and the Colonies. It is impossible to disarrange these instruments unless actually broken.

No. 9.—Negretti and Zambra's Patent Solar Radiation Registering Thermometer in Vacuum has a blackened bulb, the scale divided on its stem, a glass shield and globe surrounding the bulb and stem, from which all air is exhausted. In use it should be placed horizontally, with its bulb in the full rays of the sun, resting on grass.

No. 10.—Dry and Wet Bulb Hygrometer or Psychrometre for determining the amount of moisture in the atmosphere, an instrument of great importance, equally useful in the sick chamber, hothouses, greenhouses, and conservatories. Glaisher's Tables for showing the quantity of moisture for a difference of every tenth of a degree furnished with each instrument.

No. 11.—Negretti and Zambra's Patent Mercurial Minimum Thermometer, for deep-sea observations, the only instrument that can safely be used.

No. 12.—Standard Thermometer divided on brass into

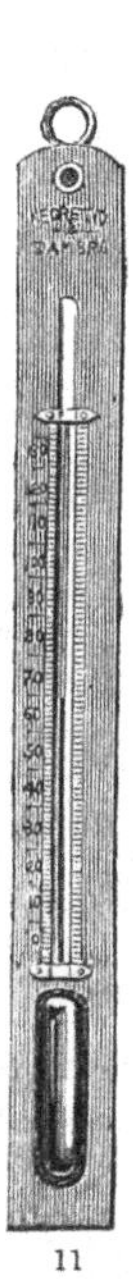

13

11 12 14 15

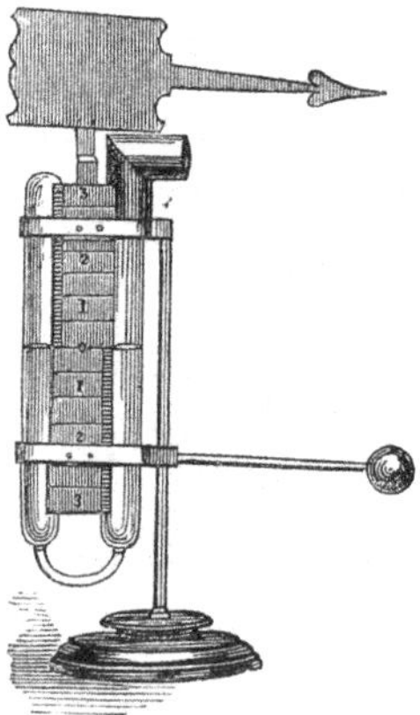

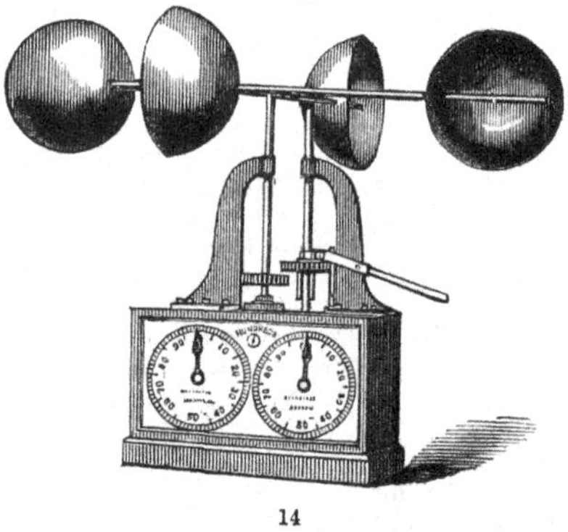

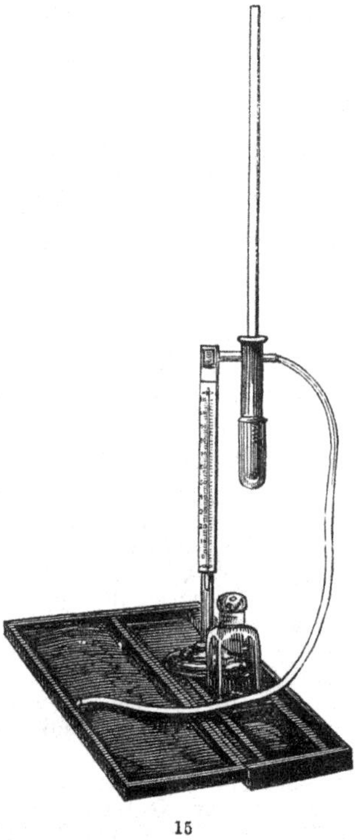

Negretti & Zambra—*continued.*

either Fahrenheit or Centigrade scale, or the divisions engraved on its stem. Negretti and Zambra's thermometers are made from selected tubes, the internal diameter of which is ascertained by very carefully conducted experiments. They are also strictly tested for index error, and a copy of the corrections furnished with each instrument.

No. 13.—Anemometer, or Wind Gauge, for showing the pressure of the Wind.

No. 14.—Anemometer for ascertaining the velocity of the wind. The readings on the dials of this Anemometer are in simple revolutions converted into actual miles.

Ostler's Self-registering Anemometer (improved by Negretti and Zambra), for showing the direction, force, and velocity of the wind; likewise the quantity of rain fallen in a given time, with clock-work and all necessary apparatus complete.

No. 15.—Regnault's Condenser Hygrometer as shown consists of a tube made of silver and glass very thin and perfectly polished; the tube is larger at one end than the other, the large part, silver, being 1·8 in depth by 8·1 in diameter; this is fitted tightly to a brass stand with a telescopic arrangement for adjustment. The more perfect form of this hygrometer has two sets of silver and glass tubes and thermometers.

Negretti & Zambra's Binoculars with Rock Crystal Lenses.

The lenses of these instruments are constructed of the finest rock crystal, cut and worked so as to insure the highest perfection both in magnifying and defining power. The superiority of crystal lenses over ordinary glass consists in the surfaces always remaining brilliant in all climates, not efflorescing or becoming dull and cloudy by exposure to the action of the atmosphere, and, from the extreme hardness of the crystal, not liable to be scratched. These instruments (the advantages of which will be appreciated by any who have had experience in the use of binocular glasses on foreign service) are offered by Negretti and Zambra at a price not much advanced on that of ordinary glasses.

For further details of prices, &c., see Negretti and Zambra's descriptive catalogue of optical, mathematical, photographic, and standard meteorological instruments, &c., illustrated by upwards of five hundred engravings. Price (post free) 2*s.* 6*d.*

Howson's Patent Long Range Barometer, made by Negretti and Zambra.

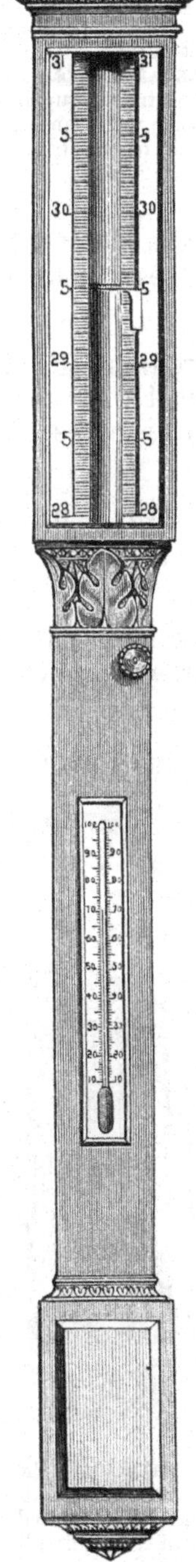

This fine instrument combines many excellences, and altogether supplies a desideratum of great importance to the progress of practical meteorology. Its action is based upon that of the Torricellian column, and is therefore of the most reliable character, but it derives its chief value from the introduction of a new principle by which a greatly extended range or length of scale is obtained.

The limits within which the ordinary barometric column oscillates are extremely narrow, and it was early felt that the public utility of the instrument would be greatly enhanced, if by any means the indications could be increased in length. This object was sought to be obtained by many devices, none of which may be said to have survived, except that of the wheel barometer, which, however, is an arrangement so inherently defective, that it has no good feature to recommend it.

The great value of the new construction consists in this, that no mechanism is employed for converting a short scale into a long one, but the mercury itself rises and falls through an extended range, naturally and in simple obedience to the varying pressure of the atmosphere. These instruments are usually made for domestic purposes with a scale from three to five, and for public use from five to eight times the scale of the ordinary standard. Their sensitiveness is consequently increased in an equal proportion, and they have the additional advantage of not being affected by variation of temperature, or from differences of level in the cistern. As regards the execution of the details and workmanship of these barometers, the makers have clearly felt that they had to deal with a valuable principle, and have endeavoured to do it full justice.

[2942]

ORCHARD, JOHN, Designer and Manufacturer, 2 *Phillimore Place, Kensington.*—Standard and compensating barometers; optical and other philosophical instruments.

A STANDARD BAROMETER in bronzed brass, constructed upon the most approved principles, so as to require as few corrections as possible, having the following additions and improvements. The cistern is made entirely of iron and glass, and is so arranged that it may be removed for cleaning without affecting the tube. The air-valve is so constructed as to open and close to the influence of the atmosphere. Two thermometers; one to take temperature of mercury in cistern, the other of that in tube.

STANDARD BAROMETERS, similar in form to the former, and constructed upon the same principles, but of less expensive workmanship; intended to meet the demand for a correct instrument at a price below that of the original standard.

COMPENSATING BAROMETERS, in bronzed brass cisterns of iron and glass, &c. The end designed in these instruments is to obtain as correct a reading as possible, without the trouble of a second adjustment, which is accomplished by displacing a quantity of mercury in the cistern equal to that taken up in the tube.

AN OXYHYDROGEN MICROSCOPE on new and improved principles, capable of illustrating to a large audience the revelations of the microscope in insect anatomy, &c., and intended to supply the want so much felt of some similar instrument to illustrate that science.

Various set of astronomical rack-slides, for illustrating the movements, &c., of the planets and other heavenly bodies.

[2943]

PARKES, JAMES, & SON, *St. Mary's Row, Birmingham.*—Microscopes; astronomical telescopes; mathematical, philosophical, and surveying instruments. (*See page* 27.)

[2944]

PASTORELLI, F., & Co., Opticians and Mathematical Instrument Makers, 208 *Piccadilly, and* 4 *Cross Street, Hatton Garden.*—Metford's theodolite; new level with micrometer for distances; standard meteorological and optical instruments. (*See pages* 28 *to* 30.)

[2945]

PILLISCHER, M., 88 *New Bond Street, W.*—Optical instruments of various kinds, principally microscopes. (*See pages* 31 *to* 33.)

[2946]

POWELL & LEALAND, 170 *Euston Road.*—Microscopes with rotating thin stage; binocular arrangement; object glasses from 2 in. to $\frac{1}{25}$th, inclusive.

[2947]

PULVERMACHER, J. L., 73 *Oxford Street.*—Galvano-piline, a self-supplying constant battery.

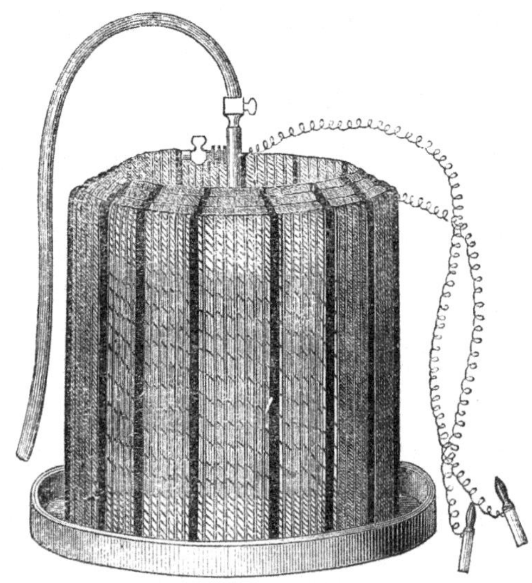

PULVERMACHER'S PATENT GALVANO-PILINE, representing a most convenient self-supplying voltaic battery of constant action, excited by one exciting liquid, as vinegar, &c., is an extremely pliable and very durable fabric, composed of galvanic metal wires and a fibrous texture, made (according to the purpose required) in any width and length. According to the size and number of its elements, it produces galvanic currents of *intensity* or *quantity* rendered constant in action by the porosity of its texture, the simultaneous contact of the atmospheric air, and the exciting liquid with the extensive surface of the galvanic metals. The Galvano-Piline is made simply flat or tubular. The first is set in action by a momentary immersion in the exciting liquid, the second by a steady self-supplying process by means of the hollow channel communicating with a small reservoir containing the exciting liquid. This simple voltaic flexible battery surpasses in power, portability, and constancy of action, all other single liquid batteries of equal size, and being always ready for use (by the simple arrangement for self-supplying the exciting liquid), its uniform generation of either continuous or intermittent currents renders it a valuable apparatus for medical cabinets, hospitals, electric baths (private or public), lectures, schools, telegraphic and other purposes. A list of apparatus for medical purposes, with prices, will be found in Class 17, page 129, of this Catalogue. These batteries can be seen in operation daily in the English Gallery, Class 13, No. 2947, and Class 17, No. 3570, and at the galvanic establishment of J. L. Pulvermacher & Co., 73 Oxford Street, London, adjoining Princess's Theatre.

PRICE LIST.

GALVANO-PILINE BATTERIES, for the instantaneous generation of constant volta-electric currents of intensity, charged with the liquid by a self-supplying arrangement.

	£	s.	d.
Galvano-piline battery, 50 elements, each element 2 square inches in surface, complete .	2	5	0
Ditto, ditto, 100 elements, each element 2 square inches in surface, complete . .	3	15	0
Ditto, ditto, 100 elements, each element 5 square inches in surface, complete . .	5	5	0
Ditto, ditto, 100 elements, 13 square inches in surface, complete	9	15	0

PARKES, JAMES, & SON, *St. Mary's Row, Birmingham.*—Microscopes; astronomical telescopes; mathematical, philosophical, and surveying instruments.

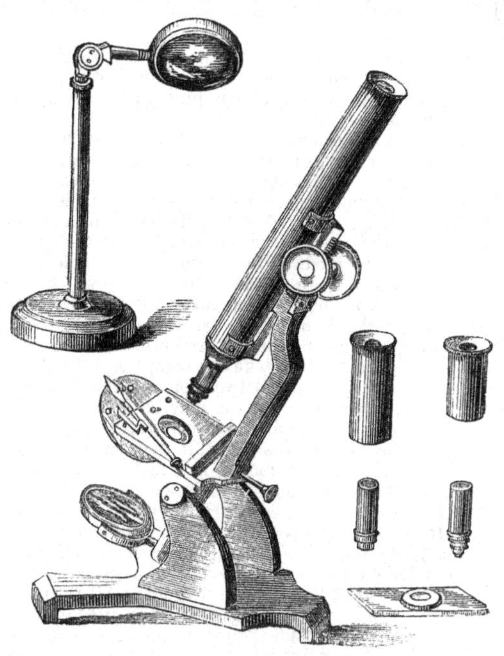

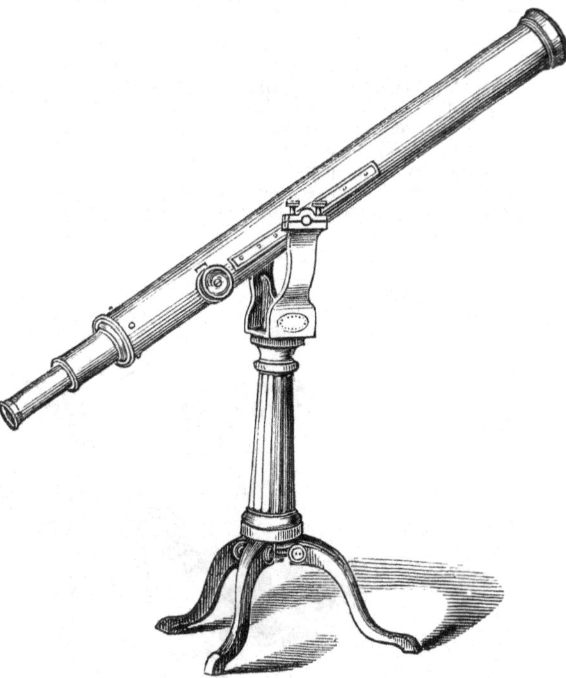

The following are exhibited :—

MICROSCOPES.

SIMPLE MICROSCOPES, brass stands, from 2*s.* to 10*s.* 6*d.*

COMPOUND MICROSCOPES (not achromatic), 4*s.* 6*d.* to 27*s.*

IMPROVED COMPOUND SCHOOL MICROSCOPE, on steady tripod, with inclining joint, with achromatic combination forming two powers, in case, price 1*l.* 1*s.*

Ditto, size larger, with stand condenser, price 2*l.* 2*s.*

* STUDENTS' MODEL MICROSCOPE (as woodcut), with two eye-pieces; 1½ inch and ¼ inch superior achromatic objectives; stand condenser, forceps, &c., complete in mahogany lock cabinet, price 3*l.* 3*s.*

* Ditto, ditto, with fine adjustment, and elongating body, 3*l.* 10s.

LARGER COMPOUND MICROSCOPES, with first-class objectives, at 5*l.* 5*s.*, 7*l.* 10*s.*, 10*l.* 10*s.*, 15*l.*, 20*l.*, 30*l.*, 40*l.*, and 50*l.*

EXHIBITION FINE ART MICROSCOPE; the most magnificent instrument ever produced, 150*l.*

MICROSCOPIC PREPARATIONS (great variety) from 1*s.* to 18*s.* per dozen.

EDUCATIONAL SERIES of ditto, with descriptive lists, in sets at 5*s.*, 10*s.*, 15*s.*, and 20*s.* per set.

POLARIZING APPARATUS, &c., may be fitted to any microscopes costing 3*l.* 3*s.* and upwards, at 25*s.* and 30*s.* each.

J. P. & S. wish to direct especial attention to their new Educational Microscope, at *five guineas*, which for quality, appearance, and price cannot be equalled. It is a full-sized instrument, with large stage, having a magnetic bar adjustment, which gives a very smooth and easy motion. It has an elongating body with coarse and fine adjustments; two eye-pieces; superior 2-inch, 1-inch, and ¼-inch objectives; condenser; all packed in mahogany cabinet. It is so constructed that all necessary apparatus can be added at any time without the instrument being returned.

* Several hundreds of these instruments have already been supplied to the medical colleges, educational establishments, &c., at home and abroad; and have met with universal approval.

ASTRONOMICAL TELESCOPES.

EDUCATIONAL TELESCOPE on improved tripod stand (as woodcut), with 2 inch objective, a terrestial and celestial eye-piece, with sun glass, in case, each 4*l.* 10*s.*

This instrument will show beautifully the lunar mountains, several planetary bodies, including Saturn's ring and one of his moons, also several double stars.

Larger ditto, ditto, with 2⅝ inch objective, 7*l.*

* SUPERIOR FOUR FEET TELESCOPE, 3 inch objective, mounted on improved equipoise garden stand 5 feet 4 inches high, with vertical rack, 3 eye-pieces, &c., 15*l.* 10*s.*

* FIVE FEET ditto, 3¾ inch objective, on large stand 6 feet high, with 4 eye-pieces, 27*l.*

* FIVE FEET SIX INCH ditto, 4¼ inch objective, 40*l.*

Larger instruments to order.

POCKET and TOURISTS' TELESCOPES; OPERAS; MARINE GLASSES.

IMPROVED PATENT DRAWING INSTRUMENTS, in cases from 2*s.* to 5*l.* each. These instruments have received the approval and recommendation of the Council of the Society of Arts, and are used in many of the Government schools.

SURVEYING INSTRUMENTS, &c.

Theodolites, levels, transit instruments, prismatic, mining, and other compasses; land chains, tape measures, air-pumps, galvanic and electrical machines, &c., &c.

N.B. By the employment of machinery, J. P. & S. have been enabled to construct many of the above instruments in a superior style, with accuracy, and at a very reduced price. They have kept in view the *educational* character of microscopes, telescopes, and drawing instruments, and have especially endeavoured to make these *complete, substantial,* and *convenient* for use.

A discount allowed to wholesale dealers and educational establishments.

For more detailed information, see wholesale illustrated catalogue, post free for 1s.

* These mounted on improved equatorial stands at 5*l.* to 15*l.* extra.

PASTORELLI, F., & Co., Opticians and Mathematical Instrument Makers, 208 *Piccadilly, and 4 Cross Street, Hatton Garden.*—Metford's theodolite; new level with micrometer for distances; and standard meteorological and optical instruments.

METFORD'S TRAVERSING THEODOLITE.—This instrument is constructed with all the important improvements introduced by Mr. W. E. Metford, C.E., the results of actual practice in the field.

The levelling gear consists of three inverted screws, having good seats fitted closely to their beds, and effectually shielded from dust by caps. To prevent the screws becoming loose, the arms that enclose them are made with sufficient spring to admit of their being slightly tightened.

The traversing stage.—The object of this is to enable the observer to shift his instrument over the exact centre, after having set it up firmly, nearly level, and approximately over the point required. The main hollow centre of the instrument carries a circular foot, which is able to travel in any direction to the extent of 1 inch from the centre, which is sufficient. The foot is properly secured by means of a three-arm pinching screw running on the hollow centre.

The check telescope, though not necessary in 5-inch instruments, or for ordinary work, is of great importance in larger ones. It lies between the traversing stage and the horizontal limb, where it can generally be used without taking advantage of its capability of sliding out; but by sliding it out a total horizontal and vertical range of 360° is obtained. The sliding horizontal and vertical motions are all fixed by one screw. Sliding the telescope out, was suggested by Mr. Newnham, C.E., of the Scinde Railway.

The horizontal limb, vernier circle, &c.—This limb is arranged to take the compass, a level with a circular bubble, and two memorandum slates on which constant errors, &c. may be recorded. The vernier plate is carried on four arms, and a diagonal brace (preventing the slightest twist in the arm) to which the tangent motion is attached. The horizontal limb has openings which enable the observer to take vertical angles to 70° in depression. Securely attached to the pivot is an arm to take the lower tangent apparatus. The brace system of tangents is adopted to prevent the loss of time occasioned by the wear of the common tangents. All the pivots have broad bearing flanges like those used in levels by Mr. Gravatt, and the pivots themselves and the bearing flanges are in one casting, thus conducing greatly to the rigidity of the whole instrument. The conical pivots fit in their sockets throughout their whole length, and not at the ends only.

The circular bubble was first used by Troughton, for the purpose of obtaining an artificial horizon, but was adapted to the theodolite by Mr. Metford. Its great advantage is, that it shows exactly the direction in which the level has been departed from, and it is thus a great aid in setting up the instrument before adjusting the traverser.

The means of supporting the upper works.—To the side of the main pivot is attached a strong curved bracket divided at the top into two arms. This bracket has a T section throughout and on the ends, and at the junction of the arms is fixed the vertical circle. The improvement is unquestionably an important one, since by it the suspension of the telescope over the axis is permitted. The use of the curved bracket is not attended by weakness, for the bracket is exceedingly stiff. It has been used by Mr. Metford for eight years with perfect success. The microscopes hang on the head of the casting, and travel far more conveniently than in the common instrument.

The telescope.—This is a "dumpy," care being taken to have all of its surface of object glass of good defining powers. The eye end passes clear over the axis, and therefore the instrument may be used as a transit. By this capability of turning over, it is of immense service in ranging railway curves, as regards accuracy in laying the tangent, and as regards time. It is also necessary in tunnelling, and in all altitude and azimuth observations, to which the instrument is perfectly adapted. A rectangular eye-piece is added to the telescope; it is taken out when the other eye-piece is used, and a stopper is inserted in its place; it is not however necessary to remove it entirely. The rays are turned with a prism, so that the loss of light is trifling.

The diaphragms.—Each diaphragm consists of two independent discs; and each takes one cobweb, and is so constructed that each web can be placed vertically or horizontally, as the case may be, and in the axis of the telescope also, independently of the other.

PASTORELLI, F., & Co.—*continued.*

Illuminating apparatus.—This consists of a small glass head placed about three quarters of an inch beyond the object glass, and just within its edge. A light thrown upon any point of its new hemisphere throws a mild faint light down the telescope.

The object glasses.—These are placed in their cases backward, so as to allow the glass surface to project beyond the brass cell. By this means rain and dust can be wiped off in the shortest time, and with the least amount of scratching, without any of the difficulty attending the same process in the case of the common deep-seated glasses. The eye-piece block—that which stops the end of the telescope barrel—pulls out, and the cobwebs and diaphragms are thus exposed.

The staff-head is made according to Mr. Froude's arrangements, having the cheeks cast to a circular plate. The leg joints resemble an inverted mortar with strong trunnions, which can be tightened in their bisected cylindrical bearings, by means of capstan-headed screws.

General summary of advantages.—From the foregoing detailed description of the construction of these instruments will be apparent their great superiority over those in ordinary use, especially as regards the great steadiness obtained by the adoption of the triple screw arrangement for levelling, in place of the ordinary parallel plates. There is also a great saving of time, when the instrument has to be set frequently in the course of the day; this is accomplished by means of the circular bubble and traversing stage, which allow of very speedy adjustment of the instrument. The various minor advantages will be best understood by the perusal of the foregoing description, which should be carefully compared by those requiring a theodolite, with the details of the ordinary maker. The great advantages of those here described will at once be apparent.

	£	s.
Seven-inch instrument, as above described, price	35	0
Five-inch ditto, but without check telescope or rectangular eye-piece	26	5

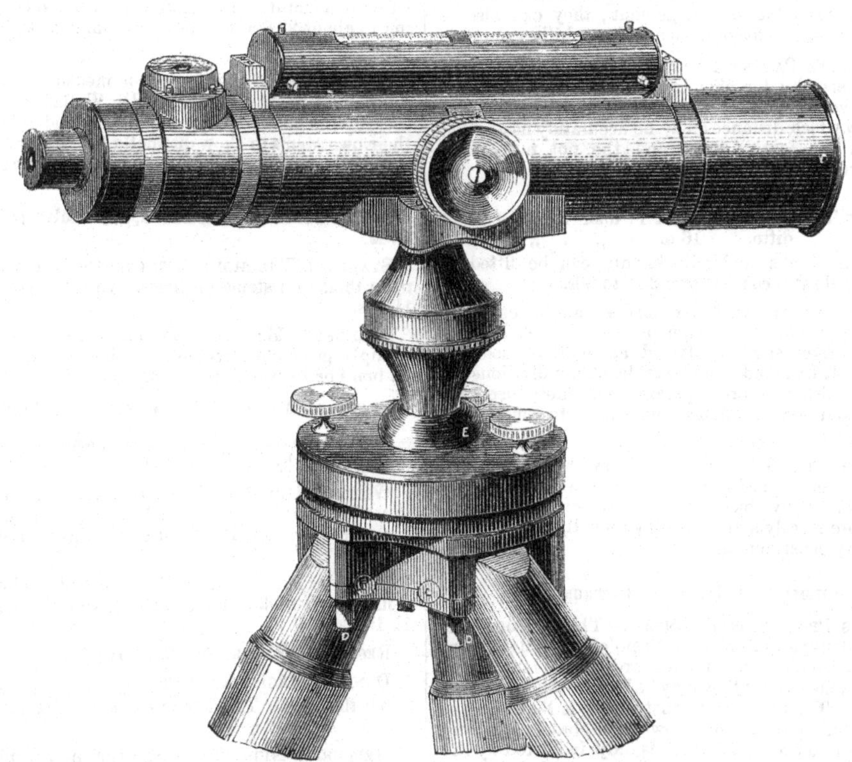

IMPROVED LEVEL.

F. PASTORELLI & CO.'S IMPROVED LEVEL.—This instrument combines several improved arrangements, giving increased facility in use, greater steadiness and freedom from vibration, more accurate adjustment, with scarcely a possibility of deranging them.

The tripod and its staff-head.—The stability of the tripod is of the utmost importance. The ordinary staff-head is defective, from the impossibility of properly tightening the joint pivots as they become worn. The new staff-head—an adoption of a plan of W. Froude, Esq., C.E.—has the cheeks cast on to a circular plate, the leg joints being similar to an inverted mortar, with strong trunnions, which can be tightened in their bisected cylindrical bearings by means of capstan-headed screws.

The ball-joint and clamp.—This ball-joint is substituted for the ordinary parallel plates, which are limited in their action, compelling the staff-head to be placed within 3° or 4° from a horizontal plane, making the tripod subservient to the level. By the ball-joint, you may set the tripod more firmly upon the ground, almost irrespective of the position of the level, the ball having a movement of nearly 20°. The instrument is approximately set by means of the circular bubble.

Mode of suspending the telescope.—The telescope is

PASTORELLI, F., & CO.—*continued.*

soldered to a saddle-piece, the base of which is made parallel to the axis. Two cylindrical gun-metal collars, very accurately turned and ground to a uniform diameter, are soldered to the telescope, and by means of these, the mechanical and optical axes of the telescope, and the line of collimation are adjusted; the spirit level being framed parallel thereto, and dead-fitted, neither admits nor requires any after adjustment. The saddle-piece is dead-fitted to the gun-metal centre or pivot, so that when the instrument is turned in any direction, the spirit bubble will be kept in the centre of its run.

The arrangement of the diaphragms.—The diaphragms are two separate discs; the horizontal web mounted on one, and the vertical web on the other. They can be moved independently of each other by means of the collimating screws. These screws are so arranged that they cannot be accidentally disturbed, being imbedded and covered by a cylindrical ring cap.

General summary of its advantages.—These instruments are less in weight and more portable than ordinary levels; they can be carried in a sling leather case, like a military telescope, with an increased stability of tripod, and capability of being adjusted with more facility and precision than the ordinary "Dumpy," and when adjusted are not liable to derangement; they can also be more readily set up for use, especially on hilly ground.

PASTORELLI'S PATENT MEASURING LEVEL.—For avoiding the necessity for chaining distances. This level embraces very important additions to the one above described, by which distances may be accurately measured; saving all the loss of time and the not unfrequent serious errors which occur when chaining them.

	£	s.
Pastorelli's improved level, 12 or 14 inch . price	14	14
Ditto, ditto, 16 or 18 „ . „	16	16

Additional webs and adjustments can be fitted to existing levels at a cost of from 20*s.* to 30*s.*

Levelling staves.—Ordinary staves can be used for ascertaining distances by means of the above measuring level, but staves slightly altered as to their mode of marking, will be found more suitable, as the divisions for estimating distances are separate from those used for levelling, and have no figures near them to confuse the observer.

PASTORELLI'S & CO.'S IMPROVED PRISMATIC COMPASS.—The prism being fixed at the back of the bisected web, the bisection of any object with the division on the silver ring, is more readily and distinctly seen than in those of the ordinary construction.

MATHEMATICAL DRAWING INSTRUMENTS.

FROUDE'S PROPORTIONAL COMPASS.—This instrument is arranged like a pair of ordinary "wholes and halves," but with these exceptions, that the legs are of equal length, one pair being jointed that they may be bent upwards towards the centre. The merit of the instrument is that it can at once be set not only to any known ratio, but to any ratio which is not known, but which is only indicated by the length of two given lines. The instrument being set to this ratio, its value can be immediately ascertained by referring to a graduated scale. It is peculiarly suitable in cases where fractional measurements are concerned and for which ordinary proportional compasses cannot be employed.

FROUDE'S IMPROVED BEAM COMPASS.—This instrument comprises many improvements over those in ordinary use, the chief being that by means of wheels fitted to each end of the beam, it is supported on the table in an upright position. By this arrangement great facility in use is obtained. Being self-supported it can be made slightly to recede from the work by a touch, and thus be always at hand. The beam also lying close to the work insures steadiness, the points being short. The pencil and ink legs are kept from the paper by means of a spring, which the slightest pressure overcomes when they are required to draw a line.

MATHEMATICAL DRAWING INSTRUMENTS, in sets or separate, to suit the requirements of military and civil engineers.

METFORD'S IMPROVED POCKET SCALES, suitable for civil engineers, architects, and land surveyors, price, in case, 2*l.* 16*s.*

STANDARD METEOROLOGICAL INSTRUMENTS WITH KEW VERIFICATIONS.

STANDARD BAROMETER mounted in brass body, ivory point in glass cistern which forms the zero of the scale, mercury boiled in the tube. The internal diameter of tube is 75, divided on silver to reach to $\frac{1}{500}$ of an inch, with French scale reading by Vernier to $\frac{1}{50}$ of a millimètre suspended by brackets to mahogany board, 21*l.*

SMALLER INSTRUMENT, internal diameter of tube ·34. In every other respect identical with the above, 8*l.* 10*s.*

STANDARD MOUNTAIN BAROMETER (upon the same principle) IMPROVED, by which the two verniers are read with great facility and exactness, divided on silver, reaching to about 22,000 feet of elevation, with tripod stand in sling case, 7*l.* 10*s.*

HYSOMETRICAL APPARATUS for measuring altitudes by the boiling point of water, on Dr. Wollaston's principle; the thermometers, engine-divided and etched upon their own stems to show distinctly the tenth of a degree; the apparatus employed is of the most portable kind, and packs in a strong leather case, price 4*l.*

METAL MARINE BAROMETER, Admiralty pattern, price 4*l.* 4*s.*

STANDARD THERMOMETERS, engine-divided and etched upon their own stems on brass. In morocco cases, price 1*l.* 15*s.*

STANDARD MAXIMUM THERMOMETER, on Professor Philip's principle, divided upon its own stem, mounted on brass or boxwood, price 15*s.*

STANDARD MINIMUM THERMOMETER, as above, price 15*s.*

SOLAR RADIATION MAXIMUM THERMOMETER, insulated as suggested by Sir John Herschel, Bart., price 18*s.* 6*d.*

WET AND DRY BULB REGISTERING MAXIMUM AND MINIMUM THERMOMETERS, divided upon their own stems, mounted on metal, and fixed to mahogny supports, price 2*l.* 5*s.*

MASON'S HYGROMETER divided and etched on the stems, upon metal stands resting upon a broad base, 1*l.* 15*s.*

REGNAULT'S HYGROMETER, 3*l.* 10*s.*

DANIEL'S HYGROMETER, 2*l.* 10*s.*

All the above Thermometers have the Kew verifications.

MERCURIAL NIGHT THERMOMETER, new form, 1*l.* 5*s.*

A SET of PORTABLE THERMOMETERS, for alpine travelling.

SYMPIESOMETER, for taking altitudes, a new form, in brass frame, most portable and light, suited for travellers, 3*l.* 3*s.*

RAIN GAUGES, from 10*s.* 6*d.* to 2*l.* 2*s.*

These can be made so portable that they may be carried in the pocket.

ANEMOMETER, a modification of Dr. Robinson's, registering the velocity of the wind in miles and furlongs, price 4*l.* 4*s.*

The exhibitors manufacture all kinds of microscopes, binocular field and opera glasses, military and naval telescopes, spectacles, &c.

PILLISCHER, M., 88 *New Bond Street, W.*—Optical instruments of various kinds, principally microscopes.

[*Prize Medals have been awarded to M. Pillischer at the Great Exhibition of* 1851, *and Paris,* 1855.]

Description of PILLISCHER'S No. 1 First Class Microscope.

Largest size Improved Compound Microscope, having a stage five-eight inch thick, and allowing of nearly one and a half inch motion in rectangular directions, sliding and rotating object plate and sliding spring holder, a secondary stage underneath the object stage, with rectangular rotatory and vertical movements for the adjustment of achromatic condenser and other apparatus, large concave and plano mirrors, diaphragm, the optical tube has coarse and fine adjustments, and draw tube with rack and pinion movement, and indicator divided into tenths to the inch for the facilitation of micrometric and erector measurements; Wenham's binocular arrangement with rack and pinion adjustment to the eye-pieces and extra single tube; Nos. 1, 2, 3, and 4, eye-pieces, glass stage, stage forceps, two live boxes, large bull's-eye condenser for opaque objects, polarizing apparatus and selenite stage, stage condenser, silver side reflector, parabolic reflector, camera lucida, compressorium, micrometers, achromatic condenser with stops and diaphragms, Brook's double nose piece, an erecting eye-piece, 2 inch, 1 inch, ½ inch, ¼ inch, and ⅛ inch object glasses of large angular aperture, and great penetrating power; the whole fitted into an elegant Spanish mahogany or walnut plate-glass case with several drawers for the different apparatus, &c. £90

A smaller Microscope having one inch motion to the stage, and similar in all other respects to the above £82

Microscropes of simpler construction than the above, but equally useful in many respects, from . . . . £20 to £40

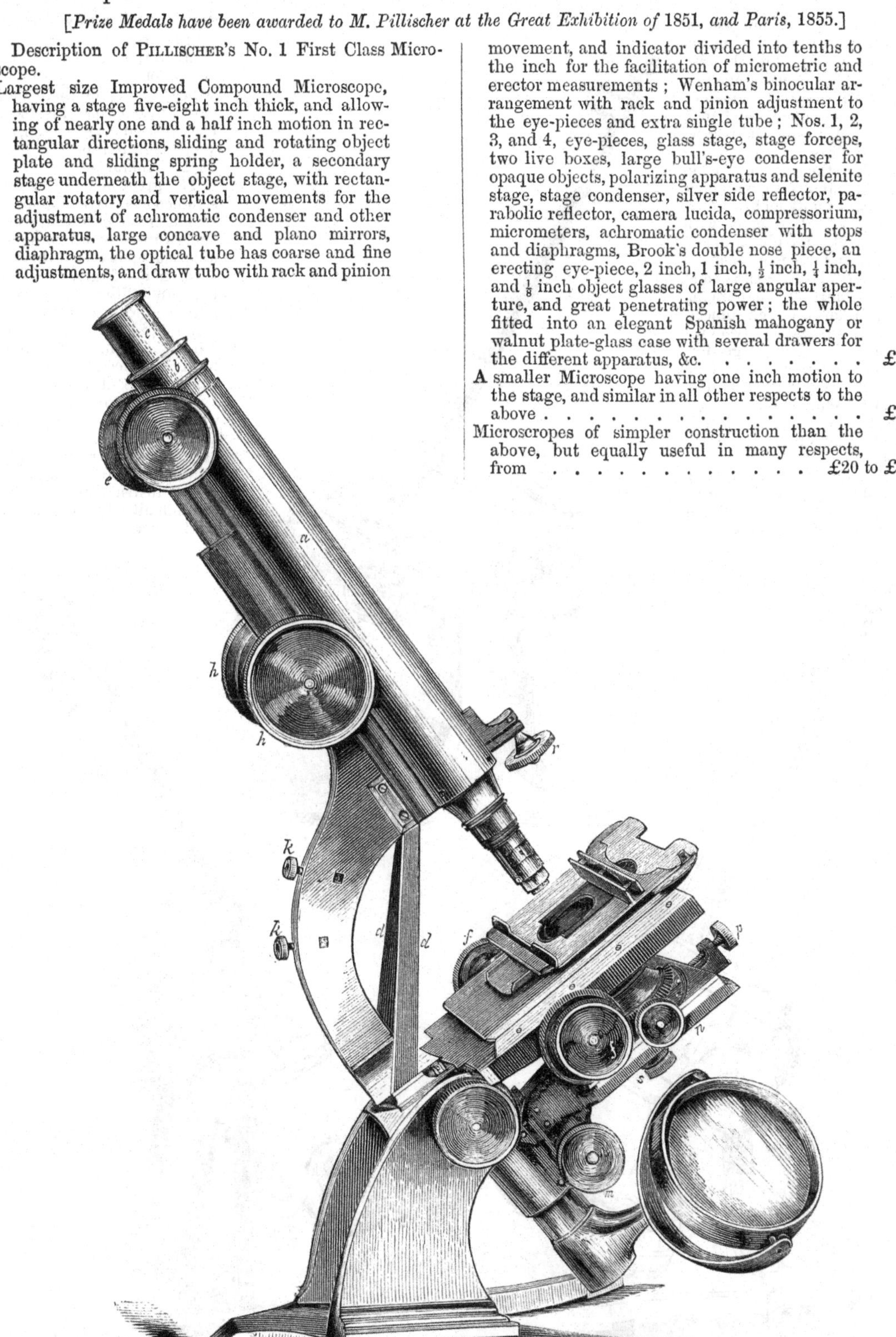

PILLISCHER'S NO. 1 FIRST-CLASS MICROSCOPE.

PILLISCHER, M.—*continued.*

PILLISCHER'S STUDENTS' MICROSCOPES.

Fig. 1. — A microscope with coarse and fine movements to the optical tube, plain stage, concave and plano mirrors, diaphragm, one eye-piece, 1-inch and $\frac{1}{4}$-inch object-glasses of the respective angle of apertures of 16° and 75°, the whole packed in a neat mahogany case, about 7 inches by $6\frac{1}{2}$ inches, 7*l.* 7*s.*

Fig. 2.—Microscope with best rack and pinion stage, coarse and fine movements to the optical tube, concave and plano mirrors, diaphragm, 1-inch and $\frac{1}{4}$-inch object-glasses 16° and 75°, two eye-pieces, live box, stage forceps, condenser for opaque objects, polarizing apparatus, and selenite parabolic reflector, and best Spanish mahogany or oak case, 10 inches by $6\frac{1}{2}$. 15*l.* 15*s.*

Fig. 3.—Pillischer's New £5 Microscope.—This microscope has been constructed with a view of supplying the public with a really good and useful instrument, equal, in many respects, to the more costly ones, the prices of which, owing to their delicate, and, in some instances, complicated construction, are very much beyond the reach of the general public.

The great merit of this instrument is its simple construction and portability, and being furnished with a moveable stage, invented by Mr. Pillischer, of great usefulness, and with a new combination object-glass, forming three different powers of 1-inch, $\frac{1}{2}$-inch, and $\frac{1}{4}$-inch focal distances. These advantages, together with the excellent form of stand, which combines lightness and stability, and to which any form of apparatus can be adapted, make it the most useful instrument ever before submitted to the public at so low a price.

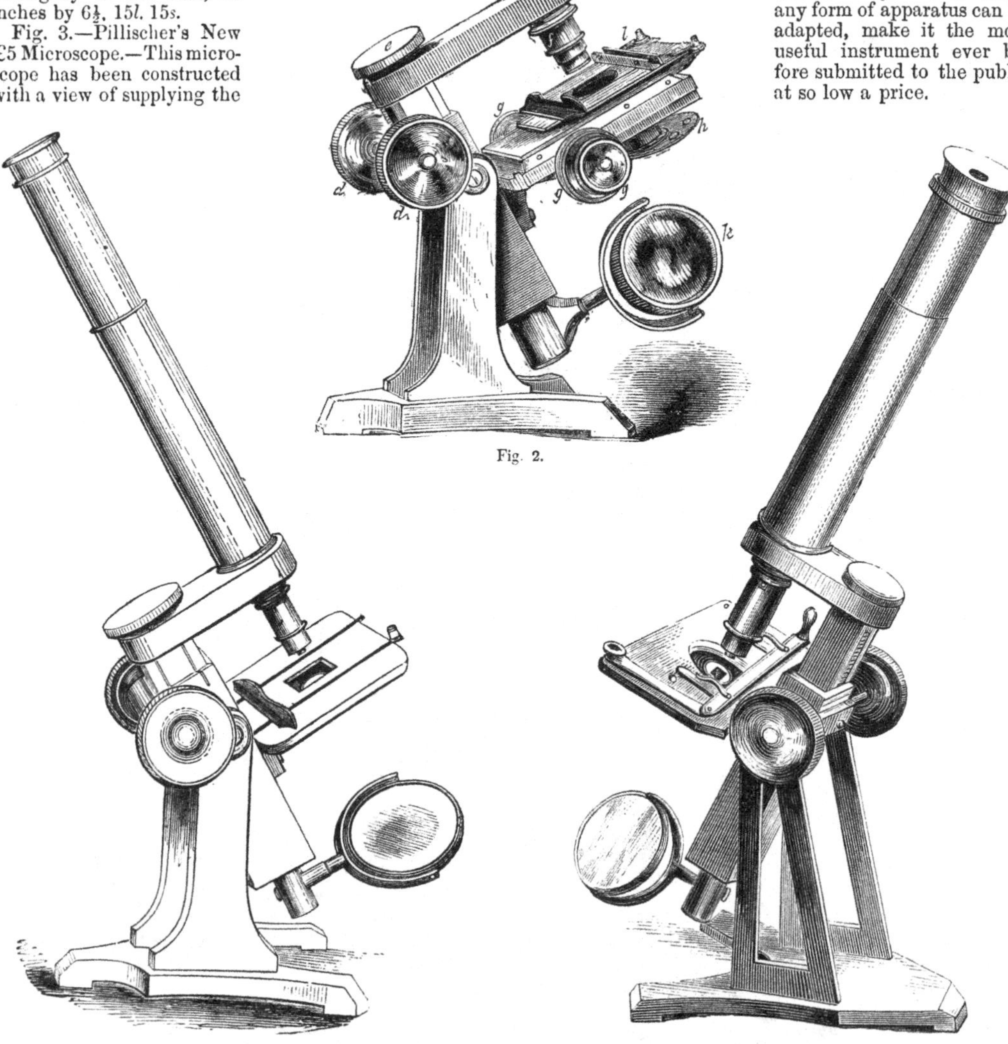

Fig. 2.

Fig. 1.

Fig. 3.

Pillischer, M.—*continued.*

Pillischer's Reading and Microscopic Lamps.

Patronized by Her Majesty, H.R.H. the Prince Consort, the Royal Family, many Scientific Societies, and most of the Nobility and Gentry of the United Kingdom.

These celebrated lamps, so well known now by the name of "Pillischer's Reading and Microscopic Lamps," combine all the advantages sought for by the scientific public, they emit a very pure white and steady light without smoke or disagreeable heat, although the common colza oil only need be used: they burn very economically, the cost of the largest size having an illuminating power equivalent to six wax candles or more, not being a halfpenny per hour. Their chief recommendation, however, is in the simplicity of the construction and management, and they can be used by any one previously unacquainted with the use of lamps.

These lamps can be used in India or any other tropical climate, since they burn the cocoa-nut oil, and can be adapted to burn perfectly steady under the Punka.

Fig. 1 represents the largest size, Queen's Pattern, so called, it being used by Her Majesty.

Fig. 2 is the same size, with an ornamental reservoir.

Fig. 3 is mounted in china, and beautifully painted with figures or flowers, and forms no unpleasant addition of a drawing-room furniture.

Fig. 3.

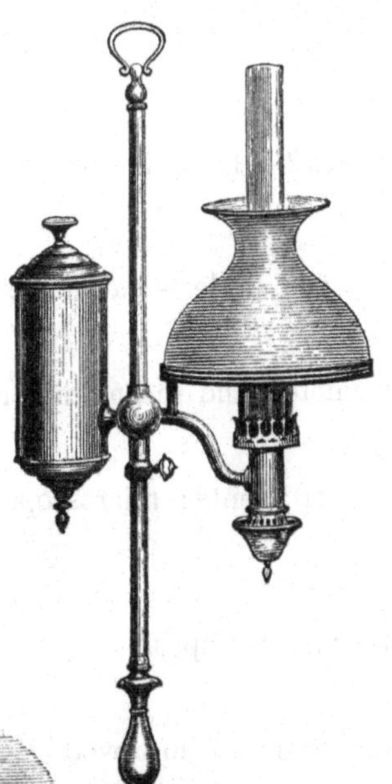

Fig. 1.

Illustrated Catalogues to be had at the Manufactory, 88 New Bond Street, W.; or, the Exhibition, Class XIII., Philosophical Instruments.

Fig. 1.

[2948]

READE, REV. J. B., F.R.S., *Ellesborough Rectory, Tring.*—Hemispherical condenser for microscopes, illustrating a new principle in microscopic illumination.

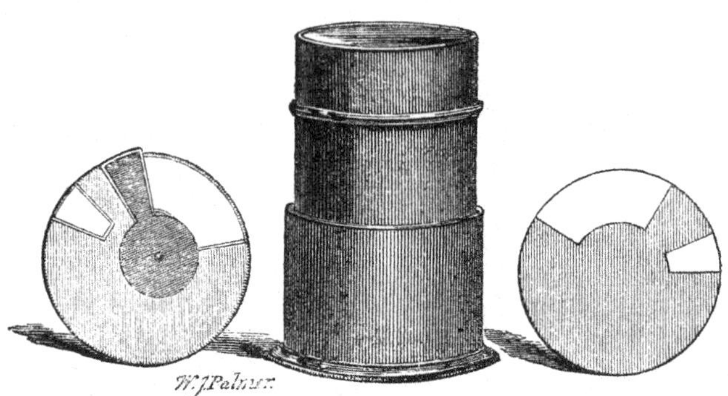

The CONDENSER consists of a hemisphere of glass, about $1\frac{3}{4}$ inches in diameter, with a new arrangement of stops, *applicable to all condensers*, for regulating the number, position, and magnitude of transmitted pencils of oblique light. The control thus obtained over the illumination of test-objects under any magnifying power, not only enables a single lens to compete with expensive achromatic condensers, but it also, for the first time, brings difficult test objects under the command of half-inch object-glasses, and thus tends to advance microscopic investigation, by saving the cost of hitherto necessary apparatus. The hemisphere is set in a thin brass ring, and screws upon a cylinder, adapted, like other fittings, to the sub-stage of the microscope. Its plain surface is covered by two similarly pierced diaphragms, shown half size in the engraving, and by the rotation of the upper diaphragm, the lower being fixed, one, two, or three apertures for transmitted light may be obtained, with distances between them varying from 30° to 120°. On taking out the eye-piece and looking down the body of the microscope, the points of light ought to be seen, and they should continue in view throughout the entire circular rotation of the sub-stage, otherwise the condenser is not truly central. The size of the apertures found best in practice is 24° at the circumference, and $\frac{4}{10}$ths of an inch in the direction of the radius; and in condensers of smaller diameter the latter dimension should never exceed half the radius. A small central aperture, closed by an eccentric shutter, is used for adjustment and for central illumination. In the application of this condenser to the resolution of lined test objects, the principle sought to be carried out is to throw the axis of the illuminating pencil in a direction at right angles to the line to be resolved. To illuminate a rectangularly marked valve, for instance, one point of light must lie over the end of the valve for bringing out the horizontal lines, and another be opposite the side of the valve to act on the longitudinal lines; and resolution into dots or squares will be readily effected by adjusting the distance of the condenser. The two apertures used in this case are necessarily 90° apart, whilst for the *Pleurosigma angulatum*, and other objects with trilinear markings, three apertures are necessary at intervals of 60°, and many bilinear oblique markings are best seen with two apertures 120° apart. Under this new arrangement of small illuminating pencils, that portion of the light of the ordinary spot lens, which really tends to obliterate the shadows by throwing them in directions opposite to each other, is stopped out,—the markings, whether elevations or depressions, are illuminated on the same side, and we preserve that uniform direction of the shadows which is the key to accurate definition. With this condenser scales of the *Podura*, and objects capable of reflecting light, can be viewed upon a black ground by bringing a single pencil of light so near to the stage of the microscope as to be beyond the angle of aperture of the object-glass. It is always desirable to use direct rather than reflected light, the source of light being placed about 10 inches from the condenser.

[2949]

REID BROTHERS, 25 *University Street, and Wharf Road, City Road.*—Electric telegraph materials.

[2950]

ROGERS, JOSEPH, 215, 216 *Gresham House, City.*—Telegraph wires and cables—patented.

[2951]

RONCHETTI, JOHN B., 9 *Cambridge Street, Golden Square.*—Hydrometers and thermometers.

[2952]

ROSS, THOMAS, 2 & 3 *Featherstone Buildings, Holborn.*—Optical instruments; microscopes; telescopes and photographic lenses, &c. (*See pages* 36 & 37.)

[2953]

SALMON, W. J., 100 *Fenchurch Street.*—Binocular and achromatic miscroscopes.

[2954]

SAX, JULIUS, 8 *Hatton Garden, E.C.*—Chemical and bullion balances on an improved style, and weights of the finest description.

The following are exhibited:—

A CHEMICAL BALANCE, in German silver, with 14 in. beam, to carry 1000 grains in each pan, and turn, when loaded, with 1000th part of a grain.

A BULLION BALANCE, with 18 inch beam, to carry 100 ounces in each pan, and turn, when loaded, with 30th part of a grain.

A set of TROY WEIGHTS in mahogany box, from 100 ounces to 1000th part of an ounce.

[2955]

SCOTT, WENTWORTH E., *Westbourne Park, London.*—1. Self-registering maximum thermometer for deep sea, &c. 2. Microscopic specimens.

[2956]

SHARPE, E. BENJAMIN, *Hanwell Park, Middlesex.*—Improvements in submarine electric telegraphs, paying-out machinery, &c. (Sharpe's patent).

[2957]

SHARP, HENRY, 38 *Bowden Street, Sheffield.*—Achromatic microscope objectives.

[2958]

SHAW, WILLIAM THOMAS, Inventor, 6 *Park Villas, Dalston, N.E.*—Stereotrope or stereoscopic thaumatrope.

[2959]

SIEMENS, HALSKE, & Co., 3 *Great George Street, Westminster.*—Telegraphic apparatus for land and submarine lines.

[2960]

SILVER, S. W., & Co., 3 *Bishopsgate Street.*—Electrical machine fitted with ebonite.

[2962]

SPENCER, BROWNING, & Co., 111 *Minories, London.*—Telescopes, Crooke's spectroscopes, pocket and improved aneroid barometers, and nautical instruments.

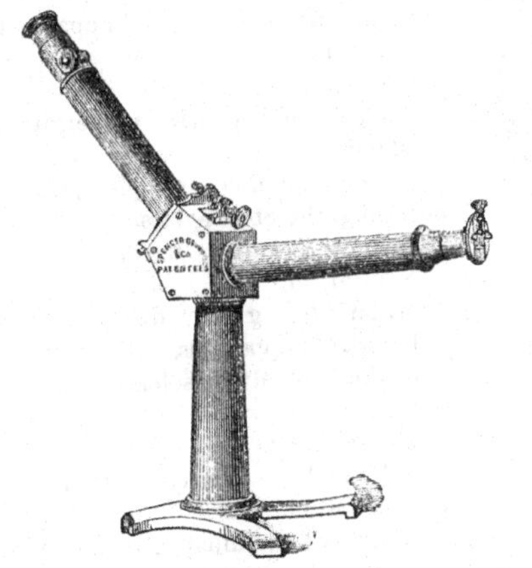

	£	s.	d.
LONG RANGE ANEROID BAROMETER, S. B. & Co.'s PATENT	2	10	0
Compensating ditto, uninfluenced by temperature	4	4	0
Long range aneroid for the waistcoat pocket, in gold case	15	15	0
Long range aneroid in silver case . . .	5	5	0
Long range aneroid, in metal case . . .	4	4	0
Crooke's pocket Spectroscope for "Spectrum analysis"	3	13	6
Crooke's large model "Spectroscope" (see illustration)	20	0	0

[2963]

SMITH, EDWARD, 16 *Queen Anne Street, London.*—Spirometer; potash-box to abstract carbonic acid during expiration.

[2964]

SMITH, BECK, & BECK, 6 *Coleman Street, E.C.*—Achromatic microscopes; objects; achromatic stereoscopes; cabinets; and other optical instruments.

[2965]

SMYTH, C. PIAZZI, *Edinburgh.*—Rotary ship clinometer; model of compound rotary apparatus; electric registering anemometer.

[2966]

SPRATT, ALICE, 118 *Camden Road Villas, London.*—Electric weather indicator.

[2967]

SPRATT, JAMES, 118 *Camden Road Villas, London.*—Lightning conductors; reproducing points; lock insulator and attachments—patented 1861.

[2968]

STANLEY, W. F., 3 *Great Turnstile, W.C.*—Mathematical and surveying instruments.

ROSS, THOMAS, 2 & 3 *Featherstone Buildings, Holborn.*—Optical instruments; microscopes; telescopes; photographic lenses, &c.

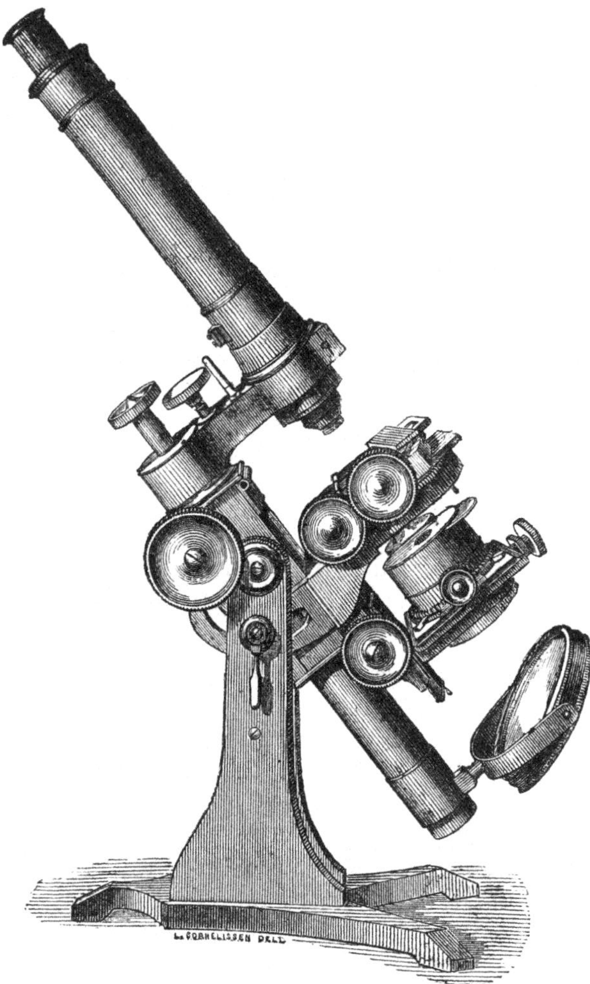

A large compound microscope, No. 1A, with Wenham's binocular arrangement; a concentric rotating stage, having one inch of motion in rectangular directions, rack and fine-screw movements to the optical part; clamping arc for fixing the instrument at any inclination; secondary stage for holding and adjusting by universal motions all the illuminating and polarizing apparatus beneath the object; flat and concave mirrors, diaphragm plate and apparatus complete *(see list)*, packed in mahogany cabinet case.

A ditto, ditto, No. 1B, with Wenham's binocular arrangement; an ordinary rotating object-plate to the stage, and apparatus in mahogany cabinet case.

A smaller binocular microscope, No. 2, having three-quarters of an inch of motion, and ordinary rotating object-plate to the stage, with apparatus, in mahogany portable case.

A smaller microscope, No. 3, having three-quarters of an inch of motion, and ordinary rotating object-plate to the stage, with apparatus, in mahogany case.

A plain microscope (basis of the above, No. 3), in mahogany cupboard case.

Apparatus for the compound microscope.

Achromatic object-glasses for microscopes, with flat field, perfect marginal definition and the maximum aperture consistent with the required performance :—

3	inches,	12	degrees	angular	aperture.
2	,,	15	,,	,,	,,
$1\frac{1}{2}$	,,	20	,,	,,	,,
1	,,	15	,,	,,	,,
1	,,	25	,,	,,	,,
$\frac{2}{3}$	,,	35	,,	,,	,,
$\frac{1}{2}$	,,	90	,,	,,	,,
$\frac{4}{10}$	,,	110	,,	,,	,,
$\frac{1}{4}$	,,	100	,,	,,	,,
$\frac{1}{4}$	,,	140	,,	,,	,,
$\frac{1}{6}$	,,	140	,,	,,	,,
$\frac{1}{8}$	,,	140	,,	,,	,,
$\frac{1}{12}$	,,	150	,,	,,	,,

A, B, C, D, E, and F, eye-pieces.
Erecting eye-piece.
Micrometer eye-piece.
Screw micrometer.
Jackson's micrometer.
Stage micrometer.
Goniometer.
Lieberkuhn's reflectors for opaque objects.
Side reflector.
Side condensing lens.
Condensing lens, with universal motions, on stand.
Brooke's double nose-piece, for rapidly changing the object-glass.
Wollaston's camera lucida.
Neutral tint do.
Millar's thin glass do.
Polarizing apparatus, with two selenites.
Darker's revolving selenite stage, and set of three selenites.
Double image prism.
Achromatic condenser for illuminating transparent objects.
Kingsley's illuminator, with diaphragms.
Reade's hemispherical condenser.
Parabolic illuminator (or paraboloid) for dark-ground illumination.
Small spotted lens for test objects.
Amici's prism, with universal motion, on stand.
Ross's centring glass.
Lister's dark wells.
Rainey's light modifier.
Plate for fixing fish, frogs, &c.
Animalculæ cages (3 sizes).
Ditto, for high powers.
Compressorium.
Wenham's ditto.
Set of animalculæ tubes in case.
Stage forceps.
Phial ditto (2 sizes).
Page's wooden ditto.
Writing and cutting diamonds.
Instrument for measuring thin glass.
Machine for cutting discs of ditto.
Paraffin lamp on adjusting stand.
Apparatus for producing photographs of enlarged microscopic objects.

ROSS, THOMAS—*continued.*

SIMPLE MICROSCOPES.

A simple microscope, with 1 in., $\frac{1}{2}$ in., $\frac{1}{4}$ in., and $\frac{1}{10}$ in., single lenses; $\frac{1}{20}$ in. Wollaston's doublet, Liberkuhn for $\frac{1}{2}$ in. and $\frac{1}{4}$ in. lenses, and stage forceps, in mahogany case.

A magnifying stand, with universal motions, and two lenses, for dissecting.

A diatom finder (field hand-microscope), with screw adjustment, cage, and single lens.

Coddington lenses and pocket magnifiers.

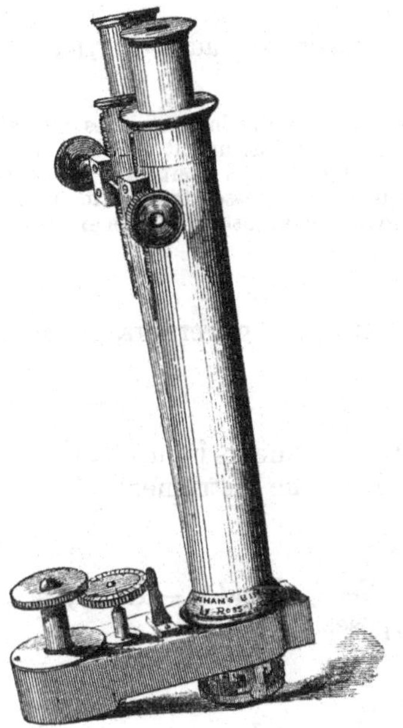

TELESCOPES.

The contact surfaces of the object-glasses of these instruments are united by a permanently transparent cement, obviating the loss of light by reflection, and preventing the decomposition of the glass.

Three portable telescopes, in German silver, brass, and aluminium mountings, opening from 5 inches to 20 inches; clear aperture $1\frac{1}{8}$ inch, magnifying power 17 times.

Three ditto, ditto, opening from $6\frac{3}{4}$ inches to 20 inches; clear aperture $1\frac{3}{8}$ inch, magnifying power 20 times.

Three ditto, ditto, opening from $8\frac{1}{2}$ inches to 28 inches; clear aperture $1\frac{1}{2}$ inch, magnifying power 20 times.

Three ditto, ditto (pancratic), opening from $12\frac{1}{2}$ inches to 43 inches; clear apertures $2\frac{1}{8}$ inches, and $2\frac{1}{4}$ inches, magnifying powers 30, 40, and 50 times.

One naval telescope, in German silver mountings, with one draw, opening from 17 inches to 23 inches; clear aperture $1\frac{1}{2}$ inch, magnifying power 14 times.

Two ditto, ditto, opening from 24 inches to $29\frac{1}{2}$ inches; clear apertures $1\frac{5}{8}$ in. and $2\frac{1}{8}$ in., magnifying power 20 times.

Two ditto, ditto (pancratic), opening from $37\frac{1}{2}$ inches to 43 inches; clear apertures $2\frac{1}{3}$ in., and $2\frac{3}{4}$ in., magnifying powers 30, 40, and 50 times.

One signal ditto, ditto (pancratic), with jointed body, opening from 55 inches to 60 inches; clear aperture $2\frac{3}{4}$ in., magnifying powers 50, 60, and 70 times. Packed in case.

One deerstalking telescope, in brass mountings, opening from $8\frac{1}{2}$ in. to $19\frac{1}{2}$ in.; clear aperture $1\frac{1}{4}$ in., magnifying power 14 times.

Three ditto, ditto, in brass and aluminium mountings, opening from 10 inches to 29 inches; clear apertures $1\frac{1}{2}$ in. and $1\frac{3}{4}$ in., magnifying power 20 times.

Three ditto, ditto (pancratic), same length and apertures as the preceding, magnifying powers 20, 25, and 30 times.

Two ditto, ditto, same length as the preceding; clear aperture $2\frac{1}{3}$ inches, magnifying power 20 times.

One 33 inch telescope, with brass tube, rack-and-pinion adjustment, and brass table-stand; clear aperture $2\frac{1}{10}$ inches. This instrument has a pancratic day eyepiece, powers 30, 40, and 50, and two astronomical eyepieces, powers 40 and 70. Packed in case.

Binocular field, race, and opera-glasses in various mountings.

Photographic lenses, various.

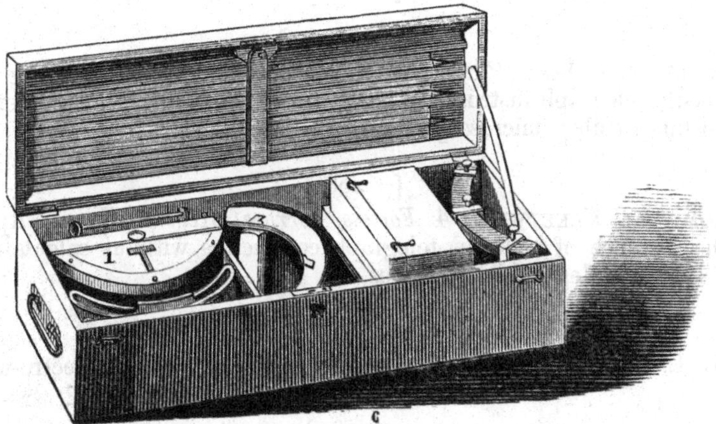

C

[2969]

STEVENSON, PETER, *Edinburgh.*—Instruments for brewers, distillers, and others; also scientific apparatus, &c.

[2970]

STEWART, BALFOUR, *Kew Observatory, Richmond, S.W.*—Self-recording magnetographs, and other philosophical instruments.

[2971]

SUBMARINE TELEGRAPH COMPANY, Chief Offices, 58 *Threadneedle Street, and* 43 *Regent Circus, Piccadilly,* L. W. COURTENAY, Secretary.—Samples of their submarine cables. (*See page* 39.)

[2972]

SUFFELL, Manufacturer, 132 *Long Acre, London, W.C.*—Improved adjusting, surveying, and drawing implements.

The exhibitor has been a manufacturer of improved adjusting, surveying, and drawing instruments, &c., for a quarter of a century. Transit theodolites, 21*l.* Everett ditto, 20*l.* 5-inch ditto, 18*l.* Improved Gravat level, 10*l.* Dumpy level, 7*l.* 10*s.* Surveying level, 4*l.* 10*s.* He also manufactures the following measures and standards of all nations; every instrument requisite for surveyors, engineers, architects, and draftsmen. Cases of drawing instruments for all classes, from 5*s.* to 5*l.*; and the improved needle socket instruments as exhibited.

[2973]

SUGG, WILLIAM, *Marsham Street, Westminster.*—Photometer, gas governors, gas meters, and other apparatus.

[2974]

SWIFT, JAMES, 3 *Matson's Terrace, Kingsland Road.*—Improvements in mechanical construction of microscope; dispensing with rackwork throughout the instrument.

[2976]

TREE, JAMES, & Co., 22 *Charlotte Street, Blackfriars Road, London.*—Rules, scales, and levels for mechanical, scientific, and agricultural purposes.

[2977]

TYER, EDWARD, 15 *Old Jewry Chambers.*—Train signalling telegraphs and electric telegraphs. (*See pages* 40 & 41.)

[2978]

UNIVERSAL PRIVATE TELEGRAPH COMPANY, 448 *Strand, London.*—Wheatstone's (magneto-alphabetic) telegraphs for railways or private use.

[2979]

VARLEY, ALFRED, 1 *Raglan Terrace, Highbury.*—Apparatus for economically heating greenhouses by gas, and regulating the temperature.

[2980]

VARLEY, CORNELIUS, 7 *York Place, Kentish Town.*—Ebonite electrifying machine; pair of single needle telegraph instruments; key and printing machine; differential galvanometer; resistance coils; microscope with lever stage.

[2981]

VARLEY, CROMWELL FLEETWOOD, 4 *Fortess Terrace, Kentish Town.*—Apparatus for indicating the distance of faults in telegraph conductors without calculation; insulators; constant batteries; telegraphic apparatus, &c.

[2982]

VULLIAMY, L. L. & H. P., *Clapham Common, S.*—Model of an electro-magnetic motive engine.

Submarine Telegraph Company, Chief Offices, 58 *Threadneedle Street, and* 43 *Regent Circus, Piccadilly*, L. W. Courtenay, Secretary.—Samples of their submarine cables.

The Submarine Telegraph Company is the only Company in exclusive communication with the Continent of Europe, the Channel Islands, Alexandria, and the East, viâ France, Belgium, Hanover, and Denmark.

The following illustrations represent the Company's cables, now in perfect working order, showing 28 lines of telegraphic communication to the Continent, making in the aggregate a submarine conductor of 2777 miles :—

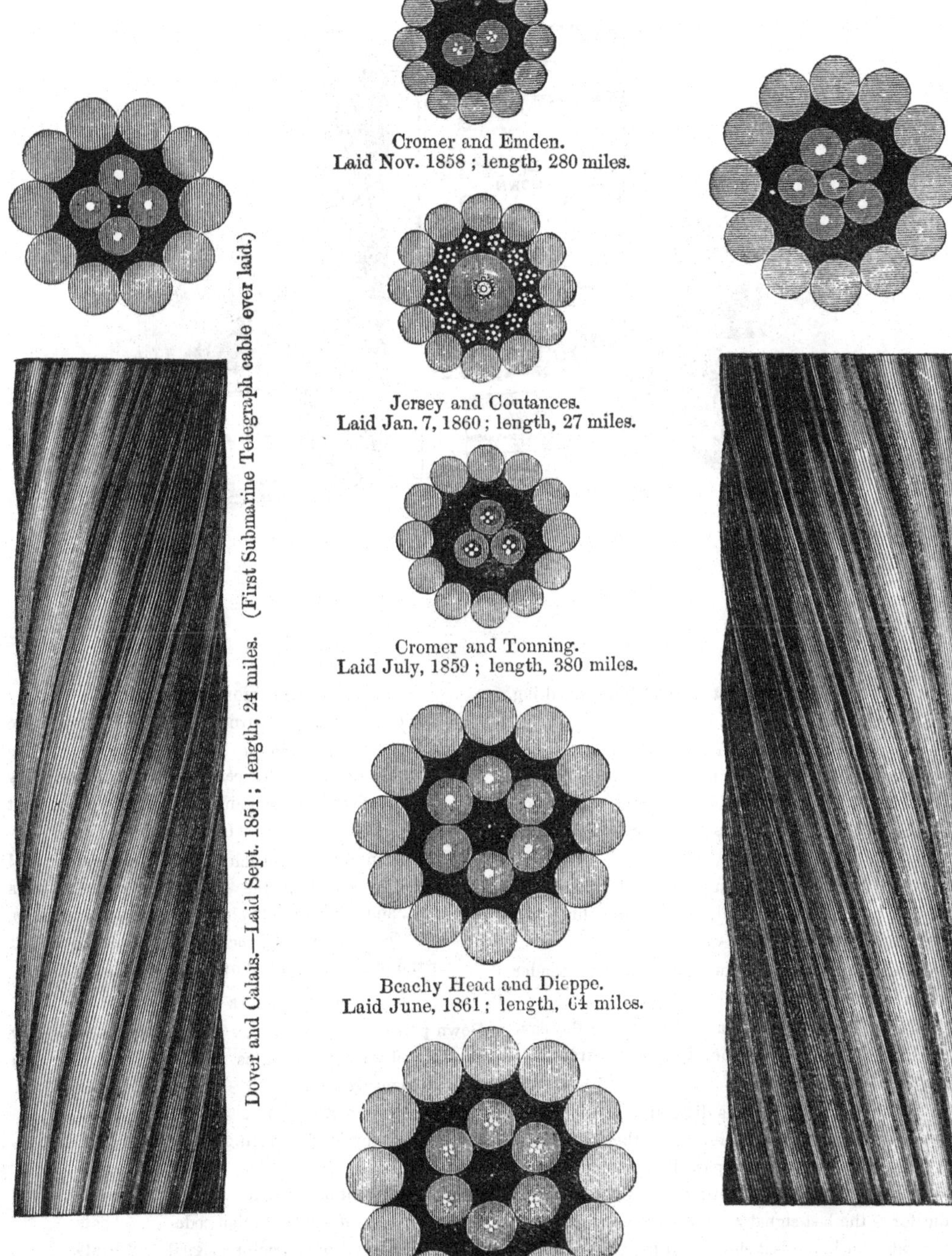

Dover and Calais.—Laid Sept. 1851; length, 24 miles. (First Submarine Telegraph cable ever laid.)

Cromer and Emden.
Laid Nov. 1858; length, 280 miles.

Jersey and Coutances.
Laid Jan. 7, 1860; length, 27 miles.

Cromer and Tonning.
Laid July, 1859; length, 380 miles.

Beachy Head and Dieppe.
Laid June, 1861; length, 64 miles.

Folkestone and Boulogne.
Laid June, 1859; length, 25 miles.

Dover and Ostend.—Laid May, 1853; length, 70 miles. (The Second Submarine cable that was laid.)

Tariff Lists and every information may be obtained at the chief offices, and at all the offices of the London District Telegraph Company.

TYER, EDWARD, 15 *Old Jewry Chambers.*—Train signalling telegraphs and electric telegraphs.

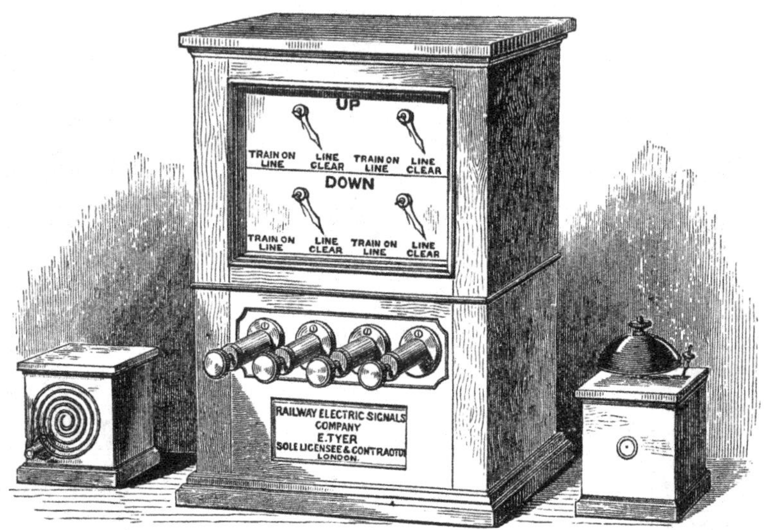

TYER'S TRAIN SIGNALLING TELEGRAPH.

System and description of Tyer's patent train signalling telegraphs :—

1. The line of railway is divided into certain portions or signal stations, and no train is to be permitted to pass one of these signal stations unless a notification of the line being clear has been received from the signal station in advance.
2. Each signal station communicates with the signal station on either side of it, so as to announce the approach and departure of every train.
3. The receiver of a signal cannot alter it; the sender alone is enabled to reverse it.
4. The signal once given remains fixed until the next signal be sent, and can therefore be referred to at any moment.
5. The dial of the instrument is divided into two parts: the UPPER part being for the *up line*, the LOWER part for the *down line;* each part has two needles or indicators, one black the other red. The *black* indicator is the last signal *received* at the station. The *red* indicator is the last signal *sent* from the station.
6. Each instrument is also furnished with a bell and gong.
7. Only two signals are used—"*Train on Line,*" and "*Line Clear.*"

OBSTRUCTION CODE.

8. In the event of any obstruction happening upon the line, the signal "*Down Train on Line,*" or "*Up Train on Line,*" as the case may be, is immediately to be given, accompanied with FIVE distinct BEATS of the "*Bell*" or "*Gong.*"
9. The station receiving such signal of obstruction will reply by sounding the "*Bell*" or "*Gong*" FIVE *times*, and immediately stop any approaching train, until the cause of obstruction has been ascertained or the indicator again shows "*Line Clear.*"

CODE FOR BELLS AND GONGS.

Down passenger trains	2 beats on *gong*.
Down goods or mineral trains . . .	3 beats on *gong*
Up passenger trains	2 beats on *bell.*
Up goods or mineral trains	3 beats on *bell.*
Inspector's signal for testing instruments	6 beats.
Acknowledgment of a signal . . .	1 beat.
Acknowledgment of obstruction code .	5 beats.
Acknowledgment of inspector's signal	6 beats.

10. Should the station to which a signal is sent, *not reply*, it must be *repeated* until such reply *is received.*
11. No signal is to be considered *complete until the reply has been received.*

TYER, EDWARD—*continued.*

TYER'S PATENT DIRECT ACTION PRINTING TELEGRAPH.

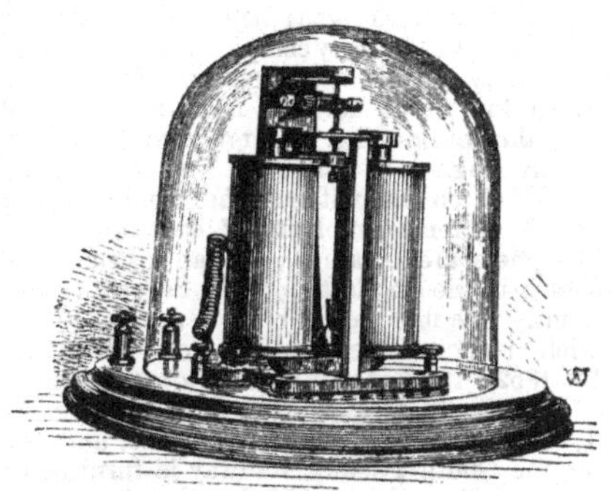

TYER'S PATENT RELAY.

TYER'S PATENT ALPHABETICAL TELEGRAPH.

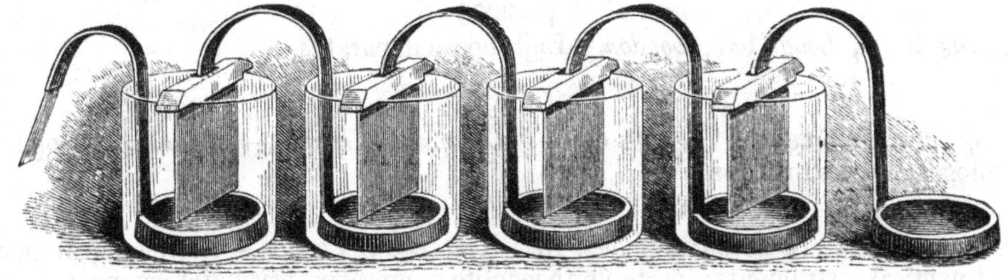

TYER'S PATENT MERCURIAL BATTERY.

[2983]

WALKER, CHARLES V., F.R.S., *Fernside, Red Hill.*—Telegraph instruments for making and for recording train-signals.

A pair of train-signal bells,—index, pecker, and platinized graphite battery complete. Employed for increasing the safety of railway travelling. The speciality of this apparatus (which does not form the subject of a patent) was the introduction of large wire, No. 16 or 18, for the electro-magnets of telegraph instruments, for actual use in direct circuit without the intervention of a relay. Hence extreme simplicity of structure and consequent cheapness are obtained; and an instrument is produced that is in the smallest possible degree liable to derangement. The first pair of bells were erected on the Greenwich viaduct of the South Eastern Railway, on January 31, 1852: the last pair on the Admiralty pier and station of the South Eastern Railway at Dover, on January 1, 1862. Every station, gate-house, and level crossing on the South Eastern Railway is furnished with one or more of these bells, 318 in all. A counting-index is attached to one or more bells, when several are in the same signal-box.

The ordinary train-signals are, one blow, two blows, or three blows; each signal being repeated back in acknowledgment of its receipt: and all signals are booked.

Plain graphite was exhibited in 1851. *Platinized* graphite, in lieu of silver, is now used extensively and with great success.

A lino-scribe, for impressing upon cotton-thread red marks when bell-signals are given; and black marks when they are received; red *and* black to mark intervals of time. Mr. D. McCallum's idea worked into form and realized by the exhibitor.

McCallum's Globotype, constructed so as to drop *red* balls when bell-signals are given, and *black* balls when they are received; *spotted* balls to mark the hours, and *blue* balls the quarters or lesser intervals. The arrangement and details are by the exhibitor.

V-trough for tunnel and subterranean wires, giving a maximum of space and protection with a minimum of material, labour, and waste. Joints of the same; and a set of joint-tools.

[2984]

WALTER, JAMES, 17 *Water Street, Liverpool.*—Barometer and weather indicator. (*See page* 43.)

[2985]

WARNER, JOHN, 72 *Fleet Street, London, E.C.*—Improved apparatus for pictorial illustration; philosophical, mathematical, and optical instruments.

[2986]

WATSON, HENRY, *Newcastle-upon-Tyne.*—Armstrong's hydro-electric machine.

[2987]

WEBB, HENRY, *George Street, Balsall Heath.*—Naturalist and preparer of microscopic objects.

[2988]

WELLS & HALL, *Mansfield Street, Southwark.*—Telegraph conductors, caoutchouc insulation; submarine cables; wires for magnetic coils. (*See page* 44.)

[2989]

WENHAM, F. H., *Effra Vale Lodge, Brixton, S.*—A binocular microscope which may be used as an ordinary instrument.

[2990]

WEST, FRANCIS LINSELL, 31 *Cockspur Street, Charing Cross.*—Self-registering mercurial and standard barometers.

[2991]

WHITEHOUSE, NATHANIEL, 2 *Cranbourne Street.*—English-made opera-glasses, and Dr. Wollaston's spectacles, &c.

[2992]

WILDE, H., *St. Anne's Square, Manchester.*—The globe telegraph, for private telegraphic communication.

[2993]

WILKINS & CO., *Long Acre, London.*—Lighthouse apparatus.

[2994]

WOOD, EDWARD GEORGE, 74 (late of 117) *Cheapside, London.*—Optical, philosophical, and photographic instruments.

[2995]

YEATES, ANDREW, 12 *Brighton Place, New Kent Road, London.*—Astronomical, geodalical, and nautical instruments; portable theodolite; improved prismatic compass.

WALTER, JAMES, 17 *Water Street, Liverpool.*—Barometer and weather indicator.

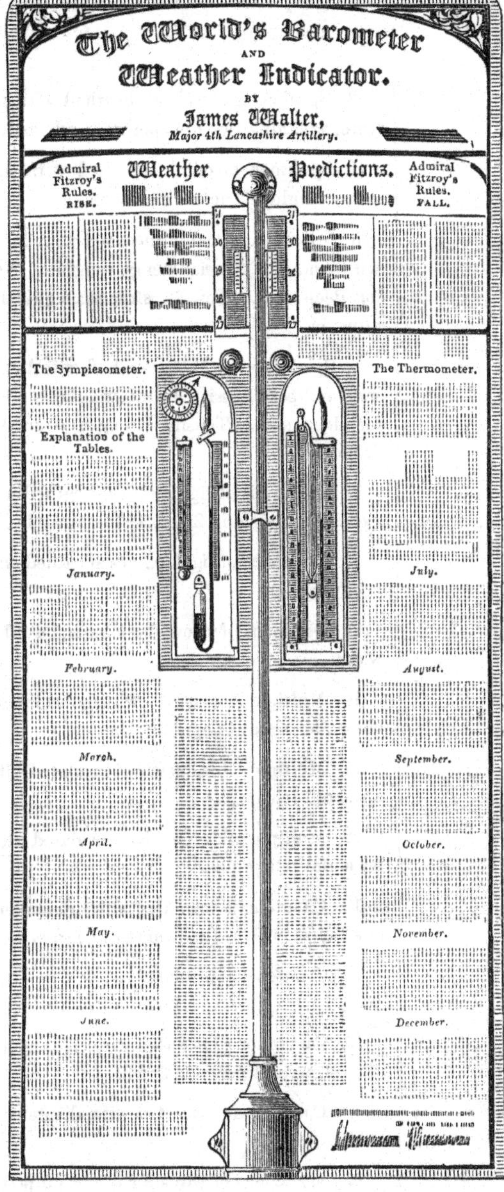

The WORLD'S BAROMETER and WEATHER INDICATOR.

Price (in mahogany frame, with Sympiesometer), five guineas.

This instrument is also made as a marine barometer, in cases specially designed and adapted for nautical purposes, and with suitable fittings for cabins.

Orders to be addressed to Wilson, Son, and Walter, Liverpool, or J. Bowden, 53 Gracechurch Street, London.

[2996]

YEATES & SON, 2 *Grafton Street, Dublin.*—Astronomical, meteorological, philosophical, and mathematical instruments.

Two equatorially mounted TELESCOPES, on iron columns, the mounting possessing many improvements in detail. The clamping circles are quite independent of the divided circle, and at the opposite ends of their respective axes, that of the polar axis being directly under the northern pivot, and that of the declination axis, close to the telescope. The clamping arrangement also differs from that in present use, being more effective and perfectly free from strain or torsion; the smaller stand is particularly adapted for those who have no convenient space to erect an observatory upon; the iron column may be permanently fixed in the open air; the equatorial arrangement packs in a small box, from which it can be lifted into its place on the top of the column in a few moments, and may be thus placed and replaced at the observer's pleasure, without its adjustments being materially affected.

YEATES & SON'S NEW ELLIPTOGRAPH. This instrument differs in three essential points from all elliptographs hitherto constructed. First. There is no limit whatever to the variation in the proportions of the ellipses formed by it. Secondly, the facility of setting it to draw any ellipse whose major and minor axes are known; and thirdly, the accuracy of the figure formed by it in all proportions.

YEATES & SON'S LARGE PUBLIC BAROMETER—dial 3 feet in diameter. This form of instrument was designed by the exhibitors in the year 1858, for the use of agricultural districts and fishing towns, in both of which situations it has been found extremely useful. The index or hand is moved by the mercurial column.

YEATES & SON'S LARGE PUBLIC THERMOMETER—dial 3 feet in diameter. This instrument was designed by them last year (1861). The index or hand is moved by a bulb of mercury one inch in diameter.

WELLS & HALL, *Mansfield Street, Southwark.*—Telegraph conductors, caoutchouc insulation; submarine cables, &c.

SUBMARINE TELEGRAPH CABLES (for deep sea) insulated with caoutchouc, having a specific gravity of 1·35, or heavier; made with Russian hemp in combination with longitudinal steel wires, thereby preventing twisting, kinking, or any perceptible elongation, when strained to, and having a tensile strength equivalent to 11,635 fathoms in sea-water.—(*Government Report on Telegraphy*, pages xxviii & 389.)

The following are exhibited :—

1. Specimen of submarine cable, sheathed with galvanized iron wire, weighing six tons per mile, for shore ends.

2. Specimens of cable sheathed with galvanized iron wire, weighing four tons per mile, for shore ends.

3. Specimens of cable sheathed with galvanized iron wire, weighing two tons per mile, for shallow waters.

4. Specimen of cable sheathed with galvanized iron wire, weighing one ton per mile.

5. Specimen of cable sheathed with longitudinal steel wires and best Russian hemp, weighing ·82 ton per mile, for deep seas. Tensile strength equivalent to 10,840 fathoms.

6. Specimen of cable sheathed with longitudinal steel wire and best Russian hemp, weighing ·72 ton per mile, for deep seas. Tensile strength equivalent to 11,635 fathoms.

7. Specimens of light cables sheathed with galvanized iron wire, weighing 226 & 44 lbs. per mile.

8. Specimens of light cables sheathed with galvanized iron wire.

9. Specimens of caoutchouc or india-rubber, insulated wires.

10. Multiple cable, 10 wires, insulated with caoutchouc, for aërial telegraphs, &c.

11. Multiple cable, 30 wires, insulated with caoutchouc, for aërial telegraphs.

12. Multiple cable, 50 wires insulated with caoutchouc, for aërial telegraphs.

13. Caoutchouc insulated wire; diameter of conductor (7 strand) ·02925, for deep-sea telegraphs.

14. Caoutchouc insulated wire, diameter of conductor ·C3721.

15. Caoutchouc insulated wire, diameter of conductor ·00871, weight per mile 3·5 lbs.

16. Caoutchouc insulated wire, diameter of conductor ·0079.

17. Specimens of wires for target purposes (grouped).

18. Specimen of Swedish wire, diameter ·02564, served with silk.

19. Specimen of Swedish wire, diameter ·0139, served with silk.

20. Specimen of Swedish wire, diameter ·00871, served with silk.

21. Specimen of Swedish wire, diameter ·0079, served with silk.

22. Specimen of Swedish wire, diameter ·0033, served with silk.

23. Specimens of Swedish wires, braided with silk, and containing two wires (diameter ·0079) and upwards.

[2997]

YOUNG, JOHN, Gas Engineer, *Dalkeith.*—Manufactured carbon for electrical batteries, and electrodes for electric lights.

[2998]

INTERNATIONAL DECIMAL ASSOCIATION, PROFESSOR LEVI, *Farrar's Buildings, Temple.*—Illustrations of the decimal and metric system of all nations.

[2999]

FIELD, R., & SON, *New Street, Birmingham.*—Microscopes, telescopes, and surveying instruments.

[3000]

GUTTA PERCHA COMPANY, 18 *Wharf Road, City Road.*—Submarine telegraph cables.

[3001]

HALL, A. J., 2 *William Street, Clerkenwell.*—Machine for describing ellipses and other oval curves.

[3002]

HALLANA, J. V., 22 *New Street, Spring Gardens.*—Steam expansion gauges.

[3003]

HUSBANDS & CLARKE, *Denmark Street, Bristol.*—Optical instruments.

[3004]

MICROSCOPICAL SOCIETY, *London.*—Peters' machine, for microscopic writing.

[3005]

NICHOLL & FOWLER, 16 *Aldersgate Street, E.C.*—Weighing and measuring apparatus.

[3006]

REGISTRAR GENERAL, *Somerset House.*—Tables calculated and stereoglyphed by the Swedish calculating machine.

[3007]

TENNANT, PROF. J., 149 *Strand, W.C.*—Models of crystals, in glass.

[3008]

TREMLETT, R., 7 *Guildford Place, Clerkenwell.*—Barometers and air-pumps.

Class XIV.

PHOTOGRAPHIC APPARATUS AND PHOTOGRAPHY.

[3029]

Adams, A., 26 *Bread Street, Aberdeen.*—Carte de visite, stereoscopic views.

[3030]

Alfieri, C., *Northwood, Hanley, Staffordshire.*—Illustrations of Welsh scenery, &c.; negatives made in field camera.

[3031]

Amateur Photographic Association, 26 *Haymarket, London.*—Photographs by the members of the Association.

[3032]

Angel, O., *High Street, Exeter.*—Photographs, enlarged by the solar camera from collodion negatives.

[3033]

Austen, W., 5 *Buxton Place, Lambeth Road.*—Presses, camera stands, head-dresses, &c.

[3034 [

Barnes, R. F., 64a *New Bond Street.*—Photographs.

[3036]

Bassano, A., 122 *Regent Street, W.*—Coloured crayon, and plain photographic portraits.

[3037]

Beard, R., 31 *King William Street, London Bridge.*—Coloured and plain photographs and microscopic portraits.

[3039]

Bedford, F., 23 *Rochester Road, Camden Road Villas.*—Photographs: landscape and architecture by the wet collodion process.

[3040]

Bennett, A. W., 5 *Bishopsgate Without, London.*—Photographs: application of photography to illustration of books.

Selection from Sedgfield's English Cathedral Views and other scenery for the stereoscope, the scrap-book, and the album—including interiors and exteriors of Beverley, Bristol, Exeter, Winchester, Salisbury, Ely, Norwich, Peterborough, Lincoln, Rochester, Canterbury, and Wells cathedrals, with others.

Price 1*s.* each for the stereoscope, or 6*d.* each for the album. The ruined castles and abbeys of Great Britain (21*s.*).

Specimen of application of Photography to the illustration of books.

[3041]

BIRD, P. H., F.R.C.S., F.L.S., 1 *Norfolk Square, W.*—Photographs of views.

[3042]

BIRNSTINGL, L., & Co., 7 *Coleman Street, E.C.*—Photographs.

[3043]

BLAND & Co., 153 *Fleet Street, London, E.C.*—Photographic cameras, materials, and apparatus.

[3045]

BOOTH, H. C., *Harrogate, Yorkshire.*—Portraits, photographed from life, on paper and ivory, plain and coloured.

[3046]

BOURNE, S., *Moore & Robinson's Bank, Nottingham.*—Photographic landscapes, by the Fothergill dry process.

[3047]

BOURQUIN & Co., 13 *Newman Street, Oxford Street.*—Photographic materials, albums, &c.

The object of BOURQUIN & Co., has always been to manufacture frames suitable to the various kinds of photographic portraits, landscapes, such as passe-partout, gilt frames, fancy frames, fancy cases, show cases, and mounts; the patent mosaic albums, &c. Specimens of these goods are exhibited. The exhibitors are also manufacturers of albumenized paper.

[3048]

BOWERS, H. T., *Gloucester.*—Photographic views, collodion and wax papers, enlarged copy of ancient print, &c.

Views of Southam de la Bere, the seat of the Earl of Ellenborough; also View of Old Chapel, Southam, which has been in ruins 500 years, and now restored by his Lordship; West Window, Gloucester Cathedral;* Views of Gardens, Alton Towers, &c. (wax paper process); Portraits, plain and coloured, in oils and water-colours; Portrait of Bishop Hooper, finished in oils upon the photograph, which is enlarged from a small print: a memorial is being erected on the spot where the Bishop suffered martyrdom (near the cathedral) in 1555. Also photographs of fresco paintings in the interior of Highham Church, near Gloucester, executed by T. G. Parry, Esq.

* Memorial of Bishop Monk. Design, "Doctrine of Baptism."

[3049]

BREESE, C. S., *Acock's Green, near Birmingham.*—Instantaneous transparent stereographs on glass.

[3051]

BROTHERS, A., *St. Ann's Square, Manchester.*—Group finished in water colours; portrait on ivory; portraits untouched.

Group of nine figures finished in water-colours.
Portrait of a lady on ivory.

Portraits (untouched) of members of the British Association.

[3052]

BROWNRIGG, S. W., 7 *Eblana Terrace, Dublin.*—Photographs.

[3053]

BULL, J. T. & G., *Great Queen Street, Lincoln's Inn.*—Photographic profiled accessories, and artistic backgrounds.

[3054]

BURNETT, C. J., 21 *Ainslie Place, Edinburgh.*—Photographic prints with uranium, copper, palladium, platinum, &c.

C. BURNETT's Illustrations of original experimental processes, from British Assciation of 1855, and Royal Scotitish Exhibtion of 1856—7 (and London Exhibition of 1859—60). I. Developments of uranium-prepared papers. 1. By red prussiate of potash (very beautiful red brown print, of March 1855). 2. By yellow prussiate (March (1855). 3 & 4. By nitrate of silver; and 5 & 6 by ammonio-nitrate, with and without gold toning, and with and without iron bath; all of 1855; and fixed in weak ammonia bath. 7. By gold (1855). 8. Same as No. 1, but toned in acid iron bath (R.S.P. Exhibition 1857). 9. By palladium. 10. Same with gold-toning. 11. By palladium with iron bath. 12. By silver, platinum-soda-toned. II. 1. Dark iron-toned copper print, image being *copper* and iron, with ferrocyanogen (R.S.P. Exhibition, 1857). 2. Manganese print. 3. Actions of light on nitro-prusside of sodium, with and without ferric salts. 4 Silver print, toned in alkaline platinum bath.

[3055]

BURTON, J., & PATESON, R., 28 *Avenham Lane, Preston, Lancashire.*—Landscapes and buildings.

[3056]

CADE, R., 10 *Orwell Place, Ipswich.*—Machinery and architecture illustrated; also views and portraiture.

[3057]

CAMPBELL, D., *Cromwell Place, Ayr.*—Large views: Land of Burns.

[3058]

CAITHNESS, EARL OF, 17 *Hill Street, W.*—Photographic views.

[3060]

CLAUDET, A., 107 *Regent Street.*—Photographic portraits; stereoscopic and visiting-cards, enlarged to the natural size. (*See pages* 50 & 51.)

[3061]

COLNAGHI, P. & D., SCOTT, & Co., 13 *and* 14 *Pall Mall East.*—Photographs from ancient and modern pictures, portraits, &c.

[3062]

CONTENCIN, J., 4 *White Cottages, Grosvenor Street, Camberwell.*—Various photographs from drawings, &c.

[3063]

CORDINGLEY, W., 14 *Wells Street, St. Helen's, Ipswich.*—Camera stand.

[3064]

COX, F. J., 22 *Skinner Street, London.*—Lenses, cameras, portable field apparatus, and instantaneous shutters.

CAMERA SHIELD for producing four carte de visite or sixteen medallion portraits on one plate and with one lens. The dark slide revolves around the axis of the lens, but the plate is always kept vertical, therefore the drainage from flowing back over the surface is prevented.

Specimens taken by the camera shield.

STEREOSCOPIC CAMERA, with three double backs, holding six plates. The weight is 3¾lbs.; it requires no unpacking, and has no loose pieces liable to be lost in use.

STEREOSCOPIC CAMERA, fitted with instantaneous shutter affixed to the central diaphragm.

CENTRAL DIAPHRAGM of a lens, showing an instantaneous shutter working without vibration. It can be opened or closed with any degree of rapidity desired.

REFLECTING STEREOSCOPE for pictures four inches square.

Field box containing chemicals for a day's use; it also forms a developing box.

PORTRAIT CAMERA, and lens with swing back.

[3065]

CRAMB BROTHERS, *Glasgow.*—Photographs on ivory; views in Palestine; half life-size portraits, not enlarged.

[3066]

CRITCHETT, C., 11 *Woburn Square.*—Photographs.

CLAUDET, A., 107 *Regent Street.*—Photographic portraits; stereoscopic and visiting-cards, enlarged to the natural size.

Mr. Claudet has obtained the following Medals:—

Council Medal, Great Exhibition of London, 1851; *First Class Medal, Great Exhibition of Paris,* 1855; *Silver Medal, Exhibition of Amsterdam,* 1855; *Bronze Medal, Exhibition of Bruxelles,* 1856; *Silver Medal, Photographic Exhibition of Scotland,* 1860; *Silver Medal, Photographic Exhibition of Birmingham,* 1861; *Society of Arts Medal presented by Prince Albert in* 1853, *for Mr. Claudet's paper on the stereoscope and its application to photograhy.*

Abstract of an address read the 9th April, 1861, before the Photographic Society of Scotland, by Sir David Brewster, K.H., F.R.S.L. and E., president, in presenting Mr. Claudet with the medal of the Society for the best portrait exhibited at their photographic exhibition of 1861. *See 'Journal of the Photographic Society' for May,* 1861, *page* 181.

"In awarding a medal for the best work of art in a photographic exhibition, it is not often, if it has ever occurred at all, that the successful competitor is distinguished by his discoveries and inventions in the art which he practises.

"These different accomplishments, however, are possessed by Mr. Claudet, to whom the Council have adjudged our silver medal for the best portrait in the Exhibition; and though they were not taken into account, yet I feel it a duty, as well as a privilege, in presenting this medal to Mr. Claudet, to lay before you a brief notice of those discoveries and inventions by which he has achieved the high reputation that he so justly enjoys in the photographic world.

"When an art has arrived at such a degree of perfection, it is a useful as well as an agreeable task to retrace the steps by which it advanced, and to do honour to the men to whom we owe them.

"When Sennebier found that muriate of silver was darkened by light, and became darker in the *violet* than in the *red* rays, he made the first step in photography. A German chemist, M. Ritter, advanced another step when he found that the muriate of silver was most powerfully blackened in the invisible rays beyond the violet; but it is to Mr. Thomas Wedgwood that we owe the application of these facts to photographic purposes, but he failed in every attempt to prevent the white parts of his pictures from being blackened by light. The experiments of Wedgwood seem to have been unknown in France, when two ingenious individuals, M. Nièpce and M. Daguerre, discovered two entirely different processes of fixing photographic pictures. The result of their joint labours was the daguerreotype, that beautiful art with which we are all acquainted, and which Mr. Claudet has done more than any other individual to bring to its present state of perfection.

"In the time of Daguerre, from 20 to 25 minutes were required to take the photograph of a landscape by this process, and nearly 10 minutes to take a portrait. In this imperfect state of the art Mr. Claudet discovered, and communicated to the Royal Society of London, in 1841, an easy and certain method of accelerating the action of light upon the film of iodine, and thus greatly shortening the process; by this means he obtained in 10 seconds pictures which would have required 4 or 5 minutes by the preparation of Daguerre. So sensitive, indeed, was this new process, that Mr. Claudet was enabled to take portraits by the oxyhydrogen light in 15 or 20 seconds, with an object-glass of short focus. He obtained, also, impressions of black lace by the light of the full moon in 2 minutes, and by the light of the stars in 15 minutes. He likewise obtained in 4 seconds an image of the moon, in which the shadowed parts were visible. In 15 minutes he obtained the image of an alabaster figure by the light of a candle, and in 5 minutes a similar image from an argand lamp.

"Next in importance to the acceleration of the photographic process, is the perfection of the image which is formed upon the sensitive plate—not of the visible image which is received and seen upon the gray glass, but of the invisible image, which is formed by the photogenic or chemical rays. In studying the subject, Mr. Claudet discovered that the chemical and visual foci were not coincident. He recommended that the rays of the photogenic spectrum should be united in one focus, even at the sacrifice of the achromatism of the lens. 'As the photogenic focus will change its place with the colour and the intensity of the light and with the distance of the object, the photographer should determine experimentally its place in relation to these varying influences.' In order to do this Mr. Claudet invented the *Focimeter*, an instrument 'for finding the difference between the two foci.'

"In the year 1847 Mr. Claudet communicated to the Royal Society an important paper 'On different Properties of Solar Radiation in Photographic operations;' and in 1848 he submitted to the Academy of Sciences, in Paris, his interesting 'Researches on the Theory of the principal Phenomena of Photography in the Daguerreotype Process,' in which he describes a new and ingenious instrument, which he calls a *Photographometer*, from its measuring the intensity of the photogenic rays, and comparing the degree of sensitiveness of various preparations.

"Another instrument of Mr. Claudet's invention is the *Dynactinometer*, for comparing the photogenic power of object-glasses, and measuring the intensity of photogenic light.

"In 1853 Mr. Claudet communicated to the Society of Arts in London a valuable paper 'On the application of the Stereoscope to Photography.' This paper was thought worthy of the Society's Medal, which was presented to him by Prince Albert.

"In a brief sketch like this of the labours of Mr. Claudet, it would be impossible to give an intelligible account of the various improvements which he has made in photography, and of the numerous papers which he communicated to the British Association at eleven of its meetings, between 1849 and 1861, and which have been published in their annual Reports and in other periodical works. There is one of his inventions, however, so remarkable that we cannot pass it by without a special notice.

"In 1857 Mr. Claudet communicated to the Royal Society a paper 'On the Phenomenon of Relief of the Image formed on the Ground Glass of the Camera Obscura;' and in the following year he described a new instrument, called a *Stereomonoscope*, founded on the principles described in that paper, and exhibiting in

CLAUDET, A.—*continued.*

relief a single image consisting of two stereoscopic images combined.

"But while Mr. Claudet has devoted so much of his leisure to the theory of photography, to the improvement of its processes, and the invention of instruments for aiding the photographer in the practice of his art, he has himself produced works of singular beauty, and placed himself at the head of British artists in the department of photographic portraiture. In proof of this we have only to state that the jury of the Great Exhibition of 1851, of which I had the honour to be chairman, awarded to him their Council Medal for the works which he then exhibited, and that the Jury of the Paris Universal Exhibition of 1855 voted to him their medal of the First Class.

"In this brief and imperfect notice of Mr. Claudet's labours I have referred only to the Daguerreotype process; but while Niepce and Daguerre were privately engaged in perfecting the art of photography on metal, Mr. Talbot was occupied with the sister art upon paper. These two rival arts long struggled for the pre-eminence; but since the discovery of collodion by Mr. Archer, the Talbotype promises to supersede the Daguerreotype, and to become the true type of the photographic art.

"Though long occupied with the practice of the daguerreotype, Mr. Claudet has pursued the Talbotype with equal success; and it is for one of his portraits, taken upon paper, that the Medal of this Society has been adjudged to him.

"I am sure that the distinguished artists with whose works those of Mr. Claudet have come into competition will not disapprove of the decision of the Council, and will welcome the distinguished stranger, who does the Society honour by his presence, and who has made such valuable contributions to its transactions."

[3067]

CRUTTENDEN, J., *Week Street, Maidstone.*—Photographs.

[3068]

CUNDALL, DOWNES, & Co., 168 *New Bond Street, and* 10 *Bedford Place, Kensington.*—Photographs from nature and from drawings.

MESSRS. CUNDALL, DOWNES, & Co., undertake to copy pictures and other works of art, maps, engineering plans and drawings, in any size from 5 in. by 4 in. up to 30 in. by 24 in., and to reproduce photographs and daguerreotypes; they are also prepared to photograph country houses, interiors of mansions, churches, works in progress, &c., and to perform every other description of work of which photography is capable.

Portraits are taken daily on the following terms:—

	£	s.	d.
Portrait (visite) and six copies	0	10	6
Portrait (visite) and twenty copies	1	1	0
Six extra copies	0	6	0
Portrait (visite) and six copies tinted	1	1	0
Portrait (visite) and twelve copies tinted	1	15	0
Portrait (visite) and one copy fully coloured	0	10	6
Portrait (visite) and three copies fully coloured	1	1	0

Terms for larger portraits and for any of the above work can be had on application.

[3069]

DALLMEYER, J. H., 19 *Bloomsbury Street, W.C.*—Photographic lenses, cameras, apparatus, &c.

[3070]

DANCER, J. B., 43 *Cross Street, Manchester.*—Microscopic photographs; landscapes and portraits.

The exhibitor can supply the trade and the public with a variety of minute photographs for the microscope; dissolving-view lanterns, with chromatic lenses, for photographic transparencies, and a variety of photographic views for the lantern; and also with a new form of the dissolving-view apparatus, for the use of schools. The lanterns with achromatic lenses were originally made by Mr. Dancer for the Manchester Mechanics' Institute.

[3071]

DAVIS, T. S., 3 *Stanley Terrace, Stockwell, S.*—Photographic manipulating camera.

[3072]

DOLAMORE & BULLOCK, 30 *Regent Street, Waterloo Place, S.W.*—Photographs.

ARCHITECTURAL AND LANDSCAPE PHOTOGRAPHY. Engineering and other works in progress.

"The Blind Beggar," from Dyckman's picture in the National Gallery.

"Early Flowers," a companion, from the original, by W. Maw Egley.

Price 10*s.* 6*d.* each.

[3073]

EASTHAM, J., 22 *St. Anne's Square, Manchester.*—French and English Treaty of Commerce, opal portraits.

[3074]

FENTON, R., 2 *Albert Terrace.*—Photographs.

[3075]

FIELD, J., *Dornden, Tonbridge Wells.*—Specimens of photolithography; plates engraved on stone by the sun.

[3077]

GANDY, T., 40 *South Street, Grosvenor Square.*—Portraits.

[3078]

GORDON, R. M., 38 *Alpha Road, St. John's Wood.*—Photographs of Madeira.

[3079]

GORDON, R., *Bembridge, Isle of Wight.*—Isle of Wight scenery.

[3080]

GRAHAM, J., *Surrey Lodge, Lambeth.*—Photographic panoramic views of Jerusalem, Syria, Naples, and Pompeii.

[3081]

GREEN, B. R., 41 *Fitzroy Square.*—Coloured photographs.

[3083]

GRISDALE, J. E., 73 *Oxford Street, W.*—Photographic camera.

[3084]

GUSH & FERGUSON, 179 *Regent Street.*—Photographic miniatures, collodion process.

[3085]

HAMILTON, A. R., *Maple Road, Surbiton, S.W.*—Photographs of the Waterloo medal, by B. Pistrucci.

[3086]

HARE, G., 140 *Pentonville Road, N.*—Photographic portrait, landscape, stereoscopic, and carte de visite cameras.

Portrait and Landscape, Stereoscopic, Carte de visite, and every other form of camera are made to order in the best possible manner, and with all the latest improvements, by the exhibitor.

[3087]

HARMER, R., 131 *Shoreditch.*—Photographs illustrating a new method of printing, adapted for book illustration.

[3088]

HART, F. W., 13 *Newman Street, Oxford Street, London.*—Life-size photograph by exhibitor's apparatus.

The following are exhibited:—

Photograph, enlarged, by F. W. Hart's Patent Instruments, complete, with lens and dark chamber, from 8*l* 10*s*. to 25*l*.

Also F. W. Hart's Patented Printing Frames for light and shade backgrounds: from 1*l*. 1*s*. to 2*l*. 2*s*.

Exhibited on Messrs. Bourquin & Co.'s stand. Class 14.

[3090]

HEATH & BEAU, 283 *Regent Street, W.*—Miniatures and photographs.

[3091]

HEATH, VERNON, 43 *Piccadilly.*—Various portraits, English and Scottish landscapes.

[3092]

HEMPHILL, W. D., M.D., *Clonmel.*—Photographs of antiquities, &c., at Cashel and Cahir, Co. Tipperary, Ireland.

[3093]

HENNAH, T. H., 108 *King's Road, Brighton.*—Collodion photographs.

[3094]

HERING, H., 137 *Regent Street, London.*—Frames of plain and coloured photographs, portraits and views.

[3095]

HIGHLEY, SAMUEL, Philosophical Instrument Maker, and Private Teacher of Photography, &c., 70 *Dean Street, Soho, London, W.*—Photographic apparatus.

Operators' actinometer, for ascertaining the amount of exposure necessary at a given time, with any light, by any lens and process. A photo-micrographic camera, with appliances for taking photographs of mounted or living microscopic objects. Photographers' travelling lamp. Highley's dropping bottle for nitrate of silver solution. Improved pneumatic plate-holder. Levelling stand. Operating room camera stand. Photo-micrographs, and photographs of scientific and artistic subjects for the magic lantern. Books illustrated by photography.

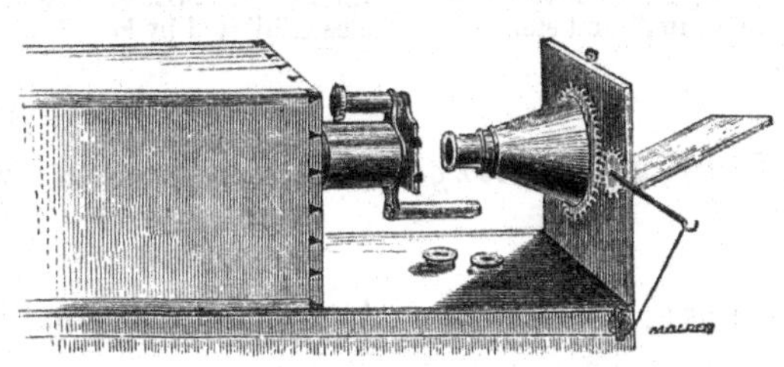

[3096]

HILL, D. O., *Edinburgh.*—Photographs.

[3097]

HOCKIN & WILSON, 38 *Duke Street, Manchester Square, W.*—Photographic set, and tent; collodion, &c., in hermetically sealed tubes.

[3098]

HOLDEN, REV. DR., *Durham.*—Photographs of cathedrals and abbeys.

[3099]

HOPKIN & WILLIAMS, 5 *New Cavendish Street.*—Photographic chemicals.

[3100]

HORNE & THORNETHWAITE, 123 *Newgate Street.*—Photographic lenses, cameras, apparatus, and chemicals.

[3101]

JAMES, COLONEL SIR H., R.E., *Ordnance Survey Office, Southampton.*—Plans reduced by photography, photozincographs, and photopapyrographs.

[3102]

JEFFREY, W., 114 *Great Russell Street, Bloomsbury, W.C.*—Photographs from busts of Alfred Tennyson, William Fairbairn, &c.

[3103]

JEFFERY, W. (SHEPHERD & Co.), 97 *Farringdon Street, E.C.*—Photographic tent, 14 lbs. weight.

[3104]

JONES, B., *Selkirk Villa, Cheltenham.*—Photographic pictures from glass negatives.

[3105]

JOUBERT, F., 36 *Porchester Terrace, W.*—Photographs in vitrifiable colour, burnt in on glass; collodion photographs, and phototypes.

[3106]

KATER, E., 46 *Sussex Gardens, Hyde Park.*—Ancient armour from Mr. Meyrick's collection.

[3107]

KEENE, R., *All Saints, Derby.*—Photographs illustrating scenery and antiquities of Derbyshire.

[3108]

KILBURN, W. E., 222 *Regent Street.*—Photographic portraits.

[3109]

KING, H. N., 42½ *Milsom Street, Bath.*—Cartes de visite; portraits of celebrities; views and stereoscopic slides.

Duplicates of the large portraits of celebrities, "Carte de Visite," and stereoscope slides exhibited by Mr. H. N. King, may be had, post free, from the Exhibitor, or from Mr. Poulton, 352 Strand, London.

[3110]

LAMB, J., 191 *George Street, Aberdeen.*—Views or portraits, or both.

[3111]

LEAKE, J. C., *Poplar, London.*—Photographic operating tent.

[3113]

LICKLEY, A., *Allhallowgate, Ripon, Yorkshire.*—Camera, with shade and shutter; positive collodion photographs.

[3115]

LOCK & WHITFIELD, 178 *Regent Street.*—Photographic miniatures.

This collection of miniatures includes portraits of the Countess Spencer, Viscountess Clifden, Lady Raglan, Lady Burghley, Duke of Sutherland, Earl of Malmesbury, Earl of Harewood, Earl Strathmore, Lord Burghley, Lord Dunkellin, Lord Colville, &c., &c.

These portraits are painted in water-colours (and not in oil or any other opaque medium), thus retaining all the correct drawing and delicacy of modelling which a good photograph possesses, combined with the transparency and finish of the best ivory miniatures.

[3116]

LONDON SCHOOL OF PHOTOGRAPHY, 103 *Newgate Street.*—The collodion knapsack, for out-door photography; illustrations of the applications of photography.

[3117]

LONDON STEREOSCOPIC COMPANY, 54 *Cheapside, E.C.*—Instantaneous stereoscopic views, large views, and portraits.

[3118]

MACDONALD, SIR A. K., BART., *Woolmer, Liphook, Hants.*—Photographic views.

[3119]

MACKENZIE, W., *Paternoster Row.*—Photographic illustrations for the Queen's Bible, by Frith.

A specimen of this superb edition of the Scriptures is exhibited in Class 28. One hundred and seventy copies only will be printed. The list of subscribers will be printed in the order received.

[3120]

M'LEAN, MELHUISH, & HAES, 26 *Haymarket.*—Photographic apparatus. (*See page* 55.)

[3121]

MARRIOTT, M., *Montpelier Square, London, S.W.*—Panoramic camera; portable stereoscopic cameras for dry processes.

[3122]

MAULL & POLYBLANK, 187A *Piccadilly.*—Photographs.

M'LEAN, MELHUISH, & HAES, 26 *Haymarket.*—Photographic apparatus; untouched and coloured photographs; universal objectives; the simultaneous camera, and highly-finished photographic miniatures.

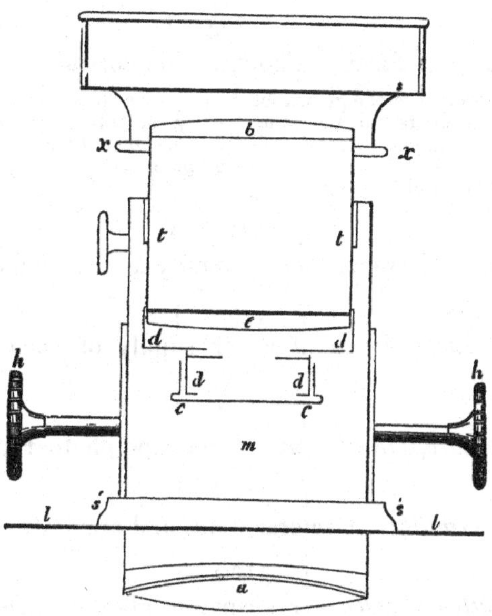

OPINIONS OF THE PRESS.

[*Reprinted from* "THE PHOTOGRAPHIC NEWS," *July* 13*th* 1860.]

These lenses have been constructed by a photographer of long practical experience, with the view of obviating the inconveniences presented by other lenses, and supplying their deficiencies. They aim at universal applicability; and it is believed that this desirable result is obtained by the peculiar construction and arrangement adopted. In the various combinations, each affords a picture equally well defined, from the centre to the margin, without any other advantages whatever being sacrificed. These results are obtained without any increase of complexity. there are no extra portions to add to the necessary incumbrances of the photographer's baggage, or to perplex and baffle him by their loss. The greatest possible simplicity, combined with universal applicability to the various requirements of the photographer, both in portraiture and landscape, were sought by the inventor in these combinations, and it is hoped they have been successfully attained.

Description.—The front lens, *b*, is of the ordinary size, but the back lens, *a*, is one size larger, thus enabling the operator to throw an equal body of light over the entire surface of the plate, and to take in a larger angle.

The diaphragm may be changed without the trouble of unscrewing any portion of the arrangement, by merely withdrawing the tube, *t*, thus the disadvantage attendant upon the cut in the middle of the brass mounting, as in Waterhouse's stop, is avoided.

The lens has a double pinion to the rack, *h*, *h*, so that either hand may be employed in focussing.

The whole may be packed as compactly and in as small a compass as other compound objectives, as shown in the above diagram.

Combination 1.—The same as shown in the figure, only with the diaphragm portion of cap, *d*, *c*, removed. It forms a portrait lens of short focus, and with the diaphragm, *d*, *in situ*, is suited for enlarging and copying.

Combination 2.—As a portrait lens, adapted for pictures up to 6½ by 4¾. It is produced by removing *c d*, and extra lens *e*.

Combination 3.—Is a portrait lens, of long focus, obtained by unscrewing the shade from *x x*, and unscrewing the whole of the brass mounting from the flange *l l*, then turning it round, and screwing *x x* into the flange *l l*. The lens *a* is removed, and the shade screwed at *s' s'*. Will take a portrait 8½ by 6½.

Combination 4.—Is a landscape lens, obtained by unscrewing the shade from *x*, and the brass mounting from the flange *l*. Then withdraw the tube *t t* from the mounting *m*, and remove the extra lens *e*, but retaining the diaphragm *d*, and the small cap *c*. Screw *x x* into the flange *l l*. Yields a picture up to 10 by 8.

[*Reprinted from* "THE PHOTOGRAPHIC NEWS," *July* 20*th*, 1860.]

IN our notice of Mr. Melhuish's new lenses, of last week, we omitted to mention that the lens described was the ½-plate size; and also to add that Combination 3 has great depth of focus, and is admirably suited for interiors.

[3123]

MAYALL, J. E., 226 *Regent Street.*—Portraits of eminent personages, studies from life; a crayon machine and daguerreotypes.

[3124]

MAYER BROTHERS, 133 *Regent Street.*—Photographic portraits.

The price of the ALBUM PORTRAITS is one guinea for twenty-five copies. Extra copies ordered at the same time are charged 1s. each. No fresh order will be executed for fewer than twelve.

MESSRS. MAYER execute photographic portraits of every size, plain and coloured. They have every facility for taking groups—such as schools, corporations, &c. At their Brompton establishment they have photographic rooms on the ground-floor.

[3125]

MAYLAND, W., *Cambridge.*—Views of the University and its vicinity.

[3126]

MOENS, W. J. C., *Lewisham.*—Views of water supply of ancient Carthage; temples in Greece, and others.

[3127]

MUDD, J., 10 *St. Ann's Square, Manchester.*—Landscape photographs.

[3128]

MURRAY & HEATH, 43 *Piccadilly.*—Cameras, tent, and apparatus.

[3129]

NEGRETTI & ZAMBRA, *Hatton Garden.*—Transparent glass pictures.

[3130]

NEWCOMBE, C. T., 135 *Fenchurch Street, E.C.*—Photographs.

[3131]

NICHOLSON, A., 23 *St. Augustine Road, Camden Town.*—Photographs from plates prepared by Fothergill's process.

[3132]

OLLEY, W. H., 2 *Bolingbroke Terrace, Stoke Newington.*—Photographs from the microscope, by reflecting process.

[3133]

OTTEWILL, T., & Co., *Charlotte Terrace, Islington.*—Photographic apparatus.

[3134]

PENNY, G. S., 14 *Rodney Terrace, Cheltenham.*—Photographs by various processes.

[3135]

PIPER, J. D., *Ipswich.*—Landscapes, &c., by collodion process.

[3136]

PONTING, T. C., 32 *High Street, Bristol.*—Photographs enlarged from small negatives; iodized negative collodion, sensitive for years.

[3138]

POULTON, S., 352 *Strand, W.C.*—Stereoscopic slides. Photographs, untouched and coloured.

[3139]

POUNCY, J., *Dorchester, Dorset.*—Photographs printed in carbon.

[3140]

PRETSCH, P., 3 *Guildford Place, Foundling.*—Printed plates and blocks, produced by photography and electrotype only; photographic engraving and printing with ordinary printers' ink.

[3141]

PROUT, V., 15 *Baker Street, Portman Square.*—Reproductions of pictures—various subjects.

[3142]

PYNE, J. B., Jun., 40 *Roxburgh Terrace, Haverstock Hill, N.W.*—Photographic copies of pictures, sculpture, portraits from life, &c.

[3143]

RAMAGE, J., *Edinburgh.*—Specimens of photolithography.

[3144]

REEVES, A., 257 *Tottenham Court Road, London.*—Microscopic photographs and microscope.

[3145]

REJLANDER, O. G., 42 *Darlington Street, Wolverhampton.*—Various photographs.

[3146]

RICHARDSON, T. W., *Brede, Sussex, and Staplehurst.*—A reflecting camera.

[3147]

ROBINSON, H. P., 15 *Upper Parade, Leamington.*—Photographs.

[*Obtained the Silver Medal of the Photographic Society of Scotland,* 1860; *the Special Medal of the same Society,* 1861; *a Silver Medal of the Royal Cornwall Polytechnic Society,* 1861; *a Medal of the Birmingham Photographic Society,* 1861; *and Honourable Mention of the Belgium Industrial Exhibition,* 1861.]

A Holiday in the Woods.
The Lady of Shalott (*vide* Tennyson).
Fading away.
The Top of the Hill.
"Here they come!"
Early Spring.
Elaine with the Shield of Lancelot.
Little Red Riding Hood and the Wolf.
"She never told her love."
Album Studies.

[3148]

ROSS & THOMSON, 90 *Princes Street, Edinburgh.*—Photographs by the collodion process.

[3149]

ROSS, T. (only son and successor to the late Andrew Ross), Manufacturer of optical instruments, 2 *and* 3 *Featherstone Buildings, Holborn.*—Photographic lenses, cameras, stands, and apparatus. (*See pages* 58 & 59.)

[3150]

ROUCH, W. W. (formerly Burfield & Rouch), 180 *Strand.*—Apparatus and chemicals; photographs, taken with new binocular camera and Hardwich's bromo-iodized collodion. (*See page* 60.)

[3151]

RUSSELL, J., *East Street, Chichester.*—Ruins of Chichester Cathedral after the fall of the spire.

[3152]

SHEPHERD & Co., 97 *Farringdon Street.*—Cameras, lenses, &c.

The Portrait Lenses combine great rapidity with extreme uniformity of sharpness over the whole field, and are therefore particularly adapted for instantaneous pictures (the leading feature in the photographic art).—Price from 1*l.* 1*s.* to 20*l.*

The Landscape Lenses are remarkable for giving flatness of field and straight marginal lines, as well as rapidity of action. Price from 12*s.* to 12*l.*

Besides superiority of quality, they are the cheapest lenses in the market.

[3153]

SIDEBOTTOM, J., 19 *George Street, Manchester.*—Photographic landscapes, by the collodio-albumen process.

[3154]

SIMPSON, H., 1 *Saville Place, Lambeth.*—Photographic cabinets, forming complete operating rooms.

The PHOTOGRAPHIC CABINET when extended forms a complete operating room. When closed it has the appearance of an ordinary closet or wardrobe, and will contain all chemicals, cameras, portable tent, and other apparatus, thus avoiding all photographic litter. It can be extended in one minute, and will then form a dark chamber about four feet square, fitted with sink, drawers, and other conveniences. The sink, drawers, &c., shift, so that light may be admitted from left or right. White light can be admitted at pleasure. There is no combustion inside chamber, and abundant ventilation without draught. No dust can arise from curtains, as they are entirely superseded by india-rubber springs.

For convenience of carriage it may be constructed in parts and fitted with screws, and carefully marked, so that with the printed directions any intelligent youth may put it together. It may be had painted or unpainted, so that it may be coloured the same as the staircase or furniture. Price, unpainted, 6*l.* 10*s.*; painted, 8*l.* 10*s.*

Various modifications of the above from 3*l.* to 3*l.* 15*s.*

ROSS, T. (only son and successor to the late Andrew Ross), Manufacturer of optical instruments, 2 *and* 3 *Featherstone Buildings, Holborn.*—Photographic lenses, cameras, stands, and apparatus. (*See also* CLASS XIII.)

PANORAMIC LENSES and apparatus for pictures, including an angle of upwards of 100°. (*For the angular extent of pictures taken by the panoramic and the ordinary Landscape Lenses respectively, see illustrations.*)

	£	s.	d.
Panoramic Lens for pictures $10\frac{1}{2} \times 5$; camera with screw adjustment, plate-holder, and screen; printing press; gutta percha bath and dipper; frames for holding glass while cleaning; box for one dozen curved glasses, and tripod stand, in varnished pine case	22	0	0
Panoramic Lens for pictures, for $14 \times 6\frac{1}{2}$, with apparatus complete . . .	28	0	0
Ditto, ditto, for 17×8, with apparatus complete	38	0	0

Stereoscopic and other sizes.

LENSES.

Portrait lenses, with No. 1 and No. 2 back combinations.

Portrait lenses of extra large apertures for producing pictures of children and animals in dull weather.

Cartes de visite Lenses, adapted for operating rooms of any length.

Cartes de visite Lenses, extra quick acting.

Single combination landscape lenses, of the very best construction.

Petzval's orthographic lenses, for groups, views, and architectural subjects.

Triplet lenses, for views, architecture, and copying, giving straight marginal lines and absolute flatness of field.

Stereographic portrait, landscape, orthographic, and triplet lenses.

Focussing eye-piece, for observing the image on the grayed glass screen.

VIEW IN JERSEY.—TAKEN WITH PANORAMIC LENS, 5 INCHES FOCUS.

Ross, T.—*continued.*

VIEW IN JERSEY.—TAKEN WITH ORDINARY LENS, 5 INCHES FOCUS.

CAMERAS.

Square mahogany sliding trunk cameras.

Ditto, ditto, with swing backs, screw and rack adjustments, &c.

Plain folding cameras.

Folding and sliding trunk cameras.

Ditto, ditto, with screw adjustment.

Improved Kinnear's portable camera.

Universal portable bellows camera, with swing back and screw adjustment, on tripod stand.

Binocular and multiple cameras, of various constructions, with rising fronts, rising and dividing fronts, swing backs, screw and rack adjustments, and instantaneous shutters for stereoscopic and cartes de visite pictures.

Latimer Clarke's stereoscopic camera.

Photographic apparatus, cases of chemicals, &c.

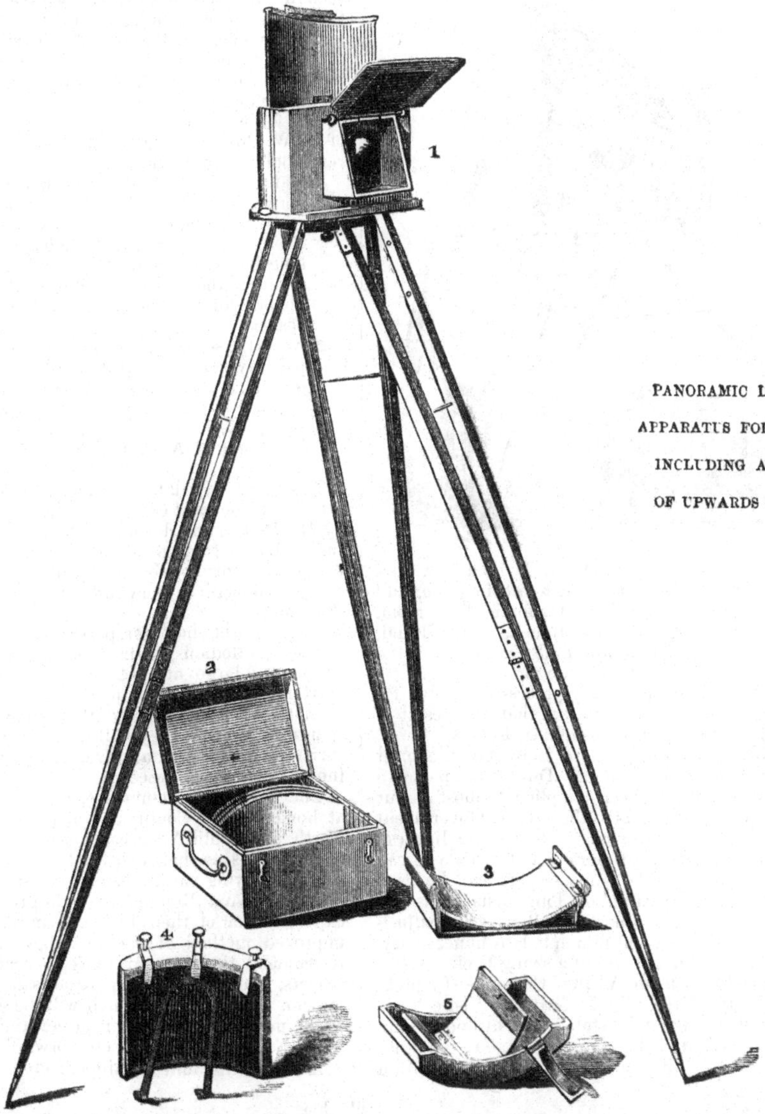

PANORAMIC LENS AND APPARATUS FOR PICTURES INCLUDING AN ANGLE OF UPWARDS OF 100°.

ROUCH, W. W. (late Burfield & Rouch), 180 *Strand, London.*—Apparatus and chemicals; photographs, taken with new binocular camera and Hardwich's bromo-iodized collodion.

From the earliest days of photographic art, the attention of this firm has been directed to the study of the various appliances required for its practice. They have introduced many of the most important improvements in apparatus, and have earned a well-merited celebrity for chemical preparations, which are unsurpassed by those of any other maker. The whole of their instruments are manufactured on their own premises, by their own workmen, and under the immediate superintendence of Mr. Rouch. They are constructed with mathematical precision, and the materials employed are the best procurable, and are carefully selected to withstand the effects of change of temperature and climate. These conditions are not made a pretext for excessive charges, as is frequently the case. The prices affixed to all their goods are strictly moderate, and as low as is compatible with first-class workmanship and material. The best proof that can be given of these statements is the well-known fact, that they have supplied many of the most important sets of apparatus yet constructed; that their apparatus has been sent to nearly every part of the world; and that in no instance has complaint of bad construction ever reached them.

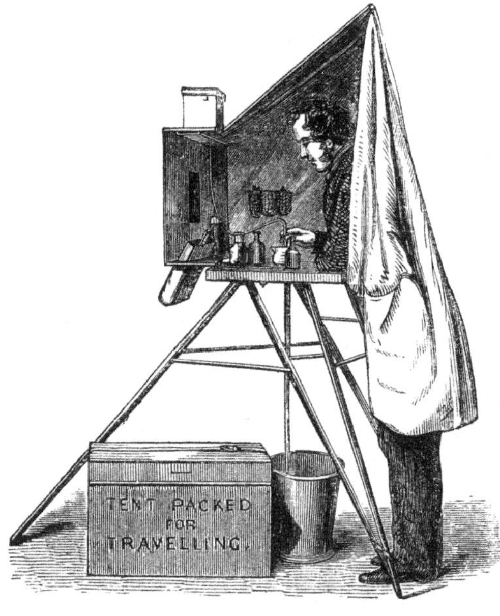

Edwards's Registered New Model Tent, a new and admirably contrived dark tent, moderate in price, and far superior to any other substitute for a dark room. Manufactured only by W. W. Rouch, 180 Strand, London. By means of this admirable contrivance, the trouble of working the wet collodion process is greatly reduced,—in fact many of the dry processes involve far more trouble, and it is well known that the result is greatly inferior. See descriptions affixed to tent.

Rouch's New Universal Camera is by far the most suitable camera for hot climates. During the past two years this instrument has been supplied to most of our best photographers, as well as to the various Government Departments, and is acknowledged to be the lightest, firmest, and most portable camera yet devised. It can be made of any size. The same camera can be used for either short-focus portrait or long-focus landscape lenses. It is fitted with a most excellent screw adjustment, thus rendering rack and pinion to lens unnecessary, and possesses all the advantages of a swing back.

Rouch's Model "Carte de Visite" Camera.—In proof of the superiority of the model "Carte de Visite" and stereoscopic camera, it may be stated that, although only recently introduced, it is at the present time in daily use at the leading metropolitan ateliers. It is fitted with a perfect rack-work adjustment, and possesses a range of focus from 3½ inches to 7 inches. The dark slide and focussing screen have each new and special advantages, the latter being permanently attached to the camera, so that it is always in its place, and cannot be mislaid or broken.

Photographic Lenses.—W. W. Rouch is the appointed agent for the most celebrated lens makers, including Ross, Dallmeyer, and Grubb, whose productions are known to possess separate and distinct advantages. Having a most intimate knowledge of the same, he will always take care to select only such as will best fulfil the requirements specified. French lenses (advertised by many as "own manufacture") at Parisian prices.

Rouch's New Registered Parallel Instantaneous Shutter and Sun Shade is acknowledged to be the only really efficient arrangement for taking children, animals, waves, &c.

Rouch's Model Operating Room Stand may be regarded as an essential requisite in every studio.

W. W. Rouch is manufacturing largely the New Binocular Camera, with movable central partition, which combines in one instrument, at a moderate cost, all the advantages of a landscape, portrait, "carte de visite," and stereoscopic camera, and producing the most charming panoramic landscapes, 12 specimens of which may be seen on the screen next the staircase, No. 263 in the Catalogue, or one can be obtained by post, with particulars of exposure, collodion, and development, on receipt of 2*s.* 8*d.* in postage-stamps. Size of picture, 7¼ by 4½, mounted on plate, India-tinted paper. Every photographer ought to possess one of these instruments.

The following are some of the important manufactures of W. W. Rouch. None are genuine unless stamped with red label and trade mark:—

WR

- Negative collodion, with usual iodizer.
- Negative collodion, with cadmium iodizer.
- Negative collodion, with bromo-iodizer. This collodion remains unchanged for a lengthened period, and with an iron developer produces the most exquisite results.
- Collodion for the Fothergill, tannin, and other dry processes.
- New extra-sensitive keeping collodion, prepared especially for portraiture and instantaneous photography, retains its sensitiveness, and is considerably improved by age. Formerly prepared by Mr. Hardwich, late Lecturer on Photography, King's College, London, and Author of "Photographic Chemistry," &c.

The laboratory, lately occupied by Mr. Hardwich, and fitted with the most complete appliances, is now occupied by Mr. Rouch, and devoted exclusively to their manufacture. Every sample is tested, and the utmost care is taken to secure perfect uniformity. Every bottle is accompanied with a new and comprehensive paper of directions.

The uniform character, persistence, and absolute purity of these collodions, their greatly increased employment by a very large number of our first professional and amateur photographers, justify the assertion that they cannot be surpassed; and that, whether for use in this country or abroad, they will be found to possess, in their various combinations, a universal applicability for any of the wet or dry processes.

The Collodion Committee, numerous correspondents at home, on the Continent, in India, China, Australia, North and South America, Egypt, &c., &c., bear ample testimony as to their excellent working qualities under the most trying variations of temperature and climate.

Burfield and Rouch have paid great attention to the export branch of their business, and having adopted an improved method of packing, they can conscientiously recommend their preparations to architects, engineers, officers, tourists, and shippers, as safe articles for transmission abroad. Mr. Rouch will be happy to advise as to the most suitable iodizer for certain climates.

Price List of chemicals etc. forwarded on application.
A liberal discount to shippers, etc.

[3155]

SKAIFE, T., 47 *Baker Street, W.*—Pistolgraph, with a selection of its productions called pistolgrams.

SKAIFE'S PATENT PISTOLGRAPH (price ten guineas), with a selection of its jewel productions, including babies from the day of birth, and album reproductions on paper, plain and coloured. A prospectus will be sent on receipt of stamped address, or a copy of Illustrated Guide (second edition) to the Pistolgraph will be forwarded on receipt of thirteen stamps.

[3156]

SMITH, L., *Cookridge Street, Leeds.*—Photographic views.

[3157]

SMYTH & BLANCHARD, *George Street, Euston Square.*—Instantaneous photographs and life-size photographs.

[3158]

SOLOMON, J., 22 *Red Lion Square.*—Photographic apparatus, &c.

[3160]

SPACKMAN, B. L., *Kensington Museum.*—Photographs of the gardens of Horticultural Society; various art reproductions; Exhibition building.

[3161]

SPENCER, J. A., 7 *Gold Hawk Terrace, Shepherd's Bush, W.*—Albumenized and other prepared photographic papers.

The articles here exhibited will possess little or no attraction for the general visitor, and will be viewed with interest only by those who are engaged in the practice of photography.

It may, however, be interesting to observe, that it is one of the productions the demand for which has arisen entirely since the former Exhibition of 1851. At that time photography was quite in its infancy, and few or no pictures were exhibited that could compare with those met with at this moment at every turn. Photographs were then obtained generally from "paper" negatives, with the exception of a few from "albumen on glass;" collodion, which is now so generally employed, only having been discovered during the time of that Exhibition, of course had not been brought to the perfection it now possesses.

The proofs from the negatives thus obtained were exclusively obtained upon what is now called "salted paper," being merely good writing-paper, saturated with a dilute solution of a soluble chloride, or brushed on one side with a stronger solution of a similar salt.

A year or two after, one of the leading French photographers, observing the universal want of sharpness upon these papers, suggested and made use of papers, the salting solution of which contained various proportions of albumen, the action of which was to keep the material employed in the production of the image upon the *surface* of the paper, instead of, as heretofore, partly penetrating into its substance, thereby insuring a sharpness and brilliancy in the proof that had not been before attainable. Recently, the enormous demand for the well-known "stereoscopic slides," and later still, of "carte de visite" or "album pictures," in which excessive sharpness is necessary, has made it requisite to increase the quantity of albumen in the preparation of the paper, till now, when pure albumen, without any dilution, is very extensively employed.

When this method of preparing paper was first employed, every photographer probably prepared paper for his own use; but experience proved, in this case as in all similar ones, division of labour to be most economical. Now, the preparation of photographic paper with salted albumen has become, in many hands, a business of itself; and some idea of the quantity used may be found in the statement, that in one establishment alone (that in which the samples exhibited were prepared) upwards of 200,000 eggs have been employed in the course of six months to furnish the requisite quantity of albumen.

[3162]

SPODE, J., *Hawkesyard Park, near Rugeley.*—Proofs from collodion negatives.

[3163]

STOVIN & Co., *Whitehead's Grove, Chelsea.*—Principal buildings, London.

[3164]

STUART WORTLEY, LIEUTENANT-COLONEL A. H. P., *Carlton Club, Pall Mall.*—Photographs of Vesuvius, during the eruption of 1861-2.

[3165]

SUTTON, E., 204 *Regent Street, W.*—Miniature photographs, plain and coloured.

[3166]

SWAN, H., 5 *Bishopsgate Without, London.*—Large (and apparently single) pictures rendered stereoscopic; new stereoscopes.

PATENT STEREOSCOPIC COMBINATIONS :—

No. 1. LARGE RECEDING PICTURE, which falls into perfect stereoscopic relief on bringing the right eye in front of the small view glass, while both eyes are still kept open. Price 21*s.*

No. 2. THE CLAIRVOYANT STEREOSCOPE. ' This instrument has the following advantages over those in common use: it suits equally for examining opaques and transparencies, paper and glass impressions; it can be used to cover plates bound in books; it adapts itself to all angles of sight and focal lengths; it is easy to hold in the hand, and admits the light with perfect freedom; it is pretty, compact, and can be put away out of sight."—*Athenæum.* Price, from 10*s.* 6*d.*

No. 3. STEREOSCOPIC SHRINE, an unpublished mode of obtaining a stereoscopic image visible from a distance as well as near.

[3167]

TALBOT, W. H. FOX, *Lacock Abbey, Wiltshire.*—Photoglyphic engravings, produced by the action of light alone.

[3168]

TELFER, W., 194 *Regent Street.*—Untouched and coloured photographs.

[3169]

THOMPSON, C. THURSTON, *South Kensington Museum.*—Photographs from the Raphael cartoons, and pictures by J. M. W. Turner.

[3170]

THOMPSON, S., 20 *Portland Road, Notting Hill, W.*—Photographs, landscapes, architectural subjects, and reproductions.

[3171]

TRAER, J. R., F.R.C.S., 47 *Hans Place, S.W.*—Photographs of microscopic objects.

[3172]

TURNER, B. B., *Haymarket.*—Photographs from paper, negatives taken by the Talbot process.

[3173]

VERSCHOYLE, LIEUT.-COLONEL, 23 *Chapel Street, Belgrave Square.*—Photographs, by wet and collodion albumen processes.

[3174]

WALKER, C., & SON, *Windsor Road, Lower Norwood.*—Carbotype photographs, unchangeable; silver printed duplicates, changeable.

[3175]

WARDLEY, G., 10 *St. Ann's Square, Manchester.*—Photographic landscapes: negatives produced by the Taupenot process.

[3176]

WARNER, W. H., *Ross, Herefordshire.*—Architectural and miscellaneous photographs from enlarged negatives.

Architectural and other photographs from enlarged negatives.

These pictures were enlarged during the month of February, 1862, from negatives of the dimensions of 3¼ by 2¾ inches. Process, wet collodion. Lens by Ross. Average exposure, four minutes. Diaphragm, eighth of an inch. For terms apply (with stamp enclosed) to the exhibitor.

[3177]

WATKINS, H., 215 *Regent Street.*—Photographic portraits.

[3178]

WATKINS, J. & C., 34 *Parliament Street, S.W.*—Portraits, plain and coloured.

[3179]

WHITE, H., 7 *Southampton Street, Bloomsbury.*—Photographic landscapes.

[3180]

WHITING, W., & SONS, *Camden Town.*—Portable developing cameras for working wet collodion in the open air.

[3181]

WILDING, W. H., 2 *Chesterfield Street, King's Cross.*—Universal eccentric camera front; instantaneous camera.

[3182]

WILLIAMS, T. R., 236 *Regent Street, W.*—Untouched and coloured photographic portraits, vignettes, cartes de visite, &c.

[3183]

WILSON, G. W., *Aberdeen.*—Views by the wet collodion process.

[3184]

WILSON, SIR T. M., *Charlton House.*—The Geysers, Iceland.

[3186]

WRIGHT, C., 235 *High Holborn.*—Photographic portraits and copies of paintings.

[3187]

WRIGHT, DR. H. G., *London.*—Portable photographic apparatus, including tent, &c.

Class XV.

HOROLOGICAL INSTRUMENTS.

[3218]

Adams, F. B., & Sons, 21 *St. John's Square, London.*—Patent reversible chronometer; duplex and lever watches.

[3219]

Agar, William, *Bury, Lancashire.*—Specimens of working men's watches and independent centre seconds, manufactured in Bury.

[3220]

Armstrong, Thomas, Inventor and Manufacturer, *Manchester.* — Armstrong's improved watchman's detector clocks, steam or speed clock, &c.

The following are exhibited:—

	£	s.	d.
Armstrong's Improved Watchman's Clock is an ordinary office Timepiece and Watchman's clock combined, for indicating punctuality and registering the neglect of it. It is entirely self-acting, portable, simple in construction, and requires little attention. Price, in plain mahogany case, dial 14 inches diameter	4	15	0
Armstrong's Improved Watchman's Clock, same as the above, but in a more highly ornamented case, dial 12 inches diameter .	5	10	0
Armstrong's Improved Steam or Speed Clock indicates the amount of work done in all establishments where steam-power is used. It also compels the engineer to work at the required pressure, by indicating any neglect. In mahogany case complete, for fixing .	2	15	0

[3221]

Aubert & Linton, 252 *Regent Street.*—Watches with recent improvements; ornamental clocks for the table and mantelpiece.

[3222]

Bailey, John, & Co., *Albion Works, Manchester.*—Improved turret clock in tower, suitable for a market-place, &c.

[3223]

Barraud & Lund, 41 *Cornhill, London.*—Chronometers and watches.

[3224]

Bayliss, William, *Finmere, Oxfordshire.*—Model of new remontoire escapement, as put up in Finmere church clock.

[3225]

Bennett, John, F.R.A.S., 64 & 65 *Cheapside, and* 62 *Cornhill.*—Marine and pocket chronometers; public and private clocks; every description of gold and silver watches.

[3226]

BENSON, J. W., *Ludgate Hill.*—Gold and silver watches and clocks, highest quality, magnificently decorated with artistic designs.

[3227]

BLACKIE, GEORGE, 24 *Amwell Street, E.C.*—A new compound balance to correct the errors in chronometers, and new auxiliary to balance.

[3228]

BROCK, JAMES, 21 *George Street, Portman Square.*—Marine chronometers.

[3229]

BROOKS, S. A., *Northampton Square, London, E.C.*—Watch jewels; sets of jewel gauges for watchmakers and material dealers. (*See page* 67.)

[3230]

CAMERER, KUSS, & Co., 2 *Broad Street, Bloomsbury.*—1. A three-part quarter skeleton on ten bells. 2. A trumpeter clock. 3. A cuckoo clock.

[3231]

CAMPBELL, ANDREW, 63 *Cheapside, E.C., and* 43 *Tottenham Court Road.*—A selection of gold and silver watches.

[3232]

CHEVALIER, BENJAMIN, 4 *Red Lion Street, Clerkenwell, E.C.*—Chronometer and watch cases.

[3233]

CLARK, DR., *Finmere House, Oxfordshire.*—Astronomical clock, impelled by gravitation, requires no oil to the escapement.

[3234]

COATHUPE, CAPTAIN H. B., 1 *Abingdon Street, Kensington.*—Everlasting shilling "silent clocks;" painting-engraving; printing-embossing; printing-painting.

[3235]

COLE, JAMES FERGUSON, 5 *Queen Square, Bloomsbury.*—Chronometers, watches, tempered springs; new horological models and descriptive treatise.

[3236]

COLE, THOMAS, 6 *Castle Street, Holborn.* — Ornamental and portable clocks of original construction and design.

[3237]

CONDLIFF, JAMES, 4 *Fraser Street, Liverpool.*—A skeleton clock.

[3238]

COOKE, THOMAS, & SONS, *Buckingham Works, York.*—Church and turret clock, astronomical clocks, and time regulators.

[3239]

CRISP, W. B., 81 *St. John Street Road.*—Chronometers.

[3240]

DAVIES, C. W., *Notting-hill.*—Clock showing time and longitude at the most important places on the globe.

BROOKS, S. A., *Northampton Square, London, E.C.*—Watch jewels; sets of jewel gauges for watchmakers and material dealers.

The exhibitor manufactures the RUBY BOTTOMED HOLES for thin cocks or plates, and is also the Inventor and sole manufacturer of the JEWEL-HOLE GAUGE, for determining the size of pivots or jewel holes. Specimens of these manufactures are shown in his case. He can supply merchants and dealers in watch-material with every description of clock, chronometer, and watch jewels, set or unset; diamond-powder, bort, rubies, sapphires, chrysolites, garnet, &c.

The Export Watch-jewel Case (see engraving), with set jewels arranged and assorted, so that any size required may be instantly found, will be of great service to watchmakers or dealers in material residing in distant parts of the world. A broken jewel can be immediately replaced from it with the greatest accuracy, at a cost of little more than one-third the usual charge, and without the risk and loss of time attending the custom of sending to London the watch or part of watch requiring the jewel.

Persons who do not wish to keep a stock of watch jewels, can be supplied with the Jewel-hole Gauge, which will enable them, by sending the measure of the hole, to obtain the jewel of the exact size required, without the necessity of sending the balance or wheels to a jeweller, whereby risk of loss or damage is incurred.

The Jewel-hole Gauges are valuable for all purposes where accuracy is required, as they neither wear nor corrode. They are arranged as follows:—

Gauge.	Holes.	
1.	1 to 12	inclusive, will gauge the escapement pivots to any modern watch.
2.	13 to 24	inclusive, will gauge the small wheel pivots of three-quarter plate and small frame watches.
3.	25 to 36	inclusive, will gauge the small wheel pivots to large frame watches.
4.	37 to 48	inclusive, will gauge centre wheel, also chronometer third and fourth pivots.
5.	49 to 60	inclusive, will gauge chronometer seconds' pivots.
6.	61 to 72	inclusive, will gauge regulator and clock-work pivots, also lower fusee and hollow centre pinions to watches.
7.	73 to 84	inclusive, will gauge small three-quarter plate fusee upper pivot.
8.	85 to 96	inclusive, will gauge large ditto ditto.
9.	97 to 108	inclusive, will gauge frame fusee upper pivot.
10.	109 to 120	inclusive, will gauge large frame fusee upper pivot.

A set of pivots, numbered to correspond in gauge with each jewel-hole can be supplied, for measuring the size of the hole required, should the balance or wheel-pivot of the watch be broken.

[3241]

Davis, W., & Sons, 84 *King William Street, City, London.*—Chronometers, watches, clocks, and specimens of horology.

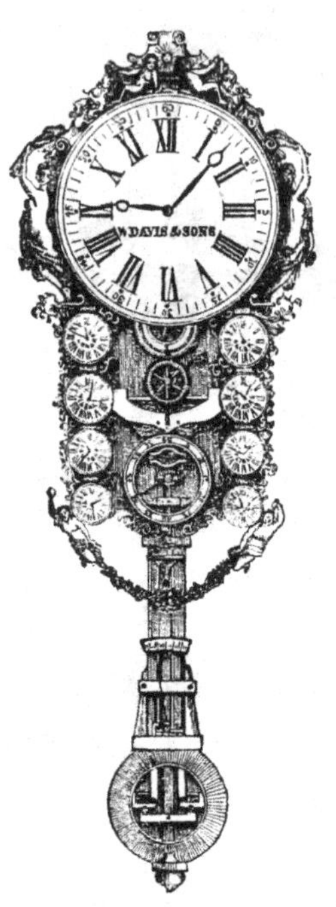

The exhibitors are manufacturers of clocks, watches, and chronometers; jewellers; dealers in diamonds and other gems, and in onyx, shell, and other cameos; and importers of Geneva watches, French clocks, bronzes, &c. They keep on hand a stock of church and turret clocks, of various sizes, and constructed on approved principles. The time is given by electricity at their city establishment, from the Royal Observatory, Greenwich.

Davis and Sons have a branch establishment at 57 New Street, Birmingham.

[3242]

Delolme, H., 48 *Rathbone Place, Oxford Street.*—Regulator, chronometers, clocks, and watches.

[3243]

Dent & Co., 61 *Strand, and* 34 & 35 *Royal Exchange.* — Chronometers, regulators, watches, and every description of time-keepers.

[3244]

Dent, M. F., Inventor and Manufacturer, 33 & 34 *Cockspur Street, Charing Cross, London, S. W.*—Watches, clocks, and chronometers; new auxiliary compensation balance. (*See pages* 69 *to* 71.)

[3245]

De Solla, J., & Son, 34 *Southampton Terrace, Waterloo Bridge.*—Original manufacturers of the royal liliputian alarm clocks.

[3246]

Dettman, Theodore, *Minories.*—Astronomical clock, constant ball escapement, compensation regulator, &c.; electro-magnetic clock, half-minute time.

[3247]

Ehrhardt, William, 26 *Augusta Street, Birmingham.*—Various kinds of watches, and instruments connected with the manufacture thereof.

[3248]

Fairer, Joseph, 188 *St. George's Street, E.*—The "Village Clock," and other turret clocks, watches, &c.

DENT, M. F., Inventor and Manufacturer, 33 & 34 *Cockspur Street, Charing Cross, London, S. W.*—Watches, clocks, and chronometers; new auxiliary compensation balance.

No. 1.—M. F. Dent's *new* compensation balance with outside auxiliary bows for extremes of temperature.

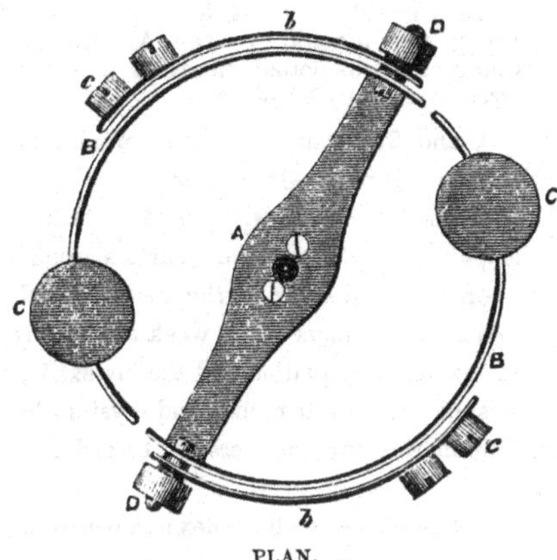

PLAN.

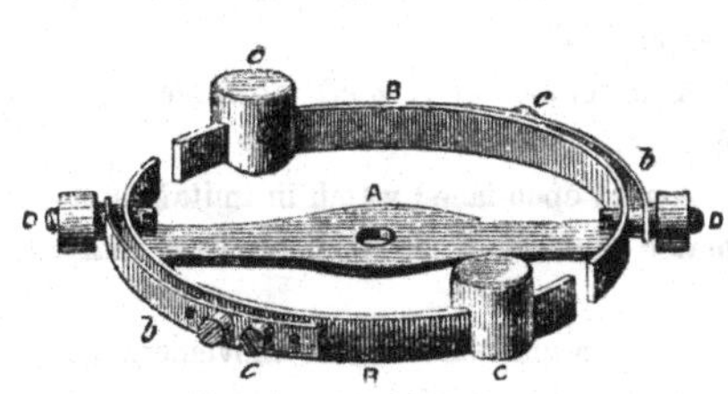

PERSPECTIVE VIEW.

A is the bar of the balance.

B B are the compensation bows, the steel within and the brass without the rim.

C C are the compensation brass weights.

These parts are the same in form as the ordinary balance, but the bows are made thinner and lighter in order that they may move through a larger space. On the studs for the timing-screws (D D) are fastened two outside auxiliary bows (*b b*), the metals being reversed, namely, brass within and steel without the rim.

These auxiliary bows are provided with a row of screw-holes, and carry one or more platina screws (*c c*), which increase or diminish the action according to the distance they are placed from the timing-screw (D).

At present the range of the ordinary compensation balance is limited, because if the brass weight (C) is placed near the extremity of the bow, the balance is, in technical terms, "over compensated;" but this is qualified by the auxiliary action, which, being reverse to that of the primary bows, adds to the error which the primary balance is intended to compensate. This enables the weight C to be placed further out on the primary bow, and by the corresponding adjustment of the platina screws on the auxiliary, a much wider range is obtained of controlling power. The general effect is that the compensation power is greatly increased in high temperatures.

This new balance is simple in construction; all its bows act perfectly free, and the correction for high and low temperature is entirely under control.

2. A marine chronometer fitted with Dent's new auxiliary compensation.

3. M. F. Dent's model watch, in gold hunting-cases, lever escapement, compensation balance, with helical pendulum spring, winding and setting at the knob, and having fusee, and swivel pendant to prevent robbery; constructed to go *two* days without winding, and having dial indicator showing the time when last wound. This watch has the "répétition à tact," whereby the time can be ascertained by an external hand; *the only kind of watch that could be used by one who is deaf and blind.*

4. Gold hunting-cased watch with independent centre seconds, lever escapement, compensation balance, Brequet pendulum spring, winding and setting hands without key.

5. Gold hunting-cased chronometer with *perpetual* calendar.

6. Gold minute repeater in hunting-cases, keyless.

7. Gold hunting-case keyless watch, with lever escapement, compensation balance, engraved case, and ornamental dial. A specimen for the Spanish market.

8. Gold observation watch; a valuable instrument for timing the transit of a star, the phase of an eclipse, or for any purpose where delicate accuracy is essential to determine the exact time of commencement, continuance, and end of any period of observation.

The effect is obtained by having the centre seconds' hand double, one part closely overlying the other, so as to give the appearance of a single hand in ordinary action.

By pressing a knob at the pendant on commencing an observation the under hand is instantly stopped; but the upper hand will continue in action until—as the moment of observation ceases—the knob is pressed a second time; both parts of the seconds' hand are then stationary, and the exact interval of time observed is seen registered on the dial. The common action of the watch is not interfered with, it continues going, and on pressing the knob a third time the two parts of the seconds' hand immediately reunite and fly to the nearest point to correspond with the minute.

DENT, M. F.—*continued.*

9. A gold hunting-cased watch with duplex escapement and cylindrical pendulum spring.

10. A specimen of a gold hunting-cased watch, lever escapement, compensation balance, helical pendulum spring, winding without a key and having fusee.

11. A specimen of a plain gold hunting lever watch with extra fine cases.

12. A specimen of a gold open-faced chronometer watch.

13. A gold open-faced watch in imitation of Brequet's celebrated flat watches with eccentric dial and solid key.

14. Under a glass-shade—the movement of a chronometer watch taken to pieces; a specimen of high manipulation, and the most approved calibre.

15 and 16. A specimen of two ladies' gold watches, engraved cases.

17, 18, and 19. Three ladies' gold watches, lever escapements, compensation balances, keyless. The cases enamelled and set with diamonds, varied designs.

20. Gold open-faced watch, silver dial, keyless "*répétition à tact.*" The bottom cover blue, enamelled with monogram in diamonds.

21. A specimen of a silver hunting-cased lever watch with compensation balance.

22. A miniature regulator with mercurial pendulum and remontoir train.

23. Gold open-face chronometer watch with patent fusee winding.

This chronometer watch is a facsimile of that made in 1859 by M. F. Dent, for Sir William Armstrong, the inventor of the Armstrong Gun, who certifies its actual variation at the end of a year to be only 45 seconds.

24, 25, 26, and 27. Four specimens of chronometer clocks, in gilt and German silver cases.

28. A chronometer clock, with patent balance for extreme temperatures, chiming the quarters upon eight bells, with perpetual calendar of the most perfect construction, indicating the days of the week and month, the phases of the moon, the equation and the bissextile, in a superbly finished case of gilt bronze and crystal glass.

29, 30. Two marine chronometers of the ordinary construction.

31, 32. Two time-pieces with duplex escapements compensated, gilt bronze cases, plain and engraved.

33. A boudoir time-piece, lever escapement, compensation balance; a gilt engraved case, silver dial.

34, 35. Two circular lever time-pieces with compensation balances, one in bronze and the other in a gilt bronze case; specially suitable as portable time-pieces.

36. A specimen of a library clock in bronze case, portable, lever escapement, compensation balance, chiming quarters.

37. A similar clock in ebony-case with improved gongs for fine tone.

The following are extracts from the reports of scientific persons as to the accuracy of Dent's horological instruments :—

Sir WILLIAM ARMSTRONG, inventor of the Armstrong Gun, says :—

"9 Hyde Park Street, W., 14*th November*, 1861.

"The chronometer watch you made for me in December 1859, has never been affected by travelling or riding; its variation at the end of a year was only 45 seconds. It has proved in every respect a most satisfactory watch.

"W. G. ARMSTRONG.

"MR. M. F. DENT,
33 Cockspur Street, Charing Cross."

The ASTRONOMER ROYAL, Greenwich Observatory, reporting in 1829 on the celebrated public trial, by order of the Lords of the Admiralty, which lasted thirteen years, during which nearly 500 chronometers were tested, says :—

"Your chronometer, No. 114, is entitled to the first premium. Actual variation in the year 54 hundredths of a second. This is superior to any other yet tried.

"J. POND, *Astronomer Royal.*

"MR. DENT.'

The RUSSIAN IMPERIAL ASTRONOMER, M. STRUVE, of St. Petersburg, reporting upon 81 chronometers tested by the Russian chronometrical expedition, in 1843, says :—

"The Dent chronometers have held first rank in a brilliant manner. They contributed, beyond dispute, the most effectually to the exactitude of the results.

"M. STRUVE."

By command of the Emperor, the Russian gold medal of the highest order of merit was presented to Mr. Dent.

G. B. AIRY, Esq., Astronomer Royal (in testimony of the excellence of Dent's turret clocks), says :—

"Royal Observatory, Greenwich, 22*nd July*, 1845.

"I believe the clock which you have constructed for the Royal Exchange to be the best in the world as regards accuracy of going and of striking.

"G. B. AIRY.

"MR. DENT.
33 Cockspur Street, Charing Cross."

Dent, M. F.—*continued.*

38, 39. Specimen drawings of heraldic and other designs, richly executed in enamel and jewels upon the cases of watches, made to special order. The following are a few selections :—

[3249]

FORREST, JOHN, 29 *Myddelton Street, E.C.*—Every description of pocket watches, various escapements and springs—London work.

The following specimens of fine London work are exhibited :—

Pocket chronometer, spiral spring.
Duplex chronometer, Brequet spring.
Lever chronometer, ditto.
Duplex ¾ jewelled tongue, ditto.
Duplex frame, ditto, capped, ditto.
Two-pin lever, ditto, ruby roller, ditto.
Best one-pin. Lever, new balance.
Do. ¾ lever with fly cap.
Independent centre seconds, double train.
Ditto ditto single train.
American block work.

Samples of watches which can be supplied by the exhibitor at the prices quoted per dozen :—

		£	s.	d.
¾-Plate gold dome hunters, compensated		..		
¾ Ditto consular, ditto		..		
¾-Plate gold hunters	from	156	0	0
¾ Ditto consulars . . .	do.	138	0	0
Gent's ditto frame hunters . .	do.	132	0	0
Ditto consulars . .	do.	126	0	0
Ladies' ¾-plate ditto . .	do.	132	0	0
Ladies' frame ditto . . .	do.	96	0	0
Silver ¾-plate hunters, compensated .				
Silver ¾-plate hunters . . .	do.	76	0	0
Ditto consulars . . .	do.	63	0	0
Silver frame hunters	do.	50	0	0
Ditto consulars . . .	do.	43	0	0

[3250]

FRODSHAM, CHARLES, 84 *Strand, London.* —New caliphers of chronometers, watches, and astronomical clocks; new equation double compensation balances. (*See page* 73.)

[3251]

FRODSHAM & BAKER, 31 *Gracechurch Street, City.*—Chronometer, watch, and clock manufacturer to the Admiralty.

[3252]

GANEVAL & CALLARD, 27 *Alfred Street, Islington.*—Watch, pendulum, spring, and wire manufacturers.

[3253]

GREENWOOD, J., & SONS, 6 *St. John's Square, E.C.*—Quarter and bracket clocks; regulators, dials, and cases.

[3254]

GUIBLET & RAMBAL, 11 *Wilmington Square, Clerkenwell, London.*—Keyless fuzee watches for scientific purposes; pocket chronometers.

[3255]

GUILLAUME, EDWARD & CHARLES, 16 *Myddelton Square, E.C.*—Watches and repeaters.

[3256]

GUMPEL, CHARLES GODFREY, 2 *Gordon Cottages, Holland Road, Brixton.*—A system of electric clocks.

[3259]

HAWLEYS, 287 *High Holborn.*—Regulator—only requires winding once in twelve months.

[ ]

HEYES, THOMAS, Manufacturer, *Appleton, Widnes, near Warrington, Lancashire.*—Steel and brass wire for pinions, clicks, screws, &c. (*See* 3295—PRESCOTT COMMITTEE.)

Pinion wire in steel and brass, round steel and other wires, and pinion wire gauges.

[3260]

HIGHFIELD BROTHERS, 5 *King Edward Terrace, Liverpool Road, N.*—Marine and pocket chronometers, duplex and lever watches, and an improved regulator.

[3261]

HILL, CHARLES JOHN, late W. H. HILL & SONS, *Chapel Fields, Coventry.*—Watches and patent pearl dials.

FRODSHAM, CHARLES, 84 *Strand, London.*—New caliphers of chronometers, watches, and astronomical clocks; new equation double compensation balances.

[*Obtained, in* 1831, *the Government Premium Prize of* £170; *in* 1848, *the Telford Medal; in* 1851, *at the Great Exhibition, the First Class Medal; in* 1855, *at the Paris Exhibition, the Gold Medal of Honour; in* 1860, *the Grand Gold Medal "Præmia Digno," from the Imperial Russian Government, for the superior performance of his Chronometers during the great Russian survey.*]

The following specimens of high-class horological workmanship are exhibited:—

Pocket chronometers, chronometer repeaters, stop-split centre seconds, and other timing and stop-watches.

Specimens of his "new series" lever chronometer watches, drawn to an entire new calipher of unrivalled timekeeping properties.

A month marine chronometer.

Large eight-day and two-day marine chronometers.

New model eight-day and small two-day marine chronometers, drawn to new and defined proportions, with important and useful changes, all founded on reliable measurements, and the result of long, accurately noted experiments.

Astronomical and sidereal clocks, with important improvements, the result of long-continued and accurate experiments.

Specimens of portable regulators, carriage clocks with compensation balances and chronometer escapements.

Carriage clocks, with lever and chronometer escapements of the highest and finest adjustments.

Small chime clocks.

Instructive specimens of new chronometer and watch movements.

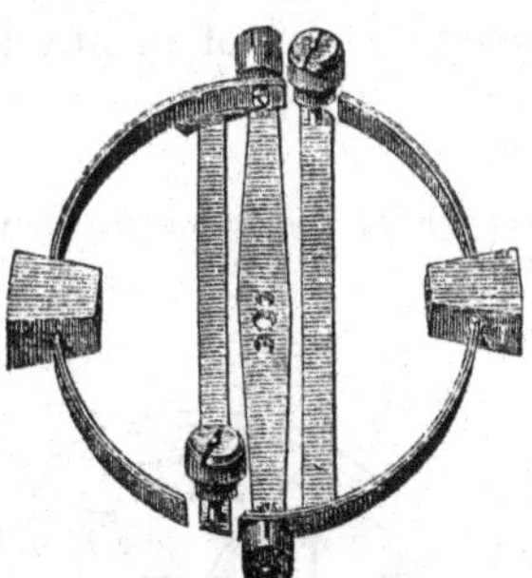

Double compound Micrometric Equation balance, invented and fecit by C. Frodsham.

Compound double inverted Differential balance, invented and fecit by C. Frodsham.

Compound triple Equation balance, invented and fecit by C. Frodsham.

An instrument to illustrate the motion of the compensation balance, showing the causes of the losing error in the extremes of temperature.

New double compensation balance of great accuracy.

New double compensation micrometric balance, extremely sensitive to sudden changes of temperature.

A model church and turret clock, constructed after designs proposed for the clock of the New Houses of Parliament, of astronomical accuracy.

C. FRODSHAM also exhibits an entirely new system of nomenclature for chronometer and watchmaking.

Tables to facilitate their construction.

New "Duo in Uno" balance springs for perfecting the adjustments of high-class watches and chronometers in their various positions.

New standard to facilitate universally the measurement of watches, with tables of comparisons and coincidences in French new and old measurements, and a work to exhibit CHARLES FRODSHAM'S system of chronometer, watch, and clock making.

He also exhibits the model of the chronometer-maker's ice-box and oven. C. F.'s new differential compensation balance, perfect for every degree of temperature, will not be ready for exhibition until August.

[3262]

HISLOP, WILLIAM, 108 *St. John Street Road, London.*—Standard or observatory clock, showing mean and sidereal time.

This clock is especially intended as a standard time-keeper. The arrangement for showing sidereal time by means of a supplementary dial may be applied to any clock, and may show mean time when the primary dial shows sidereal time. It prevents the necessity of reducing by calculation a sidereal observation to mean time, or vice versâ. The wheelwork may also be applied to show both times on a single dial. Price of clock complete, 60 guineas; in a plain case, from 35 guineas. Sidereal dial work, independent of fittings and adjustment to clock, 10*l.*

[3263]

HOLDSWORTH, SAMUEL, 220 *Upper Street, Islington, N.*—Chronometer and watch jewels: chronometer pallets and duplex rollers.

[3264]

HOLL, FREDERICK RICHARD, 284 *City Road.*—Patent non-winding chronometers and watches.

[3265]

HOLLIDAY, THOMAS, 304 (late 108) *Goswell Road, London.*—Gold, silver, and metal watch-case and dial maker.

[3266]

HOLLOWAY & CO., 128 *Minories, and New Square, London.*—Pendulum and lever clocks of the simplest construction.

[3267]

HOWARD, RAYMOND, 29 *King Square, Goswell Road.*—Sunk seconds dial maker and enamel maker.

[3268]

HOWELL, JAMES, & CO., 5, 7, 9 *Regent Street.*—Clocks, watches, &c. (*See page* 75.)

Fig. 1. English ormulu candelabra.—Gothic.
Fig. 2. Ditto ditto clock, silver dial.—Gothic.
Fig. 3. Ormolu candelabra, jewelled enrichments.—Moresque.
Fig. 4. Elaborate pierced star timepiece.—English Gothic.
Fig. 5. Pierced ormolu timepiece.—Mediæval.
Fig. 6. Ormolu travelling 8-day timepiece.—Registered padlock.
Fig. 7. Ormolu clock, jewelled enrichments, silver dial—Moresque.
Fig. 8. Ormolu 8-day travelling timepiece.—Registered horseshoe.

[3270]

HUTTON, JOHN, 10 *Mark Lane, London.* — Marine chronometers, Hartnup pocket chronometer, and other sorts; improved cheap watches.

[3271]

JACKSON, W. H. & S., 66 *Red Lion Street, Clerkenwell.*—Chronometers, day of month, keyless, and other watches.

[*Obtained a Prize Medal in Class X.*, 1851.]

The following specimens of chronometers and watches, day of month, eight day, local and mean time, keyless, solid key, with several improved modifications of the lever escapement are exhibited :—

DAY OF MONTH WATCH.

A. Two-day marine chronometer.
B. Pocket chronometer.
C. Day of month, adjustable, with ruby solid impulse lever escapement.
D. Eight-day (fuzee), with J. F. Cole's resilient pallet.
E. Watch (toothed barrel, solid key), with J. F. Cole's patent repellant lever escapement.
F. Watch showing local and mean time from one train. Hands set independently; lever escapement, with horizontal ruby pin, adjustable.
G. Various keyless and solid key watches.
H. Tool for indicating pallet angles and arc of lever escapement.

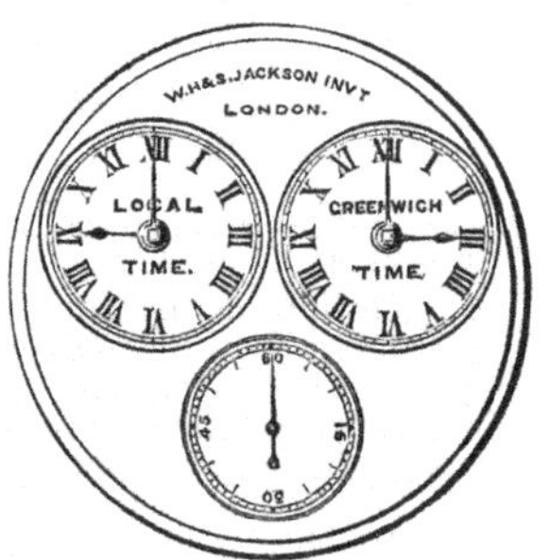

EXTERNAL ARRANGEMENT OF LOCAL AND MEAN TIME WATCH.

Howell, James, & Co., 5, 7, 9 *Regent Street.*—Clocks, watches, &c.

Fig. 1. Fig. 2. Fig. 3.

Fig. 4. Fig. 5.

Fig. 6. Fig. 7. Fig. 8.

[3272]

JOHNSON, EDWARD DANIEL, 9 *Wilmington Square, Clerkenwell, London.*—Chronometers, watches, pendulums, horological machinery, and various improvements and inventions.

MARINE CHRONOMETERS, eight-day and two-day.

HERMETIC BOX for Marine chronometer.

Magnetic Disperser for Marine chronometer. Rotating machinery in the case causing the chronometer to revolve on its own axis in 24 hours, thus dispersing the effects of local magnetism.

Surveying Chronometer. An ordinary small-sized two-day movement, fitted in a silver case as well as the ordinary gimbals, so as to make it portable in the pocket as well as suitable to the navigation of a ship. Removed from its gimbals by turning round the glass cover.

Pocket Chronometers, half plate and frame.

Duplex watches, half plate.

Lever watches, half plate and three-quarter.

Railway watches, full plate and three-quarter.

"Automaton seconds" watches. Keyless watches.

Combination of both these last.

"Universal seconds," a new watch, designed and patented especially to commemorate the Exhibition of 1862 horologically; consisting of a new caliper and train of wheels, effecting the union of "Automaton seconds" and permanent side seconds without complexity.

A new Escapement for Equatorial Telescopes, for circulating pendulums; consisting of a single crank motion, giving freedom to the motion of the pendulum, with the equable continuous rotary motion to the telescope required.

New and improved models of mercurial pendulums.

Model suspensions for pendulums.

New auxiliary compensator for wooden pendulums; can be applied to any pendulum in one minute, at a cost of 7*s.* 6*d.* It consists of a glass tube, divided into two chambers by being drawn out into an upper and a lower part, joined by a small tube; the lower, and part of the upper chamber containing mercury. This arrangement effects the transposition of small quantities of the mercury long distances, doing proportionately more work. Adjustable by a screw at the bottom.

Model of a public Timeball, discharged by electric current from Greenwich Observatory.

Groups of watch-movements, showing various constructions and qualities of workmanship.

Manufacturer of Chronometers and Watches, of which fair samples only are exhibited in Class 15. No article among those shown is made at unnatural expense on purpose to show, but each is a fair representation of his ordinary work, and his stock is manufactured of the same material and workmanship.

Inventor and Patentee of:—

The automaton-seconds watch.

The self-contained winder.

The magnetic disperser.

The hermetic box or chronometer safe.

The universal seconds watch.

And manufacturer of goods for all the foreign markets, on the models specially suited to each.

Patronized by the Admiralty.

[3273]

JONES, JOHN, 338 *Strand.*—Watches.

Case containing the following specimens of watches:—

1*st Row.*—Ladies' gold watches, with a new application of jewels in the notation of minutes on the dials, and decorated cases.

2*nd Row.*—Gold watches, with a new application of colour for the adornment of enamel dials appropriate for ladies' use.

3*rd Row.*—Specimens of the perfection of railway watches adjusted for position and temperature; also 2-day watches with correct adjustments.

4*th Row.*—Gold ¾-plate hunting levers, with specimens of the most perfectly proportioned escapements that the trade can produce, with Isochronal springs and compensated balances.

5*th Row.*—Gold hunting lever watches, with newly arranged spring caps, suitable for use in foreign climates.

6*th Row.*—Silver levers to compete with the foreign manufacturer in price while retaining the English superiority in quality, 3*l.* 3*s.* each.

[3275]

KLAFTENBERGER, CHARLES J., 157 *Regent Street.*—Minute repeaters, chronometers, &c.

[3276]

KULLBERG, V., 12 *Cloudesley Terrace, N.*—Chronometers, watches, and clocks.

[3277]

LANGE, CHRISTIAN, 9 *Salisbury Street, Strand, London.*—Watches and timepieces.

[3278]

LEONARD, G. W., 1 *Cloudesley Terrace, Liverpool Road.*—Compensation balances.

[3279]

LOSADA, JOSÉ R., 105 *Regent Street.*—Watches, marine chronometers, table clocks, turret clocks, and astronomical pendulums. (*See pages* 78 & 79.)

[3282]

MARRIOTT, BENJAMIN, 38 *Upper Street, Islington, London.*—Watches, gold chains, &c.

[3283]

MERCER, THOMAS, 45 *Spencer Street, Clerkenwell.*—Marine chronometers.

[3284]

MOORE, B. R. & J., 38 *Clerkenwell Close.*—Turret and other clocks.

[3285]

MORRIS, WILLIAM, *Blackheath, S.E.*—Electric regulator with centre seconds, and other companion clocks, all beating simultaneously.

[3286]

MUIRHEAD, JAMES, & SON, *Glasgow.*—House, turret, and railway clocks, &c.

[3287]

MURRAY, JAMES, 30 *Cornhill, London.*—Chronometers, watches, clocks, patented keyless watches, patented regulator, models, jewelry, &c.

[3288]

NEAL, JOHN, Watchmaker and Jeweller, 18 *Edgware Road, London, W.*—Jewelry; onyx clocks; duplex, lever, and chronometer watches—new construction.

[3289]

NICOLE & CAPT, 14 *Soho Square.*—Nicole's patent keyless watch and conteur.

[3290]

ORAM, GEORGE JOHN, 19 *Wilmington Square, Clerkenwell.*—Watches.

[3291]

PARKINSON & FRODSHAM, 4 *Change Alley, Cornhill, E.C.*—Chronometers, watches, regulators, astronomical clocks, &c.

[3292]

PLASKETT, WILLIAM, 12 *Alderney Road, Globe Road, Mile End, London, N.E.*—Marine chronometers with improved compensation.

These chronometers are manufactured with improved compensating auxiliary balances for correcting the difference of time occasioned by extreme temperature. They are also furnished with air-tight valves which perfectly exclude all damp from the works.

[3293]

POOLE, JOHN, 57 *Fenchurch Street, London.*—Marine and pocket chronometers and watches.

[3294]

PORTHOUSE & FRENCH, 16 *Northampton Square, Goswell Road.*—Specimens of marine chronometers and watches for home and foreign markets.

[3295]

PRESCOTT COMMITTEE FOR THE EXHIBITION OF TOOLS, HOROLOGICAL INSTRUMENTS, &c.:—

Preston, J.
Hewitt, S. & J.
Wycherley, J.
Copple, J. & W.
Scarisbruk, C.
Hunt, J., & Co.
Ford, R.
Welsby, J.
Brown, Ann.
Johnson, C. B.
Houghton, S.
Pendleton, P.
Stockley, Jas.
Taylor, Richard.
Preston, Wm.
Saggerson, E.
Molyneux, Wm.
Whitfield, J. J.
Alcock, J.
Jacques, J.
Smith, J.
Naylor, Thos.
Heyes, Thos.

[3297]

QUAIFE, THOMAS, Clockmaker, *Hawkhurst, Kent.*—Chime clock, fifty changes, in marble and gold; and chronometer.

[3300]

ROTHERHAM & SONS, *Coventry.*—Gold and silver watches, and parts of a watch in every stage of manufacture.

[3302]

RUSSEL, THOMAS, & SON, *Liverpool.*—Watches, hard tempered nickel movements, patented; especially adapted to hot climates.

LOSADA, JOSÉ R., 105 *Regent Street.*—Watches, marine chronometers, table clocks, turret clocks, and astronomical pendulums.

DE S.S. M.M. C.C., REAL FAMILIA Y ARMADA MILITAR.

[*Thrice decorated by her Catholic Majesty for merit in his art.*]

1. Astronomical pendulum, escapement jewelled, in glass case.

2. Two astronomical pendulum movements complete, escapements jewelled, dials unfinished.

3. Musical chiming clock, to strike the quarters on eight bells, and the hours on a deep gong; plays one of four different overtures at each of the hours; in rosewood case, gilt engraved dial.

4. Same as No. 3, in oak case and silvered dial.

5. Chiming quarter clock, with centre seconds and duplex escapement, compensated and adjusted, carved mahogany case, with carved dolphins as supports, gilt engraved dial.

6. Skeleton centre seconds clock, under glass shade, with chronometer and escapement compensated and adjusted, with emblem of *Fidelity.*

7. Small table chronometer with brass engraved and gilt case, and gilt engraved dial.

8. Ting tong carriage clock, with lever escapement, brass gilt case, and gilt engraved dial.

9. Binnacle clock, with lantern and reflector, lever escapement compensated and adjusted, brass bronzed case.

10. Cabin dial, with lever escapement compensated and adjusted, in black mahogany case.

11. 8-day marine chronometer.

12. 2-day ditto.

13. Two marine chronometers in construction.

14. Two silver acompañantes with mahogany case.

15. Gold hunting grand clock watch, to strike the hours and quarters, and to repeat the hours and quarters every quarter of an hour, and hours, quarters, and minutes at pleasure, showing the days of the week and month. Jewelled in 40 holes.

16. Gold hunting clock watch to strike the hours and quarters, and to repeat the hours and quarters every quarter of an hour, and at pleasure. Jewelled in 24 holes.

17. Gold hunting minute repeater. Jewelled in 24 holes. Highly ornamented.

18 Gold hunting half-quarter repeater. Jewelled in 20 holes. Highly ornamented.

19. Gold hunting duplex watch, to show 6 different meridians. Jewelled in 8 holes.

20. Gold hunting pocket chronometer, 13 jewels. Highly ornamented.

21. Gold hunting pocket chronometer, 13 jewels. Plain.

22. Gold hunting duplex, independent centre seconds. Jewelled in 20 holes. Highly ornamented.

23. Gold open face, ditto, ditto. Plain.

24. Gold hunting duplex, centre seconds. Highly ornamented.

25. Gold open face duplex, centre seconds.

26. Gold hunting duplex watch. Highly ornamented.

27. Gold demi-hunting watch. Highly ornamented.

28. Gold hunting duplex watch. Plain.

29. Gold demi hunting lady's duplex keyless watch. Highly ornamented.

30. Gold hunting lever watch. Highly ornamented.

LOSADA, JOSÉ R.—*continued.*

31. Gold hunting lever watch. Plain.

32. Gold demi hunting lever watch.

33. Gold hunting lady's lever watch. Highly ornamented.

34. Gold hunting lady's lever watch. Plain.

35. Watches in construction.

36. Silver hunting duplex watch.

37. Silver hunting lever watch.

38. Three orders, conferred by her Catholic Majesty Isabella the Second for merit in his art, viz. :—

a. Cross of Charles the Third.

b. b. Orders of Comendador de Numero of Isabella the Catholic.

39. A very elegant brooch, being the device borne on the reverse of the Mexican doubloon, and representing the secretary bird destroying a serpent; the body of the bird is composed of a very large pearl; the head, neck, wings, tail, and feet of brilliant and rose diamonds, on a spray, also of diamonds and gold, with a large single pendant, the snake of gold beautifully enamelled; the whole set in gold.

Vease anuncio en los Catálogos é Iluminado.

[3303]

SAMUEL, A., & SON, 29 *Charterhouse Square, E.C.*—Various descriptions of English watches, manufactured by exhibitors.

[3304]

SANDERS, JOHN, 15 *West Bar, Sheffield.*—Regulator, timepiece, and keyless watches.

[3305]

SCHOOF, WILLIAM GEORGE, 9 *Ashby Street, Northampton Square.*—Regulator, with detached escapement and mercurial pendulum.

[3306]

SEWILL, JOSEPH, 61 *South Castle Street, Liverpool.*—Gold and silver watches; pocket and marine chronometers.

[3307]

SHEPHERD, CHARLES, 53 *Leadenhall Street, City.*—Galvano-magnetic clocks.

[3309]

SMITH, J., & SONS, *St. John's Square, Clerkenwell, London.*—Church, turret, and house clocks, &c.; illuminated and other dials. (*See page* 81.)

[3311]

STRAM, NUMA, *Ashby Street, Northampton Square.*—Reversible and self-winding watch.

[3312]

STRATH BROTHERS, 7 *Park Terrace, Camden Town.*—Models of the English and Geneva watches.

[3313]

TANNER & SON, *Lewes.*—Clock with perpetual register of day, week, and month—requires no correction.

[3316]

THOMSON & PROFAZE, 25 *New Bond Street, W.*—Watches, clocks, timepieces, "tell-tale," and jewelry.

English skeleton clock, supported by figures emblematic of day, night, twilight, and dawn, designed and manufactured by the exhibitors. English chronometer repeating clock. The specimen of engraving on brass case is unique. Chronometer timepiece in gilt case, ornamented with river gods, &c.

English timepieces of various kinds.

Tell-tale timepiece registering time within five seconds; could be adapted for astronomical and meteorological observations. English watches with the latest improvements, winding and setting hands by pendant, the engraving and enamelling on cases of the best and most elaborate description. Gold chains, &c.

Marine set, bracelet, brooch, necklet and earrings with dolphins and shells enamelled and set with rubies, emeralds, and diamonds.

Further particulars may be learned upon application.

Smith, J., & Sons, *St. John's Square, Clerkenwell, London.*—Church, turret, and house clocks, &c.; illuminated and other dials.

1. Turret clock tower and summer house, with eight-day turret striking clock, with four faces 3 ft. 6 in. diameter, intended for illuminating. The clock is constructed on the repeating principle; has maintaining power to keep it going during winding: inside dial plate to set the four pairs of outside hands by; and various other improvements. The clock tower is surrounded by seats and bronze rail, and surmounts the summer house, which has wings that may be used for choice flowers, &c. The intention of the whole arrangement is to supersede the old custom of placing a turret clock on stables, by rendering this most useful article an ornament to the park, lawn, or ornamental garden.

2. Turret timepiece, suited for railway termini or public buildings, stables, &c.

3. Eight-day skeleton clock, strikes on cathedral tone gong, and the half-hour on bell. The decoration of this clock is of a very elaborate character.

5. Skeleton striking clock (design, Temple of Flora)

6. Ditto ditto, plain design.

7. Eight-day chiming bracket clock, in carved oak case of Old English style, introducing dolphins and acorns; chimes the quarters on eight musical bells, and strikes the hours on a gong.

8. Striking bracket clock, carved oak case (new design).

9. Ditto ditto ditto, solid mahogany carved case.

10. Ditto ditto ditto.

11. Regulator or astronomical clock, mercurial compensated pendulum, suited for a gentleman's hall, ornamental carved Spanish mahogany case.

12. Detector clock, or watchman's clock, which, in addition to forming a bracket timepiece, detects and registers neglect of duty in watchmen or night wardens.

13. Skeleton eight-day striking clock, mosque pattern.

14. Model of the turret clock tower and summer-house, erected by Messrs. J. Smith and Sons in the Eastern Annex, Class IX.

15. Various models and samples of eight-day office shop dials. Clocks for various climates, all manufactured by the exhibitors.

16. Samples of materials and tools used in the manufacture of English clocks.

17. Eight-day turret or church clock, of the same construction and material as that supplied by the exhibitors to the order of the Government Department of Science and Art, and which may be seen in the Museum South Kensington. The wheels and bosses for the pivots to act in are of gun metal, the mixture being the same as that used for the manufacture of ordnance bearings, the pinions of wrought steel, cut and finished in an engine as well as the wheels; thus securing the greatest possible accuracy. The frames are of iron, and so constructed that any part can be removed for cleaning without disturbing the remaining parts. The escapement is on the principle of Graham's dead beat, and the steel pads are made to slide in turned grooves, so as to set the pitch with the greatest exactness; they may be removed, as they are secured by screws. The striking apparatus is on the repeating principle, which prevents the possibility of striking wrong hours—a fault so common in many clocks with locking plates. The maintaining power to keep the clock going during winding is by lever and bolt; there is a small inside dial to set the hands by. The pendulum has a heavy spherical ball, and the rod, which is of prepared pine, coated with varnish and afterwards French polished, is thus secured against the action of air or damp; the pendulum is set in beat by means of a traversing screw, and the crutch has also two large screws to regulate and reduce its friction.

18. Metal drum case dial, made expressly for India, China, and tropical climates. The face of this is twelve inches diameter, though all sizes are made on the same principle and construction. The front of the case solid brass, with thick plate-glass; the movement has jointed steel chain, and neither case nor clock can be injured by climate or insects.

19. Revolving machine, strong spring movement in mahogany box, with circular plate, for the exhibition of figures in shop windows; adapted for "hairdressers," models, &c., &c.

20. Small models of office dials in oak, walnut tree, and mahogany, carved in various styles, suited for public buildings, lecture-rooms, in Elizabethan, Gothic, Grecian, Mediæval, and modern styles of architecture.

21. Illuminated dial, for outside of public buildings; the numerals, minute stops, and mouldings are of copper, and glazed with opal glass. By the construction of this dial perfect distinctness and durability are secured, and the gas light equally diffused over the surface of the clock face.

22. Eight-day school dial in solid oak case.

23. Eight-day bedroom clock with alarum.

24. Eight-day striking kitchen or country clock in long case.

WHITE, EDWARD, 20 *Cockspur Street, Pall Mall, S.W.*—Chronometers, watches, clocks, and gold chains.

No. 1. A monthly astronomical clock, with mercurial compensation pendulum, and pallets jewelled with sapphires.

CHRONOMETERS.

2. An eight-day marine chronometer.

3. A two-day do. do.

4. A ditto, with auxiliary compensation.

5. An eight-day chronometer timepiece, in plain gilt metal case, with enamel dial and engine-turned gilt dial cover.

6. A smaller do., in ornamental gilt metal case with chased columns, enriched mouldings and chased lion on top. (Registered design.)

PORTABLE CLOCKS.

7. An eight-day lever clock with compensation balance (striking hours and half-hours, and repeating the last hour on bell-spring, with very fine cathedral tone), in German silver case.

8. A do. of different pattern.

9. A do. in bronze metal case.

10. A do. chiming the quarter on four bells (Cambridge chimes) and striking the hours on bell-spring, in very handsome gilt metal case, with chased columns and figure on top. (Registered design.)

GILT DRAWING-ROOM CLOCK, "THE TRIUMPH OF NEPTUNE."

11. A gilt drawing-room clock, "The triumph of Neptune," with base and columns of Algerine onyx, designed by E. W., manufactured and registered by his agent at Paris.

White. Edward—*continued.*

GOTHIC HALL CLOCK.

Bracket Hall Clocks.

12. An eight-day clock, striking hours on bell-spring, and quarters on four bells, in richly carved oak case, with columns, roof, crockets, finials, crestings, and panels in polished brass. (Registered design.)

WHITE, EDWARD—*continued.*

13. An eight-day 3 part quarter clock, chiming the quarters on 4 bell-springs, and striking hours on large ditto. Black wood case with chased gilt metal mouldings; the cornice supported by caryatid figures of the four seasons, and with eagle on top. (Registered design.)

14. A ditto chiming quarters on four bells (Cambridge chimes), and striking hour on large bell. Carved Gothic case of various woods. (Registered design.)

15. A ditto in carved oak Gothic case with crockets, crestings, finials, and side panels in polished brass. (Registered design.)

16. A case containing eighteen specimens of monograms, arms, crests, &c., in engraving, enamelling, and precious stones.

KEYLESS WATCHES.

17. A gold hunting pocket chronometer, with two dials—one to show English and the other Turkish time—the case richly engraved with oak leaves and acorns, and with very handsome gold Albert chain to correspond.

18. A gold hunting minute repeater, with dark-blue enamel dial to show the repeating work in the centre—the case richly engraved with vine leaves and grapes, and with very handsome gold Albert chain to correspond. (See opposite page.)

19. A gold hunting quarter repeater, with duplex escapement and compensation balance—the case ornamented with "lilies of the valley," the leaves being in green enamel and the flowers in diamonds, and with brooch and chain to correspond. (See opposite page.)

20. A gold hunting duplex watch, with compensation balance; the case set with diamonds on dark-blue enamel ground, and with brooch and chain to correspond.

21. A ditto, the case set with pearls and diamonds on Maroon enamel ground.

22. A gold hunting lever watch, with compensation balance and independent seconds.

23. A gold open face "blind man's" watch.

24. A gold hunting duplex watch, with compensation balance, repeating hours and quarters.

25 A ditto, repeating half-quarters.

26. A gold open face observation watch, with double eccentric stop seconds, to register the commencement and termination of an observation without stopping the watch.

Eight other keyless watches of different patterns.

WATCHES WINDING WITH A KEY.

27. A gold hunting pocket chronometer.

28. Ditto, open face ditto.

29. A gold hunting lever watch, with compensation balance and brequet pendulum spring. Plain case, with hour circle on cover.

30. A gold open face ditto.

31. A ditto ditto, with double roller escapement.

32. A ditto ditto, with gold balance.

Six ladies' gold watches, with engraved cases and dials of different patterns.

An assortment of gold Albert and neck chains, with lockets and other pendants.

White, Edward—*continued.*

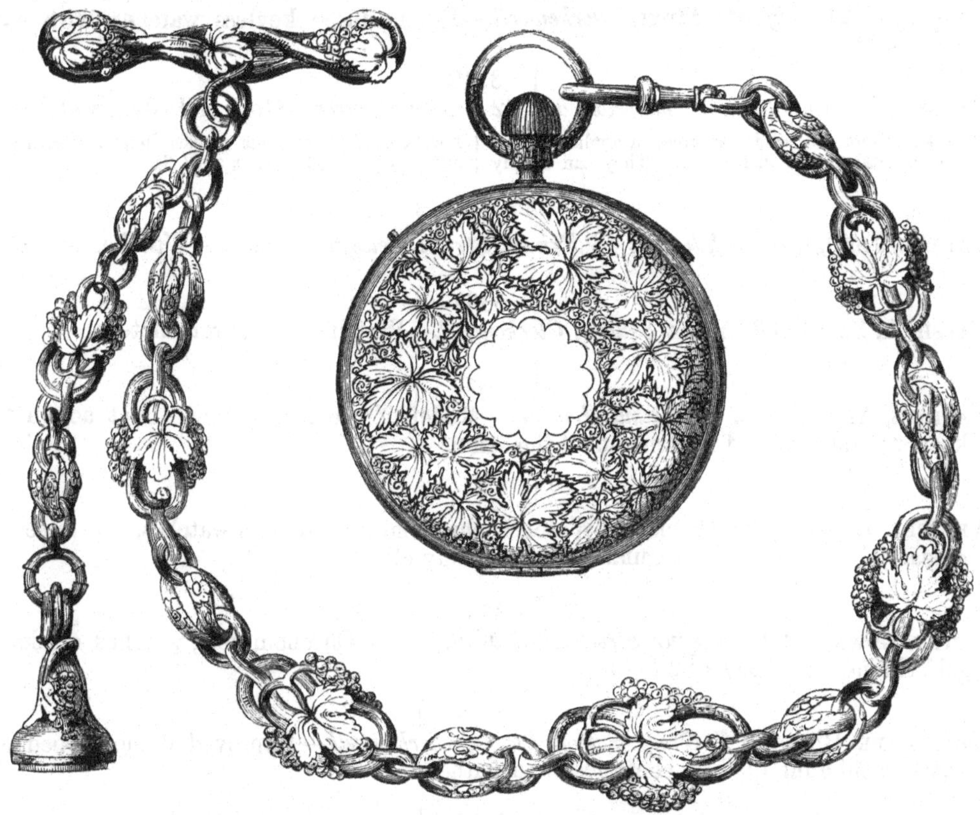

GOLD HUNTING MINUTE REPEATER.

GOLD HUNTING QUARTER REPEATER.

[3318]

VIVIER, O., 21 *Sekforde Street, Clerkenwell.*—Patent fusee keyless watches, with various movements.

[3319]

WALES & M'CULLOCH, 56 *Cheapside, and* 32 *Ludgate Street.*—Gold and silver watches.

The exhibitors will send post free, on application, an illustrated catalogue of their stock. They can supply handsome drawing-room clocks, in gilt cases, at 5*l.* 5*s.*, and in variegated marble, at 3*l.* 3*s.*

[3320]

WALKER, JOHN, 68 *Cornhill, and* 48 *Princes Street, Leicester Square.*—Watches and clocks.

[3321]

WALSH, A. P., 46 *Wilmington Square, Clerkenwell.*—Watches and chronometers.

[3322]

WATKINS, ALEXANDER, 67 *Strand, London.*—Model of the new patent direct action time-keeper; watches, and movements of the same.

[3324]

WEBSTER, RICHARD, 74 *Cornhill.*—Watches, chronometers, keyless watches, centre seconds, repeaters, touch watches, regulators, and railway clocks.

[3325]

WHITE, EDWARD, 20 *Cockspur Street, Pall Mall, S.W.*—Chronometers, watches, clocks, and gold chains. (*See pages* 82 *to* 85.)

[3326]

WHITTAKER, RICHARD, 7 *Great Sutton Street, Clerkenwell.*—Improved dome-capped lever watch, combining quality, cheapness, and flatness.

[3327]

WOOD, THOMAS JAMES, 12 *Long Lane, City.*—Black Forest clocks, with brass works, partly English manufacture.

		s.	*d.*
1. The International clock	price	5	6
Exhibited as the smallest cost at which a really durable and accurate clock has yet been produced.			
2. A small dial	,,	7	6
3. The school and workshop dial	,,	12	0
4. Clock to strike the hours	,,	12	0
5. Ditto, large size	,,	16	0
6. Striking clock, with buhl frame ..	,,	27	0
7. Ditto, large size	,,	35	0

Nos. 6 and 7 are exhibited as specimens of ornamentation.

		s.	*d.*
8. An alarum clock	price	8	6
9. Ditto, large size	,,	14	0
10. Double action alarum clock	,,	18	0
11. Alarum clock, striking the hours ..	,,	14	0
12. Ditto, large size	,,	18	0

Gravity being both the maintaining and regulating power of these clocks, they possess an accuracy of performance unsurpassed by the most costly productions.

[3329]

YOUNG, JAMES, *Knaresborough.*—Improvements in the construction of lever watches to save time in repairing, &c.

[3330]

McLENNAN, J., 6 *Park Place.*—Pocket chronometers.

[3331]

PETIT, S. A., 69 *Princes Street, Leicester Square.*—Regulators, watches, &c.

CLASS XVI.

MUSICAL INSTRUMENTS.

[3360]

ALLISON, RALPH, & SONS, *Wardour Street, W.*—Elegant oak piano, temp. Charles I., and Improved London Model. (*See page* 88.)

[3361]

ALLWOOD, THOMAS, 16 *Bow Street, Birmingham.*—Six violins and violoncello.

[3362]

BATES & SON, 6 *Ludgate Hill, London, E.C.*—Cottage pianofortes.

A semi-cottage pianoforte in Italian walnut wood case, handsomely carved; compass seven octaves; trichord, treble, &c.

[3363]

BELL, JOSEPH, *Gillygate, York.*—An harmonium with wood reeds and pedals, two octaves; also an instrument containing bassoon, oboe, and clarionet, in the shape of a violoncello, with two rows of keys and wood reeds.

[3364]

BESSON, F., Manufacturer, late of *Paris*, now of 198 *Euston Road, N.W.*—Musical instruments (brass).

Family of *transposition* instruments, enabling the player to perform the most difficult music, and to change instantly from one key to another without once removing the lips from the mouthpiece. The system may be adapted to any three-valve instrument, to which it gives the equality and almost the resources of the violin.

Family of *neoform* instruments—with moveable bells. The main advantage of this model, and which F. BESSON'S [a somewhat similar shape being made by other houses] alone possesses, is that the instruments are perfectly equipoised, and accordingly will stand upon their bells; thus rendering them commodious and less liable to injury.

Family of bugles—simple or chromatic at will.

Circular instruments (*passing over the shoulder*), very suitable for cavalry and the field, equipoised.

Usual form instruments, with F. BESSON'S latest improvements. New French horns, with and without piston attachment (2 and 3 valves); Koenig horns, pocket saloon cornets, for officers, amateurs, and for presentation. Ophicleides, trombones with double slides (only half the length of the single slide instruments with increased power of tone), new trumpets, chromatic and regulation; duty bugles, &c.

To all the above instruments the pistons *à colonne d'air pleine* may be applied, whereby the wind passages are rendered so perfectly clear and equal that freedom and softness must necessarily follow.

Clarionets, cymbals, side drums; *musical instruments in paper, linen, gutta-percha,* and various other substances,* of perfect tone, tune, and (*apparently*) metallic vibration.

A cornet in aluminium, composed of 105 pieces, each soldered, a feat hitherto deemed impracticable, and in every sense complete. Weight under 12 ounces; that of the brass instrument averages about 36 ounces.

F. BESSON is the possessor of the PROTOTYPE machinery; that is to say, the apparatus by which approved instruments may be repeated in any number with mathematical certainty. Mr. B. devoted ten years' labour exclusively to the perfecting of this machinery, and in his success achieved the greatest desideratum ever sought by the manufacturer and demanded by the patrons of brass musical instruments.

* These are exhibited to prove the all-importance of the proportions and the soundness of the acoustical principles upon which F. BESSON'S instruments are constructed.

ALLISON, RALPH, & SONS, *Wardour Street, W.*—Elegant oak piano, temp. Charles I., and Improved London Model.

ROSEWOOD PIANOFORTE.

The following are exhibited :—

Small rosewood pianoforte, known as the "London Model," suitable for the boudoir or schoolroom: exhibited to show the progress made in the manufacture of pianos by machinery, by the aid of which every part of this little instrument is made.

An elegant oak cottage piano; style, Charles I. (For detailed description, see handbills.)

An elegant walnut-tree (*Italian wood*) semi-cottage piano.

Warerooms :—108 Wardour Street.

"Steam-power Pianoforte Works,"
Werrington Street, N.W.

[3365]

BETTS, ARTHUR, 27 *Royal Exchange.*—Violins.

[3366]

BEVINGTON & SONS, 48 *Greek Street, and Rose Street, Soho, London.*—An organ, of three manuals and pedals: chancel organ, two, and five stops. (*See page* 90.)

[3367]

BOND, WILLIAM & JOHN, 44 *Norton Street, Liverpool.*—Pianoforte: construction of wrest plank on a new principle.

[3368]

BOOSEY & CHING, 24 *Holles Street, London.*—Six harmoniums — two with pedals, one having self-blowing machine. (*See pages* 92 & 93.)

[3369]

BOOSEY & SONS, 24 *Holles Street, London.*—Military band instruments, reed and brass; Pratten's perfected flutes. (*See pages* 94 & 95).

[3371]

BRINSMEAD, J., 15 *Charlotte Street, Fitzroy Square.*—Pianos. (*See page* 91.)

[3372]

BROADWOOD, JOHN, & SONS, 33 *Great Pulteney Street, London.*—Four grand pianofortes; also parts and models illustrative of construction.

[3373]

BROOKS, HENRY, & SONS, *London.*—Patent pianoforte hammer-rails, keys, actions, mouldings, fret carvings, &c.

[3374]

BUTLER, GEORGE, *Greek Street, Soho, London.*—Cornets, saxhorns, flutes, and drums.

The following brass band instruments are exhibited: Cornets in seven different models; a full set of saxhorns from soprano to bombardon; bass and side drums; improved military side drums with screws for tuning; circular vibrating horns, including every instrument from the soprano in E ♭ to the bombardon in B ♭, made in a circular form. These instruments are remarkable for the clearness and brilliancy of tone, caused by there being no impediment to the full passage of the wind. The monster bombardon is made to encircle the body, and to rest on the right shoulder, which is a great assistance in marching or riding.

[3375]

CADBY, CHARLES, 3, 33, 38, & 39 *Liquorpond Street.*—Pianofortes and harmoniums.

The instruments from these manufactories are well known throughout the United Kingdom and the Colonies for their valuable qualities. Intending purchasers, either for home use or export trade, can make their selections from a large, well-seasoned, and varied stock in the show-rooms of the exhibitors.

[3376]

CARD, E. J., 29 *St. James's Street.*—Semi-metallic and metal flutes.

[3377]

CHALLEN, CHARLES, & SON, 3 *Berners Street, Oxford Street.*—Oblique grand and cottage pianofortes.

AN OBLIQUE GRAND PIANOFORTE, in the Louis XVI. style, of walnut, inlaid with box and purple woods, and with ormolu mouldings and enrichments; chased and gilt.

Also two GRAND COTTAGE PIANOFORTES, in fine Italian walnut cases, ornamented with simple carvings in the Elizabethan style, and finished internally with patent double actions, which can be regulated to suit a dry or damp climate with the greatest facility.

These exhibitors (whose business has been established nearly sixty years) have not only an English, but a foreign reputation, and are favourably known for the general excellence of their instruments. The examples exhibited will bear the test of comparison with the workmanship of any other makers.

BEVINGTON & SONS, 48 *Greek Street, and Rose Street, Soho, London.*—An organ, of three manuals and pedals: chancel organs, two and five stops.

Builders of the great organ, Paris Exposition of 1855, which gained the first-class medal for tone and workmanship.

Also of the celebrated organs in the chapel of the Foundling Hospital, the churches of St. Martin's in the Fields, St. Gabriel's, Pimlico, St. Paul's, Covent Garden, and Dublin Exhibition of 1853.

Exhibitors of the great organ, showing the mechanism, in the Royal Horticultural Garden entrance.

Also of chancel organs, at 35 guineas and 75 guineas, being Nos. 1 and 3 on the annexed list of prices.

CHANCEL ORGAN, NO. 1.

CHANCEL ORGAN, NO. 3.

STOPS.	No. 1. PRICE 35 GUINEAS.	PIPES.
1.	Open diapason, wood bass, CC to C	49
2.	Principal, metal, CC to C	49
	Total	98

No. 2. PRICE 50 GUINEAS.

1.	Open diapason, metal, G to F	47
2.	Stop diapason / Claribel } wood, CC to F	54
3.	Principal, metal, CC to F	54
	Total	155

Size, 3 feet deep, 5 feet 9 inches wide, 10 feet 6 inches high. Octave of German Pedals.

	No. 3. PRICE 75 GUINEAS.	PIPES.
1.	Bourdon, CCC to CC, 16 feet tone, wood	13
2.	Open diapason, metal (G), wood bass, CC to F	54
3. 4.	Stop diapason / Claribel } wood, CC to F	54
5.	Dulciana, metal, C to F	42
6.	Principal, metal, CC to F	54
	Total	217

Size, 3 feet 8 inches deep, 6 feet 4 inches wide, 11 feet high. Octave of German pedals.

STOPS.	No. 4. PRICE 100 GUINEAS.	PIPES.
1.	Bourdon, CCC to CC, 16 feet tone, wood	13
2.	Open diapason, metal (FF), wood bass, CC to F	54
3. 4.	Stop diapason / Claribel } wood, CC to F	54
5.	Dulciana, metal, C to F	42
6.	Principal, metal, CC to F	54
7.	Flute, wood, C to F	42
8.	Mixture, metal (12th and 15th), CC to F	108
	Total	367

Size, 4 feet deep, 6 feet 6 inches wide, 12 feet high. Octave of German pedals.

These organs are built of the best material and workmanship; simple in construction, rich and full in tone, and have been designed by Messrs. Bevington to supply a want long felt, viz., a small church organ of architectural character, suitable for any position in the building, with the quality of tone of a large instrument. Barrel attachment, to play *eight tunes* in absence of organist, 8 guineas.

Manufactory, Greek Street, and Rose Street, Soho, London.

BRINSMEAD, JOHN, 15 *Charlotte Street, Fitzroy Square.*—Pianos.

The exhibitor's perfect Check Repeating Grand and Upright Pianos were patented by him February 1862.

The characteristics of this action is its very rapid repeat, the check acting with the slightest movement of the key, an advantage long desired, but until now unattained; the simplicity of mechanism renders these pianos most durable. The equally balanced arrangement of metal and wood in the construction of the case particularly adapts them to meet the requirements of extreme climates.

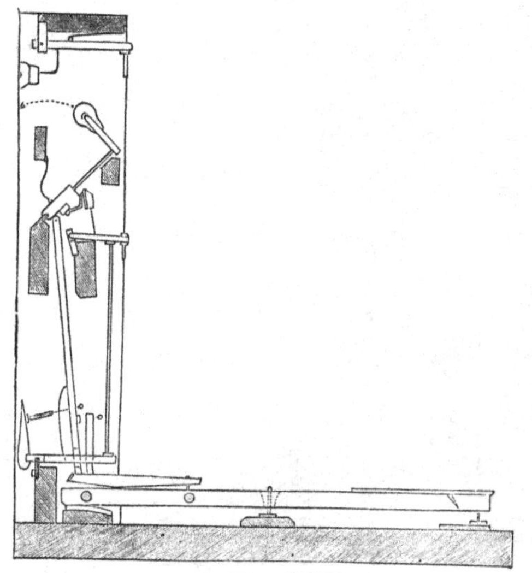

UPRIGHT ACTION.

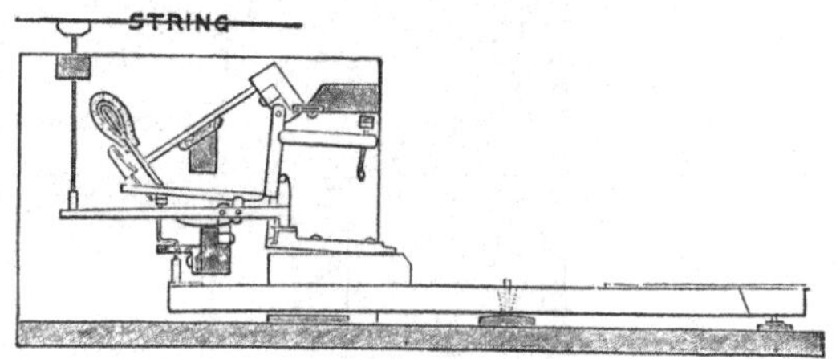

GRAND ACTION.

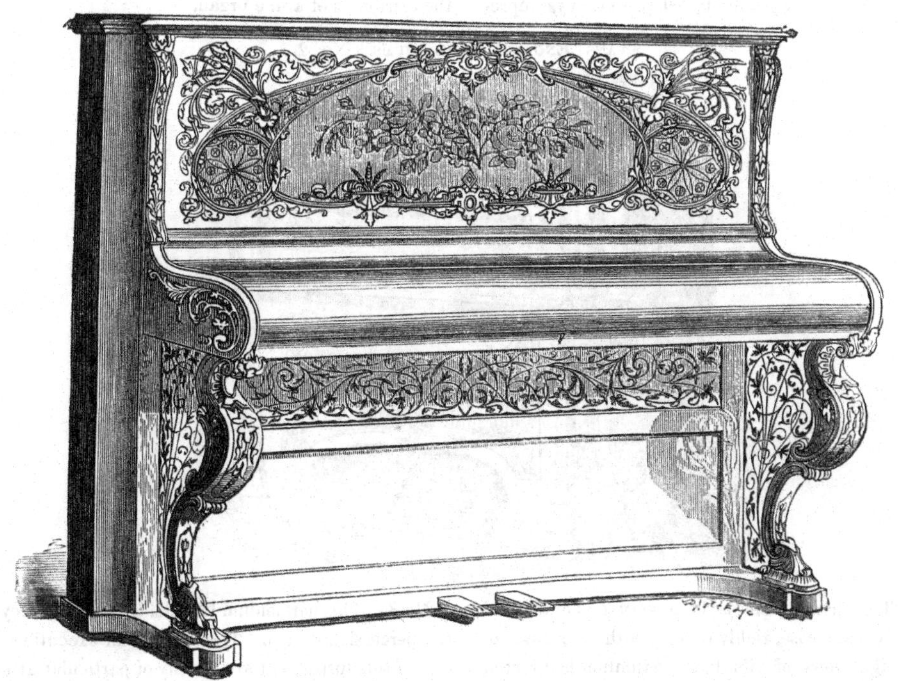

BRINSMEAD'S PATENT CHECK-REPEATING PIANO.

Boosey & Ching, 24 *Holles Street, London.*—Six harmoniums; two with pedals, one having self-blowing machine.

Evans' English Harmoniums. No. 1.

No 1.—The above is a drawing of the Organ Harmonium in a carved oak case, with two rows of keys, and two and a third octaves of pedals, with independent reeds, 32 and 16 feet scales. The upper row of keys represents the swell, and the lower row the great organ. Couplers from pedal to great, and from swell to great. This instrument has eleven rows of vibrators, and all the attributes of a fine organ.

Evans' English Harmoniums. No. 2.

No 2.—This harmonium is in a veiy elaborate and handsome walnut case, richly carved, with two rows of keys and eight rows of vibrators. Attention is directed to the great resources of this instrument, although it is of such moderate dimensions.

No 3.—This harmonium has a single row of keys and the percussion action. The design and execution of the case of this instrument are worthy of particular attention.

⁂ The cases of the above instruments are from designs by Mr. Hugh Stanus, of the Sheffield School of Art

BOOSEY & CHING—*continued.*

No. 4.—Harmonium in an American walnut case, with one row of keys and two and a fourth octaves of pedals. Attached to the seat of this instrument is the new patent self-acting blowing machine. Although many attempts have been made to manufacture a self-acting blowing machine, Boosey and Ching believe that this is the only one of the kind that has ever proved successful.

No. 5.—A specimen of the School or ten-guinea Harmonium.

No. 6.—A specimen of the cottage or six-guinea harmonium in a polished pine case. Double pedals and full compass of five octaves.

General remarks about Evans' Harmoniums.

These instruments, first introduced by Mr. Evans in 1843, were brought prominently before the public in 1859, when Messrs. Boosey undertook the full development of the plans Mr. Evans had so successfully designed. Since that period they have rapidly increased in popularity, and have been the means of dissipating the prejudice which formerly existed against the harmonium. Quickness of "speech," flute-like quality of tone, and a great combination of delicacy and power of expression, are some of the charisteristics of the English harmonium. Very beautiful effects may be produced by the combination of the harmonium with the pianoforte and chamber stringed instruments, so as to form a miniature orchestra capable of rendering the highest class of chamber music.

CASE'S PATENT CONCERTINAS, Manufactured and Exhibited by BOOSEY & CHING, 24 Holles Street, London.

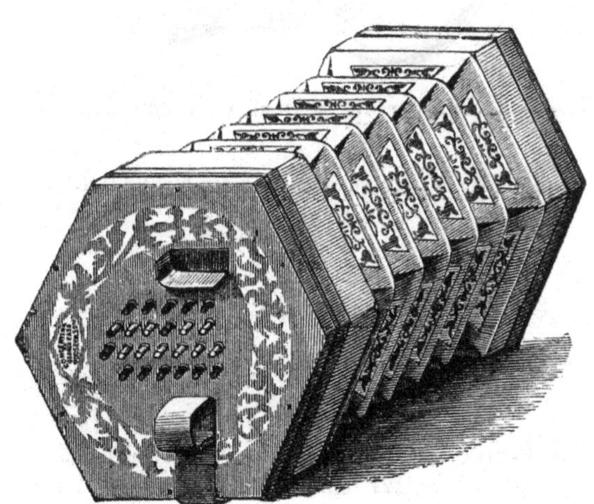

The universal popularity of the concertina may be ascribed to the many advantages which it possesses over other musical instruments. Its tones are pure, sweet, and brilliant. Its compass is greater than that of the flute, and almost equal to that of the violin. It admits of very great execution and expression. Music written for the pianoforte, violin, flute, or any other instrument, can be performed with equal effect on the concertina. By creating harmonies of any number of parts, it produces a variety of tones and effects only attempted on the pianoforte. The concertina is more easily learnt than any other instrument. It is compact and portable, and appears to equal advantage in the hands of ladies and gentlemen.

The Concertinas by Case are manufactured by BOOSEY and CHING, under the personal superintendence of Mr. GEORGE CASE, the eminent professor and performer, with the aid of experienced workmen and patent machinery. These instruments will be found to remain well in tune—an important feature peculiar to CASE'S concertinas.

The case exhibited contains specimens of treble, baritone, and bass concertinas.

BOOSEY & SONS, Manufacturers of military band instruments, 24 *Holles Street, London.*— Military band instruments, reed and brass, and Pratten's perfected flutes.

A case of reed and brass instruments containing specimens of the following :—

An Euphonion or solo bass in B flat, with four rotary cylinders.
A Bombardon in E flat, with four rotary cylinders.
An Althorn in B flat, with three ditto ditto.
A Trumpet in F, ditto ditto.
A Flugel horn in F, with four rotary cylinders.
A ditto in B flat, ditto ditto.
Two Cornet-à-pistons of the new gold metal, one with cylinders and the other with valves
A round or rotary model Cornet-à-piston.
A sterling silver presentation field bugle.
Several Clarionets in B flat and E flat.

STERLING SILVER CORNET-À-PISTON, WITH ROTARY CYLINDERS AND GILT BELL.

Particular attention is directed to the very superior workmanship displayed in the manufacture of the above instruments.

R. S. PRATTEN'S PERFECTED FLUTES, FIFES, AND PICCOLOS, Manufactured and Exhibited by BOOSEY & SONS, 24 Holles Street, W.

The CONCERT FLUTE Number 1A, is the fac-simile of that upon which Mr. Pratten plays at the Royal Italian Opera, musical festivals, &c., and is so constructed that all the keys are within the reach of the fingers whilst in the act of playing. The holes, which are extremely large, and the same size throughout the instrument, are closed with keys regulated to obey the most delicate touch, and can be fingered with perfect ease by the smallest hand, as all unnatural extension of the fingers is avoided. Thus perfect equality is obtained, and the performer can produce the most rapid passages either piano or forte, with the same facility as upon the small-holed flute, and without endangering the intonation. The fingering is the same as that of the old flute, with the addition of a perfect C♯ in the two middle octaves, fingered without the aid of the C♮ key, thus simplifying all the sharp keys where arpeggios are concerned, as also the D♭ in flat keys.

The top octave possesses advantages which facilitate the execution of passages almost insurmountable on other flutes, whilst its peculiarly convenient arrangement of the keys under the fingers renders all shakes perfectly easy to produce, without in the least altering the position of the hand.

Several of the CONCERT FLUTES as well as MILITARY FLUTES and PICCOLOS in E flat and F, and FIFES are also exhibited.

Boosey & Sons—*continued.*

MESSRS. BOOSEY AND SONS' BASSO PROFONDO (REGISTERED).

The Basso Profondo, or double slide Contra Basso Trombone in B flat; lowest note 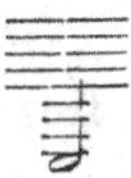

The Basso Profondo was first introduced at the Crystal Palace Brass Band contest in July, 1861, and excited the universal admiration of the judges in consequence of the depth and brilliancy of its tone, and the ease with which they were produced. Although bombardons in B flat are occasionally to be met with, their great weight, and the inability of men with ordinary lungs to perform upon them with any satisfaction, have prevented their general adoption. The basso profondo, on the contrary, weighs only eight pounds and a few ounces, and is played with as much ease as an ordinary bass trombone. Boosey and Sons have purchased of the inventor the original instrument, with the sole right of manufacture, and fully expect the basso profondo will in future be an indispensable bass instrument in every military band. Price, complete with scale and exercises, 14 guineas.

[3378]

CHAPPELL, ARTHUR, 214 *Regent Street.*—Military clarionets, bassoons, flutes, Azemar's silent practice drum, &c.

[3379]

CHAPPELL & CO., 50 *New Bond Street.*—Patent pianofortes and harmoniums, with and without pedals.

[3380]

CHIDLEY, EDWARD, 28 *Store Street, W.C.*—Treble and baritone concertinas.

[3381]

CHIDLEY, ROCK, 135 *High Holborn.*—Harmoniums and concertinas.

[3382]

CLINTON & CO., 35 *Percy Street, Tottenham Court Road.*—Wood and metal flutes of every description.

EQUISONANT FLUTE. This instrument is the only flute equal in tone and tune *throughout* The system of fingering (which is nearly the same as the ordinary flute) offers unprecedented facilities for every description of passages. Clinton and Co. now manufacture the equisonant, the Boehm, and the eight-keyed flute with the cylindrical bore and parabola head, both in metal and wood; but having discovered the means of removing the inequalities and objectionable parts of that system, their cylinder flutes will be found far superior to any others ever offered to the public. One trial will prove their superiority. They can be obtained *only* of the patentees and manufacturers, Clinton & Co., 35 Percy Street, Tottenham Court Road. Explanation gratis.

[3383]

COLLARD & COLLARD, 16 *Grosvenor Street, W., and* 26 *Cheapside, E.C.*—Four pianofortes. (*See pages* 98 *to* 101.)

[3384]

CONS, F. & F., 81 *John Street, Tottenham Court Road.*—The interior action of a piano.

[3385]

COOK, CHARLES & H. E., *Tavistock Place.*—Pianoforte silk-fronts.

[3386]

CORFE, EDWARD, 28 *Bedford Terrace, Old Ford Road, Victoria Park.*—Musical strings.

[3387]

COXHEAD, CHARLES J., *Castle Street, Shrewsbury.*—Oblique pianoforte, with new patent action.

[3388]

CROGER, THOMAS, 483 *Oxford Street, London, W.C.*—Æolian harps, educational, transposing, metallic harmonicon and metronome, for giving sixty or any number of vibrations in a minute for music, photography, or marching. (*See page* 97.)

[3389]

DAVIS, J. MOIRATO, 40 *Esher Street, Kennington Lane.*—Valves to musical instruments—action inclosed, free from dust.

[3390]

DEARLOVE, MARK WILLIAM, 156 *North Street, Leeds.*—Violins, viola, &c.—own make.

[3391]

DIMOLINE, ABRAHAM, 34 *College Green, Bristol.*—Rosewood cottage pianoforte.

CROGER, THOMAS, 483 *Oxford Street, London, W.C.*—Æolian harps, educational, transposing, metallic harmonicon and metronome, for giving sixty or any number of vibrations in a minute for music, photography, or marching.

The most delightful effect is obtained by THOMAS CROGER'S NEW PATENT ÆOLIAN HARP, which will produce music in the garden, conservatory, summer-house, on the balcony, or window-ledge, on board any vessel on the water, on the branches of a tree, or "*any other place,*" without a performer. It merely requires placing on a table or stand, or laying across the branches of a tree, or it may be suspended from one, or from any convenient place. It does not signify whether it is placed perpendicular, horizontal, or diagonally; the object is to cause the draught to pass through where the strings are, which will set them in vibration, and bring forth the most melodious sounds ever heard, far superior to anything else. At a distance the tones are truly delightful; and what renders it so amusing is, that any one not being aware of its position cannot trace from whence it proceeds; the effect is so peculiar it seems to be in every direction at once. All persons are sure to be surprised and delighted at the romantic effect; it may be used by persons totally unacquainted with music; and will produce an endless source of amusement by its various sounds. Full instruction is attached to each one.

SINGLE HARPS.	12 Strings.	24 Strings.
Plain wood	£0 14 0	£0 18 0
White varnished . . .	0 16 0	1 0 0
Amber varnished, and ornamented with black and crimson lines, best finish . .	0 18 0	1 2 0
Of Honduras Mahogany French Polished	1 4 0	1 8 0
Of choice Spanish Mahogany Rosewood, Walnut Wood, or Bird's-eye Maple, with a band round the edge, about ½ in. wide, of Zebra-wood or black ebony, beautifully French polished, very chaste, for the drawing-room.	1 10 0	1 14 0

DOUBLE HARPS.		
Plain wood	1 0 0	1 8 0
White varnished	1 4 0	1 12 0
Amber varnished, and ornamented with black and crimson lines, best finish . .	1 8 0	1 16 0
Of Honduras Mahogany French polished . .	1 16 0	2 4 0

	24 Strings.	48 Strings.
	£ s. d.	£ s. d.
Of choice Spanish Mahogany Rosewood, Walnut Wood, or Bird's-eye Maple, with a band round the edge, about 1½ in. wide, of Zebra-wood or black ebony, beautifully French polished, very chaste, for the drawing-room	2 2 0	2 10 0

The Double Harps are so contrived that they can be separated, thus forming two Single ones, for two different positions, if required.

All the above harps are 32 in. long, and may be had shorter at the same prices; but if ordered longer, they will be charged extra as follows: those at 14*s.*, 16*s.*, and 18*s.*, 6*d.* per inch for every inch beyond 32 in.; those at 20*s.*, 22*s.*, and 24*s.*, 9*d.* per inch. do; those at 28*s.*, 30*s.*, 32*s.*, 34*s.*, and 36*s.*, 1*s.* per inch do.; and those at 42*s.*, 44*s.*, and 50*s.*, 1*s.* 3*d.* per inch do.

THE NEW PATENT EDUCATIONAL TRANSPOSING METALLIC HARMONICON.—The quality of the notes or sound is the same in them all; it is the finish of the case which makes the difference in the price of any one size, for example:—

3 Octaves, with semitones	.	23*s.*, 30*s.*, 40*s.*, and 84*s.*
2½ Do.	do.	. 18*s.*, 25*s.*, 35*s.*, „ 73*s.* 6*d.*
2 Do.	do.	. 14*s.*, 20*s.*, 30*s.*, „ 63*s.*

WITH A SINGLE ROW OF NOTES.

3 Octaves, 22 notes	12*s.* and 18*s.*
2½ „ 19 „	10*s.* „ 15*s.*
2 „ 15 „	8*s.* „ 12*s.*
1½ „ 12 „	6*s.* „ 10*s.*
10 „	5*s.*
8 „	4*s.*

Musical instruments and materials of every description of the highest quality on the most advantageous terms.

Notes or vibrators, keys, pipes, stops, &c., for harmonium making or organ building.

THOMAS CROGER'S newly revised, illustrated explanatory price lists, for musical instruments of every description, with testimonials from eminent professors, amateurs, and opinions of the press, should be in the possession of every person as a book of reference, before purchasing anything whatever in the musical business, and which may be had gratis, or post free, from the manufactory as above.

COLLARD & COLLARD, 16 *Grosvenor Street, W., and* 26 *Cheapside, E.C.*—Pianos.

AN OBLIQUE GRAND PIANOFORTE ON AN EXTENDED SCALE, BY COLLARD AND COLLARD, IN WALNUT WOOD AND GOLD, IN THE LOUIS SEIZE STYLE.

The following PIANOFORTES and MODELS OF ACTIONS are exhibited by COLLARD & COLLARD:—

1.—A CONCERT GRAND PIANOFORTE of 7 octaves, A to A, with patent repetition action, in very choice walnut-wood case, with carved enrichments in the *renaissance style.*

2.—AN ELEGANT CONCERT GRAND PIANOFORTE of 7 octaves, A to A, with patent repetition action, in very choice rosewood case, with massive carved cabriole trusses.

⁂ The attention of the connoisseur is particularly directed to the mechanism of these instruments, as showing fewer centres of friction, and greater simplicity of contrivance than in any other Grand action, whereby greater durability is secured, and the operation of regulating is considerably facilitated. The power of blow in the hammer is also increased, and the delicacy of touch and unfailing promptness of repetition, so essential for the requirements of the modern school of Pianoforte playing, are secured to an extent to satisfy the most fastidious finger.

3.—AN OBLIQUE GRAND PIANOFORTE of 6⅞ octaves, C to A, with patent escapement and repetition action, in satin-wood case, with carved and gilt decorations in the *Italian style.*

4.—AN OBLIQUE GRAND PIANOFORTE, on extended scale, of 6⅞ octaves, C to A, with patent escapement and repetition action, in very choice walnut-wood case, with carved and gilt enrichments in the *Louis Seize style.*

⁂ The Oblique Grand Pianoforte (of which Nos. 3 and 4 are very unique specimens) is an instrument of comparatively recent introduction. The application of Collard and Collard's well-known and important improvements in upright pianofortes have tended in no

COLLARD & COLLARD—*continued.*

A WALNUT WOOD CONCERT GRAND PIANOFORTE BY COLLARD AND COLLARD, IN THE RENAISSANCE STYLE.

small degree to strengthen the favourable judgment which musical connoisseurs and the fashionable world have bestowed on them. Convenient and elegant in form, and effective in the highest degree, both as regard power of tone and perfection of touch, these charming instruments are found to be, for rooms of limited size, the most effective substitute for the full Grand Pianoforte, to which, in character of tone, they closely approximate.

The mechanism of these Pianofortes is exemplified by Model No. 3.

5.—A SOLID WOOD SPANISH MAHOGANY SQUARE SEMI-GRAND PIANOFORTE of 6⅞ octaves, C to A, with patent repetition action and transverse bass strings, as manufactured by Collard and Collard expressly for the East Indies and tropical climates.

*** The principle of construction of this pianoforte differs in a striking degree from that of the ordinary Grand Square. The object sought to be attained is the greatest power of resisting the destructive influence of East Indian and tropical climates, without the usual reliance on the adhesive properties of glue. This result has been most successfully accomplished, and the experience of several years has proved that under the most trying ordeals, these instruments have satisfactorily stood the test;—public opinion in India having awarded them the highest praise.

The principle of the action is illustrated by the model No. 2.

6.—A PIANINO, OR SMALL COTTAGE PIANOFORTE, in plain rosewood case, of 6⅞ octaves, C to A, O G fall; fretwork front and octagon legs.

*** This instrument is an example of the cheapest upright instrument manufactured by Messrs. COLLARD & COLLARD. Such is the popularity of these instruments that, during periods of active trade, the yearly demand reaches the large number of nearly 2000.

COLLARD & COLLARD—*continued.*

AN OBLIQUE GRAND PIANOFORTE BY COLLARD AND COLLARD, IN SATINWOOD AND GOLD, IN THE ITALIAN STYLE.

LIST OF MODELS.

No. 1.—The action of the Concert Grand Pianoforte.

No. 2.—The action of the New Square Semi-Grand Pianoforte.

No. 3.—The action of the Oblique Grand Pianoforte.

No. 4.—The action of the Cottage Pianoforte.

No. 5.—The Model of a Cottage Pianoforte, in two divisions, and extensively manufactured for the South American market. The weight of the instrument being equally divided and brought within the limit of a mule's burthen, its transport over the Andes (otherwise impossible) is thus rendered of easy accomplishment. The parts are readjusted without the smallest difficulty, and the instrument in no respect suffers from its temporary disjointment.

Dates and particulars of Patents, Registrations, &c., assigned to MESSRS. COLLARD & COLLARD, *London.*

1827.—*March 2nd.*

For "certain improvements in pianofortes, and in the mode of stringing the same;" viz., an application of the check action to the square pianoforte, thenceforward called the grand square; and a new mode of stringing,* adapted to instruments of all kinds by passing the wire round a single pin,—thus superseding the use of the noose or eye before in general use: also for a new arrangement of the damper, known as the elongated damper-head, by which the jarring consequent on the old method was entirely prevented, and more effectual damping secured.

1829—*November 2nd.*

For "improvement in upright pianofortes," viz., applying a check to the under hammer to prevent the rebound of the hammer against the string.

1835—*January 15th.*

"For improvements in the mechanism of horizontal grand and square pianofortes," consisting of an entirely new construction of the action, the escapement being placed upon the key and coming in contact with a lever or crank, and thus regulating the rise and fall of the hammer, thereby imparting greater vigour to the blow and increased durability to the touch.

* This mode of stringing has become almost universal since the expiration of the patent.

Collard & Collard—*continued.*

A SOLID WOOD MAHOGANY SQUARE SEMI-GRAND PIANOFORTE, MANUFACTURED EXPRESSLY FOR THE EAST INDIES, BY COLLARD AND COLLARD.

List of Patents, &c.—continued.

1838—*January 1st.*

The introduction of a new class of square pianoforte, entitled the "patent square semi-grand pianoforte," being a further improvement of the grand square, by which a closer approximation to the peculiarities of the grand pianoforte was attained.

1841—*November 11th.*

For "certain further improvements in the action of horizontal pianofortes," consisting of the introduction of the traversing escapement fixed upon the hammer rail, resulting in a greater amount of precision and increased vigour of action, as also the introduction of a repetition movement.

1843—*April 29th.*

For "further improvement in the action of pianofortes," viz., the application of the repetition movement to square and to vertical or upright instruments.

1844—*January.*

For the construction of a cottage pianoforte in two divisions, for the purpose of facilitating transport on the backs of mules in the mountainous districts of Central America, otherwise inaccessible by reason of weight.

1847—*October 15th.*

Registered. A new design for the shape of a square pianoforte, entitled the "symmetrical grand square," by which greater beauty of form was secured. The key-board being placed in the centre of the instrument, thus obviating the inelegant appearance of the old instruments.

1855—*May.*

Registered. "An improved key-board," the ends of the sharps being rounded for the purpose of giving to the performer increased facility for rapid execution, and imparting to the key-board a more pleasing appearance.

1857—*February.*

Patented. "Further improvement in the action of vertical pianofortes," having for its object to add increased vigour to the blow of the hammer, giving to the performer the power of a more prompt repetition, and imparting increased durability to the touch.

[3392]

DISTIN, HENRY, 9 & 10 *Great Newport Street, St. Martin's Lane.*—Military musical instruments of every description.

INTERIOR OF HENRY DISTIN AND CO.'S MUSICAL INSTRUMENT MANUFACTORY.

The exhibitors are manufacturers of musical instruments to Her Majesty's army and navy, the forces in India, volunteer corps, and the Royal and Imperial Italian Operas of London and St. Petersburgh.

Persons interested in the manufacture of musical instruments are invited to visit the above factory.

[3393]

DODD, JAMES, *Image Cottage, Holloway Road, Islington, N.*—Violin, tenor; violoncello bows; silvered music-strings; specimen of workmanship.

[3394]

DUFF, HODGSON, & Co. (late TOWNS), 20 *Oxford Street.*—Pianofortes.

The exhibitors beg to call the attention of the musical world and the public to the excellence of the improved pianofortes made by them; specimens of which may now be seen at the International Exhibition, where they have met with the unqualified approbation of some of our most distinguished pianists. These instruments, for quality and quantity of tone, delicacy of touch, and durability of construction, cannot be surpassed. They have been exported to the most trying and extreme climates, and have been found superior to most others.

The following is a descriptive and priced list of those most in demand:—

No.	Description	Price
1.	Solid walnut or mahogany boudoir, full compass	30 guineas.
2.	Elegant rosewood or zebra wood boudoir	38 "
3.	In French walnut (of great beauty).	42 "
4.	Ditto, extra elegant . . .	45 "
5.	Ditto, with carved scroll legs and plinth	50 "
6.	Trichord rosewood cottage, ditto .	55 "
7.	Ditto, walnut cottage, ditto .	60 "
8.	Rosewood cottage	45 "
9.	French walnut cottage . . .	50 "

[3395]

EAVESTAFF, WILLIAM, 17 *Sloane Street.*—A trichord walnut-wood pianoforte, seven octaves.

[3396]

EAVESTAFF, WILLIAM GLEN, 60 *Great Russell Street, Bloomsbury.*—Pianoforte.

[3397]

Fincham, John, 110 *Euston Road, London.*—Six stops of organ-metal pipes shown in skeleton organ.

[3398]

Forster & Andrews, *Hull.*—A grand church organ and a model chancel organ. (*See page* 104.)

[3399]

French, James Martin, 67 *Bull Street, Birmingham.*—Cottage grand pianoforte, with tubular braced back.

[3400]

Geary, John, *Prince of Wales Road, Kentish Town.*—A rosewood truss piccolo pianoforte; a walnut truss semi-cottage pianoforte.

[3401]

Glassbarrow, C., 104 *Great Russell Street.*—New and improved piano.

[3402]

Glen, Thomas, 2 *North Bank Street, Edinburgh.*—Highland regimental bagpipes in metal, made expressly for tropical climates.

[3403]

Greaves, Edward, 76 *Milton Street, Sheffield.*—Æolian pitch-pipes, tuning-forks and hammers, chromatic tuning-forks, portable metronomes, &c.

[3404]

Greiner & Sandilands, 1 *Golden Square, London.*—Boudoir, grand, and cottage pianofortes, with patented choir tuning.

[3405]

Hampton, Charles, 31 *Charlotte Street, Fitzroy Square.*—Improvements in the construction of first-class pianofortes. (*See page* 105.)

[3406]

Harrison, Joseph, & Co., 65 *John Street, Fitzroy Square, London, W.*—A pianoforte with patent iron clipper plates and gilt steel wire, that will not rust.

[3407]

Higham, Joseph, *Victoria Bridge, Manchester.*—Brass musical instruments.

[3408]

Hill, William Ebsworth, 192 *Waterloo Bridge Road, London.*—Gold and silver-mounted violin, &c.; bows, viola, and a violin.

[3409]

Holdernesse, W., 444 *Oxford Street, London, W.C.*—A cottage pianoforte in an elegant walnut-tree case.

[3410]

Holman, J. & E., 43 *London Street, Fitzroy Square.*—Patent model action of piano.

[3411]

Hopkins, Thomas M., *Worcester.*—Double bass, with apparatus attached, for producing enharmonic scales of harmonics.

FORSTER & ANDREWS, *Hull.*—A grand church organ and a model chancel organ.

GREAT ORGAN CC to G.

1.—Double open diapason	16 feet	...	56 pipes.
2.—Open diapason ...	8 „	...	56 „
3.—Gamba	8 „	...	56 „
4.—Hohlflöte	8 „	...	56 „
5.—Stopt diapason ...	8 „	...	56 „
6.—Principal	4 „	...	56 „
7.—Waldflöte	4 „	...	56 „
8.—Twelfth and fifteenth	3 & 2 „	...	112 „
9.—Mixture	...	...	280 „
10.—Posaune	8 „	...	56 „
11.—Trumpet	8 „	...	56 „
12.—Clarion	4 „	...	56 „
			952 „

PEDAL ORGAN CCC to F.

1.—Open diapason bass	16 feet	...	30 pipes.
2.—Stopt diapason bass	16 „	...	30 „
3.—Principal bass ...	8 „	...	30 „
4.—Stopt flute bass ...	8 „	...	30 „
5.—Trombone bass ...	16 „	...	30 „
6.—Pedal organ in octaves	16 „	...	60 „
			210 „

SWELLING ORGAN CC to G.

1.—Bourbon	16 feet	...	56 pipes.
2.—Open diapason ...	8 „	...	56 „
3.—Bell gamba ...	8 „	...	56 „
4.—Stopt diapason ...	8 „	...	56 „
5.—Principal	4 „	...	56 „
6.—Flauto traverso ...	4 „	...	56 „
7.—Mixture	...	...	280 „
8.—Double trumpet ...	16 „	...	44 „
9.—Cornopean ...	8 „	...	56 „
10.—Hautboy	8 „	...	56 „
11.—Clarion	4 „	...	56 „
			828 „

CHOIR ORGAN CC to G.

1.—Lieblich gedact ...	16 feet	...	56 pipes.
2.—Dulciana	8 „	...	56 „
3.—Stopt diapason ...	8 „	...	56 „
4.—Spitzflöte	4 „	...	56 „
5.—Dulcet flute ...	4 „	...	56 „
6.—Gemshorn ..	2 „	...	56 „
7.—Harmonic piccolo	2 „	...	56 „
8.—Clarinet	8 „	...	37 „
9.—Grand ophicleide	8 „	...	56 „
			485 „

ACCESSORY MOVEMENTS AND COUPLETS.

1—Great to pedals.
2.—Swell to pedals.
3.—Choir to pedals.
4.—Swell to great.
5.—Choir to great.
6.—Swell to choir.
7.—Sforzando pedal No. 1.
8.—Sforzando pedal No. 2.
9.—Tremulant to swell.
10.—Combination pedal.
11, 12, 13, 14, 15 & 16.—Composition pedals.

REMARKS.

The whole of the accessory movements are labelled similar to the registers. Sforzando pedal No. 1 couples the great organ to the swell. Sforzando pedal No. 2 couples the pedal organ to the great. When this pedal is down and the various couplets drawn, the full power of the instrument is concentrated on the great organ and pedals, and although forty-six pipes speak for each key pressed down, and fifty-one for each pedal, the touch remains the same as for a single pipe. The patent pneumatic combination pedal acts simultaneously on the stops in the various organs, producing eight different combinations from one pedal. INTERIOR OF ORGAN. The movements are principally direct action. Improved pneumatic movements are applied to the great and pedal organs, which also act on the whole of the couplets. The bellows are blown by Joy's patent hydraulic engines, supplying wind at four different pressures. The scales of the pipes have been arranged by Professor Töpfer of Weimar, on the proportion of $1 : \sqrt{8}$. The wood pipes from four feet C upwards are of Swiss pine. The large pedal open diapason, and the 16-feet metal double diapason have conical valves under feet. This valve was introduced by F. & A. in 1850. The organ is tuned equal temperament, and the pitch is for C, 528 vibrations in a second. The registers are arranged at an angle of 45° (first introduced by F. & A. in 1850). The pedal keys are concave and radiating. The total number of pipes is 2,475, and of registers 45.

MODEL CHANCEL ORGAN, containing—

1.—Open diapason ...	... ...	56 pipes.
2.—Stopt diapason ...	... ...	56 „
3.—Principal ...	... ...	56 „
4.—Octave couplet ...	... ...	36 „
		204 „

Oak frame, illuminated pipes.

Hampton, Charles, 31 *Charlotte Street, Fitzroy Square, London.*—Improvements in the construction of first-class pianofortes.

Height, 4 ft. 2 in.; Width, 4 ft. 6 in.

No. 6.—Three unisons; seven octaves; ivory bridge. Price 35 guineas.

The principle upon which these pianofortes are made, absolutely prevents settling in the groundwork of the instrument, the long-sought desideratum.—See "Hunt's Handbook." It also improves the tone, renders the necessity for tuning less frequent, and the pianoforte much more durable.

C. Hampton's Cottage Pianofortes are warranted to stand in tune in any climate, and are especially adapted for exportation at prices varying from 20 to 50 guineas.

C. H. begs respectfully to thank those who have so kindly given him their support from his commencement in '51, and to invite the critical attention of the scientific world to his invention of 1860, called "The Double Tension or Compressed Principle," as shown in the glass case No. 7: it will be observed that the "back," or groundwork of the instrument is simply suspended in and otherwise entirely independent of the glass case. The object being to show the construction of the back, and the time and method of applying the compression referred to.

The three tension rods or bars remaining have each a ton pressure on them; three other bars have been applied in the same direction and at the same tension in the treble part of the instrument from its commencement, till it was strung and tuned; hence it follows that six tons pressure has been equally distributed over the piano in the same direction as the strings, before the strings were applied, or even the sounding-board was fixed in its place; it must be evident, therefore, that the shrinking or settling of the groundwork by the pull of the strings, which do not exceed five tons, is obviated by the application of this principle.

These pianos are especially adapted for exportation, or exposed situations, for three reasons:—

1st. They are compensating in principle, and will not rise and fall in pitch with the alternations of temperature; the iron tubes being of the same length and in the same direction as the steel strings contract and expand in the same ratio.

2ndly. The whole of the internal mechanism having been manufactured on the premises for the last eleven years, is warranted first-class, and

3rdly. The veneering being laid in cement instead of glue, will bear an immense amount of heat or damp before it will strip from the underwood. Upwards of 300 on this principle have been sent out, and not a single complaint has been made against them.

In answer to those who think that metal should not be used in the construction of cottage pianos, C. H. begs to draw attention to the fact that our most eminent makers have hitherto taken the best prizes for grands which have contained the greatest quantity of metal in their construction; and respectfully states that his constant endeavour is to assimilate the cottage to the grand, both in its construction and tone, and leaves the public to judge how far he has succeeded.

C. Hampton's pianofortes may be purchased through any music-seller, at the same price as at the factory; but if purchased direct they will be packed and sent free to the nearest railway station in any part of England, and a warranty of three years given with each instrument.

[3412]

HOPKINSON, JOHN & JAMES, 235 *Regent Street, London.*—Patent grand and cottage pianofortes and models.

[*Obtained First Class Prize Medals at the Exhibitions of* 1851 *and* 1855.]

PATENT CONCERT GRAND PIANOFORTE. WALNUT, INLAID WITH IVORY, TULIP, BOX, AND KING WOODS.

COTTAGE GRAND PIANOFORTE, WITH CARVINGS IN THE ITALIAN STYLE.

[3413]

HUGHES, W., & CO., 148 *Drury Lane.*—Covered strings for pianofortes; copper and other music wires.

[3414]

IMHOF & MUKLE, 54 *Oxford Street.*—Orchestrion, or self-acting organ. (*See page* 108.)

[3415]

IVORY & PRANGLEY, 275 *Euston Road, London.*—Semi-cottage pianoforte with patent grand action and keys.

[3416]

JACKSON & PAINE, *Store Street, London, W.C.*—Patentee of the anti-blocking hopper for cottage-pianofortes.

[3417]

KIND, CARL, 50 *George's Grove, Holloway, N.*—Model of a grand pianoforte action, new invention—patented.

[3418]

KIRKMAN, JOSEPH, & SON, 3 *Soho Square.*—Pianofortes. (*See pages* 109 *to* 111.)

[3419]

KNOLL, CHARLES, & CO., 187 *Tottenham Court Road, W.*—Grand pianofortes; oblique grand, and cottage.

[3420]

KÖHLER, JOHN, 35 *Henrietta Street, Covent Garden.*—Brass musical instruments of every kind for military bands.

[*Obtained Prize Medal at the Exhibition of* 1851.]

The following new inventions and modifications will be found among the instruments exhibited by Mr. KÖHLER.

1. THE PATENT HARMONIC CORNOPEAN, introducing a fourth valve, by means of which an instantaneous echo can be produced.

2. The addition of a double slide to the "Harper's Slide Trumpet" rendering the chromatic scale of that instrument perfect in the *lower* as well as in the upper notes.

3. An invention to substitute the water-key in all brass instruments, preserving a perfectly level surface in the wind passage, and facilitating the discharge of the accumulated water.

[3421]

LACHENEL, LOUIS, 8 *Little James Street, Bedford Row, W.C.*—Manufacturer of English patent concertinas. (*See page* 112.)

[3422]

LOCKE, EDWARD CHARLES, 7 *Great Ducie Street, Manchester.*—The peri, campanula, or fairy bells.

[3423]

LUFF, G., & SON, 103 *Great Russell Street, Bloomsbury, W.C.*—Model piccolo piano.

[3424]

MATTHEWS, WILLIAM, & SONS, 5 *St. James's Street, Nottingham.*—Pianoforte with propeller action.

Imhof & Mukle, 547 *Oxford Street.*—Orchestrion, or self-acting organ.

ORCHESTRION, OR SELF-ACTING ORGAN.

The Orchestrion, built for the International Exhibition of 1862, is a striking example of the capabilities of mechanism for producing perfect music. On this instrument hundreds of different effects, variations, and shades of tone can be produced. The mechanism is so perfect that its action is instantaneous, and free from noise and inconvenience to the person working it. The great simplicity of its construction renders the Orchestrion a most durable instrument. As the two barrels can be conveniently removed from the front, the Orchestrion does not require more space than its width.

The deepest notes are placed in the centre of the instrument, so that the tuner can tune each and every pipe easily from the sides without removing anything. By the application of an additional fly, the speed can be regulated to the greatest nicety, so as to give detailed effects to the music in performing. In this and many other respects the Orchestrion is different and superior to other self-acting musical instruments.

Imhof & Mukle are the manufacturers of the "Flutonichorde," which can be instantly attached to any pianoforte. Subjoined is a price list of musical instruments manufactured by this firm, and also instruments for which they are agents :—

Orchestrions, 1000 Guineas and upwards.

Euterpeons, 200 to 800 guineas.

Flute instruments à la Davrainville, 30 to 400 guineas.

Self-acting organs, 24 to 60 guineas.

Musical clocks, 24 to 200 guineas.

Portable organs and pianos, 5 to 30 guineas.

German handle-organs and pianos, for schools and nurseries, 5 to 60 guineas.

Pianofortes, 1st class quality, 25 to 100 guineas.

Nicole Frères' musical boxes, 4 to 40 guineas.

All these instruments are built to stand tropical climates.

KIRKMAN, JOSEPH, & SON, 3 *Soho Square.*—Pianofortes.

CONCERT GRAND PIANOFORTE, with seven octaves, A to A, under-dampers, repetition action, and all the latest improvements, in solid rosewood case, elaborately carved. The case of this instrument was carved at Madras, East Indies; the designs and working drawings were sent from England by J. KIRKMAN & SON; the case was made, and the carvings executed, by the native workmen in the most correct manner. As a specimen of native Indian skilled labour it is interesting, as showing the ready capability of the native carvers to apply the art in which they excel to any purpose that may be required. The top of this pianoforte is made out of a solid piece of rosewood, without a joint; it is 5 feet wide, and even in India it is rare to meet with rosewood of such large dimensions.

** This piano is exhibited in the Indian Department.

KIRKMAN, JOSEPH, & SON—*continued.*

PATENT IMPROVED TRICHORD SEMI-COTTAGE PIANOFORTE, with seven octaves, A to A, and all the latest improvements, in ebony case richly carved and gilt.

CONCERT GRAND PIANOFORTE, with seven octaves, A to A, repeating action, and under-dampers, with new and improved up and down bearing bridges to preserve the sounding board in perfect equilibrium, and prevent its sinking; in English pollard oak case richly carved and gilt.

Patent Improved Trichord Semi-Cottage Pianoforte, with seven octaves, A to A; single action, English model, and all the latest improvements, in walnut, tulip-wood, and ebony case.

Oblique Grand Pianoforte, with seven octaves, A to A; grand check action, and under-dampers, with improved sounding board, in Amboyna-wood case richly carved and gilt.

Lachenal, Louis, 8 *Little James Street, Bedford Row.*—English patent concertinas.

All 48 Keys, Double Action, Iron Screwed Brass Notes, and Warranted. Compass G to C

(Instruments of a Smaller Compass made to Order only, at the usual Prices.)

Price List, 1862, of English Patent Concertinas.

		£ s. d.	Ordinary Metal Vibrators, but of the very best quality made.	Silver Vibrators, which give a a beautiful, mellow, and subdued tone, suitable for Drawing-Rooms.	Tempered Steel Vibrators, which give a very sonorous, full-bodied tone, suitable for Concert-Room.	Gold Vibrators, which give the most distinct and justly proportioned quality of tone of any Metal.
1. The People's Concertina. Mahogany, in neatly covered Box		2 2 0				
2. Rosewood. Superior tone and finish .	Mahogany Box	3 3 0				
3. Ditto, best finish, Five-fold Bellows, best finished	Mahogany Box	4 4 0				
4. Ditto, best finish, Five-fold Morocco Bellows, Moulded Edges	Rosewood Box	5 5 0				
5. Ditto, extra best finish, Five-fold Morocco Bellows, plain tops and no gilding, but with ornamented paper, German silver studs, and in every other respect as the Concert Concertina before its improvement . .	Rosewood Box	6 16 6				
			Guineas.	Guineas.	Guineas.	Guineas.
6. Ditto, Newly Improved. Ornamented throughout, with Silver Touches for Concerts, contains louder and sweeter tone than any Treble Concertina ever before produced, and is adopted by all the most eminent Professors	Rosewood Box	..	8	10	12	13
7. Ebony, Newly Improved, etc., as above, with glass studs	Rosewood Box	..	10	12	14	15
8. Amboyna Coromandel-Zebra (or any description of wood preferred), with Bellows and all pertaining to exterior finish tastefully matched, Silver Touches or Glass Studs, as preferred	Box to match	..	12	14	16	17
9. Ivory Tops, all pertaining to exterior appearance tastefully matched, Silver Touches or Glass Studs, as preferred	Box of any wood	..	13	15	17	18
Nos. 6 to 9 can also be had with Double Pans at Two Guineas each extra.						
Tenor or Baritone Concertinas.						
10. Rosewood, Ivory Keys, three octaves and three notes, sounding an octave below the full compass Treble Instruments. Finished as No. 4	Rosewood Box	9 9 0	..	..	..	..
11. Ditto, Silver Touches. Finished as No. 6	Rosewood Box	..	11	13	15	..
Ebony „ „ 7	Rosewood Box	..	13	15	17	..
Amboyna, etc. „ 8	Rosewood Box	..	15	17	19	..
Ivory Tops „ 9	Box in any kind of Wood	..	18	20	22	..
No. 11 can be had with Double Pans at Two Guineas extra.						
Small or Large Bass Concertinas.						
12. Rosewood, Ivory Keys, three octaves, and three notes. Finished as No. 4 . . .	Rosewood Box	12 12 0	..	..	..	..
13. Ditto, Silver Touches. Finished as No. 6	Rosewood Box	..	15	..	..	..
Ebony, „ „ 7	Rosewood Box	..	17	..	..	..
Amboyna, etc., „ 8	Box to match	..	19	..	..	..
Ivory Tops, „ 9	Box in any kind of Wood	..	22	..	..	..
Tuning Apparatus (Mahogany), six-sided, with screw-driver, file, etc., complete, by means of which parties at a distance from a tuner can keep their own instrument in repair		0 12 6	..	..	..	..
Ditto, ditto, Superior quality, with four holes . . .		0 18 6	..	..	..	..

[3425]

METZLER, G., & Co., *Great Marlborough Street, W.*—Brass military instruments, clarionets, &c.; specimens of printed music.

BRASS MUSICAL INSTRUMENTS, of improved circular form. Clarionets, &c., &c. Patented.

The SONOROPHONE. Invented by Mr. J. Waddell, Band-master of the First Life Guards. The excellence of these new instruments, and the marked improvement of their formation over the old system, has been admitted by the most competent judges. They are now in use in the bands of the First Life Guards, Royal Engineers, several Regiments of the Line, the Navy and the Volunteer Corps. The following drawing is intended to show the relative size of two Contre-bass Instruments in *E* flat. Fig. 1.—the old form of the Saxe-horn; Fig. 2.—the Patent Sonorophone. The advantage and portability of the new form over the old may be seen at a glance

Fig. 1. *Fig* 2

Drawings, Testimonials, and Lists of Prices of METZLER & Co.'s various new Circular Brass Instruments may be had on application as above.

[3426]

MINASI, C., 3 *St. James's Terrace, Kentish Town Road.*—Music-stool; harmonium.

[3427]

MOORE, JOHN & HENRY, 104 *Bishopsgate Street Within, City.*—Microchordon grand pianoforte.

[3428]

MURPHY, GEORGE, *Albert Street, Camden Road, and* 28 *Cheapside.*—Pianofortes.

[3429]

OATES, JOSEPH PIMLOTT, *Erdington, Birmingham.*—Cornet with equi-tritubular, or champion, pistons and improved water-exit.

[3430]

OETZMAN & PLUMB, 151 *Regent Street, London.*—Three pianofortes.

[3431]

PEACHEY, GEORGE, Pianoforte Manufacturer, 73 *Bishopsgate Street Within, E.C.*—Improved tri-chord piccolo pianofortes.

PEACHEY'S IMPROVED TRICHORD PIANOFORTES are remarkable for their durability, power, and quality of tone. They may be bought or hired, with option of purchase from the maker.

[3432]

POTTER, HENRY, 36 *Charing Cross, W.C.*—Improved flute-valve, brass instruments, and drums.

[3433]

PRIESTLEY, FREDERICK, 15 *Berners Street, Oxford Street, W.*—Small pianofortes.

PATENT SIREN PIANOFORTE. Key-board full scale. Cash price 22 Guineas.

The action of these pianofortes possess all the advantages of the "Repetition Action," while, from the simplicity of their construction, they are superior to it in durability, and can be produced at one half the price.

[3434]

ROBSON, T. J., 101 *St. Martin's Lane.*—Organ.

[3435]

RUDALL, ROSE, CARTE, & Co., 20 *Charing Cross.*—Patent clarionets, flutes, military brass instruments, drums, &c.

[3436]

RUSSELL, GEORGE, 35 *Brook Street, Euston Road, N.W.*—Rosewood grand pianoforte.

[3437]

RÜST, ROBERT ANDERSON (RÜST & Co.), 34 *Great Marlborough Street, W.*—Pianoforte with patent tubular sounding-board, and newly-constructed case.

[3438]

SCOWEN, THOMAS LAYZELL, *Allen Road, Stoke Newington, London.*—Compass for dividing circles; ocular music timekeeper.

Newly invented compass for drawing circles without centre marks, and dividing the same into any number of parts without moving the instrument.

Patent ocular demonstrator of time in music, and accentor. This instrument shows the exact duration of the various notes and rests in each bar, and is provided with a hammer so constructed as to give a louder beat at the accented parts of the different bars, and varies to any degree of quickness or slowness, if required.

[3439]

SHAW, J., & SON, *High Street, Glossop.*—Enharmonic piano.

[3440]

SIMPSON, JOHN, 266 *Regent Street.*—German concertinas, with Simpson's easy method; English concertinas, flutes, and flageolets.

[3441]

SPARKS, W. J., 13 *Eversholt Street, Oakley Square.*—Pianos.

W. J. SPARKS, inventor and manufacturer of the Trichord Cottage Piano, equal in power and quality of tone to the Horizontal Grand, price from 50 guineas. Superior cottage pianos from 25 guineas. Pianos for hire. W. J. Sparks, 13 Eversholt Street, Oakley Square, London, N.W.

[3442]

STARCK, JOHN EDWARD, 25 *Old Street, St. Luke's, E.C.*—Flutes, flageolets, clarionets, drums, fifes, &c.

[3443]

THOMPSON, H., 322 *Regent Street.*—Orchestral piano, extra pedal, producing chords and octaves.

[3444]

TURNBULL, WILLIAM, 83 *Mary Street, Hampstead Road.*—A set of pianoforte keys.

[3445]

WALKER, J. W., 27 *Francis Street, Bedford Square, W.C.*—Church and chamber organs.

[3446]

WARD, HENRY, 100 *Great Russell Street, Bloomsbury.*—Piano.

[3447]

WILLIAMS, WILLIAM, 36 *New King Street, Bath.*—Patent grand pianoforte.

[3448]

WILLIS, HENRY, *Albany Street, Regent's Park.*—An organ with four manuals and pedal organ, and 60 stops.

[3449]

WILSON, WILLIAM, *Fairbank Villa, Talfourd Road, Camberwell.* — An omnitonic flute, adjustable at will to any key.

The most important features of this flute are its perfection of intonation and capability of adjustment, combined with simplicity of manipulation. The several parts of the flute adjustable at pleasure to vary the distances between the finger-holes; by which means the relative intervals are determined with mathematical precision, and (being variable) are preserved perfectly true alike in all keys; while the fingering is reduced to the utmost simplicity. The instrument being tuned to the key of the piece to be performed, it is only necessary generally to learn one simple scale, as the scale *D* on an ordinary concert flute.

[3450]

WORNUM, R., & SONS, *Store Street, Bedford Square, London.*—Upright and horizontal pianofortes.

[3451]

BATES & SON, 6 *Ludgate Hill.*—A small organ.

[3452]

COHLMANN & SON, *Halifax.*—Grand action, oblique, and upright pianos.

[3453]

NUTTING & ADDISON, 210 *Regent Street.*—A piano.

[3454]

STIDOLPH, G. F. & J.—*Woodbridge, Suffolk.*—A small minima organ.

CLASS XVII.

SURGICAL INSTRUMENTS AND APPLIANCES.

[3482]

ARBUCKLE, JOSEPH, *South Bridge, Edinburgh.*—A hernia truss; illustrations; improvements on working tools for making trusses.

[3483]

ASH, CLAUDIUS, & SONS, 7, 8, & 9 *Broad Street, Golden Square, London.*—Artificial teeth, and dental materials.

[3484]

ATKINSON, BENJAMIN FREDERICK, 3 *Hemming's Row, Charing Cross.*—Trusses for piles, prolapsus ani and uteri, and inguinal and scrotal hernia; splint for diseased hip-joint.

[3485]

BAILEY, WILLIAM HUNTLY, 418 *Oxford Street, London.*—Trusses, elastic stockings, deformity and surgical instruments, enemas, and belts.

[3486]

BARLING, JOSEPH, 7 *High Street, Maidstone, Kent.*—Specimens of crystal gold in sponge and leaf, for dentists.

[3487]

BASSINGHAM, BENJAMIN, Manufacturer, 5 *Ruby Street, Wisbeach.*—Artificial leg upon self-acting principles, &c.

[3488]

BIGG, HENRY HEATHER, 29 *Leicester Square.*—Orthopædic and anatomical appliances for bodily deformities, weaknesses, and deficiencies. (*See page* 118.)

[3489]

BLACKWELL, W. & Co., *Cranbourne Street, and Bedford Court.*—Surgical instruments, crutches, trusses, &c.; tailors' shears, razors, crayons, cutlery, &c.

[3490]

BLUNDELL, WALTER, Dentist, 3 *Holles Street, Cavendish Square, London.*—Improved artificial teeth.

BIGG, HENRY HEATHER, Assoc. Inst. C.E., 29 *Leicester Square.*—Orthopractic and anatomical appliances for bodily deformities, weaknesses, and deficiencies.

[*Obtained Prize Medal at the Exhibition of* 1851.]

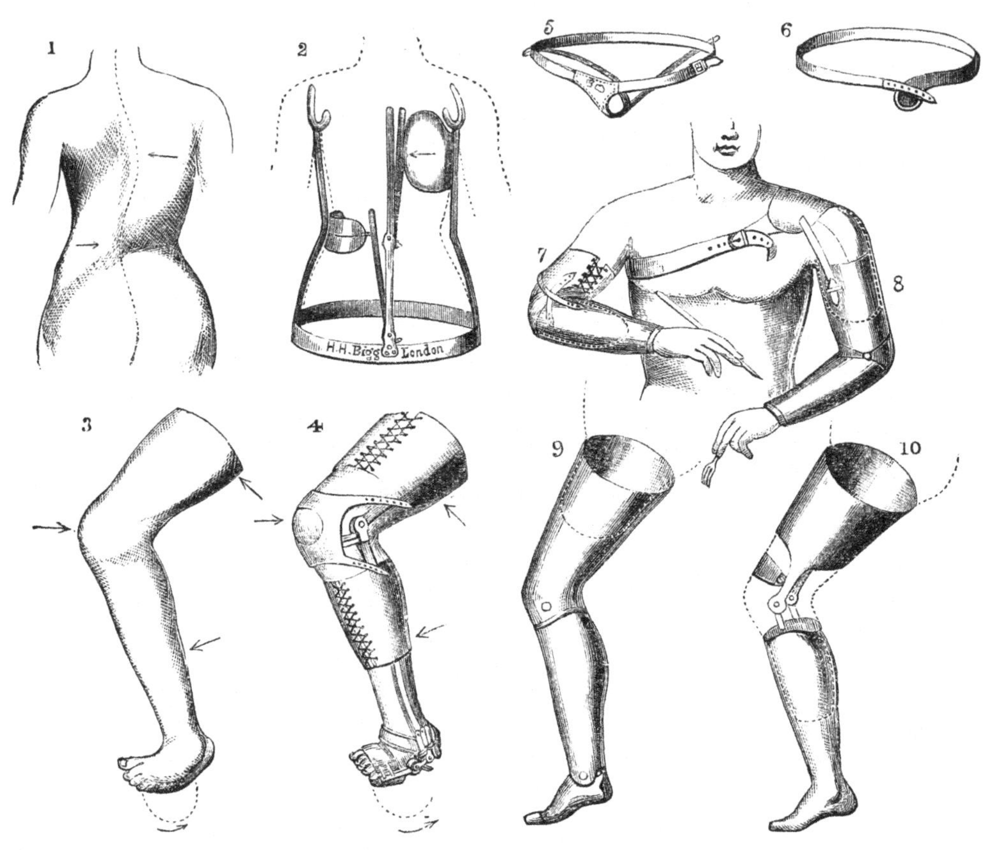

This collection of inventions and appliances is intended to illustrate the rapid progress of the new science of Mechanical Orthopraxy, of which the exhibitor is one of the most active promoters. The case contains 42 new inventions, amongst which are—

1, 2. Apparatus for treatment and cure of spinal curvature.

3, 4. Appliance for cure of contracted knee, club foot, and deformities of the foot and ankle (tibio-tarsal region).

5. Triple lever truss, for the treatment of hernia or rupture.

6. Sand pad truss, for inguinal, scrotal, and femoral hernia.

7. Artificial arm, with wrist and finger articulations, spring-thumb, &c., for use after amputation below elbow.

8. Artificial arm, with moveable elbow, wrist and finger joints, for use after amputation above elbow.

9. Artificial leg, for amputation above knee, with elastic tendons and muscles acting as they do in nature.

10. Artificial leg, for amputation below knee, with knee, ankle, and toe articulations moved by elastic springs.

[3491]

BROWN, SAMUEL SHAW, *Ellesmere Works, Runcorn.*—Flax and cotton lint, elastic stockings, abdominal belts, knee-caps, &c.

[3492]

BROWNING, EDWARD, 38 *Montague Square, W.*—Artificial teeth, &c.

[3493]

CALKIN, JOSEPH, 12 *Oakley Square, N.W.*—The "occhiombra," or patent transparent ventilating eye protector.

[3494]

CAPLIN, DR., 9 *York Place, Baker Street.*—Electro-chemical bath, for the cure of chronic diseases of all kinds.

[3495]

CAPPIE, JAMES, M.D., *Edinburgh.*—Obstetric forceps, in which the handle and blade are united by socket-joint.

[3496]

CARTE, ALEXANDER, M.D., T.C.D., F.R.C.P.I., *Royal Hospital.*—Instrument for the treatment of aneurism by compression.

[3497]

CLELAND & HILL, 146 *George Street, Glasgow.*—Artificial limbs on a new principle, strong, light, and substantial.

[3498]

CLOVER, J., 3 *Cavendish Place.*—Inhaler, chloroform, &c. Gives chloroform vapour any strength required, under 4½ per cent.

[3499]

COGHLAN, JOHN, M.D., *Wexford.*—A probe-pointed knife for dividing the neck of the womb; a drill-carrier for dentists, to be used within the mouth.

[3500]

COLES, WILLIAM, & Co., Patentees, 3 *Charing Cross.*—Patent spiral-spring trusses.

This is a novel and greatly improved truss, and is commended by the patronage of Sir Ashley Cooper, many of our most eminent surgeons, and by the adoption and recommendation of William Cobbett. It is perfectly efficacious, and at the same time agreeable to the wearer. For thirty years it has had a steadily increasing reputation. Each truss bears the name and address of the patentee.

[3501]

COLLINS, DANIEL JOSEPH, 48 *Foley Street, London, W.*—Surgical appliance; also various instruments, dental and surgical.

[3502]

COXETER, JAMES, 23 & 24 *Grafton Street East, Tottenham Court Road.*—Surgeon's instruments, including new form of lithotrite and double current catheter.

[*Obtained Prize Medal in* 1851.]

The following are exhibited, viz.:—

COXETER'S LITHOTRITE, with new movement which greatly facilitates the alternate use of *sliding action* and *screw action*, a desideratum of great importance in seizing and crushing the stone.

Coxeter's double-current catheter, with opening for inlet stream, so formed as to keep the "debris" in motion, to promote its more speedy exit.

Urethrotome and catheter combined, by H. Thompson, Esq.

Stethometer of new form, by T. Griffiths, Esq.

Coxeter's Magneto-electro machine, worked by the foot instead of the hand.

Coxeter's Spring pessary, for Prolapsus Uteri.

Coxeter's Spirometer, of new and simple form.

Coxeter's Compound Uterine Syringe.

[3503]

CRAPPER & BRIERLEY, Manufacturing Dentists, *Hanley, Staffordshire.*—Registered porcelain trays, mineral teeth, &c.

These registered impression trays are invented and manufactured by the exhibitors. They are made in every variety of shape in china, porcelain, and earthenware. Specimens of various kinds are exhibited, together with mineral teeth of extra strength for vulcanite work, tubular or flat, and a variety of materials used in dentistry. The porcelain tooth-powder may be obtained in boxes, price 1*s.* each.

[3504]

DIXON, THOMAS, 4 & 7 *St. James's Place.*—Nightingale cradle; Nightingale bed; stove for hothouses and drying-rooms.

[3505]

DURROCH, WILLIAM FRAZER, 28 *St. Thomas Street East, and* 1 *Dean Street, Borough.*—Surgical instruments, &c.

The exhibitor is Surgical Instrument Maker to the Royal Navy, Greenwich Hospital, Guy's Hospital, &c. He manufactures all descriptions of surgical instruments, and has attained a high reputation by several important improvements. He manufactures instruments to drawings; and both in making and repairs employs the best materials and skilled labour.

[3506]

ERNST, FRIEDRICH GUSTAV, 19 *Calthorpe Street, W.C.*—Orthopædic and anatomical appliances; and surgical instruments.

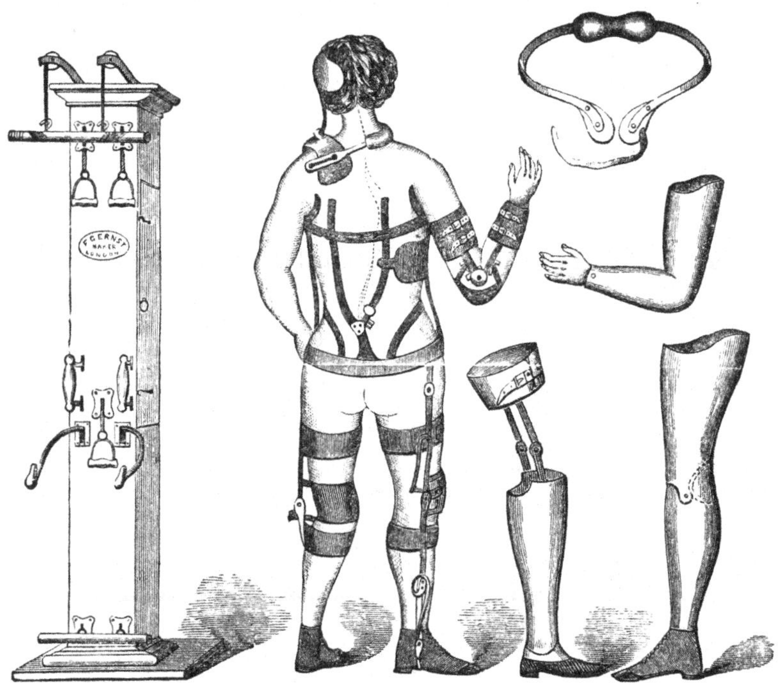

MR. ERNST is the author of the "PORTABLE GYMNASIUM," a manual of exercises for self-instruction in Home Gymnastics. He is also favourably known as a maker of surgical instruments, elastic bandages, and the various appliances necessary in cases of lost, deformed, or weak limbs, vertebral deflexions, and other local muscular relaxations

[3507]

EVANS & STEVENS, 12 *Old Fish Street, St. Paul's, London.*—A complete collection of surgical instruments.

[3508]

EVANS, CALEB, *The Hospital, Birkenhead.*—Arm splint.

[3509]

EVRARD, JOHN, 35 *Charles Street, Middlesex Hospital.*—Instrument for lithotrity; bone-cutting forceps, with parallel action; dental instrument.

[3510]

FAULKNER, HENRY, 24 *Keppel Street, Russell Square.*—Improved method of constructing artificial teeth in vulcanite.

[3511]

FAULKNER, JOHN, Practical Dentist, 2 *Mornington Crescent, Hampstead Road, N.W.*—Specimens of pink vulcanite base for artificial teeth.

[3512]

FERGUSON, J. & J., 21 *Giltspur Street, London, E.C.*—Surgical instruments.

[3513]

FITKIN, WILLIAM, 88 *Fleet Street.*—Patent safety elevator for the instantaneous and painless extraction of teeth and stumps.

The object of FITKIN'S PATENT SAFETY ELEVATOR is the extraction of teeth and stumps with greater safety and much less pain than attends the use of the ordinary instruments.

[3514]

FRANÇOIS, HENRY, 42 *Judd Street, Euston Road.*—Artificial teeth, with bases of india-rubber, coralite, gold, &c.

Various specimens of gold, vulcanized india-rubber, and coralite bases for ARTIFICIAL TEETH.

A complete set and a partial set in gold.

Complete sets and pieces of from one to ten teeth, in various kinds of vulcanized india-rubber, namely, pink, red, coralite, and black; the teeth used are the best mineral, and some have artificial mineral gums. Vulcanized india-rubber as a base for artificial teeth has nearly superseded metal, bone, &c. The advantages arising from its adoption are very numerous. All sharp edges are avoided; no springs, wires, or ligatures are required; no extraction of stumps, nor other painful operations, are necessary; a greatly increased freedom of suction is supplied; a natural elasticity, hitherto wholly unattainable, and a fit perfected with the most unerring accuracy, are secured; while, from the softness and flexibility of the agents employed, the greatest support is given to the adjoining teeth when loose or rendered tender by the absorption of the gums; the acids of the mouth exert no agency on the prepared india-rubber, all unpleasantness of taste and smell being at the same time provided against.

A complete set in gold varies from 10*l.* 10*s.* to 21*l.*; partial sets, from 10*s.* 6*d.* to 15*s.* per tooth. Sets in vulcanized india-rubber, from 5*l.* to 15*l.*; partial sets, from 5*s.* to 10*s.* 6*d.* per tooth.

[3515]

FRESCO, ANDRÉ, 7 *Grosvenor Street, Grosvenor Square.*—Artificial teeth; and tooth-powder called Fresco's odonto.

[3516]

GABRIEL, M. & A., 27 *Harley Street, and* 34 *Ludgate Hill.*—Artificial teeth, with improved air-cells and soft gums. (*See page* 122.)

[3517]

GANNON, THOMAS, Manufacturing Gas-fitter, &c., *Liquorpond Street, London.*—Improved patent self-adjusting leg and foot-rest.

[3518]

GARDEN, DR., *Edinburgh.*—New forceps and elevator, adapted for the extraction of all kinds of teeth and stumps.

GABRIEL, M. & A., 27 *Harley Street, and* 34 *Ludgate Hill.*—Artificial teeth, with improved air-cells and soft gums.

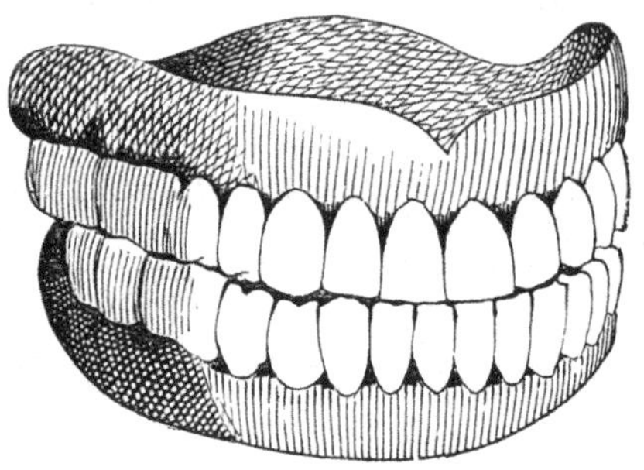

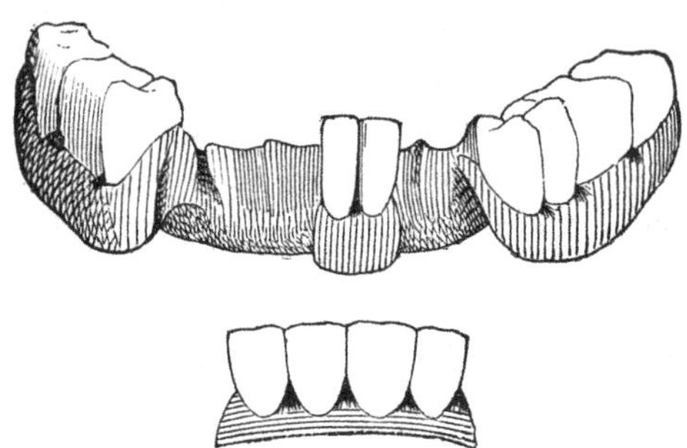

The exhibitors are the patentees and sole proprietors of the OSTEO EIDON, or artificial bone, as a base for GABRIELS' INDESTRUCTIBLE MINERAL TEETH and SELF-ADHESIVE GUMS. One set will last a lifetime, and is warranted to answer every purpose for mastication and articulation, even when all others fail. They are adjusted without springs, wires, or any unpleasant operation.

Specimens of Messrs. Gabriels' patented improvements may be seen on their stand; where also a descriptive catalogue, in French and English, with the cost of the various descriptions of artificial teeth, may be obtained gratis. Complete sets of these teeth can be made with *one visit*, where time is an object.

GABRIELS' PATENT WHITE ENAMEL, for restoring and preserving front teeth, retains its colour without injury to the enamel.

Their addresses are :—27 Harley Street, Cavendish Square, and 34 Ludgate Hill, London; 134 Duke Street, Liverpool; 65 New Street, Birmingham.

[3519]

GARRETT, JAMES ALEXANDER, 38 *Wardour Street, W.*—Trusses and surgical bandages.

[3520]

GILL, THOMAS DYKE, 84 *John Street, Tottenham Court Road.*—Improved gas vulcanizer for dentists.

[3521]

GRAY & HALFORD, 171 *Goswell Road, E.C.*—Artificial human eyes.

[3522]

GRAY, JOSEPH, & Co., 154 *Fitzwilliam Street, Sheffield.*—Surgical, dental, veterinary trusses; enema apparatus, lancets, &c.

[3523]

GRIFFITHS, RAYMOND, 2 *Duke Street, West Smithfield.*—Medicine chests and sample cases.

[3524]

GROSSMITH, WILLIAM ROBERT, 175 *Fleet Street, London.*—Patent and prize-medal artificial eyes, legs, arms, hands, &c. (*See page* 123.)

[3525]

HALLAM, F. H., 9 *Endell Street, Long Acre, W.C.*—Dental instruments.

GROSSMITH, WILLIAM ROBERT, 175 *Fleet Street, London.*—Patent and prize-medal artificial eyes, legs, arms, hands, &c.

[*Obtained Prize Medals at the Exhibitions of* 1851 *and* 1855.]

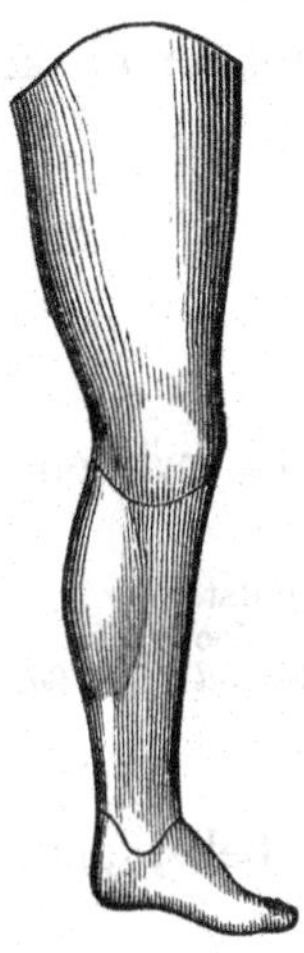

ARTIFICIAL LEGS, ARMS, HANDS, NOSES, &c., with the following newly invented improvements :—

1. A limb for contracted knee-joints (amputation below knee), giving a perfect artificial action at the knee.

2. A foot apparatus, for Symes and Chopart's operations, securing a neater appearance and firmer bearing than has been yet obtained for these cases.

3. A new limb for "Thigh amputations," containing all the advantages of the tendon action, Palmer's patent, and Grossmith's patent knee and ankle actions, with thorough durability and lightness in weight. Also a new method of making the joints of artificial limbs waterproof and noiseless in action.

4. Artificial eyes, of a new and hardened enamel, to prevent corrosion and secure a more lasting brilliancy and life-like appearance.

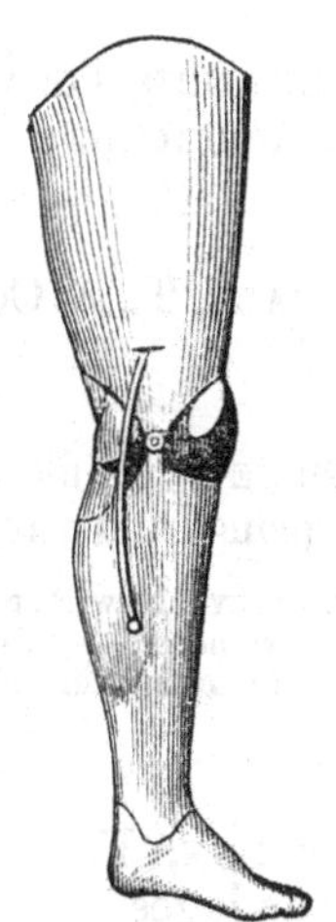

[3526]

HARNETT, WILLIAM, 12 *Panton Square, Coventry Street.*—Mineral teeth for vulcanite and gold, and all articles appertaining to dentistry.

[3528]

HAYES, GEORGE, M.D., 66 *Conduit Street, Regent Street, W.*—Mechanical dentistry fully illustrated.

[3530]

HILLIARD, WILLIAM B., 65 *Renfield Street, Glasgow.*—Surgical instruments (original inventions), artificial leg; hernia trusses; table-knife cleaner on new principle.

[3531]

HOOPER, WILLIAM, *Pall Mall East, S.W.*—Hydrostatic beds and cushions, and patent bed-lift for invalids.

[3532]

HOY, J., 6 *Pickering Place, W.*—Truss for hernia.

[3533]

HUDSON, THOMAS, Chemist, *South Shields.*—Deodorizing ashpit—preserving fertilizing elements for vegetation; improved tooth-stump instrument.

[3534]

HUXLEY, EDWARD, 12 *Old Cavendish Street, Oxford Street.*—Surgical bandages and moc-main trusses.

[3535]

JOHNSON, THOMAS, *Commercial Road East.*—Model of a portable apparatus for slinging horses whilst under surgical treatment.

[3536]

LAMBERT, PETER, 18 *Charlotte Street, Bedford Square, W.C.*—Artificial teeth.

[3537]

LAURENCE & CO., *Islington, London.*—Patent horse-hair, dry friction, and bath gloves, for promoting health.

[3538]

LAWSON, BUXTON, & Co., *Shales Moor Works, Sheffield.*—Surgical, dental, and veterinary instruments.

[3539]

LEARWOOD, THOMAS, *Fairmantle Street, Truro, Cornwall.*—Artificial limbs for all kinds of amputations; trusses without spring.

[3540]

LEMALE, T., & Co., 62 *Chandos Street, W.*—Artificial teeth and gums.

[3541]

LINDSEY, MARK JOHN, 37 *Ludgate Street, City.*—Lindsey's patent truss, without steel spring, and various other trusses, &c., with improvements.

LINDSEY'S NEW PATENT TRUSS, the most recent invention for hernia, consists of a covered plate with patent padding to support *both* hernia rings, and an elastic waist-belt: the pressure can be regulated by the patient, and the truss is perfectly easy and effective.

Prices 15*s.* 6*d.*, 21*s.* 6*d.*, 26*s.* 6*d*, 31*s.* 6*d.*

[3542]

LONGDON, F., & Co., *Derby.*—Surgical elastic stockings, knee-caps, belts, and other bandages.

[3543]

LOWS, ANDREW, 19 *Lowther Street, Carlisle.*—Specimens of dental workmanship.

[3544]

MACINTOSH, CHARLES, & Co., *Cannon Sıreet, London; and Cambridge Street, Manchester.*—Vulcanized rubber surgical and chemical apparatus.

[3545]

MACINTOSH, JOHN, 40 *North Bank, Regent's Park.*—Collodion, used as a setting for artificial teeth.

[3546]

MARSDEN, W. J., *Upper Thorpe Road, Sheffield.*—Patent respirators; registered shield chest protectors; ventilated eye-shade; animal oil wool knee-cap.

[3547]

MASTERS, MOSES, Manufacturer, 1 *Paragon Street, New Kent Road, London.*—Artificial hands, arms, legs, and crutches.

[3548]

MATTHEWS, WILLIAM, 8 *Portugal Street, Lincoln's Inn Fields, W.C.*—Surgical instruments and appliances.

[3549]

MAURICE, JOSEPH, 3 *Langham Place, W.*—Artificial teeth, showing the various applications of vulcanized india-rubber.

[3550]

MAW, S., & SON, 11 *Aldersgate Street, London.*—Surgical instruments. (*See page* 125.)

[3551]

MILLER, CLAUDIUS MONTAGUE, M.D., *Claremont Villa, Stoke Newington Road.*—Spectacles for the relief of conical cornea.

[3552]

MILLIKIN, JOHN (late BIGG & MILLIKIN), 9 *St. Thomas's Street, Borough.*—Surgeons' instruments and appliances.

MAW, S., & SON, 11 *Aldersgate Street, London.*—Surgical Instruments.

Amputating Cases, including those ordered for Army and Navy surgeons, in accordance with the latest Government regulations. Portable set of field instruments.

Abdominal Supporters.—Laced and elastic stockings, knee-caps, bandages, and suspenders.

Breast Pumps, brass and electro-plated, in mahogany and morocco cases; and a number of very useful and modern apparatus for relieving the breast.

Bougies and Catheters, in silver, German silver, electro-plated, and elastic gum.

Brass Anatomical and Ear Syringes, with ivory pipes; also a variety of patterns of syringes in glass and Britannia metal, for the eye, ear, urethra, vagina, &c.

Caustic Cases and holders, in silver, silver-gilt, platinum, ebony, &c., &c.

Cupping Apparatus, complete sets, in morocco and mahogany cases.

Dissecting and Post-Mortem Instruments, full sets, in mahogany cases. Dislocation apparatus, splints, &c.

Dentists' Instruments, including patterns of all the newest and most approved forceps, punches, scaling and stopping instruments, in handsome mahogany cases, as used by the most eminent dentists in London and the Provinces. Maw's new and improved dentists' drill, capable of being worked in any position.

Ear Instruments.—Hearing-trumpets, conversation-tubes, &c., &c.

Eye Instruments, complete set in case; also a new and elegant double-action Eye Douche, electro-plated, with glass reservoir, in morocco case; a most complete and valuable instrument.

Elastic Surgical Apparatus and appliances.

Enema Apparatus, brass and electro-plated, in mahogany and morocco cases; several specimens of the most approved kinds, both single and double action. Maw's new Enemas, with glass reservoirs, both single and double action.

Feeding Bottles for Infants. Maw's patent fountain, with German silver, electro-plated mounts, in cases, also in cheaper forms, sold at 2*s.* 6*d.* each and upwards. Maw's 1*s.* feeding-bottle, in case complete, and a variety of other patterns, with glass, metallic, and elastic tubes of the most modern and improved construction suitable for export.

Guillotine for the Tonsils (quite a new instrument).

Hernia instruments. Hydrocele instruments.

Inhalers, in Britannia-metal, glass, and earthenware. A cheap earthenware inhaler, with a new patent application and valve, by S. Maw and Son.

Lithotomy instruments. Lithotrity instruments, complete sets in cases.

Lancet Cases, silver, handsome engine-turned, engraved and chased, specimens in all sizes.

Midwifery Instruments, a complete set, in chequered ivory handles and mahogany case.

Minor Operating Instruments, a complete set, in case.

Nipple Shield, of glass, with elastic tube and teat, a perfect little instrument, in box complete, retail price, 1*s.* 6*d.*; also a variety of india-rubber teats. Glass, metallic, and india-rubber shields.

Pessaries, an assortment of india-rubber, vulcanite, and boxwood; also a beautiful specimen in thin ivory.

Pill Machines, of superior make, with marble and mahogany slabs.

Pocket Instruments, several complete sets in elegant cases, mounted in handsome engine-turned gilt handles with fluted backs; also in tortoiseshell and ivory.

Respirators, Ethereon, plated with silver and gilt; also the Ethereon Scarf Respirator.

Scissors for surgeons' and druggists use, in great variety.

Specula, an elegant assortment for the eye, ear, vagina, rectum and nose.

Stethoscopes, a variety of specimens, in ebony, ivory, cedar, &c.

Stomach-pump, in mahogany case.

Trephining Instruments, a full set in mahogany case; also Maw & Son's improved set, consisting of three Trephines, electro-plated, fitting into spring socket, and mounted in ivory.

Trusses, specimens of several kinds, improvements upon expired patents; also of the ordinary common and patent trusses, of superior make, for hospital and general use.

Urethra Instruments.—Full set of Wakley's dilating canulæ, with elastic and silver catheters, in case complete. A set of Brodie's silver catheters, in chequered ivory handles. A set of three prostate catheters, in chequered ivory handles.

Urinometers.—Urinals and uterine instruments.

Veterinary Instruments.—Complete sets of pocket and dissecting instruments, in cases; also Maw's improved veterinary enema and stomach-pump, in mahogany case.

[3554]

MOGGRIDGE & DAVIS, 18 *George Street, Hanover Square.*—Specimens in dentistry.

The PATENT PNEUMATIC PALATE, in gold, bone, and their celebrated flexible base.

Two heads illustrating the same face with and without teeth.

No. 1.—A set of mineral teeth on a flexible base, with artificial palate, cheek, shield, and uvula.

No. 2.—A set of mineral teeth on a flexible base, with an artificial palate, air chamber, and uvula.

No. 3.—A set of mineral teeth on gold plate and flexible gum.

No. 4.—A set of mineral tube teeth, with gold sockets inserted in a flexible base.

No. 5.—Specimens of mineral teeth on flexible bases, suited to various cases.

No. 6.—Two entire sets of natural teeth, socketed in a flexible base.

No. 7.—A gold articulating palate, with four mineral teeth set, and gold uvula.

No. 8.—Specimens of gold palates, with air chambers.

No. 9.—Various specimens of mineral teeth, set on gold plates.

No. 10.—Specimen of artificial teeth made from the hippopotamus' tusk.

No. 11.—A set of natural teeth on an hippopotamus palate.

No. 12.—A set of mineral teeth made in the seventeenth century.

[3555]

MORRISON, JAMES DARSIE, *Edinburgh.*—Dental appliances, processes, and products; safety couch for chloroform patients.

[3556]

MOSELEY & CO., Dentists, 30 *Berners Street, London, W.*—Different descriptions of artificial teeth and dental appliances.

[3557]

NORMAN, S., Jun., 1 *Cheltenham Place, Westminster Road, S.*—A lift for a short leg, and shell for boot: also a boot for a wooden leg.

[3559]

O'CONNELL, EDWARD, *Bury, Lancashire.*—Patent siphonia, or infant's feeding bottle; also for applying drinks to invalids and others. (*See page* 127.)

[3561]

PARSONS, JAMES, & CO., 15 *Manor Row, Bradford.*—Artificial teeth.

[3562]

PATRICK, HUGH W., 18 *Broad Street, Golden Square, W.*—Artificial palates, block and single teeth, continuous gum work; dental application of artificial ivory, coral, vulcanite, and materials in the process.

[3563]

PAUL, ANDREW, Surgeon, 27 *Mecklenburgh Square.*—Douche bath (two models), applicable in diseases requiring aspersion or percussion with water.

[3564]

PEARCE, WILLIAM, & CO., *Bridge Street, Bristol; and Brooke Street, Holborn, London.*—Surgical appliances.

PEARCE & CO.'s newly invented Truss for Hernia, is so constructed that the pressure may be increased or decreased as required, and the necessity of an under-strap is obviated. An elastic spring is introduced at the back, which yields to the motion of the body, thereby rendering the truss easy and comfortable to wear.

PEARCE & CO. are manufacturers of spine supports, umbilical belts, abdominal belts, artificial legs, &c.

PEARCE & CO.'s Improved Stethoscope consists of various kinds of wood, so joined as to render it a good conductor of sound, and making it considerably stronger and much less liable to break than the ordinary wooden ones.

[3565]

PINDAR, CHARLES, Maker and Inventor, 19 *John Street, Holland Street, Blackfriars Road.*—Pill and press, tincture press, pill machine, plaster machine, &c.

[3566]

POLLARD, CHARLES & EDWARD, *Brompton Turkish Baths, Alfred Place, Thurloe Square.*—Turkish bath, gout, invalid, and bathing sandal.

O'CONNELL, EDWARD, *Bury, Lancashire.*—Patent siphonia, or infant's feeding bottle; also for applying drinks to invalids and others.

The great facility and comfort afforded by this truly valuable invention in the rearing of infants, has elicited from parents of all classes the warmest expressions of their approval and gratitude to the original inventor of so great a boon.

No other contrivance for a similar purpose has ever before been so highly recommended by the Medical Profession, ladies of distinction, and others who have patronized its use, as this universal favourite of mothers and babes.

To mothers who, from delicacy of constitution or other causes, are unable to nurse their own infants, and who dislike the aid of "wet-nursing," the Siphonia affords the greatest assistance, by enabling them to bring up their little ones under their own care in a very healthy manner.

Nothing can be more convincing of the valuable nature of this invention, and of its near approach to the principle of the natural breast itself, than the great number of fine healthy children who have been brought up from birth on its use. There are thousands of parents throughout the country who can bear willing testimony to the truth of this statement.

It is indeed gratifying to the inventor of the Siphonia to know that his humble efforts in endeavouring to supply a want, previously felt in many families, have been so highly appreciated by parents of all classes, from the occupant of the palace to that of the humble cottage.

The inventor of the Siphonia feels it unnecessary to quote any of the numerous letters received from medical men, who have been satisfied as to the merits of the invention, and who invariably recommend it to their patients whenever such aid is required.

Fitted up in plain and elegant styles, with new patent improvements, to suit the taste and means of all classes. Price 21*s*., 10*s*. 6*d*., 5*s*., and 2*s*. 6*d*. May be had of the Inventor and Patentee, or through any of his agents.

N B.—The Siphonia Nursery Lamp is included with the Guinea Box. Price separately, 5*s*. 6*d*.

Improved Night Lamp for the sick-room, with means for giving drink to invalids without having to be raised up in bed. Price 3*s*. 6*d*. and 5*s*. 6*d*.

The Biberon, a new patent invention, adapted for a lady's travelling companion. Price 2*s*. 6*d*. plain; in neat style, 3*s*. 6*d*.

Among the many who have borne testimony to its value, the following distinguished persons have permitted the inventor of the Siphonia to make use of any extracts from their letters to him expressive of their high opinion of his useful invention :—

The Countess of Hopetoun will gladly give Mr. O'Connell permission to quote from her letters of last month, or from this one, if he prefers it :—She has experienced still greater comfort and satisfaction from his clever invention since she last wrote, as she has been ill and unable to nurse her little baby, who has consequently lived upon his Siphonia bottle and has never given an hour's trouble or suffered from the change in the least. She cannot, therefore, sufficiently praise and recommend Mr. O Connell's valuable and beautiful arrangement.—Lubenham Hall, Rugby, March, 1861.

Lady Middleton begs Mr. O'Connell will make use of her letter if it will be of any advantage to him. The Siphonia is most certainly a great boon, as her own baby is thriving well on its use. Lady Middleton takes every opportunity of recommending the Siphonia to other ladies, having herself found it so useful.—Birdsall Hall, Malton, Yorkshire, February 22, 1861.

Lady Burrard will with pleasure grant Mr. O Connell permission to add her name to the list of ladies who have experienced the comfort afforded by his charming Siphonia. She has already frequently recommended the Siphonia among her friends.—The Mount, Yarmouth, April 21, 1861.

The Hon. Mrs. Ryder, of Sendon Hall, Stone, Staffordshire, will be very happy for Mr. O'Connell to make any use he pleases of her name, being glad to bear testimony to the great value of his excellent invention the Siphonia, which has been of the greatest service to her baby.—39 Grosvenor Square, London, May 25, 1861.

Thurlam Castle, Kirby Lonsdale, Feb. 8, 1861.

Mrs. North Burton will feel obliged by Mr. O Connell sending two of his Patent Siphonias as soon as possible, which she requires for lending amongst the villagers. Mrs. North Burton considers Mr. O'Connell's invention a most invaluable one, having found it to answer so admirably with her own little babe, that she takes every opportunity of recommending it to others.

Crowsley Park, Henley-on-Thames,
June 20, 1862.

Sir,

The Siphonia which you sent me some time ago came to hand, and I am quite charmed with it, it answers the purpose so admirably. I shall have much pleasure in recommending it to my friends. I feel quite indebted to Mrs. North Burton (my sister-in-law) for mentioning the Siphonia to me.

I remain, sir, yours faithfully,
M. A. BASKERVILLE.

Mr. Edward O'Connell.

Godmersham Vicarage, Canterbury, July 20.

Mrs. Gale has much pleasure in informing Mr. O'Connel that she has experienced the greatest comfort from the use of his Siphonia, which she takes every opportunity of recommending amongst her friends. Mrs. Gale has been truly interested in a case where an infant has been deprived of the blessing of a mother. It is now being brought up entirely on the use of the Siphonia, and a more lovely or healthy child of three months old cannot be seen.

Masham, Yorkshire, January 28, 1862.

Dear Sir,

In reply to your letter duly received, I beg to assure you that I can never say too much in praise of your very happy invention. I am glad to say that my baby continues the picture of health and happiness, all of which I attribute to the use of your Siphonia.

Believe me yours sincerely,
MARGARET J. FISHER

[3567]

POWELL, SAMUEL, 2 *Surrey Cottages, Surrey Grove, Old Kent Road, S.*—Breast drawers, glass syringes, tube bottles, &c.

[3568]

PRATT, JOSEPH FRANCIS, 420 *Oxford Street, W.*—Apparatus for various deformities, and surgical instruments.

[3569]

PUCKRIDGE, F. L., 4 *York Place, Walworth.*—Liston's membrane plaster, court plasters, and gold-beaters' skins.

[3570]

PULVERMACHER, T. L., 73 *Oxford Street.*—Patent galvano-piline, a flexible galvanic constant battery for medical use, &c. (*See page* 129.)

[3571]

REDFORD, GEORGE, M.R.C.S. (late Army Medical Staff), *Cricklewood.*—1. Portable stretcher in halves fitting universally. 2. Medicine pouch.

[3572]

REIN, FREDERICK CHARLES, 108 *Strand, London.*—Surgical instruments, and acoustic appliances.

[3573]

REIN, MRS. S., 108 *Strand, London.*—On a new principle elastic abdominal supports; improved elastic stockings, knee-caps, and every support for the human body.

[3574]

REYNOLDS, JOHN, 20 *St. Anne Street, Liverpool.*—Artificial leg, with improved knee-joint and springs; trusses for hernia, and appliances for deformity.

[3575]

RIMMEL, EUGENE, 96 *Strand.*—Patent aromatic disinfector, for destroying all bad smells and purifying the air in hospital wards, dissecting rooms, dead-houses, coroners' inquests, sick-rooms, lodging-houses, ships, steamers, &c. (*See page* 130.)

[3576]

ROGERS, CHARLES, Inventor, 40 *Great Tindell Street, Birmingham.*—Either side double lever truss for single or compound hernia.

[3577]

ROGERS, MAURICE, 18 *New Burlington Street, W.*—Specimens of artificial teeth.

[3578]

ROOFF, WILLIAM B., 7 *Willow Walk, Kentish Town.*—Respirators, acoustic and medical instruments; patent safety seat.

Respirators in gold, silver, plated metal, charcoal, and aluminium. The excellence of these patent respirators is attested by the very large demand for them. Each is stamped with the maker's name.

ROOFF'S Patent Inhaler permits easy respiration, and prevents the expelled breath contaminating the in-coming vapour.

ROOFF'S Patent Tympani, or invisible sound-magnifier, has proved of great service to numbers affected with deafness.

ROOFF'S Patent Lavement apparatus is widely patronized on account of its great convenience and lightness.

These goods are sold by all chemists; descriptive catalogues may be obtained from the manufacturer, post free.

PULVERMACHER, J. L., 73 *Oxford Street.*—Patent galvano-piline, a flexible constant battery for medical use, &c.

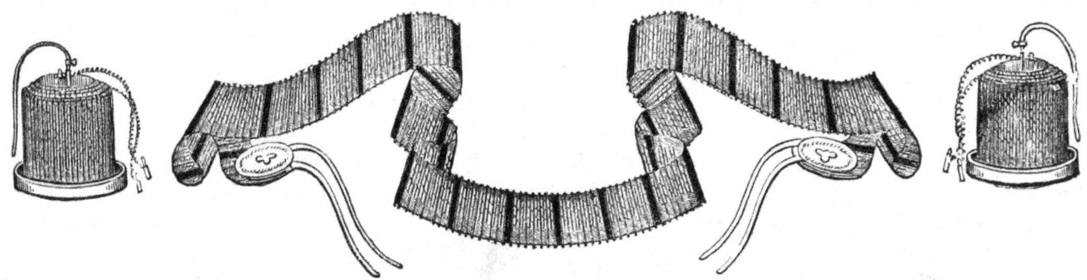

PULVERMACHER'S PATENT GALVANO-PILINE (for Medico-galvanic purposes) is a fabric composed of galvanic metal wires and a fibrous texture, representing a diminutive voltaic battery described in page 26, Class XIII. It possesses the same properties and advantages there enumerated, which, combined with its extreme pliability and durability, admirably adapt it for every imaginable mode of physiological experiment, or medical application of *intermittent* or *continuous* currents. A momentary, prolonged, localised, or diffused action can be administered by it with equal ease and comfort.

According to the mode of application required, the Galvano-piline is arranged, firstly, as a diminutive self-supplying pocket battery, for momentary operations. Secondly, in the form of bands, belts, necklaces, &c., for the prolonged application of diffused and gentle currents. These are easily worn on the part affected; infusing into the system a steady supply of gentle galvanic currents, analogous to the physiological functions of the animal economy. These batteries can be seen in operation at Messrs. J. L. PULVERMACHER & Co.'s (Galvanic Establishment), 73 Oxford Street, London, adjoining the Princess's Theatre.

PRICE LIST.

FLEXIBLE BATTERIES, manufactured from the Galvano-Piline, for the instantaneous generation of volta-electric currents of intensity, to be charged with the exciting liquid simply by immersion.

	£	s.	d.
GALVANO PILINE battery of 50 elements, each element, 2 square inches in surface, complete	1	10	0
Ditto, ditto, of 100 elements, each element 2 square inches in surface, complete	2	10	0
Ditto, ditto, of 100 elements, each element 6 square inches in surface, complete	3	10	0
Ditto, ditto, of 100 elements, each element 13 square inches in surface, complete	6	10	0

Galvano-Piline batteries can be made to order for intensive or quantitive electricity, with any number and size of elements required. Batteries can be made of zinc and silver, or zinc and platinum, platinized for obtaining a double effect with the same surface.

GALVANO-PILINE DIMINUTIVE BATTERIES for the prolonged and steady application of moderate continuous currents in form of chain-bands to be worn on the body, to be charged with the exciting liquid simply by immersion.

	£	s.	d.
No. 0, narrow, full electric power, 36 inches long, applicable for the limbs	1	2	0
No. 1, narrow, less power, 33 inches long, applicable for the limbs and stomach	0	18	0
No. 2, narrow, medium power, 24 inches long applicable for the stomach, head and face	0	15	0
No. 3, narrow, weak power, 16 inches long, applicable for the head and face	0	10	6
No. 4, narrow, weakest power, 8 inches long, applicable for the head and face	0	5	0
No. 0, broad, full power, 25 inches long, spinal band	1	2	0
No. 1, broad, less power, 21 inches long, applicable for the loins, spine, and stomach	0	18	0
No. 2, broad, medium power, 16½ inches long, applicable for the abdomen, head and face	0	15	0
No. 3, broad, less power, 12 inches long, applicable for the knee joints, and head	0	10	6
No. 4, broad, weak power, 6 inches long, short band, bracelet	0	5	0
Combined bands for acting upon the spinal column, limbs, &c., simultaneously	2	0	0

GALVANO-PILINE DIMINUTIVE FLEXIBLE BATTERIES for the prolonged and steady application of moderate continuous currents in form of chain bands, to be worn on the body, charged with the exciting fluid by a self-supplying arrangement.

	£	s.	d.
No. 0, narrow, full electric power, 36 inches long, applicable for the limbs	1	13	0
No. 1, narrow, less power, 30 inches long, applicable for the limbs and stomach	1	7	0
No. 2, narrow, medium power, 24 inches long, applicable for the abdomen, head and face	1	2	6
No. 3, narrow, weak power, 16 inches long, applicable for the head and face	0	15	0
No. 4, narrow, weakest power, 8 inches long, applicable for the head and face	0	7	6
No. 0, broad, full power, 25 inches long, spinal band	1	13	0
No. 1, broad, less power, 21 inches long, applicable for the loins, spine, and stomach	1	7	0
No. 2, broad, medium power, 16½ inches long, applicable for the abdomen, head and face	1	2	6
No. 3, broad, less power, 12 inches long, applicable for the knee-joints or head	0	15	0
No. 4, broad, weak power, 6 inches long, short band, bracelet	0	7	6
GALVANO-PILINE DIMINUTIVE BATTERY (self-supplying arrangement), in the form of a belt, 3 inches wide, complete with conducting wires, pole, plates, &c.	1	15	0

ACCESSORIES required in different modes of application of galvanic intermittent and continuous currents.

	£	s.	d.
VIBRATING INTERRUPTER to produce muscular contractions	0	2	0
Interrupting conductors	0	3	0
Interrupting clockwork	1	10	0
Galvanizing cylinders with insulating handles, per pair	0	3	6
Electro conducting spongeo-piline, per square inch	0	1	3
Electro conducting caps	1	10	0
Electro conducting brushes, according to size, from 15*s*. to	1	10	0
Electro conducting pessaries	0	15	0
Electro conducting ear-sponges	0	1	0
Electro conducting catheters	0	4	0
Electro conducting tooth-brushes, from 1*s*. to	0	5	0
Electro conducting bathing-drawers	1	10	0
Electro conducting suspenders	0	15	0
Graduated voltameter, for measuring the degree of power of the galvanic current	0	5	6
Water decomposing apparatus, from 1*s*. to	0	3	6
Pocket galvanometer	1	1	0
Flexible conducting wires, per pair	0	2	6
Electro physiological forceps	0	10	0

RIMMEL, EUGENE, 96 *Strand.*—Patent aromatic disinfector, for destroying all bad smells and purifying the air in hospital wards, dissecting rooms, dead-houses, coroners' inquests, sick-rooms, lodging-houses, ships, steamers, &c.

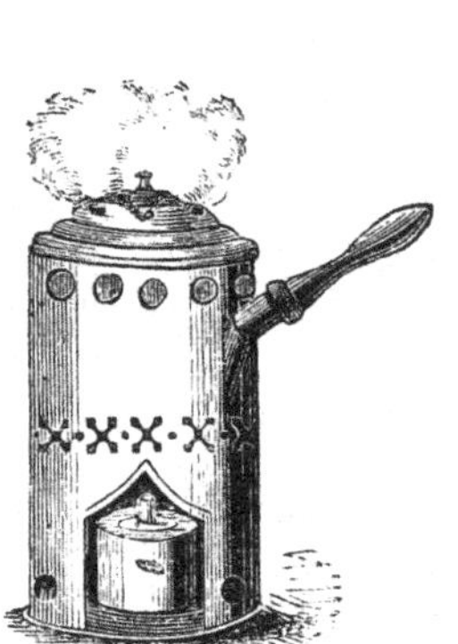

For Hospital Wards, Dissecting Rooms, &c.
Price 4*s.* 6*d.* & 10*s.* 6*d.*

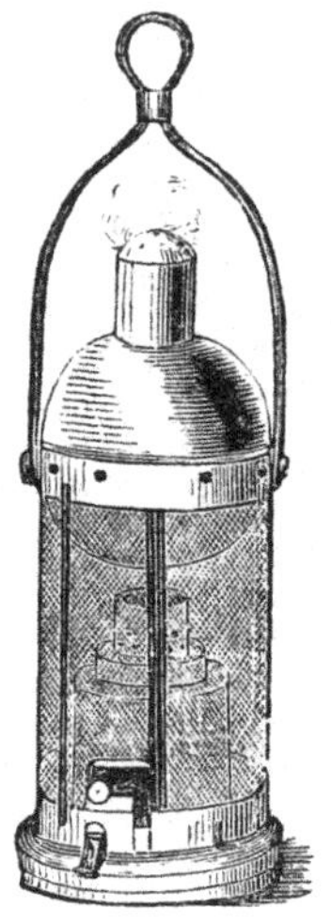

For Ships and Steamers, with safety lamps and hanging gear.
Price £1 1*s.* & £1 10*s.*

For Household use and Sick Rooms.
Price 1*s.* 6*d.*

Pocket cases for medical men, or persons visiting the sick, 1*l.* 1*s.* Aromatic Compound to be used in the disinfector, in small bottles for household use, 1*s.*; in large quantities, 16*s.* per lb.

RIMMEL's Aromatic Disinfector acts on the same principle as his Perfume Vaporizer exhibited in Class 4, and well known in fashionable circles, but it is of a more simple and economical form, and is confined to sanitary purposes. The aromatic compound prepared for it is not extracted from flowers, as that used in the Vaporizer, but from plants noted for their beneficial and prophylactic actions—such as rosemary, thyme, &c. It is therefore incapable of affecting the most nervous invalid, or of proving unpleasant even to those most averse to perfume.

It removes instantaneously all sorts of bad smells, whatever may be their nature or intensity, and substitutes a reviving and grateful atmosphere. The potency and rapidity of this system may be judged from the fact that it only takes five minutes to saturate an immense area like that of Covent Garden Theatre with fragrant vapours.

The apparatus and *modus operandi* are both very simple. The former consists in a pan heated with an oil lamp, and half filled with hot water, into which a few drops of the compound are poured; the effect is produced as soon as the water commences to boil.

Some scientific men are of opinion that aromas are not positive disinfectants, but merely cover one smell by means of another. It may be said in reply to this, that many noxious effluvia have hitherto resisted all attempts at analyzation, and that it has likewise been found impossible to ascertain the true nature of fragrant volatile emanations, their solid basis only being known; and in such a case we may admit the evidence of our senses for want of better tests, and if we find a bad smell replaced entirely by a pleasant one, we may fairly assume that the former has become neutralized. In fact, E. RIMMEL has had positive proofs of his apparatus answering when all other disinfectants had failed, which is probably to be attributed to the penetrating influence of fragrant molecules, developed *ad infinitum* by means of steam, and perhaps also to their ozonizing or oxygenating properties. Those aromatic fumigations have even been tried and found to succeed in arresting the progress of infectious diseases, and some very interesting experiments might be made in that way by medical practitioners.

RIMMEL's Disinfector has been adopted by the Royal College of Surgeons for their dissection meetings, and by many of the London hospitals to be used in the wards. It has also been tried successfully at the Amphithéâtre de Clamart in Paris, and at the principal hospitals in Vienna. It was introduced on board of Her Majesty's Steam Yacht, the Victoria and Albert, to remove the nauseous smell proceeding from the engines, and is now in use on some of the Peninsular and Oriental Company's vessels, and other steamers, where it is found most useful and agreeable to the passengers, producing a reviving atmosphere, and allaying the sufferings of sea sickness.

E. RIMMEL will be happy to present gratuitously any hospital or charitable institution with his apparatus and the necessary compound. He hopes that in return medical men when quite convinced of its efficacy in the sick-room, will do him the favour of recommending it to their patients.

N.B.—E. RIMMEL has received many letters from eminent scientific men bearing testimony to the useful qualities of his Disinfector, and is ready to show them privately to any members of the profession who may favour him with a call at No. 96 Strand.

EUGENE RIMMEL, Inventor and Patentee of the Aromatic Disinfector, and Perfume Vaporizer, 96 Strand, and 24 Cornhill, London.

[3579]

RUSSELL, CAPTAIN GODFREY, *Swan Hill, Shrewsbury.*—Improved hospital bed appliance; ditto hospital stretcher; camp hospital spring-bed or stretcher; incontinent urinal.

IMPROVED HOSPITAL BED APPLIANCE.—This model of the pattern in the Tower is shown by the kind permission of the Hon. Secretary for War, with the hope that its use may become more general, and that by thus giving it for the public benefit, and inviting competition, it may be still improved and the cost of its construction reduced. It has been in constant beneficial use in two of Her Majesty's military hospitals for more than twelve months; has undergone every test and examination by numerous boards of the highest medical authority; and has been found of great service to the medical profession in diseases and injuries of the hip joint, pelvis, spine, and all extreme cases in which absolute rest is required. Long trial has shown that by preventing painful movements, the patient is saved a good deal of exhaustion; and it may be added that a single nurse has full control over the invalid. It was given to the army and navy.

Improved HOSPITAL STRETCHER.—Accepted by Her Majesty's Service, and is shown and given to the public on the same authority and grounds as the improved bed appliance. The chief object is to remove patients to and from the operating-room, or extreme cases, as it can be removed without their feeling any motion.

RUSSELL'S CAMP HOSPITAL SPRING BED OR DHOOLEE STRETCHER.—This forms a very comfortable bed, and being on springs, prevents any shock on changing bearers. It has a sun and rain-awning, and packs up in a small compass. It was expressly made for Her Majesty's Service, has undergone examinations, and a certain number sent to the camps; but circumstances have compelled the inventor, unwillingly, to seek protection by patent. He considers that the commoner form would be of great service at hospitals and railway stations, for the easy conveyance of injured persons.

INCONTINENT URINAL, for day and night use, expressly made for Her Majesty's hospitals and invalid depôts. Considering the great importance of appliances of this kind, the inventor has shown it in order that it may be more generally adopted and improved. Many experiments have been made, by an eminent Professor of chemistry, to test the material, and that manufactured by the successors to Charles Goodyear, 11A Adam Street, Adelphi, has been found to be superior.

[3580]

SANSOM, DR. A. E., M.B., *Ashburton Villa, Lower Road, Islington.*—Apparatus for the gradual administration of chloroform.

[3581]

SAVORY & MOORE, 143 *New Bond Street.*—Portable medicine chests, &c. (*See page* 132.)

[3583]

SILLIS, FRANCIS, 2 *George Street, Euston Square, London.*—Artificial legs, hands and arms, spring crutches, and hand instruments.

[3584]

SIMPSON, HENRY, 55 *Strand, London.*—Surgical instruments (various).

[3585]

SMALE BROTHERS, 19 *Great Marlborough Street, London.*—Mineral teeth, dental implements, and appliances.

[3586]

SMITH, JOHN COX, *Week Street, Maidstone.*—Tooth instruments for the especial use of army and navy surgery.

[3587]

SMITH, WILLIAM & FRANCIS, 253 *Tottenham Court Road, London.*—Water bed, or floating mattress for invalids.

[3588]

SPARKS & SONS, 28 *Conduit Street, Hanover Square, W.*—Surgical bandages and appliances for the relief and cure of deformities, and giving support to the human frame.

[3589]

SPRATT, WILLIAM HENRY, 2 *Brook Street, Hanover Square, W.*—A collection of trusses and orthopædic instruments.

Savory and Moore, 143 *New Bond Street.*—Portable medicine chests, &c.

TOURISTS' AND SPECIAL CORRESPONDENTS' PORTABLE MEDICAL CASE.

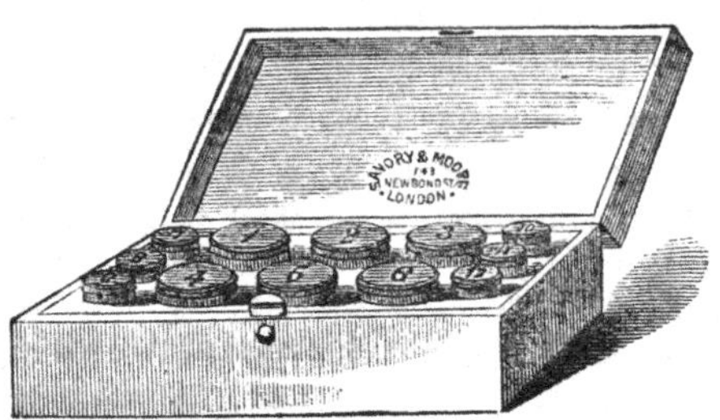

THE OFFICER'S MEDICINE CHEST USED IN THE CRIMEA, INDIA, AND CHINA.

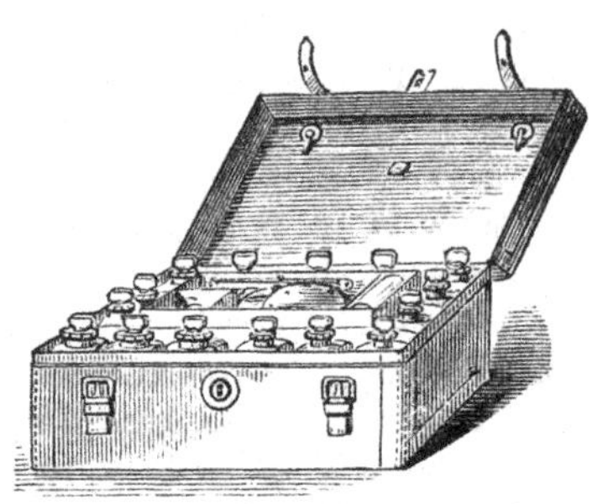

Messrs. Savory & Moore, of New Bond Street, exhibit a very complete collection of Medicine Chests, from the large box, adapted to the use of a detachment on active service, to the small, but no less efficient pocket case, that may be conveniently packed in the portmanteau of a tourist, or even conveyed in the coat-pocket in case of emergency.

These latter small travelling Medicine Chests are novel in their design, and compact in their arrangement. A strong leather case, six inches by nine, and only two and a half in thickness (in fact of the form and size of an ordinary octavo volume, and opening in much the same manner), contains a pair of scales, with the necessary weights, a small glass measure, eight small bottles adapted to receive either powders or pills, eight of larger size, stoppered for liquids, and two still larger for holding any medicines required in greater bulk. This little case, which would scarcely take up any appreciable room in the travelling bag, will contain all the Medicines required in any case of emergency. Its value to a party of tourists, or to a single traveller, removed from medical aid, can hardly be overrated. The "Special Correspondents" of our daily papers have used these cases in nearly every quarter of the globe, and have spoken most highly of their utility.

Larger Leather Cases, of the same character, containing a greater number of remedies, are also shown by the firm.

The Yacht Medicine Chest, as its title implies, is adapted for use in short sea voyages. It contains sufficient medicine for a crew of twenty persons, and in addition a few surgical appliances that may be required in an emergency, such as splints and bandages for fractures.

The most important articles exhibited by Messrs. Savory & Moore are unquestionably those valuable aids to military surgery which they have put together. Under the title of a Medical Field Companion, Savory & Moore have designed a case, weighing only ten pounds, to be carried on the march by a soldier in lieu of his rifle. This Companion contains all that could be required during a reconnaissance; such as mixture for diarrhœa, tincture of opium, chloroform, sal volatile, also packets of powder most likely to be useful in an emergency in their proper doses; several varieties of pills, and all the appliances likely to be required, as lint, bandages, plasters, splints, sheeting, tourniquet, &c., &c.

The Improved Medical Panniers, for the use of the army, are designed to convey all the appliances, both medical and surgical, that may be required by a regiment in the field and during a march. Within the compass of two panniers of ordinary size, and the regulation weight, are contained, on the one side, some thirty different drugs, with all the required accessories of scales, weights, &c., each so accessible as to be obtained in a moment; medical comforts for the sick and wounded, such as brandy, concentrated beef tea, arrowroot, &c.; a lamp with reflector and such adjustment as enables it to be used in warming a small quantity of food. In the other pannier may be found the case of operating instruments, tourniquets of different kinds for field use, bandages, plasters, sheeting, splints, and everything to hand.

The panniers may be used on or off the mule's back, and are so constructed that they can be made to form a very good and firm operating table, by placing them on the ground, throwing open the lids, and securing them in the required position. The advantage of this arrangement, when the surgeon is in the open field, far from houses, is obviously very great.

The Army Detachment Medicine Chest is a strongly-bound polished oak box, containing, in the compass of a few feet, a larger and more complete assortment of medicines and materials than would be found in most ordinary surgeries.

This Military Chest is so constructed that, by merely opening the lid, a dispensing counter of convenient height is at once formed; and, without shifting his position, the dispenser will find everything at hand, the whole being so admirably arranged that no one article has to be displaced to gain access to another.

In addition to these valuable aids to Military Surgery, Messrs. Savory & Moore also exhibit Eye and Ear Douches of improved construction. The great peculiarity of these instruments is that, in addition to the elastic bottle and tube conveying the stream of liquid to the eye or ear, there is a second tube from the glass

cup, which is placed against the affected organ; this tube conveys away the water into a basin, and so prevents it running down the face or neck, to the great discomfort of the patient.

The Enemas shown by the same firm are supplied with an elastic tube in the place of the usual inflexible bone or metallic nozzle. This affords very great facility for introduction into the bowel, and removes all risk of lacerating the lining membrane. As thus fitted, these instruments are especially adapted for the self-administration of injections, which are so valuable in the removal of habitual costiveness, without the necessity of continually having recourse to aperient medicines.

[3590]

SYKES, MARY EFLAT, 280 *Regent Street, Castle Square, Brighton.*—Corsets for pregnancy, and an abdominal bandage for after accouchement.

[3591]

THRING, CHARLES, 3 *Little Randolph Street, Camden Town.*—Inodorous commode for the sick chamber; cheap arm-sling.

[3592]

TOMPSON, W. A., 18 *Cecil Street, Strand.*—Inhaler for applying caustic solution internally in throat diseases.

[3593]

TUFNELL, JOLIFFE, *Mount Street, Dublin.*—Tubular bougies for the cure of stricture of the rectum.

[3594]

TWEEDIE, WILLIAM, 337, *Strand, London.*—The respirator, composed of ten layers of gold wire—a perfect instrument. (*See page* 134.)

[3595]

WAITE, GEORGE, 2 *Old Burlington Street.*—Surgical instruments.

[3596]

WALTERS, FREDERICK, 16 *Moorgate Street, City.*—Surgical instruments, and instruments for deformities.

[3597]

WEEDON, T., Surgeons' Instrument Maker, *Hart Street, Bloomsbury, London.*—Instruments for microscopical preparations, morbid anatomy, and animal preserving.

[3598]

WEISS, J., & SON, 62 *Strand.*—Variety of surgical instruments.

[3599]

WELLS, GEORGE S., 59 *Euston Square.*—Artificial teeth and gums.

[3600]

WELTON, THOMAS, 13 *Grafton Street, Fitzroy Square.*—A case with jointed pin-leg, artificial human leg, and others.

[3601]

WELTON & MONCKTON, 13 *Grafton Street, Fitzroy Square.*—A magnetic chain and battery for curing diseases.

[3602]

WESTBURY, ROBERT, 26 *Old Millgate, Manchester.*—Trusses and deformity instruments.

The following truss and deformity instruments are exhibited :—

No. 1. Instrument for correcting lateral curvature and torsion of spine; with eight distinct movements for adjustment.

No. 2. Instrument for a case of disease of upper cervical vertibræ close to the occiput; with seven movements for adjustment.

No. 3. Apparatus for remedying permanent contraction of fingers, after burns or other injuries. Cast No. 1 shows such a case previous to the use of this instrument, and Cast No. 2 the same after three months' treatment.

No. 4. Instrument for talipes, *equino varus*, or club foot. Case No. 3 represents such a case as the instrument is adapted for; with special arrangement for rack and pinion, and rotatory movements in the sole of the shoe.

No. 5. Truss with concave pad, for a case of irreducible femoral hernia.

No. 6. Truss with coil-spring pads, for a case of double inguinal hernia.

No. 7. Frame-work, showing construction of imperceptible curative truss; with rotatory movement for adjusting the pad.

No. 8. Truss for umbilical hernia.

No. 9. Children's trusses, single and double.

No. 10. Spinal support, for cases of slight curvature.

No. 11. Steel stays, used as a preventative in case of tendency to curvature of the spine.

Nos. 12 & 13. Instrument for genu-valgum, or knock-knees; with improvements.

TWEEDIE, WILLIAM, 337 *Strand, London.*—The respirator, composed of ten layers of gold wire—a perfect instrument.

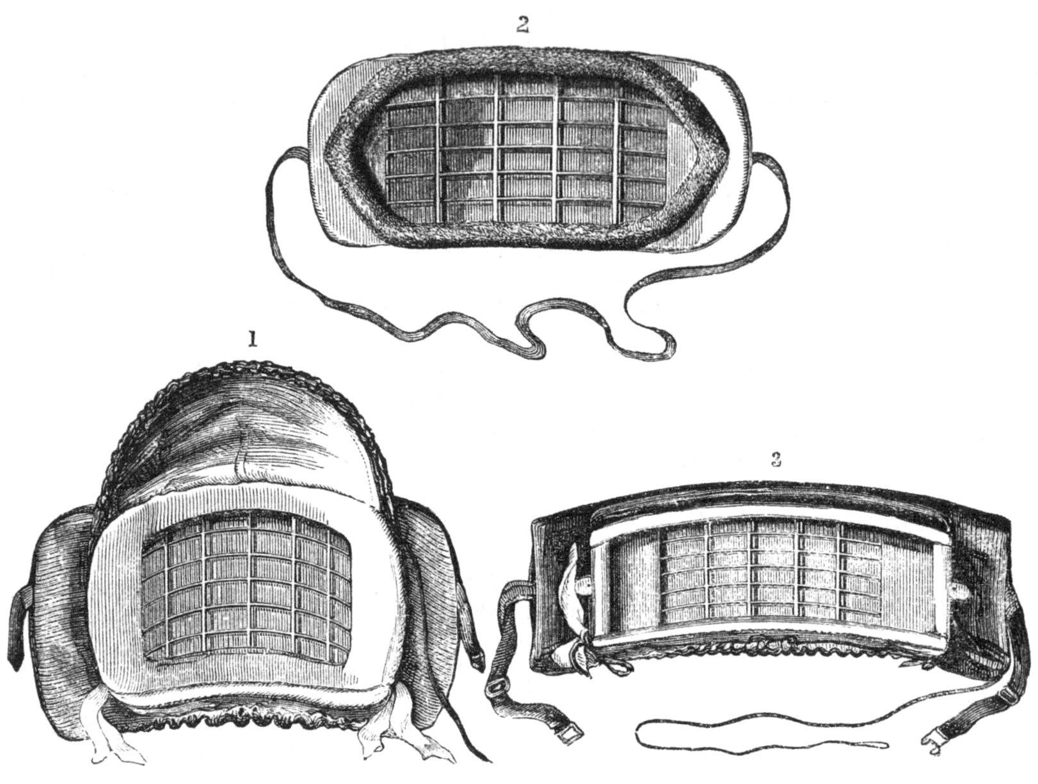

THE RESPIRATOR.

The word "RESPIRATOR" was introduced into the language twenty-six years ago by MR. JULIUS JEFFREYS, F.R.S., shortlyafter his retirement from the Indian Medical Staff. This word was chosen as an appropriate name to designate an instrument of a peculiar and elaborate metallic construction, which, when *respired* through (*i. e.* breathed through both ways in drawing and expelling breath), should have the property of promoting a free and easy respiration, by transferring warmth and moisture from each outgoing breath (the impure gases of the breath being freely voided) and imparting that warmth and moisture to each fresh-entering breath—thereby rendering it genial and soothing to irritable breath-passages.

Thus may be produced a climate for the lungs, variable at will, and fulfilling many important pathological purposes which cannot be here enumerated.

But the name RESPIRATOR has been so prostituted by its assumption for articles bearing an outward appearance to the true instruments (as a toy-watch may to the real one), that Mr. Jeffreys is very doubtful if any public object can be served by this occupying of space in the International Exhibition Catalogue with mere sketches of the different forms of the true Respirator. Figs. 1, 2, and 3 refer severally to the orinasal respirator for the mouth and nostrils, and to the dwarf (which will henceforth be discontinued) and the standard oral Respirator, instruments for the mouth alone. Besides these, there are the Nasal, an instrument for the nose only, and the Hand Respirator—held in the hand and applied to the mouth or nose according to the make of the instrument.

It is these instruments which have acquired for the name RESPIRATOR its world-wide reputation by the benefits they have conferred upon a multitude of sufferers from all varieties of pulmonary disorder—benefits which they who have recourse to trashy articles in lieu of the true instruments will never experience.

[3603]

WETHERFIELD, JOHN, *Henrietta Street, Covent Garden.*—Amadou plaster—a surgical appliance for purposes of support and defence.

[3604]

WHIBLEY, EBENEZER, 41 *Radnor Street, Chelsea.*—Surgical operating table.

[3605]

WHICKER & BLAISE (late SAVIGNY & Co.), 67 *St. James's Street, S.W.*—Surgical instruments and appliances.

[3606]

WHITE, JOHN, 228 *Piccadilly.*—White's moc-main patent lever truss; elastic surgical appliances for hernia; new patent elastic stockings.

WHITE'S MOC-MAIN PATENT LEVER TRUSS and elastic surgical appliances for hernia; new patent elastic stockings, spinal machines, spinal corsets, chest expanders, ladies and gentlemen's belts, spring trusses with spiral springs and ivory pads, improved prolapsus ani.

White's Moc-main Patent Lever Truss is allowed by 500 medical men to be the best for hernia. It consists of an elastic pad, with a lever, and instead of the usual spring, a soft band, fitting so closely as to avoid detection. A descriptive circular may be had by post.

Single, 16*s.*, 21*s.*, 26*s.* 6*d.*, & 31*s.* 6*d.*; postage, 1*s.* Double, 31*s.* 6*d.*, 42*s.*, & 52*s.* 6*d.*; postage, 1*s.* 8*d.*

[3607]

WHITING, WILLIAM, & SONS, *High Street, Camden Town.*—Improved spinal supports for lateral and angular curvatures.

[3608]

WILLIAMS, G. J., 17 *Cavendish Place, Cavendish Square, W.*—Improvements in artificial palates and teeth.

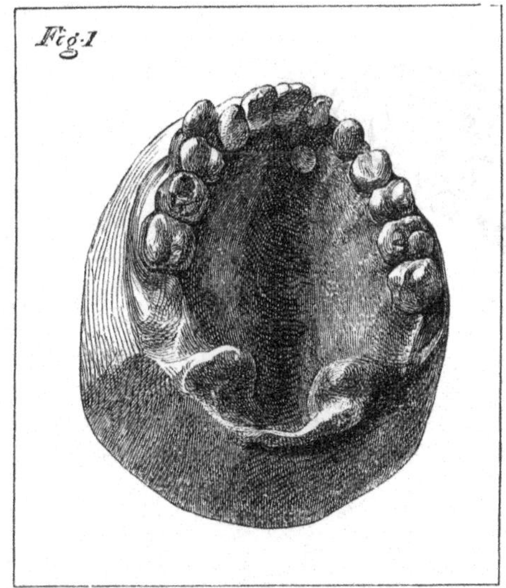

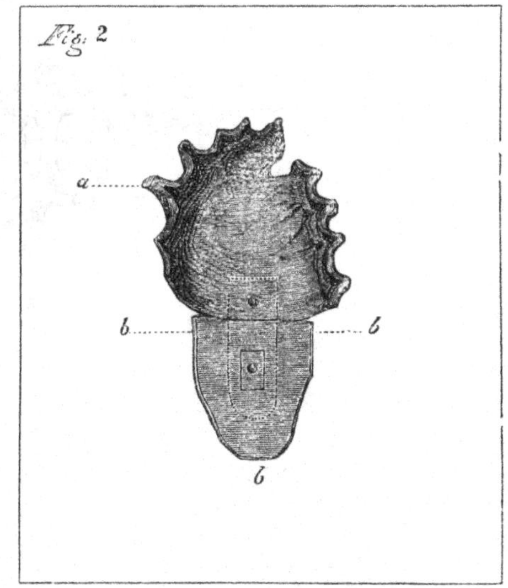

Fig 1. Case of complete fissure of the hard and soft palate, the fissure extending through the whole of the hard palate and uvula.

Fig. 2. Represents one of Mr. G. J. Williams's improved Obturators for the above case. The portion *a*, which covers the palate as far as the second molar teeth, is constructed of hard vulcanite; the velum, or soft palate, *b b b*, is formed in soft vulcanite, the two portions being united by a narrow band of elastic gold, allows the artificial velum to follow the muscular action of the palate, by which means the patient can perform the acts of deglutition and articulation with comparative ease.

[3609]

WOOD, WILLIAM ROBERT, Dentist, *Carlisle House, Brighton.*—Models presenting irregularities of teeth, and their cure; also general dentistry.

Specimens exhibiting certain means and processes employed in general dentistry, regard being had, in practice, to apply peculiar treatment to individual states of the mouth. Models of nine mouths, illustrating irregularities of teeth, with the conditions and results of REGULATION.

[3610]

Wright, Dr. Henry G., 23 *Somerset Street, Portman Square.*—A substitute for the stra waistcoat.

[3611]

Young, James Anderson, 47 *Bath Street, Glasgow, Scotland.*—Patent dental forceps, &c.

[3612]

Read, Messrs., 8 *Holles Street,* Cavendish Square.—Artificial teeth.

LONDON: PRINTED BY WILLIAM CLOWES AND SONS, STAMFORD STREET AND CHARING CROSS.

SECTION III.

Class XVIII.

COTTON.

[3640]

Ashworth, Edmund, & Sons, *Egerton Mills, Bolton.*—Sewing cotton of every description; embroidery and skein cotton.

[3641]

Auld, Berrie, & Mathieson, 111 *Union Street, Glasgow.*—Plain and fancy Scotch muslins.

[3642]

Barlow, Goody, & Jones, *Manchester.*—Toilet Marseilles quilts, counterpanes, cotton blankets, quiltings, dimities, and cotton damasks.

[3643]

Brittain, Thomas, *Manchester.*—Patent sponge cloths for cleaning machinery and fire-arms; protective garden nets.

These cloths are especially suited for use on railways, locomotives, and steam-boats. They absorb oily matter, and remove particles of dirt better than any other fabric. Price 23*s.*, 27*s.*, and 30*s.* the gross.

Protective Garden Net for the preservation of fruit trees, vines, &c., from insects and all extremes of temperature. Price, 54 inches wide, No. 1, 4*d.*; No. 2, 5*d.*; No. 3, 6*d.* the square yard.

[3644]

Brook, J., & Brothers, *Meltham Mills, Huddersfield.* (*See page* 2.)

[3645]

Brown, Sharps, & Tyas, 18 *Watling Street.*—Embroidered muslins.

[3646]

Carlile, James, Sons, & Co., *Barkend Mills, Paisley.*—Cotton and linen threads, in balls, hanks, and on reels.

[3647]

Christie, Hector, *Blackfriars Mills, Manchester.*—Samples of doubled, gassed, dyed, printed, satined, and polished yarns.

BROOK, J., & BROTHERS, Manufacturers, *Meltham Mills, Huddersfield.*—Sewing cottons, crochet, and embroidering; finished and in process.

Obtained Prize Medal at the London Exhibition, 1851, *and the First Class Prize Medal at the Paris Exhibition,* 1855.

The exhibitors are the manufacturers and winders of BROOK'S PATENT GLACÉ THREAD, nine, six, and three-cord, for hand and machine sewing, also crochet and embroidering cottons, all of guaranteed lengths, and marked with the name of 'BROOK' and a 'Goat's Head' crest.

They exhibit a Case of Threads, in every process of manufacture, in Class XVIII.; and in the Machinery Department, Class VII., they work a Self-Acting Winding Machine in motion, manufactured by Messrs. Sharp, Stewart, & Co., Manchester.

[3648]

CHRISTY, W. M., & SONS, *Fairfield, near Manchester.*—Royal Turkish towels, blankets, huckaback or Brighton towels, and patent Terry counterpanes.

[3649]

CLARK & Co., *Seedhill and Cumberland Mills, Paisley.*—Sewing, crochet, and embroidery cottons.

[3650]

CLARK, JOHN, JUN., & Co., *Mile End, Glasgow.*—White and coloured sewing thread.

[3651]

CLARKE, I. P., *Leicester.*—Sewing cottons, reels, spooling and mill bobbins.

[3652]

COATS, J. & P., *Ferguslie Works, Paisley.*—Sewing cottons.

The Agent in London for the sale of Messrs. COATS' sewing cottons is William Gilmour, 37 King Street, Cheapside. The Paris Agents are L. & B. Curtis & Co., 6 Boulevard Poissonnière.

[3653]

COATS, NEILSON, & Co., *Thorn Mill, Johnstone, Renfrewshire.*—Cotton yarns, cop warp, doubled yarns, yarns for embroidering muslins.

[3654]

COPESTAKE, MOORE, CRAMPTON, & Co., 5 *Bow Churchyard*, Sewed muslin manufacturers.

COPESTAKE, MOORE, CRAMPTON, & Co.'s first and second cases contain a quantity of Leno and harness book curtains of Scotch manufacture.

Their third case contains the following examples of muslin embroidery:—

A child's robe, flouncings and insertions, a lady's robe, chemisettes, habit shirts, collars, sleeves, and cambric handkerchiefs. Productions of Scotland and Ireland.

[3655]

CREWDSON & WORTHINGTON, *Manchester.*—Medium and fine shirtings, bleached.

[3656]

DICKINS & Co., *Spring Vale Works, Middleton.*—Dyed and polished yarns and sewings.

[3657]

ERMEN & ENGELS, *Manchester.*—Sewing and knitting cotton, and Godfrey Ermen's patented polished thread.

[3658]

EVANS, WALTER, & Co., *Derby.*—Samples of cotton threads for sewing, crochet, knitting, and embroidery.

Sewing, Crochet, Embroidering, Mending, and Knitting Cottons, Cotton Cord, &c.

Six-cord Crochet Cotton on Spools.

Patent Glacé Thread.

Sewing Cotton on Spools: two-cord, three-cord, four-cord, six-cord, and nine-cord.

Sewing Machine Cottons.

Sewing Cotton in Balls, &c.

ESTABLISHED A.D. 1783.

[3659]

FAULKNER, HENRY, 6 *Castle Court, Lawrence Lane.*—Cotton twines, run 30 per cent. longer length than hemp, same weight.

[3660]

FORD, FRANCIS, *Stanley Street Mills, Manchester.*—Sewing, crochet, knitting, embroidering, and marking cottons; spools, balls, skeins, &c.

[3661]

GILLS & HARTLEY, *Wood Street, London; Cannon Street, Manchester.*—Double and single coutils for stays.

[3662]

GOODAIR, SLATER, & SMITH, *Kent Street Mills, Preston.*—Bleached long cloths, twilled and plain shirtings.

Obtained First Class Medal at the Paris Exhibition in 1855.

EXTRACT FROM JURORS' REPORT:—

CLASS 18, No. 1264.—'The low price, quality, and regularity leave nothing to be desired.'

[3663]

GREENWOOD & WHITTAKER, 15 *Marsden Square, Manchester.*—Water twist, shirting calicoes.

[3665]

HASTINGS, WILLIAM, *Huddersfield.*—Cotton yarns for warps and wefts.

[3666]

HAWKINS, JOHN, & SONS, *Green Bank Mills, Preston, Lancashire.* *(See page 5.)*

[3667]

HAWORTH, RICHARD, & CO., *Manchester*, Spinners and manufacturers of jeannettes, India twills, silicias, casbans, and mediums.

[3668]

HOLLINS, EDWARD, *Sovereign Mills, Preston.*—Cotton shirtings and sheetings.

[3669]

HOPWOOD, ROBERT, & SON, *Nova Scotia Mills, Blackburn.*—Calicoes, and the incidental processes of manufacturing them.

[3670]

HORROCKSES, MILLER, & CO., 9 *Bread Street, London; Manchester; and Preston.*—Long cloths and twilled shirtings.

[3671]

HOULDSWORTH, THOMAS, & CO., *Manchester.*—Fine cotton yarn, single and doubled.

[3672]

HUDSON, J., & SONS, Manufacturers, *Leicester.*—Sewing cotton of various kinds on reels.

[3673]

JACK, JOHN R., 37 *Virginia Street, Glasgow.*—Jacquard muslin window-curtains.

[3674]

JOHNSON, JABEZ, & FILDES, *Spring Gardens, Manchester; Moor Mills, Bolton.*—Quiltings, quilts, counterpanes, toilette-covers, skirts, &c.

HAWKINS, JOHN, & SONS, *Green Bank Mills, Preston, Lancashire.*—Plain and twilled cloths for shirts, &c.

The exhibitors are Cotton Spinners, and Manufacturers of Power-loom Shirtings, Long Cloths, Twills, Striped Dimities, &c.

WAREHOUSES: 8 Faulkner's Street, Manchester; and 22 Lawrence Lane, London.

[3675]

JOHNSON, J. MARSHALL, *Britannia Mill, Mirfield.*—Single and double cotton yarns, dyed granderelle twists, and fancy warps.

[3676]

KENYON, JOHN THRELFALL, & Co., *White Hall Mill, Over Darwen, near Blackburn.*—Printing cloths and India and China shirtings.

[3677]

KERR & CLARK, *Linside Thread Works, Paisley; and* 88 & 90 *Reade Street, New York.* Spool cotton, enamelled and six-cord.

The Branch Establishments of KERR & CLARK are at

No. 17 Silver Street, City, London.
„ 97 Boulevard de Sebastopol, Paris.

No. 88 & 90 Reade Street, New York.
„ 5 Bank Street, Philadelphia.
„ 87 Devonshire Street, Boston.

[3678]

KESSELMEYER & MELLODEW, *Manchester*, Manufacturers and patentees of fast pile silk, imitation silk, and cotton velvets.

[3679]

LOWTHIAN, FAIRLIE, & CO., *Carlisle*, Cotton spinners, dyers, and manufacturers of ginghams, checks, stripes, drills, &c.

[3680]

MANCHESTER COTTON TWINE COMPANY, 51 *Corporation Street, Manchester*.—Cotton twine and cotton mill bands, by steam power.

These Cotton Twines, manufactured by steam power, are nearly double the length of hemp to the same weight; they are much more even, and superior in colour. The improved Sea Island Cotton Mill Bands are stretched in the single strand so as to prevent slack bands, thereby saving much waste in slack yarn.

Samples will be sent on application.

[3681]

MANLOVE, SIMEON, *Holy Moor Mills, Chesterfield*.—Sewing cotton on reels; embroidery, crochet, and knitting, in skeins.

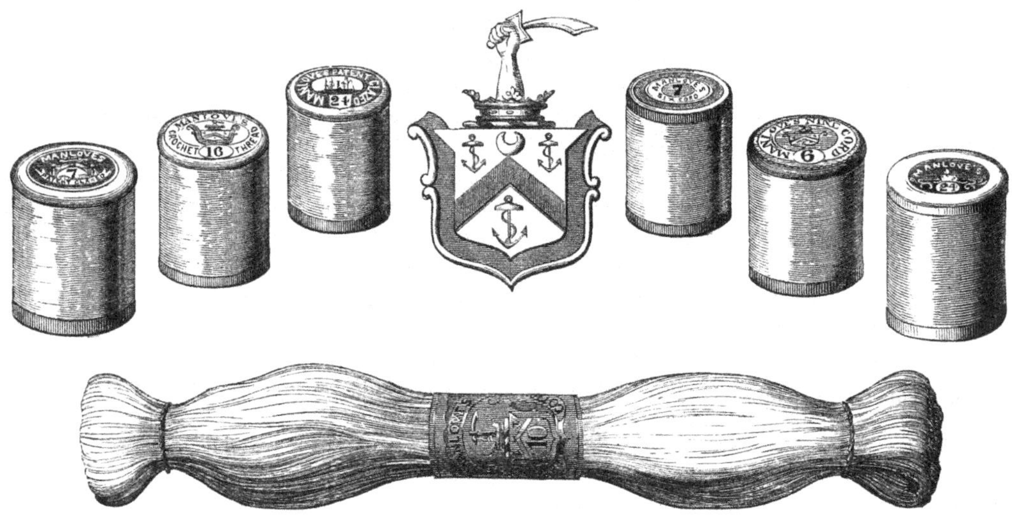

This case contains reels of Patent Glazed Thread, six and nine-cord Sewing Cottons, Crochet, Embroidery, Knitting, and Mending Cottons.

[3683]

MARTIN, JOHNSON, & JOULE, *Bolton and Manchester*.—Furniture dimities and damasks.

[3684]

MOORE, J., 33 *Piccadilly, Manchester*.—Velvet ribbons, with patent edges.

[3685]

MORGAN, JOSEPH, *Ducie Works, Manchester*.—Plaited and self-consuming wicks for hard material, tallow, mould, and dip candles.

[3686]

NORTHCOTE, S., & CO., 29 *St. Paul's Churchyard*.—Embroidered collars, sleeves, and other goods.

[3687]

OUTRAM, R., & Co., 13 *Watling Street.*—Plain and figured muslins, counterpane, quilt, &c.

[3688]

PHILLIPS, JAMES, 8 *Lawrence Lane, Cheapside, London.*—Woven fancy quiltings, and printed fancy quiltings for waistcoats.

[3689]

RAWORTH, JOHN THOMAS, Manufacturer to the Queen, *Leicester.*—Nine-cord, six-cord, and glacé sewing cotton.

[3690]

SHAW, JARDINE, & Co., *Manchester.*—Spinning and doubling, lace, sewings, and cotton crape yarns.

[3691]

SMITH, W. J., & Co., 40 *Faulkener Street, Manchester.*—Satteens, drills, royal ribs, quiltings, corset ribs.

[3693]

SWAINSON, BIRLEY, & Co., *Fishwick Mills, Preston; Portland Street, Manchester; and* 42 *Cheapside, London.*—Calicoes for shirts and ladies' under-clothing.

The calicoes manufactured by this firm for the home and export trades, are especially adapted, both in texture and finish, for shirts and ladies' and children's under-clothing.

[3694]

SYMINGTON, R. B., & Co., 9 *Cochrane Street, Glasgow.*—Harness figured muslin curtains, lappets, and linings.

[3695]

TOWNSEND, THOMAS, & SON, *Coventry.*—Grey, dyed, and dressed cotton yarns.

[3696]

WILSON, T. & D., & Co., 145 *Ingram Street, Glasgow.*—Plain and fancy muslins.

[3697]

WRIGLEY, H. & E., *Huddersfield.*—Single and double cotton yarns, grey, gassed, bleached, and coloured granderelle.

[3698]

YATES, BROWN, & HOWAT, Manufacturers, *Springfield Court, Glasgow.*—Plain and fancy muslins, Jacquard muslin curtains, &c.

[3699]

YOUNG, J. H., & Co., *Glasgow.*—Harness curtains and blinds, lappet and plain muslins, various, made by power.

[3700]

HALE & UDALE, *Manchester.*—Cotton velveteens.

Class XIX.

FLAX AND HEMP.

[3728]

Ainsworth, Thomas, *Chator Mills, Whitehaven.*—Sewing machine and other linen threads; flax yarns.

[3729]

Austin, James, *Princes Street, Finsbury.*—Imperial patent sash, blind, and picture lines; military cord; crinoline steel.

Imperial Patent Sash Lines in various qualities, Patent Lines for ships' halyards and logs, Patent Rope for turret clocks, Clock Lines, Coloured Thread Blind Lines, Worsted Blind, Curtain, Lamp, and Picture Lines.

The Patent Halyard Lines are specially adapted for ships' use, as they can be made of unlimited length, and are free from the fault of twisting, so much complained of in those commonly in use.

[3730]

Barbour, William, & Sons, *Lisburn.*—Linen, tailors', and shoe threads, various colours; yarns, flax, and linens.

[3732]

Baxter brothers & Co., *Dundee.*—Linen yarns and linen manufactures.

[3733]

Belfast Local Committee, *Belfast.*—Damasks.

[3734]

Bell, Richard, & Co., Flax Spinners and Linen Damask Manufacturers, 13 *Donegal Street, Belfast.*—Damask goods.

[3735]

Bennett & Thorn, 190 *High Street, Borough, London.*—Hemp, flax, jute, ropes, twines, lines, mats, cocoa-matting, &c.

[3736]

Bird, Robert, *Crewkerne, Somerset,* Manufacturer of linen and woollen saddlery webs, also brace; upholstery; and patentee of straining webs.

[3737]

Birrel brothers, *Dunfermline.*— Damask table-cloths, with napkins and slip-cloths to match.

[3738]

BROOK, WILLIAM, & Co., *Castleton Linen Works, Leeds.*—Bed ticks, sheetings, and drills.

[3739]

BROWN & LIDDELL, *Belfast*, Manufacturers and bleachers of table linen, bird's-eye diaper, sheetings, linen and cambric handkerchiefs.

[3740]

BROWNE, W., *Patent Rope Works, Wivenhoe, Colchester.*—Improved patent cordage, rope lines, and twines.

[3741]

BUCKINGHAM, JOHN, 33 *Broad Street, Bloomsbury, W.C.*—Twines, lines, rope, web, matting, mats, sackings.

[3742]

CARTER BROTHERS, *Oak Mills, Barnsley.*—Sheetings, towellings, huckabacks, diapers, damasks, domestic linens in general, drills, &c.

[3743]

CHARLEY, JOHN & WILLIAM, & Co., *Seymour Hill, Belfast.*—Irish linens.

The following are exhibited:—

BLEACHED LINENS (*in Case*).

12 parcels $\frac{4}{4}$ Fronting linen { First quality; one piece of these very fine, 3,200 set, value 15*s*. per yard.
6 ,, $\frac{4}{4}$ Heavy linen.
5 ,, $\frac{7}{8}$ Light fronting linen.
3 ,, $\frac{4}{4}$ Linen diaper, bird's-eye pattern.
3 ,, $\frac{3}{4}$ Lawns.
2 ,, ,, Linen sheetings, broad widths.
5 ,, ,, Ditto { Narrow widths for pillow-cases.
1 parcel $\frac{7}{8}$ Linen at width, French style.
1 ,, $\frac{4}{4}$ Grano de oro
1 ,, $\frac{4}{4}$ Britannia, or Bramante finos
1 ,, $\frac{3}{4}$ Estopilla lawn } Style for West Indies and South America.
5 parcels $\frac{4}{4}$ Fronting linen { Style for United States and North America.

BLEACHED LINENS (*on Wall*).

1 piece very fine linen sheeting, $3\frac{1}{2}$ yards wide.

UNBLEACHED GOODS.

Samples of fine cloth direct from loom.

[3744]

CLEUGH, ALEXANDER, *Imperial Mills, Bromley, London, E.*—Jute, hemp, and flax yarns, rug weft, twines, and firewood tyers.

[3745]

CLIBBORN, HILL, & Co., *Banbridge, County Down, Ireland.*—Bird's-eye diapers, bleached.

[3746]

CONNOR, FOSTER, *Linen Hall, Belfast.*—Linen drills of all classes.

[3747]

COSTERTON & NAYLER, *Flax Works, Scole, Norfolk.*—Prepared flax, tows, yarn, and waste for paper-making.

[3748]

CROGGON & Co., 2 *Dowgate Hill, E.C.* (*See page* 11.)

[3749]

DAGNALL & TILBURY, *Farm Lane, Walham Green, S.W.*—Mats, matting lines, twines, coir yarn and fibre.

[3750]

DEWAR, D., SON, & SONS, *Wood Street, London; Dunfermline, Scotland; Waringstown, Ireland*, Manufacturers of table linens, cambric handkerchiefs, sheetings, Irish linens, &c.

[3751]

DUNBAR, DICKSONS, & Co., *Belfast.*—Linens, sheetings, cambric and linen handkerchiefs, damasks, diapers, and lawns.

CROGGON & Co., 2 *Dowgate Hill, London, E.C.*—Asphalte roofing, inodorous; ship sheathing, and dry hair felts.

THE PATENT ASPHALTE FELT FOR ROOFING

IS PATRONISED BY

Her Majesty's Woods and Forests.

The War Department.

The Royal Agricultural Society of England.

The Leeds and Manchester

The London and North Western

The Liverpool and Manchester

The Chester and Holyhead

The Norfolk

And other Railways.

The Corporation of the City of Edinburgh.

The Duke of Buccleuch.

The Marquis of Anglesey.

The Marquis of Westminster.

The Birkenhead Dock Company.

The Dock Trustees of Liverpool.

Most of the Nobility, Gentry, and Agriculturists.

And many Members of the Royal Agricultural Societies of England, Scotland, and Ireland.

The FELT has been extensively used and pronounced efficient, and particularly applicable for warm climates.

1. It is a non-conductor.
2. It is portable, being packed in rolls, and not liable to damage in carriage.
3. It effects a saving of half the timber usually required.
4. It can be easily applied by any unpractised person.
5. From its lightness, weighing only about 42 lbs. to the square of 100 feet, the cost of carriage is small.

The Felt has been extensively used under slates, in church or other roofs, to regulate the temperature.

INODOROUS FELT,

For damp walls, and for damp floors, under carpets and floor-cloths; also for lining iron houses.

PRICE—One Penny per square foot.

CROGGON & CO.'S PATENT FELTED SHEATHING,

For covering ships' bottoms &c., and

DRY HAIR FELT,

For covering steam boilers, pipes, &c., preventing the radiation of heat, and saving 25 per cent. of fuel.

Samples, testimonials, and full instructions may be obtained on application to the exhibitors.

[3752]

DUNBAR, MCMASTER, & Co., *Gilford, County of Down, Ireland.*—Specimens of flax, linen yarns, and linen sewing threads.

[3753]

EDGINGTON, F., *Thomas Street, Old Kent Road.*—Marquee, tent, flag, rickcloth, sack, and tarpaulin manufacturer.

[3754]

EDWARD, A. & D., & Co., *Logie Works, Dundee.*—Linens, linen yarns, and jute fabrics.

[3755]

ELSTOB & BLINKHORN, *Spalding, Lincolnshire.*—Patent canvas folding buckets, portable water-cisterns, and seamless hose piping.

[3756]

FAULDING, STRATTON, & BROUGH, 13 *Coventry Street.*—Choice table linen, from patterns drawn by the School of Design. (*Prizes awarded.*)

[3757]

FENTON, SON, & Co., *Belfast.*—Bleached linens and damasks.

[3758]

FINLAYSON, BOUSFIELD, & Co., *Johnstone, near Glasgow.*—Bleached and coloured linen threads, shoe threads, machine threads.

Obtained a Prize Medal at the Exhibition of 1851.

The exhibitors are flax spinners and manufacturers of Bleached and Coloured Linen Threads, plain and satin finish, Shoe Threads, Saddlers' Threads, and Gilling Twines.

The 'strength, taste, and neatness' of these Threads were mentioned by the Jurors in 1851 as the ground of their award.

[3759]

FLEMING, W. & J., & Co., *Baltic Works, Glasgow.*—Jute yarns, carpetings, sackings, baggings, hessians, sacks, &c.

The following are exhibited, viz.

Jute and Tow Yarns, Jute and Tow Sackings and Sacks, Hessian or Packing Canvas, Cotton, Coffee, and Wool Baggings, Hop Pocketings, Striped Beddings, Bed Sackings and Ticks, Osnaburghs, Brown Sheetings, and Padding Canvas.

Sailcloth, Tarpaulins, Flax Seaming and Roping Twines.

Jute Carpetings, Mattings, &c.

The warehouses of the exhibitor are situated at No. 5 Ingram Street, Glasgow.

[3760]

FOX, CHARLES JAMES, Manufacturer, *Doncaster.*—Wool sheets, canvas, tarpaulins, sacks, &c.

Canvas, Wool Sheets, Carpets, Sacking, &c.

[3761]

FRASER, DOUGLAS, & SON, Manufacturers, *Arbroath.*—Sail canvas, duck, tarpaulin, &c.

[3762]

GAVIN, PETER, & SONS, *Leith Ropery, Leith.*—Power-loom sailcloth.

[3763]

GILL, JOSEPH, *Headingley, near Leeds.*—Grey, bleached, and dyed linen yarns and twines.

1. SAMPLE OF GREY OR UNBLEACHED LINEN YARNS.

2. SAMPLE OF HALF-BOILED LINEN YARNS, as used in the Manufacture of Blouse Linens.—The process to which the yarns are subjected facilitates in a remarkable degree the manufacture of the cloth, while goods woven from such yarns take a very superior finish.

3. SAMPLE OF CHANGED LINEN YARNS.—This colour is much used in the manufacture of tickings &c. where a lightish or yellow shade is required.

4. SAMPLE OF HALF-BLEACHED LINEN YARNS.—This colour is much used in the manufacture of drills &c. for the home trade, as well as for the American, Spanish, and Italian markets. Goods woven from yarns in this colour take a very superior finish in the cloth.

5. SAMPLE OF SPANISH CREAMED LINEN YARNS.—This colour is produced by a peculiar process not known to many bleachers, and is a favourite colour in the South American and Spanish linen clothing trade.

6. THREE-QUARTER BLEACHED LINEN YARNS.—A colour much used by the manufacturers of Union cloths, being wefted with coloured cotton yarns for the West Indian trade.

7. SAMPLE OF FULL-BLEACHED LINEN YARNS.—Used in the manufacture of the finest classes of linen goods. Where the quality of the yarn is sufficiently good, a beautiful pure white colour can be obtained without injuring the strength of the yarns; while the uniformity of the shade, the glossy lustre of the thread, and the superiority of the finish, recommend this colour to all manufacturers of first-class goods.

8. AN ASSORTMENT OF DYED LINEN YARN SAMPLES, in all the different shades and colours in which such goods are used.

9. AN ASSORTMENT OF LINEN TWINE SAMPLES, grey, white, and coloured, and in all qualities and thicknesses; suitable for grocers, druggists, general shopkeepers, upholsterers, Jacquard machine makers, paper makers—in fact, for all purposes to which such goods are applied.

Samples of all colours in bleached and dyed yarns, of all the regular qualities of twines, with note of prices and charges, will be supplied on application to Mr. GILL, who has also stocks of grey and bleached linen yarns in all the different qualities and counts for sale.

Yarns are bought to order on commission, and bleached at list prices by the exhibitor, from whom full particulars may be learned on application.

[3764]

GIRDWOOD, W., & Co., *Old Park, Belfast.*—Printed linens; linen lawn dresses and handkerchiefs; drills, &c.

[3765]

GRIMOND, J. & A. D., *Bow Bridge Works, Dundee.*—Jute carpeting, hessians, sacking, dyed and undyed jute yarns.

[3766]

GRIMSTON, R. & T., & Co., *Clifford Mills, near Tadcaster.*—Shoe threads &c.

[3767]

GUNDRY, JOSEPH, & Co., *Bridport, Dorsetshire.* (*See pages* 14 *and* 15.)

[3768]

HARFORD, GEORGE, *Newcastle-on-Tyne.*—An improved sailcloth in closeness of texture and strength.

Improved Sail cloth, combining closeness of texture and strength with durability.

[3769]

HARRIS, JONATHAN, & SONS, *Derwent Mills, Cockermouth.*—Samples of linen threads.

GUNDRY, JOSEPH, & Co., Manufacturers, *Bridport, Dorsetshire.* — Seines, nets, lines, and twines for fishing, &c.

SEINES.

Cod, Lance, Caplin, Mackerel, Herring, &c.

1. Best Hemp Netting for cod, barked, Newfoundland and British America.
2. Do. for lance, do. . . . do.
3. Do. for caplin, do. . . . do.

NETS.

Made by Improved Machinery.

4. Best Hemp (herring), barked, mounted complete, Newfoundland and British America.
5. Best Cotton, do. . . do. do.
6. Best Hemp Drift (mackerel, herring), not mounted, British Coast.
7. Best Cotton Drift (herring) . . . . do.

LINES.

8. Best Hemp, Deep sea.

9.	Do.	Log	30	fathoms.
10.	Do.	Hambro' . . .	,,	do.
11.	Do.	Bank	40	do.
12.	Do.	St. Peter cod . .	33	do.
13.	Do.	Long shore . . .	30	do.

LINES

14.	Best Hemp,	Pollock, brown and blue	30	fathoms.
15.	Do.	Mackerel do. .	,,	do.
16.	Do.	Squid or Jigger do. .	10	do.
17.	Do.	for Australia . .	20	do.
18.	Do.	Fine fishing . . .	10	do.

19. Best Cotton, fishing.
20. Brown Whip Cords.
21. Green do.
22. Carpenters' Chalk.
23. Masons' do.

TWINES.

Best Russia Hemp, Patent Topped.

24.	Seal	3 thread.
25.	Trawl	do.
26.	Salmon trawl	do.
27.	Salmon	do.
28.	Turtle	do.
29.	Mullet	do.
30.	Seine	do.
31.	Shad	do.
32.	Mackerel	do.
33.	Herring	do.

PYMORE MILL COMPANY, Manufacturers, *Pymore, near Bridport.*—Shoe threads, yarns, shop twines, &c.

SHOE THREADS.

No. 1.—Brown or Grey.
Fine Flax.
Fine.
No. 1 Fine.
Fine Thread.
S. C.
Best Com.
No. 1 Com.
,, 2 Com.

No. 2.—Bleached.
No. 9 Patent.
,, 12 H. B.
Fine Scotch.
No. 4 H. B.
Com. Scotch.
Patent.
No. 10 H. B.
O. W. Patent.

SHOE THREADS.

No. 3.—Yellow.
Nos. 25, 11, 3, 1, 2, and 9.

No. 4.—Slate.
No. 9 Patent.
Fine.
Best.

No. 5.—Green.
Nos. 12, 10, and 9.

No. 6.—Green Hemp.
Best Handspun.
2 Patent.
Patent.

No. 7.—H. B. Closing.
Nos. 40, 25, and 8.

No. 8.—Yellow Closing.
Nos. 40, 25, and 8.

PYMORE MILL COMPANY, Manufacturers, *Pymore, near Bridport.*—Shoe threads, yarns, shop twines, &c.

GUNDRY'S PATENT AND STANDARD THREADS.

No. 9.—Brown or Grey Patent.
,, 10.—Brown or Grey Standard.
,, 11.—Bleached Patent.
,, 12.—Bleached Standard.
,, 13.—Yellow Patent.
,, 14.—Yellow Standard.
,, 15.—Green Patent.
,, 16.—Silver Grey Patent.
,, 17.—Silver Grey Standard.
,, 18.—Bleached Closing.
,, 19.—Yellow Closing.

SHOP TWINES AND CORDS.

No. 20.—Surgeons' Twine.
,, 21.—New Zealand.
,, 22.—Sealing Twine.
,, 23.—Fine Fine.
,, 24.—Middle Thread.
,, 25.—Lay Cords.
,, 26.—Dutch and Ell Twines.
,, 27.—Box Cords.

SAIL TWINES.

No. 28.—Seaming.
,, 29.—Roping.

London Warehouse—13 & 14 Camomile Street: Hill & Hartridge, Agents.

[3770]

HAWKE, E. H., & SON, *Scorrier, Cornwall.*—Various descriptions of rope for mining, marine, and other purposes. Patent safety fuse.

[3771]

HIND, JOHN, & SONS, *Durham Street Mills, Belfast.*—Brown and bleached linens, linen and cambric yarns.

[3772]

HOLDSWORTH, WM. B., & Co., *Leeds.*—Hemp and flax yarns, sewing threads, shoe threads, twines, and netting threads.

[3773]

JAFFÉ BROTHERS, Linen Manufacturers, *Belfast.*—Linens, linen and cambric handkerchiefs.

[3774]

JOHNSTON & CARLISLE, *Brookfield, Mills, Belfast.* (*See page* 16.)

[3775]

KINNIS, WM., & Co., *Dunfermline, Fifeshire, N.B.*, Linen manufacturers, by hand and power looms, of damask and diaper table linen, huckabacks, sheetings, &c.

[3776]

LOCKHART, N. & N., *Kirkcaldy, Scotland.*— Fishing nets, hemp and cotton, mackerel, herring, and other kinds.

[3777]

LOCKHART, N., & SONS, Power and hand loom manufacturers, *Kirkcaldy, Fife.*—Ticks, sheetings, towellings, sackings, and sacks.

JOHNSTON & CARLISLE, *Brookfield Mills, Belfast.*—Flax and tow yarns, brown linens, bleached family and fronting linen.

Irish linens, brown and bleached, with specimens of the flax and yarns from which they are manufactured, viz.

Samples of flax used by exhibitors in their manufacture, grown in Ireland and Belgium.

Samples of Irish and Courtrai flax after being heckled.

Samples of linen yarns of various qualities, viz.

	Leas	Per Bundle s.	d.
Line yarns, quality No. 1. P. . . .	40	7	3
,, ,,	45	6	9
,, ,,	50	6	4½
,, ,,	55	6	1½
,, ,,	60	6	0
,, ,,	65	6	0
,, ,,	70	6	3
,, ,,	80	6	9
,, ,,	90	7	6
,, ,,	100	8	6
,, ,,	110	9	6
Line yarns, quality No. 1. . . .	45	6	0
,, ,,	50	5	7½
,, ,,	55	5	4½
,, ,,	60	5	3
,, ,,	65	5	3
,, ,,	70	5	4½
,, ,,	75	5	6
,, ,,	80	5	9
,, ,,	90	6	0
,, ,,	100	6	3
,, ,,	110	6	9
,, ,,	120	7	6
,, ,,	130	8	6
,, ,,	140	9	6
Line yarns, quality No. 2. . . .	120	6	0
,, ,,	130	6	3
,, ,,	140	6	9
,, ,,	150	7	3
,, ,,	160	8	3
,, ,,	170	9	6
Tow yarns, quality No. 1. . . .	22	6	3
,, ,,	25	6	0
,, ,,	30	5	9
,, ,,	35	5	6
,, ,,	40	5	3
Line yarns, extra quality, No. 0	50	10	9
,, ,,	70	11	3
,, ,,	100	12	0

(*The prices quoted are per bundle of* 60,000 *yards.*)

Sample bunches of line and tow yarns, boiled.

Sample bunches of line and tow yarns, bleached full white.

Samples of unbleached hand-loom linens, viz.

$\frac{7}{8}$ and $\frac{4}{4}$ heavy family linens, Nos. 1 to 15.

$\frac{7}{8}$,, $\frac{4}{4}$ medium linens, Nos. 5 to 16.

$\frac{7}{8}$,, $\frac{4}{4}$ light linens, Nos. 7 to 17.

$\frac{7}{8}$,, $\frac{4}{4}$ fine fronting linen, Nos. 5 to 17.

$\frac{7}{8}$,, $\frac{4}{4}$ superior fronting linen, Nos. 9 to 17.

Samples of unbleached power-loom linens, viz.

33 and 38 inch light linen, Nos. 10 to 16.

34 inch light linens, unboiled yarns, No. 16.

34 ,, ,, boiled yarns, No. 16.

34 ,, ,, boiled yarns, No. 20.

34 ,, linens made from tow yarns, Nos. 8, 9, and 10.

Samples of bleached linens, including all the above qualities of power-loom and hand-loom linens.

Bleached linens, unfinished.

Dry beetled linens, Nos. 8 to 16.

Dyed and finished linens and hollands.

Sheeting and pillow-case linens.

Brown and bleached linen drills.

Woven shirt-fronts.

Linen and cambric handkerchiefs.

Sample of $\frac{7}{8}$ fronting linen, exhibited as a specimen of the finest quality of linen woven in Ireland.

[3778]

MCINTYRE & PATTERSON, Flax and Yarn Merchants, *Belfast.*—Specimens of Irish flaxes and linen yarns.

[3779]

MARSHALL & Co., *Leeds.*—Linen yarn and sewing thread, and woven fabrics of linen.

[3780]

MATHEWSON, JAMES, & SON, *Dunfermline, Scotland.*—Double damask table-cloths and napkins &c.

[3781]

MATIER, HENRY, & Co., Linen and Handkerchief Manufacturers, *Belfast.*—Linens and handkerchiefs, plain, printed, and sewed.

[3782]

MILLER, O. G., Flax Spinner and Bleacher, *Dundee.*—Line, tow and jute yarns, grey, bleached, and otherwise prepared.

[3783]

MOIR, JOHN, & SON, *Dundee.*—Linens woven by power; ducks, dowlas, osnaburgs; brown, cream, and bleached sheetings, &c.

[3784]

MONCUR, A., & SON, *Dundee.*—Linens, salt and grain sacks, and sacking.

[3785]

MOORE, W. F., *Cronkbourne, Douglas, Isle of Man.*—Sailcloth and fishing nets.

[3786]

MOORE & WEINBERG, *Belfast, Ireland.*—Linen manufactures.

The following are exhibited, viz.

White Linens of every description for Home, Continental, and Transatlantic markets.

Tinted, Dyed, and Printed Linens.

Linen and Cambric Handkerchiefs, unbleached, bleached, and printed.

Table Linen, manufactured expressly for His Majesty the Emperor of all the Russias.

Crests and Arms of every description, inserted to order, in Table Cloths.

Hollands, Drills, and every kind of Linen Fabrics suitable for clothing in warm climates.

Unbleached and Bleached Linen Yarn.

[3787]

MORISON, JOHN, Government Contractor and Manufacturer, 25 *Norton Folgate, London, N.E.* Rope, twine, sack, marquee, tent, tarpaulin, rick cloth, &c.

[3788]

NORMAND, JAMES, & SONS, *Dysart, Fifeshire, Scotland.*—Linen, cotton, and union table damasks, diapers, and hucks.

[3789]

NUTT, RICHARD, 31 *Trippett, Hull.*—Oil press, hairs, and press bagging.

[3790]

PATERSON, JAMES, *Dundee.*—Hemp carpeting, Manilla and coir mattings, hearth rugs, corn sacks, &c.

[3791]

Preston, Smyth, & Co., *Belfast.*—Flax; yarns; linens, bleached and grey; handkerchiefs, linen and cambric.

[3792]

Richardson, James N., Sons, & Owden, *Belfast, Ireland,* Flax spinners; power and hand loom, linen, damask, and handkerchief manufacturers and bleachers.

[3793]

Robertson, John, Linen Manufacturer, *Middle Hills, near Coupar Angus, Perthshire.* Flax-tow sheetings, hessians, sackings, &c. Linen shirtings, ticks, paddings, canvas, &c.

[3794]

Russell, J. N., & Sons, *Lansdowne Mills, Limerick.*—Flax, yarns, linens, &c., the produce of the province of Munster.

[3795]

Samson, Hugh, & Sons, *Hill Bank, Dundee, Scotland,* Manufacturers of linens of various fabrics.

[3796]

Smieton, James, & Son, *Dundee.*—Linen and union dowlas, osnaburgs, crequelas, coletas, russias, paddings, ducks, hessians, &c.

[3797]

Smith, Thomas & William, *Royal Exchange Buildings, London; and Newcastle-upon-Tyne.*—Wire and hemp ropes.

[3798]

Stephens, J. P., & Co., and Hounsell, William, & Co., *Bridport.*—Twines, lines, nets, seines, sail canvas, &c.

[3799]

Stuart, J. & W., *Musselburgh, Scotland.*—Patent pilchard, sprat, and herring fishing nets, and fishing twines in cotton and hemp, all manufactured entirely by machinery.

These mackerel, herring, pilchard, and sprat nets, made by machinery from cotton and hemp, are very superior to hand-made goods of the same description. They are now universally adopted by the British fisheries, and generally by foreign countries. These nets are more durable and more successful in taking fish, from the exactitude of their make.

J. & W. S. are the sole patentees.

[3800]

Terrell, W., & Sons, 6 *Welsh Back, Bristol.*—Manilla and other cordage, twine cord and line, dressed flax and hemp, &c.

[3801]

Thomson, D. & J., & Co., *Seafield Works, Dundee.*—Jute carpeting, sacking, pocketing, tarpaulin, bagging, hessians, yarns, rugs, matting, and mats.

[3802]

Unite, John, 130 *Edgware Road, London, W.*—Hemp and flax in its raw state, also manufactured.

[3803]

Walker, John, & Co., Canvas Manufacturers, *Arbroath.*—Sailcloth and sail twine.

[3804]

Walker, J. & H., *Dundee.*—Jute yarns, sackings, hessians, bagging, guano bags, wool packs, &c.

[3805]

WILFORD, JOHN, & SONS, *Brompton, Northallerton, Yorkshire.*—Plain and fancy linen drills, white, padded, and printed.

[3806]

WILKS BROTHERS & SEATON, 80 *Watling Street, London.*—White, coloured, fancy and silk and wool flannels.

[3807]

WILSON BROTHERS, 29 *Lowther Street, Whitehaven.*—Double-twilled, diamond-twilled, and plain double sailcloth.

[3808]

WILSON, GEORGE, *Hutton Rudby, Cleveland, Yorkshire.*—Long flax, sailcloth.

In this case are samples of the 'CLEVELAND SAIL CLOTH,' now so extensively used, and appreciated for its strength and durability.

[3809]

YEOMAN & Co., *Osmotherley, Northallerton.*—Linen drills, ducks, huckabacks, and yarns suitable for their manufacture.

[3810]

HOUNSELL, W., & Co., *Bridport.*—Twines, lines, nets, canvas, &c.

Class XX.

SILK AND VELVET.

[3840]

Adshead, William, & Co., *Higher Fence, Macclesfield.*—Dyed silks.

[3841]

Allen, Joseph, *Spa Mills, Derby,* Silk throwster and manufacturer of elastic gusset webs, worsted braids, &c.

[3842]

Alsop, Downes, Spilsbury, & Co., *Leek, Staffordshire, and Huggin Lane, London.*—Braids, bindings, serges, buttons, sewings, twist.

[3843]

Ballance, Thomas, & Son, 13 *Spital Square.*—Rich black silks and velvets.

[3845]

Bickham, Pownall, & Co., 2 *York Street, Manchester.*—Broad silk goods.

[3846]

Birchenough, John, *Macclesfield, and* 38 *Gresham Street, London.*—Gentlemen's scarfs, neck and pocket handkerchiefs, sarcenets, &c.

[3847]

Bower, George, *Lower Temple Street, Birmingham:* London Agent, Tweedie, 337 *Strand.*—Odd Fellows' and Temperance Societies' flags.

[3848]

Brocklehurst, J. T., & Sons, *Macclesfield, and* 33 *Milk Street, London.*—Thrown silk and waste silk goods.

[3849]

Browett, Frederick, *Coventry.*—Ladies' dress trimmings, gimps, and fringes.

[3851

CAMPBELL, HARRISON, & LLOYD, 19 *Friday Street, London, E.C.*—Silk and velvet fabrics.

(*Obtained The Silver Medal of the Society of Arts,* 1849.
The Gold Medal of the Society of Arts, 1850.
The Prize Medal of the Exhibition of 1851.
The Silver Medal at the Paris Exhibition, 1855.)

[3852]

CARR, THOMAS, & Co., *Leek,* Silk throwsters, manufacturers of bindings, tailors' and machine twist, sewings, serges, &c.

[3853]

CARTER & PHILLIPS, *Coventry.*—Plain and fancy ribbons.

[3854]

CASH, J. & J., *Coventry.*—Ribbons and cambric frillings.

[3855]

CHADWICK, JOHN, 12*a Mosley Street, Manchester.*—Plain and fancy silks.

[3856]

CHADWICK, JOSEPH, 33 *Fountain Street, Manchester.*—Silks.

[3857]

CHRISTY & Co., *Stockport,* Manufacturers by power (patented) of hat plushes, velvets, and piled fabrics.

[3858]

CORNELL, LYELL, & WEBSTER, 15 *St. Paul's Churchyard.*—Ribbons and moiré antiques.

[3859]

CORNS, W. W., & Co., *Macclesfield, and No.* 1 *Wood Street, London.*—Gentlemen's scarfs, cut up cloth, dress silks, ladies' ties, shawls.

[3860]

COURTAULD, SAMUEL, & Co., 19 *Aldermanbury, London.*—Crapes and aerophanes, black and coloured.

Obtained Gold Medal at the Paris Exhibition, 1855.

Black and Coloured Crapes and Aerophanes.

As the exhibitors were Jurors for this Class in the Exhibition of 1851, they were not permitted to compete for the Prize.

The Gold Medal obtained by them at Paris was the only one awarded in this Class to a British exhibitor at that Exhibition.

[3861]

COX, R. S., & Co., *Coventry, and* 7 *St. Paul's Churchyard, London.*—Ribbons, silks, handkerchiefs, &c.

[3862]

CRITCHLEY, BRINSLEY, & Co., *Macclesfield, and* 1 *Wood Street, Cheapside, London.*—General assortment of ladies', gentlemen's, and boys' silk handkerchiefs and scarfs &c.

[3863]

CUNLIFFE, PIGGOTT, & Co., *Spring Gardens, Manchester.*—Black glacés, and other plain and fancy silks.

[3864]

DALTON & BARTON, 173 *Aldersgate Street, London, and Coventry,* Ribbon, carriage lace, and upholstery trimming manufacturers.

[3865]

Davidson & Myatt, *Leek, Staffordshire,* Manufacturers of machine twist for patent sewing machines, tailors' twist, &c.

[3866]

Elise, L. M., & Isaacson, F. W., 170 *Regent Street.*—English silks, own manufacture.

[3868]

Franklin, William, & Son, *Coventry.*—Medium qualities of plain ribbons, &c.

[3869]

Gibson, Silas, jun., *Leek.*—Machine twist, legee and silk twist, sewing silks, sewn buttons &c.

[3870]

Grant & Gask, 59 *to* 62 *Oxford Street.*—Tissue de verre.

[3871]

Grout & Co., 12 *Foster Lane, London.*—Crapes, aerophanes, and lisses.

[3872]

Hadwen, J. Wilson, *Kelroyd Mills, Halifax.*—Silk and cotton yarns and tissues.

[3873]

Hart, James, *Coventry.*—Silk ribbons made by steam power.

[3874]

Hennell & Eld, Manufacturers and Throwsters, *Coventry and Derby.*—Raw and thrown silk.

[3875]

Houldsworth, James, & Co., 23 *Portland Street, Manchester.*—Silk damasks, brocatelles, and other furniture fabrics; machine-embroideries &c.

Obtained the Honorary Certificate of the Royal Society of Dublin, 1850, *and Prize Medals at the International Exhibitions, London,* 1851; *New York,* 1853; *and Paris,* 1855.

These silk and other furniture fabrics are woven by power-looms, and are sufficiently wide for curtains without seaming. The recent application of power to their manufacture places them within the reach of large classes hitherto unable to obtain such goods. Silk brocatelles equal in durability of fabric and colour to any produced either in France or England, are here exhibited; the cost of which at any of the large upholstery and furnishing establishments in the kingdom will not exceed £8 8*s.* to £10 10*s.* per window, independent of making up, lining, &c.

The French have not yet succeeded in applying steam power to the manufacture of such goods; and as these are equal in design, quality, colour, &c., they compete successfully with French goods in the principal markets of the world.

The embroidery is done by machinery, and at a very moderate cost.

This firm makes a great variety of fancy goods in silk, worsted, cotton, &c., for the South American, African, Indian, and other markets; many of them in imitation of native productions; but these it has not been thought desirable to exhibit.

[3876]

Keith & Co., 124 *Wood Street, London.*—Furniture silks; silks for carriage lining.

[3877]

Kemp, Stone, & Co., 34 & 35 *Spital Square, N.E.*—Broad silks and velvets.

[3878]

Le Mare, Ebenezer Robert, *Manchester.*—Plain and fancy silk goods.

[3879]

Newsome, Charles, *Coventry.*—Plain and fancy ribbons of Italian and Chinese silk.

[3880]

PAYN, J. J., *Aldermanbury.*—Plain and figured silk reps, tissues, brocatelles, borders, &c.

[3881]

PEEL, GREENHALGH, & Co., Manufacturers, *Bury*; WHYATT, GEORGE, & SON, Dyers and Finishers, *Manchester.*—Silk union velvets.

[3882]

POTTS & WRIGHT, *Macclesfield.*—Sarcenet, and black handkerchiefs.

[3883]

POWNALL, STUBBS, & Co., *Leek, Staffordshire.*—Sewing silks, twists, needleworked buttons, military ornaments, machine twists, whip lashes, tassels.

[3884]

RATLIFF, JOHN, & SON, *Coventry.*—Plain and fancy ribbons.

[3885]

RUSSELL, DALGLISH, & Co., *Blackhall Factory, Paisley.*—Thrown silks, gum and soft-dyed and spooled; fringes, sewings, &c.

[3886]

SALKELD, JOHN, & Co., *Dalton, near Huddersfield.*—Silk samples, illustrative of the processes of silk throwing and spinning; also patent sewings.

[3887]

SEAMER, THOMAS, 5 *Milk Street, Cheapside.*—Moiré antiques, velvets and plain silks.

[3888]

SLATER, BUCKINGHAM, & SLATER, 35 *Wood Street, London, E.C.*—Gentlemen's silk cravats, scarfs, and ties.

[3889]

SIMPSON, M. & W., *Leek.*—Machine-sewing silk, without knots; unweighted dye; also bindings, trimmings, &c.

[3890]

SLINGSBY, HENRY, *Park Street, Coventry.*—Specimens of silk scarfs, neck-ties, badges, mantle and shawl trimmings, &c.

[3891]

SMEALE, WILLIAM, *Macclesfield, and* 20 & 41 *Gutter Lane, London.*—Sarcenets and gentlemen's scarfs.

[3893]

SUDBURY LOCAL COMMITTEE, *Sudbury.*—Specimens of silks, velvets, and brocades manufactured at Sudbury, Suffolk.

[3894]

TAYLOR, SHERBROOKE, 45 *Friday Street, London.*—Plain and fancy silks; moiré antiques, plain and figured; velvets, satins, &c.

[3895]

THOMPSON, WILLIAM, & Co., *Galgate, near Lancaster.*—Material in its several processes, from waste to yarn.

Obtained the Silver Medal at the Paris Exhibition, 1855

Specimens of waste silk in the raw material; illustrations of the same in the dressed state and carded; slubbings; thick and fine rovings; single and double spun silk yarns; lace and dyed yarns.

[3896]

THORP, JOHN & SAMUEL, Silk Manufacturers, 20 *Piccadilly, Manchester, and Macclesfield.* Galloons, doubles, bindings, ribbons, &c.

[3897]

VAVASSEUR, TAYLOR, & Co., 3 *Watling Street.*—Silk scarfs and handkerchiefs; and silk for ladies' garments.

[3898]

WALTERS, DANIEL, & SONS, 43, 44, & 45 *Newgate Street, and New Mills, Braintree, Essex.* Furniture silks.

[3899]

WANKLYN, WILLIAM, *Fountain Street, Manchester;* 42 *Cheapside, London.*—Thrown silks, printed and woven silk handkerchiefs, &c.

[3900]

WATSON & HEALEY, *Rochdale.*—Velvet and plush, made from spun silk waste.

[3901]

WINKWORTH & PROCTER, *Manchester.*—Coloured glacé; plain and figured chinés; and figured crystalline.

[3902]

WREFORD, JOHN, & Co., 17 & 18 *Aldermanbury, London.*—Sewing silks and twists, and machine-sewing silks.

[3903]

CROSS, P. R., *Sudbury, Suffolk.*—Respirator scarf.

[3904]

CLABBURN, SONS, & CRISP, *Norwich.*—Silk shawls.

Class XXI.

WOOLLEN AND WORSTED, INCLUDING MIXED FABRICS.

[3934]

Adie, Scott, 115 *Regent Street.*—Scotch clan tartans, plaids, dresses, vicuna wools, shawls, cloakings, linsey woolseys.

[3935]

Akroyd, James, & Son, *Halifax.*—Worsted yarns and fabrics, mixed fabrics, furniture and dress goods, coatings.

[3936]

Anderson, J. & A., *Princes Square, Glasgow.*—Gingham and fancy dresses.

[3937]

Andrews, Henry, & Co., *Albion Street, Leeds.* — Woollen and union cloths, mediums, meltons, and tweeds.

[3938]

Armitage brothers, Manufacturers, *Huddersfield.*—Fancy coatings and doeskins.

[3939]

Armitage, Samuel & Benjamin, *Shepley, near Huddersfield.* — Cloakings, coatings, cassimeres, fancy vestings and quiltings, Balmoral skirts.

[3940]

Barber, Joshua, & Sons, *Holmbridge Mills, near Huddersfield.*—$\frac{3}{4}$ fancy trouserings and $\frac{6}{4}$ fancy coatings.

[3941]

Barker, B., & Son, *Cookridge Street, Leeds.*—Superfine woollen cloths, beavers, and cassimeres.

[3942]

Barron, Robert, *Gillroyd Mills, Morley, and Leeds.*—Union cloths, made from Sydney wool and mungo.

[3944]

Bennett, Samuel, & Son, *Winsham, near Chard, Somerset.*—Woollen cloths, livery drabs, Kerseys, and drab Devons.

These exhibitors are manufacturers of Livery and Kersey Cloths, and spinners of Lambswool.

[3945]

BIRCHALL, J. D., *Wellington Street, Leeds.*—Woollen cloths, ladies' mantle cloths, and fancy woollen coatings.

[3946]

BIRD, OLIVER, *Southfields, Stroud, Gloucestershire,* Manufacturer of finest scarlets, blues, blacks, green billiard-cloth, doeskins, &c.

[3947]

BIRRELL, WALKER, & Co., *Glasgow.*—Fancy dress fabrics.

[3948]

BISHOP, SON, & HEWITT, *Leeds.*—Waterproof tweeds &c.

[3949]

BLAKELEY BROTHERS, *Dewsbury.*—Shoddy and mungo, made from woollen rags.

[3950]

BLISS, WM., & Co., *Chipping Norton, Oxon, and* 26 *Basinghall Street, London.*—Shawls and cloakings from a variety of furs and wools; bed coverlets; all kinds of woollens; &c.

[3952]

BOLINGBROKE, C. & T., & JONES, Manufacturers, *Norwich.*—Paramattas, poplins, poplinetts, shawls, fancy cloakings, and fancy dresses.

[3953]

BOWMAN, JAS., & SON, *Langholm, Scotland.*—Scotch tweeds.

[3954]

BRADFORD LOCAL COMMITTEE, *Bradford.*—Wools.

[3955]

———— Yarns.

[3956]

———— Alpacas and mohair goods, plain and figured.

[3957]

———— Orleans cloths, plain and figured.

[3958]

———— Cobourg, paramatta, barathea, reps, cords, cloths.

[3959]

———— Lastings, serge de Berri, crapes, stockinetts, gambroons, camlets.

[3960]

———— Italian summer cloths, Russell and mottled cords.

[3961]

———— Umbrella cloths.

[3962]

———— Mixed and mottled worsted and alpaca goods, and winseys.

[3963]

BRADFORD LOCAL COMMITTEE.—Fancy goods: alpaca, mohair, worsted, silk, &c.

[3964]

———— Worsted goods: merinos, says, shalloons, &c.

[3965]

———— Moreens.

[3966]

———— Damasks, reps, and table-covers.

[3967]

———— Wool shawls, delaines, and shawl cloths.

[3968]

BRAITHWAITE & Co., *Kendal.*—Plain and fancy woollens, coatings, coat linings, linseys, and collar checks.

Obtained a Medal at the Paris Universal Exhibition in 1855.

The Committee appointed by the Huddersfield Chamber of Commerce to report on the Paris Exhibition, state as follows:—

'The black and white tweeds from Kendal are very good in colour and mixture.' . . . 'The Kendal tweeds, shepherds' plaids, and trouserings are superior to any exhibited, in purity and firmness of colour, and fine soft woolly handle.'

[3969]

BREWIN & WHETSTONE, *Leicester.*—Worsted, lambswool, and merino yarns, for hosiery, knitting, and weaving.

[3970]

BRIGGS & SONS, 4 *Park Lane and Carlton Cross Mills, Leeds.*—Woollen shawls of various fabric and material.

[3971]

BROWN BROTHERS, Woollen Manufacturers, *Buckholm Mills, Galashiels, Scotland.*—Fancy Scotch tweeds.

[3972]

BROWN & COLLANDER, Warehouse—20 *Bread Street*; Manufactory—*Yeadon, near Leeds.*—Tweeds, meltons, and fancy cloakings.

[3973]

BROWN, J. & H., & Co., *Selkirk, N.B.*—Scotch tweeds, shawls, and mauds.

[3974]

BULL & WILSON, 52 *St. Martin's Lane, London.*—Woollen cloths and vestings.

[3975]

BURGESS, ALFRED, *Leicester.*—Samples of English and foreign and colonial wool.

[3977]

BUTTERWORTH, JAMES, & SON, *Grunbooth Mills, near Rochdale.*—Flannels and the incidental processes of manufacturing them.

[3978]

CALEY BROTHERS, *Windsor.*—Furniture, silk damasks, and satins; silk and cotton diaphane, for transparent window-blinds.

[3979]

CARR, ISAAC, & Co., *Twerton, Bath.*—Fur, Twerton and patent beavers, elastic meltons, &c.

[3980]

CARTER, WILLIAM, & GEISSLER, HERMANN, *Kirkburton.*—Cloakings, coatings, livery Valencias, shepherd checks, trouserings, &c.

[3981]

CHEETHAM, C., G., & W., *Woodbottom Mills, Horsforth, near Leeds.*—Volunteer army cloths.

[3982]

CHILD, J. & J., *Shelley, near Huddersfield.*—Waistcoats, ladies' skirts and mantles.

[3983]

CLABBURN, SONS, & CRISP, *Norwich.*—Patented figured silk shawls, paramattas, tamatives, grenadines, poplins, and fancy dresses.

[3984]

CLARK, JOHN & THOMAS, *Trowbridge, Wilts.*—Woollen cloths, various.

[3985]

CLAY, J. T., *Rastrick, near Huddersfield.*—Woollen and worsted fancy goods.

[3987]

CLOYNE, C. G., *Dewsbury.*—Woollen fabrics.

[3988]

COCHRANE, J. & W., *Mid Mill and Nethedale, Galashiels.*—Scotch tweeds.

[3989]

COGSWELL, JAMES, & CO., *Trowbridge, Wiltshire.*—Woollen doeskins, diagonals, Bedford ribbs, cross ribbs, deerskins, Venetians.

[3990]

COLLIER, HORATIO, SEN., *Crawley Mills, Witney, Oxon.*—Samples of Witney blankets.

[3991]

COMYNS, ALEXANDER, SON, & CO., 10 *College Green, Dublin.*—Finest Irish friezes and tweeds.

[3992]

COOKE, A. M., 115 *Cheapside.*—Alpacas, mohairs, &c.

[3993]

COOK, THOMAS, SON, & WORMALD, *Dewsbury Mills.*—Blankets for home and shipping trades, rugs and cloths.

[3994]

COOPER, ARTHUR, & CO., *Leeds.*—Plain and fancy woollens and unions, and cloth caps.

[3995]

COOPER, D. & J., *Leeds.*—Woollen and union cloths.

[3996]

CRAVEN, J., *Thornton, near Bradford.*—Superfine llama d'Ecosse, rep, and other shawls.

[3997]

CROMBIE, JAMES & JOHN, *Grandholme Works, Aberdeen.*—Woollen goods.

[3998]

CROSLAND, BENJAMIN, *Oaks Mill, near Huddersfield.*—Mohairs, sealskins, furs, and velvet cloths.

[3999]

CROSLAND, W. & H., *Huddersfield.*—Fancy woollen and Angola goods.

[4000]

CROSS, WILLIAM, 62 *Queen Street, Glasgow.*—Tartan and fancy woollen shawls and piece goods.

[4001]

Crowther, B., 60 *Albion Street, Leeds.*—Blankets and woollens.

[4002]

Cubitt, Wilson, & Randall, Manufacturers, 36 *King Street, Cheapside, London, and at Banbury.*—Printed mohair tapestry, Utrecht velvets, livery and other plushes.

[4003]

Dalrymple, William, *Union Mills, Isle of Man.*—Woollen goods, manufactured by the exhibitor.

[4004]

Davies, Robert S., & Sons, *Stonehouse Mills, Gloucestershire.*—Black, blue, and scarlet cloths, cassimeres, and doeskins.

[4006]

Day nephew & Co., *Dewsbury.*—Pilots, Cheviots, velvet piles, fancy and heavy woollens.

[4007]

Day & Watkinson, Woollen Manufacturers, *Huddersfield.*—Drab kerseys and Bedford cords.

[4008]

Dicksons & Laings, *Wilton Mills, Hawick,* Manufacturers of Cheviot, Australian, and Saxony wool tweeds and plaids.

[4009]

Dixon, Thomas Daniel, *Morley, near Leeds.*—Black medium union cloth, black milled union cloth, made from Sydney wool and waste.

[4010]

Dobson, J. & A., Manufacturers, *Innerleithen, N.B.*—All-wool tweeds, shirtings, shawls, cloakings, and Indian cloths, to order.

[4011]

Dobson & Riley, *Fenay Mills, Huddersfield.*—Fancy woollen manufactures.

[4012]

Dolan, J. C., *Britannia Street, Leeds, and Horsforth.*—Army, police, and export cloths.

[4013]

Drinkwater, William, *Salford Woollen Mills, Manchester.*—Woollen cords, worsted and cotton tweeds.

[4014]

Early, Edward, & Son, *West End, Witney.*—Witney blankets, tiltings, yarns, rugging, collar cloths.

[4015]

Early, John, & Co., *Witney.*—Blankets, pilot cloths, and tweeds.

[4016]

Early, Richard, jun., *Witney.*—Prince's checks, webs, kerseys, tiltings, mop yarns.

[4017]

Ecroyd, William, & Sons, *Lomeshaye Mills, Burnley.* (*See page* 32.)

[4018]

Edmonds & Co., *Bradford, Wiltshire.*—Sample of superfine blue cloth, two samples of wool dyed black.

ECROYD, WILLIAM, & SONS, *Lomeshaye Mills, Burnley.*—Worsted and mixed stuffs, dyed and mixed fabrics for printing.

The exhibitors manufacture stuffs and mixed fabrics for dyeing and printing. The following specimens of their goods are exhibited:—

1. WARP OF COTTON. WEFT OF WOOL.

Coburg Cloth, twill, heavy.
Do. do. medium.
Do. do. light.
Union Merino do. heavy.
Double Twill, twilled on both sides.
Italian Cloth, satteen twill.
Union Satteen do.
Satin de Chêne do. light and fine for linings.
Summer Rep, cord, for dresses.
Canton Cloth, plain, do.

2. WARP OF COTTON, TWO-FOLD, AND DYED BEFORE WOVEN. WEFT OF WOOL.

Paramatta, twill, for dresses.
Do. twilled on both sides.
Sicilian Cloth, cord, for dresses.
Canton Cloth, plain, do.
Union Princetta, twill.
Union Shalloon, twilled on both sides.
Italian Cloth, satteen twill.
Union Russell, do. extra heavy.
Russell Cord.

3. FABRICS MADE ENTIRELY OF WOOL.

Mousseline de Laine.
Merino, single and double warp.
Bunting, for flags and signals.
Bagging, for oil and sugar mills.

4. WARP OF SILK. WEFT OF WOOL.

Paramatta, or Henrietta Cloth, twill.
Venetian Cloth, twill, lighter.
Sultana Cloth, do. very light.
Castilian Cloth, twilled on both sides.
Canton Cloth, plain.

5. FABRICS FOR PRINTING, EXHIBITED SCOURED.

Mousseline de Laine, plain, all wool.
Do. do. cotton warp.
Orleans Cloth, plain, double warp and lustre weft.
Do. do. double warp and demi-lustre weft.
Do. do. single warp and lustre weft.
Do. do. single warp and demi-lustre weft.
Persian Cloth, plain, light, and soft.
Challi Cloth, do. very light and soft.
Twill, soft wool weft.
Sicilian Cord, lustre weft.
Do. demi-lustre weft.
Heavy Foulard, for autumn.
Light do. for spring.

The dimensions of all cloths can be varied for special orders.

WILLIAM ECROYD & SONS supply wholesale houses only. All goods sold by them, and bearing their trade mark, are their own manufacture, and no variation in dimensions, quality, or substance is made in any article, so long as it retains the same quality mark. All special orders are registered at the Works, to ensure exact reproduction.

[4019]

ELWORTHY, W. & T., *Wellington, Somersetshire.*—Serges, blankets, yarn, &c.

The following goods, of which samples are exhibited, are manufactured by this firm, viz.

Serges, estamenes, long ells for China, and swanskins for Newfoundland, &c.

Sadlers' serges, serge bandages, summer racing clothing, collar cloths and checks, and super stove cloths, &c.

Striped and check kerseys, coalpit cloths, blankets, blanketings, white flannels, and Welsh checked shirting flannels, hosiery and tweed yarns, &c.

[4023]

FIELD, RICHARD, *Skelmanthorpe, Huddersfield.*—Silk, worsted, and cotton vestings, Balmoral skirts, &c.

[4024]

FIELDING & JOHNSON, *Leicester.*—Worsted, for plain and fancy hosiery, knitting and power looms.

[4025]

FIRTH, EDWIN, & SONS, *Heckmondwike and London.*—Blankets, cloths, sealskins, mohairs, railway rugs, and horse rugs.

[4026]

FORBES & HUTCHESON, 5 *Forbes Place, Paisley.*—Shawls and dresses.

[4027]

FOX, JOHN JAMES, & SON, *Devizes.*—Broad and narrow cloth, entirely of English wool.

This case contains examples of Devon kerseys, Bedford cords, tweeds, and various fancy trouserings, coloured, drab, and undyed. These are all made entirely of British Southdown wool, and for durability are unsurpassed by any other cloths whatever.

[4028]

FRANCIS & FLINT, *Nailsworth Mills, Stroud.*—Wool-dyed black and blue cloths, and ditto ditto doeskins.

[4029]

FRY, WILLIAM, & Co., 31 *Westmoreland Street, Dublin.*—Irish poplins or tabinets, and brocatelles for curtains.

[4030]

FYFE, ALEXANDER, & Co., 77 *Queen Street, Glasgow.*—Fancy dress fabrics.

[4031]

FYFE, H., & SON, Manufacturers, *Glasgow.*—Ginghams, drugget skirts, fancy dresses.

[4032]

GARVIE & DEAS, *Perth.*—Dress, linen, and cotton fabrics; dress wincies; fingering, lambs-wool, and wheeling hosiery.

[4033]

GILL, ROBERT, & SON, Manufacturers, *Inverleithen.*—Scotch tweeds, wool tartans, &c.

[4034]

GLOYNE, C. G., *Bradford Road, Dewsbury.*—Plain and figured pilots, witneys, velvet piles, Cheviots, mantle cloths, &c.

[4035]

GODFREY, BINNS, & SONS, *Deighton, near Huddersfield.*—$\frac{6}{4}$ tweeds, diagonals, and fancy coatings of every description.

[4036]

GOTT, BENJAMIN, & SONS, *Leeds;* 37 *Milk Street, London.*—Woollen cloths, mantle cloths, blankets, woollen yarns.

[4037]

GOW, BUTLER, & Co., 55 *Wilson Street, Glasgow.*—Shawls and fancy dress fabrics.

[4038]

GREENWOOD & CARTWRIGHT, *Rawfolds, near Leeds.*—Woollen cloaks for the Eastern markets.

[4039]

GREENWOOD, JOHN, & SONS, *Dewsbury.*—Pilots, petershams, witneys, cheviots, &c.

[4040]

GRIST, SONS, & Co., *Brinscombe, Gloucestershire.*—Mattress wools, woollen and shoddy flocks made from woollen rags.

[4041]

GRIST, HENRY, SON, & TABRAM, Manufacturers, *Nailsworth, Gloucestershire.*—Shoddy, a preparation of woollen rags for remanufacture; flocks for beds and mattresses.

Case of samples of shoddy for manufacturing purposes, flocks for beds and mattress wools, with prices.

[4043]

HARGREAVE & NUSSEYS, *Farnley Low Mills, near Leeds.*—Fancy and superfine woollen cloths.

[4044]

HARTLEY, JOHN, & Co., *Low Fold Mills, Leeds*—Plain and fancy union cloths, beavers, woollens.

[4045]

HARTLEYS & HARDWICK, *Aire Street, Leeds.*—Cloths, unions, tweeds, and coatings.

[4046]

HASTINGS, ROBERT, *The Votch Mills, near Stroud.*—Superfine woollen cloths and doeskins.

[4047]

HATTERSLEY, GEORGE, & SON, *Quarmby Clough Mills, near Huddersfield.*—Fancy union cloths.

[4048]

HEMINGWAY, JOHN, *Watergate Mill, Dewsbury.*—Woollen yarns for carpets.

[4049]

HEWITT, H., & ST. J., *Heytesbury, Wilts.*—Fine woollen fabrics.

[4050]

HEY, GEO., *Kirkbarton, near Huddersfield.*—Fancy woollen cloths.

[4051]

HILL, JOHN, & Co., Manufacturers, *Banbury.*—Mohair and worsted plush, worsted webbing for braces, girths, horse clothing, &c.

[4052]

HINCHCLIFF, DANIEL, *Providence Mill, Morley, Leeds.*—Union cloths of wool and other materials, unequalled at prices.

[4053]

HINCHLIFFE, JOHN & JAMES, *Morley and Leeds.*—Union cloth made from Sydney wool, waste, and mungo.

[4054]

HINDE, F., & SON, *Norwich, and* 82 *Watling Street, London.*—Paramattas, tamatives, grenadines, poplins, &c.

[4055]

HITCHCOCK, GEORGE, & COMPANY, 72, 73, 74 *Saint Paul's Churchyard.*—Ladies' fancy dresses, in woven and printed fabrics.

[4056]

HODGES, T. W., & SONS, *Leicester;* 18 *Noble Street, London;* 13 *Lever Street, Manchester.*—Elastic webs.

[4057]

HOLDSWORTH, JOHN, & Co., *Halifax and Bradford.*—Damasks, reps, table-covers and other textile fabrics, worsted yarns, &c.

[4058]

HOLLINS, WM., & Co., *Pleasby Works, near Mansfield, Nottingham.*—Cotton, merino, silk, cashmere, hosiery.

[4059]

HOLROYD, S., & Co., *Huddersfield.*—Cotton warp and doeskin.

[4060]

HOLT, RUSSELL, & BATES, 114 *St. Martin's Lane.*—Fancy woollen goods.

[4061]

HOOPER, CHAS., & Co., *Eastington Mills, Stonehouse, Gloucester.*—Superfine black, scarlet, coloured cloth, doeskins, and elastics.

[4062]

HOWGATE, HOLT, & Co., *Dewsbury.*—Plain and fancy woollen and union cloths.

[4063]

HOWSE, MEAD, & SONS, 18 *St. Paul's Churchyard, London.*—Woollen manufactures

[4064]

HUDSON & BOUSFIELD, *Leeds.*—Plain and fancy woollen and union cloths.

[4065]

HUNT & Co., *Lodgemore and Frome Hall Mills, Stroud, Gloucestershire.* — Superfine black cloths and doeskins, scarlets, billiards, &c.

[4066]

HUNT & WINTERBOTHAM (late THOMAS HUNT & Co.), Manufacturers, *Cam and Dursley Mills, Gloucestershire.*—Superfine woollen cloths.

[4067]

IRELAND, JOHN, & Co., *Kendal.*—Railway rugs, kersies, linseys, saddle-cloths, coat linings, Prince's checks, and collar checks.

[4068]

IRWIN, EDWARD, *Leeds.*—Double-milled woollen cloths, beavers, and pilots.

[4070]

JAY, GEORGE, & SON, *Albion Mills, Norwich.*—Mohair and alpaca yarns.

[4071]

JEBSON, J. & J., *Skelmanthorpe, near Huddersfield.*—Fancy waistcoatings and skirtings.

[4072]

JENKINS, WILLIAM, *Carmarthen Road, Swansea.*—Welsh cloth and Welsh striped flannel.

[4073]

JOHNSTON, JAMES, Manufacturer, *New Mill, Elgin.*—Tweeds and mauds of Cheviot and Australian wool; ditto made from Vicugna wool.

[4074]

JONAS, SIMONSEN, & Co., *Huddersfield.*—Woollen and union goods, and patent felt carpets.

[4075]

JONES, RANDALL, & WAY, 127 *Cheapside.*—English and Scotch fancy woollen goods, original designs.

[4076]

JORDAN, JAMES, *Huddersfield.*—Fancy vestings, coatings, mantle cloths.

[4077]

KENYON, J. & T., *Huddersfield.*—Fancy woollens.

[4078]

KELSALL & KEMP, *Rochdale, Lancashire.*—White and coloured flannels, swanskins and dometts.

[4079]

KERR, SCOTT, & KILNER, 58 *Cannon Street West, London.*—Shawls.

[4080]

KOHNSTAMM, HEIMAN, 33 *Dowgate Hill, Cannon Street, London.*—Leather cloths, with materials for its manufacture.

[4081]

LAIDLAW, WILLIAM, & SONS, *Hawick*, Yarn spinners, manufacturers of Scotch hosiery, tweed trouserings and coatings.

[4082]

LAIRD & THOMSON, 69 *Ingram Street, Glasgow.*—Mixed fabrics, woollen cloth, and poncho cloth.

[4083]

LAVERTON, ABRAHAM, *Westbury, Wiltshire.*—Fancy woollen cloths, silk mixtures, Meltons, Venetians, Arctic fur, and sable beavers.

[4084]

LEACH, JOHN, & SONS, 2 *Yorkshire Street, Rochdale;* 83 *Wood Street, Cheapside, London.* Flannels, baizes, &c.

[4085]

LEES, GEORGE, *Gala Bank Mill, Galashiels.*—Saxony shawls and cloakings, Angolas or Saxony tweeds.

[4087]

LIDDELL, BENNETT, & MARTIN, *Upperhead Mills, Huddersfield.*—Fancy woollens.

[4088]

LIEBMANN, M., *Huddersfield.*—New felted carpet, called 'Airdale Felt.'

[4091]

LITTLES, LEACH, & CO., *Britannia Mills, Leeds.*—Superfine wool and piece-dyed cloths, doeskins, unions, and fancies.

[4092]

LOCKE, CROSIER, & EDWARDS, 64 *Friday Street, E.C.*—Shawls, mantle cloths.

[4094]

LOCKWOOD & KEIGHLEY, *Huddersfield.*—Patent woollen cords, all-wool cords, and woollen velveteens.

[4095]

LUPTON, WILLIAM, & CO., *Leeds, Yorkshire.*—Woollen cloths.

[4096]

MACDOUGALL & CO., *Inverness.* (*See page* 37.)

[4097]

MALLINSON, DAVID, *Lepton, near Huddersfield.*—Fancy vestings, skirts, and trouserings.

[4098]

MARLING, STRACHAN, & CO., *Ebley and Stanley Mills, Stroud, Gloucestershire.*—Wool-dyed cloths, doeskins, and cassimeres.

[4099]

MARRIOTT, THOMAS, & SON, *Wakefield, Yorkshire.*—Yarns and worsteds.

Single and double yarns and worsteds, in low and fine qualities, dyed, undyed, mixed, corded or counted, thick or small numbers.

[4100]

MELLOR, JOSEPH, & SON, 30 *New Street, Huddersfield.*—Coatings and trouserings.

MACDOUGALL & Co., Manufacturers and Warehousemen, *Inverness.* — Highland tweeds, tartans, linseys, plaids, shawls, knitted hosiery, &c.

Obtained a Prize Medal at the Exhibition of 1851.

Since the International Exhibition of 1851, the exhibitors have effected great improvement in the quality, design, and colour of their woollen manufactures. Their hand-made fabrics are largely patronised by several European courts, and the nobility and gentry of their own country. Many poor families of the remote straths and glens of the Highlands find employment in the manufacture of these goods.

The following Specimens of these well-known fabrics are exhibited:—

DEER-STALKING and SHOOTING TWEEDS.

Heather and Granite Mixt ELASTIC TWEEDS for Gentlemen's Clothing for travelling, country, and town wear, and for tropical climates.

HAND-KNITTED STOCKINGS and SOCKS in natural colours and dyes.

The HIGHLAND DRESS for Men, Youths, and Boys.

WOOL CLAN TARTANS in fine textures and brilliant colours.

LINSEY WOOLSEYS, the favourite fabric for Ladies' Dresses, in pretty natural tints.

HAND-KNIT SHETLAND SHAWLS in plain and various fancy colours.

FINE CLAN TARTAN PLAID SHAWLS.

VICUNA SHAWLS AND WRAPPERS.

Patterns sent post free on application, and goods forwarded to all parts of the world free of risk.

[4101]

MIDDLETON & ANSWORTH, *Norwich, and* 92 *Watling Street, London.*—Fabrics for dresses. Shawls and crinoline.

[4102]

MILLMAN, L., & Co., *Nind Mills, Wotton-under-Edge.*—Superfine woollen cloths and doeskins.

[4103]

MILNER & HALE, *Huddersfield.*—Fancy coatings and mantle cloths.

[4104]

MILNER & NOKES, *Thurlstone, near Penistone, Yorkshire.*—Claret, drab, grey, and fancy hairlines.

[4106]

MITCHELL & WHYTLAW, *Glasgow.*—Fancy dress fabrics.

[4107]

MORGAN, JOHN, & Co., 110 *Causeyside, Paisley,* Manufacturers of shawls of all kinds, and woollen tartans.

[4108]

NEWBY & WOODHOUSE, *Bookfoot, Brighouse, Yorkshire.*—Fancy unions.

[4109]

NOLDA, CHARLES, & Co., 2 *Church Court, Old Jewry.*—Fancy woollens.

[4111]

NORTON, JOSEPH, *Clayton West, near Huddersfield.* — Shawls, mantle cloths, coatings, rugs, dress goods.

[4112]

OATES & BLAKELEY, *Dewsbury.*—Frieze, figured pilot, and velvet cloths, Cheviot tweeds, and fancy cloakings.

[4113]

OATES, HENRY, & SON, *Heckmondwike, near Leeds,* Manufacturers of blankets, woollens, army goods, &c.

[4114]

O'REILLY, DUNNE, & Co., 30 *College Green, Dublin.*—Irish poplin and tabinets.

[4115]

PATON, J. & D., & Co., *Tillicoultry, Scotland.*—Woollen shawls and cloakings, in clan, shepherdess, and fancy patterns.

[4116]

PATON, JOHN, SON, & Co., *Kilncraig's Factory, Alloa, N.B.*—Woollen hosiery yarns for knitting and other purposes.

Woollen Yarns, for hosiery and other purposes.

[4118]

PEACE, DAVID, *Shelley, near Huddersfield.*—Waistcoatings.

[4119]

PEACE, J., & SON, *Denby Dale, near Huddersfield.*—Vestings, skirts, and dresses.

[4122]

PEASE, HENRY, & Co., *Darlington and Bradford.*—Worsted mixed fabrics (Coburgs, Henriettas, Baratheas) and worsted yarns.

Obtained Prize Medals
at the Exhibition of 1851; *at the New York Exhibition,* 1853; *and the Paris Exhibition,* 1855.

The exhibitors' London Agents are:—

For Piece Goods: THOMAS LAWES, 5 Castle Court, Lawrence Lane.

For Yarns: HENRY BATEMAN, 15 King Street, Cheapside.

[4123]

PIM BROTHERS & Co., *South Great George's Street, Dublin.*—Irish poplins of every description.

[4124]

PLAYNE, P. P. & C., *Nailsworth, near Stroud.*—Superfine woollen cloths.

[4125]

PODD, THOMAS, & Co., *Leicester.*—Worsted, woollen, and Berlin yarns; for embroidery, knitting, and weaving.

[4127]

RATCLIFFE & SONS, *Staley, Huddersfield.*—Plain and fancy flannels.

[4128]

REID & TAYLOR, *Langholm, Scotland.*—Scotch tweed and maud manufacturers.

[4129]

RHODES, DANIEL, & SONS, *Dewsbury.*—Sealskins, velvet piles, Cheviots, travelling rugs, &c.

[4130]

RIPLEY, RICHARD, *Isle Mill, Holbeck, near Leeds.*—Woollen yarn.

[4131]

ROBERTS, JOWLINGS, & Co., *Lightpill Mills, Stroud.*—Woollen cloths and doeskins.

[4132]

ROBERTSON, JOHN, Linen Manufacturer, *Middle Hills, near Coupar Angus, Perthshire.*—Fast-coloured mixtures of flax &c., with volunteer uniforms of same, costing 15*s.* only.

[4133]

SALTER, SAMUEL, & Co., *Trowbridge, Wiltshire.*—Plain and fancy woollens.

[4134]

SCHOFIELD, JOHN, & SON, *Commercial Mills, Huddersfield.*—Tweeds, wool and angola yarns, knickerbocker yarns.

[4135]

SCHWANN, KELL, & Co., *Huddersfield.*—Cloakings and coatings.

[4136]

Scott, Alexander, & Son, *Alice Valley Mills, Morley, near Leeds.*—Woollen and union cloths from 3*s.* 6*d.* to 15*s.*

[4138]

Shaw, J., *Huddersfield.*—Trouserings.

[4139]

Shaw & Beaumont, Manufacturers, *Kirkheaton and Huddersfield.*—Fancy woollen trouserings and coatings.

[4140]

Sheard, M., & Sons, *Batley, Yorkshire.*—Pilots, velvets, and reversible cloths.

[4141]

Sheppard, W. B. & G., *Frome.*—Superfine cloths and fancy woollen goods.

[4142]

Sibbald, John, & Co., *Abbot's Mill, Galashiels.*—Fancy Scotch woollen trouserings.

[4143]

Smith, William, jun., *Morley and Leeds.*—Union cloths, made from Sydney wool, machine waste, and mungo.

[4144]

Smith, Robert, & Son, *Hayford Mills, near Stirling,* Wincey, linsey, and woollen manufacturers.

[4146]

Smith, William, Son, & Co., 14 *Cookridge Street, Leeds.*—Woollen cloths, fancy meltons, naps, witneys.

[4147]

Speirs, David, & Co., 167 *George Street, Paisley.*—Paisley long and square shawls, and Scotch woollen shawls.

[4148]

Spence, James, & Co., 77 & 78 *St. Paul's Churchyard, London.*—Shawls of various textures, poplins, mohairs, challies, and alpacas.

[4149]

Stancomb, William & John, *Trowbridge, Wilts, and* 8 *Basinghall Street, London.*—Fancy woollens.

[4150]

Standen & Co., 112 *Jermyn Street, St. James's.*—Hand-knit Shetland goods.

[4151]

Stanton & Son, *Stafford Mills, near Stroud.*—Woollen cloth.

[4152]

Starkey, James & Abel, *Sheepridge, near Huddersfield.*—Woollen cords, Bedford cords, and velveteens.

[4155]

Stockdale, William, *High Burton, near Huddersfield,* Manufacturer of fancy cloakings and coatings.

[4156]

SYKES, DAVID, *Brookfield Mills, Hunslet.*—Woollen cloths.

[4157]

SYKES, GODFREY, *Dalton, Huddersfield.*—All-wool and mixed fancy coatings, trouserings, fancy vestings, and Balmoral skirts.

[4159]

TAYLOR, JOHN, & SONS, *Newsome, Huddersfield.*—Fancy woollen and silk trouserings and coatings; fancy vestings, quiltings.

[4161]

THORPE, WILLIAM, *Almondbury, near Huddersfield.*—Trouserings, coatings, and tweeds.

[4162]

THRESHER & GLENNY, 152 *Strand, London.*—Kashmir flannel, India tweed, Mudah cotton.

These exhibitors are the only manufacturers of

THRESHER'S Kashmir flannel shirts.
,, India tweed suits.
THRESHER'S Kashmir woollen socks.
,, India gauze waistcoats.

Their establishment is situated next door to Somerset House, Strand.

[4163]

TOLSON BROTHERS, *Dalton, near Huddersfield.*—Manufacture of quiltings, and other waistcoatings, dresses, &c.

[4164]

TOWLER, ROWLING, & ALLEN, 15 *Watling Street, E.C.*—Norwich paramattas, poplins, fancy dress, shawl, hosiery.

[4165]

TURNBULL, WILLIAM, & CO., 21 *Glassford Street, Glasgow.*—Scotch tweeds and hosiery.

[4166]

TURNER & MUTER, 14 *West Nile Street, Glasgow.*—Dresses in mixed fabrics of cotton, mohair, and silk.

[4168]

VICKERMANN, B., & SONS, *Huddersfield.*—Broad and narrow cloths.

[4169]

WADE, JOSEPH, & SON, *Morley, near Leeds.*—Union cloths for home and export.

[4170]

HOLROYD, J., & CO., *Leeds.*—Plain and fancy woollen and union cloths.

[4171]

WALKER, G., *Lindley, near Huddersfield.*—Coatings.

[4172]

WALL & CO., *Welshpool, North Wales.*—Her Majesty's own and other flannels, tweeds, and clothing of Welsh manufacture.

[4173]

WANDLE FELT COMPANY, *Hanover Street, Long Acre, and Royal George Mills, Manchester.* Cloths and felts for mechanical operations.

[4174]

WATSON & NAYLOR, *Pike Mills, Kidderminster.*—Samples of worsted yarns, carded and combed.

[4175]

WATSON, WM., & SONS, *Dangerfield Mills, Hawick, Scotland.*—Scotch tweeds and plaids.

[4176]

WEBB, THOMAS R., & Co., Woollen Manufacturers, *Huddersfield.*—Elastic stockingetts.

[4177]

WHEELER, THOS., & Co., *Abbey Mills, Leicester, and* 13 *Goldsmith Street, London.*— Elastic webs.

[4178]

WHEELER, W. S., & SONS, 4 *Ludgate Street, London.*—Fine woollen trouserings and coatings.

[4179]

WHITE, JOHN (Successor to W. B. MacKenzie), 12 *Frederick Street, Edinburgh.*— Shetland woollen articles, hand-knitted in Shetland.

The following are exhibited: —

Lace shawls, handkerchiefs, and neck ties, in white, black, natural colours, and dyed colours.

Veils in black, white, and natural colours.

Thick close-knit shawls and handkerchiefs, in white and natural colours.

Various articles of under-clothing, elastic, light, and warm, for children, ladies, and gentlemen.

[4180]

WHITEHEAD, EDMUND, 21 *Rook Street, Manchester, and Springwood Mills, Middleton.*— Poplins and poplinettes.

[4181]

WHITEHILL, M., & Co., *Paisley.*—Embroidered shawls and table-covers.

[4182]

WHITELEY, THOS., & SON, *Stainland.*—Tweeds.

[4185]

WILKS BROTHERS & SEATON, 80 *Watling Street, London.*—Irish linens.

[4186]

WILSON, JOHN J. & WILLIAM, *Kendal,* Manufacturers of railway wrappers, Scotch tweeds, coat linings, horse clothing, girth webs, &c.

Obtained Prize Medal in Exhibition of 1851, *and Silver Medal at the Paris Exhibition,* 1855.

The exhibitors manufacture railway wrappers, Scotch and Cheviot tweeds.

Checked serges, Prince's checks, roller and girth webs, &c.

Woollen coat linings, winceys, kersey horse clothing.

[4187]

WILSON, WALTER, *Allars Crescent, Hawick, N.B.*—Tweeds and cloakings.

[4188]

WILSON & ARMSTRONG, *Hawick, N.B.*—Tweeds and wool shawls.

[4189]

WISE & LEONARD, *Nailsworth; and Holcombe Mills, near Stroud.*—The finest superfine black and blue cloths, single kerseymere and double beaver.

[4190]

WOODHOUSE, JOHN, & Co., *Aire Street, Leeds, and New Mills, Holbeck.*—Tweeds, meltons, and union cloths.

[4193]

WRIGLEY, J. & T. C., & Co., *Dungeon Mills, near Huddersfield.*—Woollen goods of all descriptions.

[4194]

WRIGLEY, JOHN, & SONS, *Huddersfield.*—Liveries, meltons, and carriage linings.

[4195]

WURTZBURG, EDWARD, & Co., *Leeds.*—Plain and fancy woollen and union cloths.

[4196]

YUILL, JOHN, & Co., *Paisley, Glasgow, and Manchester.*—Shawls, mufflers, and dresses.

[4197]

WALKER, J., & SONS, *Lindley, near Huddersfield.*—Mohairs, Hudson bays, sealskins, cashmere, furs, brennas, shells, rugs, &c.

CLASS XXII.

CARPETS.

[4227]

BOYLE, J. W., 9 *Great Marlborough Street.*—Carpets.

[4228]

BRINTON & LEWIS, *Kidderminster.* (*See page* 46.)

[4229]

CAWLEY, JOHN, 28 *Red Lion Street, Clerkenwell, E.C.*—Adelaide mats and rugs.

Adelaide hearth-rugs and mats can be made of any size or colour to pattern, with Axminster rug or padded back. These Heald mats are very suitable for private or railway carriages, as they do not get matted with dust, and can be easily cleaned, without injury to the colours, as they are dyed ingrain.

[4230]

COOKE, HINDLEY, & LAW, 12 *Friday Street, London ;* Manufactory, *Liversedge, near Leeds.* Specimens of various descriptions of carpeting.

[4231]

CROSSLEY, JOHN, & SONS, *Halifax.*—Carpet rugs &c.

[4232]

DIXON, HENRY JEEKS, & SONS, *Long Meadow Mills, Kidderminster.*—A Brussels velvet pile carpet.

[4233]

DOWNING, GEORGE F., *Knightsbridge, London.*—Specimens of floor-cloth, ten yards wide, without seam or join.

[4234]

FAIRFAX, KELLEY, & SONS, *Heckmondwike.*— Carpets, blankets, sealskins, and carriage rugs.

BRINTON & LEWIS, *Kidderminster*, Manufacturers of velvet pile carpets made by power loom, and also of chenille and other rugs.

Mr. BRINTON, Juror, Class 22, Carpets, in the International Exhibition of 1862.

(*Holders of a Prize Medal at the Exhibition of* 1851.)

This firm, established in the late and present hands upwards of forty years, is the largest carpet manufacturing concern in the locality, and, in connection with their worsted spinning mills, affords employment to about 600 workpeople. Special attention is devoted to produce, in the higher as well as in the ordinary carpet fabrics of daily use, excellence of design and durability of wear, so as to equal the French manufacturers in style, and by adapting all the recent improvements of power-loom machinery, far to surpass them in economy of production. These exhibitors employ their own artists on the spot (some of them trained in the School of Design), and while they superintend their own designs personally, they avail themselves also of the services of other artists of celebrity. The chief fabrics manufactured by BRINTON & LEWIS are — 'Brussels and velvet pile carpets by steam power, chenille carpets and rugs, wide sofa carpets, figured and other rugs by power and hand loom.'

The patterns exhibited are as follows:—

In the Furniture Court.

No. 1. MEDALLION SMYRNA DESIGN, double breadth, in the illuminated style, in which the combination of rich colours with others more subdued imparts a mellow tone to the whole. The crimson medallions are contrasted with the rich tracery of the arabesques which surround them.

No. 2. MEDALLION TURKEY ALHAMBRA.—The figures here are disposed upon a crimson ground, covered with small tasteful objects, the colours employed being something similar to those of No. 1, but presenting quite a different effect.

No. 3. CASHMERE ARABESQUE.—A dining-room carpet, tesselated ground. The leading colours are crimson and brown, with arabesque tracery in green, crimson, and blue.

No. 4. A KNOTTED AND LOOPED RIBBON DESIGN, in self colours, grouped in trellis forms. The colours are rich crimsons upon a dark ground, with a jewel-ornamented border attached to the same. This carpet is specially intended for drawing-rooms.

On the Wall over the Grand Staircase
(Picture Gallery Entrance).

No. 5. PERSIAN CARPET.—This design is derived from manuscripts of the fifteenth and sixteenth centuries. The colouring is a combination of greens, blues, and browns, toning down the richer colours and harmonising with the full ponceau ground.

No. 6. DRAWING-ROOM CARPET, crimson centre and ribbon border. This carpet is intended to represent a veined marble ground pattern, the design coloured in shades of harmonised crimsons. The border which surrounds the crimson is upon a white ground and wreath of leaves and white roses intertwined with a ribbon; the intention of this treatment of the border being to relieve the general weight of colour in the marble within, and to add to its appearance. It should be observed that the effect of this and other carpets hung upon the walls of the building is naturally very different to what the same would be seen on floors, as intended by the producers.

No. 7. JEWEL PATTERN.—This is a parlour or small sitting-room best Brussels carpet, and, as its name implies, is in the *jewelled style.* The objects on the green ground are introduced for the purpose of giving a sparkling lively effect, and when seen in the evening by artificial light the appearance is equally pleasing as by day.

No. 8. STAIR CARPET, also a best Brussels, intended to go with the preceding Jewel pattern, the figures being arranged to work continuously forwards upon a *staircase*, and to harmonise in treatment and effect with the jewel body carpet, or others of this class.

The design and colouring of these carpets are of the advanced taste now in general preference to the gaudy and floral style of the past. The Persian objects are arranged to produce an inlaid effect. The colours are so harmonised as to enhance the general furnishing of the apartments they are intended to occupy.

BRINTON & LEWIS have contributed also a BRUSSELS CARPET POWER LOOM in the Machinery Department of the Western Annexe, in full process of weaving carpets for Messrs. Jackson & Graham.

In the South Kensington Museum, (Raw Materials section), they also exhibit, as their contributions, specimens of WOOL and WORSTED in their various stages of working, from the fleece into the finished state of carpet.

Duplicate patterns of the above may be seen at the London Warehouse, 90 Newgate Street; and at the Manufactory, Kidderminster.

[4235]

FILMER, J. H., & SON, 28, 31, & 32 *Berners Street, London, W.*—Jacquard pile carpets, designed in competition by students of the South Kensington School of Art.

[4237]

FRITH, EDWIN, & SONS, *Heckmondwike.*—Kidderminster carpets, Dutch carpets, and twilled stairs carpeting.

[4238]

GOATLEY & CHORLEY, 39 *Westminster Bridge Road,* Floor-cloth manufacturers.

[4239]

GREGORY, CHARLES, 212 *Regent Street, London.*—Two patent Axminster carpets; Brussels and velvet pile Brussels carpets; Axminster rugs.

[4240]

GREGORY, THOMSON, & CO., Carpet Manufacturers, *Kilmarnock.*—Specimens of carpeting.

[4241]

HARE & CO., Designer and Manufacturer, *Bristol.*—Floor-cloths; Corinium pavement complete, tiles and marbles, with centres &c.

Obtained Prize Medals at the Great Exhibitions of London and Paris in 1851 *and* 1855.

Perfect copy of the CORINIUM TESSELATED PAVEMENT which was discovered some years since at Cirencester, and is considered one of the most beautiful in the world.

Composition of ENCAUSTIC TILES, with centre and borders.

Composition of MARBLES and MOSAIC, with centre and borders.

Any pattern in either supplied at the ordinary price

The canvas and all the materials used in the fabric manufactured by the exhibitors.

[4242]

HARRISON, CHARLES, *Stourport, and* 59 *Skinner Street, London.*—Velvet pile and Brussels carpets.

[4243]

HAWKSWORTH, SAMUEL, Floor-Cloth Manufacturer, *Baker Street, Doncaster.*—Floor-cloth in paint imitation of Mosaic pavement.

[4244]

HEAD, JOHN, & CO., *Stourport, Worcestershire.*—Velvet pile and Brussels carpets and rugs.

[4245]

HENDERSON & CO., *Durham.*—Samples of velvet pile, Brussels, Kidderminster, and Venetian carpeting.

[4246]

JACKSON & GRAHAM, *Oxford Street, London.* (*See page* 48.)

[4247]

KINDON & POWELL, *Swan Street, Old Kent Road, S.E.*—Floor-cloths, table-covering, and stair-cloths.

[4248]

LAPWORTH BROTHERS, 22 *Old Bond Street, and Wilton, Wilts,* Carpet manufacturers and dealers; importers of Turkey and foreign carpets.

[4249]

MORTON & SONS, *Kidderminster, and* 17 *Skinner Street, London.*—Velvet and Brussels carpets.

JACKSON & GRAHAM, *Oxford Street, London.*—Patent tapestry velvet, Brussels, velvet pile, and Axminster carpets.

A PATENT AXMINSTER CARPET, 25 ft. by 18 ft., the ground of rich marone colour, with crimson rosettes, the centre with group of flowers, on white ground, su rounded by rich brown and gold ornamental framing, intersecting with the border which consists of arabesque scrollwork, on white ground, with shields and festoons of flowers at each corner. Upon the margin, outside the border, a light arabesque ornament has been introduced.

Samples of different designs, some with borders, of velvet pile carpet, woven by steam power in Jacquard loom. Sample of patent tapestry velvet carpet.

[4250]

NAIRN, MICHAEL, & Co., *Scottish Floor-Cloth Manufactory, Kirkcaldy.*—First-class printed floor-cloth.

NAIRN, MICHAEL, & Co.—First-class printed floor-cloth—*continued.*

These floor-cloths are made in sizes of twenty five yards long, and from half a yard to eight yards wide, without seam. The foundation canvas is warranted to be manufactured from flax alone; and the paints, oils, lead, and other ingredients are guaranteed to be genuine. These goods can be procured from all first-class carpet warehousemen and upholsterers in the provinces; and, in London, from J. Shoolbred & Co., Tottenham Court Road.

[4251]

PALMER BROTHERS, *Kidderminster.*—Specimens of hand-loom Brussels carpet, exhibited for quality, correctness of design, and colouring.

[4252]

ROLLS, JAMES, & SONS, *Kennington Lane, Lambeth, S.*—One piece of floor-cloth 18 feet square.

[4253]

SEWELL, HUBBARD, & BACON, *Compton House, Old Compton Road, and Frith Street, Soho.* (*See page* 51.)

[4254]

SMITH, TURBERVILLE, & Co., 9 *Great Marlborough Street, W.*—Carpets 25 feet by 18 feet square.

[4255]

SMITH & BABER, 1 *South Place, Knightsbridge, London.*—Two specimens of floor-cloth.

[4256]

SOUTHWELL, H. & M., *Bridgnorth; and* 29 *Cannon Street West, London.*—Bordered Wilton carpet.

[4257]

STEVENSON, WILLIAM, 16 *Piccadilly, London.*—Bordered pile carpet.

[4258]

SWALLOW, MICHAEL, & SONS, *Heckmondwike.*—Carpets, alpaca coat linings, &c.

[4259]

TAPLING, THOMAS, & Co., 1 *to* 8 *Gresham Street West, City.*—Axminster and Turkey carpets.

A carpet commemorative of the Treaty of Commerce between England and France, designed specially for THOMAS TAPLING & Co. by Mr. William A. Parris.

[4260]

TAYLER, HARRY, & Co., 19 *Gutter Lane.* (*See page* 52.)

[4261]

TEMPLETON, J., & Co., *Glasgow and London.* (*See page* 53.)

[4262]

TEMPLETON, J. & J. S., *Glasgow.*—Improved patent Axminster carpeting, woven by power; silk and wool curtains and covers.

[4263]

WATSON, BONTOR, & Co., 35 & 36 *Old Bond Street.* (*See page* 54.)

The design on page 54 illustrates a handsome velvet carpet, by WATSON, BONTOR, & Co., of Old Bond Street, in the Grecian style of ornament, with appropriate borders. This carpet is the finest production of the hand loom, and so constructed in breadths as to be produced in various sizes, either in velvet or Brussels, at the ordinary prices.

SEWELL, HUBBARD, & BACON, *Compton House, Old Compton Street, and Frith Street, Soho.*
Patent Axminster carpet.

Patent Axminster carpet 24 ft. × 17 ft.. The same pattern can be made to any size.

TAYLER, HARRY, & Co., 19 *Gutter Lane, Cheapside, London;* Works, *Deptford Green.*—Patent tinted ground and ordinary kamptulicon floor-cloth.

KAMPTULICON, OR ELASTIC FLOOR-CLOTH, is an article possessing advantages peculiar to itself. Composed of non-absorbent water-repellent and warm materials, the result is an invaluable covering for damp and stone floors. The materials are non-conductors of heat and electricity, are soundless, durable in themselves, and when amalgamated produce a material possessing the united advantages of carpet and oiled cloth, being warm, noiseless, impermeable by damp, impenetrable by dust, and extremely durable, thereby recommending itself for general use in all public buildings and private houses.

In designing for this material, the manufacturers have been exceedingly careful to figure the cloth with only such patterns as will leave the ground as much as possible exposed, so as not to interfere with *its warmth.* Most of the designs are of a purely architectural character.

[4264]

WAUGH & SONS, Manufacturers, 3 & 4 *Goodge Street.*—Patent Axminster and velvet pile carpets.

[4265]

WHITWELL & Co., *Dockray Hall Mills, Kendal.*—Carpeting of various descriptions.

[4266]

WHYTOCK, RICHARD, & Co., 9 *George Street, Edinburgh.* (*See page* 55.)

TEMPLETON, JAMES, & Co., *Glasgow and London.*—Carpets without seam, and in breadths; hearth-rugs, mats, &c.

No. 2650

II

The drawing No. 2650 represents a quarter of one of the carpets prepared by Messrs. Templeton for the International Exhibition of 1862. The other drawings or diagrams (on which are merely indications of pattern) are intended to represent the capabilities of making carpets in one piece to any form of room, however elaborate or simple in design.

I. Circular: on which a pattern of any style or colouring, with or without medallion, can be arranged and woven without seam.

II. Represents two rooms, separated at A by folding doors, but when the folding doors are open appear as one, with the border running round both rooms, including projections at A.

This figure may also be filled with two distinct carpets, separated by a piece of carpet similar to figure IV. between the projections A. This arrangement, however, can still be woven as one carpet, as well as in three distinct pieces.

III V

III. Shows an oblong drawing-room with two columns, one at either side towards the centre; and represents a carpet prepared for it with three medallions; the centre one (that within the pillars) being smaller than those at either end. This room has a bow window, and shows how the border is carried round the bow.

IV. Represents the piece of carpeting which divides the drawing-room from the octagon-shaped room adjoining. A complete pattern is indicated on this small piece.

V. This octagon, while it indicates the arrangement of pattern for a room of this form, at the same time shows how the borders of a carpet can be carried into bay windows, after the manner of the bow of Diagram III. or otherwise. These diagrams, from I. to V., are merely indications of how carpets can be woven in one piece to the shape of any room.

Watson, Bontor, & Co., 35 & 36 *Old Bond Street.*—Indian, Turkish, and velvet carpets.

WHYTOCK, RICHARD, & Co., 9 *George Street, Edinburgh.*—One Wilton carpet, and one Scoto-Axminster carpet.

WILTON CARPET OF THE LOUIS XVI. STYLE.

A Wilton carpet for drawing-room of the Louis XVI. style, which has recently been revived in a very pure manner in France.

This is the first attempt to introduce this style in a quality of carpet within the reach of the general public. The design presents a marked contrast to the other carpet exhibited by this firm, being an example of the shaded style of ornament.

A SCOTO-AXMINSTER CARPET, BYZANTINE STYLE.

A Scoto-Axminster carpet for dining-room, Byzantine style.

This is an example of the much-approved flat, or unshaded ornamentation, in carpet design. The warm neutral colouring is an agreeable advance upon the showy combinations of colour hitherto so much in demand.

[4267]

WIDNELL, HENRY, & SON, *Lasswade, New Edinburgh.*—Carpets, rugs, table-covers, moquettes, velvet and tapestry curtains.

[4268]

WOODWARD BROTHERS & CO., *Kidderminster.*—Brussels velvet pile carpeting and rugs.

[4269]

WOODWARD, HENRY, & SONS, Carpet and Rug Manufacturers, *Stour Vale Mills, Kidderminster.*—Brussels and Tournay velvet carpets.

[4270]

HUMPHRIES, J., & SONS, *Mile Street, Kidderminster.*—Velvet pile carpets.

Class XXIII.

WOVEN, SPUN, FELTED, AND LAID FABRICS, AS SPECIMENS OF PRINTING OR DYEING.

[4300]

Barlow, Samuel, & Company, *Stakehill, Chadderton, Manchester.*—Cotton goods bleached by Barlow's patent, dyed and finished.

[4301]

Baynes & Son, *Queen's Road, Bayswater, London;* Works, *Blackman Street, Borough, London.*—Dyed furniture hangings &c.

[4302]

Beaulieu, J., & Co., *Sloane Street, S.W.*—Dyed goods.

[4304]

Berrie, John, 13 *Oldham Street, Manchester.* (*See page* 58.)

[4305]

Black & Wingate, *Glasgow.*—Imitation French cambrics and bishop's lawns, handkerchiefs white and printed, embroideries, &c.

[4306]

Botterill, John, Stuff Dyer, *Leeds.*—Orleans, lustre, cobourg, camlet cloths, lastings, moreens, &c.

[4307]

Bradshaw, Hammond, & Co., *Levenshulme Works, and* 35 *Mosley Street, Manchester.*—Variety of calico prints produced by machine throughout.

[4308]

Butterworth & Brooks, *Manchester.*—Calico prints and mousselines de laine.

[4309]

Calder Vale Printing Company, *Calder Vale, Burnley.*—Patent furniture, damask-printed on both sides by machinery.

BERRIE, JOHN, 13 *Oldham Street, Manchester.*—Specimens illustrating new methods of dyeing and finishing silk, merinoes, &c.

This case contains the following samples, viz.

French merinoes, finished in an improved manner by the newly invented machinery of the exhibitor.

Silks dyed in the usual manner by hand, and as dyed by the exhibitor's machinery without handling. The advantage of the new process will be seen in the freedom of the machine-dyed silk from the cracked appearance common to those dyed by hand.

Satin embossed with various patterns. The material of the sample is an old plain dress.

Shawls cleaned, showing the fringes as usually done, and as cleaned by J. BERRIE's method, the latter being as good as when new.

The following machinery is used by this exhibitor, over and above the ordinary appliances of most dyers in the country:—

LARGE STEAM CLEANING MACHINE for moreens, damasks, carpets, &c.

PATENT HYDRO-EXTRACTORS, in which all goods are dried in five minutes, without wringing or pressure, thereby saving delicate fabrics, and preserving the colours.

NEW GLAZING MACHINE, for chintz curtains and covers, and the only one of its kind ever yet made.

STEAM FINISHING FRAMES, rendering it impossible for goods to be scorched.

NEW STEAM APPARATUS, for finishing French merinoes, &c., giving to the small pieces a surface equal to the large ones.

SCOTCH MUSLIN MACHINES, for finishing lace and muslin curtains.

NEW MACHINE, for taking frame marks from satins after being finished.

NEW MACHINES, for tacking the pieces of dresses together for dyeing and finishing.

The following are some of the articles which he cleans and dyes, &c.:—

DRESSES. Silks, satins, moiré antiques, velvets, poplins, French merinoes, stuffs, and all kinds of fancy dresses; children's clothing; ribbons, laces, hose, handkerchiefs, and bonnets.

SHAWLS. All kinds of Paisley, India, and China crapes, Cashmere, barège, llama, woollens, and silks.

CURTAINS. Damasks, moreens, tabarets, satins, poplins, reps, &c. &c. Lace and muslin curtains got up like new at 1½*d.* a square yard.

TRIMMINGS, such as gimps, tassels, silk and worsted fringes, &c.

CHINTZ FURNITURE, BLANKETS, GENTLEMEN'S WEARING APPAREL, GLOVES, FURS, FEATHERS, TABLE-COVERS, CARPETS, RUGS, AND DRUGGETS.

Persons having much dyeing and cleaning to be done, will find the methods employed by the exhibitor conducive to economy. Should they reside at a distance, special arrangements may be made as to carriage.

Circulars and catalogues may be obtained on application.

[4310]

CLARKSON, THOMAS, & CO., 17 *Coventry Street, Haymarket.*—Specimens of block and machine printed chintz furnitures.

[4311]

DAILY & CO., 9 & 10 *St. James's Place, Hampstead Road.*—Specimens of French dyeing and cleaning; improved method of dyeing and finishing velvets; embossing satins; re-watering moiré antiques.

[4312]

DEWHURST, SAMUEL, & CO., *Broughton Works, Manchester.*—Bookbinders' cloth, patent tracing cloth, beetled twills.

[4313]

DONOVA, R., 2 *Great Pulteney Street.*— A piece of tapestry, one part cleaned by the exhibitor's own method.

[4314]

GEORGE, T. W., & Co., *Spring Gardens Dye Works, Leeds.*—Worsted and mixed stuff goods.

[4315]

GIRDWOOD, W., & Co , *Old Park, Belfast.*—Printed linens, linen lawn dresses and hand kerchiefs, drills, &c.

[4316]

GRAFTON, F. W., & Co., *Broad Oak, Accrington, and 25 Portland Street, Manchester.*—Printed calicoes.

[4317]

HANDS, SON, & Co., Silk Dyers, *Coventry.*—Specimens of silks dyed in the skein.

[4318]

HINE, RICHARD E., & Co., 36 *York Street, Manchester.*—Silk handkerchiefs printed, of every description, and silks generally.

[4319]

HOYLE, THOMAS, & SONS, *Mayfield, Manchester.*—Printed cambrics, challis, and de laines.

[4320]

KEYMER, J., & Co., *Dartford, Kent.*—Printed silk handkerchiefs.

[4321]

LITTLEWOOD, WILSON, & Co., *Manchester, and Foxhill Bank, Accrington.*—Muslin, calico, and de laine printed.

[4322]

LOCKETT, JOSEPH, SONS, & LEAKE, *Strangeways Engraving Works, Manchester.*—Specimens of engravings for printing calicoes and other woven fabrics.

Obtained the Gold Medal at the Paris Exhibition, 1855.

[4323]

MCKINNELL & Co., *Ancoats Bridge Works, and 27 Dickinson Street, Manchester.*—Blue and madder printed cottons.

[4324]

MACNAB, JAMES, 145 *Ingram Street, Glasgow.*—Printed calicoes and muslins in dresses and handkerchiefs.

[4325]

MCNAUGHTON & THOM, 80 *Mosley Street, Manchester.*— Printed calicoes.

[4326]

MONTEITH, HENRY, & Co., *Glasgow.*—Turkey red and other dyed and printed goods, and Turkey red yarns.

[4327]

MUIR, BROWN, & Co., 29 *West George Street, Glasgow.*—Dyed and printed cotton and woollen fabrics.

[4328]

NEWTON BANK PRINTING COMPANY, 51 *Mosley Street, Manchester.*—Printed cottons.

[4329]

EWING, J. O., & Co., 32 *St. Vincent Place, Glasgow;* Works, *Alexandria and Levenfield, Dumbartonshire.*—Turkey red goods.

[4330]

ORMEROD, R., & Co., 50 *Mosley Street, Manchester.*—Patent printed ribbons.

[4331]

OXFORD, JOSEPH, & Co., 5 & 6 *Bury Court, St. Mary Axe, London, E.C.*—Printed corahs, and China crape shawls.

[4332]

PALMER & Co., *Holme Works, Carlisle,* Manufacturers of beetled silesias, taffetas, silk and cotton umbrella cloths.

[4333]

PULLAR, ROBERT, & SONS, *Perth.*—Umbrella cloths, and various descriptions of dyed cotton goods.

[4334]

RICHARDSON, BENJAMIN S., Dyer, *Priory Fields, Coventry.*—Set of silk patterns, as samples of dyeing.

[4335]

SALOMONS, A., *Old Change, London.*—Printed cottons and muslins of British manufacture.

[4336]

SIDEBOTTOM, ALFRED, Chemist, *Laboratory, Crown Street, Camberwell.*—The wool on cavalry skins dyed permanent black.

[4338]

SMITH, SAMUEL, & Co., *Horton Dye Works, Bradford.*—Specimens of dyeing and finishing goods, mixed and plain.

[4339]

STEAD, MCALPINE, & Co., *Cummersdale Print Works, Carlisle.*—Calico, and cotton damask chintz furnitures.

[4340]

STIRLING, WILLIAM, & SONS, *Glasgow.*—Turkey red plain and printed cottons.

[4341]

TATTON, SAMUEL, *Patent Dye Works, Mill Street, Leek.*—Sewing silk, and twist for the sewing machines.

[4342]

TURNER, CORNELIUS, *Airedale Felt Mills, Leeds, and* 18 *Lawrence Lane, London.*—Felt carpets and felted goods.

[4343]

TURNER, NORRIS, & TURNER, *Manchester, and Hayfield, Derbyshire.*—Printed cottons.

[4344]

TURNER, NORRIS, & TURNER, *Stockport.*—Calico prints.

[4345]

VICTORIA FELT CARPET COMPANY, *Leeds, and* 8 *Love Lane, Aldermanbury, London.*—Felt carpets, table-covers, tablings, thick felt, saddle-cloths, waddings, shoe cloth, &c.

[4346]

WALFORD, FAIRER, & HARRISON, *London, Manchester, and Glasgow.*—Silk manufacturers and printers.

[4347]

WATSON & STARK, 51 *George Street, Manchester.*—Printed vestings, trouserings, coatings, and cambrics.

[4348]

WELCH, THOMAS, *Merton, Surrey.*—Woollen cloth printed patterns, designs produced since 1851.

[4349]

WILKINSON, JOHN, SON, & CO., *Leeds.*—Felt carpets, rugs, squares, nummahs, wadding, padding, sheathing, boot felt.

Specimens of Felt Carpeting, shown for superiority of manufacture and printing.

Also Nummahs or Cavalry Blankets and Saddle-Cloths.

Boot Cloths, and Boots made therefrom.

Ship Sheathing, tarred and untarred.

[4350]

YATES, MATTHIAS WILLIAM, 2 *Wood Street, Cheapside, London, and Fountain Works, Mitcham,* Manufacturer of table-covers.

[4351]

ROSENDALE PRINTING COMPANY, 8 *Nicholas Street, Manchester.*—Prints.

[4352]

WHITWELL, BUSHER, & CO., *Kendal.*—Dyed worsted yarns.

CLASS XXIV.

TAPESTRY, LACE, AND EMBROIDERY.

[4381]

ABRAHAM, R., & SONS, 5 *Lisle Street, W.*—Cover for the Law and other articles used in Jewish synagogues; also various specimens of embroidery &c.

[4382]

ADAMS, THOMAS, & Co., *Nottingham.*—Curtains and other descriptions of lace.

[4383]

ALLEN, CHARLES, 108 *Grafton Street, Dublin.*—Irish, point, appliqué, and guipure lace.

[4384]

AUSTIN, JAMES, *Princes Street, Finsbury.*—Improved patent blind, curtain, picture, and sash-lines, chandelier rope, &c.

The following lines, of which the exhibitor is the manufacturer, are shown :—

Imperial patent flax and coloured thread blind lines. Worsted blind, curtain, lamp, and picture lines. Chandelier rope. Cotton blind and curtain lines. Silk blind, curtain, and picture lines. Metallic picture cord, sash-lines, Albert lines, &c.

Especial attention is directed to the great strength and durability of the metallic picture cord, and to the superior quality of the Albert lines.

[4385]

BAGLEY, J. W., *Nottingham.*—Patent Valenciennes, Maltese, and Honiton edges, laces, beadings, and insertions.

[4386]

BARNETT, MALTBY, & Co., *Stoney Street, Nottingham.*—Silk laces, nets, falls, black Spanish laces, shawls, &c.

[4387]

BATES, J., MISS, *Newington, Surrey.*—A lace fall and other articles worked by hand.

[4388]

BLACKBORNE, ANTHONY, 35 *South Audley Street, London.*—Irish, English, French, Spanish and Brussels laces.

[4389]

BORWICK, MISS, 2 *Henstridge Villas, St. John's Wood.*—A raised crochet counterpane.

[4390]

BRADBURY, CULLEN, & FISHER, *Broadway, Nottingham.*—Lace shawls, falls, &c.

[4392]

CARDWELL, COOPER, Lace Manufacturer, *Northampton.*—Articles of pillow lace manufacture.

[4393]

CATT (late SLOAN), 198 *Sloane Street.*—Ladies' dress trimmings and fancy needlework.

[4394]

CHAMBERS, JAMES, & Co., 4 *Upper Sackville Street, Dublin.*—Irish embroidery insewed muslins, initialled handkerchiefs, silk embroidery on cloth, &c.

[4395]

CLARKE, ESTHER, 18A *Margaret Street, Cavendish Square.*—Honiton lace, flounce, and other specimens of the same manufacture.

[4396]

COPESTAKE, MOORE, CRAMPTON, & Co., 5 *Bow Church Yard,* Lace manufacturers.

COPESTAKE, MOORE, CRAMPTON, & Co. exhibit in the stained glass gallery a collection of lace curtains, the production of their Nottingham factory.

COPESTAKE, MOORE, CRAMPTON, & Co.'s octagon case contains a collection of hand-made lace.

A dress and half shawl of Honiton lace, the production of Devonshire.

A half shawl, parasol cover, cape, &c., of Irish laccet work.

Capes, collars, and sleeves, parasol covers of crochet work, the productions of Ireland.

Laces, coiffures, barbes, falls, parasol-covers (in black and white), in Maltese lace, the productions of Buckinghamshire, Northamptonshire, and Bedfordshire.

In their second case they exhibit black pusher or royal point tunic and shawls, black Spanish laces and shawls, productions of Nottingham (machine-made).

Crochet tunic, capes, collars, and sleeves, productions of Ireland.

White lace square, half square, tunics, and flounces in tambour lace, productions of Limerick and London.

[4397]

COWAN & Co., 24 *St. Vincent Place, Glasgow.* (*See page* 65.)

[4398]

CROOME, MISS MARY ALBINIA B., *Middleton Cheney, Banbury.*—Egyptian-work dress, and lace, made at Middleton Cheney.

[4399]

DART & SON, 12 *Bedford Street, Covent Garden.*—Lace for carriages.

[4400]

DEBENHAM, SON, & FREEBODY, 42 & 44 *Wigmore Street.*—Various articles in Honiton, Buckinghamshire, and Nottingham lace.

[4401]

DE LA BRANCHARDIÈRE, MLLE. RIEGO, 1 *Princes Street, Cavendish Square, W.*—Crochet and modern point lace.

[4402]

DIXON, GEORGE, 13 *Goldsmith Street, Cheapside.* (*See page* 66.)

[4403]

DUNNICLIFF & SMITH, *Nottingham.*—Valenciennes and other lace.

COWAN & CO., 24 *St. Vincent Place, Glasgow.*—Embroidered muslins and patent household frillings.

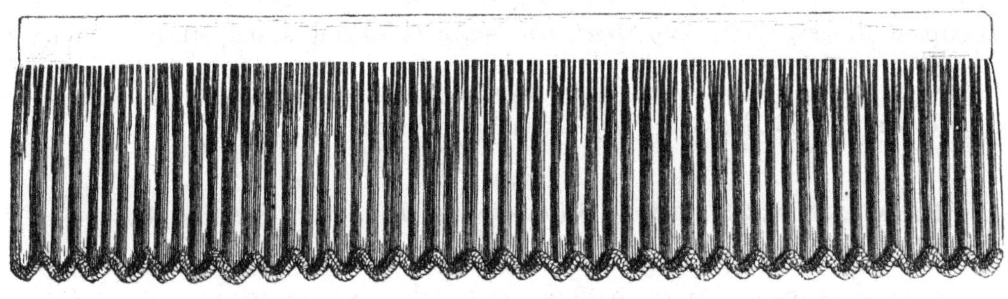

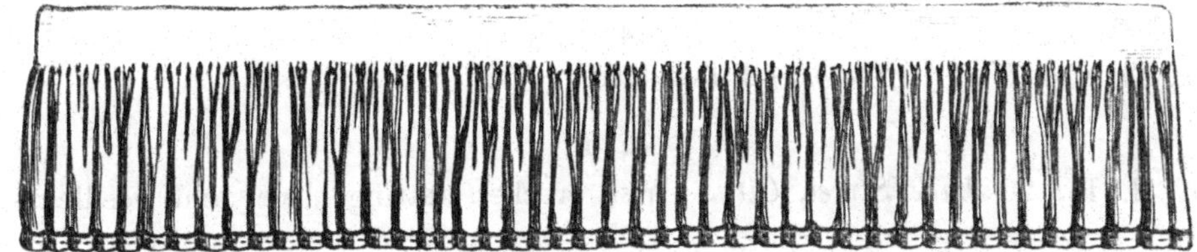

This case contains a choice assortment of useful and beautifully designed EMBROIDERED MUSLINS, the great proportion most serviceable in ornamenting ladies' and children's dresses, as well as for baby linen purposes.

Messrs. COWAN & CO. are also the Patentees of the 'PATENT HOUSEHOLD FRILLING' (see engraving), numerous specimens of which are shown in the case.

This frill, by an ingenious process in the manufacture, is attached to a double cloth band, and at the same time is made sufficiently full for goffering, thus saving ladies the trouble attending the use of a gathering thread. It is a most economical and useful frill for the purpose of trimming under-clothing, children's dresses, baby linen, &c. It is made in various widths.

DIXON, GEORGE, 13 *Goldsmith Street, Cheapside.*—Patent velvet pile cornice fringes, general upholstery trimmings, and carriage laces.

This patent has been applied to various classes of upholstery trimmings, with results which, in richness, beauty, and cheapness, far surpass those of any other process. The designer is not limited to the ordinary commonplace styles in fringes made by his patent, as it admits of the execution of the most elaborate patterns without any of the irregularities or other defects observable in fringes made by other machines. A striking merit preserved by these fringes, is, that they do not alter in shape with use.

[4404]

EHRENZELLER, FERDINAND, 15 *Gower Street, Islington; and* 35 *Cannon Street West, City.*—Needlework, lace articles, &c.

[4406]

ELLINGTON & RIDLEY, 89 *Watling Street, London, E.C.*—Fringes and other trimmings for upholstery purposes.

[4407]

ERNE, COUNTESS OF, 95 *Eaton Square.*—Irish Valenciennes lace.

[4408]

EVANS, RICHARD, & CO., 24, 25, & 74 *Watling Street, London.* (*See page* 67.)

[4409]

FORREST, JAMES, & SONS, 101 *Grafton Street, Dublin.*—No. 14. Irish laccet, point lace, tunic; No. 15. Irish, point, Brussels lace, tunic; No. 16. Irish, point lace, tunic.

[4410]

GILBERT, THOMAS, *High Wycombe.*—Pillow lace goods.

[4411]

GOBLET, H. F., 20 *Milk Street, City.*—Irish crochet, tatting, lace, and needlework.

[4412]

GODFREY, EDMOND, *Buckingham.*—Black point, ground, tunic, flounce; black guipure, tunic, flounce; veils and lappets.

[4413]

GRAHAM, ALEXANDER, 34 *York Street, Glasgow.*—Patterns on cloth for needlework; and ornamental cut velvet, silk, &c.

Evans, Richard, & Co., 24, 25, & 74 *Watling Street, London.* — **Upholstery, mantle and dress trimmings, military braids, &c.**

Obtained the Prize Medal in Class 19, 1851.

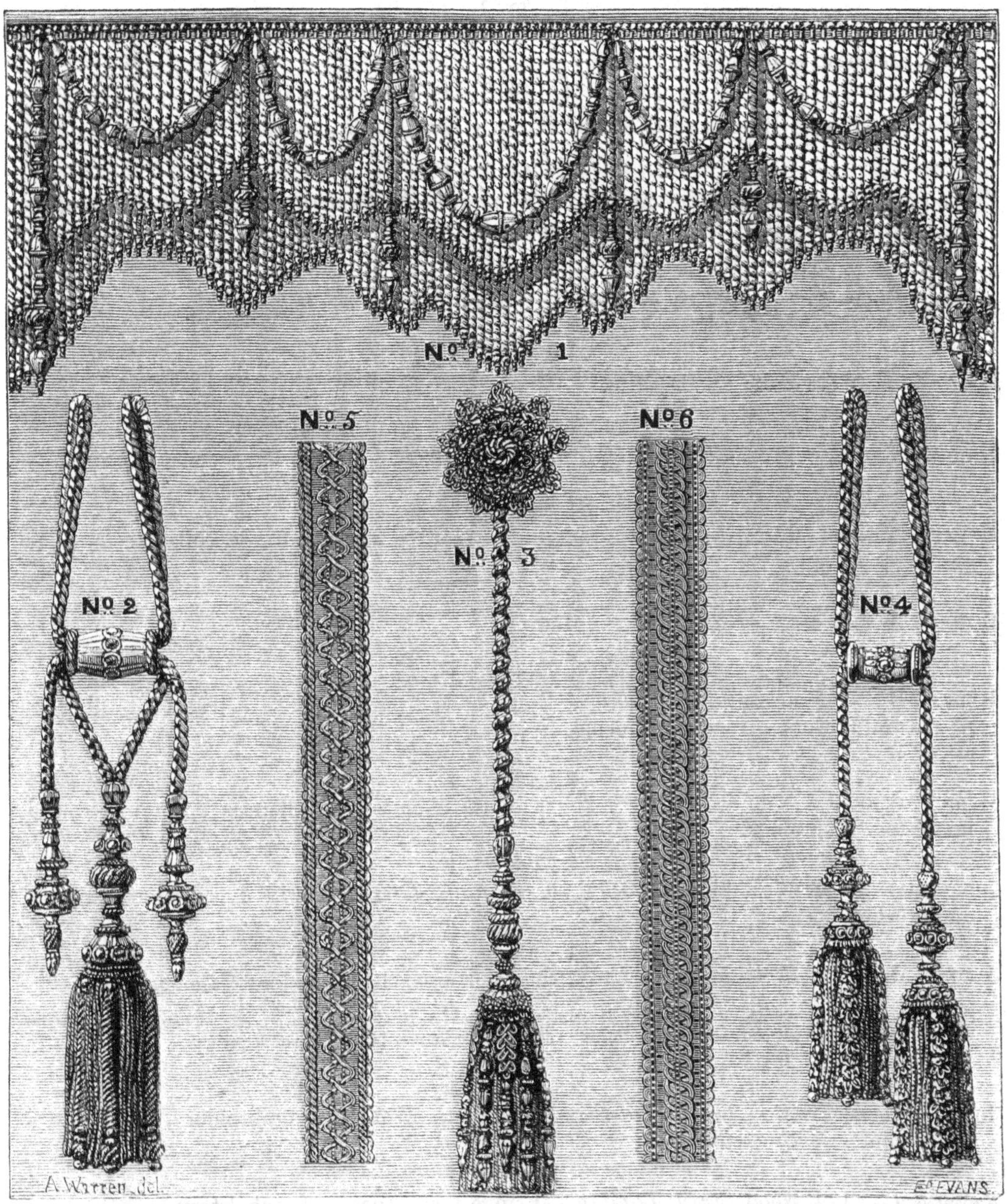

No. 1. Double-shaped Cornice Fringe, for a drawing-room window, tipped and richly ornamented.
No. 2. Curtain-Holder, in crimson and gold silk, with side ornaments.
No. 3. Bell-Rope, blue and white (made to any length).
No. 4. Double Curtain-Holder, green and gold (used in the arranging and suspending the drapery of windows).
No. 5. Figured Border Gimp, for trimming curtains and drapery, crimson and gold silk.
No. 6. Ditto ditto in green and gold.

The following are also exhibited with the above:—

Fancy Dress Gimps, Rouches, Fringes, Buttons, Mantle Ornaments and Tassels, Ottoman Tassels and Rosettes, Drapery Ornaments, Curtain Cords, &c. &c.

These exhibitors manufacture for the trade only. Their factories are situated at 197 Upper Thames Street, and 13 Garlick Hill, City.

[4414]

GREEN, ALEXANDER, 136 *Buchanan Street, Glasgow.*—Fancy needlework, materials used, velvets, scales, bullion, &c.

[4415]

HAYES, MISS E. J., 24 *Richmond Place, East Street, Walworth.*—Infant's embroidered cloak and bead embroidery.

[4416]

HAYWARDS [DANIEL BIDDLE], 81 *Oxford Street.*—Lace. (*See pages* 69 *and* 70.)

[4417]

HERBERT, THOMAS, & Co., *Houndsgate, Nottingham.*—Manufacturers of tattings, crochets, muslins, Valenciennes, laces, fringes, &c.

[4418]

HEUGH, WIGHT, & Co., 56 *Friday Street.*—Arnold's patent stitched frilling.

[4419]

HEYMANN & ALEXANDER, *Nottingham.*—Nets, laces, and curtains.

[4420]

HIGGINS, EAGLE, & HUTCHINSON, 57 *Cannon Street West.*—British lace goods.

[4421]

HORNSEY, JOSEPH, *Bedford.*—Point, ground, and Maltese laces, coiffures, collars, sets, white and black falls.

[4422]

HOWELL, JAMES, & Co., *Regent Street.*—Lace.

[4423]

HYDE, MRS., 7 *Finsbury Place South.*—Embroidery from nature on velvet &c., imitations of natural flowers, principally crochet.

[4424]

HYDE, ARCHER, & Co., 7 *Finsbury Place South, London.*—Vallences embroidered, and upholstery trimmings.

[4425]

INDUSTRIAL SOCIETY (established 1847), 76 *Grafton Street, Dublin.*—Irish point, guipure, and crochet lace.

[4426]

JACOBY, M., & Co., *Nottingham.*—Lace goods, curtains, shawls, patent Valenciennes.

The exhibitors are manufacturers of every description of cotton and silk lace goods — nets, curtains, anti-macassars, shawls, bed covers, &c. They also manufacture Valenciennes lace by a patent process.

[4427]

JONES, WILLIAM, & Co., 236 *Regent Street, London.*—Gold lace embroidery, army, navy, and volunteer accoutrements.

[4428]

KEITH & Co., 124 *Wood Street, London.*—Fringe for purposes of upholstery.

[4429]

LECHÊNE, ACHILLE, 2 *Foley Street, Portland Place.*— Chromo on velvet applied for decorative tapestry, curtains, sofa and table-covers.

[4430]

MERY, LÉON, & Co., Draftsmen, 87 *Upper Ground Street, Blackfriars.*— Patent velvet application.

[4431]

LESTER & SONS, *Bedford.*—Collars, sets, crowns, laces, falls, flouncings, lappets, coiffures, handkerchiefs.

[4432]

LONG, GEORGE, *Loudwater, Wycombe, Bucks.*—Manufacturer of hats and bonnets made on the pillow lace principle; and straw embroidery.

[4433]

MACARTHUR, D., & Co., 26 *Bothwell Street, Glasgow.*—Manufacturers of lace, embroidered muslin, fancy linen, and crape goods.

[4434]

MACDONALD, HELEN J., 1 *Stafford Street, Edinburgh.*—Silk patchwork table-cover.

[4436]

MADDERS, WILLIAM, & Co., *Leamington Place, Manchester.*—Embroidered table-cover, cradle coverlet, pair of antimacassars.

[4437]

MALLET, HENRY, *Nottingham.*—Thread and Valenciennes laces, guipures, black and white silk blonde and Spanish laces, shawls, mantillas, &c.

[4438]

MANLOVES, ALIOTT, & LIVESEY, *Bloomsgrove Works, Nottingham.*—Pile and velvet nets, falls, laces, and trimmings.

[4439]

MANLY, GEORGE N., 43 *New Finchley Road, St. John's Wood.*—Specimens of Irish lace.

[4440]

MEE, CORNELIA, 71 *Brook Street, Grosvenor Square.*—Embroidery and ornamental needlework.

[4441]

NEWMAN & PUMEY, 118 *Oxford Street, W.*, Fringe and trimming manufacturers.

[4442]

NORTHCOTE, S., & Co., 29 *St. Paul's Churchyard, E.C.*— Lace shawls, flounces, parasol-covers, &c.

[4443]

PALMER, HELEN, *Dunse, Berwickshire, Scotland.*—Specimens of embroidery in coloured silks, ditto of white embroidery.

[4444]

PALMER & COTTRELL, 87 *Blackfriars Road, S.*—Painted velvet and cloth work, with designs for embroidery and braiding.

[4447]

PULLING & MOODY, 39 *Gresham Street, E.C.*—Manufactured crape goods. (*See page* 72.)

[4448]

RADLEY, EDWARD, 20 *Lamb's Conduit Street, W.C.*—Upholsterers' trimmings.

[4449]

RECKLESS & HICKLING, *Nottingham.*—Pusher and new grenadine lace shawls; half shawls and flounces.

HAYWARDS [DANIEL BIDDLE], 81 *Oxford Street.*—Honiton, British point, guipure, black point, and other British lace.

IMPORTERS OF FOREIGN LACE AND EMBROIDERIES. MANUFACTURERS OF HONITON AND OTHER BRITISH LACES. READY-MADE LINEN FOR WEDDING AND INDIA OUTFITS. LAYETTES. BERCEAUNETTES. LACE AND EMBROIDERY ROBES, ETC. SILK AND VELVET MANTLES. PARIS MILLINERY.

The exhibitors have a large stock of lace flounces, squares, &c., especially designed for wedding orders, with every requisite in the linen-outfit department, from prices suited to the most economical.

Brussels lace squares	from 10 guineas.
Imitation ,,	,, 1 guinea.
Brussels lace tunics	,, 11 guineas.
,, double flouncings	,, 16 ,,
Imitation ,,	,, 45 shillings.
Honiton lace squares	,, 3 guineas.
,, flouncings	,, 7 ,,
Swiss lace squares	,, 3½ ,,
Black real point lace flouncings	,, 18 ,,
Imitation ,, ,,	,, 2 ,,

A large assortment of flounces, squares, &c., in Limerick, point d'Angleterre, and other inexpensive laces.

Brussels lace, sets of collars and sleeves	from 24*s.* to 20 guineas.
Honiton lace do. do.	,, 10*s.* 6*d.*
Full trimmed sets of collars and sleeves in lace and embroidery	,, 12*s.* 6*d.* to 5 guineas.
Muslin and cambric embroidered sets	,, 5*s.* 6*d.* to 30*s.*
Lace and muslin double-skirt dresses	,, 15*s.* 9*d.*
Embroidered cambric handkerchiefs	,, 3*s.* 6*d.*
Trimmed lace do.	,, 10*s.* 6*d.* to 25 guineas.
Black lace mantillas and shawls	,, 25*s.*

The ladies' and infants' READY-MADE LINEN DEPARTMENT is replete with every article of under-linen, dressing gowns, &c., at prices saving to customers all intermediate profit.

Messrs. HAYWARDS have no other establishment than 81 Oxford Street, opposite the Pantheon.

Mr. D. BIDDLE, the present head of the firm, being a Juror for lace and embroidery, no official recognition of the merit of the articles exhibited by this firm can be admitted.

HAYWARDS [DANIEL BIDDLE], 81 *Oxford Street.*—Honiton and point lace—*continued.*

HONITON GUIPURE TUNIC FLOUNCE, EXHIBITED BY HAYWARDS [D. BIDDLE].

Messrs. HAYWARDS exhibit in the British Lace Department a tunic flounce and square to match, coiffures, handkerchiefs, cape, infant's robe, &c., of Honiton lace, and a black point lace tunic flounce of Buckingham manufacture.

They also display several flounces, shawls, mantles, &c., in the Belgian and Spanish Lace Departments.

The *real ground* Brussels lace square (rare), exhibited in the case of J. Strehler, is especially worthy of attention. The process of manufacture is shown in the same case.

Near this a black real point lace shawl, of superior design, manufactured by Bruyneel ainé, will attract the notice of connoisseurs. The price of this shawl, and of many other articles exhibited by Messrs. HAYWARDS, is marked in plain figures.

Pulling & Moody, 39 *Gresham Street, E.C.*—Manufactured crape goods.

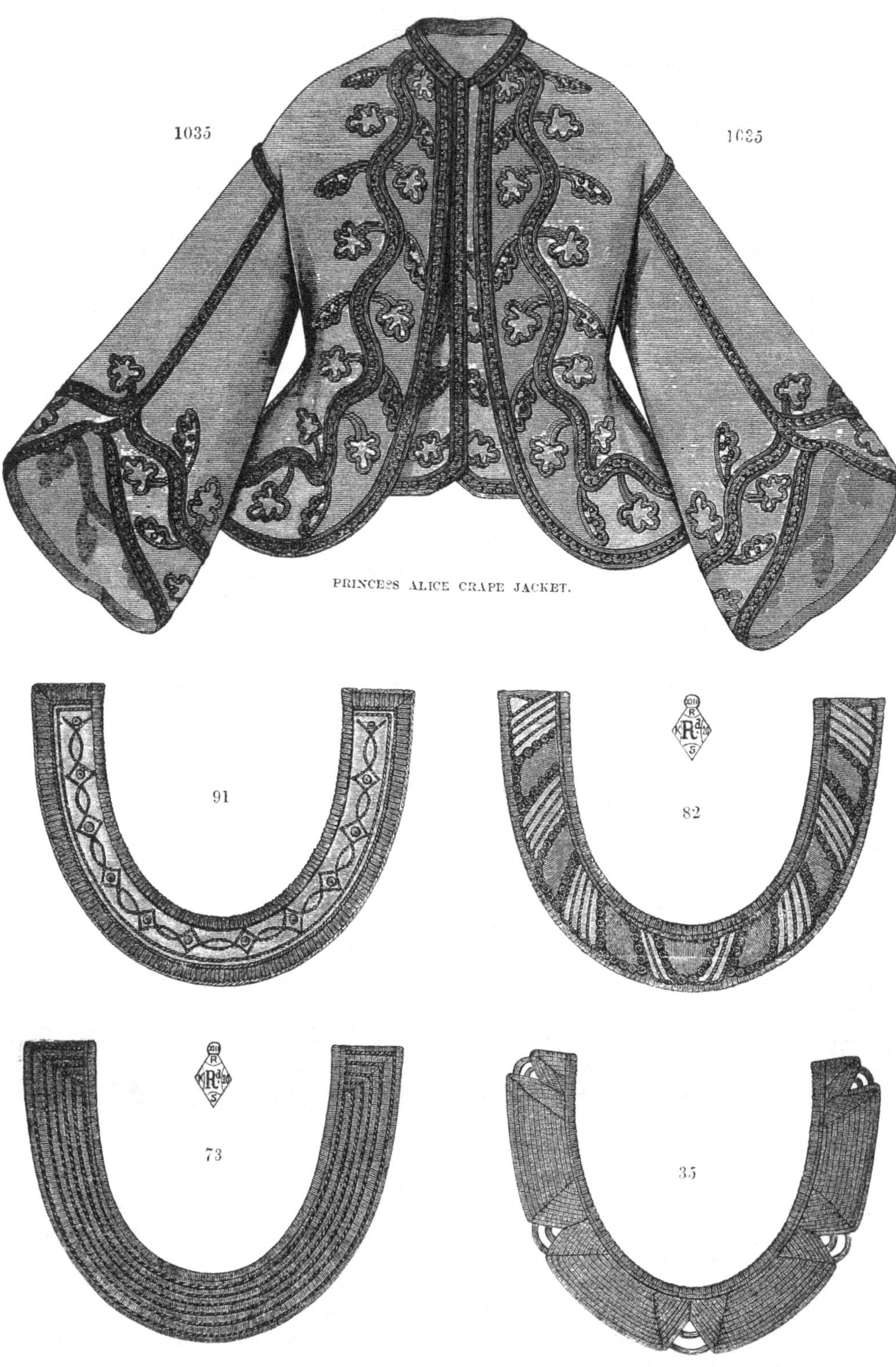

PRINCESS ALICE CRAPE JACKET.

CRAPE COLLARS.

[4450]

ROBINSON, HENRY, *Watling Street and* 42 *Cannon Street West.*—Real and imitation lace goods of British manufacture.

[4451]

SARGEANT, JOHN, *Savoy, Bedfordshire.*—Lace handkerchiefs, collars, sets, and lace.

[4454]

SMITH, SIDNEY, *Week-day Cross, Nottingham.*—Various articles in lace.

[4455]

STANDRING, JOHN, & BROTHER, *Manchester.*—Braids, cords, laces, fringes, &c.

[4456]

STEEZMANN, EDWARD, & Co., *Nottingham.*— Lace curtains, bed-covers, &c., made by machinery.

[4457]

STILLWELL, SON, & LEDGER, 25 *Barbican, City, London.*—Gold lace, embroidery, and army and navy fittings.

[4458]

URLING, G. F., 224 *Regent Street.*—Tunic; handkerchief; coiffure collar and sleeves in very fine Honiton lace.

[4459]

VERKRÜZEN & Co., 96 *Hatton Garden.*—Traced embroideries, the same worked; their new patented gold drawings on velvet, &c.

[4460]

VICCARS, RICHARD, *Padbury, Buckingham.*—Pillow lace, white Maltese laces, single flounce, collars, and sets.

[4461]

VICKERS, WILLIAM, *Nottingham.*—Black silk lace (imitation point de Chantilly) in shawls, points, and other articles.

[4462]

VOKES, FREDERICK S. T., *Royal Surrey Theatre.*—Specimen of the best novel braiding for clothes and regimentals.

[4463]

WELSTED, HONORIA, *Ballywalter, Castletownroche, County Cork.*—Pocket handkerchiefs and other articles.

[4464]

WILLS, S., & Co., *Broadway, and Mansfield Road, Nottingham.* (*See page* 73.)

[4465]

WILSON, CHARLOTTE G., *Guildhall, Broad Sanctuary, Westminster.*—Chessboard needle-work.

[4466]

WOOLCOCK, C. M. & A., 13 *Old Quebec Street, Oxford Street.*—Tapestry, drapery, portière, folding screen, &c.

Wills, S., & Co., *Broadway, and Mansfield Road, Nottingham.*—Machine lace, nets, curtains, &c.

HALF DESIGN FOR STORE LACE CURTAINS.

Wrought entirely by machinery. Manufactured by S. Wills & Co.

PRINTED FOR HER MAJESTY'S COMMISSIONERS

BY

SPOTTISWOODE AND CO., NEW-STREET SQUARE, LONDON

Class XXV.

SKINS, FUR, FEATHERS, AND HAIR.

Sub-Class A.—*Skins and Feathers.*

[4499]

Bevington & Morris, 67 *Cannon Street West.*—Manufactured furs, sheep-skin and Angora goat rugs.

[4500]

Clark, Cyrus & James, Manufacturers, *Street, near Glastonbury.*—Sheep-skin hearth, door, and carriage rugs; boots and shoes.

[4501]

Drake, Robert, 25 *Piccadilly.*—Manufactured furs, various.

[4502]

Holden, John Trippett, *Collis Works, Birmingham,* Inventor, patentee, and manufacturer of self-fasteners for victorines and mantles.

[4503]

Ince, Thomas H., & Co., Furrier, 75 *Oxford Street.*—Fashionable furs.

[4504]

Jeffs, Richard, 244 *Regent Street.*—Fur skins and manufactured ditto.

[4505]

Lillicrapp, W., 19a *Davies Street, Berkeley Square,* Real fur seal-skin cloak maker, and general furrier.—Skins dressed and mounted.

[4506]

Meyer, S. M., 71 *Cannon Street West, and Bow Lane.*—Furs, various.

[4507]

Poland, G., & Son, Fur Manufacturers, 90 *Oxford Street, W.* (*See page* 2.)

[4508]

Roberts, Edward Boyd, 239 *Regent Street.*—Manufactured furs.

POLAND, G., & SON, Fur Manufacturers, 90 *Oxford Street, W.*— Manufactured furs and specimens of skins.

The following are exhibited: —

Real fur seal cloak, trimmed with dark sable.
Velvet cloak, trimmed with chinchilla.
Real fur seal cloak, trimmed with grebe.
Fine Hudson Bay sable circular cloak.
Real ermine cloak, lined with quilted satin.
Real Russia sable muff and boa.
Fine Hudson Bay sable muff and mantilla.
Astracan muff and collaret.
Real ermine muff and mantilla.
Fine dark sable tail muff and boa.
Grebe muff and boa, lined with satin.
Gentleman's real fur seal coat and waistcoat.
Gentleman's cloth coat, lined throughout and trimmed with fur.
White Arctic-fox carriage wrapper, lined with quilted silk.
Carriage wrapper, made of the throats of the sable.
Lynx wrapper, lined with scarlet cloth.
Wolverein do. handsomely mounted.
A fine Bengal tiger, mounted with black bear, as rug.
Several specimens of fur rugs.
A variety of furs of the most recherché description.

[4509]

SENGER, A. H., 3 *Lamb Alley, Bishopsgate Street.*—Fur seal-skin in different stages of progress of manufacture.

[4510]

SMITH, GEORGE, & SONS, 9 *to* 11 *Watling Street.*—Manufactured furs.

[4511]

TUSSAUD BROTHERS, 105 *Marylebone Road.*—Method of saving skins from furs, and applying artificial pelts.

[4512]

NICHOLAY, E. J., 82 *Oxford Street.*—Furs for ornaments, carpets, rugs, &c.

SUB-CLASS B.—*Feathers.*

[4522]

DE COSTA, ANDRADE, & Co., 7 & 8 *Cripplegate Buildings, E.C.*—Ostrich feathers, in the raw and dressed state.

[4523]

SUGDEN, SON, & NEPHEW, 12 & 16 *Aldermanbury.*—Case of ostrich and fancy feathers.

SUB-CLASS C.—*Manufactures from Hair.*

[4534]

ASTON, JOHN, 20 *Dale End, Birmingham.*—Household and saddlery brushes of a very superior quality.

[4535]

BARRETT, ANDREW, 63 *Piccadilly;* 186 *Oxford Street;* 29 *St. George's Place, Knightsbridge.*—Toilet, household, and stable brushes.

[4536]

BLYTH & SONS, 4 *Chiswell Street, London; and Henry Street, Liverpool.*—Feathers, horsehair, and wools.

[4537]

BOOTH & FOX, 80 *Hatton Garden, London; and Cork,* Feather purifiers.—Patent feathers, down quilts, and petticoats.

No. 1. RUSSIAN WHITE GOOSE FEATHERS.
„ 2. IRISH WHITE.
„ 3. RUSSIAN GREY.
„ 4. IRISH GREY.
„ 5. HUDSON'S BAY.
„ 6. EIDER DOWN.
„ 7. WHITE GOOSE DOWN.
„ 8. TURKEY HACKLES.

Patent feather fur, muffs, victorines, and feather trimmings.

Eider down quilts of Arctic (goose) down in glacé silk and alpaca, alpaca tufted, chintz, damask, and Turkey red.

These Arctic down quilts are so well filled with the finest down, that one of them affords nearly as much warmth as three blankets.

Pure down contains such a large amount of the great non-conductor — atmospheric air — that it retains the natural heat with very little weight.

Down quilts afford so much comfort and warmth, that they are constantly used in winter in Germany and other cold climates of the north of Europe, and there is now a greatly increased demand for them in Great Britain and Ireland.

[4538]

BROWNE, FREDERICK, 47 *Fenchurch Street.*—Perukes, head-dresses, and ornamental hair generally, made on human hair foundations.

[4539]

CARLES, H. R., 45 *New Bond Street, W.*—Imperceptible capillamenta. (*See page* 4.)

[4540]

CHILD, W. H., 21 *Providence Row, Finsbury.*—Specimens of electro-galvanic, metallic, and other brushes, in ivory, bone, and wood.

Obtained a Prize Medal at the Paris Exhibition, 1855.

CHILD'S PATENT ELECTRO-GALVANIC AND METALLIC HAIR AND FLESH BRUSHES for preventing neuralgia, rheumatism, loss of colour, and the falling off of the hair. These are superior to the ordinary hair brushes, as they do not soften in use, have the same action as the bristle, and do not tear or injure the hair. The exhibitor manufactures all other brushes for the home or colonial markets.

[4541]

CLIFFT, JOHN, *Bristol.*—Specimens of ornamental hair in wigs, &c.; specimens of human hair dyed.

[4542]

CONDRON, T. & R., 51 *Bingfield Street, Caledonian Road, N.*—Fancy brushes.

CARLES, HYACINTH R., 45 *New Bond Street, W.*, Peruke and Hair Dresser.—Imperceptible capillamenta.

Obtained Honourable Mention in Class XVI. in 1851, *and a Medal at Paris in* 1855.

IMPERCEPTIBLE CAPILLAMENTA.

1. THE BUST OF AN OLD MAN. A study.
   The hair of the head and beard implanted on removable hair net.
4. LADIES' HEAD DRESSES.
10. GENTLEMEN'S PERUKES.

To whatever excess and absurdity fashion might have at times, and particularly during the long period of the last three centuries, carried the misapplication of of humau hair, there are evidences in the writings Xenophon, Tacitus, Suetonius, and Juvenal, that even in remote antiquity it was felt that human hair contributes to the more or less pleasing expression—to the more or less striking setting off of the character of the countenance,—and that those whom age or infirmity have deprived of it endeavoured to supply it by art, in many instances not out of vanity, to which the world is prone to attribute it, but out of the praiseworthy desire to spare to others unpleasant impressions. And as the hair is the only ornament derived direct from the human frame itself, for its embellishment it requires by its very nature not only a mechanical but also an artistical treatment, to produce the most becoming effect.

The exhibitor having always endeavoured during his long practice to combine the best workmanship which his ability afforded, with an artistical arrangement and finish, is desirous of exemplifying in the articles exhibited this twofold treatment of the hair.

Thus, considered from the material and mechanical point of view, the objects exhibited show—

Uncommon beauty and quality of the principal material, viz. the hair;

A preparation of that material, peculiar to the exhibitor, which preserves the colour, and, without making the hair stiff or brittle, causes it to retain its curl, or any form or position, unless exposed to excessive wet or too profuse perspiration;

A peculiar outline and fixing of the foundations, adapted to the shape and character of the head, securing the firmest and most comfortable fit;

Hair partings, which the exhibitor was the first to introduce, and a specimen of which was for the first time seen at the Great International Exhibition of London of 1851. The hair of which these partings are made being white and transparent, produces the imperceptibility which, until their introduction by the exhibitor, was neither known nor attainable;

Finally, skilful tressing, and most careful workmanship and finish, in which the exhibitor endeavoured not to be surpassed by any of his fellow competitors.

As to the artistical merit of these specimens, the exhibitor would beg to direct attention in the first instance to the fact, that effect is less influenced, in works made of hair, by the colour of that material, than by its properties, and chiefly depends upon the style, and the artistical touch of the comb. A glance of an experienced eye at the full-bottomed wigs of Brinette (the hair-dresser of Louis XIV. of France), or any eminent performer, well acquainted with the effects which a well-chosen coiffure produces,—at the bob wigs introduced some time after,—at the hair-bags, knots, and tails,—at the statues of antiquity (particularly when those of Nero and Marcus Aurelius are contrasted),—will easily detect this fact, corroborated even by history. For what called forth the indignation of the quiet and staid people of that time, and Mr. Prynne's quarto volume of remonstrance and condemnation of the so-called 'Love Lock' of which Charles I. set the fashion, but that air of coquettish frolicsomeness which it imparts to the wearer?

Indeed, to produce the desired effect, and give to the head-dress a style corresponding with the character, or modifying the expression, of the countenance, the hair-dresser must not only espy the means by which nature produces these effects, but thoroughly study the types of character and expression, so as to be able to adapt to them even the fantastical, sometimes absurd, modes of the fashion of the moment, to which the eye becomes so accustomed as to consider handsome and becoming what, at other times, it would pronounce improper and ugly.

The exhibitor trusts that his views will be borne out by the specimens exhibited; and if he ventures to suggest these few hints, it is because none of the treatises on head-dress known to the exhibitor, even the one of Le Févre (1778) which the French Academy selected for insertion in their 'Encyclopædia,' have ever examined the subject from an artistical point of view; and the public seems to be but imperfectly aware of the fact that not mere mechanical skill, but a sound conception of art, and a well-trained artistical taste, are required for the production of intended effects, and the proper application of the hair for the ornament of the head.

[4543]

COOPER & HOLT, 50 & 51 *Bunhill Row, Finsbury.*—Curled horsehair, bed feathers, and wool.

[4544]

DICKINSON, JONATHAN, & SON, *South Market, Meadow Lane, Leeds,* General brush manufacturers, wholesale and for exportation.

[4545]

DOHERTY, THE MISSES, *Sligo.*—Horsehair ornaments, made by peasant girls.

[4546]

DOUBBLE, THOMAS, 18 *Bartlett's Buildings.*—Brooms, brushes, and combs.

[4547]

DOUGLAS, ROBERT, 21 & 23 *New Bond Street.* — Beautiful specimens of ladies' and gentlemen's wigs, &c.

[4548]

DOW, ANDREW, 1 *Hardwick Street, Liverpool.*—Brushes for plate, watches, and jewellery, with wooden backs, filled with horse, goat, and human hair.

[4549]

ELLINGTON & RIDLEY, 89 *Watling Street, London, E.C.*—Purified bed feathers and eider down bed quilts.

[4550]

ESSEX, FREDERICK, 53 *Percival Street.*—Wool rugs, foot muffs, boots, furs, &c.

[4551]

FARRANT, RICHARD E., 16 *Queen's Row, Buckingham Gate.*—Tooth, nail, shaving, cloth, hat, and hair brushes.

[4552]

FORSTER, GEORGE, 9 *Hatton Wall, London.*—Shaving brushes, ivory and bone.

These brushes, made of Bukka or French hair, in ivory and bone rollers for dressing-cases, are sold wholesale only by the maker. As a large assortment is always kept in stock, shipping orders can be executed on the shortest notice. Every article is warranted.

[4553]

GOSNELL, JOHN, & CO., 12 *Three King Court, Lombard Street.*—Perfumery and soaps; hair, and other kinds of brushes.

[4554]

GRAY, E. M., 44 *Ebury Street, Pimlico.*—Hair flowers.

[4555]

GRAY, L., 44 *Ebury Street, Pimlico.*—Hair coronet.

[4556]

GREENWOOD, BENJAMIN, *Bond Street, Tyrrel Street, Bradford, Yorkshire.*—Improved brooms and circular brushes.

[4557]

Hastings, Stephen, *Limerick.*—Shoe, horse, cloth, and hair brushes, &c., made of oak taken from the old Cathedral of Limerick after being 700 years in use.

[4558]

Herrmann, Augustus, 4 *Oxenden Street.*—Fancy hair work on a new principle; landscapes in cork; imitation in lace work; and a crown of victory in natural leaves.

[4559]

Hewlett, Anthony H., 5 *Burlington Arcade, Piccadilly.*—Wigs, fronts, scalps, and several kinds of artificial hair; also specimens showing the effects of a new hair dye.

[4560]

Hopekirk, Walter, 88 *Westminster Bridge Road.*—Ladies' wigs and partings for thin hair; gentlemen's wigs and scalps.

Specimens are exhibited of wigs, fronts, curls, partings for thin hair, straight and ringlet hair on invisible foundations, plaits, twists, frizzettes, and plicaturas.

Nature is so closely imitated in the manufacture, that detection is almost impossible.

A large stock is always on hand, at moderate charges.

[4561]

Hovenden, R., & Sons, *Great Marlborough Street; and Crown Street, Finsbury.*—Human hair, raw and manufactured.

1. Specimens of the Raw Human Hair, of the principal descriptions in use.

2. Specimens of Human Hair, showing it in the various stages of preparation from the raw state to its completion into a gentleman's wig, a lady's band, and a lady's back plait.

i. Heads of hair washed.
ii. Do. do. drawn even at point.
iii. Do. do. further cleansed.
iv. Do. do. finished into the different lengths.
v. Do. do. curled for a lady's wig.
vi. Natural curling hair for a gentleman's wig.
vii. Craped hair for a lady's frizzette.
viii. Woven hair, ready for wig-making.
ix. Partings of various descriptions for gentlemen's wigs.
x. Partings of various descriptions for a lady's bands.
xi. A gentleman's wig complete.
xii. A lady's band complete.
xiii. Frizzettes of various descriptions.
xiv. A gentleman's wig, as a specimen of knotting upon gauze.

3. A Specimen of Hair 74 inches long (6 feet 2 inches) supposed to be the longest and most extraordinary piece of human hair in the world. It was cut from the head of an English lady.

4. A Gentleman's Wig, manufactured entirely of human hair, no other material being used in it.

5. A Foundation for a Gentleman's Wig, made in one piece, and entirely of human hair.

6. Specimens of Human Hair dyed by Batchelor's 'Instantaneous Columbian Hair Dye' — H. & Sons sole wholesale agents.

[4562]

Howard, William, 23 *Great Russell Street, Bloomsbury.*—Introduction of gutta percha for securing the hairs in brushes.

Gutta Percha Secured Toilette Brushes.—The application of gutta percha prevents the decay of the wire and consequent loosening of the hair. No extra charge is made for this improvement. These brushes are specially adapted for first-class India and colonial trades.

Painting and stable brushes are made on the same principle.

Prices can be learned by application to the exhibitor.

[4563]

Jeffcoat, Joseph, 9 *Middle Queen's Buildings, Brompton.*—Painters' brushes, the bristles being tied on self-tightening principles.

Painters' Brushes, the bristles secured on self-tightening principles. The binding string used by the exhibitor being prepared by him to resist the action of water, oil, acids, &c., will not yield or burst by the swelling of the bristles when the brush is saturated in use, a fault very common with brushes as usually manufactured. The bristles of brushes tied with binding string thus prepared do not work loose; and they are laid upon a principle which prevents their working hollow or wearing swallow-tailed; and they are effectually secured by each of the many turns of the binding string round them exercising a much tighter grip upon the whole of the bristles than can possibly be obtained by the use of metal collars or patent and other sockets and bands. The durability and working qualities of brushes made upon these principles are well vouched for in the service of the contractors of this Exhibition building, of Messrs. Cubitt, Kelk, Lucas, Myers, Crace, Kershaw, and other eminent firms. The cost of the brushes to the consumer is not increased by the application of these improvements.

[4564]

KING, GEORGE, & SON, 116 *Bunhill Row, London.*—Brushes, more especially for manufacturing purposes.

[4565]

KOLLE, H., & SON, 65 *Queen Street, Cheapside.*—Seatings, curled hair, crinolines, drawn white hair, drafts.

[4566]

LOYER & SON, 33 *Gracechurch Street.*—Patent self-supplying water brush, or carriage varnish and paint preserver.

SELF-SUPPLYING WATER-BRUSH FOR CARRIAGES.

This apparatus is simple, and cannot get out of order. One end of an india-rubber tube is attached to a water tap, and the other end to the brush, through which the water flows continuously, and at once removes each particle of dirt as soon as detached, thereby preventing the varnish or paint being sanded or scratched. Its use effects an immense saving of time and labour.

A slight pressure of water is required, so that the flow may be more abundant.

Apparatus and brush, from 25*s*.

[4567]

MARSH, J., 175 *Piccadilly.*—Manufactured hair, perfumery, and brushes.

[4568]

MASON, THOMAS, 40 *Portland Street, Leeds.*—Transparent front and parting, hair restorative, and Turkish dye.

[4569]

METHERELL, JOHN KINNEARD, 47 *Carey Street, Lincoln's Inn, W.C.*—Extra light full-bottomed wig.

[4570]

NASH, THOMAS, JUN., Brush Manufacturer, 134 *Great Dover Street, Borough, London.*—Patent paint and other brushes.

[4571]

NIGHTINGALE, WILLIAM & CHARLES, *Wardour Street.*—Bed feathers and downs; horse hair, curled and spun.

[4573]

PEMBERTON, ABRAHAM, 15 *Broad Street, Worcester.*—Saddlery and stable brushes.

[4574]

Saville, H., *Leeds.*—Scalp with mechanical movement; specimens of invisible fabric for wig-making; &c.

[4575]

Smith, Augustus, *Wentworth Street, Whitechapel.*—Brushes and specimens of piassara.

[4576]

Taylor, Robert, 3 *Brunswick Place, Brompton Square, South Kensington.*—St. Neots Church, Hunts, worked with hair on glass; ornamental designs in human hair.

[4577]

Truefitt, H. P., 20 & 21 *Burlington Arcade.*—Specimens of wig-making &c.

[4578]

Truefitt, Walter, 1 *New Bond Street.*—Perukes, ladies' head-dresses, and articles in hair.

[4579]

Unwin & Albert, 24 *Piccadilly.*—Perukes and ornamental hair, combining art and nature, on the most perfect principles.

[4580]

Vickers & Short, 12 & 13 *Boar Lane, Leeds, Yorkshire.*—Perukes, scalps, ornamental hair, &c.

The exhibitor has always a stock of the above goods on hand, and makes to order when required. His establishment, known as the 'Acme of Fashion,' was established in 1804.

[4581]

Wall, Thomas, 3 *Upper Arcade, Bristol.*—Vase of artificial flowers, size of nature, worked in human hair.

[4582]

Watkins, C. A., 10 *Greek Street, Soho Square.*—Patent wire-bound painting brushes, round and oval, graining tools, shaving brushes, &c.

Patent Wire-bound Painting Brushes, round and oval.—The bindings with which these brushes are secured are of great strength and very light; the wire being galvanised after it is wound, the binding has the properties of a metal band without a joint.

Graining Tools, comprising badger softeners, motlers, overgrainers in sable and hogs' hair, fitches, sable and camel-hair pencils, stippling brushes.

Shaving Brushes, in ivory and bone handles.

Hair Brushes.

[4583]

Webb, Edward, *Worcester.*—Plain and figured hair seating, curled hair, cider and hop cloth, crinoline, fine cloth for buttons, &c.

[4584]

Whitfield, Samuel, & Sons, *Birmingham.*—Samples of purified and unpurified bed feathers.

[4585]

Williams, John, 46 *Westminster Road.*—Improved brooms and brushes, of various descriptions, for household and toilet use.

[4586]

WINTER, WILLIAM, 205 *Oxford Street, London.*—Ornamental hair, combining lightness, durability, and elegance with good workmanship.

W. WINTER'S PILUS REDIVIVUS restores the hair in all cases of sudden baldness or bald patches, where no visible signs of roots exist. Frequently one or more bald patches make their appearance in the hair, and, if neglected, spread over the head, causing entire and permanent baldness; but by the use of the above the hair is restored to its natural colour, if quite white, and becomes as strong as on any other part of the head. Price *5s. 6d.*

In W. WINTER'S UNION HEAD DRESSES the foundation is perfectly transparent; so that the hair appears to grow from the pores of the skin, and detection is impossible. Fronts, partings, and toupées on the same novel principle.

W. WINTER'S LIQUID HAIR DYE produces natural and permanent colours, from the lightest brown to black, without any green, purple, red, or other extraordinary tints, unpleasant odour, or the least injury to the hair or skin, leaving the hair softer and more glossy than before the dye was applied. In cases at *5s. 6d.*, *10s. 6d.*, and *21s.* Each colour is quite a different preparation.

W. WINTER'S QUININE BALSAM, the original preparation, invented and made only by him. The extraordinary effect produced by its use on dry heads of hair, where there is a want of tone and deficiency of natural support in the nutriment tubes of the hair, is well known. It not only causes the young short under hair to grow up strong and prevents the hair from falling off, but also prevents it becoming grey.

As there are numerous counterfeits, purchasers should observe that the original quinine balsam bears the name and address of WM. WINTER, 205 OXFORD STREET, near Portman Square, London, W., inventor of the celebrated genuine essential botanic extract, for cleaning the hair and eradicating the dandriff. This innocent and efficacious compound has been in use now for sixty years.

[4587]

WYATT, CORNELIUS, 1 *Conduit Street, Regent Street, W.*—Specimens of ornamental hair; various shades of dyed hair.

CLASS XXVI.

LEATHER, INCLUDING SADDLERY AND HARNESS.

SUB-CLASS A.—*Leather.*

[4618]

BATTY, JOSHUA, *Tottenham.*—East India and English sheep skins and seal fleshers.

[4619]

BEVINGTON & MORRIS, 67 *Cannon Street West;* Manufactory, *Blue Anchor Road, Bermondsey.*—Leather.

[4620]

BEVINGTON & SONS, 2 *Cannon Street West, and Neckinger Mills, Bermondsey, London,* Leather manufacturers.—Leather trophy.

[4621]

BOAK, ALLAN, 59 *West Port, Edinburgh.*—Rough and curried hog skins, tanned and patent leather.

[4622]

BOWEN & ATKINS, *Shipston-on-Stour, Worcestershire,* Tanners.—Pure-oak-bark-tanned sole, gaiter, and glove leather.

[4623]

BRITISH AND FOREIGN TANNING COMPANY Limited, 37 *George Street, Bermondsey, London.*—Sole leather.

[4625]

CLARK, JOSEPH, & SONS, 76 *Dean Street, Soho Square, London.*—Leather manufactures.

[4626]

COOPER, FREDERICK EDEN, *Brunswick Court, Artillery Street, Bermondsey.*—Leather for bookbinding and upholsterers' purposes.

[4627]

DEED, JOHN S., & SONS, 461 *Oxford Street, London.* (*See page* 12.)

[4628]

DRAKE, RICHARD, *Bristol.*—Oak bark and valonia tanned leather.

DEED, JOHN S., & SONS, 451 *Oxford Street, London.*—Moroccos, roans, skivers, curried leather, enamelled hides, sheep and Angora rugs.

Obtained the Prize Medal in 1851; *Two Prize Medals at New York,* 1853; *and a Bronze Prize Medal at Paris,* 1855.

Moroccos, in various colours and styles of finishing for carriage linings, furniture, bookbinding, casemaking, &c.

Calf skins, roans, and skivers in various styles, for bookbinding, case-making, &c.

Sheep and lambskin wool rugs, colours brilliant and fast.

Angora goatskins for hearth rugs, also for muffs, trimmings, &c.

Specimens of curried calf, kip, horse, and goat, for boots and shoes.

Curried hides, enamelled cow hides, and border hides for carriage and harness purposes.

[4629]

DRAPER, HENRY, *Kenilworth.*—Two butts, tanned two years. Soles from three butts, tanned two years.

[4630]

ESSEX, WILLIAM, & SONS, 28 *Stanhope Street, Strand, London.*—Leather for carriages, saddles, harness, army accoutrements, &c.

[4631]

FISHER, N., & SONS, 31 *Maze Pond, Southwark, S.E.*—Enamelled, curried, and coloured leather.

[4632]

FLITCH, J. J., & Co., Leather Manufacturers, *Leeds.*—Various descriptions of fancy leather.

[4633]

FRANKLIN, W. & J., Curriers, *Walsall.*—Leather for bridles, saddles, and harness.

[4634]

GEORGE, CLEMENT, 102 *Dean Street, Soho Square.*—Morocco and Russia leather for furniture and dressing-case purposes.

[4635]

HEMSWORTH, LINLEY, & WILKS, 30 *West Smithfield, London.*—Leather adapted for boots and shoes.

[4636]

HEPBURN & SONS, Tanners, Curriers, and Leather Manufacturers, *Bermondsey, London.*—Rough, dressed, and machine leather.

[4637]

HOLDEN, EDWARD THOMAS, *Walsall.*—Coach, saddle, and harness leather; japanned, enamelled, and coloured leather.

[4638]

HOLMES, THOMAS, & SON, *Antaby Road Tannery, Hull.*—Patent walrus-hide belting.

These belts have been well tested for upwards of five years in all the principal manufactories and saw mills in Hull (to which reference can be made), and are in use in London, Liverpool, Manchester, Leeds, Birmingham, Newcastle, Sheffield, and other places.

For strength and pliability, as well as from their being less liable to stretch (having undergone a powerful tension in process of manufacture), they are acknowledged by those who have used them, to be superior to any other article in use. They may be had of any thickness, from $\frac{3}{8}$ths to $\frac{5}{8}$ths of an inch, and of any width to twenty-four inches.

[4639]

HUDSON, SAMUEL, 65 *Dawson Street, Dublin.*—Saddlery, harness, &c.

Obtained Three Prize Medals at the Exhibitions of the Royal Dublin Society; also Three Honorary Certificates at the Exhibitions in Dublin, London, and Paris.

The following are exhibited:—
- A set of Pair-horse Harness.
- A set of Single Harness.
- Two Hunting Saddles.
- A Lady's Saddle, with Victoria leaping-head, invented by exhibitor.
- A Windsor Cramp.
- An American style Harness Saddle.

[4640]

HYDE, ARCHER, & Co., 7 *Finsbury Place, S.*—Collection of superior coach, harness, and saddle leather.

[4641]

JONES, W. H., & SON, 179 *High Street, Borough;* Manufactory, *Russell Place, Bermondsey.* Enamelled and curried leather.

[4642]

LAMBERT, BLAKEY, & MOWBRAY, *Bermondsey New Road.*—Blocked boot fronts, white and brown tops, jockey legs, Spanish cordovan, kip butts, Memel skins, &c.

[4643]

LEVER, JOHN, *Neate Street, Coburg Road, Old Kent Road.*—Vellums and parchments.

[4644]

LIDDELL, W. H., 135 *West Port, Edinburgh.*—Hog skins, harness, bridle, and patent leathers.

[4645]

LLOYD, T., 16 *Newcastle Street, Strand.*—Parchment, vellum, forrel, &c.

[4646]

MCRAE, J. & J., 43 & 46 *Bermondsey Street, and Mitcham Common, S.*—Sole leather, buff leather for army purposes, chamois leather, enamelled and japanned leather.

[4647]

MARSHALL, WILLIAM, & SON, Tanners, *Ladyburn, near Greenock.*—Saddlers' basils.

[4648]

MATHEWS, GEORGE, Leather Manufacturer, *Market Street, Bermondsey.*—Goat, calf, sheep, and seal skins; and horse hides.

[4649]

MATTHEWS, WILLIAM, *Spa Road, Grange Road, Bermondsey.*—Enamelled and patent leather.

[4650]

MOFFAT, JOHN, Tanner, *Musselburgh.*—Sample Scotch crop hides and curried leather.

[4651]

MONTGOMERY, G. F., *Dowgate Hill, London.*—Hides preserved, prepared, and tanned by Laperouse's process.

[4652]

MUNDY, W. P., *Tyers' Gateway, Bermondsey.*—Tanned East India kips and English calf skins.

[4653]

NORRIS & Co., *Shadwell, London, E.*—Leather for machinery, belting, hose pipes, and fire buckets.

[4654]

POOLE, JOHN & CHARLES, Curriers and Boot-Top Manufacturers, *Walworth Common, London.*—Boot tops, legs, and fronts.

[4655]

PULLMAN, ROBERT & JOHN, 17 *Greek Street, Soho, and Lostiford Mills, Surrey.*—Chamois, deer, and buffalo leather.

[4656]

RICHARDSON, EDWARD & JAMES, *Newcastle-on-Tyne.*—Furniture and shoe roans, kid, calf, seal, and enamelled leathers.

[4657]

ROBERTS, DANIEL & E. W., *Page's Walk, Bermondsey*, Tanners; and morocco, patent, enamelled, kid, calf, and sheep leather manufacturers.

[4658]

SAXTON, WADDINGTON, & CAREY, 85 *to* 89 *Bartholomew Close.*—Skins, kips, fronts, shoe legs, jockey legs, cordovan, grained calf.

[4659]

SHAW & MORRIS, *Wyld's Rents, Bermondsey.*—Patent, enamelled, and harness leather in black and colours.

[4660]

SIDEBOTTOM, ALFRED, Chemist, *Crown Street, Camberwell.*—Kid skins for gloves.

[4661]

SMITH, EUSEBIUS, *Camomile Street, London.*—Boot fronts, boot tops, jockey legs, enamelled horse and cow hides.

[4662]

SOMERVELL BROTHERS, *Ne*r*therfield, Kendal.* — Shoe and harness leather. (*See pages* 15 & 16.)

[4663]

SOUTHEY & Co., 16 *Little Queen Street, Lincoln's Inn Fields, W.C.*—Manufactured hides and skins.

[4664]

STOCKIL, WILLIAM, 37 *Long Lane, and* 69 *Bermondsey New Road.*—Curried and blocked calf leather.

[4665]

SUTTON, WILLIAM, Tanner and Currier, *Scotby Works, near Carlisle.*—Boot and shoe leather in Spanish cordovan, shoe hides and kips.

[4666]

TOMLINS, WILLIAM, *Black Swan Yard, Bermondsey Street.*—Morocco and sheep leather, parchment, and vellum.

[4667]

WILSON, WALKER, & Co., *Sheepscar, Leeds.*—Fancy sheep leather and coloured calf.

[4668]

WINSOR, GEORGE, & SON, 58 *Russell Street, Bermondsey, London.*—Sheepskin wool rugs of every description.

[4669]

YORKSHIRE LEATHER Co., 482 *New Oxford Street.*—Harness, bridle, and saddle leather.

[4670]

FOORD & MOIR, *Pontsburgh Tan Works, Edinburgh.* — Tanned and curried leather grained shoe butts and calf skin, &c.

[4671]

MEGSON, E., *Dean Gate, Manchester.*—Hose pipes.

SOMERVELL BROTHERS, *Netherfield, Kendal.*—First-class shoe and harness leather, manufactured especially for foreign markets.

For outside view of the Works, see Advertisement at end of Part 9.

SHOE LEATHER.

Light English Calf Skins and Kip Butts for foreign markets, £18, £24, and £30 per dozen.

These Skins are all selected so as to be free, or as nearly so as possible, from flaws and imperfections.

In some cases half of the Skins are thrown out when they come from the tan-yard, in others one-third, all of which are disposed of for other purposes.

English Calf Butts in various weights.
Kip Butts: East India, Petersburg, and English.
Grained Kip Butts for shooting boots.
Black and Russet Grained Hides.
Spanish Cordovan Hides.
English do. do.
Spanish Colt do.
English Sole Butts, bark tanned.
Foreign do. do. do. do.
English Shoulders, light, middling, stout.
Foreign do. do.
English Belly Middles.
Do. Insole Bellies.
Foreign do. do.
Curried Welt Shoulders.
Calf Kids, light, middling, stout.
Satin Kid Calf, do. do. do.
Sumac Grained Calf.
Bark do. do.
Grained Moroccos.
Do. Persians.
Russet Grained Calf.
Do. Lining do.

COLOURED ROANS.

Geranium, Green, Amber, Bronze, Blue, Violet, Cream.

BLACK SHOE KIDS.

White, Pink, and Blue Shoe Kids.

BOOT MOROCCOS.

Violet, Marone, Green, Geranium, and Black.
Coloured Shoe Moroccos in all shades.
Blue, Pink, and Black Lambs.

Enamelled Hides.
Enamelled Seals.
Peel Boot Skins.
White Sheep and Lambs.
White Wool Lambs.
Lining Basils, glass finished.
Kid and Chevalier Bindings.
Striped Seal do.
Striped Goat do.
Do. Cordovan do.
Black Persian do.

JOCKEY TOPS.

White, Brown, Mahogany, Amber.
Single Jockey Legs.
Wellington Fronts.
Albert do.
Clarence do.

GRAFTS.

Harness, Carriage, and Army Leather.
Black Harness Hides, light, middling, stout.
Do. Backs do.
Brown do. Hides do.
Do. do. Backs do.
Stirrup Hides and Butts.
Bridle do. do.
Black and Brown Rein Hides.
Do. do. Backs.
Brown Backs for Round Reins.
Russet Bridle Bellies.
Handpart Middlings.
Brown Collar Backs.
Brown Strap Hides.
Skirt Hides.
Skirt Shoulders.
Seat Shoulders.
Hog Skins.
Covering Skirts and Flaps.
Solid do. do.
Chaise do. do.
Housings.
White Horse Hides.
Buff Army do.
White Enamelled Hides.
Chamois.
Enamelled Hides.
Patent Middlings.
Patent Horse Hides.
Do. do. Butts.
Do. Bag Hides.
Do. Splits.
Do. Sheep.
Do. Calf.
Bag Hides.
Black Chaise Hides.
Collar Basils.
Strained do.
Lining do.

For list of prices apply to SOMERVELL BROTHERS, Netherfield, Kendal; who have besides a distinct branch for the manufacture of first-class Closed Boot and Shoe Uppers and Leggings in 400 different kinds.

See Catalogue for Class XXVII., p. 58, for description and sketches.

Somervell brothers, *Netherfield, Kendal.*—Shoe and harness leather—*continued.*

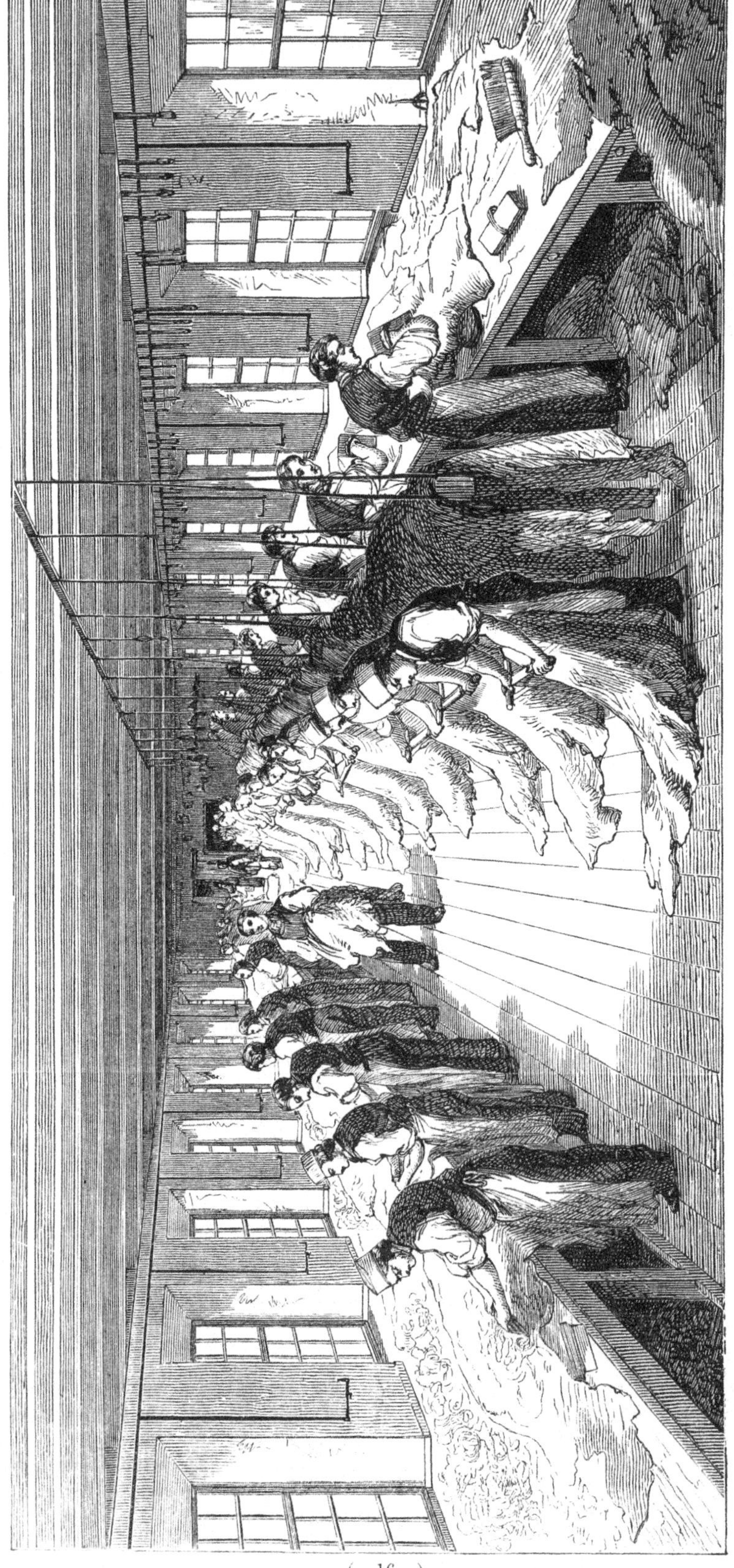

ONE OF THE LEATHER-DRESSING ROOMS, NETHERFIELD, KENDAL.

SUB-CLASS B.—*Saddlery, Harness, &c.*

[4680]

ANGUS, JOHN, & Co., 131 *Trongate, Glasgow.*—Saddle trees, wood and iron cart hames, and saddlery chains.

[4682]

BANTON, EDWARD, *Walsall.*—Saddlers' ironmongery &c., including saddles, harness, hunting bridles, breastplates, and martingales.

[4683]

BARTLEY, CHARLES ALFRED, 20B *Portman Street, Portman Square, W.* — Harness and saddlery.

[4684]

BLACKWELL, Samuel, 259 *Oxford Street.*—Saddlery and harness. (*See pages* 18 *and* 19.)

[4685]

BLYTH, ROBERT, & SONS, 4 *Park Lane, W.*—Ladies' saddle and Somerset saddle, harness pad, &c.

Prize Medal, London, 1851; *Honourable Mention, Paris,* 1855.

Lady's Saddle, with horizontal elastic seat.
Improved light Somerset Saddles.
Ladies' and Gentlemen's Riding Bridles, and Harness Pad.

[4686]

BOURNE, THOMAS, 5 *College Road, Cork.*—Set of carriage harness, with improved tug buckles.

[4687]

BRACE, HENRY, *Walsall.*—Saddles, bridles, harness, and horse appointments.

[4688]

BRILEFOLD, CHARLES, 21 *Wellington Street,*—Patent saddle trees.

[4689]

BROWN & SON, 7 *Moat Row, Birmingham,* Manufacturers of every description of saddle trees.

Awarded the Prize Medal in 1851.

[4690]

CALLOW, THOMAS, & SON, 8 *Park Lane, Hyde Park Corner.*—Whips for riding, driving, and hunting.

[4691]

CAMPBELL, JAMES, & Co., *Adams Row, Walsall.*—Ladies' and hunting saddles, deer saddles, harness, bridles, bits, stirrups and spurs, &c.

[4692]

CARTER, LIEUT.-COLONEL, *Monmouth.*—Perfected harness. (*See page* 20.)

[4693]

CLARK, WILLIAM, & SON, *Leeds* —Assortment of saddlery, harness, and horse clothing, for home and exportation.

BLACKWELL, SAMUEL, 259 *Oxford Street.* — Saddlery and harness, with four patented improvements, gutta percha crib straps, dumb jockeys, and india-rubber springs.

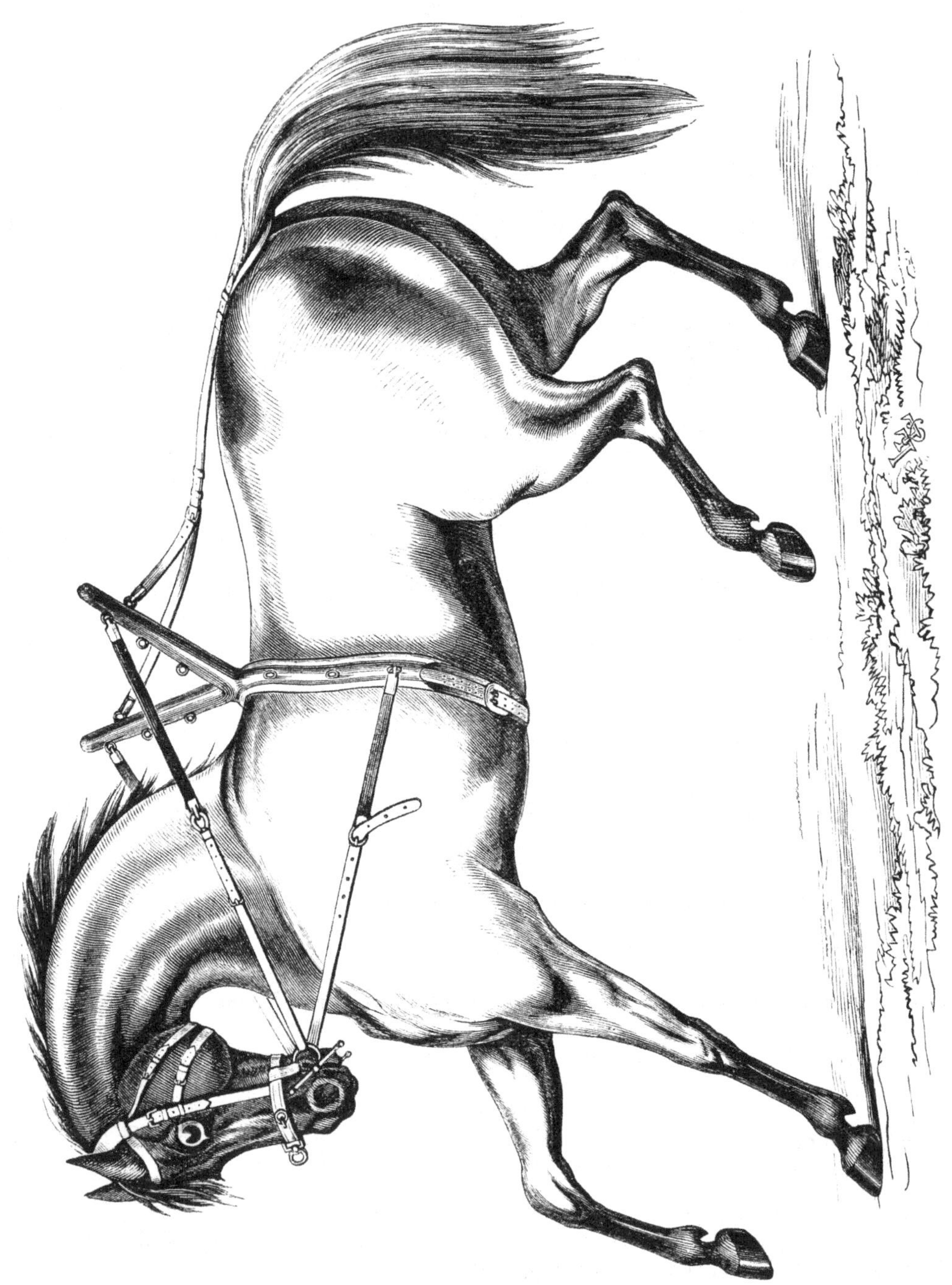

The following are exhibited:—

PATENT GUTTA-PERCHA JOCKEYS, elastic, for breaking horses easy-mouthed and temperate, and exercising in frosty weather in loose boxes and stalls, and on led horses; preventing falling and broken knees. They yield easily if a colt rolls over on his back, and are not injured as with the old wooden jockey. They are fitted with elastic vulcanised rubber springs of varied strengths to the reins, of from 3 lbs. to 10 lbs. each, making the total pull of the four springs 40 lbs. A few very violent horses require two sets of springs, making a pull of 80 lbs. These jockeys are used by the first breeders and owners of horses in the kingdom and abroad: above 3,000 are now in use. Price from 56*s.* to 60*s.*; on hire, 2*s.* per week, with option of purchase.

BLACKWELL, SAMUEL, 259 *Oxford Street.*—Saddlery and harness—*continued.*

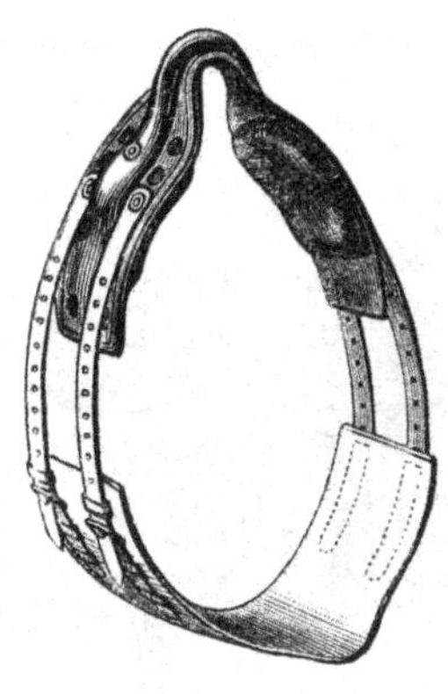

PATENT CRIB STRAP.

VULCANISED RUBBER SPRINGS for saddle straps, girths, rollers, martingales, and any part of saddlery and harness where elasticity is useful, 2*s.* each.

HOOKS (vulcanised rubber spring) in brass, iron, galvanised and German silver, of all strengths, for reins, pillar reins, chains, ropes, and where spring hooks are used, from 9*d.*

A large assortment of RIDING, MILITARY, ARTILLERY, DRIVERS', and DOG WHIPS, with springs, making dog leads and couples.

WEB, FETLOCK, SPEEDY, SPLINT, LEG, HOCK, HOOF, and STRENGTHENING BOOTS, 3*s.* to 9*s.* TRAVELLING KNEE BOOTS, 8*s.*; with patent springs, 10*s.*; and rubber knee caps, 12*s.* SANDAL-JOINTED HUNTING SHOES.

SAFETY SPRINGS, for riding and driving reins, to both bits, and one rein in hand, to act on the bradoon, and when the horse pulls hard the spring stretches, and the rein then acts on the curb. For light-mouthed horses in harness that occasionally run away, they are invaluable, as they act as two pairs of reins. Ladies and timid riders cannot use the wrong rein. 8*s.* and 12*s.* per pair. 1,000 in use. Springs to bearing, side, and gag reins, to allow the horse to lower his head.

ANTI-CRIB-BITER, elastic, of gutta percha, iron, and leather, constructed not to injure the mane. A sure preventive against crib-biting keeping a crib biter in condition. 18*s.*

Exhibited at the Museum of Patents, South Kensington.

Agents:—AULD, Saddler, Quebec; MORRIS, Saddler, Montreal; HENOQUE & VANWEARS, 56 Rue de Paradis Poissonnière, Paris; BRAY & HAYES, Merchants, Cornhill, Boston, U.S.; Mr. VERGOE, Merchant, 41 Collins Street West, Melbourne, Australia.

[4694]

COOPER, MATTHEW, 2 & 3 *Railway Street, York.*—Hunting saddles, steeple-chase saddle, side saddle, and horse clothing, &c. &c.

[4695]

COWAN, JAMES, Saddler and Harness Maker, *Barrhead, and Union Street, Glasgow.*—Scotch cart and van harness.

[4696]

CUFF & SON, 18 *Cockspur Street.*—Saddlery.

[4697]

DAVIS, ALEXANDER, 33 *Strand.*—Saddles, harness, bridles, &c. (*See pages 22 and 23.*)

CARTER, LIEUT.-COLONEL, *Monmouth.*—Harness on principles securing the steadiest draught and greatest power of management.

By Royal Letters Patent.

LIEUT.-COLONEL CARTER'S PERFECTED HARNESS.

The advantages gained by this harness are:—

1st. The chief power of control is in the hame of the collar, where the points of the shafts terminate and are attached; consequently the awkward, unsightly, and dangerous projections of the points of the shafts are got rid of.

2nd. Leather traces are dispensed with.

3rd. A small-sized pad only is necessary.

4th. The tugs are only one inch in diameter.

5th. Encircling the shaft with a portion of the back-band is not required.

6th. The straps to the pad and the surcingle are to the horse as one strap.

7th. The objectionable crupper-dock is, or can be, dispensed with.

8th. Should the horse fall, the shafts (which are the traces) are not liable to fracture.

9th. In crowded thoroughfares the shafts protect the horse from injury.

10th. The horse cannot touch the shafts, nor can he kick over them or reach the splash-board.

11th. As the pad is placed in the centre of the horse's back, the girths do not interfere with the action of the fore-legs, and the motion of the shafts is the least possible.

12th. The carriage is turned, not by a push and a strain upon the pad, but by the collar.

13th. For retarding, stopping, or backing the carriage, the pull back is as direct as the pull forward, and in its action is almost as immediate.

14th. When retarding the carriage, ventilation is obtained under the collar.

15th. The powers of the horse—as relating simply to horizontal draught—are usually taxed in five different ways:—(1) He draws the carriage with the collar; (2) he turns it by pushing with his shoulder and side, and he stops it conjointly with (3) his thighs, (4) his tail, and (5) his withers. In the 'perfected harness' the horse has but to learn two things—to pull forward with the collar, and to pull back with the breeching.

16th. The horse is put to and removed from the carriage without the necessity of either being moved from their respective positions.

17th. The twisting of the reins, when driving, is prevented by an improved terret.

[4698]

DEER, FREDERICK A., *Neath, Glamorganshire.*—Saddles, harness, and ornamental leather frame, on an improved principle.

[4699]

DOVEY, FRANCIS, 30 *Brownlow Street, Drury Lane, London.*—Saddle trees.

These saddle trees are made of well-seasoned timber, and are moderate in price. A large stock of trees of various qualities and patterns is always kept by the exhibitor.

[4700]

DUFFY, JOSEPH, Saddler, *Market Harborough, Leicestershire.*—Economical safety collar, for all purposes.

[4701]

DUNLOP, JAMES, Harness Maker, *Haddington.*—Farm harness for two horses in cart and plough, with expanding neck collar.

[4702]

ELLAM, BENJAMIN, 213 *Piccadilly.*—Whips of every description, saddles, bridles, horse clothing, harness, military appointments, &c. &c.

Among the articles in this case are a great variety of rich race prize whips, of the newest designs; ladies' riding whips, also of novel construction, with fan or sun-shade attached; ladies' and gentlemen's improved chowrie or Arab riding whips, with horse-hair plumes, especially adapted for India or other parts where horse and rider are subject to annoyance from insects; ladies' and gentlemen's riding whips of entirely new patterns and devices, and all of excellent workmanship; state carriage and postilion whips; ladies' and gentlemen's driving whips of new patterns and extraordinary finish; driving whips with horns and warning whistles in the handle; prize or gift hunting whips, with sporting devices; riding canes with novel mountings.

Saddles, bridles, horse clothing, harness, and every requisite for the stable, manufactured of the very best materials at exceedingly moderate prices. Merchants, shippers, and saddlers purchasing at the above establishment will find a great advantage.

As a proof of the superior quality of the saddlery and whips manufactured by B. ELLAM, see paragraphs on articles made for the Emperor and Empress of France in the following leading journals:—*Bell's Life, Illustrated London News, Daily News, The Era, Morning Advertiser, Sunday Times, The Atlas, Grindlay's Home News, The Globe, The Moniteur*, and most of the leading journals of France and England.

[4703]

GARDEN & SON, 200 *Piccadilly.*—Saddles and harness.

[4704]

GARNETT, WILLIAM, 4 *Bridgeman Place, Walsall.*—Four gentlemen's saddles, one lady's saddle, one set of gig harness.

[4705]

GIBSON & Co., Saddlers &c. to the Queen, 6 *New Coventry Street, Leicester Square.*—Saddlery &c.

[4706]

GORDON, ALEXANDER, 39 *Lisle Street, Leicester Square, and* 99 *Piccadilly.*—Harness and saddles.

[4707]

GRAY, EDWARD, 44 *High Street, Sheffield.*—Saddles and harness.

[4708]

GREATREX, CHARLES, & SON, *Walsall.*—Harness, saddlery, whips, &c.

DAVIS, ALEXANDER, 33 *Strand, London.*—Saddles, harness, horse clothing, bridles, ladies' saddles, &c.

1. SIDE SADDLE, embroidered in coloured silks on pig-skin, after a mediæval design.

Embroidery, as applied to saddlery, has generally been employed by letting the finished parts of embroidered work into pieces cut out of the skin. In this instance the embroidery is worked into the skin itself, and thus forms part of the material. Care has been taken to ensure its general application, at a small cost, after special designs (such as family crests, arms, &c.); and this very novel and beautiful saddle will be likely to lead to an extension, hitherto wanting, of the higher class of ornamental art to this branch of manufacture. Attention is also drawn to the fine lines of the saddle itself, and the fit, as applied from the latest practical experience and works of the best authorities.

Designs and estimates will be furnished on application.

2. HUNTING AND RIDING SADDLE. Pig-skin.

Attention is drawn to this class as exhibiting beauty of shape, especially with the object of acquiring the greatest ease and comfort to the rider, and fitting the horse with exactness in its various points of bearing. Workmanship and materials specially deserve notice.

3. PAIR-HORSE HARNESS FOR PHAETON. Furniture silver on German silver.

Attention is drawn to the very fine quality and texture of the leather, to the light and yet substantial forms of the various portions of the harness, the bridles, pads, and martingales. The bits, furniture, and ornaments have been especially designed with a view to grace and strength united, and particularly to give true and artistic forms in place of the fantastic ornament much employed. The collars, hames, and tugs are suited to give the horse the least fatigue in drawing the vehicle. Workmanship and finish are of the most superior kind.

Prices of the various descriptions of harness at the manufactory or branches.

4. SINGLE-HORSE HARNESS. Brass furniture.

Same as above in quality, and general notice of otherwise distinctive features.

5. A SUIT OF WINTER HORSE CLOTHING. Kersey, bound with cloth, stitched with silk. Initials embroidered.

This clothing, after a novel and beautiful design, is the most calculated to give warmth to the horse, fitting him comfortably and closely; also to give a beautiful appearance when clothed. The kersey is beavered, a new process giving great gloss and finish to the pattern; the roller padded to fit the horse with ease, and keep the cloth in its place.

Prices on application.

Davis, Alexander, 33 *Strand, London.*—Saddles, harness, &c.—*continued.*

6. A Suit of Summer Horse Clothing, from a new design. Newmarket kersey, bound with cloth, and stitched with silk.

This material facilitates the perspiration of the horse, and is superior to linen, which absorbs the same.

Remarks as to other distinctive features as above.

General Features.

In all manufactures emanating from this factory, there are three points to which general and particular attention is paid, viz.:—

1. Beauty of form, designers being employed to produce new shapes, and handsome forms.

2. Quality of materials employed

3. Lowness of price.

Every description of saddlery, harness, horse clothing, bridles, cart and railway harness, steeple-chase, racing and ladies' saddles, is made on the premises.

Printed price lists kept, which may be procured on application.

Especial facilities for carrying out new patterns to particular orders.

Goods shipped to all parts of the world by the cheapest and best routes. Particulars as to freight and general cost will be given to trade buyers.

Price lists on application.

On parle Français.
Deutsch gesprochen.

[4709]

Green, Robert, 8 *Edwards Street, Portman Square.*—Saddlery and harness on an improved principle, combining elegance and utility.

	£	s.	d.		£	s.	d.
Best hog-skin saddles complete .	4	0	0	Best double bridles	1	0	0
Ditto, second ,, . .	3	10	0	Best patent collars	0	18	0
Ditto, soiled ,, . .	3	3	0	Large horse blankets	0	12	0

Best Pair-Horse Silver Harness, 20 Guineas; Single ditto, 9 Guineas.

[4710]

Hargraves, John, & Son, *Carlisle.*—Horse clothing. (*See page* 24.)

[4711]

Hawkins, John V., 26 *Francis Street, Tottenham Court Road, W.C.*— Ladies', gentlemen's, and children's saddles.

[4712]

Haynes & Son, 27 & 28 *Brownlow Street, Long Acre.*—Saddle trees.

[4713]

Henton & Son, 7 *Bridge Street, Lambeth.*—Saddlery and harness. (*See page* 25.)

[4714]

Hesketh, James, Saddler, *Ashton-upon-Mersey, Cheshire.* — Four sets of harness for agriculture; van and pony harness.

[4715]

Hinkson, J., 76 *Dame Street, Dublin.*—Saddlery and harness.

[4716]

Hodder, Cæsar, 8 *Nelson Street, Bristol.*—Park and hunting saddles; saddles for single or double harness.

[4717]

Holgate, J., & Co., 33 *Dover Road, Southwark.*— Saddlery and harness.

HARGRAVES, JOHN, & SON, *Carlisle.*—New styles of horse clothing, girth, roller, brace and shoe webs.

1. The PONCHO HORSE SHEET (Registered) combines sheet and breast cloth in one garment, which is close at the breast, and draws on over the head.

For comfort to the horse, elegance of appearance, durability, and cheapness, the poncho is unequalled.

2. The DOUBLE-BREASTED NEWMARKET is close at the breast, and fastens with a sliding strap and buckle.

3. Specimens of girth webs, roller webs, brace and belt webs, machine webs, and shoe webs.

Price for the sheet as shown upon the model horse, £1 17*s.* 6*d.*; for the suit complete, £4 4*s.* Crests and initials extra.

[4718]

HOLMES, *Derby, Lichfield, and London.*—Double and single carriage harnesses, for private use.

[4719]

HOLMES, HERBERT MOUNTFORD, JUN., *London Road, Derby.*—Pillar rein and saddle drier.

[4720]

HOOD & STEPHENSON, *Dunse, Berwickshire, N. B.*—Complete set of agricultural harness for pair of horses; complete set of gig harness; riding saddle and bridles.

[4721]

HOUGHTON, GEORGE, *Tewkesbury.*—Elliptical spring-seat saddle, and tree showing action of spring.

[4722]

JACKMAN, JOHN, Saddle, Bridle, and Harness Manufacturer, 110 *Wardour Street.*—Bridles, holsters, and saddle-bags.

[4723]

LANE, HENRY WILLIAM, 3 *Little Compton Street, W.*, Saddle-tree manufacturer.

HENTON & SON, 7 *Bridge Street, Lambeth.*—Saddle and harness, patent elastic saddle.

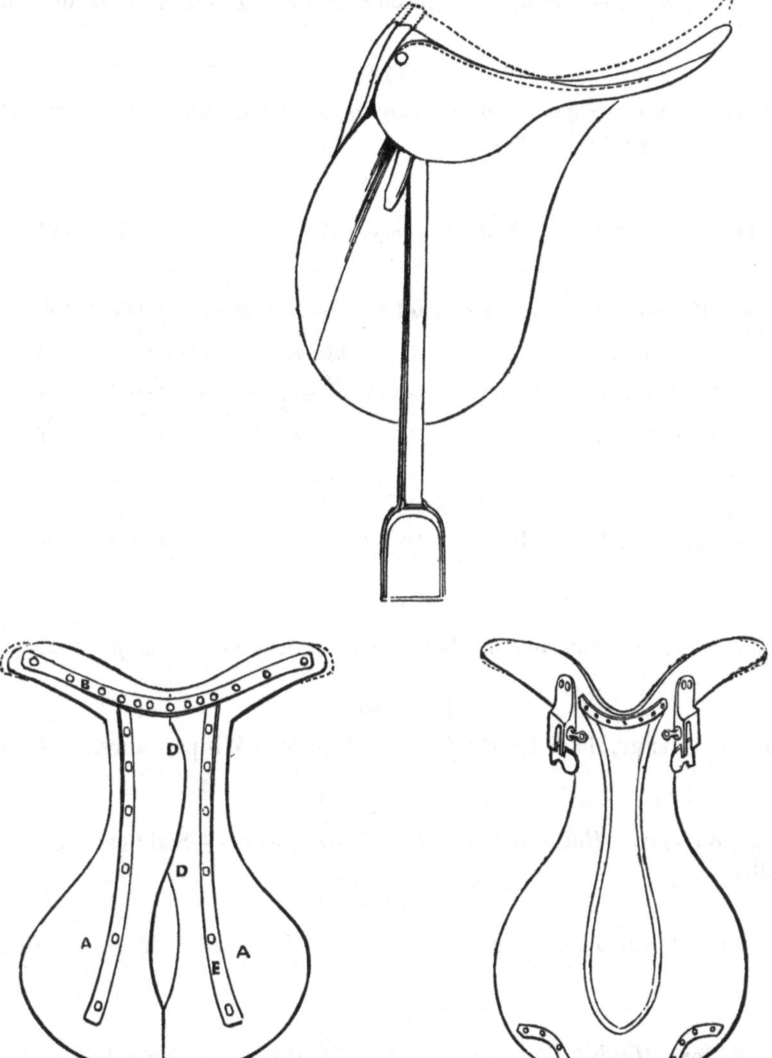

In pursuing horse exercise, it is essential that the saddle should possess such amount of elasticity as to be able to accommodate itself to the action of both the horse and the rider.

The exhibitors have therefore manufactured and patented an IMPROVED RIDING SADDLE, which combines great firmness with flexibility, and is constructed at the same time with so much simplicity as to render it well deserving the consideration of all lovers of horse exercise. HENTON & SON'S principle is to have fixed points of pressure only where actually necessary,—that is, at the head or pommel, and at the cantle or back part,—leaving the intermediate space elastic by the simple contrivance of a strong leather foundation, with two steel plates acting as springs on either side, between the head and cantle. In the specification of the patent it is stated that H. & S. dispense with the antiquated wooden trees hitherto used, and substitute an elastic foundation.

The advantages gained by this mode of construction over the old form of the stiff wooden tree, are many, and may be thus enumerated:—

1. A more accurately fitting saddle.
2. A more perfect bearing and distribution of the weight upon the back of the animal.
3. Chafing and sore back are thus to a great extent, if not entirely, avoided.
4. A firmer and more comfortable seat, with a greater power of purchase.
5. Less fatigue to the horse, arising from the elastic action of the saddle.
6. Increased ease and a more agreeable motion to the rider.

The dotted lines represent the extent of the spring.

To outward appearance, this saddle presents nothing different from the ordinary saddle. But scarcely is the rider seated and tries the various paces of the horse, when he is almost immediately aware that there is a very considerable difference in the motion; whatever the pace, whether riding, walking, or trotting, the elasticity produces an easier action for the rider, and relieves the dead weight from the horse's back, and thus prevents galling.

The price of this patent saddle is the same as the ordinary wooden-tree saddle, viz. £5 10*s.*

[4724]

LANGDON, MESSRS., 9 *Duke Street, Manchester Square, London.* — Improved side-saddles and harness.

[4725]

LEA, CORPORAL MAJOR, *Royal Horse Guards.* — Collar for prevention of crib-biting, practically proved to be effectual.

[4726]

LENNAN, WILLIAM, Saddler to Her Majesty, 29 *Dawson Street, Dublin.* — Saddlery, harness, &c.

Set of Pair-horse Harness, richly chased and plated on solid nickel silver.
Set of light Pair-horse Harness.
Set of Brougham Harness.
Lady's Saddle, quilted all over, with spring leaping-head and safety slipper.
Gentleman's Somerset Saddle, quilted all over.
Two Hunting Saddles, full shafted.
One light Steeple, or Park Saddle.
Double-rein Snaffle Bridle for preventing running away.
German Rider, with breaking tackle.
Set of Trotting Harness, complete, only 8 lbs. weight.
Race Saddle, mounted complete, only 2 lbs. weight.

[4727]

McDOUGALL, ARCHIBALD, 11 & 200 *Upper Thames Street, City.*—Van harness and cart harness.

[4728]

McNAUGHT & SMITH, *Worcester.*—One pair-horse harness &c. (*See page* 27.)

[4729]

MARTIN, WILLIAM HENRY, 64 & 65 *Burlington Arcade.*—Whips, canes, sticks, &c.

[4730]

MIDDLEMORE, WILLIAM, *Holloway Head, Birmingham.* — Saddlery &c. (*See pages* 28 *and* 29.)

[4731]

MERRY, SAMUEL, 21 *St. James's Street, London.*—Harness, bridles, saddles, and horse clothing.

[4732]

MORE, JOHN, & SON, *Market Street, Finsbury.*—Double and single harness, pads, collars, round reins, pole pieces.

[4733]

NANSON, ROBERT, *English Street, Carlisle, Cumberland.*—Ladies' and gentlemen's saddles and portmanteaus.

[4734]

NICHOLSON, WILLIAM HENRY, JUN., 57 *Market Street, Manchester.*—Lady's side saddle and gentleman's hunting saddle.

[4735]

NICKOLLS, GEORGE ALBERT, 1 *Oxford Market, London.*—Harness crupper, which prevents kicking and makes the horse carry its tail gracefully.

THE CULERON ELEVATEUR, or IMPROVED HARNESS CRUPPER DOCK.

The great advantages obtained by the use of NICKOLLS' Improved Harness Dock consist in making the horse carry its tail well, gracefully, and with perfect ease. It is also a preventive against kicking, can be attached to any crupper, cannot be seen when worn, and never galls. The invention has been pronounced by competent judges to be a perfect success. Price 12*s.* 6*d.*

McNAUGHT & SMITH, *Worcester.*—One pair-horse harness, two single ditto, one West India mule harness.

No. 1.—A LIGHT PAIR-HORSE CARRIAGE HARNESS, with swage furniture, best plated on German silver, single buckles, martingales, bearing and driving reins, and trace bearers £28

No. 2.—A HANDSOME SINGLE-HORSE BROUGHAM HARNESS, with split hip-strap, breeching, martingale, bearing and driving reins, mounted with best Chatham furniture plated on German silver, with double buckles . . . . . £12 12*s.*

No. 3. A LIGHT SINGLE-HORSE DOG CART OR PHAETON HARNESS, with long breeching and kicking strap, new patent safety tugs, martingale, bearing and driving reins, mounted with wire furniture all best plated on German silver . . £11 11*s.*

No. 4.—A STRONG WEST INDIAN MULE HARNESS, with galvanised furniture, leatherwork, copper riveted. In quantities. Per set . . . £2 10*s.*

[4736]

OERTON, FRANCIS B., *Walsall.*—Saddlery, harness, and saddlers' ironmongery.

[4737]

OLDFIELD & SON, 1 *Motcomb Street, Belgrave Square.*—Saddlery and harness.

The exhibitors have devoted great care and attention to the manufacture of their saddles, and have availed themselves of all the latest improvements to ensure ease to the rider and a perfect fit to the horse. In the harness department they exhibit sets possessing the elegance and richness required for state and town wear. In harness for general use, they have directed their efforts to combine lightness with durability.

[4738]

OWEN, JOHN A., 7 *Lisle Street, Leicester Square,* Saddler and harness maker.

[4739]

PEARL, JAMES JOHN, 2 *Friendly Place, Old Kent Road.*—Pad cloths, fronts, and rosettes.

[4740]

PEAT, HENRY, Saddler, 14 *Old Bond Street, London.*—Ladies' and gentlemen's saddles and harness.

[4741]

PETCH, THOMAS, 40 *Albert Street, Hampstead Road, London, N.W.*—Gig harness, saddles, pads, &c. Ditto tops and trees.

[4742]

RAND & BECKLEY, 297 *Oxford Street, London.*—Saddlery and harness.

[4743]

SHATTOCK, JAMES M., & Co., *Bristol.*—Gig, carriage, and American buggy harness, saddles, and saddle trees.

[4744]

SHIPLEY, JOHN GEORGE, 181 *Regent Street.*—Whips, saddlery, harness.

[4745]

SMITH, R., & Co., 1 *Beech Street, City.*—Riding and driving whips, hunting crops, and walking canes.

MIDDLEMORE, WILLIAM, *Holloway Head, Birmingham.*—Saddlery, harness, and saddlers' ironmongery.

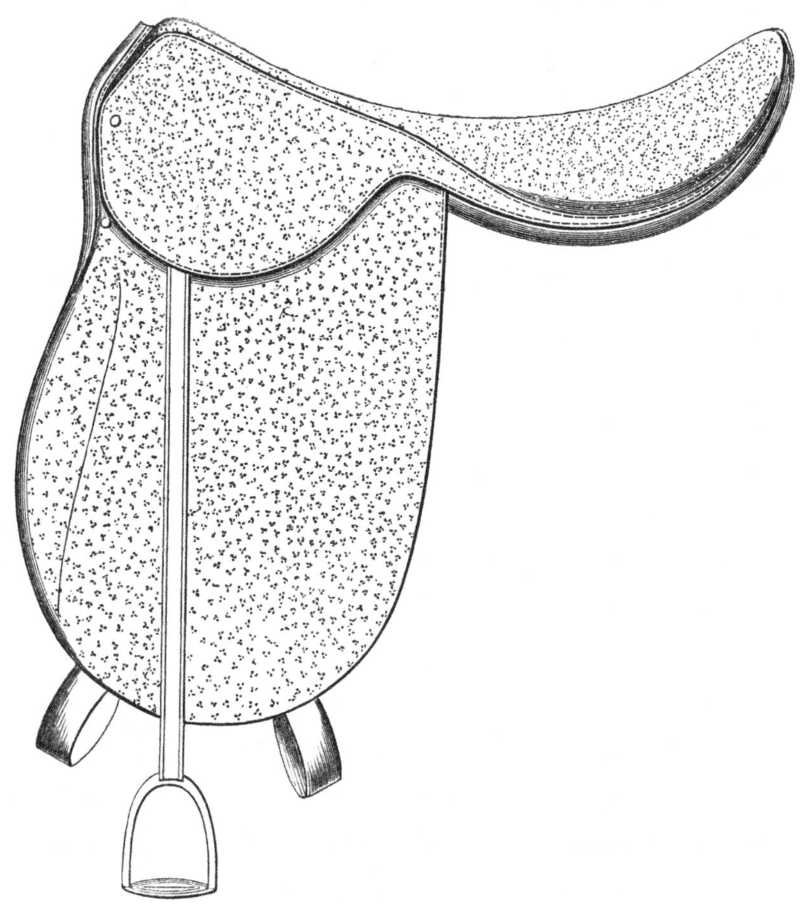

PATENT SADDLE.

MIDDLEMORE, WILLIAM, *Holloway Head, Birmingham.*—Saddlery—*continued.*

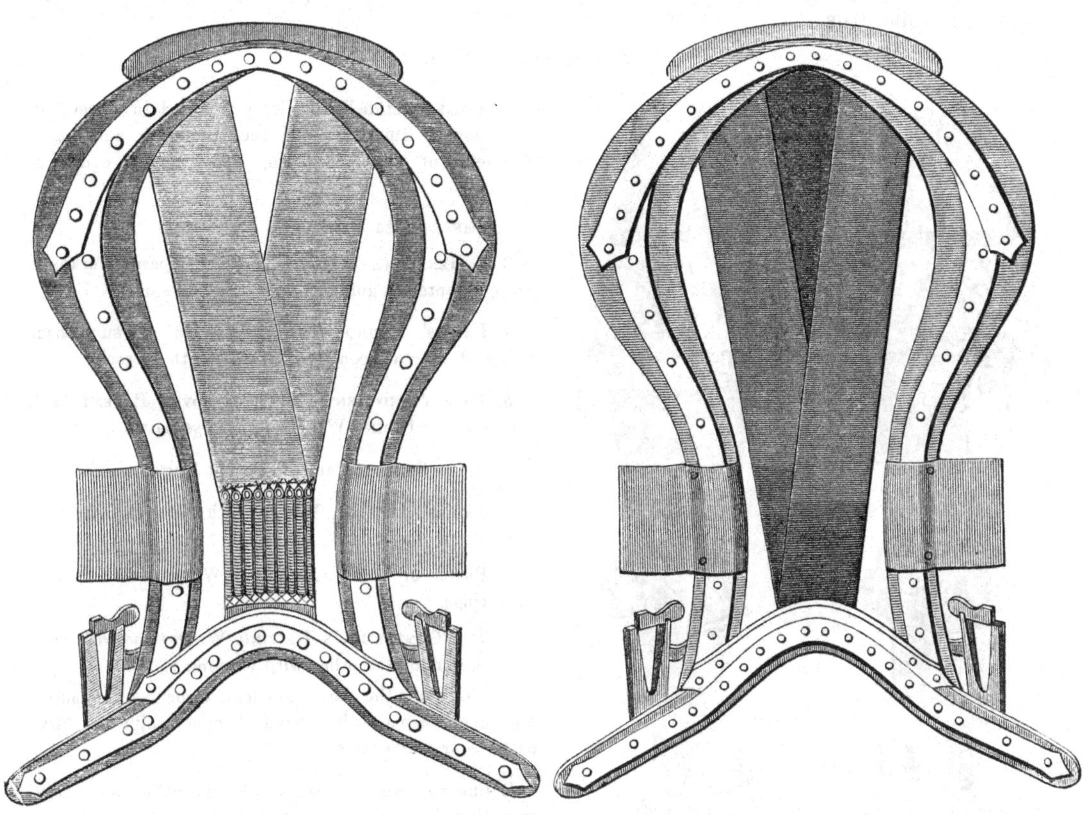

ALBERT SPRING-SEAT SADDLE TREE.

ELASTIC-SEAT SADDLE TREE.

Foundation, explaining the principle of the 'PATENT ALBERT SPRING-SEAT SADDLE,' viz. the insertion of spiral springs in the webs of the tree.

Foundation, explaining the principle of the 'PATENT ELASTIC-SEAT SADDLE,' viz. the substitution of india-rubber instead of linen web for the foundation of the tree.

Hunting saddles made up with either of the above trees.

[4746]

SWAINE & ADENEY, 185 *Piccadilly.*—Whips, thongs, and canes. (*See page* 30.)

[4747]

TIBBITS, JOHN, & SON, Manufacturers, *Walsall.*—Bits, stirrups; ladies', hunting, military, and harness bridles; heraldry and harness mountings.

[4748]

URCH & Co., 84 *Long Acre.*—Saddlery, harness, and horse clothing, &c. (*See page* 31.)

[4749]

WEIR, JOHN, Saddler and Harness Maker, *English Street, Dumfries.*—Riding saddles, cart and gig harness.

[4750]

WHILLOCK, DANIEL, 24 *Tabernacle Row, St. Luke's.*—A youth's improved saddle; a young lady's ditto, adapted for riding either side.

SWAINE & ADENEY, 185 *Piccadilly, London, W., opposite Burlington House,* Manufacturers to the Queen and Royal Family.—Whips of all descriptions, thongs, canes, and sporting apparatus.

Obtained a Prize Medal at the Exhibition of 1851.

1. PRIZE RACING WHIP, richly mounted in silver gilt; the mounts illustrative of the universal and pacific character of the Exhibition, and also of equestrian sports.

2. RIFLE PRIZE WHIP.

3. A GENTLEMAN'S RIDING WHIP, of superior workmanship, mounted in gold, set with precious stones.

4. LADIES' RIDING WHIPS, with fan or sun shade attached, of novel construction, also with parasol.

5. LADIES' AND GENTLEMEN'S IMPROVED PATENT ARAB OR CHOWRIE RIDING WHIPS, with horse-hair plumes.

6. STATE CARRIAGE and POSTILION WHIPS.

7. LADIES' DRIVING WHIPS, with parasols attached, elegantly mounted.

8. PRIZE or GIFT HUNTING WHIPS, with superb mountings.

9. TWO GENTLEMEN'S DRIVING WHIPS, one with mail horn in handle, the other with warning whistle.

In addition to the above, a general assortment of ladies' and gentlemen's riding and driving whips of new patterns or extraordinary finish.

Riding and walking canes, with beautiful and novel mountings.

Ladies' and gentlemen's driving whips, made from rhinoceros horn, handsomely mounted in gold and silver.

Rhinoceros horn riding whips and walking sticks.

Australian stock whips.

Depôt in Paris,—Messrs. DARRÉ & TEXIER,
5 Rue du Faubourg St. Honoré.

[4751]

WHIPPY, STEGGALL, & CO., *North Audley Street.*—Saddles and harness.

[4752]

WHITE, JAMES CHADNOR, *Liverpool Street, City, London.*—Harness. (*See page* 31.)

[4753]

WILKINSON & KIDD, 257 *Oxford Street.*—Saddles and harness.

[4754]

WILLIAMS, WILLIAM EVAN, *High Street, Wandsworth, Surrey.*—Elastic frame patent leather horse collars; harness and saddlery.

[4755]

WRIGHT, SAMUEL, *Stowmarket, Suffolk.*—Set of gig harness; plated furniture, light and very strong.

URCH & Co., Army, Hunting, and Colonial Saddlers and Harness Makers, 84 *Long Acre, London.*—Army and hunting saddlery, harness, and horse clothing, &c.

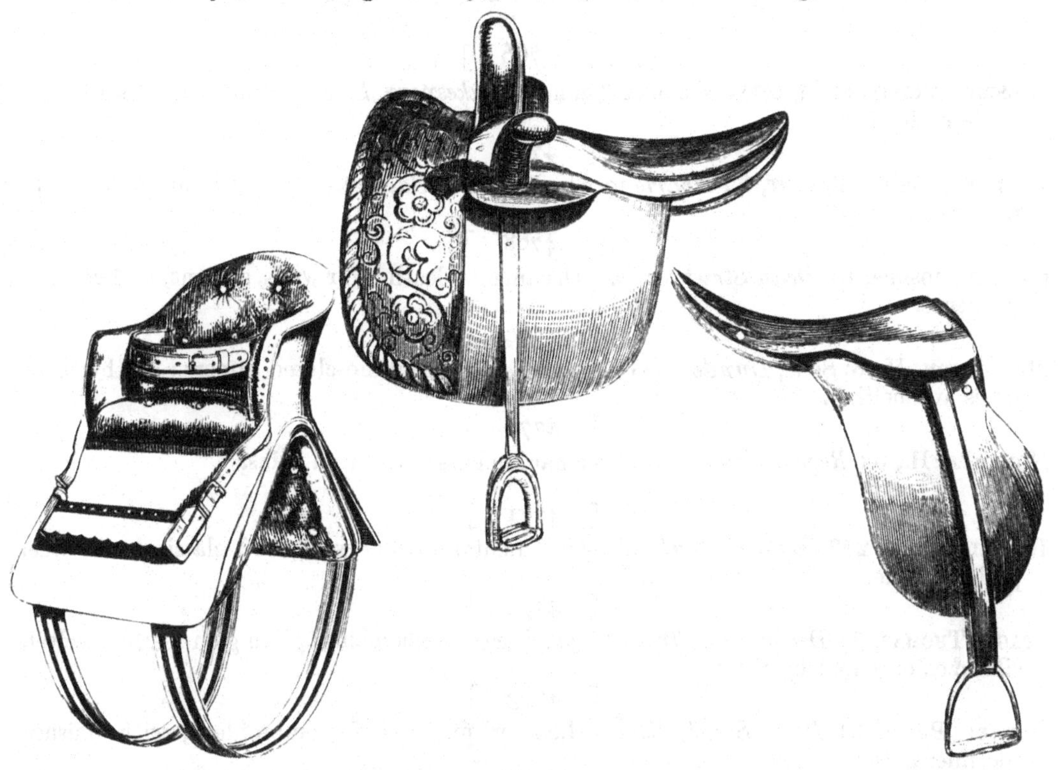

Price lists and designs of harness and saddlery may be obtained on application.

WHITE, JAMES CHADNOR, *Liverpool Street, City, London.*—Harness with improved tugs, saddles &c., wholesale and retail.

Obtained Prize Medal at the Exhibition of 1851.

Inventor of the patent tugs for traces &c., the greatest improvement ever made in harness, by the use of which many serious accidents have been prevented, and for which a prize medal has been awarded. See the differ ence between the Patent Tug and the Buckle.

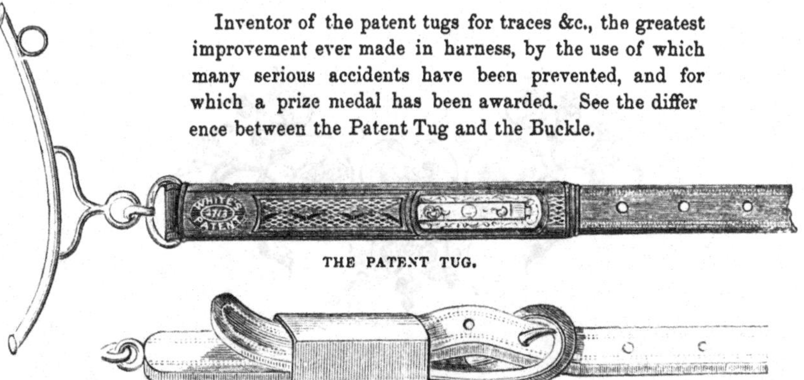

THE PATENT TUG.

THE BUCKLE TUG.

WHITE'S PATENT REIN SWIVEL, to prevent Reins twisting.

WHITE'S PATENT TANDEM BARS, a great improvement on the old style.

SUB-CLASS C.—*Manufactures generally made of Leather.*

[4766]

CROSBIE, ARCHIBALD WILLIAM GEDDES, Tanner, *Shakespear Street, Dumfries.*—Snuff boxes made of leather.

[4767]

EARRATT, JOHN & ROBERT, 25 *Henrietta Street, Covent Garden.*—Military officers' accoutrements &c.

[4768]

GEORGE, JOSEPH, 81 *Dean Street, Soho.*—Ornamental leather for walls, screens, and covering furniture.

[4769]

HENDERSON, H., & SONS, *Dundee.*—Grained leather; machine-closed uppers; leather hose-pipes and belting.

[4770]

NICHOLLS, H., 52 *Regent Street.*—Leather habiliments; prepared skins.

[4771]

REVELL, JAMES, 272 *Oxford Street, W.*—Ornamental leather work, and classical and other potichomanie.

[4772]

STAGG, THOMAS, 37 *Devonshire Street, Bloomsbury.*—Ornamental gilding on leather, velvet, silk, &c., executed by hand.

[4773]

TURNER, PETER, 31 *Dean Street, Soho.*—Leather prepared for embroidery, with finished specimens.

Class XXVII.

ARTICLES OF CLOTHING.

Sub-Class A.—*Hats and Caps.*

[4804]

Ashton, Joseph, & Sons, 54 & 55 *Cornwall Road, Waterloo Bridge, London, S.*—Gentlemen's Paris hats; and 'Prince of Wales Exhibition hat,' registered for 1862.

Black and drab beaver hats, patent granted 1813.

The 'Raglan Paris Satin Hat,' registered Dec. 13, 1854.

The 'Prince of Wales Exhibition Hat,' registered November 5, 1851.

Celebrated for the lightness and elasticity of their manufacture.

Beaver and Paris hats on gossamer and felt bodies.

[4805]

BLAIR, JOHN, & Co., *Glasgow*.—Satin hats (by royal letters patent).

The following extract from the *Practical Mechanics' Journal* of December, 1855, sufficiently describes

BLAIR'S PATENT HAT.

'The benefit of ventilation in head coverings seems now to be generally recognised, and the various manufacturers are bent on discovering the best means of securing it. Amongst these, Mr. Blair puts forth his claims in the invention which is the subject of the present patent; and he arrives at the desired result by constructing the bodies of hats in such a manner as to form a thin space between the interior surface and the outside of the hat, at the part where the hat fits upon the head, such space communicating by perforations with the interior of the hat above, or round the upper part of the head.

'Our engraving is a vertical section of a hat, as constructed according to the improved system. The cylindrical portion, A, of the hat body is composed of plies or layers of woven material, stiffened by means of shellac or other suitable preparation. These plies are formed into a single thickness from the crown of the hat, B, to the point, C, rather more than half-way down the side of the hat. The lower part, from C to D, is formed into two thicknesses in any convenient manner, as, for example, by inserting additional plies of material between the inner and outermost plies of the body. The inner thickness, E, thus formed, is slightly contracted in circumference, so as to leave a thin annular space between itself and the outer part—being joined to the body at C. The brim, F, of the hat is attached at D to the bottom of the outer thickness, C D, of the cylindrical portion of the hat body, whilst the usual leather lining, G, is attached by its outer bottom edge to the inner thickness, E. By these means an annular passage or air space is formed between the lining and the outside of the hat; and the inner thickness, E, being perforated at H, above the level of the head, whilst the crown of the hat is perforated at I, a free passage is left for the air, so that the interior of the hat is kept constantly well ventilated. It will be obvious to the practical hatter, that the same general form of hat body may be produced in a variety of ways besides that especially described herein; but the essential feature of the invention is the formation of a thin or narrow air space round the portion of the hat body which fits the head of the wearer, this space communicating with the interior of the hat for the purpose of ventilating it.

'The invention is obviously applicable to helmets and other head coverings; and in all cases of its application, its use not only induces a good system of ventilation, but also an easy and soft fit to the head.

'We are well able to give an opinion upon this effective contrivance, for we have practically tried it. In about a couple of days, the inner lining takes the exact shape of the head—there being no undue bearing upon any one part—whilst the coolness of the enveloping part is very refreshing after the obduracy of the common hat.'

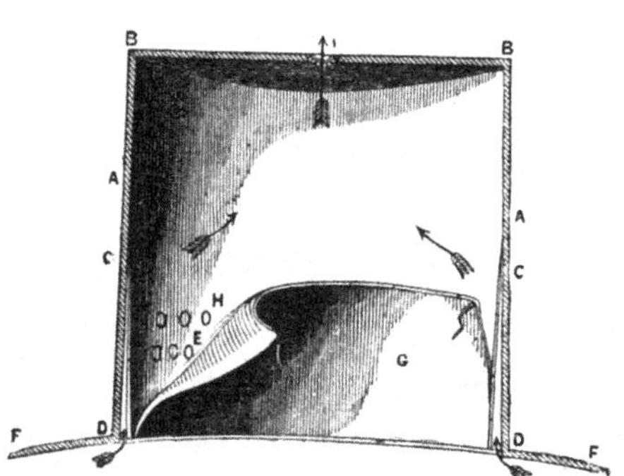

The principal advantages derived from wearing BLAIR'S PATENT HAT are—

First.—The immediate yielding of the hat to any form of head, produces at once an extremely easy and exact fit.

Second.—By the correct principle on which this hat is ventilated, the head is kept always cool and comfortable, thereby conducing to the general health of the body, and preventing premature baldness or loss of the hair.

Third.—From the construction of the frame or body of the hat, it is impossible that the brim or side crown can become saturated, or in the least destroyed by grease or perspiration.

Fourth.—The part of the hat that fits on the head being double, with a space between the two thicknesses, and the inner one only taking the shape of the head, the exterior of the hat retains its proper and original shape.

The above advantages, which are in every respect of the greatest importance, cannot be claimed for any other hat at present in use.

[4806]

BOOTH & PIKE, *Manchester.*—Hatters' trimmings, imperial plush and other materials used for hats.

[4807]

BRIGGS & PREEDY, 98 *Gracechurch Street, corner of Leadenhall Street.*—Hats, caps, felt hats, and umbrellas.

[4808]

CARRINGTON, S. & T., Hat Manufacturers, *Stockport, Cheshire.*—Felt and silk hats, patent corrugated ventiduct hat.

[4809]

CHRISTYS, MESSRS., Hatters, and Hatters' Furriers, *London and Stockport.*—Illustrations of the manufacture of felted and silk hats.

[4810]

DOUGLAS & URE, *Glasgow.*—Hand-knitted Scotch caps.

[4811]

ELLWOOD, J., & SONS, *Great Charlotte Street, Blackfriars Road.*—Wholesale hat and helmet manufacturers.

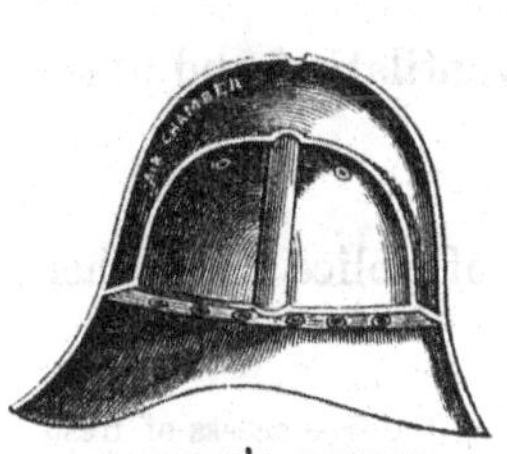

ELLWOOD'S PATENT AIR-CHAMBER HELMET. (Section.)

LION HELMET.

ELLWOOD'S PATENT AIR-CHAMBER SHOOTING HAT.

ELLWOOD'S PATENT AIR-CHAMBER HATS, HELMETS, &c., manufactured upon the only principle by which the head can be effectually protected from the heat of the sun in tropical climates. The annexed section will show that these hats are composed of two parts, the outer part forming a space or air-chamber round the inner one. The non-conducting properties of the air in this space or chamber have the effect of intercepting the rays of the sun, and prevent them from passing to the head of the wearer.

[4812]

GAIMES, SANDERS, & NICOL, 22 *Birchin Lane, Cornhill, and* 111 *Strand.*—Light ventilating hats, entirely new manufacture.

Pliant ventilating hats of very light weight, from 3½ oz., on improved porous bodies.

[4813]

GARRARD, ROBERT & JOHN, *Loman Street, Southwark.*—Japanned leather and felt hats, cap peaks, cockades, &c.

[4814]

HEATH, ROBERT, 25 *St. George's Place, Knightsbridge.*—Patented inventions in hats and umbrellas.

About the year 1850 the exhibitor directed his attention to the more careful and systematic use of straw as a material for ladies' and children's hats. Hitherto hats made from straw plaits were of the simplest forms, few possessing any feature of proportion or elegance to recommend them. They were generally sewn together row by row, according to the taste or idea of the sewer; any little incongruity perhaps being partially corrected in an after process of blocking.

The exhibitor, considering straw plait in its best qualities to be well worthy any care and attention that could be advantageously brought to bear upon it, proceeded to initiate a more scientific and painstaking regularity in the due attention to measurement and design; and after some trouble succeeded in inducing sewers to respect and observe a detailed precision, in lieu of their former uncertain practice of working by so many rows, more or less. The consequent production of shapes and styles elegant, comfortable, and becoming to the wearer, at once created a demand for the hats made from a material so well suited for this climate. Fashion acknowledged and accepted the improvement. This exhibitor was honoured with the personal commands of Her Most Gracious Majesty the Queen and the Royal Family of England, the ladies of the Court, and the nobility; and, in rapid succession, with the patronage of the Courts of France, Russia, Austria, Prussia, Hanover, Sweden, Denmark, Belgium, the Netherlands, Wurtemburg, Germany, and Naples. The good qualities of straw plait at once assumed a very increased value, and an increasing demand ensures its workers liberal prices for their labour, rendering it a valuable auxiliary employment for the females and children of the home counties of England. In this case will be observed a folding hat, or *chapeau bras*, a very ingenious adaptation, so convenient that it recommends itself to ladies as a most desirable *compagnon de voyage*, and as a protection from exposure upon leaving the opera or the ball-room. This exhibitor also displays a great number of shapes patented by him during the last seven years for the especial use of his lady patronesses, forming together an interesting illustration of how much can be produced with care and diligence from so simple a staple as the straw off the harvest field.

UMBRELLAS constructed upon a patent by Captain Francis Fowke, R.E., and upon an improved principle patented by Robert Heath.

These umbrellas are undoubtedly the most compact and the lightest of any yet manufactured; and these advantages are not gained at the expense of their durability. The simple yet scientific principles upon which they are constructed ensure far more service to the purchaser than he can hope to procure from any umbrella as usually manufactured.

These umbrellas do not exceed the dimensions of the ordinary walking cane, and in this climate of sunshine and shower they will certainly become the constant companion out of doors. The walking cane will be discarded in favour of so compact yet so desirable a protection for the promenade and the journey.

[4815]

HUSBAND, RICHARD, *Parsonage, Manchester.*—Patent spring leathered ventilating and other hats in silks and felts.

[4816]

JACOBS, WILLIAM, Hat Manufacturer, *Dorchester, Dorset.*—Specimens of police and other hats, waterproof and ventilating; registered.

These police hats have been supplied to the Dorset constabulary for the last four years, and have given entire satisfaction.

The ventilator, while admitting the free access of fresh air, effectually excludes wet.

The exhibitor is prepared to undertake contracts.

[4817]

LINCOLN & BENNETT, 1 *Sackville Street, Piccadilly.*—Hats &c.

[4818]

MELTON, HENRY, 194 *Regent Street, St. James's.*—Ladies' and gentlemen's hats.

[4819]

MOLLADY, JOHN, & E. E., Hat Manufacturers, *Denton, near Manchester.*—Felt and silk hats.

[4820]

PRITCHARD, A. & F., 31 *Stamford Street, London.*—Hats; silk, silk on cork, felt, and combined cork and felt for India.

[4821]

SIMMONS & WOODROW, *Oldham.*—Gentlemen's silk and felt hats, and ladies' and children's felt hats.

[4822]

SIMPSON, JOHN ANDERSON, 6 *St. George's Crescent, Liverpool.*—Patent easy-fitting soft-band hats.

The advantage of these hats is, that by a simple, easy, and common-sense appliance, the natural hardness of the silk hat has been completely overcome. The hat is rendered soft, and as elastic as the beaver hat of former days.

In windy weather the hat holds comfortably yet firmly to the head; and by those subject to headache this invention has, in hundreds of cases, been pronounced *invaluable.*

Retailers of hats can be supplied with the patent band on application to the patentee.

[4823]

SMITH, J., 8 *Merchant Street, Bristol.*—Hats and caps.

[4824]

TOWNEND, THOMAS, & Co., 16 & 18 *Lime Street.*—Silk and felt hats.

[4825]

TRESS & Co., 27 *Blackfriars Road, London.*—Silk, felt, and beaver hats.

[4826]

WESTLANDS, LAIDLAW, & Co., *Glasgow.*—Patent expanding hats, helmets, fancy hats, felts, caps, Scotch bonnets, &c.

[4827]

WILSON, WILLIAM, & Co., *Newcastle-on-Tyne.*—Hatters' furs, silk and felt hats and caps.

[4828]

ZOX, LAMEN, 85 *Long Acre.*—Fancy hats and caps.

SUB-CLASS B.—*Bonnets and General Millinery.*

[4840]

BORNE, C., & SON, from Paris, 11 *Queen Street, Oxford, and Regent Circus.*—Drawn front bonnet, shapes, and crowns.

[4841]

EMES, MISS, 31 *St. John's Villas, N.W.*—Dress fasteners.

[4842]

FOSTER, SON, & DUNCUM, 16 *Wigmore Street, London.*—Artificial flowers made of various materials.

[4843]

FRANCIS, MISS E., 26 *Wellington Road, Stoke Newington.*—Ladies' night-caps.

[4844]

JONES, WILLIAM, 85 *Chapel Street, Pentonville.*—Artificial May-tree, made of muslin and cambric.

[4845]

SHERRIN, S. H., 24 *Well Street, Cripplegate, London, E.C.*—Gold and silver flowers, bullion, bead, and bridal ornaments.

[4846]

STUART & TAYLOR, 37 *Old Change.*—Millinery, bonnets, caps, head-dresses, flowers, wreaths, and dress trimmings.

[4847]

TRESOLDI & BAKER, 20 & 21 *Coppice Row, Clerkenwell.*—Materials for artificial flower makers.

[4848]

VALLI, DOMINICO, *St. John's Lane, Smithfield.*—Manufactured sprays, birds, leaves, seeds, and other artificial florists' materials.

[4849]

VYSE & SONS, *London.*—Straw plaitings, straw hats and bonnets, flowers, feathers, mantles, shawls, millinery, juvenile dresses.

[4850]

WHITE, WILLIAM, 21 *Nassau Street, London, W.*—Artificial flowers; spring, summer, autumn, winter, and fancy foliage, in cases.

SUB-CLASS C.—*Hosiery, Gloves, and Clothing in general.*

[4861]

ALLEN & SOLLY, *Nottingham, Godalming, and London.* — Hosiery; also samples showing cotton-spinning from earliest date.

[4862]

ASHWELL, THOMAS, & CO., *Nottingham; firm in Chemnitz,* R. H. LOWE & THOMAS ASHWELL, *under the style of* R. H. LOWE & CO.—Hosiery, half-hose, hose, vests, and drawers.

[4863]

AUSTIN, JAMES, *Prince's Street, Finsbury, E.C.*—Crinoline steel, military and other cords, blind and picture lines.

Imperial patent Albert lines, military cords, crinoline steels; stay, skirt, bonnet, and mat cords; thread blind lines in all colours; worsted blind, curtain, lamp, and picture lines; cotton blind and curtain lines; silk blind, curtain, and picture lines. Special attention is invited to the superior quality of the Imperial patent Albert lines.

[4864]

NEWLAND, W. B., 24 *Gutter Lane, E.C.*—Ladies' collars and wrists; gentlemen's shirt collars and wrists.

[4865]

BARRS, WILLIAM, & SON, 7 *Edmund Street, Birmingham.*—Umbrella furniture.

[4866]

BIGGS, JOHN, & SONS, *Leicester.* — Plain hosiery and under-clothing, fancy hosiery, gloves, piece webs, boots, and shoes.

[4867]

BINYON, ALFRED, 37 *Eastcheap, London, E.C.* — Patent chest expander, for the cure of stooping of the shoulders and contraction of the chest.

[4868]

BLYTH, CHARLES, & Co., 4 *Cripplegate Buildings, London.* — Shirts, clothing, &c., for exportation.

The goods shown in this case are made expressly for shipment to the markets of Australia, New Zealand, the Cape, the West Indian Islands, and South America.

[4869]

BOSS, J. A., Wholesale and Export Umbrella and Parasol Manufacturer, 1 *Little Love Lane, Wood Street, Cheapside.*—Umbrellas, sunshades, and parasols.

Honourable Mention, Great Exhibition, 1851.

The exhibitor is the originator of the application of steel to the manufacture of ribs for umbrellas, sunshades, and parasols.

The following are exhibited:—

Umbrellas in plain and coloured gingham . . . } Suitable for all markets for export and home use.
„ in alpaca . . . .
„ in silk . . . .

Parasols, sunshades, frames, &c.; sticks, ribs, and furniture applicable to same; umbrellas in different stages of manufacture.

[4870]

BOWEN, BENJAMIN, Manufacturer, *Chipping Norton, Oxfordshire.* — Leggings made from pure oak bark; tanned leather, and gloves.

Black and drab enamelled and tan leather leggings, made from leather of a pure oak bark tannage.

Buck, doe, cape, and tan gloves.
Price lists will be sent free on application.

[4871]

BRIDE, JOHN HENRY, 68 *Grange Road, Bermondsey.*—Shirts, collars, and fronts.

The exhibitor manufactures shirts, collars, &c., and invites public attention, more especially that of the trade, to the specimens exhibited in his case.

[4872]

BRIE, JOSEPH, & Co., 43 *Conduit Street.*—Shirts, collars, &c. (*See page* 40.)

[4873]

BROCKSOPP, THOMAS, 114 *Wood Street, London.*—Hosiery &c.

[4874]

CARPENTER & Co., 43 *Temple Street, Birmingham.*—Braces, belts, &c. (*See page* 41.)

Brie, Joseph, & Co., 43 *Conduit Street.* — Shirts, shirt fronts, collars, under waistcoats, drawers, dressing gowns, handkerchiefs, &c.

Obtained Honourable Mention and a Prize Medal at the Exhibition of 1851.

The exhibitors devote special attention to the making of flannel shirts and under waistcoats. These are made on their premises from flannels which have been thoroughly shrunk.

[4875]

Cartwright & Warners, *Loughborough.*—Merino hosiery.

[4876]

Coles, William Fletcher, 5 *Aldermanbury Postern, E.C., and* 61 *Paul Street, Finsbury.* — Cork soles, and patent fleecy hosiery.

The following are exhibited, viz.:—

Cork soles of fifteen different kinds, covered with wool and other materials.

Patent fleecy shirts, drawers, petticoats, &c., equal to flannel in durability and warmth, and much cheaper, manufactured of the finest lambswool and cotton, one side being soft and fleecy.

Patent cork linings for boots and shoes.

Silk and cotton fleecy linings for gloves &c.

[4877]

Cooper, Thomas, & Co., 2 *South Bridge, Edinburgh.*—Braces.

[4878]

Corah, Nathaniel, & Sons, *Leicester and Birmingham.* — Hosiery and hosiery yarns.

[4879]

Desborough, S., 24 *Noble Street, E.C.*—Patent umbrella and parasol ribs and stretchers.

[4880]

Dicksons & Laings, *Hawick.* — Yarn and Cheviot and lambswool hosiery and under-clothing.

[4881]

Ellis, J., & Co., 79 *Castle Street, Bristol.*—Ladies' stays and corsets.

[4882]

Elstob, William, 19 *Woodstock Street, Oxford Street, opposite Marylebone Lane.* — Belt, breeches, and trousers.

[4883]

Ensor, Thomas, & Sons, *Milborne Port, Somerset.*—Gloves.

CARPENTER & Co., 43 *Temple Street, Birmingham.*—Webs, braces, belts, part wove; also crinoline steel &c.

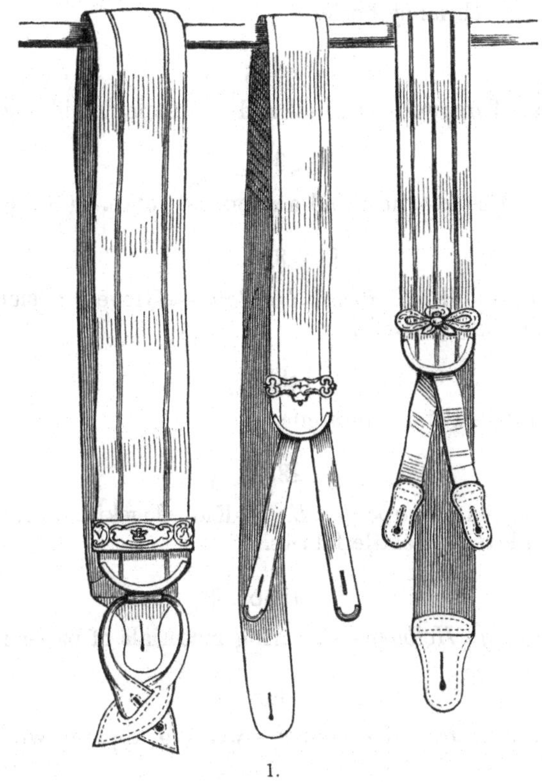

1.

2.

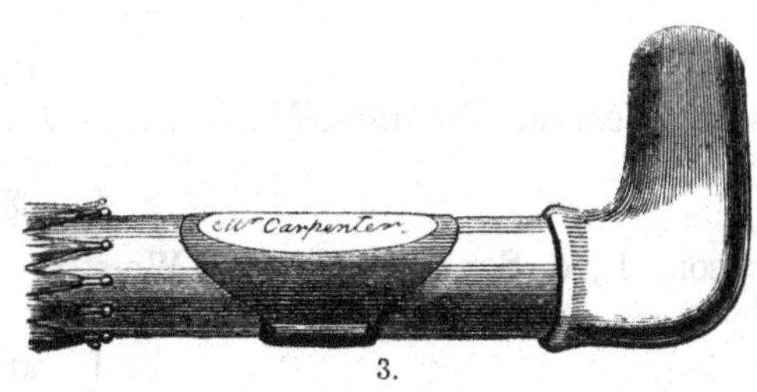

3.

No. 1. The LEVER BUCKLE, which has no prongs, moves with perfect ease in either direction, and does not tear the web or prick the fingers.

Applied to braces
,, belts
,, garters
,, waistcoat backs
&c. &c.

No. 2. The ADJUSTER, for the backs of trousers or waistcoats, which adjusts those garments with great nicety, and remains firmly fixed where it is placed.

No. 3 The UMBRELLA PROTECTOR, by which the owner can instantly attach his name and address, written by himself, to any umbrella, walking stick, whip, carpet bag, &c. &c.

[4884]

EWEN, ROBERT, Manufacturer of Hosiery, *Hawick, Scotland.* — Lambswool hosiery and under-clothing, and Scotch fancy hosiery.

[4885]

FIRKINS, JOSEPH, & Co., *Worcester.*—Ladies' and gentlemen's kid and other leather gloves.

[4886]

FOSTER & Co., *Oxford.*—Ecclesiastical and academical robes. (*See page* 43.)

[4887]

FOSTER, PORTER, & Co., 47 *Wood Street, London.* — Gloves, hosiery, bandannas, shirts, dresses, ribbons, trimmings, parasols.

[4888]

FOWKE, CAPT F., *London.*—Patent umbrella.

[4889]

FOWNES BROTHERS & Co., 41 *Cheapside, E.C.*—Kid, dogskin, and military gloves; also every description of gloves in textile fabrics.

[4890]

GRANGER, ARTHUR, 308 *High Holborn.*—Wearing materials of paper manufacture.

[4892]

HALLIDAY, THOMAS W., *Dundee.*—Gentlemen's wearing apparel, without seam or artificial joining, in felt.

[4893]

HARRIS, RICHARD, & SONS, *Leicester.*—Plain and fancy hosiery, children's socks, gloves, braces, &c.

[4894]

HARRISON, C. H., 13 *Wood Street, Cheapside, London.*—Umbrellas and parasols.

[4895]

HEPPLE, JOSEPH, *Wenlaton, Newcastle-on-Tyne.*—Cutting mensurators.

[4896]

HUDSON, J., & SONS, *Leicester.* — Worsted, lambswool, cashmere, cotton hose and half-hose.

[4897]

JOHNSON, WILLIAM GORDE, *Wheeler Gate, Nottingham.* — Gloves, hair nets, lace scarfs, hosiery, and under shirts.

[4898]

JOHNSTON, JOHN, *St. Ninian's, Stirling.*—Checked tartan hose.

[4899]

JONES, FREDERICK J., 10 *Aldermanbury, London.*—Shirts, collars, belts, braces, &c.

[4901]

JOY, STANDEN, & Co., *Oxford.*—Academical, ecclesiastical, and civil robes.

FOSTER & CO., *Oxford.*—Ecclesiastical and academical robes.

The languishing and depressed state of the Established Church and of education generally during the last century was abundantly manifest, as well in minor details as in great features. The vestments of the clergy, both as regards shape and material, as a rule, were of the meanest possible character; so that external signs indicated the existence of the necessity of reforms. These reforms, both in great matters and in small, have been gradually accomplished during the past thirty years, and Oxford has been foremost in the promotion of them.

In matters of detail, which, though some may deem them unimportant, nevertheless in their degree tend to render the public services of the Church more perfect and complete, many improvements have been effected. In the question of ecclesiastical vestments, it was the privilege of Messrs. FOSTER & Co. to lead the way as regards reform. Many years ago, when the most incorrect and ungraceful garments were furnished, with no regard to ancient precedent or modern requirements, they set themselves to discover some sound standard of authority, by which even in matters of this small moment they might be enabled to meet the demand which renewed activity in the Church, and those warmly interested in the education of the people, had with such wonderful and unexpected energy created. Assisted by the publications of the learned societies which had made this, amongst others, the subject of their investigations, as well as by that practical experience which the last twenty years had given them in the preparation of ecclesiastical and academical robes, they were enabled to accomplish this point with success. Some of their robes and vestments, prepared after ancient models, and manufactured of the best possible materials, were exhibited at the Oxford Architectural Society's meetings, where they met with the general approbation of that influential body. Independent of this, the testimony they have received from those whose practical and theoretical knowledge has enabled them to form an accurate judgment, permits their taking to themselves the credit of having assisted in improving the taste of those who created the demand, whilst they themselves were called upon to render the supply. The various specimens of robes and vestments in the present Exhibition are of themselves sufficient to illustrate what Messrs. FOSTER & Co. are now enabled to effect. This fact, more valuable than mere words, added to the above consideration, will be sufficient to point out to the public the desirability of applying to those who have thoroughly studied these matters, rather than to others who appear content to lag behind, satisfied either with indifferent imitations or in continuing to distribute only such ill-made and conventional vestments as were a disgrace to the Church in her day of neglect and inactivity, and which cannot be longer tolerated in her present season of improvement and progress.

In the subjoined list of robes &c., materials of the best quality and workmanship of the first class will be found united with charges strictly moderate.

GOWNS.

		£	s.	d.		£	s.	d.
B.A.	Mohair Gown	1	10	0	to	2	10	0
M.A.	,, ,,	1	10	0	,,	2	10	0
B.A.	Silk ,,	5	5	0	,,	8	10	0
M.A.	,, ,,	5	5	0	,,	8	10	0
Preaching	,, ,,	6	6	0	,,	8	8	0
S.C.L.	,, ,,	4	4	0	,,	5	10	0
D.C.L.	,, ,,	6	6	0	,,	8	8	0
D.C.L.	Scarlet Cloth Gown	6	6	0	,,	8	8	0
D.D.	,, ,,	10	10	0	,,	12	12	0
D.D.	Scarlet Cloth Habit	4	4	0	,,	5	5	0
Mus. Bac.	Silk Gown	6	6	0	,,	8	8	0
Mus. Doc.	Dress ,,							
Gent. Coms.	Silk ,,	5	5	0	,,	7	0	0
Undergraduate's	Stuff ,,	0	16	6	,,	1	1	0
Scholar's	,, ,,	1	5	0	,,	2	2	0

SURPLICES.

	£	s.	d.		£	s.	d.
Foster's Oxford Surplice	1	10	0	to	2	10	0
Do. Circular ditto	1	10	0	,,	2	10	0
French Cambric or Lawn ditto	2	2	0	,,	3	3	0
Embroidered Collars, in white, blue, or red, 10*s.* 6*d.* extra.							
Chorister's Surplices (Linen)	0	10	6	,,	0	16	6
Lay Clerk's ,, ,,	1	1	0	,,	1	5	0

ACADEMIC CAPS.

	£	s.	d.
Plain	0	8	6
Patent Folding Scull	0	8	6
Choristers, from	0	4	6

HOODS.

		£	s.	d.		£	s.	d.
B.A.	Silk Hood	1	0	0	to	1	10	0
M.A.	,,	1	8	0	,,	2	2	0
S.C.L.	,,	1	1	0				
B.C.L.	,,	1	5	0				
B.D.	,,	2	10	0	,,	3	3	0
D.D.	Scarlet Cloth Hood	3	3	0	,,	4	4	0
D.C.L.	,, ,,	3	3	0	,,	4	4	0
Literate's	Stuff Hood	0	7	6				
Mus. Bac.	Silk ,,	1	1	0	,,	1	15	0
Mus. Doc.	Dress ,,							

CASSOCKS.

	£	s.	d.		£	s.	d.
Mohair (English)	1	10	0	to	2	10	0
Do. (Belgian)	2	10	0	,,	3	3	0
Silk	4	4	0	,,	6	6	0

STOLES.

	£	s.	d.		£	s.	d.
Black Silk, Plain	0	10	6	to	1	1	0
Do. Embroidered in Silk or Gold	1	1	0	,,	2	2	0
Coloured Silk, Plain	0	10	0	,,	1	1	0
Do. Embroidered	1	1	0	,,	2	2	0
Chaplain's Scarf	1	1	0	,,	2	2	0

ALMS BAGS, CORPORALS, ALTAR COVERS, BANDS, SERMON CASES, &c. &c. &c.

	£	s.	d.		£	s.	d.
A SUIT OF BLACK, for ordinary parish wear, of sound good quality, and colour warranted					5	5	0
,, of extra superfine cloth, best quality made	6	6	0	to	7	7	0

Patterns and Illustrated Catalogue free by post on application.

[4902]

LAING, JOHN, Manufacturer of Hosiery, *Hawick, Scotland.*—Indian gauze, gauze, and elastic merinos, and lambswool.

[4903]

LAURENCE, FREDERICK RICHARD, 20 & 21 *Southampton Street.*—Shirt collars, shirts, and other white goods.

[4904]

LAWRENCE, WILLIAM, 2 *St. Paul's Villas, Ball's Pond.*—Chamois leather under-clothing, and travelling sheets.

[4905]

LINKLATER, ROBERT, 172 & 113 *Commercial Street, Lerwick, Shetland Isles.*—Shetland knitted shawls, veils, and hose.

[4906]

MACINTOSH, CHARLES, & Co., *Cannon Street, London; and Cambridge Street, Manchester.* Waterproof and elastic articles of clothing.

[4907]

McINTYRE, HOGG, & Co., Manufacturers, 9 & 10 *Addle Street, London;* 122 *Brunswick Street, Glasgow; and Manchester;* Manufactory, *Londonderry, Ireland.*—Shirts and collars.

Shirts, shirt fronts, and collars of different descriptions, suitable for various climates.

The shirts exhibited are specimens of superior Irish needlework, and show the high state of perfection to which this branch of manufacture has now attained.

The shirt in the centre has the arms of the City of Londonderry (at which place, and in the neighbourhood, these goods are produced) embroidered in colours on the front; and two of the fronts for insertion exhibited have designs in needlework on them. There are also shown specimens of shirts suitable for the South American trade, of the 'International' and of the 'Alma' shapes, as well as specimens of embroidered flannel drawers and under vests.

[4908]

MARION & MAITLAND, MESDAMES, 238 *Oxford Street.*—Corsets.

[4911]

MEYER, S. & M., 71 *Cannon Street West; and Bow Lane.*—Umbrellas, various.

[4912]

MEYERS, MICHAEL, 9 *Great Alie Street, Goodman's Fields, London.*—Umbrellas and parasols.

[4913]

MIDDLEMASS, JAMES, 18 *South Bridge, Edinburgh.*—Presbyterian pulpit gowns, gentlemen's clothing, and shirts.

[4914]

MONEY, HENRY KNAPP, *Woodstock, Oxfordshire.*—Leather gloves.

[4915]

MORLEY, J. & R., 18 *Wood Street, London.*—Hosiery.

[4916]

MUNDELLA, HINE B. H., & Co., *Station Street, Nottingham.*—Hosiery of all descriptions manufactured principally by power machines, under special patents.

[4917]

NEVILL, J. B. & W., & Co., 11, 12, & 13 *Gresham Street West, E.C., London.*—Hosiery.

[4918]

NEVILL, W., & Co., *Langham Factory, Godalming.*—Hosiery.

[4920]

PAYNE, THOMAS, *Hinckley, Leicestershire.*—Men's, women's, and boys' cotton stockings, men's and boys' half hose.

[4923]

REYNOLDS, G. W., & Co., 12 *Cheapside, London; and Birmingham.*—Ladies' and juvenile under-clothing.

[4924]

SALOMONS, A., *Old Change, E.C.*—Stays, crinolines, and corsets. (*See page* 46.)

[4925]

SANGSTER, WILLIAM & JOHN, 140 *Regent Street; 94 Fleet Street; 10 Royal Exchange; and 75 Cheapside.*—Umbrellas and parasols.

[4926]

SCOTT, PETER, & Co., *Edinburgh.*—Improved shirt; seamless coat.

[4927]

SHARP, PERRIN, & Co., *Old Change, London.*—Ladies' underclothing, skirts, and stays, baby linen, children's dress.

[4928]

SILVER, S. W., & Co., *Cornhill, and Bishopsgate Within.*—Shirts, caps, and waterproof garments.

[4929]

SINCLAIR, ROBERT, & Co., 80 *Wood Street, London.*—Shirts, collars, and ladies' under-clothing.

[4930]

SMYTH & Co., Original Balbriggan Hosiers, 36 & 37 *Lower Abbey Street, Dublin;* Factory, *Balbriggan.*—Hosiery.

[4931]

STEARS, SAMUEL, 36 *Briggate, Leeds.*—Parasol.

[4932]

SWEARS & WELLS, 192 *Oxford Street.*—Lilliputian hosiery.

[4933]

TAYLOR, BENJAMIN, 67, 68, & 69 *Camden Street, Birmingham.*—Braces, belts, leggings, webs, girths, bridles, accoutrements, rifle lock protector.

[4934]

THOMSON, W. T. & C. H., *Fore Street, London.*—Patent crown crinolines.

[4935]

TILLIE & HENDERSON, *Glasgow.*—Shirts and under-clothing.

SALOMONS, A., *Old Change.*— Stays and crinolines, and Smith's & Castle's patent corsets &c.

Obtained Honourable Mention at the Universal Exhibition, 1851, *and the only English Stay Maker who obtained a Prize Medal at the Paris Universal Exhibition,* 1855.

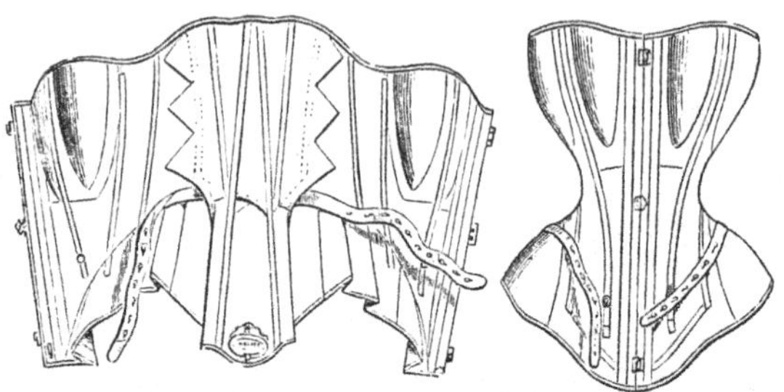

SMITH'S NEW PATENT HARMOZON SELF-ADJUSTING CORSET.

1. CORSET without laces, eyelet holes, elastic, or any mechanism likely to get out of order.

The fastenings are attached to the Corset instead of to the busk, to facilitate the removal of the busk when the stays require to be cleaned.

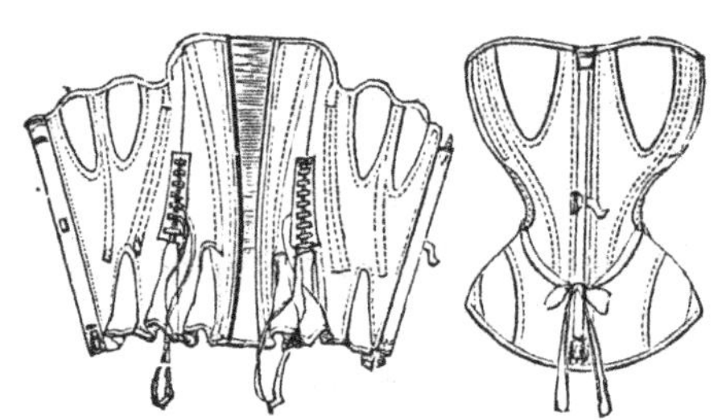

SMITH'S PATENT ROYAL SYMMETRICAL CORSETS.

2. CORSET with front fastening, elastic backs, and side lacing.

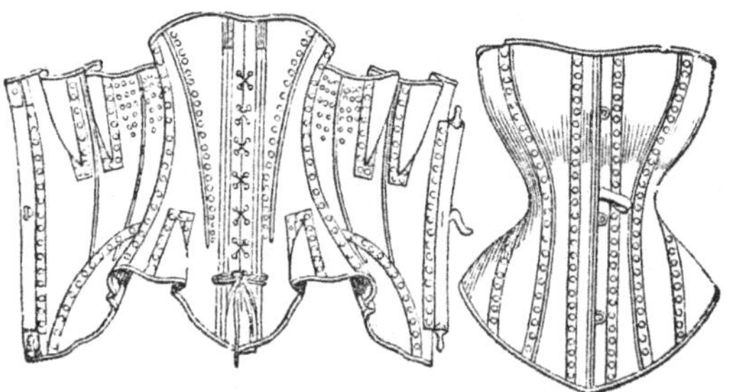

CASTLE'S PATENT VENTILATING CORSETS.

3. VENTILATING CORSETS for ball-rooms, hot climates, and equestrian exercise.

LADIES' SKIRTS.

4. The CARDINIBUS COLLAPSING JUPON (Patented).

Collapses into the smallest possible compass on the slightest pressure.

5. The IMPERIAL SYMMETRICAL CORSAGE JUPON (Patented).

This Jupon is attached to a bodice, which can be adjusted to any figure, and supports the skirt.

6. SPIRAL CRINOLINE STEEL and BRONZE for LADIES' SKIRTS (Patented).

Cannot be broken, and will fold into the smallest possible space.

[4936]

WADDINGTON & SONS, 1 *Coleman Street, London.*—Umbrellas and parasols.

[4937]

WARD, STURT, & SHARP, *Wood Street, London; and Belper, Derbyshire.*—Silk, cotton, and woollen hosiery, and gloves.

[4939]

WELCH & SONS, 44 *Gutter Lane, London.*—Straw hats, millinery, and flowers.

[4940]

WELLS, JOHN SCOTT, *Mount Street, Nottingham.*—Cotton and merino hose, vests, &c.

[4041]

WHITE, F. & W. E., *Loughborough.*—Plain and fancy worsted, woollen, and cotton hosiery.

[4942]

WHITBY BROTHERS, Manufacturers, *Yeovil, Somerset.* — Gloves and leather manufactured from foreign lamb skins.

Obtained the Prize Medal in Class XX., 1851.

LEATHER, dressed from Italian lambskins, and stained into various colours suitable for gloves.

GLOVES, of various sizes and colours, made from similar leather to that exhibited; cut on a uniform plan, which is calculated to leave the leather with that precise degree of elasticity best adapted to easily putting on the gloves. The Children's have an improved fourchette, and certain portions of the sewing are strengthened on a new principle.

The smallest hands are almost invariably the longest in proportion to their width, except in the unformed hands of children. The Exhibitors have adopted the average length for gloves of each degree of width, grouping the sizes as follows:—

FOR GIRLS.

No. 00. No. 0. No. 1. No. 2. No. 3. No. 4. No. 5. No. 6. No. 7.

FOR BOYS.

No. 1. No. 2. No. 3. No. 4. No. 5. No. 6.

FOR LADIES.

$5\frac{1}{2}$	$5\frac{3}{4}$	6	. . very long
$6\frac{1}{4}$	$6\frac{1}{2}$	$6\frac{3}{4}$	. . long
7	$7\frac{1}{4}$	$7\frac{1}{2}$	. . medium
$7\frac{3}{4}$	8		. . short
$8\frac{1}{4}$	$8\frac{1}{2}$		. . very short

FOR GENTLEMEN.

$7\frac{1}{2}$	$7\frac{3}{4}$		. . very long
8	$8\frac{1}{4}$		. . long
$8\frac{1}{2}$	$8\frac{3}{4}$	9	. . medium
$9\frac{1}{4}$	$9\frac{1}{2}$	$9\frac{3}{4}$	. . short
10	$10\frac{1}{4}$	$10\frac{1}{2}$	. . very short

[4944]

WHITEHEAD, WILLIAM, & SON, 63 *North Bridge, Edinburgh.*—Tartan hosiery and vicuna under-clothing.

[4945]

WILSON, WALTER, *Allan Crescent, Hawick, N.B.*—Hosiery in wool and merino, and under-clothing.

[4946]

WILSON & ARMSTRONG, *Drogheda Street, Balbriggan; and* 11 *Nassau Street, Dublin.* Balbriggan hosiery.

[4947]

WILSON & CO., 18 *Southampton Row, Russell Square.*— Baby's clothes protector, patent Stilla bib, &c.

[4948]

WILSON & MATHESON, *Glasgow Street, Glasgow.*—Umbrellas, Scotch bonnets, cloth caps, overshoes, &c.

[4949]

WYATT, JOHN WILLIAM, & Co., 64 & 65 *Bunhill Row, and* 154 *Old Street, Finsbury.*—Crinoline skirts.

[4950]

WILSON, C. E., & Co., *Monkwell Street, Falcon Square, E.C.*—Peglet.

SUB-CLASS D.—*Boots and Shoes.*

[4960]

ALDRED, WILLIAM, *Manchester.*—Adjustable heels for boots and shoes, with silver and other metal fastenings.

[4961]

ALLEN, CHARLES ENOS, *High Street, Haverfordwest, South Wales.*—Boots and shoes.

[4962]

ATLOFF, JEAN GEORGE, 69 *New Bond Street.*—New style of boots, showing excellence of workmanship and improvements.

THE EUGÉNIE.

The NEW ELASTIC BOOT is an elegant, graceful, and perfect-fitting boot. The elastic web is so placed as to allow the free rising of the instep, and produces that arched and *cambré* shape which is always so greatly admired.

The elastic web being ornamental, these boots are in every way adapted for dress, and do not give that gouty appearance to the ankles which has so long been complained of in the ordinary side-spring boots.

[4963]

BALL, WILLIAM, & Co., *New Weston Street, Bermondsey.*—Machine and hand-closed boot uppers.

[4964]

BARRON, W. J., & Co., 66 & 67 *Aldermanbury.*—Elastic webs, and shoe mercery.

[4965]

BAULCH, CHARLES, *Bristol.*—Improved rivet-soled boots; also registered cork clumps, and self-adjusting leather clogs.

[4968]

BIRD, WILLIAM, 86 *Oxford Street.*—Ladies' boots of elaborate and highest class design and workmanship.

[4969]

BOWLER, JAMES, 19 *Blandford Street, Manchester Square.*—Mechanical boot-stretcher, lasts, &c.

[4970.]

BOWLEY & Co., *Charing Cross.*—Boots and shoes.

HOUSEHOLD CAVALRY.

SHOOTING.

REGIMENTAL.

All kinds of military and general boots and shoes.

Spurs for service, hunting, or general use.

Shooting boots without front seams.

Flax gaiters for cover and warm climates.

Brown enamelled India boots.

FISHING.

[4971]

BRISON, ROBERT, 1 *St. Augustine's Parade.*—Anatomical lasts, made to the form of the foot.

[4972]

BROWN, EDWARD, 67 *Prince's Street, W.*—Boots, blacking, and polishers.

[4973]

CARTER, LIEUT.-COLONEL, *Monmouth.*—The 'Hythe boot.' (*See page* 50.)

[4974]

CHAPPELL, JAMES, 388 *Strand.*—The Pulvinar boot.

The new 'PULVINAR,' or CUSHION BOOT, gives a soft bed for the sole of the foot in all kinds of walking.

To gentlemen who suffer from tender feet especially, this boot will prove to be a luxury hitherto unknown by them: the comfort of it is felt until the boots are quite worn out.

[4975]

CHARLESWORTH, WILLIAM, Wholesale Manufacturer, *Stamford Street, Belvoir Street, Leicester.*—Riveted and sewed boots and shoes.

[4976]

CHRISTMAS, GEORGE, 12 *Mount Street, Westminster Road.*—The 'accelerating boot,' to assist walking and prevent splashing.

[4977]

CLARK, CYRUS & JAMES, Manufacturers, *Street, near Glastonbury.*—Ladies', gentlemen's, and children's boots, shoes, and slippers.

CARTER, LIEUT.-COLONEL, *Monmouth.*—The 'Hythe boot.'

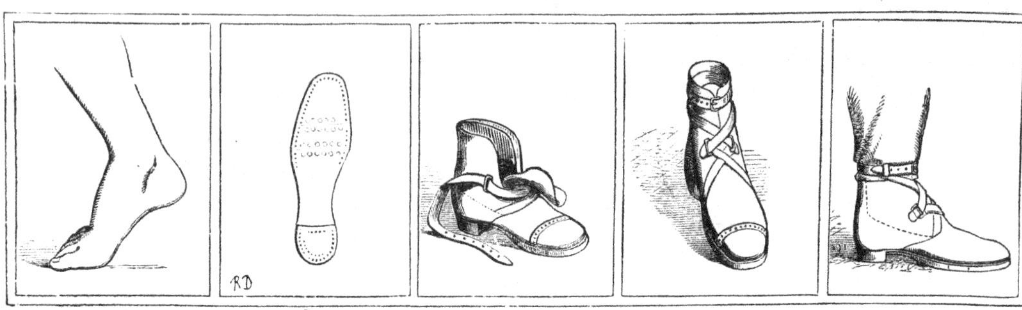

1 Shows the natural bend of the foot when walking.

2 The sole of the Hythe Boot, showing how it is divided for facilitating the natural bend of the foot.

3 Shows the large opening of the Hythe Boot.

4 Shows how that by two straps buckled, the large opening can be effectively, quickly, and neatly closed.

5 Shows how the trousers can be worn for muddy roads, the fields, and the moors.

For riflemen, the sole of the Hythe Boot, which is double or 'clumped,' and very strong, supports the weight of the body comfortably and steadily when at the 'kneeling position;' but though rigid for this purpose, it is pliant for marching.

For sportsmen and all great walkers, the Hythe Boot will be found to possess most important advantages. By the mode of fastening (two straps buckled) it is easily, comfortably, and firmly secured to the foot. While no clay soil can draw the boot off, ample room is with certainty given to the instep, as the wearer adjusts the boot to its size. The sole (all leather), however thick, is as pliant as a thin one.

For the police (town as well as county), the Hythe Boot is particularly well adapted. Sergeant No. 33 of the Monmouthshire Constabulary, who tested a pair of Hythe Boots during a part of the very wet summer of 1860, and over every kind of road, after walking in them 630 miles, stated that he found them to be the easiest and the most durable he had ever worn, impervious to wet, and considerably less fatiguing.

Fishermen will find the Hythe Boot (made easy) a convenient covering for waterproof stockings.

Youths should wear the Hythe Boot, as its increased strength does not prevent the symmetrical growth of the foot, or check muscular developement.

The Sole of the Hythe Boot can be advantageously used with *every* kind of boot. It is likewise the most effectual repair to worn boots.

The Fastening can be used with single-soled or light boots.

The Hythe Boot can be easily and expeditiously repaired.

The Hythe Boot can be made waterproof. The sole is invariably watertight.

The Hythe Boot does not creak; and if the sole is applied to a creaking boot, the creaking will cease.

Ladies will find the sole of the Hythe Boot impervious to wet; and in walking, by its thickness and pliancy will experience greater comfort and less fatigue. In the country and at the sea-side a thick-soled boot ought to be regarded as indispensable; but on *wet* ground everywhere the sole of the Hythe Boot is necessary for health and comfort. Goloshes not needed.

The principles upon which the Hythe Boot has been constructed are:—

1st. A hinge in the outer sole, to allow the foot to bend when walking. This is obtained simply by dividing the outer sole, where the foot naturally bends, in two places. Many imagine that gravel must insert itself into these divisions; but this is not possible, for the foot, when walking, is placed on the ground *flat*, so that the divisions are *closed*, and it is not until the foot is leaving the ground that the foot bends and the divisions open. Practically, the divisions, except from the comfort they afford, are not felt.

2nd. Two straps — worn after the manner of sandals — so placed that, by buckling them, a large opening is effectively closed, the size of the boot at the instep regulated, and the sole made to adhere to the foot. These straps being as durable as the boot, very easy to fasten and unfasten — so easy that either can be done in the dark — will, no doubt, supersede universally and for ever the troublesome and fragile lace.

The Hythe Boot, by giving an easy balance to the body, allowing the foot to bend when walking, freeing the instep, causing a firm adherence of the sole, and being modelled after the proper formation of the foot, reduces fatigue, when compared with ordinary thick boots, to one half. The Hythe boot is not dearer in its first cost than other strong boots.

No change of bootmaker is necessary, as licenses ar granted by the patentee gratuitously.

Manufacturers are requested to apply for further particulars to LIEUT.-COLONEL CARTER, Monmouth.

[4978]

CLARKE, EDWARD WILLIAM, 12 *Southampton Row, Russell Square.*—Boots for lame and tender feet. Volunteer gaiter boots.

The exhibitor is inventor and manufacturer of various kinds of boots, suitable for very tender and lame feet to the club-foot hospitals and surgical instrument makers.

He manufactures boots adapted for corns, bunions, weak ankles, and club and flat feet; and all descriptions of cork boots.—Established 1784.

[4979]

CREAK, JAMES, *Wisbeach.* — Ladies' and gentlemen's improved side-spring bluchers and button boots.

Obtained Honourable Mention at the Exhibition of 1851.

Specimens of ladies' and gentlemen's new and improved button and side-spring boots, selected from the exhibitor's stock.

Patterns may be obtained for thirty postage-stamps, on application.

[4980]

CREMER & Co., 126 *New Bond Street, W.*—The 'Cremerian boot,' with elastic spring in waist or arch of foot.

[4981]

CROSSDALE, JAMES, 2 *Rotherfield Street, Lower Islington.*—Ventilating boots and shoes for hot weather and damp feet.

[4982]

DAVIES, JOHN, 46 *Great Queen Street, Lincoln's Inn Fields.*—Sample of gentlemen's guinea boots.

[4983]

DENNANT, FREDERICK, 10 *Allington Street, and* 6 *Bedford Place, Vauxhall Road.*—Machine-closed boot uppers &c.

[4984]

DERHAM BROTHERS, *Bristol and London.*—Boots and shoes manufactured by machinery.

[4985]

DOWIE, JAMES, 455 *Strand.*—Elasticated leather soled boots.

[4986]

DUTTON, W. H., & SONS, *Knightsbridge, London, S.W.*—Boots and shoes.

[4987]

EAST, SAMUEL, 103 *Fore Street, Exeter.*—Boots, shoes, rifle leggings, lasts, and boot trees.

[4988]

EVANS, RICHARD, *Newtown, Montgomeryshire.*—Improvement in the manufacture of patent boots. Also specimens of superior workmanship.

[4989]

FRAMPTON, SOPHIA, 79 *Regent Street, London.*—Ladies' and children's boots and shoes.

[4990]

GARNER, DAVID, 23 *Clarence Road, Bristol.*—Lasts and boot trees; lasts for riveted boots.

Obtained Honourable Mention at the Exhibition of 1851.

INSERTED IRON-PLATED LAST FOR RIVETING PURPOSES.

Specimens of lasts for stitched and riveted or screwed boots.

Configurations of feet, classified for army purposes &c., numbered 12 to 17.

Improved portable and Balmoral boot trees.

[4991]

Glew, John Henry, 19 *Howland Street, Fitzroy Square.*—Ladies' boots and shoes.

[4992]

Gordon, Edwin, 6a *Prince's Street, Leicester Square.*—Boots, various.

[4993]

Grundy, Thomas, 44 *St. Martin's Lane.*—Easy boots.

[4994]

Gullick, Thomas, *Pall Mall.*—Boots and spurs. (*See page* 53.)

[4995]

Gundry & Sons, 1 *Soho Square.*—Variety of boots and shoes, showing improvements and inventions.

[4996]

Hall, C. G., 89 *Regent Street.*—Boots and leggings. (*See page* 54.)

[4997]

Hall, Joseph Sparkes, 308 *Regent Street.*—Boots and shoes for the rich and the poor.

[4998]

Hall & Co., *Wellington Street, Strand.* Pannus corium, or leather-cloth boots and shoes.

[4999]

Hallam, J. & E., 149 *Waterloo Road, London.*—London and Lancashire clogs.

[5000]

Hamilton, John, 4 *Diana Place, Euston Road, Fitzroy Square.*—Six pairs of ladies' boots.

[5001]

Hartley, Joshua, & Sons, 11 *King Street, St. James's.*—Top boots of English leather.

The exhibitors are Boot and Shoe Makers by special appointment to Her Majesty the Queen, H.R.H. the Duchess of Cambridge, &c. The specimens exhibited are entirely of their own manufacture.

[5002]

Heath, Austin, & Mycock, *Browning Street, Stafford.*—Ladies' boots and shoes.

[5003]

Hickson, William, & Sons, *Smithfield.*—Boots and shoes for home and colonial trade, sewed and riveted.

[5004]

Hook & Knowles, 66 *New Bond Street.*—Boots for dress, riding, and walking; costume and other shoes; over-shoes; dress and plain brogues.

[5005]

Hudson, Alfred, *Cranbrook.*—Boots and shoes with improved inner sole, for tender feet.

[5006]

Hutchings, John Thomas, 5 *Inkermann Terrace, Charlton, near Woolwich.*—Boots and shoes of every description, with composition soles.

[5007]

James, A., 2 *Trevor Square, Knightsbridge.*—Boots and shoes.

Gullick, Thomas, *Pall Mall.*—**Boots, of lac Japan leather, with spurs affixed, illustrating the 'Eclipse' box.**

BOOT-MAKER TO HIS LATE ROYAL HIGHNESS THE PRINCE CONSORT, AND TO HIS IMPERIAL MAJESTY NAPOLEON III.

Imperial Present.—A very costly scarf pin, the head of which is of the form of an imperial crown, and consists of a large globe emerald with a fine brilliant set in its centre, was presented, May 16, 1861, by the Emperor Napoleon to Messrs. Gullick, as a testimonial of approval of their Patent Eclipse Spur Box, and their newly invented Lac Japan Leather (unequalled for hunting and riding), as supplied by them to His Imperial Majesty.

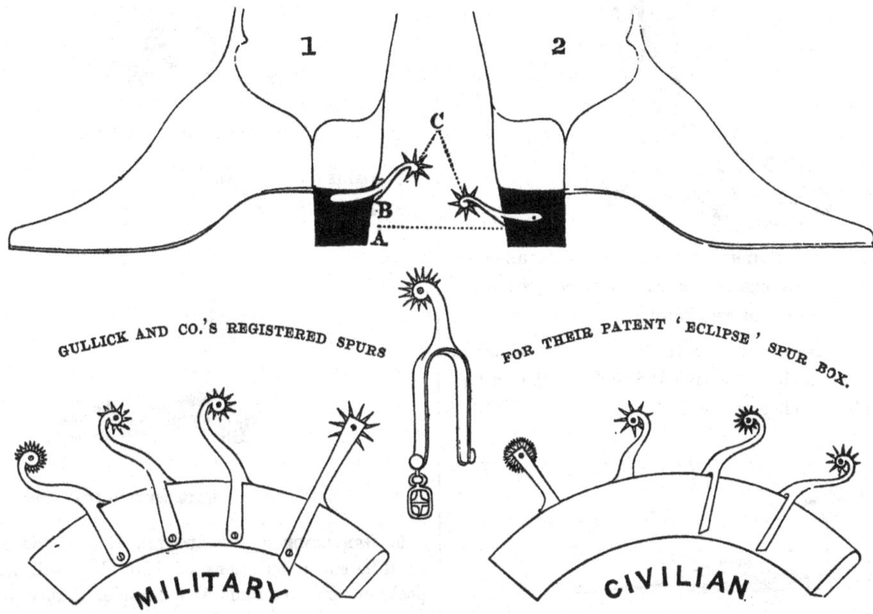

1. The Newly Invented Spur Box, by which the spur is elevated from A to B, and is half the distance to C, *i. e.* the horse's flank.

2. The Old Spur Box, in which the boot heels are unsightly, and the spur inconveniently near the ground.

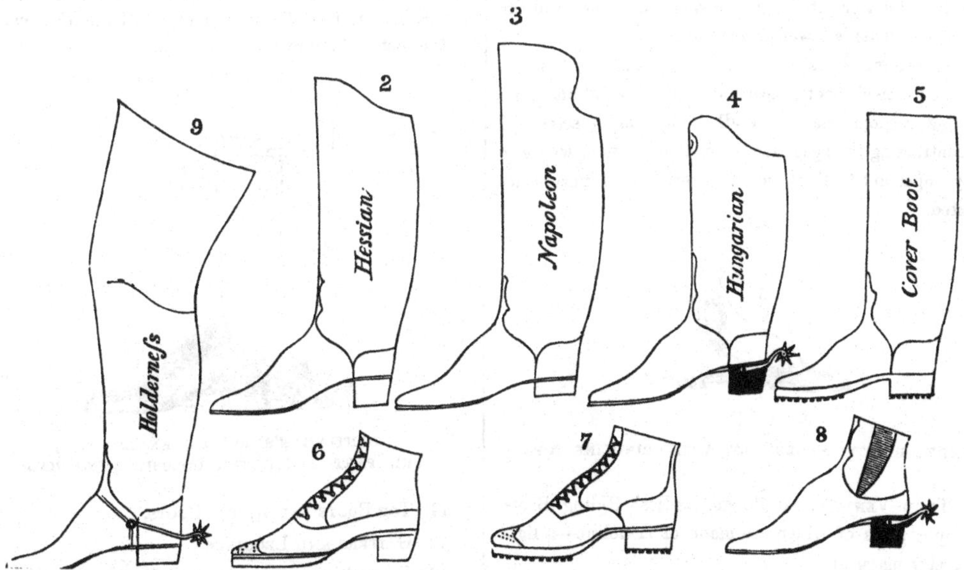

6. The Patent Impilia, invented in 1842, has an improvement in the sole which renders it warm, dry, and creakless. It is suitable for winter wear, and can be applied to all descriptions of boots.

7. The Cambrian, invented 1843, is a shooting boot: the lacing part not being confined, the feet are allowed freedom, so that the strongest leather can be worn without discomfort.

8. The Carlton, registered April 1856, suitable for uniform, having the appearance of a Wellington boot.

Hall, C. G., 89 *Regent Street.*—Boots of a novel adaptation and improved shapes, according with the requirements of nature.

SHOWS THE COMMENCEMENT OF THE INTERNAL SAPHENEOUS VEIN, WHICH COVERS THE INSTEP AND INNER SIDE OF THE FOOT.

1. The Skeleton of a Lady's Foot, the bones of which are all perfect, showing the arch of the instep, which has been well preserved.

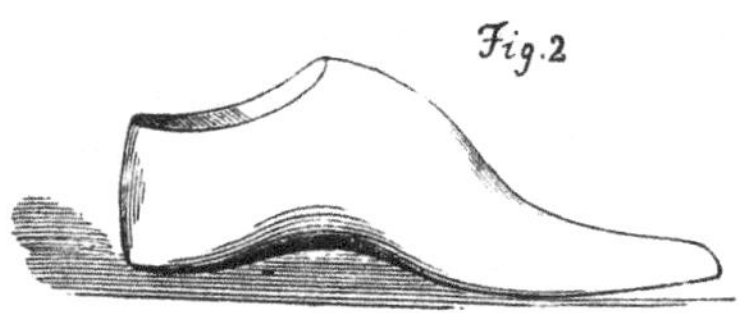

2. The Model Lasts for Ladies' Boots, when the feet are not injured by unskilful fitting.

3. Skeleton of a Gentleman's Foot. Parts of the tibia and fibula are attached, showing the position of the leg bones. This is a well-formed foot.

4. A Model Pair of Lasts. The correct form to make gentlemen's boots, according to the laws of anatomy, allowing freedom of action to all the leading muscles, and not contracting the space for the sole of the foot, where so many important tendons, veins, arteries, and nerves are situated.

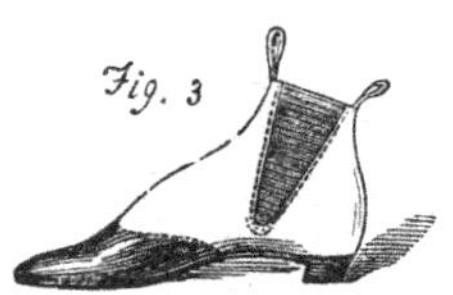

LADY'S ELASTIC ANKLE BOOT WITH CRINOLINE TOP.

5. True Ventilating Boots. Ladies' Riding Boots, the upper parts of which are made of crinoline—a light and pretty material.

6. Ladies' Elastic Ankle Boots, with crinoline tops, giving freedom to the instep and ankle, and to all the tender organs of the foot.

LADY'S VENTILATING RIDING BOOT.

7. Ladies' Crinoline Top Lace Boot.

8. Ladies' Shoes.

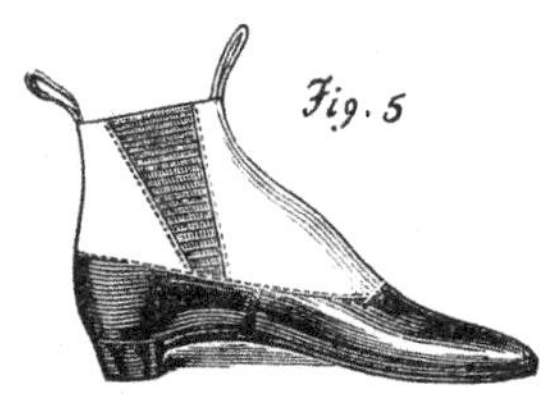

GENTLEMAN'S HAIR WELLINGTON BOOT.

9. Gentlemen's Wellington Boots, the legs made of different descriptions of hair that will keep their position, and are light and durable. Truly ventilating boots, especially adapted to all military gentlemen in the colonies.

10. Gentlemen's Elastic Ankle Boots and Shoes, the upper parts made of various descriptions of hair, admitting the free circulation of all the delicate vessels of the foot and ankle.

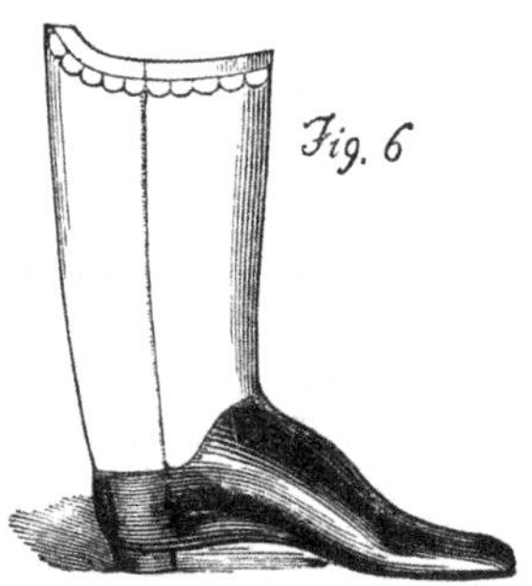

GENTLEMAN'S ELASTIC ANKLE BOOT, THE PART ABOVE THE GOLOSH BEING HAIR.

11. Top Boots and Riding Boots.

12. Gaiters and Leggings.

13. Ladies' and Gentlemen's Boots and Shoes of the improved elastic enamelled cloth.

14. Boots and Shoes for Children, made partly of hair, being light yet very durable.

[5008]

JENNETT, JOHN, 44 *Whitcomb Street, Leicester Square.*—Boot trees and lasts.

[5009]

JOSEPH, JOSHUA, & SONS, 13 *Skinner Street, Snow Hill, London.*—Ladies' and children's boots and shoes.

The exhibitors manufacture every description of boots and shoes suitable for the Australian, Indian, and home markets.

These specimens are exhibited as models of excellence of materials, correct proportions, and moderation in price, combined.

[5010]

JUDGE, CHARLES, 6 *Sion Place, East Street, Walworth, London.*—Leather buttons, laces, and leggings.

[5012]

KNIGHT & MAY, *Eagle Factory, Tewkesbury.*—Riveted boots, shoes, patented goloshes, and machine-closed uppers.

By Her Majesty's Royal Letters Patent.

THE KENSINGTON GOLOSH, or solid leather over-shoe for ladies' or gentlemen's wear.

This golosh is made with either a high back or a legging attached; and being of solid leather, is very superior to india-rubber in wear, and is easily repaired.

It is put on without the aid of a shoe-lift, and when on forms a protection for the back of the boot from the rubbing of the steel skirts.

Class 1.—Boots with extra light riveted soles for the ball-room, house, and summer wear.

Class 2.—Boots with medium soles for ordinary wear.

Class 3.—Boots with extra stout and clump soles for winter wear.

Class 4.—Machine-closed uppers.

Boxes containing sample dozens of assorted uppers to be obtained at the factory, price £3 3*s.*

[5013]

LANAGAN, 9 *Brownlow Street, Bedford Row.*—Illustrations of a principle in boot-making; instrument for measuring distorted feet.

[5014]

LANGDALE, H., 57 *Mount Street, Grosvenor Square.*—Boots and shoes, and needlework.

[5015]

LATHAM, JOHN, 214 *Oxford Street.*—Ladies' boots and shoes.

[5016]

LEPRINCE, 261 *Regent Street.*—Patented chameleon shoes, with transparent changing colour, according to the lady's taste.

[5017]

LINE, WILLIAM & JOHN, *Daventry.*—Gentlemen's boots and shoes.

[5018]

LOWLEY, JAMES, 71 *Briggate, Leeds.*—Boots and shoes, various.

[5019]

MABANE, JAMES & WILLIAM, 3 *Templer Street, Leeds.*—Boot tops and dressed leather.

[5021]

MEDWIN, JAMES, & Co., 86 *Regent Street, and* 23 *Gracechurch Street.*—Boots and shoes; royal resilient boots, perfectly elastic, without india-rubber.

[5022]

MURRAY, JAMES FRANCIS, 34 *Great Russell Street, Bloomsbury.*—Morning slippers.

[5023]

NEALE, GEORGE, 4 *Albert Place, Queen's Road, Holloway.*—Boots for persons lame or with contracted hips, ankles, &c.

[5024]

NORMAN, S. W. & E. G., 3 & 4 *Oakley Street, S.*—Box cork boot, without rand or stitch in sole; can be made any thickness for lameness; also ladies' cork heel boots and shoes.

Obtained Honourable Mention at the Exhibition of 1851.

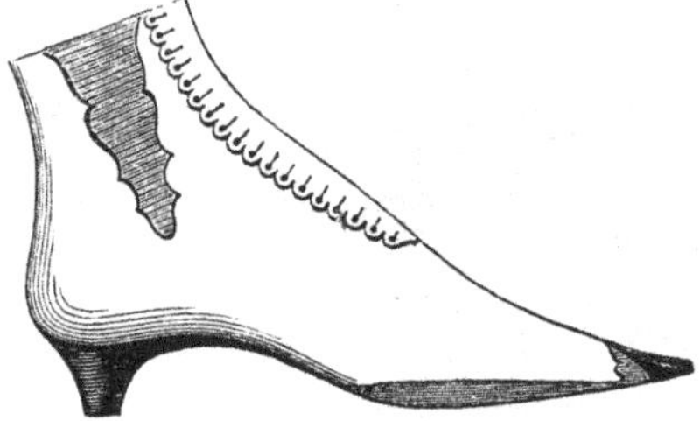

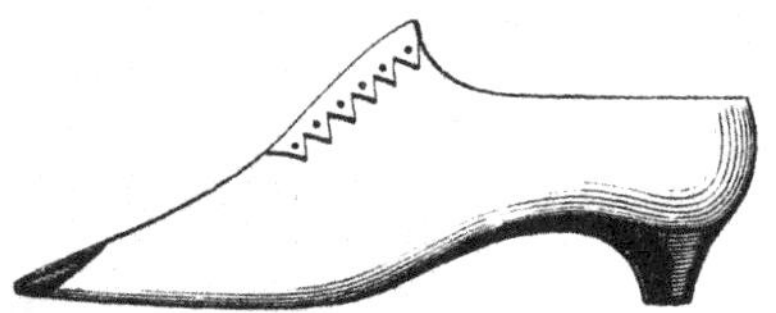

The construction of the heel of these boots is such that it will retain its shape and polish. It also supports the waist of the foot, is lighter and more durable than any other heel, and easier and quicker to manufacture.

NORMAN'S gaiter boot, requiring no spring or fastening, can be worn inside or outside of the trousers, made to measure, 35*s.*

Cork sole boots made for every description of lameness.

LIST OF PRICES.

	£	*s.*	*d.*
GENTLEMEN'S Wellington Boots	1	5	0
Side Spring Boots	1	2	0
Lace Calf Boots	1	5	0
The Best Shooting Boots	1	10	0
LADIES' French Kid Cork Heel Boots . .	0	13	6
Patent Leather, with Norman's Heel . .	0	13	6
Riding Boots	1	5	0

Cork Soles, 4*s.* extra.

An assortment of the best goods of their own manufacture kept in stock.

[5025]

OGDEN, FREDERICK, 66 *Prince's Street, Leicester Square.*—Manufacturer of boot, breeches, and glove trees, stretchers, &c.

Sample articles are kept in stock, manufactured in the best manner, and with the best materials, and include portable and solid boot and shoe trees.

His newly-invented instep-stretching boot tree; also his new instep- and toe-stretching boot tree. Instep and joint stretchers. Glove and gauntlet trees and stocking boards.

Lasts made to the feet on anatomical principles.

Boots, breeches, and gloves sent to be fitted will receive immediate attention. Price lists on application.

[5026]

PALEY, ROGER, *Leeds.*—Boots and shoes.

[5027]

PANZETTA & ANDREW, 141 *New Bond Street.*—Shooting boots.

[5028]

PARKER, WILLIAM, & SONS, *Wood Street, Northampton.*—Boots, shoes, slippers, &c.; also a small boot and shoe machine.

Obtained a Prize Medal at the Exhibition of 1851.

[5029]

PATENT PLASTIQUE LEATHER COMPANY, *Quay, Ipswich.*—Boots with sole and heel moulded, perfectly solid, damp-proof.

[5030]

PEAL, NATHANIEL, *Duke Street, Grosvenor Square, London.*—Boots, shoes, and materials, principally of Peal's waterproof leather.

This leather (of which specimens are exhibited) is waterproof throughout its entire substance; possesses great toughness and flexibility; and being exempt from the injurious influence of atmospheric changes, is much more durable than ordinary leather.

The samples of manufactured goods consist of boots for hunting, shooting, and walking, both of waterproof and common leather.

[5031]

PHIPPS, BARKER, & CO., *Cadogan House, Sloane Street, S.W.*—Boots and shoes.

[5032]

POCOCK BROTHERS, 20 *to* 23 *Southwark Bridge Road, S.E.*—Boots and shoes.

[5033]

REID, JOHN, 99 *Regent Street.*—Ladies' and gentlemen's boots and shoes, of the best quality.

[5034]

ROBERT, AUGUSTE, 26 *Change Alley, Cornhill.*—Boots.

[5035]

ROBERTS, DANIEL, 9 *New Bond Street.*—Hunting and military boots.

The boots exhibited have no extraordinary work about them, but are simply specimens of the exhibitor's usual make.

[5036]

SCARD, ANTHONY, 8 *Bow Lane, Cheapside.* (*See page* 64.)

[5037]

SEAGER, *Ipswich.*—Boots and shoes.

[5038]

SOMERVELL BROTHERS, *Netherfield, Kendal.*—Uppers and leggings. (*See pages* 58 *to* 63.)

[5039]

STAGG, ANTHONY, 34 *Little South Street, Wisbeach.*—Men's and women's lasts, boot and glove trees.

[5040]

STOKES, HENRY, 27A *Coventry Street, Haymarket.*—Patent gaiter boots.

SOMERVELL BROTHERS, *Netherfield, Kendal.*—First-class gentlemen's, ladies', and children's boot and shoe uppers and leggings.

For Outside View of the Works, see Advertisements at the end of this Part.

GENTLEMEN'S HALF BOOT UPPERS.

MEMEL LEGS.

No.
1 French Calf Goloshed, Elastic Sides
1x Do. do. do. Mock Buttoned
2 Do. do. Buttoned
2x Patent Calf do. do.
3 French Calf do. Laced Front Balmoral
3c Do. do. do. do. Calf Linings
4 Do. do. do. do. Toe Cap
4c Do. do. T. C. do. do. Calf Linings
9 Do. do. Elastic Sides, Mock Laced
9x Do. do. do. do. Toe Cap
10 Enamelled Hide Goloshed, Laced Front, Balmoral
11 Do. do. do. do. Toe Cap
95 Patent Calf do. Elastic Sides, Mock Buttoned
95x

KID LEGS.

No.
3x French Calf Goloshed, Laced Front, Balmoral
4x Do. do. do. do. Toe Cap
5 Do. do. Elastic Sides
5x Do. do. do. Mock Buttoned
5xxp Satin Finish, Leg and Golosh, Elastic Sides
5xx Do. do. do. Mk. B.
6 French Calf Goloshed, Buttoned
6x Patent Calf do. do.
7 Do. do. Elastic Sides
7x Do. do. do. Mock Buttoned
8xx Do. Toe Cap do.
80x Half Boot, Seamless do.
10x Patent Calf Goloshed, Laced Front, Balmoral
11x Do. do. do. do. Toe Cap
12 Enamelled Hide do. do. do.
12x Do. do. Elastic Sides, Mock Laced
13 Do. do. Laced Front, Balmoral, T. C.
13x Do. do. Elastic Sides, Mk., L. do.
14 Patent Calf do. do. do.
14x French Calf Goloshed, Elastic Sides, Mock Laced
15 Patent Calf do. do. do. Toe Cap.
15x French Calf do. do. do. do.
90x Enamelled hide do. do. Mk. Buttoned
90xp do. do. do.

PATENT CALF.

No.
80 Half Boot, Seamless, Elastic Sides
81 Do. Laced Front, Fancy Toe Cap
90 Do. Elastic Sides, Mock Laced
94 Do. Goloshed, do.
94x Do. do. do. Mock Buttoned

ENAMELLED HIDE.

No.
16 Half Boot, Blocked Back, Bellows Tongues, Toe Cap
85x Do. Elastic Sides, Mock Buttoned
86 Do. do. Second quality
91 Do. Laced Front do.

FRENCH CALF.

No.
60 Half Boot, Laced
60x Do. do. Second quality
0 Do. do. Third do.
80x Do. Seamless, Elastic Sides
85 Do. do. Mock Buttoned
100 Do. Laced, Bellows Tongues

WELLINGTON LEGS.

No.
92 Patent Dress Short Wellingtons
93 Do. Long do.
104 Paris Calf do. do.
104x Bordeaux Calf do. do.
105 Paris do. Short do.
105x Bordeaux do. do. do.

SHOOTING BOOTS.

No.
16x Grained Kip, Blocked Back, Bellows Tongues, Toe Cap
16xp Do. do. do.
102x Do. Goloshed do. Toe Cap
100x French Calf do. do.
102 Do. Seamless do.

MISCELLANEOUS.

No.
8 French Lasting Leg, Patent Calf Goloshed, Elastic Sides
8x Do. do. Toe Cap do.
*101 Chagriné Morocco Leg, Patent Calf Goloshed, Elastic Sides
*101x Do. do. do. do. Mock Buttoned
*101xx Do. Enamelled Hide Vamp do. do.
*103 Levant Morocco Leg, Patent Calf Goloshed, Buttoned
106 Jockey Legs and Tops, White or Brown
107 Peel Riding Boots
108 Enamelled Hide Napoleons
108x Do. Peel Top, or the 'Havelock Boot'
109 Kid Quarter Boot, French Calf Goloshed, Elastic Sides
109x Do. Patent Calf do. do.

SHOE UPPERS.

No.
17 French Calf Oxford Laced
17x Do. Buttoned
19 Do. Elastic Sides
19x Do. do. Mock Buttoned
96 Do. Seamless, Elastic Sides, Mk. B.
18 Memel Quarters, French Calf Vamp, Buttoned
18c Chagriné Morocco Quarters, French Calf Vamp, B.
18L Memel do. do. Laced
18x Memel Qrs, French Calf Vamp, Elast. S. Mk. B.
27x Do. Patent do. Laced
20 Kid Quarters do. do. Buttoned
20L Do. do. do. Laced

SOMERVELL BROTHERS, *Netherfield, Kendal.*—First-class uppers and leggings—*continued.*

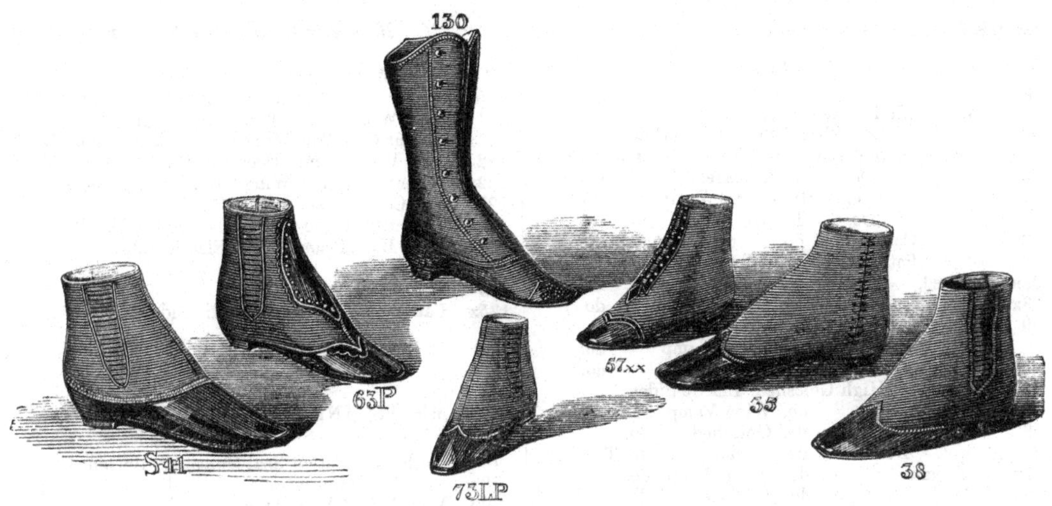

LADIES' UPPERS.

SHOE UPPERS—*continued.*

No.				
20x	Kid Quarters, Patent Calf Vamp, Elast. Sds. Mk. B.			
20xxp	Do. Satin Finish, Vamp and Qrs.	do.		
20xx	Do.	do.	do.	do. Mk. B.
24	Kid Qrs. Patent Calf Goloshed, Laced Oxford			
28	Do.	do.	Vamp, Elastic Fronts	
29	Do.	do.	Goloshed, Elast. S. Mk. Lace	
21	Enamelled Hide Oxford Laced			
21x	Do.	Elastic Sides		
26	Do.	Buttoned		
26x	Do.	Mock Buttoned		
28x	Do.	Seamless, Elastic Sides, Mk. B.		
99x	Do.	do. Laced		
22	Patent Calf Oxford	do.		
22x	Do.	Elastic Sides		
22xx	Do.	Seamless	do.	Mk. B.
*23	Morocco Qrs. Patent Calf Vamp, Buttoned			
*23x	Do.	do.	Elastic Sides, Mk. B.	
*24x	Do.	do.	Goloshed, Laced	
*97	Chagriné Morocco Qrs. Patent Calf Goloshed, Laced			
25	Tweed Qrs. Patent Calf Goloshed, Laced			
25x	Do.	do.	do.	do. Toe Cap
27	Do.	do.	Vamp	do.

BOYS' AND YOUTHS' UPPERS.

No.				
60B	French Calf Half Boot, Laced, 10-13			
60Y	Do.	do.		1-4
61B	Memel Leg, French Calf Goloshed Balmoral			10-13
61Y	Do.	do.	do.	1-4
61XB	Do.	do.	Elastic Sides	10-13
61XY	Do.	do.	do.	1-4
61XXB	Do.	do.	do. Mk. B.	10-13
61XXY	Do.	do.	do. do.	1-4
62B	Do.	do.	Buttoned	10-13
62Y	Do.	do.	do.	1-4
62XY	French Calf Oxford Shoe, Laced			1-4

LADIES' UPPERS.

CLOTH LEGS.

No.			
30	Patent Seal Toe Cap, Laced Sides		
30x	Do.	Low Straight Goloshed, Laced Sides	

LADIES' UPPERS—CLOTH LEGS—*continued.*

No.				
31	Patent Seal	Low Wing Vamp,		Laced Sides
32	Do.	High	do.	do.
32x	Cordovan	do.	do.	do.
33	Patent Seal	do. Goloshed		do.
33x	Cordovan	do.	do.	do.
42	Patent Seal	Toe Cap,		Elastic Sides
43	Do.	Low Wing Vamp		do.
44	Do.	High	do.	do.
45	Do.	do. Goloshed		do.

CASHMERE LEGS.

English.

No.				
46	Patent Calf Toe Cap, Elastic Sides			
47	Do.	Low Wing Vamp, Elastic Sides		
48	Do.	High do.	do.	
49	Do.	do. Goloshed	do.	
46x	Do.	Toe Cap	do.	Mk. B.
47x	Do.	Low Wing Vamp	do.	do.
48x	Do.	High do.	do.	do.
49x	Do.	do. Goloshed	do.	do.
46L	Do.	Toe Cap	Laced Sides	
47L	Do.	Low Wing Vamp	do.	
48L	Do.	High do.	do.	
49L	Do.	do. Goloshed	do.	
50	Patent Seal	Toe Cap	do.	Second quality
51	Do.	Low Wing Vamp	do.	do.
52	Do.	High do.	do.	do.
52x	Do.	do. Goloshed	do.	do.

French.

No.				
53	Patent Calf	Toe Cap	Laced Sides	
54	Do.	Low Wing Vamp	do.	
55	Do.	High do.	do.	
56	Do.	do. Goloshed	do.	
53x	Do.	Toe Cap	Elastic Sides	
54x	Do.	Low Wing Vamp	do.	
55x	Do.	High do.	do.	
56x	Do.	do. Goloshed	do.	
53xx	Do.	Toe Cap	do.	Mk. B.
54xx	Do.	Low Wing Vamp	do.	do.
55xx	Do.	High do.	do.	do.
56xx	Do.	do. Goloshed	do.	do.

SOMERVELL BROTHERS, *Netherfield, Kendal.*—First-class uppers and leggings—*continued.*

LADIES' UPPERS—*continued.*

KID LEGS.

No.
34 Patent Calf Toe Cap, Laced Sides
35 Do. Low Wing Vamp, Laced Sides
35B Blocked, Patent Calf Low Wing Vamp, Laced Sides
35x Do. do. Goloshed do.
36 Do. High Wing Vamp do.
37 Do. do. Goloshed do.
37x Cordovan do. do. do.
38 Patent Calf Low Wing Vamp, Elastic Sides
38B Blocked, Do. do. do. do.
38x Do. do. Goloshed do.
39 Do. Toe Cap do.
40 Do. High Wing Vamp do.
40B Blocked, Do. do. do. do.
41 Patent Calf High Goloshed, Elastic Sides
57 Do. do. Wing Vamp, Balmoral
58 Do. do. Goloshed do.
59 Do. do. do. do. Toe Cap
57x Cordovan do. Wing Vamp do.
58x Do. do. Goloshed do.
59x Do. do. do. do. Toe Cap
57xx Patent Calf Fancy Toe Cap do.
63 Do. High Wing Vamp Elastic Sides, Mk. B.
63x Do. do. do. do. Pearl
64 Do. do. Goloshed, Elastic Sides, Mk. L.
64x Blocked, Patent Calf High Wing Vamp, Buttoned
65 do. Do. Goloshed do.
65x Cordovan do. do. do.
66 Patent Calf High Wing Vamp, Balmoral
66x Do. do Elastic Sides, Mock Lace, Patent Facings, two Rows Black or Gilt Buttons
68xK Half Boot (Seamless) Elastic Sides
69 Patent Seal Vamp, Balmoral, Second quality

MISCELLANEOUS.

No.
67 Patent Calf Half Boot, Laced Front
68 Do. do. Elastic Sides
68x Do. do. Seamless Elastic Sides
74 Do. Leg and Low Wing Vamp, Elastic Sides, Mk. B.
68xM Strasbourg Morocco Half Boot, Seamless, Elastic Sides
*67x Chagriné Morocco Leg, Patent Calf High Wing Vamp, Balmoral
*74x Do. do. Elastic Sides, Mk. B.
74xx Mauve Kid Leg, Patent Calf High Wing Vamp, Elastic Sides, Pearl Mk. B.
75 Memel Goat Leg, Cordovan Goloshed, Buttoned
78 Cordovan Leg, High Wing Vamp, Laced Front
78x Do. Half Boot, Goloshed, Laced Front
79 Do. High Wing Vamp, Laced or Buttoned

Bronze or Black Glacé Kid Leg.

No.
120 High Wing Vamp, Elastic Sides, Mk. B.
120x Patent Calf High Wing Vamp, Elastic Sides, Mk. B.
121 Low Wing Vamp, Elastic Sides, Mk. B.
121x Patent Calf Low Wing Vamp, Elastic Sides, Mk. B.
122 Low Wing Vamp, Elastic Sides
122x Patent Calf Low Wing Vamp, Elastic Sides
123 Toe Cap, Elastic Sides, Mk. B.
123x Patent Calf Toe Cap, Elastic Sides, Mk. B.
124 Half Boot, Seam to Toe, Elastic Sides
124x
125 Toe Cap, Elastic Sides
125x Patent Calf Toe Cap, Elastic Sides

MISSES' AND CHILDREN'S UPPERS.

No.
70 Cloth Leg, Patent Seal Goloshed, Buttoned
70x Do. Cordovan do.
70G Memel Goat Leg, Patent Seal do.
70GX Do. Cordovan do.
70K Kid Leg, Patent Seal do.
70KX Do. Cordovan do.
71 Memel Goat Leg, Patent Seal Vamp, Laced Front
71x Do. Cordovan do. do.
71K Kid Leg, Patent Seal do. do.
71KX Do. Cordovan do. do.
72 Cordovan Half Boot, Laced Front
73 Kid Leg, Patent Calf Vamp, Elastic Sides
73x Do. do. do. Mk. B.

KNICKERBOCKER LEGGINGS, manufactured in

Black Enamelled Leather
Drab do. do.
Black Grained do.
Russet do. do.
Spanish Cordovan
Tanned Bazil
Leather Cloth, Leather lined, either Buckle, Button, or Spring.

Also Leggings for the Army.

For list of prices apply to SOMERVELL BROTHERS, Netherfield, Kendal.

For Leather List refer to Catalogue for Class XXVI., page 15.

SOMERVELL BROTHERS, *Netherfield, Kendal.*—First-class uppers and leggings—*continued.*

BALMORAL SHOE UPPER.

ELASTIC SHOE UPPER.

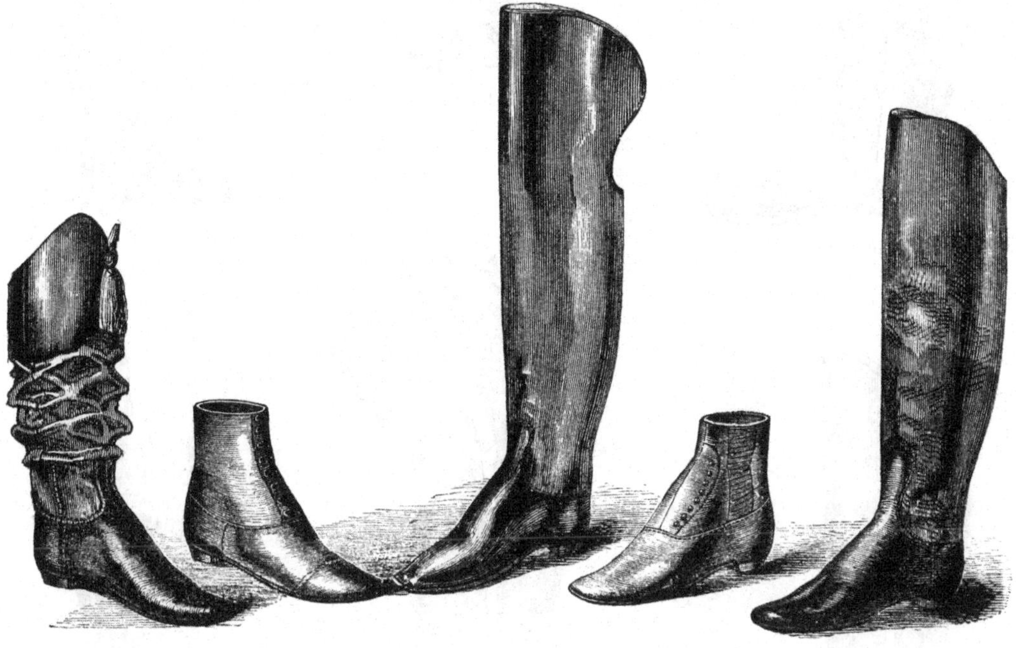

HESSIAN. SHOOTING. PEEL RIDING. ELASTIC. NAPOLEON.

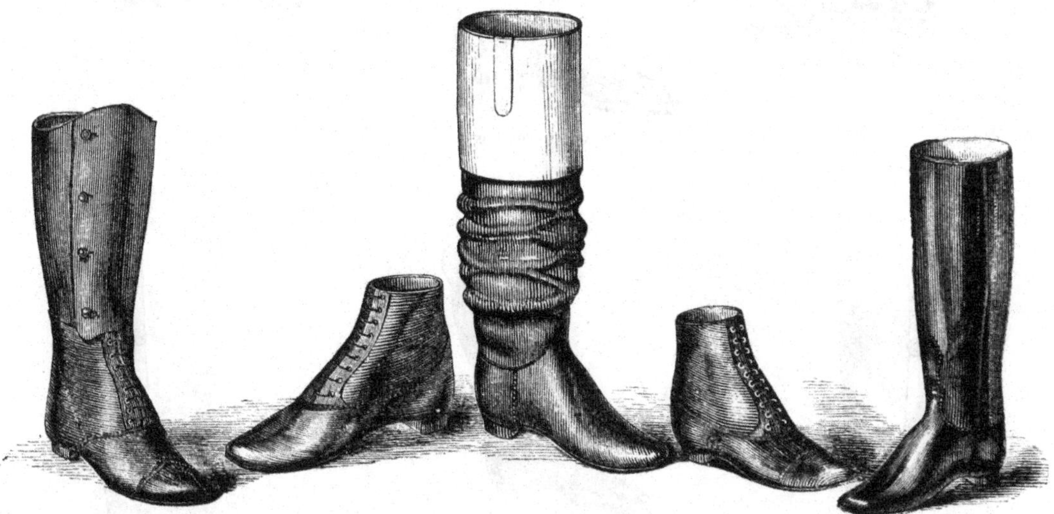

ELCHO. SEAMLESS SHOOTING. JOCKEY. BALMORAL. DRESS WELLINGTON.

SOMERVELL BROTHERS, *Netherfield, Kendal.*—First-class uppers and leggings—*continued.*

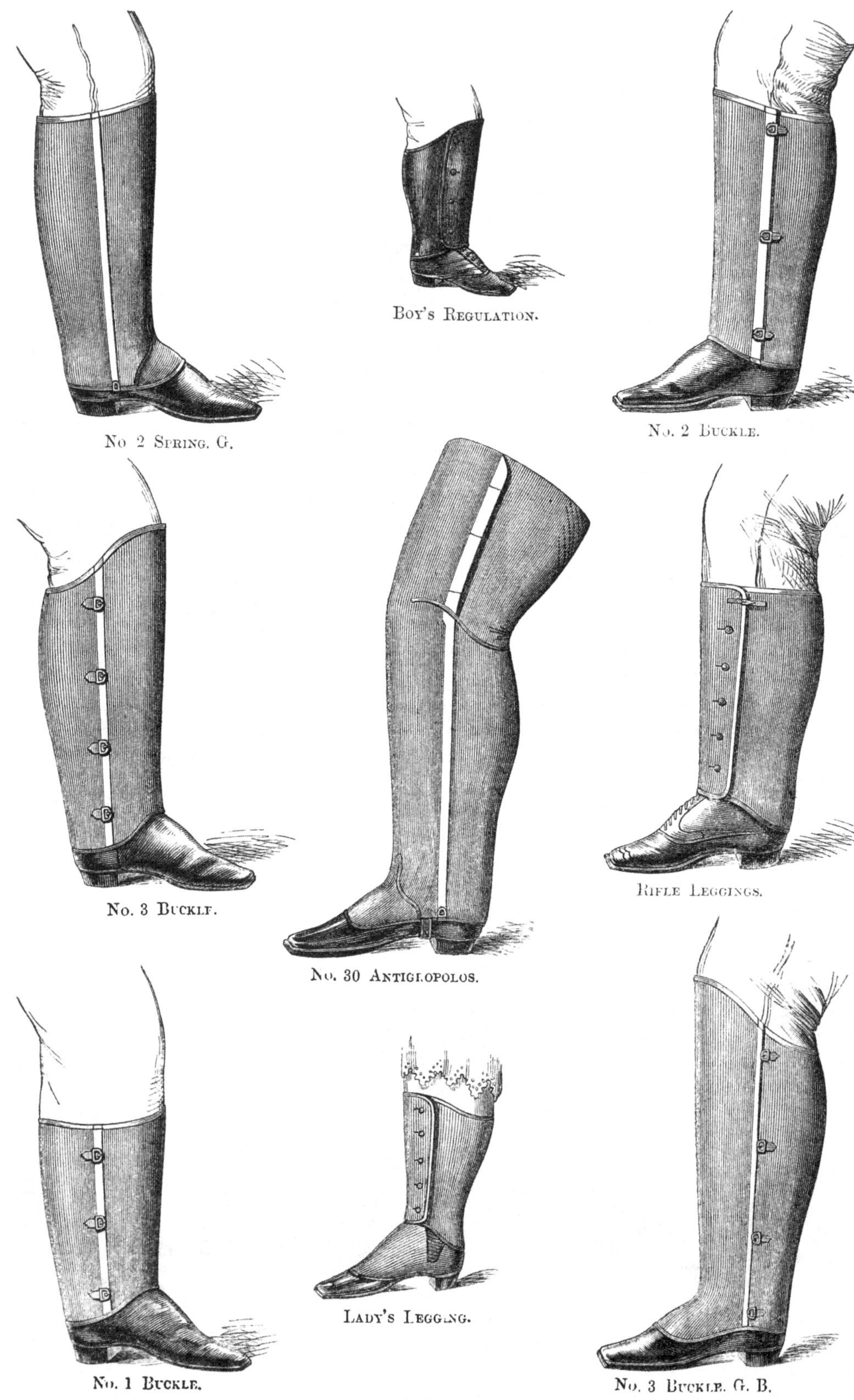

No 2 SPRING. G.

BOY'S REGULATION.

No. 2 BUCKLE.

No. 3 BUCKLE.

No. 30 ANTIGLOPOLOS.

RIFLE LEGGINGS.

No. 1 BUCKLE.

LADY'S LEGGING.

No. 3 BUCKLE. G. B.

SOMERVELL BROTHERS, *Netherfield, Kendal*—continued.

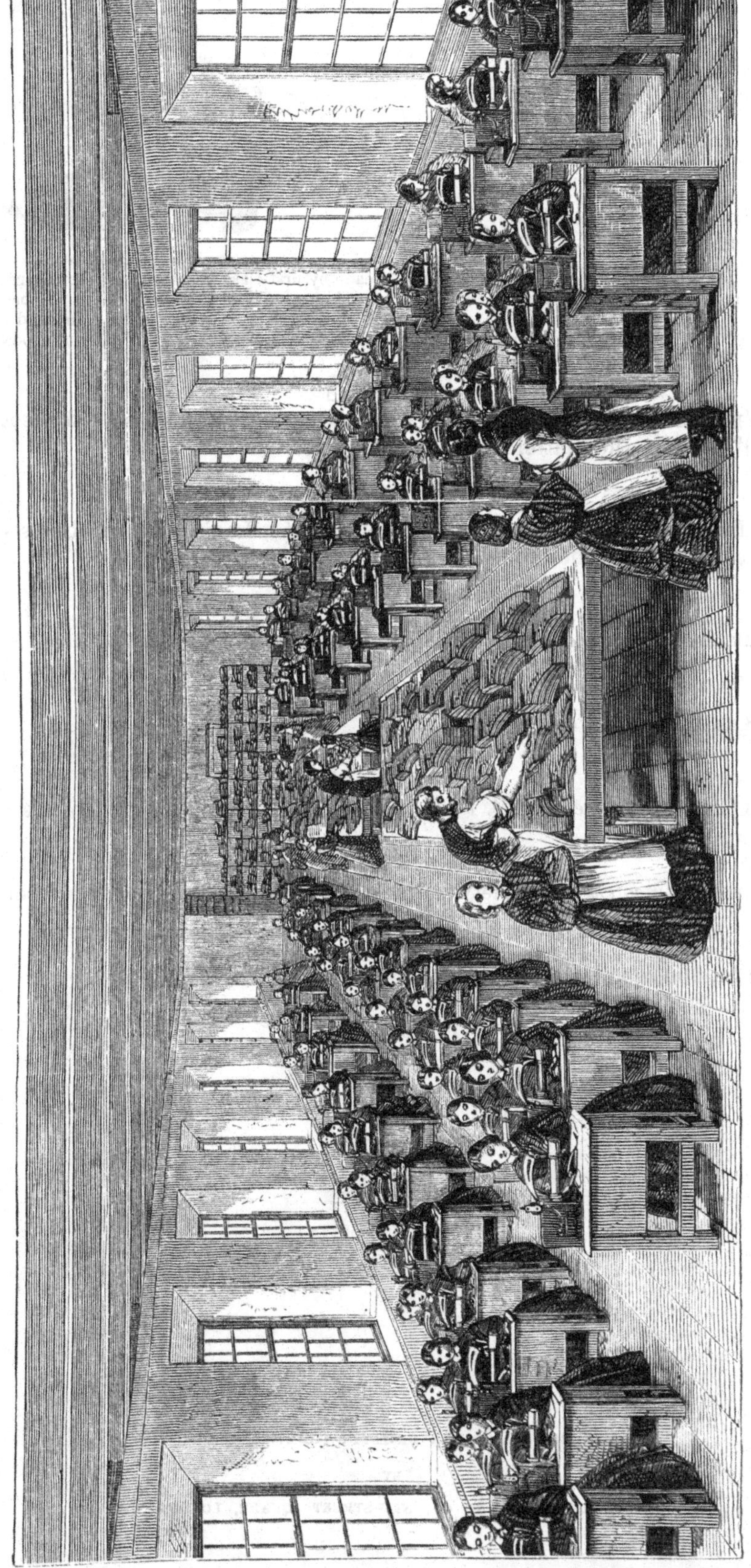

SKETCH OF THE MACHINE-ROOM, NETHERFIELD, KENDAL.

SCARD, ANTHONY, 8 *Bow Lane, Cheapside.*—Specimens of workmanship, and the process of bootmaking in various stages.

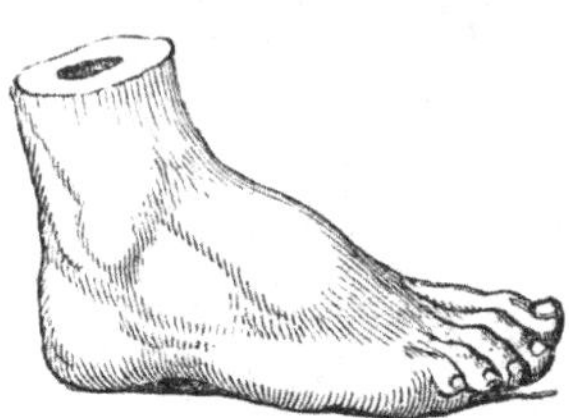

The exhibitor devotes special attention to the formation of the feet, constructing for each foot lasts adapted to its peculiarities, without any additional charge to the purchaser. He employs no machinery; every article is made by first-class workmen, so as to ensure neatness and durability. The materials used are the best procurable, the boots are at once easy and elegant, and the prices are less than those often asked for inferior goods.

[5041]

SURRIDGE, WILLIAM HENRY, Naval, Military, and General Bootmaker, 275 *Regent Street, London, W.*—Boots and matters pertaining thereto.

[5042]

TALLERMAN, REBECCA, & SON, 131 *Bishopsgate Without.*—Ladies' and children's waterproof boots and shoes of every description.

[5043]

TODD, THOMAS, 24 *Colliergate, York.*—Fancy boots in silk and leather.

[5044]

WALKER & KEMPSON, *Leicester.*—Riveted boots.

[5045]

WALSH, WILLIAM, 44 *Bolsover Street, Portland Place.*—Ladies' and gentlemen's boots.

[5047]

WINTER, CHARLES, *Norwich.*—A variety of ladies' boots, for home trade and exportation.

[5048]

YAPP, PETER S., 200 *Sloane Street, London.*—Boots and shoes.

First-class ladies' boots and shoes. First-class gentlemen's boots.

PRINTED FOR HER MAJESTY'S COMMISSIONERS

BY

SPOTTISWOODE AND CO., NEW-STREET SQUARE, LONDON

Class XXVIII.

PAPER, STATIONERY, PRINTING, AND BOOKBINDING.

Sub-Class A.—*Paper, Card, and Millboard.*

[5081]

Barling, Joseph, *Park Mill, East Malling, near Maidstone.*—Specimens of paper and millboard made from hop-bine.

[5082]

Burgess & Ward, *Mendip Paper Mills, near Wells, Somerset.*—Straw paper, with illustrations of its manufacture and applications.

[5084]

Greer, Alfred, & Co., *Dripsey and Glenville Mill, Cork.*—Writing, printing, and brown papers.

[5085]

Lamb, John, *Holborn Paper Mills, Newcastle, Staffordshire.*—Pottery tissue and other papers.

[*Prize Medal, Great Exhibition,* 1851.]

Pottery tissue for printing from copper-plates, &c., and transferring the pattern on china and earthenware; and the material from which it is manufactured.

[5087]

Routledge, Thomas, Patentee, *Eynsham Mills, Oxford.*—Paper from esparto or alfa fibre; half stuff from same.

[5088]

Saunders, Thomas Harry, *Queenhithe, London, and Dartford, Kent.*—Hand and machine made paper of every description.

[*Obtained first-class Medals at the Exhibitions of* 1851; *New York,* 1853; *and Paris,* 1855.]

Bank-note, account-book, drawing, writing, plate, printing, cheque, blotting, and other papers.

Machine Mills—Phœnix and Hawley Mills, Dartford, Kent.

Hand or Vat Mills—Darenth and Sundridge Mills, Kent; Beech Mill, Wycombe Marsh, and Rye Mill, High Wycombe, Bucks.

Town offices and depôt—Maidstone Wharf, Queenhithe, London.

[5089]

SZERELMEY, NICHOLAS CHARLES, *Laboratory, New Houses of Parliament, and* 20 *Abingdon Street, S.W., and Pannonia Building Leather Cloth Factory, Park Road, Acre Lane, Clapham.*—Arabian zopissa waterproof and paper boards processes.

No. 1. Zopissa paper boards and paper pipes, to any size and length, impregnated and prepared with Zopissa composition.

The Zopissa paper boards are much stronger than oak or other timber, they are unabsorbent, thoroughly waterproof, not liable to decay, a non-conductor, and can be used with the greatest advantage for ship building on the largest scale. Portable houses, hospitals, barracks, coach panels, railway carriages, tanks, and cisterns (particularly those intended to contain a strong acid solution) covering roofs and floors, boxes, &c. &c.

No. 2. The Zopissa paper pipes for water, are non-conducting and consequently particularly applicable for the supply pipes for houses, as the water will not freeze in them. They are not corroded or affected by acids or any other liquids or gases. They are light, and stronger than metal, and not brittle or liable to crack.

No. 3. Unexpansible rocket cases and impervious artillery cartridge cases of Zopissa paper.

No. 4. Pannonia leather cloth for boots and shoes, and leggings, carriage heads and aprons, furniture, bookbinding, travelling bags, waterproof coats and capes, &c. &c.

The Pannonia leather cloth has been privately tested for nearly two years, and found to answer much better and durable than leather for boots and shoes. Its properties are that, it is softer to the foot, is not liable to crack or shrink, is impervious to wet, permits the perspiration to pass off, is peculiarly adapted for the army, is quite equal in appearance to, and can be rendered at least 40 or 50 per cent. cheaper than leather.

[5090]

HOOK, TOWNSEND, CHARLES, & CO., *Snodland Paper Works, near Rochester, Kent.*—Writing and printing papers; webs for envelopes.

[5091]

TURNBULL, J. L. & J., *Holywell Mount, Shoreditch.*—London, Bristol, and crayon drawing-board, mounting-board, pasteboard, and address card.

[5092]

WOOLLEY & CO., 210 *High Holborn.*—Photographic, drawing, mounting, card, and paste boards; ivory and message cards.

[5093]

MOULEY, J. W., *Lower Mills, Wooburn, Bucks.*—Mill-boards.

[5094]

TOWLE & JEFFERY, *Oxford.*—Paper boards, and pipes made of straw.

SUB-CLASS B.—*Stationery.*

[5102]

ARNOLD, P. & J., 135 *Aldersgate Street.*—Writing inks and fluids.

[5104]

BANKS & CO., *Greta Pencil and Black-lead Works, Keswick.*—Pencil, penholder, leads for pencil-cases, and solid block black-lead. (*See page* 3.)

BANKS & Co., *Greta Pencil and Black-lead Works, Keswick.*—Pencil, penholder, leads, &c.

[*Obtained Exhibitor's Medal,* A.D. 1851.]

The exhibitors are manufacturers of drawing pencils, from the celebrated Borrowdale lead (obtained only in that locality), and every other description of drawing, office, pocket-book, drapers', and carpenters' pencils, penholders, &c., and also of the solid block black-lead for stoves and grates.

BANKS & Co., are the original and only manufacturers of the polished fireproof points or leads, suitable for Perry's, Lund's, Mordan's, or any other pencil-case, of any length or gauge. Pencil manufacturers to Her Majesty the late Queen Adelaide, the King of Saxony, and the King of the Belgians. Established A.D. 1833.

London Warehouse, 21 Cannon Street West.

[5105]

BARCLAY, ROBERT, 29 *Bucklersbury, E.C.*—Patent indelible paper, bank-cheques for the prevention of fraud, patent copying paper for the prevention of the fading of copies taken by machine.

[*Obtained a Prize Medal at the Exhibition of* 1851 *for account books.*]

The exhibitor is the patentee and inventor of Barclay's patent indelible bank-cheques and paper, for the prevention of fraudulent alterations of the amount, crossing, or cancelling of a bank cheque. Adopted by the Bank of Ireland and its branches.

Barclay's patent indelible paper affords an effectual protection to bankers from a description of fraud of the most dangerous character. Several cases have lately occurred of the fraudulent alteration of bank cheques by chemical means, and in every instance known this fraud has escaped detection. Any vegetable colour or printing can be applied to the production of cheques upon this paper. This invention is also applicable to the prevention of the fading of common writing ink from natural causes. Professors Brande, Miller, Warrington, and Frankland have stated that this invention affords the most perfect protection hitherto devised against this description of forgery.

Testimonial from PROFESSOR BRANDE, M.D., F.R.S., F.C.S., &c., *of the Royal Mint, Author of the well-known* "*Manual of Chemistry,*" *&c, &c.*

"Royal Mint, 5th October, 1859.

"I am unable to point out any method, which would escape detection, of discharging a portion of the writing placed upon this paper; and am of opinion that it throws greater difficulties in the way of fraudulent alterations than any other invention with which I am acquainted.

(Signed) "WILLIAM THOMAS BRANDE, M.D.,
"F.R.S., F.C.S."

Barclay's chemically prepared patent copying paper, prevents the fading of faint copies of letters taken in common copying ink. It also insures a sharper, blacker copy. A specimen is exhibited which shows that the copy is as distinct as ever, after 10 hours soaking in acidulated solution of chlorine.

Articles Exhibited.

Specimen of the removal of an ink cancelling from the signature of a bank cheque, for which the cash was afterwards obtained.

Specimens illustrating the security of the indelible paper.

Specimens of cheques on the patent paper in every variety of printing.

Specimens of relief engraving, including a trade label possessing unusual difficulties to fraudulent imitators. Executed for Messrs. Guinness of Dublin.

Bank, mercantile, and private account books.

[5106]

BAUERRICHTER & Co., 41 *Charterhouse Square, London.*—Ornamental boxes of cardboard, wood, and fancy papers.

[5107]

BAUS, HENRY, 59 *Hatton Garden, London.*—Impressions from dies and seals; models and drawings.

[5108]

BENNETT, C. W., & Co., 14 *Smith Street, E.C.*—Plume and handkerchief boxes.

[5109]

BLACKWOOD & Co., 18 *Bread Street Hill.*—Copying and writing inks, bankers' safety sealing-wax, indelible marking and stamping ink.

The following are exhibited, viz.:—

A fine black writing ink for steel pens, perfectly pure, jet black, and fluid.

A new manufacture of blue-black writing fluid, made for ledger writing, dries quickly, not apt to smear the books, flowing freely from the pen, and turning to a deep permanent jet black.

A jet black fluid possessing same qualities.

Samples of an entirely new manufacture of writing ink, being chemical solutions, without sediment, and perfectly black.

Improved mercantile machine copying ink, very powerful and fluid.

The same, double strength, to copy months after written with.

Ordinary red ink of an improved quality.

Permanent ultramarine blue ink.

A new red ink termed "rubian," perfectly clear, fine bright permanent scarlet, very difficult to obliterate from paper; valuable to merchants, bankers, and lawyers for endorsing.

Permanent marking and stamping ink for linen, &c.

Improved black bordering ink.

Various descriptions of sealing wax.

Bankers' safety sealing wax, of great strength.

These inks are contained in the patent syphon bottle, which obviates the nuisance of drawing a cork; A is the stopper, with tube divided in two channels, moving freely in the hollow cork, which with the tube is bored crosswise at C C; the air enters at E, causing the ink to flow up the other division of the tube and out at the spout D. When not in use the stopper is turned half round, keeping the contents free from air and dust.

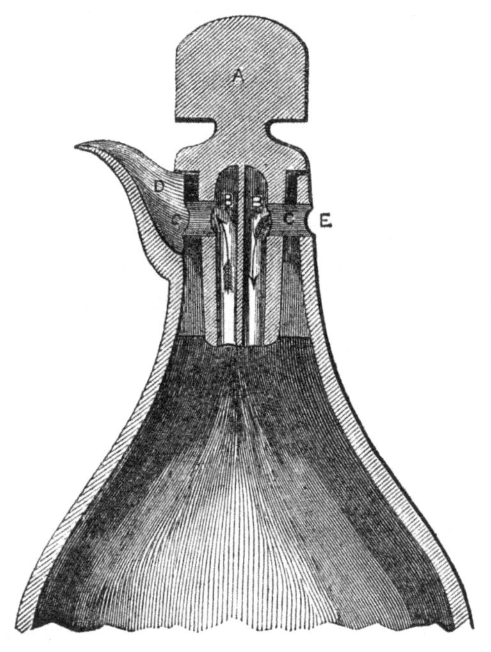

[5110]

BOUSQUET, ISAAC, 28 *Barbican, City, E.C.*—Gold, silver, and foil papers, gold paper for paper-hangings.

[5111]

BOWDEN, GEORGE, Inventor, Artists' and School Stationer, 314 *Oxford Street, W.*—Book head-bands and elastic welling-bands.

[5112]

BRETNALL, 24 *Huntley Street, Tottenham Court Road.*—Tracing-papers, cloths, &c.

[5113]

BROOKMAN & LANGDON, 28 *Great Russell Street, Bloomsbury, London, W.C.*—Pencils and patent pencil-holders.

[*Obtained a Medal at the Paris Exhibition,* 1855.]

The patent pencil-holders are the invention of the exhibitor.

[5114]

BROOKS, HENRY, & Co., *Cumberland Market, London, N.W.*—Inkstands, pen-trays, letter-seals and other articles of stationery.

[5116]

BROWN, WILLIAM, & Co., 40 & 41 *Old Broad Street, E.C.*—Banking and commercial stationery.

[5117]

CALDWELL BROTHERS, 15 *Waterloo Place, Edinburgh.*—Designs of arms, crests, monograms, for stamping on paper, envelopes, &c.

[5118]

CANTON, ROBERT, 7 *Dowgate Hill, and* 27 *College Street, E.C.*—Fancy stationery.

[5119]

CARLYLE, GEORGE, 28 *Bold Street, Liverpool.*—Manifold writer, and carbonic paper.

CARLYLE'S MANIFOLD WRITER is different from any in use. By the old method a style was the only medium by which a copy could be produced, neither the original or fac-simile being in ink. This was a serious defect, and rendered the use of manifold writers only very limited. By G. C.'s method every letter can be written with ink and a fac-simile is produced at the same time. It is simple in its application, effecting a great saving of time and trouble.

Prices, large letter size, 500 leaves, paged and index, 10*s.* 6*d.*, large note, 7*s.* 6*d.*

[5120]

CLEMENTS & NEWLING, 96 *Wood Street, London.*—Account-books in general, and special stationery for drapers.

[5121]

COCHRAN, PETER, *Liverpool.*—Improved red (scarlet) ink by new process; also writing-fluid, office and copying inks.

[5122]

COHEN, B. S., 9 *Magdalen Row, Great Prescot Street.*—Improved process of manufacturing artists' and account-book pencils. (*See page* 6.)

[5123]

COLLYER, ROBERT HANHAM, M.D., F.C.S., 8 *Alpha Road, St. John's Wood.*—Patent chemical ink pencils.

COLLYER'S EVER-POINTED DRY INK PENCIL. By Royal Letters Patent, dated 1859.

In case complete, price one penny.

This will be found to be an efficient substitute for pen and ink, being hard, of a full black colour, and perfectly indelible.

Sold also in boxes, suitable for the propelling pencil cases in ordinary use, price sixpence per box.

These pencils may be had of all Stationers in town and country, and wholesale, of Grosvenor, Chater and Co. 86 Cannon Street West, London, E.C.

[5124]

CORFIELD, JOSEPH, & SON, 7 *Farringdon Street, E.C.*—Marble papers, head-bands, specimens of book-edges, &c.

[5125]

COWAN, ALEXANDER, & SONS, 77 *Cannon Street West, and Valley-field Mills, near Edinburgh.*—Writing paper, all qualities; printing paper; parchment paper; account books, &c.

[5126]

CREESE, J., & Co., *Birmingham.*—Patent illuminated crystal and gold show and other tablets; printed mouldings, boxes, &c.

The exhibitors are the Sole Licensees, under Mr. Breese's Patent, for producing the gorgeous CRYSTAL SHOW TABLETS.

The gold and colours, as well as pearl, &c., being printed and permanently fixed and protected on the back of the glass, are not in the least affected by atmospheric changes or the sun's rays, even if placed in the hottest window or room: they are not one third the cost of those done by hand.

They are also, under the same patent, imitators of wood, for frames, boxes, imitation inlaying &c.

Invoices, account and note heads, business cards, circulars, and manufacturers' pattern books, in chalk or ink drawing and engraving, are executed in the highest style by Creese & Co., with the greatest promptitude.

COHEN, B. S., 9 *Magdalen Row, Great Prescot Street.*—Improved process of manufacturing artists' and account book pencils.

The illustration shows the workman engaged in filling in pencils with B. S. Cohen's newly invented Continuous Compressed Cumberland Lead. Slips of this material having been prepared of the exact size and degree required, the workman proceeds to fill the centre cavity of the cedar with one unbroken length of lead, thus rendering each pencil unvarying in its degree throughout, and obviating the inconveniences arising from the ordinary method of laying in several small pieces in succession, which snap at each joint and incur the further liability of having different degrees mixed in one pencil.

This lead is manufactured from the far-famed Borrowdale plumbago, which, cleansed from all impurity, rivals the original mineral in its most valued qualities, being rich in colour, remarkably smooth and tough in texture, and alone possessing the quality of rubbing out readily without leaving a trace. These pencils are adapted to every variety of climate, as the extremes of heat, cold, or damp have not the slightest effect on them, while, from the peculiar tenacity of the lead, they are not liable to break, even at the finest point. This improved mode of manufacture has secured the entire approval of the most eminent artists of this country, as the accompanying testimonials will show.

These pencils may be obtained of any respectable artists' colourman or stationer in the United Kingdom. Price 3*d.* each, or 2*s.* 6*d.* per dozen, thus supplying the want so universally felt, of a really good pencil at a moderate cost. In ordering, please be particular in asking for B. S. Cohen's Compressed Cumberland Lead.

TRADE MARK

From the Palace, Osborne.

"The Princesses find them excellent, possessing every advantage a pencil can offer."

From FREDK. TAYLER, ESQ., *President of the Old Society of Painters in Water Colours.*

"Every good quality that a pencil can possess is combined in those you have sent me.

Fredk. Tayler

From JOHN GILBERT, ESQ., *Wilmington, near Dartford.*

"They are in all respects better than any I have yet drawn with.

John Gilbert

From BIRKET FOSTER, ESQ., 12 *Carlton Hill East, St. John's Wood.*

"I find them most excellent, and can only say I wish I had had them a year or two ago.

Birket Foster.

From HENRY WARREN, ESQ., *President of the New Society of Painters in Water Colours.*

"They are without exception the best pencils I have ever used. Among their many qualities of excellence, that of unaptness to break in cutting and in use is no trifling advantage; your very hard pencils are a boon to those who, like myself, draw frequently on wood.

Henry Warren

From JAMES BRIDGES, ESQ., *Head Master of Royal Military Academy, Woolwich.*

"I have thoroughly tried each variety of mark, and find them uniform in quality and strength of colour throughout, also easy of erasure; in short, they appear excellent.

James Bridges

From the Head Master of the Training School for Masters of Schools of Art, Science and Art Department, South Kensington.

"Mr. BURCHETT has much pleasure in bearing testimony to the very excellent quality of the pencils submitted to him by Mr. COHEN; they are without doubt the best pencils he has ever tried, and merit the patronage of all draughtsmen and Schools of Art, &c. &c."

[5127]

DOBBS, KIDD, & CO., *London.*—Note-papers, envelopes, valentines, perforated and embossed goods, dies, and stamping.

Note papers in packets.
Mourning note papers and envelopes.
Hand-folded envelopes.
Wedding envelopes.
First class valentines.

Dies for embossing crests, arms, monograms, and addresses.

Specimens of stamping in relief in one or more colours.

Embossed and perforated goods.

[5129]

EDWARDS, ELIEZER, *Birmingham.*—Ink in bottles for export, glass inkstands, paper-weights, holy water fonts, &c.

[*Obtained Honourable Mention at the Exhibition of* 1851.]

Glass inkstands of various sizes, shapes, and qualities; some of which are ornamented with gilding and colour. Paper weights, holy water stoups, &c., ornamented. Ink in glass bottles suitable for exportation. The whole exhibited only as articles adapted for general sale.

[5130]

ELLIOTT, DANIEL, *York Cottage, Park Road, Old Kent Road.*—Marking-ink, linen stretcher, &c., with specimens.

[5131]

FASE & SON, *Edward Terrace, Kensington.*—Piccolo writing-cases, fairy cakes, cottage writing-cases, puzzle castles, and conundrum cubes.

Piccolo writing-case in leather, size 6 inches by 3, containing full size paper and envelopes for 30 letters, pens, holder, ink, pencil, knife, scissors, housewife (filled) blotter, &c.; from 10*s.* to 21*s.* A writing-case in shape of a cottage, with contents. A puzzle castle made from paper. A fairy surprise cake in upwards of 100 pieces; prices various, according to contents.

[5132]

FETHERSTON, JOHN J., 18 *Suffolk Street, Dublin.*—Sealing and chromo-embossing official dies, seals, stamps, &c.

Combined sealing and chromo-embossing official die and press; by which any lady or gentleman can emboss in colours, official, heraldic, or other devices, on books, documents, &c.—the die being easily detached from the press, becomes a seal for wax, &c.

Combined office or wine seal, with changeable centre, by which the expense of many seals and the liability to use a wrong one is greatly obviated.

Register, hotel, or other stamp with movable numbers, by which the articles of an individual or department can be identified.

[5133]

GILL, MISS JANE, 9 *High Street, Hastings.*—Artificial flowers in tissue paper.

[5134]

GOODALL, CHARLES, & SON, *Camden Town, N.W.*—Playing cards.

[5135]

GOODHALL & DINSDALE, *Pancras Lane, London, E.C.*—Account books, paper envelopes, and general commercial stationery.

[5136]

GRAHAM, T. & R., 10 *High Street, Paisley.*—Spool tickets.

[5137]

HARRINGTON, JOSIAH, *Lansdowne Terrace, Brixton, S.*—Specimen of apparatus containing twelve blades for pointing pencils

[5138]

HIGGINSON, MRS., *Uxbridge.*—Flowers manufactured in paper.

[5140]

HOWARD, WILLIAM, 23 *Great Russell Street, Bloomsbury.*—Specimens of tracing papers, vellums, and mounting linens.

[5141]

HYDE & CO., 61 *Fleet Street, London, E.C.*—Copying letter-book, stationery, sealing-wax, and writing inks.

[*Obtained a Prize Medal for sealing wax at the Exhibition of* 1851.]

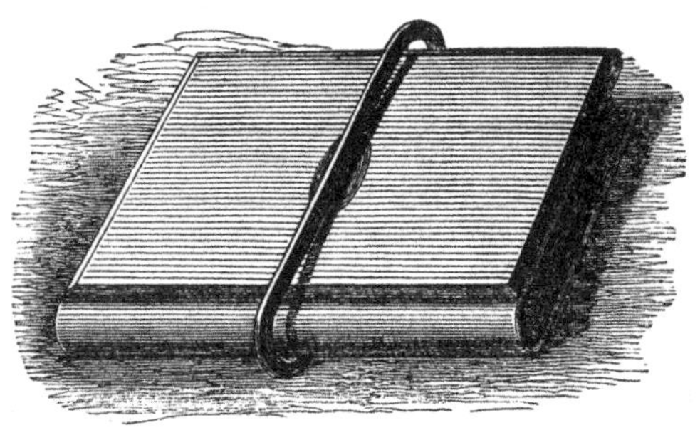

Hyde's patent portable Clamp Copying Apparatus.

This is a perfectly novel and simple method of applying great and effective pressure to the ordinary letter-copying book.

The means employed differ from any hitherto known, and give a most effective copy in less time than by the old method.

The Patentees submit the invention to the public, believing it supplies the desideratum of an effective, portable, and economical copying apparatus.

Hyde & Co.'s sealing wax is used exclusively by the British Government, and exported to every quarter of the world. They are the only makers of the well-known "Bank of England wax," and the "India wax," for hot climates.

Hyde's British Empire writing inks.

The makers have given great attention to the production of every description of writing inks, so as to combine fluidity in writing and permanency of colour with freedom from any tendency to thicken.

Hyde's Scarlet Ink, or Commercial Red Fluid, is of peculiar brilliancy and does not fade in any climate. Hyde's "India Writing Fluid," is specially made for use in hot climates and for rapid writing; it becomes a very intense and lasting black.

[5142]

IBBOTSON & LANGFORD, *Manchester.*—Plain, fancy, and tinfoil papers and pasteboard makers.

[5144]

JARROLD & SONS, 47 *St. Paul's Churchyard, London, and Norwich.*—Specimens of bookbinding.

[5145]

JOB, BROTHERS, & CO., 75 *Cannon Street West.*—Tracing papers, for strength, transparency, and colour, to stand all climates; also drawing paper, plain and mounted.

[5146]

JOHNS, GEORGE EDWARD (formerly of 4 *Falcon Street, City*), 9 *Bath Street, Newgate Street, City.*—Decorated boxes manufactured of paper and other materials.

[5147]

JOHNSON & ROWE, 17 *Warwick Square, Paternoster Row, E.C.*—Pocket-books, purses, &c. (*See page* 9.)

[5148]

JONES & CAUSTON, 47 *Eastcheap, City, London.*—Account books, stationery, and printing.

[5149]

KING, JONATHAN, 56 *Seymour Street, Euston Square, London, N.W.*—Fancy floral valentines.

[5150]

LAW & SONS, 37 *Monkwell Street, London.*—Specimens of bookbinding cloths manufacfactured by them.

JOHNSON & ROWE, 17 *Warwick Square, Paternoster Row, E.C.*—Pocket-books, purses, &c.

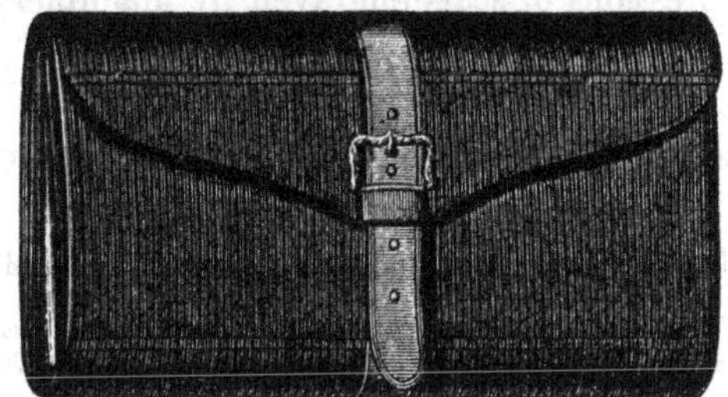

Metallic books in every style of binding, with elastic bands or clasps, ruled or plain.

Metallic books in russia or velvet, with gilt rims and locks, round corners, and lined with silk; also same size with elastic band or clasp.

Metallic wallets in numerous patterns and styles, in roan, morocco, or russia, gilt or plain.

Metallic betting books in a variety of bindings.

Metallic wallets with registered strap and buckle fastenings in all sizes—see illustration.

Instrument wallets in russia or morocco with elastic bands or locks, gilt or plain.

Gentlemen's housewifes in russia with gilt lock, places for instruments, thread, needle, &c., &c.

Patent secret purses in russia leather, various patterns, with secret pockets for gold—see illustration.

Tourists' writing-cases in russia, morocco, or roan, in a variety of patterns.

Despatch boxes in every quality and pattern.

Patent secret purses with registered fastenings.

Purses in every kind of leather, pattern, or size, with elastic bands, locks, or registered fastenings.

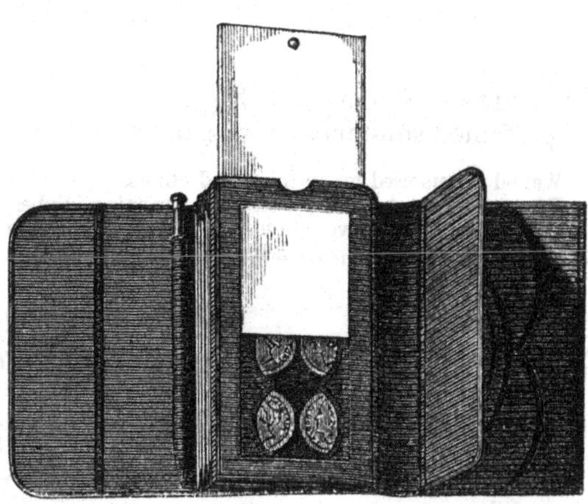

Card cases (registered) in russia or morocco, with gilt metal or silver rims, plain or engraved.

Purses in russia or velvet, with ivory leaves or books, gilt metal or silver rims and locks.

Purses in russia or morocco, with gilt metal or silver perforated mounts, in a variety of artistic designs.

Pocket ledgers in all sizes and styles of binding.

Manuscript and account books in every size, quality and style of binding.

Blotting books in roan, morocco, or russia, with or without locks.

Card cases in russia or morocco, round or square edges gilt or plain.

[5151]

LETTS, SON, & Co., 8 *Royal Exchange.*—Diaries, commercial stationery, and leather goods.

No. 1. Printers in letter-press, copper-plate, lithography, and photography, and the new chemical process, by M. Paul Pretsch.

No. 2. Engravers of bank-notes, coupons, cheques, bills, arms, crests, &c.

No. 3. Publishers of Letts's diaries, and other compilations for MS. purposes, and commercial works generally by commission.

No. 4. Stationers, account-book, and copying-machine manufacturers, wholesale and for exportation.

No. 5. Bookbinders in vellum, cloth, paper, and all kinds of leather.

No. 6. Guide, atlas, chart, and globe sellers. All the newest publications.

No. 7. Map agents, by appointment, to H.M. Board of Ordnance.

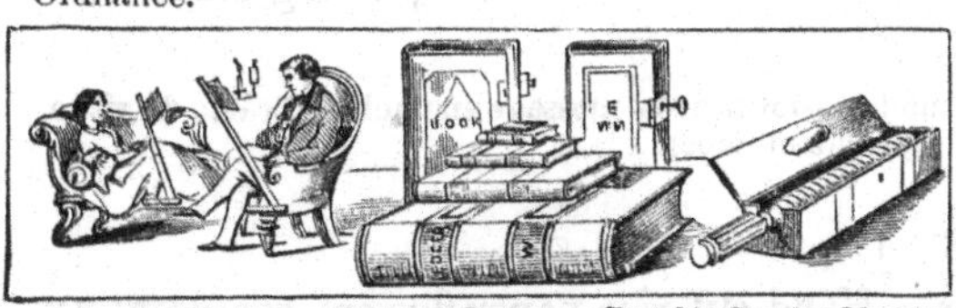

Reading Easels. Account Books. Portable Copying Machine.

LETTS'S DIARIES provide (in addition to numerous tables of reference) a ruled space, in which to enter the events, appointments, or requirements of life, for every day in the year, and as a book of direct personal appliance is adapted to the wants of every class of society by many varieties of form, size, and price; details being furnished by a small descriptive catalogue, supplied gratuitously through nearly every bookseller in the civilized world (also at the Exhibition stall). Large quantities for shippers, public companies, and the trade are supplied at wholesale prices, or (in the event of special private forms) by contract, &c. The other publications for MS. purposes embrace library catalogues, cellar, game, stable, and rent books, sermon, church, and parish registers, &c. Messrs. Letts undertake the publication of any commercial work on commission, if desired The ordnance and geological maps of England are alike valuable and moderate in price, surpassing, in both respects, anything that can be produced by private enterprise. Descriptive catalogues, specimens, and illustrations may be obtained at the Exhibition stall, or at the above address.

[5152]

LUNTLEY, JOHN, 42 *Bishopsgate Street Without.*—Ticket, receipt, and till protector.

[5153]

LYONS, WILLIAM, *Fennel Street, Manchester.*—Ink for writing and copying purposes; also brilliant red ink.

[5154]

MCGLASHAN, HENRY, 8 *Helmet Row, Old Street, St. Luke's.*—Gold, silver, and foil paper.

[5156]

MARTIN, THOMAS, *Newton Abbott, Devon.*—Wax impressions of seals engraved by machinery.

[5157]

MATTHEWS, WILLIAM, 1 *Wigmore Street.*—Waxed papers for wrapping oily, adhesive, or perfumed substances, soaps, mustard, &c.

Waxed papers used by chemists and others.

For covering ointments, plasters, &c., wrapping jujubes, scented soaps, violet powder, mustard, linseed meal, horse balls, and other greasy, perfumed, or adhesive substances, without any of the objectionable results of using tinfoil, and at half the cost.

Price per box of 50 square feet, white, 2*s.*, various tints, 2*s.* 6*d.*; per ream, white, 30*s.*, various tints, 32*s.* 6*d.*

[5158]

MEAD & POWELL, 101 *Whitechapel, and* 73 *Cheapside.*—Account books and manufactured stationery.

[5159]

MEEK, GEORGE, 2 *Crane Court, Fleet Street, E.C.*—Embossed and laced papers and envelopes.

[5161]

MORDAN, FRANCIS, *Albion Works,* 326 *City Road.*—Gold pens, pencil-cases, sealing-wax, patent purses, inks, &c.

[5162]

NICHOLSON, JOHN, 45 *Leeds Road, Bradford, Yorkshire.*—Account books, fancy stationery, and pattern cards.

[5163]

ORTNER & HOULE, 3 *St. James's Street, S.W.*—Heraldic seal and die engraving, embossing, and designing.

[5164]

PATERSON BROTHERS, *Peel Grove, Old Ford Road.*—Mechanical guard books for filing commercial papers in volumes.

[5165]

PERRY, JAMES, & CO., 37 *Red Lion Square, and* 3 *Cheapside.*—Pencils, elastic bands, and inkstands.

[5166]

POLLARD, GEORGE, *Foot's Cray Paper Mill, Kent, and* 10 *Walbrook, City, E.C.*—Paper for envelopes, and envelopes.

[5167]

REYNOLDS, JOSEPH, & SONS, *Vere Street, Lincoln's Inn Fields.*—Patent playing-cards, and cards for the blind.

Manufacturers of photograph and mounting boards, drawing and Bristol boards, message enamel and carte de visite cards; hot-pressers and glazers of paper.

[5168]

RIDDIFORD, JANE, 14 *Cowley Street, Westminster.*—Hand-cut rice-paper flowers, representing the seasons; and other specimens.

[5169]

ROBINSON, J. B., & SON, *Brampton, near Chesterfield, and* 17 *Bouverie Street, London, E.C.*—Chemists' and perfumers' cardboard boxes, &c.

Arrowroot cases. Aromatic cashoo boxes.
Dispensing powder boxes. Violet powder cases.
Lozenge boxes, various patterns. Marking-ink cases.
Fumigating pastille boxes.
Powder puff boxes with glass tops.
Cardboard pill boxes; assorted sizes, plain and shouldered.
Seidlitz and soda powder boxes.
Slide pill boxes. Tooth-powder boxes.
Cosmetique cases, and camphor ball boxes.

[5170]

ROWSELL, SAMUEL W., & SON, 31 *Cheapside, E.C.*—Account books suitable for public companies and commercial purposes.

[5171]

SHOLL, JAMES, 5A *Chapel Street, Spital Square, London.*—Patent improved writing-paper, and anticorrosive writing and copying fluid.

[5172]

SMITH, JEREMIAH, 42 *Rathbone Place, London.*—Patent tracing cloth and adhesive envelopes.

[5174]

STEAD, CHARLES, & SON, *Dalton, near Huddersfield.*—Design, or point paper.

Design paper, ruled by machine in all numbers, 30 inches square.

This design paper possesses numerous properties not possessed by any other, and is now in general use in the manufacturing districts of Great Britain. Price 22s. per quire, nett.

References are permitted to first-class designers who use this paper.

[5175]

STODART, MATILDA, 31 *Cloudesley Terrace, Islington.*—Flowers modelled in rice paper, prepared paper, paper and wax.

M. Stodart's dahlias and other flowers in rice and prepared paper, have been commended at the flower shows, and highly praised by the press. For sale in groups and sprays at the Centre Transept, Crystal Palace, Sydenham, and at the Floral Hall, Covent Garden.

[5176]

STRAKER, SAMUEL, & SONS, 26 *Leadenhall Street, and* 80 *Bishopsgate Street Within.*—Mercantile stationery.

[5177]

SUTCLIFFE, WILLIAM, 101 *Bunhill Row, London, E.C.*—Labels, and improved glove bands.

[5178]

TANNER BROTHERS, *Welsh Back, Bristol.*—Account books; specimens of rulings and bookbindings.

[5179]

THOMPSON, HENRY, *Albert Cottage, Weybridge Heath, near Chertsey, Surrey.*—Medallion or imitative cameo wafers on envelopes.

[5180]

UNWIN, GEORGE, *The Gresham Steam Press*, 31 *Bucklersbury, E.C.*—Specimens of letterpress printing, lithography, engraving, bookbinding, stationery, &c. (*See page* 12.)

[5181]

WARNER, R., 18 *Newman Street, W.*—Specimens of seal-engraving.

[5182]

WATERSTON, GEORGE, *Edinburgh, and* 3 *Queen's Head Passage, Paternoster Row, London.*—Sealing wax and wafers for home and export.

[*Obtained Prize Medal at the Exhibition of* 1851; *and at Paris in* 1855.]

[5183]

WEBSTER, HENRY, 23 *Litchfield Street, Soho, W.C.*—Patent portable travelling inkstands, and writing-cases.

UNWIN, GEORGE, *The Gresham Steam Press*, 31 *Bucklersbury, E.C.*—Specimens of letter-press printing, lithography, engraving, bookbinding, stationery, &c.

Letter Press.—Forme of Pearl type, containing 6,588 pieces.
Stereotype cast from ditto, and proof.
Forme of Gem Music type, containing 2,686 pieces.
Stereotype cast from ditto, and proof.
Raised blocks for printing in colours, and proof.
Specimens of bookwork, in ancient and modern types.
Specimens of typographic printing.
Specimen book of the various founts in use. (110 pp.)
Lithography.—Stones showing the method of printing in colours, with proofs.
Specimens of lithographic printing.
Engraving.—Copper plates for share certificates, cheques bill-heads, cards, &c., with proofs.
Raised electrotype of cheque for surface printing.
Specimens of copper-plate printing.
Account Books.—Ledger, 6 quires royal, bound white vellum, double russia bands, laced; price 63s.
Journal, 5 quires demy, bound rough calf, single russia bands, laced, lettered, and paged; price 53s.
Cash book, 5 quires fcap., bound green vellum, single russia bands, laced, lettered, and paged; price 22s.
Specimens of General Stationery, &c., &c., &c.

Bookwork, Magazines, and Newspapers.—Attention has been specially directed to this department, with a view to render it as complete as possible. The founts of modern type are numerous, and their beauty and utility cannot be surpassed. They are constantly being replaced from the best Foundries. Estimates for works of any extent, in either modern or ancient faced type, forwarded on application.

GUTTEMBERG AND FAUST'S FIRST PROOF FROM MOVEABLE TYPE.

Commercial Printing—of every description is executed with the greatest possible speed and economy. Contracts entered into with merchants, and others.

Bankers' Cheques & Notes.—By a new process, these can be printed at a very moderate rate, either in fugitive inks or on water-marked paper, secure from forgery, without extra charge, and perforated, numbered, and bound in various sizes.

The Ancient Faced Series of Type have been for ſome years in conſtant uſe. They are caſt from the original matrices, which were cut at the beginning of the laſt century. To render this ſeries more complete, a great variety of ornamental letters, and head and tail pieces, have been engraved at a great coſt, and are not to be found in any other office. The early deſigns have been followed as nearly as poſſible, thus preserving the unity of deſign, without which the ancient faced work is ſhorn of half its individuality and beauty. See Specimen Book, pp. 39 to 71.

Merchants, Public Companies, Charitable Institutions and Societies, supplied with every requisite in Stationery. Writing Paper of the finest quality, procured from the best mills. Envelopes of all sizes, adhesive, and stamped with the name of the firm on the flap, if required. Account Books of all sizes and patterns, made to order, with patent spring backs, and bound in a superior manner, on the shortest notice. Stationery exported to all parts of the world.

[5184]

WEDGWOOD, R., & SONS, 9 *Cornhill.*—Manifold writers.

[5185]

WETHERFIELD, ROSALIE, 1 *Henrietta Street, Covent Garden.*—Paper flowers.

[5186]

WILLIAMS, COOPER, & CO., 85 *West Smithfield, London.*—Paper manufactured by William Joynson and Son; rags; rags bleached and in pulp; writing-papers, note-papers, &c.

[5187]

WILSON, JAMES LEONARD, 128 *St. John Street.*—Cloth for bookbinding.

[5188]

WILSON, ROBERT, *Keswick, Cumberland.*—Black-lead pencils, and ever-pointed leads.

[5189]

WOOD, JOSEPH THOMAS, 278 *Strand.*—Engravings; perforated cards and papers; ornamental stationery.

Lace and fancy papers, cards, collars and cuffs, views of public buildings, and ornamental stationery.

[5190]

HOUGHTON, W., *New Bond Street.*—The Queen's note-paper.

[5191]

WARD, M., *Belfast.*—Ledgers, bookbinding, and illuminating.

[5192]

WARNER, G. E., *Poland Street.*—Seals.

[5193]

WYON, J. S.—Impressions from seals.

SUB-CLASS C.—*Plate, Letterpress, and other Modes of Printing.*

[5200]

ADAMS & GEE, Printers, 23 *Middle Street, West Smithfield, E.C.*—Printing on metal, as invented by them.

[5201]

ASHBY & Co., 79 *King William Street, London, E.C.*—Ornamental engraving for bank-notes, certificates, and bonds, by patented machine.

[5202]

AUSTIN, STEPHEN, *Hertford, Herts.*—Printed books in Oriental and other languages, and bookbinding.

[5203]

BAGSTER, SAMUEL, & SONS, 15 *Paternoster Row.*—Polyglot typography; binding; illumination; and gold and other mountings.

[5204]

BANK OF ENGLAND, *Threadneedle Street.*—Specimens of surface printing.

[5206]

BELL & DALDY, *Fleet Street.*—Books.

[5207]

BEMROSE, WILLIAM, & SONS, *Derby and Matlock Bath.*—Bound books; letterpress and lithographic printing.

[5208]

BESLEY, ROBERT, & Co. (CHARLES REED and B. FOX), *Fann Street, Aldersgate Street.*—Metal types and specimens.

[5209]

BISHOP, JOHN, 4 *North Audley Street, London.*—Polysciagraphic engravings by clockwork for preventing forgery.

[5210]

BLACK, A. C., *Edinburgh.*—Books.

[5212]

BOOTH, LIONEL, 307 *Regent Street, W.*—Reprint of first edition of Shakspeare. (*See page* 15.)

[5213]

BONNEWELL, WILLIAM HENRY, & Co., 76 *Smithfield, E.C.*—Wood types and materials used in printing, &c.

The exhibitors are wood letter and block cutters, engravers, manufacturers of every description of printing materials, printers' brokers, and agents for printing machines, presses, and the Caxton printing inks, &c., &c. Letter stamps, arms, &c., cut for castings, shop-fronts, soap, sacks, and for stamping, marking, architecture, &c. Specimens of these goods are exhibited.

[5214]

BRADBURY, WILKINSON, & Co., 12 & 13 *Fetter Lane, E.C.*—Bank-notes, share certificates, debentures, and documents requiring printing security.

BANK-NOTES, printed from plate, embodying every known method for the effectual prevention of forgery, especially prepared against attempts at successful imitation by the two most dangerous processes, photography and transfer. The ornamental work is produced by the most expensive and complex machinery—in combination with first class hand engraving. The vignettes are executed by the first line engravers in the country, and the whole note possessing a degree of security never before attained.

BANK-NOTES, printed from surface (as the Bank of England). These notes contain specimens of the highest and most elaborate style of work of which surface printing is capable—possessing a security second only to those printed from plate.

FOREIGN POSTAGE AND RECEIPT STAMPS.—These are produced by either of the above modes of printing, specimens of each being shown in various colours.

PATENT STEEL FACED PLATES.—This process enables the plate to yield any number of impressions without the slightest deterioration to the work, the steel surface only being renewed from time to time as required. The closest uniformity is thus secured, a matter of so much importance in Bank-note issues.

[5215]

BRANSTON, F. W., *The Grove, Southwark.*—Patent imperishable advertising tablets.

[5216]

BROOKS, VINCENT, 1 *Chandos Street, Charing Cross.*—Lithographic printing.

[5217]

CASLON, H. W., & Co., 22 & 23 *Chiswell Street.*—Types, and printed specimens of types.

[5218]

COATHUPE, CAPT. H. B., 1 *Abingdon Lane, Kensington.*—Printing on metals.

[5219]

COLLINGRIDGE, W. H., *City Press, Aldersgate Street, London.*—Specimens of printing, engraving, &c., with materials used.

[5220]

COLLINS, WILLIAM, *Glasgow.*—Specimens of binding; family Bibles; specimens of printing; New Testament on one sheet.

In Case.

1. Family Bible, with Commentary, with steel engravings from photographs, bound in real morocco, gilt edges, 32*s*.
2. Family Bible, with Commentary, with steel engravings from photographs, bound in Turkey morocco, flexible super-extra gilt edges, 40*s*.
3. Family Bible, with Commentary and Chromo-lithograph illustrations, bound in Turkey morocco, antique extra tooled edges, 52*s*. 6*d*.

The above Bibles are exhibited as combining excellence in typography, illustration, and binding with lowness of price.

On Screen.

The New Testament printed on a single sheet of paper.

[5221]

COLVILL, HENRY J. M., 52 *Queen Street, Camden Town.*—Printing in pure silver.

Booth, Lionel, 307 *Regent Street, W.*—Reprint of first edition of Shakspeare.

Mr. WILLIAM

SHAKESPEARES

COMEDIES,
HISTORIES, &
TRAGEDIES.

Publiſhed according to the True Originall Copies.

*

LONDON

Printed by Iſaac Iaggard, and Ed. Blount, 1623; and Re-Printed for Lionel Booth, 307 Regent Street, W., 1862.

* The famed Engraving by Droeſhout will be reproduced on the Title-page in the beſt way to ſecure identity of appearance with the original that preſent art can accompliſh.

This reprint compriſes *three* ſizes; one to range with all good Octavo Editions of Shakeſpeare, another to range with Knight's Pictorial, and ſimilar Editions, the third being uniform with the Original Folio.

The chief object in the reproduction of this, for all critical purpoſes, the moſt important edition of Shakeſpeare extant, has been, not mere reſemblance, but that it ſhall prove "ſo rarely and exactly wrought"—page for page, line for line, word for word, letter for letter, ornamentation for ornamentation—as to be, excepting a more convenient ſize, "one and the ſelf-ſame thing" with its prototype. That attempt has been ſucceſsful, the teſtimony of the moſt important Journals of the time has ſatisfactorily ved.

[5222]

COX, GEORGE JAMES, 46 *Stanhope Street, Hampstead Road, and Polytechnic, London.*—Impressions taken from ferns, leaves, lace, &c.

The Foliographic Press for printing from nature. A new and exceedingly simple machine for printing from fresh leaves, ferns, grasses, feathers, or lace; manufactured and sold by the inventor, and at the Polytechnic Institution, Regent Street, London.

The apparatus is not only useful to botanists, but affords an interesting employment to ladies; and is of so simple a construction that children may use it as an amusing and instructive toy.

Press, roller, tube of ink, and sheet of carbonized paper, price—small size, 5*s.* 6*d.*; large, 8*s.* 6*d.*

[5223]

CROSS, JOSEPH, & SON, 18 *Holborn Hill.*—Specimens of engraving, lithography, and printing, and labels cut by machinery.

[5224]

DAY & SON, 6 *Gate Street, Lincoln's Inn Fields, London.*—Specimens of lithography, chromo-lithography; plate printing of every description, artistic and commercial; and of illustrated and illuminated works produced as well as published by the exhibitors.

[5225]

DELACY, GEORGE, 38 *Sekforde Street, Clerkenwell.*—Specimens of tools, letters, blocks, &c., for bookbinding.

[5226]

DICKES, WILLIAM, 5 *Old Fish Street, Doctors' Commons, London.*—Specimens of engraving and oil-colour printing.

[5227]

DULAU & CO., 37 *Soho Square.*—Bound books—various.

[5228]

ELECTRO PRINTING BLOCK COMPANY (Limited), 6 & 8 *Burleigh Street, Strand.*—Enlargements and reductions from copper plates, wood blocks, lithographic stones, &c.

[5229]

EYRE & SPOTTISWOODE, Her Majesty's Printers, 43 *Fleet Street.*—Bibles, prayer books, &c. in various bindings.

[5230]

FAITHFULL, MISS, *Victoria Press, Great Coram Street.*—A specimen of printing by women; dedicated to Her Majesty by special permission.

[5231]

FALKNER, GEORGE, Lithographer, *King Street, Manchester.*—Examples of engraving upon and of printing from stone.

[5232]

FIGGINS, V. & J., *West Street, Smithfield, London.*—Specimens of new durable newspaper and book founts.

[5233]

FONTANNE, A., 5 *Bunhill Row, E.C.*—Improved cast brass type.

[5234]

GABALL, J. H., 3 *Russell Court, Brydges Street, W.C.*—Fac-simile of ancient manuscript by letterpress process.

[5235]

GARDNER, THOMAS BILSON, 45 *Greek Sreet, Soho, London.*—Improved stencil plates for marking plans, books, &c.

[5236]

GAUCI, PAUL, *London.*—Chromo-lithography. First portrait executed—1842; also specimens of his new manner—1861.

[5237]

GEORGE, BENJAMIN, *Hatton Garden.*—Patent ornamental show tablets and frames in one piece, for advertising purposes.

[5238]

GILMOUR & DEAN, *Royal Exchange Place, Glasgow.*—General lithography, engraving, die-cutting, embossing, and ornamental printing.

[5239]

GRANT & CO., *Broadway, Ludgate Hill.*—Colour printing.

[5240]

GRIFFITH & FARRAN, *Corner of St. Paul's Churchyard.*—Works for the instruction and amusement of young persons, and illustrated gift books.

[5241]

GROOM, WILKINSON, & CO., *Queen's Head Passage, Paternoster Row, London.*—Lithographs of manufacturers' patterns and show-cards.

[5242]

GUITTON & MENUEL, 2 *Bartlett's Passage, Fetter Lane.*—Brass type.

[5243]

HANHART, M. & N., 64 *Charlotte Street, Fitzroy Square.* — Lithography and chromo-lithography.

[*Obtained a Prize Medal at the Exhibition, London,* 1851 ; *and a Medal of the first class at the Paris Exhibition,* 1855.]

The following are the exhibitors' applications of the arts of lithography and chromo-lithography.

Portraits from life taken upon stone, or copied from oil, water-colour, and other drawings.

Copies from photographs executed upon stone, the size of the originals, or enlarged.

Oil paintings and water-colour drawings accurately imitated.

Landscapes and architectural drawings executed in tinted and coloured lithography.

All the works in chromo-lithography executed by M. & N. Hanhart may be had at their establishment.

[5245]

HAYMAN BROTHERS, 13 *Gough Square, Fleet Street.*—Samples of ornamental letterpress printing.

[5246]

HOME, R., & CO., *Edinburgh.*—Printed music.

[5247]

HUGHES & KIMBER, *Red Lion Passage, Fleet Street*—Copper and steel plates prepared for engraving.

[5248]

JEWELL, J. H., 104 *Great Russell Street, Bloomsbury* —Music engraving and printing.

[5249]

JOHNSON, J. M., & SON, 3 & 10 *Castle Street, Holborn.*—Improved show cards and tablets.

WINDOW TABLETS.

The elaborate care with which these tablets are executed, their artistic arrangement and superior lettering, combined with effectiveness of design and colour, have insured their general adoption by the leading manufacturers in every trade throughout the kingdom.

Manufacturers observe with satisfaction that first-class shopkeepers readily exhibit the "CHROMO-FULGENT SHOW-CARDS," because they present an additional attraction in ornamenting the windows, whilst also serving as trade advertisements. The printed matter cannot in any way be detached from the ornamental portions of the tablets; nor is it possible to apply them to the exhibition of popular engravings, portraits, &c.,—a misappropriation frequently practised when expensive show-cards are issued in frames and glasses.

For preservation *in transitu*, each tablet is enclosed in an envelope-bag; and distribution can be effected per "book-postage" (which includes printed matter of this nature) at the rate of 1*d.* for every four ounces.

[5250]

KENT, W., & Co., 23 *Paternoster Row.*—Books.

[5251]

LAVARS, T., *Broad Street Hall, Bristol.*—Chromo-lithography.

[5253]

LEFEVRE, CHARLES, 12 *Red Lion Street, Clerkenwell.*—Colours; printing on leather.

[5254]

LEIGHTON BROTHERS, *Milford House, Strand.*—Surface colour printing by machinery.

[5255]

LEIGHTON & LEIGHTON, 9 *Buckingham Street, Strand, W.C.*—Specimens of designing and engraving, and of processes connected with the production of printing surfaces.

[5256]

LINTON, WILLIAM JAMES, 85 *Hatton Garden, E.C.*—New process of engraving for surface printing (books, &c.).

[5257]

LONGMAN, GREEN, LONGMAN, & ROBERTS, *Paternoster Row, London.*—Printing and bookbinding.

[5258]

LOW, SAMPSON, SON, & Co., 47 *Ludgate Hill, E.C.*—Illustrated books. (*See pages* 20 & 21.)

[5259]

MACKENZIE, WILLIAM, *Glasgow; Paternoster Row, London.*—Bible composed by machinery, superbly illustrated by photographs.

The Queen's Bible. A superb edition of the Holy Scriptures, printed in the highest style of the art, from new types cast expressly for this Bible by Milne and Co., Edinburgh, and set up by steam-power type-composing machinery. Illustrated by magnificent photographs by Frith. (See specimens in Class 14.) Bound in morocco antique, in the most superb style, from designs by Leighton, with silver-gilt mountings, &c. Price fifty guineas.

This edition is limited to one hundred and seventy copies. Early orders are therefore necessary. The list of subscribers will be printed in the order received.

[5260]

MACLURE, MACDONALD, & MACGREGOR, Lithographers to the Queen, 37 *Walbrook, London, and at Manchester, Liverpool, and Glasgow.*—Lithography and engraving.

[5261]

MACPHERSON, D, 28 *Salisbury Street, Edinburgh.*—Stereotype plates; moulding material used for an indefinite period.

[5262]

MCQUEEN BROTHERS, 184 *Tottenham Court Road.*—Plate printing.

[5263]

MANSELL, JOSEPH, Manufacturer, *Red Lion Square.*—Oil coloured prints, ornamental papers and cards, pierced, embossed, and illuminated.

The following are the various branches of business in which this exhibitor is engaged.

Designer and manufacturer of all kinds of ornamentations printed in gold and colours, and embossed on paper for linen, cotton, and other fabrics. Show-boards of all kinds. Publisher and producer of pictures by the oil-printing process. Maker of lace and embossed paper and valentines, perforated and embossed cards and card-boards. Inventor and patentee of the process of producing damascened and pictorial subjects on the surface of papers.

[5264]

MELVILLE & Co., 68 *Marylebone Road, London.*—Patent solid indelible ink in black and colours.

[5265]

MILLER & RICHARD, Designers and Manufacturers, Foundry, *Edinburgh;* Warehouse, *Bartlett's Buildings, Holborn, London.*—Newspaper, book, old style, and other printing types. (*See pages* 22 & 23.)

[5266]

MOREL, V., & Co., General Electrotypists, 3 *Playhouse Yard, near the 'Times' Office, E.C.*—Electrotypes and surface plates produced by a new process from engraved steel and copper plates and lithographic stones.

[5267]

MOTTRAM, JOHN, 35A *Ludgate Hill.*—Medallion and engine-turned ruling.

[5268]

MUNRO, FREDERICK P., 4 *Gibson Street, Lambeth, S.*—Improved hand stamps for stamping, and sealing letters.

[5269]

MURRAY, J., *Albemarle Street.*—Books.

[5270]

NAPIER, JOHN, 13 *East Scienns Street, Edinburgh.*—A new patent method of stereotyping.

[5271]

PARSONS, FLETCHER, & Co., *Bread Street.*—Printing and lithographic inks, black and coloured—machine and press.

[5272]

PARTRIDGE, SAMUEL WILLIAM, Publisher, 9 *Paternoster Row.*—Illustrated periodicals, books, and tracts. (*See pages* 24 & 25.)

[5273]

PATENT TYPE FOUNDING COMPANY (Limited), 31 *Red Lion Square, London.*—Type cast and dressed by machinery, of extreme hardness.

[5275]

ROWNEY, GEORGE, & Co., Chromo-lithographers, Printers, and Publishers, Retail Department, 51 *and* 52 *Rathbone Place;* Wholesale and Export Department, 10 *and* 11 *Percy Street, London, W.*—Specimens of chromo-lithography applied to the production of fac-simile copies of pictures, drawings, and sketches. (*See pages* 26 & 27.)

[5276]

SCHENK, F. R., 50 *George Street, Edinburgh.*—Lithography.

[5277]

SCOTT, R. J., 8 *Whitefriars Street.*—Blocks for wood-engraving.

[5278]

SEARBY, WILLIAM, Type-founder, 2 *Crown Court, Threadneedle Street, City.*—Letter punches, and relief engraving for bank-notes, &c.

[5279]

SIDEY, CHARLES, 5 *Stephen Street, Tottenham Court Road.*—Specimens of engraved music plates, plans, &c.

[5280]

SILVERLOCK, HENRY, *Doctors' Commons, London.*—Printing, engraving, electrotyping, stereotyping, &c.

Various specimens of surface engraving, medallion ruling, and printing electrotypes and stereotypes.

[5281]

SKINNER, JAMES, 47 *Whitecross Street, Cripplegate, E.C.*—Glyphography, or surface printing-blocks from copper-plates, engraved or etched the forward way.

LOW, SAMPSON, SON, & CO., 47 *Ludgate Hill, E.C.*—Illustrated books.

THE PSALMS OF DAVID. Illustrated from designs by John Franklin. With coloured initial letters and ornamental borders. Choicely printed on toned paper, and appropriately bound. Small 4to. Bevelled boards, 1*l.* 1*s.*, or in morocco antique, bound by Hayday, 2*l.* 2*s.* Ten copies are printed on vellum for illumination, price 10 guineas each, bound in Russia, for which immediate application is requested.

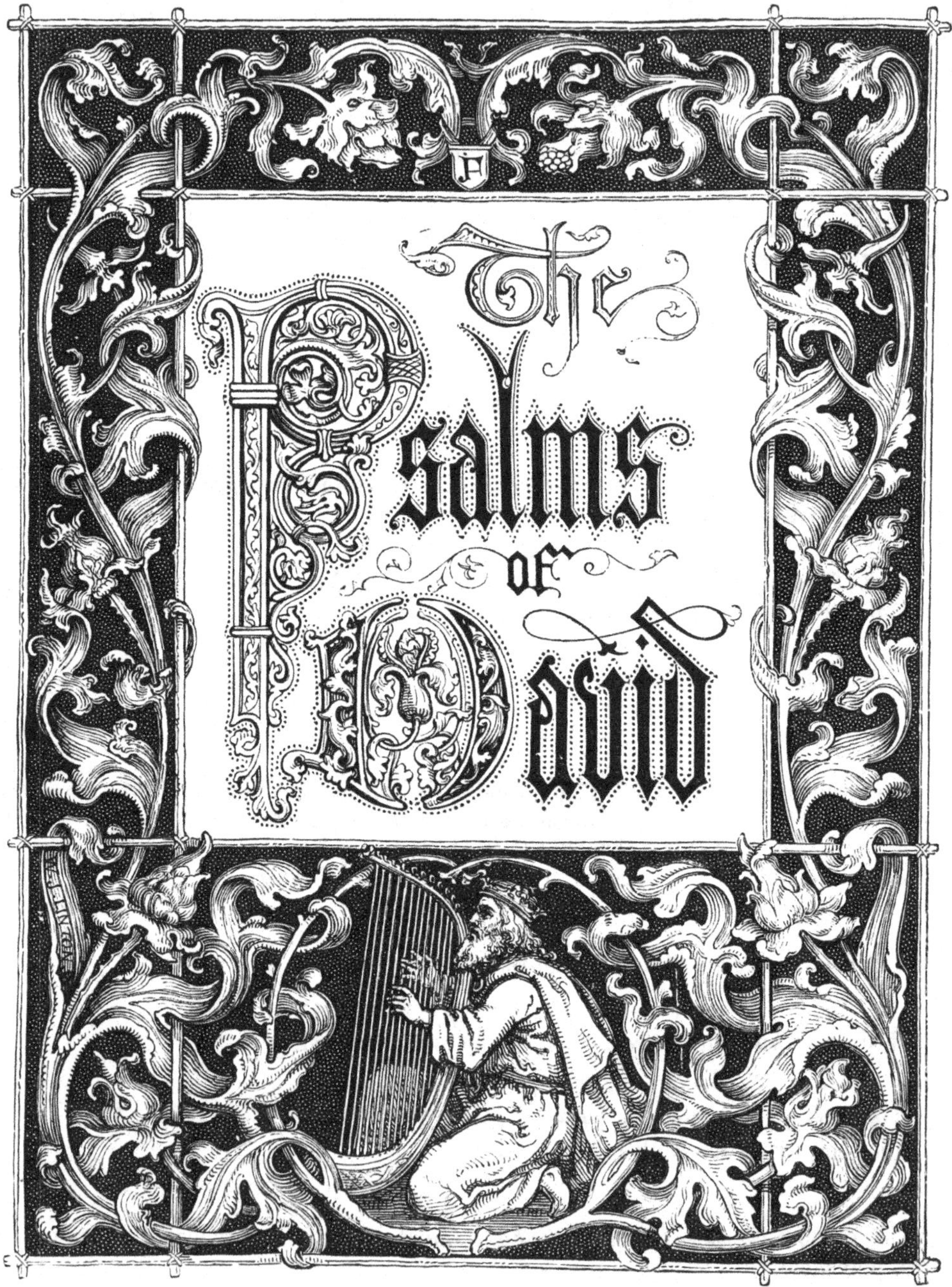

SPECIMEN OF ILLUSTRATIONS.

"This is an edition de luxe which is highly creditable to the mechanical and technical perfection of our extant topography."—*Saturday Review.*

"The manner in which classical accessories and religious treatment are blended in the composition of the subjects is remarkable, and the evident idea of the publication has been most successfully carried out."—*Illustrated London News.*

"The most handsome gift-book of the season."—*Observer.*

"One of the most beautiful gift-books of this or any season."—*Examiner.*

"A very handsome book, suited especially to the eyes and the tastes of the old. The ornamentation, moreover, though abundant, is not overwhelming in quantity, but remains in due subordination to the text."—*Guardian.*

LOW, SAMPSON, SON, & CO.—*continued.*

FAVOURITE ENGLISH POEMS OF THE LAST HUNDRED YEARS—Thomson to Tennyson, unabridged. With 200 illustrations by eminent artists. An entirely new and improved edition, handsomely bound, cloth, 1*l.* 1*s.*; morocco extra, 1*l.* 15*s.*

THE POETS OF THE ELIZABETHAN AGE: a Selection of Songs and Ballads of the Days of Queen Elizabeth. Choicely illustrated by eminent artists. Crown 8vo. Bevelled boards, 7*s.* 6*d.*; morocco, 12*s.*

THE POETRY OF NATURE. Selected and illustrated with thirty-six engravings by Harrison Weir. Small 4to., handsomely bound in cloth, gilt edges, 12*s.*; morocco, 1*l.* 1*s.*

MR. TENNYSON'S MAY QUEEN. Illustrated with thirty-five designs by E. V. B. Small 4to., cloth, bevelled boards, 7*s.* 6*d.*; or in morocco antique, bound by Hayday, 1*l.* 1*s.* Crown 8vo. edition, cloth, 5*s.*; bevelled boards, 5*s.* 6*d.*; morocco, 10*s.* 6*d.*

A New Edition of CHOICE EDITIONS OF CHOICE BOOKS. Illustrated by C. W. Cope, R.A., T. Creswick, R.A., Edward Duncan, Birket Foster, J. C. Horsley, A.R.A., George Hicks, R. Redgrave, R.A., C. Stonehouse, F. Tayler, George Thomas, H. J. Townshend, E. H. Wehnert, Harrison Weir, &c. Crown 8vo. Cloth, 5*s.* each; bevelled boards, 5*s.* 6*d.*; or, in morocco, gilt edges, 10*s.* 6*d.*

Bloomfield's Farmer's Boy.
Campbell's Pleasures of Hope.
Coleridge's Ancient Mariner.
Elizabethan Poetry.
Goldsmith's Deserted Village.
Goldsmith's Vicar of Wakefield.
Gray's Elegy in a Country Churchyard.
Keats' Eve of St. Agnes.
Milton's L'Allegro.
Tennyson's May Queen.
Warton's Hamlet.
Wordsworth's Pastoral Poems.

"Such works are a glorious beatification for a poet. Such works as these educate townsmen, who, surrounded by dead and artificial things, as country people are by life and nature, scarcely learn to look at nature till taught by these concentrated specimens of her beauty."—*Athenæum.*

ART STUDIES.—The Old Masters of Italy. By James J. Jarves, Esq. Two vols. medium 8vo., printed on toned paper, with forty-three engravings on copper (uniform style with Kügler's Work on Painting, edited by Eastlake), price 32*s.*

NEW BOOKS FOR YOUNG PEOPLE.

LIFE AMONGST THE NORTH AND SOUTH AMERICAN INDIANS; a Book for Boys. By George Catlin, author of "Notes of Travel amongst the North American Indians," &c. With illustrations. Small post 8vo., cloth, 6*s.*

"An admirable book, full of useful information, wrapt up in stories peculiarly adapted to rouse the imagination and stimulate the curiosity of boys and girls. To compare a book with 'Robinson Crusoe,' and to say that it sustains such comparison, is to give it high praise indeed."—*Athenæum,* Oct. 26.

THE BOY'S OWN BOOK OF BOATS. By W. H. G. Kingston. Illustrations by E. Weedon, engraved by W. J. Linton. Fcap. 8vo., cloth, 5*s.*

"This well-written, well-wrought book."—*Athenæum.*

"This is something better than a play-book; and it would be difficult to find a more compendious and intelligible manual about all that relates to the variety and rig of vessels and nautical implements and gear."—*Saturday Review.*

THE CHILDREN'S PICTURE BOOK OF THE SAGACITY OF ANIMALS. With numerous illustrations by Harrison Weir. Super-royal 16mo., cloth, 5*s.*; coloured, 9*s.*

"A better reading-book for the young we have not seen for many a day."—*Athenæum.*

THE CHILDREN'S PICTURE BOOK OF FABLES. Written expressly for Children, and illustrated with fifty large engravings, from Drawings by Harrison Weir. Square, cloth extra, 5*s.*; or coloured, 9*s.*

THE CHILDREN'S TREASURY OF PLEASURE BOOKS. With 140 illustrations, from drawings by John Absolon, Edward Wehnert, and Harrison Weir. Plain, 5*s.*; coloured, 9*s.*

LITTLE BIRD RED AND LITTLE BIRD BLUE: a Song of the Woods told for Little Ones at Home. With coloured illustrations and borders, by T. R. Macquoid, Esq. Beautifully printed, with coloured illustrations and borders, bevelled boards, 5*s.*

"The appearance of this little book is positively refreshing. Full of innocent fancy, and altogether childlike."—*Queen.*

"One of the most beautiful books for children we have ever seen. It is irresistible."—*Morning Herald.*

DR. WORCESTER'S NEW AND GREATLY ENLARGED DICTIONARY OF THE ENGLISH LANGUAGE. Adapted for library or college reference, comprising 40,000 words more than Johnson's dictionary, and 250 pages more than the quarto edition of Webster's dictionary. In one volume, royal 4to., 1,834 pp., price 31*s.* 6*d.* The cheapest book ever published.

"The volumes before us show a vast amount of diligence; but with Webster it is diligence in combination with fancifulness—with Worcester in combination with good sense and judgment. Worcester's is the soberer and safer book, and may be pronounced the best existing English Lexicon."—*Athenæum,* July, 13, 1861.

"We have devoted a very considerable amount of time and labour to the examination of 'Worcester's Quarto Dictionary of the English Language,' and we have risen from the task with feelings of no ordinary satisfaction at the result, and admiration of the care, scholarship, philosophical method, and honest fidelity, of which this noble work bears the impress upon every page. . . . As a complete and faithful dictionary of our language in its present state, satisfying to the full those requirements, the fulfilment of which we have laid down as essential to such a work, we know no work that can bear comparison with it."—*Literary Gazette.*

"We will now take leave of this magnificent monument of patient toil, careful research, judicious selection, and magnanimous self-denial with a hearty wish for its success."—*Critic.*

MILLER & RICHARD, Designers and Manufacturers, Foundry, *Edinburgh;* Warehouse, *Bartlett's Buildings, Holborn, London.*—Newspaper, book, old style, and other printing types.

Modern Athenian Series.

Pica, No. 28. To those Artists in Letterpress, whose genius and enterprise have, by pictorial illustration and chasteness of typography, so diffused the love of the Beautiful, that the "People's Edition," equally with the "Drawing-room Scrap Book," has become at once the educator of taste and the medium of its gratification, the series of faces here presented, which have been produced more especially for illustrated works, the density of the metal and the high finish of the type taking the strongest colour, and the lightness

Small Pica, No. 28. To those Artists in Letterpress, whose genius and enterprise have, by pictorial illustration and chasteness of typography, so diffused the love of the Beautiful, that the "People's Edition," equally with the "Drawing-room Scrap Book," has become at once the educator of taste and the medium of its gratification, the series of faces here presented, which have been produced more especially for illustrated works, the density of the metal and the high finish of the type taking the strongest colour, and the lightness of outline affording the fullest relief to the animate woodcut or elaborate ornament, whilst the careful support given to the ceriphs, secures durability with grace, are

Long Primer, No. 28. To those Artists in Letterpress, whose genius and enterprise have, by pictorial illustration and chasteness of typography, so diffused the love of the Beautiful, that the "People's Edition," equally with the "Drawing-room Scrap Book," has become at once the educator of taste and the medium of its gratification, the series of faces here presented, which have been produced more especially for illustrated works, the density of the metal and the high finish of the type taking the strongest colour, and the lightness of outline affording the fullest relief to the animate woodcut or elaborate ornament; whilst the careful support given to the ceriphs, secures durability with grace, are dedicated with the desire to aid so worthy a design, and in the

Bourgeois, No. 28. To those Artists in Letterpress, whose genius and enterprise have, by pictorial illustration and chasteness of typography, so diffused the love of the Beautiful, that the "People's Edition," equally with the "Drawing-room Scrap Book," has become at once the educator of taste and the medium of its gratification, the series of faces here presented, which have been produced more especially for illustrated works, the density of the metal and the high finish of the type taking the strongest colour, and the lightness of outline

Brevier, No. 28. To those Artists in Letterpress, whose genius and enterprise have, by pictorial illustration and chasteness of typography, so diffused the love of the Beautiful, that the "People's Edition," equally with the "Drawing-room Scrap Book," has become at once the educator of taste and the medium of its gratification, the series of faces here presented, which have been produced more especially for illustrated works, the density of the metal and the high finish of the type taking the strongest colour, and the lightness of the outline affording the fullest relief to the animate woodcut or elaborate ornament, whilst the careful sup-

Minion, No. 28. To those Artists in Letterpress, whose genius and enterprise have, by pictorial illustration and chasteness of typography, so diffused the love of the Beautiful, that the "People's Edition," equally with the "Drawing-room Scrap Book," has become at once the educator of taste and the medium of its gratification, the series of faces here presented, which have been produced more especially for illustrated works, the density of the metal and the high finish of the type taking the strong-

Nonpareil, No. 28. To those Artists in Letterpress, whose genius and enterprise have, by pictorial illustration and chasteness of typography, so diffused the love of the Beautiful, that the "People's Edition," equally with the "Drawing-room Scrap Book," has become at once the educator of taste and the medium of its gratification, the series of faces here presented, which have been produced more especially for illustrated works, the density of the metal and the high finish of the type taking the strongest colour, and the lightness of outline affording the fullest relief to the animate woodcut or elaborate ornament; whilst the

Pearl, No. 9. To those Artists in Letterpress, whose genius and enterprise have, by pictorial illustration and chasteness of typography, so diffused the love of the Beautiful, that the "People's Edition," equally with the "Drawing-room Scrap Book," has become at once the educator of taste and the medium of its gratification, the series of faces here presented, which have been produced more especially for illustrated works, the density of the metal and the high finish of the type taking the strongest colour, and the lightness of the outline affording the fullest relief to the animate woodcut or elaborate ornament: whilst the careful support given to the ceriphs, secures durability with grace, are dedicated with respect and gratitude, by their faithful Servants,

MILLER & RICHARD.

MILLER & RICHARD—*continued.*

Old Style Series.

Great Primer.

TYPE of the OLD STYLE of face is now frequently uſed—more eſpecially for the finer claſs of

Pica, Old Style.

TYPE of the OLD STYLE of face is now frequently uſed—more eſpecially for the finer claſs of book work; as, however, the faces which were cut in the early part of the laſt century are now unpleaſing both to

Small Pica, Old Style.

TYPE of the OLD STYLE of face is now frequently uſed—more eſpecially for the finer claſs of book work; as, however, the faces which were cut in the early part of the laſt century are now unpleaſing both to the eye of the critic and to the general reader, on account of their inequality of

Long Primer, Old Style.

TYPE of the OLD STYLE of face is now frequently uſed—more eſpecially for the finer claſs of book work; as, however, the faces which were cut in the early part of the laſt century are now unpleaſing both to the eye of the critic and to the general reader, on account of their inequality of *ſize* and conſequent irregu-

Bourgeois, Old Style.

TYPE of the OLD STYLE of face is now frequently uſed—more eſpecially for the finer claſs of book work; as, however, the faces which were cut in the early part of the laſt century are now unpleaſing both to the eye of the critic and to the general reader, on account of their inequality of *ſize* and conſequent irregularity of *ranging*, the

Brevier, Old Style.

TYPE of the OLD STYLE of face is now frequently uſed—more eſpecially for the finer claſs of book work; as, however, the faces which were cut in the early part of the laſt century are now unpleaſing both to the eye of the critic and to the general reader, on account of their inequality of *ſize* and conſequent irregularity of *ranging*, the Subſcribers have been induced to produce this ſeries, in which they have endeavoured to avoid the

Nonpareil, Old Style.

TYPE of the OLD STYLE of face is now frequently uſed—more eſpecially for the finer claſs of book work; as, however, the faces which were cut in the early part of the laſt century are now unpleaſing both to the eye of the critic and to the general reader, on account of their inequality of *ſize* and conſequent irregularity of *ranging*, the Subſcribers have been induced to produce this ſeries, in which they have endeavoured to avoid the objectionable peculiarities, whilſt retaining the distinctive characteriſtics of the mediæval letters.

Miller & Richard.

Nonpareil, Old Style.

Inland Bill of Exchange, Draft, or Order for Payment to the bearer, or to Order at any time otherwise than on demand.

Not exceeding	£5	1d.	Above £75 not ex.	£100	1s.
Above £5 not ex.	10	2	100	200	2
10	25	3	200	300	3
25	50	6	300	400	4
50	75	9	400	500	5

The duty increasing up to £4000.

Founts used by "The Times."

RUBY-NONPAREIL.

THE EXTRA HARD METAL made and used by the Subscribers possesses two distinct advantages—those of increased DURABILITY and DIMINISHED WEIGHT.

Of durability—One Newspaper Proprietor states that "from one Fount of MILLER & RICHARD'S EXTRA HARD METAL Type upwards of TWENTY-TWO MILLIONS of copies (of his paper) were printed;" while another says,—"After three years' wear,' Miller & Richard's fount of Extra Hard Metal is as good as new, while a new fount of the ordinary metal after the same wear is as much worn as if it had been in use ten years."

Of diminished weight—a Newspaper Proprietor writes, "Miller & Richard's fount of Extra Hard Metal set out about seven columns, while another fount of the ordinary metal barely set five."

MILLER & RICHARD.

MINION.

IN every age, and in all countries, printing denotes the state of civilization, of which books are the reflex, and the history of the human mind is written in the progress of bibliography. Thus the first printed books of Germany were almost all devoted to theology and scholastic philosophy, while at Paris ancient literature occupied an equal rank with theology; thus also at Rome, where the remembrance of ancient literature maintained a still stronger empire, printing under the guidance of the Bishops of Aleria and Seramo, principally reproduced the masterpieces of classic times. In France, however, under the

BOURGEOIS (*Leaded*).

THE art of printing is of comparatively modern origin; four hundred years have not yet elapsed since the first book was issued from the press; yet we have proofs that the principles upon which it was ultimately developed existed amongst the ancient Chaldean nations. Entire and undecayed bricks of the famed city and tower of Babylon have been found stamped with various symbolical figures and hieroglyphic characters. In this, how-

BOURGEOIS.

ever, as in every similar relic of antiquity, the object which stamped the figures was in one block or piece, and therefore could be employed only for one distinct subject. This, though a kind of printing, was totally useless for the propagation of literature, on account both of its expensiveness and tediousness. The Chinese are the only existing people who still pursue this rude mode of printing by stamping paper with blocks of wood. A great step in the science of typography was that

Partridge, Samuel William, Publisher, 9 *Paternoster Row.*—Illustrated periodicals, books, and tracts.

The British Workman, and Friend of the Sons of Toil.

This monthly penny paper, illustrated by first-class artists, is issued by the editor with the earnest desire of promoting the health, wealth, and happiness of the industrial classes.

Yearly parts :—the seven yearly parts, 1*s.* 6*d.* each, in illustrated covers.

A parlour edition for the years 1859-60-61, in crimson cloth and gilt edges, 2*s.* 6*d.* each year.

Volume :—700 illustrations, a complete edition of the "British Workman," from 1855 to 1861. Bound in plain cloth, price 10*s.* 6*d.* Crimson cloth, gilt edges (a handsome gift-book), price 12*s.*

Specimens of the above, and also of the "Mother's Picture Alphabet," may be seen at the stand.

PARTRIDGE, SAMUEL WILLIAM—*continued.*

THE MOTHER'S PICTURE ALPHABET.

Imperial Quarto, plate paper, with Twenty-seven original designs by Henry Anelay.

DEDICATED BY HER MAJESTY'S PERMISSION TO HER ROYAL HIGHNESS THE PRINCESS BEATRICE.

	s.	d.
IN ILLUSTRATED WRAPPER, PRICE	6	0
PLAIN CLOTH, RED EDGES	7	6
CLOTH, EXTRA GILT, AND GILT EDGES . .	10	6

"This book may be said to typify something of that *royal road* to learning of which one has sometimes heard; for as regards illustration, type, paper, and binding, NOTHING illustrative of the Alphabet has, we imagine, been yet produced, which will *bear the* REMOTEST *comparison* with it."—*Illustrated London News.*

"The Twenty-six letters of the Alphabet are illustrated by Anelay in his best style. It is got up in a style worthy of a royal drawing-room. We have never seen a more *exquisite* book for young children."—*Illustrated News of the World.*

"This book which has just come into our hands appears to merit *especial notice* from us, in our character of Art-Journalists."—*Art Journal.*

MONTHLY ILLUSTRATED PUBLICATONS.

THE BRITISH WORKMAN. One Penny.

THE BAND OF HOPE REVIEW. One Halfpenny.

THE CHILDREN'S FRIEND. One Penny.

The above are illustrated from drawings by Sir Edwin Landseer, J. Gilbert, Birket Foster, H. Anelay, H. Weir, and L. Huard.

BOOKS FOR PRESENTS.

160 ENGRAVINGS.—ILLUSTRATED SONGS and HYMNS. Cloth 5*s*., with coloured plates, and gilt, 7*s*. 6*d*.

700 ENGRAVINGS.—A Complete Edition of the "BRITISH WORKMAN," from 1855 to 1861. Cloth 10*s*. 6*d*., gilt, 12*s*.

700 ENGRAVINGS.—The First Ten Years of the "BAND OF HOPE REVIEW." Cloth, 10*s*., gilt, 12*s*.

THE YEARLY PARTS WTH NUMEROUS ILLUSTRATIONS.—BRITISH WORKMAN, illustrated cover 1*s*. 6*d*. cloth, gilt, 2*s*. 6*d*. BAND OF HOPE REVIEW, illustrated cover, 1*s*. cloth, gilt, 2*s*.

100 ENGRAVINGS.—The Yearly Volume of the "CHILDREN S FRIEND." Boards 1*s*. 6*d*., cloth, 2*s*., gilt, 2*s*. 6*d*.

MORNING DEW DROPS. Cloth, 3*s*. 6*d*., gilt, 4*s*. 6*d*.

ILLUSTRATED SHILLING BOOKS.

Reminiscences of the late Prince Consort.
Toil and Trust.
Confessions of a Decanter.
Voice from the Vintage.
A Mother's Stories.
Widow Green and her Three Nieces.
Two Christmas Days, &c.
Our Moral Wastes.
Wanderings of a Bible.
The History of a Shilling.

ILLUSTRATED SIXPENNY BOOKS.

"Scrub," the Workhouse Boy.
The Bible, the Book for All.
Never Give Up!
The Drunkard's Death.
The Victim.
The Warning.

ILLUSTRATED TRACTS, &C.

Publications for the Suppression of Intemperance.
Little Tracts for Little Folks.
Illustrated Hand-Bills.
Illustrated Tracts.
Illustrated Four-Page Tracts.

[5282]

SMITH, BENJAMIN, & SON, 7 *Wine Office Court, Fleet Street.*—Printing ink, and various products in its manufacture.

Every variety of black and coloured inks, of lampblack and other blacks, also of varnishes, pale and fast drying, of the best description, are manufactured by the exhibitors, and can be obtained from them at the lowest prices.

[5283]

SMITH, ELDER, & CO., 65 *Cornhill, London.*—Books.

[5284]

SPIERS & SON, *Oxford.*—Book of arms of the colleges.

The following are exhibited, viz.:—

Arms of the Colleges of Oxford; the plates richly emblazoned in gold, silver, and colours, by Henry Shaw, F.S.A.; with Historical, Biographical, Heraldic, and Antiquarian notices of each foundation, by the Rev. J. W. Burgon, M.A., of Oriel College.

A volume bound by J. and J. Leighton, from a design by John Leighton, F.S.A., and a set of the Arms framed.

"This book is a pleasant memorial for old Oxford men. Mr. Shaw, elegant and exact as ever, makes his department shine as gloriously as a herald's tabard."—*Athenæum.*

"Each plate is a perfect gem, and certainly the most magnificent art production of a local character ever attempted."—*Oxford Herald.*

[5285]

SPRAGUE, ROBERT W., & CO., 5 *Ave Maria Lane.*—Specimens of lithography, ornamental writing on vellum, and the new lithotype process.

ROWNEY, GEORGE, & Co., Chromo-lithographers, Printers, and Publishers, Retail Department 51 *and* 52 *Rathbone Place;* Wholesale and Export Department, 10 *and* 11 *Percy Street, London, W.*—Specimens of chromo-lithography applied to the production of fac-simile copies of pictures, drawings, and sketches.

Chromo-lithography has recently become one of the most popular arts in this country, from its having been adopted as a means for multiplying copies of oil paintings and water-colour drawings; and so admirably is it adapted for this purpose, that not only is each colour and gradation of light and shade rendered with remarkable accuracy, but even the very texture of the paint and the rough surface of the paper is copied with strict fidelity. Now, although this latter process may seem to the casual observer to be a matter of little moment, it is, in reality, of the greatest importance to the truthful representation of an artist's work, which, without texture, is apt to appear tame and insipid.

Previous to the discovery of chromo-lithography, copper and steel-plate engraving were the usual methods employed to reproduce the pictures of popular artists. But beautiful as are many of the fine line and mezzotint engravings, and perfect as they undoubtedly are in light and shade, they must always fail to give an accurate idea of a painter's style, owing to the absence of the colour of the original work. And when it is considered that colour is one of the greatest charms of the English school, and that, in this respect, the British artist is unrivalled, it will be readily admitted that without this new process many fine works, if published, would lose half their interest by being divested of the quality which appeals most directly to the eye, and produces that sense of pleasurable emotion so desirable when contemplating works of art.

ULYSSES DERIDING POLYPHEMUS.—AFTER J. W. M. TURNER.

The following are the titles of a few of the works already published, some of which are not exhibited for want of space.

The cave beneath the Holy Rock, Jerusalem, after a drawing by Carl Haag, executed for Her Majesty; by whose gracious permission this chromo-lithography is published. This cave has recently acquired great interest in consequence of modern investigation having pointed to it as the burial-place of Christ. A detailed account of this interesting place is given with each copy of the print.

Title	Artist	£	s.	d.
Ulysses deriding Polyphemus	after *J. W. M. Turner, R.A.*	3	3	0
Venice, The Dogana, Campanile of St. Marco " Ducal Palace, Bridge of Sighs, &c. (Canaletti painting)	*Ditto.*	3	3	0
The Canal of the Guidecca, and church of the Jesuits, Venice	*C. Stanfield, R.A.*	2	2	0
Crossing the Ford	*W. Mulready, R.A.*	1	11	6
Nuder Lahnstein and Castle of Lahneck	*T. L. Rowbotham.*	1	11	6
Oberwesel, on the Rhine	*Ditto.*	1	11	6
Antwerp Cathedral	*E. Dolby.*	1	11	6
Cathedral Porch, Evereux	*Ditto.*	1	1	0
Rouen Cathedral, West Front	*Ditto.*	1	1	0
Rouen Cathedral, South Transept	*Ditto.*	1	1	0
Macbeth and the Murderers of Banquo	*G. Cattermole.*	1	1	0
Macbeth: the Murder of Duncan	*Ditto.*	1	1	0

ROWNEY, GEORGE, & CO.—*continued.*

		£	s.	d.
A sketch of St. Paul's, from the Shot Tower	after *G. Dodgson.*	1	1	0
Sea Fog. Luccombe Bay, Isle of Wight	*T. L. Rowbotham.*	0	10	6
Italian Sketches	*F. Goodall, A.R.A.*			
Frankfort	*W. Callow.*	0	15	0
Cologne	*Ditto.*	0	15	0
The Gulf of Spezzia	*T. L. Rowbotham.*	0	15	0
The Madonna and Child, from the celebrated picture in the Dulwich Gallery	*Vandyke.*	0	15	0
Diffidence	*W. Hunt.*	0	10	6
Tower of the Church at Gorcum	*D. Roberts, R.A.*	0	10	6
Youth and Age	*F. Taylor.*	0	10	6
Fowey Castle, Cornwall.	*S. P. Jackson.*	0	7	6
Scarborough Castle, Yorkshire	*C. Bentley.*	0	7	6
Water Gate on the Rhine.	*S. Prout.*	0	7	6
Gipsy Camp. Claygate, Surrey	*R. P. Noble.*	0	5	0
Chiswick by Moonlight	*Ditto.*	0	5	0

[5286]

STANDIDGE & CO., 37 *Old Jewry, London.*—Fac-similes of documents, and other lithographic printing.

[5287]

STANESBY, SAMUEL, Illuminating Artist, 6 *St. George's Terrace, Kensington.*—Illuminated books, and various illuminations.

[5288]

STEPHENSON, BLAKE, & CO., *Sheffield.*—Specimens of printing types.

[5289]

TERRY, CHARLES, 183 *High Holborn, London.*—Artistic colour-printing for commercial purposes.

The samples of coloured labels, show cards, &c. &c., have been selected from the general bulk as supplied to the consumer.

In this particular class of work, not only is great EXACTNESS and UNIFORMITY required, but the number of colours and printings in each work will very considerably affect the price. The specimens exhibited will be found to comprise COLOURED LABELS, SHOW CARDS, BOOK COVERS, ILLUMINATED ORNAMENTS, LITHOGRAPHY, and GENERAL COLOUR PRINTING.

No. 1. SHOW CARD in seven printings.—RIMMEL'S PERFUMED ALMANACS.

No. 2. Ditto ditto DR. LOCOCK'S PULMONIC WAFERS.

No. 3. SHOW CARD in seven printings.—DR. HUGO'S MEDICAL ATOMS.

No. 4. Engravings on stone for book illustrations.

No. 5. Samples of printing in gold,

No. 6. „ book covers. 3 colours only.

No. 7. Illuminated almanack. 10 colours.

Nos. 8, 9, 10, and 11. Specimens of colour printing, used by wholesale and export houses. The large consumption of this class of work necessitates a great effect in as few colours or printings as possible.

Nos. 12 to 20. Specimens of labels and wrappers, box tops, &c., in gold and colours.

CHARLES TERRY also exhibits specimens of colour printing, showing the progress and appearance of the work after the addition of each colour.

[5290]

TRÜBNER & CO., 60 *Paternoster Row.*—Books.

[5291]

ULLMER, FREDERICK, *Old Bailey.*—Improved saw block; improvement in the manufacture of wood type and designs.

[5292]

UNDERWOOD, THOMAS, *Birmingham.*—Chromo-lithographs.

[5293]

WALLER, FREDERICK, 18 *Hatton Garden.*—Specimens of lithography.

Specimens of the following are exhibited:—Maps, plans, architectural and engineering drawings, portraits, book illustrations, manufacturers' patterns, show cards, &c. Commercial and artistic printing in colours, tints, gold, &c.

[5294]

WALLIS, GEORGE, Inventor and Patentee, 16 *Victoria Grove, West Brompton.*—Specimens of the new art of autotypography.

By this new art-process, the invention of the exhibitor, drawings can be executed on a variety of substances, in such a manner that by the aid of suitable machinery, they can be engraved upon metal plates almost instantaneously, and with the certainty that every touch of the original drawing, both washed effects and lines, will be reproduced. The process by which the drawings are executed is easy and comparatively inexpensive, whilst that of engraving, being purely mechanical, is controlled with ease and certainty. The plates can be printed from at the ordinary copper-plate printing-press; and when the style of execution in the drawing is suitable, may be used for transfer plates for lithographic purposes, and also for both "bat" and "press" transfers to porcelain and earthenware.

[5295]

WATTS, WILLIAM MAVOR, *Crown Court, Temple Bar.*—Oriental and other printing, and embossing for the blind, &c.

[5296]

WESTWOOD, PROFESSOR J. O., *Oxford.*—Volume of fac-similes of Anglo-Saxon manuscripts, with wooden covers carved with Anglo-Saxon designs.

[5297]

WHITEMAN, F. J., 19 *Little Queen Street, Lincoln's Inn Fields.*—Masterpiece specimen of plain and ornamental writing engraving.

[5299]

WINSTONE, BENJAMIN, 100 *Shoe Lane, E.C.*—Printers' ink, letterpress and lithographic, with printed specimens.

[5300]

WYATT, STEPHEN, 22 *Gerrard Street, Soho, W.*—Specimens of lithographic drawing and printing.

[5301]

CLAY, R., SON, & TAYLOR, 7 *Bread Street Hill.*—Specimens of printing.

[5302]

BELL & DALDY, *Fleet Street.*—Books.

[5303]

BERRI, D. C., 36 *High Holborn.*—Letter-stamps, as used in the Post-office.

[5304]

BRADBURY & EVANS, *Fleet Street.*—Books.

[5305]

CHAMBERS, W. & R., *Paternoster Row.*—Books.

[5306]

CIVIL ENGINEER & ARCHITECTS' JOURNAL, *Warwick Court, Gray's Inn.*—Specimens of architectural lithography.

[5307]

FRY, J., *Cosham, Bristol.*—Books.

[5308]

HOGG, J., & SONS, *London.*—Books.

[5309]

JONES, O., 9 *Argyle Place, London.*—Books.

[5310]

MCMILLAN & Co., *Henrietta Street, Covent Garden.*—Books.

SUB-CLASS D.—*Bookbinding.*

[5310]

BEDFORD, FRANCIS, 9 *Gloucester Street, Warwick Square, S.W.*—Specimens of ornamental bookbinding.

"Dresses and decorations of the Middle Ages." Bound by F. Bedford, from the designs, and under the superintendence of Henry Shaw, F.S.A., to whom the volume belongs.

[5311]

BONE, W., & SON, 76 *Fleet Street, London.*—Bookbindings.

[5312]

CHATELIN, ANTOINE, 15 *Newman Street, Oxford Street, W.*—Bookbinding, old and modern style.

[5313]

CLARK, WILLIAM, *Dunfermline.*—Specimens of bookbinding finished with hand-tooling.

[5315]

HOLLOWAY, M. M., 25 *Bedford Street, Strand, London.*—Bookbinding.

[5316]

JEFFREY, J., *Charlotte Street, Portland Place.*—Bookbinding, &c., ancient and modern styles —executed solely by the exhibitor.

[5317]

LEIGHTON, JOHN, F.S.A., 12 *Ormond Terrace, Regent's Park.*—Specimens of designs principally executed for British publishers, illustrated books, &c.

[5318]

LEIGHTON, J. & J., 40 *Brewer Street, Golden Square, W.*—Specimens of bookbinding, &c.

[5319]

LEIGHTON, SON, & HODGE, 13 *Shoe Lane, London.*—Bookbinding applicable for publishers.

[5320]

NELHAM, W., 48 *Liverpool Street, King's Cross.*—Specimens of bookbinding.

[5321]

POTTS, WATSON, & BOLTON, *Garter Court, Barbican.*—Designs on leather for upholstery, &c.

[5322]

RAINES, THOMAS, 24 *Great Ormond Street.*—Specimens of bookbinding.

[5323]

RAMAGE, JOHN, *North-Bridge, Edinburgh.*—Palæography inlaid with leather, hand-worked; various other books hand-worked.

[5324]

REYNOLDS, WILLIAM, 6 *Eldon Street, Finsbury.*—Twenty specimens of leather backs for binding volumes of printed music.

[5325]

RIVIÈRE, ROBERT, 196 *Piccadilly.*—Specimens of bookbinding.

[5326]

SETON & MACKENZIE, 80 *George Street, Edinburgh.*—Bookbindings.

[5327]

TONKINSON, J., Manufacturer, 16 *St. John's Street, Clerkenwell.*—Specimens of book-clasping, edging, and mounting in metals.

[5328]

WESTLEYS & Co., 10 *Friar Street, Doctors' Commons.*—Specimens of bookbinding.

[5329]

WRIGHT, JOHN, Trustees of the late, 14, 15, & 16 *Noel Street, Soho, W.C.*—Bookbinding.

[5330]

ZAHNSDORF, JOSEPH, 30 *Brydges Street, Covent Garden.*—Various samples of bookbinding.

Class XXIX.

EDUCATIONAL WORKS AND APPLIANCES.

Sub-Class A.—*Books, Maps, Diagrams, and Globes.*

[5361]

Albitès, Achille, LL.B., *Paris; Edgbaston Proprietary School.*—Books: 'How to speak French; or French and France.'

This book is a condensed, simplified, and progressive cyclopædia of the French language. The 'Athenæum' and other reviews have spoken of it in most favourable terms. Sixth edition, price 5*s.* 6*d.*

[5362]

Allman, T. J., 42 *Holborn Hill, London.*—Educational works, appliances, and metal corners for slates.

[5363]

Bean, J. W., Bookseller, *Leeds.*—Series of sixteen copy-books, with engraved heads, in foolscap and post.

Nos. 1, 2, 3. Initiatory series. No. 4. Large Hand. No. 5. Text hand. No. 6. Round hand. No. 7. Small hand. No. 8. Text and round hands. No. 9. Text and small hands. No. 10. Three hands. No. 11. Four hands. No. 12. Commercial hand. No. 13. Ladies' pointed hand. No. 14. Current hand. No. 15. Alphabets, German text, &c. No. 16. Introductory small hand.

Published in post and foolscap. Sold in London by Darton & Co., Holborn.

[5364]

Bell, William, Phil.D., 30 *Burton Street, W.C.*—Stream of time, or figurative representation of universal history and chronology.

[5365]

Bell & Daldy, *Fleet Street.*—Educational works.

[5366]

Berthon, Rev. E. L., M.A., *Romsey.*—Specimen of very large globes of new construction for educational purposes.

[5367]

Betts, John, 115 *Strand, London.*—Portable globe, geographical plates, maps, dissected puzzles, &c.

Betts's portable globe consists of a framework of steel wire, covered with a flexible material on which a map of the world is printed, as in an ordinary globe. It is four feet in circumference, and can be expanded or collapsed in a few seconds. Price, in a neat box, 12*s.* 6*d.*

Betts's geographical slates have two outline maps permananently engraved on each slate, price 2*s.*

Betts's educational maps, of various sizes, engraved on steel with great distinctness and effect. Also the interrogatory maps, each accompanied by a book of exercises.

[5368]

Bevan, Henry, *St. Mary's Street, Shrewsbury, Shropshire.*—Bevan's tablets for facilitating arithmetical operations.

[5369]

Bishop, T. B., *Wimbledon, Surrey.*—Two chronological charts of European history, A.D. 1400 to 1800.

[5370]

BLACK, ADAM & CHARLES, *Edinburgh.*—Books, atlases, and maps.

[*Obtained the International Jury's Medal in Class XXVI. of the Paris Exhibition,* 1855.]

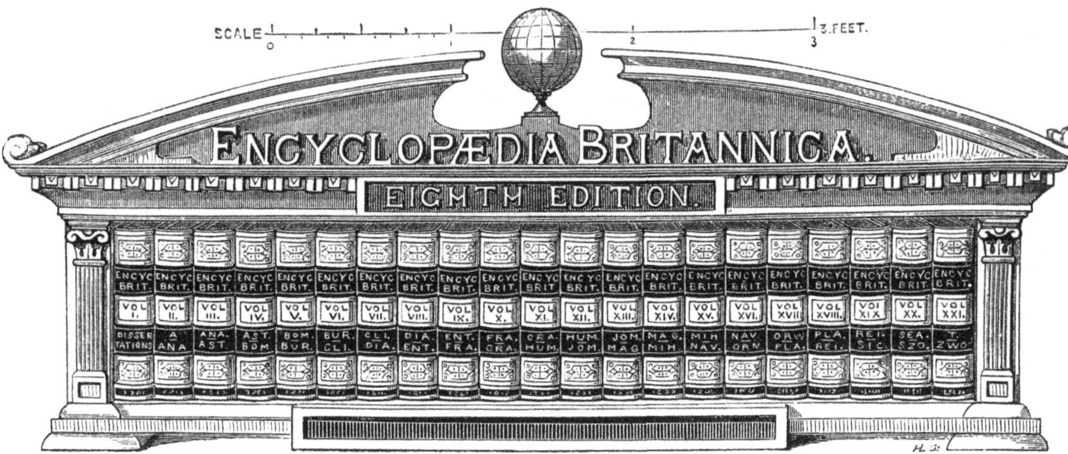

No. 1. BOOKS OF REFERENCE.

The Encyclopædia Britannica, or Dictionary of Arts Sciences, and General Literature. Twenty-one volumes quarto, with upwards of 5000 engravings on wood and steel. Elegantly bound in cloth. Price 25*l.* 4*s.*; or half-bound, Russia leather, 31*l.* 10*s.* Index separately, 8*s.*

The above diagram shows the dimensions of the work.

No. 2. ATLASES AND MAPS.

Black's General Atlas of the World, a series of 56 maps of the principal countries and divisions of the globe. Elegantly printed in colours, and containing all the latest discoveries. Strongly half-bound, in morocco leather, gilt edges. Price 3*l.*

Black's New Large Twelve Sheet Map of Scotland, on the scale of 4 miles to the inch. Divided according to the accompanying diagram, into 12 sheets. Each sheet measures 19 × 18 inches. The size of the map when complete and mounted for wall will be nearly 6 × 5 feet.

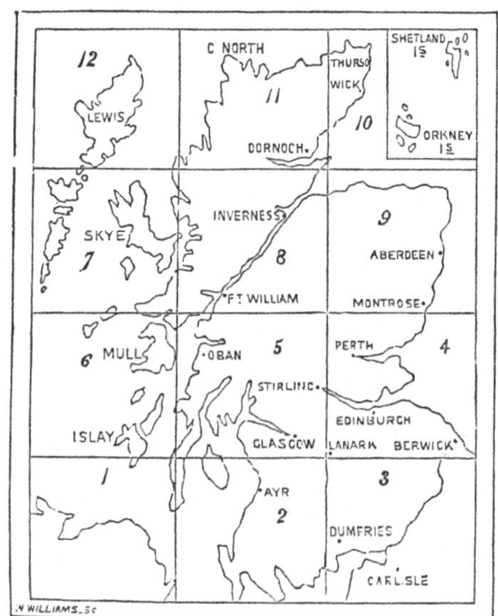

The sheets are sold separately, price 1*s.* 6*d.*, or 2*s.*
coloured.

No. 3. GUIDE-BOOKS FOR TOURISTS IN GREAT BRITAIN AND IRELAND.

Black's Picturesque Guide to England, 10*s.* 6*d.*

Black's Picturesque Tourist of Scotland, 8*s.* 6*d.*

Black's Picturesque Guide to Ireland, 5*s.*

Black's Guide to London, 4*s.* 6*d.*

Black's Plan of London, 1*s.*; Map of Environs, 1*s.*

Also to the following picturesque districts and counties :—

The lakes of Westmoreland and Cumberland, 5*s.*

Wales—North and South, 5*s.*

The Lakes of Killarney, 1*s.* 6*d.*

The south of England—Kent, Surrey, Sussex, Hants, Isle of Wight, Devon, and Cornwall.

Derbyshire, Yorkshire, Warwickshire, etc., etc.

No. 4. SCHOOL AND COLLEGE BOOKS.

History.—Kitto's History of Palestine, 3*s.* 6*d.*; Tytler's History of Scotland, 3*s.* 6*d.*; Tytler's Ancient and Modern History, 3*s.* each; Scott's History of Scotland, 6*s.*

Geography.—Black's School Atlas, 10*s.* 6*d.*; Black's Beginner's Atlas, 2*s.* 6*d.*

Literature.—Class-books of English Poetry, by Scrymgeour, 4*s.* 6*d.*; English Prose, by Demaus, 4*s.* 6*d.*; French Literature, by Masson, 4*s.* 6*d*; Introduction to English Literature, 2*s.*; Introduction to French Literature, 2*s.* 6*d.*

Students' Text-books. — Palæontology, by Professor Owen, 16*s.*; Geology, by Professor Jukes; Mineralogy, by Professor Nicol, 5*s.*; Botany, by Professor Balfour, 31*s.* 6*d.*; Medicine, by Professor Bennett; Physical Geography and Meteorology, by Sir John Herschel.

No. 5. WORKS BY POPULAR AUTHORS.

Sir Walter Scott, Bart.—Waverley Novels, &c.

Thomas de Quincey.—Confessions of an Opium Eater; Essays, &c.

Hugh Miller.—Old Red Sandstone; Autobiography, &c.

Thomas Guthrie, D.D.—Pleas for Ragged Schools, &c., &c.

Black, Adam & Charles—*continued.*

Specimen of Illustrations to Waverley Novels.

SCENE FROM 'PEVERIL OF THE PEAK,' DRAWN BY W. MULREADY, ESQ., R.A.

"There are sundry rates. Gentlemen must choose for themselves.
He asked nothing but his fees. But civility," he muttered, "must be paid for."

The Waverley Novels, by Sir Walter Scott, Bart.; Various Editions:—

No. 1. The library edition, in 25 volumes, demy 8vo., bound in cloth, extra gilt, and illustrated with two hundred and four engravings on steel, after Vandyke, Le Tocque, Wilkie, Turner, Roberts, Landseer, and others. Printed in long-primer type. Each volume contains an entire novel. Price for the set, 12*l.* 12*s.*

No. 2. The new illustrated edition, in 48 volumes, foolscap 8vo., bound in green cloth. Profusely illustrated with sixteen hundred wood-engravings, and ninety-six steel frontispieces and vignettes, after drawings by Sir W. Allan, R.A., A. E. Chalon, R.A., A. Cooper, R.A., C. W. Cope, R.A., W. P. Frith, R.A., C. R. Leslie, R.A., W. Mulready, R.A., C. Stanfield, R.A., and others. Printed in long-primer type. Each novel is divided into two volumes. Price of set, 10*l.* 10*s.*

No. 3. The edition of 1847, in 48 volumes, foolscap 8vo., bound in cloth, lettered. Each volume has a frontispiece and vignette. Price of set, 6*l.* 10*s.*

No. 4. The cabinet edition, in 25 volumes, foolscap 8vo., bound in cloth, lettered. Each volume contains an entire novel, with a frontispiece and vignette. Printed in brevier type. Price of set, 3*l.* 10*s.*

No. 5. The people's edition, in 5 volumes, royal 8vo., bound in cloth, gilt backs, and illustrated with one hundred page woodcuts, and a portrait of Scott, by Raeburn. Printed in double columns. Each volume contains five novels. Price of set, 2*l.* 2*s.*

Sir Walter Scott's entire works are contained in ninety-eight volumes, foolscap 8vo., printed in long-primer type. Price of complete set, 14*l.*

[5371]

BLACKWOOD, WILLIAM, & SONS, 45 *George Street, Edinburgh.*—The royal atlas of modern geography, and geological maps.

[5372]

BLANCHARD, REV. H. D., *Lund, Beverley.*—A school register.

[5373]

BOUVERIE, J., *Maida Hill, W.*—Books and illustrative drawings.

[5374]

BOWER, BENJAMIN, *Cheadle.*—Dial-map of the panorama of Alderley Edge, Cheshire—sketches and radii scaled.

[5375]

BRITISH AND FOREIGN BIBLE SOCIETY, 10 *Earl Street, Blackfriars.*—One hundred and ninety-one versions of the Holy Scriptures, in various languages.

[5376]

CASSELL, PETTER, & GALPIN, *La Belle Sauvage Yard, London.*—Valuable educational works for all classes.

[5377]

CHRISTIAN VERNACULAR EDUCATION SOCIETY FOR INDIA, 5 *Robert Street, Adelphi, London.*—Publications in the Indian languages.

[5379]

CRAMPTON, THOMAS, *The Butts, Brentford, W.*—School books, apparatus, school music, reading frames, satchels, &c.

[5380]

CRUCHLEY, GEORGE FREDERICK, 81 *Fleet Street.*—Reduced Ordnance, England, Wales, and other maps, atlases, globes, &c.

[5381]

CUIPERS, PETER, 32 *Castle Street, Holborn, E.C.*—Map of Central America; specimens of topography.

[5382]

CURWEN, JOHN, *Plaistow, E.*—Books and diagrams on the tonic sol-fa method. (*See page* 35.)

[5383]

DARTON & HODGE, 58 *Holborn Hill, London.*—Educational books, maps, prints, and diagrams.

Original Juvenile Library, Infant School, and Kinder Garten Repository.

Indestructible juvenile works.

English and German educational cubes.

Modern dissected maps and puzzles.

Mechanical and other educational diagrams.

NEW SERIES OF EDUCATIONAL PRINTS.

Leather and its applications.

Silk ditto.

Wool ditto.

Horn, ivory, tortoiseshell, and bone, and their applications.

Examples of the plants from which medicines are obtained.

Examples of the most useful spice plants.

Examples of the most useful palms.

Vegetable productions used for food, Part 1.

Ditto ditto, Part 2.

Ditto ditto, Part 3.

Examples of plants used for clothing and cordage.

Vegetable productions used in the manufacture of fermented liquors.

Others to follow, making a series of sixteen, and when completed a book will be published containing a description of each sheet.

Value of a dead horse. Price 1*s.*

CURWEN, JOHN, *Plaistow, E.*—Books and Diagrams on the tonic sol-fa method.

Origin and Progress.—About the year 1812, Miss Glover, the daughter of a clergyman of the Church of England, commenced teaching children to sing by means of a "musical ladder" and a simple notation of letters taken from that ladder. This system she published under the title "Scheme for rendering Psalmody congregational." At the close of A. D. 1840, Mr. Curwen taught himself to read simple music by the help of this book, and became convinced himself, by experiment and study, of its educational and scientific truth. On this system he founded, with Miss Glover's consent, the Tonic Sol-fa Method. He has endeavoured to adapt this method to the various wants of the School, the Home, and the Church, by publications; and to propagate it by lectures and the encouragement of the best teachers. The simplicity, the cheapness, and, above all, the moral purposes of this method, have won for it many devoted and most active friends. A very large number of these friends necessarily remain unknown to Mr. Curwen, but he has recently obtained "Census returns" from a thousand teachers who are now giving lessons to 47,000 children, and about the same number of adults. The progress of the method has been steady and constant under the *sole patronage* of its pupils.

The main Principle.—This method teaches the pupil to measure his intervals (not from any fixed sound in the region of absolute pitch, but) from that sound which is fixed on, for the occasion, as the Governing or Key-sound of the tune to be sung, whatever may be the place of that key-sound in absolute pitch. This is the simpler and surer plan, because there is an *invariable rule* of measurement, the same in all keys, when you start from the Governing tone,—but the measurement *differs* with every new key when you take it from a fixed sound of *absolute* pitch. The pitch of the key-tone is first found from the scale of C, and then all the other measurements are *relative* to that. This is also the *oldest* plan of teaching interval to the human voice. Only organs and pianos have tempted men away from it. It is generally allowed that there can be no true sight-singing without bearing the key-tone in mind; but, as General Thompson has said, "There wanted something to *enforce* and *necessitate* the referring of all intervals to the key-note." This key-note (or more properly tone) is called the Tonic. Hence the *Tonic* Sol-fa Method of teaching to sing.

The Modulator.—This pointing-board for teaching to sing represents the key-note of a tune and its six attendant notes, placed at their proper intervals. The side columns represent the "related keys." As the intervals of the Modulator and the syllables which represent them are unchangeable, the mind quickly associates the one with the other, just as the words of a well-known song infallibly associate themselves with their tune. Thus a system of natural Mnemonics is established.

Teaching by pattern.—In the early lessons the teacher Sol-fas a short musical phrase while he points to the notes on the Modulator,—the pupil listening the while, but not singing. This "pattern" the pupil then imitates, and his errors are corrected, not by singing with him, but by stimulating his attention, and setting him the pattern again.

The Mental Modulator.—By habit the pupil soon comes to have a Modulator "printed on his mind's eye," and as he sings his familiar exercises he sees the notes move up and down upon it.

The Letter Notation.—These letters are the initial letters of the Sol-fa syllables, and they act as pointers to the Mental Modulator. Thus the notation of "Tune" is nothing but the teacher's pointing on the Modulator *written down.* "Time" is measured by inches (so to speak) along the page. These "inches" represent the beats or "pulses" of the measure; and there are different marks for the strong, the weak, and the medium pulses. A note placed alone in a pulse fills that "beat" of the time. A horizontal stroke continues the sound through another pulse, or part of one, and so on.

Mental effect of Tones in a Key.—Great assistance in first learning to strike the tones correctly is given by the pupil's being led to observe the effect on the mind which properly belongs to each tone of a key. These effects in slow music, and apart from harmony, are proximately described by words.

Accidentals.—Sharps are represented by changing the vowel of a syllable into *e;* flats by changing the vowel into *aw.* But Transition or Modulation into a new key is represented in a truer way. See the manuals. The Tonic or key-note of the Minor Mode is *Lah.*

Harmony.—The study of Harmony is greatly facilitated by our Tonic principles, as well as by the elementary and progressive development of chords which this method adopts.

"*Easy, Cheap, and True*"—is imperatively the motto of any method for the people. Our music, printed on a single line, instead of five lines, is at least four times cheaper than that in the common notation. It supplies even the poor man with a large selection of music. Many other publishers, beside Mr. Curwen, are making good use of it.

The Established Notation.—Pupils find the Sol-fa syllables the quickest interpreters of the Established Notation. When they have once learned music itself, no difficulties of notation can hinder them. But if they begin with flats and sharps and clefs, they lose music itself in the maze of notation.

The Standard Course of Lessons (price 1s. 6d., Ward and Co, 27, Paternoster Row) is the book of practical instruction for the teacher. It contains also a list of the publications. An "Account of the Tonic Sol-fa Method" (16 pages) is sold by Messrs. Ward & Co., four copies for a penny. For information in reference to the locality of teachers apply to Mr. W. H. Thodey, Richmond House, Plaistow, London, E.

Certificates of progress—See Stall.

The Modulator,

OR POINTING BOARD FOR TEACHING TUNES.

DOH is the key-tone of a tune.

r'	s	d'		FAH'				
		t		ME'		l	r'	s
d'	f							
t	m	l		RAY'		s	d'	f
							t	m
l	r	s		DOH		f		
				TE		m	l	r
s	d	f	ta					
	t_1	m		LAH		r	s	d
f			la	se				t_1
m	l_1	r		SOH		d	f	
				bah	fe	t_1	m	l_1
r	s_1	d		FAH				
		t_1		ME		l_1	r	s_1
d	f_1		ma		re	se_1		
t_1	m_1	l_1		RAY		s_1	d	f_1
		se_1	de				t_1	m_1
l_1	r_1	s_1		DOH		f_1		
				TE_1		m_1	l_1	r_1
s_1	d_1	f_1						
	t_2	m_1		LAH_1		r_1	s_1	d_1
f_1								t_2
m_1	l_2	r_1		SOH_1		d_1	f_1	

NOTE.—Accidental sharps change the vowel into *ee,* written *e,* thus, *f fe s.* Accidental flats change the vowel into *aw,* written *a,* thus, *t ta l.* Their places show their relation to other keys.

OLD 100th. KEY A. :d |d :t_1 |l_1 :s_1 |d :— |r :— |m :— |— :m |m :m |r :d |f :— |m :— |r :— |—

 D 2

[5384]

DAY & SON, Lithographers to the Queen, 6 *Gate Street, Lincoln's Inn Fields, London.*—Coloured diagrams for educational purposes, illustrative of various branches of science; produced under the direction of the Government Department of Science and Art.

[5385]

DEAN & SON, 11 *Ludgate Hill, London.*—Maps, and other educational works; and movable children books.

[5386]

EASTON, WILLIAM, *Scudamore Schools, Hereford.*—Arithmetic for younger scholars; arithmetic and mensuration for elder scholars.

These books are drawn up chiefly with a view of facilitating the home lessons of the pupils in arithmetic and their individual work in school, and thus to help the collective explanation of principles. Care has been taken to avoid such language and expressions as are suitable only for adults. In 'Arithmetic for Younger Scholars,' the exercises are constructed on a new plan, the invention of the author, in order to render much easier the examination of the work done by the pupils.

Prices :—Arithmetic and Mensuration for elder scholars (new edition), ninepence: Arithmetic for younger scholars, sixpence.

[5387]

FLETCHER, PETER, *Clyde Street, Edinburgh.*—Globe for use of the blind. (Attended by blind man.)

[5388]

FYFE, WALLACE, *Dorchester.*—Text-book, catechism, school-calendar, diagrams of his new natural system of agricultural instruction.

[5389]

GALL & INGLIS, *Edinburgh.*—Educational charts and atlases.

Chronological and genealogical chart of Sovereigns of Great Britain to the present time, size 25 by 42 inches, coloured, on rollers, 7*s*. 6*d*.

Royal quarto school atlas, 31 maps, coloured, cloth lettered, 3*s*. 6*d*.

Chronological chart of Bible history, from the creation to the Christian Era, coloured, size 28 inches by 52 inches, on rollers, 10*s*. 6*d*.

Shilling school atlas, 11 royal 4to. maps, coloured, sewed, 1*s*.

[5390]

GILBERT, JAMES, 2 *Devonshire Grove, Old Kent Road, S.E.*—Ince and Gilbert's Outlines of English, French, and Grecian history, geography, general knowledge, and arithmetic.

No. 1. Outlines of English History. By Henry Ince, M.A., and James Gilbert. The 205th thousand.

No. 2. Outlines of French History. By Henry Ince, M.A., and James Gilbert.

No. 3. Outlines of General Knowledge. By Henry Ince, M.A., and James Gilbert.

No. 4. Outlines of Geography. By Professor Wallace and James Gilbert.

No. 5. Outlines of Grecian History. By E. Walford Esq., M.A.

No. 6. Outlines of Arithmetic. By John Box.

No. 7. English History, being an extracted edition of the "Outlines." By Ince and Gilbert.

[5391]

GORDON, JAMES, 51 *Hanover Street, Edinburgh.*—Educational class books, cards with objects, and books for school libraries.

[5392]

GOVER, EDWARD, *Princes Street, Bedford Row, London.*—Historic-geographical atlases, scripture prints, and other educational publications.

[5393]

GRIFFITH & FARRAN, *St. Paul's Churchyard.*—Educational books.

[5394]

HALL, ARTHUR, VIRTUE, & Co., 25 *Paternoster Row, London, E.C.*—Printed books.

[5396]

HOGG, J., & SONS, 9 *St. Bride's Avenue, Fleet Street.*—Educational books.

[5397]

HOPPER, ARTHUR, B.A., *Edgbaston, near Birmingham.*—Elementary lessons, and lessons on language for the deaf and dumb.

[5398]

JARROLD & SONS, 47 *St. Paul's Churchyard, London, and Norwich.*—Educational works, books, pens, and pencils.

[5399]

JONES, ALFRED (Corresponding Secretary of the United Association of Great Britain), *Shakspere Terrace, Albion Grove, Stoke Newington, N.*—Publications and proceedings of the Association of Great Britain.

[5400]

JONES, REV. CHARLES WILLIAM, *Pakenham, Suffolk.*—Three adults' reading-books: two very large type. London, Longmans.

[5401]

KNIPE, JAMES A., *Moorville, Carlisle.*—Geological map of the British Isles; geological map of England and Wales; geological map of Scotland.

[5402]

LONGMAN, GREEN, LONGMAN, & ROBERTS, *Paternoster Row, London.*—Educational works and appliances.

[5403]

LUCAS, GEORGE, 44 *Kennedy Street, Manchester.*—Terrestrial and celestial globes, for students, schools, &c.

[5404]

MACKIE, S. J., 25 *Golden Square, W.*—Scientific and educational diagrams and books.

[5405]

MACMILLAN, & Co., *Cambridge.*—Educational books.

[5406]

MAIR & Co., 34 *Bedford Street, Strand.*—Directory of 25,000 schools in United Kingdom, Continent, &c.

Mair's school list is a comprehensive guide to the public and private schools of the United Kingdom and the Continent. Price 6s. Foolscap 8vo., 850 pp.

[5407]

MARTIN, G. W., 14 & 15 *Exeter Hall, W.C.*—Books, &c., on the theory and practice of vocal music, and instruments and apparatus for musical education.

[5408]

MORRISON, LIEUTENANT, R.N., 17 *Surrey Street, Strand.*—Diagrams: sun, earth, and planets' motion through space, in cycloid curves.

DIAGRAMS SHOWING THE MOTION OF THE SUN THROUGH SPACE.

A. The *red* circles show the sun's yearly course, from east to west, on the line A B, commencing at A, 21st June, 1861, and ending at B, 21st June, 1862.

The *blue* and *white* circles show the places of the earth: 1st, at *a*, when the sun is on the north tropic, 21st June; 2nd, through *b* to *c*, when the sun is on the equator, 23rd September; 3rd, through *d e* to *f*, when the sun is on the south tropic, 21st December; 4th, through *g h* to *i*, when the sun is on the equator again, 21st March; lastly, through *j* to *k*, when the sun returns to the north tropic, completing the year. Thus the sun's course is a straight line from A to B, ever moving to the *west;* and the earth's yearly course is along the curve formed by the *red* letters, *a b c d e f g h i j k*.

Scale.—A, *a*, earth's distance from the sun, 95,000,000 miles. A B, sun's motion through space in one year, 875,694,000 miles; at the rate of 100,000 miles per hour.

B. *Two* years' motion of the earth, sun, and planet Venus, with all her phenomena. This diagram reverses all the motions, for convenience of reading.

[5409]

Newton & Son, Globe-makers, 66 *Chancery Lane.*—Improved mode of mounting globes and orreries for schools.

[5410]

Nutt, David, 270 *Strand.*—Books, chiefly educational.

[5411]

Oakey, Henry, 20 *Newman Street, Oxford Street.*—Henry Oakey's musical atlas.

[5412]

Oliver & Boyd, Publishers, *Edinburgh.*—Educational class books.

[5413]

Parker, Son, & Bourn, 445 *West Strand.*—Hullah's musical works.

[5414]

Philip, George, & Son, 32 *Fleet Street, London, and Liverpool.*—Atlases, schoolroom and commercial maps, educational works, &c.

The following are exhibited :—

No. 1. Philips' Imperial Library Atlas, a series of new and authentic maps, engraved, from original drawings, by John Bartholomew, F.R.G.S. Edited by William Hughes, F.R.G.S. Size—Imperial folio. The maps elaborately printed in colours. Publishing in monthly parts, each containing three maps, price 5*s.* The work will be accompanied by a valuable index of reference, and when complete will form one handsome volume, *half bound russia, price 5l. 5s.*

No. 2. The Family Atlas, a series of 52 maps, imperial quarto, printed in colours ; accompained by illustrative letter-press, describing the natural features, climate, productions, and political divisions of each country, with its statistics, brought down to the latest period, and a copious consulting index. Edited by W. Hughes, F.R.G.S. *Handsomely bound in cloth, gilt edges, price 1l. 15s.; or half-bound Turkey morocco, gilt edges, 2l. 2s.*

No. 3. The Library Atlas, a series of 44 maps, printed in colours, imperial quarto ; accompanied by a copious consulting index. Edited by W. Hughes, F.R.G.S. *Handsomely bound in cloth, gilt edges, price 15s.; or with the maps interleaved and elegantly half-bound Turkey morocco, cloth sides, gilt edges, 1l. 1s.*

No. 4. The Popular Atlas, a series of 39 maps, coloured, imperial quarto, with a copious consulting index. Edited by W. Hughes, F.R.G.S. *Handsomely bound, price 12s. 6d.*

No. 5. The Cabinet Atlas, a series of 33 maps, imperial quarto, coloured, with a copious consulting index. Edited by W. Hughes, F.R.G.S. *Handsomely bound, price 10s. 6d.*

No. 6. Atlas of Physical Geography, a series of maps and diagrams in illustration of the features, climates, various productions, and chief natural phenomena of the globe. Edited by W. Hughes, F.R.G.S. *Imperial quarto, handsomely bound, price 12s. 6d.*

No. 7. The Training-School Atlas, by William Hughes, F.R.G.S.; A series of maps, illustrating the physical geography of the great divisions of the globe. New edition. Coloured maps. *Medium folio, bound in cloth, price 15s.*

Series of School-room Maps, by W. Hughes, F.R.G.S. Size—5 feet 8 inches by 4 feet 6 inches; mounted on rollers and varnished.

List of Maps.

	s.	*d.*
The World, in hemispheres . .	16	0
Europe	16	0
Asia	16	0
Africa	16	0
North America	16	0
South America	16	0
Australia and New Zealand . .	16	0
England and Wales	16	0
Scotland	16	0
Ireland	16	0
Palestine	16	0
The World, on Mercator's projection	21	0
British islands	21	0
India	21	0

The publishers were led to undertake the present series from a conviction of the inadequacy of any similar maps hitherto published fully to represent the geographical knowledge of the present day, or to supply the wants of the educational community. They are content to rest their claims to notice upon their merits alone, and they invite the attention of all persons interested in education to the particular maps which the series embraces.

Catalogues of George Philip and Sons geographical and educational works may be had at their establishments in London or Liverpool, and at the Exhibition Building, Class XXIX.

[5415]

Potts, R., M.A., *Trinity College, Cambridge.*—Educational works. (*See pages* 40 & 41.)

[5416]

PROPRIETORS OF WEEKLY DISPATCH NEWSPAPER, *Fleet Street.*—Maps of London, &c., given with the paper.

[5417]

RELIGIOUS TRACT SOCIETY, 56 *Paternoster Row, and* 164 *Piccadilly, London.*—Printed books, &c.

The Religious Tract Society was instituted 1799, for the circulation of religious books and treatises throughout the British dominions and in Foreign countries.

Its publications include :—I. Works to aid in the study of the Sacred Scriptures,—Annotated Paragraph Bibles, Commentaries, Concordance, Bible Hand-book, Bible Text Cyclopedia, Bible Dictionary, Explanations of Scripture Terms, Horæ Paulinæ, Scripture Natural History, Biblical Atlas, Manners and Customs of the Jews, &c., &c.

II. Educational Works,—Historical, including volumes upon Babylon, Egypt, Greece, Nineveh, Rome, England, France, Japan, Spain, Russia, Brazil, India, &c. Scientific, including works on Botany, Conchology, Entomology, Ethnology, Geology, Natural History, Natural Philosophy, &c.

III. Works of Divinity, Theological, Practical, Experimental, and Polemical.

The Society also publishes five periodicals, The Leisure Hour, and Sunday at Home, weekly; the Cottager, the Child's Companion, and the Tract Magazine, monthly.

The donations and subscriptions are applied—I. To furnish free grants of Tracts, Handbills, &c., to the Clergy, Ministers, and other Tract distributors, and for the use of soldiers, sailors, emigrants and foreigners.

II. Libraries at reduced prices to parishes, congregations, Sunday schools, gaols, workhouses, hospitals, and asylums.

III. To assist the operations of affiliated societies throughout Europe, Asia, Africa, America, by grants of money and paper.

The publications upon the Society's catalogues, number about 2,500 Tracts and handbills in various languages, and 1,400 bound books, and numerous smaller works for children. Its grants vary from £12,000 to £14,000 per annum.

Depôts, 56 Paternoster Row, E.C., 65 St. Paul's Churchyard, and 164 Piccadilly, W.

[5418]

REYNOLDS, JAMES, 174 *Strand, London.*—Specimens of school diagrams.

[5419]

SALMON, E. W., *Nottingham.*—An engraved plan of Nottingham.

[5420]

SCOTT, DR. W. R., *Exeter.*—Reading made easy for the deaf and dumb; exercises in language for ditto; position in society of ditto; education of idiots.

The deaf and dumb, their position in society and the principles of their education considered, 5*s.*

Exercises in English composition for the deaf and dumb, being a series of lessons so graduated as to enable the deaf-mute more easily to acquire the more difficult idioms of English expression, 2*s.* 6*d.*

Reading made easy for the deaf and dumb, being a series of progressive reading lessons, illustrated by upwards of one thousand engravings, coloured, 3*s.* 6*d.*

Remarks, theoretical and practical, on the education of idiots and children of weak intellect, 1*s.* 6*d.*

[5421]

SHEAN, WILLIAM, 1 *Liverpool Street, City.*—An improved attendance register book, for the use of schools.

[5422]

SIMPKIN, MARSHALL, & CO., *Stationers' Hall Court.*—Educational works of Dr. Cornwell, F.R.G.S. (*See page* 42.)

[5423]

SMITH & SON, 172 *Strand.*—Map of London, world, Europe.

[5424]

STANFORD, EDWARD, 6 *Charing Cross, London.*—Maps, atlases, and books; specimens of map mounting.

[5425]

STEPHENSON, REV. NASH, M.A., *Shirley, near Birmingham.*—The unity and completeness of the Church catechism.

[5426]

SUNDAY SCHOOL UNION, 56 *Old Bailey, E.C.*—Books and educational appliances.

POTTS, R., M.A., *Trinity College, Cambridge.*—Educational works.

The following are published by John W. Parker, Son, and Bourn, 445 West Strand, London:—

No. 1. EUCLID'S ELEMENTS OF GEOMETRY. The University edition (the second). The first six books, and the portions of the eleventh and twelfth books, read at Cambridge, with notes, questions, and geometrical exercises, from the Senate House and College examination papers to the year 1861, with hints for the solution of the problems, &c. Enlarged and improved. 8vo., pp. 520, cloth, 10*s.* (The eighth thousand.)

⁎ In this second edition, the brief outline of the history of geometry is omitted, but reserved for the sequel, which will contain some account of the extensions of the Euclidean geometry, including the transversals, porisms, poles and polars, &c., &c., &c.

No. 2. EUCLID'S ELEMENTS. The first six books: the school edition (the fifth), with notes, questions, geometrical exercises, and hints for the solution of the problems, &c. 12mo., cloth, 4*s.* 6*d.* (The ninety-second thousand.)

"This is the best and cheapest school Euclid (fifth edition) we have ever examined, and runs to 361 pages. Here is given an abundance of exercises, selected from the best sources, to which are appended hints of the solutions.—*Educational Times.*

No. 3. EUCLID'S ELEMENTS. A supplement to the school edition, containing the portions read at Cambridge, of the eleventh and twelfth books, with notes, a selection of problems and theorems, and hints for the solutions. 12mo., 1*s.* (The sixth thousand.)

No. 4. EUCLID'S ELEMENTS. The first three books, reprinted from the school edition, with notes, questions, geometrical exercises, and hints for the solutions of the problems, &c. 12mo., 3*s.* (The sixth thousand.)

No. 5. EUCLID'S ELEMENTS. The first two books, with notes, questions, and geometrical exercises. 12mo., 1*s.* 6*d.* (The fourth thousand.)

No. 6. EUCLID'S ELEMENTS. The first book, with the notes, questions, and geometrical exercises. 12mo., 1*s.* (The sixteenth thousand.)

No. 7. EUCLID'S ELEMENTS. The definitions, postulates, axioms, and enunciations of the propositions of the first six, and of the eleventh and twelfth books. 12mo., 9*d.* (The third thousand.)

The object of these editions of Euclid has been to exhibit the Cambridge mode of teaching geometry. In addition to its use in the Universities of Oxford and Cambridge, Mr. Potts's Euclid is used at Eton College, and the chief grammar schools in England. The Council of Education at Calcutta, in 1853, was pleased to order the introduction of Mr. Potts s Euclid into the schools and colleges under their control in Bengal.

Mr. Potts's Euclid is also supplied from the Depositories of the National Society, Westminster, and of the Congregational Board of Education, Homerton College, &c., at reduced cost, for purposes of national education.

"In my opinion Mr. Potts has made a valuable addition to geometrical literature by his editions of Euclid's Elements."—*W. Whewell, D.D., Master of Trinity College, Cambridge.*

"Mr. Potts's editions of 'Euclid's Geometry' are characterized by a due appreciation of the spirit and exactness of the Greek geometry, and an acquaintance with its history, as well as by a knowledge of the modern expressions of the science. The Elements are given in such a form as to preserve entirely the spirit of the ancient reasoning, and having been extensively used in colleges and public schools, cannot fail to have the effect of keeping up the study of geometry in its original purity."—*James Challis, M.A., Plumian Professor of Astronomy and Experimental Philosophy in the University of Cambridge.*

"By the publication of these works, Mr. Potts has done very great service to the cause of geometrical science; I have adopted Mr. Potts's work as the text-book for my own lectures in geometry, and I believe that it is recommended by all the mathematical tutors and professors in this University."—*Robert Walker, M.A., F.R.S., Reader in Experimental Philosophy in the University, and Mathematical Tutor of Wadham College, Oxford.*

"The plan of this work is excellent."—*Spectator.*

"When the greater portion of this part of the course was printed, and had for some time been in use in the academy, a new edition of Euclid's Elements, by Mr. Robert Potts, M.A., of Trinity College, Cambridge which is likely to supersede most others, to the extent, at least, of the six books, was published. From the manner of arranging the demonstrations, this edition has the advantages of the symbolical form, and it is at the same time free from the manifold objections to which that form is open. The duodecimo edition of this work, comprising only the first six books of Euclid, with deductions from them, having been introduced at this institution as a text-book, now renders any other treatise on plane geometry unnecessary in our course of mathematics."—*Preface to Descriptive Geometry, &c., for the Use of the Royal Military Academy, by S. Hunter Christie, M.A., of Trinity College, Cambridge, Secretary of the Royal Society, &c., Professor of Mathematics in the Royal Military Academy, Woolwich.*

"Mr. Potts has maintained the text of Simson, and secured the very spirit of Euclid's geometry, by means which are simply mechanical. It consists in printing the syllogism in a separate paragraph, and the members of it in separate subdivisions, each, for the most part, occupying a single line. The divisions of a proposition are therefore seen at once, without requiring an instant's thought. Were this the only advantage of Mr. Potts's edition, the great convenience which it affords in tuition would give it a claim to become the geometrical text-book of England. This, however, is not its only merit." —*Philosophical Magazine*, January, 1848.

"If we may judge from the solutions we have sketched of a few of them [the geometrical exercises], we should be led to consider them admirably adapted to improve the taste as well as the skill of the student. As a series of judicious exercises, indeed, we do not think there exists one at all comparable to it in our language—viewed either in reference to the student or teacher."—*Mechanics' Magazine*, No. 1175.

"The 'hints' are not to be understood as propositions worked out at length, in the manner of Bland's Problems, or like those worthless things called 'Keys,' as generally 'forged and filed'—mere books for the dull and the lazy. In some cases references only are made to the propositions on which a solution depends; in others we have a step or two of the process indicated; in one case the analysis is briefly given to find the construction or demonstration: in another case the reverse of this. Occasionally, though seldom, the entire process is given as a model; but, most commonly, just so much is suggested as will enable a student of average ability to complete the whole solution—in short, just so much (and no more) assistance is afforded as would, and *must be*, afforded by a tutor to his pupil. Mr. Potts appears to us to have hit the 'golden mean' of geometrical tutorship."—*Mechanics' Magazine*, No. 1270.

POTTS, R., M.A.—*continued.*

"We can most conscientiously recommend it [the school edition] to our own younger readers, as the *best* edition of the *best* book on geometry with which we are acquainted."—*Mechanics' Magazine*, No. 1227.

Mr. Potts's first octavo edition of Euclid appeared in 1845, and since then has been gradually gaining ground in the estimation of our best teachers, as one of the most unexceptionable books of its class at present within the reach of the students in our schools and universities. The work is too well-known to require description, and little need be added to the notice already given in this magazine (see vol. xxxii., s. 3, p. 69), beyond the assurance that the new edition possesses all the best and most characteristic features of the old one, *minus* many of its imperfections. . . . We will merely repeat, then, that although we trust the work, considered as an *introduction* to the science of geometry, will some day be superseded, we are convinced that as a careful English reproduction of Euclid's Elements, illustrated by the notes of an able and judicious teacher, and enriched by a large collection of very useful exercises, it will long maintain its ground."—*Philosophical Magazine*, January, 1862.

No. 8. SAMMLUNG GEOMETRISCHER AUFGABEN UND LEHRSATZE FÜR DEN SCHULGEBRAUCH UND ZUM SELBSTUNTERRICHT. Aus der englischen Ausgabe des Euklides von Robert Potts, ins Deutsche übersetzt von Hans H. von Aller, oberst a. D. Ritter, &c. Mit einer Vorrede von Professor Dr. Wittstein. 8vo. 3*s*. Hanover, 1860.

No. 9. LIBER CANTABRIGIENSIS. Part I. An account of the aids, encouragements, and rewards offered to students in the University of Cambridge, to which is prefixed a collection of maxims, aphorisms, &c. Designed for the use of learners. Fcap. 8vo., pp. 570. 5*s*. 6*d*., in boards.

"A suitable book for school libraries."

"The (annual) income of the University and Colleges together, amounts to no less a sum than 209,500*l*."—*Speech of the* RIGHT HON. E. P. BOUVERIE, *in the House of Commons, on Friday*, May 30, 1856.

In the Cambridge University Act, 19 *&* 20 *Vict*, *cap.* 88, *it is stated, in section* 46:—

"From and after the first day of Michaelmas Term, 1856, it shall not be necessary for any person, on obtaining any exhibition, scholarship, or other college emolument, available for the assistance of an undergraduate student in his academical education, to make or subscribe any declaration of his religious opinion or belief, or to take any oath, any law or statute to the contrary notwithstanding."

"It was not a bad idea to prefix to the many encouragements afforded to students in the University of Cambridge, a selection of maxims, drawn from the writings of men who have shown that learning is to be judged by its fruits in social and individual life."—*The Literary Churchman.*

"A work like this was much wanted."—*Clerical Journal.*

"The book altogetner is one of merit and value."—*Guardian.*

"The several parts of this book are most interesting and instructive."—*Educational Times.*

"No doubt many will thank Mr. Potts for the very valuable information he has afforded in this laborious compilation."—*Critic.*

"A vast amount of information is compressed into a small compass, at the cost evidently of great labour and pains. The aphorisms, which form a prefix of 174 pages, may suggest useful reflections to earnest students.'—*The Patriot.*

No. 10. LIBER CANTABRIGIENSIS. Part II. Containing a full account of the recent changes in the statutes of the Colleges and University, made under the powers of the Act (19 and 20 Vict., cap. 88); and of the minor scholarships instituted and open to the competition of students before residence; with a brief account of the course of Collegiate and University studies at Cambridge. Fcap. 8vo. 2*s*. 6*d*., in boards.

The following is published by Longman & Co., London:—

No. 11. THE EVIDENCES OF CHRISTIANITY AND THE HORÆ PAULINÆ, by William Paley, D.D., formerly Fellow and Tutor of Christ's College, Cambridge. A new edition, with notes, an analysis, and a selection of questions from the Senate-House and College examination papers; designed for the use of students, by Robert Potts, M.A., Trinity College. 8vo., pp. 568. Price 10*s*. 6*d*., in cloth.

"The theological student will find this an invaluable volume. In addition to the text there are copious notes, indicative of laborious and useful research; an analysis of great ability and correctness; and a selection from the Senate-House and College examination papers, by which great help is given as to what to study and how to study it. There is nothing wanting to make this book perfect."—*Church and State Gazette.*

"The scope and contents of this new edition of Paley are pretty well expressed in the title. The analysis is intended as a guide to students not accustomed to abstract their reading, as well as an assistance to the mastery of Paley; the notes consist of original passages referred to in the text, with illustrative observations by the editor; the questions have been selected from the examinations of the last thirty years. It is an useful edition."—*Spectator.*

"Attaching, as we do, so vast a value to evidences of this nature, Mr. Potts's edition of Paley's most excellent work is hailed with no ordinary welcome—not that it almost, but that it fully answers the praiseworthy purpose for which it has been issued. In whatever light we view its importance—by whatever standard we measure its excellences—its intrinsic value is equally manifest. To these 'Evidences' the 'Horæ Paulinæ' has been added, inasmuch (we quote from the preface) 'as it forms one of the most important branches of the auxiliary evidences of Christianity.' It is further added—'To the intelligent student, no apology will be necessary for bringing here before him in connection with the 'Evidences' the 'Horæ Paulinæ'—a work which consists of an accumulation of circumstantial evidence elicited from St. Paul's Epistles and the Acts with no ordinary skill and judgment; and exhibited in a pellucid style as far removed from the unnatural as from the non-natural employment of language.' Without this volume the library of any Christian man is incomplete."—*Church of England Quarterly Review.*

"We do not hesitate to aver that Mr. Potts has doubled the value of the work by his highly important preface, in which a clear and impressive picture is drawn of the present unsettled state of opinion as to the very foundations of our faith, and the increased necessity for the old science of 'Evidences' is well expounded by his masterly analyses of Paley's two works—by his excellent notes, which consist chiefly of the full text of the passages cited by Paley, and of extracts from the best modern writers on the 'Evidences,' illustrative or corrective of Paley's statements—and by the examination papers, in which the thoughtful student will find many a suggestion of the greatest importance. We feel that this ought to be henceforth the standard edition of the 'Evidences' and 'Horæ.'"—*Biblical Review.*

"The editor has judiciously added the 'Horæ Paulinæ' as forming one of the most important branches of the auxiliary evidences. He has added many valuable notes in illustration and amplification of Paley's argument, and prefixed an excellent analysis or abstract of the whole work, which will be of great service in fixing the points of this masterly argument on the mind of the reader. Mr. Potts's is the most complete and useful edition yet published."—*Eclectic Review.*

SIMPKIN, MARSHALL, & Co., *Stationers' Hall Court.*—Educational works of Dr. Cornwell, F.R.G.S.

GEOGRAPHICAL.

Price 1*s.* 6*d.*; 2*s.* 6*d.* coloured, MAP BOOK FOR BEGINNERS: consisting of twelve pages of maps (above 70, large and small).

Also for MAP-DRAWING, price 1*s.*, BOOK OF BLANK MAPS. The above 70 maps in outline; that is, complete in everything but the names, which are to be filled in by the learner; including England, Scotland, Ireland, Russia, Switzerland, Italy, and the various British Colonies, as required at the next Oxford local examinations.

Also, price 1*s.*, BOOK OF MAP PROJECTIONS. The lines of latitude and longitude only to the above maps.

10th edition, price 1*s.*, GEOGRAPHY FOR BEGINNERS.

32nd edition, 3*s.* 6*d.*; or, with thirty maps, on steel, 5*s.* 6*d.*, SCHOOL GEOGRAPHY.

Price 2*s.* 6*d.* plain; 4*s.* coloured, SCHOOL ATLAS. This atlas consists of 30 beautifully-executed small maps on steel, in which is found every place mentioned in the author's "School Geography." It also contains a list of several hundred places, with their latitude and longitude. These names are accentuated; and, in cases of difficulty, the pronunciation is also given.

*** Recent geographical discoveries and changes are embodied in the current editions of the above works.

GRAMMATICAL, ETC.

25th edition, price 1*s.* 6*d.*, THE YOUNG COMPOSER; or, Progressive Exercises in English Composition.

Price 3*s.*, KEY TO THE YOUNG COMPOSER: with hints as to the mode of using the book.

4th edition, price 3*s.*, cloth, DR. ALLEN'S EUTROPIUS: with a complete dictionary for schools.

32nd edition, price 2*s.* red leather; 1*s.* 9*d.* cloth, ALLEN AND CORNWELL'S SCHOOL GRAMMAR: with very copious exercises, and a systematic view of the formation and derivation of Words, together with Anglo-Saxon, Latin, and Greek lists, which explain the etymology of above 7000 English Words.

38th edition, 1*s.* cloth; 9*d.* sewed, GRAMMAR FOR BEGINNERS.

11th edition, price 4*s.*, SELECT ENGLISH POETRY. For the use of schools and young persons in general. Edited by the late Dr. ALLEN.

*** This edition is got up in a superior manner, and the book is considered to be well adapted for prizes or presents.

ARITHMETICAL.

5th edition, price 1*s.* 6*d.*, ARITHMETIC FOR BEGINNERS: An introduction to Cornwell and Fitch's "Science of Arithmetic;" being a first book of practical arithmetic, with an inductive explanation of each rule, and containing numerous questions for purely mental calculation.

7th edition, price 4*s.* 6*d.*, THE SCIENCE OF ARITHMETIC A systematic course of numerical Reasoning and Computation, with very numerous exercises. By JAMES CORNWELL, Ph.D., and JOSHUA G. FITCH, M.A.

Nearly ready, KEY TO ARITHMETIC FOR BEGINNERS. In this key every question in the Arithmetic is worked in full, and practical directions accompany each rule, in addition to an introduction giving general hints for teaching arithmetic.

[5427]

THELWALL, SARA M., 9 *Stanhope Street, Bath.*—Syllabic primer and reading-book; tables of syllables to ditto.

[5428]

THIMM, F., 3 *Brook Street, Grosvenor Square.*—Grammars of the European languages.

[5429]

TILLEARD, JAMES, *Council Office, London.*—Works in music (for scholars), and in school management (for teachers).

[5430]

WALLIS, GEORGE, 16 *Victoria Grove, West Brompton.*—Wallis's drawing book for ornamental designers, artisans, and others.

[5431]

WALTON & MABERLY, *Upper Gower Street, and Ivy Lane, Paternoster Row.*—Educational and scientific works.

[5432]

WATKINS, JAMES, L.C.P., C.S.F., *The College, Dulwich, S.*—Educational diagrams to illustrate the doctrine of fractions.

[5433]

WHARTON, JAMES, 42 *Queen Square.*—Logical arithmetic, algebra, examples (graduated) in

[5434]

WHITE, GEORGE, Head Master, *Abbey Street Schools, London, N.E.*—Educational works.

[5435]

WILKINS, ERNEST P., M.D., *Newport, Isle of Wight.*—Maps, giving mountains, hills, valleys, &c., in modelling, instead of linear shading, by Brion.

[5436]

WYLD, JAMES, 12 & 13 *Charing Cross,* 457 *Strand, and* 2 *Royal Exchange.*—Geographical and topographical novelties.

[5437]

ARUNDEL SOCIETY, 24 *Bond Street.*—Publications of the Society.

[5438]

BORSCHITZKY, J. F., 32 *Tavistock Place, Tavistock Square, London, W.C.*—Materials for the international system of musical education.

[The international system owes its origin to Frobel, by whose example the author has chosen the best letters of alphabets of several languages for the nomenclature suitable for theoretical and practical purposes, instead of adopting an arbitrary system, by taking the first seven letters, as the Teutonic nations do, or the first syllables of a Latin hymn, as the Latin nations do.]

PRELIMINARY.

Thirty-one nursery and juvenile songs, with piano *ad libitum.* Book I. (or seven separate sheets), dedicated to mothers.

Forty ditto. Book II. (or seven sheets), dedicated to little children.

Kindergarten songs for piano; containing the 32 songs in "Ronge's Guide," arranged for the capacity of infant voices, with an accompaniment of a second voice, and bass *ad libitum.*

New Kindergarten songs, arranged as above.

March for piano to gymnastic exercises.

ELEMENTARY.

Twenty-two two-part school-songs, with piano or bass accompaniment *ad libitum;* arranged progressively as to intoning, time, etc., in all the major keys. Separate parts, with glossary and concise rules for the pronunciation of Italian. Pianoforte score, with the author's account of teaching singing in classes, by the new (international) system, which enables students to learn the theory of music, for sight-singing, while at the same time cultivating the voice and ear.

Forty-six nursery and juvenile songs, with piano or viola accompaniment *ad libitum.* Book III. (or eleven separate sheets), dedicated to school-children. These songs are in conformity with the school-songs, *i. e.*, the figures and initials for the modulators and minor and chromatic intervals are duly indicated in the voice-part. The purpose of these songs is to introduce now and then a sheet of them between the school-songs, according to the necessity of different schools (as very young children require several illustrations to each problem), so that children may begin to learn singing theoretically and practically as soon as they begin reading, writing, and arithmetic; the one becoming assistant to the other, and forming an agreeable variety.

ADVANCED.

Eighteen three-part school-songs, arranged as an introduction to the knowledge of harmony.

Juvenile drama
Vocal quadrille
Vocal polacca } for four treble or mixed voices.

INSTRUMENTAL.

Elementary studies and exercises for pianoforte in treble clef (with international fingering).

Violin school for joint practice of the elementary and advanced classes; containing four sheets of studies, three of chords and sequences of scales, and thirteen of exercises.

One hundred and seventeen nursery and juvenile songs (the three books in one volume). These songs, with their preludes and postludes, being written only on two staves, and marked with "the international" fingering, are intended as exercises on the piano for small hands; so that the young student, having sung the tune, or heard it sung, may, in combination with his reasoning and imitative faculties, facilitate his progress.

[5439]

BRUCCIANI, DOMENICO, 39 *Russell Street, Covent Garden.*—Casts of fruit and foliage from nature—art studies for schools. (*See page* 44.)

SUB-CLASS B.—*School Fittings and Furniture.*

[5447]

ABBATT, RICHARD, *Stoke Newington.*—1. Itinerary library regulator. 2. Orthographic projection of two hemispheres of the earth. 3. Terrestrial spinning top.

[5448]

AGAR, W. TALBOT, *Milford Lodge, Lymington, Hants.*—Pedal music desk—the foot turning over the leaves.

[5449]

ANDREWS, RICHARD, 144 *Oxford Street, Manchester.*—Guida-Mano—for giving a correct position of the hand, arm, wrist, and fingers of young children learning the pianoforte.

[5450]

ASSOCIATION FOR PROMOTING THE GENERAL WELFARE OF THE BLIND, 127 *Euston Road.*—Manufactures and educational apparatus.

[5451]

ATKINS, RICHARD, Builder, 27 *Bethel Street, Norwich.*—School desks, easels, and general fittings.

Brucciani, D., 39 *Russell Street, Covent Garden.*—Casts of fruit and foliage from nature, for drawing studies.

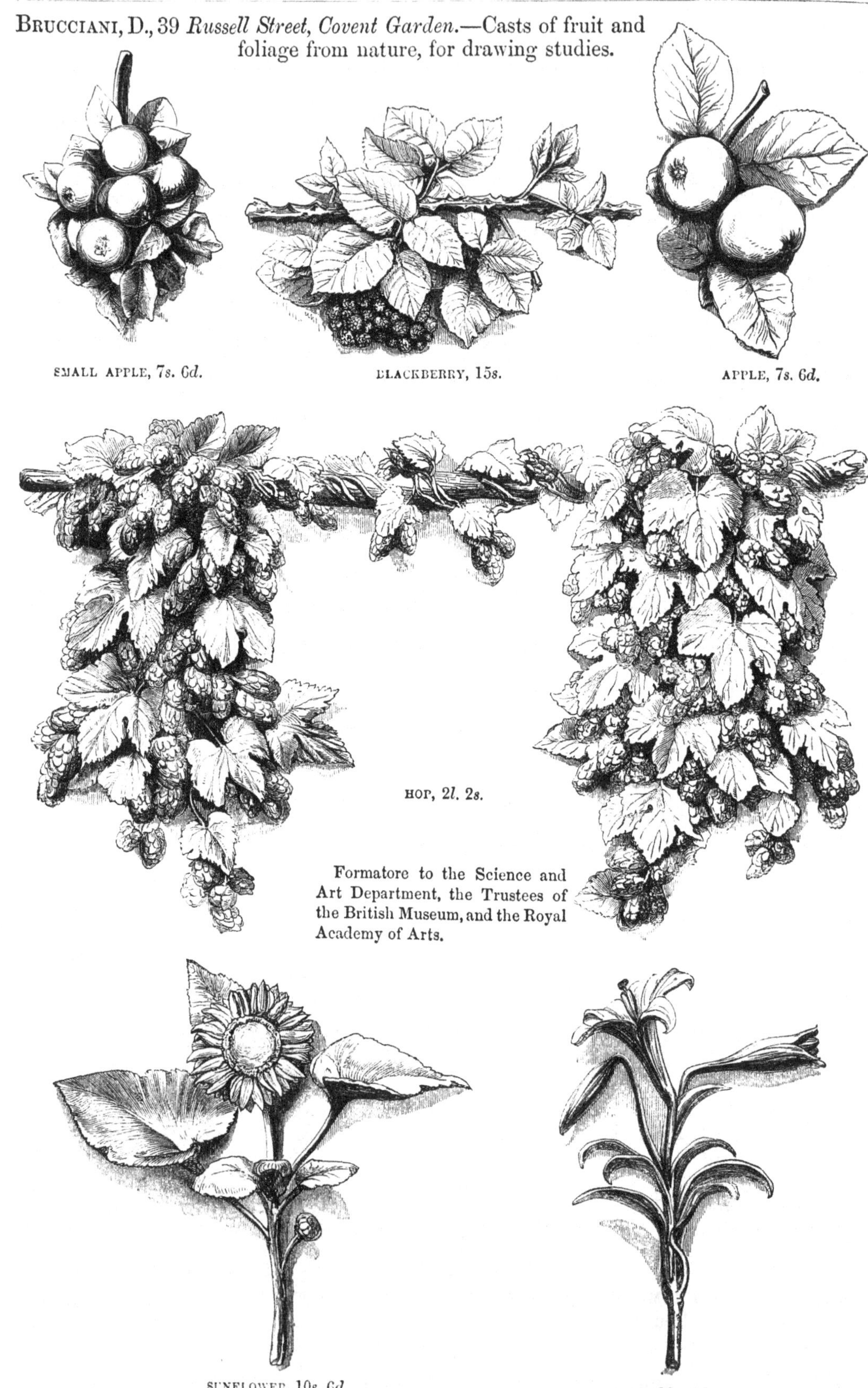

small apple, 7*s.* 6*d.*

blackberry, 15*s.*

apple, 7*s.* 6*d.*

hop, 2*l.* 2*s.*

Formatore to the Science and Art Department, the Trustees of the British Museum, and the Royal Academy of Arts.

sunflower, 10*s.* 6*d.*

lily, 10*s.* 6*d.*

[5452]

BARNARD, J., & SON, 339 *Oxford Street, and* 11 *Winsley Street, W.*—New chemical colours; improved sketching apparatus; educational diagrams.

CHEMICAL COLOURS.—Specimens in powder, and prepared in cakes, and moist for water-colour painting, and in oil.

No. 1. Rubine, a permanent colour, from naphthaline, closely allied in chemical composition to the madder colours.

No. 2. Chinese yellow, a new and permanent preparation of zinc.

No. 3. Mauve lake, the most permanent of the aniline colours.

No. 4. Extract of vermilion, a very near approach to orange.

No. 5. Real ultramarine, of different shades from lapis lazuli.

No. 6. Carmine.

No. 7. Oxide of chromium.

An improved waterproof sketching-tent, combining lightness, portability, and, when pitched, great available space: a model.

Improvements in sketching-apparatus. The diminutive colour-box, and specimens of colour-boxes, specially adapted to the various branches of art. Brushes and other materials.

Wall space.—Diagrams illustrating the permanency of different pigments used in painting, contrasts of colours, etc.

[5453]

BIGGS & SONS, 24 *Guildford Street East, W.C.*—Educational works and appliances.

1. The nursery alphabet. 2. The clock alphabet.
3. The infant teacher. 4. The picture-card alphabet.
5. The pictorial toy alphabet. 6. The infant tutor.

These toys are novel and ingenious. They help children to acquire a knowledge of letters, figures, spelling the clock face, etc., and are therefore invaluable in the nursery or schoolroom. No. 1 shows the alphabet with the faces of little children, corresponding in name with the letters exhibited. No. 2 teaches the clock face with moveable hands, and shows one by one the letters of the alphabet and figures. No. 3 forms no less than 720 (!) words and syllables, such as cat, dog, man, boy, nut, fig, etc. No. 4 is an alphabet on cards, in large capital letters, with pictures on the reverse sides. No. 5 shows the letters of the alphabet with pictures of flowers or fruits in contrast. No. 6 teaches the use of figures, and forms small sums in addition, subtraction, multiplication, and division. Price 1*s.* each.

[5454]

BLIND ASYLUM, *Bristol.*—Hearth-rugs, door-mats, baskets, brushes, knitting, and embossed books.

[5456]

BOWDEN, GEORGE, Manufacturer, 314 *Oxford Street, W.*—Assortment of sketch books, blocks, portfolios, &c., for the use of artists and schools.

[5457]

BRIGGS, R., & SON, 1 *Welbeck Street, Cavendish Square.*—Drawing materials, &c.

[5458]

BRITISH AND FOREIGN SCHOOL SOCIETY (EDWARD D. WILKS, Secretary), *Borough Road, London.*—Educational appliances.

[5459]

BROCKEDON, W., & CO., 34 *Great Ormond Yard, London.*—Compressed pure Cumberland lead, with samples, indelible lead for official documents, solid ink for linen, bone, wood, or zinc.

[5460]

CAMPBELL, THOMAS, 24 *South Richmond Street, Edinburgh.*—Original plan for making rugs, wrought in Blind Asylum, Edinburgh.

[5461]

COLSON, JOHN, *St. Swithin Street, Winchester.*—Drawings of new Training College at Winchester, Hants.

[5462]

CONGREGATIONAL BOARD OF EDUCATION.—*The College, High Street, Homerton, London, N.E.*—Model of schools, school apparatus, and results.

[5463]

CROGER, THOMAS, 483 *Oxford Street.*—Music in the garden without a performer, and the educational instrument.

[5464]

CRONMIRE, J. M. & H., 10 *Bromehead Street, Commercial Road East.*—Mathematical instruments.

[5465]

CROYDON, WILLIAM JAMES, 9 *York Place, Windsor.*—Dissecting, drawing models; each forming a complete study.

[5466]

DENTON, ALFRED, 8 *South Place, Finsbury.*—Miniature models for gold and silver pins, brooches, seals, &c.

[5467]

FETHERSTON, JOHN I., 18 *Suffolk Street, Dublin.*—Ancient and modern alphabets; illuminated colour charts and boxes.

Alphabets, Ancient, Mediæval & Modern, for Schools, Artists & Penmen. with Arabesque Elizabethan, Louis Quatorze, Floral & Foliated Borders. & Illuminated Chromatic Chart arranged on English & French Scales for mixing Colours registered designs.

Model colour boxes on new and original arrangements, with English and French chromatic charts attached, by which the inexperienced photographic, manufacturing, or other colourist can prepare every variety of shade or colour from three primaries, red, blue, and yellow, for oil or water.

[5468]

FORD, R. D., 32 *Great Carter Lane, Doctors' Commons, E.C.*—Desks and school fittings.

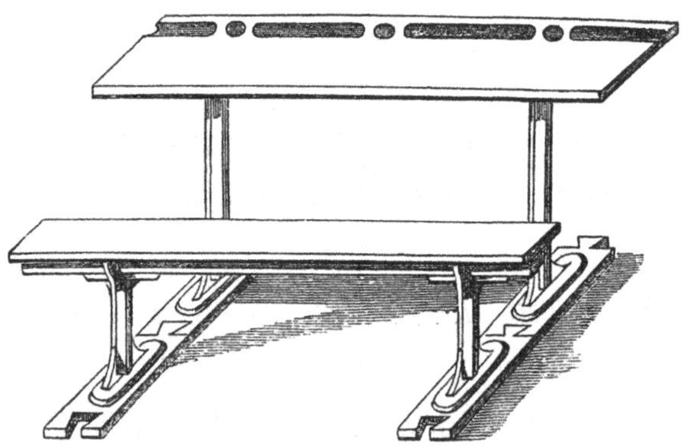

An improved desk and form extensively patronized, recommended by the Society of Arts to Messrs. Chapman and Hall, publishers, Piccadilly, and sent by them to Melbourne as the best in use.

Reference is kindly permitted to Edwin Abbott, Esq., head master of the philological school, Marylebone Road, N.W., where they have been in use six years. Frame for drawing, drawing board, and boys' satchel desk.

[5469]

GALL, JAMES, *Myrtle Bank, Trinity, Edinburgh.*—Triangular alphabet for reading and writing for the blind, the types of which may be reduced to those of school testaments.

[5470]

GLOVER, SARAH ANNA, *Great Malvern, Worcestershire.*—Sol-fa harmonicon, with rotary indexes to the diatonic scale.

[5471]

GRAY, JAMES, 33 *Richmond Place, Edinburgh.*—Raised maps for the blind, by needlework (designer blind).

[5472]

GREEN, BENJAMIN R., 41 *Fitzroy Square.*—Rustic drawing models, for preparing pupils to sketch from nature.

Series of rustic drawing models, exactly imitating the actual objects; designed to prepare pupils to sketch from nature; illustrating the rationale of perspectives, the principles of light, and shade, and colour.

First or Elementary Series, 16*s*., fitted in box with rustic figure, one guinea. Second or Advanced Series, 25*s*., fitted in box with rustic figure, one guinea and a half.

Prices of Models single.

1st Series.	*s.*	*d.*	2nd Series.	*s.*	*d.*
Roller	2	6	Well	8	6
Stile	3	0	Bridge	8	6
Pump	3	6	Water-wheel	10	6
Cottage-door	4	6	Figures, each	3	0
Hen-coop and Dove-cot	4	0			

Third Series (in progress), old timber porch, 16*s*. 6*d*. Summer-house, 15*s*. Figures, 1*s*. 6*d*.

[5473]

GREW, THOMAS, *Plaistow Park, West Ham, Essex.*—Mathematical drawing instruments.

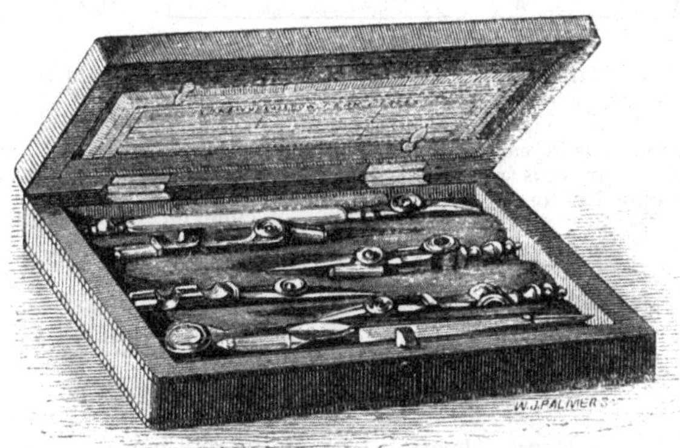

The above prize case of approved instruments, as supplied by him to the Science and Art Department at the South Kensington Museum, can be obtained for ten shillings' worth of stamps, or post-office order to the amount. Useful school set, in a mahogany case, from 4*s*.

A superior set of college instruments, mounted in best German silver, with additional joint to compass, and handsome set of marquiose scales, complete, in highly-finished mahogany case, with lock and key, 2*l*. 10*s*.

A surveyor's full set, consisting of double-pointed compass to hold needles, hair divider, ink and pencil point, bar, and two drawing pens, likewise double-jointed ink and pencil bows to hold needles, in German silver, with a set of steel spring bows to match, and superior set of green ivory rules, in 13-inch rosewood case, 4*l*. 10*s*.

[5474]

HAMMER, GEO. M., 44 *Harrington Street, London, N.W.*—Models of school fittings and educational apparatus. (*See page* 48.)

[5475]

HANCOCK, CHARLES, *Medmenham Lodge, Quadrant Road, Highbury New Park, London, N.*—Animals and figures modelled in paper.

[5477]

HASKINS, JAMES F., 14 *Victoria Street, E.C.*—Class-room desks for pupil-teachers and elementary musical works.

[5478]

HAY, JOHN HENRY, *Kennington Oval.*—Class register for 55 names; abstract for one year portfolio.

Hammer, Geo. M., 44 *Harrington Street. London, N.W.*—Models of school fittings and educational apparatus.

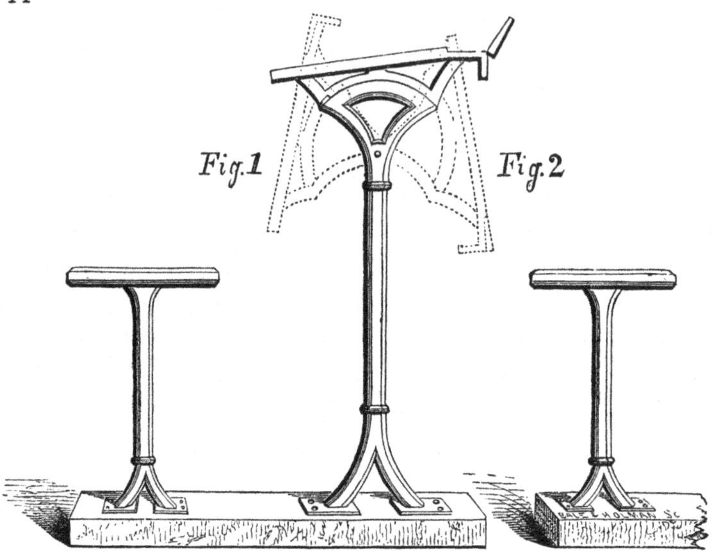

Improved desk and iron standard. The desk folds down to form back to seat (fig. 1), or to the seat in front (fig. 2). It can also be set flat, for use as a table. Price of desk, per foot, 1*s.* 9*d.* Standards 5*s.* each.

Desk, with sliding covers to ink-wells (registered), by which all are opened or closed at once, and can be secured with lock and key.

Mistresses' work and writing table. Price 4*l.* 10*s.*

Set of mechanical powers. Price 5*l.* 5*s.*

Sets of drawing models, at 10*s.* 6*d.*, 1*l.* 2*s.*, and 3*l.* 10*s.*

Sets of geometrical solids, 16*s.*, 1*l.* 11*s.* 6*d.*

Machine for illustrating centrifugal force, 12*s.*

[5479]

Herdman, William Gawin, *West Villa, Liverpool.*—The poetry of England's wealth: a scene in the midland counties.

[5480]

Holmes, Charles, *London Road, Derby.*—One improved school desk and form.

[5482]

Home and Colonial Training Institution, *King's Cross, London.*—Models of an infant school, of a school desk, &c.

[5483]

Houghton, P., & Co., 25 *Stamford Villas, Fulham.*—Drawing stumps in paper, leather, and cork; do. pins, &c.

[5484]

Hughes, George A., 47a *Edgeware Road.*—Embossed books, music, and writing apparatus for the blind.

[5486]

Jackson, Elizabeth Sarah, 3 *Sheffield Terrace, Kensington.*—Floreated motto.

[5487]

Johnston, W. & A. K., 4 *St. Andrew Square, Edinburgh.*—Specimens of maps and illustrations of physiology, printed in colours.

[5488]

Joseph, Myers, & Co., 144 *Leadenhall Street, E.C.*—Educational models and appliances.

[5489]

Leathes, Major Hill M., *St. Margarets, Herringfleet, Suffolk.*—Model picture.

[5490]

LONDON SOCIETY FOR TEACHING THE BLIND TO READ, &C., *Upper Avenue Road, N.W.*—Baskets; brushes; knitting.

[*Obtained a Medal at the Exhibition of* 1851.]

The Holy Scriptures, as embossed by the pupils.

Specimens of baskets, brushes, knitting, and embossing—the work of the blind pupils.

Similar articles, in great variety are on sale at the institution, which is open daily for the inspection of visitors from 2 to 5 o'clock, p.m., Saturdays excepted.

[5491]

MACINTOSH, C., & CO., *Cannon Street, London; Cambridge Street, Manchester.*—Portable globes and other educational requisites.

[5492]

MARTIN, JOHN, *Alfreton Road, Nottingham.*—Writing machine for the blind; enabling them to read their own writing.

[5493]

MILL, JANE, 1 *Foundling Terrace, W.C.*—Kindergarten articles used in the educational employments of children.

[5494]

MITFORD, BERTRAM, *Cheltenham.*—1. Conversational tablet. 2. Orthographical tablet for producing all words and short sentences mechanically.

[5495]

MOON, WILLIAM, F.R.G.S., *Brighton.*—Reading in several languages, maps, diagrams, &c., for instructing the blind.

The origin and success of this system of reading for the blind is somewhat remarkable; and being of so universal a character, the following outline will not fail to interest the visitors of the International Exhibition:—

In the year 1839 Mr. Moon, the inventor of the system by which the books are printed, lost his sight, and immediately commenced learning to read by a system of embossed reading invented by Mr. Frere. This accomplished, he began to seek out and teach others similarly afflicted with himself. Among this number was a poor boy, who for five years endeavoured, but without success, to learn by the plan invented by Mr. Frere. During these years Mr. Moon from time to time laid this case before the Throne of Grace, beseeching the all-wise Disposer of events to put into his mind a plan by which this poor child might be taught to read. At length his prayers were answered, and the Lord put into his mind a plan by which the poor lad was able to read easy sentences in the short period of ten days. Means were then required to carry out the plan of printing books for the use of the blind generally. At the end of two years, a gentleman gave Mr. Moon sufficient movable type to commence the work. The first publication appeared in June, 1847, in the form of a monthly magazine for the blind. This was compiled by Mr. Moon in his leisure hours. No sooner did these books begin to circulate than a demand was made for various portions of the Bible. To print the Scriptures with the small quantity of type Mr. Moon possessed would have been a work of many years, and would have required an immense capital to print even a small edition, on account of the large quantity of paper, &c., used in an embossed book. He then invented a new mode of stereotyping, which is done at a comparatively small expense. By this means, aided by subscriptions from a benevolent public, he has stereotyped the whole of the Bible in English, besides a variety of other books, and portions of Scripture in 50 other languages. At first, the work was carried on in Mr. Moon's private residence; the increase was rapid, and in 1856 it was found necessary to commence the erection of the larger and more suitable premises in the Queen's Road. The foundation-stone of this building was laid by C. H. Lowther, Esq., on the 4th September, 1856. Missionaries are continually carrying books from this establishment to foreign lands, and blind readers now rejoice in the instruction thus afforded, not only in Europe, but also in India, America, Australia, and China, while the Negro of Africa, and the Arab of Egypt, Palestine, and the desert are not forgotten, but have the word of life in their own vernacular tongue. Twenty-seven societies are established in various parts of our country for teaching the blind at their own homes, and lending them books from free libraries, and for several years past Mr. Moon has from time to time visited the Continent, for the purpose of benefiting the blind by the spread of this system, and both he and his system have met with a very cordial reception in Holland, Germany, and France.

Further particulars of this interesting work may be had free of cost, on application to Mr. Moon, at his establishment for English and foreign books, maps, &c., for the blind, 104 Queen's Road, Brighton.

Mr. Moon's publications comprise the whole of the Bible in English, and portions in 50 other languages, a good variety of other books in English, such as the Pilgrim's Progress, Biblical Dictionary, English Grammar, Geography, History of England, &c., &c. Also embossed maps, diagrams, music, pictures of animals, &c., &c., writing-frames, and many other useful things.

[5496]

MOORE, G. B., 9 *Lansdowne Terrace, Gloucester Road North, N.W.*—Crayon drawing fixed by a new method for portfolios.

[5497]

MUSSELWHITE, JOHN, *Devizes.*—Movable note music board.

[5498]

NATIONAL SOCIETY FOR THE EDUCATION OF THE POOR, *Sanctuary, Westminster.*—Educational appliances.

[5499]

NEWMAN, JAMES, 24 *Soho Square, London.*—Artists' colours, varnishes, brushes, &c.

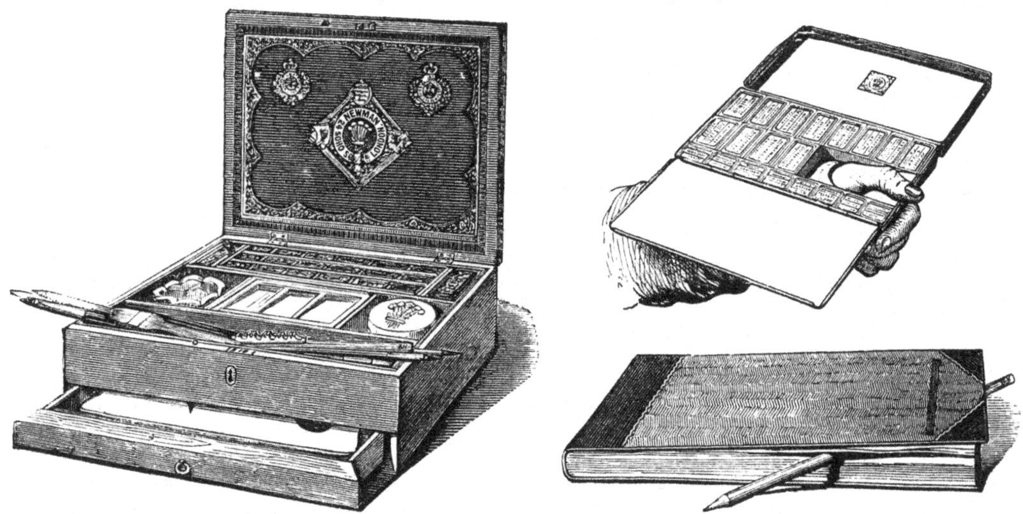

Artists' Colours, Brushes, Papers, &c., for Water-Colour Painting.

Boxes of Colours, and Materials for Oil Painting.

The New Body Colours,—solid, but not opaque,—and from a peculiarity in their manufacture, as deep and brilliant as the pure colour.

Boxes of Colours and materials for Illuminating and Missal Painting.

Photographic Colours, Varnishes, Brushes, &c.

The Improved Moist Colours, which do not become hard and useless in any climate, and are particularly brilliant and pure.

Sketching Boxes, Folding Seats, Easels, &c., and a variety of conveniences for sketching from nature.

Drawing Papers and Drawing Books in variety.

Indian Ink in great variety, some very curious.

Mathematical Instruments of superior quality.

[5500]

PEARCE, T. BLEWETT, 93 *Newman Street.*—The octave dissected; keys, &c., in music illustrated by colours.

[5501]

PEMBERTON, ROBERT, F.R.S.L., 33 *Euston Square, London.*—Science of education in various works; patent school organ. (*See page* 51.)

[5502]

PHILANTHROPIC SOCIETY'S FARM SCHOOL, *Redhill.*—Model of a school-house.

[5503]

PITMAN, ISAAC, *Bath.*—Phonetic shorthand, and printing alphabets. Uses: shorthand writing, true orthography, and easy reading.

PEMBERTON, ROBERT, F.R.S.L., 33 *Euston Square, London.*—Science of education in various works; patent school organ.

THE ATTRIBUTES OF THE SOUL FROM THE CRADLE, &c. This theory develops the false system of all scholastic education throughout the world, and demonstrates that the true basis is entirely centred in the mother, who imparts to her infant her own language in the course of the first three years; so that if the natural system of oral instruction were continued, and artistically carried out, seven languages would be produced upon the mind in the course of the first twenty-one years, the period of the natural growth of the human body, and consequently the season for the growth of all knowledge.

Specimens of the natural or oral system of teaching, for the use of the nursery and infant schools.

THE ELEMENTS OF GRAMMAR, 2*s.* 6*d.* The same on cards, with box complete 1*l.* The cards are arranged in small cases for carrying in the pocket, for the use of the mother or nurse at every opportunity, that the precious time of infant life may be continually occupied; as infant spirits imbibe and take every imprint from the sounds of the voice, its natural teacher.

THE TECHNICAL LANGUAGE OF ANATOMY, with a glossary of terms used in the work. All professional languages must be taught in infancy, as they cannot be perfectly acquired in after life, when the season for learning is passed, 3*s.* 6*d.*, on cards 1*l.* 10*s.*

THE INFANT DRAMA, a model of the true method of teaching and developing speech from the cradle, by spelling daily to the infant, and reading poetry and describing objects, aided by music; and for this purpose the nursery chromatic barrel-organ was expressly invented by the author, in order that music may form and grow upon the mind in the same way as the vernacular tongue. Addressed to the Mother, 2*s.* 6*d.*

THE HAPPY COLONY, with bird's-eye views of the Model Town and Colleges, 7*s.* 6*d.* This work develops a perfect system of emigration, to be carried out by distinct societies, and is perfectly arranged for the best civilized life, founded upon a new development of the human mind by the practical science of education, which combines the united powers of physical and mental instruction. The model town was devised expressly for a perfect arrangement for promoting human happiness, viz., perfect education, health, recreation, and occupation and mutual benefits. The form of the town is circular, taken from the prevailing forms of the creation. The inner circle contains fifty acres; upon the borders thereof are placed four colleges equi-distant, at the four cardinal points. Conservatories, workshops, swimming-baths, and riding-schools are attached to each college. Geography to be taught from the maps of the world laid out upon five acres of ground. The celestial map is laid down in the same manner, with glass balls to represent the different magnitudes of the stars, all named. This method serves two essential purposes at the same time: exercise and instruction, and, indeed, this is the secret of all education—pleasure must unite with all the processes of instruction, teaching, and training. In this circle of the maps of the world, &c., are also placed the botanic gardens, arboretum, and groves embodying history and the muses, and biography; and in the centre of the college grounds is placed the miniature farm, thus uniting all our educational forces and social wants, for perfecting the education of rising generations. The circles of streets round the college grounds are very spacious, and planted with trees. The manufactories are hid in groves of ornamental trees. The wide spaces between the backs of the houses, instead of being mere strips of garden, are laid out in beautiful orchards. The fourth circle is laid out for public gardens, in which are four churches; and the outer circle is the park, three miles in circumference, planted with groves and clumps of all kinds of valuable trees. All the cross streets verge from the park to the centre of the town, and beyond the park a public reserve of a belt of heath or forest land is provided, beyond which are the cultivated farms, which extend on the same principle. It may be presumed that this design for a town or city is the most perfect that has ever been planned.

THE SCIENCE OF MIND FORMATION, and the process of the reproduction of genius elaborated, involving the remedy for all our social evils, 1*s.* This work was produced in consequence of the state of our social evils, described by the various speakers at the inauguration of the Social Science League, held at Birmingham, by Lord Brougham and his learned coadjutors. A more valuable association has never existed, as every nation is awaiting the development of the social sciences.

THE MODEL INSTITUTION OF THE SCHOOL OF GENIUS. This work describes a royal infant opera for foreign languages, for the infants of princes, of the nobility, ambassadors, functionaries of state, and merchants. The present scientific and vast commercial age requires a free intercourse with all nations. In truth, the world requires international education. The science of developing the spoken languages has long been wanting. The absence of this science has been a serious inconvenience to the government of that immense population of her Majesty's possessions in the East: however the written languages may have reached perfection through the medium of alphabetical signs, the spoken languages have been entirely neglected, no attention whatsoever has been paid to the development of speech. The natural growth of the spoken languages is one language every two years. The natural forces of speech to be used in teaching a language are from four to seven thousand words per hour. The vernacular tongues of all nations alone possess the attribute power and forces of developing speech and imparting, transferring, or transmitting the spoken languages to infants, children, and youth. A staff of teachers of any nation can impart their own language grammatically, to a large audience of infants, from the ages of three to fourteen years, in the course of two years, far better and more correctly than any children of the age of twelve or fourteen years can acquire their own language from studying books, even when brought up in genteel life in the capital of any country. It is the abstract system of the schools which has deceived the world upon the subject of education.

London: Houlston and Wright, 65 Paternoster Row and by the author, 33 Euston Square, N.W.

By Her Majesty's royal letters patent. Pemberton's Chromatic Barrel-Organ, for the nursery and infant schools.

[5504]

RAHLES, DR. FERDINAND, 13 *Albert Street, Camden Road, N.W.*—1 and 2. alphabet and spelling game, adapted for infant-schools and nurseries.

[5505]

REEVES & SONS, 113 *Cheapside, London, E.C.*—Economic series of one shilling drawing materials for the use of students. Water-colours, artists' materials, &c.

This series comprises the following:—

One shilling box of water-colours, combining the ten colours and three brushes as selected by the Society of Arts.

One shilling case of drawing pencils, containing six pencils, india-rubber, and four drawing pins.

One shilling box of coloured chalks, &c., containing twelve coloured chalks, two stumps, porte-crayon, and charcoal.

One shilling deal drawing board and stand, forming a wooden desk 15 × 11 inches, with inclined support for holding a copy

One shilling solid drawing tablet, composed of twenty-four leaves of good drawing paper, 9 × 7½ inches, or sixteen leaves 11 × 9 inches.

One shilling T square, blade 15 inches long.

One shilling parallel rule, 10 inches long.

One shilling set of angles, with curve and protractor.

One shilling drawing portfolio, size 15 × 11 inches.

The eighteenpenny half-set of mathematical drawing instruments.

Water-colours in cakes, also moist in pans and tubes.

Boxes of water-colours of every description.

Japanned tin boxes of moist water-colours.

Water and powder colours prepared for illuminating.

Powder colours prepared for photography.

Pure Cumberland and polished and gilt drawing pencils.

Sable, Siberian, and camel hair brushes.

Mathematical drawing instruments and rules.

Solid and other sketching books.

Chalks, crayons, porte-crayons, stumps, palettes, &c., &c.

[5506]

REFORMATORY AND REFUGE UNION, 118 *Pall Mall.*—Apparatus used in reformatories, and models illustrating their operations.

[5507]

RIDLEY, REV. NICHOLAS JAMES, 10 *Paternoster Row.*—Articles and apparatus (including models) illustrating the book-hawking system.

[5508]

ROBERSON & CO., *Long Acre.*—Drawing materials, &c.

[5509]

ROWNEY, GEORGE, & CO., Manufacturing Artists' Colourmen, Retail Department, 51 *and* 52 *Rathbone Place;* Wholesale and Export Department, 10 *and* 11 *Percy Street, London, W.*—Materials used in the fine arts. (*See page* 53.)

[5510]

RUSSELL & BUGLER, *Ashford, Kent.*—Improved Rusthall school desks, with iron standard, made of different heights and lengths.

[5511]

RYFFEL, T. E., *Wimbledon.*—Calculating cubes—numbers made visible and tangible by 100 cubes, each 10 of a different colour; a multiplication table represented on the same system, without figures.

[5512]

SCHOOL FOR THE INDIGENT BLIND, *St. George's Fields, Southwark.*—Books for the blind, and goods manufactured by the blind.

Founded 1799. Supported by public subscriptions. Patron—Her Most Gracious Majesty the Queen.

The average number of pupils is 160: they are admitted for six years by election. The education consists of religious instruction, reading, writing, ciphering, industrial work, basket work of all kinds, mat making, weaving, knitting, brush making, hair work, &c., &c. Pupils having talent for it, are taught to sing, and to play the organ so as to become parochial organists.—B. G. Johns, Chaplain, Mr. Thos. Grueber, Secretary.

Embossed books and other educational appliances, as well as specimens of all their various industrial work, are exhibited.

The Society for Printing Embossed Books for the Blind (office, Blind-school, St. George's Fields) have published the following books in the Roman letter: Gospels of St. Matthew and St. John, Sunlight in the Clouds, Robinson Crusoe, The Psalms, an English History.

ROWNEY, GEORGE & Co., Manufacturing Artists' Colourmen, Retail Department, 51 *and* 52 *Rathbone Place;* Wholesale and Export Department, 10 *and* 11 *Percy Street, London, W.*—Materials used in the fine arts.

Artists' colours prepared for water-colour painting, ground by their new process, by steam-power. Manufactured in whole, half, and quarter cakes; or moist in tubes, pans, and half pans.

Oil-colours, in compressible tubes, ground extra fine by machinery; colours prepared for missal-painting, and illumination in soluble powder-colour.

Picturesque Drawing Models.—A series of objects, carefully copied from nature, adapted for instruction in sketching.

No. 1. The Windmill (see illustration), price 15*s.*

No. 2. The Boat-house, price 24*s.*

Also numerous smaller models of simple subjects, as stiles, gates, palings, sign-posts, &c., &c.

Geometric models for instruction in form, light, and shade drawing.

Perspective Models.—Showing the solid object and its appearance on the picture. These models so simplify the principles of perspective, that none can fail to comprehend them at once.

	s.	*d.*
Model No. 1. Vertical, horizontal, and oblique planes, single . .	8	0
No. 2. Cube and Pyramid . . .	10	6
No. 3. Cone, column, and prism .	8	0
No. 4. Bridge, coloured	12	6
No. 5. House and grounds, coloured	15	6

The set complete in a box, 2*l.* 15*s.*

Drawing pencils, artists' brushes, crayons, prepared canvas, easels, drawing boards, T squares, and every description of material used in the various styles of painting or drawing.

Geo. Rowney and Co.'s portable tent, for sketching tours, picnics, or summer excursions. Size when packed, 4 inches by three inches; 4 feet 2 inches long. Size when set up, 6 feet square; 7 feet high; weight, under 11 pounds, including case. Price 31*s.* 6*d.*; American leather cloth case, 3*s.* 6*d.* With extra cord for increased strength, 36*s.*, case, 4*s.*

The advantages of this tent consist in its portability and light weight when packed, and its strength and spaciousness when pitched.

Lithographic and other drawing books for students in various styles of drawing. Figure, landscape, cattle, and mechanical subjects. These specimens are exhibited on the staircase leading to the Tower.

[5513]

SCIENCE AND ART DEPARTMENT OF THE COMMITTEE OF COUNCIL ON EDUCATION, *South Kensington Museum.*—A set of drawings illustrating the stages of instruction in art afforded by the Schools of Art in connection with the Science and Art Department of the Committee of Council on Education.

The following is a list of the stages of instruction, with the names of the students whose works are exhibited as illustrations.

STAGES.	NAMES OF STUDENTS.
1. Linear drawing by aid of instruments:	
a. Linear geometry	Sarah A. Doidge.
b. Mechanical and machine drawing and details of architecture from copies	James Dundas. Wilmot Pilsbury.
c. Linear perspective	Catherine Baines.
2. Free-hand outline drawing of rigid forms, from examples or copies:	
a. Objects	
b. Ornament	Henry Mayes.
3. Free-hand outline drawing from the "round:"	
a. Models and objects	
b. Ornament	W. J. Griffiths.
4. Shading from flat examples or copies:	
a. Models and objects	
b. Ornament	Ellen Rose.
5. Shading from the round or solid forms:	
a. Models and objects	H. H. Lock.
b. Ornament	F. Hunt.
c. Time sketching and sketching from memory	
6. Drawing the human figure and animal forms, from copies:	
a. In outline	Margaret Johnson.
b. Shaded	Thomas Hampton.
7. Drawing flowers, foliage, and objects of natural history, from flat examples or copies:	
a. In outline	Joseph Platt.
b. Shaded	Mary J. Woodcock.
8. Drawing the human figure or animal forms from the round or nature:	
a. In outline from casts	Lewis Shanks.
b. Shaded	A. E. Mulready. S. George Pollard.
c. Studies of the human figure from nude model	A. E. Mulready.
d. „ draped „	
e. Time sketching and sketching from memory	
9. Anatomical studies:	
a. Of the human figure	Henry Hancock.
b. Of animal forms	Alfred T. Elwes.
c. Of either modelled	
10. Drawing flowers, foliage, landscape details, and objects of natural history from nature:	
a. In outline	Edward T. Haynes.
b. Shaded	Fanny Thorpe.
11. Painting ornament from the flat, or copies:	
a. In monochrome, either in water-colour, tempera, or oil.	Robert Fielding.
b. In colours, either in water-colour, tempera, or oil.	William Reid.
12. Painting ornament from the cast, &c.:	
a. In monochrome, either in water-colour, oil, or tempera	John Willshaw.
13. Painting (general) from flat examples or copies flowers, still life, &c.:	
a. Flowers or natural objects, in water-colours, in oil, or in tempera	T. Clack. S. E. Peal.
b. Landscapes	P. B. Brophy.
14. Painting (general) direct from nature:	
a. Flowers or still life, in water-colour, oil, or tempera	Lætitia M. Cole.
b. Landscapes	Maria A. Williams.
15. Painting groups as compositions of colour	Harriett Bradford.
a. In water-colour, oil, or tempera	
16. Painting the human figure or animals in monochrome, from casts	John Rennison.
a. In oil, water-colour, or tempera	
17. Painting the human figure or animals in colour:	
a. From the flat or copies	Helena Wilson.
b. From nature, nude or draped	W. S. Cosbie.
c. Time sketches and compositions	William Clews.
18. Modelling ornament:	
a. Elementary, from casts	
b. Advanced, from casts	
c. From drawings	
d. Time sketches from examples and from memory.	
19. Modelling the human figure or animals:	
a. Elementary, from casts of hands, feet, masks, &c.	
b. Advanced, from casts or solid examples	E. Bale.
c. From drawings	
d. From nature, nude or draped	A. B. Joy.
20. Modelling fruits, flowers, foliage, and objects of natural history from nature	James Marsh.
21. Time sketches in clay of the human figure or animals from nature.	
22. Elementary design:	
a. Studies treating natural objects ornamentally	H. Johnson.
b. Ornamental arrangements to fill given spaces, in monochrome	J. G. Woodward.
c. Ornamental arrangements to fill given spaces in colour.	Louisa A. Crawshaw.
d. Studies of historic styles of ornament, drawn or modelled	George Rhead.
23. Applied designs, technical or miscellaneous studies:	Eliza M. Bryant.
a. Machine and mechanical drawing, plan drawing, mapping, and surveys done from actual measurement	William Cairns.
b. Architectural design	
c. Surface design	Harriett E. Harman.
d. Plastic design	
e. Moulding, casting, and chasing	
f. Lithography	
g. Wood engraving	
h. Porcelain painting	

SCIENCE AND ART DEPARTMENT—*continued.*

The works to which medals have been awarded at the various local Schools of Art throughout the United Kingdom, are annually brought together to compete for national medallions and Queen's prizes awarded by the President and two members of the Royal Academy, and the drawings now exhibited are, for the most part, those which have been selected for reward at the national competition of this year, representing the best work that has been done in the Schools of Art during the year.

The school in which the student whose work has been selected for reward receives works of art or publications to the value of 10*l.*, and therefore benefits in proportion to the advance of the students, and it is by means of the presentation of these works of art or publications that the nucleus of a local museum or library may be established.

An annual examination of each school takes place, when the works in the various stages of drawing, painting, modelling, and design are examined in competition with each other, and medals awarded to the best works in each stage; and for every local medal awarded, examples or books to the value of 10*s.* are presented to the school.

At the time when the inspector makes his annual visit to the school to award these medals, a public examination in drawing, open to all persons, is held by him, when prizes, consisting of boxes of instruments, &c., are given to all who reach a certain standard in freehand and model drawing, and practical geometry and perspective.

Ninety Schools of Art are now established in various parts of the United Kingdom. The following table shows the name of the locality and the number of students and children of poor schools taught drawing by the agency of each School of Art:—

NAME OF PLACE.	No. of Scholars taught in Public and other schools according to the last returns.	No. of Students in Schools of Art according to the last returns.	Total Number under instruction.
ABERDEEN	1310	266	1576
ANDOVER	385	76	461
BATH	974	95	1069
BASINGSTOKE	See Andover.		
BIRKENHEAD	485	158	643
BIRMINGHAM, with branch at Spon Lane, *Smethwick*	1158	902	2060
BOLTON	1030	129	1159
BOSTON	552	106	658
BRIDGENORTH	65	57	122
BRIDGEWATER	150	113	263
BRIGHTON	1523	198	1721
BRISTOL	1705	296	2001
BROMSGROVE	127	100	227
BURNLEY	668	153	821
BURSLEM			
CAMBRIDGE	563	180	743
CARLISLE	No return.		
CARMARTHEN and SWANSEA	1108	45	1153
CARNARVON, with branches at *Bangor* and *Portmadoc*	924	89	1013
CHELTENHAM	480	278	758
CHESTER, branches at *Crewe* and *Wrexham*	2971	195	3166
CIRENCESTER	483	149	632
CLONMEL	160	66	226
COALBROOKDALE, with branches at *Madeley* and *Broseley*	763	103	866
CORK	209	178	387
COVENTRY	673	101	774
DARLINGTON	805	166	971
DEVONPORT	No return.		
DUBLIN (Royal Dublin Society)	330	396	726
DUDLEY	1063	79	1142
DUNDEE	2062	496	2558
DURHAM	384	118	502
EDINBURGH (Bd of Manufactures)	1814	495	2309
EXETER	882	245	1127
GLASGOW	2261	827	3088
GLOUCESTER	350	129	479
GREENOCK	546	105	651
GUILDFORD	570	30	600
HANLEY	654	98	752
HALIFAX	379	160	539
HEREFORD	354	29	383
HUDDERSFIELD	See Leeds.		
HULL	580	128	708
IPSWICH	846	258	1104
LANCASTER	820	192	1012

NAME OF PLACE.	No. of Scholars taught in Public and other schools according to the last returns.	No. of Students in Schools of Art according to the last returns.	Total Number under instruction.
LEEDS	3488	205	3693
LIMERICK	326	112	438
LIVERPOOL—			
North District School, (Collegiate Institution) Shaw Street	1523	551	2074
South District School, (Mechanics' Institution) Mount Street	1735	283	2018
LLANELLY	742	49	791
MACCLESFIELD	901	137	1038
MANCHESTER (Royal Institution)	3330	376	3706
METROPOLITAN District Schools:			
Bloomsbury, 43 Queen Square (Female School)	7	133	140
Finsbury		149	149
Lambeth	1560	128	1688
Saint Martin's	973	377	1350
Saint Thomas Charterhouse	1029	129	1158
Hampstead, *Rotherhithe*, *Saint George's-in-the-East*, *South Kensington*, *Spitalfields*, *Westminster*	11499	1033	12532
NEWCASTLE-UNDER-LYME	360	79	439
NEWCASTLE-UPON-TYNE	927	391	1318
NORWICH	749	217	966
NOTTINGHAM	1950	230	2180
PAISLEY	840	124	964
PENZANCE	874	119	993
PRESTON	1088	208	1296
SHEFFIELD	750	342	1092
SOUTHAMPTON, with branches at *Romsey* and *Ringwood*	743	229	972
STIRLING	650	126	776
STOKE-UPON-TRENT	170	187	357
STOURBRIDGE	576	130	706
STROUD	230	106	336
SUNDERLAND	510	104	614
TAUNTON	478	205	683
TRURO	490	36	526
WARRINGTON	983	147	1130
WATERFORD	205	106	311
WENLOCK	See Coalbrookdale.		
WOLVERHAMPTON	965	159	1124
WORCESTER	596	248	844
YARMOUTH, Great	845	127	972
YORK	831	177	1008

Total 86 Schools, and including the 10 Branch Schools, 96 Schools. Number of Persons under Art Instruction in Public and other Schools, 76,303; in Central Schools, 15,533; Total, 91,836 persons.

SCIENCE AND ART DEPARTMENT—*continued*

The master of a School of Art is appointed by a local committee responsible for its proper government, on the recommendation of the Science and Art Department. To meet the demand for masters, a training school is established at South Kensington, to which male and female students are admitted when properly qualified. They receive instruction from competent masters in free-hand, mechanical, and architectural drawing, practical geometry and perspective, painting in oil, tempera, and water-colours, modelling, moulding, and casting. Examinations are held twice annually, and certificates of competency to teach granted, in respect of which payments are made by the State on certain conditions.

A School of Art can be formed in any locality where the public provides and maintains at its own liability suitable premises. When it is necessary that new premises be erected, the department is authorized to make a grant in aid of the cost. The local committee is assisted in purchasing copies, models, and examples at a reduction on the net cost; and also in the purchase of objects of art which can be dispensed with at the Art Museum at South Kensington. A collection of specimens of ornamental art is also made from time to time from the objects in the Art Museum, for the purpose of being exhibited in the local schools.

This travelling collection, to which the local possessors of works of art generally make valuable additions, lent for the time that the collection is in the locality, is regarded with much interest by the local schools, and is visited annually by a large number of persons. During the past year this collection has been sent to 6 Schools of Art, and has been visited by 221,281 persons.

Books, drawings, and prints are circulated among the Schools of Art from the library at South Kensington, and as far as possible, consistently with the security of the objects, works of art of all kinds deposited in the Art Museum are lent to the Schools of Art.

SOUTH KENSINGTON MUSEUM.

This museum was commenced with the erection of an iron structure of the cheapest character in 1856. It was built under the superintendence of Sir W. Cubitt, and when completed was passed over by the Commissioners of 1851 to the Science and Art Department. Since that period permanent brick structures, erected under the superintendence of Capt. Fowke, R.E., have been added, which with the new courts now form the greater part of the museum

IN THE PERMANENT BRICK BUILDINGS, INCLUDING THE COURTS.

1. The Art Museum.
2. The collection of British Pictures is contained in four rooms of the upper portion of the brick building.
3. The Art Library.
4. The National Gallery, British School.

IN THE IRON BUILDING.

5. The Museum of Patented Inventions.
6. The Architectural section of the Art Museum.
7. The Educational Museum.
8. The collections of materials used in architectural construction.
9. The collections of animal products and food.

ADMISSION.

The National Gallery, British School, is open every day and is entered through the Museum.

The Museum is open free on Mondays, Tuesdays, and Saturdays. The student's days are Wednesdays, Thursdays, and Fridays, when the public are admitted on payment of 6*d.* each person. The hours on Mondays, Tuesdays, and Saturdays, are from 10 A.M. to 10 P.M., on Wednesdays, Thursdays, and Fridays, from 10 A.M. till 4, 5, or 6 P.M., according to the season. The entrances are in Cromwell Road, and in the Exhibition Road, opposite to the East entrance to the Exhibition.

Tickets of admission to the Museum, including the Art Library and Educational Reading-room, are issued at the following rates: weekly, 6*d.*; monthly, 1*s.* 6*d.*; quarterly, 3*s.*; half-yearly, 6*s.*; yearly, 10*s.* The public paying the entrance fee of 6*d.* on the students days have the privilege of both the Art and Educational Libraries for the purpose of study.

Yearly tickets are also issued to any school at 1*l.*, which will admit all the pupils of such school on all students' days. They may be obtained at the catalogue sale-stall of the Museum.

ART LIBRARY.

The Library of Works on Art is open during the same hours as the Museum. It is a special library, and contains works on art, original drawings and engravings of ornament, illuminations, and photographs.

Copying objects in the Museum is permitted on the students' days, on application to the keepers of the respective divisions.

Application for copying the pictures must be made in writing, addressed to the secretary, and if necessary specimens of competency must be forwarded.

Application to copy works of living artists must be accompanied by the permission of the artist. Copying in oil is not permitted.

The Museum was opened on the 22nd of June, 1857, and to the 20th June, 1862, was visited by 2,815,511 persons.

The evening attendance was 1,067,707.

Numerous catalogues are published.

THE SCIENCE DIVISION.

The Science Division of the Science and Art Department is constituted to encourage the teaching of science throughout the United Kingdom.

The branches of science thus aided are divided into seven heads or subjects, and each of these into two subdivisions, except the first, which is divided into three subdivisions.

I. Practical Plane and descriptive Geometry, with Mechanical and Machine Drawing and Building Construction, or Naval Architecture.
II. Mechanical Physics.
III. Experimental Physics.
IV. Chemistry.
V. Geology and Mineralogy.
VI. Animal Physiology and Zoology.
VII. Vegetable Physiology, Economic and Systematic Botany.

Assistance towards instruction in these sciences is afforded in four different forms, viz :—

A. Allowances to teachers on their certificates.

B. Public examinations, in which Queen's medals and prizes are awarded to all successful candidates, whether taught by a certificated teacher or not, held at all places complying with certain conditions. On the results of these examinations certificate allowances and payment on results are made to the teachers.

C. Payments on prizes to certificated teachers.

D. Grants towards the purchase of apparatus, &c.

1. *Certificate allowances to certified teachers.*

In November of each year the department of Science and Art holds an examination at South Kensington in all the above-mentioned subjects. Any one may attend this examination without payment of fees by sending in his name to the Secretary, Science and Art Department, in September, and may take up any one or more of the subjects or subdivisions at one time.

Certificates of three grades are given for success in these examinations, entitling the holder to payments under certain conditions.

At the first examination for teachers in November, 1859, shortly after the publication of the first Minute, 57 candidates came up, of whom 43 were successful, taking 65 subdivisional certificates. The next year, 1860, 89 candidates came up, 75 were successful, and 121 subdivisional certificates taken. This last November, 103

SCIENCE AND ART DEPARTMENT—*continued.*

candidates came up, 97 were successful, and 175 subdivisional certificates taken. There are now 168 teachers certificated under this system.

The teacher obtains the certificate payments in the following manner:—The classes are examined once a year (see below, Public Examinations), and then for every pupil of the artisan class who *passes* such an examination as will qualify the examiner in reporting that his instruction has been sound, and that he has benefited by it, the teacher receives 4*l.* of his certificate allowance. The artisan class is broadly defined as including all who are in the receipt of weekly wages, and their children. A pupil on account of whom payment is claimed, must have received forty lessons at least from the teacher since the last examination at which payments were claimed on his account.

A committee must be formed of at least five well-known persons in the neighbourhood, who have to give the necessary vouchers that these conditions have been strictly complied with.

2. *Public examinations.*

In order to test the efficiency of the instruction, on the proof of which alone the payments are made to the teacher, an annual examination is held in May simultaneously all over the kingdom, an evening being fixed for the examination in each subdivision of the seven subjects.

It is conducted by the Committee previously mentioned, to whom the examination papers for the pupils in each particular subdivision are sent.

The results of these examinations are classified.

1. All those who have *passed* in each subdivision of a subject. The standard of attainment required being low, and only such as will justify the examiner in reporting that the instruction has been sound, and that the students have benefited by it.

2. From among those who *passed*, those who attained a degree of proficiency qualifying them for the 1st, 2nd, or 3rd class Queen's prize.

3. The six most successful candidates in each subject throughout the United Kingdom, if the degree of proficiency attained be sufficiently high to warrant their being recommended for Queen's medals.

The Queen's prizes consist of books to be chosen by the candidates from lists furnished for that purpose, and are unlimited in number.

The Queen's medals are one gold, two silver, and three bronze, in each subject for competition throughout the United Kingdom.

At the last examination in May there were just 1000 papers, and of these 725 were *passed*, and would qualify the teacher for payment if they were of the industrial classes and had received forty lessons; 310 of these received Queen's prizes, 59 1st class, 100 2nd, and 151 3rd class, while 4 gold, 21 silver, and 16 bronze medals were awarded.

Although payments to the teacher are made only on pupils of the industrial classes, others are not excluded from examination. Any person may present himself or herself; but the local committee is permitted to charge a fee not exceeding 2*s.* 6*d.* to cover the expense of gas, &c. Such candidates are eligible to receive Queen's prizes and medals.

3. *Payments on prizes to certified teachers.*

Besides the above payment on certificates to the teachers, there are other payments which are not limited in amount as in the case of the certificate allowances.

For every pupil of the industrial classes who obtains a Queen's prize, the teacher, if he is certificated and has given the pupil forty lessons, receives a payment of 3*l.* if the pupil obtain a first grade Queen's prize; 2*l.* if a second, and 1*l.* if a third.

This amount is not limited, and at the last May examination many teachers obtained 30*l.* or 40*l.* in addition to their certificate allowance from this source.

4. *Grants towards the purchase of apparatus.*

A grant of 50 per cent. on the cost of apparatus, diagrams, &c., necessary for the instruction of the class is made. These grants are limited to 10*l.* to schools taught by a master who is not certificated.

5. *Aid to Instruction in Navigation.*

Examinations are held in January and July of each year for Navigation Certificates in the following groups:—

I. General acquaintance with Mathematics.
II. General Navigation and Nautical Astronomy.
III. Adjustment and skilful handling of instruments.
IV. Physical Geography.

On the results of these examinations, augmentation grants of three classes are made. The grade of the augmentation grant allowed to the master in each of the three classes of head master, first assistant, and second assistant, depends on the manner in which he passes his examination, and the certificates he takes. Pupil-teachers are allowed at the rate of one to thirty boys in the school, and receive 20*l.* per annum each, and the master 6*l.*, per annum for their instruction on condition of their general progress being favourably reported on by the Inspector.

Any master certificated in navigation as above, earns the certificate allowance according to the grades there fixed in the following way:—

He will receive payment,—1st, for boys who having been taught by him for not less than one year are examined by the inspector for navigation schools, and finally proved to have been apprenticed on board a merchant vessel, or entered on board a man-of-war; 2nd, for seamen, masters, or mates, who having received twenty lessons at least from the master, pass an examination at the local Marine Board for a higher certificate.

The payments under the first head will be at the rate of 6*l.*, 4*l.*, or 2*l.*, for each boy, according to the result of the examination. The payments under the second head will be at the rate of 1*l.* per head.

But the total payments under these heads can never exceed the amount which the teacher is qualified to earn by the grade of his certificates.

A local committee of not less than five responsible persons must be formed to give the teacher the necessary vouchers.

A room with firing, lighting, etc., must be provided to give instruction in.

Evening Navigation Classes may be established independent of (day) Navigation Schools.

An examination is held once a year at South Kensington, at which any person may present himself by registering his name to be examined for an evening Navigation School Certificate, or if he prefers it, he may present himself for examination by the Inspector of Navigation Schools when he visits any of the existing Schools on his tour of inspection.

The certificate is of three grades, qualifying the holder to earn 20*l.*, 15*l.*, or 10*l.*, according to the grade of the Certificate. And further payments on results dependent on the results of instruction only.

The certificate allowance is dependent on the average number of bonâ fide Sailors—Seamen and Apprentices—who attend during 200 evenings in the year, and is paid at the rate of 10*s.* per head of the average up to the maximum which the teacher is qualified to earn by the grade of his certificate. The payments on results, which are unlimited, are dependent on the number of prizes taken by the pupils when examined by the Inspector, and are at the rate of 5*s.*, 10*s.*, and 1*l.*, according to the grade of the prize.

A responsible Committee of not less than five persons must be formed to give the master the requisite vouchers. A proper room, with firing, lighting, etc., must be provided for giving instruction in. When these conditions are fulfilled, any persons holding the Evening Navigation Class Certificate may obtain the above payments.

[5514]

SEATON, JOHN LOUIS, 3 & 4 *Frederick Place, Hampstead Road.*—Improved desks; forming tables (if required) and seats.

AS A SEAT WITH BACK. AS A DESK AND FORM. AS A DESK AND FORM. AS A TABLE.

Sectional models of improved desks and forms convertible into seats with backs and book boards, adapted for schoolrooms, used occasionally for lectures or public meetings. Price in lengths of six feet and upwards, stained deal 5*s.* 6*d.*, birch 7*s.* 3*d.*, per foot run.

Improved desk and form with reversible back, to be used either flat or sloped, when placed back to back, they form dining and tea tables. Price in lengths of six feet and upwards, stained deal 7*s.*, birch 8*s.* 6*d.* per foot run. Without the back to seat, 1*s.* per foot less.

[5515]

SHARP, GEORGE, 16 *Wentworth Place, Dublin.*—Models for teaching elementary drawing.

[5516]

SHERRATT, THOMAS, Jun., 5 *Westmoreland Place, Westbourne Grove North, Bayswater, W.*—Time globe, or planetary clock. (*See page* 59.)

[5517]

SOCIETY FOR PROMOTING CHRISTIAN KNOWLEDGE, *Great Queen Street, Lincoln's Inn Fields.*—Books, maps, prints, &c.

[5518]

SPENCER, WILLIAM, *Beverley.*—1500 arithmetical exercises on 277 cards:—practice, fractions, decimals, per centages, and government questions.

[5519]

STANLEY, WILLIAM F., 3 *Great Turnstile, Holborn, London.*—Mathematical instruments, case in aluminium, engine divided scales, &c.

[5520]

STEPHENS, HENRY, 18 *St. Martin's-le-Grand, London.*—Specimens of inks, and of papier-mâché, slates, &c. (*See page* 59.)

[5521]

SUNDAY SCHOOL INSTITUTE, 41 *Ludgate Hill, London.*—Material for organization and management of Sunday schools.

[5522]

WEDGWOOD & SONS, 9 *Cornhill, London.*—Patent manifold writers, and writing machine, for the blind.

Sherratt, Thomas, Jun., 5 *Westmoreland Place, Westbourne Grove North, Bayswater, W.*—Time globe, or planetary clock.

Sherratt's Time Globe, or Planetary Clock.

At a period like the present, when the locomotive, the steamer, and the electric telegraph are receiving every improvement and extension that the united mental and mechanical genius of this and other countries can give them, it must be obvious that whatever conduces to even a superficial knowledge of the situation and extent of the vast regions of this planet—explored and traversed by these aids of commerce and civilization, and peopled by daily arriving crowds of enterprising emigrants—must be a boon to those who, from whatever cause, have not studied matters of that kind.

There are few now among us who have not friends and relations living in remote dependencies and colonies of their native land. To these, and, in fact, to all who value an object of the most essentially domestic use, my *Time Globe* or *Keeper* must recommend itself; it being a terrestrial globe, rotating on its axis simultaneously with the pointer or hand, together making one revolution before the 24-hour clock-face in that time. Every hour is divided into twelve parts of five minutes each,—the quarters and halves being more distinctly marked. The several portions of the day are also inscribed and coloured thereon, and the cardinal points given. In a right line with the pointer, a black meridian passes over whatever place on the globe the time may be set for,—showing it not only there, but at all other places under the said meridian. On turning a button, which projects through the circular opening of the glass cover, it will cause the white meridian, attached at the south pole of the globe, to move over any place at which the time is required to be known; and the index fixed thereto, on passing before the hour circle on the clock-face, will show the time at that and all other places under such meridian.

We all know how often the eye is turned to the clock: with how much more interest, then, will the glance be directed, and with how much more entertainment and instruction will that glance be repaid, when it rests upon an object such as this, either fixed, level with the sight, to the wall, or standing on the mantelpiece,—instead of the comparatively meaningless faces that characterize the clocks in general use. It is worthy of remark that the price of this ingenious timepiece does not exceed that of others.

The Inventor therefore anticipates, that when the elementary astronomical and geographical advantages possessed by these time globes are taken into consideration, combined as they are with the usefulness and cheapness of the common clock, they will become the means of stimulating inquiry and encouraging thought; in short, that they may of such things make

"Those think who never thought before,
And those who have thought make them think the more."

Stephens, Henry, 18 *St. Martin's-le-Grand, London.*—Specimens of inks, and of papier-maché, slates, &c.

The long and persevering attention which the exhibitor has paid to the various combination of colouring matters, enable him to prepare writing fluids upon the best principles.

These fluids consist of—

The original blue-black writing fluid, which is a blue fluid, changing to an intense black colour.

The following letter expresses the pleasure which the easy flow of the article universally gives:—

"10 Pittville Lawn, Cheltenham,
Dec. 26th, 1861.

"The Rev. H. Philipps wishes to express to Mr. Stephens his gratification at having met with his blue black writing fluid. Having for several years endeavoured to obtain a good ink, he considers himself fortunate in having at length met with this, manufactured by Mr. Stephens."

The patent unchangeable blue writing fluids, remaining a deep blue colour. Two sorts are prepared, a light and a dark blue.

The black inks:—

A superior black ink, of common character, but more fluid.

Instantaneous black ink, which writes black at once.

Record and registration ink. Practical proofs of its superior durability over inks made with galls has been shown by its use on garden-tickets, for flowers and plants.

The copying inks:—

A very superior black copying ink, which gives one or more strong copies with the machine, or one copy long after writing.

The blue-black copying ink. This article is very fluid.

A brilliant red, for contrast writing. This is a very beautiful and vivid colour.

Purple, green, violet, crimson, and yellow inks of very fine tints.

Stephens's concentrated and soluble ink powders.—These articles contain the constitutent parts of ink in a dry state, yet so readily soluble, that by the addition of water only, an ink fit for use is quickly formed. These powders are manufactured to make a black ink, the blue black, the unchangeable blue, and the brilliant red writing fluids.

Sold in packages to make a quarter-pint, half-pint, a pint, and a quart.

A new material as a substitute for slate.—Stephens's papier-mâché tablets, manufactured into the form of slates, pocket-books, drawing and diagram boards, diary tablets, and pocket memorandum books.

The advantages of this material are its extreme lightness and portability, strong visible contrast in the writing, pleasing to the eye and touch, rubbing out without the use of sponge or wetting; for this a cloth or handkerchief merely will suffice.

Manufactured and sold by Henry Stephens, 18 St. Martin's-le-Grand, London, E.C., late 54 Stamford Street.

Prospectuses and detailed descriptions can be had on application, or free by post.

[5523]
WEDLAKE, 58 *Warren Street.*—Choir organ; pedal organ; choir organ with knee pedal to swell.

[5525]
WOLFF, E., & SON, 23 *Church Street, Spitalfields.*—Drawing pencils, black-lead, coloured crayons, creta lævis, or indelible chalks in cedar, solid ink pencils, &c.

[5526]
CRANBOURNE, THE VISCOUNT, 20 *Arlington Street.*—Books for the blind.

[5527]
HOWARD, *Colchester.*—Plans of school-buildings.

[5528]
JOHNSON, E. C., *Savile Row.*—Books for the blind.

[5529]
WILLIAMS, A., *Windsor.*—Model of improved school desk and form.

SUB-CLASS C.—*Toys and Games.*

[5538]
ASSER & SHERWIN, 81 *Strand, London, W.C.*—Various parlour games and evening amusements.

[5539]
BAER, JOHN, 399A *Oxford Street.*—Scientific toys and wood carvings.

[5541]
BURLEY, GEORGE, 28 *George Street, Blackfriars Road, Southwark.*—Dolls of a novel de scription.

[5542]
CAMP, WILLIAM, 81 *Tottenham Court Road.*—Cricket stumps, skittle, lawn billiard, American, and bowling-green balls.

[5543]
CREMER & SON, 27 *New Bond Street, and Regent Street.*—Toys and games.

[5544]
DARK, M., & SONS, *Lord's Cricket Ground, St. Marylebone, London, N.W.*—Cricket-bats and wickets.

[5545]
DARK, ROBERT, *Tennis Court, Lord's Cricket Ground.*—Cricket-balls, india-rubber gloves leg-guards, and gauntlets. (*See page* 61.)

[5546]
DUKE & SON, *Penshurst, Kent.*—Cricket balls, bats, stumps, leg-guards, gloves, &c.

[5547]
FELTHAM, JOSEPH, & CO., 2 *Barbican, London.*—Cricketing, archery, and fencing requisites.

[5548]
GILBERT, WILLIAM, Foot-ball Manufacturer, *Rugby.*—Foot-balls and foot-ball shoes, repre senting the game as played at Rugby school.

DARK, ROBERT, *Tennis Court, Lord's Cricket Ground.*—Cricket-balls, india-rubber gloves, leg-guards, and gauntlets.

[*Prize Medal for Gauntlets, Leg-guards, Spiked-soles, Balls, and other implements used in the game of Cricket. —Vide Report of the Juries.*]

The exhibitor is the inventor and original maker of the tubular india-rubber gloves, improved leg guards, and wicket-keeping gauntlets. These, and also his celebrated cricket balls, are warranted made by the best workmen and of the best materials. They may be purchased from the proprietors of cricket grounds, and at every respectable shop in the kingdom where articles of this kind are sold.

Agencies exist for the sale of R. Dark's goods, at Calcutta, Madras, Bombay, Quebec, New York, Adelaide, Sydney, Melbourne, &c.

The trade can be supplied direct, by forwarding business-card.

The exhibitor has manufactured some hundreds of dozens of cricket balls, and has exercised constant care in the selection of materials, and a watchful supervision over the workmanship. As these balls are the only ones that have been used by the Marylebone club for many years, and as they are in high favour at Oxford, Cambridge, Eton, Harrow, Winchester, he feels confident that they will give universal satisfaction in the cricket-field.

Secretaries of cricket clubs, merchants, shippers, and other large consumers, can be supplied immediately with Dark's cricket balls, gloves, leg-guards, &c., on liberal terms, all manufactured ready for inspection, and may be despatched to any destination on the shortest notice.

India and Colonial orders carefully packed and shipped.

Lists of prices forwarded on application at the Tennis Court, Lord's Cricket Ground, London, where also may be obtained the M.C.C. laws of cricket.

The exhibitor had the honour to be appointed manufacturer of cricketing implements to his late Royal Highness the Prince Consort.

The implements and appliances in the nature of gloves, guards, &c., which the present mode of playing the game, and especially the practice of swift overhand bowling, has brought into use, are so various, that there is more room for ingenuity in the manufacture of them than might at first sight appear.

The articles contributed by Robert Dark (198) have great merit. The gauntlets which he exhibits, as designed to guard the wrist from the blow of the ball, are lined with slips of cane, and are thereby lighter than those which are thickly wadded, or lined with india-rubber; whilst they furnish a stiffer and more effectual defence against the ball.

So also in regard to the gloves, which protect the fingers of the hand that holds the bat, by a tube fixed along the back of each finger. An improvement is apparent in those exhibited by R. Dark, inasmuch as a second and smaller tube is fixed within the first, thereby materially increasing the resistance to a blow from the ball, without sensibly adding to the weight or diminishing the pliancy of the glove.

[5549]

HOFFMAN, HEINWICK (pupil of Frederick Frobel), 19 *Norland Square, Notting-hill.*—The kindergarten illustrated by models and designs.

[5550]

JAQUES, JOHN, & SON, 102 *Hatton Garden, London.*—Billiard-balls, chessmen, games, &c.

[5551]

JEFFERIES & MALINGS, *Wood Street, Woolwich.*—Racket bats, fives and racket balls, presses, and patent soled racket shoes.

[5552]

JOHNSON, SIDNEY, 6 *Heathpole Street, Paddington.*—A model, or doll's house made of white wood, carved ivory, and ebony.

[5553]

KEEN, THOMAS EDWARD, Depôt, Messrs. Myers, 144 *Leadenhall Street.*—The cork model maker: a scientific toy.

[5554]

KENNEDY & Co., 15 *Westbourne Grove Terrace, Bayswater.*—A patent mechanical trotting pony, and spring bassinet.

[5555]

LILLYWHITE, JOHN, 5 *Seymour Street, Euston Square, N.W.*—Articles connected with cricket.

The exhibitor has always a large stock of cricketing requisites on hand.

Cane-handle bats, each, 15*s.*, 17*s.* 6*d.*, and 21*s.*

Best match bats, each, 8*s.* 6*d.*, 10*s.* 6*d.*, and 12*s.* 6*d.*

Ebony-top stumps, per set, 10*s.* 6*d.*

Best match balls (warranted), per dozen, 72*s.*

Newly-invented leg-guards, per pair, 12*s.* 6*d.*, and 15*s.*

Tubular india-rubber gloves, per pair, 8*s.* 6*d.* and 10*s.* 6*d.*

Wicket-keeping gloves, per pair, 10*s.* 6*d.*

Every article connected with cricket may be obtained from J. Lillywhite.

The newly-invented "balista" (a great improvement on the catapulta), price 6*l.*

Boxing-gloves, foot-balls, quoits, and every article connected with British sports.

Illustrated lists of prices post free.

[5556]

LOYSEL, EDWARD, C.E., 92 *Cannon Street, London, E.C.*—Loysel's patent chivalric game of a combination of chance and skill. (*See page* 63.)

[5557]

MATHEWS, CAROLINE ELEANOR, Inventor, *Oatlands Park, Surrey.*—Giuoco di Clio; a miscellaneous game.

[5558]

MEAD & POWELL, 73 *Cheapside* (from *London Bridge*).—Improved rocking-horses, children's repose perambulators, baby-jumpers, model carts, &c.

[5559]

MONTANARI, AUGUSTA, 198 *Oxford Street, London.*—Model wax dolls, with all modern improvements, and model rag dolls.

[5560]

MOORE, JOSEPH LYNN, *West Street, Dorking, Surrey.*—Set of improved cricket stumps.

[5561]

NICHOLSON, HAMLET, Inventor, *Rochdale.*—Compound cricket and playing balls. Patent life-buoys.

The patentee claims for his ball a superiority over the leather ball, on the following grounds:—

1. The compound ball is cheaper than the leather ball.
2. It is more durable.
3. It is a true sphere, or circle, in every direction, which no leather ball can be; consequently it flies more accurately, handles better, and is easier to field, and to bowl.
4. It cannot absorb moisture, and therefore never varies in weight.
5. After being immersed in water, it is not as slippery as a leather ball, consequently, when the ground is wet, it is easier to hold, field, and bowl with precision.
6. In consequence of the inner substance being soft fibrous material, it is less liable to break the handle, and drives easier off the face of the bat.
7. It never loses its shape. The balls are exact facsimiles of each other, and the exact weight and size prescribed by the laws of cricket.

Many very satisfactory testimonials have been received from various cricket clubs and public schools.

LOYSEL, EDWARD, C.E., 92 *Cannon Street, London, E.C.*—Loysel's patent chivalric game of Tournoy, a combination of chance and skill.

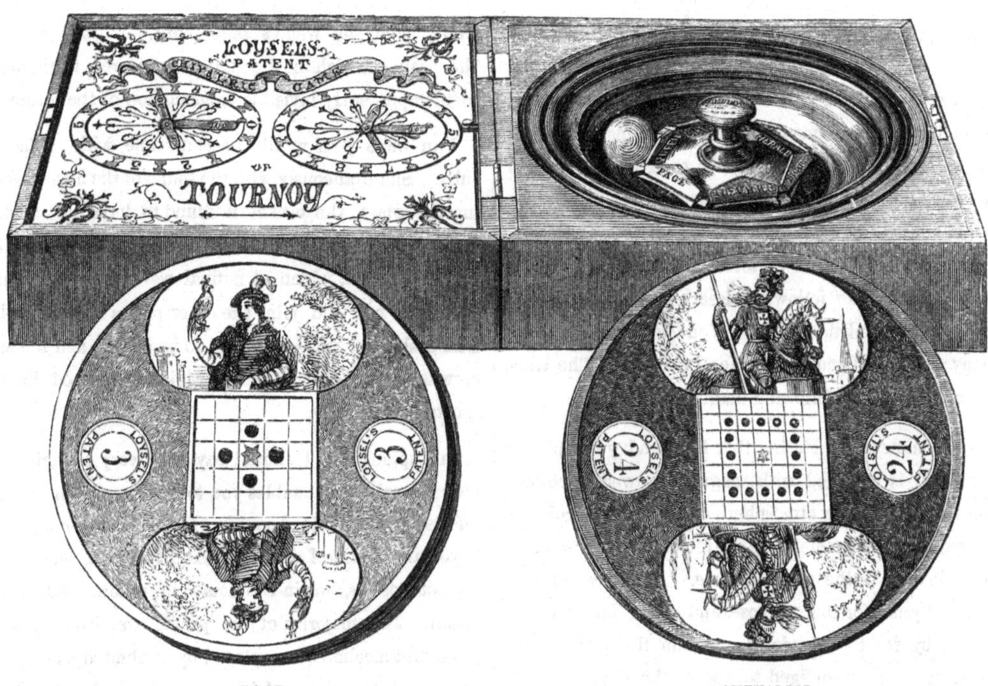

PAGE. CHEVALIER.

The Game of Tournoy is played either by two persons on a board of 36 squares, or by four persons (placed as at whist) on a board of 81 squares. Each player has six pieces which are denominated—QUEEN, CHEVALIER, CONNETABLE, HERALD, PAGE, AND DWARF. The personage of each piece is represented in colours upon a round block of porcelain, in the style of the two accompanying engravings, which represent the Page and the Chevalier. The geometric diagram in the centre indicates the power of the piece; to understand which the player must for the moment suppose, that instead of playing on the large board, he is playing on the small one, which is figured on the piece; and that the piece itself is placed in the centre square (where there is a star); then the piece can move or take in all the squares in which dots are marked: for example, the *Page* can move or take one square, front and back, right and left; and the *Chevalier* can move or take in any direction, at a distance of two squares.

The *novel* and *distinctive* feature in the Game of *Tournoy*, being the *introduction* of Chance to *govern the moves of the pieces*, each player must draw in turn the name of the piece he is to move: this is done by means of a *Roulette*, on the circumference of which are six recesses with hinged flaps, on which the names of the six pieces are engraved. By spinning the *Roulette* the *ivory ball* will fall into one of the recesses; and the player must exercise his judgment in moving to the best advantage, the piece thus indicated.

An illustrated book containing a minute description of the Game of Tournoy, the power of the pieces, their arrangement on the board, the etiquette, the laws of the game, written in both French and English by the inventor, accompanies each game, which can be so readily understood, that even a child will be able to play in ten minutes.

Each piece having a numeric value, and the game being generally played in a given number of points, two dials or point markers are provided as shown above.

The Game of Tournoy is made so that it may be used for either two or four players, and each box contains, 1st, a board (which can be expanded at pleasure to 36 or 81 squares); 2nd, 24 pieces in four sets, red, blue, green, and black; 3rd, the roulette, ivory ball and cup; 4th, the dials; 5th, book of rules. The whole is enclosed in a compact box, which, when shut, measures only 8 inches square and 4 inches thick: this box, while the game is being played, lies open on the table, as shown above.

The Game of Tournoy will be found interesting by persons of every age; and its varied combinations will never create *ennui*.

The Inventor in introducing this new game has had no idea of rivalling or trying to improve on the noble game of chess; but has merely sought to present the public with a recreation of quite a different order.

[5562]

NOVRA, HENRY, 95 *Regent Street, London, W.*—Collection of magical apparatus for amateur, juvenile, and adult drawing-room and fireside entertainment.

The apparatus selected for exhibition are principally remarkable for the superiority of their deceptive powers, the elegance and high finish of their manufacture, and the ingenious concealment of their mechanism, so as to render detection next to impossible, even when placed in the hands of the most scrutinizing observer. As space will not admit of a detailed account of each of the experiments which may be performed by their aid, the exhibitor subjoins the following description of one of the newest tricks entitled—"The Wonders of Modern Alchemy; or, the Wand of the Metal Seeker." The trick is performed in the following manner:—

You commence by stating that the wand you hold in your hand is endowed with magnetic power to collect from the surrounding atmosphere (in the form of coins of the realm) various metals, which recent discoveries have shown to exist in the gaseous state. As the wand you professedly use for this purpose, and submit for examination, is by far too slender to contain the money, the spectator is either obliged to accept the above preposterous explanation, or to attribute to you an unusual amount of skill and dexterity.

You then retire to a distance, and slowly wave your wand in circles, when, in accordance with what you have stated, a glittering substance forms at its extremity the size and shape of a half-crown, which you immediately remove and cast upon the table, to dispel all doubts as to its reality.

The process is repeated again and again, until a pile of sufficient magnitude is obtained to satisfy the spectator.

Having collected the money, you now state that you will further prove the magnetic power of the wand.

Accordingly, desiring some person present to privately mark four of the coins, you at the same time place in the hands of another an elegant jewel-case of velvet and gold. Taking the money from the first person, you openly deposit it in the casket, desiring the holder firmly to close the lid. Now placing a tumbler, which you show to be empty, at some distance from the casket, you take up a position midway: slowly waving your wand as before, you command one of the half-crowns to leave the casket and deposit itself on its point; as soon as it appears you remove it, and, opening your hand, launch it invisibly ut audibly into the tumbler; the distant sound, your empty hand, the wand's bare point, and last, not least, the casket, which now contains but three, testifying as to the apparently invisible transmission of the money.

You wave your wand a second time, with similar results. Simultaneously with the sound of the money reaching the glass, the casket is found to have lost another coin. You continue; attract a third piece, leaving in the casket but *one*, which at a final wave from your wand, leaves the box and joins the other pieces in the tumbler. The casket of course is found to be empty, and the half-crowns in the glass—those originally marked by the spectator.

In explanation, I should say that very cleverly concealed mechanism enables you to produce the marvellous effects above described, causing the spectator to attribute to you an amount of skill and dexterity which you not only do not possess, but in the course of nature could never attain. The majority of the public are little aware of the secret mechanism employed by celebrated public practitioners even in those tricks professed to be played entirely without apparatus.

The exhibitor (Mr. Henry Novra) has made this branch of amusement his peculiar study. The Magical Repository (95 Regent Street), of which he is the proprietor, has been established since 1844. Since his accession to the sole management, he has devoted so much of his time to the perfecting of the articles, and the production of novelties, that he may safely state that in no town in Europe can be found so large a stock or so complete a variety; from the simplest tricks for the child just beginning to understand, to the most complicated and scientific for the most ambitious of amateurs, who fain would rival (as indeed they can) the far-famed Robert-Houdin in the elegance and completeness of their entertainments.

The establishment (95 Regent Street) is fitted up exclusively for the sale of these articles, and affords every facility for the thorough instruction of customers in the manipulation of tricks of which they become the possessors.

Catalogues, with lists of prices, forwarded on application at the Repository or the Stand in the Exhibition, Class XXIX. No. 5562.

Branch establishment for wholesale and shipping orders, No. 2 Piccadilly Place, Piccadilly, W., opposite the St. James's Rectory.

NOVRA, HENRY—*continued.*

MECHANICAL CONJURING TABLE AND APPARATUS.

[5563]

NORMAND, G. B., 54 *Old Compton Street, Soho.*—India-rubber balls and balloons.

[5564]

PAGE, EDWARD J., 6 *Kennington Row, Kennington Park, S.*—Articles required in the game of cricket.

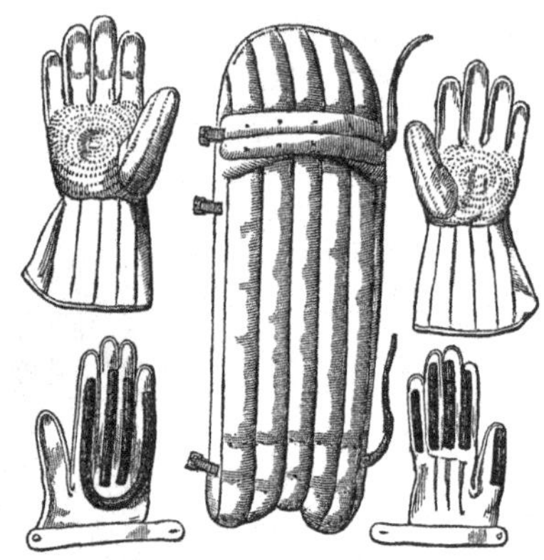

By appointment to the Surrey County Cricket Club.

The exhibitor's "superior cane handle," from the peculiarity of its construction, has been found impossible to break; and having stood the test for upwards of five years, is most confidently warranted. It is particularly adapted for warm climates.

He has also a "single cane handle," which is strongly recommended, and will be found more durable than the whalebone now in use, and which it will probably eventually supersede.

From always keeping a large and well-seasoned stock, he is able to supply "match bats," which are almost universally used, and are admitted by cricketers to be unsurpassed in the trade. They are strongly recommended for the use of clubs.

The "match balls" manufactured by him, from the care exercised in the selection of the materials, and the skill employed in the workmanship, can be confidently recommended. A sufficient guarantee of their excellence is the fact that no other balls are ever used in the great matches by the Surrey Club.

The best "match stumps" are made of lance-wood (the most durable wood in use), brass ferruled, with ebony tops. E. J. Page has always a large stock of these for sale.

He is the inventor of the "improved double protecting leg-guard," and "ventilating gauntlet," and manufactures Albert batting-gloves, and steel spikes (which can be fixed to the ordinary boot or shoe), telegraphs, club-boxes, and tr--elling-bags. These goods are used by members of many clubs in the United Kingdom, and are in general use in America, Australia, and India.

In E. J. Page's manufactory, every article required in the game can be seen in the process of manufacture, and purchasers are invited to inspect the same.

A general price-list will be sent free on application. Post-office orders should be made payable at Kennington Cross.

[5565]

PEACOCK, THOMAS, 515 *New Oxford Street, London.*—Model and composition dolls.

The exhibitor manufactures all kinds of wax model and composition dolls. He has always on hand a stock of a thousand, dressed and undressed, to select from. He supplies single dolls at the wholesale prices—ranging from 1*l.* to 5*l.*

[5566]

PIEROTTI, CELIU, 13 *Mortimer Street, Oxford Street.*—Foreign and English toys.

[5567]

PIEROTTI, HENRY, 13 *Mortimer Street, Oxford Street.*—Wax model dolls, with inserted hair; dolls and wax figures.

[5568]

PRAETORIUS & WARNER, 32 *Tavistock Place, London, W.C.*—Kindergarten materials and diagrams.

[5569]

PRINCE, MISS ABELINDE, 29 *Norfolk Crescent, Hyde Park.*—The English Pinakothek, a new artistic recreation for 1862.

[5570]

Pritchard, W. H., Press Japanner, 29 *Peartree Street, St. Luke's.*—Nursery friend and walking assistant.

[5571]

Rich, William, 14 *Great Russell Street, Bloomsbury.*—Wax figures illustrating the natives of various countries, wax fruit, botanical dissections, model wax and rag dolls, and wax and its appliances to the fine arts.

[5572]

Roth, M., Physician, *Old Cavendish Street, London.*—Gymnastic figures, diagrams, &c. (*See pages* 68 & 69.)

[5573]

Spratt, Isaac, 1 *Brook Street, Hanover Square, W.*—Collection of games and toys.

[5574]

Van Noorden, P. E., 115 *Great Russell Street, Bedford Square, W.C.*—Van Noorden's musical games, combining amusement with instruction.

These amusing and instructive games are founded upon and illustrate various branches of musical science. Their prices range from eighteenpence to one guinea.

[5575]

Wilson, George, *Castle Street, Shrewsbury.*—Chess-tables, nacre and pearl specimen cribbage board, alliance vase, and ebony.

[5576]

Woodman, William, 13 *Three Colt Court, Worship Street.*—Morocco backgammon table.

[5577]

Zimmerman, William, 18 *Bishopsgate Street Without, London.*—Toys, games, &c., for England, India, Australia, and America.

Sub-Class D.—*Illustrations of Elementary Science.*

[5588]

Ashmead, G. B., 10 *Duke Street, Grosvenor Square.*—Selections from a collection of British small birds—nests and eggs.

[5589]

Bartlett, A. D., & Son, *Zoological Gardens, Regent's Park.*—Preserved and mounted birds, &c.

[5590]

Bohn, James, 45 *Essex Street, Strand.*—Marine and fresh water vivaria, and all requisite implements.

[5591]

Cutler, H. G., 8 *Earl Street, Southwark Bridge Road, S.E.*—A superior drawing-room aquarium.

These superior aquaria are made with brass frames, and lined with white metal, to prevent verdigris injuring the fish, plants, &c. Every aquarium made by Mr. Cutler is warranted water-tight. Fish, aquatic insects, and plants, with every other requisite for the aquarium, may be had of the exhibitor.

[5592]

Damon, Robert, *Weymouth.*—Sample elementary and other named collections of foreign shells, British shells, and fossils.

[5594]

Elliott Brothers, 30 *Strand, London.*—Sectional and other models.

ROTH, M., Physician, 16A *Old Cavendish Street, London.*—Gymnastic figures, diagrams, &c.

Dr. Roth's Collection * of original high-relief models of gymnastic positions and movements for educational, sanitary, and military purposes, as well as for the physical development of the blind, deaf, and dumb.

Figs. 1, 2, 3, 4, 5, 6, 7, 8, are some of the positions in which the elementary movements are done; 9, 10, 11, 12, elementary, and 13, 14, combined head-movements; 15, 16, 17, 18, 19, 20, 21, forearm and arm movements; 22, 23, 24, 25, 26, are elementary movements of the trunk with various positions of the arms; 27, 28, combined trunk movements; 29, 30, 31, 32, 33, 34, 35, 36, elementary foot and leg movements, some of them combined with arm movements.

* Modelled, under Dr. Roth's superintendence, by Mr. Megret, drawn by Mr. Boehm, and engraved by Mr. E. Hewitt.

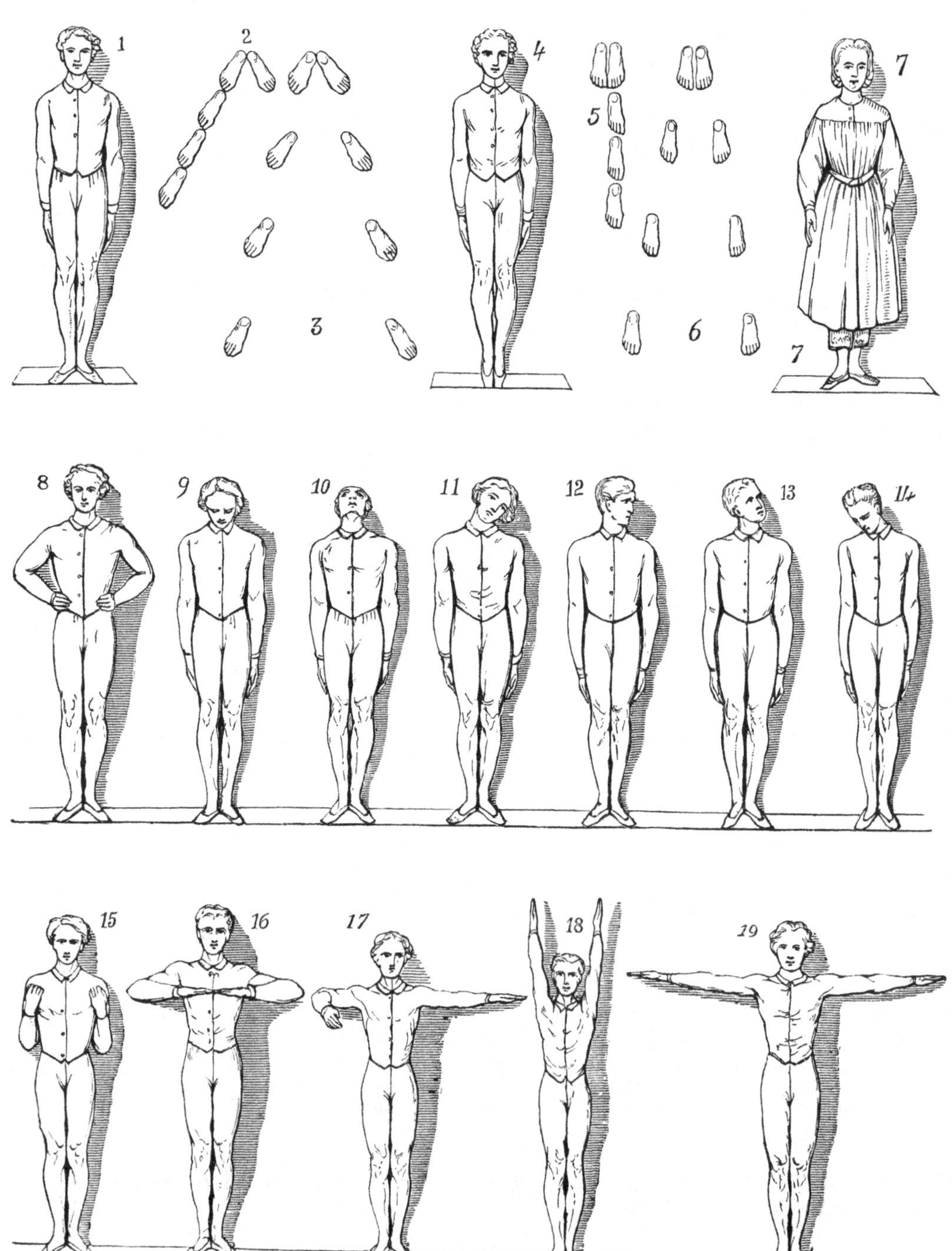

Roth M.—*continued.*

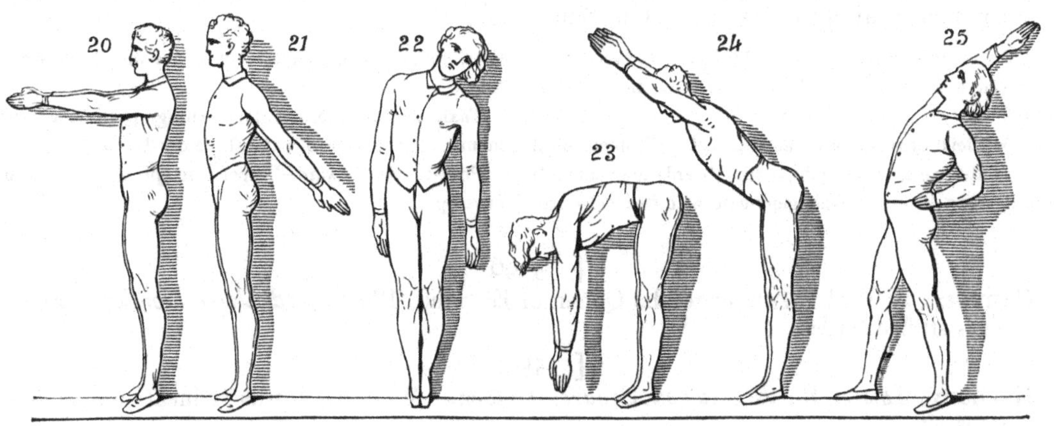

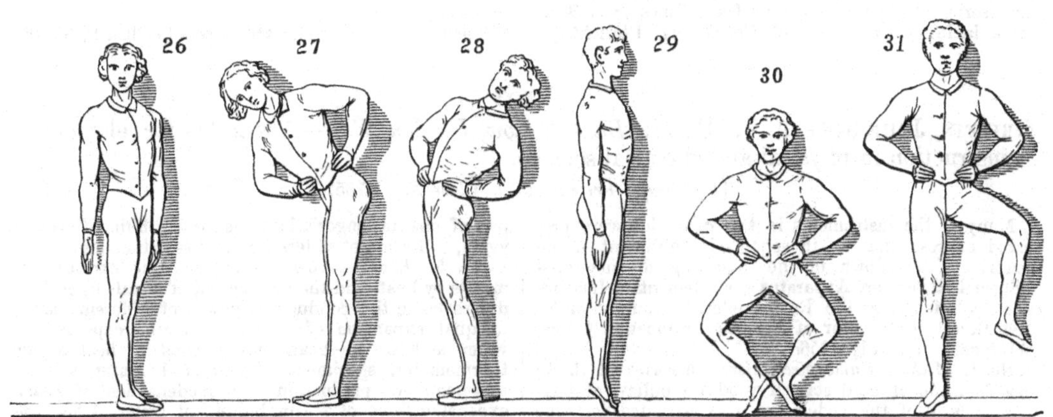

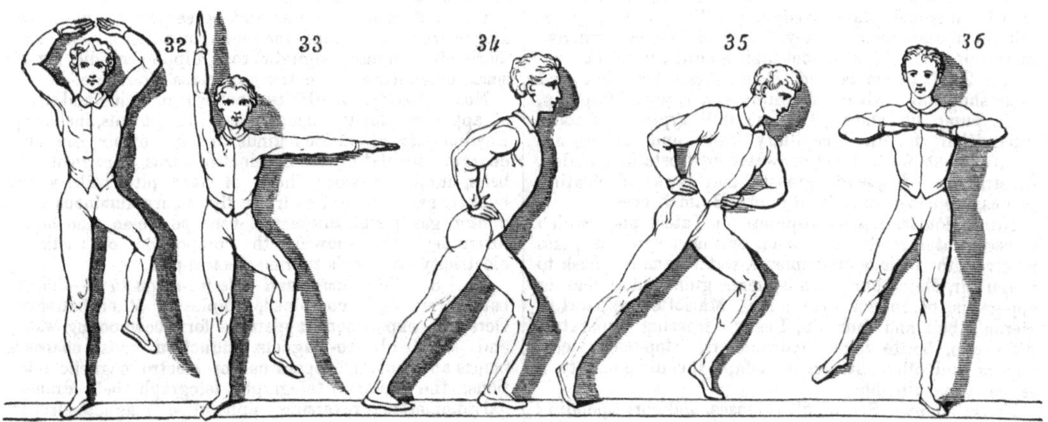

The object of this collection (which is a part of a larger one) is the diffusion of the knowledge of scientific physical education.

Copies of these figures in plaster, terra-cotta, papier-maché, stearin, and other materials will be distributed at the cost price when wanted for asylums for the blind, deaf-and-dumb, for ragged, national, parish, and training schools, or for any philanthropic institution. Applications to be made to Dr. Roth's Institution, 16A Old Cavendish Street, London, or 21 Gloucester Place, Brighton, where further information can be obtained.

[5595]

GARDNER, JAMES, 292 *Oxford Street, and* 52 *High Holborn.*—Stuffed birds; artificial eyes; entomological apparatus and publications.

Ornamental groups of foreign birds under glass shades, at prices varying from three guineas to one hundred guineas.

Artificial eyes for birds and animals. Entomological apparatus, nets, boxes, pins, setting-boards and cabinets always on hand. A mahogany case with two locks and keys, containing a complete outfit for an entomologist for £3.

Taxidermy or the art of collecting, preserving, and mounting specimens of natural history 1*s.* 6*d.*

Priced lists forwarded upon receipt of a stamped envelope.

[5596]

GARDNER, J., Bird Preserver to the Queen of England, 426 *Oxford Street, London.*—First-class stuffed birds.

[5597]

GREGORY, JAMES R., 25 *Golden Square, London.*—Minerals, fossils, educational collections, &c.

Educational and other collections of minerals, fossils, and rocks for students, schools, colleges, &c., and a beautiful series of old red-sandstone fossil fishes from Scotland, including specimens of Coccosteus, Pterichthys, Cheirolepis, Diplacanthus, Osteolepis, Diplopterus, Holoptychius, &c. Samples of minerals for chemical purposes and analysis.

The fossil fishes are now transferred to Class I., No. 131.

[5598]

GRIFFIN, JOHN JOSEPH, F.C.S., 119 *Bunhill Row, London, E.C.*—Apparatus for elementary instruction in physical and chemical science.

[*Obtained a Prize Medal in Class X., in* 1851.]

Many of the instruments in this collection were prepared expressly for use in schools, in fulfilment of the conditions laid down in the following document:—"Special Report on Apparatus for Elementary Instruction in Science, by the Rev. Frederick Temple, M.A., Her Majesty's Inspector of Schools. Education Office, Whitehall, 11th July, 1856."

No. 1. *Mechanics and Mechanism.*—A series of thirty models made of hard stained wood, the pulleys of box-wood. Most of the models are so constructed as to be fixable on the schoolroom black-board, by means of a rail which accompanies them. They include two levers, four kinds of pulleys, centre of gravity piece, simple and compound wheel-and-axle, tilt-hammer, screw, train of wheels, inclined plane, wedge, parallel motion, intermittent motion, lock and key, four models of constructive mechanics, with blocks, hooks, pins, cords, weights, &c.

No. 2. *Hydrostatics and Hydraulics.*—Overshot and undershot water-wheels, Archimedean screw, lift pump, force pump, fire engine, Barker's mill, syphon fountain, intermittent fountain, capillary tubes and plates, apparatus to show the level of water, hydrostatic paradox, illustrations of specific gravity and laws of floating bodies, Appold's centrifugal pump, Bramah press.

No. 3. *Pneumatics.*—Air-pump on Tate's plan, which freezes water readily in vacuo, syringes, syphons, glass receivers for various experiments, water hammer, flask to weigh air, Magdeburg hemispheres, guinea and feather apparatus, barometer experiment, Mariotte's apparatus, Heron's ball and fountain, Leslie's freezing apparatus, filter cup, bottle imps, hydrometers, stop-cocks, correctors, and other fittings to adapt the air-pump to a series of experiments.

No. 4. *Optics.*—Set of glass lenses, and corresponding half-lenses, magnifying and multiplying glasses, simple and compound microscopes, prism, camera obscura, apparatus and objects for experiments on polarized light, pair of phantasmagoria lanterns for exhibiting dissolving views, collection of sliders for the dissolving views.

No. 5. *Heat.*p—Apparatus for showing expansion of metals by heat, and the contraction of metals by cooling, difference in the conducting power of different metals, unequal expansion of different metals, expansion of water and air by heat, measurement of heat by the thermometer, specimens of mercurial, water, and air thermometers, production and condensation of steam, expansive force of steam, boiling of water at low temperatures in vacuo, the pulse glass, tension of vapours, illustrations of latent heat and of high-pressure steam, reflection and absorption of heat by different surfaces and colours, radiation of heat.

No. 6. *Magnetism.*—Bar and horse-shoe magnets, unmagnetized iron bars, magnetic needle, miners' and mariners' compass, magnetic toys, dipping needle, box of small magnetic and electric apparatus after Tate.

No. 7. *Electricity.*—Plate electrical machine, and series of apparatus for the usual class experiments, including Leyden jars of various kinds, battery of six jars, discharger, insulating stool, electrophorus, electrometers, bells, luminous wood, head of hair, pith balls, water bucket, swan, spider, whirl or fly, orrery, luminous conductor, gas pistol, thunder house, sportsmen, and other instrument for showing the proportion of frictional electricity, Melloni's thermo-electric pile.

No. 8. *Galvanism and electro-magnetism.*—Smee's battery, six cells combined, samples of other batteries, Oersted's experiment, apparatus for decomposing water and salts, electro-magnets, induction coils, charcoal points and holder, apparatus for electro-magnetic rotations, the electric telegraph, telegraph bell, cannon, galvanometers, reversing apparatus, magneto-electric machine.

A priced catalogue of the above may be had gratis at 119 Bunhill Row.

[5599]

HENSON, ELIZA MARIA, 113A *Strand.*—Educational, geological, and mineralogical collections; boxes for preserving specimens of natural history.

[5600]

HIGHLEY, SAMUEL, F.G.S., F.C.S., &c., Scientific Educationalist, and Private Teacher of Elementary Science, 70 *Dean Street, Soho, London, W.*—Educational apparatus, specimens, &c.

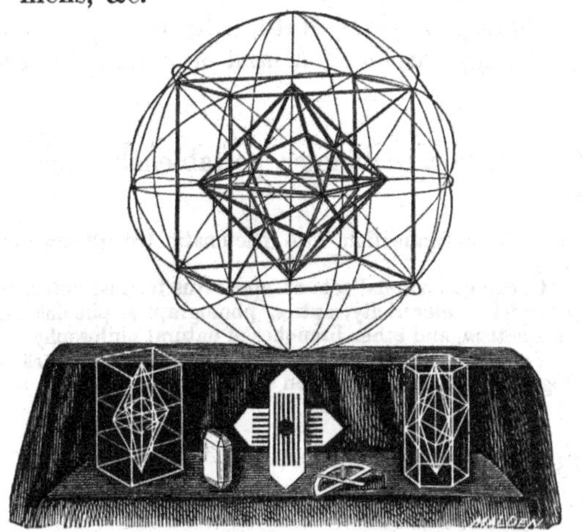

Collections of models in glass, wood, wire, &c., to illustrate crystallography, including the Rev. Walter Mitchell's "Crystallographic Armillary Sphere."

Apparatus, specimens, models, &c., to illustrate the morphological, chemical, optical, thermotic, electrical, magnetic, organoleptic, and crystallogenetic characters of minerals.

A collection of typical minerals, rocks, and fossils selected for lectures or self-instruction, and of sizes suitable for the student, lecturer, or museum.

Casts of rare characteristic fossils.

Collections of specimens and models to illustrate the British Association's lists of typical species in zoology.

Typical skeletons and skulls (according to Huxley and Owen) of mammalia, aves, reptilia, amphibia, pisces.

Naturalists' collecting appliances, as dredges, drags, nets, travelling-cases, geological hammers, &c.

Microscopes, and collections of apparatus and instruments used in connection with the microscope.

Lecturers' demonstrating lantern for optical experiments, &c., with stand, and various sources of light for the same. Portable oxy-hydrogen lantern, with combined gasometer and stand.

"Highley's Science and Art Photographs for the Magic Lantern."

[5601]

HOLT, EDWARD, 24 *White Rock, Hastings.*—Collection of Hastings seaweeds, with their botanical names.

[5602]

LA TOUCHE, Rev. JAMES D., *Vicarage, Stokesay, Newton, Salop.*—An orrery for the use of schools.

An orrery, to exhibit the motions of the earth and moon round the sun. It shows the inclination of the earth's axis to the plane of its obit, its diurnal rotation, the motion of the moon round the earth, including the inclination of the plane of its orbit to that of the earth, and therefore the nodes, and their progress through the ecliptic, and the regular recurrence of eclipses every eighteen years. It is of simple and cheap construction, and adapted for the use of schools.

[5603]

LAWRENCE, EDWARD, 10 *King Street, Lambeth Walk.*—Magic lantern and sliders.

[5604]

LLOYD, WILLIAM ALFORD, 19 & 20 *Portland Road, W.*—A tank containing sea-water, and various living marine animals.

[5605]

MAJOR, ROBERT, 2 *Sussex Terrace, Old Brompton.*—Stuffed birds and animals.

[5606]

PINNELL, THOMAS, 30A *Thomas Street, Oxford Street.*—Fern frame and aquarium combined.

[5607]

RICKMAN & HOBBS, 17 *Grove Place, Brixton Road.*—Cyrena Dulwichiensis, pitharella Rickmani, fossils discovered at Peckham and Dulwich, 1860—61.

[5608]

RIGG, ARTHUR & JAMES, *Chester.*—Apparatus for teaching motions and principles of machinery, mechanism, and engineering.

[5609]

ROBERTSON, CHARLES, 13 *Queen Street, Oxford.*—Elementary zoological series for schools.

[5610]

SHARP, CAROLINE, 248 *Bright Street, Sheffield.*—Educational series of microscopic objects.

[5611]

SHORT, WILLIAM, Naturalist, 50 *Praed Street, Paddington, W.*—Natural history specimens of kestrel, hawks on prey, kingfishers and young, &c., stuffed and mounted by exbibitor.

[5612]

STATHAM, WILLIAM EDWARD, 111 *Strand, W.C.*—Chemical cabinets; portable laboratories; cheap educational sets of scientific apparatus.

Cabinets of chemicals and apparatus, and portable laboratories (twenty varieties), adapted for youths, students, schoolmasters, lecturers, agriculturists, &c.

Hydro-pneumatic apparatus, combining in one a pneumatic trough (with large tray for holding gas jars), an hydraulic blow-pipe (complete with glass-blowers, lamp, and tongs), and a gasometer for storing gases till required for use (when they can readily be transferred to air-jars, in connection with pneumatic trough arrangement).

Cheap educational sets of apparatus for instruction in chemistry, electricity, optics, photography, pneumàtics, magnetism, and other branches of natural philosophy.

For prices and further particulars, see illustrated catalogue, which may be had on application as above.

[5613]

SUTTON, CHARLES, 2 *Hampstead Street, Fitzroy Square.*—Dissolving view lanterns, and oxyhydrogen or limelight apparatus.

[5615]

WARD, E. HENRY, Naturalist, 2 *Vere Street, London, W.*—Specimens of natural history and high art in taxidermy.

Divers specimens of birds and quadrupeds, prepared and stuffed by this firm are exhibited in different cases in the Exhibition building. These illustrate in a remarkable manner the great progress made in the art of taxidermy. The exhibitor buys, sells, and renovates zoological collections or single specimens; and is the inventor of the "permanent and chromatic models of wild and domestic animals' heads.

[5616]

WILKINSON, CHARLES WILLIAM, 8 *Linton Street, Islington, W.*—Improved aquarium, combining fern case and bulb growing.

Victorian aquarium and fern-case, of novel ornamental form, designed to obviate the objection of naturalists to the angular shape, as being often injurious to the fish, and to the globular, on account of their constant liability to entire destruction by slight accidents. The combination of a fern-case, the facility of growing flowering bulbs (in the water), are great additions to its attractions.

[5617]

WILSON, FREDERICK WILLIAM, 1 *Myrtle Terrace, Paxton Park, Sydenham, S.E.*—Specimens of mounted animals and birds, &c.

[5618]

WRIGHT, BRYCE, 36 *Great Russell Street.*—Elementary collections of minerals, fossils, and shells, beryls, and iceland spar.

Educational collections of minerals, fossils, and shells, with descriptive catalogues, in neat cabinets, from 1*l.* to 10*l.* Very large and fine collections of minerals, fossils, and shells, from which single specimens may be selected, may be seen at the exhibitor's establishment.

LONDON: PRINTED BY WILLIAM CLOWES AND SONS, STAMFORD STREET AND CHARING CROSS.

Class XXX.

FURNITURE, PAPER-HANGING, AND DECORATION.

Sub-Class A.—*Furniture and Upholstery.*

[5651]

Andrews, William Smith, 6 *King's Row, Walworth.*—Flower vases and ornaments of wood and glass.

[5652]

Annoot, Charles, 16 *Old Bond Street.*—Carved walnut tree cabinet, four carved and gilt chairs, and carved and gilt centre table.

[5653]

Arrowsmith, Arthur John, 80 *New Bond Street, London, W.*—Solid parqueteries, patented for borders round carpets, for halls, dados, panelling, altar floors, &c.

[5654]

Aspinwall, William, 70 *Grosvenor Street, Grosvenor Square.*—Furniture.

[5655]

Asser & Sharwin, 81 *Strand, W.C., London.*—Bagatelle table, economic adaptation for conversion into billiard table.

[5656]

Avery, Joseph, 81 *Great Portland Street.*—Window blinds for the inside or outside of windows. (*See page* 2.)

[5657]

Ayckbourn, Frederick, 17 *Bond Street, Vauxhall Cross.*—Patent incomparable bed, with air-tubes in tick case.

[5658]

Ayres, William Mountford, 59 *St. Anne Street, Liverpool.*—Models of furniture and appliances.

[5659]

Baker, Rev. Robert Sibley, *Hargrave Rectory, Kimbolton.*—Carved oak eagle lectern.

AVERY, JOSEPH, 81 *Great Portland Street.*—Window blinds for the inside or outside of windows.

The exhibitor's patent spring blind roller cannot be put out of order. He manufactures every description of window blinds.

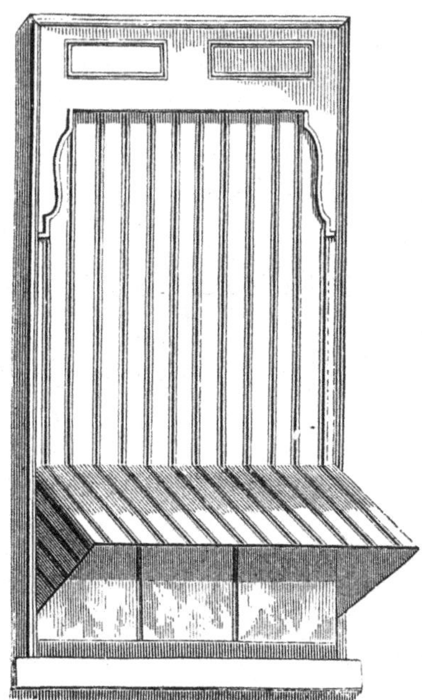

WINDOW BLINDS FOR THE INSIDE OR OUTSIDE OF WINDOWS.

[5660]

BALDWIN, CHARLES, inlayer, *Amersham, Bucks.*—Specimens of ornamental woodwork, for table-tops and house decorations.

[5662]

BAYLIS, WILLIAM HENRY, 69 *Judd Street, London.*—Jewel-casket in boxwood frame, carved in lime-tree flowers.

Obtained the Extra First Prize of the Architectural Society, South Kensington Museum, 1859.

A jewel casket, carved in box-wood; original design of the Renaissance period. Simple outline. At each corner a puma's head, with winged body, conventionalised, and swags of natural flowers depending from the mouth, over the sides. The ground work is enriched with soft foliage. The principal feature of the top is a bird in the folds of a snake. A portrait or panel frame of natural flowers in lime-tree. A new design.

[5663]

BEARD, JOHN, *Stonehouse, near Stroud.*—Improved sofa bed.

[5664]

BEDFORD, EDWARD, 42 *King Street Terrace, Islington, N.*—Oval and carved looking-glass, in carved (from Nature) gilt frame. (*See page* 3.)

[5665]

BELLERBY, WILLIAM, 10 *Bootham, York.*—Oak reading-stands, with carved panels on burnt wood.

BEDFORD, EDWARD, 42 *King Street Terrace, Islington, N.*—Oval and carved looking-glass, in carved (from Nature) gilt frame.

LOOKING GLASS AND CARVED FRAME.

The carving by E. Bedford, the gilding and glass by A. Jenkins. London Looking Glass Company, 167, Fleet Street, where all orders should be sent.

[5666]

BETTRIDGE, JOHN, & Co., *Constitution Hill, Birmingham.*—Tea-trays, tables, chairs, caddies, inkstands, writing-desks, &c. Iron japanned tea-trays, &c.

[5668]

BIELEFELD, CHARLES, 21 *Wellington Street, Strand.*—Papier-mâché decorations, patent silicious panels, *basso-relievo*, and other works of art.

[5669]

BIGGS & SON, 31 *Conduit Street, Bond Street.*—Picture frame for portrait, with armorial bearings.

[5670]

Bird & Hull (late Doveston, Bird, and Hull), 106 *King Street, Manchester.*—Decorative furniture, and an ebony cabinet.

AN EBONY CABINET AND OAK CANDELABRA.

An ebony cabinet, ornamented with white wood, in the Italian style.

A pair of oak candelabra with bronze lights.

A suite of bed-room furniture, consisting of bedstead, wardrobe, toilet and wash-hand tables executed in sycamore, and relieved with other native woods; the silk manufactured by Messrs. Houldsworth, of Manchester.

[5671]

BORNEMANN, A. F. G., *Bath.*—Ebonised furniture, original designs, bas-relief finish, solely introduced and popularised by exhibitor.

[5672]

BRADLEY, JOHN, 129 *Fore Street, Exeter.*—Two slate tables, painted in imitation of the madrepores of Devon, and six slate slabs, painted in imitation of the marbles of Devon.

[5673]

BRIDGES, HENRY, 406 *Oxford Street, London, W., near Soho Square.*—Kitchen tables, plain oak furniture, and dairy articles.

[5674]

BROOKS, HENRY, & Co., *London.*—Music stools, with patent self-fixing screws, and Canterbury what-not.

[5675]

BROWN, BROTHERS, 165 *Piccadilly.*—Patent portable easy chairs and folding bedstead.

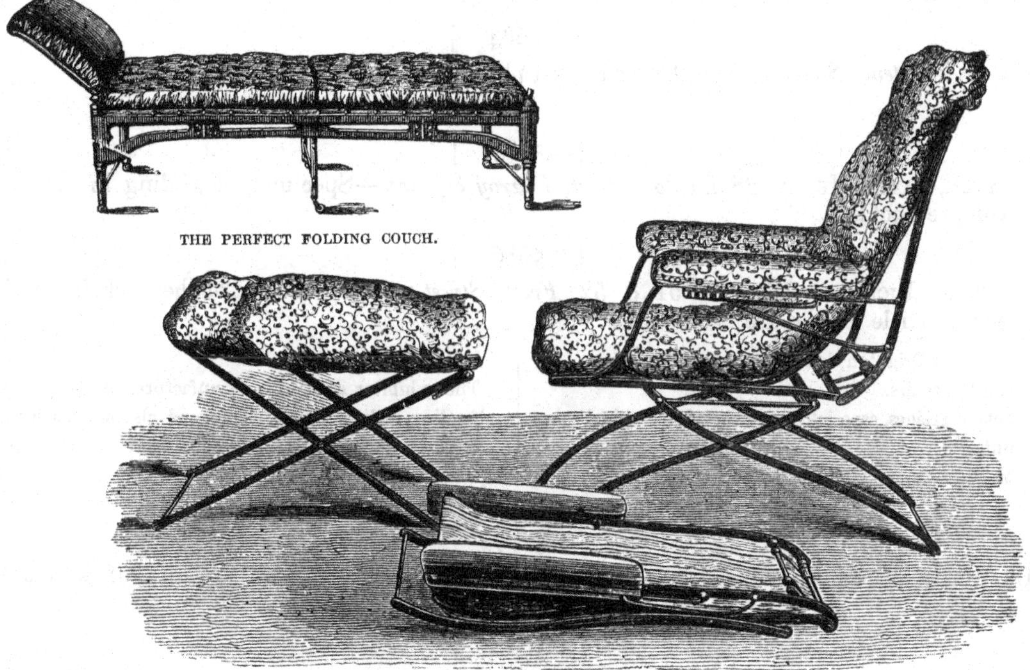

THE PERFECT FOLDING COUCH.

THE MARVELLOUS SPRING CHAIR, WITH REST. SAME FOLDED.

[5676]

BROWN, G. & A., 25 *Newman Street, W.*—Composition ornamented furniture for gilding, and examples of interior decorations.

[5677]

BRUNSWICK, M., 26 *Newman Street, W.*—Marqueterie cabinet, buhl cabinet, marqueterie commode, work table, and card table.

[5678]

BRYER, WILLIAM, *Southampton.*—Specimens of wood carving, spaniel in alabaster.

1. The "Moment of Victory," from Alexander Fraser's picture, carved in box-wood, in various degrees of relief, and highly elaborated; size, 9 in. by 6 in.
2. An altar-piece, from Rubens' "Crucifixion," carved in English oak, in bold relief; size, 5 ft. 6 in. by 3 ft. 2 in.
3. Model for a stand for a lamp or compass of a yacht. The figure of Atlas carved in mahogany.
4. Lectern. Golden eagle carved in oak; life size.
5. Yorkshire canary in full song; a life-size portrait in boxwood.
6. King Charles spaniel, in alabaster; from the life.

[5679]

Burges, William, 15 *Buckingham Street, Strand.*—Painted and decorative furniture, designed by W. Burges; decorations by Harland and Fisger.

[5680]

Burridge, Henry, 15 *Grenville Street, Brunswick Square.*—Granite and marble papers, for ornamental and decorative purposes.

[5681]

Burroughs & Watts, *Soho Square, London.*—An oak billiard table; Gothic style, time of Henry VIII.

[5683]

Caldecott, William, 53 & 54 *Great Russell Street, Bloomsbury.*—A carved English oak sideboard.

[5684]

Casey, *Academy Street, Cork.*—Carved wood plate chest.

[5685]

Chance, James Henry, 28 *London Street, Fitzroy Square.*—Specimen of gilding in various coloured golds.

[5686]

Chaplin, Richard Parsons, 51 & 52 *Frith Street, Soho, W.*—Couches, chairs, and every article for the use of invalids.

Bath and Brighton three and four-wheel chairs.

Garden chairs.

Sofa carriages, especially constructed for spinal complaints.

Merlin self-propelling chairs.

Spinal and invalid couches.

The exhibitor is the sole manufacturer of the patent self-acting reclining easy chairs, and also manufactures first-class perambulators for children and adults; velocipedes, &c. &c.

[5687]

Charlton, Thomas & Co., 128 *Mount Street, Grosvenor Square.*—Suite of bedroom furniture.

[5688]

Collmann, Leonard W., 53 *George Street, Portman Square.*—Sideboard in oak, designed and manufactured by the exhibitor.

[5689]

Cooke, Rev. R. H., *Cheltenham.*—Lectern for parish church.

[5690]

Cox & Son, 28 and 29 *Southampton Street, Strand.*—Church furniture, embroidery, wood and stone carvings, Gothic metal work. (*See page* 7.)

[5691]

Crace, John G., 14 *Wigmore Street, W.*—Decorations and cabinet furniture.

[5692]

Cremmin, Jeremiah, *Killarney.*—Fancy cabinet articles in bog oak, &c.

COX & SON, 28 and 29 *Southampton Street, Strand*, exhibit in the Mediæval Court church furniture, embroidery, wood and stone carvings, Gothic metal work, pulpit, reredos, &c.

SPECIMEN OF DECORATIVE ART.

ECCLESIASTICAL WAREHOUSES, 28 and 29 Southampton Street, Strand, London.

PAINTED GLASS WORKS, 43 and 44 Maiden Lane, adjoining Southampton Street.

CARVING AND GOTHIC METAL WORKS, Belvedere Road, Lambeth.

The drawings and prices of articles exhibited sent free, or an illustrated catalogue for six stamps.

[5693]

CRISWICK & DOLMAN, 6 *New Compton Street, Soho.*—Imitation carved bedstead, book-cases, candelabra.

[5694]

DEAR, WILLIAM, 30 and 31 *St. George's Place, Hyde-park Corner.*—Marqueterie and mosaic tables, draperies, &c.

[5695]

DEXHEIMER, CH., 27 *Connaught Terrace, Edgware Road.*—Marqueterie table, buhl cabinet.

[5696]

DRAPER, FRANCIS, 26 *Old Cavendish Street, and* 70 *Great Titchfield Street, London.*—Gilt picture frames, with spandril and an oval frame.

[5697]

DREW, JOHN, 2 *Great Warner Street, Clerkenwell.*—Portable house furniture for exportation and military use.

[5698]

DYER & WATTS, 1 *Northampton Street, Islington.*—Suite of bedroom furniture, ornamented pine imitation marqueterie and inlays.

The exhibitors are manufacturers of japanned and polished deal furniture, and patentees of the POLISHED PINE ORNAMENTAL FURNITURE.

WARDROBE IN POLISHED PINE.

ECCLESIOLOGICAL SOCIETY, 78 *New Bond Street.*—Ecclesiastical art-products, comprising embroidery, ceramics, metal-work, stone-carving, and wood-carving.

The following form only a portion of the subjects exhibited in the Mediæval Court, Class XXX., under the superintendence of this Society. The combination of various exhibitors into this Court was only decided on subsequently to the Catalogue going to press:—

[5699]

BLENCOWE, MISS, (*Ladies' Embroidery Society*), *West Walton*, *Wisbeach.*—Altar cloths of silk for Peterborough Cathedral and for Clehonger Church, Herefordshire; the latter from a design by Mr. Preedy.

[5701]

BRANDON, R., 17 *Clement's Inn, W.C.*—Model of roof, executed on a scale of half an inch to the foot.

[5702]

NICHOLLS, T., *Hercules Buildings, Lambeth.*—Effigy and altar tomb; model of the reredos for Waltham Abbey; one figure of same completed in alabaster, and painted: all designed by W. Burges, Esq., F.R.I.B.A.

[5703]

CLAYTON & BELL, 311 *Regent Street, W.*—Circular incised group, on stone—a replica of one of the medallions for the pavement of Lichfield Cathedral.

[5704]

EARP, T., *Kennington Road, Lambeth.*—Reredos and lectern.

[5705]

FORSYTH, J., 8 *Edward Street, Hampstead Road, N.W.*—Study, in plaster, of the head of our Lord, in full relief; model of front panel of pulpit of Limerick Cathedral; two mural tablets.

[5706]

KEITH, J.—Case of church plate.

[5707]

NORTON, J., F.R.I.B.A., 24 *Old Bond Street.*—Cast of life-size statue of Edward III., at High Cross, Bristol; font, in alabaster and marble; two casts of panels for pulpits; cast of the sculptures of reredos at St. John, Bedminster.

ECCLESIOLOGICAL SOCIETY—*continued.*

[5707*]

O'CONNOR, Messrs. M. & A.—Triptych, wooden and painted.

[5708]

PARRY, T. GAMBIER, *Highnam Court, Gloucester.*—Circular panel to illustrate system of painting on damp walls.

[5709]

PHILIP, J. B., *Roehampton Place, Vauxhall Bridge Road.*—Effigy of the late Rev. W. H. Mill, D.D.—a fac-simile of the original monument in Ely Cathedral; the architectural portion designed by G. G. Scott, R.A.

[5710]

PRICHARD & SEDDON, F.R.I.B.A., 6 *Whitehall, S. W.*—Two decorated organs (works by Gray and Davison).—Nine subjects in encaustic tiles.—Mediæval furniture, various.

[5711]

REDFERN, J., 29 *Clipstone Street, Fitzroy Square, W.*—Cast of bas-relief of the Resurrection, from the Digby Mortuary Chapel, Sherborne. (*See page* 9.)

[5712]

SHAW, N., *London.*—Book-case, executed by J. Forsyth.

[5713]

SLATER, W., F.R.I.B.A., 4 *Carlton Chambers, S. W.*—Portion of mural monument for Limerick Cathedral. Only one of the three canopied arches, which form the monument, is exhibited: subject, the Annunciation, by Mr. Redfern.

[5714]

STREET, G. E., F.R.I.B.A., 33 *Montague Place, Bedford Square, W.*—Iron font cover and standards for lights; executed by Mr. Leavers.

[5715]

TEULON, S. S., F.R.I.B.A., *Craig's Court, S. W.*—Reredos in alabaster and marble, with mosaics: subject of the centre space, the Last Supper; sculptor, Mr. Earp: marble font.

[5716]

WHITE, W., F.R.I.B.A., 30A *Wimpole Street.*—Case of inexpensive altar plate, for colonial use, executed by Messrs. Benham; reredos for Claydon Church, Oxfordshire, with panels on enamelled slate.

[ ]

HALSTEAD, MESSRS., *Chichester.*—Grills for Chichester Choir, from design by W. Slater, Esq.

[ ]

CARPENTER, late R. C., F.R.I.B.A.—Model of tomb of Marshal and Lady Beresford, at Kilndown, Kent; executed by Mr. Myers: tiles for wall decoration; executed by Messrs. Minton.

[5716A]

GODWIN, W., *Lugwardine, Hereford.*—Encaustic tiles.

[5716B]

GRAY & DAVISON, 370 *Euston Road.*—An organ.

[5717]

EGAN, JAMES, *Arbutus Factory, Killarney.*—Fancy cabinet table, &c.

[5718]

ELLIOTT, HENRY, 6 *Vere Street, Oxford Street, London.*—Satin wood wardrobe, and octagon revolving pedestal.

[5719]

ELLIS, CHARLES, 21 *Bedford Street, Covent Garden.*—A wainscot sideboard; dinner wagon, to correspond; dining-room chair; dining table; a book-case of walnut and ebony.

[5720]

FAULDING, JOSEPH, 338 *Euston Road.*—Specimens of fret-cutting and curvilinear sawing.

[5721]

FILMER & SON, T. H., 28, 31, 32, 34 *Berners Street, London, W.*—Patent dining table, convertible ottomans and chairs, bedroom furniture, chintzes, and wall papers, of same design, from prize competition drawings. (*See pages* 12, 13.)

[5722]

FOLSCH, FREDERICK WILLIAM, *Long Buckby, Northamptonshire.*—A jewel-box and dressing-case, on stand, inlaid with 17,000 pieces.

[5723]

FORSYTH, JAMES, 8 *Edward Street, Hampstead Road, London.*—Book-case, wood-carvings, stone-sculpture, models, &c., ecclesiastical and monumental. (*See page* 14.)

[5724]

FOX, THOMAS, 93 *Bishopsgate Street Within.*—A suite of drawing-room furniture in walnut and holly.

[5725]

FREYBERG, JAMES, *Grosvenor Street West, Belgravia.*—Furniture of a lady's boudoir, chintz, and decorations.

[5726]

GANN, MARY C., 32 *Dorset Square.*—Drawing-room tree-stand; new tile, natural, with colours burnt and unburnt.

Drawing-room tree-stand, movable by the slightest touch, with the heaviest weight upon it. A new design of a mosaic tile, in simple earth colours, burnt and unburnt—red, buff, and blue joined in black. Mounted in walnut wood, with ivory castors.

[5727]

GARDNER, JOHN HENRY, *Poppin's Court, Fleet Street.*—Mahogany dressing-table and wash-stand combined, and a mahogany cheval-glass.

[5728]

GARROOD, ROBERT ELMY, *Chelmsford, Essex.*—Mitre-box for mitreing German and other mouldings for picture-frames.

FILMER & SON, T. H., 28, 31, 32, 34 *Berners Street, London, W.*—Patent dining-table, convertible ottomans and chairs, bedroom furniture, chintzes, and wall-papers, of same design, from prize competition drawings of students of the South Kensington School of Art.

See also Class XXII.

Circular extending dining-table, of the finest English pollard oak, the framework of Italian design, with scroll supports, ornamented with festoons of fruit, &c. Manufactured on a novel plan, to open to an increased diameter by an extension of the framework, the top being preserved entire, and quarter-circle leaves introduced in several series round the circumference, thus preserving at all sizes the perfect circle.

The movement of this table combines extreme simplicity with the utmost certainty of action; it is fitted with a screw and cog mechanism (by Hawkins), by the operation of which the whole framework is expanded simultaneously; the leaves being placed, are all fastened by the same means at once, thus enabling one person without assistance, and in a few minutes, to arrange a table sufficiently large to dine thirty or forty people.

Two pollard oak dining-room chairs, of Italian design, with stuffed panelled backs, covered with morocco leather to accord with the table.

An easy chair, to match with the above.

Easy chair, the frame carved in walnut wood; the back arranged with an oscillating spring, combining the comfort of an ordinary lounge with the pleasant motion of a rocking-chair.

Movable convertible ottoman for centre of rooms, richly carved in the style of Louis XVI., the groundwork finished in white enamel, with mat and burnished gold relief; covered with rich figured silk. This ottoman is made in four separate parts, forming two easy chairs and two settees, which are constructed to fit together, and can be instantaneously formed into a complete and elegant centre seat.

Suite of furniture for a bedchamber, of novel forms, consists of a washing-stand, with closets under, and marble top, made of fine figured walnut wood, relieved by carved mouldings, &c., in white woods and ebony.

A winged wardrobe, with circular ends.

A lady's toilet, with numerous drawers and full-length glass attached.

The four following designs are the result of a competition among the students of the Government Schools of Art, for prizes given by this firm. Two are from the Male School, at South Kensington, and two from the Female School, at Queen Square.

Competition patterns from the Government School of Art, South Kensington. Conventional treatment of flowers in their natural colours. Six printings,

FILMER & SON—*continued.*

NO. 1. PERSIAN.

Block-printed chintz furniture and wall paper.

Designed by H. H. Lock.

Obtained First Prize in competition.

NO. 2. PASSION-FLOWER.

Block-printed chintz furniture and wall paper.
Designed by J. Randall.
Obtained Second Prize in competition.

Competition patterns from Female School of Art, Queen Square. Natural treatment of flowers in their proper colours. Seventeen printings.

NO. 4. CONVOLVULUS.

Designed by Miss Charlotte James.
Obtained First Prize in competition.

We have carried out all four permeated designs in various methods of block-printing, both with wood and copper blocks, suited to the different fabrics of cotton, cotton velvet, and worsted, for curtains; and also on a silk ground, a novel application for furniture purposes, the effect of which is remarkably soft and rich, and much more economical than an ordinary wove coloured silk.

NO. 5. BLACKBERRY.

Designed by Miss Mary Julyan.
Obtained Second Prize in competition.

Each pattern has also been reproduced on paper for walls, in order that, where it is desirable to have the whole decorations of a room *en suite*—a taste that is now much cultivated—the same designs may be repeated in the hangings both of the walls and windows.

Forsyth, James, *Sculptor*, 8 *Edward Street, Hampstead Road, London.*—Book-case, wood-carvings, stone-sculpture, models, &c., ecclesiastical and monumental.

OAK BOOK-CASE, WITH WRITING TABLE.

1. Oak book-case with writing table, inlaid with various woods, designed by R. Norman Shaw. Price £80.

2. New woodwork for Chichester Cathedral, executed from the designs of Mr. William Slater, architect. Various parts of the new stalls for Chichester Cathedral. In the ends are carved representations of various plants, herbs, &c., mentioned in Scripture. The carved upright panels and most of the discs in the fronts are designed upon the same principle; but some of the latter are made purely conventional, so as to contrast with the natural foliage.

3. Two Caen stone figures, representing St. Peter and St. Paul, executed for Lord Crewe.

4. Adoration of the Magi, designed for a reredos panel.

5. Four panels, being the working models of sculpture on Lichfield Cathedral font, the gift of the Hon. Mrs. Howard. Mr. William Slater, architect.

6. Series of four panels, executed in oak, for the pulpit in Witley Church. The gift of the Earl of Dudley. Mr. S. W. Dankes, architect.

7. Group in plaster (to be executed in marble) of the Departure of Hagar and Ishmael.

[5730]

GEORGE, CLEMENT, & SON, 16 *Berners Street.*—Oak sideboard, in the Italian style, and chairs.

[5731]

GILLOW & Co., 176 *Oxford Street.*—A walnut sideboard, Renaissance style; an inlaid cabinet, Italian style; richly carved chairs.

[5732]

GOSLETT, ALFRED, 26 *Soho Square.*—Looking-glasses, picture-frames, cornices, console-tables, and girandoles.

[5733]

GOW, JOHN, 13 *Argyle Street, King's Cross.*—Moulds for casting composition and papier mâché ornaments from.

[5734]

GRIFFITHS, JOHN, 89A *Ratcliffe Terrace, Rathbone Street, Liverpool.* — Miniature frame, carved in box-wood.

[5735]

HARDING, MADDOX, & BIRD, 70 *Fore Street, Finsbury.*—Louis XVI.'s bedstead, in white and gold, silk hangings.

[5736]

HARLAND & FISHER, 33 *Southampton Street, Strand.*—Specimens of ecclesiastic and domestic decorative art.

[5737]

HARROLD, C. & G., *Hinckley.*—Figured oak table; ditto ditto top and panels.

[5738]

HATCHWELL, H. & S. B., *Newton Abbot, Devonshire,* 4 *Langham Street, Langham Place, London.*—Patent revolving church stool, one side being stuffed for kneeling on, and the opposite side for placing the feet on, or foot stool.

[5739]

HAWKINS, S., 54 *Bishopsgate Street.*—Model of dining-table.

[5740]

HEAL & SON, 196 *Tottenham Court Road, London.* —Bedsteads, wardrobe, dressing table and glass. (*See pages* 16, 17, 18.)

[5741]

HERBERT, MRS., *Royal Avenue Terrace, Chelsea.*—Paper transparencies.

[5742]

HERMANN, FREDERICK, 54 *Devonshire Street, Portland Place.*—Library cabinet in buhl.

[5743]

HERRING, SON, & CLARK, 109 *Fleet Street, E.C.*—Sideboard and chairs, and dinner wagon.

HEAL & SON, 196 *Tottenham Court Road, London, W.*—Bedroom furniture, wardrobe, dressing table and glass.

BEDSTEAD.

Bedstead in the style of Louis Seize, of mahogany, with enamelled white surface, carved and gilded. The furniture and eider-down quilt are of cerise-colour silk The curtains are lined with white silk. The head-cloth is of white silk, embroidered in cerise colour, and the valance is trimmed with white silk.

HEAL & SON—*continued.*

WARDROBE.

Wardrobe in the style of Louis Seize, to correspond with the bedstead on the previous page, of mahogany, with enamelled white surface, carved and gilded.

Heal & Son's Illustrated Catalogue of Bedsteads and Bedroom Furniture, and Priced List of Bedding, sent free by post on application to the factory, 196, Tottenham Court Road, London, W.

Heal & Son—*continued.*

TOILET TABLE.

Toilet table with glass, in the style of Louis Seize (to correspond with bedstead and wardrobe on pages 16 and 17), of mahogany, with enamelled white surface, carved and gilded, and china ornaments on the pedestals.

The best talent available has been engaged in the decorative part of the work, the object being to produce articles of domestic furniture which should combine elegance of design and the highest class of art manufacture with a simplicity of style and effect which should render them suitable for a nobleman's or gentleman's mansion.

Heal & Son's Illustrated Catalogue of Bedsteads and Bedroom Furniture, and Priced List of Bedding, sent free by post on application to the factory, 196, Tottenham Court Road, London, W.

Howard & Sons, Manufacturers, 26 *Berners Street.*—Book-case fittings, library table, and seats (style, Pompeian).

The subject engraved on this page is a piece of furniture from the set of library fittings exhibited by Howard and Sons. The style is Pompeian. The surface decoration is carved in the wood below its general surface, thus preserving it from the effects of friction The more prominent ornaments are gilt bronzes of English manufacture. The workmanship is of the highest class. It may be tested by the drawers within the pedestal being turned in any way, when they will be found to fit with the nicest accuracy.

[5744]

Hindley & Sons, Charles, 134 *Oxford Street.* — Cabinet and other furniture, silk hangings, &c.

HINDLEY & SONS, CHARLES—*continued.*

[5745]

HOLLAMBY, HENRY, *Parade, and Frant Road, Tunbridge Wells.*—Mosaics in natural coloured woods.

[5746]

HOLLAND, WILLIAM, *Stained Glass and Decorative Works, St. John's, Warwick.*—Decorations for the interior of rooms.

[5747]

HOLLAND & SONS, 23 *Mount Street, Grosvenor Square.*—Furniture and decoration.

JACKSON & GRAHAM, *Oxford Street, London.*—Oak sideboard, cabinet, pianoforte, chimney-piece; other articles and decorations.

A CABINET OF EBONY, inlaid with ivory, the centre inclosed by doors, with oval medallions of hymeneal subjects in bronze, highly chased and gilt, the ends open and rounded; the plinth, columns, and frieze all enriched with mouldings and ornaments of bronze, finely chased and gilt, and surmounted by a slab of the finest Algerian onyx.

JACKSON & GRAHAM—*continued.*

A BUFFET or SIDEBOARD, 10 ft. long, and 12 ft. 6 in. high, of pollard oak of rare beauty, enriched by carvings in brown English oak. The doors of the pedestals are niched, with figures of boys gathering grapes, and reaping, and the friezes of them ornamented with the hop plant springing from shields. The frieze of the centre division has a richly carved shield, with barley springing from each side, the shield itself being enriched with fruit. The angles of the pedestals are canted to form a background for richly carved chimeras with leopards' heads, which appear to support the slab. Above the slab, and over the pedestals, are plinths, enriched with carved panels, the one representing a wine cup entwined by the vine, and the other a tankard surrounded by the hop plant. Upon these plinths are placed two caryatides, the one with attributes of the field and forest, the other of the ocean and river. These figures support the cornice and pediment, which has a boldly carved shield in the centre, with festoons of fruit hanging gracefully from it, and partly resting upon the cornice. The caryatides are flanked by richly carved pilasters, the one representing game, surmounted by the head of a retriever, and the other various denizens of sea and stream surmounted by the head of an otter. The centre and side panels of the upper part are filled with silvered glass.

A FINE WALNUT-WOOD WARDROBE, 9 ft. long, in three divisions. The plinth, cornice, and end panels inlaid with lines, and ornamental corners of Amboyna, purple-wood, and holly. The centre door has a panel of silvered glass, and on each side are pilasters, richly inlaid with various woods, the caps and bases finely carved, which support the cornice and pediment; the latter has a shield in the centre, from which spring rich festoons of flowers in marqueterie work. The corners of the plinth and cornice are rounded, and a hollow worked upon the angles of the wardrobe to receive columns inlaid and carved *en suite* with the pilasters, to complete the support of the cornice. The doors on each side of the centre have small oval mirrors, surrounded by rich floral marqueterie work.

DECORATION FOR THE SIDE OR END OF A DRAWING ROOM, in the style of Louis XVI., panelled and enriched with mouldings, and relief ornament, gilt. The centre panel fitted with silvered plate glass, and the side panels with rich crimson silk, of thrice the ordinary width, designed and manufactured expressly for exhibition.

A CHIMNEY-PIECE OF ALGERIAN ONYX, the pilasters and frieze enriched with bas-reliefs of bronze, chased and richly gilt.

A PIANO (the interior by Messrs. Erard), the case of fine Amboyna wood, richly inlaid in various ornamental devices, musical trophies, and flowers in marqueterie work; the front, above the fall, of very finely perforated purple wood, in which are framed three highly finished paintings, on porcelain, that in the centre representing a group of children playing upon musical instruments, upon the left of which is a medallion, with a boy playing the pandean pipes, and on the right another medallion, with a boy playing cymbals.

A SMALL DRAWING-ROOM CHAIR, very finely carved and richly gilt, in the style of Louis XVI.

AN ETAGERE OF EBONY, inlaid with ivory.

AN EBONISED BOOK-CASE, with engraved lines and ornaments or pilasters, panels, and framing, carved shields in centre of panels, carved shield and scroll on pediment, and carved astragals.

A LADY'S TOILETTE TABLE, with cheval glass in centre, and pedestals with drawers on each side, of fine satin wood, inlaid with tulip wood.

[5748]

HOWARD & SONS, 26 *Berners Street.*—Book-case fittings, library table, and seats (style, Pompeian). (*See page* 19.)

[5749]

HUMBLE, G. & J., *Kelso.*—Antique arm chair, suitable for an entrance hall.

[5750]

INGLEDEW, CHARLES, 17 *Market Row, Oxford Market.*—Dining-room and library chairs.

[5751]

JACKSON & GRAHAM, *Oxford Street, London.*—Oak sideboard, cabinet, pianoforte, chimney-piece; other articles and decorations. (*For Illustration, see page* 22.)

[5752]

JACKSON, GEORGE, & SONS, 49 *Rathbone Place, Oxford Street.*—Ornamental furniture, in carton-pierre and papier-mâché. (*See page* 24.)

[5753]

JEFFERSON, JOSEPH HARDEY, 46 *College Green, Bristol.*—Richly gilt picture and miniature frames.

[5754]

JENNER & KNEWSTUB, 33 *St. James's Street, & 66 Jermyn Street.*—Fancy cabinets, dressing cases, and work table.

[5755]

JOHNSTON, MRS., *Ashley, Newmarket.*—Oval frames in leather, &c.

[5756]

JOHNSTONE & JEANES, 67 *New Bond Street.*—Cabinet of various woods, ormolu mountings, and glass over.

Jackson, George, & Sons, 49 *Rathbone Place, Oxford Street.*—Ornamental furniture, in carton-pierre and papier-mâché.

Various specimens of furniture for gilding, including candelabra tables, table legs, jardinières, girandoles, &c.

Portion of a panel in Louis XVI. style.
Ornamental panel, executed in carton-pierre.
Ornamental room cornices, executed in carton-pierre.

[5758]

Jones & Co. (late Robson and Jones), 214 *Piccadilly, London, W.*—Decoration and furniture.

[5759]

Jones & Willis, *Temple Row, Birmingham.*—Church furniture and decoration.

[5760]

Joubert, Amedee, 18 *Maddox Street.*—A drawing-room cabinet, with jardinières; a newly-invented music chair; an easy chair covered in morocco; an easy chair in canvas, &c.

[5761]

Kendall, Thomas Henry, *Chapel Street, Warwick.*—Articles of furniture, and specimens of wood-carving.

[5762]

KIRK & PARRY, *Sleaford.*—Ancaster stone font, with English oak cover, designed and executed by Kirk & Parry, Sleaford.

[5763]

KNIGHT, THOMAS, *George Street, Bath.*—Library table, with cabinet, containing desk and stationery; ebony chair.

LIBRARY TABLE.

A library table of oak and ebony, surmounted by a cabinet occupying its whole length, containing drawers on either side, and a desk in the centre, which, when shut, forms an ornamental panel with malachite, lapis lazuli, cornelian, and serpentine; and when open, discloses an arrangement of the materials necessary for correspondence.

A drawing-room chair in ebony, covered with silk; style of Louis XV.

[5764]

LAKE & SON, *Old Kent Road.*—Mahogany telescope; bedstead pillar, for secret and secure fastening; and models of wrought-iron bedsteads.

[5765]

LAMB, JAMES, 29 *John Dalton Street, Manchester.*—Sideboard in English oak.

A sideboard in English pollard oak, walnut, ebony, and gold. In the upper part are two life-size figures of Vintage and Harvest; between them an oval frame containing a mirror, but designed to receive a painting (if required), trophies of fruit, corn, &c., are introduced. The lower part is supported by figures of boys on pedestals; the central panels are arranged to form one connected *reliero* of game, fish, &c. Groups of fruit and vegetables fill the curved end panels. Designed by W. J. Estall; modelled by Hugues Protat.

Dining-room chairs in embossed and gilt morocco leather.

A marqueterie, Louis Seize, cabinet of Thurgau and other woods. A card table.

[5766]

LAUNSPACH, L., 9 *Upper Berkeley Street West, Hyde Park.*—Marqueterie cabinet, with plate glass back, and buhl cigar cabinet.

[5767]

LAWFORD, 89 *Newman Street.*—Reclining arm-chairs, table bedsteads, &c.; concentration unequalled, thirty inches long, three inches diameter.

Self-acting flap dining table.

Chair, stuffed seat and back, the frame in one piece, to fold into the small compass of 18 inches long by 2½ inches in diameter.

Reclining arm-chair bed, to concentrate into the small space of 30 inches long by three inches diameter, in one piece. Full-size when expanded.

Bed, to answer the purpose of table, 6 feet long by 41 inches wide. To pack in a roll 30 inches long by 30 inches in diameter.

[5768]

LAWRIE, WILLIAM, *Downham Market, Norfolk.*—Monument, reredos, monumental cross.

[5769]

LECAND, SAMUEL, 246 *Tottenham Court Road.*—Carved and gilt console frame and table, with glass; also a pair of carved and gilt Italian tripod stands.

[5770]

LENZBERG & WALTON, 492 *New Oxford Street, London.*—Manufacturers of cornice poles, cornices, window blinds, all novelties. (*See page* 27.)

[5771]

LEVIEN, J. M., 10 *Davies Street, Grosvenor Square.*—Cabinets, tables, and sideboard.

Obtained Prize Medals at the Exhibitions of 1851 *and* 1855, *and a Testimonial from the Society of Arts.*

Ebony cabinet, in the Pompeian style.

Cabinet in the Renaissance style, inlaid with marqueterie, and plate glass back.

Circular wood mosaic table.

Small cabinet table, with plaques of Sevres china, and ornamented with ormolu.

Sideboard of New Zealand wood, richly carved, is exhibited in the Colonial Department.

[5772]

LITCHFIELD & RADCLYFFE, 30 *Hanway Street, Oxford Street.*—Marqueterie furniture. (*See page* 28.)

[5773]

LOTH, JOHN THOMAS, *Carlisle.*—Marble chess-table and urn-stand (imitation in mosaic), papeterie jewel-box.

[5774]

LOVEGROVE, J. I., *Isleworth.*—Decorative church writing.

[5775]

LOWSON, GEORGE, *Broughty-Ferry, near Dundee.*—Inlaid chess table.

[5776]

M'CALLUM & HODSON, *Summer Row, near the Town Hall, Birmingham.*—Papier mâché and japan goods.

[5777]

M'DONALD, DAVID, *Melrose, Roxburghshire.*—Oak burr; carved Davenport desk.

Lenzberg & Walton, 492 *New Oxford Street, London.*—Manufacturers of cornice poles, cornices, and window blinds.

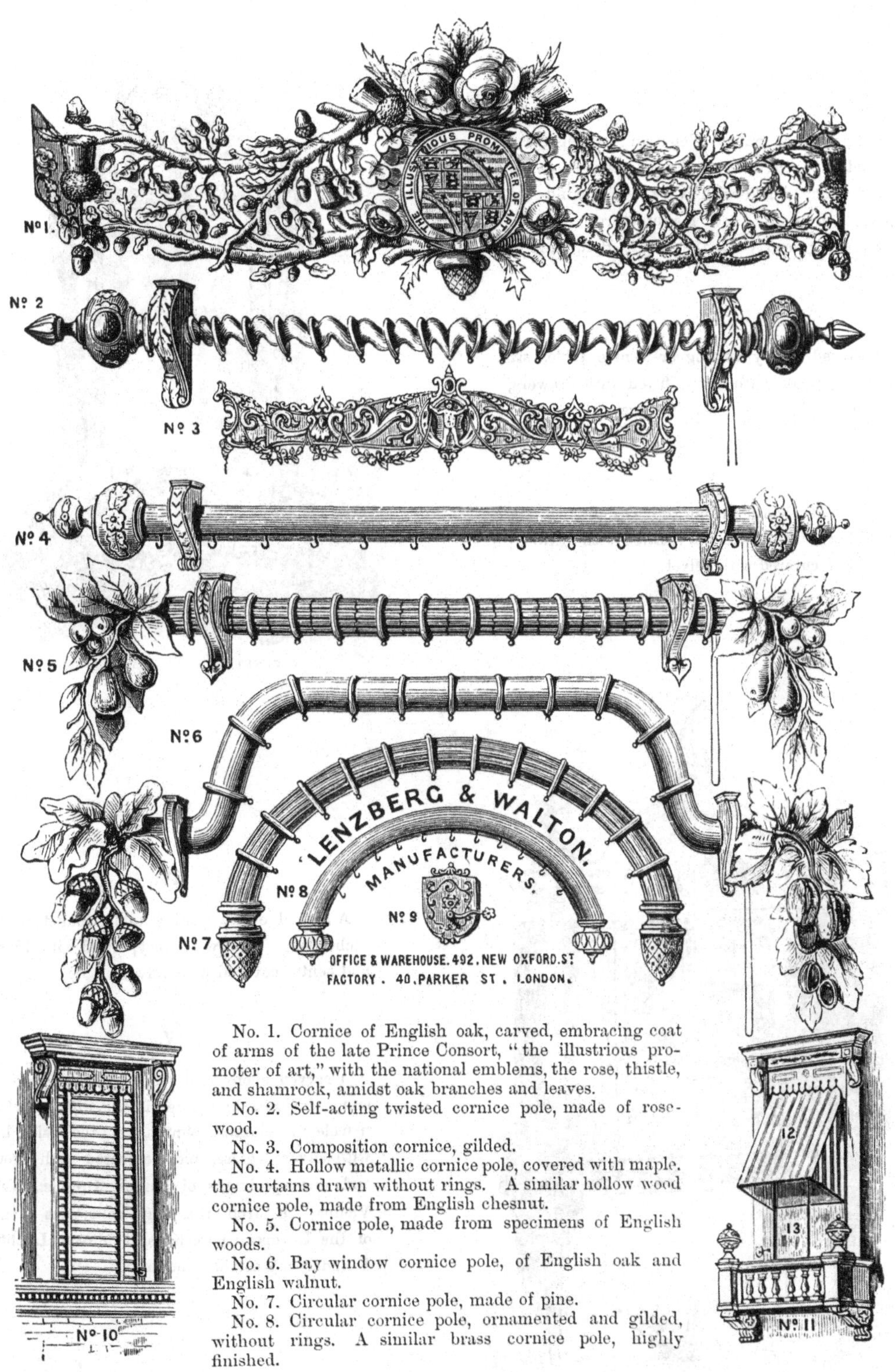

No. 1. Cornice of English oak, carved, embracing coat of arms of the late Prince Consort, "the illustrious promoter of art," with the national emblems, the rose, thistle, and shamrock, amidst oak branches and leaves.

No. 2. Self-acting twisted cornice pole, made of rosewood.

No. 3. Composition cornice, gilded.

No. 4. Hollow metallic cornice pole, covered with maple. the curtains drawn without rings. A similar hollow wood cornice pole, made from English chesnut.

No. 5. Cornice pole, made from specimens of English woods.

No. 6. Bay window cornice pole, of English oak and English walnut.

No. 7. Circular cornice pole, made of pine.

No. 8. Circular cornice pole, ornamented and gilded, without rings. A similar brass cornice pole, highly finished.

All the above cornices and cornice poles are arranged with pulleys on a new and simple plan, to draw the curtains.

No. 9. Lenzberg & Walton's new invention, the Window Blind Regulator, for elevating and lowering Venetian and Outside Blinds of any weight, as well as other blinds.

No. 10. Venetian blind, fitted with the Window Blind Regulator.

No. 11. An outside blind, marked 12, and a roller blind, 13, fitted with the Window Blind Regulator.

LITCHFIELD & RADCLYFFE, 30 *Hanway Street, Oxford Street, and* 19 *Green Street, Leicester Square.*—Marqueterie furniture.

I.

A fine old engraved Venetian looking-glass, adapted as a cheval glass by a stand of richly carved ebony, inlaid with ivory, introducing the cipher and crest, and manufactured to the order, and property of, the Right Hon. Earl of Craven.

II.

A carved ebony writing or centre table, six feet long, inlaid with ivory, fitted with drawers, and centre of top covered with velvet.

III.

A carved ebony writing table, four feet six inches long, inlaid with ivory, fitted with drawers, and centre covered with velvet.

A VENETIAN CHEVAL GLASS.

A CABINET FOR EXHIBITING FINE PORCELAINS.

IV.

A carved ebony writing table, four feet three inches long, inlaid with ivory, fitted with drawers, and centre covered with velvet.

V.

A carved ebony and ivory upright cabinet on stand, inclosed by doors, inlaid with cornelians, marble in columns, plinths, &c., the inside lined with silvered glass, and decorated with groups, and fine specimens of Sèvres, Dresden, Berlin, Vienna, &c., from the large collections of most of the European porcelains at Messrs. Litchfield & Radclyffe's establishments.

[5778]

M'FARLANE, WALTER, *Saracen Foundry, Glasgow.*—Oak book case, with bronze mountings.

BOOK-CASE—part of a library suite in solid oak. This piece of furniture is designed with a boldly projecting base, above which it is divided into five compartments by moulded haffets, enriched with a band of holly and berry. Over each division shields of varied designs support semicircular panelled tablets in line with and breaking upon the cornice. These are to be surmounted with bronze busts of celebrated men; the tablets contain the name, whilst the date of birth and death are inscribed on the shields below; the two centre shields bear family monograms. All the letters are of bronze, and in design characteristic of the different periods in which they lived. A curved pediment crowns the centre compartment, the tympanum of which is occupied by the exhibitor's crest in bronze. A bust of the Queen is to be placed on the pediment; and above the cornice is a light bronze cresting. The doors are filled in with glass, and have crank handles of an elaborate spiral form, with serpents entwined. The hinges have projecting barrels of similar design, all in bronze. The details have been studied with due regard to the nature of the materials, the object aimed at being to combine truth, fitness, and beauty with a certain degree of expressiveness and individuality.

The wood work by Robert White and Son, Glasgow. The bronze crest and letters by Elkington, Birmingham. The other bronze work is of Glasgow manufacture.

[5779]

M'LAUCHLAN, D. J., & SON, *Water Lane, Blackfriars, London, E.C.*—Gilt carved console-table and frame, with looking-glasses.

Carved oak sideboard, gilt in parts, surmounted with looking-glasses, in carved and gilt frame, with bold canopy and triple branches on either side for gas lights; or these can be adapted for oil lamps.

[5780]

MADDOX, GEORGE, 21 *Baker Street, Portman Square, London.*—Bed-room furniture in deal, inlaid and French polished.

A wardrobe, chest of drawers, washstand, and dressing table, in polished deal, with imitative marqueterie enrichments; a patented invention for ornamenting common woods, and one-half the cost of anything yet attempted.

[5781]

MARGETTS & EYLES, 127 *High Street, Oxford.*—Chimney glass, carved walnut-wood frame. Style: Louis XIV.

[5782]

MASSEY, THOMAS, *City Walls, Chester.*—Circular table, curiously inlaid with 1,000 specimens of foreign hard woods.

[5783]

MORRIS, MARSHALL, FAULKNER, and Co., 8 *Red Lion Square.*—Decorated furniture, tapestries, &c.

[5786]

NORTH, BENJAMIN, *West Wycombe, Bucks.*—Specimens of fancy chairs.

[5787]

NOSOTTI, CHARLES, 398 *and* 399 *Oxford Street.*—Looking-glasses, gilt furniture, and decorations. (*See page* 30.)

[5788]

NUTCHEY, JAMES, D.M., 5 *West Street, Soho.*—Reading-table, supported on columns canted by turning-lathe.

[5789]

OGDEN, HENRY, *Manchester.*—Sideboard in oak, Renaissance style; and drawing-room settee and chairs in walnut-wood.

Pedestal sideboard, in the Renaissance style, 12 feet wide, and 13 feet high, constructed of Dantzic and English oak, by Henry Ogden, of Manchester, as a specimen of design and workmanship by the hands and artists regularly employed in his establishment.

Drawing-room settee, lounge chair, companion chair, and single chair, in walnut-wood, of the Louis XVI. style. The carving is elaborately executed, and is entirely hand work. The upholstering is of a novel description, and the covering of green English silk.

NOSOTTI, CHARLES, 398 *and* 399 *Oxford Street.*—Looking-glasses, gilt furniture, and decorations.

An enriched gilt cabinet, with panels containing cameo drawings.

A looking-glass over ditto, in three compartments, with best French plate glass, containing cameo drawings, fitted with side branches for gas; the whole finished in pure gold.

A panel decoration, in peach colour and white and gold, with medallion painting after Watteau.

A gilt window cornice, with medallion painting, and rich draperies of violet velvet and muslin.

A richly designed and gilt flower basket.

[5790]

PAGE, HARCOURT MASTER, 23 *Coventry Street.*—Gilt console-table and glass, girandole, cheval screen, &c. (*For Illustration, see page* 32.)

No. 1. Console-table.
No. 2. Console-glass.
No. 3. Girandole.
No. 4. Set of shelves for china.
No. 5. Combination dressing glass.
No. 6. Cheval screen, with painting on velvet, by Madame de Fleury.

No. 7. Pair of pole screens, with paintings on velvet, by Madame de Fleury.

[5791]

PALMER, HENRY, 7 *St. Michael's Place, Bath.*—Drawing tables.

[5792]

PARKER, JOHN, *Woodstock, Oxon.*—Buckhorn hall furniture.

This furniture consists of hat, coat, and umbrella stand; hall chairs, gong stands, and candelabra and chandeliers for any description of lights. They are made to suit both large and small houses, in various styles, and mounted either in light-coloured wood or fine dark oak.

The above articles may be purchased in suites, or separately.

The exhibitor manufactures also revolving trays for the centre of dinner or breakfast tables.

Lithographs and testimonials may be obtained on application.

[5794]

PASHLEY, JOHN, 19 *Red Lion Square, Holborn.*—Louis XVI. console-table and glass.

[5795]

PATERSON, T., 15 *Rupert Street, Haymarket.*—Carvings and cabinets.

[5796]

PERRY, W., 5 *North Audley Street.*—Specimens of wood carving.

[5797]

PHILLIPS, THADDEUS, 10 *Park Street, Bristol.*—Looking-glass.

[5798]

POOLE & MACGILLIVRAY, 24 & 25 *Princes Street, Cavendish Square.*—Jewel-stand and two chairs.

[5800]

RICHARDSON, THOMAS, 9 *Swift's Row, Carlisle.*—Devonport desk of old oak, taken from Carlisle Cathedral.

[5801]

RIVETT, WILLIAM & SAMUEL, 50 *Crown Street, Finsbury Square.*—A mahogany sideboard, with plate-glass back.

[5802]

ROGERS, GEORGE ALFRED, 21 *Soho Square.*—Wood carvings, brackets, frames, toilette glasses, &c.

[5803]

ROGERS, WILLIAM GIBBS, 21B *Soho Square, W.*—Wood carvings.

PAGE, HARCOURT MASTER, 23 *Coventry Street.*—Gilt console-table and glass, girandole, cheval screen, &c.

[5804]

RORKE, J., 75 *Oakley Street, Westminster Road, Lambeth.*—Projecting letters for shop fronts.

[5805]

ROSS & CO., *Ellis's Quay, Dublin.*—Portable furniture.

[5806]

ROWLEY, CHARLES, *Bond Street, Great Ancoats Street, Manchester.*—Patterns of picture-frames, and imitation ormolu frames.

[5807]

SANDERS, WILLIAM COOMBE, 59 *Queen Anne Street, London.*—Leather carving.

[5808]

SANDEMAN, R., *Edinburgh.*—Mirror tables, in walnut and Quebec ash.

[5809]

SANDFIELD, JOHN, 38 *Newman Street, Oxford Street.*—Imitation ormolu metal miniature frames.

[5812]

SCOTT, HENRY D., *Boston, Lincolnshire.*—Frame and birds from Nature, carved in wood.

Circular frame: Hops and Convolvulus, carved in lime-wood, the property of Joseph Wren, Esq., Boston.

Golden Plover, carved from Nature in lime-wood; price £10. Robin, carved from Nature in lime-wood; price £4.

[5813]

SCOTT, JAMES & THOMAS, 10 *George Street, Edinburgh.*—Cabinet in the style of Louis XVI.

[5814]

SCOWEN, THOMAS L., *Allen Road, Stoke Newington.*—Patent fin-expanding canopies for carriages, boats, gardens, couches, chairs.

MAKER TO HER MAJESTY AND THE LATE PRINCE CONSORT.

Canopies fitted to Carriages as above, from 3½ guineas.
Ditto „ Waggonette, „ 3½ ditto.
Ditto „ Bath Chairs, „ 31s. 6d.
Ditto „ Perambulators, Single, 18s.; Double, 20s.

These prices include fixing within ten miles of London.

Garden canopies as above, near 10 feet long by 6 feet wide, from £3 5s. to £5 10s. This canopy folds into one-third of its length, and is taken down in one minute. Canopies can be made of any material and to any shape required. They expand with great facility, and occupy but small space when folded. Their simple and ready mode of action, utility in protecting from sun or rain, and ornamental effect, together with the moderate prices, cannot fail to recommend them.

[5815]

SEDDON, THOMAS, *London.*—Articles of furniture executed from designs of Messrs. Prichard and Seddon, Architects.

A case of shelves for drawings and prints, suitable for a library, with desk—to serve also as a portfolio rest—of oak inlaid with various woods; the panels painted by Messrs. Morris, Marshall, & Co., illustrating the fine arts as follows:—*Architecture*, by King René and his Queen, considering the design of their house; *Sculpture*, by the same, carving figures upon the house; *Painting*, by the same, decorating its walls; *Music*, by the same, singing to celebrate its completion; and in the upper panels are representations of *ornamental carving, metal work, embroidery, weaving, stained glass,* and *mosaic work.*

An escritoire of oak and pitch-pine inlaid with various woods.

A chancel organ, manufactured by Messrs. Gray & Davison, with one clavier and four stops: three in a swell-box, and two octaves of pedals in an oak case, with illuminated pipes.

An arm-chair of walnut, with panels painted with the story of Pyramus and Thisbe.

A ditto, with perforated panels, fitted with cane-work.

An oak dining-room chair, with inlaid ornament.

Also, other specimens of furniture designed to combine character with economy, so as to be suitable for houses, parsonages, &c., of the mediæval style.

[5816]

SEDLEY, ANGELO, *Regent Street.*—Furniture, novel and various.

[5817]

SHARP, D. F., *Ingram Terrace, Sleaford.*—Carved wood bracket.

[5818]

SILVER, S. W., & CO., 4 *Bishopsgate Within.*—Portable furniture; camp, cabin, and household.

[5819]

SKIDMORE ART MANUFACTURES COMPANY, *Coventry.*—Furniture in mediæval style.

[5820]

SMEE, WILLIAM, & SONS, 6, *Finsbury Pavement, London.*—Examples of modern household furniture and bedding, of improved construction. (*See pages* 35, 36, 37, 38.)

[5821]

SOUTHGATE, JOHN, 76 *Watling Street, London.*—Improved camp, barrack, and military equipage, and cabin furniture.

[5822]

SPIERS & SON, 102 & 103 *High Street*, 45 & 46 *Corn Market Street, Oxford.*—Ornamental furniture for the writing table, boudoir, toilet, &c. (*See pages* 39, 40.)

[5823]

STANTON, THOMAS, 22 *Davies Street, Berkeley Square.*—Carved walnut and marqueterie door chandelier, sideboard, and cut-glass candelabra in silver and rock crystal.

[5824]

STATHER, JOHN, *Hull.*—Manufactured photographic oak paper-hangings, washable; granite column, imitated with machine painted paper.

SPECIMEN OF PHOTOGRAPHIC OAK PAPER-HANGINGS.—GRANITE COLUMN, IMITATED WITH MACHINE-PAINTED PAPER.—In design, these papers are true to Nature, which is their great recommendation, and which has secured for the oak especially an extraordinarily large sale in this and other countries. In colour they are fast and permanent, without the expense of varnishing; in price they are within the reach of all, and may be had of almost every respectable dealer in the United Kingdom.

SMEE, WILLIAM, & SONS, 6 *Finsbury Pavement, London.*—Examples of modern household furniture and bedding, of improved construction.

A SIDEBOARD OF LIGHT OAK, banded with and relieved by mouldings and carvings of dark oak, the back of silvered plate glass, in a richly moulded frame, mounted with ornamental carvings. Size—8 ft. 4 in. long, 9 ft. 6 in. high.

Smee, William, & Sons—*continued.*

The Smees' Spring Mattress, Tucker's Patent, or Somnier Tucker, is equally luxurious with, but firmer and affording more support than, the best description of French and German spring mattresses. It has, moreover, considerable advantages over all others in

Its Simplicity.—Every part of it is exposed to view, and it is of so simple construction that its application cannot be misunderstood.

Its Cleanliness.—By merely lifting off the upper mattress, each part of it may be easily dusted and brushed, or, if necessary, washed. It affords no harbour for vermin, nor is there any canvas, hair, or other material in which moth can collect.

Its Portability.—When folded, it forms a package, as shown below, 6 ft. 6 in. long, and only 8 in. square.

It has no screws or fastenings, so that any one may fit it, or unfix it, in a few moments.

Its Cheapness.

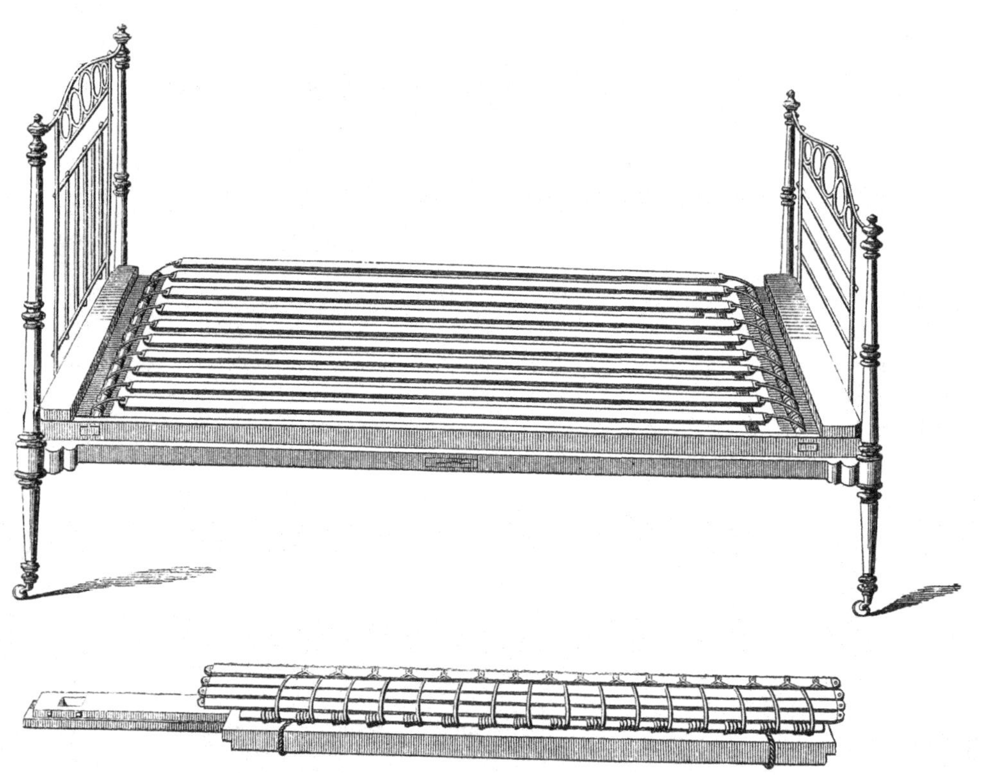

THE PRICES ARE AS UNDER :—

Size		Width	£	s.	d.
Size, No. 1. ...	for Bedstead	3ft. 0in. wide ...	£1	5	0
" " 2. ...	"	3ft. 6in. " ...	1	12	6
" " 3. ...	"	4ft. 0in. " ...	1	17	6
" " 4. ...	"	4ft. 6in. " ...	2	2	6
Size, No. 5. ...	for Bedstead	5ft. 0in. wide ...	£2	5	0
" " 5½....	"	5ft. 3in. " }	2	10	0
" " 6. ...	"	5ft. 6in. " }			
" " 7. ...	"	6ft. 0in. " ...	3	0	0

SMEE, WILLIAM, & SONS—*continued.*

A TOILET TABLE, OF FINE ITALIAN WALNUT-WOOD, supported by carved ornamental columns, upon a shaped plinth; the toilet glass suspended between carved columns.

Size—4 ft. 6 in. long, 5 ft. 9 in. high.

A LADIES' WARDROBE, OF RICHLY-FIGURED BIRCH-WOOD, banded and relieved with tulip-wood, and mounted with finely-carved ornamental work; made in three compartments: the centre one fitted with drawers and tray shelves, the wings for hanging dresses; the panel of the centre door of silvered plate glass.

Size—8 ft. long, 8 ft. 4in. high.

A TOILET TABLE, OF BIRCH-WOOD, relieved with tulip-wood, supported by pedestals of drawers, inclosed by panelled doors; the toilet glass in a carved ornamental frame, supended between two nests of jewel trays, and inclosed by doors.

Size—5 ft. long, 6 ft. high.

SMEE, WILLIAM, & SONS—*continued.*

A CHIFFONIER, OF FINE ITALIAN WALNUT-WOOD, relieved with tulip-wood; the lower part inclosed by four doors, having panels of fret-work of elaborate design, the back of plate glass; the mouldings and ornaments throughout richly carved.

Size—6 ft. 4 in. long, 10 ft. high.

[5825]

STEEVENS, JOHN, 64 *East Street, Taunton.*—Carved mahogany sideboard, representing hunting, dead game, fish, fruit, &c.

THE TAUNTON SIDEBOARD, designed and manufactured by Mr. John Steevens, is an elegant piece of furniture, ten feet long and twelve feet high, representing, in artistically carved panels, hunting the otter; the wild duck disturbed; vintage; harvest; fish; and dead game. Arranged in different brackets are goats' heads, &c. In the centre of the back is a frame containing plate glass, around which ivy is entwined; and above the frame is an ornamental cornice, with a shield and festoons of fruit surrounding the head of Bacchus. The frame under the top is fitted with drawers, richly carved with medallions, and supported by four figures, the heads of which are encircled with different devices of wheat, barley, hops, grapes, &c. The pedestals are conveniently fitted up with trays for plate, and cellarets for wine, &c., calculated for convenience as well as ornament.

SPIERS & SON, 102 & 103 *High Street, Oxford.*—Ornamental furniture for the writing-table, boudoir, toilet, &c.

[*Obtained Honourable Mention at the Great Exhibition of* 1851, *and at the Paris Exposition of* 1855; *and the Prize Medal at New York*, 1853.]

SPIERS AND SON'S CARVED OAK DISPATCH BOX.

Dispatch box in carved pollard oak. The mouldings and gilt ornaments adapted from the new Museum of the University of Oxford.

Set of writing table furniture in English pollard oak, with richly gilt mediæval mounts, comprising casket for paper and envelopes, inkstand, pen-tray, blotting-book, book-slide, letter-weigher, candlesticks, taper stick, casket for letters, date indicator, paper-weight, match-box, stamp-box, and paper-knife, all *en suite.*

Stationery case of Coromandel wood, with richly gilt Elizabethan mounts.

Tazza of Worcester honeycomb porcelain, on an ormolu stand of entwined serpents.

Set of writing table furniture, in richly gilt ormolu, inlaid with porcelain medallions enamelled by the painters of the Queen's new dessert service, and in the same style, comprising seventeen articles, *en suite.*

Lady's dressing-case in dressed shagreen, with gilt mounts. The interior lined with polished shagreen and Genoa silk velvet; containing diamond cut-glass fittings, with silver mounts, gilt and enamelled in the Moresque style, with gold initial plates; cutlery and other instruments in silver gilt, to correspond; brushes in polished shagreen handles; mirror in gilt and enamelled frame, on a new plan; and drawer fitted for jewellery, with secret divisions for gold, notes, &c.

Tazza in ormolu, of Etruscan form and ornament, inlaid with a porcelain plaque, enamelled in the Limoges style.

Silver-gilt model of an Oxford outrigger sculling-boat, on marble stand, serving as a pen-tray, and suitable for a regatta prize.

Cigar cases, similar to some made for H.R.H. the Prince of Wales; others in velvet, shagreen, &c.

SPIERS & SON, 102 & 103 *High Street*, 45 & 46 *Corn Market Street*, *Oxford—continued.* "Oxford Cyclopean" Washstands, combining the largest capacity with the smallest requirement of space.

SPIERS AND SON'S "OXFORD CYCLOPEANS."

One in polished white wood, with an eighteen and a half-inch basin, ewer holding four quarts, sponge basin, soap box, and brush tray, of best white glazed ware, price 40s.

One ditto, ditto, larger size, with a twenty-one inch basin, ewer holding six quarts, &c., with ware in coloured band and line, or printed flowing blue pattern, price 55s.

Other qualities in superior white wood, mahogany, walnut, &c., with ware of various patterns, some with plug basin and slop jar.

One with rails for towels.

One artistically carved in Elizabethan style in walnut wood, white and gold ware, price £20.

[5826]

STEVENS, GEORGE HENRY, 56 *Great Queen Street, London, W.C.*—Glass mosaic jardinières, table tops, &c. &c.

[5828]

STRAHAN, R. & CO., *Dublin.*—Carved cabinet.

[5829]

STRAPPS, MARSHALL GEORGE, *Wisbeach, Cambridgeshire.*—Carved oak chair in *alto-relievo*, by a self-taught artist.

[5830]

STRONG, WILLIAM, 137½ *New Bond Street.*—Glass frame, richly carved in walnut; a clock-case for mantelshelf.

[5831]

SUTER, ALFRED, 65 *Fenchurch Street.*—New castor for furniture, with specimens on what-not.

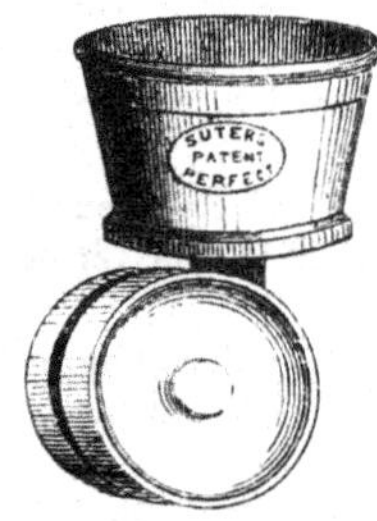

The superiority of these castors over those now used consists in the wheels being so placed that the weight of the article they have to support is very nearly central, thereby preventing all strain or leverage upon the pin; and the friction (except the floor friction) being upon a small section of the centre piece or horn, allows the wheels to rotate with perfect truth and freedom, and within a space little in excess of their diameter.

These castors have been tested, and found to perfectly realise the long-desired requisites of a castor—viz., easy and true action, simple and durable construction, and elegant appearance. They are manufactured either with a case to inclose the wheels, or without, and with sockets, plates, or screws, to suit all descriptions of furniture, musical instruments, &c.

To be had of the patentee, and of cabinet ironmongers throughout the United Kingdom.

[5832]

TAYLOR, HENRY JOSHUA, *Daisy Hill, Dewsbury.*—Drawing-room table, painted in imitation of inlaid woods.

[5833]

TAYLOR & SONS, 167 *Great Dover Street, S.E.*—Expanding dining tables and seats for ships' use.

These patent dining tables, when closed, are no larger than the ordinary tables, but possess the following great advantages:—

1. They are fixed in the centre of the cabin, instead of the sides, and have seats attached to them, with revolving backs.

2. These tables will extend fore and aft into a single range of tables the whole length of cabin, and, by the patented improvement, will also extend sideways into two distinct ranges of tables, of equal length, and always with the seats attached, thus giving a double amount of accommodation of the ordinary tables when in use, and a larger clear cabin space when closed.

3. The tables, with the seats attached, are secured to the cabin deck on sliding frames. By this means the seats always move with the tables. Any extent of table accommodation can be quickly and safely obtained, and being self-fastening, under any change of position they cannot get adrift.

The patent tables and seats are made suitable for every class of ship, steamer, or yacht.

[5833 *a*] An Arabian Bedstead of Australian cypress pine, imported by Captain Denham, R.N., from Queensland; manufactured by Taylor and Sons, 167 Great Dover Street, S.E.

[5834]

TAYLOR, JOHN, & SON, 109 *Princes Street, Edinburgh.*—Sideboard, cabinet, and sarcophagus in walnut wood.

[5835]

THOMAS, JOHN, 32 *Alpha Road.*—Inlay cabinet figure, portion of drawing-room.

[5836]

THURSTON & CO., *Catherine Street, Strand, London.*—Billiard-table carved in oak, by J. O'Shea, from designs by J. M. Allen, Esq. Model of patent combination billiard-table. (*See pages* 42 *and* 43).

[5837]

TOLLNER, J., *John Street, Tottenham Court Road.*—Inlaid wood table.

[5838]

TOMLINSON, W., *Hulme, Manchester.*—Specimens of medical shop fittings, furniture, and decorations.

[5839]

TOMS & LUSCOMBE, 103 *New Bond Street, W.*—Buhl cabinets and tables.

[5840]

TRAPNELL, C. & W., 2 *St. James's, Barton, Bristol.*—Sideboard in Riga oak, 10ft. high. Ebonised candelabrum, 11ft. high. Cabinet in walnut-wood.

[5841]

TROLLOPE, GEORGE, & SONS, 15 *Parliament Street.*—Carved chimney-piece, decorations, and cabinet furniture.

[5842]

TUCKER, JOHN, 2 *and* 3 *North Street, Finsbury.*—Fancy writing-table, with escritoire and cylinder front, in mahogany.

THURSTON & CO., *Catherine Street, Strand.*—Billiard table.

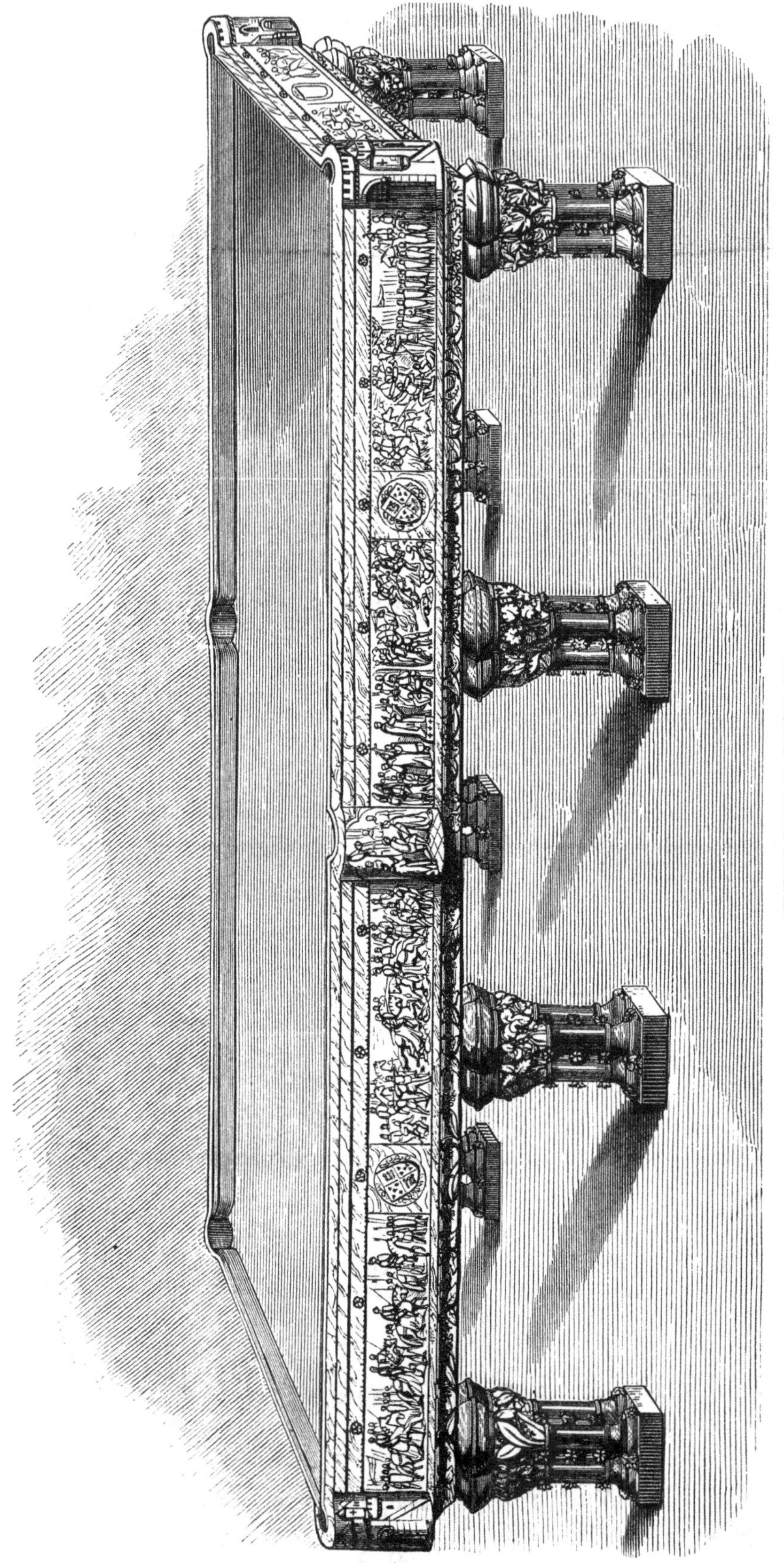

BILLIARD TABLE IN CARVED OAK.

THURSTON & Co., *Catherine Street, Strand, London.*—Billiard table carved in oak, by J. O'Shea, from designs by J. M. Allen, Esq. Model of patent combination billiard table.

Obtained the Prize Medal in Class XXVI. in 1851.

No. 1. An oak billiard table, style of the fifteenth century; the panels of the sides and ends carved in low relief, illustrating the history of the Wars of the Roses—supported on eight legs, each composed of four clustered columns, with richly foliated caps, having a central crocketed shaft, with carved spurs on square moulded base.

DESCRIPTION OF PANELS.—Departure of the Duke of York from Ludlow Castle, 1455. Battle of St. Albans, death of the Duke of Somerset, and reconciliation of the Duke of York and the Queen. Battle of Blore Heath. The Earl of Salisbury leaving Middleham Castle. (From necessity, these two subjects are chronologically reversed.) Desertion of Sir Andrew Trollope and veterans from the fortified camp at Ludlow, 1459. Somerset repulsed at Calais. Warwick's triumphant entry into London. Battle of Northampton. Desertion of Lord Grey de Ruthyn. Battle of Wakefield Green, and death of the Duke of York. Death of the Duke's son on Wakefield Bridge. Great battle at Towton, 1461, in the midst of a terrible snow storm: the Lancastrians lost 28,000 men. Battle of Mortimer Cross. Battle of Barnet, and death of the Earl of Warwick. Battle of Tewkesbury, defeat of Margaret, and death of Edward, Prince of Wales.

Marking board and cue rack *en suite.* (*See page* 42.)

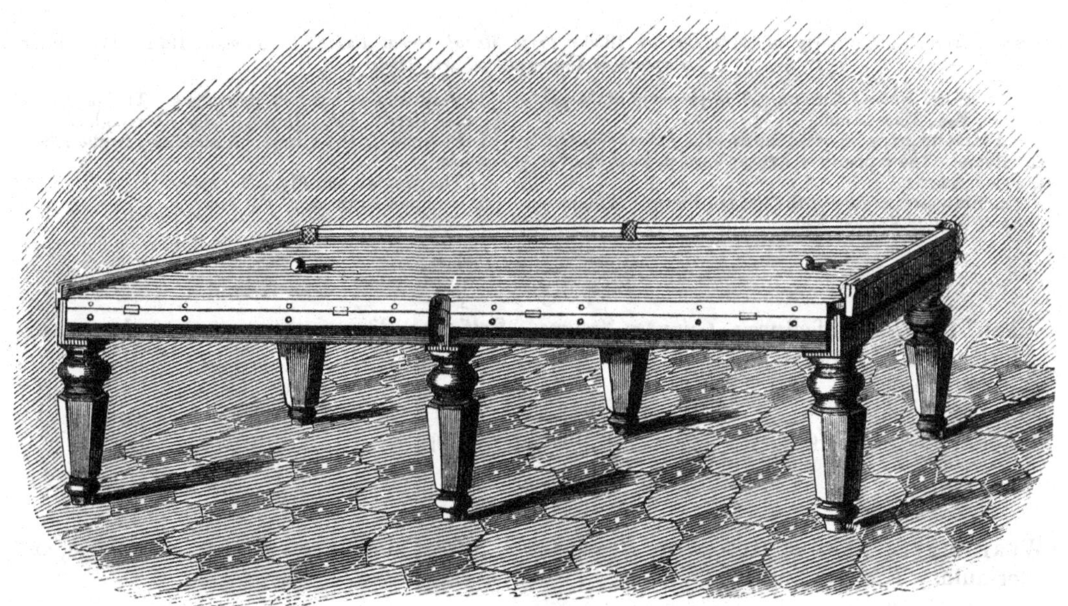

BILLIARD TABLE.

No. 2. Model of patent combination billiard table, easily convertible into a dining or supper table, the cushions being hinged, and made to turn down.

Also a complete end of a billiard table, made of fine pollard oak.

[5843]

TUDSBURY, R. J., *Edwinstowe, Ollerton, Notts.*—Carvings from Nature, in lime-wood: dead game, flowers, foliage, &c.

[5844]

TURLEY, RICHARD, 381 *Summer Lane, and* 1 & 2 *Hospital Street, Birmingham.*—Japanned and papier-mâché articles.

[5845]

TUTILL, GEORGE, 83 *City Road, London.*—Patent india-rubber preparation for banners, flags, &c., to prevent cracking, &c. (*See page* 45.)

[5846]

TWEEDY, THOMAS HALL, 44 *Grainger Street, Newcastle-on-Tyne.*—Sideboard, illustrated from Shakespeare; ditto, illustrated from "Robinson Crusoe."

[5847]

VOKINS, JOHN & WILLIAM, 14 & 16 *Great Portland Street.*—Portfolio frames; ormolu frames; imitation ormolu frames; frames for pictures and drawings.

[5848]

WALKER, JOHN, 3 *Kensington Place, Silver Street, Notting Hill.*—Carving in lime-tree—Spring.

[5849]

WALLIS, THOMAS WILKINSON, *Louth, Lincolnshire.*—Wood-carving of dead game and wild herbage, from Nature.

Large Silver Medal, Society of Arts, London, 1850; *Prize Medal, Great Exhibition, London,* 1851; *Prize Medal Exposition Universelle, Paris,* 1855.

1. A Rustic Nook—Autumn. The property of James Carlton, Esq., Manchester.

2. Statuette group of dead birds—the Snipe, Wagtail, and Robin. The property of Arthur Vardon, Esq., Worth, Crawley, Sussex.

2 A. Statuette carving of the Golden Plover. The property of Arthur Vardon, Esq.

3. Dead Game—brace of partridges. The property of Marshall Carritt, Esq., Shirley Lodge, Birmingham.

4. Promise for the Future. A medallion in alto-relievo. For sale.

5. The Wagtail and Fly. For sale.

6. Group of Fruit, &c. For sale.

[5850]

WARD, JOHN, Manufacturer, *Leicester Square.*—Self-propelling and patent recumbent chairs for the use of invalids.

[5851]

WEBB, JOHN, 22, *Cork Street, W.*—An inlaid table top.

[5852]

WERTHEIMER, SAMSON, 154 *New Bond Street, W.*—Cabinets, etageres, council table, and ormulu. (*See page* 46.)

[5853]

WESTRUP, CHARLES, 83 *Old Street Road, E.C.*—Six fancy willow seat, walnut, and sycamore occasional chairs. (*See page* 47.)

[5854]

WHITE, J., *Shrewsbury.*—Gilt fire screen.

[5855]

WHYTOCK, R. & CO., 9, *George Street, Edinburgh.*—Pollard oak sideboard.

[5856]

WILKIE, JOHN, 1 *Addington Place, Lambeth.*—Figure of our Lord, in carved canopied niche, in oak.

TUTILL, GEORGE, 83 *City Road*, *London.*—Patent india-rubber preparation for silk banners, flags, &c., to prevent cracking, &c.

GROUP OF FLAGS.

The exhibitor is the sole manufacturer of the patent india-rubber silk flags, banners, &c., for the use of the Army, Navy, and Volunteers, Ecclesiastical and Friendly Societies. These banners are remarkably soft and pliable. GEORGE TUTILL makes to order sashes, aprons, medals, regalias, &c.

Wertheimer, Samson, 154 *New Bond Street, W.*—Cabinets, etagères, council table, and ormolu.

WESTRUP, CHARLES, 83 *Old Street Road, E.C.*—Six fancy willow seat, walnut, and sycamore occasional chairs.

1. Walnut wood fancy occasional chair, with willow seat, carved back, feet, and rails to correspond; centre of top inlaid with buhl work; centre of stay, a shield for crest or initials, ornamented with gold and colours.

2. Walnut ditto ditto ditto, carved back, rails, and feet.

3. Walnut ditto ditto ditto, carved back, rails, and feet, ornamented with gold and green colour.

4. White sycamore fancy occasional chair, with willow seat, carved, with Prince Albert's portrait let in top, ornamented with gold and blue colour.

5. Sycamore ditto ditto, carved back, rails, and feet.

6. Sycamore ditto ditto, turned feet and rails, with carving.

WALNUT WOOD FANCY CHAIR.

[5857]

WILKINSON, C., & SON, 8 *Old Bond Street.*—Various specimens of decorative furniture. (*For Illustration, see page* 48.)

1. A drawing-room cabinet, with large glass and frame above, of fine walnut wood, with all the ornaments either in relief or inlaid, of white holly, transparent glass in doors below, and the top a fine specimen of jaune Fleuri marble; the carvings emblematic of the Four Seasons.

2. A washstand, on pedestals of Savannah pitch pine, with mouldings and ornaments in relief of purple wood, with porcelain tile top, made in one piece, and decorated border *en suite*, with the chamber ware.

3. A lady's toilet table, and glass *en suite* with the above.

4. A walnut drawing-room chair.

5. A gilt occasional ditto, Louis XVI.

6. A library chair, stuffed back and seat, in morocco.

7. An oak sideboard, with enrichments of Potts's patent electro bronzes.

[5858]

WILSON, J. W., & CO., 18 *Wigmore Street, W.*—A patent brass Arabian bedstead; a walnut drawing-room bagatelle and chess table; a spring stuffed couch; an easy chair.

[5859]

WINFIELD, WILLIAM, 22 *Upper Charlton Street, Fitzroy Square.*—Flowers carved in walnut.

[5860]

WOOLVERTON, CHARLES, *South Quay, Great Yarmouth.*—Improved window blind for drawing-rooms, offices, &c.

[5861]

WRIGHT, WILLIAM, 27 *Smith Street, St. George's, Birmingham.*—A pearl table and chimney glasses, with pearl frames, &c.

[5862]

WRIGHT & MANSFIELD, 3 *Great Portland Street, W.*—Piano, and decorative furniture.

WILKINSON, C., & SON, 8 *Old Bond Street.*—Various specimens of decorative furniture.

[5863]

FOXLEY, G., *Welwyn, Herts.*—Carved wood spoons, forks, &c.

[5864]

HALSTEAD & SON, *Chichester.*—Grille, for choir of Chichester Cathedral.

[5865]

HOPE, BERESFORD.—Model of a monument to the Viscountess Beresford.

[5866]

OWEN, J., *Sheffield.*—Iron bedstead, with patent mattress.

[5867]

RICHARDSON, —, *Stamford.*—Carved chairs.

SUB-CLASS B.—*Paper Hanging and General Decoration.*

[5877]

ARTHUR, THOMAS, 3 *Sackville Street, London, W.*—Ornamented panel and pilaster drawing-room decorations.

[5878]

BLACKMORE, CHARLES, 372 *Euston Road, N. W.*—Specimens of illuminated lettering and mediæval decorations.

[5879]

BUCHAN, & SON, *Southampton.*—Artistic decoration for drawing-room, shewing effect of one side of room.

[5880]

CARLISLE AND CLEGG, 81 *Queen Street, Cheapside, and* 31 *Essex Street, Islington.*—Decorative paper hangings.

[5881]

COOKE, WILLIAM, *Grove Works, Leeds.*—Paper hangings.

[5882]

COTTERELL BROTHERS, *Bristol and Bath.*—Specimens of panel decorations and paper hangings.

[5883]

COULTON, ISAAC LOVE, 7 *Robert Street, Hampstead Road, N. W.*—Allegorical arabesque decoration, in the style of Louis XVI.

[5884]

DOW, ROBERT, *Painter*, 59 *George Street, Perth.*—Imitations of the finer woods, and imitations of mouldings.

[5885]

EARLE, JAMES HOWARD, 28 *Howland Street, Fitzroy Square.*—Drawing-room decoration.

[5886]

GIRARDET, F., 4 *Charles Street, Manchester Square.*—Specimens of graining and marbling for house decorations.

[5887]

GODDARD, WILLIAM EDWARD, *Hull, Yorkshire.*—Pyrography, or carving upon charred wood, as adapted to ornamental furniture, &c.

[5888]

GRANT, W. H., 81 *King Street, Camden Town.*—Imitations of woods and marbles.

[5889]

GREEN & KING, Decorators, 23 *Baker Street, W.*—Painted washable wall decorations of moderate cost.

SPECIMENS OF WALL DECORATION OF MODERATE COST.—This material consists of paper prepared in a peculiar manner, and finished in oil paint. It is exceedingly desirable, and can be washed; a single coat of paint will at any time render it as good as new. When carried out in one colour, it is particularly suited for a background to paintings, &c., as it then combines the most delicate neutral colouring, with great richness of texture.

[5890]

GRIFFIN, J., 7 *Nauslin Street, East India Road, Poplar.*—Imitation of woods.

[5891]

HASWELL, D. O., 49 *Greek Street, Soho.*—Specimens of writing for signs and tablets.

[5892]

HAWTHORNE, JAMES, 98 *St. John Street, Clerkenwell.*—Writing inks of various kinds.

[5893]

HAYWARD & SON, 88 *Newgate Street, London.*—Church and domestic decorations.

[5894]

HEYWOOD, HIGGINBOTTOM, SMITH, & CO., *Manchester, London, and Glasgow.*—Machine-made paper hangings.

[5895]

HODSON, GEORGE GOODWIN, 92 *Drummond Street, Euston Square, N. W.*—Ecclesiastical decorations and writings.

[5896]

HORNE, ROBERT, 41 *Gracechurch Street.*—Block-printed paper hangings.

[5897]

HUMMERSTON BROTHERS, *Leeds, Yorkshire.*—Painted imitations of woods and marbles.

[5898]

HUNT, CHARLES, 40 *Spring Street, Paddington.*—Imitations of woods and marbles; an inlaid table top.

[5899]

HURWITZ, BENJAMIN, 9 *Southampton Street, Strand, W.C.*—Specimens of interior decorations, furniture, &c.

[5900]

JEFFREY & CO., 115 *Whitechapel, and* 500 *Oxford Street, London.*—Specimens of paper hangings.

[5901]

JONES & CO., *Arlington Street, New North Road, Islington.*—Paper hangings.

[5902]

KENSETT, JOHN, 18 *Southampton Street, Strand.*—Imitations of wood and marble.

[5903]

KERSHAW, THOMAS, House Painter, 38 *Baker Street, Portman Square.*—Painted house decorations for walls and wood work.

[5904]

LAINSON, GEORGE, 1 *Henry Place, Clapham.*—Specimens of wall decorations; pilasters painted on satin.

[5905]

LAMAMY, AUGUSTE, 3 *Percy Street, Bedford Square.*—Tableaux in marqueterie and wood mosaic.

[5906]

LEA, CHARLES JAMES, *High Street, Lutterworth.*—Decoration for wall of a church in Lancashire.

[5907]

M'LACHLAN, JAMES, 35 *St. James's Street, Piccadilly, S.W., and Clapham, S.*—Artistic drawing and dining-room decorations, and imitations of woods and marbles.

[5908]

MASLIN, W., & CO., 32 *Foley Street, W.*—Imitations of British and foreign marbles and serpentines on paper.

Their marble papers comprise designs in imitation of every description of British and foreign marbles, serpentines, &c.; their truthful character and colouring being an especial feature.

[5909]

MORANT, BOYD, & MORANT, 91 *New Bond Street, London.*—Decoration and articles of furniture.

[5910]

NAYLOR, WILLIAM, *No.* 4A *James Street, Oxford Street.*—Patterns of staining deal to imitate all kinds of woods; flooring of rooms, pannelling, and other decorative purposes; patterns of enamel painting, &c.

[5911]

NEWMAN & LAWDER, Decorators, 20 *Poland Street, W.*—Designs and specimens of ecclesiastical and domestic decorations.

[5912]

OWEN, A. I., 249 *Oxford Street, London.*—Interior decorations and furniture, &c.

[5913]

PEARSE, JOSEPH SALTER, 8 *Barnsbury Street, Islington.*—White and coloured enamel centre, as applied to walls or cabinet, &c.

[5914]

PITMAN, WILLIAM, 210 *Euston Road, Euston Square, London.*—Mediæval paintings and designs.

[5915]

PURDIE, BONNAR, & CARFRAE, 77 *George Street, Edinburgh.*—Decoration for a drawing-room in French style.

[5916]

PURDIE, COWTAN, & Co. (late Duppas), 314 *Oxford Street.*—Dining-room decorative imitation woods, and paintings in water glass; boudoir decoration and curtains.

1. Ceiling painted in encaustic. In the centre an Aurora, and in the surrounding panels medallions and ornaments emblematic of the Seasons, &c.

2. Decoration for a dining-room, showing on one side of the room the panelling and framework, painted in imitation of marbles, ebony, walnut, and tulip woods. The three pictures, copied from the originals in Hampton Court, are painted—that in the centre panel over the chimney in oil, the two in the side panels in fresco, in the water-glass or sterochromic method. Style, Renaissance.

3. Cabinet in carved wood, finished in enamel, white and gold, with figure of Plenty in centre, and arabesques in surrounding panels. Style, Louis XVI.

4. Pilasters to match cabinet.

5. Embroidered window-curtain; cornice in same style.

6 Richly carved wardrobe in maple wood, inlaid with purple and tulip woods.

[5917]

READ, WILLIAM, 153 *Marylebone Road.*—Imitations of woods and marbles.

[5918]

RODGERS, JOHN & JOSEPH, *Sheffield.*—Painted wall and wood-work decorations.

[5919]

SCHISCHKAR, EDWARD, *Leightcliffe, Yorkshire.*—"Marmography," produced by chemical means on glass, and other transparent articles, for decorative purposes.

[5920]

SCOTT, CUTHBERTSON & Co., *Whitelands, Chelsea.*—Specimen of block-printed paper-hangings.—*For Illustrations, see pages* 54, 55.

Specimen of Italian decoration. This is an entirely new process with block-printing, the design being considerably "raised" and printed in dead gold, the etching in bright gold, producing an effect perfectly unique. This decoration can be printed on a white or any delicate tinted ground suitable for drawing-rooms, saloons, &c. &c. The accompanying plate represents this design.

Specimen of Italian decoration by block-printing, suitable for drawing-rooms, saloons, &c.

Specimen of decorative paper hangings by block printing in the mediæval, Gothic, and Indian styles, &c. &c.

Specimen of block-printed paper hangings in "relievo," adapted for drawing-rooms, dining-rooms, libraries, halls, and staircases, painted either in oil or distemper colouring.

[5921]

SIBTHORPE, H., & SON, *Dublin.*—Specimens of internal decorations, in three styles.

[5922]

SIMPSON, W. B., & SONS, 456 *West Strand.*—Painted wall decoration.

[5923]

SMITH, CHARLES, 43 *Upper Baker Street.*—Basso-relievo decorations, with imitations of inlaid marbles, &c.; all in paint.

[5924]

SMITH, GEORGE THOMAS, 1 *Wenlock Road, City Road, London.*—Ornamental wood-work, printed by agency of heat.

[5925]

SOUTHALL, CHARLES, & Co., 157 *Kingsland Road.*—Grained woods and marble on paper.

[5926]

SPRECKLEY, JAMES, *Castle Gate, Newark-on-Trent, Notts.*—Specimens of painting and graining.

[5927]

STENITZ, CHARLES, *London Parquetry Works, Grove Lane, Camberwell, S.*—Parquetry floors, patent veneered wall panellings and ceilings.

[5928]

STEPHENS, HENRY, 18 *St. Martin's-le-Grand, London.*—Specimens of wood, stained, as a substitute for paint.

SUBSTITUTE FOR PAINT.—Stephens' stains for wood imitate oak, mahogany, rosewood, wainscot, walnut, and satinwood. These preparations exhibit to great advantage the beautiful variations of the natural grain, so that the cheapest descriptions of wood, when stained, sized, and varnished, far surpass paint in beauty of appearance; while the rapidity of the process, its great economy, and the absence of every disagreeable and unwholesome smell, always gain the approbation of those who use them. The durability of these stains, when used for interior decoration, is at least three times as much as that of paint, suffering but little even after an interval of fifteen years.

Mr. Stephens is kindly permitted to publish the following extract from a letter, addressed by the Rev. R. H. Chichester, of Chittlehampton, near South Morton:—"The effect produced by the staining fluid and varnish has given such entire satisfaction, that the parishioners have requested me to procure five times the quantity now paid for, in order to finish the church."

For the roof timbers in churches, and for boarded ceilings, boiled oil may be used as a varnish with very good effect, and will considerably diminish the expense. The more carefully wood is selected and prepared, the more beautiful the appearance; but in cases where very little care has been taken in the selection of the wood, it nevertheless surpasses cheap painting. The stains drying almost immediately, the work can be sized and varnished shortly after; one process has not to wait for the other, as in painting, so that the whole interior of a house which would take six weeks to paint may be (if stained) finished in one week.

In the fitting up of churches, chapels, halls, &c., where economy of expenditure is important, they are of great advantage, and are largely used.

The exterior woodwork of the Exhibition of 1851 was stained by Mr. Stephens with his stains.

The liquid stains are sold at 8s. per gallon, and sold also in powder, at 8s. per lb. (which makes one gallon of liquid stain).

These dye powders will be found most convenient articles for use. One pound makes one gallon of liquid stain of the deepest tint, which will *cover more than* 100 *square yards.* One gallon of liquid colour is thus condensed into one pound, a most material saving in the cost and convenience of carriage.

Prospectuses and small pieces of stained deal (as specimens) will be sent free by post, on application.

[5929]

TAYLOR, JOHN, 5 *Compton Street.*—Imitations of woods and marbles.

[5930]

TURNER & OWST, *Elizabeth Street, Pimlico.*—Paper-hangings, with frieze and pilaster, representing the Four Seasons (block-printed).

Paper-hangings entirely block-printed; a frieze running at the top, representing Spring, Summer, Autumn, and Winter, supported by pilasters, at the base of which is the Globe, with Products of the Seasons represented by the rose, the grape, the wheat, and the ivy, etched with bronze. On the side-screens, specimens of various styles of paper-hangings are exhibited.

[5931]

WARNE, STANNARD, 4 *Bruton Street, Berkeley Square.*—Furniture decorations and paper hangings.

[5932]

WARNE, EDMUND, & SON, 31 *Soho Square, W.*—Decorations and fittings.

Scott, Cuthbertson, & Co., Paper Stainers, *Whitelands, Chelsea.*—Specimen of block-printed paper hangings.

Scott, Cuthbertson, & Co.—*continued.*

[5933]

White & Parlby, 49 *and* 50 *Great Marylebone Street, London.*—Architectural decorations in relief; decorative furniture for gilding.

The above illustration shows the application of White & Parlby's new cement to the formation of curved and ornamental surfaces of buildings, by which are produced large curved ornamental forms of every description.

It is especially adapted for ornamental coves, gallery

fronts, domes, ceilings (flat, pendented, or domed), and for all large complicated and elaborate curved ornamental surfaces. The ribs, bays, and relief ornament of ceilings, domes, &c., are produced complete, and form, without plastering, a perfect covering, which is fixed at once to the rafters or other construction of the roof. The material is perfectly dry and durable.

It has been successfully applied in numerous private mansions, and in the following public buildings:—Wright's Bank, Britannia Theatre, and in the Oxford, Weston's, Middlesex, and Wilton's Music Halls.

White and Parlby further exhibit a ceiling in their new cement, a cabinet and frame console-table and frame, toilet glasses, girandoles, candelabrum, &c. (See opposite page.)

Louis XVI.'s drawing-room door and architrave.

WHITE & PARLBY.—*continued.*

DRAWING-ROOM DOOR AND ARCHITRAVE

[5934]

WILLIAMS, COOPERS, & Co., Manufacturers, 85 *West Smithfield, London.*—Wall decorations, in Italian and other styles.

[5935]

WILSHERE & RABBETH, *Great Western Road, Paddington.*—Varnishes, colours, and wood stains.

Extra pale body varnish, for whites, &c., 28s. per gallon.
Pale body ,, 26s., 24s., 22s. ,,
,, carriage ,, 18s., 16s., 14s. ,,
Super Japan ,, 18s. ,,
Pale Copal ,, 18s., 16s. ,,

Oak stain, 6s. per gallon.
Wainscote stain, 5s. ,,
Mahogany and walnut stain, 8s. ,,
Church varnish, 12s., 10s. ,,
Pale oak varnish, 12s., 10s. ,,

[5936]

WOOLLAMS, WM., & Co., 110 *High Street, near Manchester Square, W.*—Paper-hangings.

Obtained Silver Medal of the Society of Arts, 1849.

Greek decorations.
Louis XVI. do.
Gothic papers, designed by Mr. Robinson.
Paper hangings, designed by pupils of the Female School of Art, Queen Square.

Flower border, designed by Mr. J. Aumonier, late pupil of South Kensington Government School.

The specimens of green paper-hangings are NON-ARSENICAL.

[5937]

WOOLLAMS, JOHN, & Co., 69 *Marylebone Lane, London.*—Paper hangings and decorations by block-printing; machine paper hangings.

[5938]

ASPREY, C., 166 *New Bond Street.*—A Davenport.

[5939]

BURNS & LAMBERT, *Great Portland Street.*—Mediæval paper-hangings.

[5940]

OWEN, J., *Sheffield.*—Springs and attachments for mattress frames, bedsteads, sofas, &c.

[5941]

GUSHLOW, G., 34 *Newman Street, Oxford Street.*—Plaster casts and composition imitation of bronzes, &c.

PRINTED BY PETTER AND GALPIN, BELLE SAUVAGE WORKS, LUDGATE HILL, LONDON, E.C.

CLASS XXXI.

IRON AND GENERAL HARDWARE.

SUB-CLASS A.—*Iron Manufactures.*

[5968]

ABBOTT, WILLIAM, 3 *Royal Terrace, Richmond, S.W.*—Perforated cages for birds, &c.

[5969]

ADAMS, WILLIAM S., & SON, 57 *Haymarket*, 14 *Norris Street, and* 54 *Whitcomb Street, Pall Mall East, S.W.*—Improved cooking apparatus for large kitchens.

The exhibitors are manufacturers of first-class kitchen fittings and cooking apparatus, and hold the appointment of ironmongers to the Queen, and also to the principal London clubs.

THEY EXHIBIT THE FOLLOWING :—

1. THE LONDON ROASTING RANGE, of new construction, with cast-iron chimney-piece ; the back and sides composed of fire lumps ; one-half of the fire may be used without the other ; the bars are so constructed as to admit of being removed at any time without displacing the brickwork, and allowance is made for their expansion. This range gives out great heat with a small consumption of fuel, being very narrow from back to front.
2. AN IMPROVED SMOKE JACK, with double action outside movement, constructed to roast both horizontally and vertically ; with a newly invented lever apparatus for throwing the several movements in and out of gear.
3. A HOT PLATE AND BROILING STOVE, for stewing and grilling, having a large pastry oven at end, with improved sliding doors running upon gun-metal wheels. It has also a large hot closet under the oven, the whole being heated by one moderate fire at the broiling stove.
4. A STEAM BAIN-MARIE, and set of copper stewpans, soup-pots, &c. heated by steam, for keeping gravies, soups, sauces, &c. hot, without the slightest risk of their being burnt or spoilt.
5. A GAS STOVE for stewing, with newly invented burners, which produce a blue flame ; the gas being mixed with common air gives out great heat without smoke, and will not soil the cooking utensils placed over it.

An improved charcoal stove, formed of Stourbridge fire clay, in lieu of iron, as hitherto used, is placed behind the gas stove.

6. A LARGE HOT CLOSET, highly finished, with double-panelled door ; the shelves inside are made of wrought-iron, lap-welded by a new process, and will not crack, a fault to which the cast-iron shelves are always liable. They may be heated either by steam or hot water.

The above apparatus is suitable for a first-class kitchen, and is exhibited as a specimen of improved construction and superior workmanship.

[5970]

ADCOCK, RICHARD CASSWELL, 4 *Halkin Street West.*—Bolt for room or closet door, indicating engaged or disengaged.

[5971]

ADDIS, WILLIAM, 6, 7, *and* 15 *Leicester Street, Leicester Square.*—Cundy's patent brick oven kitchen range; cottager's cooking stove; pedestal stove.

CUNDY'S PATENT PNEUMATIC WARM-AIR VENTILATING STOVE, for warming churches, halls, staircases, and public buildings.

For this invention Mr. Cundy received the Society of Arts medal, and also a medal at the International Exhibition of 1851.

This is the only patented stove the inside of which is entirely constructed of fire-clay tiles.

CUNDY'S PATENT BRICK OVEN WINDING-CHEEK OPEN FIRE AND SEMI-CLOSE RANGE.

4 ft. range from £13 to	£16	0
A larger class of range from £24 to	30	0
Ditto, open fire, from £11 to	24	0

CUNDY'S PATENT ECONOMIC COTTAGER'S COOKING STOVE.

24 in.	£3	3
30 in.	4	4

These goods are manufactured and sold by William Addis, wholesale ironmonger and stove-grate manufacturer.

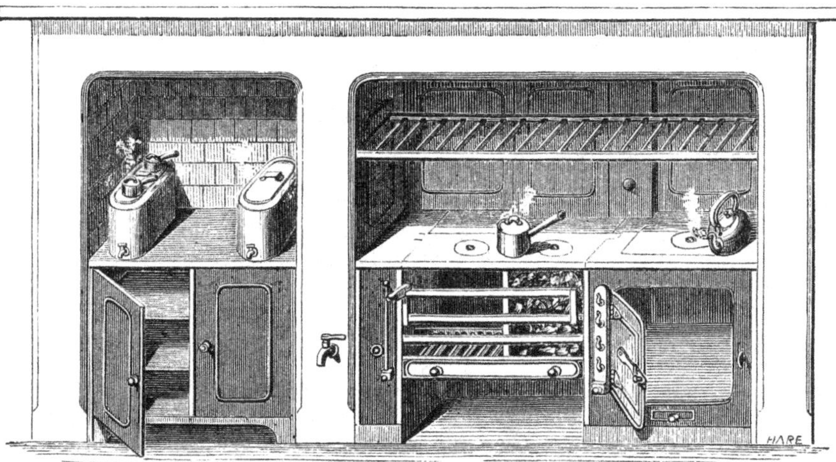

CUNDY'S PATENT BRICK OVEN WINDING-CHEEK SEMI-CLOSE RANGE.

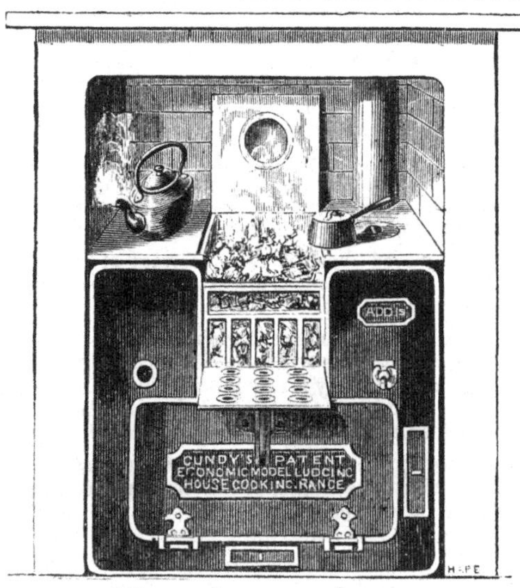

When used as an open-fire range.

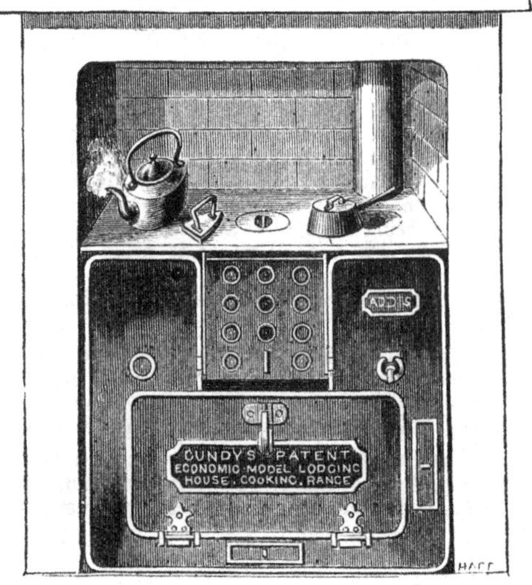

When used as a close range, for baking.

CUNDY'S PATENT ECONOMIC COTTAGER'S COOKING STOVE.

The ranges can be seen in operation every day at the exhibitor's ware rooms. A prospectus will be forwarded on application to the above address.

[5973]

ALLEN, THOMAS, *Clifton, and Hotwells, Bristol.*—Patent metallic tubular bedsteads for general use, military and portable.

PATENT METALLIC TUBULAR BEDSTEAD.

The advantages of the patent bedstead are:—

Its strength, being formed entirely of tubes and malleable iron.

Its durability and simplicity in construction. It can be put up and taken down in five minutes, without the use of tools of any kind.

It is guaranteed insect-proof.

Palliasses can be dispensed with; the sackin being equal to any spring mattress.

The prices vary from 30*s.* to £50.

TESTIMONIALS.

"*Chew Magna, June* 17, 1862.

"DEAR SIR,—I have seven of your patent tubular bedsteads, and have now had two years' experience of them, and I have much pleasure in adding my testimony to their general excellence, both as regards construction, material, and design. I have tried many other iron bedsteads, but I greatly prefer yours, being quite free from the common defect of working loose in the joints. The bedsteads I got from you are of various patterns, three are of what you term your "hospital pattern," and this is, from the moderate price, great strength, and very neat appearance for general purposes, in my opinion, the best bedstead out. I have also one of your camp bedsteads, to the excellence of which military men can better speak than I can, but I find it particularly useful to have in the house, as it is so very easily and quickly moved from one room to another, as occasion may require. Wishing you the success which I think you well deserve for having brought out a really good, useful article.

"I am, dear sir, yours faithfully,

"MR. THOMAS ALLEN, "CONWAY L. ROSE.

"*Hotwell Road Iron Works, Bristol.*"

From MAJOR BUSH, 100th Regiment.

"*Gibraltar, June* 9, 1862.

"SIR,—Having now tried your patent portable iron bedsteads in use for three years, whilst serving in the West Indies, Mediterranean, and Aldershot, I have great pleasure in stating that in my own opinion (and that of several military officers who have seen it), your bedstead far surpasses anything of the kind in present use, owing to the facility with which it is put together and small space it occupies when taken to pieces, also its strength, comfort, and neat appearance.

"I have the honor to be your most obedient servant,

"H. S. BUSH,

"MR. T. ALLEN." *Major,* 100*th Regiment.*

From W. BRUCE GINGELL, Esq. Architect, Bristol General Hospital.

"*Bristol, October* 26, 1856.

"DEAR SIR,—I beg to inform you that your patent tubular bedsteads were this day selected by the committee and faculty, as the best and most perfect submitted to them. I have no hesitation in stating that they must supersede the ordinary iron bedsteads, either for public institutions or for private houses.

"I am, dear sir, yours truly,

"MR. T. ALLEN." "W. BRUCE GINGELL.

[5974]

ASHWELL, JAMES CHARLES, 28 *Dorchester Street, Hoxton.*—Roasting jack without spring, the weight of meat the motive power.

[5975]

AVERY, W. & T., *Digbeth, Birmingham.*—Scales and weighing machines.

[5976]

BACKHOUSE, WILLIAM N., 46 *Westgate Street, Ipswich.*—The improved kitchen range; an economical ditto for cottages.

[5977]

BAILY, WILLIAM, & SONS, 71 *Gracechurch Street, E.C.*—Ornamental iron work, gates, staircase-work, stoves, &c.

[5978]

BAMBER, W. C., 12 *Little College Street, S.W.*—Mortise balance night bolt, and an improved night-latch.

[5979]

BARLOW, JAMES, 14 *King William Street, City.*—Patent cask tilt, no sediment disturbed; improved roasting-jack screen, &c.

[5980]

BARNARD, BISHOP, & BARNARDS, *Norwich.*—Park entrance gates in ornamental wrought-iron, designed by Thomas Jeckell.

This beautiful specimen of iron-work comprises a pair of principal gates and two side gates. The piers and lintels, with the griffins surmounting them, are of cast-iron. The tracery and panels of the gates and piers, with the foliage, &c. over the lintels, are of wrought-iron.

[5981]

BARRETT, R., & SON, *Beech Street, Barbican.*—Chimney-sweeping and drain machinery.

R. Barrett & Son are the only manufacturers of the patent chimney-sweeping and drain-cleaning machinery. The general adoption of these apparatus, and the satisfaction they have given, are sufficient guarantees of their excellence. The exhibitors execute all descriptions of metal castings, and are prepared to furnish estimates, upon application.

[5982]

BARTLETT, J., & SON, *Welsh Back, Bristol.*—Railway and road weighbridges and weighing machines for general purposes. (*See page* 5.)

[5983]

BARTON, JAMES, 370 *Oxford Street.*—Patent stable fittings and enamelled mangers; harness-room fittings.

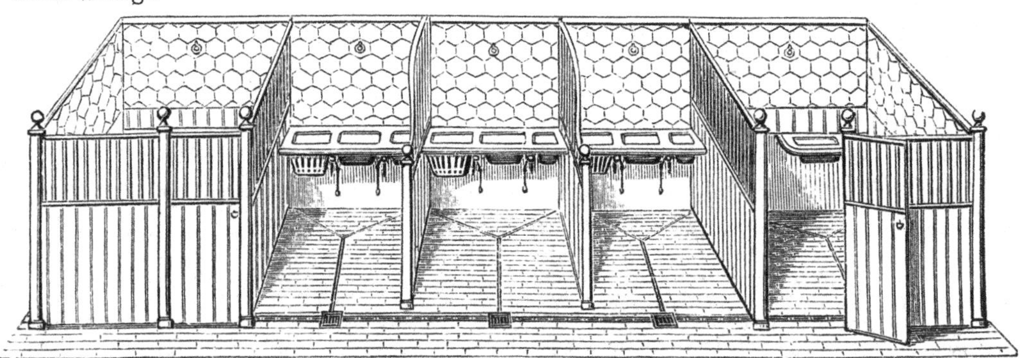

PATENT STABLE FITTINGS.

BARTON'S PATENT STABLE FITTINGS, AND ENAMELLED MANGERS.

These superior-class stable fittings are patronised by the principal nobility, and adopted by many of the first architects and builders of the United Kingdom.

The above arrangement consists of three stalls and two loose boxes, fitted with improved ventilating divisions, patent enamelled fittings, and all the latest improvements.

A stable, newly erected, with full-sized stalls and loose boxes, may be seen at the manufactory, where an extensive assortment of stable fittings upon the most modern and improved principles, together with a large collection of harness-room fittings, for single and double harness, ladies' and gentlemen's riding saddles, brackets, &c. are also on view.

The exhibitor's new Exhibition Catalogue, containing numerous illustrations of the improved method of fitting up stables, will be forwarded on receipt of four postage stamps.

BARTLETT, J., & SON, *Welsh Back, Bristol.*—Railway and road weighbridges and weighing machines for general purposes.

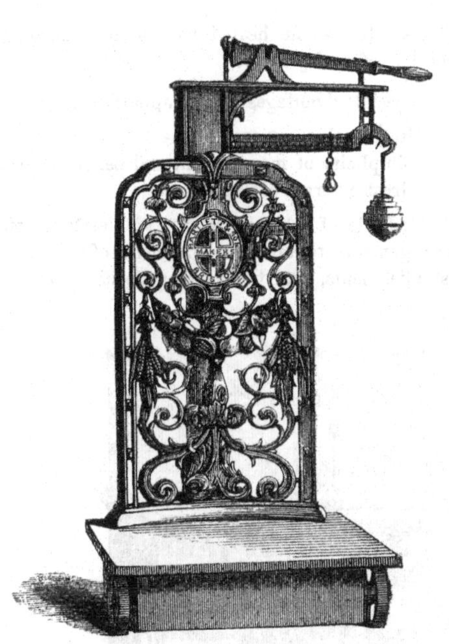

LEVER WEIGHING MACHINE.

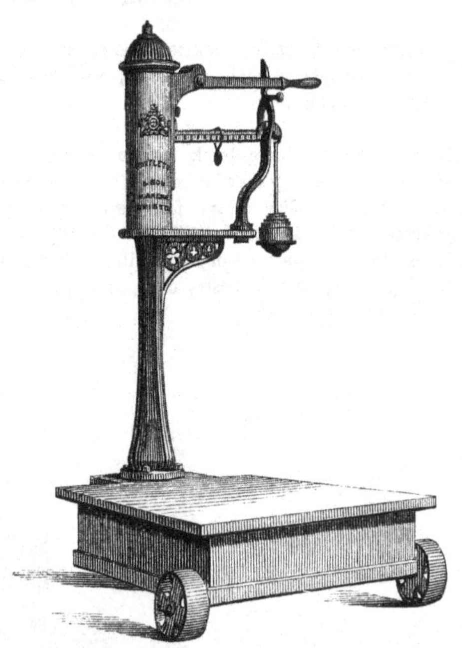

LEVER WEIGHING MACHINE.

WEIGHING MACHINE.

LEVER WEIGHING MACHINE to weigh from ¼ lb. to 1 ton £9 10 0

LEVER WEIGHING MACHINE with back iron, to weigh from ¼ lb. to 4 cwt. £3 10 0

LEVER WEIGHING MACHINE to weigh from ¼ lb. to 12 cwt. £6 5 0

LEVER WEIGHING MACHINE with dial indicator, to weigh 1 cwt. £6 0 0

WEIGHING MACHINE for equal weights . . £3 10 0

WEIGHING MACHINE fitted with chain and improved indicator.

BENHAM & SONS, 19, 20, 21 *Wigmore Street, W.*—Ornamental metal work; stoves, fenders, kitchen fittings; patent cooking apparatus.

Obtained Prize Medal in Class 22, *in* 1851; *Bronze Medal in Paris,* 1855.

1. BENHAM'S PATENT COOKING APPARATUS for large establishments, schools, hospitals, workhouses, barracks, and ships.

It consists of a large brick roasting oven, in which also bread or pastry may be baked; a hot-water boiler, which also supplies steam for steaming vegetables, puddings, &c.; a second boiler for the supply of hot baths; two or more meat or soup boilers, a hot-plate and broiling stove, an iron pastry oven, and a hot closet for plates; the whole heated by one fire, burning about 200 lbs. of coal per day.

The special advantages of this apparatus are—

Remarkable economy of fuel;
Simplicity of management and perfect control;
Great external coolness.

It admits of various modifications of form, size, and arrangements, to suit the requirements of public or private establishments, and the varying positions of fireplaces

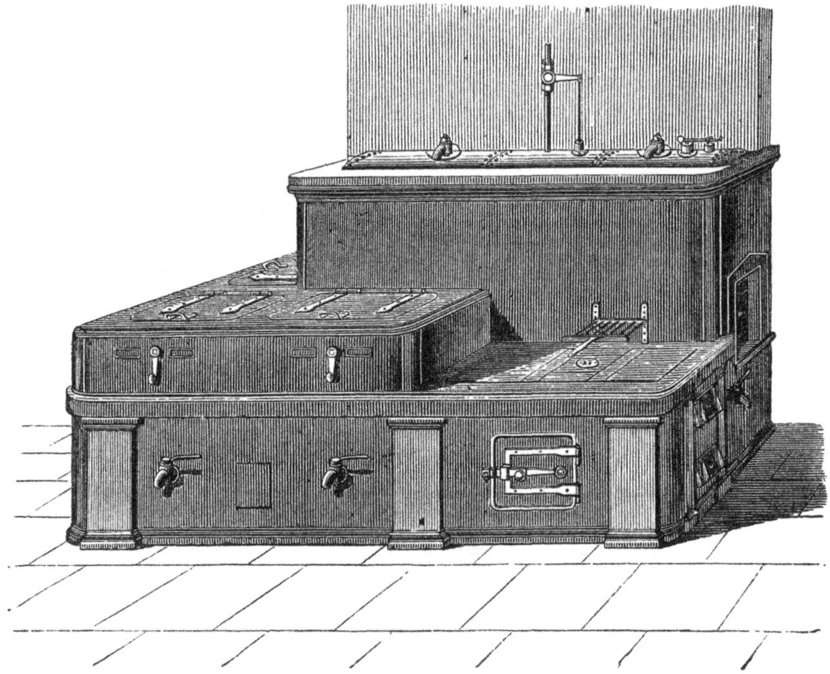

No. 1. PATENT COOKING APPARATUS.

and flues; but the following examples, amongst others, may be referred to in proof of its efficiency:—

NEW ROYAL MARINE INFIRMARY, Woolwich.
ROYAL MEDICAL BENEVOLENT COLLEGE, Epsom.
MESSRS. COOK, SON, & CO., St. Paul's Churchyard, London.
WEST LONDON UNION WORKHOUSE, West Street, Smithfield, London.
WAREHAM UNION WORKHOUSE, Dorset.
LEIGHTON BUZZARD UNION WORKHOUSE, Bedfordshire.
PENINSULAR & ORIENTAL STEAM NAVIGATION CO.'S STEAMER "Mooltan."
PENINSULAR & ORIENTAL STEAM NAVIGATION CO.'S STEAMER "Ripon." (2.)

It has also been adopted by the War Department at the following Barracks in consequence of the favourable report of the "Barrack and Hospital Improvement Commissioners:"—

ROYAL ARTILLERY BARRACKS, Woolwich. (2.)
WATERLOO BARRACKS, Tower of London.
PERMANENT BARRACKS, Aldershot.
EDINBURGH CASTLE BARRACKS.
STIRLING CASTLE BARRACKS.
GIBRALTAR BARRACKS. (2.)

2. RANGE, WITH OVEN, for married soldiers' quarters.

3. RANGE, WITH BOILER, for officers' servants' rooms.

These two ranges are fitted with hollow fire-lump backs, and louvred ventilators for the admission of warm air to the rooms. They are adopted by the War Department in all the new barrack buildings.

BENHAM & SONS, *continued.*

4. A FIRST-CLASS KITCHEN, with appropriate fittings, consisting of

The OXFORD ROASTING RANGE, the first specimen of which, in London, was introduced by Benham & Sons at the celebrated kitchen of the Reform Club, and which, with the whole of the cooking apparatus, was fitted up by them under the superintendence of the late M. Soyer.

It has an open fire, with vertical bars, but its peculiar excellence consists in the intense heat radiated from it, and its great economy of fuel—the space from the bars to the back being only one-half the usual depth. The back is formed of Stourbridge fire clay. The size of the fire may be increased or diminished at pleasure, and the whole of the front opens on hinges, like a gate, so as to give ready access for the removal of cinders, &c. The boiler for hot water is placed behind the back.

No. 4. FIRST-CLASS KITCHEN, FITTED WITH COOKING APPARATUS.

IMPROVED SMOKE JACK, with double movements, dangle-spits, and universal joints. The apparatus is kept in motion solely by the upward current of air in the chimney, without springs or weights.

STEWING STOVES AND STOCKPOT STOVE, heated by charcoal or gas, and therefore requiring no flue. The lower stove at the end is for large stockpots.

BAIN-MARIE PAN, for keeping sauces, soups, gravies, &c. always hot and ready for use without the slightest risk of burning or spoiling.

COOK'S SINK, with water for cooking purposes, &c. laid on.

HOT-PLATE AND BROILING STOVE, with movable gridiron, to which can be also added the oven on the top, as shown in the drawing, all heated by one fire.

HOT CLOSET, heated by steam or hot water, for keeping silver and china hot; also for receiving the different courses of a dinner after being dished up.

STEAM TABLE for dishing up.

Benham & Sons, *continued.*

Ovens with a separate furnace underneath.

Steam Kettles, of copper or block tin, for boiling meat, vegetables, puddings, &c.

Dinner Lift for conveying the dinner to the floor above. In large establishments, coals, &c. are carried up to the top of the building in this manner, which effects a great saving of labour.

No. 5. KITCHEN FITTED WITH COOKING APPARATUS.

5. The above engraving represents a very complete and efficient Cooking Apparatus, which Benham & Sons can confidently recommend for London or country houses, although on a much smaller scale than that which is represented in the preceding engraving.

Benham & Sons' Improved London-pattern Kitchen Range, with oven and boiler; all heated by one fire. The oven can be thoroughly depended upon for acting properly, and the boilers can be arranged to supply hot water for a bath, or for nursery and bedroom use (in addition to the kitchen), as well as to heat a hot closet and steam kettles, as represented in the engraving.

Improved Smoke Jack, with single movements, dangle-spit, and universal joint.

Hot-plate and Broiling Stove, with iron front and top, movable gridiron, &c.

Stewing Stoves, heated by charcoal or gas.

Hot Closet, heated by steam, for keeping silver and china hot and ready for use; also for receiving the different courses of a dinner after being dished up, and until taken to the dining room, without any possibility of their being scorched or dried up.

Ovens, with a separate furnace underneath.

Steam Kettles, for boiling meat, puddings, vegetables, &c.

BENHAM & SONS, *continued.*

6, 7. TWO POLISHED STEEL DRAWING-ROOM STOVES, with ormolu and porcelain enrichments.

In the adjoining court are exhibited four specimens of Mr. John Billing's PATENT DOUBLE-DRAUGHT OPEN FIRE STOVES (for curing smoke, &c.) of which Benham & Sons are the sole manufacturers.

NO. 8. INDEPENDENT STOVE WITH DOGS.

8. The DOG STOVE represented above, and which is adapted either for coals or wood, is exhibited in the South East Transept in the Metropolitan Trophy for Class XXXI.

9. There are exhibited with it other specimens of mediæval metal work, in the style of the 12th and 13th centuries, designed and manufactured by Benham & Sons; amongst the rest—

A chancel screen of hammered iron-work.

A pair of rood-screen gates in hammered iron and brass.

A brass eagle lectern or reading desk.

A standard chancel light.

Various wall and bracket lights.

Communion plate in latten and electro-plate, &c.

A small collection of church plate manufactured by Benham & Sons from the designs of W. White, Esq. is exhibited in the Ecclesiological Court in Class 30.

[5984]

BAYLISS, SIMPSON, & JONES, 43 *Fish Street Hill, Victoria Works, Wolverhampton.*—Chain cables, railway fastenings, iron hurdles, fencing, &c.

[5985]

BAYMAN, HENRY, 1 *Johnson Street, Old Gravel Lane, E.*—Double and single lifting jacks; set of iron blocks; improved ship's hearth and single winch.

[5986]

BENHAM & SONS, 19, 20, 21 *Wigmore Street, W.*—Ornamental metal work; stoves, fenders, kitchen fittings; patent cooking apparatus. (*See pages* 6 *to* 9.)

[5987]

BENNETT, WILLIAM, *Sir Thomas's Buildings and Sale Rooms, St. George's Place, Lime Street, Liverpool.*—Kitchen cooking ranges for coal and gas; smoke jacks; improved stoves and grates.

DR. ARNOTT'S SMOKELESS REGISTER GRATE, black finished, with patented improvements.

These grates will contain fuel for a whole day's consumption, are easily regulated, and well adapted for house and office purposes, and specially suited for chimneys with bad draughts.

Price of the one exhibited £11 0

DR. ARNOTT'S SMOKELESS REGISTER GRATE, best bright finished, with patented improvements, as above.

Price of the one exhibited £22 0

BENNETT'S SMOKE-CURER AND WARM-AIR GRATE.

These grates give an equal temperature throughout a room, perfect control over combustion, are very economical in use, and are manufactured to suit all classes of property.

Price of the one exhibited £12 0

BENNETT'S PATENT SMOKELESS STOVE, on Dr. Arnott's principles.

Price of the one exhibited £25 0

BENNETT'S LIVERPOOL KITCHEN RANGE, with modern improvements.

This range can be made in all sizes, and adapted to suit either large or small establishments. The boilers are constructed so as to heat baths, supply wash-houses, and for other domestic uses.

Price of the one exhibited £16 0

KING'S LIVERPOOL GAS-COOKING RANGES, made for cooking for from ten to five hundred persons.

These apparatus are most cleanly and economical in use, and perform all cooking operations in a superior manner.

Price of the one exhibited £10 10

[5988]

BERRY, GEORGE, 19 *Buttesland Street, N.*—Locks with crypted guards, not tentable by instrument or true key.

[5989]

BILLING, JOHN, *Westminster.*—Patented stove for more effectual combustion, and its regulation, and reducing smoke annoyances.

[5990]

BILLINGE, JAMES, *Ashton, near Wigan.*—Wrought-iron hinges, and locks of various sorts.

[5991]

BINKS, BROTHERS, *Millwall, Poplar.*—Round and flat wire ropes, conductors, fencing, strand sash line.

[5992]

BISSELL, WILLIAM, *Union Street, Wolverhampton.*—Rim and mortise locks, upon improved equi-action principles.

[5993]

BLACKETT, FRANK W., 31 *West Smithfield, London.*—The inaccessible lock.

A lock for the protection of a safe or strong room should possess two conditions of security :

1. It must be so constructed that it cannot be charged with gunpowder so as to be blown off the door, it must also defy all attempts by picking either by keys or other instruments.

2. It must be so placed that it cannot be reached by boring or cutting holes through the door of the safe so as to be taken out.

A perfect lock should defy the skill of the scientific thief to pick it, and the violence of the burglar to destroy it. The so-called "unpickable" locks (not yet picked is all that can be said of the best), as though to vaunt their security and challenge attack, are placed on the door in front of the safe, where they afford the expert thief every chance of tampering and trying them with success. The best locks thus violate the first condition of security in spite of the ingenuity of their construction.

The second condition is still farther from being complied with. Thick iron doors with steel plates or pegs over the lock, or even doors made altogether of case-hardened iron afford no efficient security. These can be cut or bored through, the lock removed or destroyed, and the safe door opened.

Within the last two or three years a patent safe with inch doors had 3 in. holes put through it in one short summer's night. The exhibitor has drilled through a case-hardened door ⅝ in. thick with portable tools, laying bare the lock and so opening the door without noise in the short space of forty minutes.

It is not the construction so much as the position of locks that is faulty. Making clever locks and strong doors and then putting the locks in the most convenient place for attack, is like carefully corking and labelling a bottle "poison," and then leaving it within reach of everybody.

This patent proposes no alteration in the principle or construction of locks, but places them in such a position as to increase the security of all, even the best.

A slight examination will show that by this patent a lock secures the two conditions of security requisite in a safe. It is placed at the back of the safe, where it can neither be blown off nor cut out, and where the lock-picker has no chance to exert his skill. The lock, of any construction, is placed at the back of the safe, its bolt is hinged to a lever or levers in the lower casing of the safe. This lever has teeth at its front end, which when the bolt is shot by the key, rock upwards and fit into corresponding recesses along the whole width if necessary of the door of the safe, fastening it much in the same way as in an ordinary safe. Unlocking of course removes these teeth from the door.

The length of handle necessary for a key to reach the back of a safe would be very objectionable, but this has been removed. The only real or effectual part of a key is that which acts upon the works of the lock. The handle in this patent remains always in the lock and adds to its security. It is a fixture in the keyhole and can only be drawn out far enough to attach the true key to it. The lock is not only more difficult to pick from its distance from the keyhole, but all access to it is absolutely cut off as the sole entrance to it is permanently occupied by the handle of the key.

This handle might be dispensed with by complicated machinery for conveying the key to the lock, but it is preferred to exhibit the safe in its simplest construction.

The advantages offered by this patent are the following :

The lock being absolutely out of reach places every conceivable difficulty in the way of the lock-picker or burglar. The handle of the key being left at all times in the keyhole to prevent access to the lock, it may be made so strong as to turn with ease a lock so heavy, that if a picker attempted it, it would refuse to answer any trying or give any hint as to the principle of its construction.

The lock cannot be got at by boring or cutting. The only points of attack are the levers. The position of these cannot be ascertained, and those at the bottom of the safe, as well as the lock itself, are absolutely beyond attack, as the safe, either from its weight or from being fastened to the wall or floor from the inside, is immovable. Gunpowder if it could be introduced might blow off the casing of the lock and expend its force on the inside of the safe, but the levers would still hold the front of the door.

It does not interfere with the fire-proof principle of any safe, or the mechanical principle of its lock fastening.

The lock is secure from the action of the atmosphere. A simple and inexpensive lock might be used, which, two feet out of the reach of the picker, would be safer than the best in its present usual position.

All these advantages may be gained with a key so small as to be at all times, day and night, in the custody of its owner, with far less trouble and inconvenience than those now in use.

It is to be observed that all these advantages are not in exchange but absolutely in addition to those offered by the very best safe hitherto constructed.

[5994]

BLOOMER, CALEB, & CO., *Golds Hill, West Bromwich.*—Cable chains, common and patent anchors, and railway fastenings.

[5995]

Bolton, Thomas, & Sons, *Birmingham, and Oakamoor, Staffordshire.*—Rolled metals, brass and copper wire and tubes, and calico rollers.

[5996]

Boobbyer, Joseph Hurst, 14 *Stanhope Street, Newcastle Street, Strand.*—Locks, furniture, bolts and hinges for buildings.

[5997]

Bracher & Gripper, 11 *Cannon Street West, and Vulcan Safe Works, Skin Yard, Bankside, Southwark.*—Fireproof safes, doors, deed boxes, and other fireproof articles. (*See page* 13.)

[5998]

Bramah & Co., 124 *Piccadilly.*—Patent locks, iron safes, cash, jewel, and despatch boxes, with the lock applied.

[5999]

Brierley & Geering, *Birmingham.*—Specimens of iron ornamented with their patent enamel.

[6000]

Brown, J., & Co., *Glasgow.*—Gill air warmers, hot air stoves, &c.

[6001]

Brown, Brothers, *Lyme Regis, and* 43 *Cranbourne Street, Leicester Square, London.*—Patent cooking ranges for mansions, villas, and cottages. (*See page* 16.)

[6002]

Brown & Green, *George Street, Luton; London Warehouse,* 81 *Bishopsgate Street Within.*—Improved patent kitchen range for economising fuel, and cure of smoky chimneys. (*See pages* 14 *and* 15.)

[6003]

Brown, Lenox, & Co., *Millwall, Poplar.*—Screw bench with panelled vice with adjusting jaw.

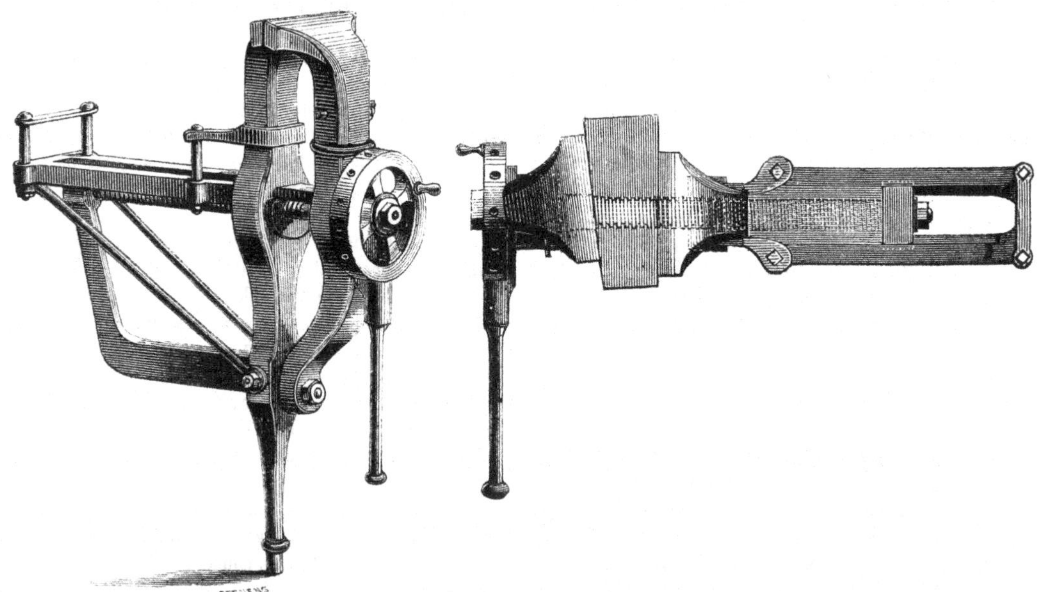

M. Roberts' Patent Parallel Vice, with adjusting jaws fitted to vice bench, complete.

The advantages of this vice are—it will take in larger work than other vices, will adapt itself to taper work, and is quicker in action, being wound out or in by the wheel, and the lever used to give the nip or let it go.

Manufactured by Brown, Lenox, & Co. Millwall, Poplar, London.

BRACHER & GRIPPER, 11 *Cannon Street West, and Vulcan Safe Works, Skin Yard, Bankside, Southwark.*—Fireproof safes, doors, deed boxes, and other fireproof articles.

Obtained a Prize Medal at the Exhibition of 1851.

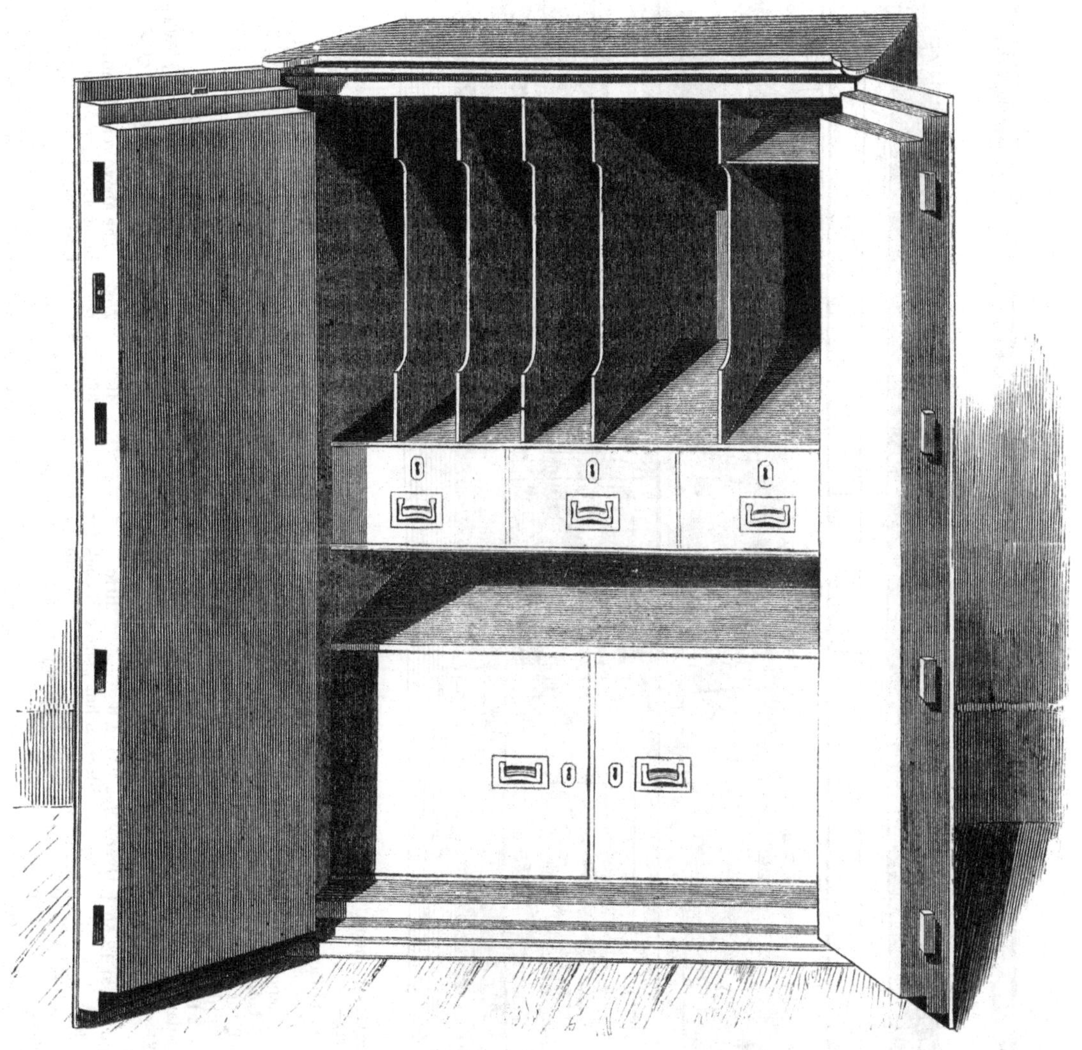

FIREPROOF SAFE.

The exhibitors are inventors, patentees, and manufacturers of wrought-iron fire and thief proof safes, chests, doors, and strong rooms; patentees of the double-security detector locks for banks, treasure rooms, &c.; and makers of cash boxes, deed boxes, &c.

They have supplied the new Houses of Parliament, Her Majesty's honourable Board of Ordnance, the National Debt Office, &c.

BROWN & GREEN, *George Street, Luton; London Warehouse,* 81 *Bishopsgate Street Within.*—New patent kitchen range for economising fuel and cure of smoky chimneys.

This range has been tested by order of the Government. The following is a copy of the official report:—

"Report of a trial of a cooking apparatus, manufactured by Messrs. Brown & Green.

"The apparatus is of the nature of a kitchener, with an open fire for roasting; it consists of 3 boilers and 6 ovens, with hot-water cistern. Two of the boilers, holding 44 gallons, are for steam; and the third, holding 18 gallons, is for hot water. The ovens are 2 ft. square; the range 18 ft. long. The trial, the particulars of which are subjoined, was exceedingly satisfactory:

	H. M.
The fire was lighted with 1 lb. of wood . . .	at 12 50
Ovens Nos. 1, 2, 3, and 4 attained 270 deg. Faht. in 35 min.	at 1 25
Ovens Nos. 5 and 6 attained 260 deg. Faht. in 59 min.	at 1 49
The steam boilers boiled in 38 min.	at 1 28
The hot-water cistern (25 gallons) boiled in 70 min.	at 2 0

One pint of water boiled on hot plate over fire in 7 min.
One ditto ditto ditto side boiler in 9 min.
One ditto ditto ditto No. 1 oven in 14 min.
One ditto ditto ditto No. 3 oven in 25 min.
One ditto ditto ditto No. 5 oven in 38 min.

	MEN
Ovens Nos. 1, 2, 3 and 4 will bake (meat with potatoes under it) for	1200
Ovens Nos. 5 and 6 ditto	400
The six boilers on hot plate will cook for . . .	360
Six stew pans ditto ditto . . .	120
The fire will roast three joints for	56
Total	2136

The steam boilers will steam potatoes for 1000 men.

The consumption of coal ("Inland," of inferior quality) was—

For the first hour	56 lbs.
For the second hour	30 lbs.
For the third hour . . . . (nil)	—
Total	86 lbs.

"*February* 18*th*, 1862.
(Signed) "G. WARRINER,
"*Instructor of Cookery to the Army.*"

Note.—The above report shows a consumption of only three-fifths of an ounce of coal per head. The whole of the apparatus remained in full action at the end of the third hour. The patentees are compelled to exhibit this range minus the two end ovens, not having a sufficient allotment of space for its entire length.

BROWN & GREEN, *continued.*

This range is constructed on a new principle (patented January, 1862), which prevents the great waste of heat, and therefore of fuel, hitherto unavoidable in all ranges from which an abundant supply of hot water or steam is required.

The method in such ranges has uniformly been to place the ovens on each side of the fire, and the boilers at the back. The heating of the boilers has always been accomplished by a flue formed under them, and carried at once into the chimney; and as the hot draught from a fire will invariably take the shortest course, the effect of the above arrangement has been to cause a great and wasteful rush of heat under the boilers, directly into the chimney, instead of its being carried round the ovens. The boilers and ovens could not, therefore, be worked simultaneously without a very much larger fire than is needful in a range constructed upon the new principle. Moreover, the old plan, with all its wasteful expenditure of fuel, fails to heat the boilers effectually, whilst it so restricts their capacity and power that the supply of hot water and steam for baths, lavatories, hot closets, steamers, &c. is far too limited for the requirements of hotels and other large establishments.

By the new patent arrangement all these disadvantages are removed.

At the side of the fire is a boiler which projects forward under the hot plate to the front of the range, and forms the side of the fire-place; and beyond this boiler is placed one or more spacious ovens.

The heat from the top of the fire is carried over the top of the side boiler, between it and the hot plate, and is then passed completely round the ovens before it obtains an outlet into the chimney. From the bottom of the fire another distinct current of heat passes through an arched flue under the same boiler and is carried up on the other side, between it and the first oven, and, uniting with the draught from the top of the fire, is also passed entirely round the ovens before it can escape.

Where an apparatus of larger dimensions is required, the same arrangement is repeated on the opposite side of the fire; whilst at the back is placed another large boiler, the lower part of which projects forward into the middle of the fire, the under surface forming a half arch which corresponds with the arched flues under the side boilers, thus affording effectual means of heating it without any escape of hot air.

The position and shape of the boilers and direction of the flues, form peculiar and most important features in this invention. The boilers being surrounded by the full heat of the fire, become the most powerful, either for steaming, circulating hot water, or any other purpose that can be desired in a kitchen range; and as the whole of that heat is under entire control and may afterwards be carried round any or all of the ovens before passing into the chimney, the desired temperature is obtained in each.

This patent principle is applicable to ranges of every size, and its power is demonstrated by the large one on view at the International Exhibition, which was made eighteen feet in length, having six ovens and three boilers, all worked by a fire only fourteen inches wide, those ovens most distant from the fire being not merely hot closets, but effective ovens; a result never before obtained with any quantity of fuel. The patentees are compelled to exhibit it minus the two end ovens—not having a sufficient allotment of space for its entire length.

This range, in its full original size, has been put into practical operation, and every part of it has been found to work most efficiently. It has been fully tested on behalf of the Government by Mr. Warriner, Inspector of Cookery and Cooking Apparatus for the Army, whose report is annexed.

Brown & Green's patent ranges are a certain cure for smoky chimneys, and possess the following special advantages:—

1. The means of roasting meat perfectly in front of the fire at the same time that the oven or ovens, boilers and hot plate are kept in full action—the movable iron plate which encloses the upper half of the fire front, coming into immediate contact with the fuel, becomes red hot; thus that portion of the joint exposed to it is equally well roasted with the part acted upon by the oven fire below, whilst the use of the roasting plate secures a full heat over the top of the range and in the ovens. For these reasons it possesses decided advantages over the ordinary door, to open which, for the purpose of roasting, seriously lessens the temperature of the ovens and hot plate.

2. The roasting plate above mentioned is perforated with a row of small holes, through which a current of oxygen is directed over the fire, causing the combustion of a portion of the smoke, and diminishing the expenditure of fuel and the frequency of cleansing the flues. This plate can also be easily removed and a set of bars slipped in its place, by which means the fire is made entirely open in front when desired.

3. The ventilating arrangement in the upper part, which is simple, never requiring attention, removes that close heat and the smell of cooking which is complained of in other kitcheners.

4. The facility with which fresh fuel can be put on the fire—the sliding top of this range being more easy to manage than any other top plate, whilst it avoids needless lifting of covers on the part of the servants and the risk of breakage.

5. Whilst an average width of fire front is retained for the convenience of roasting, the depth of the fire from the bars to the back is very much less than usual, so that a furnace heat and consequent self-destruction of the range and much waste of fuel are avoided.

In the above important respects, Brown & Green's kitchen ranges differ from and surpass all others; in many other details also they are more convenient and complete. They are well adapted for private families, and the ovens are well ventilated, and perfect either as roasters, or for the baking of bread and pastry. The larger sizes, fitted with steaming and bath apparatus, hot closets, and other appliances, form the most complete appointment for clubs, hotels, public institutions, and other large establishments.

These ranges are made from 3 ft. to 24 ft. in width.

Prospectuses, prices, references, and designs may be had on application.

BROWN BROTHERS, *Lyme Regis, and* 43 *Cranbourne Street, Leicester Square, London.*— Patent cooking ranges for mansions, villas, and cottages.

BROWN BROTHERS' UNIVERSAL KITCHENERS.

The above-named kitcheners have justly obtained a world-wide celebrity for their ingenuity, safety, durability, elegance, and economy of fuel, time, and labour. Since receiving the prize medal at the Great Exhibition of 1851, and other certificates of merit from scientific and agricultural societies, many striking additions and improvements have been made, and these kitcheners, adapted to the use of the mansion or the cottage, are now so perfectly

UNIVERSAL KITCHENER.

arranged that by a single small fire every culinary operation can be effectually carried on at the same time, and while roasting, baking, broiling, stewing, steaming, frying, and boiling are being efficiently performed, hot water may be obtained in abundance for baths, heating conservatories, shops and offices of any description, and by an ingenious application linen may be aired without danger from smoke, fire, or dust. These kitcheners are so constructed that they can be fixed without difficulty, cannot be damaged by carelessness or neglect, and prove an infallible cure for smoky chimneys.

The automaton roaster supplied with the various sizes is an invention unique in itself, and by the application of a simple principle becomes the nearest approach to perpetual motion, and is used without the trouble and inconvenience of every other kind of roasting apparatus.

[6004]

BRYON, THOMAS, *Salop Street, Wolverhampton.*—A bedstead in the Elizabethan style, with improved sacking, registered.

[6005]

BUIST, GEORGE, 70 *St. Mary's Wynd, Edinburgh.*—Lightning conductors, and metallic cords.

WIRE STRAND for fencing and signal cords.

Gauge of Strand.	Per cwt.	Gauge of Strand.	Per cwt.
0. . . .	£1 4 0	4	£1 8 0
1. . . .	1 5 6	5	1 9 6
2. . . .	1 6 0	6	1 12 6
3. . . .	1 7 0		

COPPER WIRE-ROPE LIGHTNING CONDUCTORS, $\frac{3}{8}$ in. diam. and upwards, from 8*d.* per foot; with fittings, complete, from 10*d.* per foot.

COPPER AND GALVANIZED IRON CORDS for sash lines, greenhouses, turret clocks, &c.; galvanized cords from 6*s.* and copper cords from 10*s.* per 100 ft.

These cords, when put up with proper weights and pullies, are cheaper and more durable than any other material.

GALVANIZED METALLIC CORDS for clothes lines, from 4*s.* 6*d.* per 100 ft. and upwards.

GILT AND SILVER-PLATED PICTURE CORDS from 3*s.* per 100 ft.

STEEL WIRE CORD FOR CRINOLINES from 12*s.* per 100 ft.

Price lists and samples will be forwarded on application.

[6006]

BULLOCK, THOMAS, & SON, *Cliveland Street, Birmingham.*—Ivory, bone, wood, and horn buttons of every description.

[6007]

BURCHFIELD, T., & SON, 8 *West Smithfield.*—Chaff-cutting machines; oat-bruisers, and weighing machines.

[6008]

BURNEY & BELLAMY, *Millwall, Poplar, E.*—Iron tanks and cisterns; navy, house, and farm patent ventilators.

[6009]

BUTLER, J., & SONS, 4 *Elm Street, Gray's Inn Lane, W.C.*—Brass, copper, and iron wove wire.

[6010]

CARPENTER & TILDESLEY, *Somerford Works, Willenhall.*—Patent rim, hall door, dead mortise, and stock locks; curry combs, horse scrapers, &c.

[6011]

CARRINGTON, JAMES, 4 *Queen's Mews, Queen's Gate, Kensington.*—Model of a horse stall, and new system of bitting horses. (*See page* 18.)

[6012]

CARRON COMPANY, *Warehouses,* 15 *Upper Thames Street, London;* 30 *Red Cross Street, Liverpool; and* 123 *Buchanan Street, Glasgow; Works, Falkirk, N.B.*—Sugar pan, bright range, stoves, &c. (*See page* 19.)

[6013]

CASEY, W. F., 10 *Raven Row, Stepney, London.*—Models of scales and beam, complete, used for weighing of bullion.

[6014]

CHAMBERS, WILLIAM, *Oozell Street, Birmingham.*—Metallic bedstead, pillars and rails.

[6015]

CHATWOOD & DAWS, *Bow Street, Bolton.*—Patent locks, gunpowder escapement, bankers' safes for valuables and parchment documents.

[6016]

CHILLINGTON IRON COMPANY, THE, *Wolverhampton.*—Burden's patent machine-made improved horse shoes.

[6017]

CHUBB & SON, 57 *St. Paul's Churchyard.*—Patent detector locks, &c. (*See pages* 20 *and* 21.)

CARRINGTON, JAMES, 4 *Queen's Mews, Queen's Gate, Kensington.*—Improved horse stall, and new system of bitting horses.

The stall or box is removable without damaging either the stall, the box itself, or the building in which it is placed, the fittings not being a fixture. The horse cannot in any way injure himself. The fittings are also a pre-

IMPROVED HORSE STALL.

ventive to the horse obtaining the habit of crib-biting. By the improved system of drainage introduced the stable is kept perfectly free from any ill effects of ammonia.

These fittings also include a new system of bitting horses, whereby the horse makes his own mouth, and by so doing makes his own temper.

Carron Company, *Carron Warehouses*, 15 *Upper Thames Street, London;* 30 *Red Cross Street, Liverpool;* *and* 123 *Buchanan Street, Glasgow; Works, Carron, N.B.*—Sugar pan, bright range, stoves, &c.

Fig. 1. PARLOUR REGISTER STOVE.

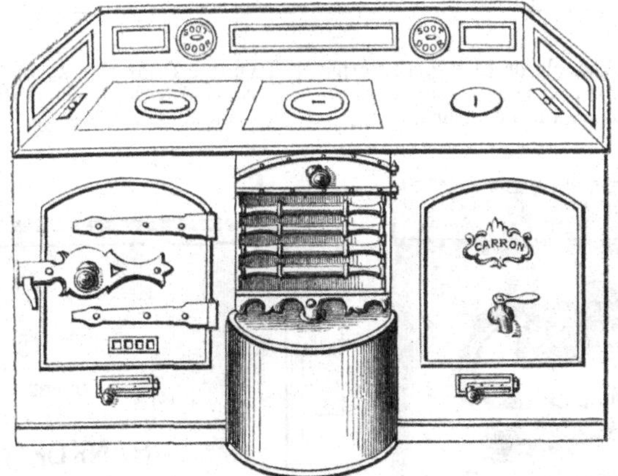

Fig. 2. KITCHEN RANGE.

1. A Parlour Register Stove, with fire-brick back, and cast-iron ornamental chimney-piece (fig. 1).
2. Various Parlour and other Register Stoves.
3. Kitchen Range, with bright fittings (fig. 2).
4. Large 350-gallon Sugar Pan, best cast-iron.
5. Sand Boiler, 50 gallons. FB. Pot, 60 gallons.
6. Cabinet Stove, with Ash Pan.
7. Umbrella Stand, and Garden Chairs.
8. Balcony Panels, different patterns.
9. Newals for iron balustrading, Flower-pot Rail, &c.
10. Sections of Carron Company's Pig Iron, showing fracture.
11. Box Bushes for colonial and other waggon axles.
12. Ornamental Door-porters, Scrapers, &c.
13. Sad Irons, various.
14. Sundries, including cast-iron boot jacks, match and candle brackets, &c.

Fig. 1. shows register stove with fire-brick back, ornamental cast-iron chimney piece, and ash-pan complete.

CHUBB & SON, 57 *St. Paul's Churchyard.*—Patent detector locks, fire-proof and thief-proof safes, strong-room doors.

Obtained a Prize Medal with "special approbation" at Great Exhibition in 1851, *and First-Class Medal at Paris Exhibition in* 1855.

CHUBB'S PATENT DETECTOR LOCKS of various sizes, and for all purposes to which locks can be applied. An illustrated price list may be obtained gratis and post-free.

WHEEL LOCK for doors of strong rooms and safes, throwing any requisite number of bolts all round the door, the whole being secured by four gunpowder-proof locks, each with distinct key.

SUITE OF TWELVE MORTISE LOCKS for room doors, each having its own key opening that lock only, and with the following sub-master keys, viz.: One key to open Nos. 1 and 2 only, one key to open Nos. 1 to 3, one key to open Nos. 1 to 4, and so on up to one opening Nos. 1 to 12. Also a master key to open all, and to double-lock and thereby shut out any and all of the other keys.

DOOR LOCK in walnut-wood stock or casing, the ornamental front, as above engraved, being wrought from a single plate of steel, hardened and burnished.

LARGER DOOR LOCK, the case being of polished steel covered with an elaborate mediæval design in ormolu open work, and the key wrought in corresponding style.

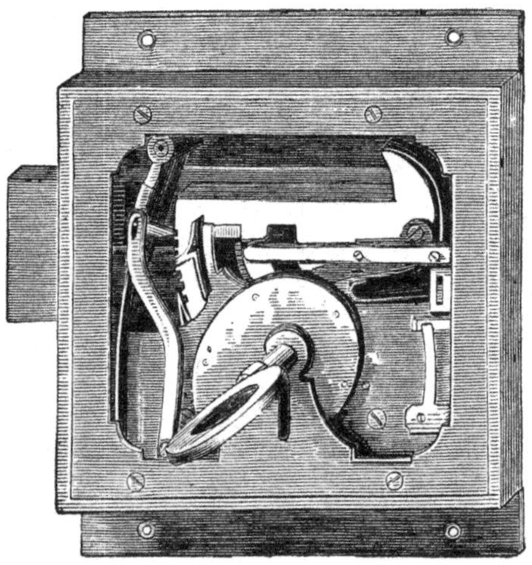

CHUBB'S BANK LOCK for special security of iron safes and doors.

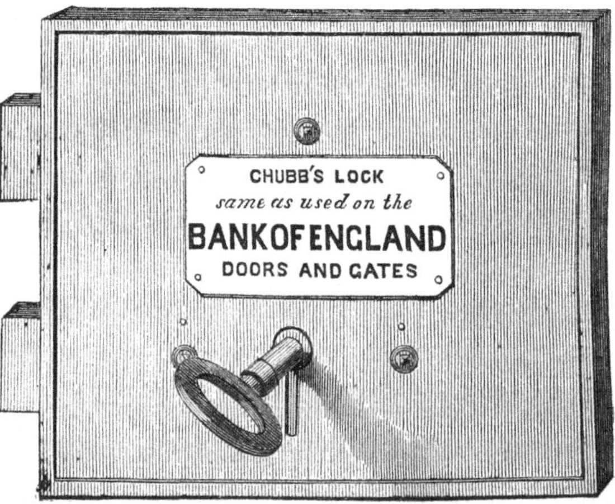

BANK OF ENGLAND LOCK, manufactured by Chubb & Son.

It will be observed that no locks of inferior quality are made by Chubb & Son. The whole of their locks sold to the public at large are exactly the same in security and excellence of workmanship as those supplied to Her Majesty, the Government offices, and other public establishments. The prices are from 10*s.* each upwards.

Chubb & Son, *continued.*

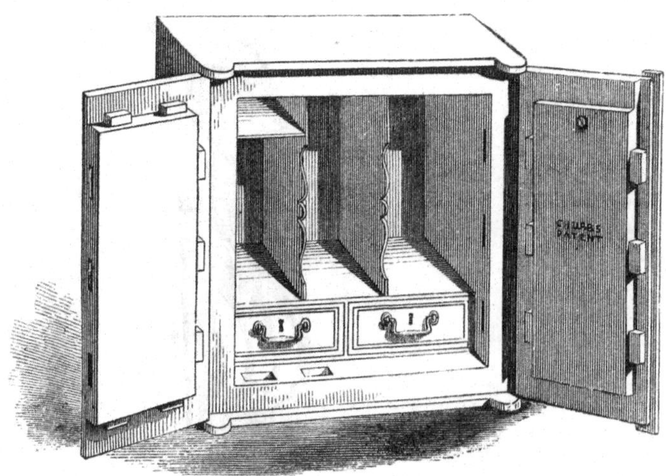

Chubb's Safe No. 29. (See price list).

Chubb's Patent Jewel Safe.

Chubb's Patent Wrought-iron Fire-proof Steel-plated Safes and Strong Room Doors, with gunpowder-proof locks.

1. Jewel Safe (see engraving) with ornamental door and sides. The design on the door is executed in a mixture of dead and burnished steel, inlaid gilt scrolls in the corners, and ormolu mountings. The interior fitted up in ornamental wood, for the reception of jewellery. The door secured by Chubb's patent wheel lock throwing bolts all round.

2. Another Jewel Safe with folding doors of dead steel, with inlaid gilt scrolls and ormolu mouldings.

3. Very large Banker's Safe, weighing about four tons, the interior fitted with drawers, cupboards, and partitions for books. The outer folding doors made of wrought-iron plates and hardened steel, combined in the most effective manner into a solid mass or plate. The doors secured by two gunpowder-proof wheel locks, throwing thirty-one bolts all round, and the main keyholes covered with case-hardened iron scutcheon locks opened by a small gold key set in a finger ring.

4. Another Banker's Safe having the above-named system of combined iron and hard steel applied throughout its entire casing.

5. Specimens of Chubb's safes and chests of various dimensions, full particulars of which will be found in their complete illustrated price list, which will be forwarded gratis and post-free.

6. Wrought-iron fireproof doors and frames of various dimensions, for strong rooms.

[6018]

Clark, T. & C., & Co., *Wolverhampton.*—Enamelled and tinned cast-iron hollow ware, and general casting.

Fig. A.

Fig. A. Cast-iron Saucepan lined either with enamel or tin.

Fig. F.

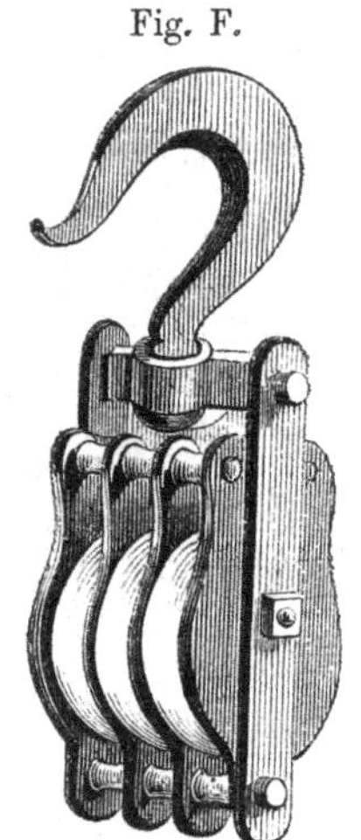

Fig. F. Wrought-iron Pulley Block, with cast-brass or iron sheaves.

Fig. B. Fig. C. Fig. D.

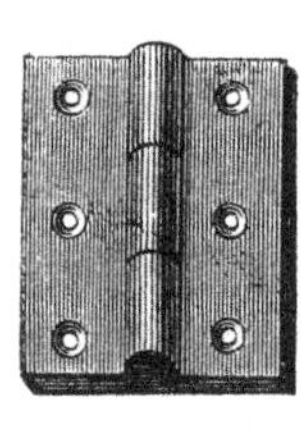

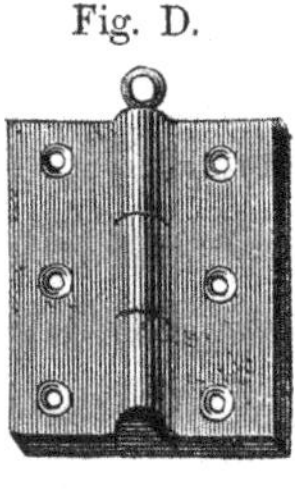

Fig. B. Fast Joint Cast-iron Hinge.
Fig. C. Loose Joint ditto.
Fig. D. Loose Pin ditto.

Fig. G.

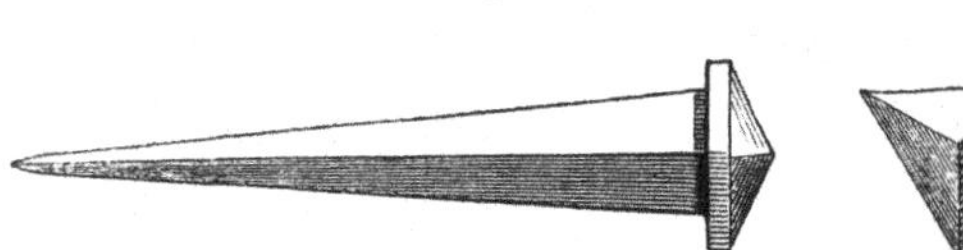

Fig. G. Carron's Patent Triangular Cast-iron Wall or Lath Nail.

Fig. E.

Fig. E. Enamelled Cast-iron Wash-hand Bowl, with plug hole.

Fig. H.

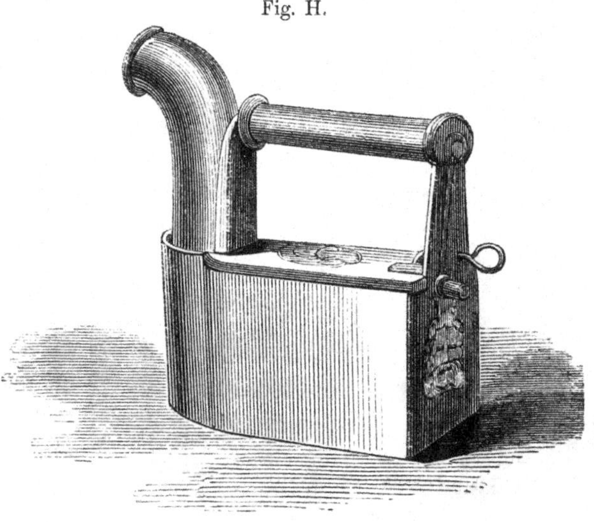

Fig. H. American Charcoal Box Iron.

[6019]

COALBROOKDALE COMPANY, THE, *Coalbrookdale, Shropshire.*—Plain and ornamental ironwork.

[6020]

COLLINS & GREEN, 7 & 8 *Albion Place, Blackfriars.*—Sculpture and marble work, marble chimney-pieces, and stoves.

[6021]

COOLEY & FOWKE, *Castle Street, Wolverhampton.*—General hardware and saddlery.

[6022]

COOPER, G. B., 121 *Drury Lane.*—A new application of ornamental tomb railing.

[6023]

CORMELL, JOHN, *Lansdowne Iron Works, Cheltenham.*—Improved wrought-iron tanks, cisterns, and cattle troughs, coated inside.

[6024]

CORNFORTH, JOHN, *Berkley Street Mills, Birmingham.*—Steel and iron wire of all kinds, and all sorts of wire nails.

[6025]

COTTAM & CO., 2 *Winsley Street, London.*—Conservatory, stable fittings, verandah staircases, tomb railings, and ornamental iron work. (*See pages* 24 *to* 26.)

[6026]

COTTRILL, EDWIN, *Vittoria Street, Birmingham.*—Metallic stationery, copying, and embossing presses; dies, detector locks, &c.

[6027]

COX, SAMUEL, *Walsall.*—Every description of saddlers' ironmongery and harness mountings.

[6028]

CRICHLEY, HENRY, *Sheffield Place, Birmingham.*—Patent enamelled stove grates, mantel-pieces, hall stoves, hat stands, and fenders.

[6029]

DAVIES, EDWARD, *Galvanized Iron Works, Snow Hill, Wolverhampton.*—Galvanized corrugated iron roofing sheets; galvanized cisterns, scoops, buckets, patent pumps, water spouts, and models, &c.

[6030]

DAWBARN, ROBERT, *The Brink, Wisbech.*—Clamp for instantaneous stoppage of leaks in fire-engine and other flexible hose.

[6031]

DAY & MILLWARD, *Birmingham.*—Patent platform and registered weighing machines, scales, scale beams, steelyards, &c. (*See page* 27.)

[6032]

DEANE, EDWARD, 1 *Arthur Street East, London Bridge, E.C.*—Patent duplex range, patent steel ovens, patent steel boiler, patent roasting apparatus. (*See page* 28.)

[6033]

DEELEY, ABEL SMITH, 27 *Brasshouse Passage, Birmingham.*—Wrought-iron shoe heels and toe tips of every description.

[6034]

DEELEY, G. H., & CO., *Campbell Street, Dudley.*—Flat and round chains for mining and other purposes.

[6035]

DIXON, ADAM, *Birmingham.*—Knife and fork cleaners, twine or string boxes, and boot or shoe warmers.

COTTAM & CO., 2 *Winsley Street, London.*—Conservatory, stable fittings, verandah staircases, tomb railings, and ornamental iron work.

MODEL OF STABLE FITTINGS, in two stalls and one loose box, to a scale of one-quarter the full size, showing the wainscot partitions and doors, wrought-iron ventilating division railing and ramps, with iron heel posts, surface gutter with movable safety covers, sanitary traps, improved registered rack, manger and water trough of enamelled iron, patent guide and halter strap with registered noiseless shackle combined, ventilating safety manger guard, enamelled head-stall plates, &c.

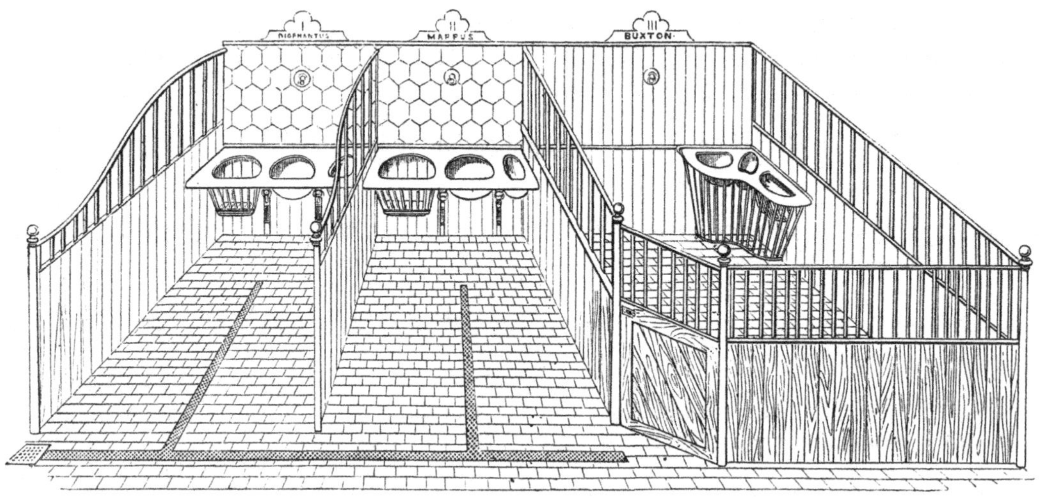

STABLE FITTINGS.

The improvements in these fittings are numerous, such as increased capacity, a better formation, no projections, the patent halter guide and noiseless swing manger shackle, patent portable seed box for saving the hay seeds for agricultural purposes, &c. drop cover for water trough, registered loose box ventilating guard to prevent the horse getting his head under the fittings, gutter to prevent the horse getting the caulkings of his shoes fixed, and numerous other additions.

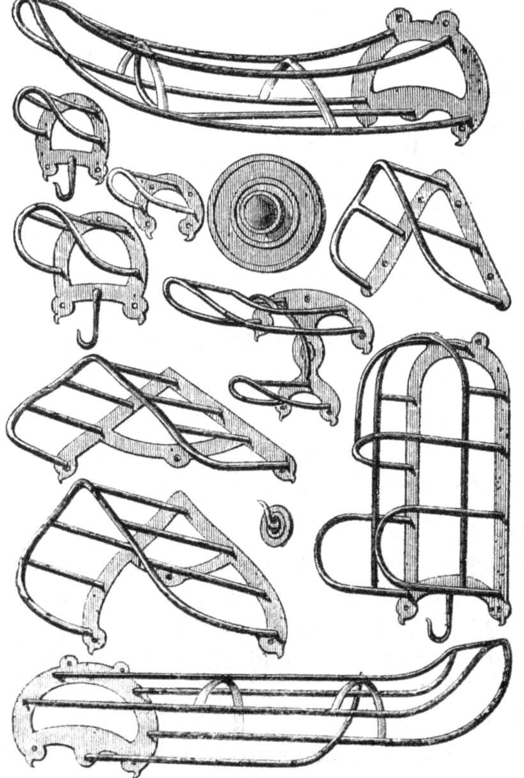

WROUGHT-IRON VENTILATING BRACKETS, for hanging saddles and harness upon.

The improvements in these are their being made in sets and of the shape of the harness, which retains its natural form when suspended and preserves the leather from cracking; the openings between the bars allowing a free admission of air to dry the under parts of the saddle, collar, or harness pads.

SAMPLES OF WROUGHT-IRON CEILING HOOKS, for cleaning harness upon.

COTTAM & CO., *continued.*

SAMPLES OF IMPROVED RACK AND PILLAR CHAINS.

WROUGHT-IRON BRUSH DRAINER for drying the cleaning brushes, &c. after use.

AN IRON FORK RACK, to hang the stable fork upon when not in use.

SAMPLES OF SLIDING HEAD-STALL PLATES for the names of horses.

PORTION OF A VERANDAH.

A portion of a VERANDAH OR COVERED WAY; having cast-iron columns for the support of the roof, which may be covered either with zinc, copper, or glass; the spaces between the columns are filled in with ornamental spandrils, and a perforated frieze above.

COTTAM & CO., *continued.*

CAST-IRON STEPS, straight and spiral, with iron railings of various designs. These stairs are suitable for the interior of buildings, also for external purposes, as into gardens from balconies, &c.

VARIOUS PATTERNS OF ORNAMENTAL BALUSTER BARS AND PANELS OF WROUGHT AND CAST IRON, applicable for stairs, galleries, communions, tombs, &c. also an example of a screen of wrought iron, grilles, lectern hinges, &c.

A PAIR OF GATES of Italian character, of wrought and cast iron, suitable for the entrance to a park, public building, and many other purposes.

DAY & MILLWARD, *Birmingham.*—Prize metal patent and platform weighing machines, of all descriptions, also manufacturers of scales, scale beams, steelyards, &c.

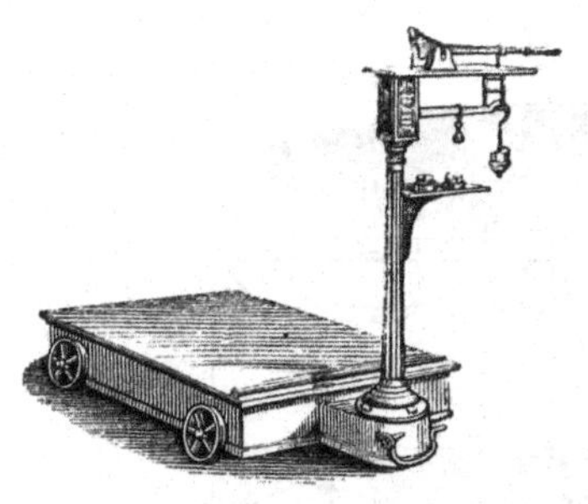

WEIGHING MACHINE.

Sole patentees of the prize medal patent weighing machines, adapted for railway stations, 118, Suffolk Street, Gee Street, Birmingham.

WEIGHING MACHINE.

These machines may be graduated to the English or foreign standard.

[6036]

DOBSON, ELIZABETH & WILLIAM, 24 *Fieldgate Street, Whitechapel.*—Specimens of branding irons.

[6037]

DOCKER & ONIONS, *Thorp Street, Birmingham.*—Smiths' bellows, portable forges, anvils, vices, &c.

[6038]

DOLLAR, THOMAS AITKEN, 56 *New Bond Street.*—Improved methods of horse-shoeing.

[6039]

DOWLER, GEORGE, *Great Charles Street, Birmingham.*—Wax vestas and boxes; hearth brushes, inkstands, bells, corkscrews, toasting-forks, candle-shades, &c.

[6040]

DOWLING, EDWARD, 2 *Little Queen Street, Holborn.*—Scales, weights, and mills, and weighbridges of every description.

[6041]

DUGARD, WILLIAM, JUN., *Newton Street Works, Birmingham.*—Carriage and railway lamps coach and harness furniture.

[6042]

DULEY & SONS, *Northampton.*—Kitchen ranges and patent bushes for axles.

[6043]

DYKE & CO., 15 *Aston Place, Holloway Road, N.*—Improved ice closet and chests.

These goods are made of any shape or size, and of the best materials. They afford a perfect safeguard against heat or dust.

DEANE, EDWARD, 1 *Arthur Street East, London Bridge, E.C.*—Patent duplex range, patent steel ovens, patent steel boiler, patent roasting apparatus.

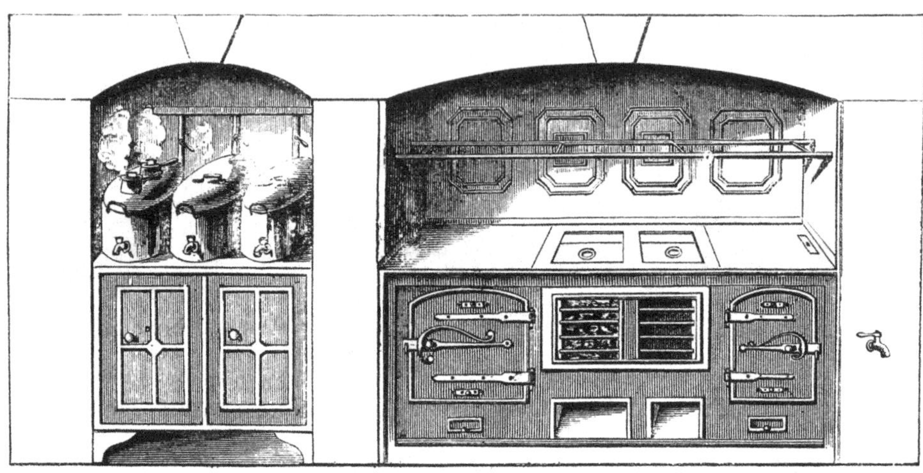

PATENT DUPLEX RANGE.

The PATENT DUPLEX RANGE is not surpassed by any range yet invonted, in effectiveness and economy, having two fires which can be regulated to any size. One or both can be used, the boiler coming in the centre. A range of 6 ft. 6 in. long, with 24 in. opening admitting the boiler front, will, with the patent revolving roasting apparatus in front, roast with ease 12 joints.

PATENT REVOLVING ROASTING APPARATUS.

PATENT STEEL OVEN.

PATENT STEEL BOILER.

The PATENT REVOLVING ROASTING APPARATUS is intended to supersede the objectionable smoke and bottle jack, made in various sizes.

The PATENT STEEL OVENS are made for Government for baking for 50, 100, and 250 men, lined with fire clay.

The PATENT STEEL BOILERS are lined with very peculiar fire clay; and with only 9 lbs. of coals they have boiled 20 gallons of water for breakfast, 20 gallons of soup for dinner, and 20 gallons of water for tea, thereby providing for 75 men.

Testimonials sent.

[6044]

EASTHOPE, WILLIAM, *Wyle Cop, Shrewsbury.*—Cooking apparatus, with open boiler for bath.

[6045]

EDELSTEN & WILLIAMS, *Newhall Works, George Street, Birmingham.*—Iron wire, pearl buttons, patent toilet and entomological solid-headed pins.

Edelsten & Williams (late D. F. Tayler & Co.) are manufacturers of iron wire, pearl buttons, patent toilet and entomological solid-headed pins, by special appointment to Her Majesty the Queen.

[6046]

EDGE & SON, *Coalport, Ironbridge, Shropshire.*—Flat gatten chains, cables, wire ropes.

[6047]

EDWARDS, ELIEZER, *Birmingham.*—Glass finger plates, lock furniture, drawer handles, bell pulls, &c., with metal mountings.

Obtained Honourable Mention at the Exhibition of 1851.

Finger plates, lock handles, key-hole plates, bell pulls, &c. *en suite*, in various styles and colours.

Drawer handles with screws complete formed entirely of glass.

Drawer handles, cupboard-turns, &c. with the iron shank firmly embedded in the glass while in a molten state.

Any of these articles can be adapted to the special requirements of foreign markets.

[6048]

EDWARDS, FREDERICK, & SON, 49 *Great Marlborough Street, London, W.*—Porcelain-tile grates, fire-brick grates, improved kitcheners. (*See page* 30.)

[6049]

EDWARDS, WILLIAM, 84 *Wellington Road, Edgbaston, Birmingham.*—Crinoline fire-protectors.

[6050]

ELIOT, EDWARD J., C.E., 7 *Southampton Row, Russell Square.*—An improved cooking apparatus.

[6051]

ELLIOTT, JOHN, 67 *Division Street, Sheffield.*—Quadrant weighing machines, adapted to English and French weights.

[6052]

ELLIOTT'S PATENT SHEATHING AND METAL COMPANY, *Newhall, Birmingham.*—Rolled metals, wire, bolts, spikes, nails, &c.

[6053]

ELLIS, ELIZABETH, *Perseverance Works, Sheffield.*—White metal buttons.

[6054]

ELLIS, G. H., *Grantham, Lincolnshire.*—Boot, knife, and fork cleaners; self-acting game, rat, and mouse traps; washing machines, &c.

[6055]

EVANS, GEORGE, 27 *St. Paul's Street, Walsall.*—Fine welded dog-chains and collars, links, Albert chains, &c.

[6056]

EVANS, JEREMIAH, SON, & Co., 33 *and* 34 *King William Street, London Bridge, E.C.*—Stoves, cooking apparatus, and lamps. (*See page* 32.)

[6057]

EYLAND, MOSES, & SONS, *Walsall.*—Spectacles and eye-glasses of every description, buckles for braces, belts, &c.

EDWARDS, FREDERICK, & SON, 49 *Great Marlborough Street, London, W.*—Porcelain-tile grates, fire-brick grates, improved kitcheners.

DRAWING-ROOM DOG GRATE, No. 7.

All the grates exhibited are of Edwards & Son's own manufacture and design. Most of these are ornamented with porcelain tiles and slabs, which have been made to Edwards & Son's designs by Mr. W. T. Copeland and Messrs. Minton & Co.

1. A LARGE GOTHIC HALL GRATE, with ormolu mouldings and handsome porcelain slabs.

2. A GOTHIC DRAWING-ROOM GRATE in polished steel, with ormolu mouldings and porcelain slabs.

3. A LIBRARY GRATE, with polished steel front and electro-bronzed Grecian mouldings; the porcelain slabs of Grecian design.

4. A BOUDOIR GRATE AND FENDER in polished steel, with ormolu mouldings and richly decorated porcelain slabs.

5. A CIRCULAR DINING-ROOM GRATE, with richly chased ormolu mouldings and porcelain slabs of Italian design.

6. A PEDESTAL HALL STOVE in polished iron, and richly ornamented with ormolu mouldings and electro-bronzed panels and ornaments. The design on the large panels is symbolical of heat. The chained figures in the lower part are intended to represent the subjection of fire to the intelligence of man. Two cherubs are shown above, nestling in foliage and enjoying the genial warmth of a vase of burning fuel. The centre of the top of the stove represents the sun. Around are figures and flowers representing the four seasons.

7. A RICHLY-DECORATED DRAWING-ROOM DOG GRATE, with fender and fire irons; the dogs and fender in ormolu, electro-gilt, and chased by gold chasers; the sides of the grate in polished steel, with porcelain tiles in white and gold, and electro-gilt ormolu mouldings.

8, 9, 10. FIRE-LUMP GRATES in one piece, made in three sizes, the fire bars of wrought-iron. These grates give a large amount of heat with a small consumption of coal, and are of very moderate price.

11. A PEDESTAL HALL STOVE in polished iron, with ormolu mouldings and porcelain panels.

12. A GOTHIC DOG GRATE, with porcelain panels at sides, and a porcelain tile hearth with fender.

13. A GRATE on the smoke-consuming principle, with porcelain slabs.

14, 15. DRAWING-ROOM GRATES, with ormolu mouldings and decorated porcelain slabs.

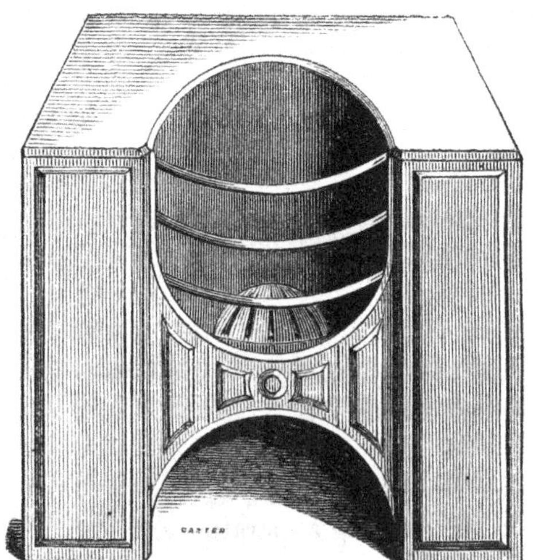

FIRE-LUMP GRATE, No. 8.

16, 17, 18, 19. GRATES of an inexpensive character, with porcelain tiles, ormolu mouldings, and fire-lump backs.

EDWARDS, WILLIAM, 24 *Wellington Road, Edgbaston, Birmingham.*—Patent improved fire-screens or guards.

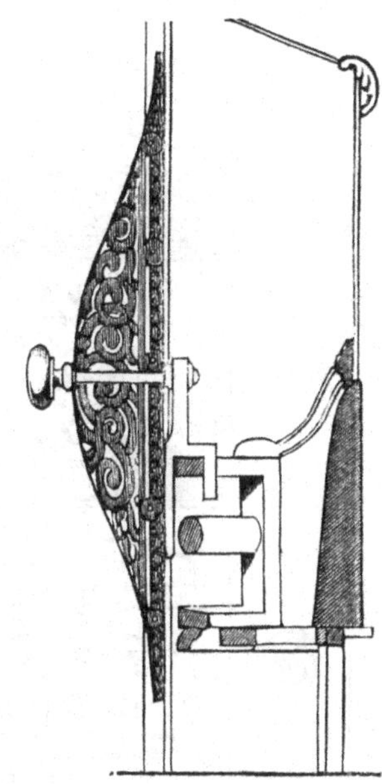

These guards are ornamental in appearance, simple in construction, and will fold or spread out with as much ease as a lady's fan. Are made in any metal, adapted for grates of all shapes and sizes, do not interfere with the cheerful appearance of the fire or the diffusion of heat in the room, do not require to be removed from the grate when the fire is wanted or needs replenishing with fuel. By simply turning the handle in front, the guard can be folded leaf over leaf into the space of one, and so give free access to the fire.

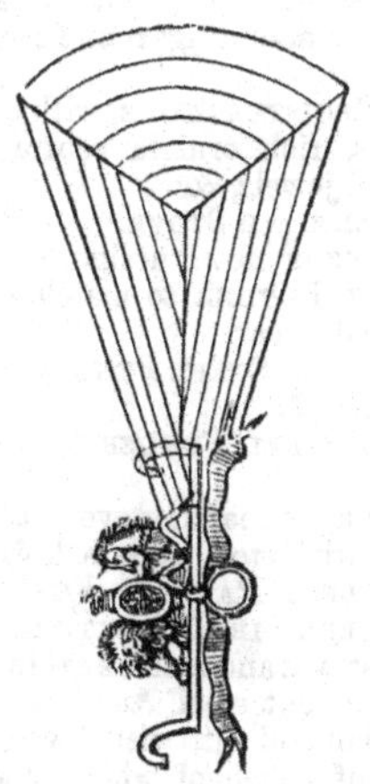

FIRE GUARD, CLOSED.

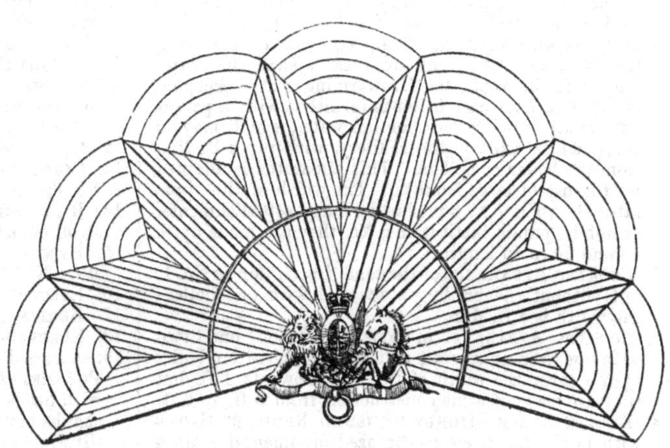

FIRE GUARD, OPEN.

Licensed manufacturers—

William Burgess, Holloway Head, Birmingham.

Henry Crichley, Coventry Road, Birmingham.

Samuel Robotham, Bradford St. Birmingham (in wire).

William Soutter, New Market Street, Birmingham.

Drawings with prices will be forwarded on application.

EVANS, JEREMIAH, SON, & CO., 33 *and* 34 *King William Street, London Bridge, E.C.*—Stoves, cooking apparatus, and lamps.

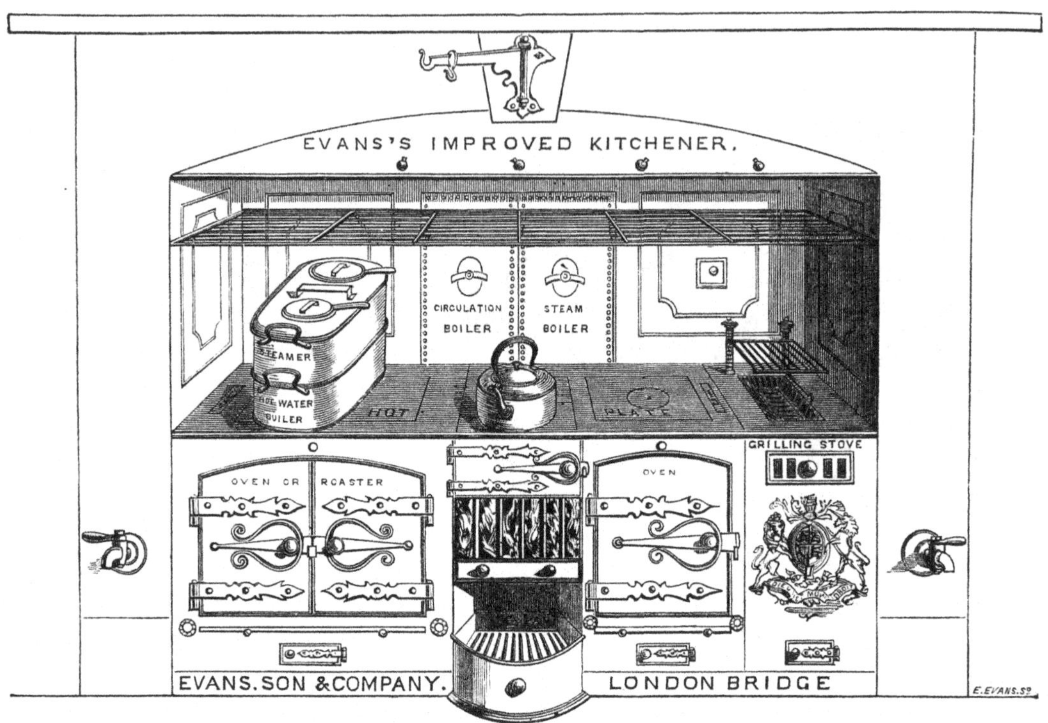

1. IMPROVED EVANS' KITCHENER, with hot-plate top, with loose plates, large wrought-iron oven on one side, and roaster on the other, fitted with shelves, best scroll spring latches and hinges; a broiling stove at one end, two capacious best hot rivetted wrought-iron boilers at back, one for steaming purposes if required, the other a sealed or pressure boiler for furnishing a constant supply of hot water all over a house for a bath, for wash-hand basons, or for the use of housemaids on the various landings, as may be found convenient, the back and sides fitted with metal covings or plates with regulating dampers, doors for cleansing the flues, and a wrought-iron rack or shelf the entire length of the apparatus for warming plates, keeping dinner hot, &c.; each boiler is furnished with a patent gun-metal draw-off cock for hot water, which may, if preferred, be fixed in the scullery, with pipes leading from the boiler or boilers. This apparatus is adapted for the use of large families. The width of the apparatus shown is 7 ft. 6 in. but for large establishments, as hotels, public institutions, &c. they are made up to 18 and 20 ft. in length.
2. COOKING APPARATUS of the same description as No. 1, but without the broiling stove and with one boiler; the width is 5 ft. 6 in.
3. COOKING APPARATUS of same description, but fitted with one oven or roaster, and one wrought-iron boiler at the side; the width of the one shown is 4 ft. but they are made of same construction from 3 ft. to 6 ft.
4. BERLIN BLACK HIGHLY-FINISHED REGISTER STOVE with bright front, electro-bronzed ornamental margin round ditto, and two ormolu twisted beads. A steel gadrone ashes pan fitted to ditto.
5. DEAD SPRUNG ARCH-FRONTED REGISTER STOVE, with burnished steel twisted moulding and lift-off ormolu ornaments, fire-brick back, &c.
6. BERLIN BLACK RADIATING REGISTER STOVE, with bevelled arch front, ornamental leafage moulding, and polished steel cable moulding. An improved radiating ashes pan fitted to ditto.
7. ELEGANT BURNISHED STEEL DRAWING-ROOM REGISTER STOVE, with highly chased enriched cable ormolu moulding, and centre and pendant ornaments; also steel cable moulding to inner front, bold bright bars with ormolu feet, banister bar, cut steel spikes, &c.
8. NEAT DRAWING-ROOM REGISTER STOVE, with burnished steel slip and bead round inner edge, and ormolu lift-off ornaments, porcelain cheeks, &c.
9. NEAT DINING-ROOM REGISTER STOVE, with splayed arch front with ormolu beaded mouldings round panels, of classic design.
10. BRIGHT DRAWING-ROOM REGISTER STOVE, with burnished steel ogee mouldings, rich ormolu centre and corner ornaments, with steel jewels, &c.
11. ROUND ORNAMENTAL WARM-AIR STOVE.
12. OCTAGON WARM-AIR STOVE of novel design.
13. CIRCULAR BERLIN BLACK FENDER to match stove No. 6, with polished steel cable bar.
14. RICH ORMOLU AND POLISHED STEEL FENDER, with scroll ends to match stove No. 7.
15. HANDSOME ORMOLU AND STEEL FENDER to match stove No. 8.
16. ELECTRO-BRONZED FENDER, to match stove No. 9.
17. BERLIN BLACK FENDER, with steel cable rod, &c.
18. ELECTRO-BRONZED FERN-LEAF PATTERN FENDER.
19. A varied assortment of STEEL FIRE FURNITURE AND POKERETTES to match the stoves and fenders exhibited, with heads of ormolu, bronze, cut steel, &c.; and with shanks plain, octagon cut, diamond cut, twisted, &c.; and illustrating the perfection of polish of which steel is susceptible.
20. BRASS CRINOLINE GUARD of improved construction.
21. JASPER MARBLE MANTEL-PIECE, very rich in colour, with arched opening; fitted to stove No. 6.
22. BOLD POVONAZZI MARBLE MANTEL-PIECE.
23. SIENNA MARBLE MANTEL-PIECE, with moulded shelf, bold columns at sides, &c.; adapted to stove No. 9.
24. PAIR OF SOLID POLISHED BRASS FIRE DOGS.

[6058]

FEETHAM, MILLER, & SAYER, 9 *Clifford Street, London.*—Ornamental iron and brass work, stoves, grates, and fenders, &c.

GROUP OF STOVES, FIRE DOGS, WROUGHT-IRON GATES, &c.

[6059]

FIELD, WILLIAM, & SON, 224 *Oxford Street.*—Patent and other horse shoes as used in England.

[6060]

FIELDHOUSE, GEORGE, & CO., 3 *Poulteney Street, Wolverhampton.*—Steel coffee and other mills.

The great superiority and cheapness of the exhibitor's best quality steel mills arises from the fact of their having introduced machinery in their production, by which means they are enabled to make each part to a standard size, so that it may be replaced (in case of loss or breakage) without the expense of carriage of the whole mill. The teeth of the grinding parts are made to one uniform angle and shape, which they have proved from considerable experience to be the best to ensure their grinding easily and quickly.

[6061]

FINCH, JOHN, *Priory Street Works, Dudley.*—Fenders, fire-irons, hat and umbrella stands, garden seats, and bedstead castings.

[6062]

FINLAY, JOHN, *Glasgow.*—Patent grates, exhibiting the most perfect central oven combination, with powerful radiation.

[6063]

FIRMAN & SONS, 153 *Strand, London, and* 2 *Dawson Street, Dublin.*—Military ornaments.

[6064]

FITZWYGRAM, LIEUTENANT-COLONEL, 15*th Hussars, Dublin.*—Improved horse shoes.

[6065]

FLAVEL, SIDNEY, & CO., *Eagle Foundry, Leamington.*—Improved kitchener.

[6066]

FRANCIS, EDWARD, *Camden Place, Dublin.*—Specimens of horse shoes for diseased and healthy feet, shod hoofs, &c.

Has obtained Medals and Honorary Certificate of the Royal Dublin Society.

This exhibitor holds the appointment of farrier to Her Majesty, the Lord Lieutenant, the officers of the staff, the metropolitan police, &c.

[6067]

FREARSON, JOHN, 10 *and* 11 *Clement Street, Birmingham.*—Patent hooks and eyes for ladies' garments.

[6068]

FULLER, WILLIAM, 60 *Jermyn Street, London.*—Improved patent freezer for making cream and water ices.

[6069]

GALE, SAMUEL, 320 *Oxford Street, W.*—An arrangement of bell-wires to prevent friction or enlargement; a register for chimney; a curious lock made by an amateur locksmith 70 years ago.

The improved register exhibited is cheap, and can be applied to any stove. It effectually prevents the smoke from other chimneys entering a room where there is no fire.

[6070]

GEDDES, JOHN, 4 *Cateaton Street, Manchester.*—Ornamental wire plant-stands, model rosery, and verandah.

[6071]

GENERAL IRON FOUNDRY COMPANY, *Upper Thames Street, London.*—Stoves, mantels, bronzes, &c., cooking apparatus, coil cases, castings.

TABLE RANGE OR COOKING APPARATUS for centre of kitchen, contains two large and powerful wrought-iron ovens and two wrought-iron roasters, with wrought boilers capable of supplying very extensive steam and hot water apparatus, and a hot plate containing about 40 ft. of cooking surface. The smoke would be conducted away by means of underground flues to any available shaft.

This range is of extraordinary power, with one fire and an exceedingly small consumption of fuel, an immense amount of cookery in every possible variety can be conducted; it is estimated that dinners could be supplied for 3,000 persons in one day from this single apparatus, while it is equally adaptable for cooking a dinner for a dozen persons.

CLOSE-FIRE COOKING RANGE, 6 ft. with hot plate, two wrought-iron roasters convertible into ovens, and wrought-iron circulating boiler capable of heating 200 gallons of water.

This range is adapted for a large private family or small hotel; but is capable of unlimited extension; it is thoroughly effective as well as economical in the working.

COIL CASE for hot-water heating, of cast-iron bronzed, with white marble top, coil of pipes enclosed.

LARGE BLACK MARBLE MANTEL-PIECE, with sculptural features in fine bronzes (Potts's patent.)

STOVE for ditto, ventilating and smoke consuming (Taylor's patent).

FENDER AND FIRE IRONS for ditto.

CAEN STONE MANTEL-PIECE, of ecclesiastical character, with fine panel (the Sermon on the Mount) and other enrichments in bronze (Potts's patent).

STOVE for ditto, ventilating and smoke consuming (Taylor's patent).

FENDER AND FIRE IRONS for ditto.

STATUARY MARBLE MANTEL-PIECE, with electro-gilt metal enrichments (Potts's patent).

STOVE for ditto in burnished steel.

FENDER AND FIRE IRONS for ditto.

SIENNA MARBLE MANTEL-PIECE, with enrichments in oxidized silver metal work (Potts's patent).

STOVE for ditto, ventilating and smoke consuming (Taylor's patent).

FENDER AND FIRE IRONS for ditto.

Four WROUGHT-IRON WINDOWS (Moline's patent), extensively used for warehouses, wharfs, railway stations, and other public buildings.

STOVE fitted to show the action and mode of fixing of Billing's patent throats and air apparatus for the prevention of smoke.

ALTAR RAIL in cast-iron of gothic character. The design taken from Mr. Digby Wyatt's "Metal Work and its Artistic Design."

Series of FINE ARTISTIC BRONZES, suitable for architectural, cabinet, and other decorations, produced by Mr. William Potts of Birmingham.

Three TABLETS or MURAL MONUMENTS, in bronze, produced by Mr. W. Potts of Birmingham.

Fourteen BALUSTERS for staircases of various designs, in cast-iron.

Four GRATINGS for heating or ventilating purposes.

CASTINGS, various.

[6072]

GIBBONS, JAMES, *St. John's Lock Manufactory, Wolverhampton.*—Ornamental locks, keys, and hinges, general ironmongery.

[6073]

GIBBONS & WHITE, 345 *Oxford Street.*—Wrought-iron weather-tight casements; Gibbon's patent lock furniture.

[6074]

GIBSON, THOMAS, *Cape Works, Birmingham.*—Specimens of springs, axletrees, and carriage iron work, patent and otherwise.

[6075]

GILLETT, WILLIAM, 18 *Back Street, Bristol.*—Two improved bottling machines.

[6076]

GINGELL, WILLIAM JAMES, *Bristol.*—Model of a uniform corn-meter.

[6077]

GLASS, ELLIOTT, & CO., 10 *Cannon Street, London; Manufactory Cardiff.*—Iron and steel wire ropes.

[6078]

GODDARD, *Nottingham.*—New patent economical cooking apparatus either for a close or open fire.

[6079]

GOLLOP, EMILIA, *Charles Street, City Road, London.*—Redmund & Gollop's patent floor springs, rising and not rising hinges, gate hinges, &c.

[6080]

GRAY, A., & SON, 9 & 11 *Weaman Street, Birmingham.*—Fire-irons, &c.

[6081]

GRAY, JAMES, & SON, 85 *George Street, Edinburgh.*—Stove with ormolu pillars.

JAMES GRAY & SON are stove grate makers to her Majesty, and manufacturers of kitchen ranges, bankers' safes, locks, &c. They exhibit a drawing-room stove with ormolu pillars, china tile covings, and fender to correspond.

[6082]

GREEN, JOSEPH, 134 *Irving Street, Birmingham.*—Builders' iron work, and other articles suitable for domestic purposes.

[6083]

GREENING & CO., *Manchester.*—Wire park fencing, manufactured of unusual strength and height by patent machinery.

[6084]

GREENING, N., & SONS, *Warrington, Lancashire.*—Wire cloth woven by steam power, of extraordinary width and strength.

[6085]

GRIFFITHS & BROWETT, *Birmingham, and* 8 *Broad Street Buildings, London.*—Wrought-iron tinned, japanned, and enamelled wares; tin-plate wares. (*See page* 37.)

[6086]

GROUT, ABRAHAM, 8 *Shephard Street, Spitalfields.*—Models of flower stands, summer houses, pheasantries, and ornamental fences in wire.

[6087]

GUY, S., 3 *Haunch of Venison Yard, Brook Street, New Bond Street.*—Specimens of horse-shoeing.

[6088]

HAGUE, THOMAS, *Bridge Street, Sheffield.*—Fire irons, with or-molu, bronze, and steel heads.

Griffiths & Browett, *Birmingham, and* 8 *Broad Street Buildings, London.*—Wrought-iron tinned, japanned, and enamelled wares; tin-plate wares.

Obtained Prize Medals at the International Exhibitions of 1851 *and* 1855.

Griffiths & Browett are general iron and tin-plate workers, japanners, manufacturers of tinned and enamelled wrought-iron hollow ware, Loysel's patent hydrostatic urns, Vose's patent hydropult, and Keevil's patent cheese-making apparatus.

GROUP OF WROUGHT-IRON TINNED AND JAPANNED WARES.

They exhibit:

A set of papier-maché trays, ornamented in the moresque style.

A set of papier-maché trays, ornamented in the Indian style.

Patent raised hot-water dishes and covers, soup and vegetable dishes, and soup tureens.

Patent tea and coffee pots, sugar basins, and cream jugs.

Papier-maché folios ornamented by a patent process.

Toilette sets, grocers' furniture, Persian coal vase.

A variety of curious and novel specimens of wrought-iron work raised from flat sheets of metal, without seam or brazing.

Paris patent enamelled ware, plain and printed.

[6089]

Hale, James, *Hatherton Works, Walsall.*—Spring hooks, curb chains, pole chains, South American bits, &c.

[6090]

Hall, Robert, 4 *Laurie Street, Leith, Scotland.*—Malleable iron branding stamps. Impressions of ditto on wood.

[6091]

HALLEN & HALLEN, 76 *Oxford Street.*—Stable fittings.

[6092]

HAMILTON & Co., 3 *Royal Exchange, E.C.*—Patent locks and safes. (*See page* 39.)

[6093]

HAMMOND, TURNER, & SONS, *Birmingham.*—Buttons, military ornaments, and fancy dress fasteners.

[6094]

HANDYSIDE, ANDREW, & Co., *Britannia Foundry, Derby.*—Fountains and vases.

Obtained a Prize Medal at the Exhibition of 1851.

FOUNTAIN, 6 ft. dia. 12 ft. 9 in. high.
Ditto 3 ft. 6 in. dia. 5 ft. 6 in. high.
Ditto 3 ft. dia. 5 ft. high.

VASE AND PEDESTAL 2 ft. 6 in. dia. 8 ft. 6 in. high.
LAMP PILLAR, with drinking fountain, 13 ft. high.

[6095]

HARLEY, GEORGE, 43 *Warwick Street, Wolverhampton.*—Patent lock and night latches.

[6096]

HARLOW & Co., *Smethwick, near Birmingham.*—Metallic bedsteads.

[6097]

HAWKINS, JOHN, & Co., 38 *Lisle Street, Leicester Square, and* 16 *Station Street, Walsall.*—Bits, stirrups, spurs, &c.

[6098]

HAYWARD, BROTHERS, 117 *Union Street, Southwark.*—Patent kitchen ranges, ventilators, coal-hole plates, lock furniture. (*See page* 40.)

[6099]

HEATON, RALPH, & SONS, *The Mint, Birmingham.*—Coins complete, and the same in progress of manufacture.

COINS complete, and the same in progress of manufacture, made by Messrs. Heaton for the English, French, Indian, Italian, and other governments.

Heaton & Sons furnish estimates for complete coinages, and execute them either in England or abroad.

[6100]

HENN, ISAAC, *Rea Street Works, Birmingham.*—Taper-pointed wood screws in iron and brass; also coach screws.

[6101]

HEWENS, RICHARD, 120 *Warwick Street, Leamington Priors.*—Improved Leamington kitchener, with Hewen's patent regulator.

[6102]

HIATT & Co., 26 *Masshouse Lane, Birmingham.*—Police handcuffs, leg-irons, padlocks curb, dog-collars, &c.

[6103]

HILL & SMITH, *Brierley Hill, Staffordshire.*—Specimens of forged iron work, railway and cart axletrees, &c.

HAMILTON & Co., 3 *Royal Exchange, E.C.*—Patent locks and safes.

Silver Medal, Society of Arts, 1859.

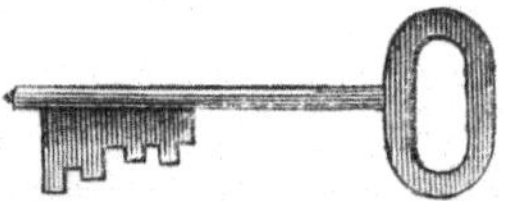

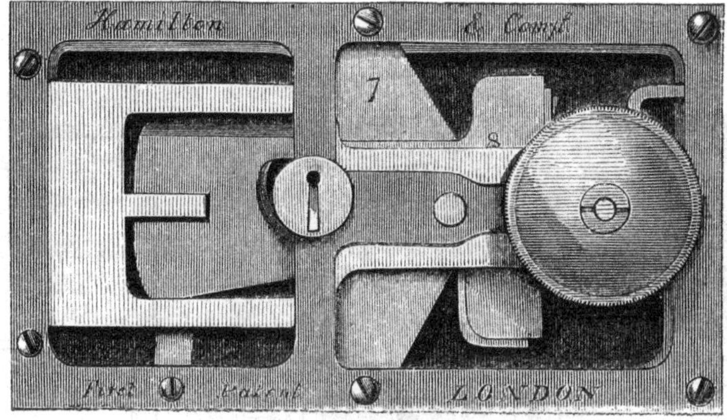

FIRST PATENT LOCK.

The FIRST PATENT LOCK possesses the following advantages:—

1. A very small key.—In locking, the bolt is shot by simply turning the knob. In unlocking, the key, which may be very small and light, is inserted and turned round once, raising the levers with a very gentle touch to the proper position, and is then taken out, and the bolt is withdrawn by turning the knob back again.

2. It cannot be picked, because the only time when the key-hole is open, is when the stump of the bolt is at a distance from the levers, and any instrument inserted through the key-hole holds the bolt fast and prevents it from being forced back against the levers in order to feel for the gratings. If the instrument be removed, the bolt can then be forced back, but the same action completely closes the key-hole.

3. It cannot be deranged, the levers being completely under control: if they are thrown down, the key raises them; if forced up too high they can be depressed by means of the handle.

4. Excludes air and damp.—When the lock is open the key-hole is closed. When locked, the key-hole may also be closed by bringing the bolt back a short distance; this excludes air and damp.

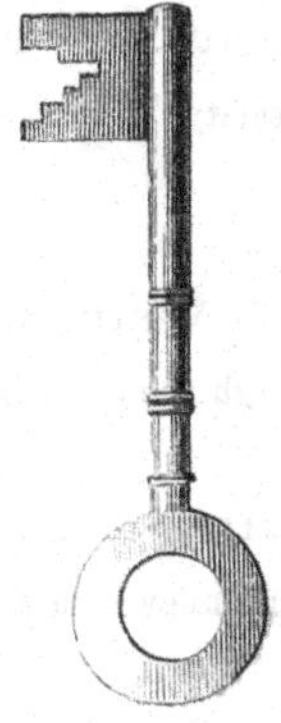

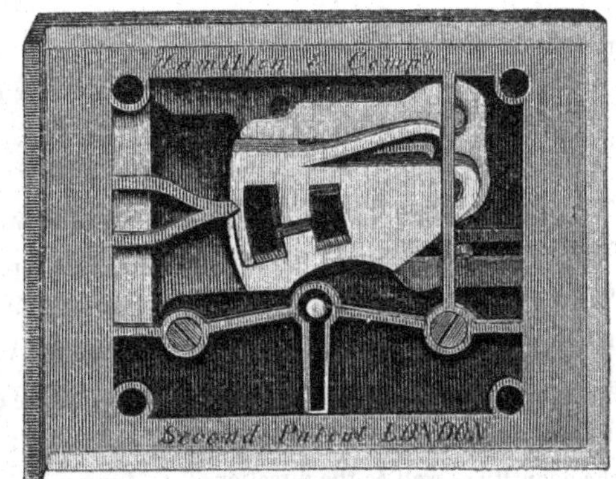

SECOND PATENT LOCK.

The peculiarity of Hamilton & Co.'s SECOND PATENT LOCK, which cannot be picked, consists in the tumblers being secured on a movable axis fixed on the tail of a bell crank lever, which when pressure is applied falls into a notch in the bolt, and on the pressure being continued the tumblers recede farther and leave a space between them and the stump of the bolt, so that the tumblers are always free.

They manufacture also lever and other locks, safes. deed and cash boxes, &c.

Hayward, Brothers, 117 *and* 118 *Union Street, Borough, London, S.E.*—Patent kitchen ranges, ventilators, coal-hole plates, lock furniture.

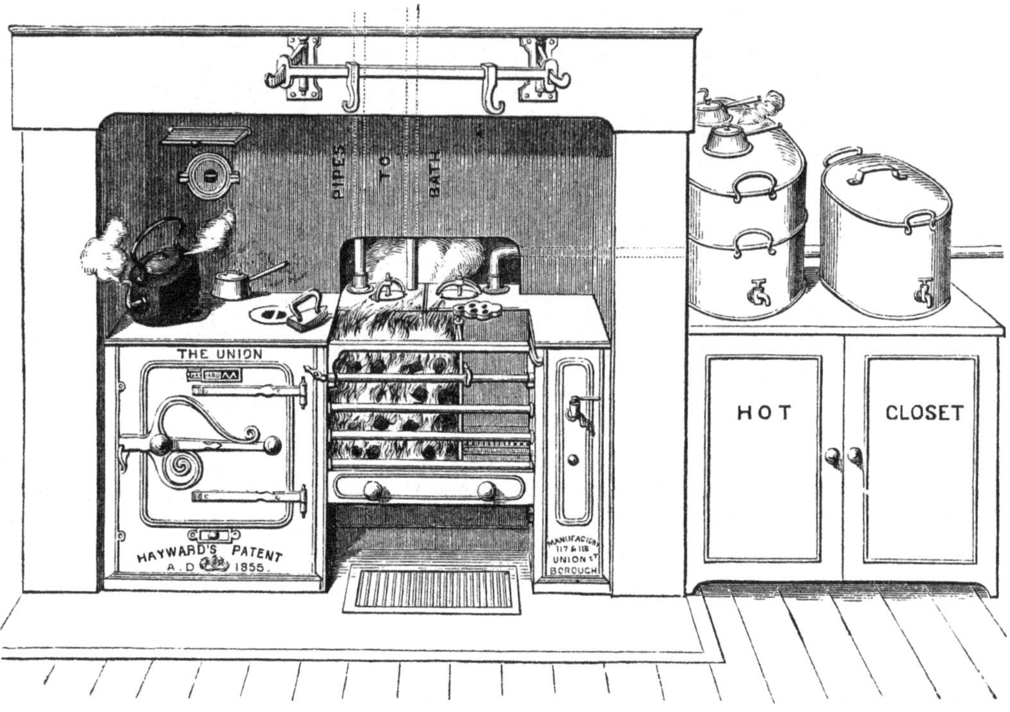

"THE UNION" KITCHEN RANGE.

1. "The Union" Kitchen Range, with open fire.

In this range the whole of the hob above the oven is a boiling surface or hot plate, and it embraces all the advantages of an enclosed cooking apparatus without its offensive smell, imperfect ventilation, &c.

The oven may be kept "slow" or raised at pleasure for baking bread, &c.; or to a quick and scorching heat for roasting meat. When baking pastry the heat can be passed to the top of the oven to raise the crust, and then equally distributed; the quantity as well as the direction of the heat being entirely under control.

The boiler is adapted for heating a large supply of water to any part of the house, for baths, &c. A second boiler can be added for steaming, if required.

2. Improved Coal-hole Plates and Pavement Lights, for safety, light, ventilation, and prevention of accident.

3. Sheringham's Ventilators, for the admission of fresh air through the external walls by day and night.

4. Arnott's Valves, for the extraction of vitiated air through the chimney breast.

5. Circular Iron Staircases on an improved principle, which renders them very strong and firm.

Price lists of the above, and estimates for hot-water work, will be forwarded on application.

[6104]

CHAPMAN, THOMAS, late **HILLIARD & CHAPMAN**, 56 *Buchanan Street, Glasgow.*—Patent knife cleaners, knife sharpeners, and lockfast table knives.

Obtained Prize Medal at the Exhibition of 1851.

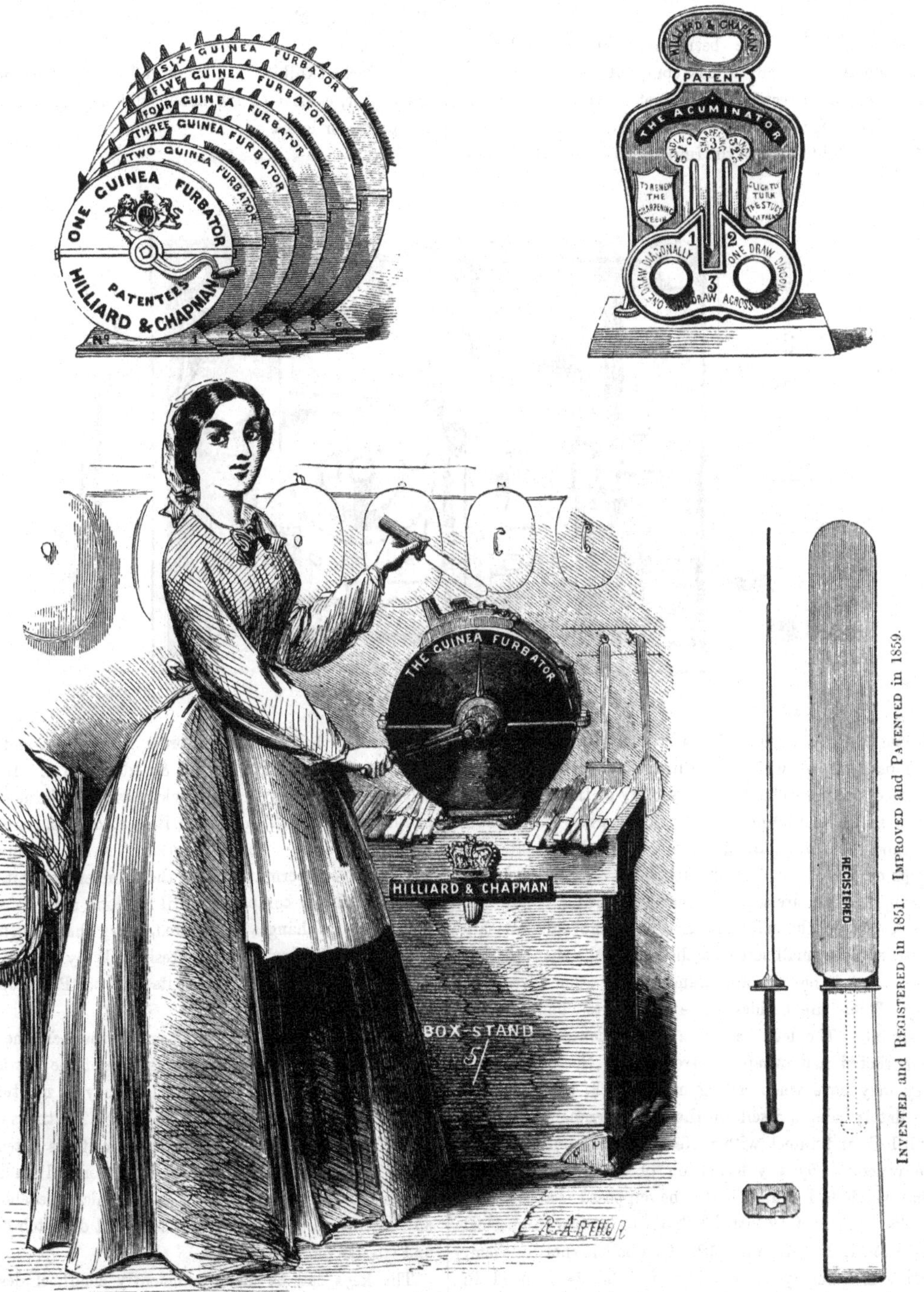

PATENT FURBATORS (improved knife cleaners), six different sizes, with pillar stand, box stand, and bracket for holding the same.

PATENT ACUMINATORS (improved knife sharpeners), various patterns and sizes.

PATENT LOCK-FAST TABLE KNIVES, various patterns, with section of handle showing the principle.

PORTABLE FORK CLEANER.

[6105]

Hobbs & Co., 76 *Cheapside, E.C.*—Locks, patent and machine made, and door lock fastenings.

The exhibitors are inventors, patentees, and manufacturers of bank, protector, and other locks, and the lock-making steam machinery.

Hobbs's locks have been awarded the following testimonials in their various competitions :—The Prize Medal of the Great Exhibition of London, 1851 ; the First-Class Medal of the Imperial Exposition of Paris, 1855 ; the Gold Medal of the Imperial National Mechanics' Institute of Vienna. In addition to these, are two gold and three silver medals from various Associations for the Promotion of Mechanical Science in the United States of America.

The locks exhibited on the stand to the right hand of the visitor, consist of the changeable-key bank lock and the protector locks.

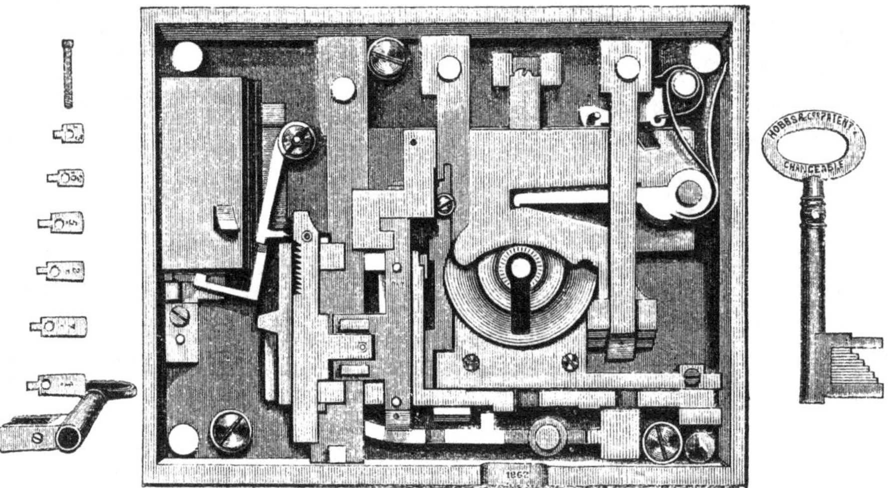

Hobbs & Co.'s Patent Parautoptic, or Bank Lock.

This lock, of which an illustration is subjoined, is deemed unapproachable as a security of the repositories of treasure, and impregnable against every practicable method of picking, fraud, or violence. The "bits" or steps on the "web" of the key, that act on the levers inside the lock, are separate, instead of being, as in other keys, cut on the solid metal. These movable bits are fastened by a small screw on the end of the shank of the key, when it has the appearance of any other lever-lock key. There are besides, spare "bits" to change, when desirable. The lock has three sets of levers, and is so constructed that, whatever arrangement the bits on the key may have when acting on the lock, the latter immediately adapts itself to the same arrangement, and will lock and unlock with perfect facility ; but it cannot be unlocked by any formation of the "bits" except that which locked it. Let it be supposed that the lock works with a "12-bitted" key, in proper numerical order, as 1, 2, 3, &c. up to 12. The bolt is shot by them, and will open by them ; but if a bit is changed in its place, the lock will remain locked, because, by the alteration, the key has become also changed in its action. to which change the levers will not answer. To re-lock in another form :—Suppose that, instead of the bits being arranged as 1, 2, 3, &c. the order is reversed, and they are screwed on as 12, 11, 10, &c down to 1. By the self-changing principle of the lock, it assumes the new form of the key, and will work with it as readily and securely as it did before. The same results can be obtained by any and every permutation of the number of "bits" of which the key is composed, until millions, and thousands of millions of changes are worked, every change virtually converting the lock into a fresh lock by this simple transposition of the key. Hence its name of "Parautoptic," or changeable.

The illustration represents a view of the lock, the key, and the spare "bits." To give an idea of the number of times this lock can be transposed, it may be mentioned, that a key of only six bits can be altered 720 times ; and if two sets of bits are used, the transpositions extend to many thousands. The price of locks for bullion safes, and the doors of strong rooms, &c. of which the above is an illustration, is £20 ; and for cash and despatch boxes, and similar purposes, £10.

The keys can be made sufficiently small, if desired, either for the waistcoat-pocket or the travelling-case. It is claimed for both locks and keys that they illustrate the highest degree of scientific and mechanical skill in the locksmith's art.

HOBBS & CO., *continued.*

HOBBS & CO.'S PATENT PROTECTOR SOLID KEY AND INDEX LOCKS.

The patent protector locks are exhibited as possessing absolute security against picking by any method at present known. The key is what is called "solid," that is, that the "bits" or "steps" are cut on the solid metal of the "web," and, therefore, not changeable. They are specially adapted for places where the most ample security against lock-picking is required.

The "protectors" of this lock consist of a peculiar arrangement of certain parts behind the bolt and levers, unreachable by any lock-picking instrument whatever. When any tampering is attempted on the lock, by pressure on the bolt through the key-hole, to discover the opening position of each lever, the bolt protector comes into action, preventing the pressure affecting the levers in any way, thus holding them clear, and thereby frustrating the calculations of the thief. This principle was first introduced in locks at the memorable Exhibition of 1851, and forms the foundation of a new security. The key and bolt fraud-protector is a movable nozzle, now first introduced. These two protectors combined are offered to the public, as the two essentials of security—protection against picking, and protection against fraud. Specimens are shown illustrating the action of the protectors. There is also a model showing the arrangement of the bolts and locks as fixed on a strong-room door. The protector locks are sold, retail, at prices varying from 10*s*. to 40*s*.

HOBBS & CO.'S PATENT LOCK INDICATOR.

This is a method of locking the doors of iron safes, strong rooms, customs stores, bonded vaults, prison cells, corridors, &c. by means of the handle, without a key, and showing to what extent the bolt has been shot. It may consist of the upper half of a dial, upon which are the words, "Open," "Shut," "Locked." When the door stands

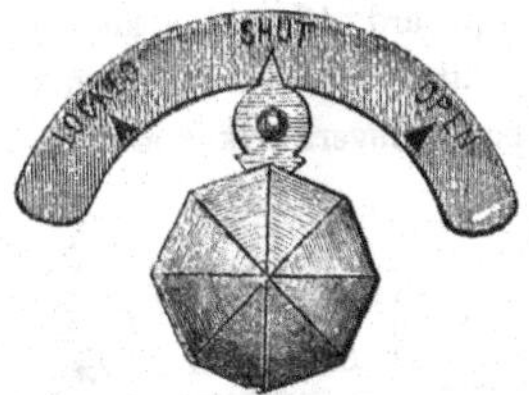

merely closed to, the index finger rests on "Open." This finger is fixed to the handle that works the lock, and therefore, whichever way the handle moves, the finger must move with it. Turn the handle, and fasten the door by the first movement of the bolt, the finger will point to "Shut." A second motion of the handle, and the bolt shoots out beyond its reach, the finger, at the same moment, resting on "Locked." The lock can only be opened by the key, because, at the second turn, the handle loses its control of the bolt. The action of the bolt returning into the lock, or unlocking, takes the index finger back to "Open," re-setting it again. The advantages of this index in dock yards, shipbuilders' stores, dock warehouses, prisons, &c. where certain officers are limited to departments of the premises, by day or night, must be of the highest importance. The superior officer would be able, by its use, to see in an instant what condition the bolts of the locks were in,

without "trying" his keys, as he passed along a corridor, or by a range of rooms. Again, if the door of a safe or strong-room was closed tight, there would be no danger of leaving it unlocked by neglect, as a glance at the index would show whether it had been locked or not. Specimens are also exhibited, showing the application of the "lock-index" principle to street-door latches, and convict and other prison cells.

In the case on the right hand is a model, showing how applied to bankers' strong-room doors, of which an illustration is here given. This arrangement gives quadruple security against violence, which is obtained by levers, eccentrics, and other means.

In the centre of the stand is a first-class strong-room door. It is made of the best iron-plate, back and front, the interior being lined with slabs of hardened steel. In this door the bolts are of the usual arrangement—they are thrown by the knob, an examination of which will show the great security attained by Hobbs and Co.'s patented method for security. It will be seen to consist of triple security against violence of all kinds, while the lock is peculiarly constructed and is powder-proof, holding less than twenty grains of gunpowder, which is totally insufficient to blow it off.

HOBBS & CO., *continued.*

HOBBS & CO.'S MACHINE-MADE LOCKS.

This group of locks is in the stand on the left of the spectator. They are constructed in all sizes and varieties, suitable for every purpose for which locks are used. In consequence of their manufacture by steam-machinery, in a manner previously unknown in the trade, they exhibit the first important step of progress in this country, in the economy of lock-producing. By means of the different machines used in the making, a faultless accuracy is arrived at, which, by hand, would be quite impossible.

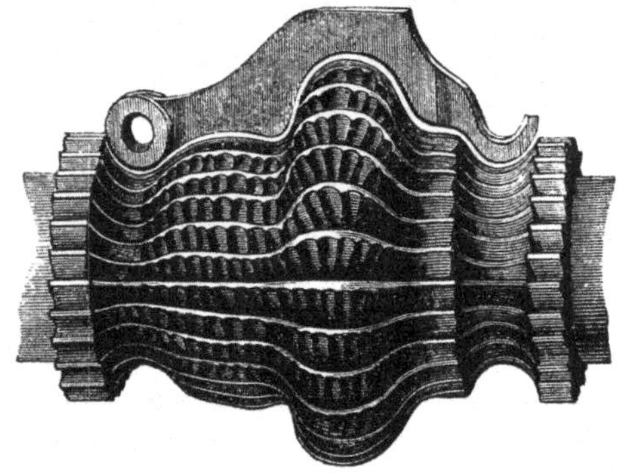

Annexed are illustrations of cutters and dies by which such accuracy is obtained :—

These locks possess a strong recommendatory quality to the owners and occupiers of house property, in their comparative cheapness to hand-made locks, being made of superior materials, combined with the nicety of the finish of the working parts, effected by the machinery, and their universal adaptability. References can be given to most of the Government offices, as to their durability and security; and also to nearly all the metropolitan and many of the country banks, as well as to architects, engineers, and builders of the highest standing. They are made from two up to five levers, and the retail price ranges from 2*s.* upwards. For cottages, mansions, warehouses, &c. these locks will be found most desirable. They are very extensively used by builders, for doors, closets, cupboards, &c. and by cabinet makers for sideboards, wardrobes, desks, drawers, dressing-cases, and all kinds of cabinet work.

The woodcut shows a mortise lock, adapted for room doors, price 8*s.* The security consists in a series of levers being raised to unequal positions by the bits of the key before the bolt can pass.

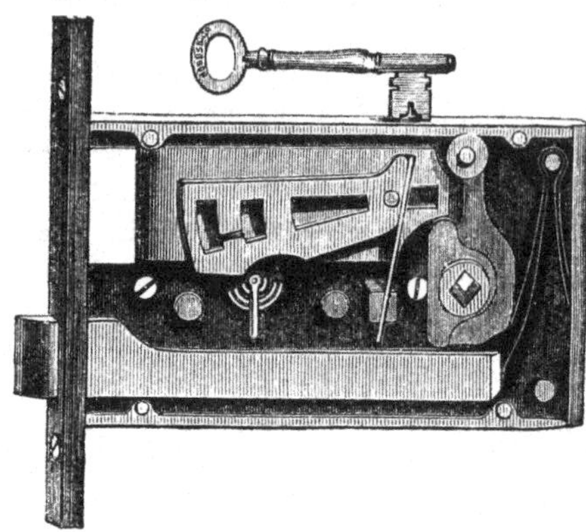

Both the protectors and the machine-made lever locks can be fitted in suite for master-keys to pass a given number of locks, that otherwise open with a different key each. The convenience afforded by a master-key in premises divided among subordinates is very great. All the locks of Messrs. Hobbs & Co. can be obtained from every respectable ironmonger in Great Britain and Ireland. Merchants and shippers are also supplied, at wholesale prices, for exportation.

DOOR LOCKS.

	s.	d.
6 in. 1 bolt mortise, 2 levers, 1 key each . .	5	0
7 in. 2 bolt ditto 4 ditto ditto . .	9	6
6 in. 1 bolt rim 2 ditto ditto . .	4	0
6 in. 2 bolts ditto 2 ditto ditto . .	5	6
7 in. 2 bolts ditto 4 ditto ditto . .	10	0

CABINET LOCKS.

	s.	d.
2¾ in. till, 4 levers, 1 key each	2	0
3 in. till, 4 ditto ditto	2	6
3 in. cupboard, 4 levers, 1 key each . . .	2	6
4 in. ditto 2 levers, japanned. . . .	2	0
2½ in. box, 2 levers, 1 key each	2	0

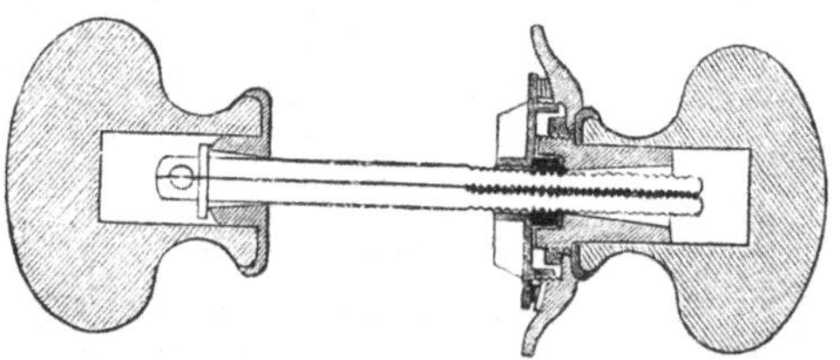

HOBBS & CO.'S PATENT LOCK FURNITURE.

This lock furniture is so planned that the most unskilful workman cannot fail to fix it properly. It is self-adjusting; the handle will not bind at whatever angle the lock may be mortised. It is very moderate in price, will endure hard and lengthened wear, and acts without friction.

Manufactory, Arlington Street, Britannia Fields, N.

[6106]

HOOD, SAMUEL, & SON, 68 *Upper Thames Street, London.*—Stable fittings.

SET OF STABLE FITTINGS, consisting of improved hay rack with patent spring top, enamelled manger, enamelled water cistern with removable cover, and noiseless tying apparatus with the halter and head piece.

SET OF LOOSE-BOX FITTINGS, consisting of hay rack, manger, and water cistern with improvements as above.

STALL DIVISION, consisting of wrought-iron heel post, cast-iron ramp, middle rail, panel, and sill.

DIVISION FOR LOOSE BOX, consisting of wrought-iron door post and shutting post, door with improved hinges, top rail, middle rail, panel, and sill.

SURFACE DRAIN and drain trap of improved construction.

IMPROVED SADDLE BRACKET.

[6107]

HOOD, WILLIAM, 12 *Upper Thames Street, E.C.*—Drinking and garden fountains, lamp-posts, lamps, and specimen castings.

[6108]

HOOLE, HENRY E., *Green Lane Works, Sheffield.*—Grates, fenders, fire irons.

[6109]

HOPKINS, J. H., & SONS, *Granville Street, Birmingham.*—Block tin, stamped, tinned iron, and japanned articles. (*See page* 47.)

[6110]

HULSE & HAINES, *Ichnield Street West, Birmingham.*—Brass and iron bedsteads.

[6111]

HURST, C. H., *Royal Road, Kennington Park, S.*—Patent wrench and mallet to save all taps from damage, and infallible transparent cement to repair china, glass, &c.

The tap wrench and mallet is an ingenious contrivance for preventing damage to wine and other taps. Price 2*s.* complete. Lever tap wrench alone, 1*s.*

The cement will effectually repair broken glass or china. It is transparent, and will bear washing in hot water. Price 1*s.* and 2*s.* per box. Post-free for 14 or 28 stamps.

[6112]

ILES, CHARLES, *Peel Works, Birmingham.*—Hooks and eyes, thimbles, pins, needles, hair pins, &c.

Obtained a Prize Medal at the Exhibition of 1851.

Specimens of hooks and eyes, thimbles, patent enamel-lined thimbles, solid-headed pins, hair pins, and fancy boxes, and articles for containing and connected with the above manufactures.

[6113]

ILIFFE, & PLAYER BROTHERS, *Birmingham and London.*—Buttons, medals, military ornaments, and patent umbrellas.

[6114]

INGRAM, GEORGE WELLS, 1 *Lombard Street, Birmingham.*—Powder flasks, shot pouches; crimping and goffering machines.

[6115]

ISMAY, THOMAS, & CO., *Dover.* — Improved close-fire ranges for large kitchens. (*See page* 48.)

[6116]

JAMES FOUNDRY COMPANY, THE, *Walsall.*—Iron and brass, and builders' ironmongery.

[6117]

JAMES & SONS, *King's Norton, and Bradford Street, Birmingham.*—Patent self-boring wood-screws.

[6118]

JAMES, J., & SONS, *Victoria Works, Redditch.*—Needles and fish-hooks.

[6119]

JEAKES, C., & CO., 5 *Great Russell Street, Bloomsbury, London.*—Kitchen range and fittings; grates, brass and iron work. (*See page* 48.)

[6120]

JEAVONS, I. & D., *Petit Street Works, Wolverhampton.*—Wrought-iron hollow-ware, &c.

[6121]

JEFFREY & JAFFRAY, 2 *Allen's Court,* 387 *Oxford Street, London.*—Wire work.

[6122]

JENKINS, HILL, & JENKINS, *Milton Works, Birmingham.*—Wire iron, iron and steel wires, &c.

[6123]

JONES, J., *Swansea.*—Flat chain.

[6124]

JONES & ROWE, *Worcester.*—Patent range, comprising fire, 4 ovens, 2 closets, boilers, and steam closet. (*See page* 49.)

[6125]

JONES, T. F., & SONS, *Soho Works, Cecil Street, Birmingham.*—Stoves, grates, fenders, fire irons, and light steel toys.

[6126]

KEITH, GEORGE, 55 *Great Russell Street, Bloomsbury.*—Ice machines; freezing powder and apparatus for hot climates. (*See page* 50.)

[6127]

KENNARD, R. W., & CO., 67 *Upper Thames Street, and Falkirk, N.B.*—Ornamental castings in iron. (*See page* 51.)

[6128]

KENRICK, ARCHIBALD, & SONS, *West Bromwich.*—Patent cast-iron tinned and enamelled hollow ware, &c.

[6129]

KENT, GEORGE, 199 *High Holborn, and Strand, London.*—Knife-cleaning machines, and other inventions for promoting domestic economy. (*See pages* 52 *and* 53.)

[6130]

KINGTON & TROWBRIDGE, 116 *Aldersgate Street.*—Platform and every description of weighing machines, scales, &c. (*See page* 54.)

HOPKINS, J. H., & SONS, *Granville Street, Birmingham.*—Block tin, stamped, tinned iron, and japanned articles.

Obtained a Silver Medal in Class 16, *in Exposition Universelle, Paris,* 1855.

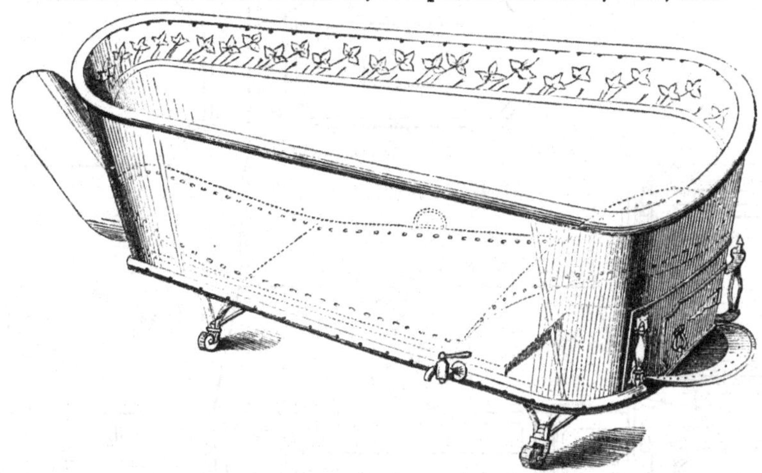

PORTABLE HOT-WATER BATH.

Portable hot-water bath, with shower bath attached, combining perfect portability, completeness, rapidity of heating, cheapness of action, and lowness of cost, J. Wilson's patent. Messrs. Peck & Co. agents in Amsterdam.

Block tin hot-water venison dish and cover, 24 in. melon pattern.

Block tin dish covers, melon pattern, complete set, viz. 1 each, 9, 10, 11, 12, 14, 16, 18, 20, 22 in.

Block tin hot-water meat dish, with cover, 18 in. fluted pattern.

Block tin hot-water dish for jugged hare, hash, or steak, fluted pattern.

Block tin soup tureens, plain oval.

Ditto vegetable or side dishes, oval, with or without hot-water pan.

Block tin vegetable or side dishes, oblong.

Ditto chop plate for hot water, with fluted cover.

Ditto hot-water plates, cheap.

Ditto ditto with earthenware plate.

Ditto plate covers only.

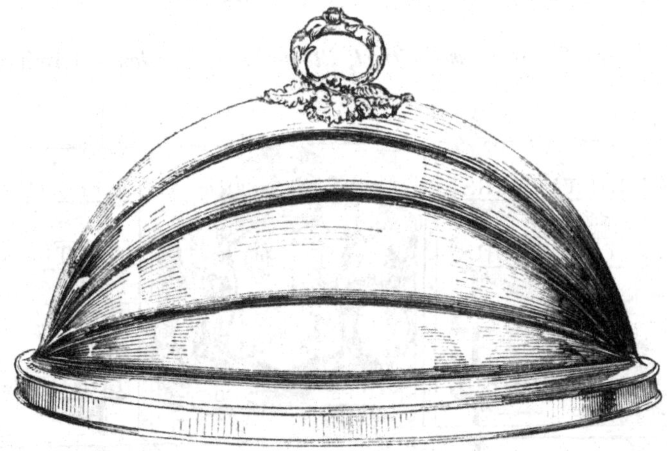

VENISON DISH AND COVER.

Block tin tea and coffee set, cheap, with stamped tinned iron tea and coffee cups and saucers.

Ditto tea and coffee set, middle quality.

Ditto ditto best quality, plain.

Ditto ditto ditto embossed.

Block tin and plate-glass lantern for Price's candle, brass bails and very strong.

Block tin and cylinder glass lantern for Price's candle, brass bails and very strong.

Block tin cottage hastener, with patent roasting jack, complete.

Stand of stamped and tinned iron bowls, 9 to 30 in. dia.

Ditto ditto pudding pans, 4 to 20 in.

Ditto ditto milk pans 16 to 24 in.

Ditto ditto baking dishes, 9 to 18 in.

Ditto ditto bowls, 4½ to 11½ in.

Set of japanned iron tea trays, 16, 24, 30 in.

Japanned tinned iron toilet set for bath-room use, consisting of foot bath, hot-water jug, and waste-water pail.

Japanned tinned iron sponging bath, stamped from one sheet of iron without seam or join.

Japanned tinned iron toilet set for bedroom use, consisting of wash-hand basins and ewers, vase, sponge tray, soap boxes, and brush trays.

Japanned tinned iron strong cash boxes, 10, 12, 16 in.

Ditto iron coal vases, with brass mountings.

Ditto tin railway lamps, including buffer head, porters', guards', and engine lamps.

Japanned tinned iron canteen, containing the most useful articles in a small compass.

Board of brass furniture for baths, lamps, &c.

ISMAY, THOMAS, & Co., *Dover.*—Improved close-fire ranges for large kitchens.

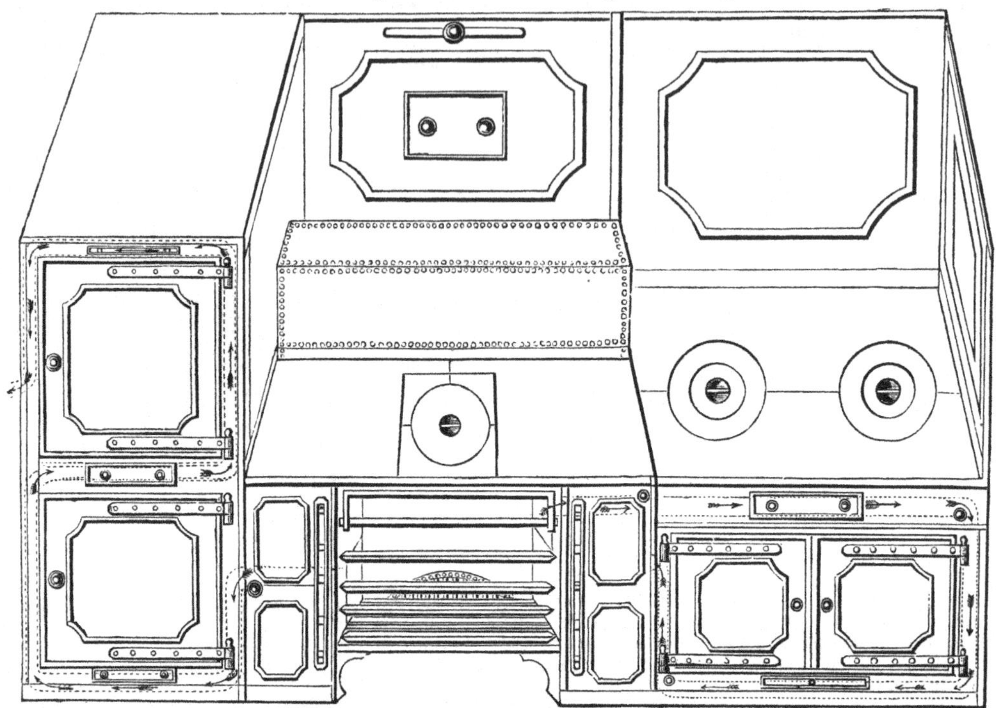

ISMAY & Co. are engineers, smiths, and iron merchants, and manufacturers of large ranges, steam closets and counters, drying closets, laundry steam apparatus, kitchen steam apparatus, dinner and coal lifts for hotels, mansions, or public buildings.

The exhibitors' workmen sent to all parts of England to fix ranges, &c.

Architects' and contractors' communications will be answered, and drawings, plans, and estimates furnished on application.

JEAKES, C., & Co., 51 *Great Russell Street, Bloomsbury, London.*—Kitchen range, &c.

The following articles are exhibited by C. JEAKES & Co.:—

1. KITCHEN RANGE, smoke jack, hot plate, with oven, hot closet, and charcoal stove.
2. PATENT SMOKELESS GRATE, will burn from 24 to 36 hours without replenishing, consuming under 1 lb. of fuel per hour, and may be seen in action at 51 Great Russell Street, Bloomsbury.
3. GRATE (circular headed), for drawing room, Italian in character, with painted tiles, and ormolu mountings.
4. GRATE, French in design, bright covings in panels, with ormolu mountings, and richly ornamented back.
5. LARGE GRATE, renaissance in character, with handsome brass dogs, rich back, and diaper covings, with monogram of International Exhibition, 1862.
6. LARGE MEDIÆVAL DOG GRATE.
7. SMALL MEDIÆVAL DOG GRATE, with diaper back, and English badge in circle of back.
8. DOGS for grates, designed and modelled especially for the Exhibition of 1862.
9. GOTHIC GRATE, with tiles and metal mountings.
10. ARCHITECTURAL TRUSSES OF FRUIT, one as a specimen of casting, the other finished for decorative purposes.
11. CHIMERA.
12. LAMP AND PILLAR, adapted for stone pedestal.
13. GOTHIC HINGES, for ecclesiastical purposes.
14. SPECIMEN OF BRASS WORK, for door furniture, for domestic purposes. Designed by T. H. Wyatt, Esq.
15. BELL PULLS (selections of).
16. WROUGHT-IRON GABLE TERMINALS AND RIDGES, for roofs.

JONES & ROWE, *Worcester.*—Patent Worcestershire range, comprising 2 large meat roasters, 2 pastry ovens, 2 large hot closets, 1 grilling stove, 2 boilers, one for steam and one for circulating hot water to any part of the house, treated by one fire, and capable of cooking sufficient for 200 persons.

JONES & ROWE'S PATENT RANGE.

"A more economical arrangement for fuel, in the accomplishment of a great deal by small means, or a more compact contrivance for cooking at once all the courses necessary for a dinner, was probably never seen."—*Worcester Herald.*

"A valuable peculiarity of Messrs. Jones & Rowe's range, is the placing of the oven and roaster, or the two ovens, one above the other, instead of side by side, the heat being made to pass, by means of flues, beneath the entire surface of the range. The roasting oven is constantly replenished with fresh air, and though the top of the range is an iron platform, enclosing the fire, a joint can be cooked at the open front in the old way."—*Daily Telegraph,* May 19, 1862.

One side of the range is occupied by a wrought-iron boiler, holding fourteen gallons; on the opposite side are two ovens, one above the other; the one for roasting meat, and the other for pastry, or for baking bread. The whole top of the range is a flat iron platform, which may be covered with vessels for boiling, stewing, &c. In front of the fire a large roasting joint may be cooked. The advantages of the patent are, that by one moderate fire hot air is generated, which, by means of flues, is made to pass beneath the entire surface of the range, whereby the ovens, boiler, and platform are sufficiently heated for all purposes of cooking; and one of the ovens, or meat roaster, is constantly ventilated with fresh air, which prevents the unpleasant flavour sometimes imparted to baked meats. The boiler is constantly kept boiling from the same fire; and a boiler might be introduced at the back of the fire, for supplying hot water to a bath or cistern, to any part of the house.

This range having now become familiarly known, is universally admitted to be the best yet introduced for economy, durability, cleanliness, and convenience. No kitchen range made can surpass it.

J. & R. have received most valuable testimonials to the excellence of the above ranges. Ranges delivered carriage free; books of illustrations, with prices and testimonials, will be forwarded on application, and estimates and plans given for fitting up large kitchens with J. & R.'s patent ranges, and supplying hot water to baths, steam kettles, or to any part of the house, to any proposel arrangement.

J. & R. wish it to be understood there is no other range made, price and size compared, that will cook for so many persons as the above.

KEITH, GEORGE, 55 *Great Russell Street, Bloomsbury.*—Ice machines; freezing powder and apparatus for hot climates.

KEITH'S IMPROVED LING'S PATENT ICE SAFE.

This invention shows the application of ice for the perfect preservation of meat, poultry, fish, and all other edibles, without destroying the original flavour or coming in contact with the ice, fitted with arrangements for icing wines, spring water, &c. at a very small daily consumption of ice. Especially adapted for the use of clubs, hotels, butchers, poulterers, and large establishments.

Size 6 ft. high, 5 ft. wide, 2 ft. 8 in. deep. Price £60.

[6131]

KNIGHT MERRY, & Co., 131 *Bradford Street, Birmingham.*—General tin-plate articles. (*See page* 55.)

[6132]

LAMBERT, BROTHERS, *Walsall, Staffordshire.*—Wrought-iron welded tubes; iron and brass fittings; chandeliers; metallic tubular bedsteads. (*See pages* 56 *to* 58.)

[6133]

LANE, HENRY, *Wednesfield, near Wolverhampton.*—Every description of wild beast, game, and vermin traps.

[6134]

LEADBEATER, JOHN, & Co., 125 *Aldersgate Street.*—Wrought-iron fire and thief proof safes.

[6135]

LEIGHTON, JOHN, 40 *Brewer Street, Golden Square.*—Reserve stoves to prevent smoke formation; Maltese chimney caps.

[6136]

LESLIE, GEORGE, *Upper Mall, Hammersmith, W.*—Patent self-acting valve for preserving brewers' casks from becoming mouldy.

[6137]

LEWIS, WILLIAM, 6 *New Westgate Buildings, Bath.*—A gas cooking stove, and a confectioner's tartlet warmer.

[6138]

LINES, W. D., & PALMER, W., 1 *Marlborough Road, St. John's Wood.*—Horse shoes suited for all purposes.

[6139]

LINLEY, THOMAS, & SONS, *Stanley Street, Sheffield.*—Patent double-blast bellows; patent portable forges; portable vice benches, &c.

KENNARD, R. W., & CO., 67 *Upper Thames St., and Falkirk, N.B.*—Ornamental castings in iron.

CAST-IRON ORNAMENTAL ENTRANCE GATES AND RAILING, manufactured by Messrs. Kennard, at the Falkirk Iron Works, N.B. for the Vista Alegre Palace, lately purchased by His Excellency Don José de Salamanca from Her Majesty the Queen of Spain. Also exhibitors of cast-iron verandah, with vases and other ornamental castings, as well as drawings of various bridges erected by them in Spain, India, and Italy, and also of the celebrated viaduct at Crumlin, Monmouthshire.

CASTINGS of every description, in loam or sand, to order or model, for engineers, builders, and machinists, gas and water works. Stoves, ranges, sugar pans, teaches, or boilers to any pattern or make.

Kent, George, 199 *High Holborn, and Strand, London.*—Knife-cleaning machines, and other inventions promoting domestic economy.

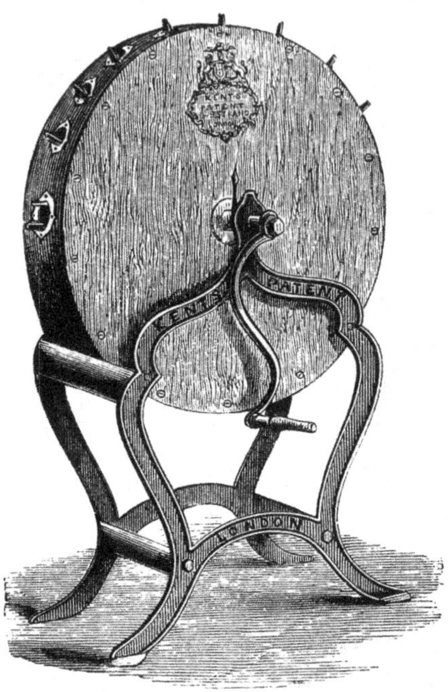

Kent's Patent Rotary Knife-cleaning Machine.

A prize medal was awarded to this invention at the Great Exhibition of 1851; since that period a second patent has been granted to G. Kent for certain improvements, which have greatly enhanced its value, not only in general efficiency and durability, but also as a sharpener of table cutlery. The unparalleled success and high reputation gained by this machine throughout the world, has tempted some unscrupulous persons to put forth spurious imitations, but as a second patent protects the construction of the most essential parts of Kent's machine, it remains unapproached in its efficiency and durability.

Made in eight sizes, from 3 to 14 guineas, to clean from 3 to 9 knives at a time.

Kent's Washing Apparatus.

A very simple, economical and effectual mode of cleansing linen, requiring comparatively no hand rubbing, and dispensing with boiling altogether.

Price from £3 10

Kent's Folding Clothes Dryer is intended to supersede clothes posts and lines, and consists of an upright standard from 10 to 13 ft. high, supporting five ribs or arms. These arms, which expand and fold like an umbrella, contain clothes lines, affording from 120 to 150 ft. of hanging space. It revolves with the wind, and may be raised or lowered as desired.

Price from £1 5

Kent's Double-action Box Mangle.

In general appearance this mangle resembles the old kind of box mangle, but has some very important advantages, viz. the backward and forward motions are obtained by turning the handle always in one direction. It is much lighter than any other mangle on this principle, more easy and rapid in working, and is perfect in general manufacture.

Price from £9 0

Kent's Self-heating Box Iron.

This iron is intended for all the purposes to which the old box and flat iron are applied. It may be heated at pleasure in three minutes, without any fire, and will remain hot at a nominal cost for any length of time.

Price from . . . . 5*s.* 6*d.*

Kent's Patent Rotary Cinder Sifter. Extensively used also for mixing and sifting guano and other artificial manures.

The object of this invention is to render a disagreeable duty as little objectionable as possible. Its operation is most certain and effectual; the unsifted cinders being thrown into the upper part of the machine, a few turns of the handle separate the ashes from the cinders in the most perfect manner, without the least dust or dirt escaping from the sifter; the refuse falls into a movable box, and the cinders are actually deposited in the coal-scuttle without the possibility of loss by mixing with the ashes.

In houses limited for room, and especially those without yards or gardens attached, and in situations where cellars become the repositories of ashes and refuse, the machine becomes invaluable. The accumulation of rubbish under such circumstances is both disagreeable and unhealthy, and its removal to the dust-cart a source of considerable annoyance. The patent cinder sifter, however, happily removes these objections; it is a compact dust-bin in itself, and to remove the refuse it is only necessary to take away the box part of the machine which contains it: there is no dirt thereby occasioned, nor can any effluvium possibly arise.

Prices and dimensions:

No. 1. 2 ft. 7 in. long, 1 ft. 3 in. wide, 3 ft. 3 in. high £3 3

No. 2. 2 ft. 10 in. long, 1 ft. 5 in. wide, 3 ft. 6 in. high £4 4

No. 3. 3 ft. 3 in. long, 1 ft. 7 in. wide, 3 ft. 9 in. high £5 5

No. 4. 5 ft. long, 2 ft. 4 in. wide, 2 ft. 7 in. high, Price £7 7

No. 4 consists of the upper portion or sifter only, expressly made for the use of large establishments, and is intended to stand on an ordinary dust-bin; it has, therefore, no ash-box or scuttle, the cinders falling on one side, and the dust on the other side of the bin.

Carpet Sweeper. George Kent, wholesale agent to the patentee.

Consists of a neat japanned iron case or box, 12 in. long, having recesses for the dust, and a patent spiral self-adjusting brush. The dust, lint, and even hairs, pins, needles, &c. are taken up directly into the box and there retained as the sweeper moves along, instead of being accumulated, driven over the entire surface, and forced into the grain of the carpet, as is usual with ordinary brooms. It will sweep cleaner than brooms, with less injury to carpets, and without raising any lint or dust.

Price . . . . 15*s.*

KENT, GEORGE, *continued.*

KENT'S PATENT TRITURATING STRAINER.

The smallest size is about 13 in. by 9 in. and 12 in. deep, the upper part has a curvilinear bottom of white metal very finely perforated, and a brush with a lever handle working in centres, and made to traverse to and fro over this metal bottom, and by continuing this for a few minutes the whole of the ingredients for making soups, sauces, purees, gravies, jams, &c. are reduced to a fine pulp or liquid, and at the same time strained into a white earthen vessel which constitutes the lower part of the apparatus, thus superseding the tedious, troublesome, dirty, and expensive process with the hair sieve and tammy cloth, while the whole of the virtues of the ingredients employed are completely extracted, and brought to a superior consistency at a much less cost and in one-tenth the time usually occupied by those very primitive means.

Size for families, 27*s.* 6*d.*; for hotels, 37*s.* 6*d.*

KENT'S ROTARY POTATO MASHER.

With this simple contrivance from 1 to 6 lbs. of potatoes can, by a few turns of the handle, be mashed more finely and perfectly than by any other means, and in less time than this brief description can be read. It is also adapted for grating bread with equal perfection and rapidity, as well as most other materials for culinary preparations generally.

Price from . . . . 7*s.* 6*d.*

MEAT MINCING AND SAUSAGE-MAKING MACHINE, for mincing any kind or quantity of raw or cooked meat, and making sausages at one operation. Price, from 21*s.*

EGG BEATER (Monroe's patent), George Kent sole manufacturer.

This is on a somewhat similar principle to Griffith's whisk described above. By it small quantities of eggs, all kinds of egg mixtures, and batters, may in a few minutes be wrought up to a degree of lightness very far superior to anything that can be produced by the ordinary hand whisk.

Price . . . . 5*s.*

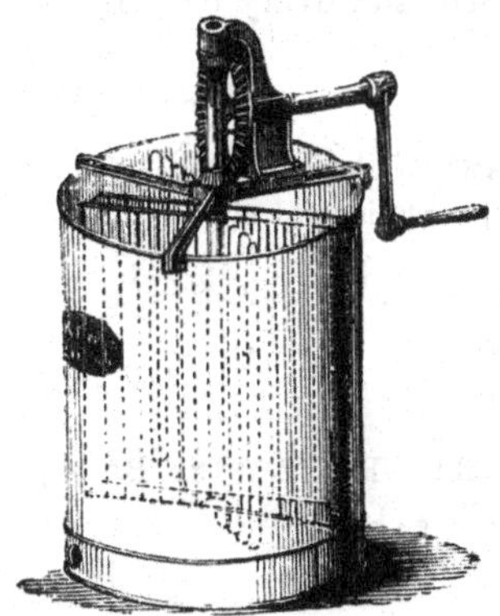

GRIFFITH'S PATENT WHISK AND MIXING MACHINE, George Kent sole manufacturer.

This machine has two metal frames with a number of wire projections thereon which are made to rotate rapidly by suitable gearing in opposite directions around one centre. These frames work in a round vessel and are put in motion by a crank handle, and thus produce a greater amount of agitation than can be produced by the means hitherto employed. It was originally designed for cooks' and confectioners' purposes in beating up eggs and batters, which it does to the highest perfection in a very few minutes; it is however adapted and extensively used for a variety of other purposes as an agitator and mixer. These machines are made in various sizes, according to the purposes for which they are required, at prices varying from 21*s.* to £5.

CHURN (Griffith's patent). This is on the same principle as the whisk, and will bring butter quicker than any other churn extant.

Price from . . . . 27*s.* 6*d.*

APPLE PARER, CORER, AND SLICER, for simultaneously paring, coring, and slicing apples.

By a simple adjustment it may be made to pare potatoes with great economy.

Price . . . . 8*s.* 6*d.*

WATER FILTER (Danchell's patent), in various plain and ornamental designs, for house, ship, agricultural, and other purposes. G. Kent sole manufacturer.

These filters are on an entirely new principle, and possess the advantage of purifying as well as brightening the water. They also obviate the great difficulty experienced with all other filters, of cleansing when becoming foul or clogged, the filtering medium being contained in an earthen cylinder, which any servant may remove, purify, and replace in its original position in a few minutes. They are capable of yielding according to size from one to ten gallons per minute.

Price from . . . . 8*s.* 6*d.*

CISTERN FILTERS on the principle described above, capable of filtering and purifying from 1 to 10 gallons per minute. Price from . . . . 25*s.*

WATER TEST (Danchell's patent), George Kent sole manufacturer.

With this test any person without a knowledge of chemistry may detect the presence of any deleterious matter or impurities in water, and whatever it is found to contain the effect may be neutralized by the adoption of Danchell's patent filters.

Price with book of instructions . . . . 10*s.* 6*d.*

WATER SOFTENING APPARATUS (Danchell's patent), G. Kent sole manufacturer, may be applied to any cistern. It is self-acting, and renders the hardest water in its course from the service pipe perfectly soft.

Price £2 2

KINGTON & TROWBRIDGE, 116 *Aldersgate Street, Corner of Long Lane, London, E.C.*—Platform and every description of weighing machines, scales, &c.

By appointment to Her Majesty's Honourable Board of War.

GILT BEAMS, of the best quality, 1*s.* 3*d.* per lb. from 30 lb. upwards, if fitted with boards and ropes or chains.

20 cwt.	£10 0 0
10 cwt.	7 0 0
5 cwt.	4 10 0

Second quality—RED PAINTED BEAMS, at 1*s.* per lb. from 25 lb. and upwards.

SOLID BRASS BELL WEIGHTS, stamped.

1 lb. to ¼ oz.	£0 5 6
2 lb. to ¼ oz.	0 8 6
4 lb. to ¼ oz.	0 15 0
7 lb. to ¼ oz.	1 6 0
14 lb. to ¼ oz.	2 10 0
Above 14 lbs. per lb.	0 1 6

N WEIGHTS, japanned and gilt, in sets, stamped.

4 lb. to ¼ lb.	£0 3 0
7 lb. to ¼ lb.	0 5 0
14 lb. to ¼ lb.	0 6 6
28 lb. to ¼ lb.	0 10 0
56 lb. to ¼ lb.	0 15 0

COPPER SCOOP MACHINES.

1, with scales, 7 in. wide	£0 10 6
2, ditto 8½ in. wide	0 13 0
3, ditto 10 in. wide	0 15 6
4, ditto 11 in. wide	0 18 6
5, ditto 12 in. wide	1 5 0

FLOUR OR POTATOE MACHINE.

3 ft. high, to weigh in sacks, 3 cwt.	£2 10 0
2 ft. 10 in. high, ditto 2½ cwt.	2 2 0
2 ft. high ditto 2 cwt.	1 15 0

PORTABLE MACHINE, mounted on wheels.

TO WEIGH.	PLATFORM.	
3 cwt.	22 by 20 in.	£2 18
4 cwt.	22 by 22 in.	3 8
5 cwt.	26 by 22 in.	3 18
7 cwt.	26 by 26 in.	4 15
10 cwt.	30 by 30 in.	5 10
12 cwt.	31 by 31 in.	6 10
15 cwt.	36 by 36 in.	7 10
20 cwt.	36 by 36 in.	10 15
30 cwt.	38 by 34 in.	12 10
40 cwt.	38 by 40 in.	15 0

MACHINE LEVEL WITH THE FLOOR, for warehouses.

TO WEIGH.	PLATFORM.	
20 cwt.	38 by 30 in.	£10 15
30 cwt.	38 by 34 in.	12 10
40 cwt.	38 by 40 in.	15 0

DOUBLE WEIGHING MACHINE, for corn, &c. to weigh off a man's back or barrow.

Best quality	£4 8 0
Second quality	3 10 0

MACHINE for seeds, hops, &c.

To weigh 2 cwt.	£2 10 0
To weigh 4 cwt.	3 15 0

This is one of the best and most useful machines made, and strongly recommended.

IRON FRAME MACHINE, for grocers, tallow chandlers, &c.

To weigh 3 cwt.	£3 5 0

SHIP SCALES to weigh coals,

Complete with weights	£13 5 0

WEIGHING MACHINES for family use, with oblong tin scale.

7 lb.	£0 9 6
14 lb.	0 10 6
28 lb.	0 12 6
40 lb.	0 14 6
56 lb.	0 18 6

REGISTERED FAMILY WEIGHING MACHINES, with weights complete.

No. 1	£0 16 6
No. 2	1 0 0
No. 3	1 6 0

FLOUR SCALES.

½ peck	£0 10 6
¾ peck	0 12 6
1 peck	0 15 6
½ bushel	1 2 0
1 bushel	2 0 0

DOUGH SCALES, 9*s.* 6*d.* per pair.

A list of every description of scales, weights, and weighing machines will b3 forwarde1 on the receipt of directed envelop to 116 Aldersgate Street, Corner of Long Lane, London, E.C.

KNIGHT, MERRY, & Co., 131 *Bradford Street, Birmingham.*—General tin-plate articles.

B 23.

B 33.

B 12.

B 44.

BEST BLOCK-TIN DISH COVERS—SILVER PATTERNS.

The UNIQUE COFFEE AND TEA URNS are strongly recommended to the public as possessing advantages over anything of the kind ever introduced.

The mechanical construction is so arranged that it is impossible for them to get out of order, and the various parts being movable ensure perfect cleanliness, and consequently purity of the article infused.

THE UNIQUE TEA AND COFFEE URNS.

The water being equally distributed over the tea or coffee, obtains perfect saturation and abstraction, requiring a smaller quantity than any other apparatus, and making the same better and in less time.

The price brings them within the means of every one.

LAMBERT, BROTHERS, *Walsall, Staffordshire.*—Wrought-iron welded tubes; stocks, taps, and dies; iron and brass fittings; steam coils; sluice valves; chandeliers; metallic tubular bedsteads.

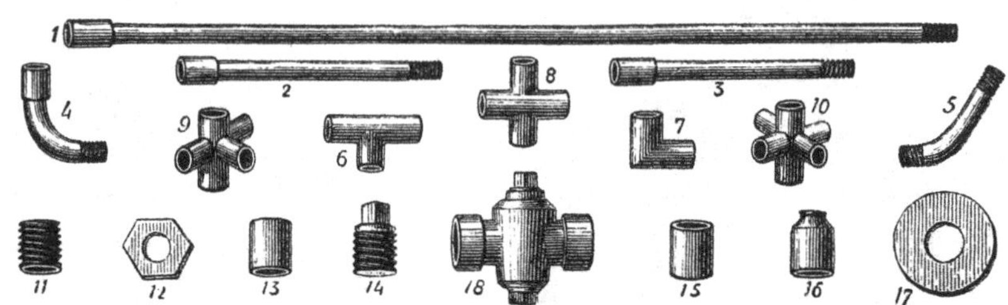

PATENT WELDED WROUGHT-IRON TUBES AND FITTINGS.

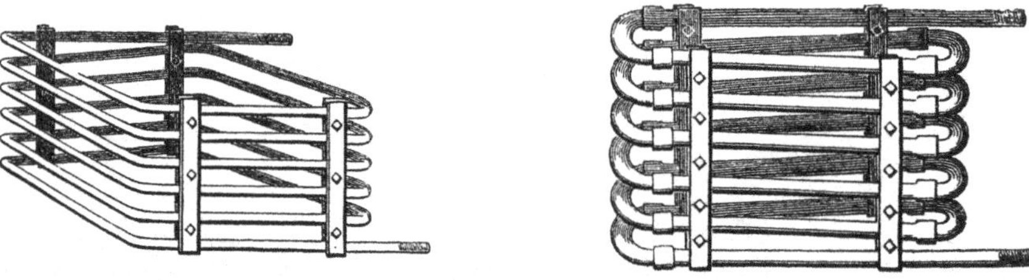

WROUGHT-IRON STEAM COILS.

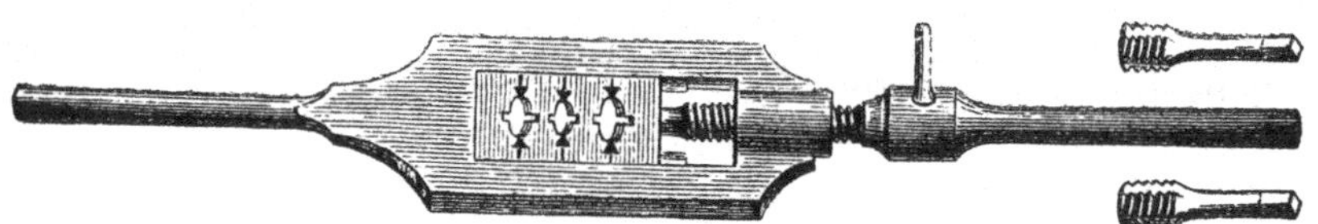

SET OF STOCKS, TAPS, AND DIES FOR IRON PIPE.

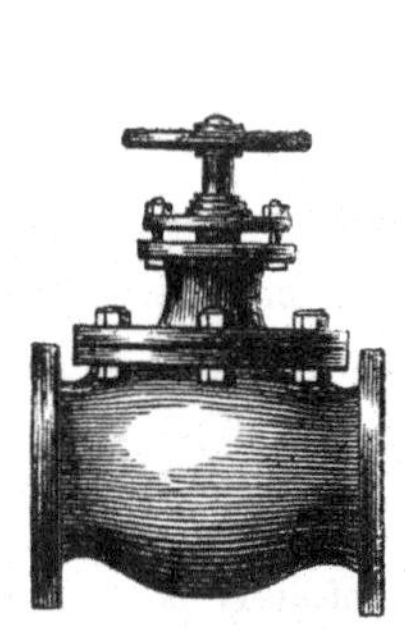

GLOBULAR STEAM VALVE.

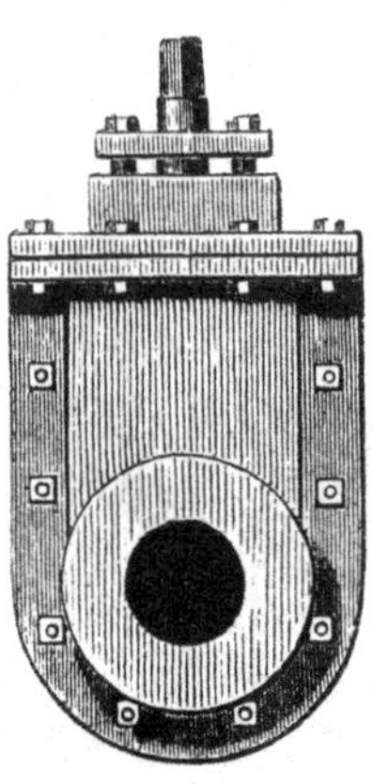

SLUICE VALVE.

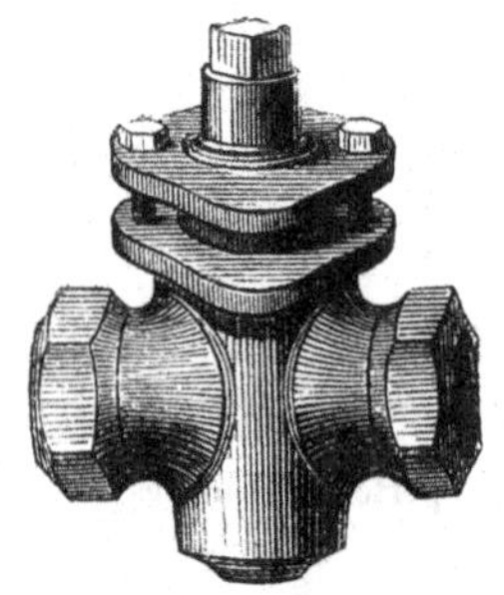

GLAND STUFFING-BOX COCK.

LAMBERT, BROTHERS, *continued.*

SINGLE GAS BRACKET.

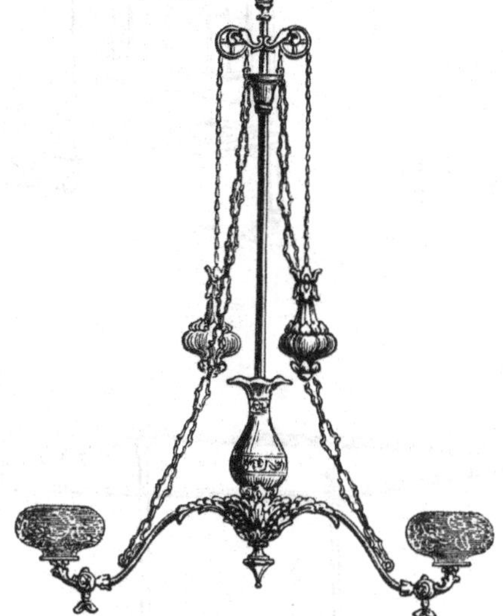

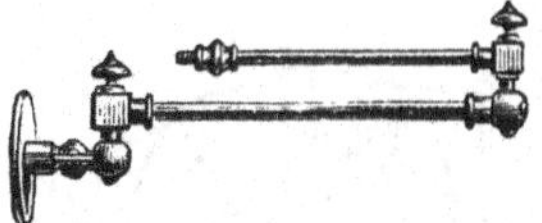

DOUBLE GAS BRACKET.

BRACKET BACK.

Two-light Hydraulic Pendant, £1 15*s*.

DOUBLE SWIVEL, WITH COCK JOINTS.

ELBOW LANTERN COCK.

SCREW-DOWN BIB COCK.

SCREW-DOWN STOP COCK.

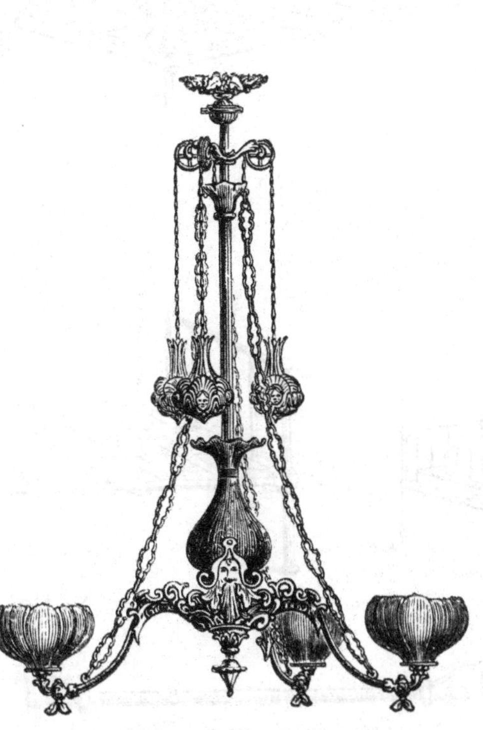

Two-light Hydraulic Pendant, £2 2*s*.
Three-light ditto . . . . . £3 0*s*.

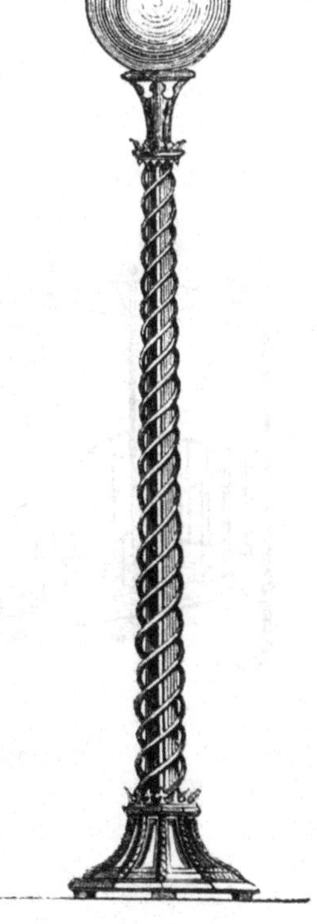

LAMP COLUMN, WITH SPIRAL TUBES.

LAMBERT, BROTHERS, *continued.*

No. 130.

No. 270.

No. 290.

No. 132.

No. 146.

PATENT EXPANDING STRETCHER.

[6140]

LLOYD, MARTIN, *Charles Henry Street, Birmingham.*—Malleable nails.

[6141]

LLOYD, THOMAS, & SONS, 15 *Old Street Road, Shoreditch, London.*—Steel mills.

Obtained First-Class Silver Medals in Boulogne, Vienna, and Amsterdam, 1856 *and* 1857.

Improved prize hand flour mills and dressing machines, to grind and dress at the same time, by hand.
No. 1 . . . . £6 15
No. 2 . . . . 8 15

Improved prize flour mills and dressing machines, to grind and dress at the same time, for horse or steam power.
No. 3 . . . . £12 0
No. 4 . . . . 15 0
No. 5 . . . . 21 0

Improved corn grinding mills, for grinding barley, beans, peas, or oats, into fine meal, by hand. 4*l.* 5*l.* 6*l.* and £7 0

Improved barley mills, for grinding any kind of grain into fine meal, for horse or steam power. 10*l.* 12*l.* and £15 0

Improved corn crushers, for crushing oats and splitting beans or peas, 4*l.* 5*l.* and £6 0

Improved corn crushers, for horse or steam power, 8*l.* to 12*l.*

Drug mill, for grinding seeds for horse or cattle medicine, 1*l.* 15*s.* to £7 10

Handsome bronzed 4 pillar frame coffee mill in double bearings, with brass hopper, and 2 fly wheels, 14*l.* 20*l.* and £24 0

Handsome bronzed coffee mill, with brass hoppers, 2*l.* 2*s.* to £14 0

Pepper and spice mills, 1*l.* 1*s.* to 9 0

Improved sugar mills, 3*l.* 5*s.* to 6 10

Improved cocoa mills, 1*l.* 10*s.* to 20 0

Improved patent currant dressing and cleansing machine £4 10

Improved patent tea mills, 2*l.* 10*s.*, 4*l.* and . . 6 0

Patent sugar choppers, 16*s.*, 1*l.* 2*s.*, 1*l.* 8*s.*, 2*l.* and 2 4

Patent sausage machines, 1*l.* 1*s.*, 1*l.* 10*s.*, 2*l.* 2*s.*, and 3 3

Patent sausage machines, for sausage makers, asylums, public institutions, for mincing a large quantity. £7 7

[6142]

LONGDEN & CO., *Phœnix Foundry, Sheffield.*—Cooking apparatus; mediæval fire-place; hot-air stove; stair balusters, railing, &c. (*See page* 60.)

[6143]

LYON, A., 32 *Windmill Street, Finsbury, London.*—Sausage-making and general mincing machines, and small mincers to assist digestion.

Obtained two Medals at the Exhibition in Paris, 1855.

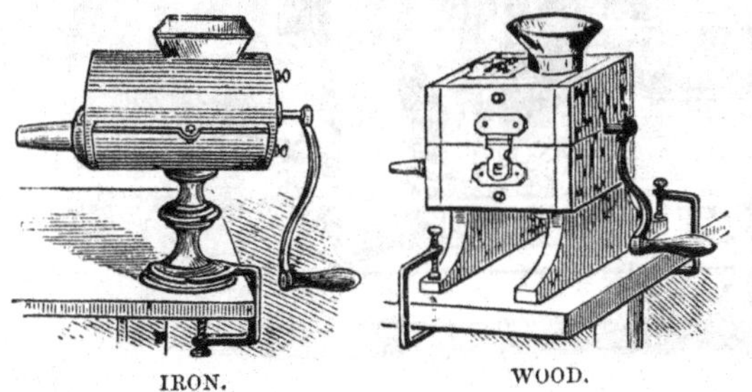

IRON. WOOD.

Machine for family use, for making sausages, and general mincing purposes. Price £1 10

Machine for small shops, pastry cooks, &c. Price . £2 0

Machine for shops, hotels, pastry cooks, &c. . . 3 10

Machine for shops, hotels, institutions, &c. . . 6 6

Small mincer for the table, to assist digestion . 1 1

Small mincer with hot-water bottles, to keep food hot while being cut up to assist digestion. Price . £2 2

Machine for cutting meat for pies, &c. Price . 2 0

Machine for pulping vegetables for poultry . . 2 5

Machine for cutting French beans, vegetables, &c. 1 5

Improved knife and board for cutting bread for large establishments, schools, asylums, &c. Price 16/0.

Improved suet chopping knife and board. 4/0 to 8/6.

[6144]

MC CONNEL, ROBERT, *Glasgow.*—Improved locks, latches, and fastenings, for security; and others for common purposes, patented.

[6145]

MAC MORRAN, JOSEPH, 33 *Leicester Square, London.*—General mincing and sausage machines, and mills of various sizes.

[6146]

MANDER, WEAVER, & CO., *Wolverhampton.*—Aluminium casket, and aluminium in various forms.

[6147]

MANGER, JAMES, *Russell Street, Liverpool.*—One pair 30-inch double-action suction bellows with frame, complete.

[6148]

MAPPLEBECK & LOWE, *Birmingham.*—Kitchen ranges, stove grates, fenders, and fire-irons.

LONGDEN & CO., *Phœnix Foundry, Sheffield.*—Cooking apparatus; mediæval fire-place; hot-air stove; stair balusters, railing, &c.

DINING-ROOM FIRE-PLACE, being an adaptation of Early Pointed art to modern requirements, consisting of register grate of cast-iron electro-bronzed with copper, with brass ornaments and glass mirrors, ash-pan and fender of electro-bronzed cast-iron and steel, fire-irons of steel with electro-bronzed handles, and mantel-piece of Devonshire and serpentine marbles, designed by Messrs. Walton & Robson, architects, London and Durham.

DINING-ROOM FIRE-PLACE.

FRENCH RENAISSANCE HOT-AIR STOVE.

ORNAMENTAL PEDESTAL for enclosing hot-water pipes.

GOTHIC BED-ROOM GRATE with fire-lump sides and back, designed by Messrs. Walton & Robson.

A selection of STAIR BALUSTERS, TOMB RAILING, AND BALCONY RAILING in various styles.

OPEN-FIRE COOKING RANGE, with a raised cast-iron oven for baking or roasting, a wrought-iron lower oven for light baking, hot-hearth for boiling over the lower oven, plate-warmer over the upper oven, welded and galvanized wrought-iron boiler for circulation of hot water for a bath; open roasting fire, with sliding spit, racks, and polished kitchen fender.

[6149]

MARTINEAU, F. E., & CO., *Cleveland Street, Birmingham.*—Wrought iron and brass hinges.

[6150]

MATHEWS, WILLIAM, 9 *Mount Street, Berkeley Square*—Horse shoes, and shoeing hammer.

[6151]

MAXWELL, H. & Co., 161 *Piccadilly, W.*—Spurs and spur sockets.

[6152]

MAY, ALFRED, 259 *High Holborn.*—Gas roasting and baking oven, ranges, stewing stoves, linen drying closet, &c.

The exhibitor manufactures hydraulic rams, pumps, gas works, gas fittings, warming by hot water and hot air, cooking apparatus of every description. He also undertakes bell-hanging, and the erection and fitting of baths, improved closets for drying linen, hot and cold water for lavatories, dressing rooms, &c.

Mr. May has been extensively employed under Sir Joseph Paxton and other eminent engineers and architects in first-class gentlemen's mansions and public institutions in England, France, Germany, and Switzerland; and from his long practical experience of upwards of 30 years, is enabled to prepare plans and estimates suitable for public or private institutions. He will guarantee the efficiency of all works intrusted to him for execution. A large stock of apparatus is always kept on hand.

[6153]

MEDHURST, THOMAS, 465 *Oxford Street, London.*—Weighing machines.

[6154]

MILLS, JOSEPH, 40 *Great Russell Street, W.C.*—Register stove with patent door; range with shifting bars and improved dampers.

[6155]

MORETON, JOHN, & Co., *Wolverhampton and London.*—Foreign and colonial hardware. (*See pages* 62 *to* 64.)

[6156]

MOREWOOD & Co., *Dowgate Dock, London, and Lion Works, Birmingham.*—Galvanized iron, manufactured and in sheets.

MOREWOOD & Co. manufacture the following, of which specimens are exhibited :—

Patent galvanized tinned iron, and galvanized iron, plain or corrugated, curved, and in tiles, of all gauges.

Black or painted corrugated iron, galvanized or black-cast gutters, pipe, &c. all of which are kept in stock.

Galvanized water and gas tubing, stamped and moulded gutters, wire, wire netting, nails, rivets, garden chairs, pails, &c.

Estimates given for roofs, and every description of galvanized buildings, at the offices and warehouse.

Morewood's patent continuous galvanized iron roofing is cheaper than felt. Full particulars may be learned on application.

MORETON, JOHN, & Co. (late Moreton & Langley), *Wolverhampton, and 22 Bush Lane, Cannon Street, City, London, E.C.*—Foreign and colonial hardware.

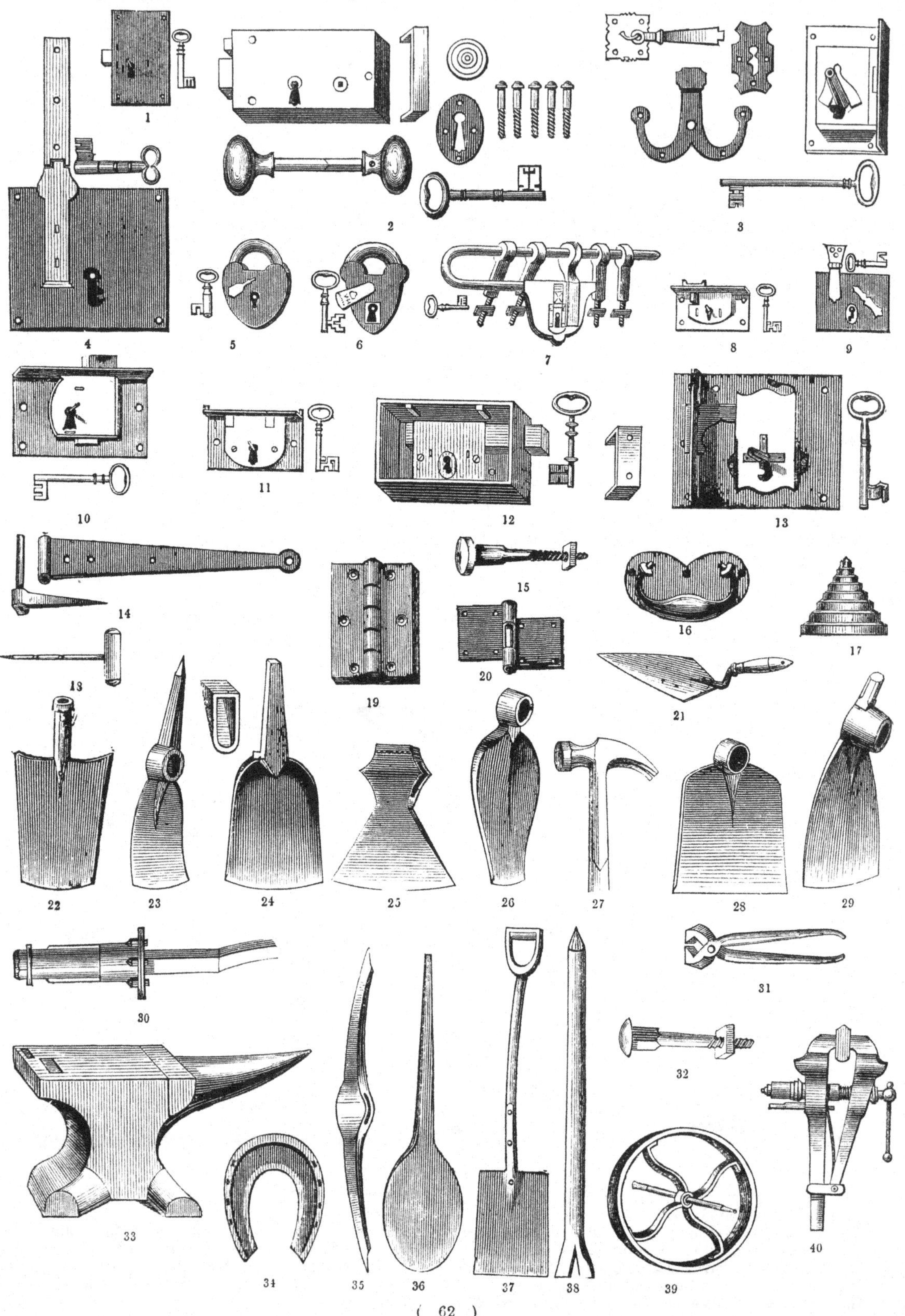

MORETON, JOHN, & CO., *continued.*

41 42 43 44 45 46 47 48 49 50 51 52 53 54 55 56 57 58 59 60 61 62 63 64 65 66 67 68 69 70 71 72 73 74 75 76

MORETON, JOHN, & CO., *continued.*

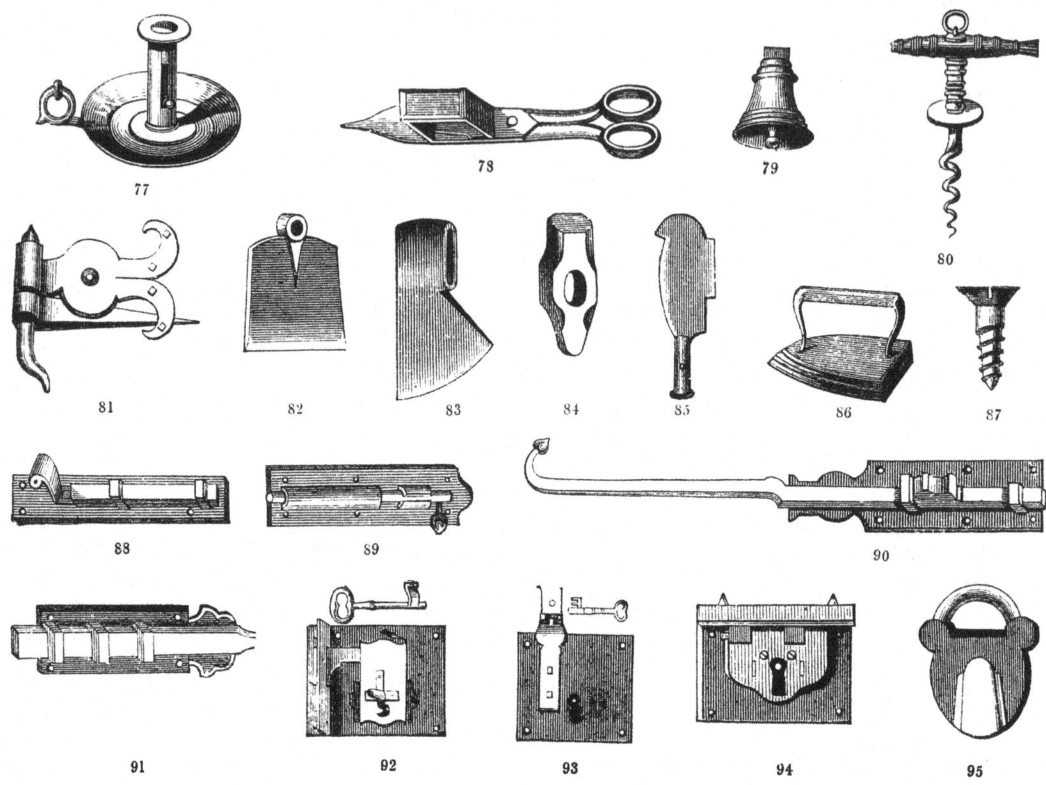

Anvils and vices.
Augers.
Awl blades.
Axes—every description.
Axles.
Balances, Salter's.
Bedsteads—iron and brass.
Bellows—house and smiths'.
Bells—dinner and tea.
,, house and yard.
Box irons and heaters.
Braces and bits.
Brass foundry.
,, cocks.
Britannia metal goods.
British plate goods.
Buckles.
Bullet moulds.
Butts—iron and brass.
Candlesticks—brass, &c.
Cash boxes.
Castors—iron and brass.
Chains and chain cables.
Chest handles.
Coach wrenches and screws.
,, bolts and nuts.
Coffin furniture.
Cooks' ladles.
Corkscrews.
Counter machines.
Cruet frames.
Curbs and bitts.
Curry combs.
Curtain rings.
Cut brads and tacks.
Cutlery—all descriptions.
Dog collars and chains.
Door springs and bolts.
Ewers and basins.
Fenders.
Fire irons.
Fish hooks.
Frying pans.
Galvanized iron
,, scoops and basins.
,, buckets, &c.
Garden rakes and tools.
German silver goods.
Gimblets.
Girths—saddle.
Gridirons.
Gun wadding.
,, implements.
Guns and pistols.
Hames.
Hammers.
Harness furniture.
Hat and coat hooks.
Head collars.
Helved hatchets.
Hoes—every description.
Hollow-ware—tinned.
,, enamelled.
Hooks and hinges.
Horse and mule shoes.
Iron mane combs.
,, pots and camp ovens.
,, dutch stoves.
,, safes.
Japanned goods.
Kaffir picks.
Kettles—sheet iron, &c.
Key rings.
Keys and blanks.
Lamps.
Lanterns.
Latches—Suffolk.
,, Norfolk.
,, bow and rim.
,, mortise.
,, night.
Lead ladles.
Leather goods.
Locks—dead.
,, mortise.
,, rim.
,, plate.
,, pad.
,, till.
,, chest.
,, cupboard.
,, trunk.
Malleable tacks.
Matchets.
Measuring tapes.
Mills—coffee, post, &c.
Nails—iron and brass.
Needles.
Percussion caps.
Pewter measures.
Pitch ladles.
Planes.
Platform machines.
Powder flasks.
Pulley blocks.
Pullies—frame and axle.
Rat and rabbit traps.
Rivets—iron and copper.
Rules—wood and ivory.
Sad irons.
Saddles and saddlery.
Sausage machines.
Scotch T hinges.
Swords and cutlasses.
Screws—iron and brass.
Shackles for chain cables.
Sheet brass and zinc.
Sheathing nails.
Ship scrapers.
Shot belts and pouches.
Singeing lamps.
Skewers.
Snuffers and trays.
Sofa springs.
Solder—tin and brass.
Spades and shovels.
Spittoons.
Spoons—all descriptions.
Spurs and spur rowels.
Stand scales.
Steelyards.
Stocks and dies.
Teapots.
Tin goods.
Toilet sets.
Traces—plough and cart.
Trays and waiters.
Trowels—garden and brick.
Wad punches.
Washers—iron.
Weights—iron and brass.
Wire goods.
Whips.
&c. &c. &c.

[6157]

MORRISON, D., & Co., *Birmingham.*—Metallic furniture.

[6158]

MORTON, JOSEPH, & SON, *Bellfield Works, Sheffield.*—Stove grates, fenders, and fire-irons.

[6159]

MUSGRAVE BROTHERS, *High Street, Belfast.*—Patent slow-combustion stoves; grates, patent iron fittings for stables, cowhouses, and piggeries. (*See page* 66.)

[6160]

NASH, RICHARD, *Ludgate Hill Passage, Birmingham.*—Presses, lathes, dies, tools, &c.

[6161]

NASH, SWAN, 253 *Oxford Street.*—Ranges, patent stoves and fuel.

JOYCE'S PATENT STOVE, manufactured by the exhibitor, is the only one that works without a flue. Price from 12/0 upwards. The prepared fuel for use with it, 2/3 per bushel.

The exhibitor's stock comprises: moderator lamps in great variety; a choice and elegant assortment of stove grates, fenders, and fire irons; an assortment of kitchen ranges and hot plates, with all the newest improvements, unsurpassed for lowness of price and excellence of quality; cutlery, electro-plated goods, gas chandeliers, and every description of furnishing ironmongery of the best quality, at the lowest prices.

[6162]

NASH & HULL, 202 *Holborn, W.C.*—Crystal glass, wood, and brass letters; stencil-plates.

Samples of the following are exhibited:—

WOOD LETTERS, gilt and painted, for facias, fronts of houses, public buildings, &c.

DECORATED GLASS LETTERS, for affixing on shop and office windows, show cases, doors, tablets, &c.

BRASS LETTERS in various patterns for the same purposes.

BRASS LETTERS for casters' patterns, monuments, tombs, &c.

LETTERING of various descriptions.

STENCIL PLATES for marking linen, packages, surveyors' and engineers' plans.

[6163]

NETTLEFOLD & CHAMBERLAIN, *Broad Street, Birmingham.*—Improvements in wood and metal screws, locks, and general iron work, introduced since 1851.

[6164]

NETTLETON, JOSHUA, 4 *Sloane Square, Chelsea.*—Open-fire ventilating stove and pan.

[6165]

NEVE, JOHN, & Co., *Union Works, Horseley Fields, Wolverhampton.*—Cut nails, shoe bills, heel and toe tips, washers, &c.

MUSGRAVE BROTHERS, *Ann Street Iron Works, Belfast.*—Patent slow-combustion stoves, patent stable fittings, cow-house fittings and iron piggeries.

MUSGRAVE'S PATENT SLOW-COMBUSTION STOVE, exhibited Class XXXI. No. 6159 in Catalogue, is the nearest approaoh to heating by hot water, and a certain and economical means of procuring a genial and steady heat.

The power of burning day and night throughout the winter, at a uniform temperature, has caused this stove to be extensively used for entrance halls, schools, libraries, &c.

It can be fixed in churches with either upright or underground flue, and is so simple that an inexperienced person can manage it.

The interior of the stove is furnished with hot-air chambers, which draw the fresh air from outside the building, and thus secure perfect ventilation.

There is no oppressive smell, nor does it form those explosive gases so much complained of in other stoves.

PATENT SLOW-COMBUSTION STOVE.

MUSGRAVE'S PATENT STABLE FITTINGS AND HARMLESS LOOSE BOXES, exhibited Class XXXI. No. 6159 in Catalogue, with tumbling manger and water pot, falling grid to prevent waste of hay, improved ventilator, sliding "barrier" to confine each horse to his stall in the event of breaking loose, and many other improvements deserving of inspection.

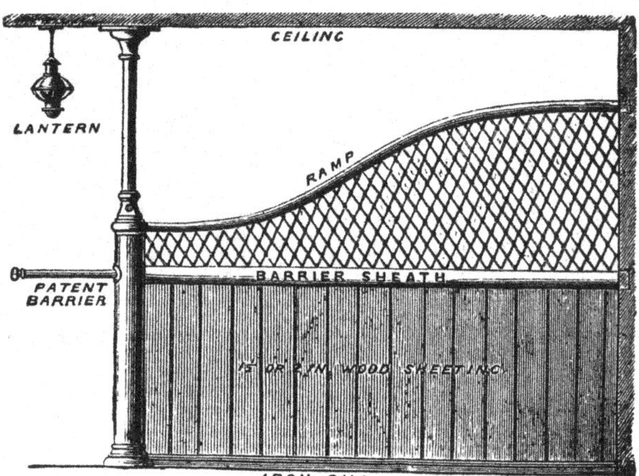

A PART OF THE PATENT STABLE FITTINGS.

MUSGRAVE'S PATENT IRON STALLS for cattle and IRON PIGGERIES and DOG KENNELS, exhibited Class IX. No. 2156 in Catalogue.

MUSGRAVE Brothers received, last season, for the foregoing inventions, the silver medals of the Royal Agricultural Societies of England and of Ireland, and the first prize at every competition where they have been exhibited.

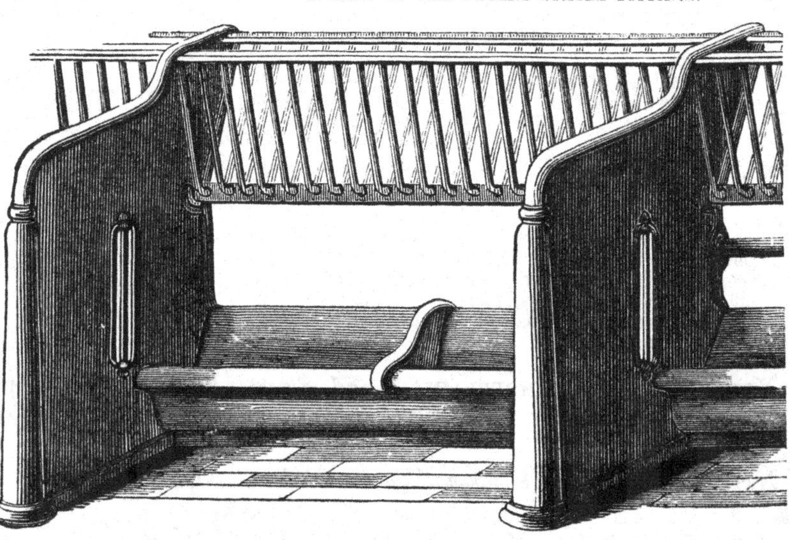

PATENT IRON STALLS FOR CATTLE..

[6166]

NEWTON, THOMAS, *Walsall, and* 84 *Long Acre, London.*—Steel bits, stirrups, spurs, chains, saddle harness, and carriage ironmongery.

[6167]

NICHOLAS, RICHARD, 32 *Water Street, Birmingham.*—Improved roasting-jack with key attached.

[6168]

NICHOLSON, WILLIAM NEWZAM, *Trent Iron Works, Newark.*—Cooking range, cottage stoves and fittings, and decorative iron work.

[6169]

NOCK & PRICE, 9 *Union Passage, Birmingham.*—Improved gas cooking range. (*See page* 68.)

[6170]

NYE, S., & Co., *Wardour Street, Soho.*—Patent mincing sausage machines, masticators, and coffee mills. (*See page* 69.)

[6171]

ONIONS, JOHN C., *Bradford Street, Birmingham.*—Portable forge; smith's, house, and fancy bellows; anvil and fire irons.

[6172]

OTTLEY, THOMAS, 59 *Spencer Street, Birmingham.*—Gold, silver, and bronze medals.

[6173]

OWEN, WILLIAM (late Sandford & Owen), *Phœnix Works, Rotherham.*—Bradley's patent kitchener; improved registered stable fittings.

PATENT KITCHENER.

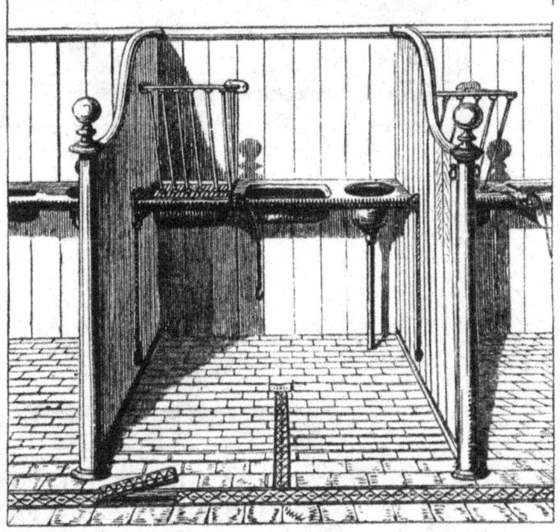

STABLE FITTINGS.

BRADLEY'S PATENT KITCHENER, which possesses all the advantages of the ordinary close-fire kitchener, in having excellent hot plates on top of oven or ovens, for boiling, steaming, and stewing, with the additional advantage of having a thoroughly effective and deep open fire for roasting purposes.

The principle of heating is self-acting, and a great saving of fuel is effected.

References in testimony of its advantages can be given.

IMPROVED REGISTERED STABLE FITTINGS, consisting of hay rack and seed box, tubular tying apparatus, manger, water-pot, &c.

These fittings have been designed expressly to meet the following requirements—

1. Of uniting strength with lightness of appearance.
2. Of preventing possibility of accident to the animal placed in them.
3. Of avoiding all waste in food.
4. Of cultivating cleanliness and comfort.
5. Of being simply and inexpensively fixed.

[6174]

PALMER, JOHN, & SONS, *Beech Lanes, near Birmingham.*—Screw railway wrenches, screw and fixed spanners, hammers, &c.

[6175]

PATENT ENAMEL COMPANY, 288 *Bradford Street, Birmingham.*—Glass enamelled hollow ware and patent tablets for street names, &c. (*See page* 70.)

[6176]

PERRY, THOMAS, & SON, *Bilston.*—Metallic bedsteads, fire-proof and thief-resisting safes.

NOCK & PRICE, 9 *Union Passage, Birmingham.*—Improved gas cooking range.

GAS COOKING RANGE.

HEATING STOVE.

IMPROVED GAS COOKING AND HEATING STOVES.

Meat cooked by gas takes less time than by ordinary fire, browns beautifully, and requires no attention. It is more nutritious, full of gravy, tender, and of superior flavour. Pastry can be baked by the same means in a superior manner, at a saving of 40 per cent. in cost and labour.

The exhibitors manufacture every description of gas cooking and heating stoves. Drawings and price lists will be sent on application.

NYE, S., & CO., *Wardour Street, Soho.*—Patent mincing sausage machines, masticators, and coffee mills.

Obtained a Prize Medal at the Paris Exhibition, 1855.

These machines are intended for mincing and mixing various substances—meat or vegetables for soups, &c.; fruit for mincemeat; suet for pastry; and for making potted or forced meats; also for preparing a great variety of dishes.

For sausages, they mince, mix, and force into the skins at the same time, and are admirably adapted for reducing meats for soups, according to the mode recommended by Professor Liebig, in his work on "The Chemistry of Food."

THE SAUSAGE MACHINE.

Their economy and efficiency are so great, and they are so well adapted for the kitchen, that they only require to be known to secure their general adoption.

"Among other objects in the show worthy of special notice, we may mention the very ingenious mincing machine, exhibited by Nye & Co. It is extremely clever, and for the mechanical skill which it displays, is eclipsed by nothing in the whole show."—*Times, July* 14, 1853.

"This is a little thing every husband ought to carry home to his wife, who, we are satisfied, will turn it to the best account, and save the price."—*Mark Lane Express, August* 15, 1854.

Price £1 10*s.* £2 2*s.* £3 3*s.* and £7 7*s.*

A SMALL MINCER for the dinner table, to assist digestion, loss of teeth, &c. Price £1 10

This machine is very neatly got up, and may be screwed on the dining table without even injuring the cloth. It is intended for mincing food for persons who cannot masticate properly. It is made hot, and meat, &c. is rapidly minced. To invalids, and to those who in order to preserve health are obliged to have their food thoroughly minced, this machine is invaluable.

"12, *Norfolk Villas, Westbourne Grove, Bayswater.*

"I have had one of your mincing machines for the dinner table in use for some time, and find it everything that could be wished. We have had masticators containing several knife-blades in one handle, but your invention is vastly superior. I recommend it to all my friends who suffer from indigestion. Yours obediently,

"Messrs. Nye & Co." "S. SAUNDERS.

By Her Majesty's Royal Letters Patent.

NYE'S PATENT IMPROVED MILLS, for coffee, pepper, spices, &c.

These mills are most conveniently arranged for domestic use, being provided with a cramp, by which they are fixed to the table or any other convenient place in an instant, and as quickly removed. By a nice and safe arrangement the grinding surfaces cannot possibly touch each other, being provided with a regulating screw, by means of which they are set to grind fine or coarse, as desired. They grind very rapidly, and are the most convenient mills ever offered to the public. Families using these mills avoid adulteration, and secure a genuine article.

Prices: No. 1, 8*s.*; No. 2, 10*s.*; No. 3, 14*s.*; No. 4, 20*s.*

TESTIMONIALS.

"GENTLEMEN,—It affords me great pleasure to add my testimony to the merits of your truly useful mincing machine, which I have now had in constant use for the last twelve months. Its performance surpasses all my expectations of it, and its great utility is only exceeded by its simplicity.

"I have already recommended the machine to many of my friends, and it will afford me much pleasure to satisfy any person as to its great efficacy, and you have my full permission to refer any one to me for that purpose.

"CHARLES GURDEN,
"*Chief Cook to the Honourable Society of the Middle Temple.*"

"60, *Tower Street, Westminster Road.*

"GENTLEMEN,—I have had your mincing machine for the last sixteen months, doing all the sausage-making for my business as a pork butcher, and am glad to say I cannot praise it too highly; it is a great saving of time, and has given me the greatest satisfaction.

"I am, gentlemen, yours, &c.

"To Messrs. Nye & Co." "J. WILSON.

"*Agricultural Department, Baker Street Bazaar.*

"SIR,—In reply to your inquiry respecting the character of your machine for mincing and sausage making, I beg to say that in all the quantity I have sold, I have never had a complaint of any kind, but in every instance in which I have made the inquiry I find they have given the greatest satisfaction; the simplicity of construction, superior workmanship, and, above all, the material of which they are composed, render them particularly clean and wholesome, and not liable to derangement.

"I am, yours respectfully,

"To Messrs. Nye & Co." "M. MEDWORTH, *Manager.*

"11, *Castlenau Villas, Barnes, Nov.* 18, 1854.

"GENTLEMEN,—In reply to your note of the 11th inst. I have great pleasure in saying that after two months' trial of your excellent patent mincing machine, I can confidently recommend it, fully answering, as it does, all the purposes you describe in your prospectuses; and I feel assured, were it more generally known, few families would be without one. Wishing you all the success the invention merits,

"I am, gentlemen, your obedient servant,

"To Messrs. Nye & Co." "R. WEDGEWOOD.

"*Refreshment Department, Western Area, Messrs. F. E. Morrish & Co. Contractors, Exhibition Building, South Kensington, May* 24, 1862.

"GENTLEMEN,—So far as our experience extends as to the utility of your Patent Mincing Machines, of which we have two in use, we have no hesitation in stating that we find them to answer the purpose in every respect.

"We are, gentlemen, yours obediently,

"To Messrs. S. Nye & Co." "F. E. MORRISH & CO.

PATENT ENAMEL COMPANY, 288 *Bradford Street, Birmingham.*—Glass enamelled hollow ware and patent tablets for street names, &c.

GRATE FRONT.

1. GRATE FRONT, in Limousine enamelled iron.

2. ALTAR FRONTAL, in Limousine enamelled iron, style of the 15th century.

3. PANELS AND FRIEZE, in Limousine enamelled iron, for exterior decorations of buildings.

4. ORNAMENTAL PANEL of Limousine enamelled iron, for bedsteads.

5. PANELLING of Limousine enamelled iron, for state cabin, &c.

6. Indestructible tablets for shop signs, door numbers, street nameplates, &c.

ALTAR FRONTAL.

7. SPECIMENS OF TABLE WARE, made of wrought-iron enamelled and printed in imitation of earthenware.

8. UTENSILS for culinary and domestic use.

9. CAST-IRON TUBES, for conveying gas or water, rendered incorrodible by enamel.

The whole of these specimens are made of wrought-iron rendered incorrodible by the process of enamelling, by which they are covered with a hard, vitreous surface, into which the colours are fused or burnt at a high temperature, so as to be perfectly unalterable by exposure to the atmosphere or weather for any length of time.

[6177]

PEYTON & PEYTON, *Bordesley Works, Birmingham; London Warehouse,* 49 *Long Acre, W.C.; City Office,* 46 *Moorgate Street, E.C.*—Metallic bedsteads; hat and umbrella stands of wrought and cast iron combined.

PATENT IRON HALF-TESTER BEDSTEAD. (No. 5.)

1. CHILD'S COT, japanned.

IRON BEDSTEADS.

2. PATENT SOLID IRON FRENCH BEDSTEAD, japanned and relieved, with improved dovetail joints to tighten the lath bottom; size, 3 ft. by 6 ft. 6 in.
3. PATENT IRON HALF-TESTER BEDSTEAD, japanned and relieved, improved dovetail joints to tighten the lath bottom; the pillars of parallel tube and ornamental castings; size, 3 ft. 6 in. by 6 ft. 6 in.
4. PATENT IRON HALF-TESTER BEDSTEAD, richly japanned and relieved with gold, improved dovetail joints (the same as in No. 3), the pillars composed of taper tube and massive ornamental castings; size, 5 ft. by 6 ft. 6 in.
5. PATENT IRON HALF-TESTER BEDSTEAD, the same as No. 4, but japanned to imitate walnut wood.

IRON BEDSTEADS, BRASS MOUNTED.

6. PATENT IRON FRENCH BEDSTEAD, japanned, the pillars of parallel tube, mounted with brass, the head and foot rails with brass ornaments to correspond, improved dovetail joints (the same as in No. 3), size, 3 ft. 6 in. by 6 ft. 6 in.
7. PATENT IRON HALF-TESTER BEDSTEAD, japanned, improved dovetail joints (the same as in No. 3), the pillars of taper tube, mounted with brass, the head and foot rails with brass ornaments to correspond; size, 4 ft. 6 in. by 6 ft. 6 in.
8. PATENT IRON TESTER BEDSTEAD, japanned, the pillars of parallel tube, mounted with brass, the head and foot rails with brass ornaments to correspond, improved dovetail joints (the same as in No. 3), size, 3 ft. 6 in. by 6 ft. 6 in.

BRASS BEDSTEADS.

9. PATENT BRASS HALF-TESTER BEDSTEAD, the pillars composed of patent taper tube, and with the rails and cornice of elaborate cast work, wrought and burnished; new patent elastic bottom; size, 5 ft. by 6 ft. 6 in.
10. PATENT BRASS FOUR-POST BEDSTEAD, with canopy and coronet, the pillars of parallel tube, improved dovetail joints (the same as in No. 3); size, 4 ft. 6 in. by 6 ft. 6 in.
11. PATENT BRASS FOUR-POST BEDSTEAD, with canopy and coronet, pillars of patent taper tube, with richly wrought ornaments, the head and foot rails and coronet to match; size, 5 ft. by 6 ft. 6 in.
12. BRASS STRETCHER OR CAMP BEDSTEAD; width, 2 ft. 10 in.

IMPROVED HAT STAND, of wrought and cast iron combined.

[6178]

PHILLIPS, THOMAS, 55 *Skinner Street, Snow Hill, London.*—Gas bath, gas cooking apparatus, stoves, &c.

[6179]

PHILLIPS, G., 27 *Featherstone Street, City Road.*—Locks and fire-proof depositories, cheap, safe, and elegant.

[6180]

PIERCE, WILLIAM, 5 *Jermyn Street, London.*—Ornamental stove grates and fenders for drawing rooms, &c.; pyro-pneumatic stove grate for churches, &c.; fire-lump grates for cottages.

[6181]

PIGOTT & CO., *London, and St. Paul's Square, Birmingham.*—Buttons, ornaments, medals, shirt studs, sleeve links, clasps, solitaires, &c.

[6182]

PLIMLEY, J. T., & CO., *Wolverhampton.*—Locks and general ironmongery.

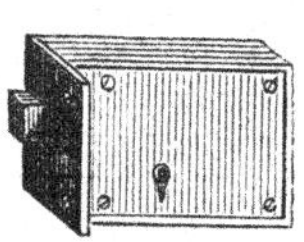

RIM DEAD TO LOCK ON BOTH SIDES.

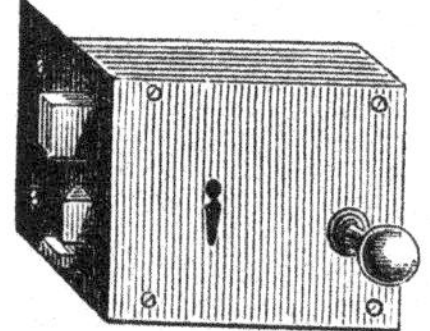

THREE-BOLT HALL DOOR.

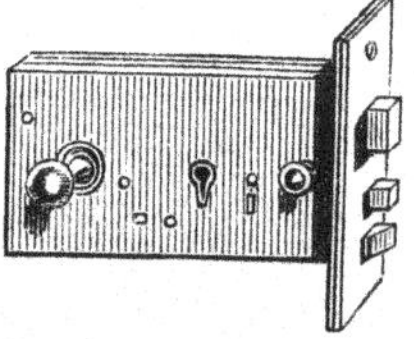

MORTISE THREE-BOLT.

TILL OR DRAWER.

STRAIGHT CUPBOARD.

BOX, CHEST, OR SLOPING DESK.

PAD.

Patterns may be seen and prices furnished for all kinds of hardware suitable for the East and West Indies, North and South America, Africa, and other foreign markets, as also a general assortment for the home and colonial trades, including anchors, anvils, augers, axles, battery pans, bells, boilers, bolts, brass work, brushes, bellows, cast-iron goods, coffee mills, chain, coach screws, copper goods, cart springs, crowbars, cannon, elliptic springs, fuze, frying pans, fenders, fire irons, files, flesh forks, guns, grindstones, gridirons, hammers, hinges of all kinds, hames, harness, jacks, keys, kibbling mills, locks of all kinds, ladles, latches of all kinds, matchets, muskets, nails (cut and wrought), picks, pots, (three-legged), planes, pistols, rivets, spades and shovels, screws, sad irons, traps, tue irons, tin goods, wire of all kinds, weights, and every other article connected with the hardware trade.

[6183]

POTTER, THOMAS, 44 *South Molton Street.*—Bronzed castings, and wrought-iron work.

[6184]

POUPARD, WILLIAM, *Blackfriars Road, London.*—Patent curvilinear beam weighing machine; imperial machine, spirometer balance, and safety wheel skid, &c.

The following are exhibited:—

1. IMPERIAL WEIGHING MACHINE TRANSVERSE COMPOUND LEVER, showing action, for weighing coals or any stipulated weight where portability or space is required.
2. Compensated curvilinear beam WEIGHING MACHINE.
3. ditto ditto ditto VAN ditto
4. Compensated curvilinear beam RETAIL COAL MACHINE.
5. ditto ditto ditto ditto (part inverted).
6. SPIROMETER BALANCE in connexion with Dr. Hutchinson's spirometer.
7. Patent safety curvilinear WHEEL SKIDS.

[6185]

PRICE, CYRUS, & CO., *Hursley Fields, Wolverhampton.*—Double-action detector locks; gunpowder fire-proof safes.

The exhibitors are the patentees and manufacturers of the gunpowder-proof double-action detector locks, and gunpowder, fire, and thief proof safes.

[6186]

PRICE, GEORGE, *Cleveland Works, Wolverhampton.*—Wrought-iron fire-resisting safes, chests and doors; cabinet, rim, and mortise locks. (*See pages* 74 *to* 77.)

[6187]

PULLINGER, COLIN, *Selsey, Sussex.*—Traps for mice, rats, &c.; each one caught resetting the trap.

[6188]

RADCLYFFE, THOMAS, *Leamington.*—Kitchen ranges; smokeless feeding screw for ditto.

[6189]

RAWLINS, EDWARD, 27 *Whittall Street, Aston, Birmingham; Works, Thimble Mill Lane Aston.*—Stampings and pressings of iron and steel for a variety of purposes.

[6190]

REDMAYNE & CO., *Wheathill Foundry, Rotherham.*—Stove grates, fenders, hat and umbrella stands, and fountains.

[6191]

REYNOLDS, JOHN, *Crown and Phœnix Works, Birmingham.*—Nails, tacks, brads, bills, washers, brackets, hooks, &c.

Obtained a Prize Medal at the Exhibition of 1851.

These Works have been established nearly half a century, for the manufacture of every description of patent cut nails, tacks, brads, and shoe bills, of best make and material, in copper, brass, zinc, and iron; also cornice fasteners and brackets, pressed hinges, washers, &c.

[6192]

REYNOLDS, JOHN, 57 *New Compton Street, W.C.*—Wire work, useful and ornamental, and patent metallic netting.

[6193]

RHODER, WILLIAM, *Westgate, Bradford.*—Indestructible fireproof safe.

[6194]

RICHARDS, HENRY, 36 *St. James' Place, Liverpool.*—A metallic meat safe.

[6195]

RICHARDS, W. & CO., *Imperial Wire Works,* 370 *Oxford Street.*—Wire work.

The exhibitors manufacture all kinds of ornamental wire-work.

They exhibit the following specimens:—

Pheasant aviaries.

Rose temple flower baskets.

Flower stands.

Garden seats.

Flower boxes, &c.

[6196]

RICKETS & HAMMOND, 5 *Agar Street, Strand.*—Gas range and gas stoves, gas globe light, and ventilator.

[6197]

RIDDELL, JOSEPH HADLEY, 155 *Cheapside, London.*—Patent slow-combustion boiler for heating by the circulation of hot water.

Price, George, *Cleveland Safe and Lock Works, Wolverhampton.* London Agents: McNeil & Moody, *Stationers, 23 Moorgate Street, Bank.*—Wrought-iron fire-resisting safes, chests, and doors; cabinet, rim, and mortise locks.

Author of a "Treatise on Fire and Thief proof Depositories and Locks and Keys," and a "Treatise on Gunpowder-proof Locks, Drill-proof Safes," &c.

MERCHANTS' OR BANKERS' BOOK AND CASH SAFE.

No. 1. Merchants' or Bankers' Book and Cash Safe, with 3-in. fire-proof composition chambers, and fitted with 38 compartments for books, formed by 32 movable partitions; 4 drawers for day use for coin, notes, and bills of exchange, with 3 distinct safes at the bottom, made of $\frac{1}{2}$-in. boiler plates with case-hardened drill-proof doors, for the additional safety of cash and securities at night. All the doors are fitted with George Price's double patent "ne plus ultra" unpickable and gunpowder-proof lock, each lock different, with a master key to pass all. Size, outside measure, (exclusive of plinth and cornice) 8 ft. 6 in. high, by 6 ft. wide, by 2 ft. 6 in. deep . . £300 0

No. 2. Single-door Precious Stone or Cash Safe, made of $\frac{5}{8}$-in. solid boiler plates, with 1-in. solid door. The whole body and door case-hardened. With 2 drawers. Size, 33 by 25 by 25 in. £40 0

No. 3. Single-door Safe with 2 drawers and case-hardened drill-proof door. The lock chamber of this is unscrewed to show the construction of the locks. Size, 26 by 20 by 20 in. Quality, 201 C . £14 2 6

PRICE, GEORGE, *continued.*

No. 4. WATCHMAKERS' AND JEWELLERS' SAFE. Not fire-proof. Made of ½-in. boiler plates with case-hardened drill-proof door, and fitted with one shelf. Size, 48 by 30 by 26 in. £39 7 6

This safe represents the strength and quality of the patentee's foreign bankers' bullion rooms.

No. 5. SINGLE-DOOR BOOK OR CASH SAFE, with 2 drawers. Size, 33 by 25 by 25 in. Quality, 201 C £18 0

No. 6. DOUBLE-DOOR BOOK OR CASH SAFE, with 2 drawers. Size, 30 by 30 by 24 in. Quality, 202 C £22 0

No. 7. MERCHANTS' COUNTING-HOUSE SAFE, with 3-in. fire-proof composition chambers, and fitted with 3 drawers, cupboards, and compartments for books and papers. The partitions are movable. The locks on the drawers and cupboards are all different, with a master key to pass. Size, 66 by 52 by 30 in. Quality, 202 D £80 0

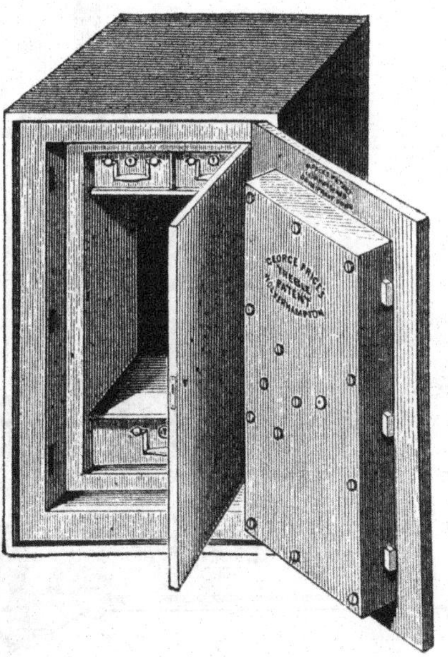

No. 8. BANKERS' SAFE, with 5-in. fire-proof composition chambers, and fitted with 3 drawers and an inner door, secured by a mortise spring lock. The outer door is rendered drill-proof by being case-hardened, and is hung on inside centres. Size, 48 by 30 by 30 in. Quality, 206 D £50 0

The left-hand cut shows the style of the safe when the door is closed, and also represents the Burnley safe, No. 2.

PRICE, GEORGE, *continued.*

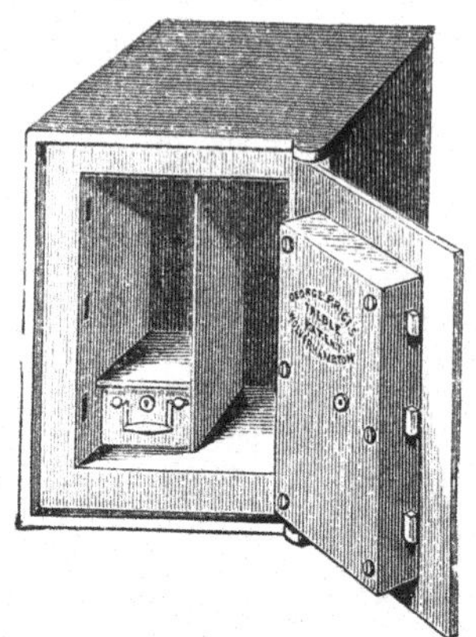

No. 9. SINGLE-DOOR SAFE, with one drawer half across and upright partition. Size, 24 by 18 by 18 in. Quality, 201 B £9 0

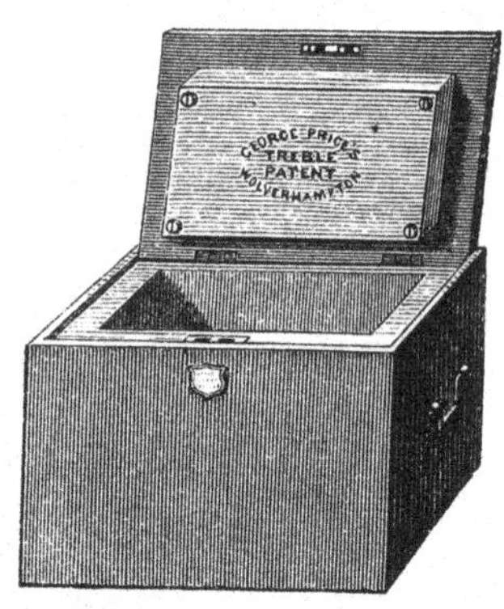

No. 10. FIRE-PROOF DEED CHEST. Size, 18 by 13 by 13 in. Quality, No. 120 . . . . £5 0

No. 11. DOUBLE CASH AND BOOK SAFE, each compartment being in itself distinct and equally secure. Size, 63 by 48 by 30 in. Quality, 204 D, with case-hardened doors. Price £100 0

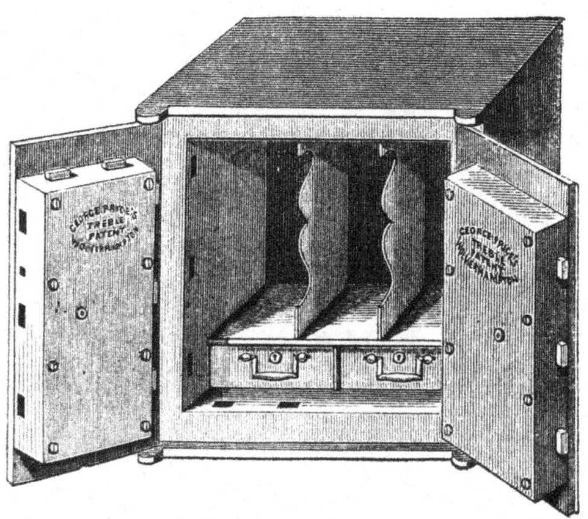

No. 12. DOUBLE-DOOR BOOK OR CASH SAFE, with 2 drawers, and 2 movable partitions. Size, 33 by 33 by 25 in. Quality, 202 B £24 0

No. 13. DOUBLE CASH AND BOOK SAFE, of a similar construction to No. 11. Size, 36 by 36 by 25 in. Quality, 202 C £32 10

PRICE, GEORGE, *continued.*

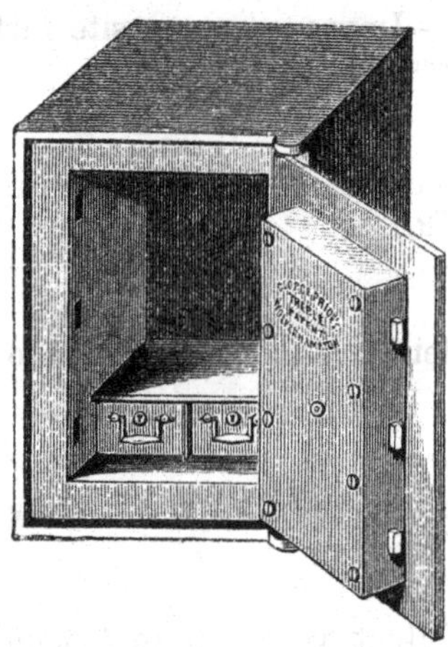

No. 14. SINGLE-DOOR SAFE with 2 drawers. Size, 28 by 22 by 22 in. Quality, 201 B £12 0

No. 15. SINGLE-DOOR SAFE with 2 drawers. Size, 28 by 22 by 22 in. Quality, 201 C £14 0

No. 16. SINGLE-DOOR SAFE with 2 drawers. Size, 24 by 17 by 15 in. Quality, the People's safe . . . £7 0

No. 17. DOUBLE-DOOR BOOK SAFE with 2 drawers. Size, 30 by 30 by 24 in. Quality, 202 B £20 0

No. 18. FIRE-PROOF DOOR AND FRAME. Size, 6 ft. by 2 ft. 6 in. Quality, 130 C £20 0

No. 19. DOOR AND FRAME NOT FIRE-PROOF, opens inwards. Size 6 ft. by 2 ft. 6 in. Quality, 130 C £20 0

No. 20. FIRE-PROOF DOOR AND FRAME. Size, 6 ft. 6 in. by 3 ft. Quality, 130 D £32 0

No. 21. FIRE-PROOF DOOR AND FRAME. Size, 6 ft. 6 in. by 3 ft. Quality, 130 D, with sunk panels . £40 0

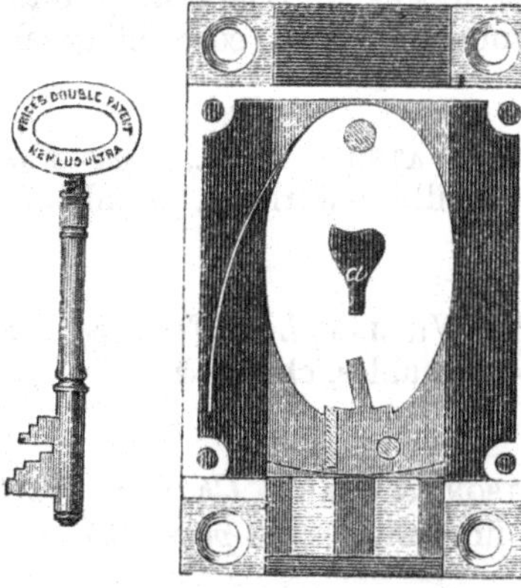

Nos. 33 to 40. PRICE'S DOUBLE PATENT "NE PLUS ULTRA" UNPICKABLE AND GUNPOWDER-PROOF LOCK, with small hardened pin keys. *a* shows the open space in which the key works, being the only cavity into which gunpowder can be forced through the keyhole. The white part represents the levers or tumblers.

Nos. 35 and 36. The same, after being tested by repeated explosions of gunpowder.

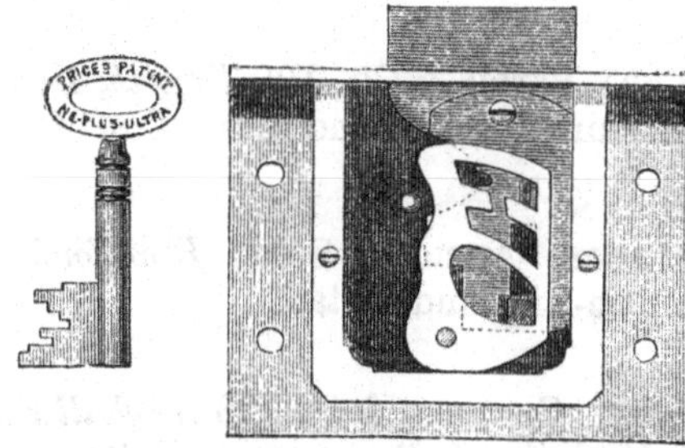

Price's patent "ne plus ultra" lock is made for all purposes and of every size.

Nos. 45 and upwards, represent specimens of the following kinds :—

Till or drawer, 1 to 4 in. from 9*s.*
Straight cupboard, 1 to 4 in. from 9*s.*
Cut cupboard, 1 to 4 in. from 9*s.*
Box, chest, and sloping desk ; mortise camp desk ; pedestal or sideboard ; link plate cupboard ; travelling desk ; and mortise box, 1 to 4 in. from 10*s.*
Cash box with fixed nosle, 2½ to 4 in. from 10*s.* 6*d.*
Pad, 1 to 5 in. from 10*s.*
Silver pad, ½ in. 30*s.*
Portfolio and writing case, 10*s.*
Trunk and portmanteau, 3 to 4½ in. from 14*s.*
Book-edge or ledger lock, from 16*s.*
Letter bag, 12*s.*
Round escutcheon lock, for locking up the keyholes of other locks, 20*s.*
Flush night latch, 3½ to 6 in. from 14*s.* 6*d.*
Drawback rim night latch, 4 to 6 in. from 16*s.*
Mortise night latch, from 3 to 4½ in. from 23*s.*
Rim dead to lock on one side only, 4 to 12 in. from 16*s.*
Rim dead to lock on both sides, 4 to 12 in. from 19*s.*
Spring lock for front doors, 6 to 10 in. from 28*s.*
Mortise 1-bolt dead, 3 to 7 in. from 20*s.*
Mortise 2-bolt, 5 in. 33*s.*; 6 in. 35*s.*; 7 in. 38*s.*
Mortise 3-bolt, 6 in. 37*s.*; 7 in. 40*s.*
Mortise hall-door lock, 5 in. 30*s.*; 6 in. 32*s.*; 7 in. 35*s.*

The above prices include two hardened keys to each lock.

Ornamental key handles, 40 specimens.
Price's patent door spindle.
Large specimen lock (24 in. dead) to show the principle of construction.
Japanned cash and deed boxes.

[6198]

RITCHIE, JAMES, 22 *South B. of Canongate, Edinburgh.*—Improved composite metallic cord for counter weights, cords of gaseliers, hanging pictures, and sash line.

[6199]

RITCHIE, WATSON, & CO., *Etna Foundry, Glasgow.*—Kitcheners, cabooses, grates, mantel-pieces, gill air warmers, plumbers' goods. (*See page* 79.)

[6200]

ROBERTS, WILLIAM, *Lion Foundry, Northampton.*—Register stoves, kitchen ranges, ornamental cast-iron tables, chairs, &c.

[6201]

ROBERTSON & CARR, *Chantry Works, Sheffield.*—Register grates, hot air stoves, fenders and fire irons. (*See page* 80.)

[6202]

ROBOTHAM, SAMUEL, *Bradford Street, Birmingham.*—Patent woven wire fencing, guards, cages, wire, and general wire work.

[6203]

ROCKE, WILLIAM, *Phœnix Foundry, Wolverhampton.*—Samples of machinery from refined pig-iron; ditto refined wrought-iron.

[6204]

ROGERS, PETER, & CO., 106 *Digbeth, Birmingham.*—Steelyards, scale beams, scales, weighing machines, stocks, and dies.

[6205]

ROLLASON, ABEL, & SONS, *Bromford Mills, Erdington, near Birmingham.*—Patent steel music-wire and metals.

[6206]

ROWLEY, CHARLES, & CO., 23 *Newhall Street, Birmingham;* 49 *Aldermanbury, London; and* 1 *High Street, Manchester.*—Buttons, ornaments, bill files, and fancy goods. (*See page* 81.)

[6207]

ROWLEY, S. A., 63 *Clement Street, Birmingham.*—Pearl buttons and studs.

[6208]

RYFFEL, I. E., *Wimbledon.*—Hygeian stove, the most effective, economical, healthy, cleanly and safe stove ever invented.

[6209]

ST. PANCRAS IRON WORK COMPANY, THE, *Old St. Pancras Road, London, N.W.*—Interior of stable, and ornamental gates.

[6210]

SCOTT, J. W., *Sidbury Works, Worcester.*—Patent solid leather buttons; patent valve leather gun wads; patent leather washers, &c. (*See page* 82.)

[6211]

SHERWIN, JOSEPH, *Tabernacle Walk, Finsbury.*—Economic kitchen range for baking, boiling, steaming, roasting, and improved supply.

[6212]

SMITH, FREDERICK, & CO., *Halifax.*—Bar-iron, and wire in various stages to finest sizes.

[6213]

SMITH, THOMAS, 27 *St. John's Square, Wolverhampton, and* 18 *St. Mary Axe, London.*—Hardware and cutlery.

RITCHIE, WATSON, & Co., *Etna Foundry, Glasgow.*—Kitcheners, cabooses, grates, mantel-pieces, gill air warmers, plumbers' goods.

RITCHIE, WATSON, & Co. are ironfounders and sole manufacturers of the patent Etnean kitcheners, open-fire kitchen ranges, stove grates, gill air warmers, hot air and gas stoves, hot-water apparatus, rain-water goods, patent mangles, patent stable fittings, &c.

1. AN ORNAMENTAL IRON CHIMNEY PIECE (cast in one piece), with a circular opening and arched slab. The slab surmounted by an ornamental stand for clock, bust, or vase. The chimney piece is fitted with an ornamental register stove grate, with fire-clay linings, radio-protectio ashes pan, circular fender, steel fire-irons, &c. Style, Italian.
2. AN ORNAMENTAL GILL AIR WARMER, with radiating surfaces and two fires, suitable for the aisle of a church. This stove is also made with one fire. The same pattern may be had with four different sizes of gills. Style, Gothic.
3. AN ORNAMENTAL GILL AIR WARMER, suitable for a large room, hall, or lobby, made also with two fires. Style, Renaissance.

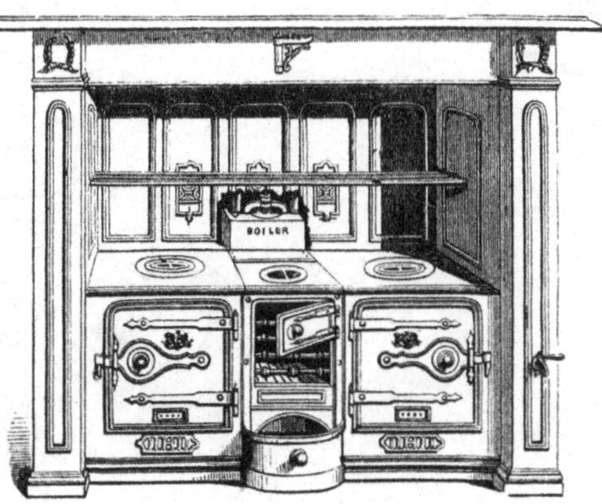

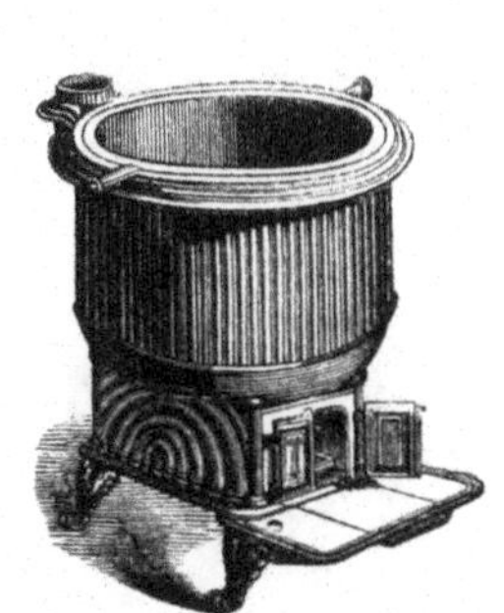

4. AN IRON CHIMNEY-PIECE, fitted with a register stove grate, &c. Style, mediæval.
5. AN ORNAMENTAL BALUSTRADE.
6. AN ETNEAN KITCHENER, with two roasters, a wrought-iron welded boiler, iron chimney piece, plate rack, &c. These kitcheners are made any size, from 3 ft. to 30 ft. and with any desirable number of roasters, brick or iron ovens, boilers, steam closets, close or open fire, &c. the manufacturers having had fully 40 years' experience in this department of manufacture.
7. A SMALL PORTABLE FARM BOILER, with fire-clay linings, made in five sizes.
8. AN ECONOMIC COOKING STOVE, on the American principle — Uncle Sam by name — made in three different sizes, and with or without boilers, hot-water apparatus, &c.
9. A miscellaneous assortment of RAIN-WATER GOODS, PUMPS, PATENT STABLE FITTINGS, HOT-WATER APPARATUS, and other goods suitable for ironmongers, plumbers, builders, &c.

ROBERTSON & CARR, *Chantrey Works, Sheffield.*—Register grates, hot-air stoves, fenders and fire irons.

A PILLAR GRATE IN THE TUDOR STYLE.

A PILLAR GRATE in the Tudor style, adapted for a baronial mansion. The body, bars, and dogs are cast-iron. The balls and jewelled ornamentation are malleable iron, being susceptible of a high polish, contrasting with the deep black of the body of grate.

Size, extreme width, 4 ft. 1 in.; ditto depth, 2ft. 3 in. but can be made in other sizes.

ROWLEY, CHARLES, & CO., 23 *Newhall Street, Birmingham;* 49 *Aldermanbury, London; and* 1 *High Street, Manchester.*—Buttons, ornaments, pins, bill files, and fancy goods.

CHARLES ROWLEY & CO. are general button manufacturers, stampers, piercers, tool makers, and patentees of the safety pins, so admirably adapted for children's under-clothing, ladies' shawls, &c. Stay, and other eyelets, military ornaments, belt, garter and other clasps; and the sole manufacturers of the Albert self-adjusting brace slide, protected by patent and registration.

This adjuster for gentlemen's braces, will be found the most simple, efficient, and economic adjuster ever offered to the public.

Open. Shut.

PATENT PINS.

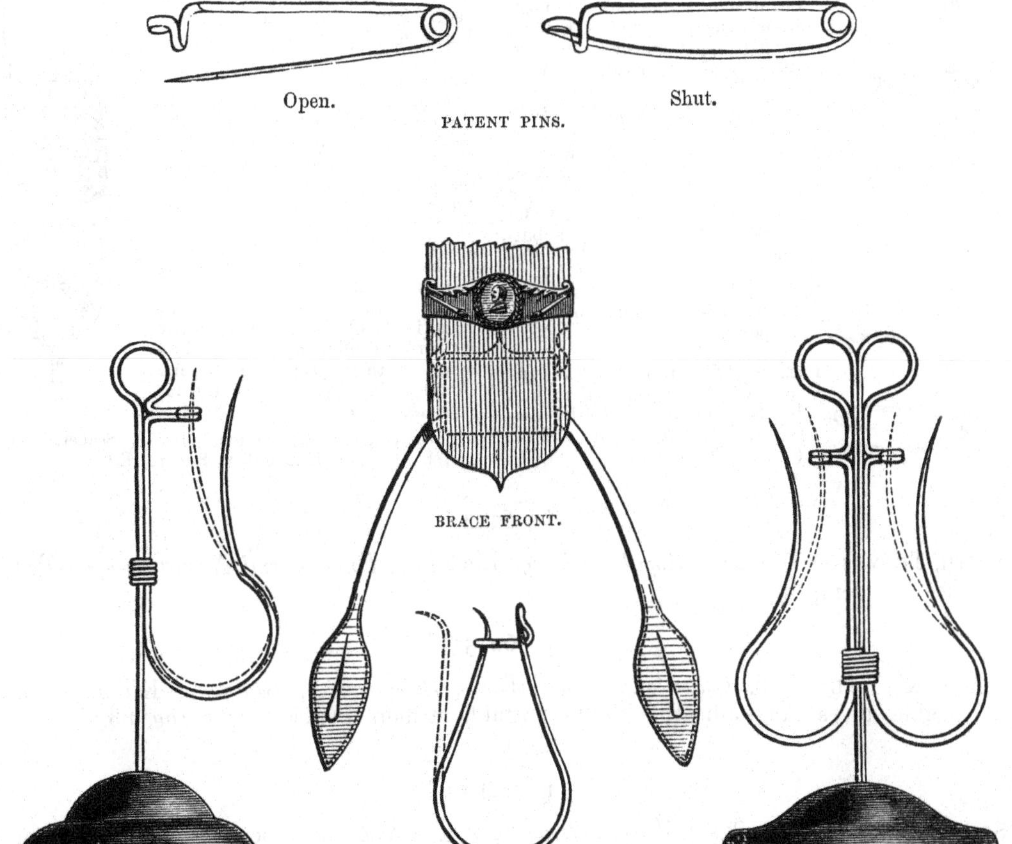

BRACE FRONT.

TICKET HOLDER OR BILL FILE. BILL FILE.

C. R. & Co. are also patentees and sole manufacturers of the universal mercantile and domestic bill file, which being made of stout or fine wire, admits of being adapted for a variety of useful purposes; the point in every instance being held and protected, the articles filed will be secured, and yet present every facility for removal or examination, and are also adapted for suspending prints, drawings, and other like articles. These useful appendages to the office, retail shop, or private house, may be made for suspension, or fitted with a stand as here illustrated. To be had retail at all respectable stationers, ironmongers, and dealers in small wares. They are, likewise, general manufacturers of brace and other buckles, naval, military, crest, and other buttons, brass and iron ship thimbles, weavers' mails, and a variety of other such-like small articles.

SCOTT, J. W., *Sidbury Works, Worcester.*—Patent solid leather buttons; patent valve leather gun wads; patent leather washers, &c.

The exhibitor is patentee of the following:—

The PATENT SOLID LEATHER BUTTONS are manufactured with leather shanks, also metal, cord, gut, &c. and in designs suitable for boots, gaiters, upholstery, dresses, and vests; they are finished in bronzes and enamelled colours, and inlaid work.

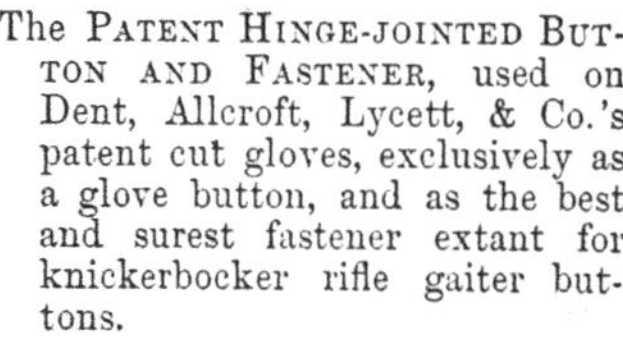

The REGISTERED STUD-SHANK PATENT LEATHER BUTTON is a ready and efficacious method of attaching buttons to boots for ornamental purposes, without sewing, instantaneous in application, and appear riveted on.

The PATENT HINGE-JOINTED BUTTON AND FASTENER, used on Dent, Allcroft, Lycett, & Co.'s patent cut gloves, exclusively as a glove button, and as the best and surest fastener extant for knickerbocker rifle gaiter buttons.

SOLID LEATHER NUTS AND BUTTONS for pianoforte and organ purposes.

PATENT CUT LEATHER WASHERS in any substance of calf or buff leather, cut to sixteenths diameter, from ¼ in. centres to 2 in. (or larger to order).

PATENT CUT LEATHER LACES, with flat heads.

SCOTT'S NEW PATENT EXPANDING VALVE, and compressed leather concave compound gun wads.

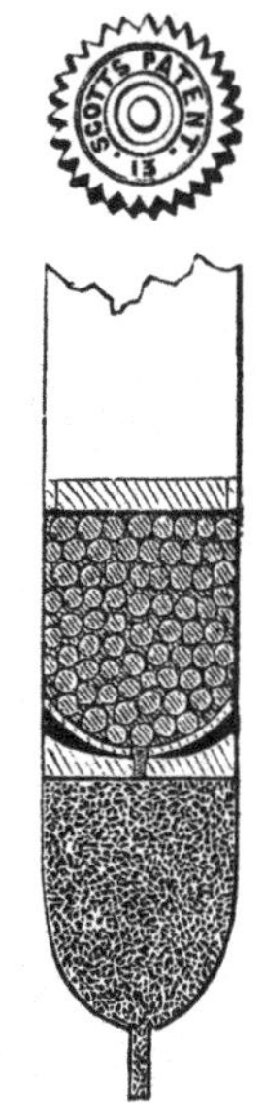

These patent wads are a scientific novelty of very peculiar construction. The compound which secures the effects hereinafter set forth, consists of a concave cup compressed in solid butt leather at certain angles, and also of a resilient valve made of leather. This inner cup or valve exactly fits the wad just described, and is riveted or firmly attached thereto at the centre only, leaving the sides free, but somewhat larger, so that it shall rest in the barrel rather in excess of the bore.

Whilst the wad is being rammed down upon the powder, the valve sufficiently collapses to allow it to pass easily down the barrel; but on its discharge, the reverse action takes place—the projecting edge of the expanding valve becomes pressed against the circumference of the barrel—effectually rendering it perfectly air-tight, and thus secures to the explosive gases the retention and resistance necessary for perfect combustion, and development of their propelling powers. Power is in proportion to the resistance.

Special agent in London, Mr. T. Seaber, colonial merchant, 21 and 22 Falcon Square, E.C.

[6214]

SMITH, THOMAS & WILLIAM, *Royal Exchange Buildings, London, and Newcastle-upon-Tyne.*—Wire and hemp ropes.

[6215]

SMITH & WELLSTOOD, *Columbian Stove Works, Bonnybridge, Glasgow.*—Kitchen cooking portable stoves, ranges, heating stoves, portable farmers' boilers. (*See page* 83.)

[6216]

SPOKES, JOSEPH, *North Street Mews, Fitzroy Square.*—Wood meat-screens and refrigerators.

[6217]

STANDING, THOMAS, *Preston, Lancashire.*—Galvanized wire netting, fencing staples, patent size, colour, and liquid agitator. (*See page* 83.)

[6218]

STANDLEY, WILLIAM, 38 *Park Street, Walsall.*—Bits, spurs, stirrups, bridles, reins, bombilloes, lasso rings, cruppers, and cavesons.

[6219]

STANLEY, JOHN M., & Co., *Midland Works, Sheffield.*—Gill air warmers, kitchen ranges, cooking apparatus, and stove grates.

Smith & Wellstood, *Columbian Stove Works, Bonnybridge, Glasgow.*—Kitchen cooking portable stoves, ranges, heating stoves, portable farmers' boilers.

No. 1. Family or Kitchen Stove, with hot-water attachment. All sizes and styles of these portable kitchen cooking stoves are made, ranging in price from
£3 3*s.* upwards.

HALL STOVE.

No. 2. Portable Laundry or Farmers' Boiler, made in 5 sizes, holding from 15 to 60 gallons, can be conveniently used in any position, and they are made to run on wheels or not as may be desirable. Prices from
£2 15*s.* upwards.

The exhibitors are manufacturers of American stoves and ranges.

Standing, Thomas, *Preston, Lancashire.*—Galvanized wire netting, fencing staples, patent size, colour, and liquid agitator.

Standing's Patent Sun and Planet Motion Agitator, and models of cisterns for mixing liquids, size, colours, starch, &c. for cotton-manufacturing, calico-printing, dyeing, and brewing purposes.

This apparatus is now in use in numerous large and well-known establishments; and is acknowledged to be the most efficient and complete invention for preparing size, colours, liquids, &c. yet extant; whilst the thorough agitating motion is not equalled by anything in the United Kingdom.

These machines can be made to any given size. Estimates on application. References to a large number of English firms using them.

The sun and planet motion consists of dashers, revolving round a fixed wheel, and at the same time turning upon their own axis in different directions. By this arrangement, the greatest possible agitation is produced, and from the very bottom to the top of the vessel the whole of the liquid is agitated to a complete foam.

For Colours and Starch.

A strong copper pan, with wrought-iron steam-jacket, fitted up with three dashers, revolving upon their own axes as well as round the vessel at the same time. Steam is introduced into the cavity between the iron jacket and copper pan, by which the liquid in the pan is kept boiling continually whilst the agitation is going on. These pans can be made to any size, but the sizes usually made are 100, 150, and 200 gallons.

For Size-preparing.

A strong cistern, fitted up with several sets of agitators, all turned from one cross shaft. These are made in sizes adapted for mixing at one time 10, 15, 20, or 30 sacks of flour, with the necessary quantity of water. In the cistern the flour and water is agitated so as to form a fine liquid; an improved pump is attached by which the liquid is raised to a copper boiling pan with iron steam-jacket. In this pan the liquid is boiled and agitated at the same time, from which it is conveyed by a pump adapted for the purpose, to the dressing or sizing frames.

For Brewing.

The patent apparatus is fitted up to large vats. The peculiar motion causes every particle of the compound to be disturbed, the whole being equally agitated from top to bottom.

Standing's Machine-made Galvanized Wire Netting and Fencing, for protecting young plantations, shrubberies, pleasure grounds, parks, gardens, wheat, barley, and general crops, against hares and rabbits; for sheep-fencings, and for flower, fruit, and vine-training, &c.

Black and Galvanized Strained Wire Fencing.

Galvanized Wire Strand Fencing.

Black and Galvanized Self-boring Staples, made by machinery, for railways, telegraph companies, &c. for fixing fencing wire and strand fencing. Made all sizes suitable for any purpose.

G 2

[6220]

STARK, JOHN C., 13 *Strand, and Swan Street, Torquay.*—Grates and kitchen ranges.

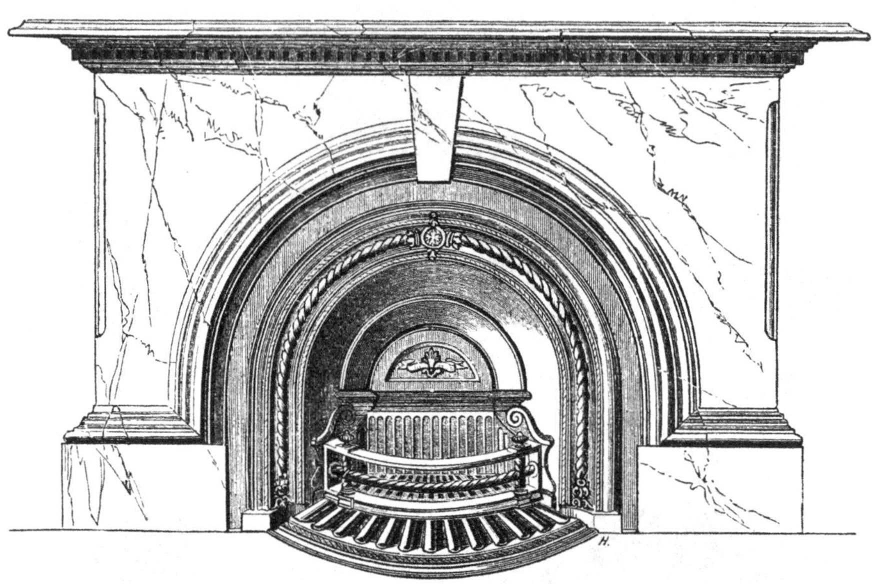

The exhibitor is a wholesale ironmonger, and manufacturer of marble chimney pieces, stoves, kitchen ranges, and all kinds of iron work.

He executes marble chimney pieces to any design, and to architects' own drawings.

Specimens of Devonshire marbles can be had on application, also see Class I. Eastern Annex.

[6221]

STEEL & GARLAND, *Wharncliffe Works, Sheffield.*—Stoves, grates, and fenders.

[6222]

STEPHENSON, JAMES, & Co., *Mill Wall Telegraph Works, Poplar, E.*—Wire rope for ships' standing rigging.

[6223]

STUART & SMITH, *Roscoe Place, Sheffield.*—Stoves, grates, fenders, fire irons, and castings.

[6224]

STUBBS, WILLIAM HENRY, *Park Crescent, London, N.W.*—Instruments for cleaning forks.

[6225]

TALBOT, C. & S., *Marylebone Iron Works, Great Titchfield Street.*—Cooking apparatus and utensils.

[6226]

TANN, JOHN, 30 *Walbrook, E.C.*—Patent reliance and other locks, fire-proof safes, iron doors, cash and deed boxes.

IRON DEED BOX.

Manufacturers to Her Majesty's Government. Established 1795.

JOHN TANN'S PATENT RELIANCE LOCKS AND FIREPROOF SAFES are the best and cheapest safeguard against fire and robbery.
John Tann's fireproof iron door for strong rooms and party walls.
John Tann's fireproof room and bullion chests.
John Tann's cash and deed boxes.
John Tann's patent Reliance cylinder locks.
John Tann's improved lever locks, street-door latches.

Patent Reliance Lock and Safe Warehouse, 30 Walbrook, London.

PATENT LOCK.

[6227]

TAYLOR, JOHN, JUN., Architect, 53 *Parliament Street.*—Smoke consuming and ventilating grates, apparatus, &c.

The Inventor during his professional practice as an architect has had his attention particularly directed to some of the drawbacks to the Englishman's enjoyment of his fire-side. These may be enumerated under the four following heads:

1. Smoke, undeprived of its carbon, or soot (which is fuel), contaminates the atmosphere, disfigures and decays buildings, and an incalculable amount of annoyance arises from smoke and smoky chimneys. Mechanical contrivances to effect the combustion of smoke will be ineffective when left, as they must often be, to the care of ordinary domestic servants.

2. The annual loss in London alone of the 75 per cent. of heat (an acknowledged fact), which escapes up the chimney without adding to the warmth of the apartment. The calculation of this does not afford any consolation in his discomfort.

3. The air necessary to support combustion makes its way to the fire from door or window, chink or crevice, and visits his back with hurtful draughts in proportion to the warmth he is receiving in front.

4. A frequent cause of smoky chimneys is that the chimney becomes filled with the air from the apartment, which, rushing in above the fire, lowers the temperature, and renders the flue incapable of acting as a sufficiently rarefied ventilating shaft, and often incapable of even conveying away the smoke at all.

VIEW WITH FRONT REMOVED, SHOWING THE ACTION.

FRONT VIEW OF THE SMOKE CONSUMING AND VENTILATING GRATE.

The peculiar features in these grates, which have been invented to cure the above evils, are:—

1. When the register is closed, the smoke, having ascended from the fuel and become mixed with atmospheric air, descends, and passes through the hottest part of the fire, where the carbon, or soot, in the smoke, is consumed as fuel.

2. The heat which would have rushed up the chimney, calculated at 75 per cent. passes down and around the hollow fire lumps of which the grate is formed, and the external air with which these communicate enters the apartment in a large body, moderately warmed, not heated.

3. The apartment is supplied with moderately warmed, instead of cold air, from door or window, and thus thorough ventilation is effected.

4. The cold air cannot rush up the chimney: the flue is, therefore, rendered a powerful extracting shaft for ventilation, and a highly rarefied, and consequently effectual, passage for the products of combustion, thus obviating another fruitful source of smoky chimneys.

Should the warmth be too great, the register can be opened, and the action of an ordinary grate will take place.

In addition to these advantages, the fire is always under perfect control, and may at all times be brought to any degree of brightness.

Its combustion of fuel is so perfect that what remains in the ash-drawer after the day's consumption, if proper attention has been paid, might be taken away in the palm of the hand.

The occupation of the sweep will be nearly dispensed with.

These grates are manufactured in designs suitable for every class or style of building, at prices ranging from 3 to 50 guineas.

Further information may be obtained by applying at the Offices, 53 Parliament Street, where also all orders are received.

[6228]

TAYLOR, WILLIAM, 11 *Sheepcote Street, Birmingham.*—Improved shutter bars, door springs, bell springs, and kneeling frame.

[6229]

THOMAS, W. H., 6 *Sloane Street.*—Fire screen for dining-room; door porters.

[6230]

TITFORD, R., VANDOME, & CO., 117 *Leadenhall Street, E.C.*—Scales, weighing machines, weights, &c.

[6231]

TONKS, SAMUEL, *Great Hampton Street, Birmingham.*—Galvanized iron and japanned goods.

[6232]

TOOVEY, EDWARD & CHARLES, *Wolverhampton.*—Ironmongery and general hardware, for home, foreign, and colonial markets.

Messrs. TOOVEY are exporters of hardwares, including locks, bolts, and all other fittings used in buildings and furniture; brass foundry of every kind, wrought and cast iron goods, edge and other tools for the use of carpenters and others, cutlery of every variety, steel toys, japanned and tin wares, including tea-trays, &c. iron coffee and other mills, and kitchen utensils of every description.

Catalogues of all articles of ironmongery manufactured at Wolverhampton suitable for continental, foreign, or colonial markets, will be forwarded upon application.

[6233]

TUCKER & REEVES, 181 *Fleet Street, London, E.C.*—Locks, tin boxes, and fire-proof safes.

[6234]

TURNER, G., 13 *Rose Terrace, Fulham Road, Brompton.*—Machinery and articles for domestic use.

[6235]

TYLOR & PACE, 5 *Queen Street, Cheapside, London.*—Metallic bedsteads, window blinds, and perforated metals.

[6236]

UPFILL, THOMAS, & SONS, *Birmingham.*—Metallic bedsteads, carriage axletrees, van arms, wrought-iron hurdles, and fencing.

[6237]

VINCENT, ROBERT, *St. George's Place, Camberwell.*—Improved smoke-resisting stove door.

[6238]

WALKER & CLARK, 6 *Cardington Street, N.W.*—Wire cloths and wire work for manufacturing and ornamental purposes.

[6239]

WALKER, THOMAS, & SON, *Oxford Street, Birmingham.*—Stoves for warming buildings.

[6241]

WARDEN, JOSEPH, & SONS, *Railway Iron Works, Edgbaston Street, Birmingham.*—Railway screw bolts, nuts, and railway appliances.

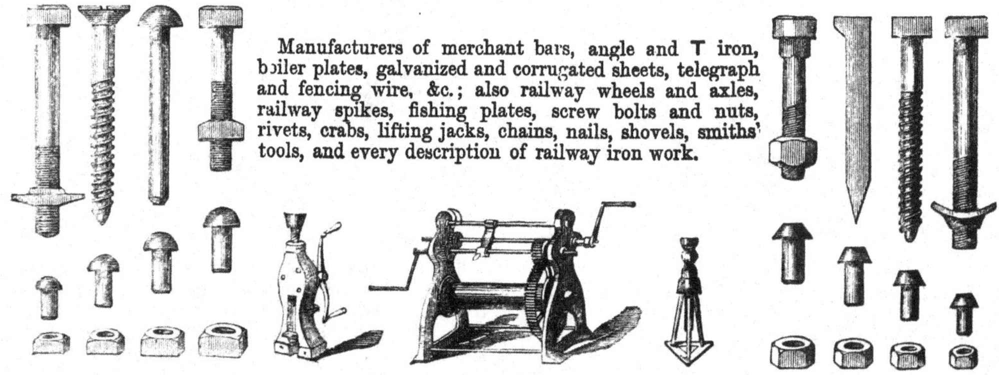

[6242]

WATKIN, WILLIAM, & Co., *High Street, Stourbridge.*—Spades, shovels, anvils, and vices.

[6243]

WATKINS & KEENE, *London Works, Birmingham.*—Bolts, nuts, couplings, tie rods, &c.

[6244]

WEBSTER & HORSFALL, *Birmingham.*—Patent steel music rope, and other wires.

[6245]

WELDON, C. & J., *Cheapside, London, and Loveday Street, Birmingham.*—Covered and other buttons.

[6246]

WENHAM LAKE ICE COMPANY, THE, 140 *Strand.*—Refrigerators, ice-cream machines, freezing powders, patent soda-water machines.

REFRIGERATORS, unequalled for preserving ice and provisions, and cooling wine, water, butter, cream, jellies, &c.

Machines for making and moulding ices, carafe freezers, improved freezing powders, and everything connected with freezing of the best, cheapest, most durable and reliable character. Patent soda water and bottling machines.

[6247]

WEST, HARRIET, 344 *Euston Road.*—Ornamental iron and wire works.

[6248]

WHALEY, BURROWS, & FENTON, *Queen's Ferry Wire and Wire Rope Works, near Flint.*—Burrows' patent conical winding drums; patent horizontal incline drums; wire rope, and telegraph cables. (*See page* 88.)

[6249]

WHITE, THOMAS, *Thorpe Hesley, near Rotheram.*—Improved screw bolts and nails, machine forged files, steel pruning hooks, gas hooks.

[6250]

WHITFIELD, SAMUEL, & SONS, *Birmingham.*—Iron bedsteads, japanned and gilt; wrought-iron fire and thief proof safes.

Obtained a Prize Medal for safes at the Exhibition of 1851.

[6251]

WHITFIELD, THOMAS, & Co., *Birmingham.*—Frying pans, wrought hollow ware and general iron-plate goods.

[6252]

WHITLEY, JOHN, *Ashton, near Warrington.*—Cathedral hinge and handles, and other wrought-iron hinges.

[6253]

WILDS, WILLIAM, *Hertford, Herts.*—A ceiling ventilator (patent); model of stove for domestic warming invented by the exhibitor.

[6254]

WILKINS & WEATHERLY, 39 *Wapping, London.*—Specimens of rope made of steel, iron, and copper wire.

[6255]

WILLS BROTHERS, *Metropolitan Drinking Fountain Depôt,* 12 *Euston Road.*—Art drinking fountains in bronzed iron.

[6256]

WINCHESTER, GRAVELEY, & SAGER, 40 *to* 42 *Upper East Smithfield.*—Patent sea water distilling and cooking apparatus.

Whaley, Burrows, & Fenton, *Queen's Ferry Wire and Wire Rope Works, near Flint.*—Burrows' patent conical winding drums; patent horizontal incline drums; wire, rope, and telegraph cables.

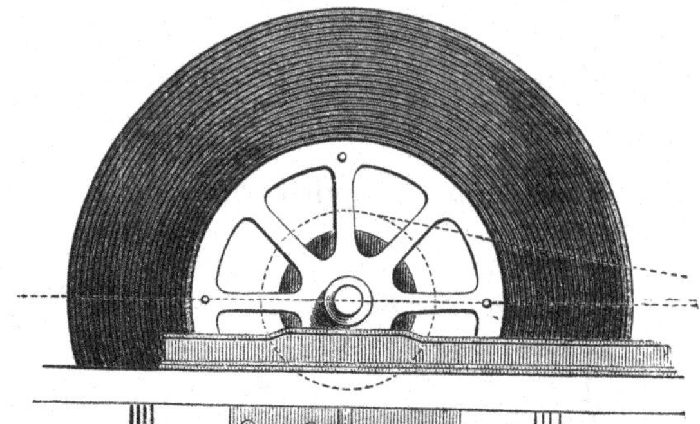

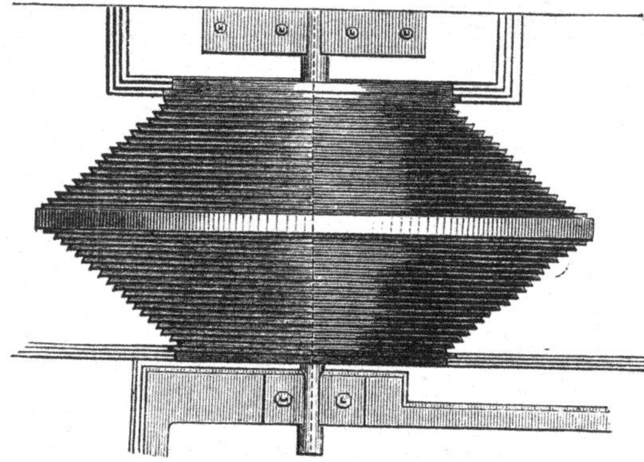

PART SECTION AND ELEVATION,
BURROWS' PATENT COMPENSATING WINDING DRUM.

Specimens of Patent Wire Rope, for collieries, mines, inclined planes, ships' standing rigging; copper rope for lightning conductors, wire cables for submarine and other telegraphs, wire strands for fencing, railway signal lines.

Wire Rope applied to winding purposes by means of Burrows' Patent Compensating Winding Drum.

The drum admits of the ropes and cages being balanced at all points in the pit; therefore, when the pit is of such a depth that the weight of one rope is equal to the weight of coals to be lifted at one time, an engine of half the usual power, or steam at half the usual pressure, will suffice, or a greater load of coals may be raised.

When starting the load from the bottom, it is raised without the violent concussion which is experienced with the usual form of drum.

It has been found in practice that it is scarcely possible to overwind with this drum.

The ropes last much longer on this drum.

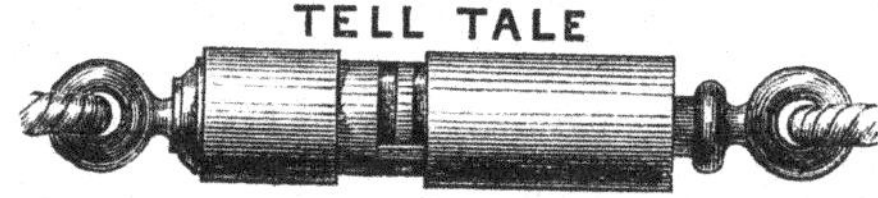

STUBBS AND FENTON'S TELL-TALE.

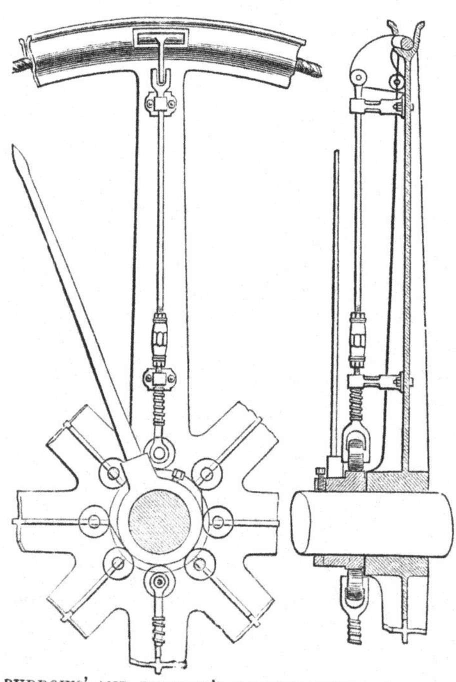

BURROWS' AND DOUGAN'S PATENT CLUTCH PULLEY.

Wire Ropes in connexion with Burrows' and Dougan's "Clutch Drum," and Stubbs' and Fenton's "Tell-tale."

The use of the clutch pulley facilitates the substitution of wire ropes in place of chains for working inclines, &c. Half a turn round the pulley is sufficient for holding, as all slip is prevented by the grippers, and wear and tear consequently reduced to a minimum.

By means of these pulleys and ropes, power may be easily transmitted to almost any distance, and in any direction with the greatest facility.

The use of the "tell-tale" diminishes the strain upon and consequently the wear and tear of the ropes. The weight of the load is registered and rendered evident at a glance. It is simple, cheap, and strong.

[6257]

Winfield, Robert Walter, & Son, *Birmingham, and* 141 *Fleet Street, London.*—Tubes, bedsteads, gas fittings, brass foundry, &c. (*See pages* 90 *to* 93.)

[6258]

Winter, Henry, 3 *Paragon Road, Hackney.*—Patent lifting and weighing machine, greatly economising labour and expense.

[6259]

Withers, George, & Sons, *Park Works, West Bromwich.*—Patent fire-proof and thief-resisting safes and money chests.

[6260]

Wood, Brothers, *Stourbridge.*—Chain cables, anchors, anvils, vices, &c.

[6261]

Woodin, Dennis, 2 *Upper Park Place, Dorset Square.*—Horse shoes which prevent slipping on stones or ice.

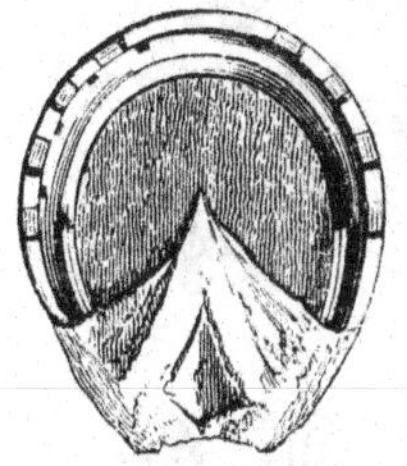

PATENT HORSE SHOES.

Woodin's Patent Horse Shoes, for preventing horses slipping on stone pavements or ice or other surfaces, have now stood the test of years, and are pronounced by competent judges to be the best ever offered to the public. They give a strong firm hold, and a level tread, prevent cutting or "clacking," are one-third lighter than the common shoe, and at the same time equally durable, and are put on at the same price.

Woodin's Elastic Prepared Padding prevents and cures contraction, corns, sandcracks, thrushes, and concussion of the joints. It gives an equal bearing to all parts of the feet, and keeps them always moist and cool. Price 6*s.* 12*s.* and £1 4*s.* per case.

Woodin's Golden Stimulating Absorbing Ointment, supersedes the firing-iron for the cure of spavins, ring-bones, curbs, splints, and enlargements of the joints, &c. It reduces them in a surprisingly short time without leaving any blemish, and frequently without interfering with work. It is sold only by the exhibitor. Prices, from 5*s.* upwards, according to size of pot. Proper directions for its application are forwarded with every pot.

[6262]

Wright, George, & Co., *Burton Weir, Sheffield.*—Stoves, grates, fenders, kitcheners, umbrella stands, chairs, tables, &c.

[6263]

Wright, Peter, *Constitution Hill Works, Dudley.*—Patent vices and anvils of various descriptions, cramps, &c.

[6264]

Wright & North, *Monmore and Cleveland Iron and Steel Works, Wolverhampton.*—Specimens of iron and steel, and Stocker's patent combined metal tyre-bars.

The following are exhibited :—

Specimens of boiler plate.
Ditto cast-steel ditto.
Ditto sheet iron.
Ditto galvanized iron.
Ditto corrugated iron.
Ditto hoop iron.
Ditto bar iron.
Ditto spring steel.
Ditto boiler rivets, stamped.

Specimens of ditto rolled, under Arrowsmith's patent. Sole manufacturers.

Specimens of combined metal for wheel tyres and horse shoes, rolled, under Stocker's patent. Sole manufacturers.

Specimens of horse shoes made from the patent combined metal bars. Trade mark, "Mon-moor."

WINFIELD, ROBERT WALTER, & SON, *Birmingham, and* 141 *Fleet Street, London.*—Gas fittings, bedsteads, brass foundry, tubes, &c.

CHANDELIERS, GAS STANDS FOR SIDEBOARDS, CHIMNEY PIECES AND TABLES, ETC.

WINFIELD, ROBERT WALTER, & SON, *continued.*

BRASS ARABIAN AND FRENCH BEDSTEADS, SWINING COT, CHAIR AND TABLE, ETC.

Winfield, Robert Walter, & Son, *continued.*

CORNICE POLE, CEILING ROSES, METAL FRAME LOOKING GLASS, TUBES.

WINFIELD, ROBERT WALTER, & SON, *continued.*

DRAWING-ROOM CHANDELIER FOR GAS, in silver bronze and relief, outline formed from ornamental panelled angular tubes, Hunt's patent.

DRAWING-ROOM CHANDELIER FOR GAS, in bright gold, fitted with an internal sliding apparatus, Hunt's patent.

CHANDELIER FOR GAS, in bronze frame-work, formed of angular tubes, filled in with rich ornamental panels, and fitted with an internal sliding apparatus, Hunt's patent.

CHANDELIERS FOR GAS. (See illustrations.)

GAS STANDS, in bright gold and bronze, for sideboards and chimney pieces, also fitted with flexible tube, as portable lights for tables, &c.

SMOKE-ABSORBING GAS SHADES, Hunt's patent.

GAS BURNERS of various kinds.

IMPROVED PATENT BRASS FOUR-POST BEDSTEAD, with massive pillars, mountings, "hop foliage" cornice, and head and foot rails to correspond; circular dove-tail joints, china bowl castors. (No. 3661.)

IMPROVED PATENT BRASS FOUR-POST BEDSTEAD, with pavilion tester, tapered pillars, made of the patent embossed metal, mountings and ornaments in arabesque style. (No. 3687.)

IMPROVED PATENT BRASS TENT BEDSTEAD, with ornamental mountings and head and foot rails. (No. 3572.)

HALF-TESTER, OR ARABIAN BEDSTEAD, of brass, with pillars of patent tubing, wrought mountings, and elaborate cast ornaments on the head and foot rails. (No. 3644.) See illustration.

HALF-TESTER, OR ARABIAN BEDSTEAD, with plain pillars, the vases and mountings in imitation of precious metals and stones. (No. 3633.)
Others, with the head portions of iron, with mountings of ruby (No. 3761) and opal (No. 3749).

FRENCH BEDSTEADS, in brass, one with panelled head and foot rails, and parallel twisted pillar (No. 3779); another with ornamental head and foot rails (blackberry and poppy), and mountings on plain pillars, (No. 3775); a third with ornamental head and foot rails (birds and oakleaves), the vases and mountings to the pillars in imitation of precious metal and precious stone (No. 3776). See illustration.

BRASS BEDSTEAD, for travellers or military officers. (No. 3055.)

SWINGING COTS, with ornamental stands (No. 3499). See illustration.

CHILDREN'S CRIBS, several examples, some with patent brass pillars.

CENTRE TABLE, with marble top (No. 3713). See illustration.

RECUMBENT CHAIRS, frames of ornamental tubing. (Nos. 3006—8.)

ELEGANT DRAWING-ROOM CHAIR, in brass, richly chased, with white satin seat. (No. 3019.) See illustration.

REGISTERED FIRE GUARDS, OR DRESS PROTECTORS, in brass. (No. 3723—24.)

REGISTERED FIRE GUARDS, OR DRESS PROTECTORS, in iron. (No. 3718—19.)

IRON BEDSTEADS, of various shapes, japanned in the ordinary manner, and likewise by the patent processes of ornamenting—viz. the pattern produced in various colours at one operation, and the combination of gold and colours.

FOLDING CHAIR BEDSTEAD, with patent double-action head piece, forming when necessary a bed rest for an invalid. (No. 3731.)

PORTABLE CHAIR, for overland or sea journeys. (No. 3014.)

N. B.—Short side rails are inserted in some instances to allow of the exhibition of a greater number of patterns, the bedsteads being in all cases made of the ordinary length. The stability of these bedsteads is guaranteed, the continuous tube pillar, patented 22d December, 1831, is used in combination with circular dovetail joints; white china bowl castors are generally applied to best bedsteads, many of the designs are registered under the copyright Act, 5 & 6 Vic. cap. 100.

LARGE OVAL GLASS, brass frame, composite ornament, with figure and candle branches, &c. (See illustration.)

Stamped window cornices, cornice poles of various kinds, stamped brass foundry, ceiling roses enamelled white and other colours, mantel-piece banner arms, screen poles, balustrades, drapers' brackets and shop-window fittings, sash bars, name plates, bed rings, stair rods, and weights of every ornamental character.

Examples of the following articles, for the use of brass founders, engineers, &c. Brass tubes for locomotives brazed joints, brass and copper tubes plain and ornamental, including twisted, reeded, fluted, and others, indented by a patent process. (See illustration.) Iron tubes coated with brass, parallel, twisted, reeded, or taper; tin and zinc tube, brass wire for pin makers and wire workers; copper wire for electric telegraph cables, plain, or coated with brass or tin, African rods in copper or brass, brass and copper piston rods, sheet brass of all sizes and gauges, spelter solder, common stair rods, beadings, and clips, &c.

[6265]

Yates, Haywood, & Drabble, *Rotherham and London.*—Stoves, fenders, and other ornamental furniture in cast-iron.

Register stoves, warm air stoves, fenders, cooking ranges, fire-irons, tables, hat and umbrella stands, gates, fencing, balusters, and various other similar goods in cast-iron, steel, and ormolu, shown as specimens of the ordinary manufactures of the house.

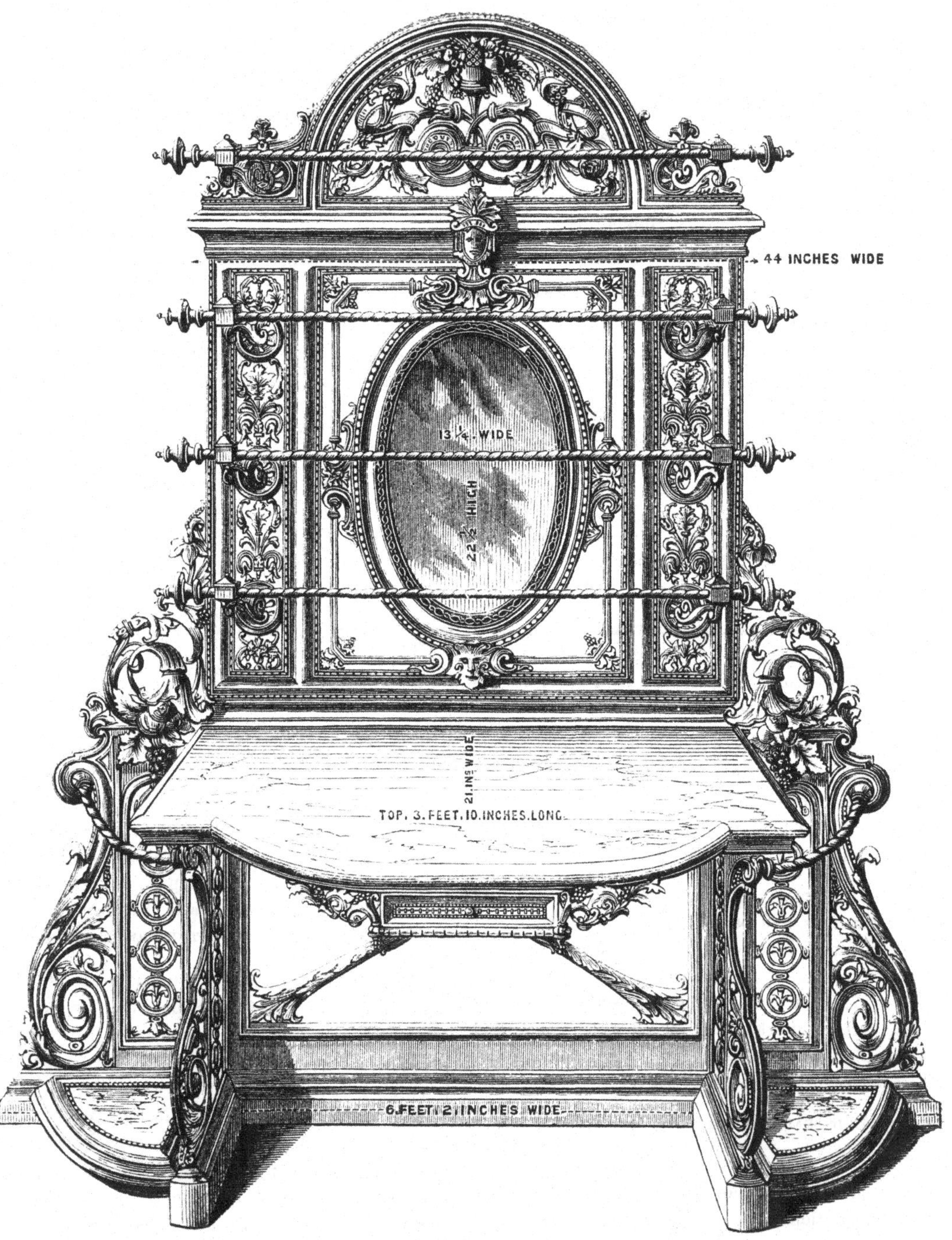

TABLE AND STAND FOR HATS AND UMBRELLAS

YATES, HAYWOOD, & DRABBLE, *continued.*

STATUARY MARBLE CHIMNEY PIECE AND STEEL GRATE, with ormolu enrichments.

WARM AIR STOVE, bright, with electro-bronzed ornaments, selected from a very copious stock of designs fit for general use.

CIRCULAR WARM AIR STOVE electro-bronzed.

YORK, SAMUEL, & CO., *Wolverhampton.*—Hardware goods made at Wolverhampton, suitable for every foreign market.

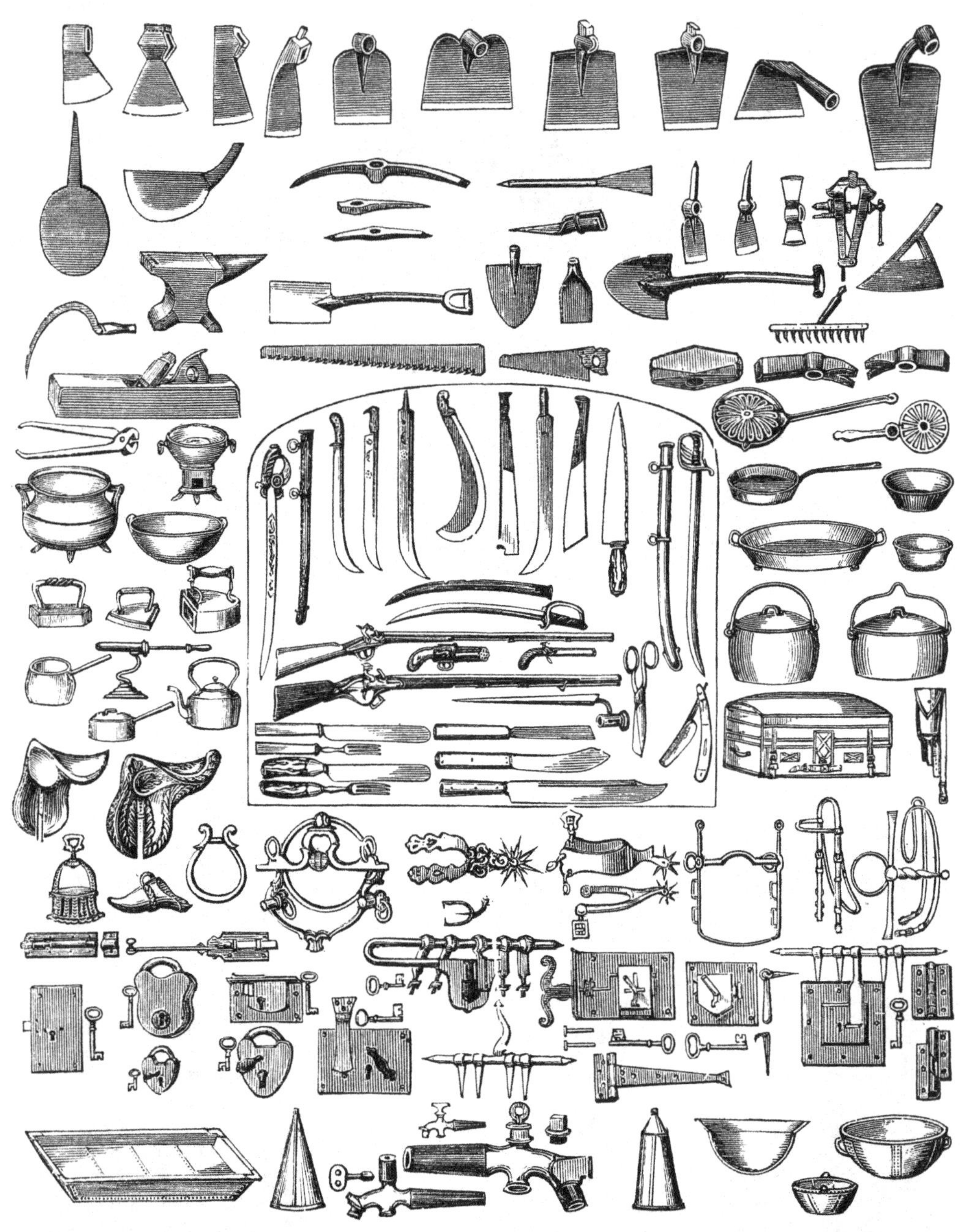

SPECIMENS OF A FEW ARTICLES OF HARDWARE, being the production of Wolverhampton and its adjacent districts, and such as are supplied for shipment to foreign markets. London agent, Wm. Roberts, 50 Little Britain.

YOUNG, WILLIAM, 33 *and* 34 *Queen Street, Cheapside.*—Spirit and oil lamps; gas-burners; smokeless grates and furnaces.

SMOKELESS GRATE.

THE SMOKELESS GRATE.

This patent smokeless grate consists of an ornamental trough being fixed at the lower portion of the grate, in which is placed a right and left handed screw connected with a ratchet at the side, which together with the screw is moved by the poker.

When the fire requires feeding, the coals are deposited in the trough, and by the revolution of the screw the burning fuel is raised up and the fresh coal conveyed into the cavity underneath. By this simple means the whole of the gases given off are burned, a great saving effected, and greater heat obtained combined with cleanliness.

This patent is also applied for kitcheners, furnaces, &c.

VESTA LAMP.—BIN-OXIDISED.

THE VESTA LAMP BIN-OXIDISED, constructed to burn rectified spirits of turpentine or other hydro-carbons.

This, the well known Vesta lamp, is now improved by the introduction of an apparatus termed a Bin-oxidiser, which apparatus is placed above the flame, and which causes a current of air to descend on the top part of same independent of the ordinary column of air passing through the burner, thus increasing the power and economy of light.

THE VESTA GAS BURNER AND PENDANT.

W. Young's Vesta gas burner consists of two or more deflectors being placed in the interior of the flame in such a manner that the column of air on its passage through the burner is taken up by the said deflectors and directed in separate currents upon the flame, consequently a more perfect combustion is obtained, and a great saving effected.

The above inventions may be seen in operation at W. Young's warehouse as per above address.

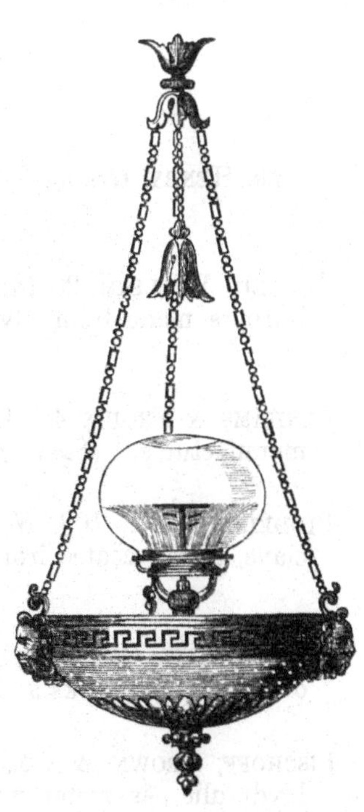

VESTA GAS BURNERS.

[6266]

YATES, HENRY, & SONS, *Merridale Lock Works, Wolverhampton.*—Hand-made and machine-made locks, with specimens of ornamental keys.

[6267]

YORK, SAMUEL, & Co., *Wolverhampton.*—Hardware goods suitable for every foreign market. (*See page* 96.)

[6268]

YOUNG, WILLIAM, 33 *and* 34 *Queen Street, Cheapside.*—Spirit and oil lamps, gas-burners, and smokeless grates and furnaces. (*See page* 97.)

[6269]

AUBIN, C., *Wolverhampton.*—Nettlefold's guardian locks, and fancy keys.

[6270]

COTTRILL, E., *St. Paul's Square, Birmingham.*—Copying presses and dies.

[6271]

KENNARD, R. W., & Co., 67 *Upper Thames Street, London.*—Cast-iron park gates, and railing; verandahs, garden seats, &c.

[6272]

MOORE, J., *Birmingham.*—Medals.

BETTRIDGE, J., & Co., have been removed to Class XXX.

SUB-CLASS B.—*Manufactures in Brass and Copper.*

[6277]

ALDER, HENRY, *Grange Works, Edinburgh.*—Gas meters.

[6278]

ALDRED, WILLIAM, 28 *Pall Mall, Manchester.*—Non-corrosive ordinary and economical gas burners made from silver and other sheet metals.

[6280]

BENHAMS & FROUD, 40, 41, & 42 *Chandos Street, Charing Cross.*—Copper, zinc, and brass manufactures. (*See page* 99.)

[6281]

BIDDELL & Co., 108 *New Street, Birmingham.*—Enamelled, inlaid, ornamental metals, clays, glass; coated iron shot; thief detectors; alarms.

[6282]

BIDDLE, ELIZABETH EMILY, *Victoria Street, Birmingham.*—Book clasps; gilt rims and ornaments for books and cabinets.

[6283]

BISCHOFF, BROWN, & Co., *Langham Works, George Street, Great Portland Street, London.*—Hydraulic gas meter, with floating measuring chamber. (*See page* 102.)

BENHAMS & FROUD, 40, 41, & 42 *Chandos Street, Charing Cross, London, W.C.*—Copper, brass, and zinc manufactures.

WEATHER VANES.

Copper, zinc, and iron weather-vanes of various styles. The two shown above are of wrought copper, from designs by Mr. S. J. Nicholl, architect.
Copper and zinc casement frames, plain and ornamental.
Bronzed copper glass-case frames.
Copper and zinc lanterns.
Copper architectural enrichments, for the exterior of houses or public buildings.
Specimens of zinc covering for roofs.
Ditto of copper ditto, recommended on account of lightness and durability.
Copper lightning conductors, solid, tubular, and rope, with platina tips.
Copper and zinc guttering, rain-water pipes, cistern-heads, &c.
Ornamental copper clock hands.

COPPER BATH.

Tinned copper bath, enamelled in imitation of marble, inside and out, designed for fixing without wood casing.
A 5 ft. 6 in. enamelled copper oblong bath, for fixing in wood casing, with set of cocks, lever handles, engraved brass plates, &c.
A full size polished zinc bath.
Copper shower baths.
Set of three tinned copper oblong steamers, for meat, fish, and vegetables.
Tinned copper stewpans, stockpots, saucepans, &c.

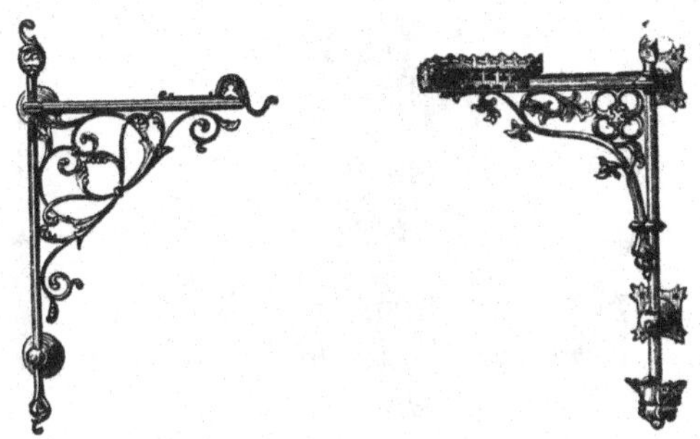

BRASS BRACKETS.

Tinned copper jelly and cake moulds, shown in various stages of the manufacture.
Copper and brass coal scoops.
Brass and copper mullers for publicans' use.
Copper saddle boiler.

ALTAR RAILS. TOMB RAILS.

Brass candlesticks, alms dishes, &c.
Ornamental polished brass brackets.
Brass altar and tomb rails.

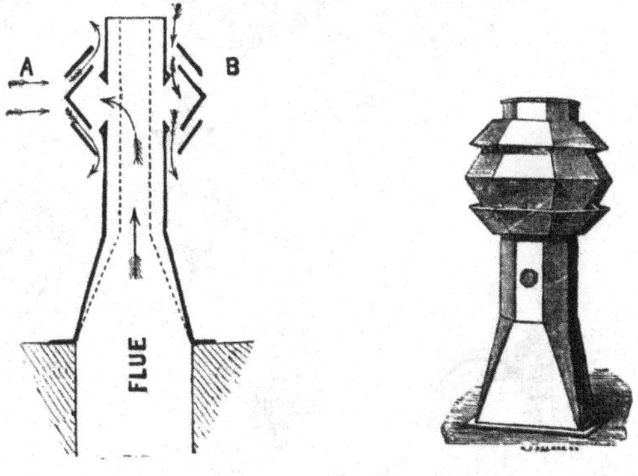

OCTAGON CHIMNEY HEAD.

The new patent octagon chimney head.
Chimney pipes of various descriptions.

[6284]

Blews & Sons, 9, 10, 11, *and* 12 *Bartholomew Street, Birmingham.*—Chandelier, gas-fitting, lamp, standard weight, and measure manufacturers; bells and brass foundry articles.

No. 820. No. 476. No. 850. No. 456. No. 480. No. 465.

BLEWS & SONS, *continued.* (In Birmingham Court.)

Established A.D. 1782. *Prize Medal awarded at Exhibition,* 1851.

A set of imperial standard measures, from a bushel down to ½ gill.

A set of imperial standard bell-shape brass weights, from 56 lbs. down to ¼ oz.

A set of imperial standard spherical-shape brass weights, from 56 lbs. down to 1 dram.

A set of imperial standard decimal troy weights, from 500 oz. down to $\frac{1}{1000}$ oz.

A set of cental weights, 50 lbs. down to ¼ oz.

A set of troy cup weights, each 256 oz. 128 oz. 64 oz. 48 oz. 32 oz. 24 oz. 16 oz. 12 oz. 8 oz. 4 oz. 2 oz. and 1 oz. down to ¼ oz.

A set of Spanish covered cup weights, each 8 lb. 4 lb. 2 lb. 1 lb. and ½ lb. down to ¼ oz.; making 16 lb. 8 lb. 4 lb. 2 lb. and 1 lb. in the whole.

A set of flat brass weights, each 7 lb. 4 lb. 2 lb. 1 lb. down to ¼ oz.

A set of bevilled-edge weights, each 4, 2, 1 lb. down to ¼ oz.

A set of registered knob weights, each 7, 4, 2, 1 lb. down to ¼ oz.

A set of frame weights, each 4, 2, 1 lb. down to ¼ oz.

A set of porcelain bell weights, 7 lb. down to ½ oz.

A turned and polished ship's bell, 22 inch diameter.

A turned and polished ship's bell, in brass frame, 7½ inch.

A railway station hand bell, turned and polished, with ebony handle.

A set of dinner bells, 3 inch to 6½ inch, turned edge and crown, with hardwood handles.

A set of dinner bells, 3 inch to 6½ inch, turned and lacquered, with ebony handles.

House bells, 6 to 24 oz. turned edges, and turned and lacquered.

Sheep bells, plain, and turned and lacquered.

Australian bullock bells, with mottoes, "Advance Australia," and "Success to Bullock Drivers."

Australian horse bells, with motto, "Success to Horse Teams."

METAL CHANDELIER,	3 light	No. 455A	£3 5 0
ditto	3 ,,	456	4 15 0
ditto	3 ,,	476	5 5 0
ditto	3 ,,	480	7 0 0
ditto	5 ,,	465	8 8 0
GLASS CHANDELIER,	3 ,,	800, plain, Queen's drops	4 15 0
ditto	3 ,,	800, cut, ditto	5 0 0
ditto	3 ,,	801, plain, ditto	4 10 0
ditto	3 ,,	801, cut, ditto	4 15 0
ditto	3 ,,	802, cut, ditto	5 10 0
ditto	3 ,,	803, cut, Queen's drops	5 10 0
ditto	3 ,,	804, plain, ditto	4 4 0
ditto	3 ,,	804, cut, ditto	4 10 0
ditto	5 ,,	805, cut, ditto	8 8 0
ditto	5 ,,	820, cut, Albert drops	9 9 0
ditto	5 ,,	850, cut, Queen's drops	9 9 0
ditto	5 ,,	860, cut, ditto	20 0 0
ditto	7 ,,	900, plain, ditto	11 11 0

The chandeliers are hung in the Court in the South-Eastern Transept, opposite Naylor, Vickers & Co.'s steel trophy.

Pattern books and price lists supplied on application.

Bischoff, Brown, & Co., *Langham Works, George Street, Great Portland Street, London.*—Hydraulic gas meter, with floating measuring chamber.

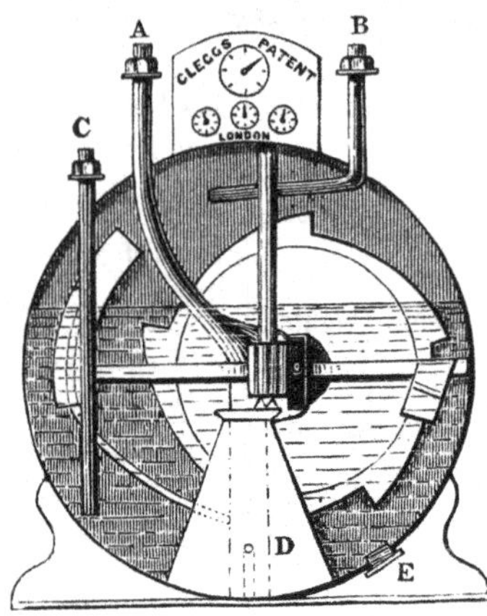

Clegg's New Patent Hydraulic Gas Meter.

The drum floats by means of a central air vessel, and thus renders the measuring capacity independent of any reduction of the water level by means of evaporation. The meter registers correctly under all variations of pressure, or increase and diminution of the number of lights.

[6285]

Bright, Richard (Successor to Argand & Co.), 37 *Bruton Street.*—Argand and indicator lamps and wicks.

The Patent Indicator Lamp has the following improvements:—

The position of a small bead shows when the plunger requires winding up. In filling the lamp, a bubbling shows when there is sufficient oil. A patent stiffened wick can be momentarily applied without a stick. The flame is regulated by raising or lowering the chimney without removing the globe. A double extinguisher renders it unnecessary to blow out the flame as requisite in moderator lamps.

[6286]

Cartwright, Sambidge, & Knight, *Lombard Street, Birmingham, and Castle Street, Holborn, London.*—Chandeliers, brackets, gas fittings, &c.

The exhibitors manufacture every description of gas fittings, gaseliers, &c. In the specimens exhibited, they have endeavoured to combine beauty of design with economy of cost, by producing such articles as come within the reach of the many rather than elaborate productions, which from their costliness could only be obtained by the few.

[6287]

Chambers & Co., 216 *Bradford Street, Birmingham.*—Railway, ship, and carriage lamps, coach furniture, &c.

[6288]

Cowan, W. & B., *Buccleugh Street Works, Edinburgh.*—Wet and dry gas meters.

[6289]

Croll, Rait, & Co., *Kingsland Road, N.E.*—Croll's patent improved dry gas meter, and gas apparatus. (*See page* 103.)

[6290]

Dale, Richard, & Son, 195 *Upper Thames Street.*—Copper boilers, copper baths, copper kitchen furniture.

CROLL, RAIT, & CO., *Kingsland Road, N.E.*—Croll's patent improved dry gas meter, and gas apparatus.

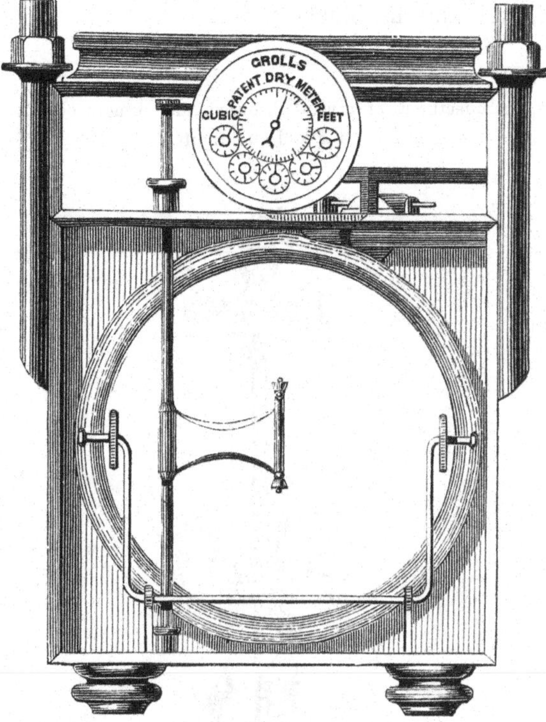

FRONT VIEW.

DRY GAS METER.

Extract from "Rutter on Gas-lighting."

"The dry meter I shall here endeavour to describe has been in use about fifteen years. During that period neither capital nor skill have been spared by its inventor and patentee, Mr. A. A. Croll, in his efforts to make it what it has now become—an accurate measurer, and a durable machine. This meter is not liable to be affected by sudden or extreme variations of temperature. It may be fixed in almost any part of the consumer's premises, either above or below the level of the entrance to the fittings, and it requires no adjustment to insure correctness."

The following articles are exhibited :—

1. Improved dry gas meter for 100 lights, made with glass front, sides, &c. to show the working.
2. Model of dry gas meter.
3. Parts of dry gas meter.
4. Model of a testing gas holder.
5. 20-light consumer's governor in glass, to show its action.
6. Public lamp meter in cast-iron case.

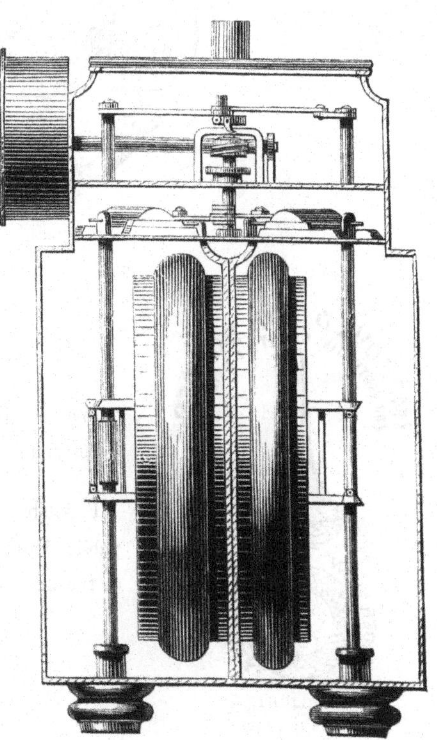

SIDE VIEW.

[6291]

DEFRIES, J., & SONS, *Works, London and Birmingham; Show Rooms,* 147 *Houndsditch.*—Brass chandeliers, bronzes, hall lanterns, brackets, &c.

J. DEFRIES & SONS, manufacturers of crystal, bronzed, and ormolu chandeliers; improved crystal star and sun lights.

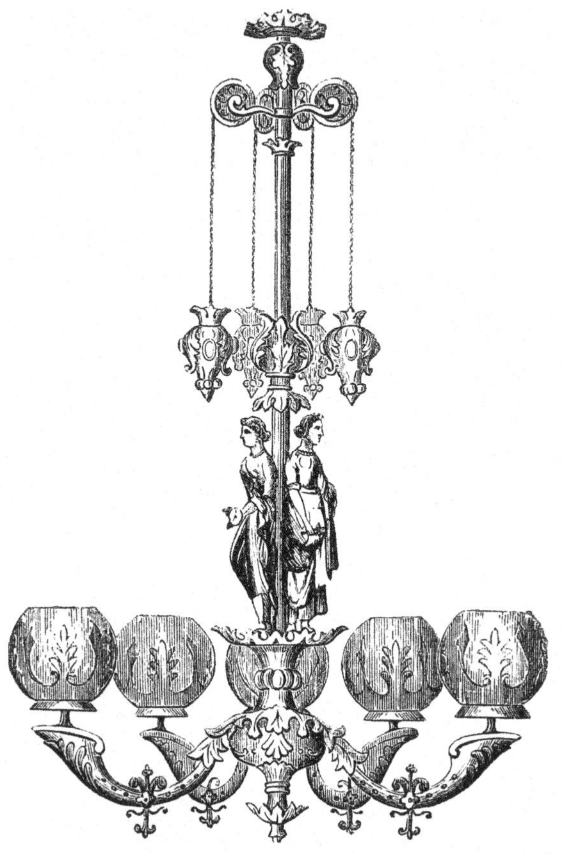

CHANDELIER.

SUN LIGHT.

Works: London and Birmingham. Principal depôt and show rooms, 147 Houndsditch, City.

Estimates and designs for lighting theatres, public buildings, &c. can be had by applying at the above address.

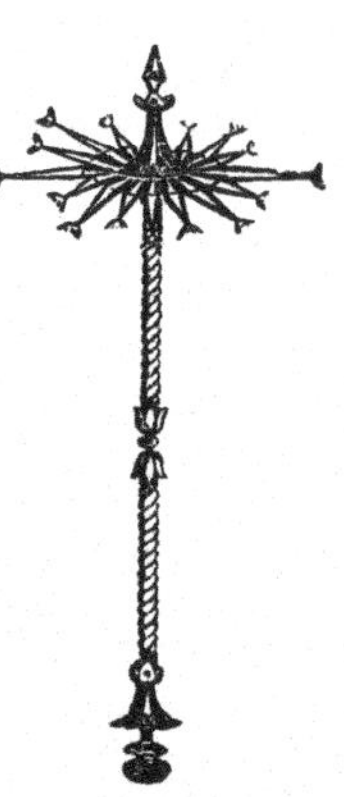

NEW STAR LIGHT.

[6292]

DICKIE, CHARLES, *Dundee.*—Specimens of wire work; working models of improvements in bell hanging.

ORNAMENTAL WIRE AVIARY.

1. ORNAMENTAL WIRE AVIARY for birds, on ornamental wire table, frame of mahogany, embellished with carved figures, and four views on glass, representing the seasons, introduced in panels. The table has vases for four flower pots, and two hanging baskets, and is ornamented with bronzed figures and ornaments. Size of aviary 3 ft. 8 in. by 22 in. by 2 ft. 6 in. high. Height of table 2 ft. 9 in. represented in accompanying illustration. Designed and manufactured by C. Dickie.
2. GARDEN STOOL, handsome design in ornamental cast-iron and wire-work, with griffins and dolphins as supporters. Price . . . . . 12*s.*
3. ORNAMENTAL WIRE GARDEN CHAIR, strong, light, and elegant, and when inverted by a simple and easy movement, becomes a flight of four steps, very suitable for a conservatory. Price £1 5
4. ORNAMENTAL WIRE ARCH, for garden walk, gothic design.
5. HANGING FLOWER BASKET with drop.
6. WIRE FLOWER TRAINER.
7. ORNAMENTAL PANEL WIRE FENCING.

BELL-HANGING.

8. INDEX DIAL BELL, by which one bell only is required for any number of apartments.
9. MANIFOLD BELL PULL, by which one pull only is made to ring any number of bells.
10. ORNAMENTAL DESIGN BELL LEVER, constructed upon an entirely new principle, so as to ring three different bells.
11. New designs in DOOR BELL PULLS.

[6293]

DRURY, FRANCIS, 10 *Duke Street, Grosvenor Square, W.*—The Campanelian and musical inventions applicable to clocks.

[6294]

DUCKHAM, H. A. F., 44 *Clerkenwell Green.*—Compensating and regulating gas meters, wet and dry self-acting gas regulators.

[6295]

EDGE, THOMAS, *Great Peter Street, Westminster.*—One each wet and dry gas-meter in glass cases.

[6296]

EVERED, RICHARD, & SON, *Bartholomew Street, Birmingham, and Drury Lane, London.*—General brass foundry articles.

[6297]

FELE, J., & Co., *St. James' Square, Wolverhampton.*—Brass chandeliers, gas fittings, &c.

[6298]

FORREST, GEORGE, & SON, *Nevill's Court, New Street Square.*—Candelabrum and 32-light gas chandelier.

[6299]

GARDNER, HENRY & JOHN, 453 *Strand, Charing Cross,* 4 *and* 5 *Duncannon Street, and* 63 *Strand.*—Lamps, chandelier, candelabra, gas fittings and apparatus. (*See page* 106.)

[6300]

GLOVER, GEORGE, & Co., *Ranelagh Works, Pimlico.*—Dry gas meters, pneumatometers, and photometers. (*See page* 108.)

GARDNER, HENRY & JOHN, 453 *Strand, Charing Cross,* 4 & 5 *Duncannon Street, and* 63 *Strand.* —Lamps, chandelier, candelabra, gas fittings, and apparatus.

***Obtained a Prize Medal at the Exhibition of* 1851.**

STEVENS.

The exhibitors, whose business has been established for more than a century, hold a warrant of appointment as lamp manufacturers to the Queen. They manufacture and supply the following goods, of which specimens are exhibited:—

Lamps for India, of an improved construction, with punkah protectors, table lamps, chandeliers, candelabra, wall branches, hall lanterns, and passage lamps, arranged for both oil, gas, and candles, to suit any style of decoration, from the richest to the most moderate.

A considerable saving effected by using Gardner's improved gas regulator. Fittings, and gas apparatus of every description.

[6301]

GLOVER, THOMAS, *Suffolk Street, Clerkenwell Green, E.C.*—Dry gas meters, and gas holders.

[6302]

GRAY, BAILEY, & BARTLET, *Berkley Street, Birmingham.*—Gaseliers, tea trays, coal vases, &c.

[6303]

GREENWAY, WILLIAM, *Princip Street, Birmingham.*—Locks, bolts, latches, door springs, fastenings, weavers' mails, patent wrought-iron hinges, &c.

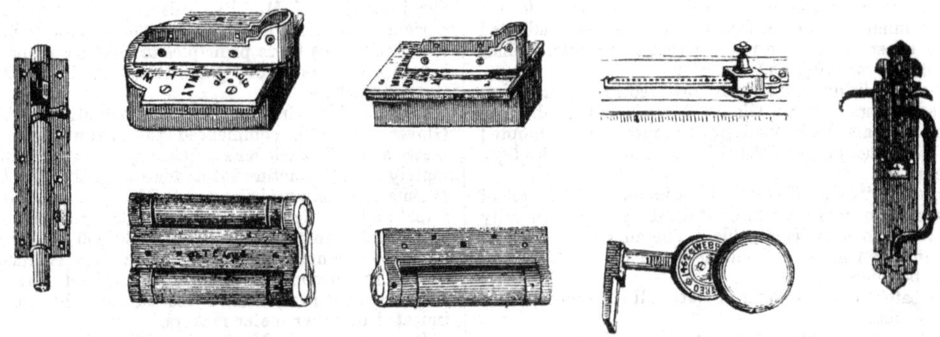

The exhibitor is a manufacturer of—

Mortise, rim, drawback, dead, and pad locks.

Drawer, cupboard, chest, box, sideboard, and every description of cabinet locks.

Copper, brass, and gun-metal locks, suitable for iron and wood ships, or powder magazines.

Iron safe and prison door locks.

Winkler's patent secure safety locks.

Eastman's patent bolt and door fastener.

Greenway's patent barrel bolts.

Greenway's patent casement stays.

Greenway's patent cupboard turns.

Greenway's patent door springs.

Cartland's ditto ditto.

Dilkes & Co.'s ditto ditto.

Greenway's Gothic and other new pattern door latches.

Greenway's improved shutter bars and window fasteners.

Greenway's patent wrought-iron hinges.

Espagnolette and casement fastenings.

Copper, brass, and steel weavers' mails.

Lingoes and umbrella furniture.

Wrought, pressed, and cast-iron metal work, and general cast-brass work.

[6304]

GUEST & CHRIMES, *Rotherham, and* 37 *Southampton Street, Strand, London.*—Water-works articles. (*See page* 109.)

[6305]

HARDMAN, JOHN, & CO., 166 *Great Charles Street, Birmingham ;* 13 *King William Street, Strand, London, W.C. ;* 1 *Upper Camden Street, Dublin.*—Mediæval metal manufactures. (*See page* 110.)

[6306]

HARROW, WILLIAM, & SON, 14 *Portland Street, Soho, W.*—Specimens of manufacture; chandeliers, tripods, brackets, &c.

AN 8-LIGHT CHANDELIER for gas, in brass, lacquered, designed and modelled by Mr. W. G. Rogers.

A 6-LIGHT CHANDELIER for gas. Fret ornament.

A LANTERN for gas. Fret panels and ornament.

A PAIR OF STANDARDS with six lights each for gas, in brass, bronzed.

BRACKETS, PILLARS, &c. in brass, bronzed and lacquered.

[6307]

HART & SON, *Wych Street and Cockspur Street.*—Ecclesiastical and domestic metal work (*See pages* 112 *to* 117.)

GLOVER, GEORGE, & CO., *Ranelagh Works, Ranelagh Road, Pimlico; Offices,* 22 *Parliament Street, Westminster, and* 15, *Market Street, Manchester.*—Dry gas meters, pneumatometers, and photometers.

The construction of a good and durable dry gas meter involves a multiplicity of chemical and mechanical considerations, to each of which its due weight must be assigned.

A subtle, invisible, elastic, aëriform body, very complex in its chemical constitution, susceptible of change in condition and volume from slight variations in temperature and pressure, has to be accurately measured; and the result of that measurement must be correctly recorded.

The instrument must be self-acting, and must act continuously or at intervals, requiring no adjustment or interference of any sort.

All its parts which come in contact with gas must be made of anti-corrosive metal; while the materials, forms, and combinations of its different parts must be so adapted to each other, that, when put together as a whole, it shall work easily, steadily, and correctly.

These conditions are strictly observed in the manufacture of Messrs. George Glover & Co.'s patent dry gas meter; the same high standard of accuracy being adopted in its construction as in that of the national standard gas holders.

Since the "Sales of Gas Act," a much closer degree of accuracy in meters than that attained by those generally used hitherto is indispensable. The admitted range of error in wet meters of from 30 to 40 per cent. can no longer be tolerated.

The patent dry gas meter obviates all the objections to the wet meter.

1. It measures accurately, and does not vary in its registration.

2. It does not cause jumping or sudden extinction of the lights, the former a common source of annoyance, the latter not free from danger, especially in large assemblies and on railway lines, where signal lights are used.

3. The dry gas meter does not require to be opened that water may be put into it; thus escapes of gas from the plugs being carelessly left open, always offensive, and occasionally producing explosions, are averted.

4. It cannot be tampered with without showing distinct evidence of having been so; and it is thus free from the many temptations and facilities to fraud which are characteristic of the wet meter.

5. The dry meter does not allow the gas to pass without being registered, a source of much greater loss to gas companies than is commonly supposed, and caused by the water level falling to a point at which the gas passes unregistered.

6. The frequent supply of water now rendered necessary by the small range of error allowed by the Act, the vigilant attention required to prevent fraud, and to ascertain when the gas is passing unregistered, needs three times the number of Inspectors requisite where dry meters are used; whilst, in testing meters, the expense of Inspectors and instruments is three times larger with the wet than the dry meter, which thus effects a great saving to Gas Companies and local authorities.

7. It does not require to be placed in the basement or lower part of the house, but may be put anywhere. The attempt with wet meters to prevent jumping of the lights by giving all the pipes a gradual ascent from the meter, so as to admit of the water trickling back into it, besides being impracticable, is expensive and detrimental to house property.

8. The dry meter works with less pressure than the wet. Not only is a saving of gas thus effected, but in large cities when, during the winter season, dense fogs occur, and the low pressure in the mains during the day is not adequate to move the wet meters so as to supply enough of gas for the burners, and only small smoky flames can be obtained from them, with the dry meter there is sufficiency of light. Thus interruptions to business, occasioning considerable loss to the owners of large warehouses, mills, and factories, are averted.

9. The action of the dry meter cannot, like that of the wet, be arrested by frost, causing the total extinction of the lights. This makes the dry meter especially advantageous on railway lines, precluding as it does the necessity of keeping up large fires near the meter during a severe and protracted frost.

10. Made of anti-corrosive metal, and not subject to the corrosive power of the chemical constituents of coal-gas and water, the patent dry gas meter is a much more durable instrument than the wet.

The dry gas meter has been brought to its present condition of excellence by successive stages. The essential improvements, invented by Mr. William Richards, and patented by Messrs. Croll & Richards in 1844, consisted in the introduction of the diaphragm and the direct action of the disc. The theoretical accuracy of the principles which the invention of Mr. Richards involved time and experience have fully established. The patentees, however, failed to reduce those principles to practice in producing a good and durable dry gas meter, and they abandoned its manufacture.

Mr. Croll having secured the patent, Mr. Thomas Glover, in 1845, commenced the manufacture of the meter as Croll & Glover's patent dry gas meter, and ultimately he manufactured it as his own. To him belongs the merit of having imparted to Mr. Richards' invention a real and practical value by the production of a correct and durable instrument. Some 200,000 of his meters have been manufactured; the medal was awarded them at the Exhibitions of London, Paris, and New York; they are now in extensive use all over the world, and imitated by other meter makers.

Improvements in Messrs. Geo. Glover & Co.'s meter:—

It has a large and distinct dial, which shows at a glance the number of cubic feet of gas passed, the number of the meter's capacity per hour and per revolution, and the number of identity, all of which the "Sales of Gas Act" requires, the maker's name, and the date of its manufacture. These points of information are inscribed on an enamelled dial in characters easily read and indelible; and they are necessary for reference, especially when disputes arise between buyer and seller, in which case the marks of identity and capacity are essential. These ought not to be entrusted to flimsy badges of thin metallic substances, which become tarnished and illegible, accidentally fall off, and can easily be transposed for the purposes of fraud.

A slot is introduced and a pin which connects the valve and valve rod. This facilitates the adjustment of the two sets of valves necessary to the uniform flow of gas, without which steady lights cannot be obtained. The attempt to adjust the position of the valve pin by giving a curvature to the valve arm is very objectionable. The valve arm has to be made soft so as to admit of this finger and thumb adjustment, its protracted immersion in gas rendering it still softer. The result is that the rod becomes more or less curved during the action of the meter as it transmits force, in the direction of its length, as a thrust or as a pull alternately.

A slot is introduced in the tangent of the meter, and a shoulder or rest is placed on the tangent pin, the flat surface of which rests on the upper surface of the tangent. The pin is secured in its place by a screw from below, the flattened head of which fixes it firmly at any desired point of the slot. This arrangement keeps the pin in a perfectly vertical position, and admits of the meter being registered with ease and precision.

Messrs. George Glover & Co. have made a modification of their meter, adapting it to the photometer, which shows its extreme accuracy in measuring the most minute quantities of gas. All the wheel-work, the spindle, and the worm are removed. The dial is placed on the top of the meter, a pointer is fixed in connexion with the crank rod, and the measurement of the gas is taken directly from the revolution of the crank. This pointer passes round a disc 6 inches in diameter, the scale of which is so divided that the $\frac{1}{600}$th part of a cubic foot of gas can be indicated with precision each second. This is the most severe test to which the steadiness, uniformity, and accuracy of a meter can be subjected. (In Class X. No. 2291, and Class XXXI. No. 6300, International Exhibition, it is seen at work.)

GUEST & CHRIMES, *Rotherham, and* 37 *Southampton Street, Strand, London.*—Water-works articles.

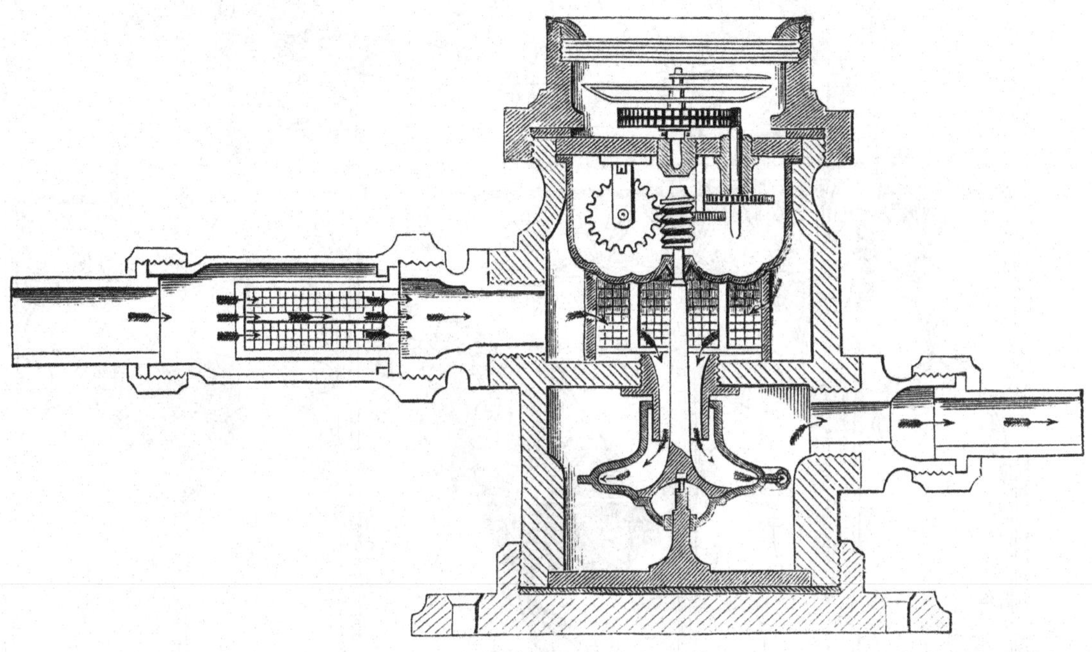

SIEMEN'S AND ADAMSON'S PATENT WATER METER.

The following is a list of the manufactures of GUEST & CHRIMES:—

Bateman & Moore's & Chrimes' patent hydrants or fire cocks, improved sluice cocks and gas valves.

Chrimes' patent high-pressure single and double loose valve and screw-down cocks.

Pilbrow's patent water-waste preventer; patent absolute water-waste preventer.

Siemens' patent balance water meter.

Bell & Chrimes' patent service box valve; improved self-acting and pull water-closet.

Eskholme's patent pneumatic regulator and valve closet.

Lowe's patent effluvia traps, Beggs' improved.

Fire-extinguishing apparatus.

Galvanized iron tubes and fittings.

Gas chandeliers, brackets, and fittings; glass chandeliers and brackets.

Crossley & Goldsmith's patent wet gas meter.

All the above articles may be seen on application to THOMAS BEGGS, 37 Southampton Street, Strand, London; or to the exhibitors.

They also execute plumbers' and gas fitters' brass-work of every description.

Agents for Scotland, Donaldson & Hume, 23 St. Enoch Square, Glasgow. Agents at Hamburgh, Messrs. Alfred Barber & Co.

Drawings, descriptions, prices, and testimonials, will be forwarded per post on application to Mr. BEGGS, 37 Southampton Street, Strand, or to the Works, Rotherham.

[6308]

HARVEY, THOMAS, 13 *Bradford Street, Walsall.*—Brass and plated harness furniture; South American spurs, stirrups, &c.

HARDMAN, JOHN, & Co., 166 *Great Charles Street, Birmingham;* 13 *King William Street, Strand, London, W.C.;* 1 *Upper Camden Street, Dublin.*—Mediæval metal manufactures.

[6309]

HICKLING & COX, *Birmingham.*—Copper and iron boat nails, rivets, and washers; cut copper and zinc tacks, screws, &c.

[6310]

HICKMAN, JOHN, 34 *William Street North, Birmingham.*—Brass cocks.

[6311]

HILL, JOSEPH, 18 *Broad Street, Birmingham.*—Chandeliers, gas fittings, stampings for metallic bedsteads, &c. (*See page* 118.)

[6312]

HIND, JAMES, 118 *Kingsland Road, London.*—Engraved and inlaid metals in door-plates and monumental brasses.

[6313]

HINKS, JAMES, & SON, *Crystal Lamp Works, Birmingham.*—Lamps for burning hydro-carbon oils.

[6314]

HOLDEN, H. A., *Bingley Works, Birmingham.*—Railway lamps and signals, and railway carriage fittings.

[6315]

HORSEY & BAKER, *Worcester Street, Southwark.*—Tea urns, coal-scoops, washing and brewing coppers, copper cooking utensils, patent wine and beer taps with steel protectors, white-metal taps with patent brace, preserving them from injury and breakage.

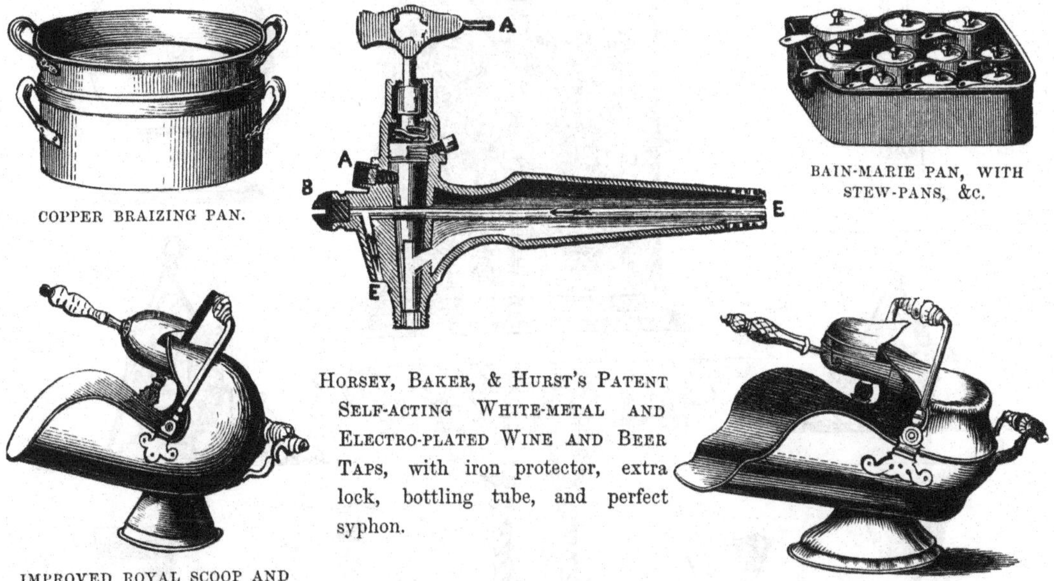

COPPER BRAIZING PAN.

BAIN-MARIE PAN, WITH STEW-PANS, &c.

HORSEY, BAKER, & HURST'S PATENT SELF-ACTING WHITE-METAL AND ELECTRO-PLATED WINE AND BEER TAPS, with iron protector, extra lock, bottling tube, and perfect syphon.

IMPROVED ROYAL SCOOP AND SHOVEL.

HAVELOCK SCOOP AND SHOVEL.

The exhibitors keep a large and well-assorted stock of tea urns, copper coal scoops, bath tubs, and other articles for domestic use of improved shapes and constructions.

They are patentees of a tap, fitted with a steel protector, and having a double lock, for wine, beer, cider, &c.

[6316]

HUGHES, RICHARD HUGH, *Atlas Works, Hatton Garden.*—Patent safety Atlas indicating chandeliers.

Hart & Son, *Wych Street and Cockspur Street.*—Ecclesiastical and domestic metal work, iron, brass, silver.

Manufacturers of every description of silver, brass, and wrought-iron metal work in the mediæval style, both for ecclesiastical and domestic uses, comprising chalices, patens, monstrances, pixes, ciboriums, flagons, tabernacles, benediction crowns and branches, sanctus bells, thuribles, croziers; altar, processional, oratory, and

Hart & Son, *continued.*

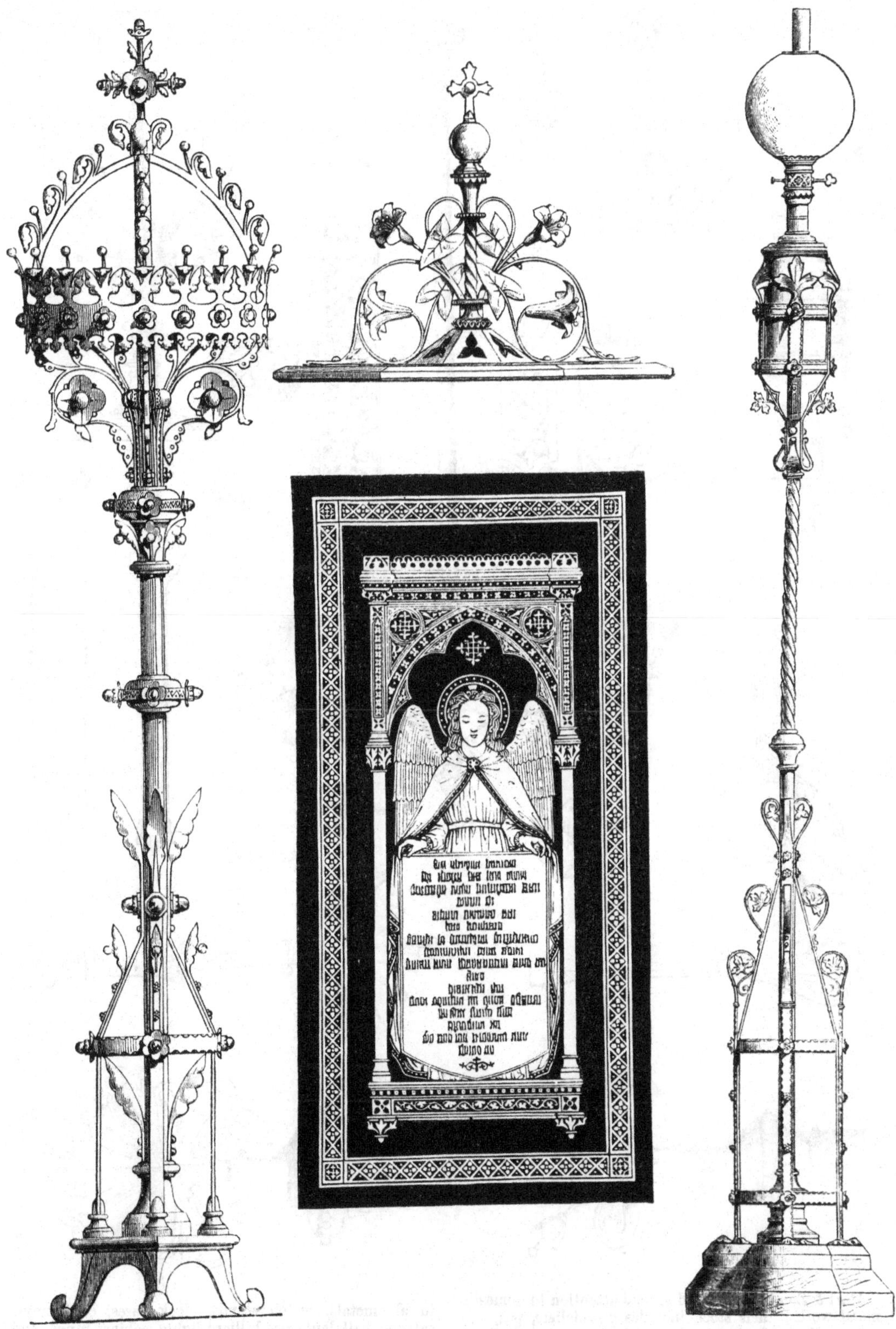

pectoral crosses; altar, gospel, and elevation candlesticks; flower vases, portable communion services, alms basins and boxes, font covers, altar and font rails and standards, chancel chairs in oak and metal, lecterns and book rests, metal screen work, and every description of ecclesiastical furniture.

Engravers of monumental brasses and heraldic devices.

HART & SON, *continued.*

Hart & Son have devoted special attention to domestic metal work; their stock includes chandeliers, gas, oil, and candle pendants; candelabra in brass, electro-silver, and gilt; gas, oil, and candle standards for halls, corridors, stairs, &c. pastile burners, call bells, inkstands in all metals, watch stands, clock cases, card trays, salvers, hall lanterns, billiard lights, cabinet hinges and fittings, door hinges, knobs, and plates, bell pulls and levers, cornice poles, ends, and brackets; tea and coffee services in silver and electro-plate.

HART & SON, *continued.*

Hot-air stoves, register and dog grates, fenders and fire irons, wrought gates and grilles, gable and spire terminals and vanes, boundary and tomb railing, mortuary crosses, out-door lamps, brackets and posts, &c.

Specimens may be seen in the South Court, and at the West Side of London Metal Workers' Trophy.

Hart & Son, *continued.*

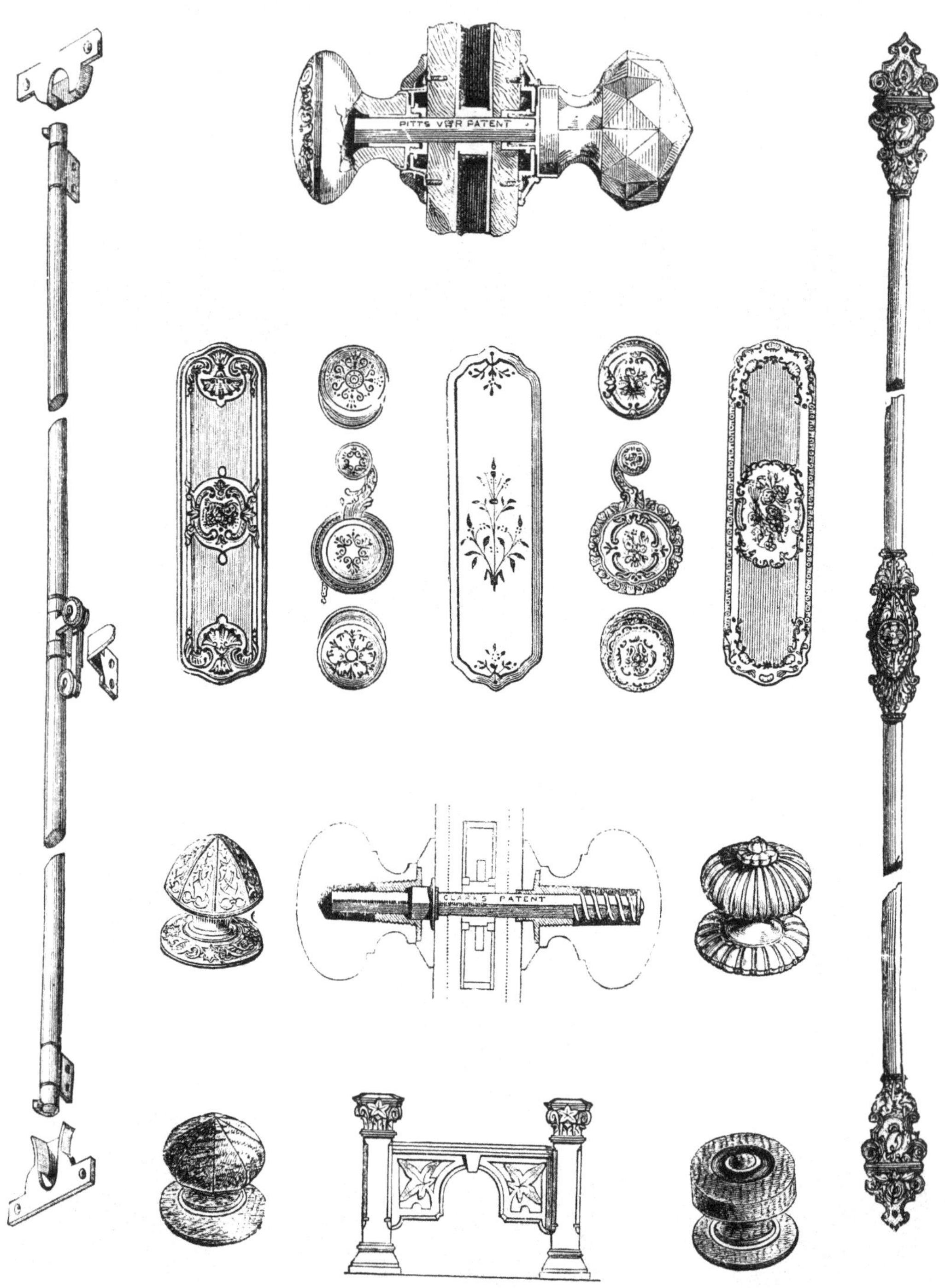

Manufacturers of Pitt's patent self-adjusting, and Clark's patent spindled door furniture, in china, glass, brass, and wood; and finger plates, bell pulls, shutter knobs, &c. en suite.

Manufacturers of espagniolette and double-action bolts, weather bars, stays, and fastenings, suitable for every description of casements.

A large assortment of general ironmongery, suitable for the colonies.

Hart & Son, *continued.*

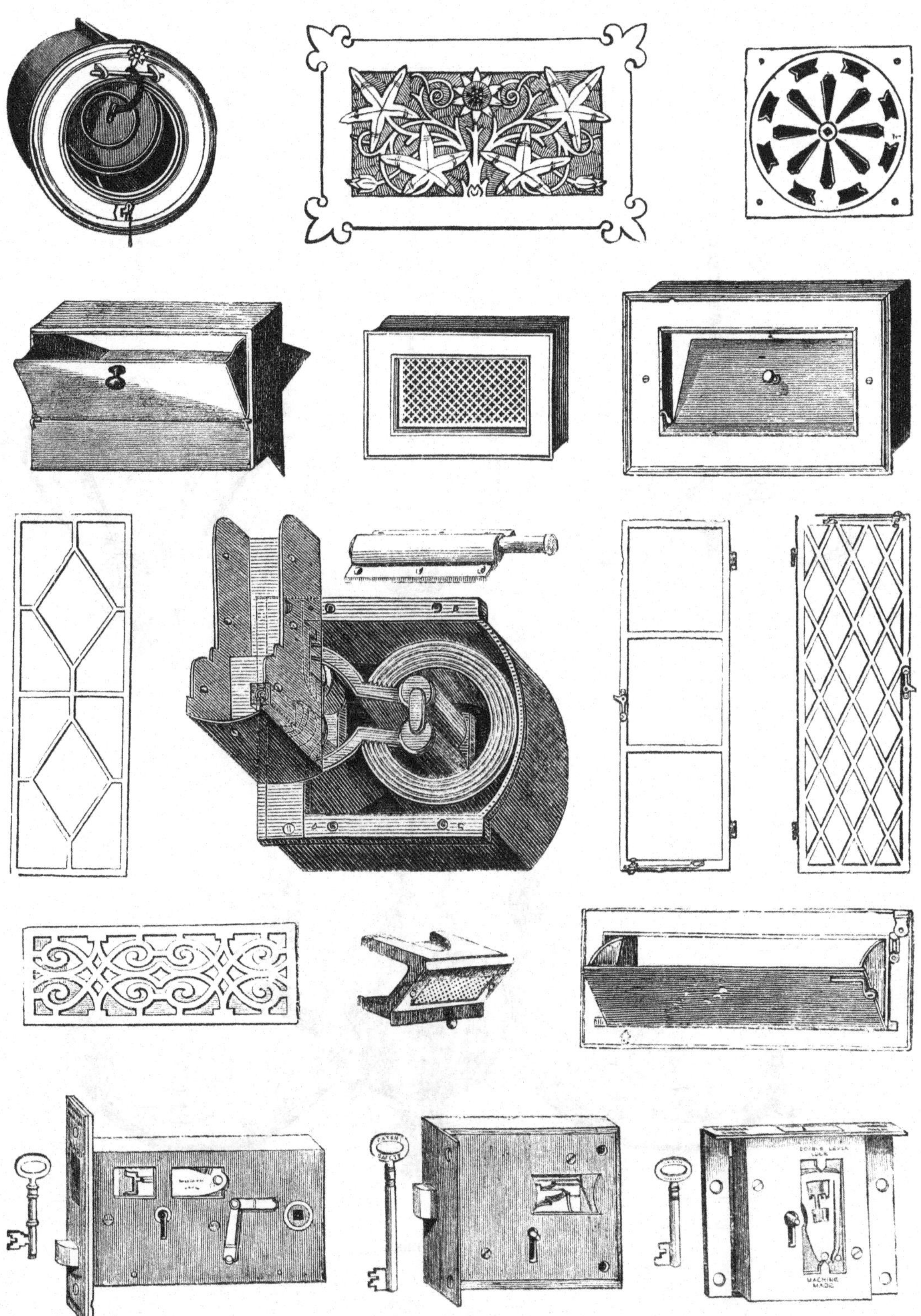

Manufacturers of Arnott's and Sheringham's ventilators, sliding and revolving ventilators of every description, ornamental air bricks, improved floor centres for swing doors and buffer springs; wrought-iron casements suitable for cottages, levered, mortise, rim, and till locks, latches, &c. and every description of general ironmongery.

HILL, JOSEPH, 18 *Broad Street, Birmingham.*— Stampings, chandeliers, gas fittings, metallic bedsteads, &c.

CHANDELIERS.

ORNAMENTAL STAMPINGS for lamps, chandeliers, and general gas-fittings.
HUSKS an l VASES for metallic bedsteads.

CEILING ROSES in various styles and sizes, finished in white and gold and other colours.

[6317]

HULETT, D., & Co., 55 & 56 *High Holborn.*—Gas-fittings.

A GASELIER IN THE RENAISSANCE STYLE.

GASELIERS, HALL LANTERNS, &c. in glass, ormolu, and bronze.

Hulett & Co. manufacture improved gas meters, station and experimental meters, governors, pressure registers, gauges, &c.

D. Hulett's improved service cleanser, for clearing out services, gas fittings, &c.

The above engraving is a representation of a gaselier in the Renaissance style, designed and modelled by the exhibitors.

Inventors and patentees of the mercurial gas regulator, sole manufacturers of Church & Mann's improved photometer, the registered convex silvered glass reflecting light, Arnott's improved ventilators, Carter's valves, high pressure cocks, and all kinds of gas and steam fittings, and every description of gas apparatus.

Prospectuses may be obtained on application at the Manufactory, 55 & 56 High Holborn.

[6318]

Jenn, Joseph, Jun., 38 *Whittlebury Street, Euston Square.*—Moulds for pound-cakes.

[6319]

Johnston, Brothers, 190 *High Holborn, London.*—A standard gaselier for a cathedral, designed by Mr. G. Trufitt.

[6320]

Lambert, Thomas, & Son, *Short Street, New Cut, Lambeth.*—High-pressure valve cocks, &c. (*See page* 121.)

[6321]

Leale, A., 4 *Litchfield Street, W.C.*—Various designs of copper vanes, weathercocks, and of cake and jelly moulds.

[6322]

Leoni, S., 34 *St. Paul's Street, N.*—Knife handles, castor-bowls, taps, gas burners, ornamental wares, of adamas, resisting wear, acids, and heat, and of great durability.

[6323]

Loysel, Edward, C. E. 92 *Cannon Street, London, E.C.*—Loysel's hydrostatic percolator, keyless locks, portable hot-air ovens, reflecto-culinarium. (*See pages* 122 *and* 123.)

[6324]

Mackey, Charles, *Great Hampton Row, Birmingham.*—Brass knobs, vases, furniture ornaments, &c., for various purposes.

[6325]

Marrian, James Pratt, *Birmingham.*—Lamps and various goods in brass for the fitting-up of ships.

[6326]

Matthews, E., 377 *Oxford Street.*—Engraving in metal for ecclesiastical decoration and other purposes.

[6327]

Messenger & Sons, *Broad Street, Birmingham.*—Chandeliers, candelabra, and general gas fittings, &c.

[6328]

Midwinter, Edward, & Co., 68 *Snow Hill, London.*—Brown and bright copper tea-urns, tea-kettles, coal-scoops, &c.

[6329]

Naylor, James, *Radnor Street, ulme, Manchester.*—Lamps for pillars, brackets, and for suspending from ceilings.

[6330]

North, E. P., 6 *Exeter Row, Birmingham.*—Ornamental metallic panelling.

[6331]

Nunn, William, 179 *St. George Street, E.*—Signal lanterns and lenses. (*See pages* 124 *and* 125.)

[6332]

Oerton, Francis B., *Walsall.*—Carriage lamps, axles, springs, handles, and fittings of all kinds.

LAMBERT, THOMAS, & SON, *Short Street, New Cut, Lambeth.*—High-pressure valve cocks; and plumbers', gas fitters', ironmongers', and engineers' furnishings; Carter's patent safety gas valves.

Society of Arts Medal, 1847; *Prize Medal,* 1851; *Bronze Medal, Amsterdam,* 1854.

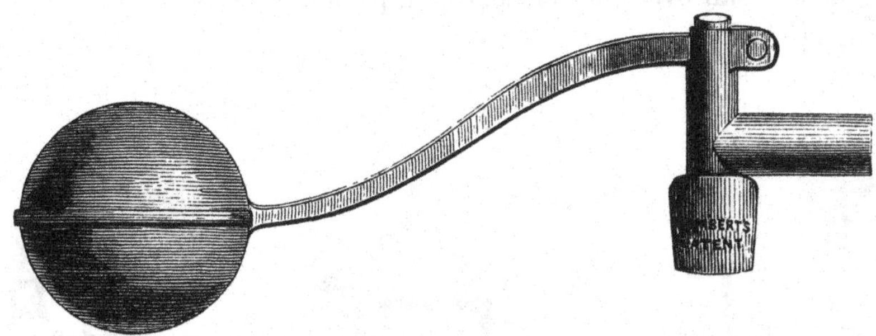

LAMBERT'S EQUILIBRIUM BALL VALVE.

This ball valve is equally adapted for high or low pressure. It runs full bore until the cistern is within two inches of being filled, an important advantage, especially where the supply is intermittent; and it is cheaper than the common ball cock.

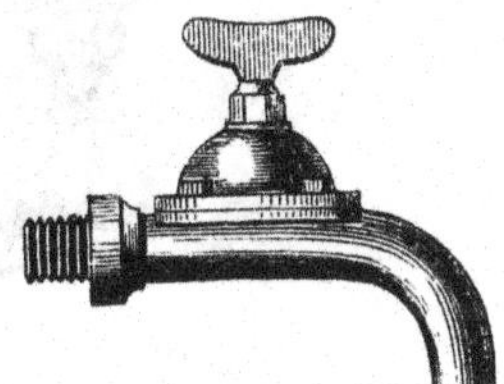

LAMBERT'S HIGH-PRESSURE BIB VALVE.

COMMON BIB COCK, CRUTCH KEY.

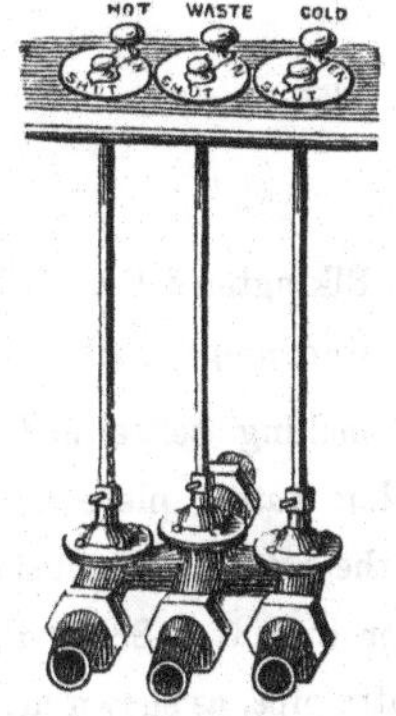

LAMBERT'S BATH VALVES.

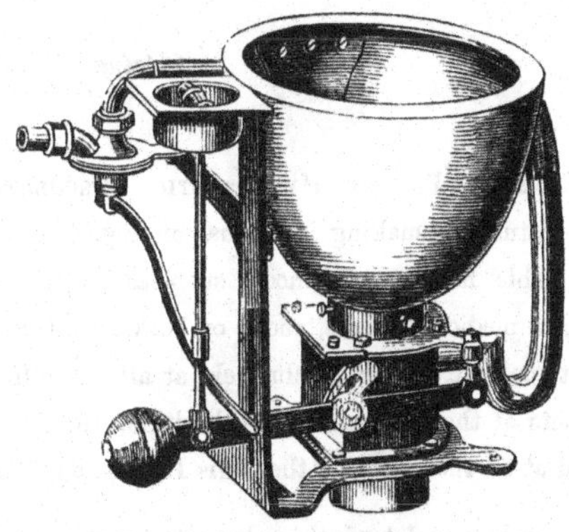

VALVE CLOSET WITH LAMBERT'S PATENT REGULATING VALVE COCK, WHICH DELIVERS A GIVEN QUANTITY AT EACH ACTION.

Section of Valve closed.

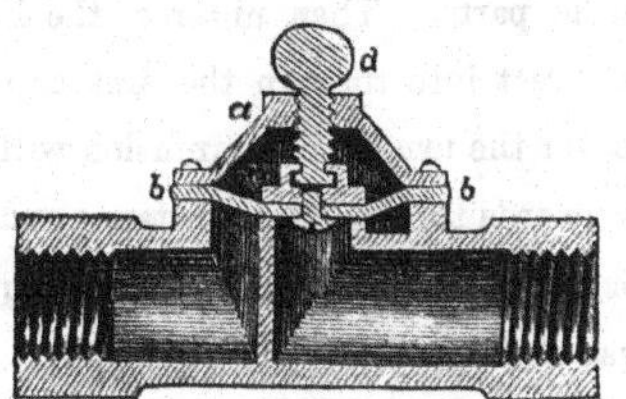

Section of Valve open.

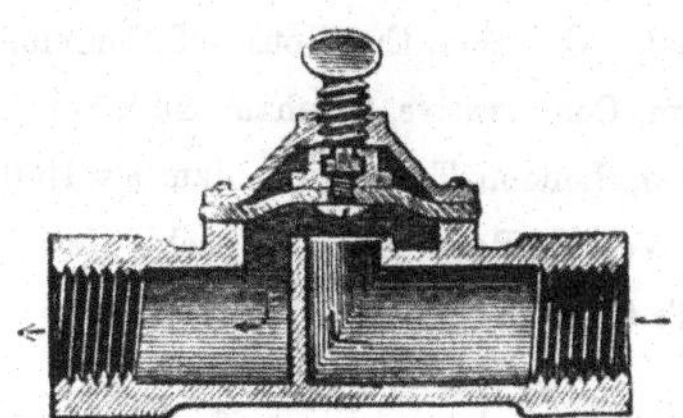

CARTER'S SAFETY GAS VALVES.

These valve cocks are made to suit every description of fittings, never leak, and never set fast, give great facility for regulating the flow of gas, and are exceedingly durable.

Illustrated catalogues post-free.

LOYSEL, EDWARD, C.E., 92 *Cannon Street, London, E.C.*—1. Loysel's patent hydrostatic percolator for extracts of vegetable substances, dye-woods, tea, coffee, &c. 2. Loysel's patent keyless locks and padlocks, safes, bags, cash and deed boxes, &c. 3. Loysel's patent portable hot-air ovens, for bread, coffee, potatoes, chestnuts, &c. 4. Loysel's patent reflecto-culinarium for broth.

IN ACTION.

HYDROSTATIC PERCOLATORS.

Obtained the Medal at the Paris Exhibition, 1855.

LOYSEL'S PATENT HYDROSTATIC PERCOLATOR, an apparatus for making infusions, or liquid extracts, of vegetable fibrous substances, such as coffee, tea, dye woods, medicinal herbs, roots, or barks, malt and hops, beet root, &c. is used exclusively at all the refreshment buffets of the International Exhibition, 1862; and was used at all the buffets of the Paris Exhibition, 1855.

Loysel's percolator is the only perfect system for making tea and coffee, either for families or hotels, refreshment rooms, schools, ships, &c. It is used already by upwards of 120,000 families, and at most large establishments in the United Kingdom, such as the buffets of the International Exhibition, the House of Commons, the Clubs (Reform, Conservative, Gresham, &c.), South Kensington Museum, London Tavern, St. James's Hall, Cremorne, Great Western Royal Hotel, Breach's New Palace Hotel, Queen's (Bull and Mouth) Hotel, &c.

Loysel's percolator is manufactured in tin, copper, bronze, britannia, electro-plate, and silver, by seven of the largest and best manufacturers in the United Kingdom, including Messrs. Elkington & Co. It is sold as low as *5s.* by all respectable ironmongers, silversmiths, &c.

Directions for making coffee and tea:—Warm the percolator with hot water; place the ground coffee or triturated tea on the bottom filter inside the urn; place the movable filter over it. Screw the inverted lid or funnel on the centre pipe, as shown here above, pour into the reservoir so formed the boiling water necessary for the quantity of infusion required. This water will instantaneously go down the centre pipe and percolate upwards through the substance under the powerful action of hydrostatic pressure, and thereby extract at once all the useful and aromatic parts. Then unscrew the lid or funnel, pour back direct into the urn the first cup drawn from the tap; cover the urn, and the infusion will at once percolate downwards through the substance and fill the cup with tea or coffee, all of uniform strength, exquisite aroma, and unrivalled brightness.

London depôts, retail and wholesale: City, 92 Cannon Street, E.C.; and West End, 309 Regent Street, W.

LOYSEL, EDWARD, *continued.*

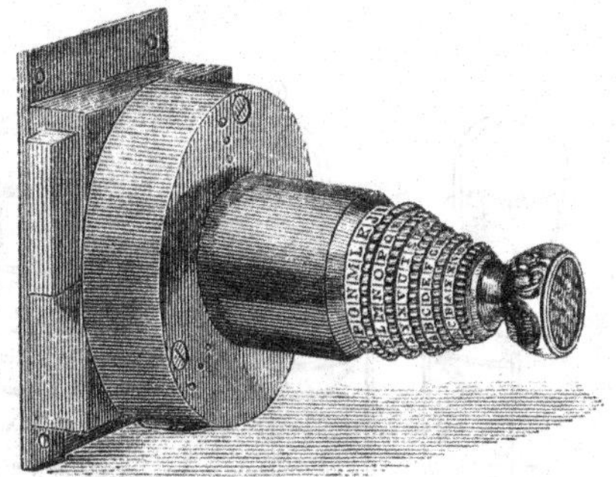

LOYSEL'S PATENT KEYLESS LOCK.

Properly speaking, the "keyless lock" may be said to be composed of two parts, the hinder part, which is the lock and contains the bolt, and the fore part, which is, if it may be so termed, a scientific fixed key, which is formed of concentric cylinders, each of which is divided at the middle into two parts, and traversed by a spindle, which is to act on the bolt for shutting and opening the lock.

The outer edges or faces of the concentric cylinders are impressed with alphabets of 24 or less letters, and it is only when a pre-determined combination is brought into coincidence, that the spindle can be brought in a position to work the bolt.

It should be observed that, owing to the division of the cylinders into two parts, the owner of a lock can instantly change the combination on which it opens, without pulling the lock to pieces. These changes in the combination may be made from the outside.

There being no key-hole, no instrument whatever can be introduced to try to pick the lock or injure it, and as to introducing gunpowder, it is an utter impossibility; and even if the fore part of the lock, or rather, scientific key, could be broken by extreme violence, the result would be merely to take away all chances of ever opening the lock, as the lock itself, and the hinder part of the key, would remain as an impenetrable block.

All the parts of the lock being hardened, it is drill proof.

In short, the lock defies violence as well as skill, as there is no possibility whatever of opening it by hearing or feeling, or by any pressure either gentle or rough, as besides the impossibility of bringing the pins opposite the holes, a screening plate has been introduced between the holes and the pins, which renders even trying an utter impossibility. Therefore, the only means of opening this lock is by finding out the combination on which it opens.

Now, a lock with 5 cylinders, of 24 letters each, gives 7,962,624 combinations; ditto 6 cylinders, of 24 letters each, gives 491,102,796 combinations; ditto 7 cylinders, of 24 letters each, gives 4,586,471,424 combinations; and it is calculated that working assiduously 10 hours a day, it would take about 2,000 years for ringing all the changes this last lock is susceptible of.

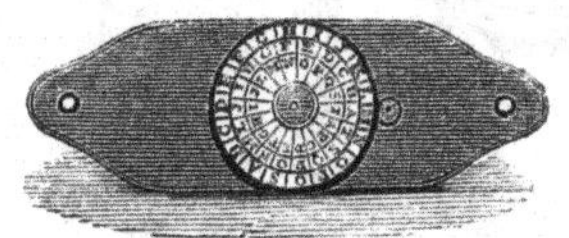

KEYLESS LOCK FOR BAGS, CASH BOXES, &C.

The keyless lock is made so as to be adapted to travelling bags and cash boxes, and will prove of great convenience, as keys are a great and acknowledged nuisance.

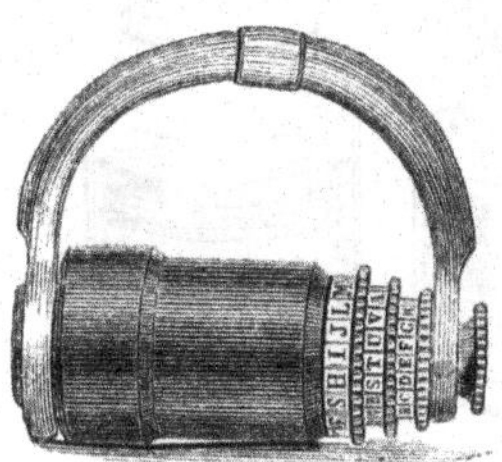

KEYLESS PADLOCK.

Padlocks are also constructed on this principle, which cannot be forced open by pressure, and in which the combination is altered at pleasure without pulling the padlock to pieces, as in the ordinary letter padlocks. The lock will also be adapted to most purposes for which locks in general are used.

As to the cost, it will be, if anything, less than locks of any other system or corresponding quality, as every part is made by machinery, which generally combines precision with cheapness.

The keyless lock, originally the invention of Viscount de Kersolon, who has worked at it for 20 years, has been improved and patented in England, by E. Loysel, C.E. Assoc. Inst. C.E.

London depôts for Loysel's patent keyless locks: City, 92 Cannon Street, E.C.; West End, 309 Regent Street, W.

NUNN, WILLIAM, 179 *St. George Street, E.*—Patent signal lanterns, lenses, and reflectors.

By Royal Letters Patent.

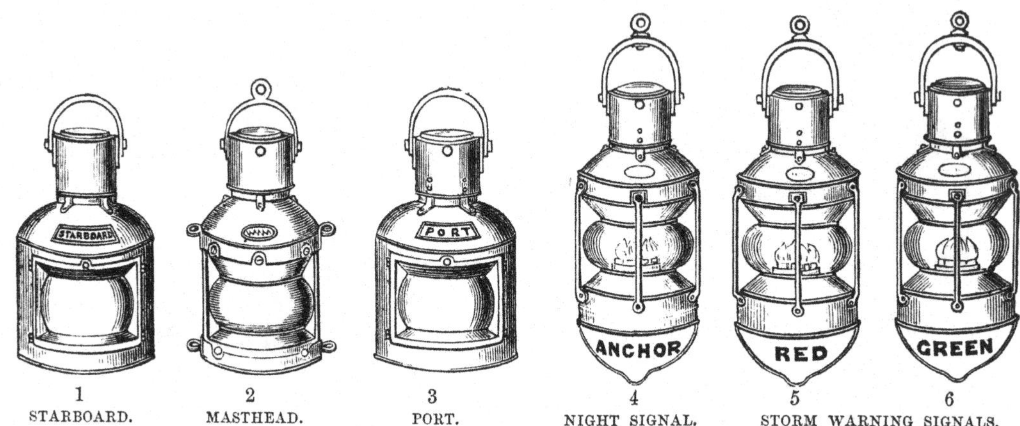

1 STARBOARD. 2 MASTHEAD. 3 PORT. 4 NIGHT SIGNAL. 5, 6 STORM WARNING SIGNALS.

The above are fitted with powerful reflecting lenses, tested, approved, and adopted in the Royal Navy, also constructed so that a new lens can be replaced in three minutes by any person on board when the old lens is broken.

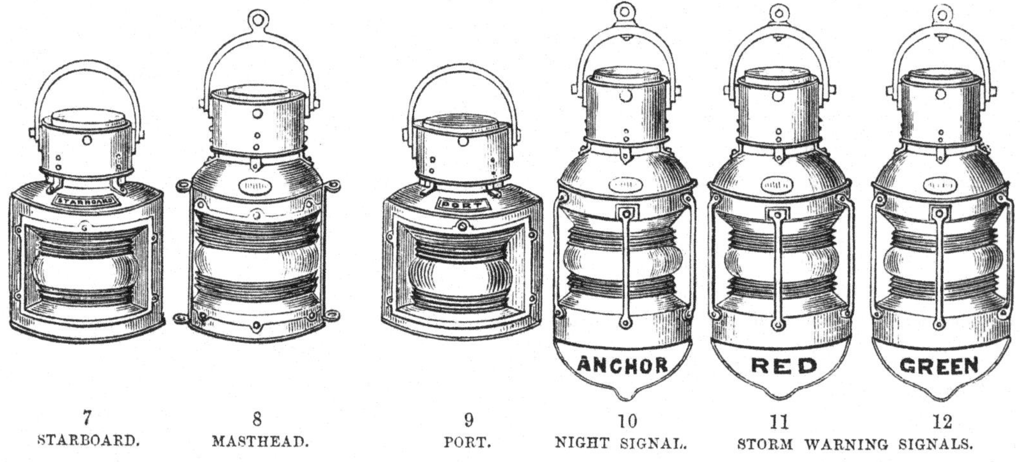

7 STARBOARD. 8 MASTHEAD. 9 PORT. 10 NIGHT SIGNAL. 11, 12 STORM WARNING SIGNALS.

These lanterns are fitted with patent *dioptric* lenses, also constructed for the refitting of a new lens in the place of one broken in three minutes, and can be made to burn oil or stearine candles, as may be required

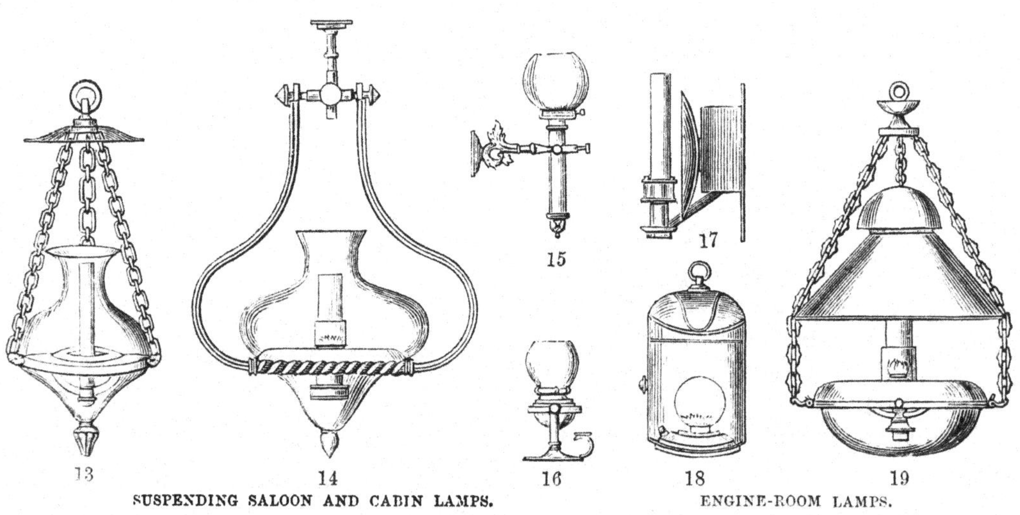

13, 14, 15, 16 SUSPENDING SALOON AND CABIN LAMPS. 17, 18, 19 ENGINE-ROOM LAMPS.

NUNN, WILLIAM, *continued.*

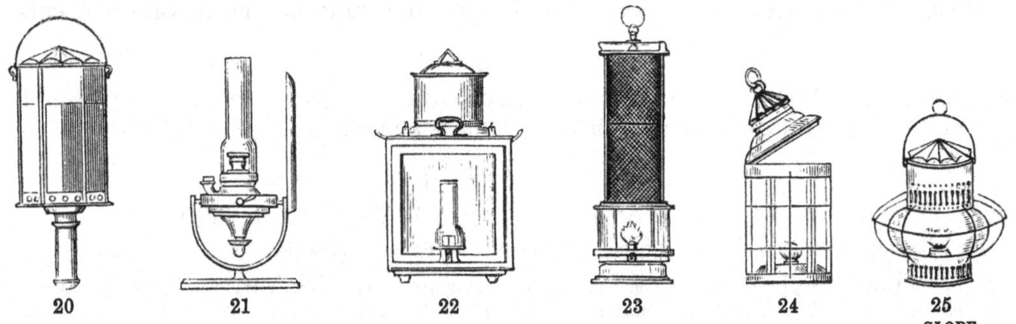

Railway lamps and reflectors, ship and railway lenses in ruby, green, and white.

In forwarding orders for lamps, &c. the tonnage of vessels in which they are to be used should be given.

***Some of these lamps and lenses are exhibited in Class* XII.**

[6333]

OLIVER, GEORGE & JOS., 286 *Wapping, E.*—Dioptric ships' signal lamps and buoy.

[6335]

PHILLP, CHARLES J., 20 *Caroline Street, and* 29 *Mary Street, Birmingham.*—Gaseliers, brackets, and gas fixtures generally.

[6336]

PONTIFEX, HENRY, & SONS, 55 *Shoe Lane, Holborn, London.*—Brewing and distilling apparatus.

[6337]

PONTIFEX, R., & SON, 14 *Upper St. Martin's Lane, London.*—Copper, brass, and steel plates.

[6338]

PROSSER, W., & H. J. STANDLY, 24 *Dorset Place, and* 20 *Cockspur Street.*—Improved lamps for lime lights.

[6339]

PYRKE, J. S., & SONS, *Dorrington Street, London.*—Bronzed tea and coffee urns, and swing kettles.

[6340]

REID, JOHN, *Edinburgh.*—Patent gas-saturator for preventing evaporation of water from gas meters.

[6341]

RENNIE & ADCOCK, *Easy Row Works, Birmingham.*—Chandeliers, candelabra, bronzes, mirrors, and works of art.

[6342]

RICHARDS, W., *Crawford Passage, Clerkenwell.*—Models of gas meters in progress of making.

1. An ordinary gasholder, converted into a gas measure.
2. Gasholder, as constructed by Mr. Clegg, supposed to have suggested the present meter.
3. An ordinary gas meter wheel, revolving in its case.
4. A wet gas meter as constructed by exhibitor, and patented by him in 1858.
5. A transverse section of a wet gas meter as constructed by exhibitor.
6. Section of a gas meter wheel as constructed by exhibitor.
7. Clegg's inferential dry gas meter.
8. Meter made by Dry Meter Company, in 1835-6.
9. Meter patented by G. Sullivan, 1837.
10. Meter patented by N. Defries and N. F. Taylor, 1843.
11. Dry meter invented and manufactured by exhibitor in 1844.
12. Dry meter invented by the exhibitor, and now manufactured by him.
13. Lowe's motive power meter.
14. An exhaust and pressure tell-tale indicator.

[6343]

SARSON, THOMAS FREDERIC, *Leicester.*—Lamp upon a new construction, that can be repaired in a few minutes.

[6344]

SINGER, JOHN W., *Frome, Somerset.*—Brass lectern, altar rails, and mediæval ornaments.

[6345]

SKIDMORE ART MANUFACTURES COMPANY, THE, *Great Coventry.*—Screens for Hereford, Ely and Lichfield cathedrals; gas corona, pendants, standards, &c.

[6346]

SOUTTER, WILLIAM, *New Market Street, Birmingham.*—Tea-urns and kettles.

The following goods, suitable for the home and export trades, are exhibited :—

1. COPPER BRONZE TEA URNS, showing the early designs.
2. COPPER BRONZE TEA URNS, most modern designs, showing the various modes of heating by spirits of wine, iron heater, and charcoal.
3. COPPER BRONZE SWING KETTLE, used for same purpose as the urn, showing latest improvements in spirit lamp.
4. BRASS URN OR SAMAVOIR, as used in Russia, heated by charcoal.
5. BRASS COFFEE URN AND BASIN, as used in Turkey, heated by charcoal.
6. BRASS DUTCH TEA URN, heated by charcoal.
7. BRASS TRAY, used in Turkey.
8. COPPER BOX IRON, heated by charcoal; used in the cape trade for ironing purposes; two patterns, showing the latest improvements.
9. LARGE COPPER SOUP OR STOCK POT, raised from the sheet metal without seam, by an entirely new process.
10. Assortment of COPPER COOKING UTENSILS.

[6347]

STEER, JOSEPH, 44 *Weaman Street, Birmingham.*—Cornices, cornice poles, ends, rings, brackets, and curtain bands.

[6348]

STONE, JOSIAH, *Deptford, London, S.E.*—Copper and cast composition boat and ship nails, &c.

[6349]

STRODE, WILLIAM, 16 *St. Martin's-le-Grand, London, E.C.*—Improved sun burner, with valve, and a bronze valve candelabra.

[6350]

SUGG, W., *Marsham Street, Westminster.*—Gas meters; governors and pressure gauges; lava burners; and public lamp governors.

[6351]

TAYLOR, JOHN, & CO., *Loughborough.*—A large bell, three tons weight, note B, suspended on frame, with hammer for striking.

Obtained the Prize Medal, Great Exhibition, 1851, *and the special approbation of the iurors.*

[6352]

THOMASON, THOMAS, & CO., 30 *St. Paul's Square, Birmingham.*—Ecclesiastical and domestic Gothic metal work.

[6353]

TILLMAN, GEORGE, 5 *Ashley Crescent, City Road.*—Carriage lamps.

[6354]

TONKS, WILLIAM, & SONS, *Moseley Street, Birmingham.*—Brass work for builders, cabinet-makers, and upholsterers.

Obtained Medals at the Exhibitions of London, 1851, *and Paris,* 1855.

Hinges, bolts, sash, casement, and espagniolette fasteners, knobs, handles, door porters and knockers, brackets, shop-window fittings, organ, pew, and desk railings, mouldings, sideboard edgings, picture and stair rods, bell pulls and levers, upright and mantel-piece screens, sconces, candlesticks, ventilators, &c.

[6355]

UNDERHAY, F. G., *Crawford Passage, Clerkenwell, London, E.C.*—Patent direct-action compensating gas meter.

[6356]

VERITY, B., & SONS, 31 & 32 *King Street, Covent Garden.*—Gaseliers and brass works.

The following chandeliers are exhibited :—

1. LARGE EIGHT-LIGHT GASELIER, designed by the late John Thomas. Price £95 0
2. EIGHT-LIGHT POLISHED GASELIER (Old English). Price £30 0
3. FIVE-LIGHT GASELIER, in bronze (Grecian design). Price £12 12
4. ORIENTAL HALL LAMP, designed by George Somers Clarke. Price £28 0
5. FIVE-LIGHT GASELIER, designed by the late John Thomas. Price £21 0
6. TAZZA GASELIER, for three lights, suitable for a boudoir. Price £3 10

And several other gaseliers, all new designs, suitable for drawing and other rooms.

A great variety of designs for bracket-lights, standards, and vases, all for gas, are likewise exhibited in their case.

VERITY & SONS manufacture and fit gas baths, and gas and hot-water apparatus. They make gaseliers and every kind of brass work to any design required, and erect private gas works. A working model of one of the latter is exhibited at the factory.

A large assortment of gaseliers, brackets, &c. of their own manufacture is always kept in stock at their show-rooms. Factory, Hart Street, Covent Garden; West End branch, Charles Street, Westbourne Terrace.

[6358]

WARNER, JOHN, & SONS, *Crescent, Cripplegate, London.*—Bells, urns, baths, lamps, braziery, weights, and measures. (*See pages* 128 *to* 132)

[6359]

WEST & GREGSON, *Oldham.*—Model gas (station) meter, with its appurtenances.

[6360]

WOOTTON & POWELL, *Parade Works, Birmingham.*—Gas chandeliers and wall brackets.

[6361]

WYATT, ALFRED, 22 *Gerrard Street, Soho, W.*—Silver-plated state carriage lamps.

[6362]

YOUNG, JOHN, & SON, 46 *Cranbourn Street, London.*—Weighing machines, with multiplying power, for persons or goods.

[6363]

CARTER & HACKS, *West Middlesex Waterworks, Kensington Reservoir.*—Screw cocks, with and without packing, under pressure.

[6364]

RICKETS & HAMMOND, *Agar Street, Strand.*—Globe-light chandeliers, and ventilating globe lights.

WARNER, JOHN, & SONS, *Crescent, Cripplegate, London.*—Bells, urns, baths, lamps, braziery weights and measures.

Prize Medal awarded 1851, *for bells.*

No. 8. A 12-IN. TURNED AND POLISHED SHIP'S BELL, in bronzed cast-iron dolphin frame.
Other sized bells can be had when required.

No. 9. A GONG, OR CALL BELL, forming an appropriate ornament for the halls of mansions.

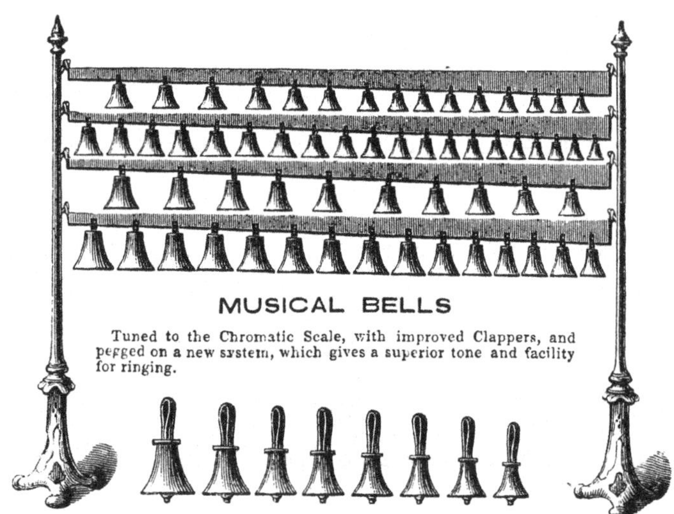

MUSICAL BELLS

Tuned to the Chromatic Scale, with improved Clappers, and pegged on a new system, which gives a superior tone and facility for ringing.

Prices of peals of hand bells :

	A peal of 15.	A peal of 12.	A peal of 10.	A peal of 8.
No. 22 size, in C	£8 10 0	£7 14 0	£7 0 0	£6 7 0
No. 21 ditto D	7 9 0	7 0 0	6 7 0	5 16 0
No. 20 ditto E	7 0 0	6 2 0	5 16 0	5 9 0
No. 19 ditto F	6 7 0	5 16 0	5 12 0	5 6 0
No. 18 ditto G	5 16 0	5 6 0	4 14 0	5 0 0
No. 17 ditto A	5 12 0	4 19 0	4 11 0	4 5 0
No. 16 ditto B	5 6 0	4 14 0	4 7 0	3 19 0
No. 15 ditto C	5 0 0	4 11 0	4 5 0	3 17 0

An extensive stock kept of—

House bells ; ditto, turned and lacquered ; ditto, with springs for shutters ; bells for ships, yachts, steamers, &c.; horse bells ; sheep bells ; dog bells ; clang bells for cattle ; ferret bells ; squirrel bells ; dinner bells ; tea bells ; bellman's bells ; self-acting alarm bells ; small clock bells.

Warner, John, & Sons, *continued.*

Obtained a Prize Medal in 1851.

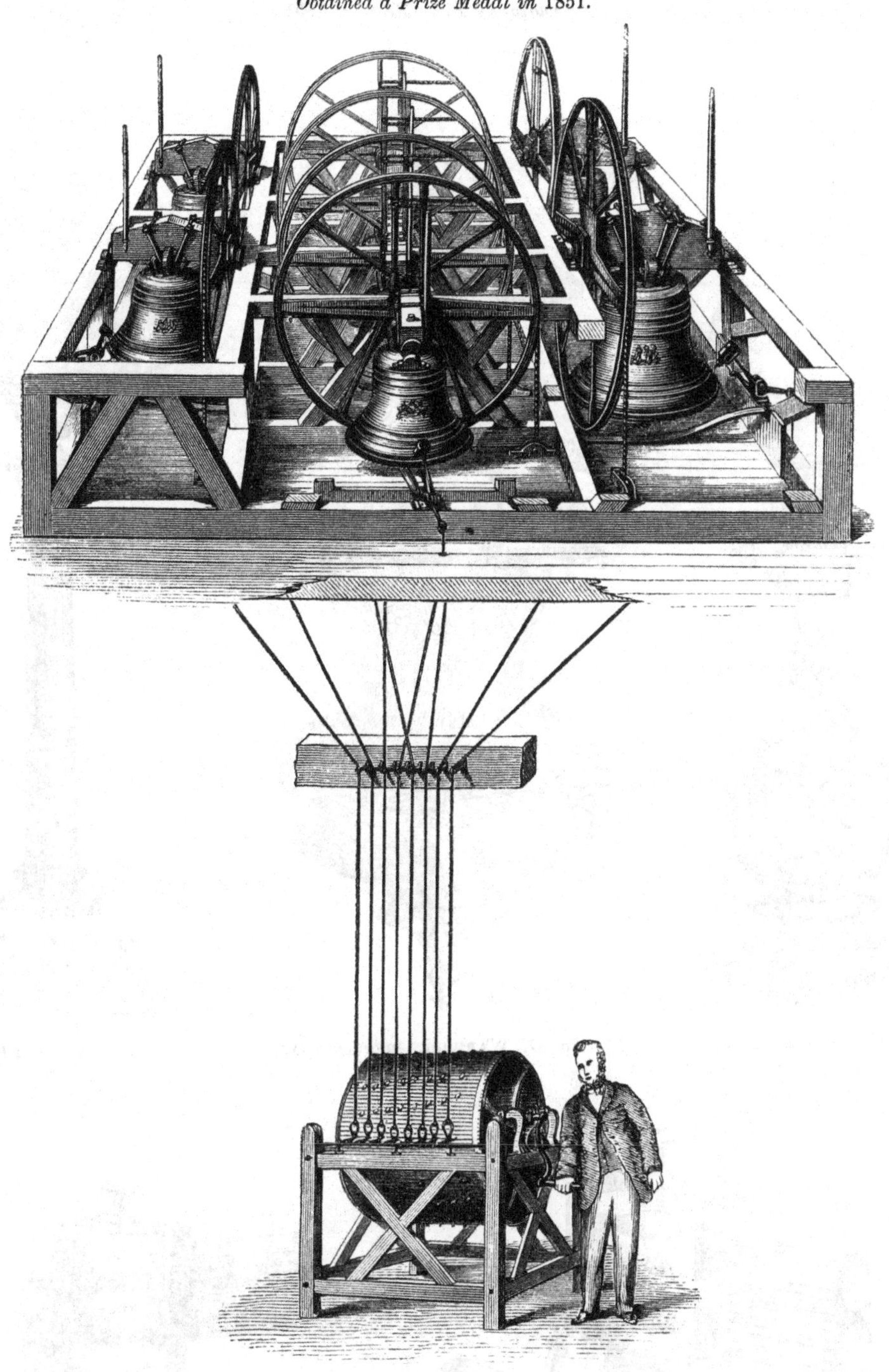

PEAL OF EIGHT BELLS.

Drawing showing a peal of eight bells fixed in a frame for ringing. In connexion with it, is shown one of J. Warner & Sons' Improved Chiming Machines, by which a lad entirely unaccustomed to music may correctly chime a whole peal. In parts of the country where no good ringers are to be obtained, this simple machine will be found invaluable. John Warner & Sons continue to supply estimates for bells of all sizes, singly or in peals, as well as for recasting broken or bad bells, new oak frames and fittings, and contract to hang bells.

The original Big Ben, the largest bell ever cast in England, the present quarter bells in the clock tower of the Houses of Parliament, the bells at Her Majesty's palace at Balmoral, the hour bell at the Leeds Town Hall, the peal of eight at Doncaster Cathedral, and the bells exhibited in connexion with Mr. Dent's large clock, were cast by John Warner & Sons.

Warner, John, & Sons, *continued.*

Obtained a Prize Medal in 1851.

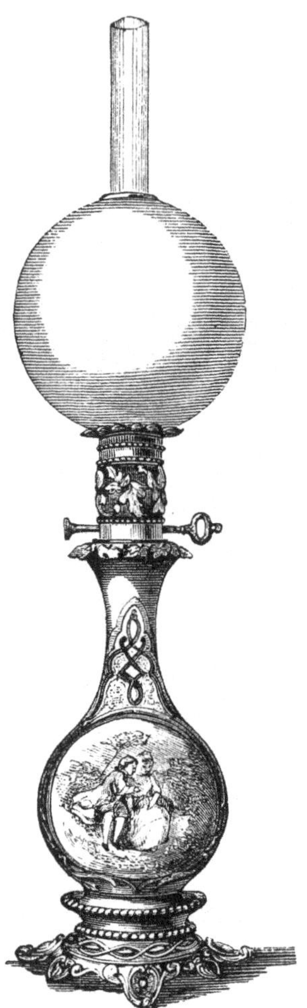

No. 56. Moderator Lamp.

No. 87. Warwick Grecian Lamp.

No. 196. Crescent Oil Lamp

No. 1. Goffering Machine.

No. 1, 2. Crimping Machine.

WARNER, JOHN, & SONS, *continued.*

Obtained the Prize Medal in 1851.

No. 296. A 4-QUART URN.

No. 297. A 5-QUART URN.

No. 1. SWING KETTLE ON BROWN STAND, with lamp or heater, several sizes.

No. 3. SWING KETTLE ON BLACK STAND, with lamp or heater, several sizes.

Urns and kettles of various patterns, and every variety of braziery goods, kept in stock.

WARNER, JOHN, & SONS, *continued.*

Obtained the Prize Medal in 1851.

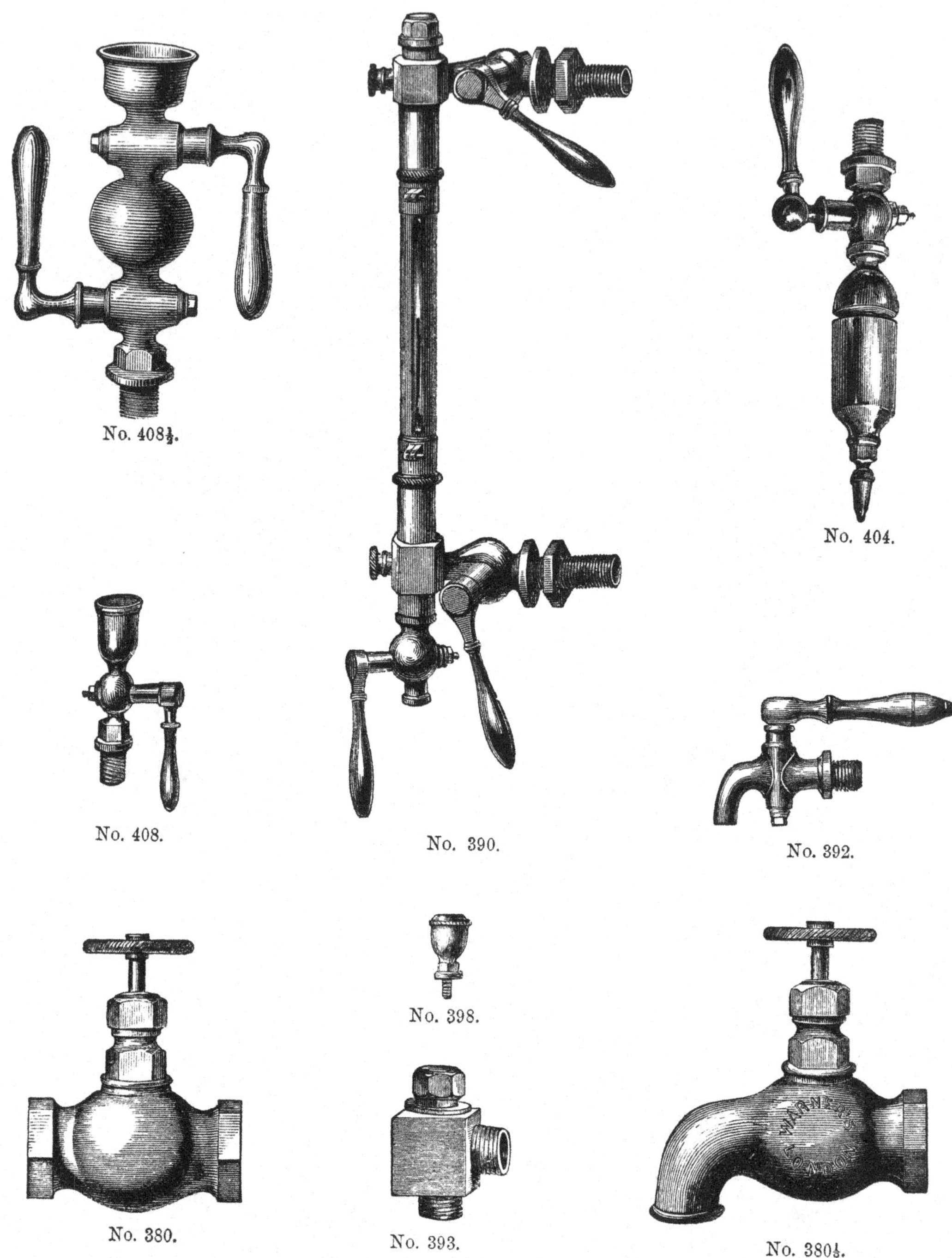

No. 408½. No. 390. No. 404. No. 408. No. 392. No. 398. No. 380. No. 393. No. 380½.

GUN-METAL STEAM GAUGE WORK, for portable or fixed engines :—

No. 408½. GUN-METAL DOUBLE GREASE COCK.
No. 390. GUN-METAL WATER GAUGE.
No. 404. GUN-METAL STEAM WHISTLE, of all sizes.
No. 408. GUN-METAL GREASE COCK.
No. 392. GUN-METAL FULLWAY GAUGE COCK.
No. 398. GUN-METAL SYPHON OIL CUP.
No. 393. GUN-METAL VALVE BOX.
No. 380. GUN-METAL VALVE, for steam, water, or gas.
No. 380½. CAST-IRON VALVE, for steam, water, or gas.

Illustrated and priced catalogues can be had on application.

SUB-CLASS C.—*Manufactures in Tin, Lead, Zinc, Pewter, and General Braziery.*

[6373]

AZULAY, BONDY, *Rotherhithe, Surrey.*—Heat-retaining vessels for boiling-water, &c.

[6374]

BEARD & DENT, 21 *Newcastle Street, Strand.*—Plumbers' appliances.

[6375]

BRABY, FREDERICK, & Co., *Fitzroy Works, Euston Road, London.*—Galvanized zinc; galvanized iron; roofing felt; perforated metals. (*See page* 134.)

[6376]

CHATTERTON, JOHN, *Wharf Road, City Road, London.*—Specimens of lead, block tin, and composition pipe.

Specimens of improved lead pipe, pure block-tin pipe, composition gas-tube, and also of lead pipe, coated internally with tin, and the patent compound tube, or lead pipe, lined with gutta percha, for use in localities where water acts upon lead.

[6377]

COOKSEY, HECTOR RICHARD, 148 *Bordesley, Birmingham.*—Coffin plates, handles, and ornaments.

[6378]

DIXON, JAMES, & SONS, *Sheffield.*—Britannia metal wares.

[6379]

ELLIS, JOSEPH, 136 *King's Road, Brighton.*—The Elutriator, for decanting wine or other liquids successfully, but without additional care or trouble.

[6380]

EWART, HENRIETTA, 346 *Euston Road, N.W.*—Baths, washstands, flower-boxes, meat-safes, spirometer, flues, mouldings, and other zinc goods. (*See page* 136.)

[6381]

FOXALL, SAMUEL, 52 *William Street, Regent's Park, N.W.*—Confectioners' moulds, piecer, &c.

[6382]

GILBERT, JOHN A., & Co., *Clerkenwell, London.*—Mills, scales, canisters, and shop fittings used by grocers.

[6383]

HICKMAN & CLIVE, *William Street North, Birmingham.*—Coffin furniture.

[6384]

LOVEGROVE, JOHN JAMES, 6 *Pembroke Place, Spring Grove, Isleworth, Middlesex.*—Specimens of plumbing, from 14th century to present time.

BRABY, FREDERICK, & Co., *Fitzroy Works, Euston Road, London.*—Zinc; galvanized iron roofing felt; perforated metals.

PERFORATED ZINC. No. 12 B.

PERFORATED ZINC. No. 25.

Perforated zinc in various designs and sizes of holes, for ventilations, sieves, window blinds, larders, meat safes, dairy windows, &c.

Zinc friezes and frets for verandahs, lamps, and decorative purposes.

Zinc saws for cutting salt.

Sheet zinc, and zinc nails.

Zinc tubing for bell hanging.

A zinc meat safe.

A perforated zinc window blind.

Pierced tin plates and percolators.

Perforated galvanized iron.

Stabbed iron for malt-kiln plates.

ROOFING FELT, 1*d.* per square ft.

Corrugated galvanized iron for roofing.

Malleable galvanized iron sheets of superb quality.

Wire netting, galvanized and japanned.

Galvanized iron coal-scoops, buckets and basins, turnip skeps, oval pans, &c.

Galvanized iron furnace pans, or wash boilers.

Portable folding galvanized wire garden-stools, 4*s.* 6*d.* each.

Ditto, chairs, 7*s.* 6*d.* each.

Patent top for curing smoky chimneys.

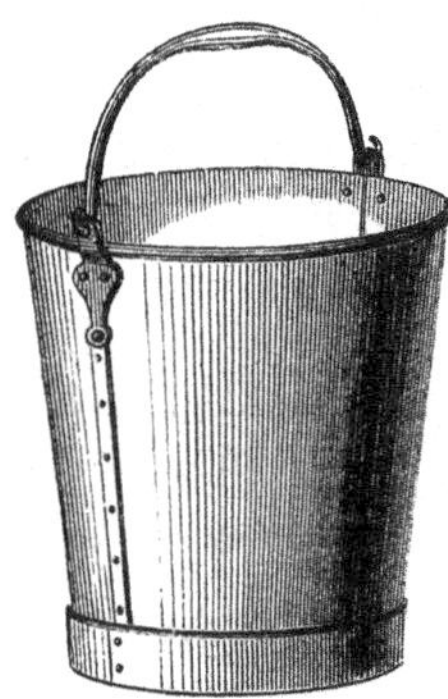

GALVANIZED IRON PAIL.

Galvanized iron basins for use of schools.

Perforated card-board and cards, fancy.

Zinc mouldings, flower baskets, stands, &c.

Spelter for brass-founders.

[6385]

Loveridge, Henry, & Co., *Wolverhampton.*—Papier-maché trays, wares, &c.

Beart & Platow's coffee pots and urns, of which Messrs. Loveridge & Co. are the patentees and manufacturers.

Dish covers.

Victoria Regia sponge and patent hip-baths.

Patent hip-bath with jointed covers.

The patent folding roasting-jack screen.

Patent Persian coal-scoops.

Albert and Windsor coal-scoops, and coal vases of every description.

Iron and patent paper and fine papier-maché tea-trays.

COAL VASE—ELIZABETHAN STYLE.

The coal vase here represented will be found in the Wolverhampton Court.

This vase is a composition after the Elizabethan period, designed by Remmett, and is the first instance with which we are acquainted of a successful union of a bright metallic surface (silver-plated or ormolu) with japanned ware.

The metal handles and pillars may be taken off for repairs and cleaning.

The plating is protected by a white lacquer from the effects of the atmosphere.

The objection to silver-plating, viz. its liability to tarnish, is thus entirely obviated.

This vase may be fitted up as a Canterbury, a cellarette, or with a loose lining as a coal vase; it was for the last purpose more particularly designed.

EWART, HENRIETTA, 346 *Euston Road, N.W.*—Baths, washstands, flower-boxes, meat-safe, spirometer, flues, mouldings, and other zinc goods.

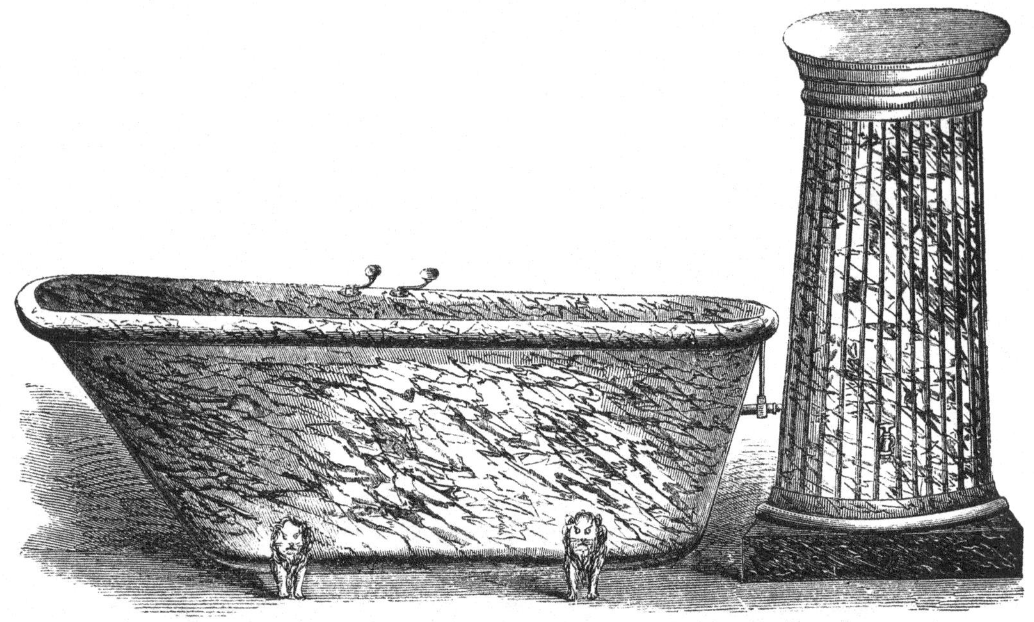

ROMAN BATH.

1. ROMAN BATH, with a new arrangement for a supply of hot water. Can be adapted for either gas or coke, and does not require any fire-place in the bath-room. It is also so constructed that hot water can be drawn off for cleaning or other purposes, as well as for the bath. The same system of heating can be applied to all kinds of baths, and will be found speedy and economical.

Price, complete, including cocks, levers, and handles £12 12

Baths of a less costly description with the same arrangement, from £8 8

DEEP WASH-HAND BASIN.

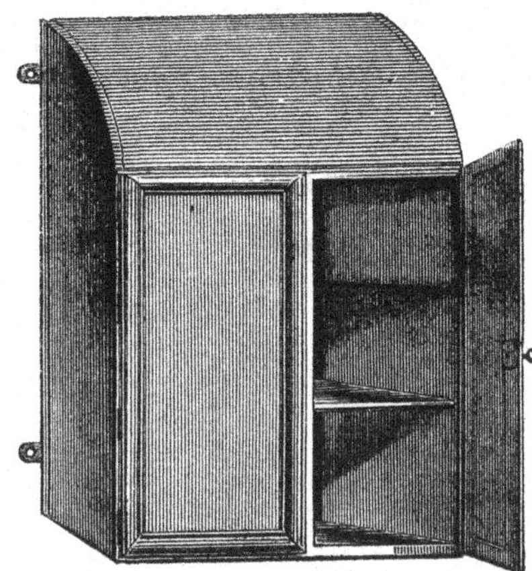

ZINC MEAT SAFE.

2. DEEP WASH-HAND BASIN, for gentlemen, 24 in. diameter, 12 in. deep, on stand 3 ft. high, with castors. The basin is provided with a plug, by means of which it can be emptied, after use, into the bath beneath without removing either.

Price, including the water can, which contains about 4 gallons £2 15

The same pattern can be made any size, or japanned any colour, at proportionate prices.

3. SPIROMETER, an instrument made of zinc, for testing the capacity of the lungs. Price £3 10

4. ZINC MEAT SAFE, with hollow shelf, which can be filled with warm water in winter, and with ice in summer. Price £4 4

EWART, HENRIETTA, *continued.*

FERN CASE ON STAND.

5. FERN CASE on stand, with sliding door at each end. Price, complete £3 3

6. MODELS OF FLUES, COWLS, and various contrivances made of zinc, for curing smoky chimneys and preventing down-draughts.

7. Specimens of zinc mouldings, rain-water pipes, gutters, and heads.

8. Specimens of zinc sash-bars, tubes, astragles, &c.

9. FLOWER BOX for windows, made of zinc, mounted with tiles. Size, 8 in. by 8 in. Price 4*s.* 6*d.* per ft.

10. FLOWER BOX for windows, made of zinc, mounted with tiles. Size, 10 in. by 10 in. With mouldings, price 6*s.* 6*d.* per foot.

11. FLOWER BOX for balconies, all of zinc. Size, 11 in. by 12 in. Price 5*s.* 6*d.* to 7*s.* 6*d.* per ft.

[6386]

MARSTON, JOHN, *London Works, Bilston, Staffordshire.*—Trays, waiters, coal vases, toilette ware, and other japanned goods.

[6387]

PERRY, EDWARD, *Jeddo Works, Wolverhampton.*—Japan and tin wares.

[6388]

TYLOR, J., & SONS, *Warwick Lane, Newgate Street, E.C.*—Baths for private dwellings. (*See page* 138.)

[6390]

WATTS & HARTON, 61 *Shoe Lane, Holborn Hill.*—Pewter articles of every description.

[6391]

WILSON, R. & W., *London.*—Baths, various, and pedestal rotary plate warmer.

[6392]

WOLVERHAMPTON ELECTRO-PLATE COMPANY, THE, *Peel Works, Wolverhampton.*—Silver-plated wares, tea and coffee services, &c. (*See page* 139.)

TYLOR, J., & SONS, *Warwick Lane, Newgate Street, London, E.C.*—Baths for private dwellings.

J. TYLOR & SONS' PATENT BATH, sienna marbled inside, verdantique outside. Taps and safe fitted. No wood casing required.

J. TYLOR & SONS' BATH, white marbled inside for fitting in wood casing. These baths are made both in copper and galvanized tinned iron, 5 ft. to 5 ft. 6 in. long.

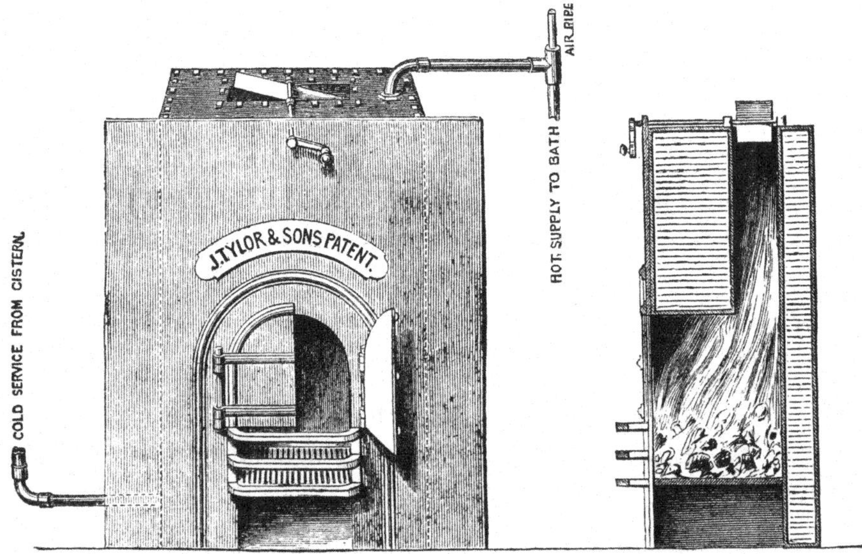

J. TYLOR & SONS' PATENT BATH BOILER, which may be fixed in any fire-place, and will serve a bath with hot water in any apartment below the level of the cold-water cistern.

Catalogues containing illustrations of nine methods of fixing a warm bath, on application.

Wolverhampton Electro-plate Company, The, *Peel Works, Wolverhampton.*—Silver-plated wares, tea and coffee services, cruet frames, spoons, forks, &c. &c.

Breakfast or Dinner Cruet, also made as an inkstand—Neptune driving through the sea. No. 1226.

[6393]

Zobel, Julius, 139 *Euston Road.*—Geometrical works; flower ornaments for gas and water, &c.

Class XXXII.

STEEL CUTLERY AND EDGE TOOLS.

Sub-Class A.—*Steel Manufactures.*

[6425]

Acadian Charcoal Iron Company (Limited), 17 *New Church Street, Sheffield.*—Pig and bar iron, steel, and steel tools and cutlery.

[6426]

Allcock, Samuel, & Co., *Unicorn Works, Redditch, and* 121 *King Street, Toronto.*—Needles, fish hooks, and fishing tackle.

[6427]

Bessemer & Longsdon, 4 *Queen Street Place, New Cannon Street.*—Various specimens of Bessemer iron and steel. (*See page* 142.)

[6428]

Boulton, William, & Son, *Redditch.*—Needles for plain and ornamental work, fish-hooks for sea and river.

[6429]

Brandauer, C., & Co., 407 *New John Street West.*—Steel pens and pen holders.

[6430]

Brown, J., & Co., *Atlas Steel and Iron Works, Sheffield.*—Cast-steel rails, bent and twisted cold.

[6431]

Caldwell, Brothers, 15 *Waterloo Place, Edinburgh.*—Serpentine pen, overcomes greasiness of paper, marks easily, quill-like, and durable.

[6432]

Cammell, Charles, & Co., *Cyclop Works, Sheffield.*—Iron, steel, files, springs, forgings, railway materials.

[6433]

Dewsnap, J., 10 *St. Thomas Street, Sheffield.*—Leather and cabinet goods, dressing cases, &c.

BESSEMER & LONGSDON, 4 *Queen Street Place, New Cannon Street.*—Various specimens of Bessemer iron and steel.

SPECIMENS OF STEEL, ILLUSTRATIVE OF ITS APPLICATION TO VARIOUS PURPOSES.

The whole of the cast-steel employed in the manufacture of the various specimens exhibited was made by the Bessemer process, at the Works of Messrs. Henry Bessemer and Co. Sheffield, with the exception of the locomotive engine tyres, which were made by the same process direct from the fluid iron as it leaves the blast furnace, by M. F. Göranson, of Gefle, Sweden.

Cast-steel is a material possessing greater strength and elasticity than any other known metal, while its power to resist wear and abrasion, and its perfectly homogeneous character, render it greatly superior to wrought-iron for nearly every purpose to which that metal is now applied.

The cost of cast-steel as ordinarily made by melting blister steel or puddled steel in crucibles, is so great as to have hitherto confined its use within very narrow limits, although enough has been done to show its great superiority over wrought-iron.

All the cast-steel made in this country, as well as that made in France and Prussia, has after its original conversion into steel by a series of laborious and expensive processes, still to be melted in clay crucibles in quantities varying from 40 to 50 lbs. in weight, and thus it is only by the simultaneous fusion of hundreds of such crucibles of steel, and the skilful organization of a numerous staff of workmen, that the molten steel can be rapidly collected and conveyed from the numerous furnaces employed for its fusion, and be poured from the separate crucibles in an unbroken stream into the mould.

The Bessemer process, instead of requiring blister steel or puddled steel as the raw material or basis of its manufacture, operates at once upon molten pig-iron, and thus entirely dispenses with the whole of the engine power, skilled labour, and fuel expended in the several processes now employed in making blister steel or puddled steel.

The Bessemer process produces from the crude molten pig-iron in a single vessel, several tons of cast-steel in a period of 20 or 25 minutes, wholly without the employment of skilled labour or any species of manipulation, or the expenditure of any fuel.

The great changes wrought in the character and properties of the crude metal in this short interval, is simply the result of forcing numerous streams of air upwards through the fluid metal, whereby the oxygen contained in the atmosphere is brought in contact with the excess of carbon present in pig-iron, producing an intense combustion and an increase of heat beyond that which has ever been obtained in furnaces employing fuel. The perfect malleability of the metal so produced, will be at once perceived by an examination of the various specimens exhibited, many of which have been bent or twisted cold, in order to show the extreme toughness of the metal, and to what extent it will suffer a change of form without fracture.

The more prominent advantages of the Bessemer process may be briefly stated as follows :—

Masses of tough cast-steel from 10 to 30 or more tons in weight, can be made in half an hour from molten pig-iron.

Large marine engine cranks, shafts, ship's plates, rifled ordnance beams, and other massive parts of machinery, may be made in one piece of cast-steel, without weld or joint.

The tensile strength of this steel varies with the degree of carburation, and ranges between 40 and 70 tons per square inch, the tough qualities most suitable for engineering purposes being about 40 to 48 tons as against 21 tons for common wrought-iron, and 26 tons for the celebrated irons of Yorkshire.

The Bessemer steel is produced in a perfectly fluid state, and admits of being cast into various forms, such as heavy spur wheels, metal rolls, guns, mortars, projectiles, screw propellers, railway wheels, marine and other engine framing, hammer blocks, &c.

The apparatus now employed in the manufacture of Bessemer steel, is rendered almost self-acting under the control of one directing hand, who applies hydraulic force to effect every movement.

The cost of the complete apparatus for carrying on this process, including steam and blast engines, is considerably less than the mere furnaces required to melt an equal quantity of blister or puddled steel.

Apparatus for carrying out the new process on an extensive scale, is now in course of erection in different parts of England, in Scotland, France, Belgium, Prussia, Sweden, and the East Indies.

[6434]

GILLOTT, JOSEPH, *London and Birmingham.*—Specimens of metallic pens and penholders.

[6435]

GOODMAN, GEORGE, 82 *Caroline Street, Birmingham.*—Patent elastic pins and needles.

[6436]

HADFIELD & SHIPMAN, *Attercliffe Steel Wire Mills, Sheffield.*—Steel wire for crinoline umbrella ribs, ropes, fish-hooks, springs, &c.

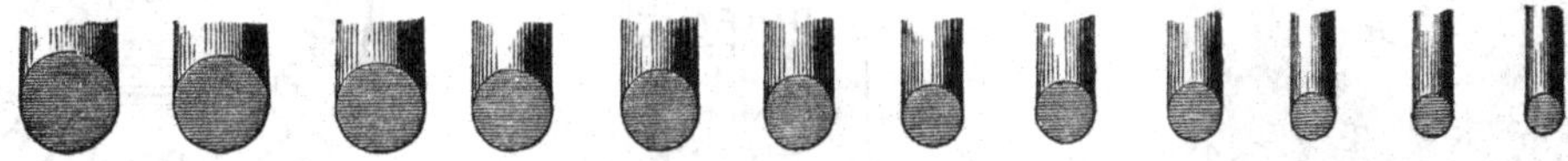

CAST-STEEL WIRE IN RINGS, IRON WIRE SIZES.

CAST-STEEL WIRE IN RINGS, IRON WIRE SIZES.

NEEDLE WIRE SIZES.

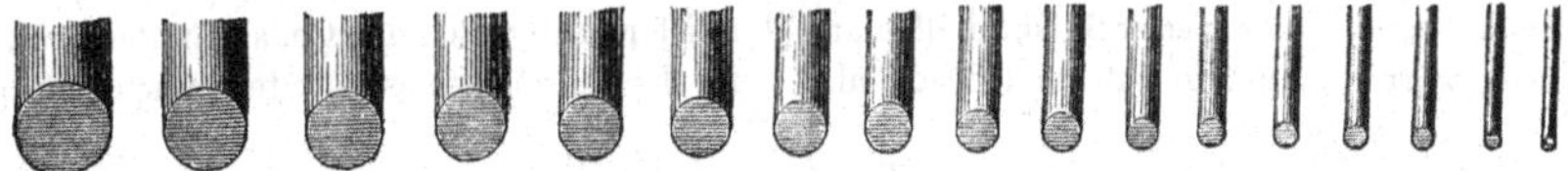

ROPE WIRE SIZES.

Hadfield & Shipman are general merchants and manufacturers of crinoline, hardened and tempered, and all sorts of cast-steel wire for needles, pins, fish-hooks, hackle pins, spiral springs, watch springs, &c.; also hardened and tempered cast-steel wire for parasol and umbrella ribs; ropes for deep pits, telegraphs, cables, &c.

[6437]

HINKS, WELLS, & CO., *Birmingham.*—Steel pens and penholders.

[6438]

HOEY, THOMAS, & CO., 25 *New Row West, Dublin.*—Pins and hair pins.

[6439]

HUTCHINSON, P., & SON, *Kendal.*—Fish hooks and fishing tackle.

[6440]

KIRBY, BEARD, & CO., 62 *Cannon Street West, E.C.*—Pins, needles, fish hooks, sewing cotton, and general warehousemen, &c. (*See page* 144.)

[6441]

KNIGHTS, WILLIAM, & CO., *Shaksperean Works, Stratford-on-Avon.*—Needles and mourning hatbands.

These drilled and egg-eyed needles are among the best that are manufactured. They possess the combined advantages of a brilliant polish, great elasticity, and an extreme smoothness of the eye which effectually prevents the cutting of the thread. Knights & Co. are the inventors of a registered needle box.

[6442]

LEWIS, HENRY, & SON, *Church Green East, and Queen Street, Redditch.*—Sewing needles and fish hooks.

KIRBY, BEARD, & CO., 62 *Cannon Street West, E.C.*—Pins, needles, fish hooks, sewing cotton, and general warehousemen.

Obtained the Prize Medal, London, 1851; *Paris,* 1855.

PINS.

NEEDLES.

SEWING COTTON.

TRADE MARKS.

PINS of very superior finish; solid heads and adamantine points, stuck on paper, loose in lbs., and in boxes.

SEWING COTTON, double spun, celebrated for smooth finish and softness, warranted the lengths marked on the reels, and the sizes are apportioned to the needles in the following order, viz.

Needles, No. 1 to 5, 6, 7, 8, 9, 10, 11, 12.
Cotton, No. 12 to 16, $\frac{20}{24}$, $\frac{30}{36}$, $\frac{40}{45}$, $\frac{50}{60}$, $\frac{70}{80}$, $\frac{90}{100}$, $\frac{120}{150}$.

NEEDLES of high temper and superior finish, drilled and burnished eyes, warranted not to cut the thread; also their celebrated egg-eyed needles, large convenient eyes, also needles for every kind of fancy work.

FANCY BOXES OF PINS, 4-paper boxes, in excellent photographic and other designs, and 1 oz. and 2 oz. boxes. Needles in rich fancy-paper boxes, and elegant morocco and silk velvet portfolios.

SEWING-COTTON BOXES, in photographic and other designs, 1 and 2 dozen boxes.

To ensure the articles being genuine, purchasers should ask for Kirby, Beard & Co.'s manufactures, as their trade marks are sometimes imitated on spurious goods.

[6443]

MILWARD, HENRY, & SONS, *Redditch.*—Needles and fish hooks; extra quality needles, specially manufactured.

Obtained a silver Medal at the Paris Exhibition, 1855, *and a First-Class Medal at the New York Exhibition.*

The exhibitors are patentees of the patent method of wrapping needles, and manufacturers of the registered needle case.

Amongst the specimens will be found the needles saleable in each quarter, and all the different countries of the globe, and in addition to the ordinary sewing needles, needles for tailors, milliners, saddlers, harness makers, stay and mattress makers, sail makers, sack makers, needles for surgeons and veterinary surgeons; needles for knitting, netting, darning, worsted darning, crochet, embroidery, chenille and tambour work; needles for carpets, and carpet needles, and every description of needles for sewing machines.

There are also fish-hooks and fishing tackle for all waters, at home and abroad.

The interest that has always attached itself to the manufacture of needles, has induced Messrs. Milward & Sons to exhibit beautifully finished models of the whole in the Process Court, of the machinery required, from which it will be easy to obtain an idea of each process through which the needle passes. Amongst the most interesting is the pointing, both on account of the great danger, indeed certain death, formerly attending it, and of the simple machine called the "fan," by which this has been overcome, and by which the dangerous particles of steel formerly inhaled are drawn away from the "pointer." Attention may be also called to the ingenious counting machine, for the use of which the exhibitors are licensed by the patentees, by which a great saving of time is effected.

For further particulars, reference should be made to the Process Court, and the South Kensington Museum.

[6444]

MITCHELL, WILLIAM, *Washington Works, Birmingham, and* 74 *Cannon Street West, London.*—Case of metallic pens and penholders.

[6445]

MITCHELL, W., 41 *London Street, Fitzroy Square.*—Springs.

[6446]

MOGG, JOSEPH, & CO., *Adelaide Works, Redditch.*—Needles, fish hooks, and fishing tackle

[6447]

Myers & Son, *Charlotte Street Steel Pen Works, Birmingham.*—Steel pens, holders, letter clips, paper knives, drapers' ticket suspenders, &c.

Obtained Medals at the Great Exhibition, 1851, *and at the Paris Exhibition*, 1855.

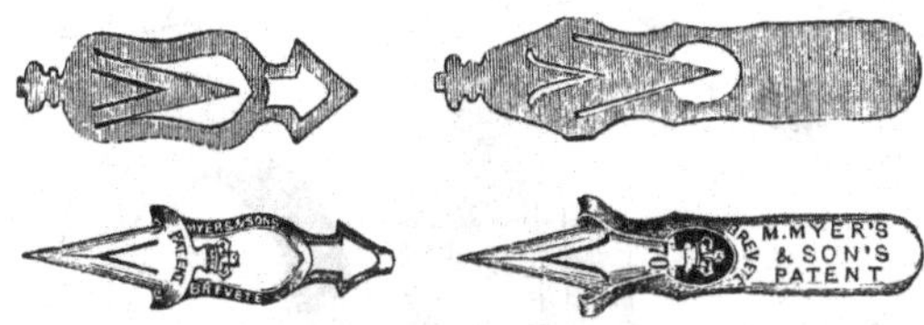

STEEL PENS.

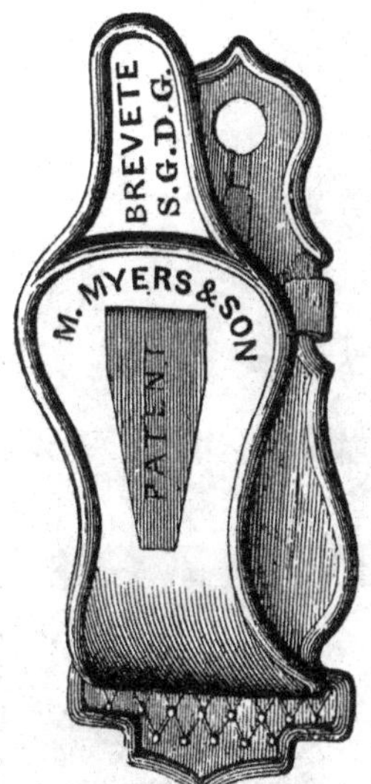

LETTER CLIP.

M. Myers & Son beg respectfully to call the attention of the public to their recently patented novelties in paper knives, book markers, and letter clips, which, for adaptation, utility, and elegance, stand prominently forward as a great step in advance of any that have yet been before the public, and at a price so exceedingly moderate, that when known, will claim general adoption.

M. Myers & Son would also take the present opportunity of sincerely thanking the public for the very flattering preference they have given to their galvanized pens, and beg to assure them that they still continue to manufacture them with the same care and attention, through their patented process, which has secured for them the enviable distinction of being a reliable pen, as, in their freedom of action they glide over the surface of the paper with that smoothness which is so desirable to the general writer, and, at the same time, resists for a much longer period the acidity of the ink. They confidently recommend them to the commercial world, and the public generally. Drapers, haberdashers, &c., would do well to try our price ticket suspenders, as, by an ingenious but simple contrivance, they can be attached and detached instantly without the least injury to the most delicate article. Secured by letters patent.

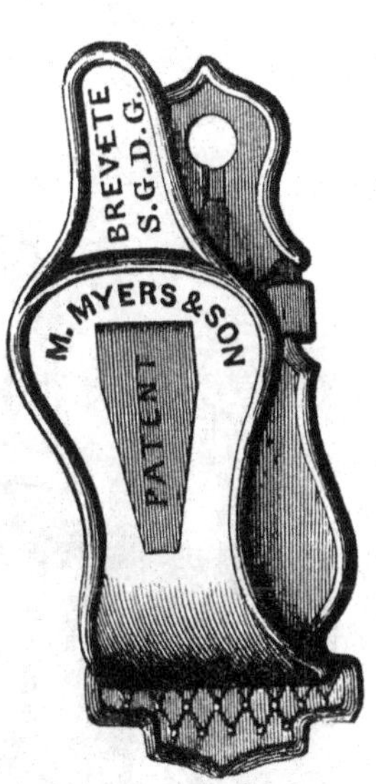

LETTER CLIP.

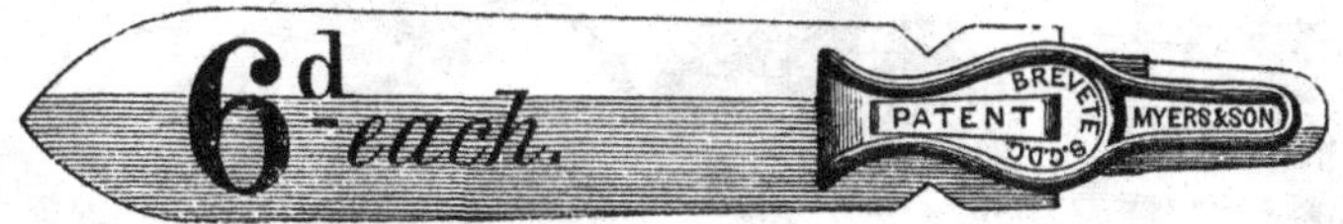

PAPER KNIFE.

No. 1. Metallic Pens, various.

No. 2. Patent Axissary Pens.

The point of this pen is formed in the body and twisted over, by which means an axis is formed on each side of the pen, upon which the point works, and a new and an agreeable elasticity is produced.

No. 3. Pen Holders, various.

No. 4. Skeleton Points, for quills.

By this patented invention, all the superior qualities of the steel pen are secured, combined with the acknowledged advantages which are only to be found in the natural quill.

No. 5. Gilt, Silvered, and Galvanized Pens.

No. 6. The Patented Gauge Pens.

The point and the body of this pen are made separately, and so connected, that the one slides within the other, and by moving the point to a given gauge, any required elasticity is attained.

No. 7. Paper Knives and Book Markers, elegance, simplicity, and utility.

No. 8. Letter Clips for adaptation, utility, and price, stand prominently forward as a great step in advance of any before the public.

No. 9. Price-ticket Suspenders, for drapers, haberdashers, &c. By an ingenious yet simple contrivance they can be attached or detached instantly, without the least injury to the most delicate fabric.

No. 10. Paper Holders, for stationers, &c.

No. 11. Railway Safety Ticket Holders.

No. 12. Crinoline Fasteners.

[6448]

NAYLOR, VICKERS, & CO., *Sheffield.*—Cast-steel disc wheels, tyres, crank and straight axles, castings to pattern.

STEEL CASTINGS AND FORGINGS.

Makers of all descriptions of bar, rod, and sheet steel, and of wrought cast-steel railroad tyres without weld.

Patent corrugation cast-steel disc wheels, with tyres in one solid piece.

Cast-steel crank and straight axles.

Cast-steel ordnance; cast-steel boiler plates.

Heavy cast-steel forgings and castings in steel to patterns.

For price lists, apply to Naylor, Vickers, & Co. River Don Works, Sheffield; 80 Lombard Street, London; 4 Cook Street, Liverpool; 99 John Street, New York; 80 State Street, Boston, U.S.; 425 Commerce Street, Philadelphia.

NAYLOR, VICKERS, & CO., *continued.*

CAST-STEEL BELLS.

Cast-steel bells are stronger and more durable than bronze bells. Cast-steel bells of the same note and volume of sound, are only two-thirds the weight, and one-third the cost, of bronze bells. For example, the peal of eight cast-steel bells (shown in the above sketch), tenor 54 in. diameter, key E, weighs about 8,000 lbs.

Price £300

The largest bell exhibited by Naylor, Vickers, & Co. is 7 ft. 6 in. diameter, note G, 9,000 lbs. Price . £300

A list of the peals of cast-steel bells in use in Great Britain, with testimonials; also estimates for peals or single bells, may be obtained by applying to Naylor, Vickers, & Co. River Don Works, Sheffield; 80 Lombard Street, London; 4 Cook Street, Liverpool; 99 John Street, New York; 80 State Street, Boston, U.S. 425 Commerce Street, Philadelphia.

THOMAS, SAMUEL, & SONS, *British Needle Mills, Redditch.*—Improved spring steel needles and fish hooks.

A VISIT TO THE BRITISH NEEDLE MILLS, REDDITCH.

WHY are needles made at Redditch? Why should a beautiful and secluded part of the county of Worcester, many miles distant from what are termed the "manufacturing districts," contain a village, whose inhabitants, one and all, live directly or indirectly by making these little steel implements? The fact is demonstrable, but the reason is not. The good housewife who mends her child's pinafore, the milliner who decks out a lady in her delicate attire, the hard-working sempstress who supplies "made-up goods" to the shops, the school girl who works her sampler—all, however little they may be aware of the fact, are dependent principally on a Worcestershire village for the supply of their needles. Their "Whitechapel needles" are no longer made at Whitechapel, even if they ever were; and though they may in some cases seem to emanate from London manufacturers, the chances are that they were made at Redditch. Not that other towns are without indications of this branch of manufacture; but in them it is merely an isolated feature, while at Redditch, as we shall presently see, needle-making is the staple, the all-in-all, without which, almost every house in the place would probably be shut up; for although there is a fair sprinkling of the usual kind of workmen, shopkeepers, dealers, &c. these are only such as are necessary for supplying the wants of the needle-making population. It is a strange thing that the Redditch manufacturers themselves seem scarcely able to assign a reason why this branch of industry has centred there, or to name the period of its commencement. Indeed, the early history of the needle-trade is very indistinctly recorded. Stow tells us while speaking of the kind of shops found in Cheapside and other busy streets of London, that needles were not sold in Cheapside until the reign of Queen Mary, and that they were at that time made by a Spanish negro, who refused to discover the secret of his art. Another authority states that "needles were first made in England by a native of India, in 1545, but the art was lost at his death; it was, however, recovered in 1650, by Christopher Greening, who settled with his three children at Long Crendon, in Buckinghamshire." Whether the negro in one of these accounts is the same individual as the native of India mentioned in the other, cannot now be determined, nor is it more clear at what period Redditch became the centre of the manufacture. There are slight indications of Redditch needle-making for a period of two centuries, but beyond that all is blank.

A reader, who associates the potteries with the clay districts of North Staffordshire, and the smelting works with the coal and iron districts of South Staffordshire, will naturally seek to know whether any features distinguish Redditch which will enable us to assign a probable origin for the needle-manufacture there. A visitor, in any degree accustomed to watch the progress

THOMAS, SAMUEL, & SONS, *continued.*

of manufactures, looks around him to seek for any indications whence he may account for the location of needle-making; he looks for a stream or canal, or something which may be to the manufacture in the relation of cause to effect; but very little of the kind is seen. Needle-making is nearly all the result of manual dexterity, requiring little aid from water or steam power. There are, it is true, a few water wheels employed for pointing and scouring the needles, but Redditch presents no other facilities for this purpose than such as are presented by a thousand other places in the kingdom. In short, there seems to be no other mode of accounting for the settlement of the needle-manufacture in this spot, than that which may be urged in reference to watchmaking in Clerkenwell, or coach-making in Long Acre. A needle-maker we will suppose—say two centuries ago—settled at Redditch, and gradually accumulated round him a body of workmen. A supply of skilled labour having been thus secured, another person set up in the same line. In time, the workmen's children learned the occupation carried on by their parents, and thus furnished an increased supply of labour, which in its turn, led to the establishment of other manufacturing firms. By degrees so many needles were made at Redditch, that the village acquired a reputation throughout the length and breadth of the land for this branch of manufacture, and hence it became a positive advantage for a maker to be able to say that his needles were "Redditch needles." This train of surmises may perhaps approach pretty nearly to the truth.

Let us, however, leave conjecture and proceed to facts. There are in Redditch about half-a-dozen manufacturers who conduct the needle-manufacture on a large scale, and employ a considerable number of persons. Some work in factories built by and conducted under the superintendence of the master manufacturers; while others work at their own homes. In no occupation, perhaps, is the division of labour more strictly carried out than in needle-making; for the man who cuts the wire does not point, nor does the pointer make the eyes or polish the needles. Both within and without the factory the same system of division is kept up; for a cottager who procures work from a needle-manufacturer does not undertake the making of a needle, but only one particular department, for which he is paid at certain recognised prices. Many of the workpeople live a few miles distant and come with their work at intervals of a few days, a plan which can be adopted without much inconvenience, since a considerable quantity of these little articles may be packed in a small space. It is, we believe, estimated that the number of operatives in Redditch is about three thousand, and in the whole district of which Redditch is the centre, six or seven thousand, of whom a considerable number are females.

The general name of "mills" is given to the needle-factories, each one having some distinctive name whereby it may be indicated. Thus the establishment which we have been obligingly permitted to visit, and the arrangements of which will be here described, is called the "British Needle Mills." To the British Needle Mills of S. Thomas & Sons, then, our visit is directed.

This factory has been recently constructed, and is situated at one extremity of the village. It consists of a number of court-yards or quadrangles, each surrounded by buildings wherein the manufacture is carried on. The object of this arrangement seems to be to obtain as much light as possible in the wo kshops, since most of the departments of needle-making require a good light. Some of the rooms in the factory are small, containing only three or four men; while others contain a great many workmen, according to the requirements of the several processes of the manufacture. From the upper rooms of the factory, the surrounding hilly districts of Worcestershire are seen over a wide extent, wholly uninterrupted by any indications of manufacture or town bustle; and it is while glancing over this prospect that one wonders how on earth needle-making came to speckle such a scene.

The sub-divisions of the factory correspond with those in the routine of manufacture, and we accordingly find that, while some of the shops are occupied by men, others contain only females, and others again furnish employment chiefly for boys. We should surprise many a reader were we to enumerate all the processes incident to the manufacture of a needle, giving to each the technical name applied to it in the factory. The number would amount to somewhere about thirty, but it will be more in accordance with our object to dispense with such an enumeration, and to present the details of manufacture in certain groups, without adhering to a strictly technical arrangement.

First, then, for the material. It is scarcely necessary to say that needles are made of steel, and that the steel is brought into the state of wire before it can assume the form of needles. The needle-makers are not wire-drawers; they do not prepare their own wire, but purchase it in sizes varying with the kind of needles which they are about to make. We will suppose, therefore, that the wire is brought to the needle factory and deposited in a store-room. This room is kept warmed by hot air to an equable temperature, in order that the steel may be preserved free from damp or other sources of injury. Around the walls are wooden bars or racks, on which are hung the hoops of wire. Each hoop contains what is called a packet, the length varying according to the diameter. Perhaps it may be convenient to take some particular size of needle and make it our standard of comparison during the details of the process. The usual sizes of sewing needles are from No. 1, of which twenty-two thicknesses make an inch, to No. 12, of which there are a hundred to an inch. Supposing that the manufacturer is about to make sewing needles of that size known as No. 6, then the coil of wire is about two feet in diameter; it weighs about 13 lbs.; the length of wire is about a mile and a quarter; and it will produce forty or fifty thousand needles. The manufacturer has a gauge, consisting of a small piece of steel, perforated at the edge with eighteen or twenty small slits, all of different sizes, and each having a particular number attached to it. By this gauge the diameter of every coil of wire is tested, and by the number every diameter of wire is known.

A coil of wire when about to be operated upon, is carried to the "cutting shop," where it is cut into pieces equal to the length of two needles. Fixed up against the wall of the shop is a ponderous pair of shears, with the blades uppermost. The workman takes probably a hundred wires at once, grasps them between his hands, rests them against a gauge to determine the length to which they are to be cut, places them between the blades of the shears, and cuts them by pressing his body or thigh against one of the handles of the shears. The coil is thus reduced to twenty or thirty thousand pieces, each about three inches long, and as each piece had formed a portion of a curve two feet in diameter, it is easy to see that it must necessarily deviate somewhat from the straight line. This straightness must be rigorously given to the wire before the needle-making is commenced, and the mode by which it is effected is one of the most remarkable in the whole manufacture. Around the walls of the shop we see a number of iron rings hung up, each from three or four to six or seven inches in diameter, and a quarter or half an inch in thickness. Two of these rings are placed upright on their edges at a little distance apart, and within them are placed many thousands of wires, which are kept in a group by resting on the interior edges of the two rings. In this state they are placed on a shelf in a small furnace, and there kept till red hot. On being taken out at a glowing heat, they are placed on an iron plate, the wires being horizontal and the rings in which they are inserted being vertical. The process of "rubbing" (the technical name for the straightening to which we allude then) commences. The workman, as here represented, takes a long piece of iron, and inserting it between the two rings, rubs the wires backwards and forwards, causing

THOMAS, SAMUEL, & SONS, *continued.*

THE PROCESS OF "RUBBING."

each to roll over on its own axis, and also over and under those by which it is surrounded. The noise emitted by this process is just that of filing; but no filing takes place, for the rubber is smooth, and the sound arises from the rolling of one wire against another. The rationale of the process is this:—the action of one wire on another brings them all to a perfectly straight form, because any convexity or curvature in one wire would be pressed out by the close contact of the adjoining ones. The heating of the wires facilitates this process, and the workman knows by the change of sound, when all the wires have been "rubbed" straight.

Our needles have now assumed the form of perfectly straight pieces of wire, say a little more than 3 in. in length, blunt at both ends, and dulled at the surface by exposure to the fire. Each of these pieces is to make two needles, the two ends constituting the points; and both points are made before the piece of wire is divided in two. The pointing immediately succeeds the rubbing, and consists in grinding down each end of the wire till it is perfectly sharp. The workman sits on a stool or "horse" a few inches distant from the stone, and bends over it during his work. He takes fifty or a hundred wires in his hand at once, and holds them in a peculiar manner. He places the fingers and palm of one hand diagonally over those of the other, and grasps the wires between them, all the wires being parallel. The thumb of the left hand comes over the back of the fingers of the right, and the different knuckles and joints are so arranged, that every wire can be made to rotate on its own axis, by a slight movement of the hand, without any one wire being allowed to roll over the others. He grasps them so that the end of the wires (one end of each) projects a small distance beyond the edge of the hand and fingers, and these ends he applies to the grindstone in the proper position for grinding them down to a point. It will easily be seen, that if the wires were held fixedly, the ends would merely be bevelled off, in the manner of a graver, and would not give a symmetrical point; but by causing each wire to rotate while actually in contact with the grindstone, the pointer works equally on all sides of the wire, and brings the point in the axis of the wire. At intervals of every few seconds, he adjusts the wires to a proper position against an iron plate, and dips their ends in a little trough of water between him and the grindstone. Each wire sends out its own stream of sparks, which ascends diagonally in a direction opposite to that at which the workman is placed. So rapid are his movements, that he will point seventy or a hundred needles, forming one hand-grasp, in half a minute, thus getting through ten thousand in an hour.

The reader will bear in mind, that the state of our embryo needle is simply that of a piece of dull straight wire, about 3 in. long (supposing 6's to be the size), and pointed at both ends. The next process is one of a series by which two eyes or holes are pierced through the wire, near the centre of its length, to form the eyes of the two needles which are to be fashioned from the piece of wire. A number of very curious operations are connected with this process, involving mechanical and manipulative arrangements of great nicety. Those who are learned in the qualities of needles—as that they will not "cut in the eye" and so forth—will be prepared to expect that much delicate workmanship is involved in the production of the eyes, and they will not be in error in so supposing. Most of the improvements which have from time to time been introduced in needle-making, relate more or less to the production of the eye. In the commoner kinds of needles, many processes are omitted which are essential to the production of the finer qualities; but it will show the whole nature of the operations better for us to take the case of those which involve all the various processes.

After being examined, when the pointer has done his portion of the work to them (an examination which is undergone after every single process throughout the manufacture), the wires are taken to the "stamping shop," where the first germ of an eye is given to each half of every wire. The stamping machine consists of a heavy block of stone, supporting on its upper surface a bed of iron, and on this bed is placed the under half of a die or stamp. Above this is suspended a hammer, weighing about 30 lbs. which has on its lower surface the other half of the die or impress. The hammer is governed by a lever moved by the foot, so that it can be brought down exactly upon the iron bed. The form of the die or stamp may be best explained by stating the work which it is to perform. It is to produce the "gutter" or channel in which the eye of the needle is situated, and which is to guide the thread in the process of threading a needle.

But besides the two channels or gutters, the stampers make a perforation partly through the wires, as a means of marking exactly where the eye is to be. The device on the two halves of the die is consequently a raised one, since it is to produce depressions in the wire. The workman holding in his hand several wires, drops one at a time on the bed-iron of the machine, adjusts it to the die, brings down the upper die upon it by the action of the foot, and allows it to fall into a little dish when done. This he does with such rapidity that one stamper can stamp 4,000 wires, equivalent to 8,000 needles, in an hour, although he has to adjust each needle separately to the die.

To this process succeeds another, in which the eye of the needle is pierced through. This is effected by boys, each of whom works at a small hand-press, and the operation is at once a minute and ingenious one. The boy takes up a number of needles or wires, and spreads them out like a fan. He lays them flat on a small iron bed or slab, holding one end of each wire in his left hand, and bringing the middle of the wire to the middle of the press. To the upper arm of the press are affixed two hardened steel points or cutters, being in size and shape exactly corresponding with the eyes which they are to form. Both of these points are to pass through each wire, very nearly together, and at a small distance on either side of the exact centre of the wire. The wire being placed beneath the points, the press is moved by hand, the points descend, and two little bits of steel are cut out of the wire, thereby forming the eyes for two needles. As each wire becomes thus pierced, the boy

THOMAS, SAMUEL, & SONS, *continued.*

shifts the fan-like array of wires until another one comes under the piercers, and so on throughout. The press has to be worked by the right hand for piercing each wire, and the head of the boy is held down pretty closely to his work, in order that he may see to "eye" the needles properly. Were not the wires previously prepared by the stamper, it would be impossible thus to guide the piercers to the proper point, but this being effected, patience, good eye-sight, and a steady hand effect the rest.

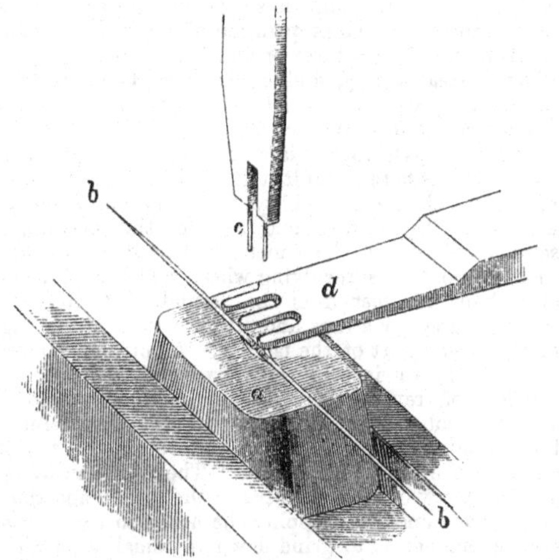

(*a* is the lower die on which the needles *b* are placed, to be pierced by the points *c*, guided by the apparatus *d.*)

There are several processes about this stage which are effected by boys; some of these little labourers take the needles when they have been "eyed" and proceed to "spit" them, that is, to pass a wire through the eye of every needle. Two pieces of fine wire, perhaps three or four inches in length, are prepared, the diameter corresponding exactly with the size of the needle eye. These two pieces of wire are held in the right hand, parallel, and at a distance apart equal to the distance between the two eyes in each needle-wire. The pierced needles, being held in the left hand, are successively threaded upon the two pieces of smaller wire, till, by the time the whole is filled, the assemblage has something the appearance of a fine toothed comb. A workman then files down the bur or protuberances left on each side of the eye by the stamper.

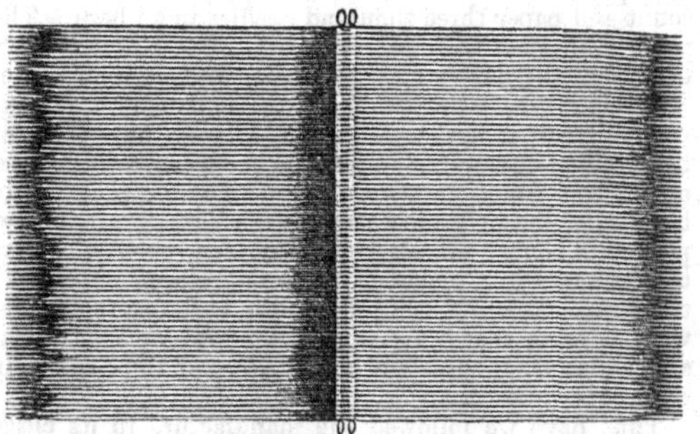

THE WIRE "SPITTED."

It must be borne in mind that throughout all these operations the needles are double; that is, that the piece of wire three inches in length, which is to produce two needles an inch and a half long each, is still whole and undivided, the two eyes being nearly close together in the centre, and the two points being at the ends. Now, however, the separation is to take place. The filer, after he has brought down the protuberances on each wire, but before he has laid the comb of wires out of his hand, bends and works the comb in a peculiar way until he has broken the comb into two halves, each half "spitted by one of the fine wires. The needles have arrived at something like their destined shape and size, for they are of the proper length and have eyes and points. In the annexed cut we can trace the wire through the processes of change hitherto undergone.

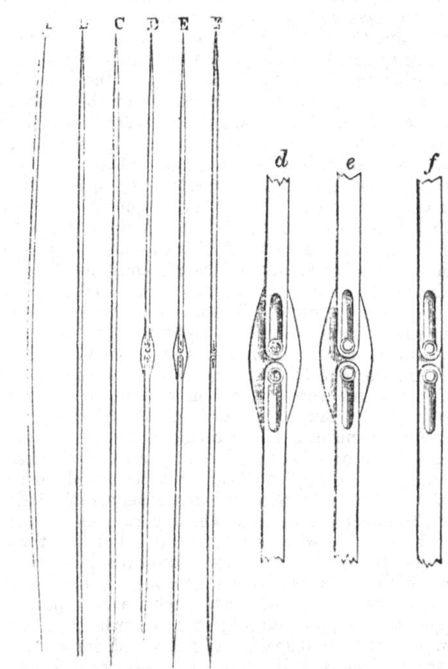

(*A* the wire for two needles; *B* the same, pointed at one end; *C* pointed at both ends; *D* the stamped impress for the eyes; *E* the eyes pierced; *F* the needles just before separation; *d, e, f,* enlargements of *D, E, F.*)

But although we have now little bits of steel which might by courtesy be called needles, they have very many processes to undergo before they are deemed finished, especially if, in accordance with our previous supposition, they are of the finer quality.

The needles are by this time pointed and eyed, but before they can be brought to that beautifully finished state with which we are all familiar, it is necessary that they should be "hardened" and "tempered" by a peculiar application of heat. After being examined to see that the preceding processes are fitly performed, the needles are taken to a shop provided with ovens or furnaces. They are laid down on a bench, and by means of two trowel-like instruments spread in regular thick layers on narrow plates or trays of iron. In this way they are placed on a shelf or grating in a heated furnace. When the proper degree of heating has been effected, the door is opened and the needles are shifted from the iron tray into a sort of colander or perforated vessel immersed in water or oil. When they are quite cooled the hardening is completed, and if it has been effected in water the needles are simply dried; but if in oil, they are well washed in an alkaline liquor to free them from the oil. Then ensues the tempering process. The needles are placed on an iron plate, heated from beneath and moved about with two little trowels until every needle has been gradually brought to a certain desired temperature.

We now leave the furnace-room and proceed to one of the upper rooms of the factory, where a multitude of minor operations are conducted. The needles have become slightly distorted in shape by the action of the heat in the processes just described, and to rectify this they

THOMAS, SAMUEL, & SONS, *continued.*

undergo the operation of "hammer straightening." A number of females are seen seated at a long bench, each with a tiny hammer, giving a number of light blows to the needles; the needles being placed on a small steel block with a very smooth upper surface. This is rather a tedious part of the manufacture, the workwomen not being able to straighten more than five hundred needles in an hour, a degree of quickness much less than that which we have had hitherto to notice.

We leave the tinkling hammers and follow the needles to the only part of the manufacture which involves apparatus other than of a small size. This is the "scouring" process. In one of the lower rooms of the factory are machines looking like mangles, or, perhaps, more correctly like marble polishing machines, a square slab or rubber working to and fro on a long bench. The object of this process is to rub the needles one against another for a very long period, till the surfaces of all have become perfectly smooth, clean, and true. This is effected in a curious manner. A strip of thick canvas is laid open in a small hollow tray, and on this a heap of needles is laid, all the needles being parallel one with another, and with the length of the cloth. The needles are then, with soft soap, emery, and oil, tied up tightly in the canvas, the whole forming a compact roll about two feet long and three inches in thickness; these are placed under the runners of the scouring machines, two rolls to each machine. A steam engine gives to the runners, by connected mechanism, a reciprocating or backward and forward motion, pressing heavily on the rolls of needles, and causing all the needles of each bundle to roll one over another. By this action an intense degree of friction is exerted among the needles, whereby each one is rubbed smooth by those which surround it. For eight hours uninterruptedly this rubbing or scouring is carried on, after which the needles are taken out, washed in suds, placed in new pieces of canvas, with a new portion of soap, emery, and oil, and subjected to another eight hours' friction. Again and again is this repeated, insomuch that for the best needles the process is performed five or six times over, each time during eight hours' continuance. This is one of the points in which the difference is shown between various qualities of needles, the length of the scouring being correspondent with the excellence of the production.

Again we accompany the needles to another part of the factory, being that which is technically termed the "bright shop," in which many processes are carried on in reference to the finishing of needles. The needles are examined after being scoured and are placed in a small tin tray, where, by shaking and vibrating in a curious manner, they are all brought into parallel arrangement. From thence they are removed into flat paper trays in long rows or heaps, and passed on to the "header," generally a little girl, whose office is to turn all the heads one way, and all the points the other. This is one among the many simple but curious processes involved in this very curious manufacture, which surprise us by the rapidity and neatness of execution. The girl sits with her face towards the window, and has the needles ranged in a row or layer before her, the needles being parallel with the window. She draws out laterally to the right those which have their eyes on the right hand, into one heap, and to the left those which have their eyes in that direction, in another heap.

About this time, too, the needles are examined one by one, to remove those which have been broken or injured in the long process of scouring; for it sometimes happens that as many as eight or ten thousand, out of fifty thousand, are spoiled during this operation. Most ladies are conversant with the merits of "drilled-eyed needles" warranted "not to cut the thread." These are produced by a modern improvement, whereby the eye, produced by the stamping and piercing processes before described, is drilled with a very fine instrument, by which its margin becomes as perfectly smooth and brilliant as any other part of the needle. To effect this, the needle is first "blued;" that is, the head is heated so as to give it the proper temper for working. Next comes the drilling. Seated at a long bench are a number of men and boys, with small drills working horizontally with great rapidity. The workman takes up a few needles between the finger and thumb of his left hand, spreads them out like a fan, with the eyes uppermost, brings them one at a time opposite the point of the drill, and drills the eye, which is equivalent to making it even, smooth and polished. He moves the thumb and finger, so as to bring the opposite side of the needles, in succession, under the action of the drill; and thus gets through his work with much rapidity. The preparation of the drills, which are small pieces of steel three or four inches long, is a matter of very great nicety, and on it depends much of that beauty of production which constitutes the pride of a modern needle-manufacturer.

We next pass into a large room (see illustration on page 148), where a multitude of little wheels are revolving with great rapidity, some intended for what is termed "grinding" and setting the needles, and some for polishing. The men are seated on low stools, each in front of a revolving wheel, which is at a height of perhaps two feet from the ground. All the wheels are connected by straps and bands with a steam-engine in the lower part of the factory. A constant humming noise is heard in the room, arising from the great rapidity of revolution among a number of wheels; and it is not difficult for the ear to detect a difference of tone or pitch among the associated sounds, due to differences in the rate of movement. The workman takes up a layer or row of needles, between the fingers and thumbs of the two hands, and applies the heads to the stones in such a manner as to grind down any small asperities on the surface. As the small grindstones are revolving three thousand times in a minute, it is plain that the steel may soon be sufficiently worn away by a slight contact with the periphery of the stone.

The grinders and the polishers sit near together, so that the latter take up the series of operations as soon as the former have finished. The polishing wheels consist of wood coated with buff leather, whose surface is slightly touched with polishing paste. Against these wheels the polishers hold the needles, applying every part of the cylindrical surface in succession; first holding them by the pointed end, and then by the eye end. About a thousand in an hour can thus be polished by each man; and, when they leave his hands, the needles are finished.

We have still to see the needles papered. In one of the rooms a number of females are cutting the papers, separating the needles into groups of twenty-five each, and folding them into the neat oblong form so well known to all users of a "paper of needles." So expert does practice render the workwomen, that each one can count and paper three thousand needles in an hour. The papered needles then pass to another room, where boys paste on the labels bearing the manufacturer's name. Even here there are sundry little contrivances for expediting the process, which would scarcely be looked for by common observers. When the papers have been dried on an iron frame, in a warm room, they are packed into bundles of ten or twenty papers each; which are further packed in square parcels containing ten, twenty, or fifty thousand needles, inclosed, if for exportation, in soldered tin cases. As a means of judging the bulk of the needles, we may state that ten thousand 6's form a packet about six inches long, three and a half wide, and under two in thickness.

Thus have we followed the manufacture to its close. None but the best needles undergo the whole of the processes enumerated; but we have wished to give them as a means of estimating the complexity of the manufacture of an article apparently so humble.

The arrangements of the "British Needle Mills," as to apparatus, &c. are adapted to the production of two hundred millions of best needles per annum. These are startling results, and show that, in considering the seats of manufacture in England, we must not forget to include the remarkable Worcestershire village of Redditch.

[6450]

PAGE, W. & J., 70 & 71 *Mott Street, Birmingham.*—Specimens of corkscrews and other steel toys.

[6451]

PEACE, JOSEPH, & Co., *Sheffield.*—Saws; bright rolled steel, and saws from same; crinoline; steel busks.

[6452]

PERRY, JAMES, & Co., 37 *Red Lion Square, and* 3 *Cheapside.*—Steel and gold pens, penholders, and stationers' sundries.

[6453]

REYNOLDS, G. W., & Co., 12 *Cheapside, London; and Birmingham.*—Steel wire, crinoline steel, umbrella frames, &c.

[6454]

ROWELL, JEREMIAH, 7 *St. Alban's Row, Carlisle.*—Artificial flies and fish hooks.

[6455]

SCHELHORN, G., 158 *Hockley Hill, Birmingham.*—Penholders of every description.

[6456]

SHORTRIDGE, HOWELL, & Co., *Hartford Street Works, Sheffield.*— Steel files and general articles.

SHORTRIDGE, HOWELL, & Co., merchants, steel converters and refiners, forge, tilts, and rolling mills.

Description of steel, &c. exhibited—

Best cast-steel in bars; cast-steel rolls; Howell's patent homogeneous metal in bars and sheets. Homogeneous metal tubes for boilers, and samples of same bent in various forms, showing the extreme ductility of the metal; also bars of the same metal bent cold in various forms; glass case containing specimens of the ends of cast-steel ingots, showing the fracture up to 10 inches square; and glass case containing samples of best cast-steel files for engineers' and machinists' purposes; cast-steel shell to be charged with molten iron on Martin's principle; Howell's patent coupling chains for railway carriages, &c. in one continuous coil.

London Offices: 39 Bedford Street, Strand; New York Store, 24 Cliffe Street.

[6457]

SMITH, JOHN WRIGHT, *High Cross Street, Leicester.*—Patent self-acting and other hosiery needles.

[6458]

SMITH & HOUGHTON, *Warrington.*—Superior qualities of pianoforte wire, pinion and round steel wire, &c.

[6459]

SOMMERVILLE, A., & Co., *Birmingham.*—Best-class carbonized patent regulator, enamelled, gilt-pointed, and other steel pens.

SLIDE UP—very flexible.

SLIDE IN THE MIDDLE—medium flexibility.

SLIDE DOWN—hard.

Registered metal spring pattern card, showing 708 different numbers of steel pens, patent and other steel pen boxes and cases.

Best class carbonized steel pens. Celebrated gilt pointed steel pens on white and blue steel.

A large model, moved by machinery, of A. SOMMERVILLE & Co.'s patent regulating spring slide pens. By moving the spring slide up or down the pen, every degree of flexibility is obtained.

TURTON, THOMAS, & SONS, *Sheaf and Spring Works, Sheffield;* 17 *King William Street, City, London;* 10 *Rue du Grand Chantier, Paris;* 83 *John Street, New York;* 3 *North Street, Fifth Street, Philadelphia.*—Steel files, edge tools, railways springs, &c.

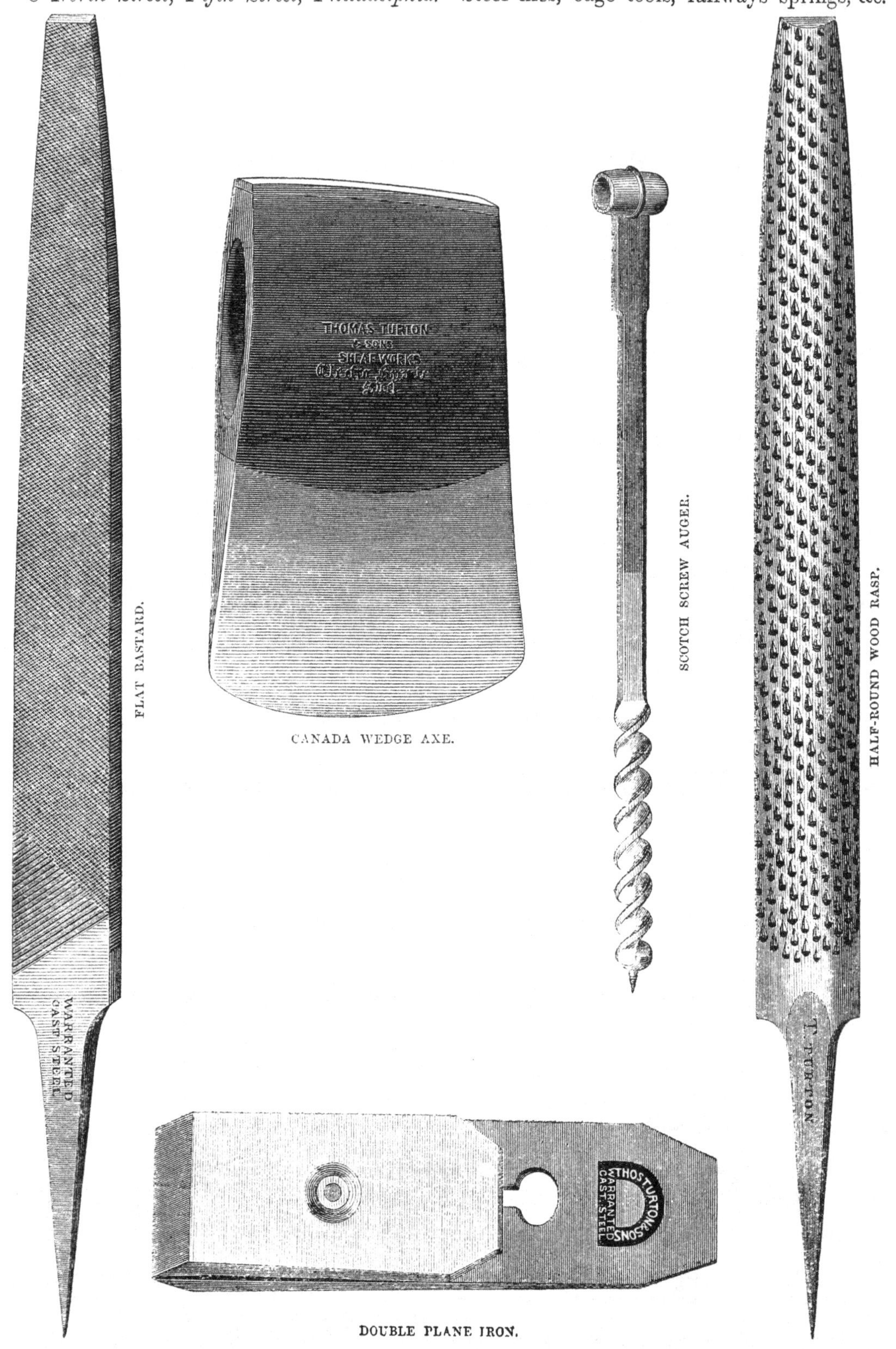

FLAT BASTARD.

CANADA WEDGE AXE.

SCOTCH SCREW AUGER.

HALF-ROUND WOOD RASP.

DOUBLE PLANE IRON.

TURTON, THOMAS & SONS, *continued.*

CAST-STEEL SLIDE BAR.

CAST-STEEL SLIDE BAR.

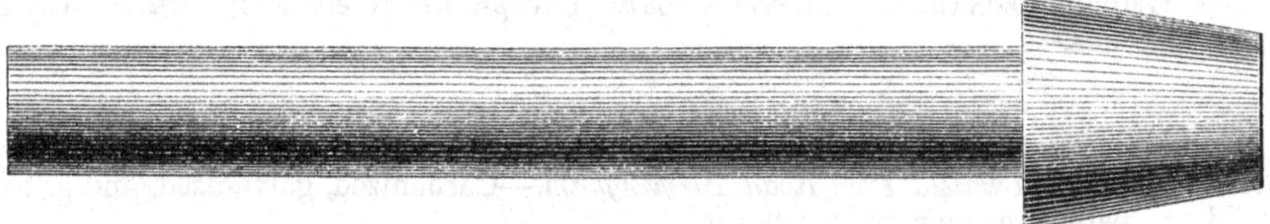

CAST-STEEL PISTON ROD.

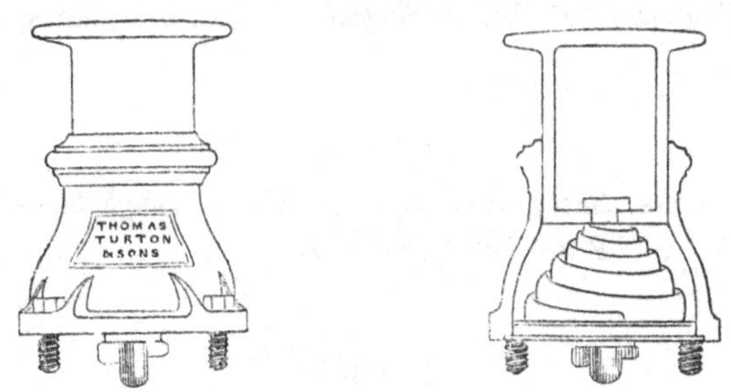

RAILWAY BUFFER WITH PATENT CONICAL SPRING.

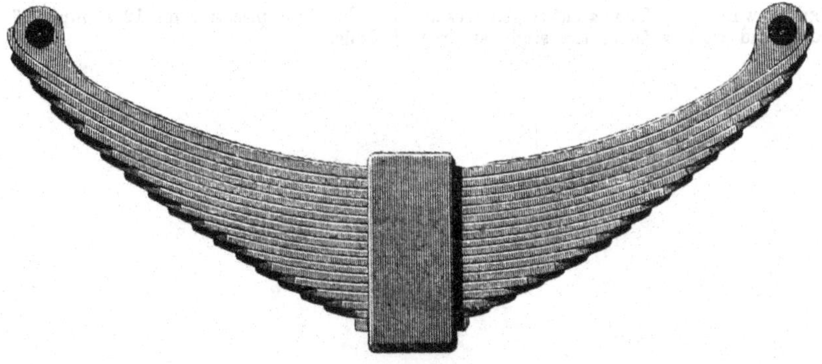

LOCOMOTIVE SPRING.

[6460]

SPENCER, JOHN, & SONS, *Newcastle-on-Tyne.*—Cast-steel tyres, volute spring buffers, springs, steel, and files.

[6461]

THOMAS, SAMUEL, & SONS, *British Needle Mills, Redditch.*—Improved steel spring needles and fish hooks. (*See pages* 148 *to* 152.)

[6462]

TOWNSEND, GEORGE, & CO., *Redditch and London.*—Machine and other needles, and tools for making them.

[6463]

TURNER, R., & CO., *London; and Old Factory, Redditch.*—Sewing machine and other needles, and tools for making them.

[6464]

TURNEY, GEORGE LEONARD, 20 *Lawrence Lane, Cheapside.*—Needles in cylinder cases and pins in books.

[6465]

TURNOR, M., & CO., *Icknield Port Road, Birmingham.*—Carbonized, galvanized, and gilt pens; holders; patent rectangular pen-boxes.

[6466]

TURTON, BROTHERS, *Phœnix Steel Works, Sheffield.*—Steel files, saws, engineering tools, &c.

[6467]

TURTON, THOMAS, & SONS, *Sheaf and Spring Works, Sheffield.*—Steel files, edge tools, railway springs, &c. (*See pages* 154 *and* 155.)

[6468]

WALKER, HENRY, 47 *Gresham Street, and Alcester.*—Patent ridged and other needles, fish hooks, &c.

H. WALKER's new needles. The patent ridged eyes are easily threaded, and work without the slightest drag. A hundred post-free for 12 stamps, of any respectabl dealer.

SUB-CLASS B.—*Cutlery and Edge Tools.*

[6480]

ADDIS, JAMES BACON, 159 *Waterloo Road.*—Carvers', carpenters', print cutters', engravers' masons', plasterers', turning sculptors', and geological screw tools and saws.

[6481]

ADDIS, SAMUEL JOSEPH, 49 *and* 50 *Worship Street, Shoreditch.*—Assortment of carvers' and general edge tools.

[6482]

ALLARTON, THOS., & POWELL, *Birmingham.*—Awls, and sewing-machine needles, of every size and shape.

[6483]

BADGER, CHARLES, 1 *Stangate, Lambeth, S.*—Planes in iron and gun-metal, for joiners, cabinet makers, &c.

[6484]

BAKER, W., *Pembroke Street, London, N.*—Awls, trade bodkins, and needles.

[6485]

BARKER, ROBERT, & SON, *Easingwold.*—Butchers' and table steels, manufactured from the best refined cast-steel.

[6486]

BEACH, WILLIAM, *Salisbury.*—Case of assorted cutlery of Salisbury manufacture.

This case contains carving, table, sportsmen's, pocket, pen, pruning, and paper knives; scissors, and cases of ditto; razors, daggers, &c. These goods are forged from the best cast-steel, and ground and fitted by the exhibitor.

[6487]

BEARDSHAW, GEORGE, *Tomcrop Lane, Sheffield.*—Table, dessert, and carving knives and forks, palette, pruning, farriers', butchers', and shoe knives, bowie and bread knives, &c.

[6488]

BOLSOVER, THOMAS, *Ford, Ridgeway, Sheffield.*—Scythes, sickles, and reaping hooks, suitable for all parts of the world.

[6489]

BOND, WM. J., *Bethnal Green Road, London.*—Saw, cabinet, bench, hand screw, and mechanical tools.

The exhibitor is a manufacturer of cabinet work, benches with double screw, also with single screw, and double slide for keeping the chop at all times parallel, and with end or cramp screw.

He also makes every description of hand screw, bench screws and chops, carvers' wood vices, &c. He supplies chests of tools fitted complete, from 18*s*.

[6490]

BOOTH, HENRY E., & CO., *Norfolk Works, Norfolk Lane, Sheffield.*—Table knives and forks, spear, butcher, bowie, and dagger knives.

Table knives and forks, spear point, palette, and butchers' knives, bowie, dagger, and hunting knives, also plated desserts, razors, and general cutlery, suitable for home and export trade.

[6491]

BROOKES & CROOKES, *Atlantic Works, Sheffield.*—Fine pen and sportsmen's knives, razors, and dressing case fittings.

[6492]

BROWN, HENRY, & SONS, *Western Works,* 108 *Rockingham Street, Sheffield.*—Braces, bits, joiners' tools, augers, gimblets, skates, tool-chests.

[6493]

BUCK, JOSEPH, *Newgate Street, and Waterloo Road, London.*—Mechanical tools for engineers, carpenters, &c.

The following goods are manufactured by this exhibitor, and supplied wholesale, retail, and for exportation.

Manufacturer of every description of saws, planes, and tools for engineers, carpenters, cabinet makers, joiners, coachmakers, carvers, wheelwrights, coopers, plumbers, &c.

Lathes, and every description of tools adapted for turning.

Circular saws from 1 to 60 in. diameter.

Mill, veneer, and endless band saws.

Cutlery of the best quality.

Fret-cutting machines.

Buhl saw-frames, and tools for buhl work and ornamental carving.

[6494]

BUXTON, E. J., & CO., *Duke Place, Sheffield.*—Electro-plated goods upon Britannia metal.

[6495]

CHAMPION & CO., 169 *Broad Lane, Sheffield.*—Fine scissors.

[6496]

COCKBAIN, JOHN, *Portland Place, Carlisle.*—Joiners' and cabinet tools.

[6497]

DIGGINS, GEORGE, 20 *Bessborough Place, Pimlico.*—Iron and metal planes; castings, rough and machine planed.

[6498]

DRABBLE, JAMES, & CO., *Orchard Works, Sheffield.*—Table knives and forks; spear, butchers', and dagger knives.

These exhibitors manufacture knives, forks, &c. entirely by machinery, and are thus enabled to supply goods more uniform in appearance, and as good in quality, at less cost than where hand labour is employed. The small case exhibited by them contains a fair sample of their productions.

[6499]

EADON, MOSES, & SONS, *President Works, Sheffield.*—Steel saws, files, machine knives, hammers, &c.

[6500]

EASTWOOD, GEORGE, 31 *Walmgate, York.*—Assortment of planes, with modern improvements, suitable for joiners, cabinet makers, &c.

[6501]

FIRTH, THOMAS, & SONS, *Norfolk Works, Sheffield.*—Files, saws, and edge tools; cast steel; large forgings in cast steel.

[6502]

FLETCHER, JOHN CARR, *Crown Works, Sheffield.*—Chisels, plane irons, axes, adzes, hatchets; augers, hammers, compasses, pliers.

[6503]

FULLER, JOHN H., 70 *Hatton Garden.*—Patent tube cutters, stocks and dies, taps, &c.

[6504]

GALLIENNE, GEORGE, 138 *Goswell Street.*—A general assortment of cutlery; boar spears, hunting knives, &c.; a very large bread knife, and trowel slicer.

[6505]

GIBBINS, J., & SONS. *Sheffield.*—Scissors; nail and champagne nippers; pruning shears and pocket cutlery.

[6506]

GILBERT, BROTHERS, 60 *St. Philip's Road, Sheffield.*—Razors; pen, pocket, and sportsmen's knives of all kinds.

[6507]

GILPIN, WILLIAM, SEN., & CO., *Wedges Mills, Cannock, Staffordshire.*—Edge tools, patent augers, matchets, iron and steel.

[6508]

GORRILL, ROBERT, & SON, 159 *Eyre Street, Sheffield.*—Fine scissors.

[6509]

GRAY, JOHN H., *Pelham Street, Nottingham.*—Improved skates.

[6510]

GREENSLADE, E. A. & W., *Thomas Street, Bristol.*—Planes.

[6511]

GREER, JAMES, 90 *Newgate Street.*—Specimens of London made table cutlery; also knives used in various trades.

The best plain ivory handle table knives, 30/0 per dozen.

Ditto, dessert knives, 24/0 per dozen.

Ditto, carvers, 10/0 per pair.

Ditto, with ornamented shoulders, 36/0, 30/0.

Ditto, carvers, 12/0 per pair.

The best electro silver-plated table spoons and forks, 40/0 per dozen.

Ditto, dessert spoons and forks, 30/0 per dozen.

Tea, salt, egg, and mustard spoons, 16/0 per dozen.

Gravy spoons, 7/0 each.

Soup ladles, 12/0 each.

Carved wood bread trays and butter dishes, from 5/0 each.

Bread knives, table steels, corkscrews, razors, scissors, pocket knives, needles.

Knives made expressly for use in all the various trades.

[6512]

HANNAH, ALEXANDER, *Calton, Glasgow.*—Screw-augers, braces, bracebits, and all kinds of tools for boring wood.

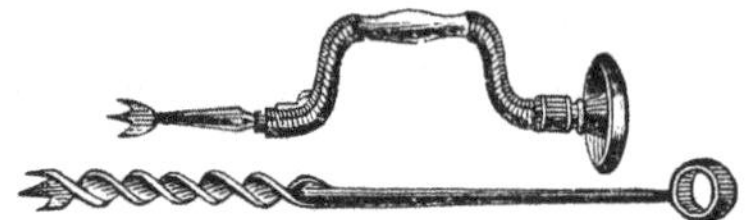

The braces, bits, augers, &c. made by A. Hannah, bear the stamp, "Thompson, Glasgow."

[6513]

HARDY, THOMAS, 44 *Milton Street, Sheffield.*—Button-hooks, nail files, corkscrews, stilettoes, tweezers, nut picks, &c.

[6514]

HARGREAVES, SMITH, & CO., *Eyre Lane, Sheffield.*—Sheffield cutlery and hardware. (*See page* 161.)

[6515]

HASLAM, JOHN, & SONS, *Ridgeway, near Sheffield.*—Scythes, sickles, and reaping hooks.

[6516]

HAWCROFT, WILLIAM, & SONS, *Bath Works*, 53 *Bath Street, Sheffield.*—Razors of superior quality in great variety of pattern and mounting.

[6517]

HAYWOOD, JOSEPH, & CO., *Sheffield.*—Pruning knife, and cutlery in general.

[6518]

HEATH, SIMEON, *Union Place Paddock, Walsall.*—Improved spring splitting machine, and general assortment of saddlers' tools.

[6519]

HILL, J. V., 5 *Gray's Inn Road, King's Cross.*—Samples of London-manufactured saws.

[6520]

HOWARTH, JAMES, *Broomspring Works, Sheffield.*—Edge tools; joiners', engravers', carvers' and turners' tools; augers, skates, &c.

Obtained Prize Medals from the London, 1851, *and Paris,* 1855, *Exhibitions, and Special Medal of Honour from the Society of Arts and Industry,* 1856, *for superior quality of goods exhibited.*

[6521]

JACKSON, NEWTON, & CO., *Sheaf Island Works, Sheffield.*—Steel files, saws, edge tools, cutlery, sheep shears.

[6522]

JOLLEY, JOHN & THOMAS, *Excelsior Works, Warrington.*—Files, railway-ticket nippers, telegraph vices, and engineers' tools.

The following are the manufactures of Messrs. JOLLEY, of which specimens are exhibited :—

Files, telegraph vices and nippers, railway-ticket nippers, stock taps and dies, wrenches, spanners, ratchet drills, cast-steel hammers, saws for buhl, iron, and steel, bench vices, hand vices, screw-plates, plain, wing, rack and millwrights' compasses, squares, spring dividers, index, plain, wing, rack in and out, pocket, and spring callipers, handcuffs, &c.

HARGREAVES, SMITH, & CO., *Eyre Lane, Sheffield.*—Sheffield cutlery and hardware.

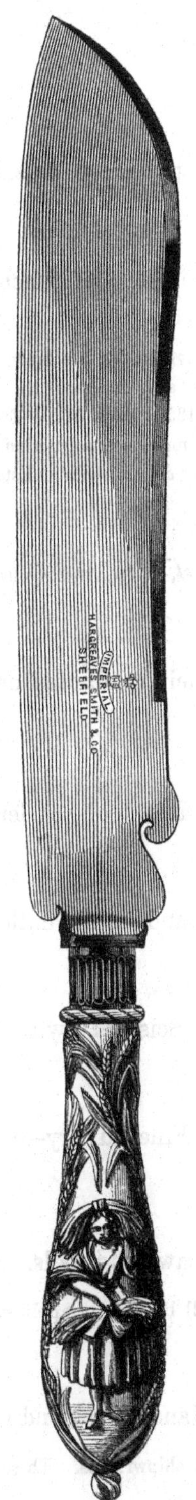

BREAD KNIFE.

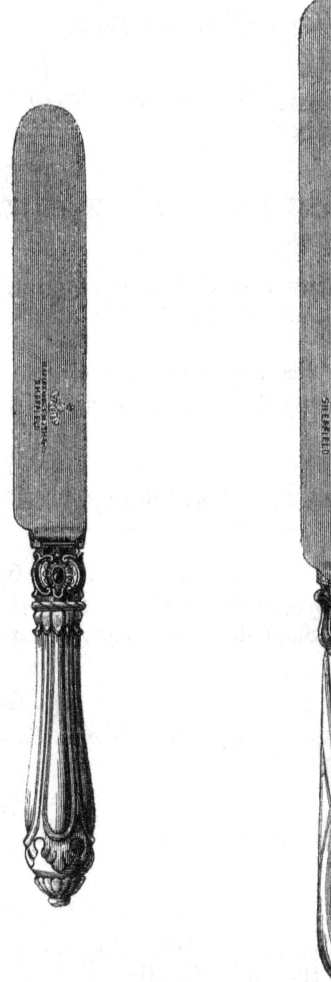

TABLE AND DESSERT KNIVES.

HARGREAVES, SMITH, & CO. exhibit—

1. One dozen carved ivory-handle table and dessert knives in a case of coromandel wood.

2. A carved ivory-handle bread knife.

HARGREAVES, SMITH, & CO. Manufacturers, Sheffield.

[6523]

JOWITT, THOMAS, & SON, *Sheffield.*—Specimens of manufactured steel, and files for engineers and exportation.

[6525]

KING & PEACH, *Hull.*—Planes, various.

[6526]

KINGSBURY, THOMAS, 9 *New Bond Street.*—Cutlery; cutlery applied to dressing-cases of a new construction.

[6527]

LINNEKER, RAVEL & JAMES, *Cobnar Works, Sheffield.*—Scythes, sickles, chaff machine knives, and straw knives.

REAPING AND STRAW-CUTTING MACHINE KNIVES, SCYTHES, SICKLES, AND HOOKS.

The temper of the edge is produced by a new process, ensuring perfect regularity; also, the back part of machine knives, scythes, &c. are made spring temper to prevent breakage and increase the strength (see *Engineer*, July 16, 1861, page 31). Every article is of the best and most modern construction, and supplied in patterns suitable for all countries. Established 1768.

[6528]

MAPPIN BROTHERS, 222 *Regent Street,* 67 *&* 68 *King William Street, City, and Queen's Cutlery Works, Sheffield.*—Cutlery.

[6529]

MAPPIN & CO., *opposite the Pantheon, Oxford Street, London.*—Their celebrated cutlery from their works at Sheffield. (*See pages* 164 *and* 165.)

[6530]

MARSH, BROTHERS, & CO., *Ponds Works, Sheffield.*—Steel files, saws, tools, cutlery, railway and carriage springs.

[6531]

MATTHEWMAN, B., & SONS, *Milton Works, Sheffield.*—Pocket and table cutlery, razors, scissors, &c.

[6532]

MATTHEWMAN, BENJAMIN, JUN., 80 *Milton Street, Sheffield.*—Scissors, with miniature photograph of the Royal Family.

[6533]

MECHI & BAZIN, 4 *Leadenhall Street, and* 112 *Regent Street.*—Fine cutlery—razors, table knives, scissors, sporting and pen knives.

[6534]

MITCHELL, J. W., 1 *Bridge House Place, Newington Causeway.*—Saws and tools.

These tools are of the best materials, and moderate prices. Lists will be sent on application.

[6535]

MITCHELL, WILLIAM HENRY, 3 *Britannia Place, Limehouse, E.*—Hand, panel, and tenon saws.

London-made saws, frames, and other tools for cabinet makers, carpenters, and shipwrights. These goods are of excellent quality, and their prices are moderate.

[6536]

MOLYNEUX, W., *Prescot.*—Varieties of Lancashire files.

[6537]

MONK, THOMAS, 74 *Edward Street, Birmingham.*—Moulders, plasterers', and stonemasons tools.

[6538]

MOSELEY, JOHN, & SON, 54 *Broad Street, Bloomsbury, London.*—Planes and other joiners' tools.

[6539]

MOSELEY, JOHN, & SON, 27 *Bedford Street, and* 17 & 18 *King Street, Covent Garden.*—Cutlery and tools.

[6540]

MUSHET, ROBERT, & CO., *Coleford and Sheffield.*—Cutlery of all kinds; edge tools, and samples of steel.

[6541]

NURSE, C., *Mill Street, Maidstone.*—Carpenters' planes.

[6542]

PARKES, FRANCIS, & CO., *Sutton Works, Birmingham.*—Cast-steel forks, spades, draining, edge, and plantation tools.

CAST-STEEL FORKS, &c.

Cast-steel forks, four to twelve prongs, for lifting tan, coke, malt, chaff, and other light substances.

Cast-steel forks, three to eight prongs, for dung, stones, and bulbous roots.

Cast-steel forks, three to six prongs, for digging, subsoiling, and clearing land.

Cast-steel forks, two to three prongs, for harvesting.

Solid cast-steel spades and shovels.

Solid cast-steel draining tools.

Draining tools partly of steel.

Spades and shovels, the surface of which is plated with cast-steel.

Cast-steel hoes, for gardening and plantation.

Cast-steel axes, hatchets, pickaxes, and mattocks, hedging bills, &c.

Cast-steel ploughshares.

[6543]

PARKIN, JOHN, *Steel Works, Harvest Lane, Sheffield.*—Saws, files, machine knives, papermakers' bars and tools.

[6544]

PEACE, WARD, & CO., *Agenoria Steel Works, Sheffield.*—Steel files, tools, saws, hammers, machine springs; cutlery.

[6545]

RODGERS, JOSEPH, & SONS, 6 *Norfolk Street, Sheffield.*—Pocket, pen, and sportsmen's knives, table cutlery, razors, scissors, fish carvers.

Mappin & Company, 77 & 78 *Oxford Street, London, opposite to the Pantheon.*—Celebrated cutlery.

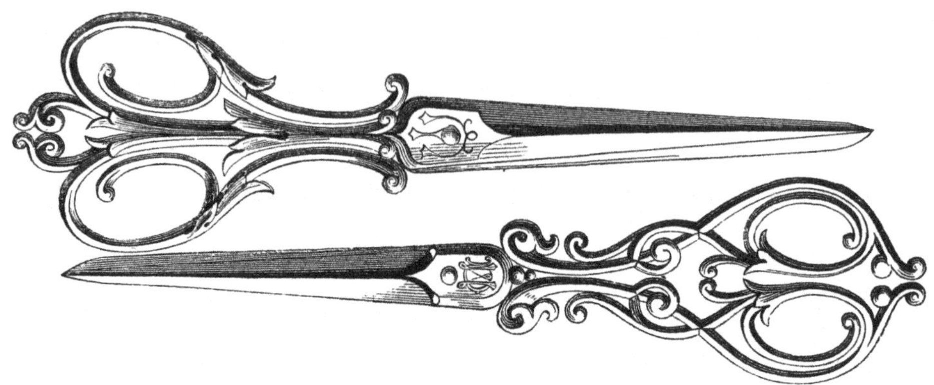

Mappin & Co.'s scissors have long been famous for their exquisite quality and finish. They can be had as low in price as 1/0 per pair, and a set of three in a handsome case for 5/0.

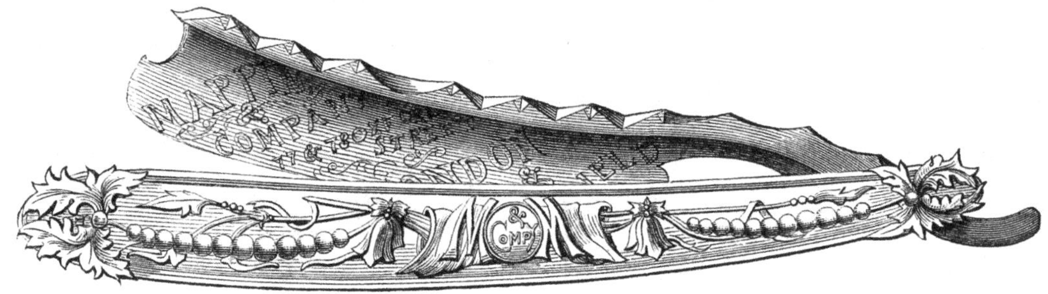

Mappin & Co.'s razors, well known for their great keenness, are well represented in the Exhibition. Their 1/0 razor has a world-wide reputation, and an enormous sale.

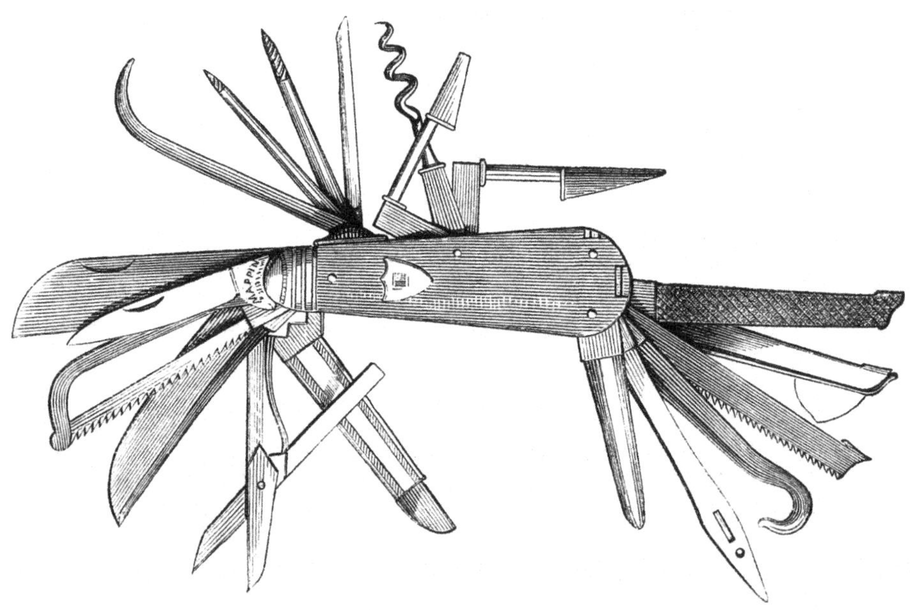

Mappin & Co.'s penknives, and sportsmen's knives for hunting, angling, &c. are of the finest quality, and of perfect mechanism; the specimens exhibited are beautifully fluted, as specimens of what can be done with steel for ornamental cutlery.

Mappin & Company, *continued.*

Table knives with secure ivory handles, balanced, from 13/0 per dozen.

All Mappin & Co.'s table cutlery is made of the best double shear steel, and warranted.

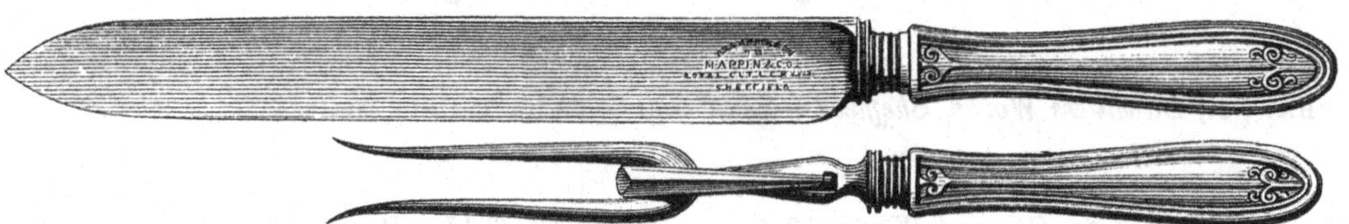

Carvers, with secure ivory handles, 4/6 per pair.

Bread knives with beautifully carved wood and ivory handles, the quality of steel warranted.

The above is one of the best designs exhibited for a dessert knife, being artistic in form, and of a most convenient shape for use. It is an admirable specimen of Mappin & Co.'s manufactures in this department, in which they greatly excel.

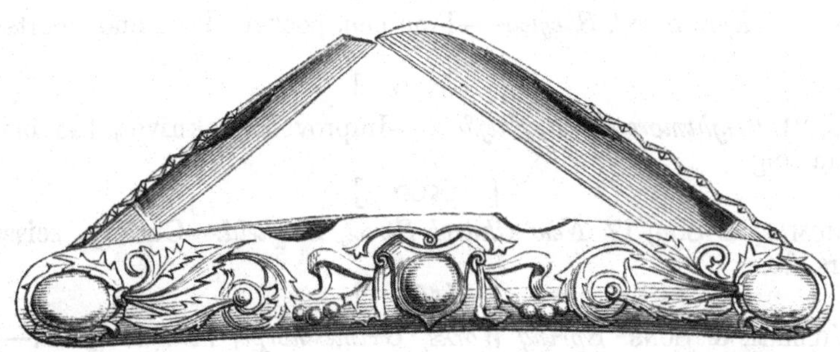

Mappin & Co.'s pen and pocket knives are unequalled for the excellence of the steel and other material used, and the exquisite finish with which they are put together; the prices also at which they are produced are very low—a most serviceable knife with ivory handle and two blades of the very best steel being sold for 1/0; selected beautiful knives with pearl and tortoiseshell handles elaborately fluted, four best steel blades, to 10/6 each.

[6546]

RUSSELL, THOMAS, & Co., *Canada Works,* 38 *Charles Street, Sheffield.*—Saws.

The exhibitors are proprietors of the marks "Russell & Horsfield," "John Sanderson," and the corporate mark,

Their case contains specimens of back saws with electro-plated back, hand saws with etched blade and Albert shield on handle, and other varieties of saws.

[6547]

SAGGERSON, E., *Prescot.*—Clock, watch, and jewellers' and various other Lancashire files.

[6548]

SAYNOR & COOKE, *Paxton Works, Sheffield.*—Pruning, budding knives, scissors, &c.

[6549]

SHAW, J., *Dauntless Works, Sheffield.*—Saws and tools for various trades.

[6550]

SHIRLEY, WILLIAM, *Crescent Works,* 19 *Carver Street, Sheffield.*—Pen, pocket, sportsman, dirk, and bowie knives.

[6551]

STEER & WEBSTER, *Castle Hill Works, Sheffield.*—Scissors, razors, knives, table cutlery, &c.

[6552]

SUTTON, W., & SONS, *New Town Row, Birmingham.*—Awl blades.

[6553]

TAYLOR, H., 105 *Fitzwilliam Street, Sheffield.*—Various trades' tools.

[6554]

THOMAS, RICHARD, *Icknield Edge-Tool Works, Birmingham.*—Edge tools for home and foreign markets.

[6555]

TUTON, MATHEW, *Scarbro' Road, Driffield.*—One stand forks and hedge tools.

[6556]

UNWIN & ROGERS, *Rockingham Street, Sheffield.*—Table and spring knives, razors, and every description of cutlery.

[6557]

WALDROW, WILLIAM, & SONS, *Bellbroughton, Stourbridge.*—Scythe, hay and chaff knife, hook, and edge tools.

[6558]

WARD, GEORGE, 171 *Eyre Street, Sheffield.*—Fine pen, pocket, desk, and sportsmen's knives, lancets, &c.

[6559]

WARD, THOMAS, 31 *Brightmore Street, Sheffield.*—Improved penknives, the blades cannot be injured in shutting.

[6560]

WILKINSON, THOMAS, & SON, 17 *New Church Street, Sheffield.*—Cutlery, scissors, improved tailors' shears, &c.

[6561]

WILKINSON, WILLIAM, & SONS, *Spring Works, Grimesthorpe, near Sheffield.*—Sheep, horse, glovers', thatchers', and other shears.

[6562]

WINKS, B., & SONS, 55 *Earl Street, Sheffield.*—Razors, table knives, and scalping blades.

[6563]

WOSTENHOLM, GEORGE, & SON, *Washington Works, Sheffield.*—Pen, pocket, table, bowie, and sportsmen's knives; razors and scissors.

Obtained the Prize Medal at the Exhibition in 1851, *and the large Gold Medal at the Paris Exhibition*, 1855.

Prize Medal, Exhibition of 1851.

Paris Exhibition, 1855.

CORPORATE MARK.

Prize Medal, Exhibition of 1851.

The only Gold Medal awarded for English Cutlery.

THE DOUBLY-CARBONIZED I X L RAZOR.

GEORGE WOSTENHOLM & SON are the sole manufacturers of the IXL table, pen, pocket, and bowie knives, dirks, scissors, and razors. These goods, for quality and workmanship, are unsurpassed. The gold medal obtained at the Paris Exhibition by the IXL cutlery, was the only one awarded to English cutlers. The exhibitors confidently recommend their registered doubly-carbonized IXL razors, which, for chasteness of design, exquisite finish, and fine shaving properties, are unsurpassed. These razors are hardened and tempered by a peculiar and secret process.

George Wostenholm & Son are the sole manufacturers of the genuine congruent and fine pipe razors. These are really good and useful articles.

LONDON:
R. CLAY, SON, AND TAYLOR, PRINTERS,
BREAD STREET HILL.

CLASS XXXIII.

WORKS IN PRECIOUS METALS, AND THEIR IMITATIONS, AND JEWELLERY.

[6595]

ADAMS, GEORGE WILLIAM, *Hosier Lane, London.*—Knives, forks, spoons, and various articles of new design.

[6596]

ADKINS, HENRY, & SONS, 22 *Weaman's Row, Birmingham; 4 Thavies Inn, London.*—Electro-plated goods, epergnes, cruet, liquor, and egg frames, &c.

[6597]

ANGELL, JOSEPH, 10 *Strand, and* 25 *Panton Street, Haymarket.*—Jewellery, gold and silver plate.

[6598]

ASTON, THOMAS, & SON, *Regent Place, Birmingham.*—Jewellery, goldsmith's and silversmith's work.

[6599]

ATKIN, BROTHERS, *Sheffield, and* 39 *Ely Place, London.*—Electro-plate, Britannia metal, silver and plated cutlery.

[6600]

ATTENBOROUGH, RICHARD, 19 *Piccadilly, London.*—Silver cups, ebony and silver casket jewellery, watches.

[6601]

BALLENY, JOHN, 74 *Hatton Garden, and* 44 *St. Paul's Square, Birmingham.*—Gold chains, and gold, plated gold, gilt, and black jewellery, &c.

[6602]

BARKER, WILLIAM, 42 *&* 43 *Paradise Street, Birmingham.*—Silver, plated, Nickel silver, and Britannia metal wares.

[6603]

BARRY, WALTER E., *Egyptian Hall, Piccadilly.*—Gilt metal work, applied to the mounting of artistic productions.

[6604]

BELL, J., & Co., *Newcastle-on-Tyne.*—Groups in aluminium.

[6605]

BENSON, J. M., *Cornhill.*—Argentine and electro-plate dinner and tea services, of rich and elegant design.

[6606]

BIRMINGHAM COMMITTEE, THE, *Birmingham.*—Gold and silver jewellery, gold and silver plated jewellery, chains, &c.

EXHIBITORS:—

C. Green.
G. Hazleton.
F. A. Harrison.
Hilliard & Thomason.
J. Lees.
R. A. Loach.
Manton & Mole.
J. Russell.
C. T. Shaw.
B. W. Westwood.
W. Spencer.

[6607]

BIDEN, JOHN, 37 *Cheapside.*—Gold seals, signet rings, stone and other engravings; specimens of moss-agate stones.

[6608]

BRAGG, THOMAS & JOHN, *Vittoria Street, Birmingham.*—Gold bracelets, brooches, earrings, pins, studs, and links.

[6609]

BRYAN, CHARLES, *West Cliff and Bartergate, Whitby.*—Brooches, bracelets, earrings, coronets, necklets and necklace, hair pins, &c.

[6610]

BRYDONE, J., & SONS, 29 *Princes Street, Edinburgh.*—Devices with human hair for brooches, lockets, pictures; also gold-mounted hair jewellery.

The art of working in human hair has been brought to its present state of perfection by the Messrs. Brydone, and so skilful have they become, that any length or quality of hair can be manufactured into some ornament. Gold-mounted rings, 5*s.* to £5; brooches, 20*s.* to £15; bracelets, 5*s.* to £20; scarf pins, 6*s.* to 40*s.*; chains, 15*s.* to £5; lockets, 10*s.* to £10; necklets, &c.

Illustrated catalogues free by post.

[6611]

COLLIS, GEORGE RICHMOND, & Co., 130 *Regent Street, London, and Church Street Works, Birmingham,*—Silver and electro-plated services.

G. R. Collis & Co. exhibit specimens of their manufactures, showing the application of electro-silver plating and gilding to objects of art and domestic utility.

They are makers of breakfast, dinner, and tea services in silver and electro-plate; and their establishments are replete with every novelty of each season.

These exhibitors succeeded Sir Edward Thomason in the business of silversmiths and its kindred trades. Their works are situated in Church Street, Birmingham, and their goods may be seen at the Crystal Palace, Sydenham, as well as at their establishment in Regent Street.

[6612]

DERRY & JONES, *Great Hampton Street, Birmingham.*—Plated table service, spoons, forks, tobacco and vesuvian boxes, pencils, &c.

The following are exhibited:—Plated tea, coffee, table, and dessert services, snuff boxes, card cases, vesuvian boxes, and pipe mounts, pencil cases, penholders, and gold pens.

Derry & Jones are the inventors and sole makers of the *New Spanish Silver Spoons and Forks*, the whitest substitute for silver ever made.

[6613]

DIXON, JOHN, 95 *Lillington Street, Pimlico, S.W.*—A collection of bronze medals.

[6614]

DIXON, JAMES, & SONS, *Sheffield.*—Best Sheffield and electro-plate.

Obtained Two Medals at the 1851 *Exhibition; One Medal at Paris,* 1855, *Exhibition; One Medal for Plated Ware in the* 1862 *International Exhibition; and another for Britannia Metal Goods in Class* 31.

1. Coffee and tea service, in Grecian style, with coffee tray.
2. Wine cooler, in Roman style.
3. Dinner service, in Flemish style.
4. Claret jug, Louis Quatorze.

[6615]

DODD, P. G., & SON, 45 *Cornhill, London.*—Artistic works in the precious metals, &c.

[6616]

DONNE, W., & SONS, 5 *Great Vine Street, and* 51 *Cheapside.*—Engravings on precious metals.

[6617]

DUCLOS, L. D., 13 *Whiskin Street, Clerkenwell.*—Cameos, &c.

[6618]

DUNCAN, J., 4 *St. Nicholas Street, Aberdeen.*—Granite jewellery, &c.

[6619]

ELKINGTON & CO., *Newhall Street, Birmingham;* 20 *and* 22 *Regent Street, S.W., and* 43 *Moorgate Street, E.C., London;* 25 *Church Street, Liverpool; and* 29 *College Green, Dublin.*—Manufactures in silver, electro-plate, and bronze. (*See pages* 4 *to* 8.)

[6620]

ELLIS, BROTHERS, *Exeter.*—Brooch, and bracelet of Sidmouth pebbles.

[6619]

ELKINGTON & Co., *Newhall Street, Birmingham; 20 and 22 Regent Street, S.W., and 45 Moorgate Street, E.C., London; 25 Church Street, Liverpool; and 29 College Green, Dublin.*—Manufacturers of artistic works in silver, bronze, and other metals, by special appointment, to Her Majesty the Queen.

SILVER REPOUSSÉ TABLE.

The ornamental portions designed and executed by Morel Ladeuil (one of the artists in the employment of Elkington and Co.) This engraving only shows the top (or upper surface) of the table.

The subject of this design is intended to represent the dreams of three figures—a minstrel, a soldier, and a husbandman—who sleep at the base, they being under the influence of the goddess who floats over the centre, strewing poppies around.

The execution of this work occupied nearly three years.

The stem and base of the table are given upon the following page.

Elkington & Co., *continued.*

STEM AND BASE OF SILVER REPOUSSE TABLE.

Elkington & Co., *continued.*

SILVER VASE AND ROSEWATER DISH.

Designed by A. Willms.

The vase is adorned with allegorical figures and reliefs, emblematical of day.

The dish is divided into four panels containing reliefs emblematic of the elements, which are treated in an original manner. The vase and dish were designed by the principal artist of this firm; and who is also the designer of the silver enamelled dessert service, and of several other important works exhibited by Elkington & Co.

SILVER ENAMELLED CENTRE-PIECE, DESIGNED BY A. WILLMS.

This centre-piece, in the Pompeian style, forms part of a complete service intended for dessert. It is enriched by enamel and gold. One of the figures represents a Priestess of the Temple of Peace; the others typify Agriculture and Commerce, as votaries of Peace.

Elkington & Co., *continued.*

PIANO CANDLESTICK.

PIANO CANDLESTICK.

SILVER ENAMELLED CANDELABRUM AND ENAMELLED PIANO CANDLESTICKS.

Designed by A. Willms.

ELKINGTON & CO., *continued.*

Extract from an Article "ON OUR WAY THROUGH THE EXHIBITION." *By* BLANCHARD JERROLD, ESQ.

Our way shall be from Minton and Co.'s St. George's fountain, with its winged Victories holding aloft St. George and the Dragon. We saunter, sloping, like great Orion,

Slowly to the west.

The way is past British furniture trophies, the model of the *Warrior*, the granite obelisk from the Cheesewing quarries, Bevington and Sons' leather trophy, Nicholay's furs—a mere shop-show from Oxford Street--the British porcelain trophies; the food trophy, a strange mixture; the shows of Hunt and Roskell, Harry Emanuel, and the great display of Messrs. Elkington and Co.

Before this last trophy we will bid the reader rest awhile, while we carry him far away to the great scene of human activity whence these fine works of science and skill proceed. It is by the studying and mastering of the difficulties that have to be overcome before the smallest statuette or the poorest plate can be fashioned, that visitors to the Exhibition may derive solid advantage out of their wanderings. . . . . To appreciate fairly all the useful and ornamental developments of which the electro process has proved itself capable, it is necessary to spend a few days in an establishment like that of Messrs. Elkington and Co., Newhall Street, Birmingham. Here may be seen in perfection every variety of the electro process; in one part are the mighty figures modelled by Durham for his Great Exhibition Memorial, lying in their copper baths; in another are the frailest brooches receiving their coat of gold. Wandering from room to room, from shed to shed, from courtyard to courtyard, the visitor is bewildered with the constant variety of skill and ingenuity brought to bear upon common objects of daily life.

ROSEWATER TAZZA, BY MOREL LADEUIL.

The art-department of the establishment merits distinct attention. Why the presiding intellects here are importations, it matters not, in this place, to discuss. The fret and toil that are going on to produce splendours for Elkington's part in the Exhibition are wonderful. Here is the most wonderful of all spirit-rapping; for, by the help of magic taps, given by little hammers, behold the magic dream which M. Morel-Ladeuil reveals upon this silver table. The design is exquisitely poetic and happily appropriate—"The Three Dreamers." At the base of the table are the Troubadour, the Warrior, and the Agriculturist. Above these sleeping figures float—aye, float as in the air, so

ELKINGTON & CO., *continued.*

exquisite is the rendering—the dreams natural to each. Mark the minstrel's dream, it is of Music, Pleasure, Fortune, Love, speaking in the lithe forms of women and children. Happy dreamer! The warrior's vision is grim. Here is laurelled glory; here victory and renown, with their fitting emblems. The agriculturist—honest tiller of the soil—dreams of abundance. Women and children, bearing fruit, flowers, corn, and grapes—bringing the brilliant and glorious treasures of mother Earth. This is all gracefully conceived, and the execution matchless. The centre figure crowns the whole idea. The exquisite draped figure is Gentle four childrens' heads represent the four winds. Jupiter's wing is the capital bit. This elaborate and artistically-studied work will delight connoisseurs. We may point to other productions from the graceful pencil of M. Willms and the skilful hammer of M. Morel-Ladeuil. Here, for instance, are the Four Seasons, represented by four masks, encompassed by seasonable flowers and plants; the flagon is a mass of flowers and butterflies fantastically grouped. M. Willms's silver tankard has theatrical art for its subject. It is in the Greek style. The bas-reliefs and figures are in ivory. These bas-reliefs represent Comedy, Tragedy, Song, and Dancing.

THE ELEMENTS DISH.

The perspective view of this is shown, with the vase standing on it, at page 6.

Sleep, scattering her poppies, and so dominating the sleepers.

We now look upon a silver flagon and dish, upon which the four elements are developed. This remarkable work is designed by M. Willms, the directing artist of this house. The dish or plateau is from bas-reliefs, representing Earth, Water, Air, and Fire. These are encompassed with ornamental work and flowers. A border with lions' heads frames the plateau. The flagon, a bold and graceful design, is decorated with two bas-reliefs, upon one of which Apollo appears, encompassed by the signs of the zodiac, while the other is occupied by Diana, surrounded by the principal planets. Beneath, A child beating time crowns the lid. The names of ancient authors are worked in letters of gold upon enamel over each bas-relief. Nor should those perfume-burners in enamelled copper be passed over. The silver enamelled dessert service, in the Pompeian style, is a splendid specimen of art-workmanship; and so are the great Indian vase, the Mauresque epergne, the centre piece discovering "Æsop's Fables," and an inkstand in the style of Louis XIV. The Indian and Renaissance tea-sets are also in the highest style of art.

We pass from before those noble works of thought and skill, fascinated by the beauty of form and brilliancy of material here offered to the luxurious and wealthy.

Emanuel, E. & E., Jewellers and Silversmiths to the Queen, H.R.H. the Prince of Wales, and the principal Courts of Europe, 101 *High Street, Portsmouth.*—Works of art, &c., in the precious metals; jewels; horological machinery.

NO. 1.—SILVER CANDELABRUM.

Emanuel, E. & E., *continued.*

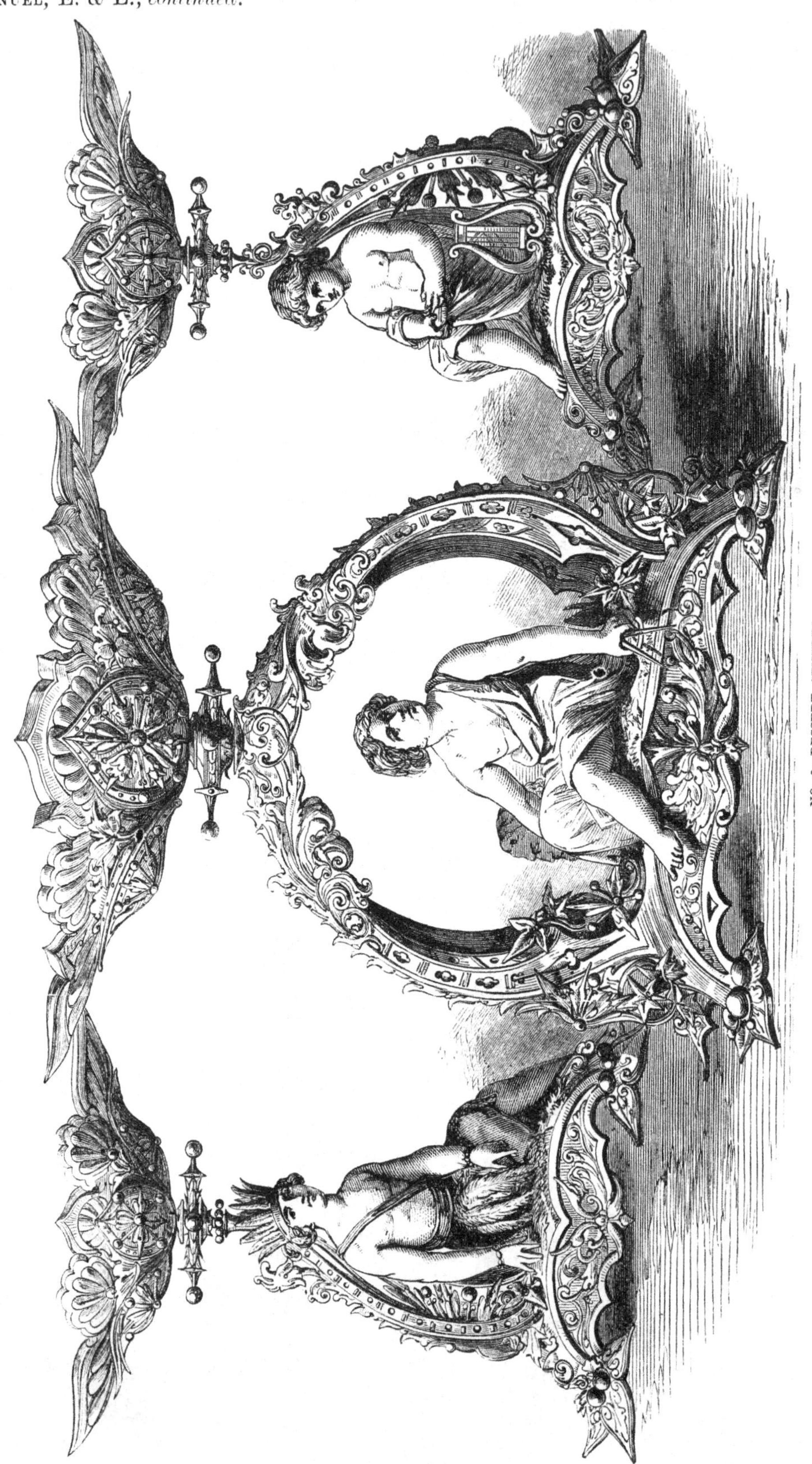

No. 2.—Dessert Centres.

EMANUEL, E. & E., *continued.*

EMANUEL, E. & E., *continued.*

A silver shield, the outer circle of which is surrounded, in relief, with a frieze of horses and warriors, taken from the celebrated Elgin marbles.

The centre of this shield is in alto-relievo, and is a copy of the classic cameo by Athenion, in the Royal Museum at Naples, illustrating "the Conquest of the Titans by Jupiter," described in the following terms, by Horace (Ode iii. 4—42):—

" . . . Scimus ut impios
Titanas, immanemque turmam
Fulmine sustulerit caduco.
Qui terram inertem, qui mare temperat
Ventosum, et urbes, regnaque tristia,
Divosque, mortalesque turmas
Imperio regit unus æquo."

Jupiter is represented in his car, drawn by four horses, and preparing to hurl his thunderbolts at the giants, who, in accordance with Ovid, are depicted by the artist as men of great stature, having serpents in the place of legs:—

"Sphingaque et Harpyias serpentipedesque Gigantas."
OVID, Tristia iv. 7, 17.

Designed and modelled by Mr. Henry Morrell, in the manufactory of the Exhibitors. The property of the Right Hon. the Earl of Lonsdale, by whom it has been kindly lent for exhibition.

A dessert service in silver, *allegorical of the International Exhibition*, for which occasion it has been expressly designed and manufactured.

The service consists of:—

A silver candelabrum, with basket at top for fruit or flowers. As portions of the branches can be removed, it is made applicable as an epergne. On the base is a large group, representing Britannia seated on a lion, and attended by an allegorical figure of Industry, distributing rewards to the representatives of the various nations of the globe, who are represented laying the products of their several countries at her feet. The decorations of the whole of this service, which are of a floral character, are of an entirely novel design. The allegory is carried out in—

Two epergnes, with figures of the Arts and Sciences; and

Four dessert centres, containing figures, severally representing, "Europe," "Asia," "Africa," and "America," containing an allusion to the four quarters of the globe being represented at the Exhibition.

In each of the figures the artist has succeeded in producing the acknowledged type of countenance and dress. Designed and modelled by Mr. E. W. Clarke, and produced in the manufactory of the Exhibitors. The accompanying plates, No. 1 and 2, represent portions of this service.

A classic group in silver, representing "Thetis bringing to her son Achilles the Armour forged by Vulcan."

"Tu vero a Vulcano allata inclyta arma accipe,
Pulchra admodum, qualia nondum quisquam vir humeris gestavit.
Sic sane locuta Dea armo deposuit
Ante Achillem: illa vero sonitum-edidere facta-artificiose omnia."
ILIAD, lib. xix.

Designed and modelled by Mr. Henry Morrell, in the manufactory of the Exhibitors.

Emanuel, E. & E., *continued.*

A vase in silver (parcel gilt), dedicated to Tasso; the outline in the Italian style, the ornaments of the Cinque-cento period.

The large group in alto-relievo, illustrates "The Combat of Clorinda and Prince Tancred," pourtrayed in Tasso's "Jerusalem Delivered," Canto iii. stanza 21:

"Meanwhile Clorinda rushes to assail
The Prince, and level lays her spear renowned:
Both lances strike, and on the barred ventayle
In shivers fly, and she remains discrowned;
For, burst its silver rivets, to the ground
Her helmet leaped (incomparable blow!)
And by the rudeness of the shock unbound,
Her sex to all the world emblazoning so,
Loose to the charmed winds her golden tresses flow."

* * * * *

One bas-relief represents the First Interview of Prince Tancred and Clorinda at a Fountain, (Canto i. stanza 47):

"To the same warbling of fresh waters drew,
Armed, but unhelmed and unforeseen, a maid:
She was a Pagan, and came thither too,
To quench her thirst beneath the pleasant shade;
Her beautiful fair aspect, thus displayed,
He sees; admires; and, touched to transport, glows
With passion—'tis strange how quick the feeling grows:
Scarce born, its power in him no cool, calm medium knows."

The second bas-relief is descriptive of the scene in which the Prince rescues Clorinda from an attack of one of his followers, (Canto iii. stanza 29):

"One base pursuer saw Clorinda stand,
Her rich locks spread like sunbeams on the wind,
And raised his arm, in passing, from behind,
To stab secure the undefended maid;
But Tancred, conscious of the blow designed,
Shrieked out, 'Beware!' to warn the unconscious maid,
And with his own good sword bore off the hostile blade."

Designed and modelled by Mr. Henry Morrell, in the manufactory of the Exhibitors. Exhibited by kind permission of Captain Alexander, Belgrave-square.

A silver vase, in the Grecian style, the bowl enriched with festoons of vine-leaves and grapes in alto-relievo, with handles formed of winged horses; the base ornamented with rich foliage, and Bacchanalian heads.

The first group in alto-relievo, on the base, represents the attack of the Trojans, Æneas and Pandarus, on Diomed, resulting in the death of Pandarus, the rescue of Æneas by Venus, and the capture of Æneas' celebrated horses. (Iliad, Book V.):

"Thus while they spoke, the foe came furious on,
And stern Lycaon's warlike race begun:
Prince, thou art met. Though late in vain assail'd
The spear may enter where the arrow fail'd."

The second alto-relievo represents the Horses being unloosed from Juno's Car by the attendant Hours. (Iliad, Book viii.):

"She spoke, and backward turn'd her steeds of light,
Adorn'd with manes of gold, and heavenly bright.
The Hours unloosed them, panting as they stood,
And heap'd their mangers with ambrosial food."

Designed and modelled by Mr. Henry Morrell, in the manufactory of the Exhibitors. The property of R. Ten Broeck, Esq., by whose kind permission it is exhibited.

A suite of articles in silver for the writing table, of entirely novel design, consisting of inkstand, blotting book, envelope case, match box, date indicator, pair of candle sticks, penholder, seal, &c.

The bodies of engraved silver, and the wire mounts in silver gilt.

EMANUEL, E. & E., *continued.*

A silver tazza. Bassi-relievi around the body, representing "The Battle between Richard Cœur de Lion and the Saracens."

The group on the summit represents "The Friendly Meeting of Richard Cœur de Lion and Saladin." Designed and modelled by Mr. W. Clarke, at the Exhibitors' manufactory.

A silver candelabrum and epergne, with allegorical figures, severally representing "Comedy," "Dancing," and "Music."

This piece of plate was presented to Edward Weston, Esq., by whom it has been lent for exhibition. Designed by Mr. Emanuel, Jun., and produced in the manufactory of the Exhibitors.

A silver vase, Louis Quatorze style. Subject on summit: "Perseus Slaying the Dragon, and Rescuing Andromeda."

The bowl is richly embossed. There are figures in silver of rampant horses, at each corner of the base. Designed and modelled by Mr. H. Morrell, in the Exhibitors' manufactory.

A silver "vase irregulier."

Jug and pedestal with rampant horses at base, and bassi-relievi groups of wild horses.

A large silver group in alto-relievo, representing "an Episode in a Steeple-chase—"Jumping a Stone Wall."

One of the competitors is represented as having "come to grief," while the others are gallantly charging the wall. Designed by Mr. W. E. Clark, and produced in the manufactory of the Exhibitors.

A silver tazza, dedicated to Homer. A group in alto-relievo, at top, represents Achilles in his chariot. The stem and base is composed of copies of ancient armour, shields, &c.

The body of the tazza is surrounded with illustrations of the following subjects pourtayed in the Iliad:—"Diomed casting his Spear at Mars;" "The Hours taking the Horses from Juno's Car;" "The Gods descending to Battle;" "Hector's Body dragged at the Car of Achilles." Designed and modelled by Mr. Henry Morrell, in the Exhibitors' manufactory.

[6622]

EMANUEL, HARRY, 70 *Brook Street, and Hanover Square.*—Original and artistic articles in precious metals and jewels. (*See pages* 18 *to* 32.)

[6623]

FORRER, ANTONI, 2 *Hanover Street, Hanover Square.*—Hair jewellery, brooches, bracelets, chains, rings, pins, studs, necklaces, earrings, pencil-cases, lockets, &c., &c.

Obtained the Prize Medal at the Exhibition of **1851.**

The exhibitor was appointed "Artist in Hair Jewellery to Her Majesty," in 1845; and it is in great measure owing to his efforts that this art has attained its present popularity. He designs and manufactures ornaments and bijouterie of every description in hair, and will send drawings by post for inspection. He has no connection with his late establishment in Regent Street. This fact is mentioned, as, from his name remaining still on his late premises, mistakes might easily occur.

[6624]

FRANCIS, WILLIAM, 13 *Hemingford Road, Islington.*—Patent regulating pencils, adapted to take all and any sized leads.

[6625]

GARRARD, R. & S., & Co., 25 *Haymarket, London.*—Works of art in silver, plate, and jewellery.

[6626]

GOGGIN, J., 74 *Grafton Street, Dublin.*—Ornamental jewellery in bog oak, &c.

EMANUEL, HARRY, 70 *Brook Street, and Hanover Square.*—Original and artistic articles in precious metals and jewels.

An equestrian statuette of H.R.H. the Prince of Wales, by Marshall Wood.

The likeness is most accurate, the Prince having honoured Mr. Wood by a sitting.

His Royal Highness is represented in the uniform of Colonel of the 100th (Canadian) Regiment, acknowledging a salute.

The horse is very life-like, and the delicacy of the chasing produces all the details with a marvellous fidelity. The different textures of cloth, lace, &c., and the flesh and hair, will bear minute examination.

EMANUEL, HARRY, *continued.*

The female figure in the accompanying illustration deserves particular attention, both from its intrinsic merits as a work of sculpture, and from the very great size of the block of ivory from which it is carved; it being of one piece, with the exception of the arms, which are added. It stands two feet high; and the

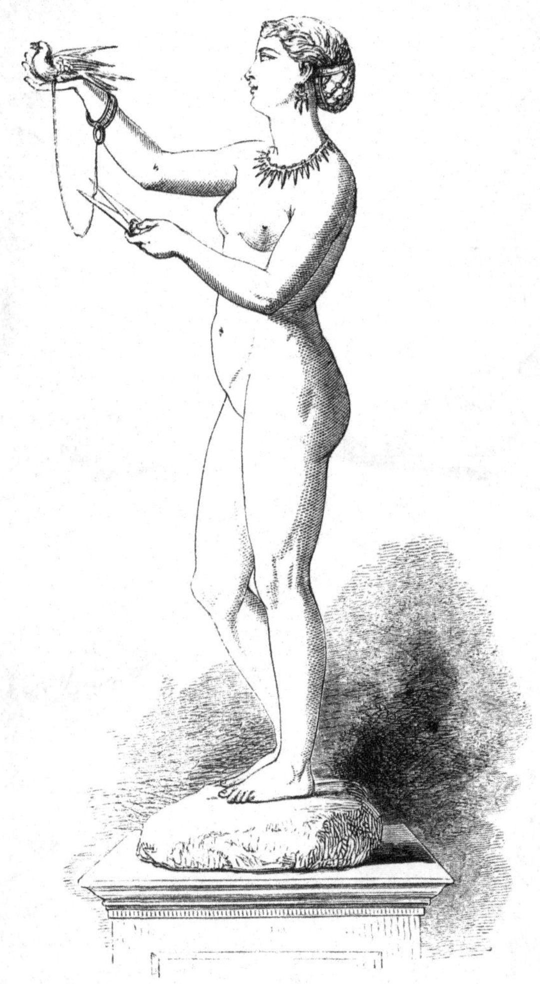

IVORY FIGURE.

lines, a translation from the Greek, from which the artist, Mr. Chesneau, has drawn his inspiration, will best convey the meaning:

> Ὡς ἐγώ, εν φυλακαῖς λυπῆχρυσεαισι δεθεῖσα
> Ἥλικας, ὦ ορνις, σόυς τε ποθοῦσα δομους
> Τέμνω τῦν κατέχοντα σ' ὅμος αεκουσαν ἵμαντα
> Ουρανού εις νεφέλας νῦν ἴθ', ἐγὼ δὲ μενω.
>
> In gilded cage like me you pine,
> And long for home and playmates dear;
> I cut the chain which makes thee mine—
> Go, thou art free; I must stay here.

The graceful attitude, the sad and pensive expression in the face of the slave girl, who, while liberating her bird, regrets her home and freedom, render most aptly the sentiment of the poet; and the delicate colour of the ivory, which of all substances comes nearest to the natural colour of flesh, and which is set off by the contrast of the jewels with which the girl is decked, explains the reason of the fondness of the ancients for this material for their works of art.

Emanuel, Harry, *continued.*

These dessert stands represent the "Seasons," and consist of children bearing the attributes of each season; thus, Spring is a flower-girl, Summer a boy reaper, Autumn a vintner, and Winter a skater.

SPRING. SUMMER. AUTUMN. WINTER.

The spirit and vigour with which these little figures are modelled and finished, have received the approbation of all connoisseurs. All these pieces were modelled by Mr. Chesneau.

Emanuel, Harry, *continued.*

This illustration represents a dessert service, consisting of seven pieces, made expressly for the Exhibition. The models are by Chesneau. The subject of the centre-piece, which forms either an epergne or a candelabrum, is the "Love-Letter." A peasant-girl is reading a letter, whilst a youth looks over her shoulder in anxious expectation.

The candelabra are supported by two trees entwined, and are ornamented by the novel introduction of crystal prisms, which add considerably to the effect when lighted. The figures are chased with much delicacy, and great attention has been paid to the perfect representation of the various textures introduced.

The two oval dessert stands, destined for either end of the table, carry out the same history; in the one we have the same maiden crossing a stile, whilst the youth, doubtful no longer, assists her descent with his arm around her waist; in the other, which presents the sequel to the homely romance, the female, more mature in appearance, holds up a child to greet his father, who, returning from work, advances with outstretched arms to receive it.

The illustration on the next page represents a fountain for the dinner-table, in lapis-lazuli and silver, forming also a centre-piece for fruit or flowers. This piece of plate, which is chased throughout with the minuteness of a piece of jewellery, and which, for finish and originality, is unequalled, stands in a circular plateau made to serve as a jardinière. The base is square, and at each side are figures of children holding urns, from which tiny jets gush forth into crystal shells, whilst underneath a dome, supported by lapis-lazuli and silver columns, is a fountain, sending its stream aloft. On the dome is seated a Venus who is being disrobed by Cupids, who, flying up, support a vase for flowers.

This work of art is remarkable alike for novelty of purpose and of design. The aërial manner in which the Cupids fly away with the drapery of Venus is marvellously light for metal-work. The introduction of lapis-lazuli, as an adjunct to plate, although common enough in mediæval works of art, is of very rare occurrence at the present day, and, no doubt, will now become again popular.

Emanuel, Harry, *continued.*

A FOUNTAIN FOR THE DINNER-TABLE.

EMANUEL, HARRY, *continued.*

SILVER EWER.

Emanuel, Harry, *continued.*

The silver ewer in the accompanying illustration is of very large size, and entirely covered with the most elaborate *repoussé* work, pourtraying the history of Undine. The body is divided into compartments, on which are represented the most salient incidents of the touching romance of Lamotte Fouqué. The handle is composed of a crane and a recumbent female figure; around the base are grouped the otter, the polar bear, the seal, &c., and every attribute and decoration is suggestive of, and has some connection with, the water.

The figure of Hildebrand on horseback, trampling through the water, is treated in a masterly manner, and the chasing and modelling of this important work are distinguished by a grace and elegance rarely met with in metal.

Five diamonds, mounted as stars, of enormous size, valued at £20,000.

More than fifteen hundred different aticles of jewellery, for personal ornament.

Various ornaments made of rock crystal, engraved in intaglio, and enamelled pâte tendre, producing the effect of carving in relief, and imitating nature marvellously.

Watches on the patent winding principle, with all the latest improvements, by Harry Emanuel, from the size of a sixpence upwards; chronometers, racing, and stop watches.

COLOURED ILLUSTRATIONS.

Gold (Perseus and Andromeda) Cup.

A brown topaz, or cairngorm, of very large size, carved in the form of a cup of early date, and hollowed with great skill, has here been mounted into a gold vase of the very highest pretensions as a work of art. It is made of pure gold, *repoussé* throughout, and partially enamelled. The grace of its composition, the spirit and truth of the modelling, and the harmonious blending of the colours, render it one of the most important specimens of goldsmith's work that has ever been produced in this country. It represents the history of Perseus and Andromeda.

Andromeda is chained to the rocks which form the base of the cup, whilst the Dragon has climbed up to attack Perseus, who, armed with the sword of Vulcan and Minerva's shield, and mounted on Pegasus, surmounts the cover. The novel design and execution are the work of Mr. Chesneau.

Emerald Brooch and Pearl Necklace.

This illustration represents a brooch, consisting of an enormous emerald, which once formed part of the Braganza jewels. It is of very fine colour, and singularly free from flaws and defects, and is mounted in a ribbon of large brilliants. This stone weighs 156 carats. It has for pendant a pearl pendeloque, two inches long, of matchless size and beauty. This pearl formed part of the French booty from the Summer Palace at Pekin, and is one of the very largest known. The mounting and diamond setting is light and graceful.

A necklace of Oriental pearls, of stupendous size and extraordinary beauty and lustre, with a brilliant snap.

Patented Ivory Jewellery.

Various ornaments in ivory and gold, inlaid with different gems.

This style is perfectly novel, and from the semi-transparency, *durability*, and beauty of the material, is eminently calculated to show the jewels and the design in the most favourable light; whilst the neutral colour of the ground admits of its employment with almost any toilet, and contrasts admirably with the gold and gems employed; in the hair, also, it has a very good effect. It is capable of adaptation in numberless ways, and from the success it has met with, and the approbation which has been universally expressed, it seems likely to enjoy a lasting popularity.

Pink Pearl-shell Jewellery.

The accompanying illustration represents various articles of jewellery, in which this novel substance has been introduced, in conjunction with various jewels. It is cut from a rare shell, found in the West Indies, and its delicacy of colour far surpasses the best and purest coral, while its brilliancy equals that of many gems. From its hardness, it is susceptible of a very high degree of polish, and it seems eminently adapted for use in jewellery, to which purpose it has not been before applied.

It has not been thought necessary to patent this beautiful addition to the material of the jeweller, as, from the practical difficulty of employing it on the part of the goldsmith, and the rarity of the material, it is not likely to be pirated.

A. CHESNEAU, Fecit. Vincent Brooks, Imp.

GOLD (PERSEUS AND ANDROMEDA) CUP.

Manufactured by Harry Emanuel, 70 and 71, Brook Street, Hanover Square.

EMERALD BROOCH AND PEARL NECKLACE.

Vincent Brooks, Imp

PATENTED IVORY JEWELLERY.

PINK PEARL SHELL JEWELLERY.

[illegible] 70 and 71, Brook Street, Hanover Square.

[6627]

GREEN, C., 48 *Augusta Street, Birmingham.*—Signet rings.

[6628]

GREEN, RICHARD A., 82 *Strand.*—Artistic jewellery and art, under £20 in value.

[6629]

HANCOCK, C. F., *Bruton Street, Bond Street.*—Jeweller and silversmith to the Queen and Courts of Europe. (*See pages* 26 *to* 32.)

[6630]

HARRISON, W. W., *Montgomery Works, Fargate, Sheffield.*—Electro-silver plate. (*See page* 41.)

[6631]

HAZLETON, 45 *Northampton Street, Birmingham.*—Filagree iewellery.

[6632]

HILLIARD & THOMASON, *Spencer Street, Birmingham.*—Silver fancy goods.

[6633]

HOWELL, JAMES, & Co., 5, 7, 9 *Regent Street.*—Goldsmiths, silversmiths, &c. (*See pages* 42 *to* 45.)

[6634]

HUNT & ROSKELL, 156 *New Bond Street.*—Artistic works in gold and silver, watches and clocks, pearls, precious stones, &c. &c. (*See pages* 46 *to* 53.)

[6635]

JAMIESON, GEORGE, Jeweller to the Queen, 107 *Union Street, Aberdeen.*—Granite and pebble ornaments.

[6636]

JENNER & KNEWSTUB, 33 *St. James's Street, and* 66 *Jermyn Street.*—Gold and silversmiths, and metal work.

[6637]

JOHNSON, JOSEPH, 22 *Suffolk Street, Dublin.* — Oak ornaments. Patronized by Her Majesty.

[6638]

KEITH, JOHN, 41 *Westmoreland Place, City Road.*—Church plate of every description.

[6639]

KONINGH, HENRIK DE, 79 *Dean Street, Soho, London.*—An ormolu enamelled clock, and specimens of enamelling.

[6640]

LAMBERT & Co., *Coventry Street, Piccadilly.*—Chased shield, large cistern, pot-pourri jars, tall beakers, antique figures, cups, chalices, monstrances, centres, &c., &c., &c.

[6641]

LA ROCHE, MISS E., 21 *Noel Street, St. James's.*—Specimens of piercing for jewellers and other workers in metal.

THE VASE OF SHAKSPEARE.

THE POETRY OF GREAT BRITAIN, A GROUP IN SILVER.

Manufactured by C. F. Hancock, Jeweller and Silversmith to the principal Sovereigns and Courts of Europe, expressly for the Exhibition. Designed and modelled by Signor Monti.

This group is intended to illustrate and embody some of the greatest and best known creations of the British poets. It consists of a central vase, dedicated to Shakspeare, supported by two loving cups and two tazzas, respectively dedicated to Milton, Byron, Moore, and Burns. Thus embracing illustrations of the dramatic, classical, romantic, lyrical, and popular poetry of the British Isles.

THE VASE OF SHAKSPEARE.

This vase is surmounted by the figure of THE POET, designed in such an inspired attitude as that described in Ben Jonson's lines,—

Thus while I wond'ring pause o'er Shakspeare's page,
I mark in visions of delight, the Sage,
High o'er the works of man, who stands sublime,
* * * * * *
Majestic 'mid the solitude of time.

The allegorical winged figures, which in this Vase occupy the place of handles, represent TRAGEDY and COMEDY.

TRAGEDY, closely draped, her snake-like ringlets encircled by a royal band, plunged in deep thought, holds the dagger, half concealed, under her cloak.

COMEDY, in loose garments, crowned with ivy and vine, the shepherd's staff and the histrionic mask in her hand, looks smiling towards the image of the Poet.

The subjects treated by Shakspeare are rendered in the decorations of the Vase in the following manner:—

Beneath the crowning figure of the Poet.

A Four Female Heads, on shields, surrounded by festoons of laurel, represent,—

SILVIA, in the "*Two Gentlemen of Verona.*"
. . Silvia is excelling,
She excels each mortal thing.

VIOLA, in "*Twelfth Night; or, What You Will.*"
Conceal me what I am, and be my aid,
For such disguise as, hap'ly shall become
The form of my intent. I'll serve this Duke.

ISABELLA, in "*Measure for Measure.*"
O, were it but my life,
I'd throw it down for your deliverance
As frankly as a pin.

HELENA, in "*All's Well that Ends Well.*"
My friends were poor, but honest, so's my ove.
Be not offended, for it hurts not him
That he is loved by me.

B The Frieze, around the swelling of the Vase, presents the following subjects—On the front,

Scene of the *Masquerade*, in "*Romeo and Juliet.*"

ROMEO. Have not saints lips, and holy palmers too?
JULIET. Ay, Pilgrim, lips that they must use in prayers.

ANTONIO, BASSANIO, and SHYLOCK, in the "*Merchant of Venice.*"

SHYLOCK. Go with me to a notary, seal me there
Your single bond.

HELENA and HERMIA, in "*Midsummer Night's Dream.*"

HERMIA. God speed, fair Helena! whither away?
HELENA. Call you me fair? that fair again unsay,
Demetrius loves you fair: O happy fair!

On the reverse—

PETRUCHIO, KATHARINA, and HABERDASHER, in the "*Taming of the Shrew.*"

PETRUCHIO. It is a paltry cap,
A custard coffin, a bauble, a silken pie.

FALSTAFF, Mrs. FORD, and Mrs. PAGE, in the "*Merry Wives of Windsor.*"

MRS. FORD. He is too big to go in there. What shall I do?
FALSTAFF. Let me see't, let me see it, I'll in, I'll in.
MRS. PAGE. What! Sir John Falstaff? Are these your letters, knight?
FALSTAFF. I love thee, and none but thee; help me away, let me creep in here.

BENEDICK and BEATRICE, in "*Much Ado about Nothing.*"

BENEDICK. How doth your cousin?
BEATRICE. Very ill.
BENEDICK. And how do you?
BEATRICE. Very ill too.
BENEDICK. Serve God, love me, and mend. . . .

C The Centre Medallions represent—On the front

OTHELLO and DESDEMONA in the Council Chamber, in "*Othello.*"

BRABANTIO. Look to her, Moor, have a quick eye to see—
She has deceived her father, and may thee.
OTHELLO. My life upon her faith. . .

On the reverse—

FERDINAND brought by the spells of PROSPERO before MIRANDA, in the "*Tempest.*"

CALIBAN. I must obey! [*exit.*
ARIEL. (*enters as a Water Nymph, playing and singing.*)
FERDINAND. Where should this music be?
PROSPERO. The fringed curtains of thine eye advance,
And say, what thou seest yond' . . .
MIRANDA. What is't? a spirit—
Lord, how it looks about! Believe me, sir,
It carries a brave form. . . .

D In the spaces between the Centre Medallions and the Allegorical Figures, are introduced the fantastic characters of the Plays of Shakspeare, viz.:—On the front of the Vase.

The Fiends forsaking La Pucelle, of Orleans. (Act V., Scene 3, First Part of "*King Henry VI.*")

The Ghost of Hamlet's Father. (Act I., Scene 5, in "*Hamlet.*")

Julius Cæsar's Ghost in the tent of Brutus. (End of Act IV. in "*Julius Cæsar.*") On the left of the spectator.

The apparition of the *Eight Kings of Banquo's issue*, and *Banquo's Ghost*, evoked by the *Three Witches* in the presence of *Hecate.* (Act IV., "*Macbeth.*") On the right.

On the reverse—

Oberon, amongst his attendants, is shown by *Puck*, *Titania*, surrounded by her Fairies caressing *Bottom*, waited on by *Peasblossom*, *Cobweb*, and *Mustard-Seed.* (Act IV., Scene 1, "*Midsummer Night's Dream.*")

The *Nymphs*, *Iris*, *Ceres*, and *Juno*, of the Mask. (Act IV., Scene 1, "*Tempest.*")

THE CUP OF BYRON.

THE TAZZA OF BURNS.

HANCOCK, C. F., *continued.*

E Four Bas-reliefs on the Stem of the Vase record—
THE PRINCESS OF FRANCE, in "*Love's Labour Lost.*"
Tell him the daughter of the King of France
On serious business, craving quick despatch,
Importunes personal conference with his Grace.
ÆMILIA as *Abbess*, in the "*Comedy of Errors.*"
What then became of them I cannot tell;
I to the fortune that you see me in.
ROSALIND, in "*As You Like it.*"
I'll have no father, if you be not he [*to* DUKE.
I'll have no husband if you be not he [*to* ORLANDO.
Nor e'er wed woman if you be not she. [*to* PHEBE.
HERMIONE, in "*Winter's Tale.*"
LEONTES . . . Her natural posture!
Chide me, dear stone, that I may say indeed,
Thou art Hermione!

F The Foot of the Vase is surrounded by Figures in Round-relief, representing—
LEAR *in the Storm*, in "*King Lear.*"
Blow, wind, and crack your cheeks! rage! blow!
You cataracts and hurricanoes spout,
Till you have drenched our steeples.
HAMLET in the Churchyard, in "*Hamlet.*"
Alas! poor Yorick! I knew him.
OPHELIA *distributing flowers*, in "*Hamlet.*"
There's a daisy; I would give you some violets,
But they withered all, when my father died.
LADY MACBETH *during the murder of the King*, in "*Macbeth.*"
. . . Hark! Peace!
. . . He is about it!

The Base of the Stand is ornamented in the following manner:—

G By Eight Miniature Low-reliefs, in which are given Troilus, Cressida, and Pandarus, in *Troilus and Cressida;* Timon and the Steward in the Wood, in *Timon of Athens;* Coriolanus and Volumnia, in *Coriolanus;* Brutus and Portia in the Orchard, in *Julius Cæsar;* Cleopatra applying the Asp, in *Antony and Cleopatra;* Imogen as Fidele, and her brothers, in *Cymbeline;* Lavinia making known her misfortune, in *Titus Andronicus;* Marina nursing her father, in *Pericles, Prince of Tyre;* and,

H By Heads in High-relief, representing the Kings treated of in the historical pieces, viz.:—*King John, Richard II., Henry IV., Henry V., Henry VI., Richard III.*, and *Henry VIII.*, to which has been added that of *Queen Elizabeth*, as the Sovereign under whom the Poet lived and flourished, and of whom he says (in the play of *Henry VIII.*)—
. . . She shall be
* * * *
A pattern to all princes living with her,
And all that shall succeed . . .

THE CUP OF MILTON.

The Cup rests upon a Stand, decorated with Fruits and Flowers, and bearing on its Stem Heads in High-relief of SAMSON (from "*Samson Agonistes.*")
He patient, but undaunted, where they led him
Came to the place.
DALILA (from the same Poem).
With doubtful feet and wavering resolution,
I came, still dreading thy displeasure, Samson.
The *Virtuous Young Lady* (from the Sonnets).
Lady, that in the prime of earliest youth,
Wisely hast shunned the broad way and the green.
LYCIDAS (from "*Lycidas.*")
Where were ye, nymphs, when the remorseless deep
Closed o'er the head of your loved Lycidas?

The subjects of the Bas-reliefs on the Cup are:—
RAPHAEL *addressing* ADAM *and* EVE (from "*Paradise Lost.*")
To whom the Angel: Son of Heaven and Earth
Attend: that thou art happy—owe to God.
That thou continueth so—owe to thyself,
That is, to thy obedience: therein stand.
And SABRINA *rising from the Water* (*Comus*).
Sabrina, fair,
Listen, where thou art sitting,
* * * * *
Listen, and appear to us.

The Cover of the Cup is surmounted by a Figure of *Urania*, from Book VII. of the "*Paradise Lost.*")
Descend from heaven, *Urania*, . . .
. . . Up led by thee
Into the heaven of heavens I have presumed.

And besides Bas-reliefs with Figures of *Fame and Genius distributing Laurels*, it has two Medallions, representing—
The Allegro. But come, thou goddess, fair and free!
and *The Penseroso.* Hail, divinest melancholy!

THE CUP OF BYRON.

The Stand is decorated by Heads representing THE GIAOUR, ("*Giaour.*")
Dark and unearthly is the scowl
That glares beneath his dusky cowl.
MYRRHA ("*Sardanapalus.*")
My lord, I am no boaster of my love,
Nor of my attributes; I have shared your splendour
And will partake your fortunes.
ALP ("*Siege of Corinth.*")
The first and freshest of the host,
Which Stamboul's Sultan there can boast.
ZULEIKA ("*Bride of Abydos.*")
Woe to the head whose eye beheld
My child Zuleika's face unveil'd.

The Bas-reliefs on the body of the Cup illustrate "*Mazeppa.*"
Bring forth the horse! The horse was brought.
* * * * * *
They bound me on, . . .
Upon his back with many a thong;
They loosed him with a sudden lash—
Away! away!—and on we dash.
DON JUAN *and* HAÏDEE.
And slowly by his swimming eyes was seen
A lovely female face of seventeen.
'Twas bending close o'er his.

The Cover is surmounted by a figure of the Nymph EGERIA (from "*Childe Harold's Pilgrimage.*")
Egeria, sweet creation of some heart,
Which found no mortal resting-place so fair
As thine ideal breast! . .
And has four Medallions of MANFRED and the *Chamois Hunter* (from "*Manfred.*")
Chamois Hunter. Hold, madman!
Away with me.—I will not quit my hold!
And of GULNARE and CONRAD (from "*The Corsair.*")
"She gazed in wonder; 'Can he calmly sleep?'
. . . . but soft his slumber breaks—
He raised his head."
The other two being occupied by figures of Genius and Fame, holding laurels.

THE TAZZA OF BURNS.

The subjects illustrated in this Tazza are the following—
On the outside, and in the centre panels,
"*The Cottar's Saturday Night.*"
The priest-like father reads the sacred page.
And "*Tam O'Shanter's Ride Home.*"
For Nannie, far before the rest
Hard upon noble Maggie prest,
And flew at Tam wi' furious ettle,
But little wist she Maggie's mettle—
Ae spring brought off her master hale,
But left behind her ain gray tail.

On the Ornaments of the Side Panels and under the Handles, are introduced "*The Mountain Daisy,*" "*The Mouse,*" "*The Wounded Hare,*" and "*The Twa Dogs,*" favourite subjects amongst those treated by the Poet.

The Spandrils between the Bas-reliefs are decorated by the emblematic *Thistle;* and the interior of the Bowl contains a Medallion representing "*Highland Mary*"—
"O my sweet Highland Mary!"

HANCOCK, C. F., *continued.*

TESTIMONIAL IN SILVER, PRESENTED BY THE OFFICERS OF THE 1ST LIFE GUARDS TO GEN. HALL.
The three different epochs of the regiment—Charles II., George II., and Queen Victoria.

THE QUEEN'S YACHT CUP, IN SILVER.

VASE IN CELLINI STYLE, IN SILVER.

THE "CRAVEN" VASE, FOR FLOWERS OR FRUIT, IN SILVER.

THE VOLUNTEER TANKARD, "OUR HEARTHS AND HOMES." IN SILVER.

HANCOCK, C. F., *continued.*

THE TAZZA OF MOORE.

This Tazza (like that of Burns,) rises on a Stand decorated by Figures of young Genii, bearing festoons of Flowers, and standing between fanciful Griffins.

The Body of the Tazza presents in its centre Bas-reliefs. The *Peri* offering to Heaven the blood of the Slain Warrior (from "*Paradise and the Peri.*")

Be this, she cried, . . .
My welcome gift at the Gates of Light.
Though foul are the drops that oft distil
On the field of warfare, blood like this,
For liberty shed, so holy is.

And NAMOUNA, the enchantress, charming NOURMAHAL to sleep (from "*The Light of the Haram.*")

No sooner was the flowery crown
Placed on her head, than sleep came down,
Gently as nights of summer fall,
Upon the lids of *Nourmahal.*
* * * *
And now a spirit.
* * * *
Hovers around her.

On the Ornamental Panels, under the Handles, are introduced, The *Sword* and the *Laurel Wreath;* the *Pledging Cup;* the *Last Rose of Summer;* and the *Harp of the Minstrel Boy;* subjects from the "*Melodies.*" The Spandrils are decorated with the *Shamrock,* and the Medallion in the interior of the Tazza represents "*Nora Creina.*"

My gentle, bashful Nora Creina!

[6642]

LAW, JOHN, 3 & 4 *North Side, Bethnal Green.*—Gold and silver leaf, and gold-beaters' skin.

[6643]

LEE, BENJAMIN, 41 *Rathbone Place.*—Bracelets, brooches, earrings, rings, guards, pins, studs, and devices of hair.

The following articles are exhibted, viz.:—

Several suites of ladies' jewels, of new and elegant designs; Ladies' brooches, bracelets, eardrops, rings, necklaces, chatelaines, necklets, crosses, lockets, pencil cases, book markers, charms, and riding whips; Gentlemen's Albert guards and keys, guard chains, pins, rings, and studs. Miniature and device brooches. A choice collection of designs and specimens of hair devices for brooches and bracelets. A large bouquet of 80 flowers, formed of delicate shades of hair, and other novelties in the art of working human hair.

[6644]

LEES, JOSIAH, 37 *Spencer Street, Birmingham.*—Gold and plated chains, patent hooks, swivels, rings, &c.

[6645]

LISTER, W., & SONS, 12 *Mosley Street, Newcastle-on-Tyne.*—Silver plate and jewellery. (*See page* 54.)

[6646]

LOACH, A. M., *Regent Parade, Birmingham.*—Plated gold brooches, bracelets, and lockets.

[6647]

LOEWENSTARK, A. D., & SON, 1 *Devereux Court, Strand, W.C.*—Masonic jewels and paraphernalia, gold and silver military medals, friendly societies' presentation medals, &c.; filagree work.

[6648]

LONDON & RYDER, 17 *New Bond Street.*—Modern gold jewellery, specimens of diamond work, and silver plate.

[6649]

MANTON, H., 110 *Great Charles Street, Birmingham.*—Fancy silver goods.

[6650]

MAPPIN, BROTHERS, 222 *Regent Street; 67 & 68 King William Street, City; and Queen's Cutlery Works, Sheffield.*—Electro-silver plate.

[6651]

MAPPIN & CO., 77 *& 78 Oxford Street, W. (opposite the Pantheon), London; Rue de l'Ecuyer, Brussels;* 17 *Boulevard des Italiens, Paris; Manufactory, Royal Cutlery Works, Sheffield.* (*See pages* 56 *and* 57.)

HARRISON, W. W., *Montgomery Works, Fargate, Sheffield.*—Electro-silver plate.

Tea and coffee urns, or vases, for which Her Majesty's Royal Letters Patent were granted, May 8th, 1861.

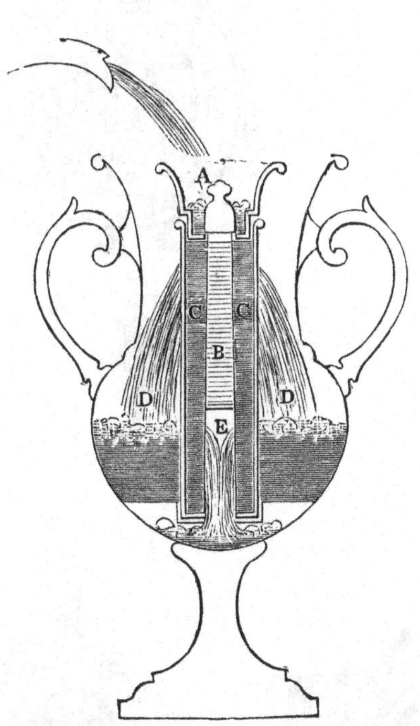

SECTION OF COFFEE URN.

This very simple invention produces the best and clearest coffee to be obtained by any known method. The ground coffee is placed in a central perforated compartment, called the Extracter, which has an inner tube, also perforated. The boiling water is poured into the receiver, A, and after passing into the inner perforated tube, B, flows out *through the coffee*, C, into the body of the vessel, D, in its course acting upon every particle of the coffee, and extracting all its useful properties. The infusion, before it can reach the outlet, must pass a *second* time through the lower portion of the ground coffee, which now acts as a filter, and renders the infusion perfectly clear.

Coffee pots on the same principal as above; Tea and Coffee services of several choice designs; Cruet stands in a variety of patterns; Breakfast stands, to contain toast, eggs, salt, and butter, with spoons, &c., complete. (Registered designs.)

Butter coolers in considerable variety; Tea caddies, neatly engraved and richly chased; Claret jugs, cut glass, with ornamental mountings; Flower vases in various styles, with coloured glass linings; Centre dish, or Salad stand, Inkstands, Tea tray, Waiters, &c., in a variety of designs; Spoons and Forks, in plain and ornamental patterns.

HOWELL, JAMES, & Co., 5, 7, 9 *Regent Street.*—Goldsmiths, silversmiths, &c.

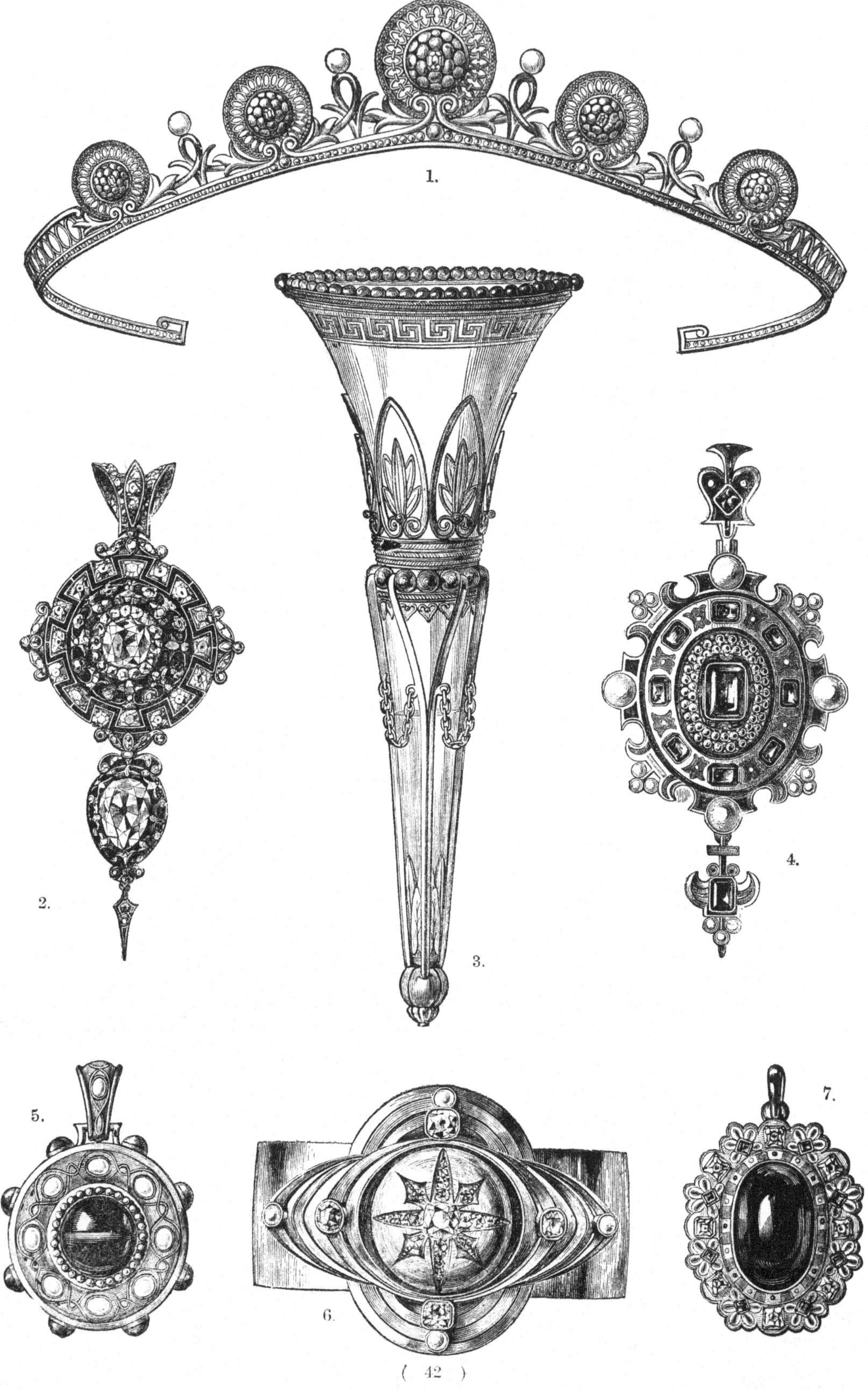

HOWELL, JAMES, & Co., *continued.*

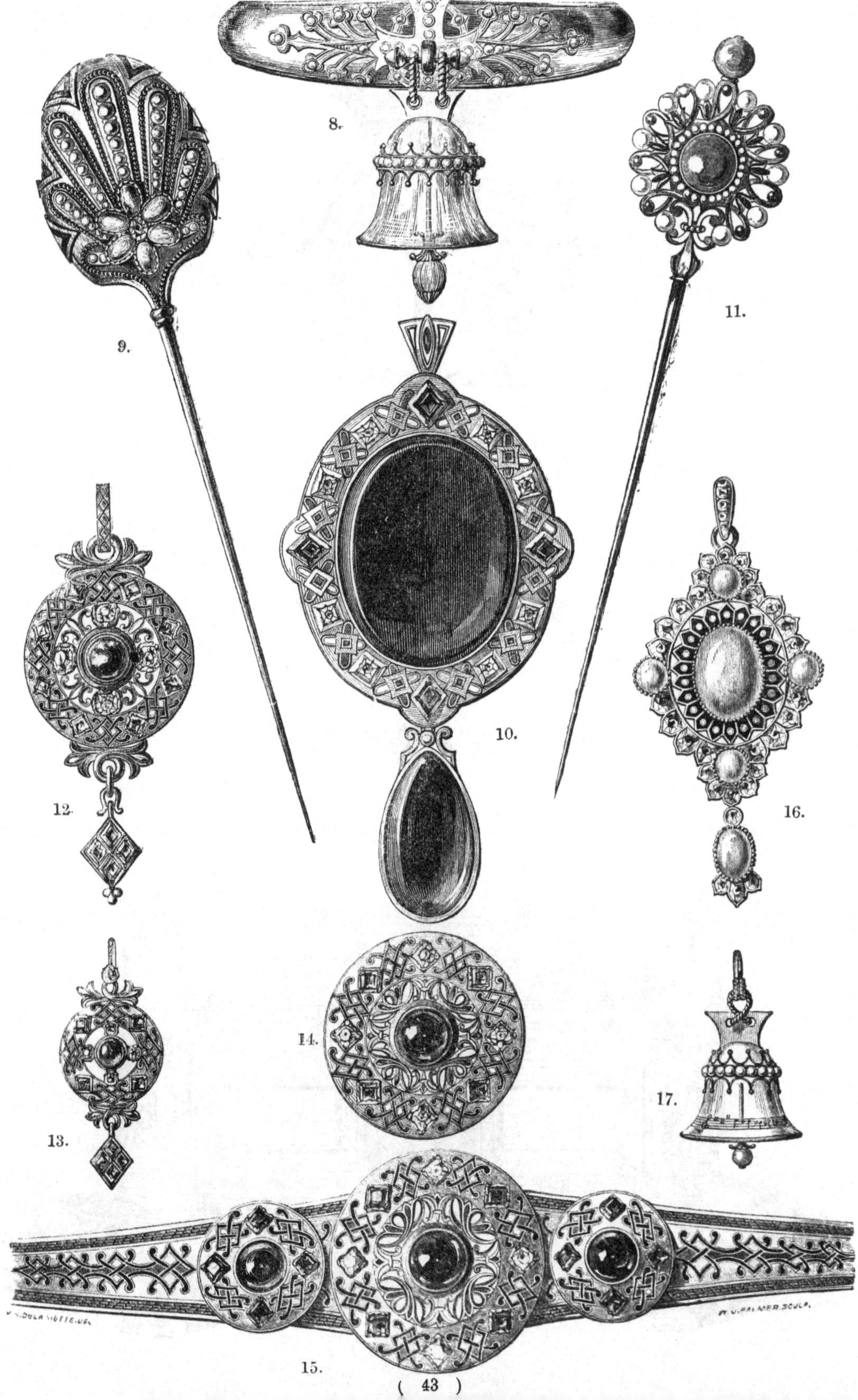

18.

19.

HOWELL, JAMES, & Co., *continued.*

1. Tiara, pierced gold, turquoise, and diamond pavés, pearl frets—Etruscan.
2. Locket, fine brilliant centre and drop, pierced open setting—Etruscan.
3. Bouquet holder, with folding cramps as flower vase, elaborate engraved gold mounts—Greek, British Museum.
4. Locket, pearl pavé, emeralds, rubies, pierced enamel—Benvenuto Cellini.
5. Locket, oriental onyx, enamel, and pearl—Anglo-Saxon.
6. Bracelet, diamonds and turquoise enamel—Celtic.
7. Locket, carbuncle, emeralds, and white enamel—Holbein.
8. Bracelet, Gothic, crystal, turquoise, and bell-pendant.
9. Shawl or hair pin, turquoise and diamond—Alhambresque.
10. Locket, very fine carbuncle, emeralds, and diamonds, to act as brooch—Holbein.
11. Shawl or hair pin, coral, enamel and pearl border, —Cellini.

12, 13, 14, 15. Carbuncle, emerald, and diamond, pierced enamelled suite, comprising brooch, ear-rings, necklace, and bracelet—Holbein.

16. Locket, pink coral and diamond, with pearl rosettes —Holbein.
17. Ear-rings, crystal bells, with pink coral coronets, engraved illuminated inscription, from Schiller's "Glocken lied"—Gothic.
18. Silver epergne and centre-piece, with pierced silver baskets for fruit, and crystal pendants for flowers; six branches for lights, with emblems of lilies and papyrus leaves. Designed by Professor Miller.
19. The Raffles Jubilee testimonial: solid silver casket, from designs and models by Professor Miller, South Kensington School of Art.

The patent dressing-case exhibited by Messrs Howell and James is a masterpiece of excellence in workmanship, great mechanical ingenuity, and good and artistic ornamentation.

The most remarkable feature in it is, undoubtedly, a clever and ingenious mechanism, by which, on the lid being raised by the hand in the usual way, the various compartments are made to open into their respective places; so that any single article that may be required for the toilet is seen at a glance, and can at once be removed without altering or disturbing any other. Those who have been accustomed to the ordinary dressing-cases, and have experienced the delay in lifting this tray, closing that drawer, and pressing the other article back into its place, will be well able to appreciate the boon now offered to them by the Inventors and Exhibitors of this case. In addition, they beg leave to call attention to the elegant and sumptuous manner in which it is fitted up throughout; it is lined with the richest silk velvet, the fittings are silver gilt, with solid gold centres, enriched with settings of pink coral. All the various instruments for the use of the toilet, with which the case is amply supplied, are of the finest cutlery, set in Russian malachite handles. The pierced ornament which surrounds the interior, and covers the backs of the ivory brushes, is also remarkable for the artistic manner in which it is treated; it is engraved to correspond with the fittings.

HOWELL, JAMES, & Co., by Appointment, 5, 7, 9 Regent Street. Goldsmiths, jewellers, silversmiths, manufacturers of dressing-cases, clocks, &c.

HUNT & ROSKELL, 156 *New Bond Street.*—Artistic works in gold and silver, watches and clocks, pearls, precious stones, &c. &c.

NO. 23.—THE BREADALBANE VASE—CANDELABRUM.

Hunt & Roskell, *continued.*

The object sought to be attained by Messrs. Hunt and Roskell in the articles which they have exhibited, is the combination of utility with the highest possible artistic excellence. They have endeavoured to make the object, whether an ornament for the table, a testimonial, or a memorial, valuable, not only for the precious metal of which it is formed, but also for the workmanship by which it is adorned; and in so doing, they believe that they are carrying out the intention which has actuated all the best workers in the precious metals. Further to forward the interests of Art, they have appended to each piece the name of the artist employed, and they are assured by the public voice that they have been so far successful that they have produced from their manufactory the finest works of art which this age can boast—masterpieces which indeed will compare with the best works in silver produced in any age or country. Two of their artists have received the Medal of Honour in this Exhibition, in addition to medals previously gained; but the fact of one of the firm having complied with a solicitation to fill the office of Juror of his Class, will explain the reason why they are otherwise unmentioned.

1. Vases, the Property of Her Majesty the Queen.

A vase in oxydized silver, damaskened. Subject: the Centaurs and Lapithæ. On the pedestal are groups and entablatures illustrative of the same subject, and enriched by the introduction of cornelian and lapis-lazuli.

A vase and pedestal in oxydized silver, marine composition. The bassi-relievi represent Venus and Adonis, and Thetis presenting to her son Achilles the armour forged by Vulcan. These highly artistic and elaborate works, by Antoine Vechte, are the property of Her Majesty the Queen, who has graciously permitted their exhibition. They were commanded by his late Royal Highness the Prince Consort.

2. An Obelisk in Silver.

This interesting memorial of the International Exhibition of 1851 was suggested by Mr. T. Hamilton. On it are engraved the statistics of the Exhibition, an explanatory, descriptive, and appropriate inscription, and mottoes in Greek and Latin.

3. The Goldsmiths' Plate. Previous to the Exhibition of 1851, the Goldsmiths' Company offered prizes for the best productions of art in silver. The works to which these prizes were awarded were not considered adapted to the tables of the Company, neither did they mark the state of art of the period. They therefore voted the sum of £5,000 to be expended in works in silver for their tables, open to competition for designs, &c., to all goldsmiths; and this resulted in the selection of the following three candelabra and two groups.

The group on the base of the grand candelabrum illustrates the granting of the Charter to the Company by Richard II., a.d. 1392. The figures represent the King delivering the Charter into the hands of the Prime Warden; a warden kneeling in front of the throne with specimens of the craft; Thomas D'Arundel, Chancellor and Archbishop of Canterbury; the Queen, Anne of Bohemia; an attendant bearing plate; William Stonden, Mayor of London, with the insignia of his office; the Chamberlain; pages playing with greyhound, &c., &c. On the base, the processes of mining, refining, and working the precious metals are illustrated. The portraits of the King and Queen are from tombs in Westminster Abbey and the painting formerly in the Star Chamber; that of Thomas D'Arundel, from a painting in the Archbishop's Palace at Lambeth.

The groups at the base of the second candelabrum represent Michael Angelo in the studio of his master, Domenico Ghirlandaio, sketching a lady who is fitting on one of the garlands from the fabrication of which, Carradi, Ghirlandaio's father, derived his name; Lorenzo de Medici, Michael Angelo's patron, is inspecting artistic works handed to him by a page. On the base are the arms of the Company. The youthful portrait of Michael Angelo is from the bust formerly in the possession of Sir T. Lawrence, and various old prints. Vasari is the authority for that of Ghirlandaio. The portrait of Lorenzo de Medici is from a terra-cotta by Michael Angelo, in the possession of the Rev. J. Sandford.

The figures on the base of the third candelabrum represent Benvenuto Cellini, George Heriot, and Sir Martin Bowes, each attended by a figure of Genius bearing emblems of the craft. The portrait of Benvenuto Cellini is from prints by Vasari; that of George Heriot, from a painting and statue in Edinburgh; that of Sir Martin Bowes, from the painting possessed by the Goldsmiths' Company.

A group in silver, illustrating the business duties or the Goldsmiths' Company. On the summit, a figure of Science, her hand resting on a crucible, points to Law upheld by Justice, in allusion to the regulation of the standard of the precious metals. On the left, a figure of Industry, with beehive and specimens of craft; Mercury, as Commerce; Plutus, the god of Wealth. On the pedestal are medallions of Edward III., Henry VII., and James I.; at the angles the arms of the Company. The portraits of Henry VII. and Edward III. are from Westminster Abbey; that of James I. from prints and painting in the British Museum.

A group in silver, illustrating the benevolence of the Goldsmiths' Company. On the summit is a figure of Prudence, and by her side Benevolence distributing to the necessitous from the Horn of Plenty. The figures beneath are a scholar and his tutor; a sick man; a widow and her orphan children, and an aged artisan about to lay aside his implements of trade—all sustained and relieved by the Goldsmiths' Company. On the pedestal are medallions of Edward III., Henry VII., and James I. At the angles are the arms of the Company. Alfred Brown, Del. et Sculp.

4. The Manchester Art Treasures Exhibition Testimonial.—Consisting of seven pieces from one model, presented to each of the members of the Executive Committee.

A figure of Genius contending with an eagle surmounts the column, around which three figures, distinguished by appropriate emblems, illustrate Painting, Sculpture, and Industrial Art. On the column are the Rose, Shamrock and Thistle, and the motto of the Exhibition—the first line of Keats's "Hyperion"—

"A thing of beauty is a joy for ever."

H. H. Armstead, Del. et Sculp.

5. The Pakington Testimonial.

A shield in oxydized silver, illustrating the public and official career of the Right Hon. Sir John Pakington, Bart., G.C.B., as First Lord of the Admiralty, Secretary of State for the Colonies, and an able advocate for general education. The alti-relievi are typical representations of events in English history.

6. The Outram Shield.—Presented to Lieut.-General Sir James Outram, Bart., G.C.B., of H.M. Bombay Army, by his friends, admirers, and brother officers.

The shield illustrates some of the most important events in the career of Sir James Outram, commencing

NO. 28.—THE DONCASTER CUP FOR 1860

Hunt & Roskell, *continued.*

with the subjugation of the Bhils, in 1822, and terminating with the Relief of Lucknow, 1857. The frame of the shield is of steel, richly damaskened with gold, and contains eight medallion portraits of Sir James's companions in the Lucknow campaign, and his companions in the Persian War. H. H. Armstead, Del. et Sculp.

7. The Lawrence Testimonial.—Presented to Sir Henry Lawrence, K.C.B., by his friends of the Punjaub.

The figure on the summit of the candelabrum typifies India. Around the shaft in bassi-relievi are five reclining deities, representing the Punjaub. The branches are richly decorated with Indian ornamentation. The palm, plantain, and fig-trees, adorn the shaft. The first group on the base is typical of the state of anarchy which existed in the Punjaub. One of Runjeet Singh's body-guard is attacked by a Hill-man; an Akalee lies dead on the ground, and above him is a dismounted Irregular horseman. The second group represents the contention with the British forces; the figures introduced are a Seikh Irregular horseman, an Artilleryman, a Seikh infantry soldier on the ground contending with a dismounted British Dragoon. The third group represents the pacification of the Punjaub. The figure of Sir Henry Lawrence is a portrait of this distinguished man; an Afghan and a Seikh chief surrender their arms to him, and accept implements of husbandry from the hands of figures allegorical of Industry and Peace. Alfred Brown, Del. et Sculp.

8. The Wellington Statuette.—Executed for the late Earl Spencer, K.G., &c., &c.

An equestrian statuette in bronze. The Orders, &c., conferred upon His Grace decorate the pedestal. H. H. Armstead, Del. et Sculp.

9. The Napier Testimonial.—Presented to Lieut.-General Sir Charles James Napier, G.C.B., Colonel of the 22nd Regiment of Foot, &c., &c., the Conqueror and late Governor of Scinde, as a memorial of their admiration, respect, and sincerest attachment, by the Officers who enjoyed the distinction of serving under him in the civil administration of Scinde.

Surmounting the base an elephant carries a rich howdah, with Beeloochee and Scindian attendants, in Ameer costume. The groups of figures, British Infantry, Zemindar and Moonshee, Water-carrier, and Fruit-merchant, severally represent the military, civil, and industrial power of Scinde. Alfred Brown, Del. et Sculp.

10. The Napier Statuette.—Presented by General Sir C. J. Napier, Commander-in-Chief in India, Colonel of the 22nd Regiment, to his brother Officers, as a mark of his friendship, and in commemoration of the conquest of Scinde, resulting from his two great victories, Meanee and Hydrabad, in which H.M. 22nd was the only European Regiment bravely led into battle, courageously followed by the native troops. Bugshai, 1850.

An equestrian statuette, in silver, of Sir Charles J. Napier, G.C.B. On the pedestal is a relievo in silver, the battle of Meanee. This statuette was selected by the officers of the 22nd in compliment to Sir C. J. Napier. Alfred Brown, Del. et Sculp.

11. The Napier Sword.—Presented to General Sir Charles James Napier, G.C.B., Colonel of the 22nd Regiment of Foot, &c., &c., by the Belooch Sirdars of Scinde, in token of the attachment and gratitude which his honourable and generous treatment of them after victory, and during a long administration of the Scinde government, has secured for him in the breasts of his former foes. Hydrabad, January 9th, 1851.

The hilt, in gold, is composed of four compartments, in which are figures of Fortitude, Truth, Justice, and Victory; the arms of Sir Charles, and the victories of Meanee, Hydrabad, and the Moodkee campaign; figures of Commerce and Industry. The sheath, of silver gilt, contains four medallions on each side, representing severally, Britannia supported by Peace, Literature, Art, and Science. The Beeloochees and Scindians casting aside their arms, and led by Victory to Civilization, Anarchy in chains. The conquest of the Hydra by Hercules. India supported by Law and Justice. Peace descending with ratified treaty to the Military, Industrial, and Commercial population of Scinde. Rebellion overthrown. Apollo's conquest of the Python. Alfred Brown, Del. et Sculp.

12. A Sword.—Presented by the Legislature of Nova Scotia to her distinguished son, Major-General Sir William Fenwick Williams, K.C.B.

This blade is manufactured of steel from the mines of Nova Scotia. On the hilt are two figures emblematical of Wisdom and Truth; on the scabbard are appropriate medallions. Thomas Brown, Del. et Sculp. Exhibited by permission of Major-General Sir W. F. Williams, K.C.B.

13. The Inglis Sword.—Presented to Major-General Sir John Eardley Wilmot Inglis, K.C.B., by the Legislature of Nova Scotia, in testimony of the admiration in which his heroic defence of Lucknow and other distinguished services, are held by the people of his native province.

The blade is manufactured of steel from the mines of Nova Scotia. The hilt, which is of silver oxydized, and partly gilt, is formed of three figures—Fame, Victory, and Justice—grouped around a shaft on which is the Royal Crown, which forms the pommel of the sword. The scabbard is divided into five entablatures, containing appropriate designs. Thomas Brown, Del. et Sculp.

14. The Richmond Testimonial.—Presented to His Grace the Duke of Richmond, Lennox and D'Aubigne, K.G., by the recipients of the War Medal, in grateful remembrance of his long and unwearied exertions in their behalf.

The group on the summit represents His Grace the late Duke of Richmond directing the attention of Britannia to the merits of her military and naval powers, illustrated by figures of Mars and Neptune. Britannia holds the Peninsula medals, which she is about to bestow. Alfred Brown, Del. et Sculp.

15. The Scarlett Testimonial.—Presented to the 5th (Princess Charlotte of Wales) Dragoon Guards, by Major-General the Hon. Sir James Yorke Scarlett, K.C.B., and Commander of the Legion of Honour, as a token of his affection for the regiment, to whose discipline at home, and gallantry before the enemy, he owes his reputation as a soldier.

Upon the shaft of this testimonial are bassi-relievi of the battles of Llerena, Salamanca, and Vittoria; and at each angle stands a trooper of the dates 1685, 1705, and 1820. The figures on the top are a mounted and unmounted vidette, in Crimean tunic. The pedestal contains three panels—Balaclava, Inkerman, and Sebastopol—and at the angles are recorded the various battles in

Hunt & Roskell, *continued*

No. 4.—The Manchester Art Testimonial.

HUNT & ROSKELL, *continued.*

which the regiment has been engaged. The panels on the plinth contain the inscription, the badge of the regiment, and the arms of the donor. Thomas Brown, Del et Sculp. Exhibited by permission of the Officers of the 5th Dragoon Guards.

16. THE LONDONDERRY TESTIMONIAL.—A memento of affection from Charles William Vane, Marquis of Londonderry, K.G., G.C.B., &c., &c., Colonel of the 2nd Life Guards, to the Regiment.

On the summit is a figure of Britannia. The bas-relief around the column represents the final charge of the Life Guards at the battle of Waterloo. On the base are figures of a mounted officer, a private, and a trumpeter of the Regiment. Alfred Brown, Del. et Sculp. Exhibited by permission of the Officers of the 2nd Life Guards.

17. A SHIELD, in silver and iron, damaskened with gold.

This *chef d'œuvre* is entirely *repoussé*, or embossed; the subjects are dedicated to Shakspeare, Milton, and Newton. This work was commenced for the Exhibition of 1851, in which the first rude sketch was shown. Antoine Vechte, Del. et Sculp.

18. THE TITAN VASE.

A vase of Etruscan form, embossed from thin sheets of silver, in the highest and lowest possible relief. The subject, which is treated in the style of Michael Angelo, is the destruction of the Titans by Jupiter, who made war upon them for having imprisoned his father Saturn. The giant sons of Cœlus and Terra, seeking to revenge the death of the Titans, are seen attacking the gods, and endeavouring to reach heaven. On the summit of the cover is Jupiter, who, with stern and angry looks, grasps thunderbolts which he hurls on the presumptuous Titans below. Antoine Vechte, Del. et Sculp.

19. A VASE and PEDESTAL in oxydized silver, by Antoine Vechte. Subject: the Centaurs and Lapithæ, same as No. 1. The property of the Right Hon. the Earl of Wemyss.

20. A VASE in oxydized silver, designed for a race prize. The subject of the design is from Homer. This was the last prize given by the late Emperor of Russia to the Ascot races. Antoine Vechte, Del. et Sculp.

21. THE KEAN TESTIMONIAL.—Presented to Charles John Kean, Esq., F.S.A., by many of his fellow Etonians, together with numerous friends and admirers among the public, March 22nd, 1862.

A vase in oxydized silver. The relievo on the body contains portrait-models of Mr. and Mrs. C. Kean, in the plays of *Lear, Macbeth, Hamlet, Richard II., Henry V., Winter's Tale, King John, Richard III., Much Ado About Nothing, Henry VIII.*, and *Merchant of Venice*, with figures of Shakspeare, Tragedy, and Comedy.

Two candelabra for five lights. Upon the bases are portraits of Shakspeare in bas-relief. The plinths are enriched with masks of Tragedy and Comedy, in bold relief.

Four dessert stands, in oxydized silver, richly decorated.

Two groups in oxydized silver. The first illustrative of Shakspeare's *Midsummer Night's Dream.* H. H. Armstead Del. et Sculp. Exhibited by permission of C. J. Kean, Esq., F.S.A.

22. A ROSE-WATER FOUNTAIN, in silver, partly gilt, presented to his Highness the Marahja Runbeer Sing, by Her Majesty Queen Victoria. A. J. Barrett, Del. et Sculp.

23. A VASE, to serve also as a candelabrum. Executed for the Most Noble the Marquis of Breadalbane, K.T., &c.

This important work forms the repository of a number of the Poniatowski gems, which are rendered translucent by interior lights. The body of the vase has upon it, in bas-relief, two figures of Venus, in keeping with the subjects of the intaglios. Upon the shoulders are those of Mars and Venus. Antoine Vechte, Del. et Sculp.

24. A GROUP OF STAGS, designed and executed for the Earl of Stamford and Warrington. The withered oaks are modelled from trees in Lord Stamford's Park, at Bradgate. Alfred Brown, Del. et Sculp.

25. THE SEYMOUR TESTIMONIAL.—Presented to H. E. Rear Admiral Sir Michael Seymour, K.C.B., Commander-in-Chief of Her Majesty's Naval Force on the East India and China Station, by the British Mercantile Community of Hong Kong, March, 1859.

A service of plate. On the base of the centre ornament are four figures, representing Britannia, China, Navigation, and Commerce; panels with views of Pekin, Canton, Victoria, dragons, an emblem of China, &c., &c. Thomas Brown, Del. et Sculp.

26. THE BRASSEY TESTIMONIAL.

A silver tazza, surmounted by a figure of Science, on a coral base with cable border. On the base are figures of a Sailor, a Navigator, a Miner, and an Engineer. The four figures surrounding the stem represent the elements. Archibald Barrett, Del. et Sculp.

A Series of Race-cups, from the year 1851 to the year 1862, including amongst them that of—

27. GOODWOOD, 1860.

A vase in silver. Subject illustrative of Chaucer's "Canterbury Pilgrimage." On the cover the Poet is seated amidst Genii. The handles are formed by figures emblematic of Spring, the poem opening at that season. By Thomas Brown.

28. THE DONCASTER CUP, 1860.

A cup in oxydized silver, illustrative of the ancient ballad of the birth and exploits of St. George. By H. H. ARMSTEAD.

22. THE GOODWOOD CUP, 1861.

A vase in silver; the form, ornament, and moulding of pure Grecian style. Two relievi illustrate the self-sacrifice of Curtius. R. H. Roskell, Del. et Sculp.

And many others.

30. Two large SILVER VASES, designed for moderator lamps. The machinery is on a new principle, for double burners. On the covers are groups of stags and hinds; on the bodies, panels containing alti-relievi of the chase; on the feet, groups of deer-stalkers.

HUNT & ROSKELL, *continued.*

A TESTMONIAL IN FORM OF A CLOCK, PRESENTED TO SIR PROBY T. CAUTLEY, K.C.B.

HUNT & ROSKELL, *continued.*

31. A grand centre MODERATOR LAMP, composed of the arms and supporters of the Earl of Dudley.

32. A MEDALLION in platinum *repoussé*, by Antoine Vechte, executed for the Department of Science and Art, South Kensington. A deposit copy of this medal is awarded to the successful students in schools of art.

33. A VASE, by Antoine Vechte, in silver *repoussé*, executed for the late Earl of Ellesmere.

On one side, a relievo represents Cupid carrying Psyche to Heaven, surrounded by the Graces. On the other, Psyche is presented to Venus by little Loves, who endeavour to appease the anger of the goddess.

34. A MISSAL-COVER (unfinished) in platinum *repoussé*, by Antoine Vechte, executed for his Royal Highness the Duke d'Aumale.

The subject of this work of art is the Assumption of the Virgin, who is represented surrounded by angels, about to place upon her head a celestial crown.

35. THE CHANDOS TESTIMONIAL.—Presented by the Edinburgh, Perth, and Dundee Railway Company, the Scottish Central Railway Company, and the Scottish North-Eastern Railway Company, to the Marquis of Chandos.

A candelabrum, or epergne, in the Greek style; the shaft supported by a figure of Science, and surmounted by that of Mercury. Two smaller candelabra, or epergnes, supported by figures of Wisdom and Justice.

36. THE WILLOUGHBY TESTIMONIAL.—Presented to J. P. Willoughby, Esq., on the occasion of his leaving Bombay, on the 3rd of May, 1851, for his philanthropic labours in the abolition of infanticide in the province of Khatiawar, &c.

A centre-piece, composed of a column of Indian character, surmounted by a basket. The two groups upon the plinth illustrate the exertions of Mr. Willoughby for the suppression of infanticide—the one, "Wisdom appealing to Natural Affection," the other descriptive of the happiness attained by the successful termination of his labours.

37. THE CONOLLY TESTIMONIAL.—Presented to John Conolly, M.D., Physician to the Hanwell Lunatic Asylum, for improving the condition of the insane. 1852.

The groups illustrate Melancholy and Raving Madness under restraint, &c.

38. THE MCCLURE TESTIMONIAL.—Presented to Captain Sir Robert J. L. Mesurier McClure, R.N., late of H.M.S. "Investigator," for penetrating through the Polar Ocean in search of Sir John Franklin, and the discovery of the North-West Passage.

A figure of Fame on a globe engraved with chart showing the North-West Passage.

39. THE SALFORD BADGE AND CHAIN.—A massive gold Corporation chain, with a richly ornamented gold and enamelled badge, with the arms, crest, and supporters of Salford.

40. THE RIPON BADGE AND CHAIN.—A massive gold badge and chain, richly enamelled with the ancient insignia, arms, and badges of the Mayors and Guilds of Ripon.

Many other works in silver.

Among the jewels exhibited by this firm are the Nassuck and Arcot diamonds, the property of the Most Noble the Marquis of Westminster, by whose permission they are exhibited.

Other fine jewels, consisting of a splendid sapphire and diamond suite, together with some fine diamonds, pearls, and opals, &c.

A Head Ornament, containing a remarkably fine and large ruby, a "*pierre d'échantillon*," set with other rubies and diamonds.

A very fine sapphire, weighing 680 grains, and a fine ruby spinelle, 323 grains; these stones are set with *entourages* of brilliants. The ruby spinelle was recently cut by Messrs. Hunt & Roskell, from a stone possessed by various kings of Delhi, similar in shape to those forming the Queen's necklace, exhibited in Messrs. Garrard's case.

A fine brilliant, weight 93½, and another weighing 95½ grains, set as stars.

A row of thirty-two extraordinary fine pearls, weighing 1254 grains, each pearl averaging upwards of 39 grains; value £8,000. This is believed to be the largest pearl necklace of so fine a quality now for sale.

To show the art of DIAMOND CUTTING, Messrs. Hunt & Roskell have erected a mill in the Western Annex, Class 7, No. 1627, where the process may be seen and clearly explained. As the experts in this art are all of the Jewish persuasion, no work will be done on Saturdays; but Mr. Auerhaan or one of his assistants will attend on those days to explain the process. The chief difficulty in diamond cutting exists in the fact that the diamond is the hardest substance known to exist, and it follows that it can only be cut by itself, and that in polishing it a great rapidity of motion is necessary. The first step taken with diamonds of ordinary size is to set them in cement on the ends of two pieces of wood, which are then held in the hands, and rubbed together; and the one diamond grinds the other away to something of the form required. The dust, which is of considerable commercial value, is carefully preserved for the purpose of polishing.

After cutting, in order to proceed with polishing the stone, the diamond is embedded in soft metal, and, by means of clamps, brought to bear upon the skieve at the proper angle.

The skieve is a horizontal plate composed of soft iron, and this plate is charged with diamond powder and oil, and has a revolution imparted to it of from 2000 to 2500 times a minute; it is about twelve inches in diameter, and the periphery, therefore, travels at the rate of about 100 miles per hour, and runs until the required facet is formed and polished; for the skieve first cuts, and then, as the powder runs fine, polishes.

On the double cut diamond there are sixty-two of these facets: namely, on the upper part thirty-three, and beneath the girdle twenty-nine. Large diamonds are not exposed to the risk of cutting, but are polished from the rough on the skieve.

In the Art Designs for Manufacturers, Class 38 A, Messrs. Hunt & Roskell have, by request, exhibited some drawings by their artists, Messrs. Armstead, Alfred Brown, Thomas Brown, A. J. Barrett, Bailey, R.A., N. R. Roskell, Antoine Vechte, and others. Also some models: the Æneas shield, by Pitts; a figure of Æsculapius, and another of Charles Kemble, in the character of Hamlet, by Chantrey, R.A.; the latter highly interesting, as being the last model made by this eminent sculptor for a work in silver: also a medallion model by Stothard.

Lister, W., & Sons, *Newcastle-on-Tyne.*—Silver plate and jewellery.

Dessert service in sliver, electro-plate, or gilt.

Registered design by Wm. Lister & Sons, Silversmiths to the Queen.

[6652]

MARSHALL, W., & Co., 24 *Princess Street, Edinburgh.*—Gold and silver enamelled jewellery, in antique style.

[6653]

MARTIN, HALL, & Co., *Sheffield.*—Silver and electro-plate, and silver-plated cutlery.

[6654]

MUIRHEAD, JAMES, & SON, *Glasgow.*—Silver and electro dessert, tea services, covers, dishes, steam-boat plate, &c., &c.

[6655]

NELIS, JOHN, *Omagh, Ireland.*—Specimens of pearls found in the river Strule, Omagh, Ireland.

[6656]

PARKER & STONE, 7 *Myddelton Street, Clerkenwell.*—Gold chains and jewellery.

[6657]

PAYNE, EDWARD ROBERT, Goldsmith to the Queen, *Bath.*—Two vases in silver, from the antique.

[6658]

PHILLIPS, ROBERT, 23 *Cockspur Street, London, N.W.*—Works in gold, silver, coral, and precious stones.

Prize Medal for Excellence in Design and Manufacture.

An enamel portrait of Shakspeare, by Essex, from the Chandos picture, mounted as brooch, of Elizabethan design, enriched with jewels, enamels, and arabesques.

Bracelet of rare pink coral, mounted in the Etruscan style, in rich massive gold, relieved by pale grey enamel; part of a *parure* made for Mrs. Thomas Fairbairn.

Mappin & Co., 77 & 78 *Oxford Street, W. (opposite the Pantheon), London; Rue de l'Ecuyer, Brussels;* 17 *Boulevard des Italiens, Paris; Manufactory, Royal Cutlery Works, Sheffield.*

Selections from the following manufactures are exhibited:—

Celebrated electro-silver plate on hard Nickel silver.

Presentation plate prizes, for Rifle and Archery matches.

Candelabra.

Epergnes.

Fruit Stands.

Flower Stands.

Plateaux and artistic plate of every description for the table.

Military messes, hotels, and large establishments supplied with complete services of plate at Sheffield prices.

Dinner Services.

Tea and Coffee Services.

Wine Coolers.

Salvers.

Tea Trays.

Cruet Frames.

Liqueur Frames.

Candlesticks.

Toast Racks.

Soy Frames.

EPERGNE.
Subject—Cupid Disarmed.

Epergnes from £11 to £100, of the most beautiful designs, in stock, ready for presentation.

FRUIT STAND.
Sea-horse supporting Shell.

INKSTAND.
Fox and the Crane.

Dessert stands from 30*s.* to £20.

Mappin and Co.'s celebrated Dressing-bags and Cases.—See page 6, Class XXXVI.

MAPPIN & Co., *continued.*

CELEBRATED ELECTRO-SILVER PLATE ON HARD NICKEL SILVER.

Mappin and Co. have a beautiful selection of Tea Services, from £4 10*s.* the complete service.

Mappin and Co.'s Dish Covers have been manufactured with special attention to beauty of form and ornament. Prices range from £10 to £35 the set of four.

CRUET STANDS,

From £0 15 0

To £10 0 0

WAITERS, from 24*s.* to £10.

BUTTER DISHES,

From £1 to £8.

Mappin and Co.'s Side Dishes, arranged so that a set of four will form eight dishes, from £9 to £30 the set.

Mappin and Co.'s celebrated Cutlery.—See pages 164, 165, Class XXXII.

[6659]

PORTLAND COMPANY (Limited), 6 *Ridinghouse Street, London.*—Manufacturers of silver and electro-plated goods, by patent machinery.

[6660]

PRIME, THOMAS, & SON, *Magneto-Plate Works, Birmingham.*—Dinner, dessert, and tea services, &c., in silver and electro-plate. (*See page* 59.)

[6661]

READING, JOHN, 82 *Spencer Street, Birmingham.*—Patent spring sleeve-links and solitaires, key-rings, hooks, swivels, &c.

[6662]

REID & SONS, *Newcastle-upon-Tyne.*—Silver and electro-plated services, articles in aluminium and aluminium-bronze, &c.

Obtained Honourable Mention at the Exhibition of 1851, *and Medal*, 1862.

Messrs. Reid and Sons have had the honour to be appointed Goldsmiths and Jewellers to Her Majesty. The following are exhibited by them, viz.:—

Silver centre piece, with figures, "Flora and Pomona."

Dessert service, Indian, 12 pieces, viz.:—centre piece, 2 assiettes montées, 6 dessert stands, and 3 plateaux.

Large inkstand, "Renaissance."

Tea and coffee service, "Columbia."

Tea and coffee service, new Vase form.

Salver, beaded, "Renaissance."

Dessert stand, "Venus rising from the Sea."

Helping spoons and ladles, Antique.

Fish carvers, ditto.

Potato spoons, with bust of Raleigh.

Alms dish, presentation trowel, watch, knives, forks, spoons, and sundry small articles, in aluminium and aluminium bronze, produced at Newcastle.

[6663]

RETTIE, MIDDLETON, & SONS, *Union Street, Aberdeen.*—Granite jewellery, and silver crest brooches.

[6664]

ROBINSON, H., 61 *Bolsover Street, Euston Road.*—Inkstands, caskets, vases, &c.

[6665]

RUSSELL, J., *Warstone Lane, Birmingham.*—Silver and gold jewellery.

[6666]

SHAW, C. T., *Great Hampton Street, Birmingham.*—Gold brooches, bracelets, rings, and chains.

[6667]

SMITH & NICHOLSON, Manufacturers, 12 *Duke Street, Lincoln's Inn Fields, London.*—Silver and electro-plated articles.

[6668]

SPENCER, WILLIAM, 33 *Regent Place, Birmingham.*—Gentlemen's rings, pins, and studs; ladies' rings, earrings, necklets, &c.

[6669]

SPURRIER, W., 5 *New Hall Street, Birmingham.*—Electro-silver tea services, entrée dishes, &c.

[6670]

TATNELL, HENRY, 120 *Salisbury Square, Fleet Street.*—Electro-silver plated on German silver wares.

PRIME, THOMAS, & SON, *Magneto-Plate Works, Birmingham.*—Dinner, dessert, and tea services, &c., in silver and electro-plate.

Obtained the Medals of the Paris Exhibition, in 1855, *and the present Exhibition.*

1. Italian epergne or centre piece, and plateau, with figures of Tragedy, Comedy, and Music.
2. Coffee and tea service, kettle and tray, designed after Greek models.
3. Coffee and tea service, kettle and tray, richly engraved: wrought by hand, without the use of dies.
5. Enriched gothic communion service.
7. Silver cruet frame, with engraved bottles; the design and ornamentation in the Moresque style.
9, 10. Engraved glass claret jugs, with plated mounts and handles.
12. Silver claret jug, richly engraved.
13. Inkstand of Moresque design—group in centre, "The Pet Lamb."
15. Presentation trowel, part oxydized, and gilt, the handle enriched with enamels.
18. Engraved glass butter dish and cover, with plated stand—Greek.
21. Pair of fish carvers, beaded handles.
22, 23. Spoons and forks, Princess and other patterns.
24, 27. Dessert, butter, and fish table knives.
28, 30. Engraved waiters, various designs.

[6671]

TENNANT, J., 149 *Strand, London.*—Stones used in jewellery, &c.

[6672]

THOMAS, 153 *New Bond Street, London.*—Elegant and useful articles in silver plate. (*See page* 61.)

[6674]

WESTWOOD, B. W., 20 *Warstone Road, Birmingham.*—Rings, brooches, pins, &c.

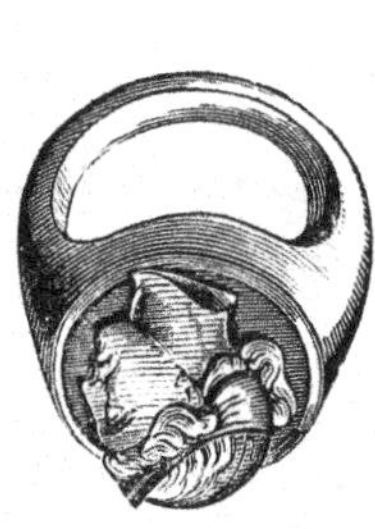

The following are shown, viz. :—

An assortment of 18 carat standard gold mourning rings, in Ecclesiastic, Egyptian, Old English, Grecian, and Arabesque styles.

An assortment of gentlemen's signet, stone, cameo, and other rings, bright and coloured, in 9 carats, 12 carats, 15 carats, 18 carats, 22 carats (or guinea gold), and 24 carats, the standard of purity, or virgin gold.

An assortment of ladies' rings in various standards, set with diamonds, rubies, emeralds, and other precious stones.

Onyx cameo brooches, with mourning and fancy borders, in 18 carat gold.

Bracelets, scarf-rings, pins, &c.

[6675]

WHEATLEY, JAMES ATKINSON, 31 *English Street, Carlisle.*—Cumberland lead, lead ore, Cumberland silver; the Cumbrian cup, jewellery, &c. (*See page* 62.)

[6676]

WIDDOWSON & VEALE, 73 *Strand.*—Silver plate and jewellery.

[6677]

WILEY, W. E., *Graham Street, Birmingham.*—Gold pens, pencil cases, &c.

THOMAS, 153 *New Bond Street, London.*—Elegant and useful articles in silver plate.

SILVER TOILET GLASS.

Mr. Thomas, of New Bond Street, exhibits many beautiful and highly-finished specimens of the silversmith's art. His attention has evidently been directed to the production of such articles as would be required in daily use.

We give an engraving of an exceedingly elegant and well-executed toilet glass, boldly chased in silver, and richly gilt.

There are also several specimens of *repoussé* work, which deserve notice on account of the general boldness of style and vigour of execution. The Bradgate Park Testimonial, a large rosewater dish or sideboard shield, the result of a penny subscription, subscribed for by the poorer inhabitants of Leicester, and presented to the Earl of Stamford and Warrington, is of a most elaborate character, and is a fine example of chasing.

The small, but prettily arranged, case of this exhibitor contains numerous pieces of plate, ably modelled, and of a very high order of merit; many admirably adapted for racing, yachting, volunteer, and other presentation prizes.

WHEATLEY, JAMES ATKINSON, 31 *English Street, Carlisle.*—Cumberland lead, lead ore, Cumberland silver; the Cumbrian cup, jewellery, &c.

Specimens from the Alston Lead Mines, Cumberland, furnished by Messrs. Shield & Dinning, Langley Smelt Mills, Haydon Bridge, viz.:—

1. Lead ore (two specimens).
2. Lead extracted from the ore.
3. Lead ore desilverized.
4. Cake of pure silver.
5. Granulated silver, lead crystals, and litharge, introduced to illustrate the process of extracting silver from lead.

Cumberland silver jewellery, designed and manufactured by the exhibitor, including the following designs:—

6. The Cumberland brooch and bracelet.
7. The Carlisle brooch and bracelet.
8. The Border brooch, bracelet, and earrings.
9. The Carlisle Castle brooch.
10. Carlisle Cathedral brooch.
11. The Musgrave brooch.
12. The Naworth brooch.
13. The Lannercost brooch.
14. The Burgh brooch, showing the monument erected to Edward I. on Burgh Marsh, erected on the spot where he died.
15. The Cumbrian cup.

The body composed of white glass of the form of a Roman urn, with Cumberland silver mounts, standing on an ebony plateau, designed to symbolize the peculiar features of Cumberland scenery.

THE CUMBRIAN CUP

16. A paper cutter, designed in the form of a Roman sword, ornamented with scroll work, and bearing on one side the words "Murus Severi," and on the other "Luguvallium," the name borne by the city of Carlisle when a Roman station.
17. A vinaigrette, the lid bearing a crown in frosted silver, surrounded by a wreath of thistles and roses entwined.
18. A miniature portrait of Napoleon I. in water colours, by Colanton, representing the Emperor in his coronation robes, and mounted in a splendid wreath of diamonds. The portrait is original, and was painted for the mother of Napoleon, and by her bequeathed to one of the old nobility of France.
19. A few fine specimens of goldsmiths' work in gold; gem and enamelled brooches, bracelets, and gem and signet rings of new designs, manufactured by eminent London firms; and scarf pins, with a new safety guard, the invention of the exhibitor, suitable to sporting or valuable gem pins, being a security against loss, and simple in operation.

[6678]

WILKINSON, H., & Co., *Sheffield, and 4 Bolt Court, Fleet Street.*—Silver and electro-plate, cutlery, &c.

[6679]

WILKINSON, THOMAS, & Co., 15 *Great Hampton Street, Birmingham.*—Best electro-plated on German silver centre piece, &c.

[6680]

WILLS BROTHERS, 12 *Euston Road, N.W.*—Useful art works in terra cotta, silver, and silver-bronze, for "Rifle Cups," &c.

[6681]

MACCARTHY, H., 21 *Lower Grosvenor Street.*—Groups in electro-silver and bronze.

[6682]

RESTELL, R., 35 *High Street, Croydon.*—Double-lock jewellery, a new mode of fastening.

[6683]

THWAITES, J. H. B., *Bristol.*—Specimens illustrative of a new method of cutting precious stones.

[6684]

HARRISON, F. A., *Birmingham.*—Jewellery.

[6685]

GOGGIN, C., 13 *Nassau Street, Dublin.*—Bog oak ornaments.

CLASS XXXIV.

GLASS, FOR DECORATIVE AND HOUSEHOLD PURPOSES.

Sub-Class A.—*Stained Glass, and Glass used in Buildings and Decorations.*

[6710]

Baillie, Thomas, & Co., 118 *Wardour Street, London.*—Specimens of stained and painted glass for windows.

[6711]

Ballantine, James, & Son, 42 *George Street, Edinburgh.*—Stained glass windows for churches, halls, and mansions. (*See page* 64.)

[6713]

Barnett, Henry Mark, *Newcastle-upon-Tyne.*—Stained glass medallion and foliage window; subject, Life of Christ.

[6716]

Chance Brothers & Co., *Glass Works, near Birmingham.*—Crown, sheet, plate, painted, and optical glasses, shades, &c. (*See pages* 66 *to* 69.)

[6717]

Claudet & Houghton, 89 *High Holborn.*—Glass shades—Painted and stained window-glass.

Window of three compartments. Subject, "The Ascension." Decorated style.

Window of one compartment. Figure of St. Peter. Perpendicular.

Window of one compartment. Figure of St. John. Decorated.

Three compartments. Specimens of geometrical windows, various styles.

[6718]

Clayton, John R., & Bell, Alfred, 311 *Regent Street.*—Stained glass.

[6719]

Cox & Son, 28 *&* 29 *Southampton Street, and* 43 *&* 44 *Maiden Lane.*—Painted glass for cathedral and church windows.

1. East window of Wimbledon Church. Subject, "The Crucifixion of our Saviour."
2. Memorial window for Worthing Church.
3. Memorial window for Haversham Church, Newport Pagnell.
4. Richly-decorated window. Subject, "Adoration of the Magi."
5. Specimens of geometric and other windows.

BALLANTINE, JAMES, & SON, 42 *George Street, Edinburgh.*—Stained glass windows for churches, halls, and mansions.

WINDOW FOR ALL SAINTS' CHURCH, KENSINGTON.

Five-light window, with tracery. Memorial to the late Thomas B. Crompton, of Farnworth, for Prestolee Church, Lancashire. Subject, "The Crucifixion."

Twin-light window, with tracery, for All Saints' Church, Kensington Park, executed for the Chevalier Burnes, K.H. Figures of St. John and the Poor Widow.

Three-light window, for Hall of South Bantaskine, where the battle of Falkirk was fought in 1745. Figures of Prince Charles Edward Stuart, Lord George Murray, and Lord John Drummond, with heraldic bearings.

Three windows, for National Bank of Scotland, Glasgow. Figures, groups, and emblems, illustrative of Commerce, Mechanics, and Agriculture.

Three-light window for Presbytery Hall of the Free Church of Scotland, Edinburgh. Medallion heads of the early and recent leaders in the Scotch Presbyterian Church—Wishart, Knox, Henderson, Erskine, Moncrieff, Thomson, McCrie, Chalmers, and Cunningham.

Vestibule window for Chevalier Burnes, K.H., Ladbroke Square, with heraldic bearings, Order of Knight of the Guelphs, &c.

These windows were designed and executed 1861—1862.

[6721]

FIELD & ALLAN, *Edinburgh and Leith.*—Painted glass window.

[6722]

FORREST, JAMES ALEXANDER & Co., (late Forrest & Bromley,) 58 *Lime Street, Liverpool.*—Two stained glass windows, intended for Glasgow Cathedral and All Saints', London.

Stained glass window for the crypt of Glasgow Cathedral, in memory of the late J. R. Nichol, Professor of Astronomy in the University of Glasgow. Subject, "The Wise Men's Journey to Bethlehem."

Stained glass window for All Saints', Islington, London, illustrating three miracles in the life of our Saviour, viz., "Making the Blind to see," "Raising of Lazarus," "Healing the Sick of the Palsy."

Price lists of plate, sheet, crown, photographic, and horticultural window glass sent free on application.

[6723]

GIBBS, ALEXANDER, 38 *Bedford Square.*—Designs for stained glass windows. (*See page* 70.)

[6724]

GIBBS, CHARLES, 148 *Marylebone Road, Regent's Park, London, N.W.*—Crucifixion window, and one from the life of David, and various examples.

[6725]

HARDMAN, JOHN, & Co., 166 *Great Charles Street, Birmingham;* 13 *King William Street, Strand, London, W C.;* 1 *Upper Camden Street, Dublin.*—Stained glass windows.

1. Canterbury Cathedral (date, close of 12th century). The subjects are the same as the window originally existing, though long since destroyed, a description of which was found in a manuscript in the library; and are in three divisions. 1st. Our Lord, with the Samaritan woman at the well, with four accompanying typical groups. 2nd. Mary and Martha, with Our Lord, and four typical groups. 3rd. Mary anointing Our Lord's feet, surrounded by four groups.

2. Worcester Cathedral (date, middle of 13th century). Memorial window of the late Colonel Unett, of the 3rd Dragoons. The subjects are taken from the life of Joshua, the great captain of the Israelites.

3. Two windows, selected from those given by His Royal Highness the late Prince Consort to the church at Whippingham, Isle of Wight. The subjects are, of the first, "The True Vine;" and of the other, the Royal Arms, those of the late Prince, and those of the Prince of Wales.

4. Harrow School Chapel (date, close of 13th century). Subject, the Transfiguration of Our Lord.

5. St. Stephen's Crypt, Westminster (date, commencement of 14th century). Subject, the Deposition of St. Stephen, being one of a series of windows illustrating the life of the Saint.

6. Danesfield (date, middle of 14th century). Subject, the Blessed Virgin, with typical subjects from the Old Testament.

7. Doncaster, St. George's Church (date, middle of 14th century), the east window. Subjects from the life and passion of Our Lord.

8. Oxford, All Souls' College (date, 15th century). "The Signing of the Servants of God," from the 7th chapter of the Apocalypse.

9. Norwich Cathedral (date, middle of 14th century). The figures of Faith, Hope, and Charity, with appropriate groups beneath.

10. Worcester Cathedral, east window of ten lancet lights (date, early 13th century). Groups from the life and passion of Our Lord.

[6726]

HARTLEY, J., & Co., *Sunderland.*—Specimens of stained glass, in the two windows terminating the nave.

Chance Brothers & Co., *Glass Works, near Birmingham.*—Crown, sheet, plate, painted, and optical glass, shades, &c.

Robin Hood's last shot.

Chance Brothers & Co., *continued.*

1. Series of specimens to illustrate the manufacture of Crown and Sheet Window Glass, the two kinds mostly used. Crown is preferred, on account of its brilliant surface, but is limited in size by the circular form and bull's-eye, for which reason sheet glass is gradually taking its place. The sheet and crown specimens are made of *extra-white* glass.

2. Sample panes, of various sizes and thicknesses, show the different qualities supplied for home trade and exportation, for glazing purposes and for prints.

3. Samples of Crystal Sheet Glass, made from sulphate. A very superior article for engravings; it does not *sweat.*

4. Chance's Patent Plate Glass is *blown* plate, obtained by grinding off the uneven surface of sheet glass, and polishing it. See samples of different colours, thicknesses, and qualities, in various sizes. Attention is particularly directed to *extra-white patent plate.*

5. Patent Rolled Rough Plate, ⅛th to ½-inch thick. An excellent glass for roofs of railway sheds, greenhouses, skylights, factories, &c. Samples of various kinds, plain, fluted, &c.

6. Coloured Window Glass, flashed colours, pot metals, and cathedral tints, in sheet, plain, and antique, for leaded windows. Rolled Plate pot metals and cathedral tints are much liked for church windows; but Antique Sheet more nearly resembles old glass in appearance and effect.

7. Hollow Lenses, and Bent Hand-lamp Glasses, red and green, of different thicknesses, for railway signal lanterns.

8. Admiralty Ship Signal-light Lenses, red and green, of various patterns.

9. Photographic Glass. — Glass plates, glass baths, dishes, and dippers. Samples of various sizes and qualities of crown plates, unpolished and polished, crystal sheet plates, purple plates (sheet and patent plate), opal plates, patent plate plates of light and extra-white colour, glass baths, greenish colour and yellow, glass dishes, glass dippers.

10. Miscellaneous Articles in Glass.—Glass tiles and slates, simple and corrugated, Ceylon pattern, &c., for roofs of barns, skylights, &c.; milkpans, propagating glasses, preserve jars, &c.; crown glass *gauge tubes* for steam boilers, much superior to those made in flint. Deck-lights and pressed tumblers of extra-white glass.

11. Crown Glass Shades, for clocks, ornaments, &c.

12. Optical Glass, for telescopes, cameras, and microscopes. Discs from 25 inches diameter downwards, and plates for cutting up, of crown and flint glass. Glass Slides, 3 by 1, of thin crown, and of patent plate, usual and white; and Thin Microscopic Glass, of three degrees of thinness, for covering objects for examination under a microscope.

13. Ornamental Window Glass.—Sheets of various enamelled stencilled patterns, and of double-etched, stained-enamelled, and embossed patterns, centre patterns, landscapes, flowers, &c.

Leaded Windows.—Two specimens, from designs of Sebastian Evans, M.A., Manager of the Ornamental Department of Chance Brothers and Co., are exhibited in the North Gallery of the East Transept. The subjects are respectively, "Robin Hood's Last Shot," and "The Madonna and Child, accompanied by the Four Archangels." A drawing of the former is given on the opposite page, and a description (from the ballad) on this page. In the latter subject, Raphael bears the pilgrim's staff and gourd, Gabriel the lily, Uriel the scroll, and Michael the sword and scales—the symbolical emblems of "The Four Angels who sustain the Throne of God." In the tracery of the window are other attendant angels, and wreaths of lilies entwined with roses.

Lenticular Lighthouse Glass is manufactured by Chance Brothers and Co., for their own lenticular apparatus. See their *First-order Revolving Light,* exhibited as a trophy in the West Nave; entered under Class 13 in the Industrial and Illustrated Catalogues, a drawing and detailed description being however, given under Class 34 in the latter, as see the following pages. A medal has been awarded to them for this Light, and for their optical glass in general.

ROBIN HOOD'S LAST SHOT. *(See opposite Page.)*

"Yet he was beguiled, I wis,
By a wicked woman,
The Prioress of Kirkleys,
That nigh was of his kin.
For the love of a knight,
Sir Roger of Doncaster,
That was her own special,
Full evil may they fare!"

* * * * * *

'"Give me my bent bow in my hand,
And a broad arrow I'll let flee,
And where this arrow is taken up,
There shall my grave digged be.
Lay me a green sod under my head,
And another at my feet;
And lay my bent bow at my side,
Which was my music sweet;
And make my grave of gravel and green,
As is most right and meet.'"

N.B.—Chance Brothers & Co. *are excluded from competition for a Medal for their Window Glass and Leaded Windows, in consequence of the senior partner of their Firm having been appointed a member of the Jury for Class* 34, *to which all the above glass belongs, with the exception of the Optical Glass* (*No.* 12), *which has been transferred by the Jury to Class* 13, *and to which a Medal has been awarded in connection with the First-order Dioptric Light.*

Chance Brothers & Co., *continued.*

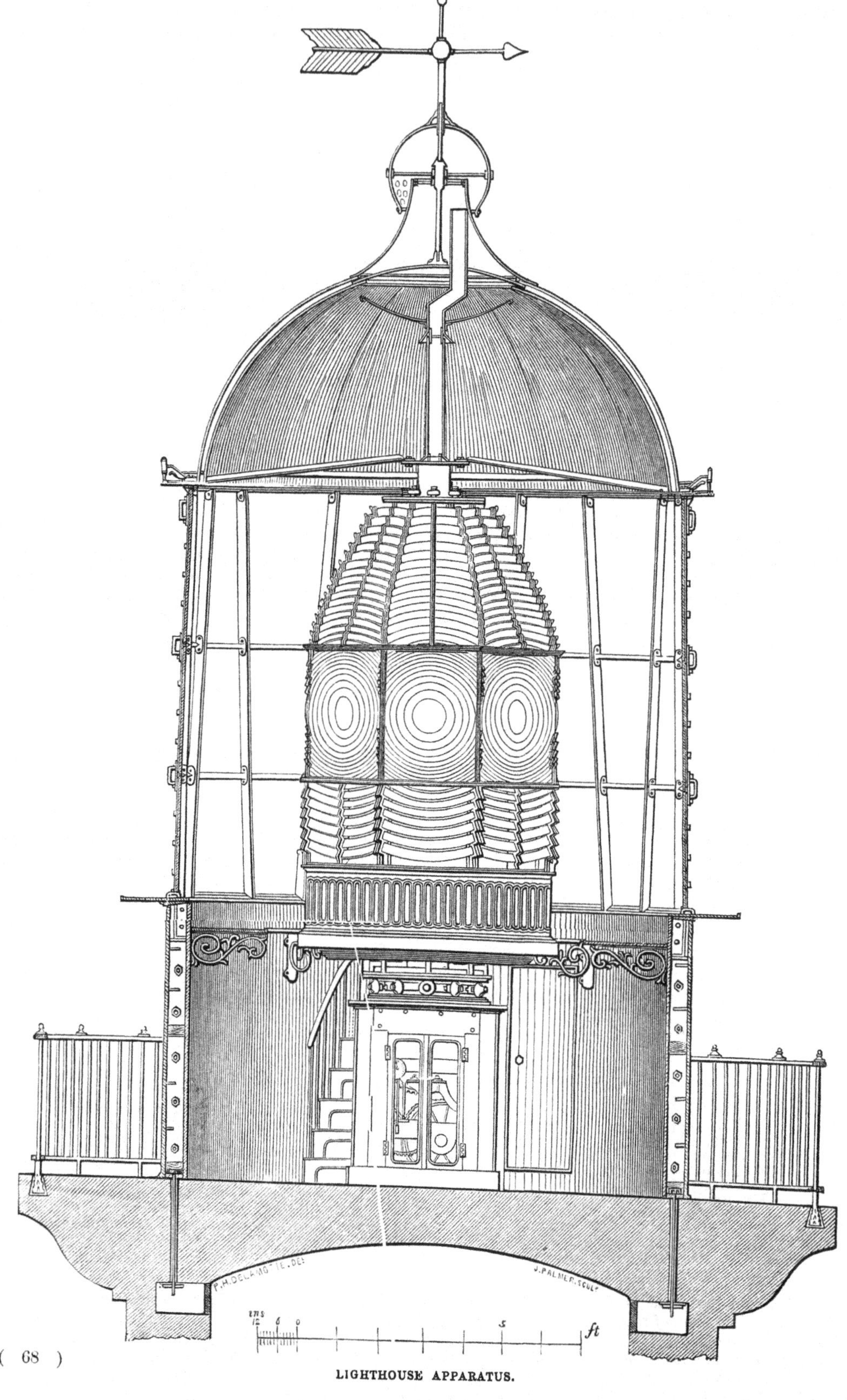

LIGHTHOUSE APPARATUS.

Chance Brothers & Co., *continued.*

Lighthouse Apparatus.

1. Dioptric Revolving Apparatus of the First Order, constructed according to Mr. Thomas Stevenson's Holophotal improvement of the system of Augustin Fresnel.

This apparatus is formed of an eight-sided frame, in the centre of which the flame is placed. Each side comprises a compound lens and a series of totally reflecting prisms both above and below the lens; all these prisms, as well as the rings of the compound lens, being concentric round a horizontal axis passing through the centre of the lens.

The result is, to condense the light proceeding from the central flame into eight beams of parallel rays, without the aid of unnecessary reflections or refractions, so as to produce the maximum effect at sea.

Light-room and Lantern.

2. The Light-room is made of cast iron; it is seven feet high, being cylindrical within, and having externally sixteen sides, which are alternately large and small, to suit the lantern which it supports. It is provided, *outside*, just beneath the lantern, with a gallery or balcony, on which the keepers can stand to clean the lantern-frames, and also with an *inside* gallery for the service of the apparatus. The inside of this light-room is lined with mahogany.

The Lantern is formed, 1st, of sixteen standards, alternately inclined to the right and left; they are made of wrought iron, covered with gun-metal facings, by which combination the greatest strength and the least interception of light are obtained, together with the usual protection from the sea air. 2nd, of gun-metal astragals, in two tiers, and of gun-metal sole-plates and sills. 3rd, of a double copper dome supported on iron rafters. The whole is surmounted by a revolving copper ball carrying a wind-vane. The panes of the lantern are purposely omitted, to facilitate inspection of the apparatus.

3. A square cast-iron Pedestal with glazed doors, containing the clockwork for imparting rotatory motion to the apparatus. By a contrivance communicated to the manufacturers by Professor Airey, and for the first time used in lighthouse machinery, the *winder* is so constructed as to maintain an uniform speed of rotation, without any check during the winding-up. In other particulars, the plan of the pedestal and of the clockwork is in accordance with the Scotch system.

A Revolving Carriage; being an arrangement of rollers and guide-rollers, to give the least possible amount of friction, whilst it maintains the perfectly vertical position of the apparatus.

A fixed cast-iron Table, on which the oil-lamp is placed, and on which the keeper stands for the service of the lamp.

4. The Oil-lamp is a novel kind of "pressure-lamp," and consists of a turned gun-metal cylinder, in which the piston that forces the oil into the burner is worked by a weight placed *outside* the cylinder, instead of *inside*, as hitherto.

Each of the four concentric wicks of the burner is supplied with oil by two independent feed-tubes communicating with the main pipe.

The Dioptric or Lenticular system of Lighthouse illumination is distinguished for its superiority to the Catoptric or Reflector system in the essential qualities of power, simplicity, durability, and economy.

The above First-order Dioptric Light is exhibited in the West Nave, near the Turkish Department. The whole of the optical apparatus, lamp, rotatory machine, lantern and light-room, has been constructed by Chance Brothers and Co., at their Glass Works, near Birmingham.

They have also constructed, under the direction of Messrs. D. and T. Stevenson, the following apparatus, exhibited by the Commissioners of Northern Lights:—

1. A Holophotal Revolving Light of the sixth order.
2. An Azimuthal Condensing Light of the sixth order.
3. A Totally-reflecting Hemispherical Glass Mirror.

And they have constructed the Fourth-order Holophotal Revolving Apparatus, in which is exhibited the Magneto-electric light of Professor Holmes, in the Western Annex. (See Class 8 in Catalogues.)

A Medal has been awarded to them in the present Exhibition, for improvements in Dioptric Lights, and great excellence in optical glass.

Chance Brothers and Co. are the only manufacturers in Great Britain of Dioptric Lighthouse Apparatus. They have constructed, within the few years during which they have pursued this branch of business, eighty-five complete Lights, of which thirty-nine are of the first and second orders; also thirty Lanterns. These have been supplied to the Lighthouse Boards of Great Britain, and of the other principal maritime countries of the world.

Their Lights embrace all the successive improvements introduced by themselves or by others, and are testified to be excellent in design, material, and workmanship.

Perfect optical adjustment of the lenses and prisms, in accordance with the local *data* of each lighthouse, is applied to all the apparatus of their construction.

GIBBS, ALEXANDER, 38 *Bedford Square.*—Designs for stained glass windows.

Note.—To gain brilliancy, very thick glass has been used in these windows, and in consequence of which they have all been leaded with the ancient round lead.

Description of the specimens of stained glass in the International Exhibition, executed by Alexander Gibbs.

Large three-light window, price £250

The principal subject in this window is taken from the early life of Christ, viz., a representation of Christ sitting in the midst of the Doctors in the Temple. Christ sitting on a throne, occupies an important place in the centre opening, having one of the Doctors in the foreground. In the right-hand opening, the principal figures are the Virgin Mary and Joseph; and those in the left-hand light, Doctors; some standing and some sitting complete the group. The whole is relieved by a diapered wall, and surmounted by rich canopies on ruby grounds.

Below the above subject are three smaller ones; the one in the centre light representing the Adoration of the Magi; in the right-hand light, the Flight of Joseph with the Infant Jesus, and Mary, his mother, into Egypt; and in the left-hand light, the Annunciation. Above these are rich canopies, and a small arcade running along the bottom forms the base of the window.

On either side of this window are four smaller lights, thus:—

Left-hand, top. Single-light early decorated window, price £20

A large figure of a Bishop occupies the greater portion of this light, the remaining portion being filled up with rich canopy and base. Behind the figure is a rich diapered screen, jewelled with ruby.

Left-hand, bottom. Single-light, semicircular head, price £35

This window, of which the accompanying engraving is a sketch, needs no description, except so far as relates to colour. The foliage is on a ruby ground, inclosed by a rich border on blue, and the subjects are richly coloured on blue backgrounds.

Right-hand, top. Single-light, early English, price £20

In this window, occupying the centre, is a figure of Christ sitting on a throne, and surrounded by a Vesica. The remaining portion is filled with geometrical interlace on a silvery ground, with an Angel top and bottom, in medallions, holding ribbons with the following texts:—

Top Angel—"I am the Resurrection and the Life."
Bottom Angel—"Behold the Lamb of God."

Right-hand, bottom. Single-light, perpendicular, price £25

Christ's Descent from the Cross is the principal feature in this light, surmounted by a rich canopy. In the base is an Angel on a blue background, holding ribbon with the following text—"It is finished;" surmounted by rich foliage.

Large two-light window, on right-hand, price . £75.

The arrangement of this window is as follows:—

The principal feature is a subject of Christ's Charge to St. Peter. In the right-hand light, Christ is in the act of giving the keys to St. Peter; and in the left-hand light, St. Peter is kneeling to receive them; and behind him, St. John. This subject is richly coloured, on a blue diapered background, broken by an apple-tree running throughout. Above this subject are canopies on ruby grounds, and below, a small arcade base. The remaining portion of the window is grisaille, with interlacing coloured bands, enriched with four Angels in medallions on blue grounds, two at top and two at bottom. The two top Angels have musical instruments, and the two bottom ribbons with the following texts:—

Left-hand Angel—"Thou knowest that I love Thee."
Right-hand Angel—"He saith unto him, Feed my sheep."

Large early English window, on left-hand, price . £75

This window, which is carried out in the early style, is arranged thus:—

There are four subjects in medallions, with blue diapered backgrounds. The bottom one represents The Last Supper; the one next above, The Agony in the Garden; the one next above that, The Crucifixion; and the top one, The Resurrection.

The ornamental portion of this window is composed of geometrical forms of rich colouring, on an interlaced background of ruby and green, the whole surrounded by a broad border on a ruby and blue ground.

[6727]

HEATON, BUTLER, & BAYNE, *Cardington Street, Euston Square, London.*—Stained windows, for St. Alban's Abbey and other places.

(*Prize Medal.*)

13th century stained window, for St. Alban's Abbey, illustrating the Baptism of Our Saviour and the Passage of the Israelites through the Red Sea.

14th century window, for Harpenden Church. (Slater, architect.) Subject, "The Six Acts of Mercy."

14th century window, illustrating the "Burial of Our Lord."

15th century window, for Skulthorpe Church, "The History of Ruth." (Seckell, architect.)

15th century window, "St. George, St. Andrew, St. Patrick, and the Royal Arms of England."

16th century window, "The Adoration."

And the east window of Langton Church, illustrating the principal events of the Life of Our Saviour.

[6730]

HOLLAND & SON, *St. John's, Warwick.*—Stained glass.

[6731]

JAMES, W. H., 37 *High Street, Camden Town.*—Enamelled window glass.

[6732]

LONG, CHARLES, 17 *Queen's Road, Bayswater.*—Specimens of ornamental, embossed, and painted window glass.

[6775]

MOORE, JOSIAH, 81 *Fleet Street.*—Patent ventilators for all buildings, patent self-shadowed glass for windows, tablets, &c.

[6734]

MORRIS, MARSHALL, FAULKNER, & Co., 8 *Red Lion Square, London.*—Stained glass windows.

[6735]

O'CONNOR, M. & A., & W. H., 4 *Berners Street, Oxford Street, London.*—Stained and painted glass windows.

Prize Medal.

1. The great west window of Aylesbury Parish Church (probable date, A.D. 1420), designed to illustrate three great epochs—viz., "The Fall of Man," "The Means of Grace," and "The Restoration of Man." This is carried out with two subjects from the Old Testament for each epoch, thus:—for "The Fall"—1. Adam and Eve eat the forbidden fruit; 2. Adam and Eve are expelled from Paradise. For "The Means of Grace"—1. The Passage of the Red Sea (type of Baptism); 2. Abraham offering Isaac (type of Holy Communion). For "The Restoration of Man"—1. The Feast of the Passover; 2. The Lifting up of the Brazen Serpent. The tracery (of a great number of openings) has, in the lower series of lights, figures of the twelve minor Prophets; above these are figures of the four major Prophets, over which again are figures of Noah, Abraham, Moses and David. Surrounding these figures are cherubim, while in the seven centre lights of the tracery are represented the seven gifts of the Holy Spirit.

2. One compartment of the west window of St. Matthias Church, Stoke Newington, London, designed and executed in the manner of the early glass stainers, with figures of Isaiah, David, and Noah.

3. Panels of heraldry in stained glass, consisting of the Royal Arms of Her Majesty the Queen, those of his late Royal Highness the Prince Consort, the arms of the Duke of Kent, and of her late Royal Highness the Duchess of Kent; being portions of a memorial window.

4. Life-sized figures of Jacob and Aaron, being a part of the series of windows for Eton College Chapel.

[6737]

POWELL, JAMES, & SONS, *Glass Works and Warehouses, Whitefriars, E.C.*—Glass for all purposes, and window work.

A

The specimens of quarries and borders are taken from stock, and can be recommended for simple glazing. They are very strong, and are found from their substance well adapted for hot or cold countries; the thickness of the glass prevents the flat effect in the cheaper kinds of painted glass. (A. B.)

Heraldry in glass, from drawings by Mr. Pollen. (C.D.) Iron casements. Wire guards.

Figures and subjects from drawings by various artists, and best painted work. (E.) Old windows restored. Geometrical patterns printed on muff glass, of good texture, with rich colours in the ornaments and borders. (F.G.)

Messrs. Powell exhibit, by permission, parts of windows executed by them; from the drawing of G. F. Street, Architect, for W. Cotton, Esq.; also from the drawing of E. B. Jones (the east window of Waltham Abbey), for the Rev. James Francis.

B

The manufacture and sale of glass was commenced at these works, A.D. 1700.

C

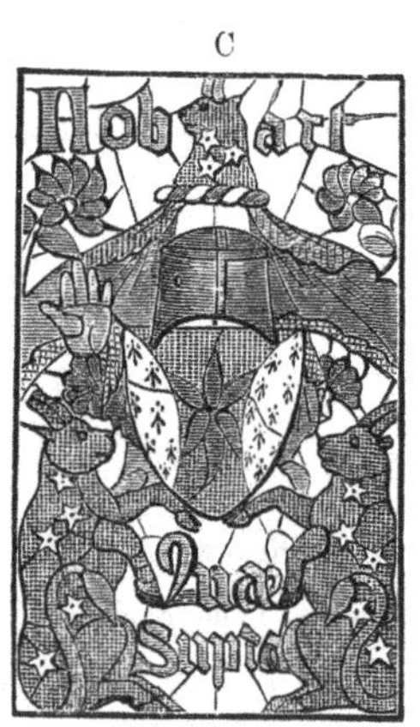

F

E

D

G

Writing on glass; Heraldry on glass for inlaying; glass mosaic.

POWELL, JAMES, & SONS, *continued.*

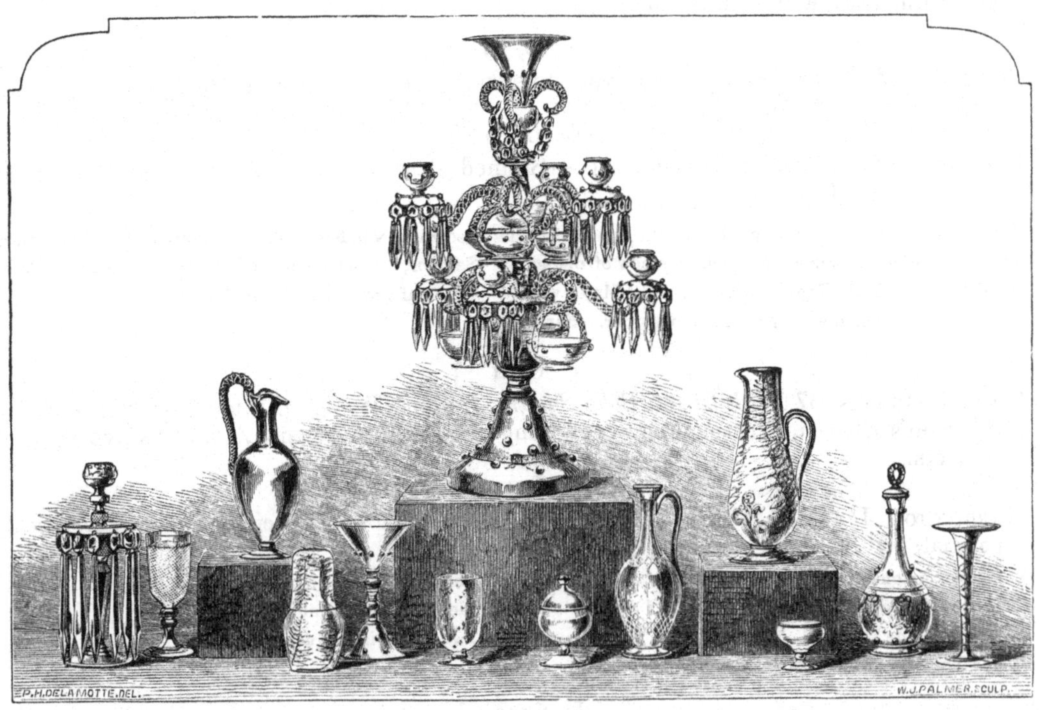

Glass made to patterns and drawings. Tube, enamels, optic-plate. Muff glass for artists, the colour and texture like old glass. Horticultural glass, glass for museums, sheet and plate glass, English and foreign glass and porcelain for chemical use. Glass water pipes. Some specimens of the above are shown, and the colours and texture of the muff glass may be seen in the painted windows exhibited.

Groups of table glass selected from stock, simple in outline and moderate in price. Cut glass and good engraving. Flower glasses, wall lights, and chandeliers.

[6738]

PREEDY, FREDERICK, 13 *York Place, Portman Square, London.*—Stained glass, English art, 13th, 14th, and 15th centuries.

[6739]

PRINCE, A., & Co., 4 *Trafalgar Square.*—Specimens of illuminated glass.

[6740]

REES & BAKER, 175 *Goswell Road, E.C.*—Stained glass windows: Norman, early English, early decorated.

Norman window, representing the Birth, Crucifixion, and Ascension of Our Lord; rich grounds of diaper and border. Early English window, "The Majesty of Our Lord;" rich diaper ground and border. Early decorated window, "The Presentation of the Infant Saviour," inserted in rich interior mosaic work on vine-leaf ground, with rich border.

[6741]

WARD & HUGHES, 67 *Frith Street, Soho Square, London, W.*—A painted glass window, for St. Ann's Church, Westminster, containing life-size "Ascension," and twelve smaller subjects.

[6742]

WARRINGTON, JAMES P., 43 *Hart Street, Bloomsbury Square.*—Specimens of ecclesiastical, palatial, and domestic stained glass.

1. Figure, Judgment. First Pointed.
2. The Annunciation. First Pointed.
3. Four Knights (Neville). Decorated.
4. The Three Beloved Disciples. Perpendicular.
5. The Magdalen at the Sepulchre. Pictorial.
6. Blessing Little Children. Pictorial.
7. Two Armorial Figures. Palatial.

[6743]

WARRINGTON, WILLIAM, JUN., 17 *Northumberland Place, W.*—Specimens of stained glass.

[6744]

WARRINGTON, WILLIAM, SEN., 35 *Connaught Terrace, W.*—Progressive examples of stained glass from the 12th century.

[6745]

LAVERS & BARRAUD, *Endell Street, W.C.*—Stained glass windows. (*See page* 75.)

The eight lower compartments of the great west window of Lavenham Church, Suffolk, illustrating the life of St. Peter. Style, circa, 15th century. Cartoon by J. M. Allen. A panel of 13th century glass. Subjects from the history of St. Alban. Cartoon by F. P. Barraud. A 14th century window. Subject, "Our Lord Sitting in Judgment." Cartoon by J. Bentley and N. H. J. Westlake. Specimens of glass for domestic buildings. Single light, "The Annunciation." Cartoon by E. B. Jones. Roundells. Scene from Enid, "Idylls of the King." Cartoon by J. M. Allen.

[6746]

HERBERT, MRS. F., 20 *Royal Avenue Terrace, Chelsea.*—Paper transparencies.

SUB-CLASS B.—*Glass for Household Use and Fancy Purposes.*

[6756]

AIRE AND CALDER GLASS BOTTLE COMPANY, 61 *King William Street, London.*—Glass bottles.

[6757]

ALEXANDER, AUSTIN, & POOLE, *Victoria Wharf, Earl Street, Blackfriars.*—Glass bottles, jars, and insulators.

Lavers & Barraud, *Endell Street, Bloomsbury, London, and 3 Oxford Street, Manchester.*

TWO LIGHTS OF THE GREAT WEST WINDOW OF LAVENHAM CHURCH, SUFFOLK, ILLUSTRATING EVENTS IN ST. PETER'S LIFE.

[6712]

BARKER, S., & Co., *Sloane Street, and Ellis Street, Chelsea, S.W.*—Ornamental window-glass, &c.

[6758]

BOWRON, BAILY, & Co., *Stockton-on-Tees.*—Glass bottles, glass for architectural, household, and horticultural uses.

[6759]

BROCKWELL, FREDERIC HENRY, 79 *&* 80 *Leather Lane, London.*—Ink glasses, flasks, mounted bottles, and case fittings.

[6760]

BROWN, MICHAEL LEWIS, 47 *St. Martin's Lane.*—Specimens of cut and engraved table glass in general use.

[6761]

CANDLISH, JOHN, *Manufactories, Leaham and Sunderland; Warehouse,* 224 *High Street, Wapping, London.*—Wine bottles.

[6762]

COPELAND, MR., 160 *New Bond Street; Manufactory, Stoke-upon-Trent.*—Examples of cut and engraved English crystal glass.

[6763]

DEFRIES, J., & SONS, 147 *Houndsditch.*—Crystal glass chandeliers, standards, lustres, and table glasses. (*See page* 77.)

[6764]

DOBSON & PEARCE, 19 *St. James's Street, Piccadilly, London.*—Ornamental table glass, lustres, and gaseliers. (*See pages* 78 *and* 79.)

[6765]

DOWLING, EDWARD, 2 *Little Queen Street, Holborn.*—Patentee and manufacturer of glass weights and glass rosettes for horses' bridles.

[6720]

EDMETT, BEEDOE, 10 *Long Acre, London, W.C.*—Specimens of plain, ornamental, and illuminated and other writing on glass.

[6766]

GARRETT, JOHN, *Arundel Place, Haymarket.*—Drinking goblet, with portable metallic handle and foot.

[6767]

GREEN, JAMES, 35 *&* 36 *Upper Thames Street, London.*—Cut and engraved table glass, chandeliers, and lustres. (*See page* 80.)

[6728]

HETLEY, HENRY, 13 *Wigmore Street, London.*—Specimens of glass shades, horticultural and window glass.

[6729]

HETLEY, JAMES, & Co, 35 *Soho Square, London, W.*—Glass shades, stained and various descriptions of window glass.

[6768]

HODGETTS, W. J., *Wordsley, near Stourbridge.*—Table and toilet glass. (*See page* 81.)

[6769]

HOOMAN & MALISKESKI, 490 *Oxford Street, London.*—Photographic portraiture for the interior of glass vases, &c. (Patented.)

DEFRIES, J., & SONS, *Works, London and Birmingham; Principal Depôt and Show Rooms, 147 Houndsditch, City.*

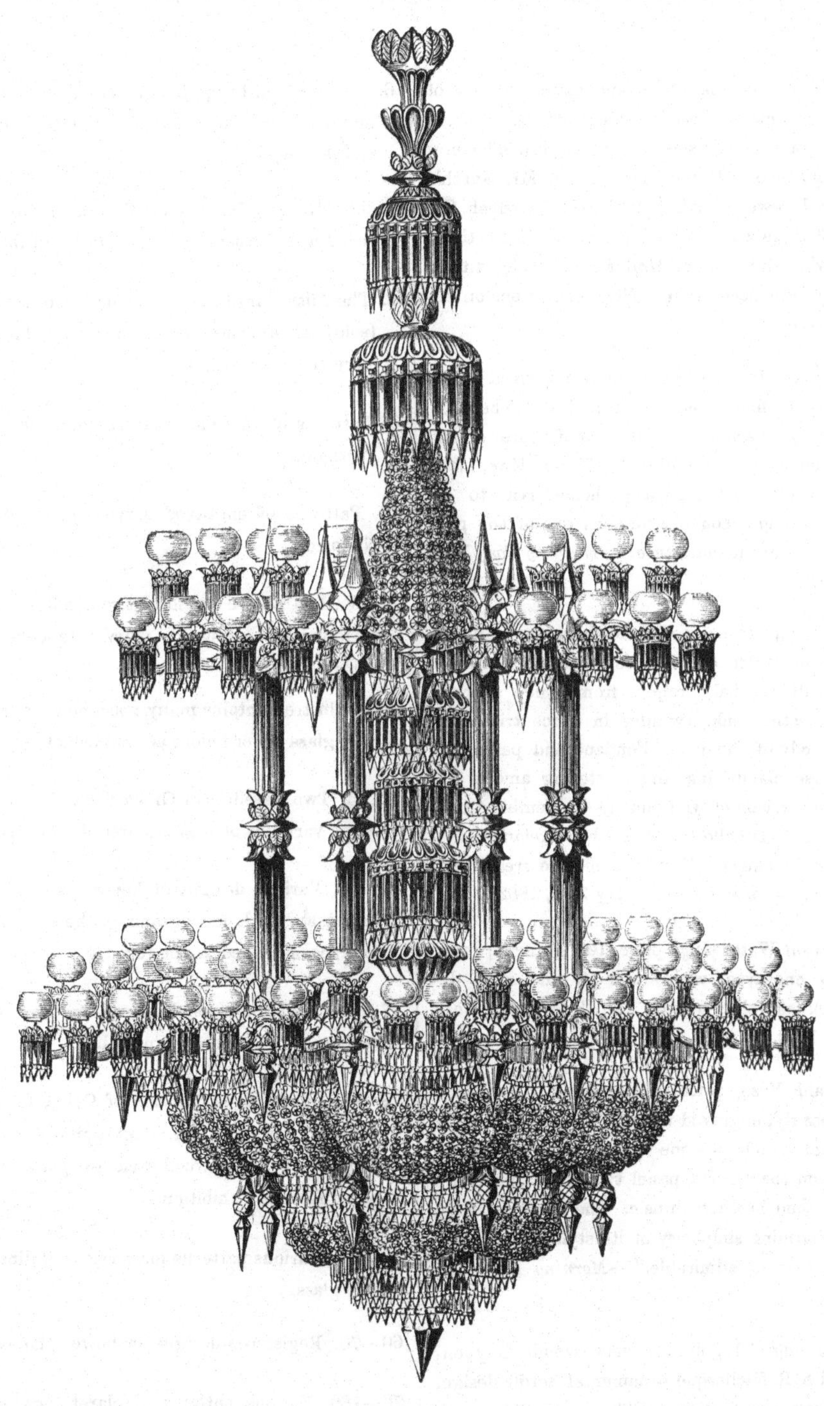

The exhibitors are manufacturers of crystal, bronzed, and ormolu chandeliers. Improved star and crystal sun-lights, and gas fittings of every description. Improved paraffin and other lamps, for India and the colonial markets in general.

Dobson & Pearce, Artists in Glass to the Queen, 19 *St. James's Street, Piccadilly, S.W.*—Ornamental table glass, lustres, and gaseliers.

Prize Medal obtained for the Exhibitions of 1851 *and* 1862.

The exhibitors are designers and manufacturers of first-class ornamental and useful household glass-work, in lustres, dinner and dessert services, gaseliers, candelabra, and flower glasses; sole manufacturers of Mr. March's much-admired flower stands (registered), for which the Special First Prize was awarded, at the Royal Horticultural Society's exhibitions of 1861 and 1862, for dinner and drawing-room decoration. (See illustrations on the opposite page.)

1. The Morrison Tazza, of very richly engraved glass, mounted in the finest gold and turquoise. "The most extraordinary specimen of art manufacture of its kind in the whole Exhibition."—*Times*, May 17th, 1862. "Price 250 guineas; henceforth to be reckoned, amongst connoisseurs, as one of the precious gems of art manufacture."—*Morning Post*, May 21st, 1862.

2. The Hamilton Vase; a flat-sided magnum claret jug, "on the centre of which is engraved a strange chimera, half cat, half dragon, from which a scroll springs on either side, twining in concentric rings over the body of the jug. Pendant and pendulous among these, clambering and clustering among the foliage, are myriads of wild animals—a reminiscence, may be, of Othello's inexplicable exclamation—Goats and monkeys! This is a picture fresh from Dreamland!'"—*Morning Post*, May 21st, 1862.

3. The Morrison Water Jug, price 100 guineas; very exquisitely engraved, with water-lily and shell designs. "A marvel of artistic skill."—*Morning Post*, May 21st, 1862.

4. The Oswald Vase, price 100 guineas. "Another thing of beauty, engraved with Pompeian designs in two circular panels, the one with Cupids playing on pipes, and on the reverse panel two other Cupids are dancing to their brothers' music. The purity of taste, and the charming simplicity of its style, must strike every beholder as admirable."—*Morning Post*, May 21st, 1862.

5. The Ailsa Jug, 50 guineas; very wonderfully engraved, with Raffaellesque ornamental scroll design, fountain, fruit, and flowers. "They are not so much engraved, as they seem to flow and ripple from the body of the vase, with an effect almost equal to an optical delusion."—*Times*, May 17th, 1862.

6. The Crawford Jug, 25 guineas (sold); very chastely engraved with flat tracery, scrolls, and figures, of Greek design.

7. The Gurney Cup: vine design, infant Bacchanals revelling amidst the vines; very beautifully engraved.

8. The Ailsa Large Claret Jug, or Jeroboam, 30 guineas, (sold): a glass claret jug, holding nine bottles of wine. Very richly engraved.

9. Patterns of cut glass service, made for the King of Portugal.

10. Patterns of engraved service, made for the Prince Napoleon.

11. A Glass Lustre for forty-two candles. "Unique in elegance and design, a marvel to connoisseurs."—*Morning Post.*

This lustre contains many specimens of art workmanship in glass never before accomplished. Price £350.

12, 13. Two Gaseliers of Greek form, for six lights each. Many varieties of this admired design may be had.

14—26. Various designs in flower glasses, in the Venetian taste, with monograms in the centre; never before accomplished at the furnace.

27, 28. Small glass candelabra, for four lights each, registered. Unique and useful.

29. The Wedgewood Service of Glass, for twenty-four persons (unique). An adaptation of Flaxman's ornament on a Wedgewood vase to glass-cutting; prepared for this Exhibition.

30—60. Various patterns for services of dinner and dessert glass.

60—70. Registered designs for flower glasses.

70—180. Various patterns of claret jugs, most beautifully engraved.

180—200. Ditto, water jugs and goblets, ditto, ditto.

Dobson & Pearce, *continued.*

Mr. March's Designs for Flower and Fruit Decorations.

Prize Epergnes, with glass stem, suitable also to drawing-rooms, 15*s.*, 18*s.*, and 22*s.* each, according to sizes.

Prize Baskets, with glass handles, turned or plain, 10*s.*, 13*s.*, and 16*s.* each, according to sizes.

GREEN, JAMES, 35 & 36 *Upper Thames Street, London.*—Cut and engraved table glass, chandeliers, and lustres.

HODGETTS, W. J., *Wordsley, near Stourbridge.*—Table and toilet glass.

RUBY SCENT JAR, CUT, WITH RAISED PILLARS.

JUG, WITH COLOURED SHIELD AND COAT OF ARMS, ENGRAVED.

Dessert service of flint glass, inlaid with a circle of ruby, consisting of decanters, carafes, finger basins, ice plates, elevated comports, jugs, goblets, champagne and other glasses.

Dessert service of rich cut flint glass, designed for six persons.

Cut flint and coloured vases, scent jars, and pieces for drawing-room ornament.

WATER JUG.

CENTRE DISH.

Water jugs and goblets cut and engraved, and with coloured shields attached.

Decanters, wines, and a variety of ornamental glass.

[6770]

JENNINGS, GEORGE, 263 *High Holborn, London.*—Specimens of writing, embossing, ornamenting, and gilding on glass.

[6771]

KILNER BROTHERS, *Dewsbury, and Brook's Wharf, Upper Thames Street, London.*—Superior glass bottles of every description. (*See page* 83.)

[6772]

LLOYD & SUMMERFIELD, *Park Glass Works, Birmingham.*—Glass show case, cut and engraved glass, glass window bars.

[6773]

MARCH, THOMAS CHARLES, *St. James's Palace.*—Examples of table decoration.

[6774]

MILLAR, JOHN, & Co., Potters to Her Majesty, *Edinburgh.*—Engraved glass and china.

[6733]

MOORE, EDWARD, & Co., *South Shields; 8 Newhall Street, Birmingham; 120 Duke Street, Liverpool; 83 Temple Street, Bristol.*—Pressed flint glass table ware (in every variety).

BUTTER COOLER, NO. 72.

COVERED SUGAR, NO. 54.

Examples are exhibited of pressed glass in flint and coloured table ware, and blown and cut goods, consisting of—

Ales, butter-coolers, water carafes and ups, cruets, custard cups, caddies, celeries, handled cans, cheese stand, candle ornaments, decanters, dishes, centre pieces, goblets, honey pots, jellies, jars, jugs, mustard pots, piano stands, paper weights, plates and bowls, pickle jars, gas reflectors, salts, salvers, sugar basins, cream jugs, covered sugars, tumblers, and wines.

Samples of the above-mentioned goods may be seen at the Works, South Shields, or at their Offices, 2, Barge Yard Chambers, Bucklersbury.

[6776]

MOTT & SONS, Wine Merchants, *Liverpool and Leicester.*—Vessels for preventing flatness in draught beer. (Patent.) (*See page* 84.)

[6777]

NAYLOR & Co., 7 *Prince's Street, Cavendish Square, London.*—Unique and classical designs, shapes, and engravings. (*See page* 84.)

Kilner Brothers, *Dewsbury, and Brook's Wharf, Upper Thames Street, London.*—Superior glass bottles of every description.

Prize Medal, the only one awarded to British Glass Bottle Manufacturers.

FOR WINE MERCHANTS.

The above are illustrations of a very superior kind of bottle recently introduced to the public.

They are in size from half-a-gallon to eight gallons each, and are made expressly for containing articles of value, such as wines, spirits, tinctures, &c., and are of especial utility when it is desirable to see the condition of the article, and draw off the contents without disturbance.

They are fitted with electro-plated or brass taps and vent-peg, and, when required, are covered with white wicker-work.

For soda-water, confectioners', drug, dispensing, and oilmen's bottles of every description, stoppered and plain, our house has long stood unrivalled; and the fact of the very great increase of late years in the demand for them, both in the home and foreign markets, is a proof of their superior quality and fitness for the purposes required.

Price lists may be had and samples seen on application.

Permanent Glass Edgings for Garden-Walks, Flower-Borders, &c.:—

No. 1.
Length 9 ins., Weight 3 lbs. to 3½ lbs.

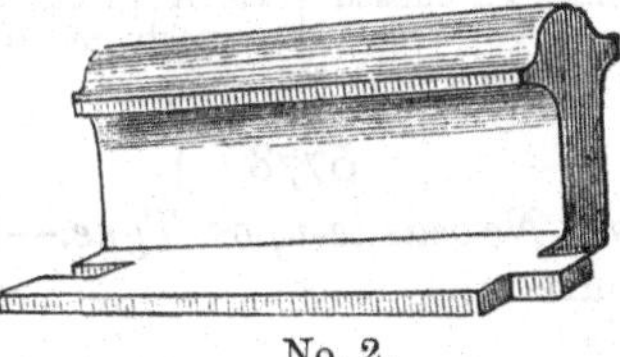

No. 2.
Length 9 ins., Weight 5 lbs.

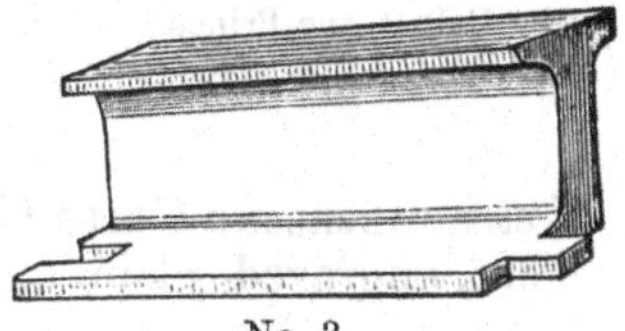

No. 3.
Length 9 ins., Weight 4 lbs.

Kilner Brothers have succeeded in making garden edgings of various patterns in glass, coloured so as to resemble dwarf box. These edgings are the most permanent that have yet been introduced, as they will resist the severest influence of the atmosphere, and are not liable to oxydization like iron, or disintegration like stone. The patterns already on sale are—No. 1, the usual upright edging with a Vandyked border; No. 2, the pattern designed by Dr. Hogg, and introduced as Hogg's Edging Tiles; and No. 3, which is a modification of No. 2.

	Per Yard—s.	d.
No. 1, 100 yards and under	1	9
No. 1, 200 yards and under	1	7½
No. 1, above 200 yards	1	6
No. 2, 100 yards and under	2	6
No. 2, 200 yards and under	2	3
No. 2, above 200 yards	2	0
No. 3, 100 yards and under	2	0
No. 3, 200 yards and under	1	10
No. 3, above 200 yards	1	9

Put on the rail at any of the stations in London, or 20*s.* per ton paid towards carriage, if delivered direct from the works.

MOTT & SONS, Wine Merchants, *Liverpool and Leicester.*—Vessels for preventing flatness in draught beer. (Patent.)

Mott's patent air-tight beer cans, for keeping ale and porter on draught out of contact with the air.

In drawing ale and porter from an ordinary cask, the air has to be admitted into the vessel, and the result is the well-known "flattening" and rapid spoiling of the beer. In the patent cans, which are made of earthenware, and are very clean and convenient, the air is admitted into a separate expanding chamber, which fills up the space from which the beer is withdrawn without suffering the air to come in contact with the liquid. Mechanical pressure being at the same time applied, the beer retains the carbonic acid gas generated in it, and grows riper instead of growing flatter while on draught. The invention, to a great extent, supersedes the necessity of bottling. The cans are made at present to hold 4½ gallons.

NAYLOR & CO., 7 *Prince's Street, Cavendish Square, London.*—Unique and classical designs, shapes, and engraving.

1. An elegant goblet, of a tall, classical shape, richly engraved. Subject, "The Lord's Supper."—"A fine work of art."
2. Companion to ditto. "The Crucifixion."
3. A flat-sided handle claret decanter, richly engraved with "Daniel and the Lion," and fine Egyptian borders.
4. A Moorish-shape claret decanter, richly engraved with Roman figures, borders, &c.
5. A unique Etruscan vase, with rope handles carried all round, exquisitely made. Richly engraved subject, "Cupids," on each side, surrounded with garlands of flowers.
6. Set of table glass, gilt, in forms of old Dutch, very chaste, with white Venetian stems.
7. Set of table glass, exquisitely engraved in rich pattern, imitation of bees' wings, &c.
8. Set of table glass, antique cut, with coloured Venetian stems, elaborately threaded.
9. Set of table glass, correct copy of old Venetian, spiral threads all over.
10. Set of table glass, finely and richly engraved borders.
11. Set of table glass, Anglo-Venetian diamond mould, of the finest metal, and antique shapes.
12. One tall, elegant, Etruscan water jug, very richly and characteristically engraved. Two goblets for ditto, to match.—"A very fine work of art." Purchased by H.R.H. the Princess Alice.
13. A magnum claret decanter, Moorish shape, richly engraved with stags, trees, water, &c.
14. A variety of claret decanters, &c., &c., different shapes, beautifully engraved.
15. One very large two-handle Exhibition tankard, for claret cup, twenty-two inches high, beautifully made.
16. One very antique jug, with ruby dabs all over; 12th century.
17. One Pompeian claret jug, with white berries all over.
18. A very beautiful and rare production in "flower glass," old Flemish, with twisted handle, and twelve tubes round to hold single flowers.
19. A Venetian flower glass, with harp-twisted stem, copied from an early date, finely engraved.
20. Three antique flower vases, in ruby; green and blue dabs all over. 12th century. The first ever made of the kind in England.
21. A pair of very classical shape Etruscan flower vases, elegantly engraved. Purchased by H.R.H. the Princess Helena.
22. A variety of flower glasses of the most unique designs, from various reigns, and of the choicest productions.
23. A group of wine glasses, richly engraved with crests, monograms, &c.

[6778]

NORTHUMBERLAND GLASS COMPANY, *Newcastle-upon-Tyne.*—Specimens of British flint glass: cut, engraved, plain, and coloured.

[6779]

OSLER, F. & C., 45 *Oxford Street, W.*—Pair of colossal candelabra, in crystal glass; glass vases, and specimens of lapidary cutting. (*See page* 85.)

[6780]

PEARCE, WILLIAM, & CO., 9 *Brooke Street, Holborn, and Bridge Street, Bristol.*—Toilets and smelling bottles.

[6781]

PELLATT & CO., Glass Manufacturers to the Queen, 58 *&* 59 *Baker Street, and Falcon Glass Works, London.*—Cut and engraved table glass and chandeliers. (*See page* 86.)

Osler, F & C., 45 *Oxford Street, W.*—Pair of colossal candelabra, in crystal glass; glass vases, and specimens of lapidary cutting.

COLOSSAL CANDELABRA, 20 FEET HIGH, IN RICHLY-CUT CRYSTAL GLASS.

PELLATT & Co., Glass Manufacturers to the Queen, 58 *&* 59 *Baker Street, and Falcon Glass Works, London.*—Cut and engraved table glass and chandeliers.

The group of glass, in the early Italian style, here illustrated and engraved, is a selection of the first prize design of the students of the Schools of Art throughout the kingdom; the prizes being awarded by Messrs. Pellatt and Co., for the best designs in engraved glass.

The group on the right is half-cased with ruby, and richly engraved and gilt in the Gothic style. Pellatt and Co. exhibit a table set of flint glass, cut in jewelled facets; various groups of engraved glass, chandeliers, lustres, gem work, medical and chemical glass, &c.

[6782]

PHILLIPS, EDWARD, *Shelton, Staffordshire.*—Glass gaseliers, chandeliers, candelabra, girandoles, and table glass, richly cut.

[6783]

PHILLIPS, W. P., & G., Designers and Producers, 359 *Oxford Street, and* 155 *New Bond Street.*—Table and dessert pieces of various kinds.

A service of crystal for table use, showing a novel application of ruby on the flint glass, consisting of port, sherry, champagne, claret, liqueur and hock glasses, decanters, carafes, and tumblers, water sets, finger basins, butter dishes, sugar vases, ice plates, salt cellars, and dessert service, with centre piece of original design.

A service of engraved table glass, complete.

A service of cut brilliant crystal.

A finely engraved water set.

Samples of jewelled glass, consisting of decanters, claret jug, water set, tankard jug, and chalices, &c., &c.

Samples of do. do., engraved in the Mediæval style.

A round table, made entirely of the purest crystal, and cut in the finest manner.

The Iliad dessert service, after Flaxman's illustrations.

Various specimens of beautiful form in plain glass.

[6785]

PRICE, JAMES, 41 *Castle Street, Leicester Square, W.C.*—Specimen of embossed and burnished gold writing on glass.

[6786]

READWIN, WILLIAM RANSOM, 44 *Warwick Street, Pimlico.*—Plain and ornamental writing on glass.

[6787]

ROYSTON, *Jesus Lane, Cambridge.*—Ornamental illuminated alphabets, in various ancient and modern styles, gilded on glass.

[6788]

Rust, Jesse, & Co., *Lambeth, S.*—Lamp glasses, globes, and perfumers' bottles; soluble glass for soap makers.

[6789]

Sharpus, T., & Cullum, W., 13 *Cockspur Street, Pall Mall.*—Unique designs of crystal table glass, &c.

[6790]

Sinclair, Charles, 5 *City Road, Finsbury Square, and* 177 *Old Street, City Road.*—Glass chandeliers and lustres, girandoles, &c.

[6791]

Spiers & Son, Glass and China Merchants, *Oxford.*—Specimens of table glass: plain, cut, and engraved.

Specimens of glass for table use; comprising decanters, claret jugs, wine glasses, water jugs and goblets, &c., of new and original forms, cut and engraved with various designs in the Etruscan, Alhambresque, and other styles, among which are the following:—

"Oxford University New Museum" claret jug, the ornament taken from various parts of the museum, and bearing the University arms.

"Christ Church" water jug and goblets, the ornament taken from stone tracery on the college, and bearing the college arms.

Decanters, wine glasses and tumblers, Egyptian shape and pattern.

Antique-shaped claret jug, with flat sides, figure subject engraved after the antique, surrounded by an Arabesque scroll border.

Tall claret jug, decorated with Alhambresque ornaments, with panel for receiving coat of arms or crest.

Claret decanters and wine glasses, old Venetian style.

Claret glasses, in form of a thistle, and many others.

Wine and Burgundy jugs, Etruscan style.

Exhibitors also, in Class 30, of the "Oxford Cyclopean Washstands, combining the largest capacity with the smallest requirement of space."

[6792]

Storey & Son, 19 *King William Street, City of London.*—Specimens of table and ornamental flint glass. (*See page* 88.)

[6793]

Toogood, William, 37 *Mount Street, Grosvenor Square, W.*—White and green glass bottles, show jars, &c.

[6794]

Westwood & Moore, *Brierley Hill.*—Glass bottles.

[6795]

Wheeler, John Jackson, 1 *Henry Place, Chelsea.*—Medical glass, glass surgical instruments, and chemical apparatus.

Storey & Son, 19 *King William Street, City of London.*—Specimens of table and ornamental flint glass.

SPECIMENS OF ORNAMENTAL TABLE GLASS.

Specimens of services of cut and engraved and etched glass, consisting of decanters, claret decanters, and water jugs, port, sherry, hock, and champagne glasses, carafes and tumblers, finger glasses and wine coolers, jellies and custards, dessert dishes, oval, round, and elevated, salts, sugar and butter vases, flower vases, cut and engraved, a fish bowl and flower vase combined, tazzas and jewel stands, claret, cider, and water sets.

[6796]

Wood, John Henry, *Cold Harbour Lane, Camberwell, S.*—Anglo-Venetian silvered ornamental mirrors.

[6797]

Chance Brothers & Co., *Glass Works, near Birmingham.*—Crown, sheet, plate, painters', and optical glass, shades, &c.

[6798]

Claudet & Houghton, 89, *High Holborn.*—Glass shades.

[6799]

Lavers & Barraud, *Endell Street, Bloomsbury.*—Painted glass.

[6800]

Powell, J., & Sons, *Whitefriars.*—Glass for all purposes.

CLASS XXXV.

POTTERY.

[6827]

Ashworth, George L., & Brothers, *Hanley, Staffordshire.*—Dinner, dessert, toilet ware, &c., in ironstone china and earthenware.

[6828]

Atkins, Thomas, & Son, Engineers, 62 *Fleet Street, London.*—Patent *moulded* carbon filters, and patent glass circulating filter fountains, supplying filtered water for public use. (*See page* 92.)

[6829]

Battam & Son, *Gough Square, London, E.C.*—Decorated Etruscan vases from the antique, and other ceramic works.

[6830]

Bell, J. & M. P., & Co., *Glasgow Pottery, Glasgow.* — Porcelain, earthenware, parian and terra cotta.

[6831]

Bevington, Messrs., & Son, *Hanley, Staffordshire.*—China, porcelain, parian, earthenware, &c. (*See page* 93.)

[6832]

Blanchard, Mark H., *Blackfriars Road, S.*—White and coloured terra cotta for architectural decoration and other purposes.

Obtained the Prize Medal at the Exhibition of 1851.

The objects here exhibited are—

1. Copy of the Western vase, from the orginal in the British Museum.

2, 3. Reduced copies of the Warwick and Albani vases.

4. Statue of Watt.

5, 6. A pair of Neapolitan vases, from the antique in the Kensington Museum, all in red terra cotta; commissioned to be executed for the Science and Art Department, South Kensington.

7. A column, as executed under the First Commissioner of Her Majesty's Parks and Buildings, for the arcades of Victoria Park.

8. A column, as executed for the arcades of the Royal Horticultural Gardens, for the Royal Commissioners.

9. A statue of Flora, from the Capitol, and a series of vases and other objects.

10. Duplicate of the Flagstaff Pedestal from St. Mark's Cross, Venice, commissioned to be executed by the Department of Science and Art, Kensington Museum.

ATKINS, THOMAS, & SON, Engineers, 62 *Fleet Street. London.*—Patent *moulded* carbon filters, and patent glass circulating filter fountains, supplying filtered water for public use.

1. Ornamental stand of filters, made in many different shapes and of various materials for portable, domestic, trade, and ship purposes. They constitute some of the most improved articles in sanitary reform. In addition to the ordinary results derived from mere mechanical percolation of impure fluids through porous substances, these filters possess the power of chemically purifying such fluids without rendering them flat, vapid, or insipid. From their peculiar construction, they impart the valuable constituents of palatable water, viz., oxygen and carbonic acid gas. It is proved from results, that both organic and inorganic substances, whether soluble or insoluble, are separated from impure water—smell and noxious vapours are dissipated, and colouring matter removed. The compound of materials used in the medium excites a galvanic action amongst the particles, and thus another powerful agent for contributing to the purity and wholesomeness of water is added. These results should be particularly noticed.

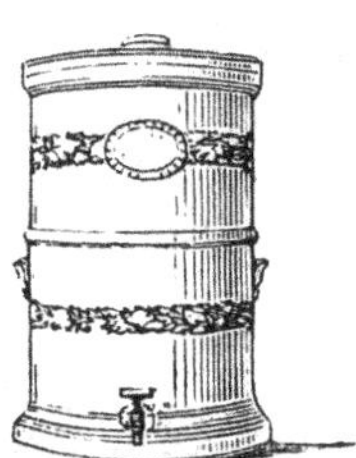

2. The media used are pure animal, vegetable, and mineral charcoals, free from other ingredients. The process of manufacture requires much skill, time, and judgment, and involves great expense. The object to be attained is the most perfect cohesion of the particles—any adulteration of the charcoal by non-absorbent or other substances not only impairs its cohesiveness, but deteriorates its imperishability. Admixture with other substances is therefore carefully avoided. Specimens of the carbon, which have been in use for many months, are exhibited, showing their perfect freedom from injury. The filters are expressly designed to facilitate cleansing, and to suit the many requirements of domestic and manufacturing establishments. Respirators and sewer ventilators, made of carbon, and which answer effectually, are also exhibited. Patent Hydro-Pneumatic Circulating Fountains are the most novel applications of hydrostatics, hydraulics, and pneumatics ever exhibited. Their utility is paramount to their beauty, being filters as well as fountains—suitable for conservatories, public exhibitions, dining rooms, saloons, &c. They are perpetual in their action, and are made in a portable shape. Designs in various materials and of various sizes kept on stock.

CISTERN FILTER.
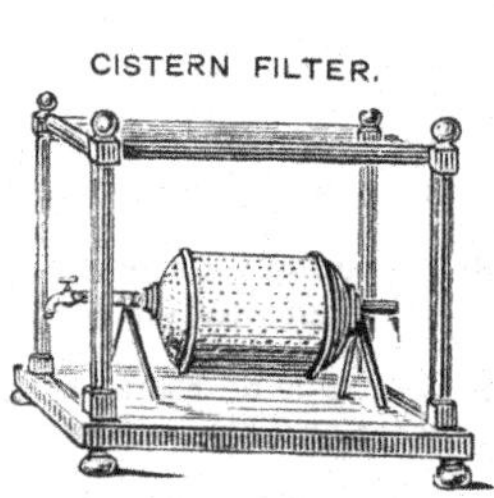
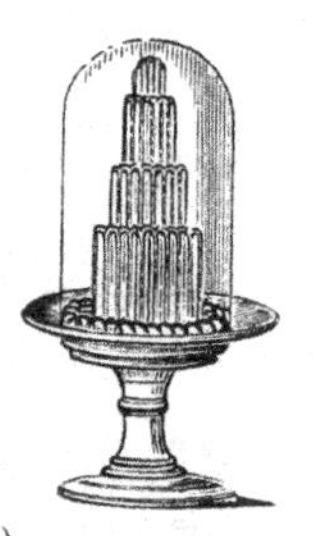

BEVINGTON, MESSRS., & SON, *Hanley, Staffordshire.*—China, porcelain, parian, earthenware, &c.

PARIAN VASES.

A pair of vases, with ornamental decoration. Likeness of Her Majesty the Queen and his late Royal Highness the Prince Consort in centre. (*See Engraving.*)

A pair of vases, marone and turquoise, richly gilt; groups of flowers in centre.

A great variety of vases.

PARIAN STATUETTES.

Leda and Swan. (*See Engraving.*)

Morning and Evening Dew (by Pradier).

Ruth and Esther (Beattie).

Equestrian Figure of Amazon (after Fouchére).

Ariadne and Bacchus (Morrey).

Flora.

Garibaldi.

Mother's Love (Morrey).

Group, representing Agriculture.

PORCELAIN.

Dessert plates, with border in mazarine blue and raised enamel and gold, groups of flowers in centre.

Also, elevated and low comports, in similar style.

Dessert plates with border in puce colour, dead and burnished gold; also, elevated and low comports in similar style.

A variety of dessert plates.

Tea service, Queen's shape. Transparent Rose du Barry and gold.

Ditto, canary colour and ditto.

Tea and coffee service, Alhambra border.

Breakfast service, Cambridge shape, Celeste and gold.

Tea and breakfast services in great variety.

[6833]

BLASHFIELD, JOHN MARRIOTT, *Stamford, Lincolnshire.*—Terra cotta and other pottery.

GROUP OF ARCHITECTURAL DETAILS.

The articles exhibited by J. M. Blashfield are made of a mixture of clays from the oolite beds of Northamptonshire and clays from the neighbourhood of Poole, in Dorsetshire, combined with flint, ground glass, sand, felspar, &c.

Moulded brick cornice, from a fragment at Villa Barberini, Rome.

Moulded brick cornice, from the convent of St. Antonio, Padua.

Pilaster of moulded bricks, after the old Italian manner.

Moulded brick details.

Tracery, designed by Owen Jones.

Old English chimney shaft, 11 feet high.

Large original model of a console, with colossal mask of a river god.

Large console, with mask of Bacchus.

Bracket, designed by Owen Jones.

Crosses for churches, designed by Hine and Evans.

Antifixæ, acroteria, balusters, trusses, string courses, &c.

Large frieze of acanthus foliage in white terra cotta.

Paving tiles, the same as those used on the footpaths of Westminster Bridge.

Details of mullions, tracery, &c., of oriel window, made for his Highness the Nizam Nawaub Moorshedabad, of Bengal.

Specimen of terra cotta cornice, 4 feet long, in one piece.

Washing trough in terra cotta.

Italian roofing tiles.

Ionic pilaster capital.

Elizabethan tracery for terraces.

BLASHFIELD, JOHN MARRIOTT, *continued.*

W. J. PALMER, SC.

STATUE OF A NYMPH, BY HALE; VASES, BUSTS, TAZZA, &c.

BUST OF SHAKSPEARE, BY HALE; VASES FROM BOLOGNA, VILLA ALBANI, &c.

Blashfield, John Marriott, *continued.*

STATUE OF TYCHO WING.

STATUE OF CERES.

Statue of Tycho Wing, the astronomer of Pickworth, in the county of Rutland, grandson of the famous astronomer, Vincent Wing, of North Luffenham, who published the "Celestial Harmony of the Visible World, 1651," "Astronomica Britannica," and several other works, and died September 20, 1668. An almanac continues to be sold, with his name prefixed as the author, by the Stationers' Company to this day.

An original design for a vase founded on the antique, the principal ornaments of which consist of four festoons, two of fruit and two of flowers, which have been designed, composed, and modelled at once upon the surface of the body, without any previous or after process, such as moulding, casting, or pressing, being used. The modelling by Henry Sibson.

A large tazza (55 inches diameter), of a plain bold character, turned and burnt without the process of moulding, on a table constructed to carry the weight of heavy masses of clay, and at the same time admit of the gradual contraction of clay in drying, without cracking.

A large tazza (48 inches diameter), after the manner of the antique, ornamented with masks of a river god. An original work, modelled and burnt without the process of moulding.

A tazza (39 inches diameter) with handles, after the antique.

A small fountain group of four boys, in red, supporting a tazza.

A large vase, with allegory of the vintage.

A vase, after the antique, in dark grey colour, with eagles drinking from the rim.

A vase, after the antique, with Greek ornament on the mouldings and modern foliage on the body.

A vase of Corinthian form, ornamented with boys and dolphins.

A wine cooler, in red, adorned with leaves and medallions of Dante and Petrarch.

A variety of vases and amphoræ, in red, copied from antique Greek outlines.

Blashfield, John Marriott, *continued.*

VASE MODELLED BY HENRY SIBSON.

A large grand bowl, designed by H. A. Darbishire, Esq.

A square vase, with consoles at corners, from which are festoons supported by boys.

A variety of flower trays, flower pots, and flower baskets.

Flower pot, with medallion portraits of Her Majesty the Queen, his late Royal Highness the Prince Consort, and their Imperial Majesties the Emperor and Empress of the French.

An original statue of Ceres, life size, modelled by Henry Hale.

An original statue of a nymph, for a fountain, modelled by Henry Hale.

Busts of Homer, Virgil, and Washington, heroic size, modelled and burnt; original works, without moulding.

Busts of Her Majesty the Queen, by Weigall.

A statuette in red, of a Roman lady reading, in the ancient fashion, from an original sketch by M. Digby Wyatt.

Original busts of Shakspeare, Locke, Newton, and Milton, modelled by H. Hale.

A very large dog, from life; an original work, modelled by Mrs. Henry Heathcote, of North Luffenham, Rutlandshire.

A pair of greyhounds, modelled by Woodington.

A pair of storks.

A crane; an original work, modelled and burnt without moulding.

Bust of Isis, after the antique, in white terra cotta, the property of R. N. Newcombe, Esq., of Stamford.

Statuettes of the Nine Muses and Apollo Musagatæ, after the antique, in red terra cotta.

A cup, after an antique, from Pompeii, with group of Cupid and Centaur.

A vase modelled by the late Mr. Nixon, with bassi relievi of the masque scene in Shakspeare's *Tempest.*

A large bowl in red terra cotta, from an antique at Bologna.

A pedestal ornamented at the base with acanthus leaves, from which ivy branches spring and entwine around it, after the antique at Villa Albani.

Vases in red terra cotta, from antiques in the British Museum, from Athens, Nola Vulci, and Cumæ.

Bust of Clyte, in red terra cotta.

Group of Hæmon and Antigone, designed and modelled by Sibson.

Large vase, from a Greek outline in Piranesi.

Bowl in red terra cotta, from a work by Cellini.

A bowl and stand for crocuses.

A statue of Triton, for a fountain.

A jardinière, with medallions emblematical of the Seasons, by John Bell.

Candelabræ brackets from the Cathedral at Sienna. Angels supporting tazzi, remodelled by Woodington.

Bust of his late Royal Highness Prince Albert, by Pitts.

Busts of children, small life, by Schultze.

Bassi relievi, after the antique at Villa Borghesi. "Bacchanalian festival."

Alti relievi of a choir of angels, life size. "*Gloria in excelsis.*"

Statuette of Victory, after Rauch.

Statuette of History, after Franz.

Statuettes of children, by John Bell. "The First Letter" and the "Boys' Own Book."

Large vase in red terra cotta, supported by four dolphins in black terra cotta.

A copy of the Medici vase.

A tazza in red, supported by four Egyptian figures in black terra cotta.

A statuette of Psyche, by Hale.

A variety of Corinthian, bell-shaped, and other vases, amphoræ, cups, jugs, water bottles, flower pots, flower trays, &c.

Pendant vases for orchidaceous plants.

[6834]

BOOTE, MESSRS., *Burslem, Staffordshire.*—Tiles and pottery.

[6835]

BOURNE, JOSEPH, & SON, *Macclesfield Street, City Basin.*—Patent vitrified stoneware, bottles, jars, insulators, and fancy articles.

[6836]

BROWN, MICHAEL LEWIS, 47 *St. Martin's Lane.*—Specimens of dinner, dessert, tea, breakfast, and toilet services.

[6837]

BROWN, WILLIAM, *Trent Pottery, Burslem, and* 8 *Old Jewry, London.*—Earthenware dinner, tea, and toilet services.

[6838]

BROWN-WESTHEAD, T. C., MOORE, & CO., *Cauldon Place, Hanley.*—Parian, china, earthenware, sanitary, and druggist ware manufacturers. (*See page* 99.)

[6839]

BROWNFIELD, WILLIAM, *Cobridge, Staffordshire Potteries.*—Dinner, dessert, and toilette services; jugs, garden seats, garden pots, &c. (*See page* 100.)

[6840]

BULLOCK, CHARLES, *Britannia China Works, Longton.*—China tea, breakfast, and dessert sets.

[6841]

CLIFF, J., & SON, *Wortley, near Leeds.*—Ornamental and useful works in terra cotta.

[6842]

COLE, HENRY, C.B., *South Kensington.*—Use of earthenware pressed Mosaics, for the exterior decoration of buildings. (*See page* 102)

[6843]

COMER, R., *Thorpe Hamlet, near Norwich.*—Pottery.

[6844]

COPELAND, W. T., 160 *New Bond Street; Manufactory, Stoke-upon-Trent.*—Examples in every class of ceramic manufacture.

[6845]

CRYSTAL PALACE ART-UNION.—Presentation pieces in ceramic statuary.

[6846]

DANIELL & CO.—(*See* ROSE & Co).

[6847]

DAVENPORT, BANKS, & CO., *Castle Field Pottery, Etruria.*—Jet, jasper, stone, red, porous, and antique wares.

[6848]

DIMMOCK, JOHN, & CO., *Hanley.*—Earthenware, plain and ornamental, more particularly in Celeste blue.

BROWN-WESTHEAD, T. C., MOORE, & Co., *Cauldon Place, Hanley.*—Parian, china, earthenware, sanitary, and druggist ware manufacturers.

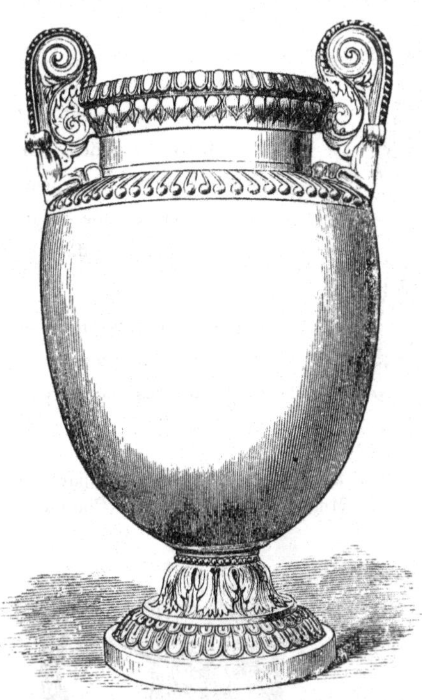

No. 1. China vase, copy of Bacchante vase in the British Museum. Height 3 ft. 2½ in.
2. China table services.
3. China dessert services.
4. China tea and déjeûner services.

No. 6. Parian bust of Apollo, from the original. Life size.
7. Parian busts of the Queen, by Durham, and the Prince Consort, by Marochetti; and group, Venus and Cupid, by Gibson.
8. Earthenware table services.
9. Earthonware toilet services.

No 5. Parian figure of Cupid, after Michael Angelo. Height 3 ft.

10. Parian figure, Crouchin Venus, from the antique. Height 2 ft. 2 in.

BROWNFIELD, WILLIAM, *Cobridge, Staffordshire Potteries.*—Dinner, dessert, and toilette services; jugs, garden seats, garden pots, &c.

Mr. Wm. Brownfield, of Cobridge, Staffordshire Potteries, exhibits many choice specimens of dinner, dessert, and toilet services, jugs, and other useful and ornamental articles in earthenware. His attention has been directed to the improvement of the quality, form, and style of decoration of this portion of the ceramic art, and he has been so successful that his earthenware is superior to a great deal of the porcelain exhibited, and has the advantage of being within the reach of all purchasers.

1. This jug is very graceful in its form; it is decorated with the fern leaf, between which hangs in a pleasing manner the wheat-head and leaves; has a rope handle, terminating with the Staffordshire knot.

2. The form of this jug is taken from the antique, the ornamentation being in the "Renaissance" style.

3. The outline of this jug is pleasing, and the decoration is very good; the ornaments are pretty and graceful; the four figures represented in compartments are—"Art," "Music," "Science," and "Commerce."

4. This flower pot and stand is useful in form and elegant in design; the decoration is easy and graceful. The trellis work, with the passion flower and foliage running through it, has a pleasing effect.

BROWNFIELD, WILLIAM, *continued.*

The tureen and vegetable dish here represented are of new design, and are not only elegant in form, but great care has been exercised to make them useful. The ornamentation is very chaste; the colours are turquoise and maroon, relieved with gold.

The dinner service, of which the engraved plate is a sample, is a novelty. The drawings are by H. K. Browne, Esq., better known as "Phiz," who is justly celebrated for his amusing sketches. The subjects in the centres of this dinner-service are of great variety, and the engravings are executed in the best style.

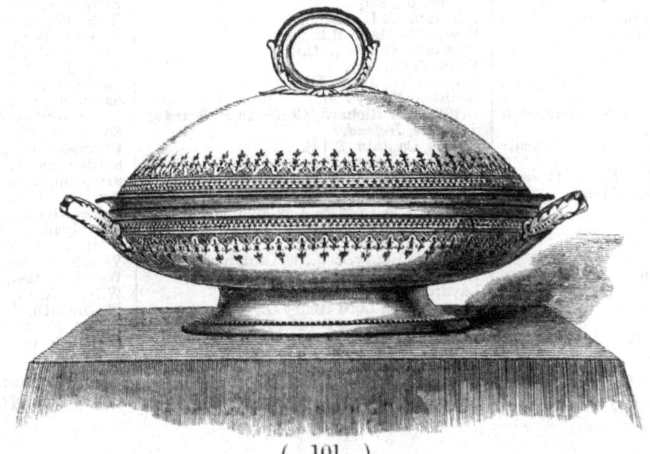

COLE, HENRY, C.B., *South Kensington.*—Use of earthenware pressed Mosaics, for the exterior decoration of buildings.

It is proposed to raise sufficient funds to execute two large Mosaic Pictures, 23 feet high by 13 feet wide, as experiments for decorating the Panels of the outside walls of the permanent Picture Galleries for International Exhibitions in Cromwell Road, South Kensington. The Mosaics will be made of Pottery in geometric forms by the pressure of dry powder. Various experiments in laying the Mosaics have been made by Messrs. Minton (Stoke-upon-Trent) with Mosaics of their own manufacture, and by Messrs. W. B. Simpson & Sons, of 456 West Strand, with Mosaics manufactured by Messrs. Maw. The experiments are very promising; and they prove that Mosaic pictures may be as easily worked and used in England, as in ancient Greece and Rome or Mediæval Italy. They will be as imperishable as the hardest and most perfect Terra cottas. They will create a new branch of industry which may be worked in any locality, and, probably, by women as well as men.

The designs will illustrate Industry, Science, and Art. Some Cartoons have been already prepared by Mr. Cope, R.A., Mr. J. C. Hook, R.A., Mr. Godfrey Sykes, and Mr. Townroe; two of these will be executed in Mosaics as soon as the funds are provided.

The ornamental borders will be designed and the Mosaics worked out under the superintendence of Mr. Godfrey Sykes and his assistants.

When two panels have been done, and all the necessary arrangements have been made after the close of the Exhibition of 1862, for filling the others, designs for other subjects will be sought from the artists named.

The following are the principal subjects which, at present, it is proposed should be executed, and the artists named are those who have already kindly consented to undertake to make designs for them, when the proper period arrives.

I. SUBJECTS ILLUSTRATING THE PRODUCTION OF RAW MATERIALS.

1. *Agriculture,* Holman Hunt.
2. *Chemistry,* W. Cave Thomas.
3. *Fishing,* J. C. Hook, R.A.
4. *Hunting,* Frederick Leighton.
5. *Metallurgy,* Eyre Crowe.
6. *Mining,* F. Barwell.
7. *Planting, &c.,* Michael Mulready.
8. *Quarrying,* G. F. Watts.
9. *Sheep Shearing,* C. W. Cope, R.A.
10. *Vintage,* F. R. Pickersgill, R.A.

II. SUBJECTS ILLUSTRATING MACHINERY.

1. *Astronomy,* S. Hart, R.A.
2. *Engineering,* (reserved).
3. *Horology,* (reserved).
4. *Mechanics,* (reserved).
5. *Navigation,* J. E. Millais, A.R.A.
6. *Railways,* R. Townroe.

III. SUBJECTS ILLUSTRATING MANUFACTURES, AND HAND LABOUR.

1. *Bricklaying,* D. Maclise, R.A.
2. *Carpentry,* R. Burchett.
3. *China Painting,* H. A. Bowler.
4. *Glass Blowing,* (reserved).
5. *Iron Forging,* Godfrey Sykes.
6. *Jewellery,* D. G. Rossetti.
7. *Lace Making,* R. Redgrave, R.A.
8. *Metal Casting,* A. Elmore, R.A.
9. *Printing,* R. Redgrave, R.A.
10. *Straw Plaiting,* C. W. Cope, R.A.
11. *Weaving,* Octavius Hudson.
12. *Pottery,* Godfrey Sykes.
13. *Stone Carving,* W. B. Scott.

IV. SUBJECTS ILLUSTRATING FINE ARTS.

1. *Architecture,* W. Mulready, R.A.
2. *Painting,* W. Mulready, R.A.
3. *Sculpture,* W. Mulready, R.A.
4. *Music,* J. C. Horsley, A.R.A.

The designs before they are executed will be approved by a Committee of the Artists.

The Marquess of Salisbury, K.G., Mr. Layard, M.P., and Mr. Cole, C.B., act as a Committee of Management for carrying out the experiments; and all communications should be addressed to G. F. Duncombe, Esq., Secretary, South Kensington Museum, London, W.

The following Noblemen and Gentlemen have subscribed in aid of the experiment. Further names may be sent to the Secretary:—

The Society of Arts.
The Earl Granville, K.G., *Lord President of the Council, and Chairman of H. M. Commissioners for the Exhibition of* 1862.
The Marquess of Salisbury, K.G.
The Lord Overstone.
The Lord Boston.
Sir Charles Eastlake, *President of the Royal Academy.*
Sydney Smirke, Esq., R.A., *Professor of Architecture in the Royal Academy.*
George Gilbert Scott, Esq., R.A., *Architect to the Dean and Chapter of Westminster.*
F. C. Penrose, Esq., *Architect to the Dean and Chapter of St. Paul's Cathedral.*
Wm. Tite, Esq., M.P., *President of the Institute of British Architects.*
A. Beresford Hope, Esq., *President of the Architectural Museum.*
The Rt. Hon. the Lord Mayor of London for 1861-62.
Rt. Hon. W. Cowper, M.P., *First Commissioner of Public Works.*
Sir C. Wentworth Dilke, Bart., *Commissioner for the Exhibition of* 1862.
Addington, S., Esq.
Alger, John, Esq.
Barker, A., Esq.
Bartley, G. C. T., Esq.
Bell, John, Esq., *Sculptor.*
Bell, Miss Margarita.
Bowler, K. A., Esq.
Bowley, R. K., Esq.
Bowring, E. A., Esq., C.B.
Brassey, Thomas.
Bury, Talbot, Esq.
Campbell, Colin M., Esq.
Chapman and Hall, Messrs.
Clutton, John, Esq.
Coalbrookdale Iron Company.
Cobden, R., Esq., M.P.
Cole, Henry, Esq., C.B.
Cole, C. A., Esq.
Coope, Octavius E.
Crace, J. G., Esq.
Donnelly, Captain, R.E.
Evans & Son, Messrs.
Fisher, R., Esq.
Fortnum, C. D., Esq.
Fowke, Captain, R.E.
Fowler, John, Esq., C.E.
Frith, J. G., Esq.
Godwin, G., Junr., Esq.
Graham, Peter, Esq.
Griffith, Sir Richard, *Chairman of Board of Works, Ireland.*
Harris, Captain, E.I.S.
Heywood, James, Esq.
Hollins, Michael, Esq.
Hope, Henry Thomas, Esq.
Hubert, S. M. (Messrs. J. Woollams and Co.)
Iselin, J. F., Esq., M.A.
James, Sir Henry, R.E.
Johnson, Henry, Esq., 39, Crutched Friars.
Kelk, John, Esq.
Layard, A. H., Esq., M.P.
Lennox, The Lord Henry G., M.P.
Lowe, Right Hon. R., M.P.
Lucas, Charles, Esq.
Lucas, Thomas, Esq.
Macdonald, J. C., Esq.
Mackrell, W. T., Esq.
Maskell, W., Esq.
Milnes, R. Monckton, Esq., M.P.
Murchison, Sir Rodk. I., G.C.S.S., *Director of the Geological Survey.*
Paxton, Sir Joseph, M.P.
Peto, Sir S. Morton, M.P.
Phillpots, Captain, R.E.
Playfair, Dr. Lyon, C.B.
Poole, E. Stanley, Esq.
Redgrave, Samuel, Esq.
Redgrave, Richard, Esq., R.A.
Roberts, David, Esq., R.A.
Rothschild, Sir Anthony, Bart.
Russell, J. Scott, Esq.
Salomons, Mr. Alderman, M.P.
Sandford, F. R., *Secretary to H. M. Commissioners for the Exhibition of* 1862.
Saunders, W. Wilson, Esq.
Sheepshanks, John, Esq.
Sladen, St. Barbe, Esq., Onslow Square.
Simpson, Messrs. W. B., and Sons.
Sopwith, T., Esq.
Sykes, Godfrey, Esq.
Trevelyan, Sir Charles, K.C.B.
Twining, T., Esq.
Veitch, J., Esq., Junior.
Webb, J., Esq.
Wilson, G. F., Esq.
Winkworth, T., Esq.
Wyatt, M. Digby, Esq., M.I.B.A.
Wylde, R. G., Esq.

Subscriptions may be paid to the account of "Mosaic Wall Pictures Fund," Messrs. Coutts, Strand, London, and to Mr. Davenport, Society of Arts, John Street, Adelphi, W.C.

G. F. DUNCOMBE, *Secretary.*

18*th March,* 1862.

[6849]

DOULTON & WATTS, *Lambeth Pottery, London.*—Articles in glazed stoneware, chemical apparatus, filters, &c.

[6850]

DUDSON, JAMES, *Hanley.*—Improved ironstone jugs and teapots with metal tops, stone candlesticks, ornamental china figures.

[6851]

DUKE, SIR JAMES, & NEPHEWS, *Hill Pottery, Burslem.*—Parian and china dessert service; china dinner, dessert, breakfast, and tea specimens; china and parian vases; parian groups and statuettes; Limoges enamels; Majolica and Palissy wares; jet and terra cotta vases; decorated and printed earthenware, dinner, and toilette ware; parian and stone jugs, &c., &c. (*See pages* 105 *to* 109.)

[6852]

FELL, THOMAS, & Co., *Newcastle-upon-Tyne.*—Dinner ware, vase, table top, lamps, chamber ware.

[6853]

GOODE, THOMAS, & Co., 19 *South Audley Street, Grosvenor Square, W.*—Fine porcelain dessert service, &c. (*See page* 104.)

[6854]

GOODWIN, E., & SON, Pipe-clay Manufacturers, *Saint Clement's, Ipswich.*—Common clay pipes of superior workmanship.

[6855]

GOSS, WILLIAM HENRY, *Stoke-upon Trent.*—Statuettes, vases, tazzi, &c., in parian and other ceramic bodies. (*See page* 110.)

[6856]

GRAINGER, GEORGE, & Co., *Worcester.*—China, chemical porcelain, dinner and dessert services, pierced parian, and busts.

Specimens are exhibited of *Worcester* china tea services; the pure chemical porcelain *(unequalled for economy and durability)*, for dinners and dessert services, chemical apparatus, telegraph insulators, &c. The exhibitors are also manufacturers of parian busts, perforated vases, toilettes, and every other variety of ornaments.

[6857]

GRIMESLEY, THOMAS, Sculptor, *Oxford.*—Sepulchral monuments in terra cotta, for churchyards.

[6858]

HOLLAND, WILLIAM THOMAS, *South Wales Pottery, Llanelly.*—Specimens of printed earthenware, table, tea, jug, and toilette.

[6859]

HYAMS, MICHAEL, *Bath Street, London.*—Registered designs and patent improvements in the manufacture of smoking pipes.

Made under Letters Patent, 2114. They possess the following advantages over other clay pipes:—Increased sweetness, porosity, colouring properties, &c.; they are also impregnated with aromatic qualities, which prevents the unpleasant taste and smell attending the smoking of ordinary washed clay pipes. A prize medal was awarded to this firm at the Exhibition of 1851. N.B. Tobacconists and shippers supplied.

[6860]

INDERWICK, J., 58, *Princes Street, Soho.*—Tobacco pipes. (*See page* 111.)

GOODE, THOMAS, & Co., 19 *South Audley Street, Grosvenor Square, W.*—Fine porcelain dessert service, &c.

DESSERT SERVICE.

The above illustrations represent a portion of a dessert service, designed by Messrs. Goode, with a view of introducing a more artistic style than hitherto produced. The service has been finished by Messrs. Minton, in the very best manner, and of their finest pâte tendre porcelain.

Messrs. Goode also exhibit another truly magnificent dessert service, which has been finished at a cost of nearly one thousand pounds, as well as an exquisite tea and coffee service, the design being pearls set in gold on a turquoise ground, with a coronet and monogram in pink roses, in the old Sèvres style. This service is one of the most expensive ever made in this country, and has been made expressly to the order of a foreign nobleman.

The pure white china figures, baskets, &c., for the dessert table are all of English design, in the old Dresden style.

All the china exhibited by Messrs. Goode has been made for them by Messrs. Minton, of whose manufactures they have an extensive collection at their warehouse.

Duke, Sir James, & Nephews, *Hill Pottery, Burslem.*—Parian and china dessert service; china dinner, dessert, breakfast, and tea specimens; china and parian vases; parian groups and statuettes; Limoges enamels; Majolica and Palissy wares; jet and terra cotta vases; decorated and printed earthenware, dinner, and toilet ware; parian and stone jugs, &c., &c.

DEATH OF MARMION.

DESSERT SERVICE, &c.

DUKE, SIR JAMES, & NEPHEWS, *continued.*

PORCELAIN VASES, TAZZAS, INKSTANDS, CARD-TRAYS, &c.

1. Pair of large vases, 39 inches high, of Grecian form, with antique subjects, representing departure of Achilles for Troy.
2. Pair of Grecian vases, representing the departure of Memnon for Troy.
3. Pair of large vases, royal blue ground, Greek tinted figures and borders, foliated enamel laurel wreath round the neck, and burnished gold.
4. Pair of Cumean pitchers, one with blue, one with black ground, representing the battle between the Greeks and Amazons.
5. Series of vases, in 3 pairs, different sizes, black, blue, and Rose du Barry grounds respectively, similarly decorated with Greek figures and ornaments.
6. Variety of vases, royal blue and other grounds, with tinted antique subjects, illustrative of Greek art.
7. Vase with perforated neck, with painted birds, pale Celeste blue ground and burnished gold.
8. Specimens of tazzas, different forms and sizes, blue and black ground, Greek figures and gold.
9. Pair of china flower pots, green ground, and painted marine views and fruits, in compartments.
10. Pair of large ewers, in compartments, royal blue and chocolate grounds, Grecian figures, and gilt.
11. Pair of ewers designed after the antique, royal blue ground.
12. Pair of ewers, Grecian tinted subjects, upon blue ground and gold.
13. Candlesticks with marone, Celeste, and blue grounds, with antique figures, and finished in burnished gold.
14. Pair of turquoise vases, flowers and fruits.
15. Pair of cigar cups, enamelled figures upon blue ground, and gilt.
16. Small inkstands, various Grecian designs, on black and blue ground.
17. Pair of bedroom candlesticks, black ground, with figures from the antique.
18. Series of small vases, 5 sizes, blue ground, ornamented with enamelled figures and gold.
19. Series of small vases, 5 sizes, black ground, similarly decorated.
20. Teapot stands, one with rich landscape and broad band of burnished gold, two Grecian style, in royal blue, and gilt.
21. Large vase, Limoges enamel, with compartments on either side, containing subjects figurative of Peace and War.
22. Pair of ewers, Limoges enamel and chased gold ornaments.
23. Tazza, Limoges enamel, representation of Neptune and Amphitrité.
24. Elevated tazza, enamel; subject, "The Triumph of Galathée."
25. Pair of pot-pourri jars, with perforated covers, painted wreaths of flowers, and gilt.
26. Pair of ewers, roses natural size, on azure ground.
27. Vase, white bisque ground, with large painted Venetian subject on one side, and landscape on the reverse, each in chased and raised gold panels.
28. Pastille burner, turquoise, painted flowers and gold ornaments, geometrically arranged.
29. Several pairs of spill pots, various designs, green, blue, and rose colour grounds.
30. Pair of vases, Chinese shape, white bisque ground, decorated with painted bouquets of flowers, mat and burnished gold.
31. Pair of large tubes, chased gold panels and Watteau vignettes.
32. Embossed porcelain ice pail, stone drab ground, with burnished gold.
33. Pair of bottles, decorated with Gothic design in blue, red, and burnished gold.
34. Small Gothic jugs, in 3 pairs, various sizes, grounds, and style of finish.
35. Large ewer, marone ground, chased gold ornaments, and four panels, enclosing paintings emblematical of the Seasons.
36. Déjeûner trays.
37. Pair of large vases, white bisque ground, with paintings of landscape, with figures on either side, in panels of raised chased gold.
38. Large porcelain jug, embossed with the flags of nations, coloured and gilt.
39. Inkstand, Rose du Barry ground, painted wreath of roses, and gilt.
40. Card basket, with perforated border, Watteau subject, turquoise and gold.
41. Pair of vases, Mazarine blue ground, body decorated by process of gold printing, painted subject in oval medallion on either side, "Summer and Winter."
42. Elaborately perforated card basket, containing study of various fruits and flowers in centre, and medallions of flowers round the border, finished in turquoise and gold.
43. An assortment of porcelain and parian jugs.

MAJOLICA, PALISSY, TERRA, AND JET VASES, PLATEAUX, &c.

44. Pair of large plateaux, 20 inches diameter, majolica, painted by G. Eyre.
45. Pair small plateaux, majolica, painted by G. re.
46. Large group, "Chamois Hunters," majolica.
47. Ewer and stand. "
48. Pair of brackets. "
49. Specimens of statuettes. "
50. Garden seat, fern leaf, and set of garden pots, majolica.
51. Pair of cups, Palissy.
52. Tazza.
53. Pair wine coolers, black-grounded terra cotta body, Greek subjects, finished in Samæan enamels.
54. Large ice pail, antique subject—"Pan performing on the double flute, in presence of Bacchus."
55. Minerva jug, terra cotta, after the antique.
56. Pair of large vases, with swan's neck handles, jet black bisque ground, Grecian figures and borders.
57. Pair of jars, jet, with solid chased gold, Grecian subjects.
58. Pair of vases with covers and handles, jet, chased silver, Greek figures.
59. Pair of ewers, jet, with Greek ornaments in red and gilt.
60. Pair of vases, jet, vermilion ground, black figures.
61. Pair of terra cotta, enamelled.

DUKE, SIR JAMES, & NEPHEWS, *continued.*

DUKE, SIR JAMES, & NEPHEWS, *continued.*

62. Pair of jet, Grecian design, upon black bisque ground.

63. Small vases in pairs, jet, antique figures, with ornaments of burnished gold.

64. Tobacco boxes, terra cotta, coloured in enamel and gilt.

65. Teapots, sugar box, and cream ewer, jet, enamelled and gilt.

66. Assortments of jugs, various shapes, jet, coloured and gilt, in different designs.

67. Various terra cotta spill pots, Samæan enamel.

68. A variety of candlesticks, jet, Greek figures and gold.

69. Pair of vases, black bisque ground, with Grecian subjects, in bright black.

70. Pair of small scent jars, jet, enamelled, with gold ornaments.

71. Pair of jet vases, the details of decoration in enamel and gold.

72. Small vases in pairs, in terra cotta, Samæan enamels.

73. Pair of jet vases, with rich burnished gold ornaments.

74. Bedroom candlesticks, jet, gilt and enamelled.

75. Water bottles and stands, decorated in a variety of designs.

76. Various specimens of porcelain and earthenware, decorated by the process of printing in gold.

PARIAN STATUARY, VASES, &c., &c.

77. Large group, "Cupid Captive," by W. C. Marshall, R.A.

78. Group, "Death of Marmion," by Bayley.

79. Group, "Dying Zouave,"

80. Group of two dogs, setter and pointer, by Bayley.

81. Group, "Ino and Bacchus," by Méli.

82. Bust, "Young Augustus," from the antique.

83. Statuettes, Venetian beauties: Modesty, Vanity, Terpsichore, Ceres, Pomona, by Méli.

84. Group, "Peace Congress," by J. Henning, after a design by A. Crowquill.

85. Statuette, late Duke of Wellington, by G. Abbott.

86. "The Hop-girl."

87. Statuette, Lord Elcho, by Beattie.

88. "Innocence Protected," bisque, by Beattie.

89. Statuette, "Flora," finished in imitation of ivory and oxydized silver.

90. Statuettes, "Venus and Shell," "Girl at Bath," "Richard Cœur de Lion," "Morning," "Evening," "Italian Improvisatore," &c.

91. Variety of small groups, dogs, &c.

92. Pair of large vases, ornamented with raised flowers.

93. Pair of large ewers, decorated with passion flowers and roses.

94. Pair of large vases, raised roses.

95. Pair of pot-pourri jars, perforated covers, one with pink ground, raised flowers on each.

96. Pair of vases, with covers and ring handles, bouquet of flowers.

97. Pair of vases, a group of raised flowers on either side.

98. Pair of ewers, with flowers in porcelain, bisque body.

99. Variety of other parian vases and ewers.

100. Parian flower pots, one pair with Celeste ground.

PORCELAIN DESSERT AND TEA SERVICES, TABLE PLATES, CUPS AND SAUCERS, &c.

101. Dessert service, comprising—
Centre-piece, 30 inches high.
Two corner-pieces, 18 do.
Four comports, 12 do.
Two cream-bowls, 10 do.; porcelain base and perforated basket, with parian figure supports, that of the centre-piece being a group of Grecian attendants on marriage; of the corner-pieces, grouped figures representing Peace, Commerce, and Industry. The porcelain portions of each article decorated in amethyst and gold, painted groups of flowers and fruits.
Twenty-four dessert plates, perforated porcelain, with borders in amethyst, gold, and groups of flowers, with painted views in centre, from Cowper's "Task."

102. Specimens of dessert plates, with borders in raised, chased, and burnished gold, with different grounds, painted, enamelled and gilt, in various designs.

103. Centre piece, elevated and low comports, each of two shapes.

104. Tea service, rose colour band, with elaborately-chased gold border.

105. Variety of cups and saucers, of diversified patterns, gilt, chased, and painted.

106. Cups and saucers, Sèvres form and style of decoration.

107. Series of cups and saucers, each with different ground, burnished gold on foot, &c.

108. Specimens of cups and saucers for more general use, assorted shapes and patterns.

109. Porcelain table plates, variously enamelled and chased gold borders, with elaborate crests and coats of arms in centre.

110. Numerous porcelain table service samples, painted, gilt, and enamelled, in different styles of ornament.

111. Breakfast cups and saucers, various shapes, grounded.

112. Porcelain soup tureen and other covered pieces.

113. Porcelain toilet service, painted wreaths of flowers, and gilt.

114. Porcelain flower pots, finished in a variety of designs in enamel and gold.

115. Specimens of printed and flowing printed table services in a variety of shapes, colours, and patterns.

116. Variety of samples of enamelled and gilt table services, different designs and style of finish.

117. Toilet service patterns, printed and flowing printed, on several shapes.

118. Various specimens of enamelled and gilt toilet ware.

119. Assortment of stone and other jugs.

120. Loaf plates, butter tubs, &c.

121. Complete assortment of chemical and apothecaries' articles.

Duke, Sir James, & Nephews, *continued.*

VASES AND STATUETTES.

VASES AND ORNAMENTS.

GOSS, WILLIAM HENRY, *Stoke-upon-Trent*.—Statuettes, vases, tazzi, &c., in parian and other ceramic bodies.

STATUETTES, VASES, TAZZI, &c.

INDERWICK, J., 58 *Princes Street, Soho.*—Tobacco pipes.

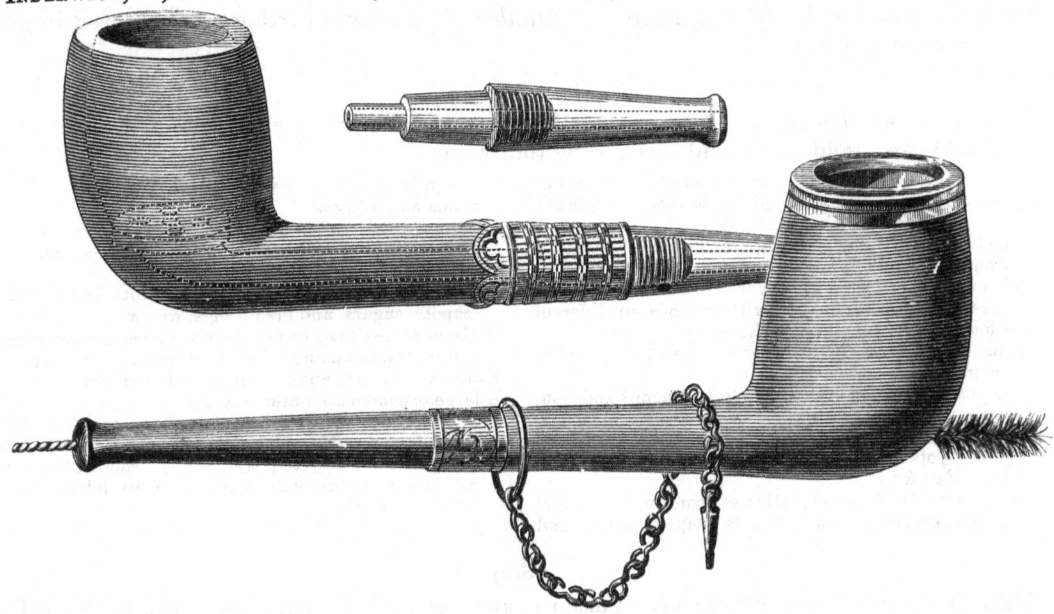

The exhibitors are importers and manufacturers of meerschaum pipes, cigar cases, snuff boxes, cigar tubes, and every article requisite to the smoker.

The following are exhibited, viz.:—Inderwick's Patent Silician Pipe, upon an entirely new construction.

Inderwick's Patent Ventilating Pipe, in meerschaum, briar-root, or clay, through which a stream of pure air is conveyed with the smoke, producing a deliciously cooling effect.

Inderwick's Patent Cigar Desideratum, an article that should be used by every smoker.

J. I. & Co. import and supply Latakia, Djbail, and all the finest Eastern tobaccos.

[6861]

JAMIESON, J., & Co., *Borrowstounness, N.B.*—Fine earthenware.

[6862]

JENNINGS, GEORGE, *Holland Street, Blackfriars.*—Drain pipes, air bricks, bending bricks, damp-proof course, invert blocks, &c.

[6863]

KERR, W. H., & Co., *Royal Porcelain Works, Worcester.*—Porcelain, parian, and stone china.

Portion of a dessert-service, manufactured by command of Her Majesty the Queen.

The decorations are in enamel, in the style of the monochromic enamels of Limoges, but on a fine turquoise ground.

The subjects are all varied, and consist of designs by Mr. Bott, and adaptations from celebrated gems. The service was painted by Mr. Bott, and the pieces were designed by Mr. Thomas Reeve.

Specimens of the Worcester enamels, in various styles, consisting of plaques, chalices, vases, tazzi, &c.

Specimens of the Worcester Raphaelesque porcelain, consisting of vases, brackets, figures, busts, tazzi, &c.

Specimens of the Worcester ivory porcelain, consisting of dessert patterns, déjeûner sets, vases, &c.

Specimens of porcelain dinner services.

„ „ dessert services.

„ „ tea and coffee services.

„ „ toilette services.

Specimens of the Worcester vitreous stoneware, in dinner services, &c., &c.

[6864]

KEY & BRIGGS, *Stoke-upon-Trent.*—Parian, majolica, mosaic, jet, and porous ware.

[6865]

KOSCH, F., *Hanley.*—China and earthenware, printed by new process in gold and colours,

[6866]

LIDDLE, ELLIOT, & SON, *Dalehall Pottery, Longport, Staffordshire.*—Earthenware dinner, tea, and toilette ware; photograph goods, and parian.

[6867]

LIVESLEY, POWELL, & Co., Manufacturers, *Hanley.*—China and earthenware, printed in gold and colours; parian, &c.

[6868]

LOCKETT, JOHN, *Longton.*—China and earthenware of all kinds; chemical earthenware and stone ware; gold lustre and black Egyptian wares.

Dessert services, various coloured grounds in Mazarine, marone, green, &c., painted in flowers, landscapes, &c., richly gilt and chased in gold.
Porcelain cups and saucers, of different coloured grounds, richly gilt and chased.
China dessert plates, in great variety of design.
China vases, painted flowers, with grounds in different colours, and richly gilt and chased.
China toilette sets, in great variety, and very chaste designs.
Gold lustre ware, in every variety, in teacups and saucers, jugs, mugs, ewers and basins, bowls and covers, bowls, vases, teapots, sugar boxes and cream jugs, garden pots and stands, toy cups and saucers, jugs, mugs, &c., &c.
Stone ware—Mortars and pestles, evaporating dishes and funnels, garden pots and stands, teapots, sugars and cream jugs, and jugs with metal covers, of different forms and designs.
Egyptian black and black lustre ware—Teapots, sugars, and creams, in oval and round shapes, of a variety of patterns.
Cane-coloured, turquoise, sage, and white earthenware teapots, sugars, and cream jugs, &c., &c.
Earthenware—Soup tureens, sauce tureens, cover dishes and plates, in enamelled pottery, printed, &c., &c.
Ewers and basins, and garden pots and stands.
A large ornamented water filter.
Jars and covers, of a variety of grounds, in blue, olive, turquoise, marone, pink, mat blue, &c., with gold lines. Tamarind, &c., jars, richly ornamented, and gold labels. Cold-cream pots and covers, labelled, &c., and diced ware.

[6869]

MELI, GIOVANNI, *Stoke-upon-Trent.*—Parian statuettes, vases, ornaments, jugs, butter-tubs, dessert pieces, &c.

The Statuettes exhibited by G. Meli are religious, classical, historical, and poetical in character. They vary in size from two to twenty-five inches in height. He exhibits also groups of figures, centre-pieces, busts of eminent men, vases, and useful ornaments. All articles are of the best workmanship and material, and are very reasonable in price.

[6870]

MID-LOTHIAN POTTERY COMPANY, *Portobello, near Edinburgh.*—Bottles, filters, jars, foot-warmers, jugs, picklers, casks, jelly-cans, &c.

[6871]

MILLAR, J., & Co., 5 *St. Andrew's Street, Edinburgh.*—Ornamental pottery, &c.

[6872]

MILLICHAMP, HENRY, *Prince's Street, Lambeth.*—Terra cotta vases, flower pots, chimney shafts, &c. (*See page* 120.)

[6873]

MINTON & Co., *Stoke-upon-Trent.*—China, earthenware, majolica, parian, tiles, &c. (*See pages* 114 *to* 117.)

[6874]

NORTHEN, WILLIAM, *Union Pottery, Vauxhall, London.*—Manufacturer of every description of stoneware, and patentee of stoneware mangers.

[6875]

OLD HALL EARTHENWARE COMPANY, THE (Limited), *Hanley, Staffordshire.*—Parian, plain and ornamental earthenware of every description.

Obtained a Prize Medal in **1851**, *and several of the Society of Art's Medals.*

Specimens of earthenware:—Dinner, dessert, tea, breakfast, and toilet services, from the most costly to the cheapest kind, both for home and foreign markets, candlesticks, jugs, and flower pots.

Parian and terra cotta clocks, vases, figures, fonts, and wine coolers.

Large vase—History of Bacchus.

Group—Prometheus bound.

Venus and Cupid, after Fraikin.

Hamlet.

Esmeralda.

Highland Mary and Glycera, with many others.

Specimens of porous goods, for hot climates.

Rockingham and green glaze ware.

Plain and decorated stone jugs, teapots, flower pots, and other articles.

Memorial jug—Prince Consort.

[6876]

PELLATT & Co., 59 *Baker Street, Portman Square, W.*—Specimens of ceramic manufacture.

[6877]

PHILLIPS, W. P., & G., Designers and Producers, 359 *Oxford Street, and* 155 *New Bond Street.*—China dessert service, vases, &c. (*See pages* 120 *and* 121.)

[6878]

POWELL, WILLIAM, & SONS, *Temple Gate Pottery, Bristol.*—Various articles in stoneware.

[6879]

PRICE, CHARLES, & JOSEPH R., *Potteries, Bristol.*—Every description of the improved stoneware.

[6881]

ROBERTS, JOHN, *Upnor, Rochester, Kent.*—Coolers, filters, stoves, grates, made of terra cotta and fire clay.

[6882]

ROSE & Co., DANIELL & Co., *Coalport, Shropshire; Wigmore Street and New Bond Street, London.*—China manufacturers.

[6883]

SHARPE, BROTHERS, *Swadlincote, Burton-on-Trent.*—Patent closet basins and specimens of Derbyshire cane and Rockingham wares.

[6884]

SHARPUS, THOMAS, & CULLUM, WILLIAM, 13 *Cockspur Street, Pall Mall.*—Unique china dessert service and other elegant designs.

Minton & Co., *Stoke-upon-Trent*.—China, earthenware, majolica, parian, tiles, &c.

MAJOLICA VASE.

MINTON & Co., *continued.*

SPRING.

AUTUMN.

Minton & Co., *continued.*

PARIAN CANDLESTICKS, DECORATED WITH GOLD.

The Majolica Fountain erected under the Eastern Dome.

Subjoined is a brief description of this magnificent work, which is 36 feet high by 39 in diameter :—At the summit there is a group, larger than life size, of St. George and the Dragon; four winged figures of Victory, holding crowns of laurel, encircle a central pavilion, on the top of which the group rests, and around which is inscribed the motto, "For England and for Victory;" underneath is a series of smaller fountains of varied shapes and sizes, which receive the water, and spread it as required. One of these, that supported by a heron, is after the model of the fountain designed for the Queen's dairy, at Windsor, by the late Mr. Thomas, under the personal superintendence of his Royal Highness the late lamented Prince Consort. The outer basin of the St. George's fountain, which is the largest, and encircles the whole work, is ornamented with an oak leaf, alternating with the Rose of England, and is divided by eight flower vases. The whole management of the heraldic colouring is considered very perfect, and the imitation of the steel armour on the arms and legs of the colossal figure is especially deserving of notice. This work was designed and modelled by the late Mr. John Thomas, sculptor, of London, a self-made artist, a man of genius and sterling merit, and who had attained to great eminence, when death removed him from the scene of his useful labours.

"The Porcelain master-piece of the Exhibition. If there were no other object in the Building but this grand work alone, it would be well worth a shilling entrance fee to see it."—*Vide Times, May* 17.

"This magnificent triumph of the Potter's Art."—*Standard, May* 19.

"The triumph and *ne plus ultra* of the Potter's Art. . . . . Every one felt, and thousands uttered the word 'success,' and the Majolica Fountain was at once installed as one of the chief 'Lions' of the Exhibition."—*Daily News, May* 26.

Parian.

A 1. Pair of candlesticks, No. 976, two Cupids adorning a pillar with garlands of laurel.

A 2. Pair of spill cases, finely pierced, No. 943, glazed Byzantine border, on gold ground.

A 4. Oval comport, No. 930, supported by two Cupids holding garlands of laurel, group of birds underneath.

A 5. Three bonbonnières, No. 906, small figures, representing Industry, Science, and Art.

A 6. Chinese lantern, No. 978, finely perforated, and ornamented in the Oriental style, on green ground, mounted in metal, with bells, &c.

A 7. Pair of vases, No. 901, four panels with Cupids, representing the Seasons, surrounded by a gold wreath of oak and laurel.

A 8. Renaissance figure, Autumn, No. 384, partially gilt.

A 9. Renaissance figure, Summer, No. 385, partially gilt.

A 10. Pair of candlesticks, No. 830, group of three Cupids holding bugles, pedestal ornamented with cameos.

A 13. Vase, No. 944, celadon ground, embossed grasses, wild flowers, flies, and birds.

A 14. Cock and Hen, No. 372, coloured naturally.

A 15. Vase, No. 944, celadon ground, ornamented with honeysuckle and wild roseberry in gold, flies.

A 21. Nautilus shell, No. 992, Marine Venus seated on the top, Cupids in relief on the body of the shell, group of two Mermen supporting it.

A 22. Pair of spill cases, finely pierced, No. 943, Byzantine border.

A 23. Pair of leafage vases, No. 687, ivory, rope, solid gilt.

A 24. Pair of vases, No. 944, Chinese, celadon ground, tied with ropes, perforated at neck and foot.

A 50. Hebe and Eagle, No. 339.

A 51. Venus, No. 386.

A 52. Season figure, No. 387, Spring.

A 53. Ditto do. No. 388, Summer.

A 54. Ditto do. No. 389, Autumn.

A 55. Ditto do. No. 390, Winter.

A 56. Amazon, No. 246.

MINTON & Co., *continued.*

A 57. Theseus, No. 254.
A 58. Cain and Abel, No. 327.
A 59. Prince Alfred and Pony, No. 357.
A 60. Prince Arthur, No. 360.
A 61. Princess Beatrice, No. 364.
A 62. Bust—Queen Victoria, No. 355.
A 63. Bust—Prince Consort, No. 356.
A 64. Bust—King of Sardinia, No. 358.
A 65. Vintagers with shell, No. 365.
A 66. Flight into Egypt, No. 223.
A 67. Undine, No. 351.
A 68. Bacchus, No. 353.
A 69. Lalage, No. 391.
A 70. Bust of Modesty, after Clodion, No. 392.
A 71. Venus and Cupid, No. 336.
A 72. Daniel Saved, No. 393.
A 73. Clodion vase, No. 647.
A 74. Roger's vase, No. 848, cast from the antique.
A 75. American Slave, No. 377.
A 76. Knitting Girl, No. 370.
A 77. Water Nymph, No. 380.
A 79. Lady Godiva, No. 383.
A 80. Prince Arthur, No. 382.
A 81. Prince Leopold, No. 381.
A 82. Princess Helena, No. 362.
A 83. Princess Louisa, No. 363.
A 91. Lady Constance Grosvenor, No. 325.
A 92. Prince of Wales, No. 206.
A 93. Princess Alice, No. 394.
A 94. Skipping Girl, No. 371.
A 95. Fanny Ellsler, No. 162.
A 98. Tazza Shell Top, No. 1003, supported by Cupid, &c.

CHINA.

B 1. Pair of vases, No. 803, Sèvres blue ground, painted Cupid subjects after Boucher.
B 2. Pair of vases, No. 974, lilac ground, painted on gold ground, with allegorical figures of Love and Spring, &c.
B 3. Pair of jars, No. 975, painted orchidaceæ and gold.
B 4. Pair of bottles (Alhambra), No. 624, painted roses.
B 5. Pair of sceaux, No. 977, painted wreath and pendants of passion flower and festoons.
B 6. Pair of vases, No. 865, French green ground, painted Cupids and landscapes and trophies.
B 7. Pair of vases, No. 467, painted festoons of various flowers and cameos, in the old Sèvres style.
B 9. Pair of vases, oval and pedestal (Queen's), No. 626, Sèvres blue ground, painted pastoral subjects, after Boucher, and trophies.
B 10. Vase No. 803, pâte tendre, turquoise ground, painted landscape and birds.
B 11. Pair of vases, No. 450, Sèvres green ground, painted landscapes, birds and fruit.
B 12. Pair of sceaux, No. 977, painted wreath of roses.
B 13. Pair of vases, No. 865, pâte tendre, turquoise and painted juvenile subjects, after Eisen.
B 14. Pair of vases, No. 840, small size, Rose du Barry ground, painted cameos on blue medallions.
B 15. Pair of vases, No. 857, Sèvres blue ground, painted subjects, Apollo and the Muses on one side, and the "competition between Apollo and Marsyus," from Bartolozzi, on the other.
B 16. Pair of sceaux, No. 383, small size, pâte tendre, turquoise ground, painted Cupids in compartments.
B 17. Pair of sceaux, No. 383, painted small bunches of roses hanging from purple ribbon, and bouquets of roses, with sprigs of cornflowers.
B 18. Pair of bottles, pilgrim, No. 680, painted Cupid and trophy in pink.
B 19. Pair of vases, No. 689, turquoise ground, painted cameos and trophies, festoons and pendants of various flowers.
B 22. Pair of vases, No. 469, turquoise ground, painted figures of the Seasons.
B 24. pair of flower pots, No. 979, painted wreath of poppies, corn, bluebells, &c., mounted in ormolu.
B 28. Comport Sevres, No. 576, Rose du Barry ground, gold wreath and bows of blue ribbon, painted group of fruit and flowers in centre.
B 30. Pair of vases, No. 662 (Rothschilds), perforated, turquoise ground, painted subjects after Boucher.
B 31. A single dejeûner set, in the Chinese style, fluted, with a wreath of fruit in gold.
B 33. Pair of sceaux, No. 383, lilac ground, painted Cupids in Camaïeu.
B 34. Vase, No. 840, pâte tendre, turquoise ground and cameos on purple ground.
B 42. Pair of vases, No. 450, Sèvres blue ground, painted pastoral subjects after Boucher, and trophies between.
B 43. Pair of bottles (Harewood), No. 299, Sèvres green, husks and gold ribbon.
B 48. Oval dejeûner set, rose embossed, coloured pink.
B 49. Pair of vases, No. 776, Sèvres blue ground, painted "The Muses," after Lesueur.
B 50. Pair of candelabra, No. 918, parian figures of piper and listener, Sèvres green ground richly gilt.
B 51. Pair of vases, No. 689, pâte tendre, Rose du Barry ground, painted cameos, trophies, and festoons of flowers in the old Sèvres style.
B 58. Flower holder, No. 935, bamboo, pink ribbon and bows, and painted small groups.
B 65. Cupid candlestick, No. 504, turquoise and gold.
B 68. Twist candlestick, No. 529, Rose du Barry, flowers and gold.
B 69. Piano candlestick, No. 794, turquoise, roses and gold.
B 70. Chinese tripod, No. 958, painted flying birds.
B 71. Pair of bottles, No. 744, hexagon pierced, painted pendants of flowers.
B 72. Pair of match pots, painted lilies-of-the-valley.
B 73. Pair of match pots, Rose du Barry ground, painted birds, flowers, &c.
B 76. Pair of match pots, blue band, painted festoons of flowers.
B 77. Tray, No. 781, turquoise band and ribbon, with painted subject in centre.
B 78. Tray, No. 781, green band and ribbon, painted fruit, and flowers in centre.
B 79. Tray, No. 781, Rose du Barry and gold, twist and painted trophy.
B 80. Duck cream, coloured naturally.
B 87. Clock piece, No. 985, green ground, Cupids holding garlands of flowers, surmounted with a Cupid holding a tazza.
B 88. Tazza, No. 987, stork support, parian, gold cord edge.
B 89. Pair of tazzas, No. 987, stork support, gold cord edge.
B 90. Pair of bottles, No. 688, oriental ornaments in turquoise and gold.
B 91. Pair of bottles, No. 984, turquoise ground, Persian coloured border.
B 92. Pair of indented bottles, No. 743, lilac ground, Chinese scrolls.
B 93. Pair of small vases, No. 988, three handles, white Greek border on red ground.
B 94. Pair of vases, square-handled, No. 470, Sevres blue ground and richly gilt.
B 97. Oval basket and pedestal, No. 989, perforated fleur-de-lis, turquoise and gold.
B 98. Chinese single dejeûner set, fluted in turquoise, painted birds on branches of hawthorn.
B 301. Oval centre piece, No. 708, Sèvres blue ground, finely perforated and richly gilt.
B 303. Pair of bottles, No. 707, pale ivory ground, gold Indian ornaments all over.

MILLICHAMP, HENRY, *Prince's Street, Lambeth.*—Terra cotta vases, flower pots, chimney shafts, &c.

Terra cotta vases, flower pots, pedestals, flowers, trusses, brackets, tracery, balusters, finials, chimney shafts, pots, &c.

TERRACE VASE.

GARDEN VASE.

RUSTIC GARDEN SEAT.

GARDEN VASE.

[6885]

Sherwin, Henry, *Wolstanton, Stoke-on-Trent.*—Dinner plates of patterns constructed of the parsley leaf, &c.

[6886]

Southorn, Edwin, *Broseley, Salop.*—The Broseley patent glazed tobacco-pipes, unequalled in purity of material.

[6887]

Southorn, William, & Co., *Broseley, Shropshire.*—The celebrated Broseley glazed tobacco-pipes.

[6888]

Stiff, James, *London Pottery, High Street, Lambeth.*—Water filters, chemical apparatus, terra cotta drain pipes, jars, bottles, &c. (*See page* 122.)

[6889]

Storey & Son, 19 *King William Street, City of London.*—Specimens of china, porcelain, and earthenware. (*See page* 123.)

[6890]

Temple, Emily, 184 *Regent Street, and Brighton.*—Dessert service, ceramic statuary figure, supporting pierced comports, richly decorated. (*See page* 124.)

[6891]

Turner, Bromley, & Hassall, *Stoke-upon-Trent.*—Parian groups, statuettes, vases, &c.

[6892]

Wathen & Lichfield, *Fenton, Staffordshire.*—Earthenware, printed, enamelled, and gilt, in dinner, tea, jugs, and toilette sets.

[6893]

Wedgwood, Josiah, & Sons, *Etruria, Staffordshire.*—Jasper, blue and white bas-relief ware; parian statuary, stone, chemical, and photographic wares; earthenware in all its branches, enamelled, printed, pearl, cream colour, green glaze, Rockingham, majolica, porous, and terra cotta. (*See pages* 125 *to* 127.)

[6894]

Wilkinson, Rickhuss, & Badock, *Hanley, Staffordshire.*—China tea, breakfast, and dessert ware, parian vases, jugs, &c.

[6895]

Ynismedew Brick and Pipe Company, *Pontardawe, Swansea.*—Sewerage pipes, vases, tazzas, chimney pots, and faced bricks.

[6896]

Robertson, Mrs. J.—Service of tartan-plaid pattern.

[6897]

Grove, R. H., *Barlaston, near Stone, Staffordshire.*—Lustre ware.

Phillips, W. P., & G., Designers and Producers, 359 *Oxford Street, and* 155 *New Bond Street.*—China dessert service, vases, &c.

Dessert service of original design, perforated border, richly gilt, consisting of centre-piece (see fig. 3), group of figures in parian, surmounted by a basket for flowers, mounted on an ornamental plinth, with elegant shields for armorial bearings, and supported on either side by a shell-shaped comportier for fruit; assiettes montées (see fig. 1), baskets with dancing-boys (see fig. 2), comportiers with figures representing the Four Seasons, ditto for corners, and cream-bowls and covers *en suite*, with plates decorated variously.

Porcelain Vases.

Pair of vases, Queen's shape, "bleu de roi" ground, finely painted subjects after Boucher, finished in the old Sèvres style.

Pair of vases, "bleu de roi" gilt, and finely painted landscape.

Pair of vases, turquoise ground, richly gilt, and very finely painted subjects—"Venus and Endymion," "Vulcan and Venus."

Pair of vases, oviform, turquoise ground, rich chased gold festoons, with beautifully painted subjects after Watteau.

Pair of vases, finished in the old Sèvres style.

Porcelain Sceaux, Trays, &c.

Large tray, Rose du Barry ground, rich raised gold borders, with very finely painted landscape in centre.

Large tray, turquoise ground, rich raised gold borders, and finely painted cameo subjects in centre, in the old Vienna style.

Chalice cup, Sèvres green ground, richly gilt, with finely painted medallion of the Amazon.

Cabinet cups with two handles, green ground, richly gilt, with finely painted medallions.

Sceaux, turquoise ground, richly gilt, and finely painted groups of flowers and fruit.

Porcelain Cups and Saucers, Plates, &c.

A variety of cups and saucers, dinner and dessert plates.

China toilet service, green band, and gold Greek border complete with every requisite for the dressing-table.

Various samples, ditto, ditto.

Enamels.

Pair of pilgrims' bottles.

A tazza with masks, &c.

A coupe and cover.

Modern Majolica Ware.

Pair of large vases, with painted Cupids in compartments, after Boucher.

A magnificent dish, with ribbon handles, black ground, with subject of "Orithya, daughter of Erectheus, King of Athens, being carried away by Boreas, King of Thrace, while crossing the Ilyssias," in white enamel.

Various samples of earthenware.

Fig. 1.

Phillips, W. P., & G., *continued.*

Fig. 2.

W.L. THOMAS. Sc. P.H.DEL. AMOTTE. Del.

Fig. 3.

STIFF, JAMES, *London Pottery, High Street, Lambeth.*—Water filters, chemical apparatus, terra cotta drain pipes, jars, bottles, &c.

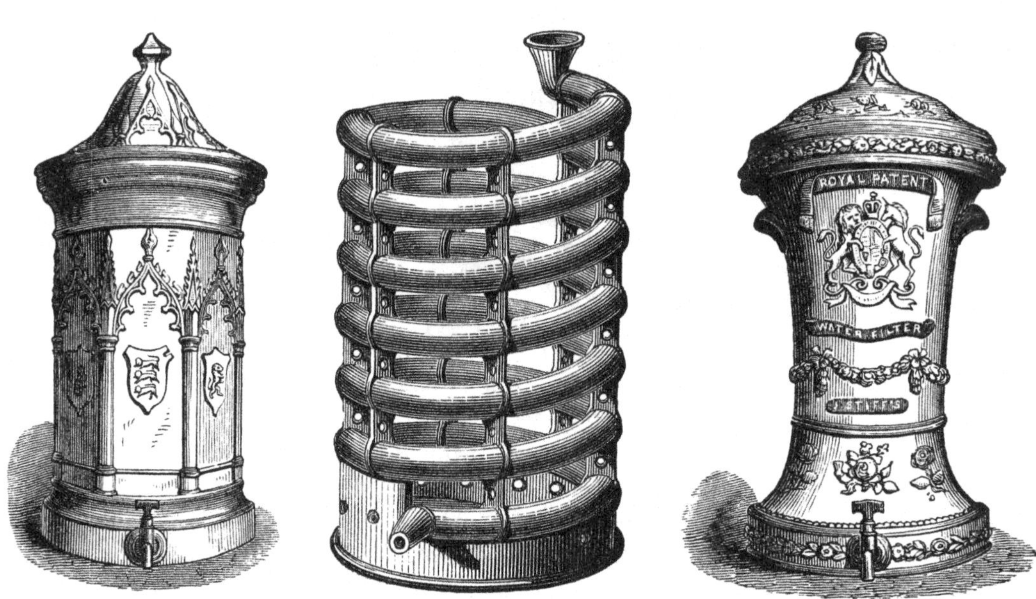

Water filters constructed on the most approved purifying principles, and of various designs. As they have been extensively used for the last twenty-five years, it would be superfluous to point out their advantages. They may also be had cased in white wicker, for ships' use.

Barrels, round and oval shape, specially adapted for spirit-merchants' use. Being vitrified, they do not absorb, therby preserving the article therein contained, for any length of time. Gilded and japanned to order.

Stiff and Sons are manufacturers also of damp-proof course, drip bands, wall copings, facing blocks, &c., according to the patents of John Taylor, Jun., Architect.

STOREY & SON, 19 *King William Street, City of London.*—Specimens of china, porcelain, and earthenware.

CHINA TEA SERVICES, &c

CHINA DESSERT SERVICES &c

Specimens of breakfast, dessert, and tea services in porcelain, painted and decorated with raised and burnished gold; dessert comportiers, with parian supports. In these the china and parian is combined, the glazed portion being decorated to match with low comportiers and plates of the dessert services, déjeûner, and luncheon sets.

Dinner and toilet services in earthenware, in a variety of colours and designs, and decorated with burnished gold.

Parian acanthus vase, with festoons of raised flowers.

" " in china, green ground, with two compartments, one painted rich groups of flowers—reverse side a landscape.

Triple violet bottle, Chinese vase and cover.

Tazzas and inkstands, flower pots, painted and gilt.

Temple, Emily, 184 *Regent Street, and Brighton.*—Dessert service, ceramic statuary figures, supporting pierced comports, richly decorated.

PORCELAIN CENTRE PIECE.

Dessert service, consisting of centre piece with figures and raised comports in statuary porcelain, elaborately ornamented in mat and chased gold, the drapery of the figures enriched with tinted borders.

The plates of fine porcelain with perforated border, and beautifully-painted festoons, baskets, and wreaths of flowers, and rich gold borders and star.

The following is a description of centre-piece and raised comports, which have been modelled by W. Beattie, of London, under the direction and from the designs suggested by Madame Temple, expressly for the Exhibition:—

Upon a base of great richness arises a column worked with ivy and scrolls, the caps are composed of rich acanthus foliage, out of which springs an elaborate perforated tazza, lined with a ruby glass, to hold fruit or flowers. Upon the base, and dancing round the column are three female figures, with various insignia, expressive of joyfulness, gratitude, and abundance. Among the foliage of the acanthus are three genii of the dance, hovering over and admiring the spirit and grace of the joyous, happy trio.

The figures of each of the different size groups are varied in composition, although emblematical of joyfulness and abundance.

The service has been manufactured exclusively for Madame Temple, by W. T. Copeland.

WEDGWOOD, JOSIAH, & SONS, *Etruria, Staffordshire.*—**Jasper, blue and white bas-relief ware; parian statuary, stone, chemical, and photographic wares; earthenware in all its branches, enamelled, printed, pearl, cream colour, green glaze, Rockingham, majolica, porous, and terra cotta.**

Received the Prize Medal in **1851**, *and the Medaille de* **1re** *Classe at Paris, in* **1855.**

A Solid jasper, that is the material now exclusively called Old Wedgwood Ware, the perfection of which depends on the smoothness of the surface and the delicacy and sharpness of the bas-reliefs. Invented by Josiah Wedgwood, F.R.S., about 1766.

1. Various vases, flower stands, &c., pale blue ground and white bas-reliefs, in the style of Louis XVI., as introduced into pottery in this material.
2. Bonbonnière and tazza, pale blue and white bas-reliefs, with cameos of various colours.
3. Various vases, black and white bas-reliefs.

B Blue and white bas-reliefs, not solid jasper.

The process of covering an inferior but more manageable material than jasper, with a dip of coloured jasper, was introduced about 1785.

4. Pair of large vases, three feet high, on pedestals, in the style of Louis XIV.; boar's-head handle and bas-reliefs, representing sacrifices.
5. Vases in dark blue, with flowers white, in alto-relievo.
6. Various vases in dark blue and white bas-reliefs, with coloured cameos.
7. The Portland and other vases, blue and white bas-reliefs.

C Black basalt: invented by Josiah Wedgwood, F.R.S., 1766.

8. Various vases on old and new models, and slabs bearing bas-reliefs representing a Roman procession and the death of a warrior.

D Black, with red encaustic paintings: invented and patented by Josiah Wedgwood, F.R.S., 1769.

9. Pair of large vases, Greek form, Leda handles; subjects from Flaxman's "Homer."
10. Various other vases, of the Greek and Campanian style in form and painting.
11. Vases in the same style, painted in encaustic colours on red, buff, and other grounds.

E Carrara or parian and terra cotta.

12. Oval vase, supported on a pedestal, consisting of two Bacchante caryatides and two satyrs, by A. Carrier, of Paris.
13. Statuettes of Faith, Hope, and Charity, by A. Carrier, of Paris.
14. Faun and Bacchante, loaded with fruit and game, running, by Clodion.
15. Oberon and Titania, by Wyon.
16. Milton and Shakspeare.
17. Candelabra, running figures, by Clodion.
18. Various other statuettes, after the antique and from models by modern artists.
19. Busts of the most distinguished warriors, writers, poets, and engineers.

F Works of art, painted by Mr. Lessore.

20. Pair of large vases, Greek form, five feet high, on pedestal. Subjects:—"Alexander and Porus," and the "Family of Darius," painted in Nevers blue.
21. One large vase in Nevers blue; allegorical subject, Her Majesty, supported by the Prince of Wales; the medallion of the Prince Consort.
22. Pair of vases, three feet high, on pedestals, in the style of Louis XIV., from a model at Versailles.
23. Pair of Greek vases, figures after Raphael.
24. Three Greek vases, figures after Raphael.
25. Pair of vases on pedestals. Subjects:—"Venus and Adonis," "Diana and Endymion," "Perseus and Andromeda."
26. Various large plateaux, framed:—"The Rape of Europa," "Massacre of the Innocents," "Emperor Charles V. at Venice," "St. Jerome," after Parmegiano, "Venus and Adonis," "Apollo," after Rubens, "Night," after Albano, "The Entombment," "The Adoration," after Titian, landscape, &c., &c.
27. Slabs framed:—"The Magdalen washing Christ's Feet," after P. Veronese, "The Bathers," after Boucher, landscapes, &c.
28. Various vases; subjects from the Stanze by Raphael, "Battle of the Amazons," after Rubens, "Rape of the Sabines," after Caravaggio; domestic subjects.
29. Tazzi; "Bacchus and Ariadne," "The Covenanters;" interiors, after Fragonati.
30. Trays, vases, &c., decorated with landscapes, pastoral and other light and pleasing subjects.

G Majolica, dark blue, and other ornamental glasses and enamels.

31. Pair of vases on pedestals, style of Louis XIV., blue glaze and gold tracery.
32. Various other vases in blue and gold.
33. Various vases in malachite.
34. Various vases, baskets, dessert, and toilet ware, in majolica.

H Green glaze, invented by Josiah Wedgwood, F.R.S., 1754; dessert ware, flower pots, &c.

I Rockingham teapots, &c.

Wedgwood, Josiah, & Sons, *continued.*

CARITAS

FIDES

W. J. LINTON

SPES

WEDGWOOD, JOSIAH, & SONS, *continued.*

J Inlaid ware, tea and toilet ware, in red, black, drab, sage, white, and brown.

K Stoneware. Drab, sage, red, black, and white, teapots, jugs, candlesticks, &c., both plain and with bas-reliefs.

L Plumbers' ware, enamelled, printed, pearl, and cream colour.

M Chemical and photographic wares; mortars, evaporating pans, crucibles, baths, and pans.

N Porous ware. Water-bottles, orange, black, white, and blue, plain and ornamented; battery-cells of all shapes.

O Enamelled dinner, dessert, and toilet ware.

P Patterns of ware printed in various colours by a new process, which obtains the effects of chromolithography, and with durable enamel colours and gold.

Q Printed dinner, dessert, and toilet ware, scale-plates and weights.

R. Pearl-white dinner and toilet ware.

S. Cream-colour, invented by Josiah Wedgwood, F.R.S., about 1750, brought into use in 1763; the first dinner-set being presented to Queen Charlotte, on her accession, 1761, and called Queen's Ware.

Dinner, tea, and toilet ware; milk pans, kitchen pans, and ware for domestic use.

LONDON:
PRINTED BY EDMUND EVANS,
RAQUET COURT, FLEET STREET.

Class XXXVI.

TOILET, TRAVELLING, AND MISCELLANEOUS ARTICLES.

Sub-Class A.—*Dressing Cases and Toilet Articles.*

[6926]

Asprey, C., 166 *Bond Street* & 22 *Albemarle Street.*—Dressing cases, travelling bags, and despatch boxes. *See pages* 2, 3, 4, *and* 5.

[6927]

Austin, T. & G., 39 *Westmoreland Street, Dublin.*—Dressing cases, despatch boxes, carriage bags.

[6928]

Betjemenn, G. & Sons, 36 *Pentonville Road, London.*—Dressing cases, writing sets, and book slides.

[6929]

Bourn, E., *College Street, Bristol.*—Russia leather travelling desks, stationery, writing and dressing cases.

[6930]

Gebhardt, Rottmann, & Co., 24 *Lawrence Lane, Cheapside.*—Dressing cases and bags, writing desks, and photograph albums.

[6931]

Howell, James, & Co , 9 *Regent Street, Pall Mall.*—Dressing cases, travelling bags, &c.

[6932]

Jenner & Knewstub, 33 *St. James's Street.*—Dressing cases, travelling bags, and despatch boxes.

[6933]

Leake, C. F., 3 *Alexander Square, Brompton.*—Relievo leather caskets, boxes, and fancy articles.

[6934]

Leuchars, W., 38 *Piccadilly.*—Dressing and writing case, travelling bag, &c.

[6935]

Mappin Bros., 222 *Regent Street.*—Dressing cases and bags.

Asprey, C., 166 *Bond Street* & 22 *Albemarle Street.*—Dressing cases, travelling bags, and despatch boxes.

An Adelaide Writing Desk, in Coromandel wood, with pierced, gilt, and engraved mounts, with china medallions, and fitted with every requisite for correspondence.

A Suite of things for the Writing Table, in plain ormolu, set with corals, and producing a very elegant and pleasing effect.

Asprey, C., 166 *Bond Street, and* 22 *Albemarle Street—continued.*

An elegant gilt and engraved Metal Tray, with Oriental Agate Centre and Inkstand, raised on elegant scroll at the back.

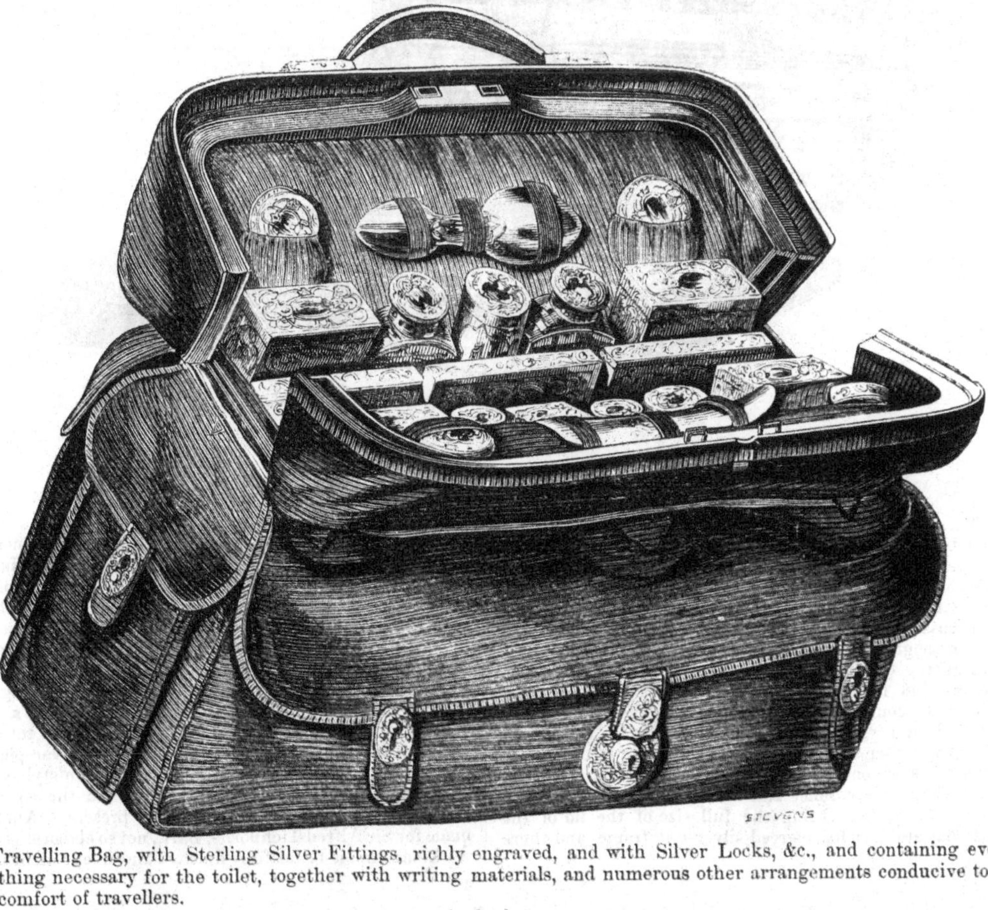

A Travelling Bag, with Sterling Silver Fittings, richly engraved, and with Silver Locks, &c., and containing everything necessary for the toilet, together with writing materials, and numerous other arrangements conducive to the comfort of travellers.

ASPREY, C., 166 *Bond Street* & 22 *Albemarle Street—continued.*

Mr. Asprey exhibits in the North Gallery, and also in two separate trophies under the Eastern Dome, a beautiful collection of first-class art manufactures, in dressing cases, travelling bags, despatch boxes, writing cases, inkstands, and other elegancies, as represented in the accompanying plates. One of the striking features being a dressing case, truly magnificent, and undoubtedly the most chaste production of the kind ever seen. It consists of a fine specimen of Coromandel wood, with brilliant cut-glass fittings, with massive silver-gilt tops, richly engraved, each top having a superb carbuncle in the centre, surrounded with pearls, and these again encircled with carbuncle drops; the whole set in fine gold, and producing a most fascinating effect. There are in the whole 872 stones used in this case. By a very ingenious arrangement, an elegant looking-glass in the top is made to turn and swing to any angle with two telescope candle branches, and two hand mirrors *en suite*. The fall front (which by a clever contrivance acts simultaneously with the lid) is fitted with a bell beautifully mounted with stones, also a vinaigrette, the mariner's compass, thermometer, whistle, magnifier, glove stretchers, shoe lift, mouth glass, set of tablets, yard measure, also a double set of highly-finished instruments of every description, and all *en suite* with the other fittings. There are cleverly arranged secret contrivances for letters, gold, and bank notes, and boxes for the different varieties of postage stamps, jewel drawer, writing slope and stationery case complete, set of four coloured drawing pencils, and other fittings too numerous to give in detail. This case is only one of a series, all being more or less beautiful, but differing materially in style and arrangement.

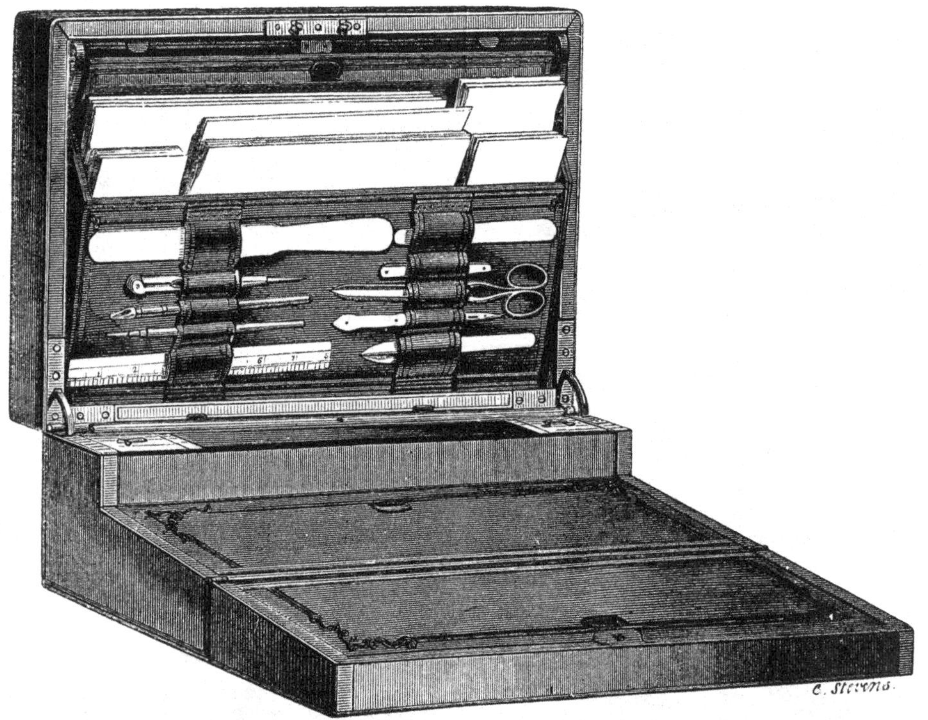

Mr. Asprey exhibits several dressing cases and travelling bags he has had the honour of making to order; one bearing the coronet and monogram of the Viscountess Lismore is a specimen of rare workmanship. The exterior is of the utmost simplicity, being the plain dead leaf russia leather, surmounted with a gold coronet and monogram. The interior fittings are very beautiful, all the tops being mounted in silver gilt of exquisite design, and finely pierced in the Moorish style, after one of the courts of the Alhambra, and over this pierced work is a crystal tablet, surmounted by a gold starlike plate, formed by two intersecting squares similar to those seen on some of the old Moorish brasses; on each plate is a coronet and monogram in blue enamel. The looking-glass (being the full size of the lid of the box) has also a richly pierced silver-gilt frame, and there is also an elegant set of instruments all *en suite*. Another, bearing the coronet and monogram of the Lady Harriet Ashley, is a splendid specimen of cabinet work, composed of fine rosewood, with kingwood cross bands with brass angles. The whole of the beautiful fittings are in massive silver gilt, with fine gold centres encircled with torquoise, all the elegant instruments being *en suite* with graduated stones, giving a very pretty effect. Another, bearing the name of Mrs. E. S. Woodhouse, is of rosewood of a very beautiful colour, and having the appearance of tortoiseshell. This is mounted outside with solid silver pierced work in very correct and tasteful Elizabethan style; the interior is fitted in the same pure style, and the case has essentially the appearance of a bridal present. Another, made for Sir Alfred Tichborne, Bart., not so elaborate in the workmanship, is beautifully finished and in excellent taste.

Aspréy, C., 166 *Bond Street* & 22 *Albemarle Street—continued.*

It is covered in green russia leather, with Mr. Asprey's improved patent handle, and his improved patent double-action Bramah lock. The fittings are of dead silver gilt, plain and massive, with gold centres, each having on it a clever monogram in green enamel. There is also a smaller sized walnut-wood case, bearing the name of Miss Aytoun, very prettily mounted outside, having a patent handle in the centre, and the fittings having silver-gilt mounts with gold centres. Also a morocco travelling bag, having engine-turned silver Bramah locks and fasteners, the interior fittings being of the same material, with writing arrangements, and every convenience for travelling, belonging to Mrs. Martin B. Stapylton.

The bag in the engraving is one selected from an immense stock of every size and style that can be imagined, and at prices to suit the most economical and luxurious taste.

The open writing case, or despatch box, forming a desk fitted with every necessary for home or travelling, is a very useful article, and may be made to any size or price desired.

The elegant writing case shut, with a slope and rounded

top, is suitable for a lady's writing table, it being both ornamental and useful. The suite of things for the writing table are plain ormolu mounted with coral drops, in very nice taste. It consists of envelope case, blotting book, inkstand, date box, candlesticks, match box, &c. &c. These sets, either larger or smaller in number of useful things, are kept in every kind of fashionable wood and other materials, elegantly mounted in the mediæval and other styles.

The beautiful inkstand in the engraving represents a style peculiar to Mr. Asprey's establishment, and which is very *recherché*. It is in richly engraved ormolu work, mounted with oriental stones, the bottom of the large tray being a beautiful *plaqué* of oriental agate.

A visit to Mr. Asprey's extensive establishment would amply repay any one, where may be seen not only a splendid stock of exquisitely-finished articles, but, what is more rare, such articles in process of manufacture on the premises.

[6936]

MAPPIN & COMPANY, 77 *and* 78 *Oxford Street, opposite the Pantheon.*—Dressing cases and bags.

MAPPIN & Co.'s Dressing and Travelling Bags and Cases are celebrated for the quality of the fittings, and their admirable arrangement to suit the many wants of modern luxury.

The most convenient ever invented for the use of travellers.

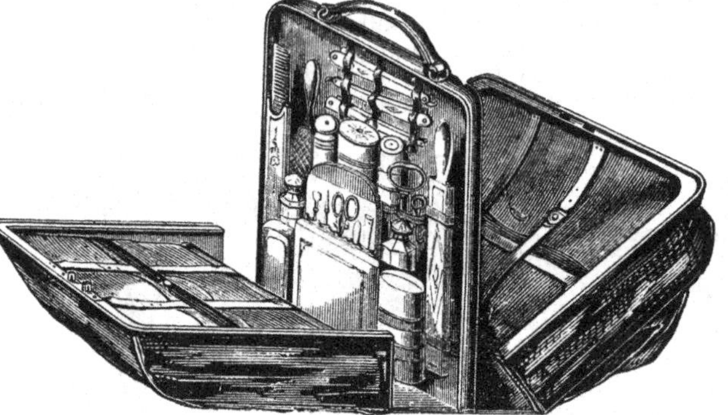

Fitted with every article for dressing and writing of the best quality.

£11 0 0

MAPPIN AND CO.'S REGISTERED "OXFORD" DRESSING AND TRAVELLING BAG

[6937]

MECHI & BAZIN, 4 *Leadenhall Street.*—Travelling dressing bags and cases, despatch boxes, &c.

[6938]

PARKINS & GOTTO, 25 *Oxford Street.*—Writing and dressing cases, bags, despatch boxes, &c.

[6939]

TOULMIN & GALE, 7 *New Bond Street.*—Despatch boxes, writing desks, and dressing cases.

[6940]

WEST, F., 1 *St. James's Street, Pall Mall.*—Dressing and writing cases, travelling bags.

SUB-CLASS B.—*Trunks and Travelling Apparatus.*

[6951]

ALLEN, J. W., 31 *Strand.*—Portmanteaus, trunks, dressing cases, &c.

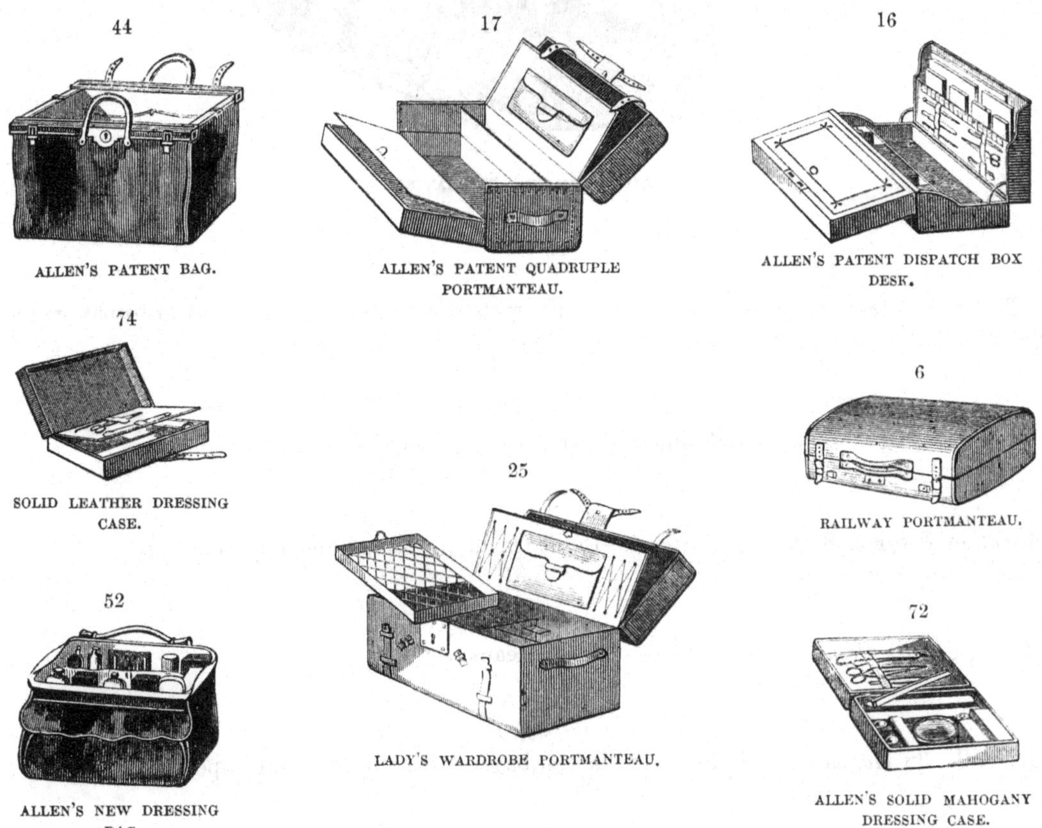

ALLEN'S PATENT BAG. ALLEN'S PATENT QUADRUPLE PORTMANTEAU. ALLEN'S PATENT DISPATCH BOX DESK. SOLID LEATHER DRESSING CASE. RAILWAY PORTMANTEAU. LADY'S WARDROBE PORTMANTEAU. ALLEN'S NEW DRESSING BAG. ALLEN'S SOLID MAHOGANY DRESSING CASE.

[6952]

BARRETT, B. BROS., 184 *Oxford Street.*—Travelling goods.—*See page* 8.

[6953]

BOURNE, T., 5 *College Road, Cork.*—Portmanteaus.

[6954]

CAVE, H. JANE, 1 *Edward Street, Portman Square.*—Waterproof dress and bonnet baskets.

[6955]

DAY, W. & SON, 353 & 378 *Strand.*—Portmanteaus and travelling requisites generally. —*See pages* 9, 10, *and* 11.

BARRETT, B. BROS., 184 *Oxford Street.*—Travelling goods.

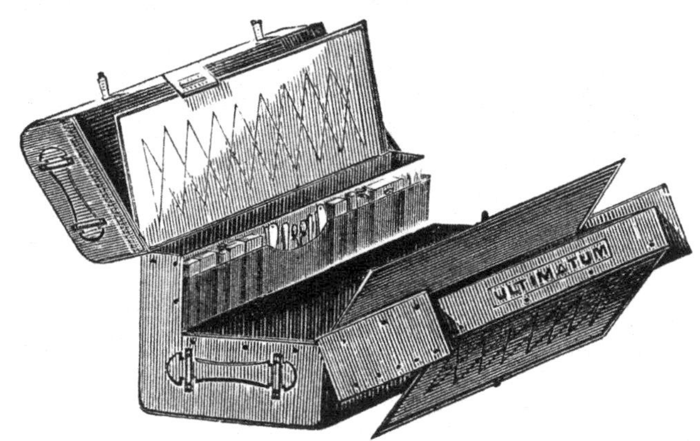

THE "ULTIMATUM" PORTMANTEAU.

Illustrated catalogues of trunks, portmanteaus, and general travelling equipage, manufactured by Barrett Brothers, may be obtained, post free on application.

[6956]

FISHER, S., 211 *Strand.*—Dressing cases and portmanteaus.

[6957]

HARROW & SON, 38 *Old Bond Street.*—Light travelling basket and imperial trunk.

[6958]

KANE, G., 70 *Dame Street, Dublin.*—Portmanteaus, &c.

[6959]

LAST, J., 38 *Haymarket.*—Trunks, hat boxes, bags, portmanteaus, and imperials.

[6960]

LAST, S., 256 *Oxford Street.*—Solid and other portmanteaus; new invented leather bag, &c.

[6961]

PRATT, H., *Chester Terrace, Eaton Square.* — Travelling wardrobe and compendium portmanteau.

[6962]

SILVER, S. W. & Co., 4 *Bishopsgate Within.*—Portmanteaus, bags, &c.

[6963]

SOUTHGATE, J., 76 *Watling Street.*—Solid leather portmanteau, travelling trunks and bags.

DAY, W. & SON, 353 *and* 378 *Strand.*—Portmanteaus and travelling requisites generally.

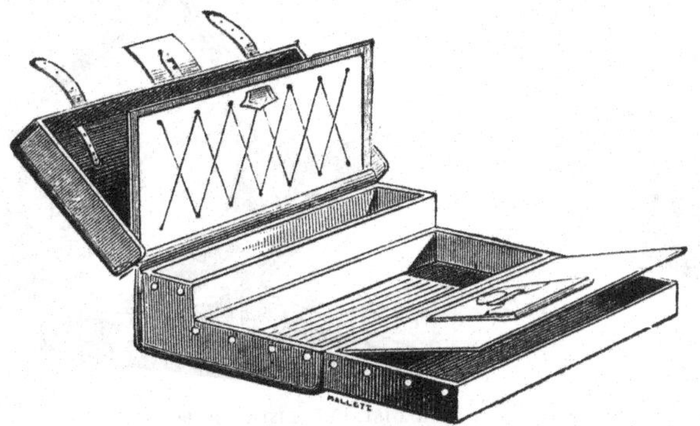

SOLID LEATHER QUADRUPLE OR WARDROBE PORTMANTEAU.

Contains four separate compartments, greatly facilitating the arrangement of a wardrobe, and keeping each article of dress distinct ; Bramah lock, with duplicate key, &c.

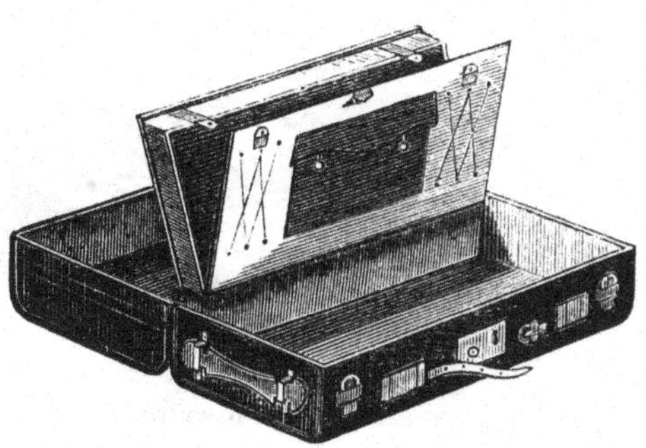

SOLID LEATHER FOLDING PORTMANTEAU.

Regulation size for the Continental Malle Poste. Contains three distinct compartments, is fitted with portfolio for papers, patent lock, with duplicate key, flush handles, &c.

DAY'S "ECLIPSE" PORTMANTEAU. BY ROYAL LETTERS PATENT.

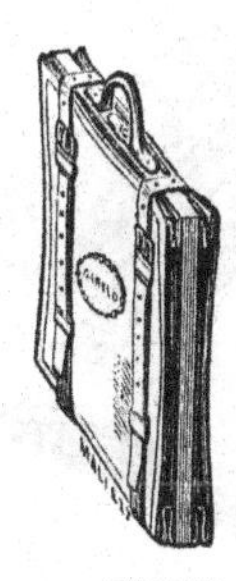

EMPTY.

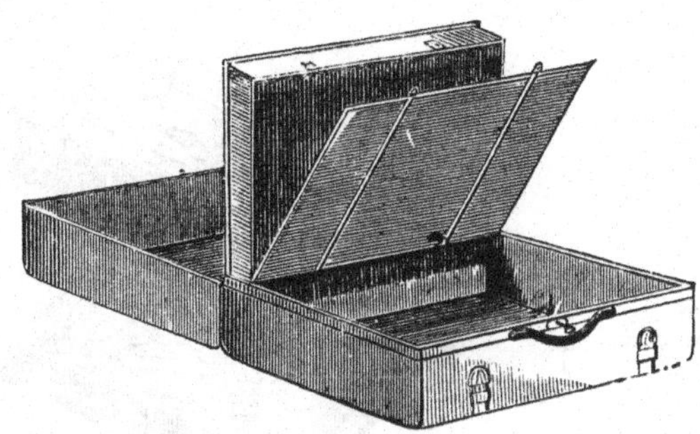

OPEN.

PACKED.

There is no hand-portmanteau superior to this. It possesses the capacity and solidity of the ordinary portmanteau, and is only half its weight, and one-fifth its size. It is made in solid and ordinary leather, black or brown, and of five different sizes.

The "Eclipse" contains three distinct compartments (one for shirts), each lying perfectly flat, and readily accessible, adapting itself in size to its contents, unaided by machinery or complication; thus insuring freedom from creasing and disarrangement, as it is always full, whether packed for a day's trip or a month's tour, to either of which the same portmanteau is applicable.

CARRIAGE OR HAND BAG.

TOURIST'S OR COURIER BAG.

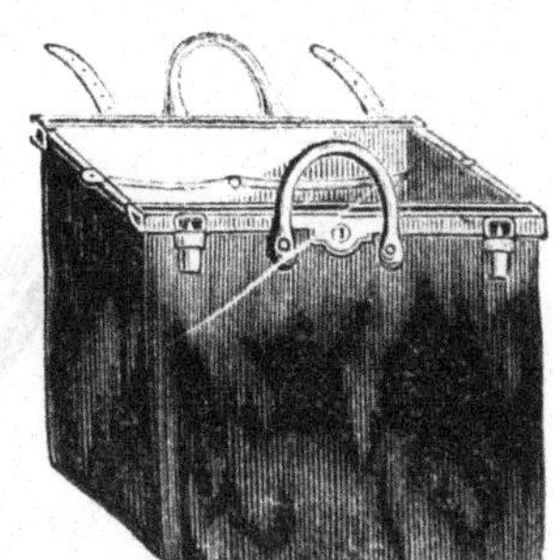

SQUARE-MOUTHED BAG.

Carriage or hand bags, of various sizes, in morocco or enamelled hide leather, with self-acting locks, &c.

Tourist's or courier bags, in morocco and other leathers, from 7s. 6d.

Square-mouthed travelling bags, of enamelled waterproof hide leather, in various sizes.

The exhibitors manufacture ladies' dress trunks and imperials, air-tight overland trunks for India, cabin portmanteaus, hat cases, haversacks, tourist's knapsacks, and travelling requisites of every known description. Illustrated catalogues will be forwarded on application.

Day, W. & Son—*continued.*

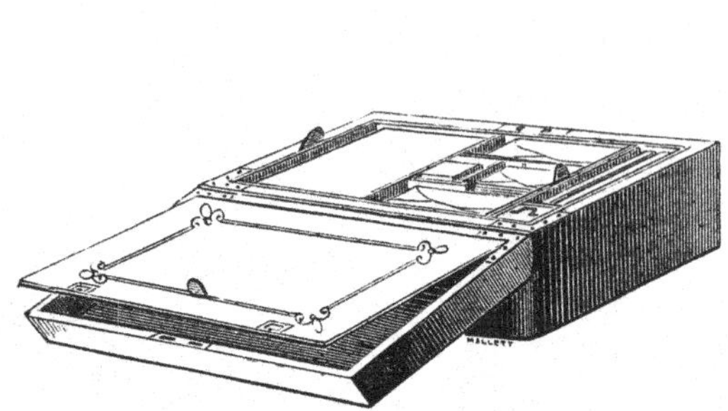

DISPATCH BOX AND WRITING CASE,

Of russia or morocco leather, with Bramah's patent lock; contains paper-folder, penknife, scissors, pen-holder, pencil, ink, and light-box; tray fitted with stationery, blotting-book, &c.

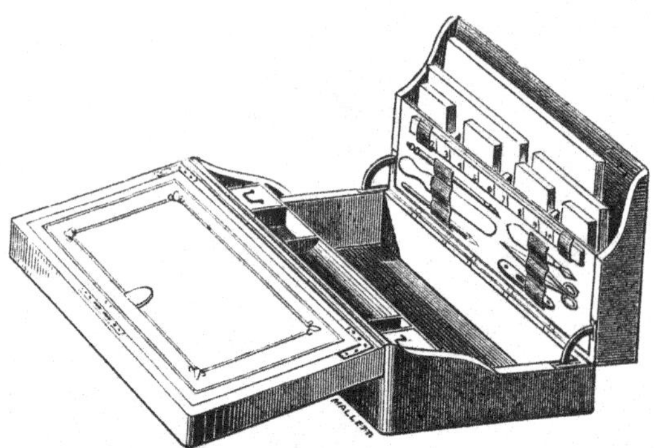

REGISTERED DISPATCH BOX.

Of russia or morocco leather, with Bramah's patent lock; a complete and convenient combination of writing desk and dispatch box, containing stationery and all the usual writing implements, so arranged that any article is instantly accessible, without disarranging the remainder. It is fitted with blotting-book, ink, light, &c. Every article warranted of the very best quality.

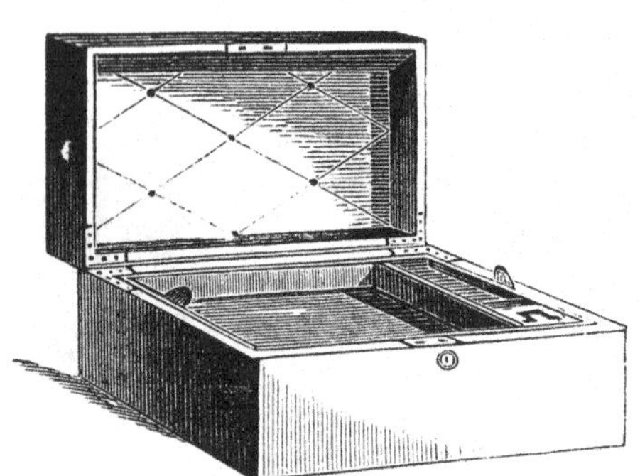

IMPROVED DISPATCH BOX,

Of morocco leather, with patent lock; fitted with ink, pen-tray, &c. In all sizes.

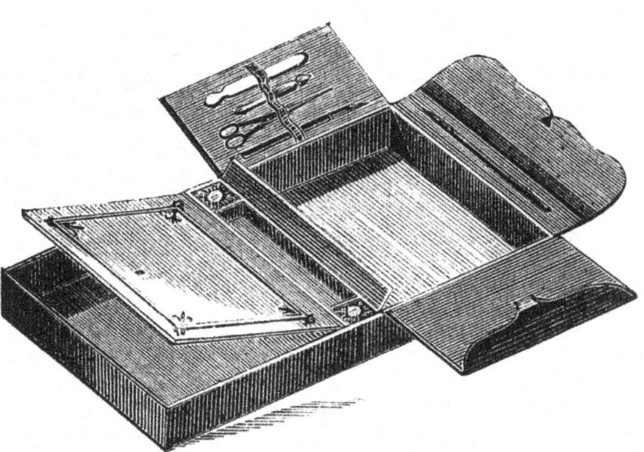

TRAVELLING WRITING DESK,

Made of imitative and of real russia or morocco leather, of various sizes, commencing at 10½ inches.

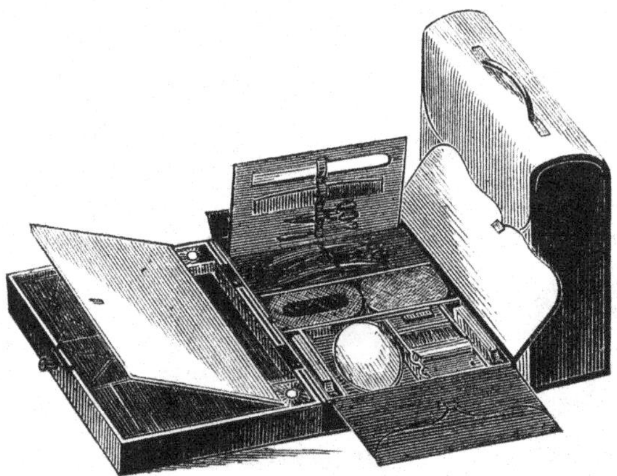

WRITING AND DRESSING CASE COMBINED.

Recommended for use where economy of space is desirable. The dressing fittings are contained in a tray, removable at pleasure, allowing the case to be used as dispatch box and writing-desk only, if desired.

Of russia leather, with Bramah lock and best quality fittings.

Illustrated catalogues forwarded on application.

Day, W. & Son—*continued.*

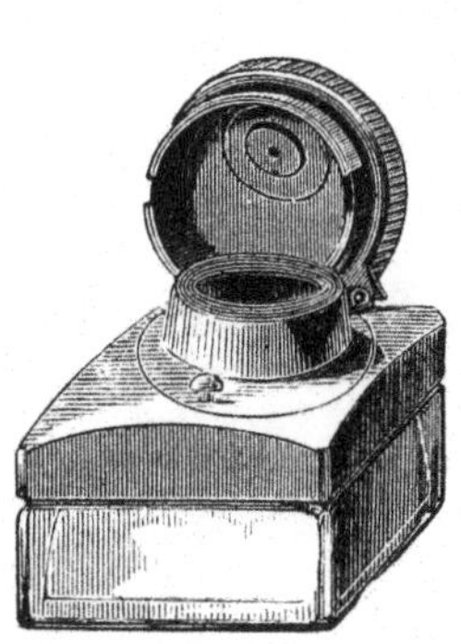

DAY'S PATENT REVOLVING-TOP INKSTAND.

DAY'S GUINEA DRESSING CASE, OF REAL RUSSIA LEATHER,

Containing cloth-brush, hair-brush, dressing-comb, shaving-brush, tooth-brush, nail-brush, shaving soap-dish, pair of razors, razor-strop, looking-glass.

The contrivances for securing portable inkstands have hitherto been confined to three: a top to pull off, a top to screw on, and a lid to close with a spring. Each of these methods is subject to objection; the pull-off top soon becomes loose and insecure; the screw-on top is troublesome and corrosive; the lid closing with a spring quickly gets out of order, and blurts the ink from the force with which it flies open.

Day's Patent Revolving-top Inkstand is submitted to the public as the simplest and most perfect inkstand that has yet appeared. It closes instantaneously, by one movement, without screw or spring; cannot corrode, or get out of order by wear; and is guaranteed to retain the ink perfectly secure, in whatever position it may be carried.

The principle of the revolving-top inkstand may be familiarly explained as the same by which the bayonet is affixed to the Enfield rifle.

Price 3s. 6d.

[6964]

Watson, C. J., 162 *Piccadilly.*—Portmanteau, with collapsing fittings for a hat.

[6965]

Wilks, E., *Cheltenham.*—Portmanteaus, &c.

PRINTED BY PETTER AND GALPIN, BELLE SAUVAGE WORKS, LUDGATE HILL, LONDON, E.C.

in the United States
nasters